Handbook of Hormones

ELSEVIER

science & technology books

ELSEVIER

Companion Web Site:

http://booksite.elsevier.com/9780128010280

Handbook of Hormones: Comparative Endocrinology for Basic and Clinical Research
Yoshio Takei, Hironori Ando, and Kazuyoshi Tsutsui, Editors

Resources for available:

- Supplementary sequence data, comparative data, figures and tables.

ELSEVIER

TOOLS FOR ALL YOUR TEACHING NEEDS
textbooks.elsevier.com

ACADEMIC PRESS

Handbook of Hormones
Comparative Endocrinology for Basic and Clinical Research

EDITED BY

Yoshio Takei
*Atmosphere and Ocean Research Institute,
The University of Tokyo, Chiba, Japan*

Hironori Ando
*Sado Marine Biological Station, Faculty of Science,
Niigata University, Niigata, Japan*

Kazuyoshi Tsutsui
*Department of Biology and Center for Medical Life Science,
Waseda University, Tokyo, Japan*

Produced in association with the *Japan Society for Comparative Endocrinology*

ELSEVIER

AMSTERDAM • BOSTON • HEIDELBERG • LONDON
NEW YORK • OXFORD • PARIS • SAN DIEGO
SAN FRANCISCO • SINGAPORE • SYDNEY • TOKYO
Academic Press is an imprint of Elsevier

Academic Press is an imprint of Elsevier
The Boulevard, Langford Lane, Kidlington, Oxford OX5 1GB
525 B Street, Suite 1800, San Diego, CA 92101−4495
225 Wyman Street, Waltham MA 02451

First edition 2016

Library of Congress Cataloging-in-Publication Data
A catalog record for this book is available from the Library of Congress

British Library Cataloguing-in-Publication Data
A catalogue record for this book is available from the British Library

ISBN 978-0-12-801028-0

For information on all publications visit
our website at http://store.elsevier.com

Typeset by MPS Limited, Chennai, India
www.adi-mps.com

**Working together
to grow libraries in
developing countries**

www.elsevier.com • www.bookaid.org

Publisher: Mica Haley
Acquisition Editor: Mara Conner
Editorial Project Manager: Jeffrey Rossetti
Production Project Manager: Melissa Read
Designer: Victoria Pearson

Printed and bound in the United States of America

Contents

Contents

Contents

Contents

Contents

Contents

PART II PEPTIDES AND PROTEINS IN INVERTEBRATES

Section II.1. Neuropeptides Related to Vertebrate Hormones

Contents

Contents

Contents

Subsection II.2.3 Regulation of Behaviors

Subsection II.2.4 Other Hormones and Neuropeptides

Contents

PART III LIPOPHILIC HORMONES IN VERTEBRATES

Contents

Masafumi Amano
School of Marine Biosciences, Kitasato University, Kanagawa, Japan

Hironori Ando
Sado Marine Biological Station, Faculty of Science, Niigata University, Niigata, Japan

Tadashi Andoh
Seikai National Fisheries Reserch Institute, Fisheries Research Agency, Nagasaki, Japan

Ivana Daubnerova
Institute of Zoology, Slovak Academy of Science, Bratislava, Slovakia

John A. Donald
School of Life and Environmental Sciences, Deakin University, Victoria, Australia

Leonard G. Forgan
School of Life and Environmental Sciences, Deakin University, Victoria, Australia

Shogo Haraguchi
Department of Biology and Center for Medical Life Science, Waseda University, Tokyo, Japan

Yoichi Hayakawa
Department of Applied Biological Sciences, Saga University, Saga, Japan

Satoshi Hirako
Department of Health and Nutrition, University of Human Arts and Sciences, Saitama, Japan

Susumu Hyodo
Laboratory of Physiology, Atmosphere and Ocean Research Institute, The University of Tokyo, Chiba, Japan

Taisen Iguchi
National Institute for Basic Biology, Okazaki Institute for Integrative Bioscience, Aichi, Japan

Akio Inui
Department of Psychosomatic Internal Medicine, Graduate School of Medical and Dental Sciences, Kagoshima University, Kagoshima, Japan

Haruaki Kageyama
Faculty of Health Care, Kiryu University, Gunma, Japan

Hiroyuki Kaiya
National Cerebral and Cardiovascular Center Research Institute, Osaka, Japan

Sho Kakizawa
Department of Biological Chemistry, Graduate School of Pharmaceutical Sciences, Kyoto University, Kyoto, Japan

Shinji Kanda
Department of Biological Sciences, Graduate School of Science, The University of Tokyo, Tokyo, Japan

Hiroyuki Kaneko
Division of Animal Sciences, National Institute of Agrobiological Sciences, Ibaraki, Japan

Hiroshi Kataoka
Graduate School of Frontier Science, The University of Tokyo, Chiba, Japan

Hidekazu Katayama
Department of Applied Biochemistry, School of Engineering, Tokai University, Kanagawa, Japan

Takashi Kato
Faculty of Education and Integrated Arts and Sciences, Waseda University, Tokyo, Japan

Yoshinao Katsu
Department of Biological Sciences, Hokkaido University, Hokkaido, Japan

Goro Katsuura
Department of Psychosomatic Internal Medicine, Graduate School of Medical and Dental Sciences, Kagoshima University, Kagoshima, Japan

Tsuyoshi Kawada
Division of Integrative Biomolecular Function, Bioorganic Research Institute, Suntory Foundation for Life Sciences, Kyoto, Japan

Atsushi P. Kimura
Department of Biological Sciences, Faculty of Science, Hokkaido University, Hokkaido, Japan

Yuki Kobayashi
Laboratory for Behavioral Neuroscience, Graduate School of Integrated Arts and Sciences, Hiroshima University, Hiroshima, Japan

Norifumi Konno
Department of Biological Science, Graduate School of Science and Engineering, University of Toyama, Toyama, Japan

Tadafumi Konogami
Graduate School of Frontier Science, The University of Tokyo, Chiba, Japan

Hiroyuki Minakata
Suntory Foundation for Life Sciences, Kyoto, Japan

Masatoshi Mita
Department of Biology, Faculty of Education, Tokyo Gakugei University, Tokyo, Japan

Shinichi Miyagawa
National Institute for Basic Biology, Okazaki Institute for Integrative Bioscience, Aichi, Japan

Mikiya Miyazato
Department of Biochemistry, National Cerebral and Cardiovascular Center Research Institute, Osaka, Japan

Akira Mizoguchi
Division of Biological Science, Graduate School of Science, Nagoya University, Aichi, Japan

Kanta Mizusawa
School of Marine Biosciences, Kitasato University, Kanagawa, Japan

Kenji Mori
Department of Biochemistry, National Cerebral and Cardiovascular Center Research Institute, Osaka, Japan

Fumihiro Morishita
Laboratory of Molecular Physiology, Department of Biological Science, Graduate School of Science, Hiroshima University, Hiroshima, Japan

Shunsuke Moriyama
School of Marine Biosciences, Kitasato University, Kanagawa, Japan

Hiroshi Nagasaki
Department of Physiology I, School of Medicine, Fujita Health University, Aichi, Japan

Shinji Nagata
Graduate School of Frontier Science, The University of Tokyo, Chiba, Japan

Yoshiaki Nakagawa
Graduate School of Agriculture, Kyoto University, Kyoto, Japan

Tomoya Nakamachi
Laboratory of Regulatory Biology, Graduate School of Science and Engineering, University of Toyama, Toyama, Japan

Teruyuki Niimi
Laboratory of Sericulture & Entomoresource, Graduate School of Bioagricultural Sciences, Nagoya University, Aichi, Japan

Yukiko Ogino
National Institute for Basic Biology, Okazaki Institute for Integrative Bioscience, Aichi, Japan

Maho Ogoshi
Graduate School of Natural Science and Technology, Okayama University, Okayama, Japan

Tsuyoshi Ohira
Department of Biological Sciences, Faculty of Science, Kanagawa University, Kanagawa, Japan

Hiroko Ohki-Hamazaki
College of Liberal Arts and Sciences, Kitasato University, Kanagawa, Japan

Hirokazu Ohtaki
Department of Anatomy, Showa University School of Medicine, Tokyo, Japan

Yoshihiko Ohyama
Department of Pharmaceutical Chemistry, Faculty of Pharmacy, Yasuda Women's University, Hiroshima, Japan

Yoshitaka Oka
Department of Biological Sciences, Graduate School of Science, The University of Tokyo, Tokyo, Japan

Naoki Okamoto
RIKEN Center for Developmental Biology, Hyogo, Japan

Tomohiro Osugi
Division of Integrative Biomolecular Function, Bioorganic Research Institute, Suntory Foundation for Life Sciences, Kyoto, Japan

Min Kyun Park
Department of Biological Sciences, Graduate School of Science, The University of Tokyo, Tokyo, Japan

Kazuki Saito
Graduate School of Frontier Science, The University of Tokyo, Chiba, Japan

Yumiko Saito
Laboratory for Behavioral Neuroscience, Graduate School of Integrated Arts and Sciences, Hiroshima University, Hiroshima, Japan

Takafumi Sakai
Graduate School of Science and Engineering, Division of Life Science, Saitama University, Saitama, Japan

Hirotaka Sakamoto
Ushimado Marine Institute, Faculty of Science, Okayama University, Okayama, Japan

Tatsuya Sakamoto
Ushimado Marine Institute, Faculty of Science, Okayama University, Okayama, Japan

Honoo Satake
Division of Integrative Biomolecular Function, Bioorganic Research Institute, Suntory Foundation for Life Sciences, Kyoto, Japan

Tomomi Sato
International Graduate School of Arts and Sciences, Yokohama City University, Kanagawa, Japan

Toshio Sekiguchi
Noto Marine Laboratory, Division of Marine Environmental Studies, Institute of Nature and Environmental Technology, Kanazawa University, Ishikawa, Japan

Munetaka Shimizu
Faculty of Fisheries Sciences, Hokkaido University, Hokkaido, Japan

Toshimasa Shinki
Promotion Center for Pharmaceutical Education, Nihon Pharmaceutical University, Saitama, Japan

Tetsuro Shinoda
National Institute of Agrobiological Sciences, Ibaraki, Japan

Haruyuki Sonobe
Konan University, Hyogo, Japan

Koichi Suzuki
Suzuki Laboratory, Center for Regional Collaboration in Research and Education, Iwate University, Iwate, Japan

Nobuo Suzuki
Noto Marine Laboratory, Division of Marine Environmental Studies, Institute of Nature and Environmental Technology, Kanazawa University, Ishikawa, Japan

List of Contributors

Tetsuya Tachibana
Department of Agrobiological Science, Faculty of Agriculture, Ehime University, Ehime, Japan

Akiyoshi Takahashi
School of Marine Biosciences, Kitasato University, Kanagawa, Japan

Toshio Takahashi
Division of Integrative Biomolecular Function, Bioorganic Research Institute, Suntory Foundation for Life Sciences, Kyoto, Japan

Yoshio Takei
Laboratory of Physiology, Atmosphere and Ocean Research Institute, The University of Tokyo, Chiba, Japan

Fumiko Takenoya
Department of Physical Education, Hoshi University School of Pharmacy and Pharmaceutical Science, Tokyo, Japan

Sakae Takeuchi
Graduate School of Natural Science and Technology, Okayama University, Okayama, Japan

Yoshiaki Tanaka
National Institute of Agrobiological Sciences, Ibaraki, Japan

Yuta Tanizaki
Department of Biology, Faculty of Education and Integrated Arts and Sciences, Waseda University, Tokyo, Japan

Takehiro Tsukada
Department of Anatomy, Jichi Medical University School of Medicine, Tochigi, Japan

Yusuke Tsukamoto
Graduate School of Frontier Science, The University of Tokyo, Chiba, Japan

Kazuyoshi Tsutsui
Department of Biology and Center for Medical Life Science, Waseda University, Tokyo, Japan

Naoaki Tsutsui
Ushimado Marine Institute, Faculty of Science, Okayama University, Okayama, Japan

Takayoshi Ubuka
Department of Biology and Center for Medical Life Science, Waseda University, Tokyo, Japan

Kazuyoshi Ukena
Graduate School of Integrated Arts and Sciences, Hiroshima University, Hiroshima, Japan

Nobuhiro Wada
Department of Anatomy, Showa University School of Medicine, Tokyo, Japan

Jun Watanabe
Center for Biotechnology, Showa University, Tokyo, Japan

Marty K.S. Wong
Laboratory of Physiology, Atmosphere and Ocean Research Institute, The University of Tokyo, Chiba, Japan

Zhifang Xu
Department of Anatomy, Showa University School of Medicine, Tokyo, Japan

Toshinobu Yaginuma
Laboratory of Sericulture & Entomoresource, Graduate School of Bioagricultural Sciences, Nagoya University, Aichi, Japan

Kiyoshi Yamauchi
Green Biology Research Division, Research Institute of Green Science and Technology, Shizuoka University, Shizuoka, Japan

Shinya Yuge
Department of Cell Biology, National Cerebral and Cardiovascular Research Institute, Osaka, Japan

Dusan Zitnan
Institute of Zoology, Slovak Academy of Science, Bratislava, Slovakia

Comparative endocrinology is a unique field of endocrine research, which aims to identify the essence of a hormone molecule and its function in the context of its evolution. This is achieved through comparisons among various species — most widely for hormones in animal systems, but a comparative approach can be applied to plant systems as well. It is increasingly evident that the comparative approach helps expand the views of all endocrinologists, including clinical researchers, and often leads to a deeper understanding of specific hormones. Thanks to the increased availability of genome databases for invertebrates and vertebrates, many hormones have been newly identified in non-mammalian species, broadening the notion of hormone families across phylogenies. Thus, there are increasing demands from endocrinologists all over the world to renew the catalog of hormones as frequently as possible.

The Japan Society for Comparative Endocrinology (JSCE) was founded in 1975, probably as the second society in the world to incorporate 'comparative endocrinology' in its name. Since its foundation, the JSCE has published useful books for the practical benefit of society members. The *Hormone Handbook* was published in 1988 (ISBN 4-524-23954-5) as a compendium of basic and important knowledge about various hormones. In response to requests from the Society's members, we revised the book in 2007, to include newly discovered hormones and published it as *Hormone Handbook, New E-book Edition* (ISBN 978-4-524-25058-5). In 2014, the new council members of JSCE gathered to discuss a revision to handle the further increase in newly-discovered hormones. It was also agreed to take this opportunity to publish the handbook in English and expand the readership to the world.

As you will find in the list of contributors, the book is written by 85 authors, almost all of whom are members of the JSCE. The number of hormones included in the handbook increased from about 120 in the first offering to more than 210 in this volume. We, as editors, recognize the rapid development of hormone knowledge and that publication of a hormone catalog within a short period is essential to provide readers with the most recent data on the hormones. We appreciate the time and efforts of all of the contributors toward the achievement of this book. The success in timely publication is solely an outcome of their tremendous efforts. Because of the quick publication, however, readers may find from place to place what we could have done better. We welcome any comments and suggestions from readers for the future revision of this book.

In addition to the contributors, we also deeply thank the many anonymous reviewers who gave useful comments and supported the publication of this book, and we thank Jeffrey Rossetti and Mara Conner of Elsevier Inc. for their invaluable help in processing this book. The value of this volume is certainly enhanced by the beautiful front cover designed by Kataaki Okubo, a member of JSCE, which is also gratefully acknowledged. Finally, the editors would like to dedicate this book, a great accomplishment by the JSCE members, to the memory of Hideshi Kobayashi, who was the founder of JSCE and who passed away on Christmas Eve of 2014 at the age of 94. He introduced *comparative endocrinology* to Japan and inspired us tremendously throughout his life. In dedicating this volume, we want to tell him that the tradition of good science that he seeded in Japan will never fade, but continue to flourish in the times to come.

Yoshio Takei
Hironori Ando
Kazuyoshi Tsutsui

BIOACTIVE MOLECULES AND TECHNICAL TERMS RELATED TO HORMONES

Abbreviations for hormones and receptors are found in full at their first appearance in each chapter.

AC	adenylate cyclase
ACh	acetylcholine
ADP	adenosine diphosphate
AhR	aryl hydrocarbon receptor
AMPK	AMP-activated protein kinase
AP-1	activator protein-1
AP-2	activator protein-2
ATP	adenosine triphosphate
bHLH	basic helix—loop—helix
C/EBP	CCAAT/enhancer-binding protein
cAMP	cyclic adenosine monophosphate
CNS	central nervous system
CoA	co-activator
CoR	co-repressor
COUP-TF	chicken ovalbumin upstream promoter-transcription factor
CRE	cAMP-response element
CREB	CRE-binding protein
CYP	cytochrome P450
DAG	diacylglycerol
DIO	deiodinase
DMSO	dimethyl sulfoxide
EC_{50}	median effective concentration
EGF	epidermal growth factor
ELISA	enzyme-linked immunosorbent assay
ERK	extracellularly regulated kinase
EST	expressed sequence tag
FAK	focal adhesion kinase
G protein	GTP-binding protein
GABA	γ-aminobutyric acid
GC	guanylate cyclase
GPCR	G protein-coupled receptor
GR	glucocorticoid receptor
HNF	hepatocyte nuclear factor 1
HPLC	high-performance liquid chromatography
HSD	hydroxysteroid dehydrogenase
IL	interleukin
IP3	inositol trisphosphate
IU	international unit

JAK	Janus kinase
LC/MS/MS	liquid chromatography—tandem mass spectrometry
MAPK	mitogen-activated protein kinase
MeOH	methanol
MR	mineralocorticoid receptor
NGFI-A	nerve growth factor-induced gene-A
NMDA	N-methyl-D-aspartate
PAS	Per-Arnt-Sim
PI3K	phosphoinositide 3-kinase
PIP2	phosphatidylinositol 4,5-bisphosphate
Pitx 1	pituitary homeobox 1
PKA	protein kinase A
PKC	protein kinase C
PKG	protein kinase G
PLC	phospholipase C
POU	Pit-Oct-Unc
PPAR	peroxisome proliferator-activated receptor
RIA	radioimmunoassay
RNAi	RNA interference
RTK	receptor tyrosine kinase
RXR	retinoid X receptor
SF-1	steroidogenic factor-1
SH	src homology
SHC	SH2 domain-containing
SNP	single nucleotide polymorphism
SP-1	specificity protein-1
StAR	steroidogenic acute regulatory protein
STAT	signal transducer and activator of transcription
TK	thymidine kinase
TLC	thin-layer chromatography
TNF	tumor necrosis factor
TRE	thyroid hormone response element
VDR	vitamin D receptor
VDRE	vitamin D response element
XRE	xenobiotic response element

PART I

Peptides and Proteins in Vertebrates

SECTION I.1

Neuropeptides

RFamide Peptide Family

Kazuyoshi Tsutsui and Takayoshi Ubuka

History

Neuropeptides that possess the $Arg-Phe-NH_2$ motif at their C-termini (i.e., RFamide peptides) have been characterized both in invertebrates and vertebrates. The first RFamide peptide to be identified was the cardioexcitatory peptide $Phe-Met-Arg-Phe-NH_2$ (FMRFamide), which was isolated from the ganglia of the Venus clam *Macrocallista nimbosa* [1]. Since then, a number of RFamide peptides have been identified in invertebrates, where these peptides seem to act as neurotransmitters and neuromodulators [2]. Subsequently, several RFamide peptides have been characterized in the brain of various vertebrates. In the past, the existence of five groups within the RFamide peptide family has been recognized in vertebrates: the gonadotropin-inhibitory hormone (GnIH) group, the kisspeptin group, the PQRFamide peptide (NPFF) group, the pyroglutamylated RFamide peptide (QRFP)/26RF amide group, and the prolactin-releasing peptide (PrRP) group (Table 1.1) [3−7]. These RFamide peptides have been shown to exert important neuroendocrine, behavioral, sensory, and autonomic functions [3−8].

Structure

Structural Features

A common structural feature of the family members is the $Arg-Phe-NH_2$ motif at their C-termini (Table 1.1) [3−9]. Human GnIH precursor mRNA encodes a polypeptide that produces two mature peptides, GnIH1 (RFRP-1) and GnIH2 (RFRP-3) [9]. Human kisspeptin precursor has several potential N-terminal cleavage sites and one C-terminal cleavage and amidation site, which generate several lengths of biologically active kisspeptins (kisspeptin-54, kisspeptin-14, kisspeptin-13, and kisspeptin-10) [10]. Human PQRFamide peptide (NPFF) precursor mRNA encodes a polypeptide that produces two mature peptides (NPFF and NPAF) [8]. Human QRFP precursor has several potential N-terminal cleavage sites and one C-terminal cleavage and amidation site, which generate several lengths of biologically active QRFPs (43RFa and 26RFa) [8]. Human PrRP precursor has several potential N-terminal cleavage sites and one C-terminal cleavage and amidation site, which generate several lengths of biologically active PrRPs (PrRP-31 and PrRP-20) [8].

Molecular Evolution of Family Members

E-Figure 1.1 shows a phylogenetic tree of the RFamide peptide family in vertebrates. The scale bar refers to a phylogenetic distance of 0.1 amino acid substitutions per site among the precursors. The phylogenetic tree of the RFamide peptide family shows that the precursors of RFamide peptides are clustered into five groups — the GnIH group, the kisspeptin group, the PQRFamide peptide (NPFF) group, the QRFP/26RFamide group, and the PrRP group — due to their structural similarities. The kisspeptin group includes two closely related peptides, Kiss1 and Kiss2, which are produced from different precursors encoded in paralogous genes of each animal [7,8,10].

Receptors

Structure and Subtypes

The cognate receptors for GnIH, NPFF, kisspeptin, QRFP/26RFamide, and PrRP are seven-transmembrane G protein-coupled receptors (GPCRs), GPR147, GPR74, GPR54, GPR103, and GPR10, respectively (Table 1.2) [7,8,10]. GPR147, GPR74, GPR103, and GPR10 belong to the family of rhodopsin β-type GPCRs. GPR54 belongs to rhodopsin γ-type GPCR family. GPR147 and GPR74 are structurally similar, and they both bind GnIH and NPFF, although GnIH and NPFF have higher affinity for GPR147 and GPR74, respectively. GPR74 also binds kisspeptin, QRFP, and PrRP with lower affinity than GPR54, GPR103, and GPR10, respectively.

Signal Transduction Pathway

GPR147, GPR74, GPR54, GPR103, and GPR10 are GPCRs, which interact with G proteins in the cell [7,8,10]. Agonist-bound GPCR activates various G proteins, such as $G\alpha_i$, $G\alpha_q$, and $G\beta_\gamma$ subunits, and regulates the activity of AC and PLC (Table 1.2). AC is a membrane associated enzyme that catalyzes synthesis of the second messenger cAMP. PLC is a membrane associated enzyme that catalyzes the synthesis of IP_3. cAMP and IP_3 can modify intracellular Ca^{2+} concentration ($[Ca^{2+}]_i$) (Table 1.2). It has been reported that GPR74, GPR103, and GPR10 can inhibit the activity of AC. GPR54 stimulates PLC and increases $[Ca^{2+}]_i$. GPR103 and GPR10 can also increase $[Ca^{2+}]_i$.

Biological Functions

Target Cells/Tissues and Functions

GnIH neurons project to the median eminence and gonadotropin-releasing hormone (GnRH) neurons [3,5]. GnIH receptor GPR147 is expressed in the gonadotropes and GnRH neurons [3,5]. Thus, GnIH may inhibit gonadotropin secretion by decreasing the activity of GnRH neurons as well as directly inhibiting gonadotropes. NPFF binds GPR74 expressed in the dorsal horn in the spinal cord and modulates pain transmission [7,8]. NPFF and NPAF attenuate the analgesic effect of morphine [7,8]. NPFF also modifies food intake [7,8]. Kisspeptin

Y. Takei, H. Ando, & K. Tsutsui (Eds): Handbook of Hormones. DOI: http://dx.doi.org/10.1016/B978-0-12-801028-0.00001-5

Table 1.1 Comparison of Amino Acid Sequences of the RFamide Peptide Family in Humans*

GnIH1	MPHSFANLPL**RF-NH$_2$**
GnIH2	VPNLPQ**RF-NH$_2$**
Kisspeptin-10	YNWNSFGL**RF-NH$_2$**
NPFF	FLFQPQ**RF-NH$_2$**
NPAF	AGEGLSSPFWSLAAPQ**RF-NH$_2$**
43RFa	QDEGSEATGFLPAAGEKTSGPLGNLAEELNGYSRKKGGFSF**RF-NH$_2$**
PrRP-31	SRTHRHSMEIRTPDINPAWYASRGIRPVG**RF-NH$_2$**

*RFamide sequences are shown in bold.

Table 1.2 Intracellular Effects of RFamide-Peptide Receptor Activation

Receptor	Intracellular Effect
GPR147	Inhibit adenylyl cyclase
GPR74	Inhibit adenylyl cyclase
GPR54	Stimulate phospholipase C Increase $[Ca^{2+}]_i$
GPR103	Inhibit adenylyl cyclase Increase $[Ca^{2+}]_i$
GPR10	Inhibit adenylyl cyclase Increase $[Ca^{2+}]$

neurons project to GnRH neurons, and GnRH neurons express GPR54 [6,8]. Kisspeptin is thought to be important for puberty onset-to-adult regulation of gonadotropin secretion and fertility. Kisspeptin also has potent anti-metastasis activity in several tumors [8]. QRFP/26RFamide and its receptor GPR103 mRNAs are expressed in the hypothalamus [7,8]. QRFP/26RFamide increases food intake, stimulates the gonadotropic axis, inhibits insulin secretion, inhibits locomotor activity, and induces analgesic effects [7,8]. PrRP receptor GPR10 mRNA is widely expressed throughout the brain, but highest in the thalamus, hypothalamus, and pituitary [7,8]. PrRP stimulates prolactin secretion, and has a variety of regulatory functions on food intake, stress, blood pressure, and pain [8].

References

1. Price DA, Greenberg MJ. Structure of a molluscan cardioexcitatory neuropeptide. *Science*. 1977;197:670−671.
2. Walker RJ, Papaioannou S, Holden-Dye L. A review of FMRFamide- and RFamide-like peptides in metazoa. *Invert Neurosci*. 2009;9:111−153.
3. Tsutsui K. A new key neurohormone controlling reproduction, gonadotropin-inhibitory hormone (GnIH): Biosynthesis, mode of action and functional significance. *Prog Neurobiol*. 2009;88:76−88.
4. Tsutsui K, Ukena K. Hypothalamic LPXRF-amide peptides in vertebrates: identification, localization and hypophysiotropic activity. *Peptides*. 2006;27:1121−1129.
5. Tsutsui K, Bentley GE, Bedecarrats G, et al. Gonadotropin-inhibitory hormone (GnIH) and its control of central and peripheral reproductive function. *Front Neuroendocrinol*. 2010;31:284−295.
6. Tsutsui K, Bentley GE, Kriegsfeld LJ, et al. Discovery and evolutionary history of gonadotrophin-inhibitory hormone and kisspeptin: new key neuropeptides controlling reproduction. *J Neuroendocrinol*. 2010;22:716−727.
7. Ukena K, Tsutsui K. A new member of the hypothalamic RF-amide peptide family, LPXRF-amide peptides: structure, localization, and function. *Mass Spectrom Rev*. 2005;24:469−486.
8. Chartrel N, Tsutsui K, Costentin, et al. Gonadotropin-inhibitory hormone. In: Kastin AJ, ed. Handbook of biologically active peptides, *Section on brain peptides*. Amsterdam: Elsevier; 2006:79−86.
9. Ubuka T, Morgan K, Pawson AJ, et al. Identification of human GnIH homologs, RFRP-1 and RFRP-3, and the cognate receptor, GPR147 in the human hypothalamic pituitary axis. *PLoS One*. 2009;4:e8400.
10. Pinilla L, Aguilar E, Dieguez C, et al. Kisspeptins and reproduction: physiological roles and regulatory mechanisms. *Physiol Rev*. 2012;92:1235−1316.

Supplemental Information Available on Companion Website

- Phylogenetic tree of the RFamide peptide family members/ E-Figure 1.1
- Phylogenetic tree of the RFamide peptide receptors/ E-Figure 1.2

Gonadotropin-Inhibitory Hormone

Kazuyoshi Tsutsui and Takayoshi Ubuka

Additional names/abbreviations: LPXRFamide peptide, RFamide-related peptide, Neuropeptide VF/GnIH, LPXRFa, RFRP, NPVF

A hypothalamic neuropeptide inhibiting gonadotropin secretion, GnIH inhibits sexual and aggressive behavior. It has an LPXRFamide (X = L or Q) sequence at its C-terminal.

Discovery

First identified in the hypothalamus of the Japanese quail in 2000 [1], GnIH's structure has been determined in the hypothalamus of various vertebrates, including humans [2].

Structure

Structural Features

Human GnIH precursor polypeptide consists of 196 amino acid residues that produce two mature GnIH peptides, human GnIH1 (RFRP-1) and GnIH2 (RFRP-3) [2]. The human GnIH1 (RFRP-1) aa sequence in the precursor polypeptide follows a basic aa lysine that is a proteolytic cleavage site preceding its N-terminal. Subsequent MPHSFANLPLRF sequence is followed by glycine as an amidation signal, and arginine as an endoproteolytic basic amino acid (Table 1A.1). The human GnIH2 (RFRP-3) aa sequence in the precursor polypeptide follows two arginine preceding its N-terminal. The subsequent VPNLPQRF sequence is followed by glycine as an amidation signal and arginine as an endoproteolytic basic aa (Table 1A.1) [2]. Most mammalian GnIH precursor polypeptides may produce two LPXRFamide (X = L or Q) peptides, whereas non-mammalian GnIH precursor polypeptides may produce three or four LPXRFamide peptides [3].

Primary Structure

Most GnIH peptides have the LPXRFamide (X = L or Q) motif at their C-termini (Table 1A.1) [3]. Although the N-terminal sequences are less conserved across vertebrate species, many identified endogenous GnIH peptides start from serine at the N-terminal (macaque RFRP-1, bovine RFRP-1, hamster RFRP-1, quail GnIH, quail GnIH-RP-2, starling GnIH, zebra finch GnIH, frog GRP, frog GRP-RP-1, newt LPXRFa-1, newt LPXRFa-3, goldfish LPXRFa-3, lamprey LPXRFa-1a, and lamprey LPXRFa-2). Other GnIH peptides start from alanine (bovine RFRP-3, rat RFRP-3, frog GRP-RP-3, newt LPXRFa-4, and lamprey LPXRFa-1b), methionine (human RFRP-1, newt

LPXRFa-2), valine (human RFRP-3), threonine (hamster RFRP-3), and tyrosine (frog GRP-RP-2) at their N-termini.

Properties

Human GnIH1 (RFRP-1): Mr 1,429, pI 10.55. Freely soluble in water. Human GnIH1 (RFRP-1) dissolved in water at 10^{-3} M is stable for more than a year at $-20°$C.

Human GnIH2 (RFRP-3): Mr 969, pI 10.55. Freely soluble in water. Human GnIH2 (RFRP-3) dissolved in water at 10^{-3} M is stable for more than a year at $-20°$C.

Synthesis and Release

Gene, mRNA, and Precursor

The human GnIH precursor gene (*NPVF*), located on chromosome 7 (p21–p15), consists of three exons. Human GnIH precursor mRNA has 1,190 bp. An orthologous GnIH precursor gene was found in sea lamprey, one of the most basal vertebrate species [4]. The gene structure and mRNA sizes are well conserved among vertebrates [3].

Distribution of mRNA

In the brain, GnIH precursor mRNA is exclusively expressed in the hypothalamus [1–5]. Transcripts are also found in the eye and the gonads [3,5].

Tissue and Plasma Concentrations

In the brain, abundant GnIH immunoreactive neuronal fibers are found in the hypothalamus followed by the midbrain [1–4]. The GnIH concentration in quail diencephalon is approximately 100 pmol/g tissue [1,6]. In the gonad, GnIH immunoreactive materials were found in ovarian thecal and granulosa cells, testicular interstitial cells, and germ cells [3]. The plasma concentration of GnIH is below its detection limit, except in the hypophysial portal system.

Regulation of Synthesis and Release

GnIH expression in the brain is regulated by nocturnal secretion of melatonin in birds [6] and mammals [3]. Melatonin may also stimulate GnIH release [3]. GnIH expression in the brain is also regulated by stress via the action of glucocorticoids [3]. Although glucocorticoid response elements (GREs) are found at $-1,665$ and $-1,530$ bp upstream of the GnIH precursor coding region in rats, the $-1,530$ GRE was identified to be critical for corticosterone responsiveness and recruitment of the

Y. Takei, H. Ando, & K. Tsutsui (Eds): Handbook of Hormones. DOI: http://dx.doi.org/10.1016/B978-0-12-801028-0.00104-5

Table 1A.1 Primary Structure of GnIH Peptides

Human GnIH1 (RFRP-1)	MPHSFAN**LPLRF**-NH$_2$
Human GnIH2 (RFRP-3)	VPN**LPQRF**-NH$_2$
Quail GnIH	SIKPSAY**LPLRF**-NH$_2$
Frog GRP	SLKPAAN**LPLRF**-NH$_2$
Goldfish LPXRFa-3	SGTGLSAT**LPQRF**-NH$_2$

Table 1A.2 Biological Actions of GnIH

Biological Actions	Animals Tested
Inhibition of gonadotropin release	quail, chicken, sparrow, rat, hamster (long-day), goldfish, zebrafish
Inhibition of gonadotropin synthesis	quail, chicken, goldfish
Inhibition of GnRH-elicited gonadotropin release	sparrow, rat, ovine, bovine, goldfish
Inhibition of GnRH-elicited gonadotropin synthesis	ovine, goldfish
Suppression of amplitude of LH pulse	ovine, bovine
Inhibition of reproductive and aggressive behaviors	quail, sparrow, rat
Stimulation of prolactin release	rat
Stimulation of gonadotropin release or synthesis	hamster (short-day), goldfish, lamprey
Stimulation of GnRH synthesis	lamprey

glucocorticoid receptor (GR). GnIH expression in the testis is also regulated by melatonin and corticosterone in birds [3].

Receptors

Structure and Subtype

Quail GnIH binds the GnIH receptor (GnIHR) with high affinity ($K_d = 0.752$ nM) [7]. GnIHR is a seven-transmembrane rhodopsin β-type G protein-coupled receptor (GPCR). GnIHR is also named GPR147 or NPFFR1 [2,5]. Mammalian GnIH (RFRP) also binds GPR74 (NPFFR2) with significantly lower affinity [3]. GPR147 (NPFFR1) and GPR74 (NPFFR2) genes are thought to be paralogous, positioned within the MetaHOX paralogons in human chromosomes 10 and 4, respectively. Human GnIHR (GPR147) has 430 aa residues, Mr 47,819. Vertebrate GnIHR has highly conserved seven-transmembrane domains as well as disulfide bridge sites between the first and second extracellular loops. Several glycosylation sites are predicted in the extracellular amino terminus and the extracellular loops.

Signal Transduction Pathway

Mammalian GnIH (RFRP) suppresses cAMP production in Chinese hamster ovarian cells transfected with GnIHR (GPR147) cDNA, suggesting that GPR147 couples to $G_{\alpha i}$ protein [5]. The cell signaling pathway by GnIH and GnIHR (GPR147) and its possible interaction with gonadotropin-releasing hormone (GnRH) signaling was investigated using a mouse gonadotrope cell line, LβT2 [8]. The results indicated that mouse GnIH peptides (mRFRPs) inhibit GnRH-induced gonadotropin subunit gene transcriptions by inhibiting AC/cAMP/PKA-dependent ERK activation in LβT2 cells [8].

Agonist

GnIHR (GPR147) binds with high affinity to GnIH peptides, which have a LPXRFamide (X = L or Q) motif at their C-termini [7]. GnIHR (GPR147) also binds neuropeptide FF (NPFF), a pain modulatory neuropeptide that has a PQRFamide motif at its C-terminus, with significantly lower affinity [3].

Antagonist

RF9 was reported as a selective and potent antagonist of GPR147 (NPFFR1) and GPR74 (NPFFR2) [9]. Central administration of RF9 evokes dose-dependent increases of LH and FSH levels in adult male and female rats, possibly by antagonizing GPR147 (NPFFR1) [3]. RF9 can block the increase in blood pressure and heart rate evoked by NPFF in rats, possibly by antagonizing GPR74 (NPFFR2) [3]. When chronically co-injected with heroin, RF9 blocks the delayed and long-lasting opioid-induced hyperalgesia and prevents the development of associated tolerance in rats [9].

Biological Functions

Target Cells/Tissues and Functions

GnIH neurons are located in the paraventricular nucleus (PVN) in birds [1], in the dorsomedial hypothalamic area (DMH) in mammals [2], and in the nucleus posterioris periventricularis in fish [3]. In many vertebrate species, GnIH neurons project to the median eminence or close to the pituitary to control anterior pituitary function [1,4]. GPR147 is expressed in the gonadotropes [2,3], and GnIH suppresses synthesis and release of gonadotropins in many vertebrate species (Table 1A.2) [2,3]. GnIH neurons also project to GnRH neurons in the preoptic area [2,3], and GnRH neurons express GPR147. Accordingly, GnIH may inhibit gonadotropin synthesis and release by decreasing the activity of GnRH neurons as well as directly acting on the gonadotropes [3]. However, GnIH can stimulate gonadotropin synthesis or release in fish and hamsters, depending on their reproductive states (Table 1A.2) [3]. GnIH also stimulates prolactin release in rats (Table 1A.2) [5]. GnIH can inhibit reproductive behavior by possibly acting within the brain (Table 1A.2) [3]. GnIH inhibits aggressive behavior of male quail by stimulating the activity of aromatase and increasing neuroestrogen concentration in the brain (Table 1A.2) [10]. GnIH and GnIHR expressed in the gonads may regulate gametogenesis and sex steroid secretion [3].

Phenotype in Gene-Modified Animals

Ablation of GPR147 elevated LH and FSH levels, and lowered the suppressive effect of stress on LH secretion in mice. There are reports showing the effect of RNA interference (RNAi) of GnIH precursor mRNA on the behavior of white-crowned sparrows and Japanese quail [10]. GnIH RNAi reduced the resting time of male and female birds, and spontaneous production of complex vocalizations, and stimulated brief agonistic vocalizations in male white-crowned sparrows. GnIH RNAi further stimulated aggressive song production of short duration in male birds when they were challenged by playbacks of novel male songs. These behaviors resembled the behavior of breeding birds during territorial defense. GnIH RNAi increased locomotor activity and aggressive behavior in the male quail [10]. These results suggest that GnIH RNAi induces arousal and aggressiveness in birds.

Pathophysiological Implications

Clinical Implications

Two endogenous human GnIH peptides, human GnIH1 (RFRP-1) and GnIH2 (RFRP-3), were identified in the human hypothalamus [2]. It was demonstrated that human RFRP-3 has a potent inhibitory action on gonadotropin secretion in ovines that has the same RFRP-3 aa sequence in its GnIH precursor polypeptide [3]. Many factors, such as stress, anorexia, diabetes, obesity, and photoperiod, inhibit gonadotropin secretion [3]. The possible role of GnIH in mediating these effects has been implicated in animal models [3,6]. Thus, GnIH has the potential of an alternative or adjunct therapeutic agent to inhibit gonadotropins and steroid hormones.

Use for Diagnosis and Treatment

The endogenous inhibitor of gonadotropin secretion, GnIH, has therapeutic potential in the treatment of gonadotropins or steroid hormone-dependent diseases, such as precocious puberty, endometriosis, uterine fibroids, benign prostatic hyperplasia, and prostatic and breast cancers. Human GnIH may also have potential as a novel contraceptive.

References

1. Tsutsui K, Saigoh E, Ukena K, et al. A novel avian hypothalamic peptide inhibiting gonadotropin release. *Biochem Biophys Res Commun.* 2000;275:661−667.
2. Ubuka T, Morgan K, Pawson AJ, et al. Identification of human GnIH homologs, RFRP-1 and RFRP-3, and the cognate receptor, GPR147 in the human hypothalamic pituitary axis. *PLoS One.* 2009;4:e8400.
3. Tsutsui K, Bentley GE, Bedecarrats G, et al. Gonadotropin-inhibitory hormone (GnIH) and its control of central and peripheral reproductive function. *Front Neuroendocrinol.* 2010;31:284−295 (Review).
4. Osugi T, Daukss D, Gazda K, et al. Evolutionary origin of the structure and function of gonadotropin-inhibitory hormone: insights from lampreys. *Endocrinology.* 2012;153:2362−2374.
5. Hinuma S, Shintani Y, Fukusumi S, et al. New neuropeptides containing carboxy-terminal RFamide and their receptor in mammals. *Nat Cell Biol.* 2000;2:703−708.
6. Ubuka T, Bentley GE, Ukena K, et al. Melatonin induces the expression of gonadotropin-inhibitory hormone in the avian brain. *Proc Natl Acad Sci USA.* 2005;102:3052−3057.
7. Yin H, Ukena K, Ubuka T, et al. A novel G protein-coupled receptor for gonadotropin-inhibitory hormone in the Japanese quail (*Coturnix japonica*): identification, expression and binding activity. *J Endocrinol.* 2005;184:257−266.
8. Son YL, Ubuka T, Millar RP, et al. Gonadotropin-inhibitory hormone inhibits GnRH-induced gonadotropin subunit gene transcriptions by inhibiting AC/cAMP/PKA-dependent ERK pathway in LβT2 cells. *Endocrinology.* 2012;153:2332−2343.
9. Simonin F, Schmitt M, Laulin JP, et al. RF9, a potent and selective neuropeptide FF receptor antagonist, prevents opioid-induced tolerance associated with hyperalgesia. *Proc Natl Acad Sci USA.* 2006;103:466−471.
10. Ubuka T, Haraguchi S, Tobari Y, et al. Hypothalamic inhibition of socio-sexual behaviour by increasing neuroestrogen synthesis. *Nat Commun.* 2014;5:3061.
11. Hirokawa T, Boon-Chieng S, Mitaku S. SOSUI: classification and secondary structure prediction system for membrane proteins. *Bioinformatics.* 1998;14:378−379.
12. Horn F, Weare J, Beukers MW, et al. GPCRDB: an information system for G protein-coupled receptors. *Nucleic Acids Res.* 1998;26:277−281.
13. Gupta R, Jung E, Brunak S. Prediction of N-glycosylation sites in human proteins. Available at: <www.cbs.dtu.dk/services/NetNGlyc/>; 2004.

Supplemental Information Available on Companion Website

- Gene, mRNA, and precursor structures of human GnIH/E-Figure 1A.1
- Precursor and mature hormone sequences of various animals/E-Figure 1A.2
- Gene, mRNA, and precursor structures of human GnIH receptor/E-Figure 1A.3 [11−13]
- Primary structure of GnIH receptor of various animals/E-Figure 1A.4
- Accession numbers of genes and cDNAs for GnIH precursor and GnIH receptor/E-Tables 1A.1, 1A.2

Kisspeptin

Shinji Kanda and Yoshitaka Oka

Additional names/abbreviation: metastin

Kisspeptin is a neuropeptide that is encoded by the kiss1/kiss2 gene in vertebrates. Kisspeptin neurons are mainly localized in the hypothalamus, and there is a growing body of evidence to support their essentiality and importance for reproduction in mammals.

Discovery

The *KISS1* gene and its product KISS1 were identified to be a natural ligand for an orphan receptor, GPR54, in humans in 2001 [1]. In addition to *Kiss1*, the existence of a highly homologous peptide was reported when the first *kiss1* gene in a non-mammalian species was found [2]. Subsequent synteny analysis as well as reporter assays strongly suggested that: (1) both peptides are kisspeptins in terms of being ligands of GPR54; (2) *kiss1* and *kiss2* genes [3,4] are suggested to have been duplicated in the early vertebrate lineage; and (3) both are conserved in the majority of vertebrates, although placental mammals and marsupials lack *Kiss2*.

Structure

Structural Features

Both of the paralogous *kiss1* and *kiss2* products (Kiss1 and Kiss2 peptides) include widely conserved important C-terminal 10 aa residues.

Primary Structure

Although longer forms have been isolated, the 10 aa residues with amidated C-terminals have been shown to activate the kisspeptin receptor Gpr54 signaling pathways. For instance, human preprokisspeptin consists of 145 aa residues and is cleaved to a shorter C-terminus that varies from 10 to 54 residues (Figure 1B.1). The C-terminal $RF/RY-NH_2$ is the most conserved motif within the RF-amide peptide family which kisspeptins belong to [5]. Native Kiss1/2 peptides have been isolated from tissues in a very limited number of species.

Synthesis and Release

Gene, mRNA, and Precursor

Ancestral kisspeptin genes have been suggested to be duplicated prior to the emergence of agnathans, due to the fact that lampreys have both *kiss1* and *kiss2* genes. Moreover, because synteny relationship is conserved between these two paralogs, it has been also suggested that these genes were

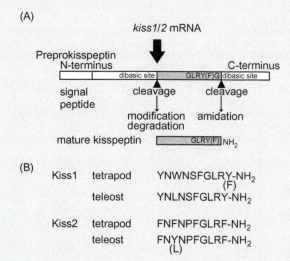

Figure 1B.1 Maturation process of the Kiss1/Kiss2 peptide and conserved peptide sequence of kisspeptins in vertebrates. (A) Translated preprokisspeptin is cleaved in its dibasic site, the C-terminus of which is amidated. After this cleavage, degradation and/or modification may occur at the N-terminus. (B) Core sequence of Kiss1 and Kiss2 peptide sequence. *Adapted with permission from Kanda and Oka, 2013 [9], © Springer.*

duplicated in either 1R or 2R whole genome duplication. On the other hand, there has been no report for the existence of two *kiss1* or two *kiss2* genes in teleosts, and it is therefore likely that one of the duplicated copies was lost immediately after the 3R whole genome duplication. Although the mechanisms have not been elucidated, some of the species or evolutionary branches lack either *kiss1* or *kiss2*; *kiss2* has been lost in placental mammals and marsupials, for instance. Interestingly, the avian species are supposed to have lost both *kiss1* and *kiss2* (Figure 1B.2). The length of kisspeptin precursors, which contain signal peptides in their N-termini, varies among species, and the precursors are cleaved into shorter mature peptides with their conserved 10 aa residues.

Distribution of mRNA

In all animals reported thus far, kisspeptin mRNA is expressed in the brain. In addition, high expression of *KISS1* in human placenta has been reported. Although kisspeptin is expressed in some other peripheral tissues [6], most research has been focused on kisspeptin expressed in the brain as a neuropeptide.

Y. Takei, H. Ando, & K. Tsutsui (Eds): Handbook of Hormones. DOI: http://dx.doi.org/10.1016/B978-0-12-801028-0.00105-7

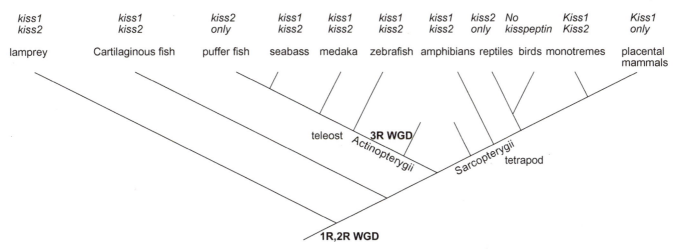

Figure 1B.2 The ancestral gene of kisspeptin has been suggested to be duplicated into *kiss1* and *kiss2* in the early vertebrate lineage, and either *kiss1* or *kiss2*, or both, are lacking in some branches. WGD, whole genome duplication.

Regulation of Synthesis and Release

In both teleosts and mammals, it has been demonstrated that some kisspeptin neurons show high steroid sensitivity in that they alter their expression in accordance with serum sex steroid concentration. This suggests that the *cis*-regulatory elements of the kisspeptin gene include estrogen responsive elements, and the presence of some transcription factors in a certain kisspeptin neuronal population determines whether positive or negative expressional regulation of kisspeptin genes occurs. Thus, kisspeptin neurons are sensitive to gonadal/breeding states, and their steroid-sensitivity of expression/release is widely conserved in vertebrates, at least in osteichthyes [7]. Recently, it was suggested that, in mouse, Kiss1 neurons in the arcuate nucleus co-express neurokinin B and dynorphin to enhance/suppress their firing activities, respectively, in autocrine/paracrine manners. Some physiological evidence further suggests a hypothesis that their interactions may underlie the pulsatile release of GnRH/LH in mammals [8].

Receptors

Structure and Subtype

Kisspeptin receptor Gpr54 is a G protein-coupled receptor. As is true with the ligands, Gpr54 also has paralogs, Gpr54-1 and -2, in vertebrates. Reporter assay studies have shown that both Kiss1 and Kiss2 activate both the Gpr54-1 and -2 signaling pathways. However, dose dependence analysis shows that Kiss1 prefers Gpr54-1, and Kiss2 prefers Gpr54-2 in many species [9].

Biological Functions

Target Cells/Tissues and Functions

In mammals, accumulating evidence strongly suggests that Kiss1, which is produced in hypothalamic Kiss1 neurons, directly and indirectly activates GnRH1 neuron firing activity and, eventually, the HPG axis. Given that mutations of the kisspeptin ligand or receptors lead to infertility, the kisspeptin system is believed to be essential for the regulation of reproduction in placental mammals. On the other hand, there is no clear evidence for kisspeptin regulation on reproduction in non-mammalian vertebrates. In addition to regulation of reproductive functions, kisspeptin is suggested to be involved

in various other functions. Some of the candidate action sites outside of the HPG axis include neurons producing oxytocin, prolactin (mammals), vasotocin, isotocin [10], and somatostatin (teleosts).

Phenotype in Gene-Modified Animals

As described above, *Kiss1* knockout mice and rats, as well as *Gpr54* knockout mice, experience drastic gonadal regression and infertility. On the other hand, in spite of intensive previous research, no one has ever identified *kiss1* or *kiss2* in avian species, although there are some genome databases available for birds. Thus, the fact that the avian species completely lack the kisspeptin gene clearly shows that kisspeptin is not essential for reproduction, at least in this group.

Pathophysiological Implications

Clinical Implications

Mutation of the *KISS1* gene or neurokinin B (*TAC3*) gene (an autocrine/paracrine neurotransmitter of arcuate kisspeptin neuron activities) causes idiopathic hypogonadotropic hypogonadism. GnRH application to patients with this disorder resulted in the partial recovery of their fertility, similar to cases in rodent and primate studies.

Use for Diagnosis and Treatment

Kisspeptin has not yet been used for diagnosis and treatment.

References

1. Ohtaki T, Shintani Y, Honda S, et al. Metastasis suppressor gene KiSS-1 encodes peptide ligand of a G-protein-coupled receptor. *Nature*. 2001;411(6837):613–617.
2. Kanda S, Akazome Y, Matsunaga T, et al. Identification of KiSS-1 product kisspeptin and steroid-sensitive sexually dimorphic kisspeptin neurons in medaka (*Oryzias latipes*). *Endocrinology*. 2008;149(5):2467–2476.
3. Kitahashi T, Ogawa S, Parhar IS. Cloning and expression of kiss2 in the zebrafish and medaka. *Endocrinology*. 2009;150(2):821–831.
4. Lee YR, Tsunekawa K, Moon MJ, et al. Molecular evolution of multiple forms of kisspeptins and GPR54 receptors in vertebrates. *Endocrinology*. 2009;150(6):2837–2846.
5. Oakley AE, Clifton DK, Steiner RA. Kisspeptin signaling in the brain. *Endocr Rev*. 2009;30(6):713–743.
6. Akazome Y, Kanda S, Okubo K, Oka Y. Functional and evolutionary insights into vertebrate kisspeptin systems from studies of fish brain. *J Fish Biol*. 2010;76(1):161–182.

7. Kanda S, Oka Y. Evolutionary insights into the steroid sensitive kiss1 and kiss2 neurons in the vertebrate brain. *Front Endocrinol*. 2012;3(28). Available from: http://dx.doi.org/10.3389/fendo.2012.00028.

8. Okamura H, Tsukamura H, Ohkura S, et al. Kisspeptin and GnRH Pulse Generation. *Adv Exp Med Biol*. 2013;784:297−323.

9. Kanda S, Oka Y. Structure, synthesis, and phylogeny of kisspeptin and its receptor. *Adv Exp Med Biol*. 2013;784:9−26.

10. Kanda S, Akazome Y, Mitani Y, et al. Neuroanatomical evidence that kisspeptin directly regulates isotocin and vasotocin neurons. *PLoS One*. 2013;8(4):e62776.

Supplemental Information Available on Companion Website

- Precursor and mature hormone sequences of various animals/E-Figure 1B.1
- Accession numbers or location of genes and cDNAs for kisspeptins and their receptor genes/E-Tables 1B.1, 1B.2

PQRFamide Peptide

Tomohiro Osugi, Takayoshi Ubuka, and Kazuyoshi Tsutsui

Abbreviation: PQRFa peptide/PQRFa
Additional names: Neuropeptide FF (NPFF)

A pain modulatory neuropeptide isolated from the medulla oblongata, PQRFamide peptide is also involved in the regulation of food intake, water balance, blood pressure, behavior, and reproduction.

Discovery

Neuropeptide FF (NPFF), a representative PQRFamide (PQRFa) peptide, was first discovered as a morphine modulating neuropeptide in 1985 [1]. NPFF was originally isolated from bovine medulla oblongata [1].

Structure

Structural Features

The PQRFa peptide precursors encode two PQRFa peptides in mammals and teleosts, and three PQRFa peptides in agnathans (Figure 1C.1) [2,3]. After cleavage of the signal peptide, biologically active PQRFa peptides, such as SQA-NPFF, SPA-NPFF, NPAF, and QFW-NPSF, are cleaved at the Arg$-$(Xaa)$_{2n}-$Lys/Arg sequence ($n = 0$, 1, or 2 in many cases) or a single basic amino acid at their N-termini (Figures 1C.1, 1C.2). NPFF and NPSF are exceptionally cleaved after alanine (Figures 1C.1, 1C.2). Hagfish possess two types of precursors that encode hagfish PQRFa, PQRFa-RP-1 and PQRFa-RP-2, or hagfish LPQRFa, PQRFa-RP-1 and PQRFa-RP-2 (Figures 1C.1, 1C.2) [3].

Primary Structure

PQRFa peptides possess a conserved C-terminal PQRFa sequence. Their N-terminal sequences are variable across different classes.

Properties

Mr 729$-$2,325. Freely soluble in water.

Synthesis and Release

Gene, mRNA and Precursor

Human PQRFa peptide gene (*NPFF*), located on chromosome 12 (q13.13), consists of three exons (E-Figure 1C.1) and has PIT1, AP2, and other elements in the promoter region [4]. Human PQRFa peptide precursor cDNA consists of 591 nt and encodes a 113-aa precursor protein [4]. PQRFa peptide genes are present in mammals, teleosts, and agnathans. In the genome database, putative PQRFa peptide genes are present in reptiles, amphibians, and cartilaginous fish. The presence of the PQRFa peptide gene in birds is not clarified.

Distribution of mRNA

PQRFa peptide mRNA is highly expressed in the dorsal spinal cord and the nucleus of the solitary tract in the rat [4]. PQRFa peptide mRNA is expressed at some level in rat hypothalamus [4]. In teleosts, PQRFa peptide mRNA is selectively expressed in the terminal nerve gonadotropin-releasing hormone (GnRH) neurons in the embryo and larva of zebrafish, and in the hypothalamus in grass puffer [5]. In agnathans, PQRFa peptide mRNA is strongly expressed around the third ventricle in the hypothalamus and is moderately expressed in the medulla oblongata [2,3].

Tissue and Plasma Concentrations

NPFF concentration is the highest (1,008 pmol/g protein) in the pituitary in the rat. In the rat central nervous system, NPFF is highly concentrated in the spinal cord (368 pmol/g protein), hypothalamus (202 pmol/g protein), and pons-medulla (136 pmol/g protein). In the plasma, NPFF concentration is 2.2 ± 0.5 pg/ml in humans.

Receptors

Structure and Subtype

In mammals, the two genes of NPFF receptors, NPFF receptor 1 (NPFFR1) and NPFF receptor 2 (NPFFR2), have been identified [4]. NPFFR1 and NPFFR2 are G protein-coupled receptors, and are also referred to as OT7TO22/GPR147/NPFF1/GnIH-R and HLWAR77/GPR74/NPFF2, respectively. NPFFR2 is considered to be the receptor for PQRFa peptides in mammals, because PQRFa peptides have higher binding affinities for NPFFR2. Human *NPFFR2* is located on chromosome 4 (q13.3) and consists of four exons (E-Figure 1C.2). In teleosts, two subtypes of the NPFFR gene (*npffr2-1* and *npffr2-2*) have been reported in zebrafish and Takifugu (E-Figure 1C.3) [6]. The fish NPFF receptors may have been duplicated by fish-specific genome duplication.

Signal Transduction Pathway

NPFFR2 is coupled to $G_{oi/o}$, and the activation of NPFFR2 inhibits forskolin stimulated adenylyl cyclase activity, intracellular cAMP levels, and phosphorylation of extracellular signal-regulated kinase 2 (ERK2) [4]. In addition, NPFFR2 and μ-opioid receptors are able to heteromerize [7]. The activation of NPFFR2 induces phosphorylation of μ-opioid receptors mediated by GPCR kinase 2 (GRK2) and results in the inactivation of μ-opioid receptors [7].

Y. Takei, H. Ando, & K. Tsutsui (Eds): Handbook of Hormones. DOI: http://dx.doi.org/10.1016/B978-0-12-801028-0.00106-9

```
                        10         20         30         40         50         60         70
Human        MDSRQAAALL VLLLLIDGG- CAEGPGGQQ- EDQLSAEEDS EPLPPQDAQT ---------- ----------
Bovine       MDARQAAALL LVLLLVTDWS HAEGPGGRDG GDQIFMEEDS GAHPAQDAQT ---------- ----------
Mouse        MDSKWAALLL LLLLLLNWG- HTEEAGSWG- EDQVFAGEDK GPHPPQYAHI PDRIQT---- ----------
Rat          MDSKWAAVLL LLLLLRNWG- HAEEAGSWG- EDQVFAEEDK GPHPSQYAHT PDRIQT---- ----------
Zebrafish    MN--GLLEDR LLVEM----- ------LRSL ---------- --LHGS---- ---------- ----------
Dwarf gourami MD--AAALVT LLALLAA--- ------TAGA SQALRSQGGL DEDDMQPGG- --AGEKVAER LLELESENTE
Lamprey      MEAKAVSAML LLALANCVLV SAARGSFSSM EEAAMPDSDS SLTKDYLAES VHEDPYRDSF DRASPDAAGS
Hagfish1     MDTKVLTALF LLILSYMAQG ATTFDATEND FKEESWEEES WSMGTAPLRG ---------- ----------
Hagfish2     MDTKVLTALF LLILSYMAQG ATTFDATEND FKEESWEEES WSMGTAPLRG ---------- ----------
```

```
                        80         90        100        110        120        130        140
Human        -------SGS LLHYLLQAME RPGR-SQAFL FQPQRFGRNT QG---SWRNEW LSPR-AGEG LNSQFWSLAA PQRFGKK-
Bovine       -------PRS LLRSLLQAMQ RPGR-SPAFL FQPQRFGRNT RG---SWSNKR LSPR-AGEG LSSPFWSLAA PQRFGKK-
Mouse        -------PGS LFRVLLQAMD TPRR-SPAFL FQPQRFGRSA WG---SWSKEQ LNPQAR---- QFWSLAA PQRFGKK-
Rat          -------PGS LMRVLLQAME RPRR-NPAFL FQPQRFGRNA WG---PWSKEQ LSPQAR---- --EFWSLAA PQRFGKK-
Zebrafish    --------- Q RYER-NPSEP HQPQRFGRGA RSG-LSTEER IQSR-DWET VPGQIWSMAV PQRFGKK-
Dwarf gourami NSIDDHLLTS VLRALLLGSQ RETR--TSVL HQPQRFGRGS RGQ-AVPEDQ LQTR-EWEA APGQIWSMAV PQRFGKK-
Lamprey      SSSEQLLLSR LARAFMHFPQ RFGRAGPSSL FQPQRFGRGS NDDEEVPPSL FYRR-SWGA PAEKFWMRAM PQRFGRKK
Hagfish1     -------IMER VVRAFSNTPQ RFGRADTSHF FQPQRFGRGI TKSDQRAEAG VAERRNSQET VPAYVWMRAF PQRFG---
Hagfish2     -------IMER VVRAFSNTPQ RFGRADTSHF FQPQRFGRGI TKSDQRAEAG VAERRNSQET VPAYVWMRAL PQRFG---
                          Peptide coding   Peptide coding                    Peptide coding
                          region A         region B                          region C
```

Figure 1C.1 Comparison of PQRFa peptide precursor amino acid sequences in representative species of vertebrates. The peptide coding regions are boxed by broken lines.

```
Encoded in the peptide coding region A
Lamprey PQRFa-RP-1                          AFMHFPQRF-NH₂
Hagfish PQRFa-RP-1                          AFSNTPQRF-NH₂
```

```
Encoded in the peptide coding region B
Human/Bovine/Mouse/Rat NPFF                 FLFQPQRF-NH₂
Human SQA-NPFF                              SQAFLFQPQRF-NH₂
Mouse SPA-NPFF                              SPAFLFQPQRF-NH₂
Rat NPA-NPFF                                NPAFLFQPQRF-NH₂
Zebrafish PQRFa-1*                          NPSVLHQPQRF-NH₂
Dwarf gourami PQRFa-1*                      TSVLHQPQRF-NH₂
Lamprey PQRFa-RP-2                          AGPSSLFQPQRF-NH₂
Hagfish PQRFa-RP-2                          ADTSHFFQPQRF-NH₂
```

```
Encoded in the peptide coding region C
Human NPAF                                  AGEGLNSQFWSLAAPQRF-NH₂
Bovine NPAF                                 AGEGLSSPFWSLAAPQRF-NH₂
Mouse QFW-NPSF                              QFWSLAAPQRF-NH₂
Zebrafish PQRFa-2*                          DWETVPGQIWSMAVPQRF-NH₂
Dwarf gourami PQRFa-1*                      EWEAAPGQIWSMAVPQRF-NH₂
Lamprey PQRFa                               SWGAPAEKFWMRAMPQRF-NH₂
Hagfish PQRFa                               NSQETVPAYVWMRAFPQRF-NH₂
Hagfish LPQRFa                              ALPQRF-NH₂
```

Figure 1C.2 Comparison of PQRFa peptides in representative species of vertebrates. The identical amino acids are shaded. * indicates putative peptides.

Agonist

[D-Tyr[1], (NMe)Phe[3]]NPFF (1DMe), which is an N-terminally modified NPFF, is broadly used as a NPFF agonist. 1DMe binds to both NPFFR1 and NPFFR2, with higher affinity for NPFFR2.

Antagonist

RF9 (2-adamantanecarbonyl−Arg−Phe−NH₂) was found by screening of small peptides that antagonize NPFFR1 and NPFFR2. RF9 displays an equally high affinity for both NPFFR1 and NPFFR2. RF9 blocks opioid-induced hyperalgesia, and prevents the development of associated opioid tolerance in the rat. RF9 also blocks the increase in blood pressure and heart rate evoked by NPFF.

Biological Functions

Target Cells/Tissues and Functions

PQRFa peptides bind to NPFFR2, which is expressed in the superficial layers of the dorsal horn in the spinal cord and modulates pain transmission at the spinal level. NPFF and NPAF attenuate the analgesic effect of morphine when administered intracerebroventricularly in rodents [4]. This anti-opioid effect of NPFF is considered to be due to the interaction between NPFFR2 and µ-opioid receptor [7]. On the other hand, NPFF also exerts an antinociceptive effect when administered intrathecally by modulating µ- and δ-opioid receptors in the rat [4]. Intrathecally injected NPFF potentiates the antinociceptive effect of morphine by modulating δ-opioid receptors [4]. NPFF and NPSF also exert pronociceptive actions by potentiating the acid sensing ion channel [4].

Moreover, NPFF inhibits food intake at low doses and stimulates food intake at high doses via opioid receptors expressed in the parabrachial nucleus in the rat [4]. Central administration of NPFF evokes an elevation in arterial blood pressure and heart rate, and stimulates water intake in the rat [4,8]. Central administration of NPFF also activates hypothalamic oxytocin neurons that are involved in the ascending visceral autonomic pathways in the rat [4,8]. In teleosts, PQRFa peptides modulate pacemaker activity of terminal nerve GnRH neurons in dwarf gourami [9]. In agnathans, one of the PQRFa peptides increases GnRH concentration in lampreys [10] and stimulates gonadotropin-β mRNA expression in hagfish [3].

Pathophysiological Implications

Clinical Implications

PQRFa peptides comprise one of the anti-opioid factors that may solve clinical problems of opiate tolerance and dependence. Thus, the antagonists of PQRFa peptide receptors, such as RF9, have the potential to be used as therapeutic agents that improve the efficacy of opioids in treatment for chronic pain in humans. It was reported that the density of NPFF fibers in the dorsal motor nucleus of vagus and the ambiguus nucleus in the medulla was significantly lower in hypertensive patients than in controls [8].

References

1. Yang HY, Fratta W, Majane EA, et al. Isolation, sequencing, synthesis, and pharmacological characterization of two brain neuropeptides that modulate the action of morphine. *Proc Natl Acad Sci USA*. 1985;82:7757−7761.
2. Osugi T, Ukena K, Sower SA, et al. Evolutionary origin and divergence of PQRFamide peptides and LPXRFamide peptides in the RFamide peptide family. Insights from novel lamprey RFamide peptides. *FEBS J*. 2006;273:1731−1743.
3. Osugi T, Uchida K, Nozaki M, et al. Characterization of novel RFamide peptides in the central nervous system of the brown hagfish: isolation, localization, and functional analysis. *Endocrinology*. 2011;152:4252−4264.
4. Yang HY, Tao T, Iadarola MJ. Modulatory role of neuropeptide FF system in nociception and opiate analgesia. *Neuropeptides*. 2008;42:1−18 (Review).
5. Ando H, Shahjahan M, Hattori A. Molecular neuroendocrine basis of lunar-related spawning in grass puffer. *Gen Comp Endocrinol*. 2013;181:211−214.
6. Zhang Y, Li S, Liu Y, et al. Structural diversity of the GnIH/GnIH receptor system in teleost: its involvement in early development and the negative control of LH release. *Peptides*. 2010;31:1034−1043.
7. Moulédous L, Froment C, Dauvillier S, et al. GRK2 protein-mediated transphosphorylation contributes to loss of function of μ-opioid receptors induced by neuropeptide FF (NPFF2) receptors. *J Biol Chem*. 2012;287:12736−12749.
8. Jhamandas JH, Goncharuk V. Role of neuropeptide FF in central cardiovascular and neuroendocrine regulation. *Front Endocrinol (Lausanne)*. 2013;4:8.
9. Saito TH, Nakane R, Akazome Y, et al. Electrophysiological analysis of the inhibitory effects of FMRFamide-like peptides on the pacemaker activity of gonadotropin-releasing hormone neurons. *J Neurophysiol*. 2010;104:3518−3529.
10. Daukss D, Gazda K, Kosugi T, et al. Effects of lamprey PQRFamide peptides on brain gonadotropin-releasing hormone concentrations and pituitary gonadotropin-β mRNA expression. *Gen Comp Endocrinol*. 2012;177:215−219.
11. Hirokawa T, Chieng SB, Mitaku S. SOSUI: classification and secondary structure prediction system for membrane proteins. *Bioinformatics*. 1998;14:378−379.

Supplemental Information Available on Companion Website

- Gene, mRNA, and precursor structure of human PQRFamide peptide /E-Figure 1C.1
- Gene, mRNA and precursor structures of human PQRFamide peptide receptor (NPFFR2)/E-Figure 1C.2
- Primary structure of PQRFamide peptide receptor (NPFFR2) of various animals /E-Figure 1C.3 [11,12]
- Accession numbers of genes and cDNAs for PQRFamide peptide precursor and PQRFamide peptide receptor/E-Tables 1C.1, 1C.2

Pyroglutamylated RFamide Peptide

Kazuyoshi Ukena and Kazuyoshi Tsutsui

Abbreviation: QRFP
Additional names: 26RFamide, 26RFa, P518, QRFP-43, QRFP-26, 43RFa

A 26-aa residue peptide originally isolated from the frog brain, QRFP is the latest member of the RFamide peptide family discovered in the hypothalamus of vertebrates.

Discovery

QRFP was independently discovered by three research groups in 2003. One group identified the 26-aa residue RFamide peptide in the brain of the European green frog (*Rana esculenta*) and designated it 26RFa [1]. Concurrently, two other research groups found QRFP precursors using a bioinformatic approach in rat, mouse, bovine, and human genomes, and paired QRFP with a previously identified orphan GPCR, GPR103, now termed QRFPR [2,3].

Structure

Structural Features

Human QRFP precursor consists of 136 aa residues, encoding one RFamide peptide with 43 aa residues. This RFamide peptide is flanked at the N-terminus by a monobasic amino acid cleavage site and at the C-terminus by a glycine amidation signal. The mature 43-aa residue RFamide peptide was identified from the culture medium of CHO cells that expressed the human peptide precursor [2]. Because the N-terminal amino acid was pyroglutamic acid, this RFamide peptide was named pyroglutamylated RFamide peptide (QRFP).

Primary Structure

The amino acid lengths of mature peptides depend on species — for example, 43 aa residues in human and rat, 27 aa residues in quail, 25 aa residues in zebra finch, and 26 aa residues in frog (Figure 1D.1) [4,5]. As there are several monobasic processing sites in the QRFP precursor protein, alternative cleavage may yield various N-terminally elongated forms of QRFP (E-Figure 1D.1). Indeed, the existence of both 26- and 43-aa residue RFamide peptide-like immunoreactivities was detected in the hypothalamus and spinal cord of humans [6]. The human and *Xenopus* QRFP precursors may also generate a 9-aa residue peptide, termed 9RFa, located upstream of QRFP (E-Figure 1D.1). However, 9RFa has not been detected in tissue extracts to date. The sequence of C-terminal octapeptide of QRFP, KGGFXFRF-NH$_2$ (X = S, A, T, or G), is highly conserved in vertebrates.

Properties

Human QRFP: Mr 4,504, pI 4.95. Moderately soluble in water. Human QRFP dissolved in water at 10^{-3} M is stable for more than a year at $-20°$C.

Synthesis and Release

Gene, mRNA, and Precursor

Human QRFP precursor gene (*QRFP*) locates on chromosome 9 (at 9q34.12). Because only the open reading frame of the precursor is deposited in the database, the exon—intron structure is obscure. Human QRFP precursor mRNA has 411 bp. The cDNAs encoding QRFP have been are identified in human, rat, mouse, quail, chicken, zebra finch, and goldfish. Furthermore, homologous sequences are listed in the genome database of reptilian (lizard), amphibian (*Xenopus*), and fish (stickleback, medaka, fugu, and zebrafish) species. These data reveal the existence of the QRFP-encoding gene in representative species of vertebrates, including fish, amphibians, reptilians, birds, and mammals.

Distribution of mRNA

The mRNA encoding QRFP is highly expressed in paraventricular and ventromedial nuclei of the hypothalamus in humans, and the dorsolateral and mediobasal hypothalamic areas in rodents. In birds, QRFP mRNA is expressed in the anterior hypothalamic nucleus in the chick brain, and in the anterior-medial hypothalamic area, the ventromedial nucleus of the hypothalamus, and the lateral hypothalamic area in the zebra finch brain. In goldfish, QRFP mRNA is expressed in the hypothalamus, optic tectum—thalamus, and testis.

Regulation of Synthesis and Release

The expression of QRFP mRNA in the hypothalamus is upregulated in fasted and genetically obese *ob/ob* and *db/db* mice [7]. In goldfish, mRNA expression is also augmented at 4 days after food deprivation.

Receptors

Structure and Subtype

In humans, QRFP is found to be an endogenous ligand of the orphan receptor GPR103, also known as AQ27 or SP9155 (now re-named QRFPR), which is a class A G protein-coupled receptor (GPCR) (E-Figure 1D.2) [2,3]. QRFPR shares relatively high sequence similarity with other RFamide receptors, notably those for neuropeptide FF (NPFF), prolactin-releasing peptide

Y. Takei, H. Ando, & K. Tsutsui (Eds): Handbook of Hormones. DOI: http://dx.doi.org/10.1016/B978-0-12-801028-0.00107-0

```
Human         <EDEGSEATGFLPAAGEKTSGPLGNLAEELNGYSRKKGGFSFRF-NH₂
Rat           <EDSGSEATGFLPTDSEKASGPLGTLAEELSSYSRRKGGFSFRF-NH₂
Quail                      GGGGTLGDLAEELNGYSRKKGGFAFRF-NH₂
Zebra finch                SGTLGNLAEEINGYNRRKGGFTFRF-NH₂
Green frog                 VGTALGSLAEELNGYNRKKGGFSFRF-NH₂
```

Figure 1D.1 Primary structure of identified QRFP.

(PrRP), kisspeptin, and gonadotropin-inhibitory hormone (GnIH). Two isoforms of the receptor for QRFP are characterized in rodents. These QRFP receptor isoforms are designated QRFPR1 and QRFPR2 in rat and mouse. In birds, the cDNAs encoding QRFPR are characterized in the brain of chicken and zebra finch. The sequence of chicken QRFPR is highly similar to those of human and rat QRFPR. QRFP increases $[Ca^{2+}]_i$ in a dose-dependent manner, with an EC_{50} value of around 40 nM [8]. The mRNA of QRFPR is widely expressed in chicken and zebra finch brains, and the highest concentration of mRNA is observed in the diencephalon. Although there are homologous sequences to QRFPR in the genome database of *Xenopus*, zebrafish, coelacanth, and lamprey, QRFPR has been studied only in mammals and birds to date (E-Figure 1D.3).

Signal Transduction Pathway

The 26- or 43-aa residue QRFP binds to QRFPR with high affinity ($EC_{50} = 3.2$ or 0.52 nM, respectively). QRFP inhibits cAMP formation with similar efficacy in QRFPR-transfected CHO cells. Furthermore, QRFP markedly increases intracellular Ca^{2+} concentration ($[Ca^{2+}]_i$) in a pertussis toxin (PTX)-independent manner. These results suggest that QRFPR couples to $G_{i/0}$ and/or G_q protein.

Agonist

A synthetic C-terminal heptapeptide of QRFP ($26RFa_{20-26}$; GGFSFRF-NH₂) is responsible for the biological activity of the peptide via QRFPR.

Biological Functions

QRFP is thought to be one of the orexigenic peptides in vertebrates. Although QRFP hardly affects food intake in normally fed rats (at least under a low-fat diet), QRFP induces a marked orexigenic effect in mice, food-restricted rats, and rats fed with a high-fat diet. In addition to its orexigenic effects, in mammals QRFP regulates various functions, including energy homeostasis, bone formation, pituitary hormone secretion, steroidogenesis, nociceptive transmission, and blood pressure. In an earlier report, intravenous administration of QRFP in rats was found to increase plasma aldosterone levels [2]. In birds, intracerebroventricular injection of QRFP stimulates feeding behavior in broiler chicks, but not in layer chicks [8]. Central injection of QRFP in free-feeding male zebra finches stimulates food intake for 24 hours, without a change in body mass. These results also indicate that QRFP exerts an orexigenic activity in various avian species. In goldfish, serum LH levels are significantly increased at 1 hour after intraperitoneal injection of QRFP. As QRFP has no effect on LH release from pituitary cells in primary culture, it is thought that, in fish, the peptide may stimulate the gonadotropic axis by acting exclusively at the hypothalamic level.

Phenotype in Gene-Modified Animals

Mice deficient in the receptor for QRFP (QRFPR1) suffer from osteopenia [9]. This observation indicates that QRFP plays a major role in bone formation, via QRFPR that is expressed in bone.

Pathophysiological Implications

Clinical Implications

Chronic intracerebroventricular injection of QRFP increases body weight and fat mass with a hyperphagic behavior in mice [10]. These data indicate that QRFP is able to cause obesity by affecting food intake and reducing energy expenditure. QRFPR1 knockout mice display osteopenia. Four single-nucleotide polymorphisms (SNPs) are identified in the promoter region of QRFP of osteoporosis-prone strain SAMP6 mice. The lower expression of QRFP mRNA in the spine and the hypothalamus of SAMP6 mice may associate with osteoporosis.

Use for Diagnosis and Treatment

QRFP/26RFa has not yet been used for diagnosis and treatment.

References

1. Chartrel N, Dujardin C, Anouar Y, et al. Identification of 26RFa, a hypothalamic neuropeptide of the RFamide peptide family with orexigenic activity. *Proc Natl Acad Sci USA*. 2003; 100:15247−15252.
2. Fukusumi S, Yoshida H, Fujii R, et al. A new peptidic ligand and its receptor regulating adrenal function in rats. *J Biol Chem*. 2003;278:46387−46395.
3. Jiang Y, Luo L, Gustafson EL, et al. Identification and characterization of a novel RF-amide peptide ligand for orphan G-protein-coupled receptor SP9155. *J Biol Chem*. 2003;278:27652−27657.
4. Chartrel N, Alonzeau J, Alexandre D, et al. The RFamide neuropeptide 26RFa and its role in the control of neuroendocrine functions. *Front Neuroendocrinol*. 2011;32:387−397 (Review).
5. Ukena K, Vaudry H, Leprince J, et al. Molecular evolution and functional characterization of the orexigenic peptide 26RFa and its receptor in vertebrates. *Cell Tissue Res*. 2011;343:475−481 (Review).
6. Bruzzone F, Lectez B, Tollemer H, et al. Anatomical distribution and biochemical characterization of the novel RFamide peptide 26RFa in the human hypothalamus and spinal cord. *J Neurochem*. 2006;99:616−627.
7. Takayasu S, Sakurai T, Iwasaki S, et al. A neuropeptide ligand of the G protein-coupled receptor GPR103 regulates feeding, behavioral arousal, and blood pressure in mice. *Proc Natl Acad Sci USA*. 2006;103:7438−7443.
8. Ukena K, Tachibana T, Iwakoshi-Ukena E, et al. Identification, localization, and function of a novel avian hypothalamic neuropeptide, 26RFa, and its cognate receptor, G protein-coupled receptor-103. *Endocrinology*. 2010;151:2255−2264.
9. Baribault H, Danao J, Gupte J, et al. The G-protein-coupled receptor GPR103 regulates bone formation. *Mol Cell Biol*. 2006;26:709−717.
10. Moriya R, Sano H, Umeda T, et al. RFamide peptide QRFP43 causes obesity with hyperphagia. *Endocrinology*. 2006;147:2916−2922.

Supplemental Information Available on Companion Website

- Precursor sequences of QRFP of various animals/E-Figure 1D.1
- Gene, mRNA and precursor structures of human QRFPR/E-Figure 1D.2
- Primary structure of QRFPR of various animals/E-Figure 1D.3

Prolactin-Releasing Peptide

Tetsuya Tachibana and Tatsuya Sakamoto

Abbreviations: PrRP, PRP, PRLH, PrRP-2, C-RFa
Additional names: Prolactin-releasing peptide (PrRP, PRP, or PRLH); Carassius RF amide (PrRP-2 or C-RFa)

Prolactin-releasing peptide, originally identified as the definitive prolactin-releasing factor, appears to be related more to the regulation of hypothalamic peptide release, feeding and energy homeostasis, and stress response. The ancestral gene of prolactin-releasing peptide also led to another bioactive peptide named prolactin-releasing peptide-2 (originally found as Carassius RFamide).

Discovery

In 1998 PrRP was identified from mammalian hypothalamus, using the reverse-pharmacological technique [1]. In the same year the isolation of PrRP2 (C-RFa) from the brain of Japanese crucian carp was reported, using the traditional bioassay system, by monitoring the contraction of the intestine [2].

Structure

Structural Features

Human proPrRP consists of 87 aa residues. From this precursor, two types of PrRPs are made: one consists of 20 amino acid residues (PrRP-20, Table 1E.1) and the other of 31 (PrRP-31) [1]. PrRP-20 is a 20-aa truncated form of the C-terminus of PrRP-31. Both PrRPs have the amidated arginine–phenylalanine motif at the C-terminus. The arginine residues at positions 26 and 30 are critical for binding to the PrRP receptor. In PrRP2, the precursor consists of 115, 108, and 108 aa residues in common carp, *Xenopus tropicalis*, and chicken, respectively [2–5]. Mature PrRP2s appear to consist of 20 aa in teleosts (Table 1E.1), while two types of PrRP-2 are produced in the chicken: one consists of 20 amino acids (PrRP2-20) and the other of 31 amino acids (PrRP2-31) [5]. Chicken PrRP2-20 is a 20-aa truncated form of the C-terminus of PrRP2-31. *Xenopus laevis* is also suggested to possess both PrRP2-20 and PrRP2-31 [4]. As in PrRP, all PrRP2 have the amidated arginine–phenylalanine motif at the C-terminus.

Primary Structure

C-terminus amino acids of PrRP and PrRP2 are well conserved among animal species. Moderate homology is also observed between PrRP and PrRP2 in each animal (for example, 70% in chicken) (Table 1E.1) [5].

Properties

Human PrFP-31, Mr 3,664.1; teleost PrRP2-20, Mr 2,350. Soluble in water.

Synthesis and Release

Gene, mRNA, and Precursor

The human PrRP gene (*PRLH*), located on chromosome 2 (2q37.3), consists of two exons and has the SP-1, AP-2, and Oct-2A binding sequence and three TATA boxes in the promoter region (E-Figure 1E.1) [3]. Human PrRP mRNA has 264 bp (E-Figure 1E.2). The PrRP gene first appeared, at least in early bony fish, by duplication of the ancestral gene, which is the same ancestor gene of PrRP2 (E-Figure 1E.3). PrRP2 genes of zebrafish, *X. tropicalis*, and chicken are located on chromosome 24, scaffold 359, and chromosome 2, respectively [6]. By a synteny analysis, the human PrRP2 gene is suggested to be located at chromosome 3, but it has not been identified [6].

Distribution of mRNA

In the rat, PrRP mRNA is located in the nucleus tractus solitarius (NTS, A2) and ventrolateral reticular nucleus (A1) in the medulla oblongata and the dorsomedial hypothalamic nucleus (DMN) [3]. PrRP is co-expressed in noradrenergic neurons in the medulla oblongata [3]. PrRP mRNA is also found in peripheral organs, such as the trachea, submandibular gland, adrenal gland, thyroid gland, pancreas, gut, and reproductive organs [3]. As well as PrRP in mammals, PrRP2 mRNA is expressed in the brain and peripheral organs in teleosts and chicken [3,6].

Tissue and Plasma Concentration

Tissue content (fmol/g tissue), PrRP: rat hypothalamus, 1800; midbrain, 720; posterior pituitary, 530; medulla oblongata, 330; adrenal gland, 62 for PrRP ($n = 6$–8) [7]. Tissue content of PrRP2 has not been investigated quantitatively.

Plasma concentration (fmol/ml), PrRP: rat (male), 0.13; (female) 0.14 [7]. Plasma concentration of PrRP2 has not been investigated in any animals.

Regulation of Synthesis and Release

PrRP mRNA expression is directly regulated by gonadal steroid hormones in the rat [3]. The mRNA level in the medulla oblongata is the highest in proestrus of the female rat. Expression in the NTS increases with the progression of pregnancy, and reaches its peak at mid-pregnancy. Food deprivation reduces PrRP mRNA expression. In addition,

Y. Takei, H. Ando, & K. Tsutsui (Eds): Handbook of Hormones. DOI: http://dx.doi.org/10.1016/B978-0-12-801028-0.00108-2

Table 1E.1 Primary Structure of PrRP-20 and PrRP2*

PrRP-20	Human	TPDINPAWYASRGIRPVGRF-NH₂
	Chicken	NPDIDPSWYTGRGIRPVGRF-NH₂
	Frog	NPDIDPSWYTGRGIRPVGRF-NH₂
	Teleost	DPNIDAMWYKDRGIRPVGRF-NH₂
PrRP2	Chicken	SPEIDPFWYVGRGVRPIGRF-NH₂
	Frog	SPEIDPYWYVGRGVRPIGRF-NH₂
	Teleost	SPEIDPFWYVGRGVRPIGRF-NH₂

*Shaded characters indicate conserved aa residues.

cholecystokin and leptin are thought to regulate the synthesis and/or release of PrRP with respect to energy maintenance [8]. Stress also affects PrRP expression in the brain [3,8]. PrRP2 mRNA expression is changed during acclimation to different environments in relation to the action of prolactin in teleosts: for example, PrRP2 mRNA expression in the brain is higher in fresh water than sea water in euryhaline teleosts [3]. In goldfish brain, PrRP2 mRNA expression is also associated with energy homeostasis because it is regulated by feeding and fasting [9]. In *X. laevis*, brain PrRP2 mRNA is increased during prometamorphosis [4].

Receptors

Structure and Subtype

The receptor of PrRP (PrRP-R, or human GPR10 and rat UHR-1) consists of about 350 amino acids (E-Figures 1E.4, 1E.5) and has lower but significant homology with the neuropeptide Y receptor [3]. In chicken and *X. tropicalis*, two subtypes of PrRP-R (PrRP-R1 and PrRP-R2) have been identified, and PrRP-R1 is thought to be a homolog of mammalian PrRP-R (E-Figure 1E.5) [6]. In chicken, PrRP-R1 and PrRP-R2 can function as common receptors for both PrRPs (PrRP and PrRP2), whereas PrRP2-R is a receptor specific to PrRP2; these three types of receptors are expected to belong to a subclass of the rhodopsin/β-adrenergic subfamily of the membrane-bound G protein-coupled receptor superfamily. PrRP2-Rs have been identified in zebrafish, pufferfish, *X. tropicalis*, and chicken, but the PrRP2-R gene has not yet been identified in mammals. PrRP2-R consists of 371, 360, and 361 amino acids in zebrafish, *X. tropicalis*, and chicken, respectively [6].

Signal Transduction Pathway

PrRP-R is coupled to G_i/G_o protein and appears to signal via multiple kinase pathways including mitogen-activated protein kinase, Jun N-terminal kinase, and serine/threonine kinase to the PRL promoter [3]. PrRP2-R also appears to be coupled to G protein and to signal via the protein kinase-A signaling pathway in chicken. The activation of PrRP2-R is also expected to trigger Ca^{2+} release from intracellular stores [6].

Agonists

Purified and recombinant human/rat PrRP, and purified and synthesized carp/tilapia/frog/chicken PrRP2, have been identified as agonists.

Antagonists

Specific antagonists of PrRP-Rs and PrRP2-R have not yet been identified.

Biological Functions

Target Cells/Tissues and Functions

PrRP was originally proposed to stimulate prolactin release from the anterior pituitary [1], but PrRP neural fibers are not found in the external layer of the median eminence where hypophysiotropic hormones are released in tetrapods [3]. In addition, the prolactin-releasing effect of PrRP is inconsistent (E-Table 1E.3), indicating that PrRP may affect prolactin release by mechanism(s) different from ordinary hypophysiotropic hormones [3]. PrRP participates in inhibition of feeding, regulation of the stress response, and stimulation of somatostatin, gonadotropin-releasing hormone (GnRH), vasopressin, and oxytocin secretions in the brain [3,8]. The physiological role of PrRP in non-mammalian vertebrates is unclear, while the role of PrRP2 has been relatively well investigated. In teleosts, PrRP2-immunoreactive neural fibers and endings are found in the pituitary rostral pars distalis, where prolactin-producing cells are localized, and PrRP2 stimulates prolactin secretion [2] and increases prolactin mRNA expression in the pituitary (E-Table 1E.3) [3]. In addition, the prolactin mRNA level in the pituitary is suppressed by PrRP2 antiserum treatment, demonstrating that PrRP2 is essential to maintain prolactin secretion in teleosts [3,10]. On the other hand, PrRP2 has less effect on prolactin release in *X. laevis* [4] and chicken [5]. PrRP2 has also been linked to the regulation of feeding, but the effects appear to differ among animal species: peripheral and central injection of PrRP2 inhibits feeding behavior in goldfish [9], while central injection of PrRP2-31 stimulates it in chicken (E-Table 1E.3) [5].

Phenotype in Gene-Modified Animals

PrRP-deficient mice show late-onset obesity, adiposity, glucose intolerance, and hyperphagia, and a low response to the anorexigenic effect of cholecystokinin and leptin. PrRP-R-deficient mice also show hyperphagia, obesity, and an increase in body fat. PrRP2-deficiency has not been reported in any animals.

Pathophysiological Implications

Clinical Implications

Disruption of the PrRP system causes an increase in body fat, obesity, and hyperphagia. There have been no clinical implications regarding PrRP2 in mammals or in non-mammals.

Use for Diagnosis and Treatment

PrRP and PrRP2 have not yet been used for diagnosis and treatment.

References

1. Hinuma S, Habata Y, Fujii R, et al. A prolactin-releasing peptide in the brain. *Nature*. 1998;393:272–276.
2. Fujimoto M, Takeshita K, Wang X, et al. Isolation and characterization of a novel bioactive peptide, Carassius RFamide (C-RFa), from the brain of the Japanese crucian carp. *Biochem Biophys Res Commun*. 1998;242:436–440.
3. Sakamoto T, Fujimoto M, Ando M. Fishy tales of prolactin-releasing peptide. *Int Rev Cytol*. 2003;225:91–130 (Review).
4. Sakamoto T, Oda A, Yamamoto K, et al. Molecular cloning and functional characterization of a prolactin-releasing peptide homolog from *Xenopus laevis*. *Peptides*. 2006;27:3347–3351.
5. Tachibana T, Moriyama S, Takahashi A, et al. Isolation and characterisation of prolactin-releasing peptide in chicks and its effect on prolactin release and feeding behaviour. *J Neuroendocrinol*. 2011;23:74–81.
6. Wang Y, Wang CY, Wu Y, et al. Identification of the receptors for prolactin-releasing peptide (PrRP) and Carassius RFamide peptide (C-RFa) in chickens. *Endocrinology*. 2012;153:1861–1874.
7. Matsumoto H, Murakami Y, Horikoshi Y, et al. Distribution and characterization of immunoreactive prolactin-releasing peptide (PrRP) in rat tissue and plasma. *Biochem Biophys Res Commun*. 1999;257:264–268.

8. Takayanagi Y, Onaka T. Roles of prolactin-releasing peptide and RFamide related peptides in the control of stress and food intake. *FEBS J.* 2010;277:4998–5005 (Review).

9. Kelly SP, Peter RE. Prolactin-releasing peptide, food intake, and hydromineral balance in goldfish. *Am J Physiol Regul Integr Comp Physiol.* 2006;291:R1474–R1481.

10. Fujimoto M, Sakamoto T, Kanetoh T, et al. Prolactin-releasing peptide is essential to maintain the prolactin level and osmotic balance in freshwater teleost fish. *Peptide.* 2006;27:1104–1109.

Supplemental Information Available on Companion Website

- Gene, mRNA and precursor structures of human PrRP/E-Figure 1E.1
- Precursor and mature PrRP and PrRP2 sequences of various animals/E-Figure 1E.2
- Phylogenetic relationship between proPrRP and proPrRP2 in vertebrates/E-Figure 1E.3
- Gene, mRNA, and precursor structures of human PrRP receptor-1/E-Figure 1E.4
- Primary structure of PrRP receptor-1, PrRP receptor-2, and PrRP2 receptor of various animals/E-Figure 1E.5
- Accession numbers of genes and cDNAs for PrRP and PrRP2 precursors and PrRP and PrRP2 receptors/E-Tables 1E.1, 1E.2
- Comparison of the effect of PrRP and PrRP2 on prolactin release and feeding behavior of various animals/E-Table 1E.3

CHAPTER 2

Corticotropin-Releasing Hormone Family

Masafumi Amano

History

A substance that stimulates "pituitary—adrenal" hormone secretion in the extract of mammalian hypothalamus was first reported in 1955, and it was named corticotropin-releasing factor (CRF) [1,2]. In 1981, a 41-aa residue peptide was identified from the ovine hypothalamus by Vale's group, and this was proved to be CRF [3,4]. CRF was subsequently identified not only in mammals such as human, mouse, rat, and pig, but also in amphibians (*Xenopus*) and in teleost fishes such as the white sucker, tilapia, and carp. CRF is now called corticotropin-releasing hormone (CRH). Subsequently, peptides structurally related to CRH have also been discovered from various vertebrate species. In amphibians, sauvagine (SVG) was isolated from the skin of the South American frog *Phyllomedusa sauvagei* in 1980 [5]. In teleosts, urotensin-I (UI), which decreases blood pressure in the rat, was reported from the urophysis of the white sucker in 1982 [6]. In 1995, another mammalian member of CRH family was discovered from rat midbrain [7]. The novel peptide, named urocortin-I (Ucn-I), showed higher amino acid similarity to UI, and higher potency than CRH at binding and activating type-2 CRH receptor (CRHR2) [7]. Subsequently, using the genome information, Ucn-II (stresscopin-related peptide) and Ucn-III (stresscopin) were identified in humans, and then in the mouse and rat, as novel CRH-related peptides that showed high affinity to the CRHR2 [8,9].

Structure

Structural Features

CRH and UI peptides isolated thus far consist of 41 aa residues with an amidated C-terminus. On the other hand, human Ucn-I, Ucn-II, and Ucn-III consist of 40, 38, and 38 aa residues, respectively, and all CRH-related peptides were similarly amidated at the C-terminus. Human CRH gene, flounder UI gene, and human Ucn-I gene consist of two exons and one intron, and the precursor-coding region exists in the second exon. Mature sequences of representative peptides for CRH, UI, Ucn-I, Ucn-II, Ucn-III, and SVG are shown in Figure 2.1.

Molecular Evolution of Family Members

Molecular phylogenetic analysis shows that the CRH family is divided into two groups: the CRH/UI/Ucn-I/SVG group and

human CRH	SEEPPISLDL	TFHLLREVLE	MARAEQLAQQ	AHSNRKLMEI	Ia	(MW 4,758)
carp U-I	NDDPPISIDL	TFHLLRNMIE	MARIENEREQ	AGLNRKYLDE	Va	(MW 4,869)
human Ucn-I	DNPSLSIDL	TFHLLRTLLE	LARTQSQRER	AEQNRIIFDS	Va	(MW 4,696)
human Ucn-II	IVLSLDV	PIGLLQILLE	QARARAAREQ	ATTNARILAR	Va	(MW 4,154)
human Ucn-III	FTLSLDV	PTNIMNLLFN	IAKAKNLRAQ	AAANAHLMAQ	Ia	(MW 4,138)
PS-sauvagine	EGPPISIDL	SLELLRKMIE	IEKQEKEKQQ	AHSNRKLMEI	Ia	(MW 4,716)

Figure 2.1 Comparison of the amino acid sequence of human CRH, carp UI, human Ucn-I, -II, -III, and PS-SVG. Amino acid residues identical to human CRH are shaded.

the Ucn-II/Ucn-III/Ucn related peptide (URP) group. The presence of a CRH-like peptide in tunicates and insects suggests that the two lineages of the CRH family are the result of a gene duplication from a single ancestral peptide gene during the early evolution of vertebrates. The phylogenetic tree of the CRH family is shown in Figure 2.2.

Receptors

Structure and Subtypes

Three types of CRH receptors were identified, all of which belong to the family of seven-transmembrane-domain GPCRs that activate adenylate cyclase (AC). The type-1 and -2 CRH receptors (CRHR1 and CRHR2) were identified in mammals, birds, amphibians, and teleosts. In mammals, CRHR2 has three splice variants: CRHR2α, CRHR2β, and CRHR2γ. The type-3 CRH receptor (CRHR3) was found in catfish. The affinities to receptors differ among ligands. CRHR1 has high affinity with CRH in all species examined. On the other hand, CRHR2 has high affinity with UI and SVG compared to CRH. Ucn-I binds to both CRHR1 and CRHR2, whereas Ucn-II and Ucn-III bind exclusively to CRHR2. In addition to the receptors, CRH binding protein (CRH-BP) has been reported in mammals and teleosts (Mozambique tilapia). CRH-BP (37 kDa) binds to CRH with high affinity and reduces CRH effects. CRHR1 is distributed mainly in the anterior pituitary and brain and is involved in adrenocorticotropin (ACTH) release from the pituitary. CRHR2 is distributed in the wider brain regions in humans.

Signal Transduction Pathway

After CRH binding, AC is activated. AC stimulates cAMP, and cAMP in turn activates PKA, which phosphorylates downstream of cytosolic and nuclear targets, such as CREB,

Y. Takei, H. Ando, & K. Tsutsui (Eds): Handbook of Hormones. DOI: http://dx.doi.org/10.1016/B978-0-12-801028-0.00002-7

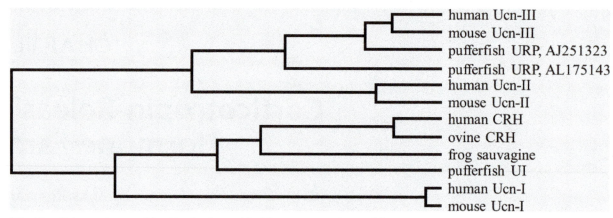

Figure 2.2 The phylogenetic tree of the CRH family, modified from Lewis *et al.*, 2001 [9].

resulting in activation of gene transcription. Ucn is considered to act mainly through both $G_{\alpha s}$ and $G_{\alpha q}$ subunits to stimulate diverse signaling cascades such as AC and PLC pathways, depending on the tissue type.

Biological Functions

Target Cells/Tissues and Functions

CRH stimulates ACTH secretion from the pituitary, resulting in stimulation of cortisol secretion from the adrenal cortex [10]. UI, Ucn-I, and SVG also stimulate ACTH secretion. CRH, UI, and Ucn-I suppress food intake. Ucn-II and Ucn-III may be responsible for the effects of stress on appetite and anxiety in the brain.

References

1. Saffran M, Schally AV. The release of corticotrophin by anterior pituitary tissue *in vitro. Can J Biochem Physiol.* 1955;33:408−415.
2. Guillemin R, Rosemberg B. Humoral hypothalamic control of anterior pituitary: a study with combined tissue cultures. *Endocrinology.* 1955;57:599−607.
3. Vale W, Spiess J, Rivier C, et al. Characterization of a 41-residue ovine hypothalamic peptide that stimulates secretion of corticotropin and beta-endorphin. *Science.* 1981;213:1394−1397.
4. Spiess J, Rivier J, Rivier C, et al. Primary structure of corticotropin-releasing factor from ovine hypothalamus. *Proc Natl Acad Sci USA.* 1981;78:6517−6521.
5. Montecucchi PC, Anastasi A, De Castiglione R, et al. Isolation and amino acid composition of sauvagine. An active polypeptide from methanol extracts of the skin of the South American frog *Phyllomedusa sauvagei. Int J Pept Protein Res.* 1980;16:191−199.
6. Lederis K, Letter A, McMaster D, et al. Complete amino acid sequence of urotensin I, a hypotensive and corticotropin-releasing neuropeptide from *Catostomus commersoni. Science.* 1982;218:162−164.
7. Vaughan J, Donaldson C, Bittencourt J, et al. Urocortin, a mammalian neuropeptide related to fish urotensin I and to corticotropin-releasing factor. *Nature.* 1995;378:287−292.
8. Reyes TM, Lewis K, Perrin MH, et al. Urocortin II: a member of the corticotropin-releasing factor (CRF) neuropeptide family that is selectively bound by type 2 CRF receptors. *Proc Natl Acad Sci USA.* 2001;98:2843−2848.
9. Lewis K, Li C, Perrin MH, et al. Identification of urocortin III, an additional member of the corticotropin-releasing factor (CRF) family with high affinity for the CRF2 receptor. *Proc Natl Acad Sci USA.* 2001;98:7570−7575.
10. Pankhurst NW. The endocrinology of stress in fish: an environmental perspective. *Gen Comp Endocrinol.* 2011;170:265−275.

Supplemental Information Available on Companion Website

- The phylogenetic tree of three types of CRH receptors/ E-Figure 2.1
- Accession numbers of genes and cDNAs for preproCRHR/ E-Table 2.1

Corticotropin-Releasing Hormone

Masafumi Amano

Abbreviation: CRH
Additional names: corticotropin-releasing factor (CRF), corticoliberin

CRH is a key activator of the hypothalamo-pituitary—adrenal (HPA) axis, in response to internal or external stresses, by stimulating adrenocorticotropic hormone (ACTH) secretion from the pituitary.

Discovery

The first evidence of a hypothalamic corticotropin-releasing substance was reported in 1955 [1,2]. The presence of a corticotropin-releasing hormone (CRH) was first reported in the ovine hypothalamus in 1981 [3,4]. Subsequently, CRH was found in the human, mouse, rat, pig, amphibian (*Xenopus*), and teleost fishes such as the white sucker, tilapia, and carp. Urotensin-I was formerly considered to be a CRH ortholog in teleosts; however, CRH has also been identified in teleosts.

Structure

Structural Features

CRH peptides of all species examined consist of 41 aa residues with an amidated C-terminus. The C-terminal region is important for physiological activity. Human, mouse, and rat CRH have identical aa sequences. CRH sequence homologies between human/mouse/rat and others are high (76—95%) (Figure 2A.1). In the elephant shark *Callorhinchus milii*, a holocephalan species, two CRH peptides were found in the genome [5].

Primary Structure and Properties

The amino acid sequence and Mr values of CRH peptides are shown in Figure 2A.1. CRH peptides are freely soluble in water. The oxidation of methionine at position 21 remarkably decreases the activity of CRH.

Synthesis and Release

Gene, mRNA, and Precursor

The human CRH gene, *CRH*, location 8q13, consists of two exons and one intron. The coding region of preproCRH exists in the second exon (Figure 2A.2). In the mouse, the CRH gene is located on chromosome 3. The basic structure of the mammalian and amphibian CRH gene is identical [6]. The CRH gene of the catfish *Ameiurus nebulosus* consists of four exons and three introns [7]. In humans, the CRH precursor consists of 196 aa residues, including a signal peptide. The CRH peptides are located in a C-terminal region of the precursor. A processing motif, Arg—X—Arg—Arg, exists at the N-terminus of CRH, while a dipeptide Gly—Lys is found at the C-terminus of precursor. After processing by an endopeptidase, the C-terminal Lys is cleaved off by an exopeptidase, and the C-terminus of CRH is amidated using glycine as an amidation donor.

Species	CRH peptide				Mr
human/mouse/rat	SEEPPISLDL	TFHLLREVLE	MARAEQLAQQ	AHSNRKLMEI Ia	4,758
pig	SEEPPISLDL	TFHLLREVLE	MARAEQLAQQ	AHSNRKLMEN Fa	4,792
ovine	SQEPPISLDL	TFHLLREVLE	MTKADQLAQQ	AHSNRKLLDI Aa	4,670
bovine	SQEPPISLDL	TFHLLREVLE	MTKADQLAQQ	AHNNRKLLDI Aa	4,697
Xenopus	AEEPPISLDL	TFHLLREVLE	MARAEQIAQQ	AHSNRKLMDI Ia	4,728
sucker I	SEEPPISLDL	TFHLLREVLE	MARAEQLAQQ	AHSNRKMMEI Fa	4,810
sucker II	SEEPPISLDL	TFHLLREVLE	MARAEQLVQQ	AHSNRKMMEI Fa	4,838
tilapia	SEDPPISLDL	TFHLLREMME	MSRAEQLAQQ	AQNNRRMMEL Fa	4,908
carp	SEEAPISLDL	TFHLLREVLE	MARAEQMAQQ	AHSNRKMMEI Fa	4,802
goldfish	SEEPPISLDL	TFHLLREVLE	MARAEQMAQQ	AHSNRKMMEI Fa	4,828
sockeye salmon	SDDPPISLDL	TFHMLRQMME	MSRAEQLQQQ	AHSNRKMMEI Fa	4,922
eel	SEEPPISLDL	TFHLLREVLE	MARAEQLAQQ	AHSNGKMMEI Fa	4,710

Figure 2A.1 Comparison of the amino acid sequence of CRH peptides. Amino acid residues identical to human/mouse/rat CRH peptide are shaded.

Y. Takei, H. Ando, & K. Tsutsui (Eds): Handbook of Hormones. DOI: http://dx.doi.org/10.1016/B978-0-12-801028-0.00109-4

Distribution of mRNA

CRH mRNA is distributed widely in the brain, especially in the paraventricular nucleus. CRH neurons in the supraoptic nucleus co-express vasopressin. In addition to the brain, CRH mRNA is widely detected in peripheral organs such as the retina, stomach, intestine, spleen, lung, and gonads.

Tissue and Plasma Concentrations

Concentrations of CRH in the brain are shown in Table 2A.1 [8–11].

Regulation of Synthesis and Release

The synthesis and release of CRH are regulated by internal or external stress that is conveyed to the brain and is integrated at the hypothalamus, where CRH neurons exist. Synthesis and release of CRH are also regulated by diurnal rhythm and the negative feedback effect of cortisol. In nocturnal rats, CRH mRNA levels are higher at midday than at midnight. CRH is released from the axon terminal depending on Ca^{2+}.

Receptors

Structure and Subtype

Three types of CRH receptors have been identified. As mentioned in Chapter 2, they belong to the family of seven-transmembrane-domain GPCRs and activate AC. Type-1 (CRHR1) and type-2 (CRHR2) receptors were identified in mammals, chicken, *Xenopus*, and teleosts. Type-3 receptor (CRHR3) was found in catfish. In mammals, the CRHR1 gene consists of 13 exons and 12 introns. The CRHR2 gene in humans consists of 12 exons and 11 introns, and 3 splice variants (CRHR2α, 2β, and 2γ) have been found; 2–12 exons are common to the 3 splice variants, and the first exon is specific for each variant. CRH binds to CRHR1 with high affinity. CRHR1 is mainly distributed in the anterior pituitary, and is involved in ACTH release. CRH also has high affinity to CRHR3. However, CRHR2 exhibits low affinity to CRH but high affinity to urotensin-I, sauvagine, and urocortins. CRHR2 is distributed in several restricted regions of the brain, but also in peripheral organs such as heart, lung, muscle, and kidney. In addition, CRH binding protein (CRH-BP) has been reported in mammals and a teleost (the Mozambique tilapia). CRH-BP (37 kDa) binds to CRH with high affinity and regulates CRH effects.

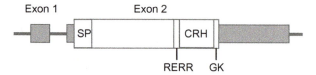

Figure 2A.2 Structure of human CRH gene and precursor.

Signal Transduction Pathway

CRH activates AC via G_s protein-coupled with CRHRs. AC stimulates cAMP production, and cAMP, in turn, activates PKA which phosphorylates downstream of cytosolic and nuclear targets, such as CREB, resulting in induced gene transcription [12]. Moreover, it is known that, in addition to signaling pathways activated by cAMP, CRHR1 and CRHR2 trigger multiple signaling pathways such as activation of p42/44 MAPK (ERK1/2) [13].

Agonist

Alpha-helical CRH is an agonist.

Antagonists

Alpha-helical CRH(9–41), [D-Phe]CRH-(12–41) and astressin are antagonists; of these, astressin is most potent in inhibiting ACTH release.

Biological Functions

Target Cells/Tissues and Functions

CRH stimulates the synthesis and processing of proopiomelanocortin (POMC) to generate ACTH in the anterior pituitary. CRH also stimulates ACTH release. ACTH is secreted into the systemic circulation, reaches the adrenal cortex, and stimulates the synthesis and release of glucocorticoids [14]. CRH also acts as a neuromodulator in the brain. CRH in the brain regulates the endocrine system, autonomic nervous system, and immune system, in response to stress. In addition to the stress response, CRH is involved in multiple physiological functions such as suppression of food intake, regulation of body temperature, growth, metabolism, metamorphosis, reproduction, and diuresis. In mammals, CRH inhibits the female reproductive axis by suppressing gonadotropin-releasing hormone (GnRH) secretion.

Phenotype in Gene-Modified Animals

CRH gene knockout mice have lower plasma corticosterone levels compared with normal littermates, and do not show increases in plasma ACTH and glucocorticoids after stress. CRH transgenic mice show elevated plasma ACTH and corticosterone levels accompanied by Cushing-like symptoms.

Pathophysiological Implications

Clinical Implications

Increased CRH production has been suggested to be associated with Alzheimer's disease and depression. CRH deficiency has been observed to induce hypoglycemia and hepatitis. CRH has been used in the diagnosis of ACTH-dependent Cushing's syndrome and Addison's disease.

Table 2A.1 Tissue Concentrations of CRH in Rat, Sheep, and Frog

Animal	Tissues				Reference(s)
	Median eminence	Hypophysial portal blood	Paraventricular nucleus	Intermediate lobe	
Rat	2.54 ± 0.41 ng/mg protein ($n = 6$)	100–1,000 fmol/ml	—	—	[8,9]
Sheep	70.3 ± 14.7 ng/mg protein ($n = 4$)	—	1.44 ± 0.42 ng/mg protein ($n = 5$)	—	[10]
Frog	1.7 ± 0.14 ng/mg protein ($n = 11$)*	—	—	12.5 ± 2.94 ng/mg protein ($n = 6$)	[11]

*Infundibulum and median eminence.

Use for Diagnosis and Treatment

CRHR1 antagonists, pexacerfont and antalarmin, are under clinical trial for the treatment of generalized anxiety disorder, and of depression and other mental disorders, respectively.

References

1. Saffran M, Schally AV. The release of corticotrophin by anterior pituitary tissue *in vitro. Can J Biochem Physiol.* 1955;33:408−415.
2. Guillemin R, Rosemberg B. Humoral hypothalamic control of anterior pituitary: a study with combined tissue cultures. *Endocrinology.* 1955;57:599−607.
3. Vale W, Spiess J, Rivier C, et al. Characterization of a 41-residue ovine hypothalamic peptide that stimulates secretion of corticotrophin and β-endorphin. *Science.* 1981;213:1394−1397.
4. Spiess J, Rivier J, Rivier C, et al. Primary structure of corticotropin-releasing factor from ovine hypothalamus. *Proc Natl Acad Sci USA.* 1981;78:6517−6521.
5. Nock TG, Chand D, Lovejoy DA. Identification of members of the gonadotropin-releasing hormone (GnRH), corticotropin-releasing factor (CRF) families in the genome of the holocephalan, *Callorhinchus milii* (elephant shark). *Gen Comp Endocrinol.* 2011;171:237−244.
6. Stenzel-Poore MP, Hedwein KA, Stenzel P, et al. Characterization of the genomic corticotropin-releasing factor (CRF) gene from *Xenopus laevis*: two members of the CRF family exist in amphibians. *Mol Endocrinol.* 1992;6:1716−1724.
7. Malagoli D, Mandrioli M, Ottaviani E. ProCRH in the teleost *Ameiurus nebulosus*: gene cloning and role in LPS-induced stress response. *Brain Behav Immunol.* 2004;18:451−457.
8. Skotfisch G, Jacobowitz DM. Distribution of corticotropin releasing factor-like immunoreactivity in the rat brain by immunohistochemistry and radioimmunoassay: comparison and characterization of ovine and rat/human CRF antisera. *Peptides.* 1985;6:319−336.
9. Crofford LJ, Sano H, Karalis K, et al. Corticotropin-releasing hormone in synovial fluids and tissues of patients with rheumatoid arthritis and osteoarthritis. *J. Immunol.* 1993;151:1587−1596.
10. Palkovits M, Brownstein MJ, Vale W. Corticotropin releasing factor (CRF) immunoreactivity in hypothalamic and extrahypothalamic nuclei of sheep brain. *Neuroendocrinology.* 1983;37:302−305.
11. Tonon MC, Burlet A, Lauber M, et al. Immunohistochemical localization and radioimmunoassay of corticotropin-releasing factor in the forebrain and hypophysis of the frog *Rana ridibunda*. *Neuroendocrinology.* 1985;40:109−119.
12. Bonfiglio JJ, Inda C, Refojo D, et al. The corticotropin-releasing hormone network and the hypothalamic−pituitary−adrenal axis: molecular and cellular mechanisms involved. *Neuroendocrinology.* 2011;94:12−20.
13. Pittman QJ, Hollenberg MD. Urotensin I−CRF−Urocortins: a mermaid's tail. *Gen Comp Endocrinol.* 2009;164:7−14.
14. Pankhurst NW. The endocrinology of stress in fish: an environmental perspective. *Gen Comp Endocrinol.* 2011;170:265−275.

Supplemental Information Available on Companion Website

- Molecular phylogenetic tree of CRHs from various animals/E-Figure 2A.1
- Accession numbers of genes and cDNAs for preproCRH/E-Table 2A.1

Urotensin-I

Masafumi Amano

Abbreviation: UI

A peptide hormone first isolated from the urophysis of the white sucker, urotensin-I has been shown to decrease blood pressure in the rat.

Discovery

The presence of UI was first reported in 1982, obtained from the urophysis of the white sucker *Catostomus commersoni* [1]. Subsequently, UI has been found not only in teleosts but also in elasmobranchii (the dogfish shark *Scyliorhinus canicula*) [2].

Structure

Structural Features

UI consists of 41 aa residues with an amidated C-terminus. Amino acid residues 4−19 are necessary for adrenocorticotropic hormone (ACTH) secretion. The C-terminus is required for complete ACTH-secreting activity.

Primary Structure

The amino acid sequence and Mr of UI are shown in Figure 2B.1. Alignment of UI shows that UI forms two structurally separate groups, one including Pleuronectiformes (flounder), and the other including Cypriniformes (white sucker and goldfish) and Salmoniformes (salmonids). The sequence homology within the group is high (90−95%), while that between groups is low (59−71%). The homology of the N-terminus of flounder UI with CRH is high (Figure 2B.1).

Properties

Freely soluble in water. Oxidation of methionine remarkably reduces UI's blood-pressure decreasing activity.

Synthesis and Release

Gene, mRNA, and Precursor

In goldfish, the UI cDNA (769 bp) encodes a 146-aa precursor that consists of a signal peptide, a cryptic region, and a 41-aa mature peptide at the C-terminal [3]. The flounder UI gene consists of two exons and one intron, and the whole precursor-coding region exists in the second exon [4]. UI is produced from the prohormone by endopeptidase. Basic amino acid residues (Lys−Arg) exist in the N-terminal of UI, for processing. The cleavage site in the C-terminal region is located at Gly−Lys; the Lys residue is cleaved off by an exopeptidase and the Gly residue is then used as an amidation donor.

Distribution of mRNA

UI mRNA is mainly detected in the urophysis rather than in the brain in carp and rainbow trout, suggesting that UI plays its major role in peripheral tissues and has a relatively small role in the brain [5]. In the goldfish brain, UI mRNA is mainly expressed in the telencephalon/preoptic area, and also in the optic tectum/thalamus, hypothalamus, cerebellum, and medulla oblongata, but not in the olfactory bulb and pituitary [3]. In the white sucker brain, UI mRNA is detected in the preoptic region and the nucleus of the lateral tuberis [6].

Tissue and Plasma Concentrations

Tissue and plasma concentrations in white sucker and rainbow trout are shown in Table 2B.1 [7,8].

Regulation of Synthesis and Release

UI mRNA is elevated in the nucleus of the lateral tuberis by removal of the urophysis in white sucker [6].

Receptors

Structure and Subtype

Receptors homologous to tetrapod type-1 and -2 CRH receptors (CRHR1 and CRHR2) have been cloned in teleosts such as rainbow trout, chum salmon, and catfish. CRHR2 has a higher affinity with UI, compared to CRH.

Biological Functions

Target Cells/Tissues and Functions

UI was first isolated from the urophysis of the white sucker and has been shown to decrease blood pressure in the rat [1]. UI modulates cortisol secretion either directly by acting on steroidogenic cells of inter-renal tissue, or indirectly via the hypothalamo-pituitary−adrenal (HPA) axis [9]. Intraperitoneal injection of UI increases plasma cortisol levels in goldfish. UI stimulates ACTH release from superfused goldfish anterior pituitary cells [10]; it also stimulates thyroid-stimulating hormone secretion from the pituitary cells of coho salmon *in vitro* [11]. UI is involved in the adaptability of fishes regarding, for example, ion and fluid equilibrium, cardiovascular activity, and glucocorticoid release. In addition to the biological function on the HPA axis, central injection of carp/goldfish UI suppresses food intake in a dose-dependent manner in goldfish [12]. Administration of UI also inhibits food intake in neonatal chicks.

Y. Takei, H. Ando, & K. Tsutsui (Eds): Handbook of Hormones. DOI: http://dx.doi.org/10.1016/B978-0-12-801028-0.00110-0

Species	UI peptide					Mr
flounder	SEDPPMSIDL	TFHMLRNMIH	MAKMEGEREQ	AQINRNLLDE	Va	4,841
sole	SEEPPMSIDL	TFHMLRNMIH	RAKMEGEREQ	ALINRNLMDE	Va	4,883
maggy sole	SEEPPMSIDL	TFHMLRNMIH	RAKMEGEREQ	ALINRNLLDE	Va	4,865
carp/goldfish	NDDPPISIDL	TFHLLRNMIE	MARNENQREQ	AGLNRKYLDE	Va	4,869
sucker	NDDPPISIDL	TFHLLRNMIE	MARIENEREQ	AGLNRKYLDE	Va	4,870
rainbow trout	NDDPPISIDL	TFHLLRNMIE	MARIESQKEQ	AELNRKYLDE	Va	4,886
fugu AL218869	PPLSIDL	TFXLLRNMMQ	RAEMEKLREQ	EKINREILEQ	Va	
shark	PAETPNSLDL	TFHLLREMIE	IAKHENQQMQ	ADSNRRIMDT	Ia	4,807
human CRH	SEEPPISLDL	TFHLLREVLE	MARAEQLAQQ	AHSNRKLMEI	Ia	4,758

Figure 2B.1 Comparison of the amino acid sequence of UI. Amino acid residues identical to flounder UI are shaded.

Table 2B.1 Tissue Concentrations of UI in White Sucker and Rainbow Trout

Animal	Tissues	Concentration	Reference
White sucker	Hypothalamus	0.9 ± 0.1 ng/mg dry weight ($n = 7$)	[7]
	Telencephalon	1.3 ± 0.2 ng/mg dry weight ($n = 7$)	[7]
	Pituitary (RPD)	$0.4-0.9$ ng/mg dry weight	[7]
	Pituitary (PPD + NIL)	$0.6-1.3$ ng/mg dry weight	[7]
	Urophysis	$1,655 \pm 227$ ng/mg dry weight ($n = 10$)	[8]
	Plasma	15.3 ± 1.9 fmol/ml ($n = 10$)	[8]
Rainbow trout	Urophysis	250 ng/mg dry weight*	[8]

*Acetone-dried.
RPD, rostral pars distalis; PPD, proximal pars distalis; NIL, neurointermediate lobe.

References

1. Lederis K, Letter A, McMaster D, et al. Complete amino acid sequence of urotensin I, a hypotensive and corticotropin-releasing neuropeptide from *Catostomus commersoni. Science.* 1982;218:162−164.
2. Waugh D, Anderson G, Armour KJ, et al. A peptide from the caudal neurosecretory system of the dogfish *Scyliorhinus canicula* that is structurally related to urotensin I. *Gen Comp Endocrinol.* 1995;99:333−339.
3. Bernier NJ, Lin X, Peter RE. Differential expression of corticotropin-releasing factor (CRF) and urotensin I precursor genes, and evidence of CRF gene expression regulated by cortisol in goldfish brain. *Gen Comp Endocrinol.* 1999;116:461−477.
4. Lu W, Dow L, Gumusgoz S, et al. Coexpression of corticotropin-releasing hormone and urotensin I precursor genes in the caudal neurosecretory system of the euryhaline flounder (*Platichthys flesus*): a possible shared role in peripheral regulation. *Endocrinology.* 2004;145:5786−5797.
5. Lovejoy DA. Structural evolution of urotensin-I: reflections of life before corticotropin releasing factor. *Gen Comp Endocrinol.* 2009;164:15−19.
6. Morley SD, Schonrock C, Richter D. Corticotropin-releasing factor (CRF) gene family in the brain of the teleost fish *Catostomus commersoni* (white sucker): molecular analysis predicts distinct precursors for two CRFs and one urotensin I peptide. *Mol Mar Biol Biotechnol.* 1991;1:48−57.
7. McMaster D, Lederis K. Urotensin I- and CRF-like peptides in *Catostomus commersoni* brain and pituitary-HPLC and RIA characterization. *Peptides.* 1988;9:1043−1048.
8. Suess U, Lawrence J, Ko D, et al. Radioimmunoassays for fish tail neuropeptides: I. Development of assay and measurement of immunoreactive urotensin I in *Catostomus commersoni* brain, pituitary, and plasma. *Pharmacol Methods.* 1986;15:335−346.
9. Pittman QJ, Hollenberg MD. Urotensin I−CRF−Urocortins: A mermaid's tail. *Gen Comp Endocrinol.* 2009;164:7−14.
10. Fryer J, Lederis K, Rivier J. Urotensin I, a CRF-like neuropeptide, stimulates ACTH release from the teleost pituitary. *Endocrinology.* 1983;113:2308−2310.
11. Larsen DA, Swanson P, Dickey JT, et al. *In vitro* thyrotropin-releasing activity of corticotropin-releasing hormone-family peptides in coho salmon, *Oncorhynchus kisutch. Gen Comp Endocrinol.* 1998;109:276−285.
12. Bernier NJ. The corticotropin-releasing factor system as a mediator of the appetite-suppressing effects of stress in fish. *Gen Comp Endocrinol.* 2006;146:45−55.

Supplemental Information Available on Companion Website

- Accession numbers of genes and cDNAs for preproUI/E-Table 2B.1

Urocortins

Masafumi Amano

Abbreviation: Ucn
Additional names: corticotensin, stresscopin, stresscopin-related peptide

The urocortins are three peptide hormones (Ucn-I, -II, and -III) that have higher affinity to type-2 corticotropin-releasing hormone (CRH) receptor (CRHR2) than does CRH itself.

Discovery

Ucn-I was first reported in 1995 from rat midbrain as a peptide having higher potency in binding and activating CRHR2 than does CRH [1]. Subsequently, using the genome information, two novel CRH-related peptides that showed high affinity to CRHR2, Ucn-II (stresscopin-related peptide) and Ucn-III (stresscopin), have been identified in mouse [2] and human/mouse [3], respectively. Ucn-related peptide (URP), which is considered to be homologous to Ucn-III, has been identified in the fugu genome [3].

Structure

Structural Features and Primary Structure

Amino acid sequences of UCNs are shown in Figure 2C.1. Mammalian (human, ovine, rat, mouse, hamster) Ucn-I consists of 40 aa residues with an amidated C-terminus. Ucn-II and Ucn-III consist of 38 aa residues (Figure 2C.1). The sequence identities of rat Ucn-I peptide with those of rat/mouse/human CRH, carp urotensin-I (UI), and sauvagine (SVG) are 45%, 63%, and 35%, respectively. Tiger puffer URP is closely related to Ucn-III (76% identity), but showed lower identity with mouse Ucn-II (37%). Tiger puffer URP-II is also more closely related to Ucn-III (53%) than to any of the other mammalian CRH family members (Figure 2C.1 [3]).

Properties

The Mr values of urocortins are shown in Figure 2C.1. Human Ucn-I is soluble in water. Human Ucn-II and Ucn-III are insoluble in water and soluble in DMSO.

Synthesis and Release

Gene, mRNA, and Precursor

Ucns are synthesized as large precursors, which are cleaved to yield the active mature peptides [4]. The human Ucn-I gene, *UCNI*, location 2p23−p21, consists of two exons and one intron (Figure 2C.2). PreproUcns are encoded in the second exon. *UCNII* is located on human chromosome 3p21.3 and on mouse chromosome 9, while *UCNIII* is located on human chromosome 10p15.1 and on mouse chromosome 5 [4]. Mature Ucn peptides are coded by the C-terminal of the precursor, and basic amino acid residues exist in the N-terminal of mature peptide for processing. In the Ucn-I precursor, consecutive basic amino acids (Arg−X−Arg−Arg) exist, and Ucn-I is processed from prohormone by endopeptidase. A dipeptide Gly−Lys is present in the C-terminal region of the mature peptide. The C-terminal lysine residue is cleaved off by an exopeptidase, while the glycine residue is used as an amidation donor. In the precursor of Ucn-II and Ucn-III, mature peptides are processed by arginine or lysine located in the N-terminal, and amidation of mature peptide occurs after cleavage of C-terminal arginine or lysine with an exopeptidase.

Distribution of mRNA

In the rat brain, Ucn-I mRNA is intensely detected in the olivary nucleus and Edinger-Westphal nucleus (EWN) [5]. Ucn-I mRNA is also detected in the anterior and intermediate part of the pituitary. In the human brain, Ucn-I mRNA is found not only in the hypothalamus and pons but also in the cerebral cortex and cerebellum, and the distribution is different from that of CRH [6]. The gastrointestinal tract and its associated macrophages constitute a major source of Ucn-I in the periphery [5]. Ucn-I mRNA is also abundantly detected in the heart of humans and rodents. Ucn-II mRNA is detected in the hypothalamic area, such as the paraventricular hypothalamic nucleus, supraoptic nucleus, arcuate nucleus, and locus nucleus in the rat [2]. Ucn-II mRNA is detected in several organs, such as the brain, heart, and adrenal gland in humans. Ucn-III mRNA is detected in the central nervous system and in almost all peripheral organs, such as the colon, intestine, muscle, stomach, thyroid, adrenal gland, pancreas, and heart.

Plasma Concentrations

Ucn-I: Healthy men, 9.6 fmol/ml; normal sheep, 15.1 fmol/ml; sheep with heart failure, 19.1 fmol/ml [7].

Regulation of Synthesis and Release

Ucn-I mRNA in the rat thymus is upregulated under stress conditions. Ucn-I mRNA in the mouse Edinger-Westphal nucleus is upregulated by stress and CRH deficiency [8].

Y. Takei, H. Ando, & K. Tsutsui (Eds): Handbook of Hormones. DOI: http://dx.doi.org/10.1016/B978-0-12-801028-0.00111-2

Species	Ucn peptide				Mr
rat/ovine/mouse Ucn-I	DDPPLSIDLT	FHLLRDLLEL	ARTQSQRERA	EQNRIIFDSVa	4,707
human Ucn-I	DNPSLSIDLT	FHLLRTLLEL	ARTQSQRERA	EQNRIIFDSVa	4,696
hamster Ucn-I	EDLPLSIDLT	FHLLRTLLEL	ARTQSQRERA	EQNRIILNAVa	4,686
human Ucn-II	IVLSLDVP	IGLLQILLEQ	ARARAAREQA	TTNARILARVa	4,154
mouse Ucn-II	VILSLDVP	IGLLRILLEQ	ARYKAARNQA	ATNAQILAHVa	4,054
rat Ucn-II	VILSLDVP	IGLLRILLEQ	ARNKAARNQA	ATNAQILARVa	4,024
human Ucn-III	FTLSLDVP	TNIMNLLFNI	AKAKNLRAQA	AANAHLMAQIa	4,138
mouse Ucn-III	FTLSLDVP	TNIMNILFNI	DKAKNLRAKA	AANAQLMAQIa	4,459
fugu URP	LTLSLDVP	TNIMNVLFDV	AKAKNLRAKA	AENARLLAHIa	4,145
fugu URP-II	FALSLDVP	TSILSVLIDL	AKNQDMRSKA	XRNAELMARIa	
human CRH	SEEPPISLDLT	FHLLREVLEM	ARAEQLAQQA	HSNRKLMEIIa	4,758
carp UI	NDDPPISIDLT	FHLLRNMIEM	ARIENEREQA	GLNRKYLDEVa	4,870
PS-SVG	EGPPISIDLS	LELLRKMIEI	EKQEKEKQQA	HSNRKLMEIIa	4,716

Figure 2C.1 **Comparison of the amino acid sequence of urocortins.** Amino acid residues identical to rat/ovine/mouse Ucn-I are shaded.

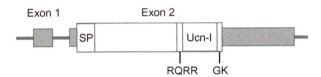

Figure 2C.2 Structure of human Ucn-I gene and precursor.

Receptors

Structure and Subtype

Ucns act through CRHR2. Ucn-I also binds to type 1 CRH receptor (CRHR1), whereas Ucn-II and Ucn-III bind exclusively to CRHR2.

Signal Transduction Pathway

Ucn-I is considered to act mainly through both $G_{\alpha s}$ and $G_{\alpha q}$ subunits to stimulate diverse signaling cascades such as AC and PLC pathways, depending on the tissue type [9].

Biological Functions

Target Cells/Tissues and Functions

Ucn-I stimulates adrenocorticotropic hormone (ACTH) secretion both *in vivo* and *in vitro*, while Ucn-II and Ucn-III do not stimulate ACTH secretion. This effect of Ucn-I on the pituitary is probably mediated by CRHR1, as is CRH. Ucn-I is more potent than CRH in suppressing food intake [10]. Ucn-specific effects are considered to be mediated by CRHR2. ICV injection of Ucn-I increases whole body oxygen consumption and body temperature in the rat [7]. Ucn-I has been reported to participate in vascular inflammation, and to function as an anxiolytic in stress response. Ucn-II and Ucn-III may be responsible for the effects of stress on appetite and anxiety in the brain. Ucn-II blocks the suppression of gastric antral contractions induced by lipopolysaccharide in freely moving conscious rats.

Phenotype in Gene-Modified Animals

Crh knockout mice show increased Ucn-I expression in the Edinger-Westphal nucleus.

Pathophysiological Implications

Clinical Implications

Ucn-I levels are high in the synovial fluid of patients with rheumatoid arthritis. Ucn-I has been linked to a reduction in inflammation and bone erosion in a mouse model of rheumatoid arthritis.

References

1. Vaughan J, Donaldson C, Bittencourt J, et al. Urocortin, a mammalian neuropeptide related to fish urotensin I and to corticotropin-releasing factor. *Nature.* 1995;378:287−292.
2. Reyes TM, Lewis K, Perrin MH, et al. Urocortin II: a member of the corticotropin-releasing factor (CRF) neuropeptide family that is selectively bound by type 2 CRF receptors. *Proc Natl Acad Sci USA.* 2001;98:2843−2848.
3. Lewis K, Li C, Perrin MH, et al. Identification of urocortin III, an additional member of the corticotropin-releasing factor (CRF) family with high affinity for the CRF2 receptor. *Proc Natl Acad Sci USA.* 2001;98:7570−7575.
4. Lovejoy DA, De Lannoy L. Evolution and phylogeny of the corticotropin-releasing factor (CRF) family of peptides: expansion and specialization in the vertebrates. *J Chem Neuroanat.* 2013;54:50−56.
5. Bittencourt JC, Vaughan J, Arias C, et al. Urocortin expression in rat brain: evidence against a pervasive relationship of urocortin-containing projections with targets bearing type 2 CRF receptors. *J Comp Neurol.* 1999;415:285−312.
6. Takahashi K, Totsune K, Sone M, et al. Regional distribution of urocortin-like immunoreactivity and expression of urocortin mRNA in the human brain. *Peptides.* 1998;19:643−647.
7. Pan W, Kastin AJ. Urocortin and the brain. *Prog Neurobiol.* 2008;84:148−156.
8. Weninger SC, Peters LL, Majzoub JA. Urocortin expression in the Edinger-Westphal nucleus is up-regulated by stress and corticotropin-releasing hormone deficiency. *Endocrinology.* 2000;141:256−263.
9. Combs CE, Fuller K, Kumar H, et al. Urocortin is a novel regulator of osteoclast differentiation and function through inhibition of a canonical transient receptor potential 1-like cation channel. *J Endocrinol.* 2012;212:187−197.
10. Tsatsanis C, Dermitzaki E, Venihaki M, et al. The corticotropin-releasing factor (CRF) family of peptides as local modulators of adrenal function. *Cell Mol Life Sci.* 2007;64:1638−1655.

Supplemental Information Available on Companion Website

- Accession numbers of genes and cDNAs for preproUcn/E-Table 2C.1

Sauvagine

Masafumi Amano

Abbreviation: SVG

Sauvagine is a peptide hormone, first isolated from the skin of the South American frog Phyllomedusa sauvagei, with adrenocorticotropic hormone (ACTH) releasing activity and antidiuretic activity.

Discovery

Sauvagine was first reported in the skin of the South American frog *Phyllomedusa sauvagei* in 1980 [1]. In 2012, a novel form of SVG was isolated in the skin of the Mexican giant leaf frog *Pachymedusa dacnicolor* [2].

Structure

Structural Features

The sauvagines of *Phyllomedusa sauvagei* (PS-SVG) and *Pachymedusa dacnicolor* (PD-SVG) comprise 40 aa and 38 aa residues, respectively, with an amidated C-terminus [1,2] (Figure 2D.1).

Primary Structure

The sequence identity of PS-SVG with human CRH is 63%. The sequence identities of PS-SVG with carp urotensin-I, human urocortin-I, human urocortin-II, and human urocortin-III are 50%, 35%, 20%, and 25%, respectively.

Properties

Mr 4,698. Soluble in water and methanol.

Synthesis and Release

Gene, mRNA, and Precursor

PD-SVG cDNA encodes 80-aa precursors. Prepro PD-SVG consists of a signal peptide, a cryptic region, and a mature peptide of 38 aa residues at the C-terminus [2].

Receptors

Structure and Subtype

Two CRH receptors, type-1 and -2 CRH receptors (CRHR1 and CRHR2), have been identified in *Xenopus laevis*. PS-SVG has higher affinity to CRHR2 ($K_d = 0.9$ nM) than to CRHR1 ($K_d = 51.4$ nM) [3].

Agonists

[Tyr^0, Gln^1, Leu^{17}]SVG (YQL-SVG) and [Tyr^0, Gln^1, Bpa^{17}]SVG (YQB-SVG) have been identified as agonists.

Biological Functions

Target Cells/Tissues and Functions

SVG stimulates ACTH release from the pituitary of rats and goldfish. In *Xenopus*, SVG stimulates α-melanocyte-stimulating hormone and β-endorphin release from the pars intermedia of the pituitary. SVG in the skin is considered to be involved in chemical defense.

References

1. Montecucchi PC, Anastasi A, De Castiglione R, et al. Isolation and amino acid composition of sauvagine, an active polypeptide from methanol extracts of the skin of the South American frog *Phyllomedusa sauvagei. Int J Pept Protein Res.* 1980;16:191−199.
2. Zhou Y, Jiang Y, Wang R, et al. PD-sauvagine: a novel sauvagine/corticotropin releasing factor analogue from the skin secretion of the Mexican giant leaf frog, *Pachymedusa dacnicolor. Amino Acids.* 2012;43:1147−1156.
3. Dautozenberg FM, Dietrich K, Palchaudhuri MR, et al. Identification of two corticotropin-releasing factor receptors from *Xenopus laevis* with high ligand selectivity: unusual pharmacology of the type 1 receptor. *J Neurochem.* 1997;69:1640−1649.

Supplemental Information Available on Companion Website

- Accession numbers of genes and cDNAs for preproSVG/E-Table 2D.1

PS–SVG   pEGPPISIDLS LELLRKMIEI EKQEKEKQQA HSNRKLMEIIa   (Mr 4,698)
PD–SVG   QGTSLDLT FDLLRHNLEI AKQEALKKQA AKNRLLLDTIa   (Mr 4,301)

Figure 2D.1   Comparison of the amino acid sequence of SVG. Amino acid residues identical to PS-SVG are shaded.

Y. Takei, H. Ando, & K. Tsutsui (Eds): Handbook of Hormones. DOI: http://dx.doi.org/10.1016/B978-0-12-801028-0.00112-4

CHAPTER 3

Gonadotropin-Releasing Hormone

Hironori Ando

Abbreviation: GnRH

Additional names: luteinizing hormone-releasing hormone (LHRH), gonadoliberin, gonadorelin, luliberin

A decapeptide neurohormone that is produced in neurosecretory cells within the hypothalamus, GnRH stimulates the synthesis and release of luteinizing hormone (LH) and follicle-stimulating hormone (FSH) from the anterior pituitary and also controls reproductive behavior, thus serving as a central regulator of reproductive function. It is used not only as a fertility drug but also as an antifertility drug.

Discovery

Andrew V. Schally's group first isolated and determined the chemical structure of porcine GnRH in 1971 [1]. In non-mammalian vertebrates, GnRH was isolated from the hypothalamus of chicken in 1982 and that of salmon in 1983. To date, 15 different isoforms of GnRHs have been identified in vertebrates. In invertebrates, multiple GnRH-like peptides were first identified in the tunicate *Chelyosoma productum* (see Chapter 38) [2].

Structure

Structural Features

Vertebrate GnRH isoforms consist of 10 aa residues with a pyroglutamate at the N-terminus and an amidated Gly at the C-terminus, while GnRH-like peptides of protostomes consist of 12 aa residues with insertions of 2 aa residues at the site between positions 1 and 2 (Figure 3.1). Based on phylgenetic and genomic synteny analyses, multiple GnRH forms are classified into four groups: namely GnRH1, GnRH2, GnRH3, and GnRH4 [3]. The peptides are bent with a β-turn around Gly^6, and the N-terminal and C-terminal aa residues are important for binding to the GnRH receptor (GnRH-R) [4].

Primary Structure

The primary structure of GnRH is highly conserved in the N-terminus (positions 1−4) and in the C-terminus (positions 9−10) (Figure 3.1). Position 8 is most variable.

Properties

Mr 1,050−1,250. GnRH is inactivated in 0.9-M HCl at 100°C for 60 minutes, and is also inactivated by endopeptidases such as chymotrypsin and papain. mGnRH and cGnRHI are soluble in water; cGnRHII and sGnRH are insoluble in water but soluble in 10-mM acetic acid.

Synthesis and Release

Gene, mRNA, and Precursor

GnRH peptide is synthesized from a precursor that consists of a signal peptide followed by the GnRH decapeptide, a Gly−Lys−Arg amidation cleavage site, and GnRH-associated peptide (E-Figure 3.1). The human GnRH1 gene, *GNRH1*, location 8p21−11.2, consists of four exons; the human GnRH2 gene, *GNRH2*, location 20p13, also consists of four exons.

Distribution of mRNA

Multiple GnRH isoform genes are expressed in different neuronal groups in the brain. Hypophysiotropic GnRH1 neurons are present in the ventral forebrain−preoptic area (POA)−basal hypothalamus. GnRH2 neurons are present in the midbrain tegmental area. GnRH3 neurons are present in the terminal nerve ganglion−POA in certain teleosts.

Tissue and Plasma Concentrations

Tissue: Ovine GnRH1, 20−80 (ng/hypothalamus); cockerel GnRH1, 6.55 (pg/μg protein) in mediobasal hypothalamus, 0.95 in medial preoptic region; edible frog GnRH1, 2.12 (ng/brain), GnRH2, 0.97; goldfish GnRH2, 4.1 (ng/mg tissue) in medulla oblongata, GnRH3, 50 in the olfactory bulb, 18 in the telencephalon.

Plasma: Rhesus monkey GnRH1 in median eminence perfusates during the periovulatory period, 21 (pg/ml); rabbit GnRH1 in pituitary stalk plasma after stimulation of gonadotropin secretion, 30−45 (pg/ml).

Regulation of Synthesis and Release

The regulation of GnRH secretion by internal and environmental factors, such as growth, energy condition, light, photoperiod, temperature, and social status, is critical for reproductive success and rhythmicity. Regulatory mechanisms of GnRH1 neurons involve many stimulatory and inhibitory factors, including gonadal steroids, neuropeptides, and neurotransmitters (GABA, glutamate, norepinephrine) [5]. Kisspeptin stimulates GnRH1 secretion and has key roles in transmission of the negative and positive feedback effects of gonadal steroids, in metabolic regulation, and in the photoperiodic control of reproduction [6]. Gonadotropin-inhibitory hormone (GnIH) has inhibitory action on GnRH1 neurons [7].

Y. Takei, H. Ando, & K. Tsutsui (Eds): Handbook of Hormones. DOI: http://dx.doi.org/10.1016/B978-0-12-801028-0.00003-9

```
GnRH                    Type   12345678910
Mammalian (mGnRH)       GnRH1  pEHWSYGLRPGa
Guinea pig (gpGnRH)     GnRH1  pEYWSYGVRPGa
Chicken I (cGnRH)       GnRH1  pEHWSYGLQPGa
Frog (frGnRH)           GnRH1  pEHWSYGLWPGa
Seabream (sbGnRH)       GnRH1  pEHWSYGLSPGa
Salmon (sGnRH)          GnRH3  pEHWSYGWLPGa
Whitefish (wfGnRH)      GnRH1  pEHWSYGMNPGa
Medaka (mdGnRH)         GnRH1  pEHWSFGLSPGa
Catfish (cfGnRH)        GnRH1  pEHWSHGLNPGa
Herring (hrGnRH)        GnRH1  pEHWSHGLSPGa
Dogfish (dfGnRH)        GnRH1  pEHWSHGWLPGa
Chicken II (cGnRHII)    GnRH2  pEHWSHGWYPGa
Lamprey II (lGnRHII)    GnRH2  pEHWSHGWFPGa
Lamprey III (lGnRHIII)  GnRH4  pEHWSHDWKPGa
Lamprey I (lGnRHI)      GnRH4  pEHYSLEWKPGa

Tunicate I (tGnRHI)            pEHWSDYFKPGa
Tunicate II (tGnRHII)          pEHWSLCHAPGa
Octopus (octGnRH)        pENYHFSNGWHPGa
(pE: pyroglutamic acid, a:amide)
```

Figure 3.1 Primary structure of the GnRH isoforms in vertebrates and invertebrates. pE, pyroglutamic acid; a, C-terminal amide.

Receptors

Structure and Subtype

GnRH-R is a membrane-bound GPCR belonging to the Class A (rhodopsin-like) subfamily. Three major types of GnRH-Rs — type I (mammalian and non-mammalian), type II, and type III — have been identified in vertebrates (E-Figure 3.4) [8]. In mammals, type I GnRH-R shows a strong preference for GnRH1, while type II GnRH-R shows a preference for GnRH2. In some mammalian species, including humans, the type II GnRH-R is a non-functional receptor encoded by a pseudogene. Mammalian type I GnRH-R lacks a C-terminal intracellular domain, which is responsible for rapid desensitization. Human type I GnRH-R consists of 328 aa residues (E-Figure 3.3).

Signal Transduction Pathway

GnRH action is mediated via a membrane receptor (GnRH-R) mainly coupled to $G_{q/11}$ protein. GnRH induces mobilization of intracellular Ca^{2+} and activation of PKC in target cells.

Agonists

Lamprey GnRHIII (FSH-releasing peptide), (Des-Gly^{10}, Pro-NHEt^9)-LHRH (fertirelin), (Des-Gly^{10},N-im-benzyl-D-His^6,Pro-NHEt^9)-LHRH (histrelin), (Des-Gly^{10},D-Leu^6,Pro-NHEt^9)-LHRH (leurolide), (Des-Gly^{10},D-Trp^6,Pro-NHEt^9)-LHRH (deslorelin), (D-Trp^6)-LHRH (triplorelin), (D-Ser(tBu)^6,Azagly^{10})-LHRH (goserelin), and gonadorelin have been identified as agonists.

Antagonists

Antide, acetyl-(3,4-dehydro-Pro^1,4-fluoro-D-Phe^2,D-Trp^{3,6})-LHRH, and (D-Phe^{2,6},Pro^3)-LHRH are antagonists.

Biological Functions

Target Cells/Tissues and Functions

GnRH1 acts on the anterior pituitary as a stimulator of the synthesis and release of luteinizing hormone (LH) and follicle-stimulating hormone (FSH). In mammals, GnRH1 is released into the pituitary portal circulation in a pulsatile manner, which is essential for a LH surge prior to stimulation of ovulation. In teleosts, GnRH stimulates secretion of other pituitary hormones, including growth hormone (GH), prolactin (PRL), and somatolactin (SL). GnRH2 and GnRH3 neurons

send fibers throughout the brain, and both hormones have neuromodulatory roles [9]. GnRH2 has been implicated in the regulation of reproductive and feeding behavior. GnRH3 is also involved in reproductive behavior in teleosts.

Phenotype in Gene-Modified Animals

Kallmann syndrome (KS) is a rare genetic disorder comprising lack of olfactory senses and hypogonadotropic hypogonadism due to GnRH deficiency. KS is caused by defective migration of GnRH neurons from the olfactory placode to the hypothalamus during fetal development. Mutations in *GNRH-R* result in hypogonadotropic hypogonadism in humans. Type I GnRH-R knockout mice, normally lacking a type II GnRH-R, show severe reproductive defects such as small sexual organs; low levels of FSH, LH, and steroid hormones; failure of sexual maturation; infertility; and inability to respond to exogenous GnRH [10].

Pathophysiological Implications

Clinical Implications

Since mGnRH has a short half-life of several minutes, a number of GnRH analogs with high potency and long half-life have been synthesized. The effects of GnRH and its analogs are dependent on the dose and method of administration. Low doses of GnRH delivered in a pulsatile fashion restore fertility in hypogonadal patients. High doses of GnRH or continuous administration first results in an increase in LH and FSH secretion, followed by a decrease in LH and FSH levels due to desensitization, and by a decline in gonadal steroid levels. The administration of antagonists interrupts GnRH-dependent LH and FSH secretion through competition with endogenous GnRH, but the doses required are much higher than the desensitizing agonists.

Use for Diagnosis and Treatment

mGnRH is available as gonadorelin for veterinary use. Various GnRH agonists are used in the treatment of hormone-responsive cancers such as prostate and breast cancers, estrogen-dependent conditions such as endometriosis, and precocious puberty. They are also widely used in ART (assisted reproductive technology including IVF-ET, in vitro fertilization, and embryo transfer) to block the endogenous LH surge in controlled ovarian stimulation.

References

1. Matsuo H, Baba Y, Nair RMG, et al. Structure of the porcine LH- and FSH-releasing hormone. I. The proposed amino acid sequence. *Biochem Biophys Res Commun.* 1971;43:1334–1339.
2. Powell JFF, Reska-Skinner SM, Prakash MO, et al. Two new forms of gonadotropin-releasing hormone in a protochordate and the evolutionary implications. *Proc Natl Acad Sci USA.* 1996;93:10461–10464.
3. Tostivint H. Evolution of the gonadotropin-releasing hormone (GnRH) gene family in relation to vertebrate tetraploidizations. *Gen Comp Endocrinol.* 2011;170:575–581.
4. Millar RP, Lu Z-L, Pawson AJ, et al. Gonadotropin-releasing hormone receptors. *Endocr Rev.* 2004;25:235–275.
5. Parhar I, Ogawa S, Kitahashi T. RFamide peptides as mediators in environmental control of GnRH neurons. *Prog Neurobiol.* 2012;98:176–196.
6. Tena-Sempere M, Felip A, Gómez A, et al. Comparative insights of the kisspeptin/kisspeptin receptor system: lessons from non-mammalian vertebrates. *Gen Comp Endocrinol.* 2012;175:234–243.
7. Tsutsui K. A new key neurohormone controlling reproduction, gonadotropin-inhibitory hormone (GnIH): Biosynthesis, mode of action and functional significance. *Prog Neurobiol.* 2009;88:76–88.

8. Sower SA, Decatur WA, Joseph NT, et al. Evolution of vertebrate GnRH receptors from the perspective of a basal vertebrate. *Front Endocrinol*. 2012;3:140.
9. Oka Y. Three types of gonadotrophin-releasing hormone neurones and steroid-sensitive sexually dimorphic kisspeptin neurones in teleosts. *J Neuroendocrinol*. 2009;21:334–338.
10. Wu S, Wilson MD, Busby ER, et al. Disruption of the single copy gonadotropin-releasing hormone receptor in mice by gene trap: severe reduction of reproductive organs and functions in developing and adult mice. *Endocrinology*. 2010;151:1142–1152.
11. Jones DT, Taylor WR, Thornton JM. The rapid generation of mutation data matrices from protein sequences. *Comput Appl Biosci*. 1992;8:275–282.
12 Tamura K, Stecher G, Peterson D, et al. MEGA6: Molecular evolutionary genetics analysis version 6.0. *Mol Biol Evol*. 2013;30:2725–2729.

Supplemental Information Available on Companion Website

- Structure of GnRH precursor genes in vertebrates/E-Figure 3.1
- Phylogenetic tree of GnRH-Rs/E-Figure 3.2
- Primary structure of GnRH-Rs of various animals/E-Figure 3.3 [9,10]
- Primary structure of GnRH precursors of various animals/E-Figure 3.4
- Accession numbers of genes and cDNAs for GnRH and GnRH-R/E-Tables 3.1, 3.2

CHAPTER 4

Thyrotropin-Releasing Hormone

Hironori Ando

Abbreviation: TRH
Additional names: thyrotropin-releasing factor (TRF), thyro-liberin, pyroglutamylhistidylprolinamide, protirelin, 5-oxo-L-prolyl-L-histidyl-L-prolinamide

The first hypothalamic hypophysiotropic neurohormone identified, TRH consists of the tripeptide pGlu—His—Pro—NH$_2$. It stimulates the secretion of thyroid-stimulating hormone (TSH), prolactin (PRL), and growth hormone (GH), and also functions as a neurotransmitter and neuromodulator.

Discovery

TRH was first isolated and characterized in 1969 by Roger Guillemin and Andrew V. Schally, who shared the Nobel Prize in Physiology or Medicine in 1977 "for their discoveries concerning the peptide hormone production of the brain" [1,2]. Biosynthesis of TRH from a precursor molecule was first clarified in 1984 by isolation of a preproTRH cDNA from the skin of *Xenopus laevis* [3]. The structure of the TRH receptor (TRH-R) was first deduced from a cDNA isolated from the mouse pituitary in 1990 [4].

Structure

Structural Features

TRH consists of three aa residues with a pyroglutamate at the N-terminus and an amidated proline at the C-terminus (Figure 4.1).

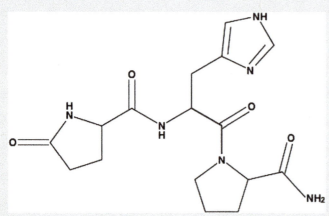

Figure 4.1 Structure of TRH.

Primary Structure

The sequence pGlu—His—Pro—NH$_2$ is fully conserved in vertebrates.

Properties

Mr 362. Soluble in water, methanol, and ethanol; partially insoluble in chloroform; completely insoluble in ether and pyridine. Stable in solution at <15°C for more than a year; partially (1%) degraded at 40°C for 6 months. Resistant to proteolytic enzymes. Inactivated by diazotized sulfanilic acid (Pauly reagent). Plasma half-life is 2—6 minutes.

Synthesis and Release

Gene, mRNA, and Precursor

TRH is synthesized from a precursor that contains multiple copies of the TRH progenitor sequence Gln—His—Pro—Gly, which is flanked by dibasic cleavage sites at its N- and C-termini (Figure 4.2). The number of progenitor sequences in a precursor is diversified: six in humans, five in rats, four in chicken, seven in frogs, six to eight in fish (E-Figure 4.1) [5]. Human preproTRH gene, *TRH*, location 3q13.3—q21, consists of three exons.

Distribution of mRNA

The preproTRH gene is widely expressed in various brain regions, including the hypothalamus, thalamus, and lower brainstem regions, and in the spinal cord. In the hypothalamus, preproTRH mRNA-expressing neurons are concentrated in the parvocellular division of the paraventricular nucleus (PVN). In tetrapods, the hypophysiotropic TRH neurons send fibers to the external zone of the median eminence and TRH is released into the hypophysial portal circulation, while in teleosts TRH neurons directly innervate the pars distalis of the pituitary.

Tissue and Plasma Concentrations

Tissue: Concentrations of TRH in various vertebrate species are shown in Table 4.1. TRH is also present in other peripheral organs, such as the gastrointestinal tract, pancreas, placenta, and testis.

Plasma: Rat <30 pg/ml; chicken, 0.3 ng/ml; leopard frog >100 ng/ml.

Regulation of Synthesis and Release

TRH secretion is regulated by norepinephrine, histamine, dopamine, and serotonin. Cold-induced secretion of TSH from the rat anterior pituitary involves α-adrenergic

Y. Takei, H. Ando, & K. Tsutsui (Eds): Handbook of Hormones. DOI: http://dx.doi.org/10.1016/B978-0-12-801028-0.00004-0

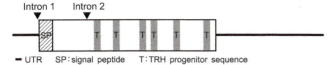

Figure 4.2 Structure of the human preproTRH mRNA.

Table 4.1 Tissue Concentrations of TRH

Animal	Tissues (ng/g Wet Weight)			
	Hypothalamus	Median Eminence	Pituitary	Skin
Rat	280	3,570	155 (neurohypophysis)	–
Chicken	41	–	168 (neurohypophysis)	–
Leopard frog	2,270	–	>5,000	4,100
Atlantic salmon	235	–	150	–

regulation of TRH secretion. TRH secretion is negatively regulated by thyroid hormones through a feedback mechanism. This feedback regulation by thyroid hormones is restricted to the hypophysiotropic TRH neurons in the PVN. Locally produced T_3 is taken by these neurons to regulate transcription, posttranslational modification, and degradation of TRH [6].

Receptors

Structure and Subtype

TRH-R is a seven-transmembrane-domain GPCR. Two major types of TRH-Rs (type I receptor including TRH-R1 and TRH-R3, and type II receptor [TRH-R2]), have been identified in vertebrates (E-Figure 4.3). In mammals, TRH-R1 and TRH-R2 have been identified in rat and mouse, while only TRH-R1 has been identified in human, pig, sheep, and cow. Three and four subtypes of TRH-Rs have been identified in *Xenopus laevis* and teleost species, respectively [7].

Signal Transduction Pathway

TRH action is mediated via a membrane receptor mainly coupled to $G_{q/11}$ protein. TRH induces mobilization of intracellular Ca^{2+} and activation of PKC in target cells.

Biological Functions

Target Cells/Tissues and Functions

In mammals, TRH-R1 and TRH-R2 are differently distributed in the brain and in the pituitary. TRH-R1 is strongly expressed in the anterior pituitary and mediates the hypophysiotropic actions of TRH — i.e., stimulation of the release of TSH, PRL, and GH. In fish, TRH stimulates GH and PRL release but does not stimulate TSH release. In amphibians and teleosts, TRH also stimulates the release of α-melanocyte-stimulating hormone (MSH) from melanotropes in the intermediate lobe of the pituitary [7]. In addition, TRH-R1 is widely expressed in the hypothalamic nuclei, brainstem regions, and spinal cord. TRH-R2 is also expressed throughout the brain. TRH-Rs in the neuronal tissues mediate multiple neurobehavioral roles of TRH in sleep, anxiety, depression, learning, and memory.

Phenotype in Gene-Modified Animals

TRH knockout mice show normal development, but exhibit tertiary hypothyroidism and hyperglycemia due to a diminished insulin secretion [8].

Pathophysiological Implications

Clinical Implications

TRH (200–500 μg) administered intravenously to normal subjects causes a rise in TSH levels within 15 to 30 minutes, resulting in an increase in T_3 levels within 90–150 minutes [9]. In primary hypothyroidism, TSH hyper-response to TRH occurs, with a typical elevation in the basal TSH levels. In secondary hypothyroidism, an impaired TSH response to TRH occurs, whereas in tertiary hypothyroidism normal or increased TSH response to TRH occurs. Isolated central hypothyroidism was reported in a patient with inactivating mutations in the TRH-R gene [10].

Use for Diagnosis and Treatment

Protirelin is used to test the response of the anterior pituitary to TRH in people who may have medical conditions of thyroid function, including hyperthyroidism, Graves' disease, and hypothyroidism.

References

1. Burgus R, Dunn T, Desiderio D, et al. Molecular structure of the hypothalamic hypophysiotropic TRF factor of ovine origin: mass spectrometry demonstration of the PCA–His–Pro–NH$_2$ sequence. *C R Acad Sci.* 1969;269:1870–1873.
2. Bøler J, Enzmann F, Folkers K, et al. The identity of chemical and hormonal properties of the thyrotropin releasing hormone and pyroglutamyl-histidyl-proline amide. *Biochem Biophys Res Commun.* 1969;37:705–710.
3. Richter K, Kawashima E, Egger R, et al. Biosynthesis of thyrotropin releasing hormone in the skin of *Xenopus laevis*: partial sequence of the precursor deduced from cloned cDNA. *EMBO J.* 1984;3:617–621.
4. Straub RE, Frech GC, Joho RH, et al. Expression cloning of a cDNA encoding the mouse pituitary thyrotropin-releasing hormone receptor. *Proc Natl Acad Sci USA.* 1990;87:9514–9518.
5. Wallis M. Molecular evolution of the thyrotrophin-releasing hormone precursor in vertebrates: insights from comparative genomics. *J Neuroendocrinol.* 2010;22:608–619.
6. Vella KR, Hollenberg AN. The ups and downs of thyrotropin-releasing hormone. *Endocrinology.* 2009;150:2021–2023.
7. Galas L, Raoult E, Tonon M-C. TRH acts as a multifunctional hypophysiotropic factor in vertebrates. *Gen Comp Endocrinol.* 2009;164:40–50.
8. Yamada M, Saga Y, Shibusawa N. Tertiary hypothyroidism and hyperglycemia in mice with targeted disruption of the thyrotropin-releasing hormone gene. *Proc Natl Acad Sci USA.* 1997;94:10862–10867.
9. Jackson IMD. Thyrotropin-releasing hormone. *N Engl J Med.* 1982;306:145–155.
10. Collu R, Tang J, Castagné J, et al. A novel mechanism for isolated central hypothyroidism: inactivating mutations in the thyrotropin-releasing hormone receptor gene. *J Clin Endocrinol Metab.* 1997;82:1561–1565.
11. Jones DT, Taylor WR, Thornton JM. The rapid generation of mutation data matrices from protein sequences. *Comput Appl Biosci.* 1992;8:275–282.
12 Tamura K, Stecher G, Peterson D, et al. MEGA6: Molecular evolutionary genetics analysis version 6.0. *Mol Biol Evol.* 2013;30:2725–2729.

Supplemental Information Available on Companion Website

- Primary structure of preproTRHs of various animals/E-Figure 4.1
- Phylogenetic tree of TRH-Rs/E-Figure 4.2 [7,8]
- Primary structure of TRH-Rs of various animals/E-Figure 4.3
- Accession numbers of genes and cDNAs for preproTRH and TRH-R/E-Tables 4.1, 4.2

CHAPTER 5

Somatostatin

Hironori Ando

Abbreviation: SS

Additional names: somatotropin-release inhibiting factor (SRIF), growth hormone-inhibiting hormone (GHIH)

A tetradecapeptide exerting a growth hormone (GH) release-inhibiting activity, SS also has a large variety of neuro-modulatory and gastrointestinal actions, acting mostly as an inhibitory hormone.

Discovery

Somatostatin of 14 aa residues (SS-14) was first isolated from ovine hypothalamic extracts by Roger Guillemin's group in 1973 and was originally named GH inhibiting factor [1]. An N-terminally extended form of 28 aa residues (SS-28) was isolated later from porcine gut. SS cDNA that encodes a precursor for SS-14 and SS-28, a product of the somatostatin 1 (SS1) gene (approved symbol SST), was first cloned from the anglerfish pancreas in 1980 [2]. In this report, another SS cDNA encoding a different precursor (formerly called SSII), a product of the somatostain 3 (SS3) gene, was also isolated. Subsequently, cDNAs and genes encoding SS precursors have been identified in many animals by molecular cloning and bioinformatic analyses, and so far six paralogous genes (SS1, SS2, SS3, SS4, SS5, and SS6) have been identified in vertebrates [3,4]. Accordingly, two SS-related peptides, cortistatin (CST) [5] and neuronostatin [6], have been identified.

Structure

Structural Features

SS-14 and SS-28 contain a disulfide bridge and have a cyclic structure (Figure 5.1). CST, a product of the SS2/CST gene, is 14–17 aa residues in length, depending on the species, and contains the SS2 signature, a proline residue at position 2. CST in placental mammals exhibits an additional lysine residue at its C-terminal extremity. Neuronostatin is a 13-residue amidated peptide that is flanked with the signal peptide of the SS1 precursor (Figure 5.2), and is an acyclic peptide.

Primary Structure

The aa sequence of SS-14 is fully conserved in vertebrates. SS-28 in mammals that contains the SS-14 moiety at its C-terminal shares 40–66% sequence identity with its counterparts in fish.

Properties

Mr 1,638 (SS-14), 3,149 (SS-28). Soluble in water, acid, and methanol. Stable in solution at −80°C for more than a year. Plasma half-life is <3 minutes.

Synthesis and Release

Gene, mRNA, and Precursor

Human preproSS gene (SS1), *SST*, location 3q28, consists of two exons. Mammalian SS-14 and SS-28 are derived from preproSS1 of 116 aa residues by specific proprotein convertases (PCs) through tissue-specific posttranslational processing. In vertebrates, six SS genes have been identified, and all these paralogs are present in teleost fish, while only SS1 and SS2/CST are present in tetrapods. Phylogenetic and comparative genomic analyses showed that whole-genome duplications, local duplications, and gene losses contribute to the divergent evolution of SS genes [3,4].

Distribution of mRNA

SS-14 is the predominant form in the brain and is responsible for the regulation of the GH axis, whereas SS-28 is mainly produced by intestinal enteroendocrine cells. In the hypothalamus, SS-immunoreactive (ir) cells are concentrated in the periventricular region, paraventricular nucleus, anterior hypothalamic nucleus, and perifornical region of the lateral hypothalamus. SS-ir cells are also widely distributed in extrahypothalamic regions, including the neocortex, caudate nucleus, putamen, piriform cortex, amygdala, hippocampus, septum, and brain stem. The SS2/CST gene is expressed in the cortex and hippocampus. In the gastrointestinal system, SS is localized in the epithelial endocrine D cells in the mucosal layer of the gastrointestinal tract and in neurons of the submucosal and myenteric plexus. Neuronostatin is localized in parietal cells of the gastric mucosa.

Tissue and Plasma Concentrations

Tissue: Male rat brain, cerebral cortex 6.7 (ng/mg protein), thalamus 2.9, corpus striatum 2.3, medulla oblongata and pons 3.3, cerebellum 0.5. Tissue concentrations of SS in various vertebrate species are shown in Table 5.1.

Plasma: Human, 2–30 pg/ml. There is almost no variation depending on age and sex.

Regulation of Synthesis and Release

SS secretion in the gastrointestinal tract is regulated by the autonomous nervous system and various gut regulatory peptides including gastrin, cholecystokinin (CCK), and substance P.

Y. Takei, H. Ando, & K. Tsutsui (Eds): Handbook of Hormones. DOI: http://dx.doi.org/10.1016/B978-0-12-801028-0.00005-2

SS1/SS-14	All vertebrates	AGCKNFFWKTFTSC
SS2/CST	Human	DRMPCRNFFWKTFSSCK
SS2/CST	Rat, Mouse	PCKNFFWKTFSSCK
SS2/CST	Frog	APCKNFFWKTFTMC
SS2/CST	Lungfish, Goldfish, Sturgeon	APCKNFFWKTFTSC
SS3	Anglerfish, Salmon, Flounder	AGCKNFYWKGFTSC
SS3	Tilapia	AGCKNFYWKGLTSC
SS3	Eel, *Takifugu*	AGCKNFYWKGPTSC
SS3	Goldfish	EGCKNFYWKGFTSC
SS3	Medaka	DGCKNFYWKGFTSC
SS3	Spotted gar	AGCRNFYWKTFTSC
Neuronostatin	Pig	LRQFLQKSLAAAAa

Figure 5.1 Primary structure of SS-related peptides. Conserved residues are shaded. a, C-terminal amide

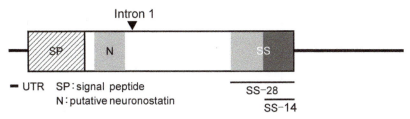

Figure 5.2 Structure of the human *SST* mRNA.

Table 5.1 Tissue Concentrations of SS

Animal	Tissues (µg/mg protein)			
	Hypothalamus	Extrahypothalamic Brain Region	Pancreas	Stomach
Shark	0.037	0.345	0.291	0.455
Tilapia	<0.01	0.027	–	<0.01
X. laevis	0.550	<0.01	0.277	0.850
Turtle	0.036	0.087	<0.01	1.020
Pigeon	0.018	0.215	1.036	0.027
Rat	0.109	0.036	0.055	0.218

The synthesis and release of hypothalamic SS are regulated by GH, growth hormone-releasing hormone (GHRH), and glucose.

Receptors

Structure and Subtype, Signal Transduction Pathway, Agonist, and Antagonist

SS receptors belong to the family of seven-transmembrane-domain GPCRs. There are five subtypes (sst_1–sst_5), and all these receptors bind to SS and CST with high affinity [7]. Structural and functional characteristics of these receptors in humans, including signal transduction pathways, agonists, and antagonists, are shown in Table 5.2.

Biological Functions

Target Cells/Tissues and Functions

Target tissues with SS receptor expression are shown in Table 5.2. In the anterior pituitary, SS inhibits the release of GH and thyroid-stimulating hormone (TSH). Pulsatile GH secretion reflects the pulsatile release of both SS and GHRH in a reciprocal fashion. In fish, SS-14 inhibits the release of GH, prolactin, and insulin. In the brain, SS has a variety of neuromodulatory roles in learning, cognitive functions, locomotor activity, anxiety, and depression. CST has physiological functions, such as depression of neuronal activity and induction of slow-wave sleep [5]. Moreover, SS exerts inhibitory effects on various gastrointestinal functions, including gastric acid secretion, gastric emptying, intestinal motility, and release of insulin, glucagon, and various gastrointestinal hormones.

Phenotype in Gene-Modified Animals

sst knockout mice have shown increased plasma GH levels, but there was no significant increase in body length compared with controls in both sexes [8]. Sexual dimorphism in GH pulsatile secretion and GH-regulated hepatic gene expression disappeared in the *sst*-deficient mice. In sst_2 knockout mice, the binding of radio-labeled SS-14 was greatly reduced in almost all brain structures, indicating that sst_2 accounts for most SS binding in mouse brain [9]. sst_2 knockout mice show increased anxiety-related behavior, and decreased locomotor and exploratory activity.

Pathophysiological Implications

Clinical Implications

Somatostatinoma is a malignant tumor that arises from transformed D cells in the pancreatic islets or duodenum. Somatostatinomas are associated with malabsorption, diabetes mellitus, steatorrhea, and cholelithiasis. In gastroenteropancreatic tumors high levels of SS receptor expression have been found, and specifically designed analogs are used for tumor imaging and radiotherapy. SS deficiency causes persistent *Helicobacter pylori* infection in the patient with chronic gastritis.

Use for Diagnosis and Treatment

Extensive SS analogs with improved pharmacokinetics, bioavailability, and receptor subtype selectivity have been

Table 5.2 Characteristics of SS Receptor Subtypes in Humans

Subtypes	sst_1	sst_2	sst_3	sst_4	sst_5
Gene	*SSTR1*	*SSTR2*	*SSTR3*	*SSTR4*	*SSTR5*
Gene location	14q13	17q24	22q13.1	20q11.2	16q13.3
Amino acid residues	391	369 (sst_{2A}), 356 (sst_{2B})	418	388	364
G protein-coupling	G_i/G_o	G_i/G_o	G_i/G_o	G_i/G_o	G_i/G_o
Selective agonist	CH275, L797591	BIM23027, L05422, MK-678	L796778	NNC269100, L803087	BIM23268, BIM23052, L817818
Selective antagonist	NVP-SRA880	CYN154806, PRL2903	NVP-ACQ090	–	BIM23056
Tissue distribution	Brain (cortex, hypothalamus, cerebellum)	Brain	Brain (cerebellum)	Brain (hypocampus)	V
	Pituitary, islets, stomach	Pituitary, islets, stomach	Pituitary, islets, stomach	Pituitary, islets, stomach	Pituitary, islets, stomach
	Liver, kidney, spinal cord	Adrenals, kidney	–	Adrenals, liver, eyes, lung, placenta	Adrenals

developed. These include non-peptidergic analogs (shown in Table 5.2) and octapeptides such as octreotide and lanreotide. Octreotide and lanreotide are long-acting sst_2-preferring agonists, and are used for the treatment of acromegaly, gastroenteropancreatic tumors, neuroendocrine tumors, and other gastrointestinal disorders such as secretory diarrhea and gastrointestinal bleeding.

References

1. Brazeau P, Vale W, Burgus R, et al. Hypothalamic polypeptide that inhibits the secretion of immunoreactive pituitary growth hormone. *Science*. 1973;179:77–79.
2. Hobart P, Crawford R, Shen L, et al. Cloning and sequence analysis of cDNAs encoding two distinct somatostatin precursors found in the endocrine pancreas of anglerfish. *Nature*. 1980;288:137–141.
3. Liu Y, Lu D, Zhang Y, et al. The evolution of somatostatin in vertebrates. *Gene*. 2010;463:21–28.
4. Tostivint H, Quan FB, Bougerol M, et al. Impact of gene/genome duplications on the evolution of the urotensin II and somatostatin families. *Gen Comp Endocrinol*. 2013;188:110–117.
5. de Lecea L, Criado JR, Prospero-Garcia O, et al. A cortical neuropeptide with neuronal depressant and sleep-modulating properties. *Nature*. 1996;381:242–245.
6. Samson WK, Zhang JV, Avsian-Kretchmer O, et al. Neuronostatin encoded by the somatostatin gene regulates neuronal, cardiovascular, and metabolic functions. *J Biol Chem*. 2008;283:31949–31959.
7. Tostivint H, Ocampo Daza D, Bergqvist CA, et al. Molecular evolution of GPCRs: Somatostatin/urotensin II receptors. *J Mol Endocrinol*. 2014;52:T61–T86.
8. Low MJ, Otero-Corchon V, Parlow AF, et al. Somatostatin is required for masculinization of growth hormone-regulated hepatic gene expression but not of somatic growth. *J Clin Invest*. 2001;107:1571–1580.
9. Viollet C, Vaillend C, Videau C, et al. Involvement of sst2 somatostatin receptor in locomotor, exploratory activity and emotional reactivity in mice. *Eur J Neurosci*. 2000;12:3761–3770.

Supplemental Information Available on Companion Website

- Primary structure of preproSS1s of various animals/E-Figure 5.1
- Primary structure of preproSS2/CSTs of various animals/E-Figure 5.2
- Primary structure of preproSS3s of various animals/E-Figure 5.3
- Primary structure of human SSTRs/E-Figure 5.4
- Accession numbers of genes and cDNAs for preproSS and SS receptors/E-Tables 5.1, 5.2

CHAPTER 6

Neurohypophysial Hormone Family

Susumu Hyodo

History

The vertebrate neurohypophysial hormone family is composed of two subfamilies called the vasopressin (VP) and oxytocin (OT) families. The pressor activity in the bovine pituitary was first found in 1895, followed by discovery of other major activities of this hormone family, including uterine contraction and antidiuresis. After these discoveries, two physiologically active principles, arginine vasopressin (AVP) and OT, were isolated from the mammalian neurohypophysis and chemically synthesized [1]. Thereafter, extensive searches in vertebrates revealed that all classes of jawed vertebrates possess both VP- and OT-family peptides (Table 6.1) [2]. In cyclostomes, only vasotocin (VT), a non-mammalian ortholog of AVP, has been found. Related peptides have also been reported for various invertebrate classes, such as Urochordata, Mollusca, Annelida, and Arthropoda. Their respective precursor and receptor structures have revealed that the invertebrate peptides are orthologs of the vertebrate neurohypophysial hormones. Neurohypophysial hormones are produced in the hypothalamic neurosecretory neurons, and are mainly released from the axon terminals in the neurohypophysis into the general circulation.

Structure

Structural Features

Neurohypophysial hormones are produced as larger precursor proteins that are subsequently processed to give rise to mature nonapeptides. A common structural feature of the mature peptides is a conserved intramolecular ring consisting of 6 aa residues and a C-terminal extension of 3 aa residues that characterizes VP (basic) and OT (neutral) family peptides (Figure 6.1). The C-termini of neurohypophysial hormones are amidated. It is noteworthy that, within each of the two subfamilies, peptides having different amino acid sequences are assigned separate names. This nomenclature is independent of other hormone family assignments; for example, hagfish VT and avian VT have identical amino acid sequences. In amphibians, two types of C-terminally extended peptides have been found; these are called hydrins.

Molecular Evolution of Family Members

As mentioned above, VP-related and OT-related peptides exist in invertebrates. However, so far, the evolutionary story connecting invertebrate and vertebrate peptides (genes) is still veiled in mystery. The ancestral molecule of the vertebrate neurohypophysial hormones is most probably VT, because cyclostomes only have a VT gene. In the course of evolution to the gnathostomes, the ancestral VT gene was duplicated to form VP- and OT-family genes (Table 6.1). Indeed, VP- and OT-family genes are closely linked on the same chromosome in tail-to-head orientation (non-mammalian vertebrates) or tail-to-tail orientation (eutherian mammals) [3]. All non-mammalian vertebrates have VT as the VP family peptide, while for OT family peptides, isotocin was isolated from bony fish and mesotocin was identified from amphibians, reptiles, and birds.

Receptors

Structure and Subtypes

Receptors for the neurohypophysial hormones belong to the G protein-coupled receptor superfamily (class A, or rhodopsin-type, GPCRs). For some time, it was thought that three VP receptors (V1aR, V1bR, V2R; V1aR and V1bR are also known as V1R and V3R, respectively) and one OT receptor arose in vertebrates as a consequence of gene duplication. Recently, a novel fifth receptor has been discovered [4,5]. This receptor is structurally related to V2R, but is functionally similar to V1 receptors (see below). Molecular phylogenetic and synteny analyses revealed that the fifth receptor arose by gene duplication of ancestral V2R. After this discovery, the conventional V2R was renamed V2aR, while the fifth receptor is referred to as V2bR [4,5]. V2bR is widely distributed throughout the jawed vertebrates, while V2aR is not found in cartilaginous fish [5]. The V2bR gene has not been found in most mammalian genomes, and thus might have lost functionality in mammals.

Signal Transduction Pathway

Following ligand binding, V1aR, V1bR, and OTR activate phospholipase C/protein kinase C pathways. V2aR is coupled with the adenylyl cyclase/protein kinase A pathways. Since V2bR is structurally related to V2aR, V2bR was initially considered to activate the cAMP signaling pathway; however, all V2bR examined (cartilaginous fish, teleost fish, birds) activated Ca^{2+} signaling pathways [5]. Therefore, the ancestral neurohypophysial hormone receptor most probably used Ca^{2+} as an intracellular second messenger. This is a reasonable suggestion because the related receptors of invertebrates

Y. Takei, H. Ando, & K. Tsutsui (Eds): Handbook of Hormones. DOI: http://dx.doi.org/10.1016/B978-0-12-801028-0.00006-4

Table 6.1 Distribution of Neurohypophysial Hormones in Vertebrates

Cyclostomes	Cartilaginous Fishes	Ray-Finned Fishes	Non-Mammalian Tetrapods	Mammals
Vasotocin	Vasotocin Various peptides	Vasotocin Isotocin	Vasotocin Mesotocin	Vasopressin Oxytocin

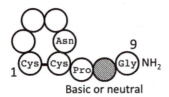

Figure 6.1 Mature sequence of neurohypophysial hormone.

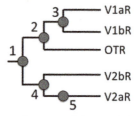

1: Duplication of V1R and V2R
2: Duplication of V1R and OTR
3: Duplication of V1aR and V1bR
4: Duplication of V2aR and V2bR
5: Switching to cAMP signaling

Figure 6.2 Evolutionary history of neurohypophysial hormone receptors.

are also coupled with Ca^{2+} signaling pathways. After the divergence from the V2bR lineage, V2aR switched to using cAMP as a second messenger, while the V2bRs retained the original Ca^{2+} signaling system (Figure 6.2) [5].

Biological Functions

Target Cells/Tissues and Functions

As described above under "History," the main peripheral actions are body fluid regulation, cardiovascular regulation, and regulation of reproductive events. In particular, renal antidiuresis [6] and milk ejection [7] are impaired by deficiencies of VP/V2aR and OT/OTR, respectively. V1bR mediates stress responses to VP/VT; in many species, VP/VT synergistically stimulates adrenocorticotropic hormone (ACTH) release with corticotropin-releasing hormone (CRH). Recently, a growing body of evidence is demonstrating involvement of neurohypophysial hormones in a variety of social and sexual behaviors, including social recognition and interactions, pair bonding, parental care, aggression, and anxiety. This is in agreement with the wide distribution of receptors and of projections of neurohypophysial hormone-producing neurons in the brain. Involvement of OT in neuropsychiatric disorders in humans has also been implicated [8].

References

1. Walter R, Rudinger J, Schwartz IL. Chemistry and structure—activity relations of the antidiuretic hormones. *Am J Med*. 1967;42:653—677.
2. Acher R. Neurohypophysial peptide systems: processing machinery, hydroosmotic regulation, adaptation and evolution. *Regul Pept*. 1993;45:1—13.
3. Gwee P-C, Tay B-H, Brenner S, et al. Characterization of the neurohypophysial hormone gene loci in elephant shark and the Japanese lamprey: origin of the vertebrate neurohypophysial hormone genes. *BMC Evol Biol*. 2009;9:47.
4. Ocampo Daza D, Lewicka M, Larhammar D. The oxytocin/vasopressin receptor family has at least five members in the gnathostome lineage, including two distinct V2 subtypes. *Gen Comp Endocrinol*. 2012;175:135—143.
5. Yamaguchi Y, Kaiya H, Konno N, et al. The fifth neurohypophysial hormone receptor is structurally related to the V2-type receptor but functionally similar to V1-type receptor. *Gen Comp Endocrinol*. 2012;178:519—528.
6. Moeller H, Rittig S, Fenton RA. Nephrogenic diabetes insipidus: essential insights into the molecular background and potential therapies for treatment. *Endocrine Rev*. 2013;34:278—301.
7. Takayanagi Y, Yoshida M, Bielsky IF, et al. Pervasive social deficits, but normal parturition, in oxytocin receptor-deficient mice. *Proc Natl Acad Sci USA*. 2005;102:16096—16101.
8. Lee H-J, Macbeth AH, Pagani JH, et al. Oxytocin: the great facilitator of life. *Prog Neurobiol*. 2009;88:127—151.

Vasopressin

Susumu Hyodo

Abbreviation: VP
Additional name: antidiuretic hormone (ADH)

Vasopressin is a mammalian VP family peptide with vasopressor and antidiuretic actions. Disorders of VP and V2aR cause central and nephrogenic diabetes insipidus, respectively.

Discovery

Vasopressor and antidiuretic effects of VP were first reported in 1895 and 1913, respectively. Arginine vasopressin (AVP) was isolated in 1951, and chemically synthesized in 1955 [1].

Structure

Structural Features

The N-terminal six aa residues of the nonapeptides are flanked by two cysteine residues (positions 1 and 6) forming an intramolecular ring, while the C-terminal extension with arginine or lysine residue confers a basic peptide property. The C-terminus is amidated. VPs are found only in mammals.

Primary Structure

Three peptides have been found (Figure 6A.1). Most mammals, including humans, have AVP, while lysine-VP (LVP) and phenypressin were discovered in pig and marsupials, respectively.

Properties

Mr 1,084 (AVP), 1,056 (LVP), 1,068 (phenypressin); pI, 10.9 (AVP). The peptides are freely soluble in water. AVP solution in water at $>10^{-4}$ M is stable for more than a year at $-20°C$, with increased stability under acidic conditions using acetic acid.

Synthesis and Release

Gene, mRNA, and Precursor

The human AVP gene located on the short arm of chromosome 20 (20p13), consists of three exons and two introns [2]. The oxytocin (OT) gene is located in the vicinity of the AVP gene in a tail-to-tail configuration, and they are transcribed in opposite directions. Human AVP mRNA has 595 bp encoding a signal peptide, a mature AVP with processing and amidation sites (Gly−Lys−Arg), a neurophysin, and a glycoprotein (copeptin) (Figure 6A.2). Processing of the prohormone occurs in the acidic pH of large dense core vesicles.

The neurophysin non-covalently binds with mature peptide in the vesicles and functions as a intravesicular chaperone and carrier protein, while the function of copeptin is still unknown. The AVP gene has glucocorticoid and cAMP responsive elements, and AP-1/2 regulatory elements in the promoter region. A downstream region of the AVP gene is also important as an enhancer.

Distribution of mRNA

Vasopressin is most abundantly produced in magnocellular neurosecretory neurons in the supraoptic and paraventricular (PVN) nuclei, transported to terminals in the neurohypophysis, and released into the general circulation. VP mRNA is also found in parvocellular neurons in the PVN; VP produced in these neurons is transported to terminals in the external layer of the median eminence, from which VP is released into the hypophysial portal system. AVP is also produced in neurons in the suprachiasmatic nucleus, the bed nucleus of the stria terminalis (BNST), and the medial amygdala in rodents. In peripheral organs, AVP mRNA has been detected in the adrenals, thymus, and gonads.

Tissue and Plasma Concentrations

Concentrations in rat tissues are shown in Table 6A.1 [3−5]. Plasma concentration is 1−5 fmol/ml. AVP content in the pituitary is higher in daytime, while the plasma level at night is twice as high as that during the day. In urine, the AVP concentration is approximately 100 times higher than that in plasma.

Regulation of Synthesis and Release

Synthesis and release of AVP are stimulated by elevation of plasma osmolality and decreases in blood volume and blood pressure. Magnocellular AVP neurons receive signals of various neurotransmitters and neuromodulators from the circumventricular organs and medulla, including GABA, noradrenaline, dopamine, glutamic acid, atrial natriuretic peptide, and angiotensin II. In the parvocellular PVN, adrenalectomy increases AVP synthesis, while elevation is restored by dexamethasone administration. Direct regulation by androgen and estrogen is shown in the BNST and amygdala. Various peptide hormones (such as galanin, enkephalin, neuropeptide Y, vasoactive intestinal peptide, dynorphin) are co-localized in AVP neurons, and are involved in the regulation of AVP and OT release.

Y. Takei, H. Ando, & K. Tsutsui (Eds): Handbook of Hormones. DOI: http://dx.doi.org/10.1016/B978-0-12-801028-0.00113-6

Arg-Vasopressin CYFQNCPRGa
Lys-Vasopressin CFYQNCPKGa
Phenypressin   CFFQNCPRGa

Figure 6A.1 Primary structure of VPs.

Table 6A.1 Tissue Concentration of AVP

Species	Tissues	Concentration	Reference
Rat	Hypothalamus	$1.58 \pm 0.28\ \mu g/g$ ($n = 4$)	[3]
	Pituitary	$130 \pm 22\ \mu g/g$ ($n = 3$)	
	Adrenal	$11.5\ ng/g$	
	Urine	$120\ pg/ml$	
Human	Urine	$50\ pg/ml$	[4]
Rat brain	Neurohypophysis	(pg/μg protein; $n = 5–7$)	[5]
	Normal	$11{,}537 \pm 1{,}247$	
	7 days' salt loading	774 ± 233	
	Median eminence		
	Normal	227 ± 37	
	2 days' salt loading	414 ± 40	
	Supraoptic nucleus		
	Normal	126 ± 17	
	2 days' salt loading	155 ± 30	

Receptors

Structure and Subtype

Three functional receptors have been identified (V1aR, V1bR, and V2aR) [6,7]. In the human, the length (aa residues), chromosomal location, and number of exons are as follows: V1aR, 418, 12q14−15, two exons; V1bR, 424, 1q32, two exons; V2aR, 371, Xq28, three exons. AVP binds to three receptors with high affinity ($K_d < 2\ nM$). GPCRs including VP and OT receptors may also function as homo/hetero-dimers/oligomers.

Signal Transduction Pathway

V1aR and V1bR couple to G_q and PLC pathways, while V2aR couples to G_s and AC pathways. In addition, a single receptor may activate multiple signaling pathways by coupling to different G proteins and/or other signaling intermediates.

Agonists

A number of peptide and orally active non-peptide agonists and antagonists have been investigated [8]. Recently, new research to find coupling-selective ligands and to design bivalent ligands has been conducted. Potent selective agonists are [Phe[2], Orn[8]]VP and F-180 (V1a agonists), d[Cha[4]]AVP (V1b agonist), and dDAVP and dVDAVP (V2a agonists). The selectivity varies between human and rat receptors.

Antagonists

Potent selective antagonists are d(CH$_2$)$_5$[Tyr(Me)[2]]AVP and SR49059 (V1a antagonists); SSR149415 (V1b antagonist); and OPC31260, OPC41061, and SR121463 (V2 antagonists).

Biological Functions

Target Cells/Tissues and Functions

The most prominent function of AVP is its antidiuretic action through V2aR/aquaporin-2 (AQP2) in the inner medullary collecting duct of the kidney. AVP administration also induces a large increase in blood pressure. This effect is caused by vascular contraction, basoreflex, renin production, and aldosterone and glucocorticoid releases from the adrenal gland, all

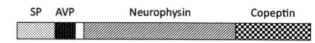

Figure 6A.2 Precursor structure of AVP.

of which are mediated via V1aR. In addition, adrenocorticotropic hormone (ACTH) and catecholamine release via V1bR and body fluid retention via V2aR contribute to blood pressure regulation. AVP also regulates metabolic function via V1aR. AVP affects fatty acid release from adipose tissue, while in the liver AVP enhances glycogenolysis to increase circulating glucose levels. AVP stimulates insulin and glucagon release from the pancreas, and these effects are mediated by V1bR. In the central nervous system, AVP is involved in social behaviors, learning and memory, aggression, anxiety and depression, water and food intake, circadian rhythm, and thermoregulation [9]. In the pituitary, AVP synergistically controls the secretion of ACTH with corticotropin-releasing hormone (CRH).

Phenotype in Gene-Modified Animals

Knockouts (KOs) of V1aR and V1bR confirm their contributions to the biological functions mentioned above [6]. V1aR KO animals showed impairments of spatial memory and social recognition, reduced anxiety-like behavior, hypotension, decreased susceptibility to salt-induced hypertension, renal tubular acidosis, decreased urine osmolality, enhanced lipid metabolism, and reduced circadian rhythmicity of locomotor activities. V1bR KO animals showed reduced aggressive behavior and social recognition memory, and decreased fasting-induced plasma glucose, insulin, and glucagon levels. Although there was no significant difference in the basal plasma levels of ACTH and corticoids, the ACTH response during chronic stress was impaired in V1bR KO mice. The Brattleboro rat is a natural KO model of AVP [10]. Although the coding region for AVP is normal, a single nucleotide deletion results in synthesis of an altered AVP precursor. This strain exhibits central diabetes insipidus, showing polyuria (inability to concentrate urine) and polydypsia (increased fluid intake). On the other hand, a loss-of-function mutation of V2aR causes X-linked congenital nephrogenic diabetes insipidus (X-CNDI) [7]. More than 200 different V2aR mutations known to cause X-CNDI have been reported, and these are classified into five types.

Pathophysiological Implications

Clinical Implications

Approximately 90% of CNDI patients are males with mutations in the V2aR gene on the X-chromosome. Besides CNDI, gain-of-function mutations in V2aR cause nephrogenic syndrome of inappropriate antidiuresis (NSIAD), characterized by excessive sodium excretion, low plasma sodium concentration, and low plasma osmolality [11]. The syndrome of inappropriate antidiuretic hormone secretion (SIADH) is typically associated with measurably elevated AVP levels and consequent V2aR hyperactivity [11].

Use for Diagnosis and Treatment

dDAVP (1-desamino-[D-Arg[8]]VP, desmopressin) is widely used clinically as a long-lasting and selective V2aR agonist for the treatment of central diabetes insipidus. For X-CNDI, the use of non-peptide antagonists (pharmacological chaperones) and non-peptide agonists (vaptans) has been tested [11]. These cell-permeable reagents stabilize ER-retained misfolded

V2aR mutants, allowing the receptors to reach the plasma membrane, or to initiate a cAMP response, leading to AQP2 translocation. Strategies bypassing non-functional or mislocalized V2aR to induce AQP2 translocation and phosphorylation have also been tested [11]. OPC41061 (V2aR antagonist, tolvaptan) has been approved in the USA and Europe for the treatment of hyponatremia in SIADH.

References

1. Walter R, Rudinger J, Schwartz IL. Chemistry and structure—activity relations of the antidiuretic hormones. *Am J Med.* 1967;42:653−677.
2. Burbach JPH, Luckman SM, Murphy D, Gainer H. Gene regulation in the magnocellular hypothalamo-neurohypophysial system. *Physiol Rev.* 2001;81:1197−1267.
3. Nussey SS, Ang VT, Jenkins JS, Chowdrey HS, Bisset GW. Brattleboro rat adrenal contains vasopressin. *Nature.* 1984;310:64−66.
4. Bankir L. Antidiuretic action of vasopressin: quantitative aspects and interaction between V1a and V2 receptor-mediated effects. *Cardiovasc Res.* 2001;51:372−390.
5. Zerbe RL, Palkovits M. Changes in the vasopressin content of discrete brain regions in response to stimuli for vasopressin secretion. *Neuroendocrinology.* 1984;38:285−289.
6. Koshimizu T, Nakamura K, Egashira N, et al. Vasopressin V1a and V1b receptors: from molecules to physiological systems. *Physiol Rev.* 2012;92:1813−1864.
7. Moeller H, Rittig S, Fenton RA. Nephrogenic diabetes insipidus: essential insights into the molecular background and potential therapies for treatment. *Endocrine Rev.* 2013;34:278−301.
8. Manning M, Stoev S, Chini B, et al. Peptide and non-peptide agonists and antagonists for the vasopressin and oxytocin V1a, V1b, V2 and OT receptors: research tools and potential therapeutic agents. *Prog Brain Res.* 2008;170:473−512.
9. Caldwell HK, Lee H-J, Macbeth AH, Young III WS. Vasopressin: behavioral roles of an "original" neuropeptide. *Prog Neurobiol.* 2008;84:1−24.
10. Bohus B, de Wied D. The vasopressin deficient Brattleboro rats: a natural knockout model used in the search for CNS effects of vasopressin. *Prog Brain Res.* 1998;119:555−573.
11. Swaab DF. The human hypothalamo-neurohypophysial system in health and disease. *Prog Brain Res.* 1998;119:577−618.

Supplemental Information Available on Companion Website

- Gene and precursor structures of human AVP/E-Figure 6A.1
- Precursor and mature hormone sequences of human AVP/E-Figure 6A.2
- Gene structures of human AVP receptors/E-Figure 6A.3
- Primary sequences of the human V1a, V1b, and V2a receptors/E-Figure 6A.4
- Accession numbers of genes and cDNAs for VPs and VP receptors/E-Table 6A.1

Vasotocin

Susumu Hyodo

Abbreviation: VT
Additional names: arginine vasotocin (AVT), antidiuretic hormone (ADH)

VT is a vasopressin family peptide of all non-mammalian vertebrates; roles in body fluid homeostasis, reproduction, behavior, and stress response have been implicated.

Discovery

Vasotocin was chemically synthesized in 1958 as an analog of neurohypophysial hormones containing the ring of oxytocin and the side chain of vasopressin. Subsequently, VT has been isolated from various vertebrate species. All non-mammalian vertebrates have VT as a VP family peptide. Since cyclostomes contain only VT, the VT gene is considered to be an ancestral gene for all vertebrate neurohypophysial hormones.

Structure

Structural Features and Primary Structure

The intramolecular ring structure of VT is identical to that of OT, while the eighth aa residue in the C-terminal extension is arginine (Figure 6B.1). Therefore, VT is a basic VP family peptide. The C-terminus of VT is amidated. In anuran amphibians, two types of C-terminally extended peptides called hydrins have been found.

Properties

Mr 1,050. VT is freely soluble in water, with increased stability under acidic conditions using acetic acid.

Synthesis and Release

Gene, mRNA, and Precursor

From cartilaginous fish to birds, the VT gene is located on the same chromosome with the respective oxytocin (OT) family genes in tail-to-head orientation [1]. Existence of *cis*-regulatory elements that mediate neuron-specific expression was suggested for the pufferfish fugu. Similarly to VP precursors, VT precursors are composed of a signal peptide, mature peptide, processing and amidation motifs, neurophysin, and copeptin moieties (see Subchapter 6A). The copeptins of tetrapods and cartilaginous fish are considered to be glycopeptides, while no glycosylation site is found in teleosts and cyclostomes.

Distribution of mRNA

VT is most abundantly produced in hypothalamic magnocellular neurosecretory neurons, and is released into the general circulation from the neurohypophysis. VT is also produced in hypothalamic and extra-hypothalamic parvocellular neurons, such as the preoptic area, bed nucleus of stria terminalis (BNST), septum, and amygdala. Co-localization of VT and corticotropin-releasing hormone (CRH) has been demonstrated. In birds, these nuclei are sexually dimorphic; VT neurons are more abundant in adult males. Male-specific expression of VT is also found in several hypothalamic nuclei in medaka. In peripheral organs, VT mRNAs are found in the kidney, ovary, uterus (shell gland), intestine, rectal gland, and several other organs, but distributions vary among species.

Tissue and Plasma Concentrations

Tissue and plasma concentrations measured by radioimmunoassay (RIA) are shown in Table 6B.1 [2–8]. VT levels are measured also by ELISA or HPLC coupled to several detection methods, and the values measured by these methods seem to be higher than those resulting from use of RIA methods.

Regulation of Synthesis and Release

In birds, the synthesis and release of VT are increased by water deprivation, drinking hypertonic saline, and hemorrhage. Involvement of a transcription factor, TonEBP, is implicated in enhanced VT synthesis after salt loading. Osmotic manipulation increases the synthesis and release of VT in reptiles, amphibians, many teleosts, and sharks, too. In birds, plasma VT levels transiently increase at the time of oviposition and decrease within 30 minutes after egg-laying, which is correlated with increases in uterine contractility. Direct and indirect regulation of VT synthesis by androgen and estrogen is shown in the sexually dimorphic BNST and amygdala. VT release from the neurohypophysis is stimulated by steroid hormones and other hormones, including angiotensin II and urotensin I. Brain VT content is high in aggressive stickleback males that take care of eggs.

Receptors

Structure and Subtype

Currently, four specific receptors have been identified in non-mammalian vertebrates [9]. Of these, V2bR is a newly discovered receptor and is either a pseudogene or has disappeared in mammals, while the other three receptors (V1aR, V1bR, and V2aR) are orthologous to mammalian VP receptors. Specifically, and only in birds, VT receptors have been named

Y. Takei, H. Ando, & K. Tsutsui (Eds): Handbook of Hormones. DOI: http://dx.doi.org/10.1016/B978-0-12-801028-0.00114-8

SUBCHAPTER 6B Vasotocin

CYIQNCPRGa

Figure 6B.1 Primary sequence of VT.

Table 6B.1 Tissue and Plasma Concentration of VT

Animals and Tissues	Concentration	Ref.
Chicken		
Ovary	2.5 ± 0.6 ng/tissue ($n = 5-6$)	[2]
Plasma (control)	25.54 ± 1.59 pg/ml ($n = 5-7$)	[3]
Plasm (1.5M NaCl injection	68.64 ± 6.84 pg/ml ($n = 5-7$)	
Plasma (oviposition)	>150 pg/ml	
Crab-eating frog		
Plasma (control)	29.8 ± 2.3 fmol/ml ($n = 6$)	[4]
Plasma (dehydrated)	57.9 ± 15.7 fmol/ml ($n = 8$)	
Plasma (80% SW)	110.4 ± 41.8 fmol/ml ($n = 10$)	
Flounder		
Pituitary (SW)	1.13 ± 0.25 nmol/g ($n = 8$)	[5]
Plasma (SW)	6.1 ± 2.7 fmol/ml ($n = 9$)	
Plasma (day)	4.4 ± 0.8 fmol/ml ($n = 13$)	[6]
Plasma (night)	1.5 ± 0.4 fmol/ml ($n = 13$)	
Shark		
Plasma (control)	82.6 ± 18.3 fmol/ml ($n = 8$)	[7]
Plasma (60% SW)	51.9 ± 9.3 fmol/ml ($n = 8$)	
Lamprey		
Plasma (freshwater, FW)	1.93 ± 0.17 pg/ml ($n = 5$)	[8]
Plasma (50% SW)	1.70 ± 0.21 pg/ml ($n = 11$)	
Pituitary (FW)	20.79 ± 6.14 ng/gland ($n = 5$)	
Pituitary (50% SW)	3.40 ± 1.02 ng/gland ($n = 5$)	

differently as VT1R (V2bR), VT2R (V1bR), VT3R (mesotocin receptor, MTR), and VT4R (V1aR) [10]. V2bR is further duplicated into type 1 (found only in fishes) and type 2 (tetrapods and some fishes) [9]. V2aR has not been found in birds and cartilaginous fishes, while the existence of V1bR in teleosts has not been proven. In teleosts, further duplications of receptors have been found.

Signal Transduction Pathway

Similarly to mammalian VP receptors, V1aR and V1bR couple to Ca^{2+} signaling pathways, while V2aR couples to the adenylate cyclase pathway. V2bRs, including chicken VT1R, are structurally related to the conventional V2 receptor (V2aR), but couple to Ca^{2+} signaling pathways.

Agonists and Antagonists

Only a few studies have tested the selectivity and potency of agonists and antagonists to non-mammalian receptors. The following antagonists for mammalian receptors have been used: V1a antagonists, OPC-21268, SR-49059, d(CH$_2$)$_5$[Tyr(Me)^2]AVP (Manning compound); V1b antagonist, SSR-149415; V2a antagonists, OPC-31260, d(CH$_2$)$_5$[D-Ile^2, Ile^4, Ala-NH$_2$]AVP.

Biological Functions

Target Cells/Tissues and Functions

VT is important for osmoregulation and ionoregulation in vertebrate groups from fishes to birds. The antidiuretic effect is particularly important in terrestrial species. VT is also implicated in cardiovascular and glycogenolytic effects, circadian and seasonal biology, uterine contraction, stress response, and social and reproductive behaviors [10–12].

Body-Fluid Regulation

In the kidney, VT constricts the preglomerular afferent arteries to reduce the glomerular filtration rate, while VT increases water permeability of the collecting duct through a VT-dependent aquaporin-2 (AQP2). Because cartilaginous fishes do not have V2aR, and since AQP2 does not appear to be present in cartilaginous fishes and ray-finned fishes, it is most probable that the VT/V2aR/AQP2 axis first appeared in lobe-finned fishes [13]. In amphibians, the urinary bladder and the highly vascularized ventral skin are also the targets of VT. VT promotes Na$^+$, urea, and water reabsorption from the bladder, and cutaneous drinking by enhancing water permeability, blood flow, and Na$^+$ transport in the skin. Hydrins increase water absorption across the skin and the bladder. In teleosts, V2aR and newly discovered V2bR are expressed in the kidney; however, their contribution to renal tubule function is yet to be clarified. The glomerular effect is most likely predominant in fishes. Indeed, localization of V1aR in the kidney of saltwater (SW) flounder implies the glomerular antidiuretic action of VT [12]. Effects of VT on ionoregulation in gills and intestine are implicated. VT may also play an important role in oocyte hydration.

Reproduction and Stress

VT is a key regulator of oviposition in birds. Both VT1R (V2bR) and VT3R (MTR) are expressed in the shell gland [10]. In catfish, VT is involved in ovarian function via the regulation of ovarian steroidogenesis. A stress response is also a common physiological effect of VT. In birds, VT and CRH synergistically induce adrenocorticotropic hormone (ACTH) and corticosterone release. Both VT2R (V1bR) and VT4R (V1aR) are expressed in the pituitary corticotrophs, and expression of VT2R (V1bR) is increased by acute stress. In teleosts, both V1aR and V2aR are expressed in many tissues, including the brain, pituitary, gills, liver, kidney, and gonads, but their distributions vary among species. Since V1bR has not been found in teleosts, V1aR and/or V2aR may mediate the ACTH release caused by VT. In the newt, only V1bR was found in the pituitary.

Behavior

In the brain, VT is important for social and sexual behaviors. In birds, VT exerts sex-specific effects on gregariousness, pair bonding, grouping, aggression, parental care, song, and sexual behaviors [14]. Testosterone treatment increases V1aR mRNA levels in the sexually dimorphic nuclei, and affects both V1aR and MTR in nuclei of the song system. In amphibians, effects of VT on calling, receptivity in female frogs, and clasping and courtship behaviors have been suggested [15]. In courtship behavior, VT enhances the incidence and frequency of these behaviors, induces the release of pheromone by contracting the abdominal gland, and induces spermatophore deposition. In teleost fishes, VT affects behaviors related to aggression, anxiety, cooperation, territory, courtship, and social subordinance, but patterns vary across species [16]. VT projection and V1aR showed widespread distribution in fish brain, from the olfactory bulb to the hindbrain. Experiments using antagonists suggest that the behavioral effects of VT in non-mammalian vertebrates are mostly mediated by V1aR.

Phenotype in Gene-Modified Animals

Antisense knockdown of VT production in finches reduces gregariousness, exerts sex-specific effects on aggression, and reduces courtship [17].

References

1. Gwee P-C, Tay B-H, Brenner S, Venkatesh B. Characterization of the neurohypophysial hormone gene loci in elephant shark and the Japanese lamprey: origin of the vertebrate neurohypophysial hormone genes. *BMC Evol Biol.* 2009;9:47.
2. Saito N, Kinzler S, Koike TI. Arginine vasotocin and mesotocin levels in theca and granulosa layers of the ovary during the oviposition cycle in hens (*Gallus domesticus*). *Gen Comp Endocrinol.* 1990;79:54–63.

3. Saito N, Furuse M, Sasaki T, Arakawa K, Shimada K. Effects of naloxone on neurohypophyseal peptide release by hypertonic stimulation in chicks. *Gen Comp Endocrinol.* 1999;115:228−235.

4. Uchiyama M, Maejima S, Wong MK, et al. Changes in plasma angiotensin II, aldosterone, arginine vasotocin, corticosterone, and electrolyte concentrations during acclimation to dry condition and seawater in the crab-eating frog. *Gen Comp Endocrinol.* 2014;195:40−46.

5. Bond H, Winter MJ, Warne JM, McCrohan CR, Balment RJ. Plasma concentrations of arginine vasotocin and urotensin II are reduced following transfer of the euryhaline flounder (*Platichthys flesus*) from seawater to fresh water. *Gen Comp Endocrinol.* 2002;125:113−120.

6. Kulczykowska E, Warne JM, Balment RJ. Day−night variations in plasma melatonin and arginine vasotocin concentrations in chronically cannulated flounder (*Platichthys flesus*). *Comp Biochem Physiol A.* 2001;130:827−834.

7. Hyodo S, Tsukada T, Takei Y. Neurohypophysial hormones of dogfish, *Triakis scyllium*: structures and salinity-dependent secretion. *Gen Comp Endocrinol.* 2004;138:97−104.

8. Uchiyama M, Saito N, Shimada K, Murakami T. Pituitary and plasma arginine vasotocin levels in the lamprey, *Lampetra japonica*. *Comp Biochem Physiol A.* 1994;107:23−26.

9. Yamaguchi Y, Kaiya H, Konno N, et al. The fifth neurohypophysial hormone receptor is structurally related to the V2-type receptor but functionally similar to V1-type receptor. *Gen Comp Endocrinol.* 2012;178:519−528.

10. Baeyens DA, Cornett LE. The cloned avian neurohypophysial hormone receptors. *Comp Biochem Physiol B.* 2006;143:12−19.

11. Uchiyama M, Konno N. Hormonal regulation of ion and water transport in anuran amphibians. *Gen Comp Endocrinol.* 2006;147:54−61.

12. Balment RJ, Lu W, Weybourne E, et al. Arginine vasotocin, a key hormone in fish physiology and behavior: a review with insights from mammalian models. *Gen Comp Endocrinol.* 2006;147:9−16.

13. Konno N, Hyodo S, Yamaguchi Y, et al. Vasotocin/V2-type receptor/aquaporin axis exists in African lungfish kidney but is functional only in terrestrial condition. *Endocrinology.* 2010;151:1089−1096.

14. Goodson JL, Kelly AM, Kingsbury MA. Evolving nonapeptide mechanisms of gregariousness and social diversity in birds. *Hormones Behav.* 2012;61:239−250.

15. Moore FL, Boyd SK, Kelley DB. Historical perspective: hormonal regulation of behaviors in amphibians. *Hormones Behav.* 2005;48:373−383.

16. Godwin J, Thompson R. Nonapeptides and social behavior in fishes. *Hormones Behav.* 2012;61:230−238.

17. Kelly AM, Goodson JL. Hypothalamic oxytocin and vasopressin neurons exert sex-specific effects on pair bonding, gregariousness, and aggression in finches. *Proc Natl Acad Sci USA.* 2014;111:6069−6074.

Supplemental Information Available on Companion Website

- Organization of VP- and OT-family genes in vertebrates/E-Figure 6B.1
- Gene and precursor structures of the elephant fish VT/E-Figure 6B.2
- Precursor and mature hormone sequences of the chicken, lungfish, chum salmon, houndshark, and hagfish VT/E-Figure 6B.3
- Primary sequences of the elephant fish V1a, V1b, and V2b receptors/E-Figure 6B.4
- Molecular phylogenetic tree of vertebrate VP family receptors/E-Figure 6B.5
- Accession numbers of genes and cDNAs for VT and VT receptors/E-Table 6B.1

Oxytocin

Susumu Hyodo

Abbreviation: OT, OXT

OT is a mammalian oxytocin family peptide stimulating uterine contraction and milk ejection. Recently, its central effects have attracted attention and it is considered a "great facilitator of life."

Discovery

The stimulatory activity on uterine contraction was reported in 1906. Oxytocin (OT) was isolated in 1949 and chemically synthesized in 1953 [1]. OT is the mammalian OT family peptide, but it is also found in holocephalan cartilaginous fishes.

Structure

Structural Features

The N-terminal 6 aa residues of OT are flanked by two Cys residues forming an intramolecular ring, while a C-terminal extension with leucine residue characterizes the neutral peptide property (Figure 6C.1). The C-terminal is amidated.

Properties

Mr 1,007; pI 7.7. Freely soluble in water. 1 mg of OT is equivalent to approximately 500 IU.

Synthesis and Release

Gene, mRNA, and Precursor

The human OT gene, located on the short arm of chromosome 20 (20p12.2) in the vicinity of the AVP gene, consists of three exons and two introns [2]. Human OT mRNA has 493 bp that encode a signal peptide, a mature OT with processing and amidation sites (Gly–Lys–Arg), and a neurophysin (Figure 6C.2). OT precursor lacks a glycoprotein (copeptin) moiety. In vesicles, the processed OT forms a complex with neurophysin in a 1:1 ratio. Cys^1 and Tyr^2 of OT are the principal neurophysin binding residues. The binding is strong in the acidic environment within the neurosecretory vesicles, while dissociation of the complex is facilitated in the neutral condition in plasma. OT promoter in the 5′-flanking region contains a POU-homeodomain binding region and composite hormone response elements (cHREs). The cHREs are composed of multiple-hormone responsive motifs, such as for interaction with estrogen receptors, thyroid hormone receptors, retinoic acid receptors, and an orphan receptor COUP-TF1. Interactions with multiple receptors results in the intricate positive and negative regulation of OT gene expression in different physiological conditions. Enhancers are found downstream of the OT gene.

Distribution of mRNA

Oxytocin is most abundantly produced in magnocellular neurosecretory neurons in the supraoptic (SON) and paraventricular (PVN) nuclei. OT mRNA is also found in parvocellular neurons in the PVN, and the bed nucleus of stria terminalis (BNST), medial preoptic area, and lateral amygdala. In peripheral organs, OT mRNA has been detected in the ovary, uterus, placenta, testis, prostate gland, adrenal, and thymus, depending on species.

Tissue and Plasma Concentrations

Concentrations in rat brain are shown in Table 6C.1 [3–5]. Concentration in the sheep corpus luteum is 2.6 μg/g. OT content in the pituitary is high in the morning of proestrus. Plasma concentration is 10–20 fmol/ml. The plasma level is high in proestrus.

Regulation of Synthesis and Release

Synthesis and release of OT changes according to sexual maturation, ovarian cycle, pregnancy, parturition, and lactation [6]. Plasma concentrations vary during the ovarian cycle in the rat and human, with the highest levels corresponding to peak estrogen levels. In the rat SON, OT mRNA levels are high during estrous. OT is accumulated in magnocellular neurons during pregnancy, and is released at the time of parturition. Manipulation of estrogen and progesterone in ovariectomized virgin rats to mimic the changes during pregnancy enhances OT mRNA levels in the SON and PVN. OT gene expression falls postpartum, and then increases during early lactation. The OT mRNA level during early lactation is maintained by the suckling stimulus. The suckling stimulus by infants immediately promotes the release of OT from the neurohypophysis. Expression of OT mRNA in magnocellular neurons is also increased by osmotic stimulus.

Receptors

Structure and Subtype

One specific OT receptor (OTR) has been identified [7]. In humans, the OTR gene is located on chromosome 3 at locus 3p25−26.2, and it encodes a protein of 389 aa residues. In most mammals the OTR gene consists of three exons, while in the human and mouse an additional intron interrupts the first exon (consequently yielding four exons). Estrogen stimulates OTR expression in the uterus. Complete and/or half-palindromic EREs are found in the promoters of OTR genes. Estrogen-induced

Y. Takei, H. Ando, & K. Tsutsui (Eds): Handbook of Hormones. DOI: http://dx.doi.org/10.1016/B978-0-12-801028-0.00115-X

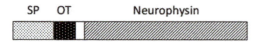

Figure 6C.1 Primary sequence of OT.

| SP | OT | Neurophysin |

Figure 6C.2 Precursor structure of OT.

Table 6C.1 Concentration of OT in the Rat Brain

Tissues	Concentration	Ref.
	(pg/mg protein)	
Hypothalamus	842.5 ± 133.2 (*n* = 8)	[3]
Supraoptic nucleus (SON)	678.3 ± 145.5 (*n* = 5)	
Paraventricular nucleus (PVN)	512.0 ± 170.0 (*n* = 5)	
Medial septum	63.2 ± 6.0 (*n* = 6)	
Locus coeruleus	174.7 ± 63.1 (*n* = 6)	
Nucleus of solitary tract	111.5 ± 37.2 (*n* = 6)	
	(pg/μg protein) (*n* = 10)	[4]
SON	71.8 ± 1.8	
PVN	34.4 ± 5.7	
Median eminence	416.7 ± 99.0	
Neurohypophysis	2.5 ± 0.3 μg/organ	
OT content in the PVN during estrous cycle in virgin rats		[5]
	(pg/μg protein)	
Proestrus	10.2 ± 1.0 (*n* = 12)	
Estrus	9.8. ± 0.7 (*n* = 18)	
Metestrus	14.4 ± 1.1 (*n* = 18)	
Diestrus	18.7 ± 1.7 (*n* = 10)	

OTR expression in the brain is abolished in ER-α knockout (KO) mice, whereas the basal OTR expression of the KO mice is similar to that in controls. One OTR has also been found in the holocephalan elephant fish [8]. In elephant fish, high expression of OTR is found in the muscle and uterus.

Signal Transduction Pathways

OTR couples to $G_{q/11}$ and phospholipase C pathways [7,8]. In addition, OTR activation leads to phosphorylation and activation of the MAPK pathway in uterine myometrial cells. OTR also activates the RhoA–Rho–kinase cascade, which in turn leads to inhibition of myosin phosphatase. Stimulation of nitric oxide production represents an additional signaling pathway to mediate vasodilation, natriuresis, and ANP release.

Agonists

[Thr^4, Gly^7]OT and HO[Thr^4, Gly^7]OT are potent and selective OTR agonists exhibiting negligible vasopressor (V1aR) and antidiuretic (V2aR) activities in rats [9]. In rats, [Thr^4, Gly^7]OT is more selective to OTR than OT. In humans, HO[Thr^4]OT exhibits high affinity and selectivity to OTR, but [Thr^4, Gly^7] OT is not highly selective.

Antagonists

Atosiban (d[D-Tyr(Et)^2, Thr^4, Orn^8]vasotocin) is a well-known peptidic OTR antagonist, but also has an affinity for V1aR [9]. The effectiveness of atosiban in delaying premature labor has been demonstrated in many species, including humans. The non-peptide antagonist SSR126768 has high specificity to OTR.

Biological Functions

Target Cells/Tissues and Functions

OT is involved in parturition and lactation [6]. OT induces milk production through stimulating prolactin release. OT also stimulates contraction of the myoepithelial cells surrounding the alveoli of the mammary gland, resulting in milk ejection. In addition to these peripheral effects, central effects of OT have attracted attention both in laboratory animals [10] and in humans [11]. OT is produced in parvocellular neurons in the PVN, BNST, medial preoptic area, and lateral amygdala. Projections of OT neurons and OTR are widely distributed throughout the brain: in rodents, for example, they are found in the olfactory bulb, cortex, hippocampus, amygdala, septum, BNST, medulla oblongata, and spinal cord. In the brain, OT is involved in the regulation of a wide variety of behaviors, including social recognition, sexual behavior, paternal and maternal behaviors, aggressive behavior, spatial and non-spatial memory, anxiety, and depression. For example, OT is implicated in the formation of social bonding between individuals, for both partners and pups; for pups, OT primarily acts in females. In humans, OT also facilitates love and pair bonding, sexual behavior, mother–infant bonding, and trust, and reduces fear and anxiety behavior; it is thus considered to enhance well-being.

Phenotype in Gene-Modified Animals

Mice lacking OT have no obvious deficits in fertility or reproduction, including during gestation and parturition. However, all offspring die shortly after birth because of the mother's inability to nurse, showing that an essential role of OT is milk ejection [12]. The OT-KO mice also have deficits in social recognition and maternal behavior, as well as increased aggression. OTR-KO mice display autistic-like deficits in social behaviors, increased aggression, and aberration in mother–offspring interaction [13]. Other physiological defects, including obesity and dysfunction in body temperature, are also shown in OTR-KO mice.

Pathophysiological Implications
Clinical Implications

Several neuropsychiatric disorders, including obsessive-compulsive disorder, autism, eating disorders, addiction, schizophrenia, and post-traumatic stress disorder (PTSD), show OT abnormalities [10,11]. A positive correlation has been indicated between OT levels and happiness, while a negative correlation with pain, stress, and depression has been observed. Lower concentrations of OT in the cerebrospinal fluid have been reported in otherwise healthy women with exposure to childhood abuse or neglect. Single nucleotide polymorphisms in the OT and OTR genes are thought possibly to be linked with autism spectrum disorders.

Use for Diagnosis and Treatment

Synthetic OT (pitocin) is used to induce labor and to help milk production. Atosiban, a peptidic antagonist, is approved in many countries under the name of Tractocile for the treatment of preterm labor. OT treatment of adults with autism or Asperger's disorder significantly reduces both the number and the severity of repetitive behaviors, and increases the ability to comprehend and remember the affective component of spoken words [10,11]. Since OT is degraded in the gastrointestinal tract, it must be administered by injection or as nasal spray.

References

1. Walter R, Rudinger J, Schwartz IL. Chemistry and structure–activity relations of the antidiuretic hormones. *Am J Med.* 1967;42:653–677.
2. Burbach JPH, Luckman SM, Murphy D, Gainer H. Gene regulation in the magnocellular hypothalamo-neurohypophysial system. *Physiol Rev.* 2001;81:1197–1267.

3. Hawthorn J, Ang VT, Jenkins JS. Comparison of the distribution of oxytocin and vasopressin in the rat brain. *Brain Res.* 1984;307:289–294.

4. Jackson R, George JM. Oxytocin in microdissected hypothalamic nuclei. Significant differences between prepubertal and sexually mature female rats. *Neuroendocrinology.* 1980;31:158–160.

5. Greer ER, Caldwell JD, Johnson MF, Prange Jr AJ, Pedersen CA. Variations in concentration of oxytocin and vasopressin in the paraventricular nucleus of the hypothalamus during the estrous cycle in rats. *Life Sci.* 1986;38:2311–2318.

6. Gimpl G, Fahrenholz F. The oxytocin receptor system: structure, function, and regulation. *Physiol Rev.* 2001;81:629–683.

7. Zingg HH, Laporte SA. The oxytocin receptor. *Trends Endocrinol Metab.* 2003;14:222–227.

8. Yamaguchi Y, Kaiya H, Konno N, et al. The fifth neurohypophysial hormone receptor is structurally related to the V2-type receptor but functionally similar to V1-type receptor. *Gen Comp Endocrinol.* 2012;178:519–528.

9. Manning M, Stoev S, Chini B, et al. Peptide and non-peptide agonists and antagonists for the vasopressin and oxytocin V1a, V1b, V2 and OT receptors: research tools and potential therapeutic agents. *Prog Brain Res.* 2008;170:473–512.

10. Ishak WW, Kahloon M, Fakhry H. Oxytocin role in enhancing well-being: a literature review. *J Affect Disord.* 2011;130:1–9.

11. Lee H-J, Macbeth AH, Pagani JH, Young III WS. Oxytocin: the great facilitator of life. *Prog Neurobiol.* 2009;88:127–151.

12. Takayanagi Y, Yoshida M, Bielsky IF, et al. Pervasive social deficits, but normal parturition, in oxytocin receptor-deficient mice. *Proc Natl Acad Sci USA.* 2005;102:16096–16101.

13. Pobbe RLH, Pearson BL, Defensor EB, et al. Oxytocin receptor knockout mice display deficits in the expression of autism-related behaviors. *Horm Behav.* 2012;61:436–444.

Supplemental Information Available on Companion Website

- Gene and precursor structures of the human OT/E-Figure 6C.1
- Precursor and mature hormone sequences of human OT/E-Figure 6C.2
- Gene structure of the human OT receptor/E-Figure 6C.3
- Primary sequence of the human OT receptor/E-Figure 6C.4
- Accession numbers of genes and cDNAs for OT and OT receptor/E-Table 6C.1

Non-Mammalian Oxytocin Family Peptides

Susumu Hyodo

Additional Names/Abbreviations: Mesotocin (MT), isotocin (IT), glumitocin ([Ser^4][Glu^8]oxytocin), valitocin ([Val^8]oxytocin), aspargtocin ([Asn^4]oxytocin), asvatocin ([Asn^4][Val^8]oxytocin), phasitocin ([Phe^3][Asn^4][Ile^8]oxytocin), phasvatocin ([Phe^3][Asn^4][Val^8]oxytocin). Nomenclature is based on the amino acid sequence of each peptide.

Oxytocin family peptides of non-mammalian vertebrates. Although their physiological importance is not clear yet, their contributions to social and sexual behaviors have been implicated.

Discovery

In the 1960s, many of the oxytocin (OT) family peptides were chemically isolated [1]. Initially, peptides were distinguished pharmacologically based on their relative activities in highly sensitive biological assay systems. However, the pharmacological approach sometimes gave erroneous or ambiguous results. With the exception of cyclostomes and sharks, all non-mammalian vertebrates have one OT family peptide. Cyclostomes do not have an OT family peptide, while sharks have two distinct peptides. Some marsupials have MT in addition to OT.

Structure

Primary Sequences and Structural Features

All of the OT family nonapeptides have a six-membered intramolecular ring and a C-terminal extension with a neutral amino acid residue at position 8. The C-termini of the peptides are amidated. All non-mammalian tetrapods (birds, reptiles, amphibians), lungfishes, and coelacanths have MT, while ray-finned fishes have IT (Figure 6D.1). Among cartilaginous fishes, holocephalans and batoids have OT and glumitocin, respectively. In sharks, the OT family peptides are species-dependent in their presence: aspargtocin and valitocin in the spiny dogfish *Squalus acanthias*, asvatocin and phasvatocin in the lesser-spotted dogfish *Scyliorhinus canicula*, and asvatocin and phasitocin in the houndshark *Triakis scyllium* [2]. In Australian lungfish, the OT family peptide is [Phe^2]mesotocin.

Properties

Mr values of MT and IT are 1,007 and 966, respectively. Solubility of IT in water is low, but is increased after acidification with acetic acid. Other peptides are freely soluble in water.

Synthesis and Release

Gene, mRNA, and Precursor

From cartilaginous fishes to birds, OT family peptide genes are located on the same chromosome with the VT gene in tail-to-head orientation [3]. Extensive search of the lamprey genome supported the claim that cyclostomes do not have an OT family peptide. Peptide precursors are composed of a signal peptide, an OT family peptide, processing and amidation motifs, and a neurophysin (Figure 6D.2). IT neurophysins of ray-finned fishes are extended in length and contain a copeptin moiety, similar to VT precursors. Proteolytic cleavage between IT neurophysin and copeptin moieties does not occur.

Distribution of mRNA

In contrast to VT neurons, OT family peptide-producing neurons are distributed in restricted areas of the brain. In birds, MT neurons are predominantly localized in the parvocellular and magnocellular subgroups of the paraventricular nucleus, and only scattered MT neurons were found in the supraoptic nucleus. MT immunoreactive fibers are found widely in the brain. In amphibians, MT neurons are found in the bed nucleus of the stria terminalis (BNST), preoptic area, amygdala, and ventral thalamus in addition to the magnocellular and parvocellular preoptic nuclei; however, substantial variation occurs among species. IT neurons are found in parvocellular, magnocellular, and gigantocellular parts of the preoptic nucleus in teleosts. IT fibers densely innervate the ventral telencephalon and numerous areas in the hypothalamus and brainstem.

Tissue and Plasma Concentrations

Tissue and plasma concentrations in various species are shown in Table 6D.1 [4–12]. In the chicken, plasma MT levels are inversely related to plasma osmolality. No circadian variation is observed in plasma IT levels in trout.

Regulation of Synthesis and Release

Hyperosmotic stimulation did not affect plasma MT, brain IT mRNA, or plasma IT levels in all species examined. In masu salmon, seasonal variations are found for IT mRNA and IT immunoreactivity, and the observed changes were accompanied by increases in plasma testosterone and estrogen levels. Brain IT levels are higher in aggressive dominant male

Y. Takei, H. Ando, & K. Tsutsui (Eds): Handbook of Hormones. DOI: http://dx.doi.org/10.1016/B978-0-12-801028-0.00116-1

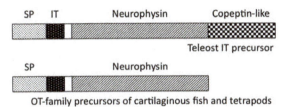

Mesotocin	CYIQNCPIGa
[Phe²]MT	CFIQNCPIGa
Isotocin	CYISNCPIGa
Glumitocin	CYISNCPEGa
Aspargtocin	CYINNCPLGa
Valitocin	CYIQNCPVGa
Asvatocin	CYINNCPVGa
Phasvatocin	CYFNNCPVGa
Phasitocin	CYFNNCPIGa

Figure 6D.1 Primary sequences of OT family peptides.

SP IT Neurophysin Copeptin-like

Teleost IT precursor

SP Neurophysin

OT-family precursors of cartilaginous fish and tetrapods

Figure 6D.2 Structure of teleost IT precursor and other OT family peptide precursors.

sticklebacks that defend their territory, but no difference in IT mRNA levels was observed between dominant and subordinate medaka males. In medaka, IT mRNA levels in the brain are higher in females. IT release from the sea bream pituitary is decreased by cortisol treatment.

Receptors

Structure and Subtype

A single receptor that belongs to the G protein-coupled receptor superfamily has been identified. Together with the mammalian OT receptors (OTRs), the receptors of OT family peptides form a single lineage in the molecular phylogenetic tree of neurohypophysial hormone receptors [13].

Signal Transduction Pathway

Similarly to V1aR, V1bR, V2bR and mammalian OTR, OT family peptide receptors couple to Ca^{2+} signaling pathways.

Agonists and Antagonists

The following antagonists to mammalian OTR have been used for MT and IT studies: d(CH₂)₅[Tyr(Me)²,Thr⁴,Orn⁸,desGly⁹-NH₂]VT, d(CH₂)₅[Tyr(Me)²,Thr⁴,Tyr⁹-NH₂]OVT (OTA), and d(CH₂)₅[D-Tyr²,Thr⁴]OVT. Potencies and selectivities of native nonapeptides, and agonists and antagonists for mammalian vasopressin and OT, have been tested for several MT, IT, and VT receptors, with the results suggesting cautious interpretation of agonist and antagonist actions.

Biological Functions

Target Cells/Tissues and Functions

Peripheral Actions

MT exhibits an ability to stimulate water and Na^+ transport *in vitro* across amphibian skin and urinary bladder, but the effect is less potent than that of VT. MT receptor (MTR) mRNA is found in the brain, pituitary, urinary bladder, kidney, adrenal, muscle, and testis of anuran amphibians [14]. In possum, MT may be important for regulating prostate growth and regression. The IT receptor (ITR) gene in sea bream is not sensitive to variations of external salinity, whereas in

Table 6D.1 Tissue and Plasma Concentrations of OT Family Peptides

Animals and Tissues	Concentration	Reference
Brushtail possum (MT)		
Corpus luteum	8.7 pmol/g	[4]
Hypothalamus	17.6 ± 0.6 ng/tissue ($n = 7$)	[5]
Pituitary	3.9 ± 0.2 µg/tissue ($n = 7$)	
Plasma (anesthetized male)	39.7 ± 9.7 pg/ml ($n = 11$)	
Plasma (conscious catheterized)	9.4 ± 6.3 pg/ml ($n = 4$)	
Tammer wallaby (MT)		
Plasma (basal)	11.6 ± 6.6 fmol/ml ($n = 12$)	[6]
Plasma (peak at birth)	546.6 ± 44.1 fmol/ml ($n = 6$)	
Chicken (MT)		
Plasma	16.8 ± 4.2 pg/ml ($n = 8$)	[7]
Plasma (0.15-M NaCl injection)	101.20 ± 8.81 pg/ml ($n = 5–7$)	[8]
Plasma (1.5-M NaCl injection)	89.45 ± 4.78 pg/ml ($n = 5–7$)	
Ovary (theca layer)	106–277 pg/tissue	[9]
Trout (IT by ELISA)		
Plasma (FW)	5.5 ± 1.3 pmol/ml ($n = 16$)	[10]
Plasma (SW)	5.3 ± 1.1 pmol/ml ($n = 13$)	
Pituitary (FW)	0.46 ± 0.10 nmol/tissue ($n = 5$)	
Pituitary (SW)	0.49 ± 0.60 nmol/tissue ($n = 4$)	
Trout (IT by HPLC)		
Pituitary (night)	0.83 ng/µg protein	[11]
Plasma	50–75 fmol/ml	[12]

pupfish, hyperosmotic stimulation decreased the expression of ITR mRNA in the gills [15]. IT exerts a relaxing effect on the upper esophageal sphincter muscle of eels, which is important for drinking behavior in fish acclimated to seawater. ITR mRNA is found in the brain, pituitary, gills, kidney, bladder, muscle, body fat, and gonads in teleosts.

Central Effects

Similarly to mammalian OT, contributions of MT and IT to social and sexual behaviors have been implicated. Central administration of MT promotes flocking behavior in the highly gregarious zebra finch [16,17]. MTR mRNA is found widely in the brain, including the lateral septum, habenula, dorsal thalamus, and cerebellum in birds [18]. MTR blockade using OT antagonists impairs the formation of pair bonds, and reduces gregariousness in finches, particularly in females. MT has also been suggested to be involved in maternal care. Testosterone treatment affects MTR mRNA expression in several brain regions. MT stimulates the biosynthesis of neurosteroids. In amphibians, MTR mRNA is found widely in the brain, including the pallium, amygdala, septum, preoptic area, preoptic nucleus, thalamus, ventromedial hypothalamus, optic tectum, and pituitary [14]. In the teleost midshipman, territorial males decrease vocalization in response to VT, while females and sneaker males respond to IT with decreases in vocalization [19]. Such contrasting effects of VT and IT are also found for social approach behavior in male goldfish. The number of IT neurons decreased over the course of female-to-male sex change in goby. IT also promotes parental care in a monogamous cichlid fish.

Phenotype in Gene-Modified Animals

Knockdown of MT in the paraventricular nucleus produces female-specific deficits in gregariousness, pair bonding, and nest cup ownership in finches. The knockdown also modulates stress-coping, while inducing hyperphagia in males [17].

References

1. Acher R. Neurohypophysial peptide system: processing machinery, hydroosmotic regulation, adaptation and evolution. *Regul Pept.* 1993;45:1–13.
2. Hyodo S, Tsukada T, Takei Y. Neurohypophysial hormones of dogfish, *Triakis scyllium*: structures and salinity-dependent secretion. *Gen Comp Endocrinol.* 2004;138:97–104.
3. Gwee P-C, Tay B-H, Brenner S, Venkatesh B. Characterization of the neurohypophysial hormone gene loci in elephant shark and the Japanese lamprey: origin of the vertebrate neurohypophysial hormone genes. *BMC Evol Biol.* 2009;9:47.
4. Sernia C, Bathgate RA, Gemmell RT. Mesotocin and arginine-vasopressin in the corpus luteum of an Australian marsupial, the brushtail possum (*Trichosurus vulpecula*). *Gen Comp Endocrinol.* 1994;93:197–204.
5. Bathgate RA, Sernia C, Gemmell RT. Mesotocin in the brain and plasma of an Australian marsupial, the brushtail possum (*Trichosurus vulpecula*). *Neuropeptides.* 1990;16:121–127.
6. Parry LJ, Guymer FJ, Fletcher TP, Renfree MB. Release of an oxytocic peptide at parturition in the marsupial, *Macropus eugenii*. *J Reprod Fertil.* 1996;107:191–198.
7. Koike TI, Neldon HL, McKay DW, Rayford PL. An antiserum that recognizes mesotocin and isotocin: development of a homologous radioimmunoassay for plasma mesotocin in chickens (*Gallus domesticus*). *Gen Comp Endocrinol.* 1986;63:93–103.
8. Saito N, Furuse M, Sasaki T, Arakawa K, Shimada K. Effects of naloxone on neurohypophyseal peptide release by hypertonic stimulation in chicks. *Gen Comp Endocrinol.* 1999;115:228–235.
9. Saito N, Kinzler S, Koike TI. Arginine vasotocin and mesotocin levels in theca and granulosa layers of the ovary during the oviposition cycle in hens (*Gallus domesticus*). *Gen Comp Endocrinol.* 1990;79:54–63.
10. Pierson PM, Guibbolini ME, Mayer-Gostan N, Lahlou B. ELISA measurements of vasotocin and isotocin in plasma and pituitary of the rainbow trout: effect of salinity. *Peptides.* 1995;16:859–865.
11. Rodriguez-Illamola A, Lopez Patino MA, Soengas JL, Ceinos RM, Miguez JM. Diurnal rhythms in hypothalamic/pituitary AVT synthesis and secretion in rainbow trout: evidence for a circadian regulation. *Gen Comp Endocrinol.* 2011;170:541–549.
12. Kulczykowska E, Stolarski J. Diurnal changes in plasma arginine vasotocin and isotocin in rainbow trout adapted to fresh water and brackish Baltic water. *Gen Comp Endocrinol.* 1996;104:197–202.
13. Yamaguchi Y, Kaiya H, Konno N, et al. The fifth neurohypophysial hormone receptor is structurally related to the V2-type receptor but functionally similar to V1-type receptor. *Gen Comp Endocrinol.* 2012;178:519–528.
14. Acharjee S, Do-Rego J-L, Oh DY, et al. Molecular cloning, pharmacological characterization, and histochemical distribution of frog vasotocin and mesotocin receptors. *J Mol Endocrinol.* 2004;33:293–313.
15. Lema SC. Identification of multiple vasotocin receptor cDNAs in teleost fish: sequences, phylogenetic analysis, sites of expression, and regulation in the hypothalamus and gill in response to hyperosmotic challenge. *Mol Cell Endocrinol.* 2010;321:215–230.
16. Goodson JL, Kelly AM, Kingsbury MA. Evolving nonapeptide mechanisms of gregariousness and social diversity in birds. *Hormones Behav.* 2012;61:239–250.
17. Kelly AM, Goodson JL. Hypothalamic oxytocin and vasopressin neurons exert sex-specific effects on pair bonding, gregariousness, and aggression in finches. *Proc Natl Acad Sci USA.* 2014;111:6069–6074.
18. Leung CH, Abebe DF, Earp SE, et al. Neural distribution of vasotocin receptor mRNA in two species of songbird. *Endocrinology.* 2011;152:4865–4881.
19. Goodwin J, Thompson R. Nonapeptides and social behavior in fishes. *Hormones Behav.* 2012;61:230–238.

Supplemental Information Available on Companion Website

- Organization of VP- and OT-family genes in vertebrates/E-Figure 6D.1
- Precursor structures of tetrapod, teleost and cartilaginous fish OT family peptides/E-Figure 6D.2
- Precursor and mature hormone sequences of the newt, lungfish, chum salmon, houndshark and elephant fish OT family peptides/E-Figure 6D.3
- Primary sequences of the newt MTR, medaka ITR and elephant fish OTR/E-Figure 6D.4
- Molecular phylogenetic tree of vertebrate OT family receptors/E-Figure 6D.5
- Accession numbers of genes and cDNAs for OT family peptides and their receptors/E-Table 6D.1

CHAPTER 7

Opioid Peptide Family

Akiyoshi Takahashi

History

Met-enkephalin, Leu-enkephalin, β-endorphin, α-neo-endorphin, dynorphin, and nociceptin/orphanin FQ (N/OFQ) are members of opioid peptides (Figure 7.1). Met-enkephalin and Leu-enkephalin were first isolated from the porcine brain in 1975. β-Endorphin was first isolated from the camel pituitary in 1976 (details for β-endorphin are described in Chapter 16). Dynorphin was first isolated from the porcine pituitary in 1979. α-Neo-endorphin was first isolated from the porcine hypothalamus in 1979. In 1995, orphanin (OFQ) was isolated from the rat and nociceptin (N) was isolated from the pig, but as the sequences for OFQ and N were identical, they are now collectively termed nociceptin/orphanin FQ (N/OFQ). All of these opioid peptides are derived from their respective precursors. Met-enkephalin and Leu-enkephalin are derived from proenkephalin (PENK, alternatively designated pro-enkephalin-A), the structure of which was elucidated in 1982; dynorphin and other related peptides are derived from prodynorphin (PDYN; alternatively, proenkephalin-B), the structure of which was elucidated in 1982; and N/OFQ is derived from pronociceptin (PNOC; alternatively, N/OFQ precursor), the structure of which was elucidated in 1996.

Structure

Structural Features

Both Met-enkephalin and Leu-enkephalin, which are derived from PENK, are composed of five aa residues; the amino acid sequence of the former is Tyr—Gly—Gly—Phe—Met and that of the latter is Tyr—Gly—Gly—Phe—Leu. Dynorphin, α-neo-endorphin, and other peptides derived from PDYN have the Leu-enkephalin sequence in their N-termini. In mammals, including humans, N/OFQ is a 17-aa peptide with a phenylalanine at the N-terminus and a glutamine at the C-terminus, and the peptide was originally termed OFQ because of this structural characteristic. The other original name N was derived from its physiological function. The N terminal 4-aa residue sequence of N/OFQ (Phe—Gly—Gly—Phe) is conserved between mammals and fish.

Molecular Evolution of Family Members

It is plausible that PENK, PDYN, PNOC, and proopiomelanocortin (POMC) originated from a common ancestor. In one scenario, PENK and POMC diverged from an ancestral opioid gene by an initial genome duplication event. During the second genome duplication event, when gnathostomes evolved from agnathans, PENK was duplicated to form PENK and PDYN, and then PNOC was derived from POMC (Figure 7.2).

The opioid core sequence repeats in each precursor, indicating that these evolved by duplication and insertion of the core sequence.

Receptors

Structure and Subtypes

Receptors for opioid peptides belong to the superfamily of GPCRs. These are classified into four subtypes: δ opioid receptor (DOR), κ opioid receptor (KOR), μ opioid receptor (MOR), and N/OFQ receptor (NOP).

Signal Transduction Pathway

The following signaling process is common to all four opioid receptor subtypes, DOR, KOR, MOR, and NOP. Opioid receptors couple to pertussis toxin-sensitive G proteins, including $G_{\alpha i}$, which inhibits cAMP formation [1]. After being activated by an agonist, the G_{α} subunit dissociates from the $G_{\beta \gamma}$ subunits and interacts with the G protein-coupled inwardly-rectifying potassium channel, Kir3. A series of processes causes cellular hyperpolarization and inhibits tonic neural activity. All four opioids also cause a reduction in Ca^{2+} currents. Opioid receptor-induced inhibition of calcium conductance is mediated by binding of the dissociated $G_{\beta \gamma}$ subunit directly to the channel.

Heteromer-Directed Signal Specificity

Within GPCR receptor—receptor interactions, or heteromerization, one protomer is likely to induce novel pharmacological properties of its interaction partner [2]. Receptor heteromerization leads to new binding properties: DOR—KOR heteromers gain active conformations that are absent in their homomeric counterparts. Heteromerization alters receptor activity and G protein activation; MOR associated with DOR decreases MOR activity in response to selective agonists. Heteromerization can lead to a switch in G protein coupling: MOR and DOR both couple to $G_{\alpha i}$ when individually expressed, whereas the heteromer selectively couples to $G_{\alpha z}$.

Biological Functions

Target Cells/Tissues and Functions

Opioid receptors are widespread in the central nervous system and nerves of the gastrointestinal tract. These receptors mediate diverse functions, including analgesia, rewards, autonomic reflexes, endocrine, and immune regulation, of endogenous opioid peptides such as enkephalins and β-endorphin.

Y. Takei, H. Ando & K. Tsutsui (Eds): Handbook of Hormones. DOI: http://dx.doi.org/10.1016/B978-0-12-801028-0.00007-6

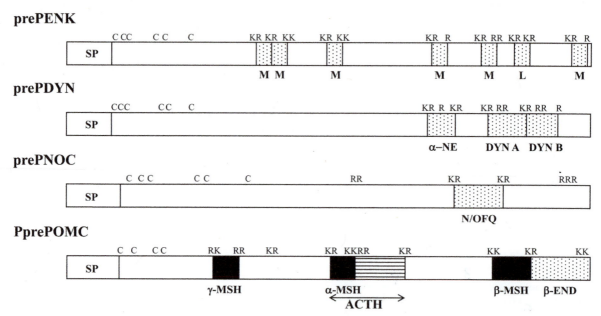

Figure 7.1 Schematic diagram of precursors for opioid peptides. M, Met-ENK; L, Leu-ENK; SP, signal peptide. C, K, and R are amino acids cysteine, lysine, and arginine, respectively.

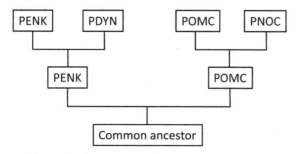

Figure 7.2 Molecular evolution of opioid gene family, based on the hypothesis proposed by Dores and Baron, 2011 [3].

In pain perception processes, these peptides increase the nociceptive threshold. In contrast, N/OFQ was first identified as a peptide exhibiting hyperalgesic activity — i.e., causing a decrease in the nociceptive threshold.

References

1. Al-Hasani R, Bruchas MR. Molecular mechanisms of opioid receptor-dependent signaling and behavior. *Anesthesiology.* 2011;115:1363–1381.
2. Rozenfeld R, Devi LA. Exploring a role for heteromerization in GPCR signalling specificity. *Biochem J.* 2011;433:11–18.
3. Dores RM, Baron AJ. Evolution of POMC: origin, phylogeny, post-translational processing, and the melanocortins. *Ann NY Acad Sci.* 2011;1220:34–48.

Enkephalin

Akiyoshi Takahashi

Abbreviations: ENK, Enk

Endogenous morphines first isolated from the brain, enkephalin pentapeptides are associated with nociception by analgesic functions.

Discovery

There are two types of ENK peptides: methionine ENK (Met-ENK, [Met^5]-ENK) and leucine ENK (Leu-ENK, [Leu^5]-ENK). Isolation of these peptides from the porcine brain was first reported in 1975.

Structure

Structural Features

The amino acid sequences of Met-ENK (Tyr—Gly—Gly—Phe—Met) and Leu-ENK (Tyr—Gly—Gly—Phe—Leu) are common in mammals. These are generated from the common precursor proenkephalin (PENK or, alternatively, proenkephalin-A [PENK-A]) by proteolytic cleavage.

Primary Structure

In addition to Met-ENK and Leu-ENK, several distinct PENK-derived peptides with C-terminal or N-terminal extensions are present in the bovine brain (Figure 7A.1). In mammals, including humans, PENK contains six copies of Met-ENK and one copy of Leu-ENK (Table 7A.1), whereas in amphibians, lungfish, and shark, PENK contains seven Met-ENK sequences. In contrast, zebrafish PENK contains five copies of Met-ENK and one copy of Leu-ENK.

Properties

Met-ENK, Mr 574; Leu-ENK, Mr 556.

Synthesis and Release

Gene, mRNA, and Precursor

The human PENK gene, *PENK*, location 8q23—q24, consists of three exons and has transcription elements such as ENKCRE-1 and ENKCRE-2. The latter acts as an enhancer. A glucocorticoid response element is also present. Human prePENK is composed of 267 aa residues. A signal peptide consisting of 24 aa residues is followed by a cysteine-containing N-terminal sequence and the region containing the repeated ENK sequences [1].

Distribution of mRNA

In the rat, neurons containing *Penk* mRNA have been found in a wide region of the brain and spinal cord. The adrenal medulla is also a major source of *Penk* mRNA, and its *Penk* mRNA content is higher than that in the pituitary and other brain regions, suggesting that the Met-ENK found in blood originates from the adrenal medulla [2].

Regulation of Synthesis and Release

In the rat, glucocorticoid is required for the maintenance of basal *Penk* mRNA expression in the forebrain, and insulin injection induces an increase in mRNA levels in the adrenal medulla. In humans, as in rats, Met-ENK that is probably derived from the adrenal medulla is present in the circulation. Acupuncture relief of chronic pain syndrome correlates with increased plasma Met-ENK concentration, whereas the level of plasma Met-ENK in chronic alcoholics is reduced [3,4].

Tissue and Plasma Concentrations

Tissue

Rat brain: Met-ENK 33.8 ng/g; Leu-ENK, 12.2 ng/g wet weight.
Met-ENK in adrenal (a) or adrenal medulla (am): human, 2,360 pg/mg (am); bovine, 4,660 pg/mg (am); dog, 36,300 pg/mg (am); guinea pig, 24.2 pg/mg (a); rat, 3.2 pg/mg (a). Leu-ENK: human, 1,924 pg/mg (am); bovine, 3,440 pg/mg (am); dog, 4,660 pg/mg (am); guinea pig, 18.1 pg/mg (a); rat, <0.2 pg/ml (a).

Plasma

Human: Met-ENK, 69.3 pg/ml; Leu-ENK, 54.0 pg/ml.

Receptors

Structure

Met-ENK and Leu-ENK are agonists for the δ-opioid receptor (DOR, also known as δ receptor, DOR-1, OP$_1$, etc.) or μ-opioid receptor (MOR; also known as μ receptor, MOR-1, OP$_3$, etc.), both of which are the subtypes of opioid receptors, which belong to the GPCR family. Human *DOR* is located on chromosome 1 (1p36.1—p34.3). Met-ENK also interacts with a non-classical opioid receptor called the opioid growth factor receptor [5].

Signal Transduction Pathway

See Chapter 7.

Y. Takei, H. Ando, & K. Tsutsui (Eds): Handbook of Hormones. DOI: http://dx.doi.org/10.1016/B978-0-12-801028-0.00117-3

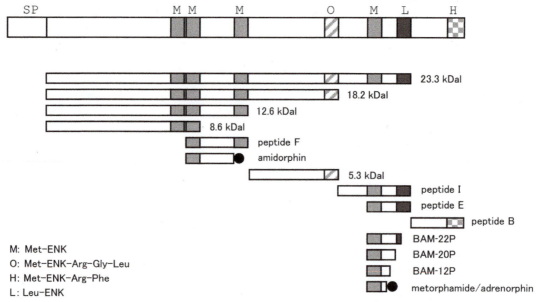

M: Met–ENK
O: Met–ENK–Arg–Gly–Leu
H: Met–ENK–Arg–Phe
L: Leu–ENK

Figure 7A.1 Location of PENK-derived peptides on the precursor in bovines.

Table 7A.1 Amino Acid Sequence of PENK-Derived Peptides in Bovines*

Leu–ENK	YGGFL	Mr 556
Met–ENK	YGGFM	Mr 574
Met–ENK–RF	YGGFMRF	Mr 877
Met–ENK–RGL	YGGFMRGL	Mr 900
Metorphamide	YGGFMRRVa	Mr 984
BAM-12P	YGGFMRRVGRPE	Mr 1,425
BAM-20P	YGGFMRRVGRPEWWMDYQKR	Mr 2,619
BAM-22P	YGGFMRRVGRPEWWMDYQKRYG	Mr 2,839
Peptide E	YGGFMRRVGRPEWWMDYQKRYGGFL	Mr 3,157
Amidorphin	YGGFMKKMDELYPLEVEEEANGGEVLa	Mr 2,947
Peptide F	YGGFMKKMDELYPLEVEEEANGGEVLGKRYGGFM	Mr 3,845
Peptide B	FAEPLPSEEEGESYSKEVPEMEKRYGGFMRF	Mr 3,657
Peptide I	SPHLEDETKELQKRYGGFMRRVGRPEWWMDYQKRYGGFL	Mr 4,848
5.3-kDa peptide	DAEEDDGLGNSSNLLKELLGAGDQREGSLHQEGSDAEDVSKRYGGFMRGL	Mr 5,324

*ENK sequences are underlined; lower case letter "a" is amide.

Agonists

DSLET, diprenorphine, DADLE, (−)-bremazocine, DPDPE, nalmefene, hydromorphone, and morphine are agonists.

Antagonists

Naltriben, naltrindole, naltrexone, quadazocine, alvimopan, and naloxone are antagonists.

Biological Functions

Target Cells/Tissues and Function

The effects of morphine represent the functions of endogenous opioid peptides via DOR, KOR, and MOR. In the central nervous system, morphine causes analgesia, euphoria, sedation, miosis (constriction of the pupils), truncal rigidity, nausea, and vomiting, and decreases the rate of respiration and the cough reflex. In the gastrointestinal system, morphine causes constipation, constriction of biliary smooth muscle, and esophageal reflux, and reduces gastric motility, digestion in the small intestine, and peristaltic waves in the colon. In other smooth muscle, morphine causes urinary retention, depresses renal function, and decreases uterine tone. In the skin, morphine causes itching, sweating, and flushing of the face, neck, and thorax. In the cardiovascular system, morphine decreases blood pressure and the heart rate if the cardiovascular system is stressed. In the immune system, morphine decreases the cytotoxic activity of natural killer cells and the formation of rosettes by human lymphocytes. Morphine also induces behavioral restlessness. Pharmacological studies using both delta agonists and delta antagonists in rodents show that the anxiolytic activity of the opioid tone is mediated by DOR. Although Met-ENK was originally identified as a neuromodulator that interacts with DOR, this peptide was subsequently revealed to be a tonically active regulator of cell proliferation as well [5–7].

Phenotype in Gene-Modified Animals

In mice lacking one of the three opioid receptors (DOR, KOR, or MOR), opioids bind to the two remaining receptors. In all mutants, the anatomical distribution of the remaining opioid receptor sites was unchanged. The absence of one opioid receptor does not drastically alter the expression of the other opioid receptors. As demonstrated in double and triple mutants, the receptors are not crucial for development in mice. Regarding pain perception, DOR- and KOR-deficient mice did not exhibit any alteration in the perception of heat, although PDYN mutant mice showed an increased pain response in the tail flick test [8].

Pathophysiological Implications

Clinical Implications

Although MOR agonists are the most commonly used drugs for the treatment of pain, mu agonists show variable efficacy in the treatment of chronic pain, partly because of the development of tolerance. Delta-opioid agonists have been shown to have beneficial effects in chronic pain and emotional disorders, and may potentially be used for treatment in these symptoms [9]. Acute alcohol intoxication stimulates the release of the endogenous opioid peptides β-endorphin, enkephalin, and dynorphin, and non-selective antagonists for opioid receptors reduce alcohol consumption in humans, and alcohol consumption and self-administration in rats. Selective antagonists of MOR and DOR have been shown to reduce alcohol self-administration. Thus, MOR and DOR are viable targets for reducing the positive reinforcing effects of alcohol in non-dependent cohorts [10].

Use for Diagnosis and Treatment

Nalmefene reduces alcohol consumption in adults with alcohol dependence. Morphine and hydromorphone are used in the treatment and management of severe pain. Naltrexone is used alongside behavioral therapy both for opiate addiction and for alcohol dependency. Naloxone is used to reverse narcotic depression.

References

1. Weisinger G. The transcriptional regulation of the preproenkephalin gene. *Biochem J*. 1995;307:617−629.
2. Pittius CW, Kley N, Loeffler JP, et al. Quantitation of proenkephalin A messenger RNA in bovine brain, pituitary and adrenal medulla: correlation between mRNA and peptide levels. *EMBO J*. 1985;4:1257−1260.
3. Kiser RS, Khatami M, Gatchel RJ, et al. Acupuncture relief of chronic pain syndrome correlates with increased plasma met-enkephalin concentrations. *Lancet*. 1983;322:1394−1396.
4. Govoni S, Bosio A, Di Monda E, et al. Immunoreactive met-enkephalin plasma concentrations in chronic alcoholics and in children born from alcoholic mothers. *Life Sci*. 1983;33:1581−1586.
5. Zagon IS, Verderame MF, McLaughlin PJ. The biology of the opioid growth factor receptor (OGFr). *Brain Res Rev*. 2002;38:351−376.
6. Al-Hasani R, Bruchas MR. Molecular mechanisms of opioid receptor-dependent signaling and behavior. *Anesthesiology*. 2011;115:1363−1381.
7. Chung PCS, Kieffer BL. Delta opioid receptors in brain function and diseases. *Pharmacol Ther*. 2013;140:112−120.
8. Kieffer BL, Gavériaux-Ruff C. Exploring the opioid system by gene knockout. *Prog Neurobiol*. 2002;66:285−306.
9. Pradhan AA, Befort K, Nozaki C, et al. The delta opioid receptor: an evolving target for the treatment of brain disorders. *Trends Pharmacol Sci*. 2011;32:581−590.
10. Walker BM, Valdez GR, McLaughlin JP, et al. Targeting dynorphin/kappa opioid receptor systems to treat alcohol abuse and dependence. *Alcohol*. 2012;46:359−370.

Supplemental Information Available on Companion Website

- Phylogenetic tree of PENK/E-Figure 7A.1
- Phylogenetic tree of DOR/E-Figure 7A.2
- Accession numbers of genes and cDNAs for PENK and DOR/E-Tables 7A.1, 7A.2

Dynorphin/α-Neo-endorphin

Akiyoshi Takahashi

Abbreviation: Dyn, Dyn A/αNE
Additional name: Dynorphin A (1-17), Dynorphin (1-17), Dyn A(1-17)/α-Neoendorphin

Dynorphin is an extraordinarily potent opioid peptide. To denote its powerful potency, the peptide was named "dynorphin" ("dyn" from the Greek dynamis = power and "orphin" for endogenous morphine peptide). α-Neo-endorphin has been identified as a "big" leucine enkephalin (Leu-ENK) with potent opiate activity.

Discovery

Both Dyn and αNE are heptapeptide sequences that contain Leu-ENK (Tyr−Gly−Gly−Phe−Leu) in their N-termini. Isolation of these peptides from the porcine pituitary and hypothalamus was first reported in 1979.

Structure

Structural Features

Dyn A, Dyn B, and αNE are generated from a common precursor prodynorphin (PDYN, also known as proenkephalin-B, PENK-B) by proteolytic cleavage. Human PDYN contains three copies of Leu-ENK, which are flanked by dibasic amino acid residues. Other mammalian PDYN has a similar structure. In amphibians and fishes such as lungfish and shark, PDYN contains two copies of ENKs, Leu-ENK or Met-ENK, in addition to Dyn A, Dyn B, and αNE [1]. Teleost PDYN also contains five ENKs, although two of them are unusual; one is present as a relic sequence and the other is Tyr−Gly−Gly−Phe−Ile. Thus, PDYN in the common ancestor may also have had five copies of ENKs.

Primary Structure

Several peptides with differential length are produced from PDYN (Table 7B.1). Figure 7B.1 shows the location of each peptide on PDYN.

Synthesis and Release

Gene, mRNA, and Precursor

The human PDYN gene, *PDYN*, location 20p13, consists of four exons. Four cAMP responsive element (CRE)-like sites (DYNCRE1, 2, 3, and 4) are important for *PDYN* expression, and an AP-1 site, representing a specific target for Jun/Fos, and an SP-1-like domain, targeted by NGFI-A and a single AP-2 consensus site, have been proposed as promoter elements [2]. Human PDYN contains 254 aa residues and comprises a signal peptide of 20 aa residues followed by a cysteine-containing N-terminal sequence and a region containing the repeated ENK sequences.

Distribution of mRNA

In the rat brain, the PDYN mRNA-containing region includes the hypothalamus, striatum, hippocampus, midbrain, nucleus tractus solitarius in the brainstem, cerebral cortex, thalamus, and cerebellum. In addition, PDYN mRNA has been detected in the adrenal gland, spinal cord, testis, and anterior pituitary [3].

Tissue Content

PDYN-derived peptides are distributed widely in the brain [4]. The highest contents in the rat central nervous system are 673.8 (fmol/mg protein) for Dyn A(1−8) in the substantia nigra, 186.1 for Dyn A(1−17) in the median eminence, 1106.2 for Dyn B in the substantia nigra, 1692.1 for αNE in the substantia nigra, and 341.4 for βNE in the median eminence.

Regulation of Synthesis and Release

In the rat, the level of PDYN mRNA expression has been shown to increase in the hypothalamus in response to osmotic challenge [5]. In addition, seven daily injections of a dopamine agonist, apomorphine, caused an increase in the level of PDYN mRNA expression in the striatum [6]. Electrical stimulation *in vivo* decreased the expression of PDYN mRNA in the dentate gyrus in the hippocampus.

Receptors

Structure

PDYN-derived peptides such as Dyn and αNE are agonists for KOR (also known as κ receptor, KOR-1, OP_2, etc.), which is a subtype of opioid receptors belonging to the G protein-coupled receptor superfamily. Human *KOR* is located on chromosome 8 (8q11.2). Dyn-A (1−17) is considered to be an endogenous ligand for KOR.

Table 7B.1 Amino Acid Sequence of PDYN-Derived Peptides in Humans*

Dynorphin A (1−8)	YGGFLRRI	Mr 981
Dynorphin A (1−17)	YGGFLRRIRPKLKWDNQ	Mr 2,148
Bigdynorphin	YGGFLRRIRPKLKWDNQKRYGGFLRRQFKVVT	Mr 3,985
Dynorphin B (aka rimorphin)	YGGFLRRQFKVVT	Mr 1,571
Leumorphin	YGGFLRRQFKVVTRSQEDPNAYSGELFDA	Mr 3,352
α-Neo-endorphin	YGGFLRKYPK	Mr 1,228
β-Neo-endorphin	YGGFLRKYP	Mr 1,100

*Leu-ENK sequences are underlined.

Y. Takei, H. Ando, & K. Tsutsui (Eds): Handbook of Hormones. DOI: http://dx.doi.org/10.1016/B978-0-12-801028-0.00118-5

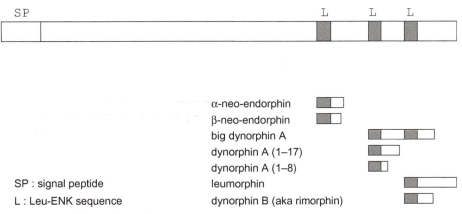

Figure 7B.1 Location of PDYN-derived peptides on the precursor.

Signal Transduction

See Subchapter 7A (Enkephalins).

Agonists

Nalfurafine, ethylketocyclazocine, hydromorphone, nalbuphine, morphine, fentanyl, etc., are agonists for human KOR.

Antagonists

Buprenorphine, nalmefene, naltrexone, naloxone, alvimopan, etc., are antagonists for human KOR.

Biological Functions

Based on the localization of KOR expression, dynorphins are involved in learning and memory, emotional control, stress response, and pain [2]. Administration of high-efficacy KOR agonist caused depressant-like effects and anhedonia in rodents [7]. See also Subchapter 7A.

Phenotype in Gene-Modified Animals

Absence of KOR does not modify expression of the other components of the opioid system, and behavioral tests indicate that spontaneous activity is not affected in mutant mice. KOR is implicated in the perception of visceral chemical pain. KOR is critical for mediating the hypolocomotor, analgesic, and aversive actions of a KOR agonist. This receptor does not contribute to morphine analgesia and reward, but participates in the expression of morphine abstinence [8]. See also Subchapter 7B.

Pathophysiological Implications

Clinical Implications

Acute alcohol intoxication stimulates the release of Dyn. Dyn/KOR systems contribute to the negative-reinforcing effects of alcohol based on the prodepressive effects of Dyn/KOR system activation, the antidepressant properties of KOR antagonists, and the involvement of KOR with the dysphoria produced by stress [9]. The Dyn/KOR system is related to the pathogenesis and pathophysiology of several psychiatric disorders. Dyn/KOR systems are recruited by various stimuli and act to shape neuronal activity, alter presynaptic neurotransmitter release, and decrease neuronal excitability. Changes in this system may contribute to symptom clusters that are shared by various psychiatric disorders [10].

Use for Diagnosis and Treatment

Hydromorphone, morphine, fentanyl, etc., are used to treat moderate to severe pain. The antagonist buprenorphine is used to treat severe pain, and to ameliorate opioid dependence. Naltrexone is used alongside behavioral therapy both for opiate addiction and for alcohol dependency. Naloxone is used to reverse narcotic depression.

References

1. Komorowski LK, Lecaude SG, Westring CG, et al. Evolution of gnathostome prodynorphin and proenkephalin: characterization of a shark proenkephalin and prodynorphin cDNAs. *Gen Comp Endocrinol.* 2012;177:353−364.
2. Schwarzer C. 30 years of dynorphins − new insights on their functions in neuropsychiatric diseases. *Pharmacol Ther.* 2009;123:353−370.
3. Civelli O, Douglass J, Goldstein A, et al. Sequence and expression of the rat prodynorphin gene. *Proc Natl Acad Sci USA.* 1985;82:4291−4295.
4. Zamir N, Weber E, Palkovits M, et al. Differential processing of prodynorphin and proenkephalin in specific regions of the rat brain. *Proc Natl Acad Sci USA.* 1984;81:6886−6889.
5. Sherman TG, Day R, Civelli O, et al. Regulation of hypothalamic magnocellular neuropeptides and their mRNAs in the Brattleboro rat: coordinate responses to further osmotic challenge. *J Neurosci.* 1988;8:3785−3796.
6. Li SJ, Sivam SP, McGinty JF, et al. Regulation of the metabolism of striatal dynorphin by the dopaminergic system. *J Pharmacol Exp Ther.* 1988;246:403−408.
7. Carlezon Jr WA, Béguin C, DiNieri JA, et al. Depressive-like effects of the κ-opioid receptor agonist salvinorin A on behavior and neurochemistry in rats. *J Pharmacol Exp Ther.* 2006;316:440−447.
8. Simonin F, Valverde O, Smadja C, et al. Disruption of the κ-opioid receptor gene in mice enhances sensitivity to chemical visceral pain, impairs pharmacological actions of the selective κ-agonist U-50,488H and attenuates morphine withdrawal. *EMBO J.* 1998;17:886−897.
9. Walker BM, Valdez GR, McLaughlin JP, et al. Targeting dynorphin/kappa opioid receptor systems to treat alcohol abuse and dependence. *Alcohol.* 2012;46:359−370.
10. Tejeda HA, Shippenberg TS, Henriksson R. The dynorphin/κ-opioid receptor system and its role in psychiatric disorders. *Cell Mol Life Sci.* 2012;69:857−896.

Supplemental Information Available on Companion Website

- Phylogenetic tree of PDYN/E-Figure 7B.1
- Phylogenetic tree of KOR/E-Figure 7B.2
- Accession numbers of genes and cDNAs for PDYN and KOR/E-Tables 7B.1, 7B.2

Nociceptin/Orphanin FQ

Akiyoshi Takahashi

Abbreviations: N/OFQ
Additional names: nociceptin (N), orphanin FQ (OFQ)

An opioid peptide that is evolutionarily related to dynorphin and enkephalin, N/OFQ is a heptadecapeptide exhibiting hyperalgesic action – i.e., a decrease in the nociceptive threshold.

Discovery

Nociceptin/orphanin FQ was first identified as an endogenous ligand for a GPCR that is homologous to the opioid receptors. A 17-aa peptide that decreased forskolin-stimulated cAMP production *in vitro* was isolated from the rat brain in 1995 and was named N, based on its ability to increase reactivity to pain [1]. Isolation of a peptide with an identical sequence from the porcine brain was simultaneously reported, and this peptide was named OFQ. The term orphanin refers to the affinity of the peptide for an orphan opioid receptor, and F and Q reflect the N- and C-terminal amino acid residues, respectively [2].

Structure

Structural Features

N/OFQ is generated from the precursor pronociceptin by proteolytic cleavage. The amino acid sequence of N/OFQ is common among human, rat, and mouse. The N-terminal pentapeptide sequence of N/OFQ, Phe–Gly–Gly–Phe–Thr, is related to those of Met-enkephalin (Tyr–Gly–Gly–Phe–Met) and Leu-enkephalin (Tyr–Gly–Gly–Phe–Leu).

Primary Structure

The amino acid sequences of N/OFQ in humans, the Carolina anole, cane toad, white sturgeon, and American eel are illustrated in Table 7C.1.

Synthesis and Release

Gene, mRNA, and Precursor

The human prepronociceptin gene, *PNOC*, location 8q21, consists of four exons, and two cAMP responsive elements are involved in the regulation of gene expression. Human preproN/OFQ (ppN/PFQ) is composed of 167 aa residues, with a signal peptide consisting of 19 aa residues followed by a cysteine-containing N-terminal sequence and the region containing N/OFQ [1–3].

Distribution of mRNA

Human *PNOC* is mainly expressed in the amygdala and the subthalamic nuclei. Moderate expression is found in the hypothalamus, substantia nigra, and thalamus [4]. In the rat, *Pnoc* is expressed in the hypothalamus, striatum, and spinal cord, as well as in the ovary [1].

Tissue Content

Plasma N/OFQ levels in humans: 8.86 pg/ml [5]. Level of N/OFQ immunoreactivities in different brain regions of naïve rats, as measured by RIA (N/OFQ-IR $\pm$ SEM, fmol/50 μl tissue preparation, $n = 6$): hypothalamus, 25.59 ± 3.76; midbrain, 15.17 ± 2.16; thalamus, 12.65 ± 0.84; brainstem, 9.25 ± 0.60; prefrontal cortex, 6.46 ± 0.48; parietal cortex, 6.42 ± 0.62, hippocampus, 5.88 ± 0.58, striatum, 3.09 ± 0.28; cerebellum, 2.46 ± 0.32 [6].

Regulation of Synthesis and Release

A TATA-box motif upstream of the human *PNOC* gene displays weak promoter activity. An increase of cellular cAMP levels by forskolin treatment upregulates *PNOC* transcription. Estrogen also upregulates *PNOC* transcription, whereas glucocorticoid downregulates transcription [7].

Receptors

Structure

N/OFQ is an agonist for N/OFQ receptors (NOP, also known as nociceptin receptor, opiate receptor-like 1 [OPRL1], ORL receptor, ORL-1, N/OFQ receptor, etc.), which are a subtype of opioid receptors belonging to the GPCR family. Human *OPRL1* is located on chromosome 20 (20q13.33).

Signal Transduction

The transducer is a member of the G_i/G_0 family, which inhibits cAMP (see also Subchapter 7A).

Agonists

UFP-102, Ro64-6198, and Ac-RYYRIK-NH$_2$ are agonists.

Antagonists

UFP-101, SB 612111, J-113397, and JTC-801 are antagonists.

Biological Functions

Behavioral Actions

The first behavioral effect of N/OFQ to be described was a hyperalgesic response (lowering of nociceptive thresholds).

Y.Takei, H. Ando, & K.Tsutsui (Eds): Handbook of Hormones. DOI: http://dx.doi.org/10.1016/B978-0-12-801028-0.00119-7

Table 7C.1 Amino Acid Sequence of N/OFQ*

Human	FGGFTGARKSARKLANQ	Mr 1,809
Anolis carolinensis	FGGFIGVRKSARKWHNQ	Mr 1,988
Bufo marinus	YGGFIGVRKSARKWNNQ	Mr 1,981
Acipenser transmontanus	YGGFIGIRKSARKWNNP	Mr 1,964
Anguilla rostrata	YGGFIGVRKSARKWNNQ	Mr 1,981

*Conserved amino acid residues are shaded.

However, the effect of N/OFQ on nociceptive responses is complicated, as various studies report hyperalgesia, blockade of hyperalgesia, analgesia, blockade of analgesia, no effect, or blockade of allodynia in response to N/OFQ. N/OFQ is also associated with locomotion, feeding, anxiety, spatial attention/learning, endocrine effects in the hypothalamus, reward, and opiate tolerance and dependence [8].

Cellular Actions

N/OFQ reduces dopamine levels in the nucleus accumbens in rats, whereas it increases dopamine release in the striatum. N/OFQ also inhibits the release of acetylcholine from the guinea pig trachea, substance P and calcitonin gene-related peptide from sensory nerve endings, and glutamate from cerebrocortical slices in rats. N/OFQ stimulates prolactin release in both sexes, but the effect is much greater in females. Conversely, growth hormone release is stimulated only in male rats. The peptide inhibits electrically induced contractions of both the guinea pig ileum and the mouse vas deferens. N/OFQ causes hypotension and bradycardia in anesthetized rats, whereas it increases blood pressure and heart rate in unanesthetized ewes [8].

Phenotype in Gene-Modified Animals

N/OFQ-deficient mice display behavioral, sensory, and endocrine symptoms of an increased susceptibility to stress. Adaptive responses to repeat stress are impaired in the mutant animals. However, mice lacking *Pnoc* do not always show phenotypic changes in anxiety-related behavior. In contrast, *Pnoc*-deficient mice display improved spatial attention and memory. Mice lacking the receptor fail to develop morphine tolerance [9].

Pathophysiological Implications

Clinical Implications

The endogenous N/OFQ system has a physiological role in mediating or regulating behavioral responses to alcohol, and the activation of NOP suppresses ongoing alcohol consumption or the reinstatement of responding to alcohol. These findings encourage the development of therapies targeted at the N/OFQ system for the treatment of alcoholism in humans [10].

References

1. Meunier JC, Mollereau C, Toll L, et al. Isolation and structure of the endogenous agonist of opioid receptor-like ORL₁ receptor. *Nature*. 1995;377:532−535.
2. Reinscheid RK, Nothacker HP, Bourson A, et al. Orphanin FQ: a neuropeptide that activates an opioidlike G protein-coupled receptor. *Science*. 1995;270:792−794.
3. Mollereau C, Simons MJ, Soularue P, et al. Structure, tissue distribution, and chromosomal localization of the prepronociceptin gene. *Proc Natl Acad Sci USA*. 1996;93:8666−8670.
4. Nothacker HP, Reinscheid RK, Mansour A, et al. Primary structure and tissue distribution of the orphanin FQ precursor. *Proc Natl Acad Sci USA*. 1996;93:8677−8682.
5. Csobay-Novák C, Sótonyi P, Krepuska M, et al. Decreased plasma nociceptin/orphanin FQ levels after acute coronary syndromes. *Acta Physiol Hung*. 2012;99:99−110.
6. Lutfy K, Lam H, Narayanan S. Alterations in the level of OFQ/N-IR in rat brain regions by cocaine. *Neuropharmacology*. 2008;55:198−203.
7. Xie G, Ito E, Maruyama K, et al. The promoter region of human prepro-nociceptin gene and its regulation by cyclic AMP and steroid hormones. *Gene*. 1999;238:427−436.
8. Harrison LM, Grandy DK. Opiate modulating properties of nociceptin/orphanin FQ. *Peptides*. 2000;21:151−172.
9. Reinscheid RK, Nothacker HP, Civelli O. The orphanin FQ/nociceptin gene: structure, tissue distribution of expression and functional implications obtained from knockout mice. *Peptides*. 2000;21:901−906.
10. Murphy NP. The nociceptin/orphanin FQ system as a target for treating alcoholism. *CNS Neurol Disord Drug Targets*. 2010;9:87−93.

Supplemental Information Available on Companion Website

- Phylogenetic tree of PNOC/E-Figure 7C.1
- Phylogenetic tree of NOP/E-Figure 7C.2
- Accession numbers of genes and cDNAs for N/OFQ and NOP/E-Tables 7C.1, 7C.2

Endomorphin

Kanta Mizusawa

Abbreviation: EM
Additional names: L-tyrosyl-L-prolyl-L-tryptophyl-L-phenyl-alaninamide (endomorphin-1), L-tyrosyl-L-prolyl-L-phenylala-nyl-L-phenylalaninamide (endmorphin-2)

The endomorphins are endogenous peptides with high affinity and remarkable selectivity for the μ-opioid receptor.

Discovery

Endomorphin-1 (EM-1) and endomorphin-2 (EM-2) were first isolated from bovine brain, and subsequently from human cortex [1,2].

Structure

Structural Features

The N-terminal message sequence is distinct from those of endorphins, enkephalins, and dynorphins. The amino and phenolic groups of Tyr^1, and the aromatic ring of Trp^3 (in EM-1) and Phe^3 (in EM-2), are required for μ-opioid receptor recognition [3]. As a stereochemical spacer, Pro^2 directs Trp^3 toward a μ-opioid receptor selectivity region in EM-1 [4].

Primary Structure

The amino acid sequences of EM-1 and EM-2 are shown in Figure 7D.1.

Properties

Mr 611 (EM-1), 572 (EM-2); pI 8.3−8.9 [5]. Soluble in water (up to 2 mg/ml). Endomorphins should be stored at −20°C. They are also hygroscopic, and must be protected from light.

Synthesis and Release

Gene, mRNA, and Precursor

No gene has been identified [6].

EM-1 YPWF–NH₂
EM-2 YPFF–NH₂

Figure 7D.1 Amino acid sequence of EM-1 and EM-2. Conserved amino acid residues are shaded.

Tissue and Plasma Concentrations

Tissue: total EMs, human brain, 151 pmol/g tissue; bovine brain, 2.1 pmol/g tissue [1,2].
  Plasma: EM-1, rat (male) mature, 221 pg/ml; EM-2, rat (male) mature, 178 pg/ml [7].

Regulation of Synthesis and Release

Electrical stimulation of the dorsal roots was shown to promote the release of EM-2 from dense-cored vesicles situated in dorsal horn axons [8,9].

Receptors

Structure

EMs are selective agonists for the μ-opioid receptor [10], which is one of the subtypes of opioid receptors belonging to the GPCR superfamily. EM-1 and EM-2 produce their biological effects by stimulating functionally diverse subtypes of μ-opioid receptors, μ1 and μ2: μ2-opioid receptors would be stimulated by both EM-1 and EM-2, whereas μ1-opioid receptors would be stimulated only by EM-2 [3].

Signal Transduction

See Subchapter 7A (Enkephalins).

Biological Functions

Target Cells/Tissues and Functions

Radioimmunological and immunocytochemical analyses have revealed that EM immunoreactivities are distributed throughout human, bovine, and rodent central nervous systems [3]. EMs are involved in many major biological processes, including perception of pain, responses related to stress, and complex functions such as reward, arousal, and vigilance, as well as autonomic, cognitive, neuroendocrine, and limbic homeostasis [3].

Pathophysiological Implications

Clinical Implications

EM has beneficial effects in various pathological conditions, such as pain, mood, and feeling disorders, related to reward and cardiorespiratory function. Thus, EM motivates the development of useful ligands as analgesic, anxiolytic, or antidepressant agents, with high μ-opioid receptor affinity and good resistance to proteolytic degradation [3].

Y. Takei, H. Ando, & K. Tsutsui (Eds): Handbook of Hormones. DOI: http://dx.doi.org/10.1016/B978-0-12-801028-0.00120-3

References

1. Zadina JE, Hackler L, Ge L-J, et al. A potent and selective endogenous agonist for the μ-opiate receptor. *Nature*. 1997;386:499−502.
2. Hackler L, Zadina JE, Ge L-J, et al. Isolation of relatively large amounts of endomorphin-1 and endomorphin-2 from human brain cortex. *Peptides*. 1997;18:1635−1639.
3. Fichna J, Janecka A, Costentin J, et al. The endomorphin system and its evolving neurophysiological role. *Pharmacol Rev*. 2007;59:88−123.
4. Paterlini MG, Avitabile F, Ostrowski BG, et al. Stereochemical requirements for receptor recognition of the mu-opioid peptide endomorphin-1. *Biophys J*. 2000;78:590−599.
5. Németh K, Mallareddy JR, Domonkos C, et al. Stereoselective analysis of endomorphin diastereomers: resolution of biologically active analogues by capillary electrophoresis applying cyclodextrins as mobile phase additives. *J Pharm Biomed Anal*. 2012;70:32−39.
6. Terskiy A, Wannemacher KM, Yadav PN, et al. Search of the human proteome for endomorphin-1 and endomorphin-2 precursor proteins. *Life Sci*. 2007;81:1593−1601.
7. Coventry TL, Jessop DS, Finn DP, et al. Endomorphins and activation of the hypothalamo-pituitary−adrenal axis. *J Endocrinol*. 2001;169:185−193.
8. Williams CA, Wu SY, Dun SL, et al. Release of endomorphin-2 like substances from the rat spinal cord. *Neurosci Lett*. 1999;273:25−28.
9. Wang Q-P, Zadina JE, Guan J-L, et al. Morphological studies of the endomorphinergic neurons in the central nervous system. *Jpn J Pharmacol*. 2002;89:209−215.
10. Ide S, Sakano K, Seki T, et al. Endomorphin-1 discriminates the mu-opioid receptor from the delta- and kappa-opioid receptors by recognizing the difference in multiple regions. *Jpn J Pharmacol*. 2000;83:306−311.

Dermorphin

Kanta Mizusawa

Abbreviations: DM, DER
Additional name: L-tyrosyl-D-alanyl-L-phenylalanyl-L-glycyl-L-tyrosyl-L-prolyl-L- serinamide

A heptapeptide isolated from amphibian skin, dermorphin contains D-alanine (a D-isomer amino acid) and shows very high selectivity for the μ-opioid receptor.

Discovery

Between the 1960s and 1980s, a number of active peptides were identified in the skin of the South American leapfrog belonging to the genus *Phyllomedusa*. DM was isolated from the skin of *P. sauvagei* as a potent opiate-like peptide [1]. At present, seven endogenous analogs (DM, [Hyp⁶]DM, [Lys⁷] DM, [Lys⁷]DM-OH, [Trp⁴,Ans⁷]DM, [Trp⁴,Ans⁷]DM-OH, and [Trp⁴,Ans⁵]DM(1–5)-OH) have been identified [2]. DM is not found in mammals.

Structure

Structural Features

Dermorphin does not contain the common N-terminal sequence for the traditional endogenous opioid peptides (Tyr—Gly—Gly—Phe), and its sequence is completely different from those of the endomorphins, which were identified as endogenous μ-opioid peptides that also do not contain the common sequence. Interestingly, DM contains D-alanine (a D-isomer amino acid) in its sequence [2].

Primary Structure

The primary structure of dermorphin is YDAFGYPS-NH$_2$.

Properties

Mr 803. Soluble in water (up to 2 mg/ml) and acetonitrile. Dermorphin should be stored at −20°C. It is also hygroscopic and must be protected from light.

Synthesis and Release

Gene, mRNA, and Precursor

Precursor mRNA has 700 bp that encode five aa residues of DM (Figure 7E.1). In these cloned cDNAs, the alanine codon GCG occurred at the position where D-alanine is present in the end product [3], suggesting posttranslational conversion of an L-amino acid to its D-isomer carried out by an amino acid isomerase [4].

Distribution of mRNA

The precursor gene is expressed in amphibian skin.

Tissue Content

The tissue content in *P. sauvagei* skin is 50–60 g/g skin [1]

Receptors

Structure

DM shows a very high affinity for μ-opioid receptors and a low affinity for δ-opioid receptors [5], which are subtypes of opioid receptors belonging to the GPCR superfamily.

Signal Transduction Pathway

See Chapter 7A (Enkephalins).

Biological Functions

Target Cells/Tissues and Functions

No endogenous function of DM in the frog has been reported.

Pathophysiological Implications

Clinical Implications

Intracerebroventricular administration of DM produces a long-lasting antinociceptive activity, with more than 200 times the potency of morphine [6]. [Lys⁷]DM also exhibits a long-lasting antinociceptive potency, exceeding that of morphine by 25- to 30-fold, by peripheral administration with a high penetration into the blood—brain barrier [7]. Within synthetic DM analogs, some of the N-terminal tetrapeptide analogs containing D-Arg² have antinociceptive profiles that are distinct from those of traditional μ-opioid receptor agonists [2]. A few studies have reported the effects of DM on endocrine systems in humans. Intravenous infusion (5.5 μg/kg per minute for 30 minutes) of DM significantly increased plasma levels of prolactin, growth hormone, thyrotropin, and renin activity, but decreased plasma levels of cortisol [8]. In fertile women, DM (5.5 μg/kg per minute for 30 minutes) decreases plasma levels of luteinizing hormone, but not of follicle-stimulating hormone [9].

Y. Takei, H. Ando, & K. Tsutsui (Eds): Handbook of Hormones. DOI: http://dx.doi.org/10.1016/B978-0-12-801028-0.00121-5

```
                    10        20        30        40        50        60
P. sauvagei 1   MSFLKKSLLLILFLGLVSLSVCKEEKRETEEENENEE-NHEEGSEMKRYMFHLMDGEAK-   58
P. sauvagei 2   MSFLKKSLLLILFLGLVSLSVCKEEKRVSEEENENEE-NHEEGSEMKRYAFGYPSGEAKK   59
P. nordestina   MSFLRNRLSLT-FPWMVSLSICKE-KRENEEENENEE-NHEEESEMKRYASGYPSGEAKK   57
P. bicolor      MSFLKKSLLLVLFLGLVSHSVCKEEKRETEEENENEEENHEVGSEMKRYAFWYPNRDTEE   60
                    70        80        90       100       110       120
P. sauvagei 1   -------------------------KRDSEE-NEIEEN-HEEGSEMKRYAFGYPSGEAKKIK   93
P. sauvagei 2   I------------------------KRESEEEKEIEEN-HEEGSEMKRYAFGYPSGEAKKIK   96
P. nordestina   I------------------------KRETEEENENEEEDHEEESEMKRYAFGYPSGEAKKKK   95
P. bicolor      KNENEEENQEEGSEMKRYAFGYPKREPEEENENEEEENHEEGSEMKRYAFEVVGGEAKKMK  120
                   130       140       150       160       170       180
P. sauvagei 1   RVSEEENENEE-NHEEGSEMKRYAFGYPSGEAKKIKRESEE-EKEIEENHEEGSEMKRYA  151
P. sauvagei 2   RESEEENENEE-NHEEGSEMKRYAFGYPSGEAKKIKRESEE-EKEIEENHEEGSEMKRYA  154
P. nordestina   RETEEENENEE-NHDEGSEMKRYAFGYPSGEAKKIKRETEE-NENDEEDHEEESEMKRYA  153
P. bicolor      REPEEENENEEEENHEEGSEMKRYAFDVVGGEAKKMKREPEEENENEEEENHEEGSEMKRYA 180
                   190       200       210       220
P. sauvagei 1   FGYPSGEAKKIKRESEEENENEE-NHEEGSEMKRYAFGYPSGEAKKM  197
P. sauvagei 2   FGYPSGEAKKIKRESEEENENEE-NHEEGSEMKRYAFGYPSGEAK   198
P. nordestina   FGYPSGEAKKMKRETEEENENEE-NHDEGSEMKRYAFGYPSGEAK   197
P. bicolor      FDVVGGEAKKMKREPEEENENEEEENHEEGSEMKRYAFDVVGGEAKKM  227
```

Figure 7E.1   Alignment of the amino acid sequences of the DM and dertrophin precursor in leapfrogs (*Phyllomedusa*). Conserved amino acid residues are indicated by shadow boxes, and the sequences of predicted peptides (DM, "YAFGYPS"; [Lys[7]]DM, "YAFGYPK"; [Trp[4],Asn[7]] DM, "YAFWYPN"; deltorphin, "YMFHLMD"; deltorphin I, "YAFDVVG"; deltorphin II, "YAFEVVG") are emphasized by underlining. Accession numbers: *P. sauvagei* 1, M18031.1; *P. sauvagei* 2, M18030.1; *P. nordestina*, JX127157.1; *P. bicolor*, M34560.1.

References

1. Montecucchi PC, De Castiglione R, Piani S, et al. Amino acid composition and sequence of dermorphin, a novel opiate-like peptide from the skin of *Phyllomedusa sauvagei*. *Int J Pept Protein Res*. 1981;17:275–283.
2. Mizoguchi H, Bagetta G, Sakurada T, et al. Dermorphin tetrapeptide analogs as potent and long-lasting analgesics with pharmacological profiles distinct from morphine. *Peptides*. 2011;32:421–427.
3. Richter K, Egger R, Kreil G. D-alanine in the frog skin peptide dermorphin is derived from L-alanine in the precursor. *Science*. 1987;23:200–202.
4. Heck SD, Faraci WS, Kelbaugh PR, et al. Posttranslational amino acid epimerization: enzyme-catalyzed isomerization of amino acid residues in peptide chains. *Proc Natl Acad Sci USA*. 1996;93:4036–4039.
5. Krumins SA. Characterization of dermorphin binding to membranes of rat brain and heart. *Neuropeptides*. 1987;9:93–102.
6. Kisara K, Sakurada S, Sakurada T, et al. Dermorphin analogues containing D-kyotorphin: structure-antinociceptive relationships in mice. *Br J Pharmacol*. 1986;87:183–189.
7. Negri L, Lattanzi R, Melchiorri P. Production of antinociception by peripheral administration of [Lys[7]]dermorphin, a naturally occurring peptide with high affinity for mu-opioid receptors. *Br J Pharmacol*. 1995;114:57–66.
8. Deqli Ubertia EC, Trasforini G, Salvadori S, et al. The effects of dermorphin on the endocrine system in man. *Peptides*. 1985;6:171–175.
9. Petraglia F, Deqli Uberti EC, Trasforini G, et al. Dermorphin decreases plasma LH levels in human: evidence for a modulatory role of gonadal steroids. *Peptides*. 1985;6:869–872.

CHAPTER 8

Agouti Family

Sakae Takeuchi

History

Agouti-signaling protein (ASIP) and agouti-related protein (AGRP) are endogenous antagonists of melanocortin receptors (MCRs), and play crucial roles in the regulation of pigmentation and energy balance, respectively. Identification of these two signaling molecules came from studies on a mouse genetic locus affecting coat color, the *agouti* locus on chromosome 2. The *agouti* locus acts within the hair follicle to control relative amounts of eumelanins and pheomelanins, which give rise to the wild-type coat color of most mammals arising from hairs with a subapical yellow band on an otherwise black or brown background. In 1992/1993, the *agouti* was demonstrated to encode a paracrine signaling molecule ASIP that promotes the synthesis of pheomelanin by melanocytes [1,2]. Expression of ASIP was normally limited to the skin; however, the mice with the lethal yellow (A^y) allele on the *agouti* locus showed deregulated ubiquitous expression of ASIP, giving rise to a pleiotropic phenotype characterized by a uniformly yellow coat color and late-onset obesity associated with hyperphagia. These findings suggested the existence of an ASIP-like protein in the brain that would stimulate feeding behavior. In 1997, two groups cloned, independently, the mouse and human ASIP homologs, called AGRP, from brain EST clones, based on its homology to ASIP. *AGRP* was found to be expressed exclusively in the arcuate nucleus of the hypothalamus in mice, and ubiquitous expression of human *AGRP* cDNA in transgenic mice caused obesity [3,4]. The human ASIP gene was cloned using linkage groups conserved between mice and humans or an interspecies DNA-hybridization approach [5,6]. Expression studies revealed that the gene is expressed in adipose tissues and testis as well as in the ovary, heart, liver, kidney, and foreskin. Orthologs of ASIP and AGRP have been identified in other vertebrates, including birds, teleosts, and cartilaginous fishes.

Structure

Structural Features

Both human AGRP and ASIP are 132-aa residue proteins with cysteine-rich C-terminal domains, which are shown to be sufficient for potent antagonist function and adopt an inhibitor cysteine knot (ICK or knottin) fold stabilized by five disulfide bonds (Figure 8.1). The C-terminal domains consist of three polypeptide loops, referred to as the N-terminal loop, the active loop, and the C-terminal loop. The active loop contains the Arg−Phe−Phe (RFF) triplet that is essential for binding and antagonist function at their cognate receptors. By contrast, the N-terminal domains of AGRP and ASIP exhibit little sequence similarity, reflecting different physiological roles, a prodomain to be removed by prohormone convertases, and a ligand for an accessory receptor attraction, respectively [7].

Molecular Evolution of Family Members

ASIP and AGRP are thought to have arisen early in vertebrate evolution by gene duplication. Distinct physiological functions of these proteins are likely to have arisen through subfunctionalization based on the specialized expression patterns of both proteins and their ability to cross-react with the other's receptor *in vitro*. Strong evolutionary constraint is observed in the C-terminal domains of both proteins and the ASIP N-terminal domain, but not in the AGRP N-terminal domain [7]. Teleost-specific members of this family, *asip2* and *agrp2*, are thought to have evolved by duplication of the *ASIP* gene (Figure 8.2) [8].

Receptors

ASIP and AGRP are endogenous antagonists of distinct subsets of the MCR family. Since MCR agonists are melanocortins, including α-melanocyte-stimulating hormone (MSH) and adrenocorticotropic hormone, details of MCRs are described in Chapter 16. ASIP acts normally as a competitive antagonist at MC1R, and AGRP at both MC3R and MC4R. ASIP can also bind to MC4R and MC3R. Beyond competitive antagonism, *in vitro* assays show that these proteins act as inverse agonists to decrease basal receptor activity in the absence of agonists.

Biological Functions

Target Cells/Tissues and Functions

ASIP prevents α-MSH from binding to MC1R on melanocytes to inhibit α-MSH-mediated elevation of cAMP levels leading to eumelanin production, which shifts the production of eumelanins to pheomelanins. ASIP also has a role in regulating lipid metabolism in adipocytes in humans and chickens. AGRP acts in the hypothalamus to control body-weight regulation and metabolism [4].

Y. Takei, H. Ando, & K. Tsutsui (Eds): Handbook of Hormones. DOI: http://dx.doi.org/10.1016/B978-0-12-801028-0.00008-8

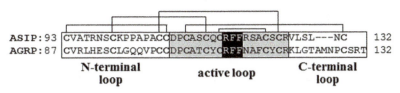

```
ASIP:93 CVATRNSCKPPAPACCDPCASCQQRFFRSACSCRVLSL---NC 132
AGRP:87 CVRLHESCLGQQVPCCDPCATCYCRFFNAFCYCRKLGTAMNPCSRT 132
```

N-terminal   active loop   C-terminal
loop   loop

Figure 8.1 Structure of the C-terminal domains of human ASIP and AGRP.

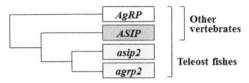

Figure 8.2 Teleost-specific members of the Agouti family, *asip2* and *agrp2*, are thought to have evolved by duplication of the *ASIP* gene.

References

1. Bultman SJ, Michaud EJ, Woychik RP. Molecular characterization of the mouse agouti locus. *Cell*. 1992;71:1195–1204.
2. Miller MW, Duhl DM, Vrieling H, et al. Cloning of the mouse agouti gene predicts a secreted protein ubiquitously expressed in mice carrying the lethal yellow mutation. *Genes Dev*. 1993;7:454–467.
3. Shutter JR, Graham M, Kinsey AC, et al. Hypothalamic expression of ART, a novel gene related to agouti, is up-regulated in obese and diabetic mutant mice. *Genes Dev*. 1997;11:593–602.
4. Ollmann MM, Wilson BD, Yang YK, et al. Antagonism of central melanocortin receptors *in vitro* and *in vivo* by agouti-related protein. *Science*. 1997;278:135–138.
5. Kwon HY, Bultman SJ, Loffler C, et al. Molecular structure and chromosomal mapping of the human homolog of the agouti gene. *Proc Natl Acad Sci USA*. 1994;91:9760–9764.
6. Wilson BD, Ollmann MM, Kang L, et al. Structure and function of ASP, the human homolog of the mouse agouti gene. *Hum Mol Genet*. 1995;4:223–230.
7. Jackson PJ, Douglas NR, Chai B, et al. Structural and molecular evolutionary analysis of Agouti and Agouti-related proteins. *Chem Biol*. 2006;13:1297–1305.
8. Vastermark A, Krishnan A, Houle ME, et al. Identification of distant Agouti-like sequences and re-evaluation of the evolutionary history of the Agouti-related peptide (AgRP). *PLoS One*. 2012;7:e40982.

Agouti-Signaling Protein

Sakae Takeuchi

Abbreviations: ASIP, ASP
Additional names: agouti-signaling peptide, agouti protein, agouti switch protein, agouti coat color protein

A paracrine signaling molecule involved in the regulation of melanogenesis, ASIP functions as an inverse agonist of melanocortin 1 receptor (MC1R) and blocks cAMP production, leading to a switch in melanogenesis from eumelanogenesis to pheomelanogenesis in melanocytes.

Discovery

In 1992/1993, the murine ASIP was reported to be identified as a paracrine signaling molecule encoded by the *agouti* locus on chromosome 2 [1]. The human ASIP gene was cloned in 1994/1995 using linkage groups conserved between mice and humans or an interspecies DNA-hybridization approach [1].

Structure

Structural Features

ASIP is a glycoprotein with a cysteine-rich C-terminal domain, which adopts an inhibitor cysteine knot (ICK or knottin). The C-terminal domains consist of three polypeptide loops: the N-terminal loop, the active loop, and the C-terminal loop. The active loop contains the Arg−Phe−Phe triplet that is essential for binding and antagonist function at MC1R. The N-terminal domain of ASIP acts as a ligand for an accessory receptor, Attractin [2].

Primary Structure

The primary structure is illustrated in Figure 8A.1, which shows the potential signal peptide, and the potential N-linked glycosylation site and cysteine residues forming five disulfide bonds (93−108, 100−114, 107−125, 111−132, 116−123).

Synthesis and Release

Gene, mRNA, and Precursor

The human ASIP gene, *ASIP*, location 20q11.22 (chromosome 20: 32,782,375−32,857,150 in Ensembl databases), consists of three exons. *ASIP* mRNA has two splice variants of 610 bp and 586 bp. In mice and rabbits, the Asip gene has two different promoters, the proximal hair cycle-specific promoter and the distal ventral-specific promoter [3], and produces multiple forms of mRNAs by alternative promoter usage and alternative splicing. This is true for chickens, in which three different promoters and at least seven mRNA isoforms have been identified [4].

Distribution of mRNA

In humans, the ASIP gene is expressed at substantial levels in adipose tissue, testis, ovary, and heart, and at low levels in the liver, kidney, and foreskin [1]. The murine Asip gene is predominantly expressed in the skin. Widespread expression was also reported in bovine and chicken ASIP genes [4].

Tissue Content

Human: subcutaneous mature adipocytes and preadipocytes in culture, 19.18 and 4.07 pg/μg protein, respectively [5].

Regulation of Synthesis

Body mass index (BMI) and ASIP gene expression in isolated omental and subcutaneous abdominal adipocytes were reported to be negatively correlated in men, whereas a positive relationship was observed in women [6]. Sexual dimorphism in the expression of the ASIP gene exists in feather follicles, which creates estrogen-dependent sexual plumage dichromatism in chickens [7].

Receptors

ASIP is an endogenous antagonist of MC1R. Since melanocortin receptor (MCR) agonists are melanocortins including α-melanocyte stimulating hormone (MSH) and adrenocorticotropic hormone, details of MCRs are described in Chapter 16. ASIP acts normally as a competitive antagonist at MC1R. ASIP can also bind to MC4R and MC3R. Beyond competitive antagonism, *in vitro* assays show that ASIP acts as an inverse agonist to decrease basal receptor activity in the absence of agonists.

Biological Functions

Target Cells/Tissues and Functions

ASIP prevents α-MSH from binding to MC1R on melanocytes to inhibit α-MSH-mediated elevation of cAMP levels leading to eumelanin production, which shifts the production of eumelanins to pheomelanins. In wild-type mice, the hair cycle-specific promoter of *Asip* acts in hair follicles during the mid-portion of hair growth to produce black hair with a sub-apical yellow band, which creates the overall appearance of a mottled brown hair coat that provides adaptive coloration in the natural environment. The ventral-specific promoter of the gene, on the other hand, acts in hair follicles throughout the hair cycle but only on the ventral side of the body, to produce entirely yellow or cream-colored ventral hairs [1]. Similarly,

Y. Takei, H. Ando, & K. Tsutsui (Eds): Handbook of Hormones. DOI: http://dx.doi.org/10.1016/B978-0-12-801028-0.00122-7

```
          10        20        30        40        50        60
┌─────────────────────────────────┐
│MDVTRLLLATLLVFLCFFTANS│HLPPEEKLRDDRSLRSNSSVNLLDVPSVSIVALNKKSK
└─────────────────────────────────┘        ‾
          70        80        90        100       110       120
QIGRKAAEKKRSSKKEASMKKVVRPRTPLSAPCVATRNSCKPPAPACCDPCASCQCRFFR
          130
SACSCRVLSLNC
```

Figure 8A.1 Primary structure of human ASIP. The potential signal peptide is shown by a box. The potential N-linked glycosylation site and the cysteine residues forming five disulfide bonds (93–108, 100–114, 107–125, 111–132, 116–123) are underlined and shaded, respectively.

two promoters of the ASIP gene produce a light-bellied agouti pattern in rabbits [3]. The distal ASIP gene promoter produces countershading in chicks and adult female chickens, similarly to the ventral-specific *ASIP* promoter in mammals [7]. In addition, the promoter plays an important role for creating sexual plumage dichromatism controlled by estrogen [7]. In teleost fishes, ASIP has been shown to establish the dorsal–ventral pigmentation in goldfish by being mainly expressed in ventral skin, where it inhibits melanophore differentiation and/or proliferation but promotes iridophore differentiation and/or proliferation [8]. ASIP also has a role in regulating lipid metabolism in adipocytes in humans, bovines, and chickens [5,9]. Furthermore, ASIP is shown to stimulate Ca^{2+} signaling and insulin release in pancreatic β-cells in humans [5].

Phenotypes in Gene-Modified Animals

Several dominant mutations causing ectopic expression of ASIP in mice result in a "yellow obese phenotype" characterized by yellow hair, adult-onset obesity and diabetes, increased linear growth and skeletal mass, and increased susceptibility to tumors.

Pathophysiological Implications

Clinical Implications

Analysis of SNP in forensic studies suggests that regulation of the expression of ASIP may underlie pigmentation variation. In primary human melanocyte cultures, the MC1R antagonists ASIP and human beta defensin 3 (HBD3) inhibit MC1R activity by blocking its activation by α-MSH, thus suggesting it affects human pigmentation and the responses to UV [10].

References

1. Dinulescu DM, Cone RD. Agouti and agouti-related protein: analogies and contrasts. *J Biol Chem*. 2000;275:6695–6698.
2. Jackson PJ, Douglas NR, Chai B, et al. Structural and molecular evolutionary analysis of Agouti and Agouti-related proteins. *Chem Biol*. 2006;13:1297–1305.
3. Fontanesi L, Forestier L, Allain D, et al. Characterization of the rabbit agouti signaling protein (ASIP) gene: transcripts and phylogenetic analyses and identification of the causative mutation of the nonagouti black coat colour. *Genomics*. 2010;95:166–175.
4. Yoshihara C, Fukao A, Ando K, et al. Elaborate color patterns of individual chicken feathers may be formed by the agouti signaling protein. *Gen Comp Endocrinol*. 2012;175:495–499.
5. Xue B, Zemel MB. Relationship between human adipose tissue agouti and fatty acid synthase (FAS). *J Nutr*. 2000;130:2478–2481.
6. Voisey J, Imbeault P, Hutley L, et al. Body mass index-related human adipocyte agouti expression is sex-specific but not depot-specific. *Obes Res*. 2002;10:447–452.
7. Oribe E, Fukao A, Yoshihara C, et al. Conserved distal promoter of the agouti signaling protein (ASIP) gene controls sexual dichromatism in chickens. *Gen Comp Endocrinol*. 2012;177:231–237.
8. Cerda-Reverter JM, Haitina T, Schioth HB, et al. Gene structure of the goldfish agouti-signaling protein: a putative role in the dorsal-ventral pigment pattern of fish. *Endocrinology*. 2005;146:1597–1610.
9. Albrecht E, Komolka K, Kuzinski J, et al. Agouti revisited: transcript quantification of the ASIP gene in bovine tissues related to protein expression and localization. *PLoS One*. 2012;7:e35282.
10. Swope V, Jameson J, McFarland K, et al. Defining MC1R regulation in human melanocytes by its agonist α-melanocortin and antagonists agouti signaling protein and β-defensin 3. *J Invest Dermatol*. 2012;132:2255–2262.

Supplemental Information Available on Companion Website

- List of useful web resources for ASIP, including alignment of primary structure of ASIP of various animals and phylogenetic tree of ASIP/E-Box 8A.1

Agouti-Related Protein

Sakae Takeuchi

Abbreviations: AGRP, AgRP
Additional name: agouti-related peptide

A neuropeptide involved in the regulation of energy balance, agouti-related protein functions as a competitive antagonist of α-melanocyte-stimulating hormone (MSH) signaling via melanocortin receptors MC3R and MC4R within the hypothalamus by inhibiting cAMP production, leading to the increase of appetite and the decrease of metabolism and energy expenditure. It is one of the most potent and long-lasting appetite stimulators.

Discovery

In 1997, the mouse and human AGRP were independently identified by two groups from brain EST clones based on their sequence similarity with agouti-signaling protein (ASIP), a protein that is synthesized in the skin and controls hair pigmentation [1].

Structure

Structural Features

AGRP contains a signal peptide and a cysteine-rich C-terminal domain, which adopts an inhibitor cysteine knot (ICK or knottin). The C-terminal domains consist of three polypeptide loops, referred to as the N-terminal loop, the active loop, and the C-terminal loop. The active loop contains the Arg−Phe−Phe triplet that is essential for binding and antagonist function at MC3R and MC4R [2]. The N-terminal domain of AGRP acts as a prodomain to suppress antagonist activity of the biologically active C-terminal domain until it is removed by digestion with prohormone convertase [2,3].

Primary Structure

The primary structure is illustrated in Figure 8B.1, which shows the potential signal peptide, and the amino acid residues of mature protein and cysteine residues forming five disulfide bonds (87−102, 94−108, 101−119, 105−129, 110−117).

Synthesis and Release

Gene, mRNA, and Precursor

The human AGRP gene, *AGRP*, location 16q22 (chromosome 16: 67,516,474−67,517,716 in Ensembl databases), contains three coding exons and has one transcript of 764 bp.

Distribution of mRNA

In humans, the AGRP gene is expressed at substantial levels in the adrenal gland and hypothalamus, and at lower levels in the testis, kidneys, and lungs [1]. In chickens, the agrp gene is expressed ubiquitously and *agrp* mRNA has been detected in all tissues tested [4].

Regulation of Synthesis and Release

Leptin secreted by adipocytes acts in the arcuate nucleus of the hypothalamus to inhibit the AGRP/neuropeptide Y (NPY) neurons from releasing AGRP and NPY [5]. Ghrelin stimulates the secretion of AGRP and NPY to increase appetite [3]. The expression of AGRP in the adrenal gland is regulated by glucocorticoids [6].

Receptors

AGRP is an endogenous antagonist of the melanocortin receptors MC3R and MC4R. Since MCR agonists are the melanocortin α-melanocyte-stimulating hormone (MSH) and adrenocorticotropic hormone, details of MCRs are described in Chapter 16. AGRP acts normally as a competitive antagonist at MC3R and MC4R. AGRP can also bind to MC5R with lower affinity. Beyond competitive antagonism, *in vitro* assays show that AGRP acts as an inverse agonist to decrease basal receptor activity in the absence of agonists.

Biological Functions

Target Cells/Tissues and Functions

AGRP prevents α-MSH from binding to MC3R or MC4R in the hypothalamus to inhibit α-MSH-mediated elevation of cAMP levels, leading to appetite stimulation and the inhibition of energy expenditure. AGRP expressed in the hypothalamus also imposes an inhibitory effect on puberty by regulating TAC2/neurokinin B (NKB) neurons in the arcuate nucleus of the hypothalamus [7]. In the adrenal gland, AGRP blocks α-MSH-induced secretion of corticosterone [6].

Phenotypes in Gene-Modified Animals

Overexpression of AGRP in transgenic mice causes severe hyperphagia and obesity [1]. However, mice lacking the Agrp gene do not exhibit alterations in feeding or body weight [8].

Y. Takei, H. Ando, & K. Tsutsui (Eds): Handbook of Hormones. DOI: http://dx.doi.org/10.1016/B978-0-12-801028-0.00123-9

```
        10        20        30        40        50        60
MLTAALLSCALLLALPATRG AQMGLAPMEGIRRPDQALLPELPGLGLRAPLKKTTAEQAE
        70        80        90        100       110       120
EDLLQEAQALAEVLDLQDRE PRSSRRCVRLHESCLGQQVPCCDPCATCYCRFFNAFCYCR
        130
KLGTAMNPCSRT
```

Figure 8B.1 Primary structure of human AGRP. The potential signal peptide is shown by a box. The amino acid residues of mature protein and cysteine residues forming five disulfide bonds (87–102, 94–108, 101–119, 105–129, 110–117) are underlined and shaded, respectively.

Pathophysiological Implications

Clinical Implications

It has been shown that polymorphisms in the AGRP gene are associated with susceptibility for anorexia nervosa [9] as well as obesity [10].

References

1. Dinulescu DM, Cone RD. Agouti and agouti-related protein: analogies and contrasts. *J Biol Chem.* 2000;275:6695–6698.
2. Jackson PJ, Douglas NR, Chai B, et al. Structural and molecular evolutionary analysis of Agouti and Agouti-related proteins. *Chem Biol.* 2006;13:1297–1305.
3. Creemers JW, Pritchard LE, Gyte A, et al. Agouti-related protein is posttranslationally cleaved by proprotein convertase 1 to generate agouti-related protein (AGRP)83–132: interaction between AGRP83–132 and melanocortin receptors cannot be influenced by syndecan-3. *Endocrinology.* 2006;147:1621–1631.
4. Takeuchi S, Teshigawara K, Takahashi S. Widespread expression of Agouti-related protein (AGRP) in the chicken: a possible involvement of AGRP in regulating peripheral melanocortin systems in the chicken. *BBA-Mol Cell Res.* 2000;1496:261–269.
5. Enriori PJ, Evans AE, Sinnayah P, et al. Diet-induced obesity causes severe but reversible leptin resistance in arcuate melanocortin neurons. *Cell Metab.* 2007;5:181–194.
6. Dhillo WS, Small CJ, Gardiner JV, et al. Agouti-related protein has an inhibitory paracrine role in the rat adrenal gland. *Biochem Bioph Res Commun.* 2003;301:102–107.
7. Sheffer-Babila S, Sun Y, Israel DD, et al. Agouti-related peptide plays a critical role in leptin's effects on female puberty and reproduction. *Am J Physiol Endocrinol Metab.* 2013;305:E1512–E1520.
8. Qian S, Chen H, Weingarth D, et al. Neither agouti-related protein nor neuropeptide Y is critically required for the regulation of energy homeostasis in mice. *Mol Cell Biol.* 2002;22:5027–5035.
9. Vink T, Hinney A, van Elburg AA, et al. Association between an agouti-related protein gene polymorphism and anorexia nervosa. *Mol Psychiatr.* 2001;6:325–328.
10. Argyropoulos G, Rankinen T, Neufeld DR, et al. A polymorphism in the human agouti-related protein is associated with late-onset obesity. *J Clin Endocr Metab.* 2002;87:4198–4202.

Supplemental Information Available on Companion Website

- List of useful web resources for AGRP, including alignment of primary structure of AGRP of various animals and phylogenetic tree of AGRP/E-Box 8B.1

CHAPTER 9

Tachykinin Family

Honoo Satake

History

In 1962, Erspamer and Falconieri Erspamer reported isolation and sequence elucidation of eledoisin from the posterior salivary gland of the Mediterranean octopus *Eledone moschata* [1]. This was the first identification of tachykinin (TK) in an organism. In 1964, the same group also purified physalaemin from the skin of the South American frog *Physalaemus biligonigerus*, as the first vertebrate TK [2]. The first elucidation of the primary sequence of a mammalian TK, substance P (SP), was reported by Chang and colleagues in 1971 [3]. Thereafter, a great number of structurally and/or functionally related peptides were characterized from the brain, hypothalamus, and gut of other chordates and amphibian skins, including neurokinin A and B (NKA and NKB), and hemokinin/endokinin (HK-1/EK) [4−7]. Consequently, the mammalian TK family is believed to consist of four peptides: SP, NKA, NKB, and HK/EK [5−7]. SP, NKA, and NKB have also been identified in non-mammalian gnathostomates as brain/gut peptides [5−8]. Frogs have been found to possess skin-specific TKs in addition to SP, NKA, and NKB [5−7]. In teleosts, two NKB subtype (NKBa and NKBb) and the species-specific gene-related peptide, neurokinin F (NKF), were characterized. In a protochordate, *Ciona intestinalis*, an authentic TK, Ci-TK, was identified in the central nervous system and digestive tract as the sole TK family representative [9]. In consequence, it has now been established that the TK family is conserved as brain/gut peptides not only in "vertebrates" but also in "chordates" [5−9].

Structure

Structural Features and Primary Structure

The TK family peptides feature the C-terminal consensus motif Phe−X−Gly−Leu−Met−NH$_2$. Putative EKC and EKD bear Phe−X−Gly−Leu−Leu−NH$_2$. Most TKs, including piscine-specific NKF, consist of 9 to 12 amino acids (aa), whereas two N-terminally extended forms of NKA, neuropeptide K (21 aa) and neuropeptide-γ (36 aa) are also detected. In teleosts, NKBb is composed of 24 aa.

Properties

Mr 900−4,000; soluble in water, physiological saline solution, aqueous acetonitrile, methanol.

Synthesis and Release

Gene, mRNA, and Precursor

Mammalian TKs are encoded by three TK genes, *tac1* (*pptA*), *tac3* (*pptB*), and *tac4* (*pptC*) [4−7]. Four splicing variants, α, β, γ, and δ TAC1, are generated from *tac1* in a tissue-specific manner. α-TAC1 and δ-*TAC1* encode only SP, whereas SP and NKA sequences are present in β- and γ-TAC1 precursors. Neuropeptides K and -γ are yielded as N-terminally extended forms of NKA from β- and γ-TAC1, respectively [4−7]. Tetrapod *tac3* encodes only NKB, although various splicing variants are generated [4−7]. Moreover, the teleost homolog generates NKB and NKF [8]. *tac4* generates various splicing variants, while only one EK/HK is bioactive [4−7]. Frog skin TK precursors encode a single copy of TK, whereas an additional TK-like sequence is found in several species [6,10]. The Ci-TK precursor encodes two TK sequences, whereas no splicing variant occurs [9].

Distribution of mRNA

In mammals, TAC1 mRNA is present in the brain (including the spinal cord), gastrointestinal tract, salivary gland, heart, skin, spleen, adrenal gland, uterus, testis (including Leydig cells), ovary immune cells, and pulmonary artery, and in small amounts in the kidney and thyroid gland, while α-TAC1 is exclusively present in the brain [4−7]. TAC3 mRNA is expressed mainly in the hypothalamus, the dorsal horn of the spinal cord, the kidney, testis, placenta, uterus, and bronchus [4−7]. The TAC4 gene is expressed primarily in the heart, muscle, thyroid, skin, and adrenal gland, whereas the mouse counterpart is detected in the hematopoietic cells, bone marrow, thymus, uterus, lymph node, lung, spleen, and eye, but only faintly in the central nervous system [4−7]. In teleosts, *tac1* expression is detected in wide regions of the central nervous system, and *tac 3a* mRNA is detected abundantly in the mid- and hindbrain, pituitary, and (to a lesser extent) in the forebrain, testis, and embryos at 6−12 days' postfertilization, whereas *tac3b* is expressed mainly in the forebrain and ovary [8]. *ci-tk* is expressed in the brain, digestive tract, and endostyle of an acidian [9].

Receptors

Structure and Subtype

All TK receptors belong to the class A G protein-coupled receptor family. Three TK receptors, NK1 (TACR1), NK2 (TACR2),

Y. Takei, H. Ando, & K. Tsutsui (Eds): Handbook of Hormones. DOI: http://dx.doi.org/10.1016/B978-0-12-801028-0.00009-X

and NK3 (TACR3) have so far been identified in vertebrates [4–7]. Two subtypes of NK1 and NK3 are present, TACR1a and TACR1b, and TACR3a and TACR3b in zebrafish, respectively [8]. In *C. intestinalis*, Ci-TK-R is the sole Ci-TK receptor [9]. NK1, -2, and, -3 display ligand selectivity: for NK1, SP > NKA > NKB; for NK2, NKA > NKB > SP; for NK3, NKB > NKA > SP [4–7]. EK/HK exhibits the highest affinity to NK1 [4–7]. Activation of TACR3a and -b by NKF is almost comparable to that by NKBa, whereas the activity of NK3b is approximately 100-fold less potent than that of NKBa [8].

Signal Transduction Pathway

All vertebrate TK receptors, expressed in *Xenopus* oocytes or cultured cells, trigger intracellular calcium mobilization and cAMP production in response to the cognate TKs, followed by PKC activation, ERK phosphorylation, or NF-κB induction in a cell- or tissue-specific manner [4–9].

Biological Functions

Target Cells/Tissues and Functions

TKs participate in a wide range of biological functions and pathogenic processes, including neural transmission, nociception, inflammation, smooth muscle contraction, and neurodegeneration [4–10].

References

1. Erspamer V, Falconieri-Erspamer GF. Pharmacological actions of eledoisin on extravascular smooth muscle. *Br J Pharmacol.* 1962;19:337–354.
2. Anastas A, Erspamer V, Cei JM. Isolation and amino acid sequence of physalaemin, the main active polypeptide of the skin of *Physalaemus fuscomaculatus. Arch Biochem Biophys.* 1964;108:341–348.
3. Chang MM, Leeman SE, Niall HD. Amino-acid sequence of substance P. *Nature.* 1971;232:86–87.
4. Severini C, Improta G, Falconieri-Erspamer G, et al. The tachykinin peptide family. *Pharmacol Rev.* 2002;54:285–322.
5. Satake H, Kawada T. Overview of the primary structure, tissue distribution, and functions of tachykinins and their receptors. *Curr Drug Targets.* 2006;7:963–974.
6. Satake H, Aoyama M, Sekiguchi T, et al. Insight into molecular and functional diversity of tachykinins and their receptors. *Prot Pept Lett.* 2013;20:615–627.
7. Steinhoff MS, von Mentzer B, Geppetti P, et al. Tachykinins and their receptors: contributions to physiological control and the mechanisms of disease. *Physiol Rev.* 2014;94:265–301.
8. Biran J, Palevitch O, Ben-Dor S, et al. Neurokinin Bs and neurokinin B receptors in zebrafish-potential role in controlling fish reproduction. *Proc Natl Acad Sci USA.* 2012;109:10269–10274.
9. Satake H, Ogasawara M, Kawada T, et al. Tachykinin and tachykinin receptor of an ascidian, *Ciona intestinalis*: evolutionary origin of the vertebrate tachykinin family. *J Biol Chem.* 2004;279:53798–53805.
10. López-Bellido R, Barreto-Valer K, Rodríguez RE. Expression of tachykinin receptors (tacr1a and tacr1b) in zebrafish: influence of cocaine and opioid receptors. *J Mol Endocrinol.* 2012;50:115–129.

Supplemental Information Available on Companion Website

- Amino acid sequences of typical vertebrate tachykinins/E-Table 9.1
- Accession numbers for cDNAs encoding chordate tachykinins and their receptors/E-Table 9.2
- Schematic representatives of tachykinin precursors/E-Figure 9.1

Substance P/Neurokinin A

Honoo Satake

Abbreviation: SP/NKA
Additional names (neurokinin A): substance K (SK), neurokinin α (NKα), neuromedin L (NKL)

SP and NKA are vertebrate tachykinins (TKs) encoded by the TAC1 gene. These peptides participate in the regulation of multiple biological functions.

Discovery

In 1971, SP was isolated from the horse intestine [1]. In 1983, NKA was purified from porcine spinal cord [2]. Since the discovery in mammals, SP and NKA have been characterized in all gnathostomate species [3–6]. In addition, N-terminally extended forms of NKA, neuropeptide (NP) K and NP-γ, were also purified from mammals [3–6].

Structure

Structural Features and Primary Structure

In all species SP and NKA are composed of 10–11 aa, with the exception of NPK (36 aa) and NP-γ (19 aa), and conserve the C-terminal Phe–X–Gly–Leu–Met–NH$_2$ TK sequence (Table 9A.1). SP family peptides contain an aromatic aa (Phe or Tyr) at position X, whereas a branched aliphatic residue (Ile or Val) is located at the corresponding position of NKA family peptides (Table 9A.1). Moreover, a neutral or basic aa is located at the seventh position from the C-terminus of SP family peptides and an acidic residue is found at this position in NKA family peptides, respectively (Table 9A.1). These motifs and residues have been shown to play crucial roles in receptor activation and binding selectivity [3–5].

Synthesis and Release

Gene, mRNA, and Precursor

SP and NKA are encoded in *tac1* (or *pptA*). The tac1 gene is organized by seven exons [3–6]. Human and rat SP and NKA are present as single copies in the third and sixth exons, respectively [3–6]. To date, four splicing variants have been shown to be generated from *tac1* in mammals: α-, β-, γ-, and δ-TAC (Figure 9A.1) [3–6]. α-and δ-TAC lack the sixth exon, and thus only SP is produced from these precursors [3–6]. β-TAC is organized by all of the exons, and γ-TAC lacks only the fourth exon of the seven exons. Consequently, both of them yield SP and NKA [3–6]. In addition, NPK and NP-γ are generated exclusively from β-TAC and γ-TAC, respectively (Figure 9A.1), due to the absence of cleavage at the endoproteolytic site located prior to the N-terminus of the NKA sequence [3–5]. In mice, *tac1* is composed of six exons, and fails to liberate NPK [3–6]. Non-mammalian tac1 is essentially organized as in the mammalian counterpart, but neither homologs thereof nor other gene-related peptides have ever been identified [3–5,7].

Distribution of mRNA

α-TAC1 mRNA is localized in numerous neurons of the brain. β-TAC1 (10–20%) and γ-TAC1 (80–90%) are expressed predominantly in intrinsic enteric neurons and sensory neurons [3–5]. Moreover, SP and tac1 expression are observed in diverse peripheral neurons, including capsaicin-sensitive and resistant neurons [3–6]. In peripheral non-neural tissues, the expression of tac1, SP, and NKA was found to be in abundance in the salivary gland, heart, skin, spleen, adrenal gland, uterus, testis (including Leydig cells), ovary, placenta, several immune cells, and the pulmonary artery, and in small amounts in the kidney and thyroid gland [3–5]. In zebrafish, *tac1* is expressed in the telencephalon, preoptic region, hypothalamus, mesencephalon, and olfactory bulb [7]. In the embryonic stage, *tac1* expression elevates progressively at 8 hours post-fertilization (hpf) to the end of the embryogenesis at 72 hpf, and also is prominent in the spinal cord in embryos at 20–30 hpf, and mainly in the telencephalon, diencephalon, hypothalamus, rhombomeres, epiphysis, heart, and somites in 36, 42, and 48 hpf embryos [8].

Receptors

Structure and Subtype

TK receptors, namely NK1 (TACR1), -2 (TACR2), and -3 (TACR3), belong to the class A G protein-coupled receptor family (E-Figure 9A.1) [3–6]. SP and NKA display selective affinity to NK1 and NK2, respectively [3–6]. NPK and NP-γ are also selective to NK2 [3–6]. NK1 homologs have also been characterized in several tetrapods. In zebrafish, *tacr1a* and *tacr1b* are present as duplicates [9].

Signal Transduction Pathway

NK1 and NK2, expressed in in *Xenopus* oocytes or cultured cells, trigger intracellular calcium mobilization and cAMP production in response to their selective ligands, followed by activation of PKC, MAPK, JAK/STAT, ERK phosphorylation, p38, and NF-κB induction in a cell- or tissue-specific manner [3–9].

Agonists

More than 30 mammalian NK1- or NK2-agonists, such as [Sar^9, Met(O$_2$)^{11}]SP (NK1-specific) and GR 64349 (NK2-specific), are commercially available.

Y. Takei, H. Ando, & K. Tsutsui (Eds): Handbook of Hormones. DOI: http://dx.doi.org/10.1016/B978-0-12-801028-0.00124-0

Table 9A.1 Primary Sequences of SP and NKA Family

Species	Peptide	Sequence
Mammals	Substance P	RPKPQQFFGLM-NH$_2$
	Neurokinin A	HKTDSFVGLM-NH$_2$
Chicken (*Gallus gallus*)	Substance P	RPRPQQFFGLM-NH$_2$
	Neurokinin A	HKTDSFVGLM-NH$_2$
Frog		
Bufo marinus	Bufokinin	KPRPDQFYGLM-NH$_2$
Rana ridibunda	Ranakinin	HKLDSFIGLM-NH$_2$
Xenopus tropicalis	Substance P	KPRPDQFYGLM-NH$_2$
	Neurokinin	YKSGSFFGLM-NH$_2$
Dogfish (*Scyliorhinus canicula*)	Scyliorhinin	AKFDKFYGLM-NH$_2$
Zebrafish (*Danio rerio*)	Substance P	RPRPHQFIGLM-NH$_2$
	Neurokinin	HKINSFVGLM-NH$_2$
Medaka (*Oryzias latipes*)	Substance P	KPRPHQFIGLM-NH$_2$
	Neurokinin A	HKVNSFVGLM-NH$_2$

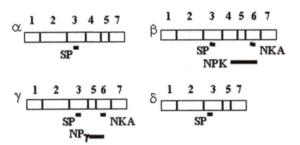

Figure 9A.1 TAC1 precursor. Digits indicate exon numbers.

Antagonists

A wide variety of NK1- or NK2- competitive antagonists such as [D-Arg^1, D-Phe^5, D-Trp^{4,6}, Leu^{11}]-SP (NK1-specific) and L-659,877 (NK2-specific) are commercially available. Some of the NK1 antagonists are also candidates for anti-tumor, anti-inflammatory, and antidepressive drugs [4,6]. It should be noteworthy that almost no evidence has been provided regarding the pharmacological effects of agonists or antagonists for mammalian NK1 or NK2 on non-vertebrate homologs.

Biological Functions

Target Cells/Tissues and Functions

In mammals, NK1 is widely distributed in the neurons, cardiovascular system, muscles, gastrointestinal tract, genitourinary tract, pulmonary system, genital organs, and immune cells, whereas NK2 is expressed predominantly in the prostate, cerebellum, lung, trachea, uterus, and bronchus [3—6]. In sperm, NK1 is distributed throughout the acrosomal region, midpiece, and flagellum; NK2 is present on the acrosomal region, the middle region of the head, and the end of the flagellum [5]. In zebrafish, *tacr1* is also expressed during embryonic development [10]. In mammals, SP and NKA have been shown to participate in a wide range of biological functions, including nociception, vasodilation, salivation, muscle contraction, inflammation, and neurodegeneration [3—6]. In a typical nociceptive process, peripheral nociceptors, including capsaicin-sensitive sensory neurons expressing TRP and/or cytokine receptors, release SP in the dorsal horn and spinal cord in response to various stimuli such as pressure, temperature, other transmitters, or cytokines, and then SP activates at NK1 on spinal neurons [3—6]. Likewise, SP and NKA participate in diverse inflammatory processes in the respiratory, gastrointestinal, neurogenic, and musculoskeletal systems by induction of inflammatory mediators such as cytokines,

prostaglandins, histamine, and oxygen radicals, frequently with calcitonin gene-related peptides [3—6]. SP and NKA also stimulate sperm motility mainly via NK1 and NK2 [5].

Phenotype in Gene-Modified Animals

TAC1 or NK1-knockout mice show a prominent deficiency in amplification and intensity of nociceptive refluxes by SP [3—6]. Elevation of morphine analgesia and reduction of side effects are also found in *tac1*-knockout mice [6]. Moreover, anxiety- and depression-related behaviors are suppressed in *tac1*-knockout mice [6]. *tacr1*-deficient mice also display a decreased preference for sexual pheromones [10].

Pathophysiological Implications

Clinical Implications

SP and NKA are likely to be involved in prevalent symptoms and diseases including cough, diarrhea, asthma, and arthritis [3—6]. Furthermore, SP and NKA are pathologically implicated in neural, colon, breast, lung, and skin cancers [6]. For instance, expression of *tac1*, *tacr1*, and *tacr2* is markedly elevated in malignant biopsies, and NK1 and NK2 antagonists suppress growth of human breast cancer cells [6].

Use for Diagnosis and Treatment

To date, a wide variety of specific agonists and antagonists targeted for NK1 and NK2 have been examined for clinical trials. The NK1 antagonist, aprepitant (EMEND™), and its prodrug form fosaprepitant (IVEMEND™), have been approved for treatment of chemotherapy-induced or postoperative nausea and vomiting.

References

1. Chang MM, Leeman SE, Niall HD. Amino-acid sequence of substance P. *Nature*. 1971;232:86—87.
2. Kimura S, Okada M, Sugita Y, et al. Novel neuropeptides, neurokinin a and b, isolated from porcine spinal cord. *Proc Jpn Acad B*. 1983;59:101—104.
3. Severini C, Improta G, Falconieri-Erspamer G, et al. The tachykinin peptide family. *Pharmacol Rev*. 2002;54:285—322.
4. Satake H, Kawada T. Overview of the primary structure, tissue distribution, and functions of tachykinins and their receptors. *Curr Drug Target*. 2006;7:963—974.
5. Satake H, Aoyama M, Sekiguchi T, et al. Insight into molecular and functional diversity of tachykinins and their receptors. *Prot Pept Lett*. 2013;20:615—627.
6. Steinhoff MS, von Mentzer B, Geppetti P, et al. Tachykinins and their receptors: contributions to physiological control and the mechanisms of disease. *Physiol Rev*. 2014;94:265—301.
7. Ogawa S, Ramadasan PN, Goschorska M, et al. Cloning and expression of tachykinins and their association with kisspeptins in the brains of zebrafish. *J Comp Neurol*. 2012;520:2991—3012.
8. López-Bellido R, Barreto-Valer K, Rodríguez RE. Substance P mRNA expression during zebrafish development: influence of mu opioid receptor and cocaine. *Neuroscience*. 2013;242:53—68.
9. López-Bellido R, Barreto-Valer K, Rodríguez RE. Expression of tachykinin receptors (tacr1a and tacr1b) in zebrafish: influence of cocaine and opioid receptors. *J Mol Endocrinol*. 2013;50:115—129.
10. Berger A, Tran AH, Dida J, et al. Diminished pheromone-induced sexual behavior in neurokinin-1 receptor deficient (TACR1 (−/−)) mice. *Genes Brain Behav*. 2012;11:568—576.

Supplemental Information Available on Companion Website

- Accession numbers for cDNAs encoding SP and NKA/E-Table 9A.1
- Phylogenetic tree of TK receptors/E-Figure 9A.1

Neurokinin B

Honoo Satake

Abbreviation: NKB
Additional name: neuromedin K (NMK)

Neurokinin B is a vertebrate tachykinin encoded by the TAC3 gene (Tac2 in rodents). In teleosts, an additional tachykinin, neurokinin F (NKF), is also encoded by TAC3.

Discovery

In 1983, neurokinin B (NKB) was isolated from the porcine spinal cord [1]; this was followed by characterization of other tetrapod NKBs [2–4]. In 2012, teleost NKB and its gene-related peptide, neurokinin F (NKF), were identified [5,6].

Structure

Structural Features and Primary Structure

Tetrapod NKBs are composed of 10 aa. In several teleosts, such as zebrafish and salmon, NKBa consists of 10 aa whereas NKBb, another zebrafish NKB subtype, comprises 24 aa. NKF is composed of 13 aa. All *tac3*-derived peptides share the C-terminal Phe–X–Gly–Leu–Met–NH_2 tachykinin (TK) sequence (Table 9B.1) [2–4]. In particular, NKBs and NKFs almost exclusively contain Val at the "X" position, as seen in NKA family peptides [2–4]. Compared with NKA, NKB possesses no basic amino acid (Lys or Arg). These motifs and residues have been shown to play crucial roles in receptor activation and binding selectivity [2–4].

Synthesis and Release

Gene, mRNA, and Precursor

NKB is encoded in *tac3* (or *pptB*). Vertebrate *tac3* is organized by seven or eight exons [2–6]. Most tetrapod NKB is present as a single copy in the fifth exon [2–6]. To date, several splicing variants have been shown to be generated from *tac3* in mammals, while only NKB is produced from these precursors (Figure 9B.1) [2–6]. In several teleosts, including zebrafish and salmon, two *tac3*, *tac3a* and *-3b*, are present. For instance, zebrafish *tac3a* encodes one 10-amino acid NKBa and one NKF, whereas one 24-amino acid NKBb and one additional NKF, where only the second Asn of *tac3a*-derived NKF is replaced with Asp, are encoded by *tac3b* [5,6].

Distribution of mRNA

tac3 is expressed predominantly in the hypothalamus, kidney, testis, placenta, uterus, bronchus, and the equatorial segment and post-equatorial region of the sperm head in mammals [2–4]. In the hypothalamus, NKB is produced in various neurons, including KNDy arcuate neurons that co-express NKB, kisspeptin and dynorphin [7]. In zebrafish, *tac3a* is expressed in the habenula, periventricular hypothalamus, periventricular nucleus of the posterior tuberculum, preoptic region, and posterior tuberal nucleus [5,6]. *tac3b* is distributed predominantly in the dorsal telencephalon area [5]. However, no obvious KNDy neurons have been observed in the zebrafish hypothalamus [5]. *tac3a* expression has also been detected in the right-habenula nuclei of zebrafish [6].

Receptors

Structure and Subtype

TK receptors, namely NK1 (TACR1), -2 (TACR2), and -3 (TACR3), belong to the class A G protein-coupled receptor family [2–4]. NKB displays selective affinity to NK3 [2–4]. In most teleosts, two subtypes, namely, *tac3ra* and *-b*, are conserved [5,6]. NKF exhibited activity at human NK3 and zebrafish TACR3a and -b comparable to mammalian NKB or zebrafish NKBa, whereas zebrafish NKBb is 100- to 1000-fold less potent than these NKB family peptides [5,6].

Signal Transduction Pathway

Mammalian and zebrafish NK3, expressed in *Xenopus* oocytes or cultured cells, triggers intracellular calcium mobilization and cAMP production in response to their selective ligands, followed by activation of PKC, MAPK, and ERK phosphorylation in a cell- or tissue-specific manner [2–4].

Agonists

A number of mammalian NK3 agonists, such as senktide and [MePhe^7]NKB, are commercially available.

Antagonists

A wide variety of NK3-competitive antagonists, such as osanetant and talnetant, are commercially available [6,8]. It should be noteworthy that almost no evidence has been provided for the pharmacological effects of agonists or antagonists for mammalian NK3 on non-vertebrate homologs.

Biological Functions

Target Cells/Tissues and Functions

In mammals, NK3 is distributed in the brain, hypothalamus (including the median eminence), kidney, lung, placenta, prostate, testis, muscle, intestine, and uterus, in the sperm midpiece, and weakly in the proximal region of the sperm

Y. Takei, H. Ando, & K. Tsutsui (Eds): Handbook of Hormones. DOI: http://dx.doi.org/10.1016/B978-0-12-801028-0.00125-2

Table 9B.1 Primary Sequences of NKB

Species	Peptide	Sequence
Mammals	Neurokinin B	DMHDF FVGLM-NH$_2$
Amphibians		
Pelophylax ridibundus	Neurokinin B	DMHDF FVGLM-NH$_2$
Xenopus tropicalis	Neurokinin B	EMNDF FVGLM-NH$_2$
Fish	Neurokinin Ba	EMHDI FVGLM-NH$_2$
Danio rerio	Neurokinin Bb	STGINREAHLPFRNMNDI FVGLL-NH$_2$
	Neurokinin F	YNDIDYDS FVGLM-NH$_2$
Oryzias latipes	Neurokinin B	DMDDI FVGLM-NH$_2$

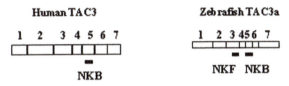

Figure 9B.1 Human and zebrafish TAC3 precursors. Digits indicate exon numbers.

flagellum [2−4]. In zebrafish, *tacr3a* is expressed mainly in the pituitary also during embryonic development. Low *tacr3b* expression was detected in the midbrain and testis [6]. In mammals, circulating NKB via enhancement of expression of the tac3 gene was elevated in the placenta of pre-eclampsia women, but not in the placenta of women in normotensive pregnancies [3,4,8]. Moreover, NKB participates in the regulation of GnRH pulse generation in the mammalian hypothalamus−pituitary−gonadal axis [9]. This GnRH release system is in part regulated by NKB in concert with kisspeptin yielded by KNDy neurons [9], whereas the involvement of non-KNDy NKB neurons or other kisspeptin-free NKB neurons in the stimulation of GnRH release is also reported [10]. In prepubertal zebrafish, estradiol treatment upregulates *tac3a* gene expression as well as *kiss1*, *kiss2*, and *gnrh3* [6]. Furthermore, in sexually mature zebrafish, intraperitoneal injection of NKBa and NKF resulted in secretion of luteinizing hormone [6].

Phenotype in Gene-Modified Animals

A homozygous point mutation of Gly^{93} to Asp or Pro^{355} to Ser in NK3 has been detected in several individuals of familial hypogonadotropic hypogonadism [3,4,9].

Pathophysiological Implications

Clinical Implications

NKB is highly likely to be involved in pre-eclampsia in the placenta, and familial hypogonadotropic hypogonadism [3,4,8,9].

References

1. Kangawa K, Minamino N, Fukuda A, et al. Neuromedin K: a novel mammalian tachykinin identified in porcine spinal cord. *Biochem Biophys Res Commun.* 1983;114:533−540.
2. Satake H, Kawada T. Overview of the primary structure, tissue distribution, and functions of tachykinins and their receptors. *Curr Drug Target.* 2006;7:963−974.
3. Satake H, Aoyama M, Sekiguchi T, et al. Insight into molecular and functional diversity of tachykinins and their receptors. *Prot Pept Lett.* 2013;20:615−627.
4. Steinhoff MS, von Mentzer B, Geppetti P, et al. Tachykinins and their receptors: contributions to physiological control and the mechanisms of disease. *Physiol Rev.* 2014;94:265−301.
5. Ogawa S, Ramadasan PN, Goschorska M, et al. Cloning and expression of tachykinins and their association with kisspeptins in the brains of zebrafish. *J Comp Neurol.* 2012;520:2991−3012.
6. Biran J, Palevitch O, Ben-Dor S, et al. Neurokinin Bs and neurokinin B receptors in zebrafish-potential role in controlling fish reproduction. *Proc Natl Acad Sci USA.* 2012;109:10269−10274.
7. Rance NE, Dacks PA, Mittelman-Smith MA, et al. Modulation of body temperature and LH secretion by hypothalamic KNDy (kisspeptin, neurokinin B and dynorphin) neurons: a novel hypothesis on the mechanism of hot flushes. *Front Neuroendocrinol.* 2013;34:211−227.
8. Page NM, Neurokinin B. and pre-eclampsia: a decade of discovery. *Reprod Biol Endocrinol.* 2010;8:4.
9. Topaloglu AK, Semple RK. Neurokinin B signalling in the human reproductive axis. *Mol Cell Endocrinol.* 2011;346:57−64.
10. Gaskins GT, Glanowska KM, Moenter SM. Activation of neurokinin 3 receptors stimulates GnRH release in a location-dependent but kisspeptin-independent manner in adult mice. *Endocrinology.* 2013;154:3984−3989.

Supplemental Information Available on Companion Website

- Accession numbers for cDNAs encoding NKB and NKF/E-Table 9B.1

CHAPTER 10

Appetite-Regulating Peptides

Yoshio Takei and Akiyoshi Takahashi

History

The hypothalamus has been implicated in appetite regulation since the middle of the 20th century, when the lateral hypothalamic area (LHA) was suggested as a "hunger center" and the ventromedial nucleus, dorsomedial nucleus, and paraventricular nucleus (PVN) as "satiety centers" by ablation experiments [1]. These regulatory centers were initially thought to be target sites for glucose and fatty acids, but thanks to the growing interest of endocrinologists in obesity and metabolic syndrome, many new appetite-regulating hormones have been identified in the periphery and the CNS [2]. In the periphery, various appetite-regulating hormones are secreted from the gastrointestinal tracts (ghrelin, cholecystokinin, glucagon-like peptide-1, peptide YY, etc.), including the pancreas (insulin, pancreatic polypeptides, etc.) and adipose tissue (leptin, adiponectin, etc.), which act on the CNS regions that have insufficient blood—brain barrier, such as the arcuate nucleus (ARC) of the hypothalamus and area postrema of the medulla oblongata, to modulate neural circuitry for appetite regulation (Figure 10.1) [2,3]. Among the peripheral hormones, ghrelin is the most potent orexigenic (stimulatory) hormone and leptin is the most potent anorexigenic (inhibitory) hormone. Readers are directed to subsequent chapters of respective hormones dealing with individual hormones regarding the appetite-regulating function (Subchapters 17C for glucagon-like peptide-1, 19A for insulin, 20B for cholecystokinin, 21A for ghrelin, 25A for pancreatic polypeptide, 25C for peptide YY, 34A for leptin and 34B for adiponectin). In this chapter, therefore, the major topic will be on neuropeptides in the hypothalamic neurons that play pivotal roles in appetite regulation. Such major neuropeptides include orexigenic neuropeptide Y (NPY) and agouti-related peptide (AgRP), and anorexigenic α-melanophore-stimulating hormone (α-MSH) and cocaine-amphetamine regulated transcript (CART) in the ARC; melanin-concentrating hormone (MCH) and orexin (or hypocretin) in the LHA; and corticotropin-releasing hormone (CRH) and thyrotropin-releasing hormone (TRH) in the PVN (Figure 10.1) [4]. NPY and AgRP, or α-MSH and CART, co-localize in the same neurons in the ARC. Among them, MCH, ORX, and CART are the neuropeptides discussed here, while NPY, α-MSH, AgRP, and CRH are described in Subchapters 25B, 16B, 8B, and 2A, respectively (21A and 34A respectively). Furthermore, several neuropeptides, such as neuromedin U/S (Chapter 13) and apelin (Chapter 31) are also involved in fine-tuning of appetite regulation, and their actions are described to some extent in each chapter.

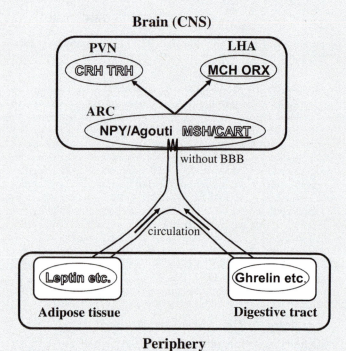

Figure 10.1 Central circuitry of appetite regulation by neuropeptides. Interaction with peripheral hormones at the arcuate nucleus (ARC) is also shown. Hormones and neuropeptides shown by black characters are orexigenic, and those shown by outline characters are anorexigenic. Underlined neuropeptides are described in this chapter. For other hormones and neuropeptides, refer to the text. For abbreviations, see text.

Structures and Biological Functions

There are no common structural features of the hormones and their receptors included in this chapter, and readers can obtain structural information in detail in the chapters dealing with individual hormones. Thus a brief account of each hormone underlined in Figure 10.1 will be made for comparison. MCH was originally discovered in fish as a neurohypophysial hormone that induces aggregation of melanin granules in the skin of chum salmon [5]. Teleost MCH has 17 aa residues with a disulfide bond between Cys^7 and Cys^{16}, while mammalian MCH has 19 aa residues including an extra aa in the

Y. Takei, H. Ando, & K. Tsutsui (Eds): Handbook of Hormones. DOI: http://dx.doi.org/10.1016/B978-0-12-801028-0.00010-6

N-terminus outside the ring. MCH neurons are present in the hypothalamus of all vertebrates from lampreys to humans, and they send axons to other parts of the brain. However, there seem to be some differences in the projection of MCH neurons between mammals and teleosts, as the mammalian MCH axons do not send axons abundantly to the neurohypophysis. MCH neurons, as well as orexin neurons, are considered to be the most downstream peptidergic neurons involved in the chain of hypothalamic signaling that stimulate food intake [5]. Therefore, NPY/AgRP neurons and α-MSH/CART neurons in the ARC are called primary appetite-regulating neurons, while MCH and orexin neurons in the LHA and CRH and TRH neurons in the PVN are designated as secondary neurons in the appetite-regulating neural circuitry. Orexin A and orexin B were originally identified as endogenous ligands for two orphan GPCRs [6]. These peptides are also known as hypocretins [7], and were independently identified as putative peptides encoded by a hypothalamus-specific transcript [7]. They were initially recognized as regulators of feeding behavior because of their exclusive production in the LHA, and their orexigenic activity when administered into the brain. Subsequently, however, the importance of orexins in the maintenance of consolidated sleep/wake states has been demonstrated by the fact that the sleep disorder narcolepsy is caused by orexin deficiency in humans and other animals [8]. CART was first discovered as an mRNA in the striatum of the rat brain, and its expression is regulated by cocaine and amphetamine [9]. Subsequently, a peptide predicted to be derived from the mRNA was identified. CART mRNA and the peptide are widely distributed in the hypothalamic area, including the ARC and retrochiasmatic area. Most CART-containing neurons in these hypothalamic areas also contain proopiomelanocortin (POMC) mRNA that produces α-MSH as a part of translated peptides [10]. CART and α-MSH are produced in the same neuron and are major anorexigenic peptides that inhibit feeding and cause high-energy expenditure. Indeed, expression of the CART and POMC genes in such hypothalamic neurons is upregulated by the adipocyte-derived hormone leptin and other peripheral anorexigenic hormones, while their expression is downregulated by ghrelin and other peripheral orexigenic hormones (Figure 10.1). Thus CART/POMC neurons serve as one of primary neurons that link the peripheral information to the distal neurons such as MCH, orexin, CRH and TRH neurons in the appetite-regulating circuitry.

References

1. Kalra SP, Dube MG, Pu S, et al. Interacting appetite-regulating pathways in the hypothalamic regulation of body weight. *Endocrine Rev*. 1999;20:68−100.
2. Hussain SS, Bloom SR. The regulation of food intake by the gut−brain axis: implications for obesity. *Int J Obesity*. 2013;37:625−633.
3. Field BCT, Chaudhri OB, Bloom SR. Bowels control brain: gut hormones and obesity. *Nat Rev Endocrinol*. 2010;6:444−453.
4. Schwartz MW, Woods SC, Porte D, et al. Central nervous system control of food intake. *Nature*. 2000;404:661−671.
5. Kawauchi H. Functions of melanin-concentrating hormone in fish. *J Exp Zool*. 2006;305A:751−760.
6. Sakurai T, Amemiya A, Ishii M, et al. Orexins and orexin receptors: a family of hypothalamic neuropeptides and G protein-coupled receptors that regulate feeding behavior. *Cell*. 1998;92:573−585.
7. de Lecea L, Kilduff TS, Peyron C. The hypocretins: hypothalamus-specific peptides with neuroexcitatory activity. *Proc Natl Acad Sci USA*. 1998;95:322−327.
8. Lin L, Faraco J, Li R, et al. The sleep disorder canine narcolepsy is caused by a mutation in the *hypocretin (orexin) receptor 2* gene. *Cell*. 1999;98:365−376.
9. Douglass J, McKinzie AA, Couceyro P. PCR differential display identifies a rat brain mRNA that is transcriptionally regulated by cocaine and amphetamine. *J Neurosci*. 1995;15:2471−2481.
10. Elias CF, Lee C, Kelly J, et al. Leptin activates hypothalamic CART neurons projecting to the spinal cord. *Neuron*. 1998;21:1375−1385.

Melanin-Concentrating Hormone

Hiroshi Nagasaki and Yumiko Saito

Abbreviation: MCH

Melanin-concentrating hormone is an orexigenic peptide produced in the lateral hypothalamic area and zona incerta of the mammalian brain. MCH neurons constitute a powerful regulatory system with wide and divergent projection, associated with food intake, energy expenditure, mood, and REM sleep.

Discovery

It was realized in 1928 that administration of teleost pituitary gland extract led to concentrated pigment in fish scales. In 1983, Kawauchi purified and sequenced the factor, named melanin-concentrating hormone (MCH), for the first time from the salmon pituitary. In mammals, the primary structure of rat MCH was determined by purification methods in 1989, and that of the human was deduced from the cDNA sequence in 1990 [1].

Structure

Structual Features

Human proMCH consists of 165 aa residues with mature MCH at the C-terminus. Mature MCH is cleaved off at the KR sequence either by prohormone convertase PC1/3 or PC2. MCH is a cyclic peptide flanked by a disulfide bridge, and comprises 17 aa in teleost fishes and 19 aa in mammals (Figures 10A.1, 10A.2).

Primary Structure

The primary structure is well conserved; human MCH is identical with that of rodents and highly similar to that of teleost fishes. The central ring sequence between two cysteines is the most significant factor for melanin-concentrating activity.

Properties

Mr 18,482; pI, >9.5 (mouse, rat, human MCH). MCH is freely soluble in water, ethanol and 70% acetone; insoluble in acetone, benzene, chloroform, and ether.

Synthesis and Release

Gene, mRNA, and Precursor

The human MCH gene (*PMCH*), located on chromosome 12 (12q 23−q24), consists of three exons [2]. Exon 2 encodes neuropeptide G-E (NGE) and neuropeptide E-I (NEI), and exons 2 and 3 encode MCH (Figure 10A.2). Two variant PMCH genes are also identified, on 5p14 (*PMCHL1*) and

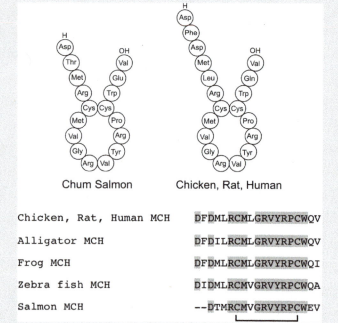

Figure 10A.1 Alignment of mature MCH sequences in vertebrates. Conserved amino acids are shaded and the disulfide bond is bracketed.

5q12−q13 (*PMCHL2*), but only a single locus is found in the rodent species. Human PMCH mRNA has 752 nucleotides, and preproMCH is 165 aa long [3].

Distribution of mRNA

In teleost fishes, MCH is expressed in two groups of neurons. One population of large MCH neurons is located in the nucleus tuberis and sends projections to the neurohypophysis, and another consists of smaller neurons in the dorsal hypothalamus projecting to various brain regions. In mammals, MCH mRNA is detected most abundantly in the lateral hypothalamic area and zona incerta. Orexin neurons are also located in the same region, but the location is distinct and not overlapping. Although the function is not consistent, MCH mRNA or immunoreactivity is identified in diverse peripheral tissues, including the nodosa ganglion of the vagus nerve, pancreatic islets, and the nerve terminals of the colon, skin, and immune and epithelial cells of various tissues.

Y. Takei, H. Ando, & K. Tsutsui (Eds): Handbook of Hormones. DOI: http://dx.doi.org/10.1016/B978-0-12-801028-0.00126-4

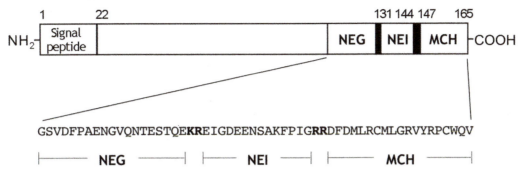

Figure 10A.2 Pro-MCH structure and the amino acid sequence of NGE, NEI, and MCH.

Table 10A.1 MCH Content of Dorsal and Ventral Hypothalamus, Pituitary (pmol/g), and Plasma (fmol/ml) of Trout Reared in Black or White Tanks [1]

Reared Backgrounds	Dorsal Thalamus	Ventral Hypothalamus	Pituitary	Plasma
Black	1.52 ± 0.52	4.0 ± 0.05	113 ± 24	13 ± 3
White	4.57 ± 0.05	147.0 ± 18.60	500 ± 72	49 ± 10

Table 10A.2 MCH Effects in CNS

Species	Function	Reference
Rat	Promotes feeding behavior after ICV administration	[10]
Mouse	Increases cocaine reward in nucleus accumbens shell	[8]
Rat	Increases REM sleep after dorsal raphe nucleus	[11]
Rat	Induces depressive effect after dorsal raphe nucleus	[12]
Rat	Stimulates GnRH release on the gonadotropin-releasing neurons	[13]
Rat	Suppresses TRH in primary culture and TSH release *in vivo*	[14]
Rat	Stimulates CRH release and ACTH secretion after PVN injection	[15]
Rat	Regulates beat frequency of ependymal cilia	[16]

Tissue and Plasma Concentrations

MCH is abundant in the pituitary of a teleost reared in a black tank (113 pmol/pituitary), and its content increased five-fold when reared in a white tank (Table 10A.1) [1]. Compared to the fish, MCH content in the mammalian pituitary is only trivial (0.26 pmol/rat pituitary and 1.2 pmol/ human pituitary). The MCH level in human plasma is around 5 pmol/l at fasting in the morning, and it correlates positively to body fat composition [4]. The plasma value is again much smaller compared with that of teleosts (Table 10A.1).

Regulation of Synthesis and Release

Synthesis

There are a couple of AP-1 motifs in the 5′ flanking region of the MCH coding sequence. A downstream target of insulin signaling, Foxa2, regulates the expression of MCH and orexin [5]. Foxa2 binds to MCH and orexin promoters and stimulates their expression during fasting. In fed animals, insulin excludes Foxa2 from the nucleus and the MCH and orexin gene expression are downregulated.

Release

MCH neurons generally do not fire spontaneously. Neurotransmitters such as norepinephrine, serotonin, and the cholinergic agonists muscarine and carbachol all hyperpolarize the MCH neurons. Neuropeptide Y also inhibits MCH neurons via pre- and postsynaptic mechanisms. By contrast, orexin activates MCH neurons. Increasing concentrations of glucose stimulate MCH neurons while exerting the opposite effect on the orexin neurons, although both are orexigenic hormones.

Receptors

Structure and Subtypes

MCH acts via two G protein-coupled receptors, MCHR1 and MCHR2, of which functional MCHR2 is not present in rodents [6]. MCHR1 is a glycoprotein consisting of 353 aa in the rat/human, and maps to chromosome 22, q13.3 in humans. Strong labeling of MCHR1 is detected in several limbic structures and anatomical areas implicated in the control of olfaction in rats. Human MCHR2 is also highly expressed in the brain.

Signal Transduction Pathway

In recombinant cell lines, rat, mouse, and human MCHR1s are promiscuous, and couple to various GTP binding proteins such as G_i, G_o, and G_q proteins. In contrast to human MCHR1, however, human MCHR2 exclusively couples to G_q protein.

Agonists

None is thus far known.

Antagonists

T-226296, SNAP-7941, SNAP-94847, ATC0065, GW803430 (=GW3430), and TPI1361-17 are being developed as anti-obesity drugs.

Biological Functions

Target Cells/Tissues and Functions

In teleost fishes, MCH acts as a pituitary hormone on the melanophore via MCHR1. In mammals, MCH mainly functions as a neurotransmitter in the CNS, where MCH fibers are distributed diversely. Besides feeding, MCH affects various brain functions, both higher (memory, cognition, mood, sleep [7], motivation [8]) and lower or autonomic (reproduction, stress, olfaction and energy expenditure) functions [9] (Table 10A.2 [8,10–16]). Although these functions are not yet fully established, the MCH system is also identified in some peripheral tissues, including the pancreatic islets, epithelium of the colon, and trachea (Table 10A.3 [17–19]).

Phenotype in Gene-Modified Animals

The transgenic mice overexpressing the MCH gene are obese and insulin-resistant [20]; mice lacking the MCH gene are hypophagic and lean. The MCH gene is upregulated in leptin-deficient ob/ob mice. MCH1R$^{-/-}$ mice are hyperphasic but lean because of hyperactivity and altered metabolism.

Table 10A.3 MCH Effects in Peripheral Tissues

Species	Function	Reference
Salmon	Accumulation of pigment in melanophore cells	[17]
Mouse	Promotes growth of pancreatic islet beta cells	[18]
Mouse	Mediates intestinal inflammation in colon epithelial cells	[19]

Pathophysiological Implications

Clinical Implications

So far, no hereditary diseases have been associated with MCH, MCHR1, or MCHR2 genes. The association of obesity with SNP in PMCH, MCH1R, and MCH2R has been investigated, but the results were not consistent. A case having autoimmunity to MCHR1 is reported to have developed vitiligo.

References

1. Baker BI. Melanin-concentrating hormone: a general vertebrate neuropeptide. *Int Rev Cytol.* 1991;126:1–47.
2. Nahon JL, Joly C, Levan G, et al. Pro-melanin-concentrating hormone gene (PMCH) is localized on human chromosome 12q and rat chromosome 7. *Genomics.* 1992;12:846–848.
3. Pedeutour F, Szpirer C, Nahon JL, et al. Assignment of the human pro-melanin-concentrating hormone gene (PMCH) to Chromosome 12q23–q24 and two variant genes (PMCHL1and PMCHL2) to Chromosome 5p14 and 5q12–q13. *Genomics.* 1994;19:31–37.
4. Gavrila A, Chan JL, Miller LC, et al. Serum adiponectin levels are inversely associated with overall and central fat distribution but are not directly regulated by acute fasting or leptin administration in humans: cross-sectional and interventional studies. *J Clin Endocrinol Metab.* 2005;90:1047–1054.
5. Silva JP, von Meyenn F, Howell J. Regulation of adaptive behaviour during fasting by hypothalamic Foxa2. *Nature.* 2009;462:646–650.
6. Saito Y, Nothacker HP, Wang Z, et al. Molecular characterization of the melanin-concentrating-hormone receptor. *Nature.* 1999;400:265–269.
7. Torterolo P, Lagos P, Monti JM. Melanin-concentrating hormone: a new sleep factor? *Front Neurology.* 2011;2:1–12.
8. Chung S, Woodward HF, Nagasaki H, et al. The melanin-concentrating hormone system modulates cocaine reward. *Proc Natl Acad Sci USA.* 2009;106:6772–6777.
9. Pissios P, Maratos-Flier E. Expanding the scales: the multiple roles of MCH in regulating energy balance and other biological functions. *Endocrine Rev.* 2006;27:606–620.
10. Qu D, Ludwig DS, Gannektift S, et al. A role for melanin-concentrating hormone in the central regulation of feeding behaviour. *Nature.* 1996;380:243–247.
11. Verret L, Goutagny R, Fort P, et al. A role of melanin-concentrating hormone producing neurons in the central regulation of paradoxical sleep. *BMC Neurosci.* 2003;4:19–28.
12. Lagos P, Torterolo P, Jantos H, et al. Immunoneutralization of melanin-concentrating hormone (MCH) in the dorsal raphe nucleus: effects on sleep and wakefulness. *Brain Res.* 2011;1265:103–110.
13. Murray JF, Adan RA, Walker R, et al. Melanin-concentrating hormone, melanocortin receptors and regulation of luteinizing hormone release. *J Neuroendo.* 2000;12:217–223.
14. Kennedy AR, Todd JF, Stanley SA, et al. Melanin-concentrating hormone (MCH) suppresses thyroid stimulating hormone (TSH) release, *in vivo* and *in vitro*, via the hypothalamus and the pituitary. *Endocrinology.* 2000;142:3265–3268.
15. Kennedy AR, Todd JF, Dhillo WS, et al. Effect of direct injection of melanin-concentrating hormone into the paraventricular nucleus: further evidence for a stimulatory role in the adrenal axis via SLC-1. *J Neuroendocrinol.* 2003;15:268–272.
16. Conductier G, Brau F, Viola A, et al. Melanin-concentrating hormone regulates beat frequency of ependymal cilia and ventricular volume. *Nat Neurosci.* 2013;16:845–847.
17. Rance T, Baker BI. The teleost melanin-concentrating hormone – A pituitary hormone of hypothalamic origin. *Gen Comp Endocrinol.* 1979;37:64–73.
18. Pissios P, Ozcan U, Kokkotou E, et al. Melanin concentrating hormone is a novel regulator of islet function and growth. *Diabetes.* 2007;56:311–319.
19. Kokkotou E, Moss AC, Torres D, et al. Melanin-concentrating hormone as a mediator of intestinal inflammation. *PNAS.* 2008;105:10613–10618.
20. Ludwig DS, Tritos NA, Mastaitis JW, et al. Melanin-concentrating hormone overexpression in transgenic mice leads to obesity and insulin resistance. *J Clin Invest.* 2001;107:379–386.

Supplemental Information Available on Companion Website

- Gene, mRNA, and precursor structures of the MCH /E-Figure 10A.1
- Primary structure of human MCHR1 and MCHR2 /E-Figure 10A.2
- Accession numbers of genes and mRNA for MCH precursor and MCH receptor/E-Table 10A.1

Orexin

Tomoya Nakamachi

Abbreviations: ORX, OX
Additional name: hypocretin (HCRT)

Orexin is a hypothalamic neuropeptide consisting of 33 or 28 aa residues. Orexin has a pleiotropic function regulating feeding behavior, the sleep—wake cycle, metabolism, locomotor activity, and the endocrine system through orexin receptors.

Discovery

Orexin (also known as hypocretin) was discovered by reverse pharmacology as an endogenous ligand for two orphan GPCRs in 1998 [1].

Structure

Structural Features

Orexin exists as two molecular forms, orexin-A and orexin-B, derived from the same 130-aa residue precursor (prepro-orexin). In mammals, orexin-A is a 33-aa residue peptide with an N-terminal glutamine cyclized to pyroglutamate, and two intrachain disulfide bonds that are fully conserved among tetrapods. Orexin-B is a linear 28-aa residue peptide. Orexin-B is deduced to consist of two α-helices connected with a short linker. The C-termini of both orexins are amidated. Human orexin-A and orexin-B have 46% homology (13/28), with their C-terminal side in particular being conserved (Figure 10B.1).

Although it was thought originally that orexin has no functional or structural identity with any other known regulatory peptide, it is now considered to belong to the incretin gene family of peptides, including members of the secretin-glucagon superfamily such as growth hormone releasing hormone, pituitary adenylate cyclase-activating polypeptide, vasoactive intestinal peptide, and glucagon-like peptides [2]. Molecular structures of orexins are relatively well conserved among vertebrates, especially in the disulfide bond regions in orexin-A and the C-terminal side in both orexins (Figure 10B.2) [3]. Although tetrapod orexin-As comprise 33 aa residues, teleost orexin-A is longer in length, at 47 aa residues in goldfish and zebrafish, due to an additional spacer sequence.

```
human orexin A    pEPLPDCCRQKTCSCRLYELLHGAGNHAAGILTL-NH2
human orexin B    RSGPPGLQGRLQRLLQASGNHAAGILTM-NH2
```

Figure 10B.1 Sequence comparison of human orexin A and B.

Properties

Human orexin-A, Mr 3,562; human orexin-B, Mr 1,937. Freely soluble in water and ethanol.

Synthesis and Release

Gene, mRNA, and Precursor

The human prepro-orexin gene (*HCRT*), located on chromosome 17 (17q21), consists of two exons and one intron (E-Figure 10B.1). Orexin-A and -B are coded tandemly in the prepro-orexin mRNA flanked by basic amino acid residues. To date, only vertebrate orexins have been reported while invertebrates seem to lack orexin-like sequences [4].

Distribution and Concentration

In mammals, orexin-containing neurons are located in the lateral hypothalamus, which is known as the orexigenic center, and the nerve fibers are widely distributed in various regions, including the cerebral cortex, hippocampus, limbic system, and brainstem [5]. In peripheral tissues, orexin immunoreactivity has been observed in myenteric and submucosal plexus neurons in the gut and pancreatic neurons. Orexin-like immunoreactivity was also observed in the hypothalamic area of the chicken, turtle, African clawed frog, zebrafish, and goldfish [3].

Regulation of Synthesis and Release

Orexin-A and orexin-B are synthesized in neurons. Food restriction increases orexin mRNA and peptide content in the brain. In cultured hypothalamic orexin neurons, low glucose triggers synthesis of orexins.

Receptors

Structure and Subtype

The orexin receptor was found to have two forms: orexin receptor-1 and -2 (OX1R and OX2R). The orexin receptors are GPCRs with seven-transmembrane domains. The amino acid sequence homology of human orexin receptors is 64%. Orexin-A binds to OX1R and OX2R with a high affinity, whereas orexin-B selectively binds to OX2R with a similar high affinity [1]. The human OX1R gene is located on chromosome 1 (1p35), and the human OX2R gene is located on chromosome 6 (6q11).

Signal Transduction Pathway

OX1R and OX2R activate G_q pathways to produce IP3 and an intracellular rise in Ca^{2+}, or to stimulate of the MAPK pathway [6].

Y. Takei, H. Ando, & K. Tsutsui (Eds): Handbook of Hormones. DOI: http://dx.doi.org/10.1016/B978-0-12-801028-0.00127-6

```
                                                                      orexin A                                    orexin B
                  10        20        30        40        50        60        70        80        90       100
Human           MNLPSTKVSWAAVTLLLLLLLLPPALLSSGAAAQPLPDCCRQKTCSCRLYELLHGAG----------------NHAAGILTLGKRRSGPPGLQGRLQRLL
Dog             MNPPSTKVPWAAVT-LLLLLLLPPALLSPGAAAQPLPDCCRQKTCSCRLYELLHGAG----------------NHAAGILTLGKRRPGPPGLQGRLQRLL
Mouse           MNFPSTKVPWAAVT-LLLLLLLPPALLSLGVDAQPLPDCCRQKTCSCRLYELLHGAG----------------NHAAGILTLGKRRPGPPGLQGRLQRLL
Rat             MNLPSTKVPWAAVT-LLLLLLLPPALLSLGVDAQPLPDCCRQKTCSCRLYELLHGAG----------------NHAAGILTLGKRRPGPPGLQGRLQRLL
Chiken          MEVPNAKLQRSACLLLLLLLLLCS-----LAGGRQSLPECCRQKTCSCRIYDLLHGMG----------------NHAAGILTLGKRKSIPPAFQSRLYRLL
African clawed frog -------VHKRHCW-LFLVLLCS-----LISTSHGAPDCCRQKTCSCRIYDILRGTG----------------NHAAGILTLGKRRSDFQTMQSRLQRLL
Goldfish        ------CTAKRVQLLLFMALLAH-----LARDAEGVATCCSSASRSCKLYEICRAGRRNDTSIARHIGRFNNDAAVGILTLGKRKVGERRVQDRLQQLL
Zebrafish       ----MDCTAKKLQVLVFMALLAH-----LARDAEGVASCCARAPGSCKLYEMLCRAGRRNDSSVARHLVHLNNDAAVGILTLGKRKVGESRVHDRLQQLL
                                          + + **      ***++++++**  *   +  +  *            ++**********  +  ** +++
```

```
                  110       120       130       140       150       160       170
Human           QASGNHAAGILTMGRR---------------AGAEPAPRPCLG------RRCSAPAAA-SVAPGGQSGI-----
Dog             QASGNHAAGILTMGRR---------------AGAEPAPRPCPG------RRCPVVAVP-SAAPGGRSGV-----
Mouse           QANGNHAAGILTMGRR---------------AGAELEPHPCSG------RGCPTVTTT-ALAPRGGSGV-----
Rat             QANGNHAAGILTMGRR---------------AGAELEPYPCPG------RGCPTATAT-ALAPRGGSRV-----
Chiken          HGSGNHAAGILTIGKREERPGTACRDALSCAAGT--QPTVTPRGTAASPRECQEHAEK-DLTKGWA--AAKSFY
African clawed frog QGSGNHAAGILTMGRRSQDKVETN-----CINGLMGSSSTSSSLSLLT-LLCPTAPEPLNASKGFGCQQDPSM-
Goldfish        HGSRNQAAGILTVGKR---------------LEDPLQDLMPR-----------PPEDLDAYETR---------
Zebrafish       HNSRNQAAGILTMGKR---------------LEEPAKFLIPT-----------VPQDVDSYEKR---------
                ++*******  *  *                              +
```

Figure 10B.2 Comparison of prepro-orexin sequences in representative species.

Furthermore, OXR2 is able to modulate the intracellular concentration of cAMP by coupling to G_s and G_i proteins [4].

Agonists

[Ala^{11}, D-Leu^{15}]-Orexin B and SB-668875 are OX2 specific agonists.

Antagonists

Among orexin receptor antagonists, ACT-335827, SB-334867, and SB-408124 are OX1-specific antagonists; TCS-OX2-29, EMPA, and JNJ 10397049 are OX2-specific antagonists; and ACT-078573 (almorexant), MK-4305 (suvorexant), MK-6096, SB-649868, and TCS 1102 are dual antagonists. Further information regarding orexin receptor antagonists can be found in the review by Lebold and colleagues [7].

Biological Functions

Target Cells/Tissues and Functions

Although orexin was first reported as appetite-stimulating hormone in mammals, further studies have revealed its multifunction status [8]. An important function of orexin is as a regulator of the sleep–wake cycle. Orexin affects monoaminergic neurons in the locus ceruleus, raphe nucleus, and ventral tegmental area to maintain wakefulness. Orexin systems have roles in regulating drinking behavior, locomotor activity, and sympathetic nervous system activation. Orexin contributes to regulation of the endocrine system, elevating blood corticosterone levels, decreasing blood prolactin levels, and suppressing the gonadotropin-producing cells in the hypothalamus. In teleosts, intracerebroventricular injection of orexin A stimulates food consumption and locomotor activity in goldfish and zebrafish [9].

Phenotype in Gene-Modified Animals

Prepro-orexin or OX2R-deficient mice exhibit a phenotype similar to the human sleep disorder, narcolepsy. In contrast, the sleep–awake cycle is normal in OX1R-deficient mice. Mutated dogs with their exon-skipping mutation in the OX2R gene also show narcolepsy symptoms. Prepro-orexin knockout mice show hypophagia and obesity when fed a high-fat diet. Orexin neuron-deficient mice, and orexin/ataxin-3 transgenic mice, show hypophagia, metabolic abnormality, and obesity.

Pathophysiological Implications

Several pharmaceutical companies have developed non-peptide orexin receptor antagonists, and clinical trials for sleep disorders have begun [10].

References

1. Sakurai T, Amemiya A, Ishii M, et al. Orexins and orexin receptors: a family of hypothalamic neuropeptides and G protein-coupled receptors that regulate feeding behavior. *Cell*. 1998;92:573–585.
2. Tam JK, Lau KW, Lee LT, et al. Origin of secretin receptor precedes the advent of tetrapoda: evidence on the separated origins of secretin and orexin. *PLoS One*. 2011;6:e19384.
3. Wong KK, Ng SY, Lee LT, et al. Orexins and their receptors from fish to mammals: a comparative approach. *Gen Comp Endocrinol*. 2011;171:124–130.
4. Scammell TE, Winrow CJ. Orexin receptors: pharmacology and therapeutic opportunities. *Annu Rev Pharmacol Toxicol*. 2011;51:243–266.
5. Sakurai T, Mieda M. Connectomics of orexin-producing neurons: interface of systems of emotion, energy homeostasis and arousal. *Trends Pharmacol Sci*. 2011;32:451–462.
6. Ramanjaneya M, Conner AC, Chen J, et al. Orexin-stimulated MAP kinase cascades are activated through multiple G-protein signalling pathways in human H295R adrenocortical cells: diverse roles for orexins A and B. *J Endocrinol*. 2009;202:249–261.
7. Lebold TP, Bonaventure P, Shireman BT. Selective orexin receptor antagonists. *Bioorg Med Chem Lett*. 2013;23:4761–4769.
8. Mieda M, Sakurai T. Overview of orexin/hypocretin system. *Prog Brain Res*. 2012;198:5–14.
9. Matsuda K, Azuma M, Kang KS. Orexin system in teleost fish. *Vitam Horm*. 2012;89:341–361.
10. Mieda M, Sakurai T. Orexin (hypocretin) receptor agonists and antagonists for treatment of sleep disorders. Rationale for development and current status. *CNS Drugs*. 2013;27:83–90.

Supplemental Information Available on Companion Website

- Gene, mRNA, and peptide organization of the human orexin /E-Figure 10B.1
- Accession numbers of genes and cDNAs for orexin, OX1R, and OX2R/E-Tables 10B.1–10B.3

Cocaine- and Amphetamine-Regulated Transcript

Hiroshi Nagasaki, Yuki Kobayashi, and Yumiko Saito

Abbreviation: CART

CART, so named because of upregulation by cocaine and amphetamine, is thought to be involved in the regulation of feeding and stress. However, its receptor has not yet been identified.

Discovery

In the 1980s, a fragment of CART peptide was identified in an extract of ovine hypothalamus. Fifteen years later, the level of CART mRNA was found to increase in the rat striatum after acute administration of cocaine and amphetamine [1]. Subsequently, in 1999, CART peptides were extracted and sequenced from the rat, and this showed that two different forms of CART (CART55−102 and CART62−102) were present [2].

Structure

Structural Features

CART peptides of different length have been found in various tissues [3]. The rat central nervous system contains CART55−102 and CART62−102 fragments. In the periphery, longer products are generated in addition to CART55−102. Adrenal glands in the rat produced two peptides, CART1−89 and CART10−89. The most widely studied peptides are CART55−102 and CART62−102 (Figure 10C.1). Both peptides contain a cysteine-knot motif, which is critical for the biological activity of the hormone.

Primary Structure

Human CART42−89 corresponds to rodent CART55−102 (Figures 10C.1, 10C.2). Genome studies have revealed that fish have multiple CART genes. Two different goldfish CART55−102 (Goldfish I and II) present a high homology with their mammalian counterparts in their C-terminal end region (Figure 10C.1).

Properties

Human CART42−89: Mr 5233.188, pI 8.36.

Synthesis and Release

Gene, mRNA, and Precursor

Due to alternative splicing, rodent CART mRNA has two spliced variants as long and short forms of proCART. The mRNA that encodes the long one is translated into a 102-aa sequence, and the short one into a 89-aa sequence [3]. Only the latter has been found in humans. The Human CART gene, *CARTPT*, location 5q13.2, consists of two introns and three exons [4].

Distribution of mRNA

Strong mRNA was found in the nucleus accumbens [1], olfactory turbercle, piriform and somatosensory cortex, amygdala complex, and hypothalamus nuclei, including the arcuate, lateral, nucleus, paraventricular, and supraoptic nuclei. Outside the nervous system, CART is also expressed in gut, pituitary endocrine cells, adrenomedullary cells, and islet somatostatin cells.

Tissue and Plasma Concentrations

Tissue: Radioimmunoassay (RIA) directed to CART-like 79−102 (pmol/g): rat pituitary (male), 373.0 ± 55.2; rat hypothalamus (male), 50.6 ± 4.4; rat duodenum (male), 26.1 ± 4.2 [5].
    Plasma: RIA directed to CART-like 55−102 (fmol/ml): 49 male 72 ± 2.8 and 49 female 74 ± 3.8 [6].

Regulation of Synthesis and Release

In hypothalamic explants, neuropeptide Y (NPY) significantly increased the release of CART(55−102)-immunoreactivity. *In vivo* experiments showed that leptin administration induces Fos expression in hypothalamic CART neurons. NPY and CART are co-expressed in the same neurons in the dorsomedial hypothalamus in chronic obesity. These neurons are activated by peripheral leptin treatment in diet-induced obesity. Thus, CART peptides are anorexigenic and closely associated with leptin and neuropeptide Y, two important food intake regulators.
    CART is also released in response to repeated dopamine release in the nucleus accumbens.

Receptors

Structure and Subtype

Although CART-responsive cell lines or neurons were reported [7,8], the putative receptor for CART has not yet been identified. Because the two active CART peptides (CART55−102 and CART62−102) show different relative features in *in vivo* activity, it could be speculated that there are multiple CART receptors.

Y.Takei, H. Ando, & K.Tsutsui (Eds): Handbook of Hormones. DOI: http://dx.doi.org/10.1016/B978-0-12-801028-0.00128-8

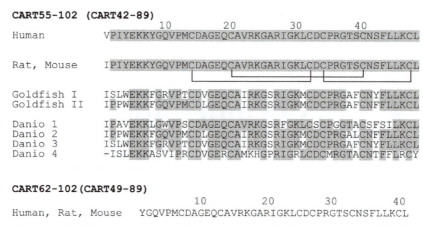

CART55-102 (CART42-89)

```
                    10        20        30        40
Human          VPIYEKKYGQVPMCDAGEQCAVRKGARIGKLCDCPRGTSCNSFLLKCL

Rat, Mouse     IPIYEKKYGQVPMCDAGEQCAVRKGARIGKLCDCPRGTSCNSFLLKCL


Goldfish I     ISLWEKKFGRVPTCDVGEQCAIRKGSRIGKMCDCPRGAFCNYFLLKCL
Goldfish II    IPPWEKKFGQVPMCDLGEQCAIRKGSRIGKMCDCPRGAFCNFFLLKCL

Danio 1        IPAVEKKLGWVPSCDAGEQCAVRKGSRFGKLCSCPGGTACSFSILKCL
Danio 2        IPPWEKKFGQVPMCDLGEQCAIRKGSRIGKMCDCPRGALCNFFLLKCL
Danio 3        ISLWEKKFGRVPTCDVGEQCAIRKGSRIGKMCDCPRGAFCNYFLLKCL
Danio 4        -ISLEKKASVIPRCDVGERCAMKHGPRIGRLCDCMRGTACNTFFLRCY
```

CART62-102 (CART49-89)

```
                         10        20        30        40
Human, Rat, Mouse   YGQVPMCDAGEQCAVRKGARIGKLCDCPRGTSCNSFLLKCL
```

Figure 10C.1 Alignment of mature CART sequences in vertebrates. Conserved amino acids are shaded and the proposed disulfide bond is bracketed.

```
Rat       MESSRLRLLPVLGAALLLLLPLLGAGA

Human     MESSRVRLLPLLGAALLLMLPLLGTRA

                   Signal peptide

Rat       QEDAELQPRALDIYSAVDDASHEKELPRRQLRAPGAVLQIEALQEVLKKLKSKR

Human     QEDAELQPRALDIYSAVDDASHEKEL-------------IEALQEVLKKLKSKR


Rat       IPIYEKKYGQVPMCDAGEQCAVRKGARIGKLCDCPRGTSCNSFLLKCL (1-102)

Human     VPIYEKKYGQVPMCDAGEQCAVRKGARIGKLCDCPRGTSCNSFLLKCL (1-89)
                          CART55-102/CART42-89
```

Figure 10C.2 Amino acid sequences of CART precursors in rat and human. The CART precursor found in humans lacks 13 amino acid residues compared to the murine precursor. Therefore, CART55−102 in rodents corresponds to CART42−89 in humans.

Signal Transduction Pathway

The signal transduction pathway is predicted to be mainly coupled to $G_{i/o}$ protein [7,8].

Biological Functions

Target Cells/Tissues and Functions

In the goldfish, central administration of human CART has been shown to decrease food intake. In mammals, CART modulates various physiological processes such as feeding, energy expenditure, stress control, and endocrine secretion (Table 10C.1 [9−18]).

Phenotype in Gene-Modified Animals

1. CART knockout mice weigh more than wild type-mice when fed with a high caloric diet [19]. Some difference has been observed in other CART knockout mice with a different genetic background.
2. Viral delivery of CART mRNA via ICV suppressed weight gain in diet-induced obese rats.
3. Knockdown of CART in the nucleus accumbens induces overeating in fed mice and inhibits stimulation of 5-HT4R (serotonin receptor)- and MDMA(3,4-N-methylene dioxymethamphetamine)-induced anorexia.
4. CART knockout mice display altered psychostimulant locomotor activation after psychostimulants, sensitization, and self-administration.

Table 10C.1 *In Vivo* Activity of CART in the Central Nervous System and Peripheral Tissues

Species	Function	Reference
Goldfish	CART55−102, ICV: Inhibits spontaneous feeding behavior	[9]
Rat	CART55−102, ICV: Inhibits spontaneous and fast-induced feeding	[10]
Mouse	CART55−102, ICV: Inhibits fast-induced feeding	[11]
Rat	CART55−102, intra-VTA: Increases spontaneous locomotor activity	[12]
Rat	CART55−102, intra-NAc: Inhibits cocaine-induced locomotor activity	[13]
Mouse	CART55−102, ICV: Increases hot-plate latency and prepulse inhibition	[14]
Mouse	CART55−102, ICV: Increases anxiety	[15]
Rat	CART55−102: Regulates islet hormone secretion in INS(832/13) and isolated islets	[16]
Mouse	CART62−102, ICV: Increases spontaneous locomotor activity	[14]
Mouse	CART62−102, ICV: Inhibits fast-induced feeding	[11]
Mouse	CART62−102, ICV: Increases hot-plate latency and prepulse inhibition	[14]
Rat	CART62−76, ICV: Increases anxiety	[17]
Rat	CART62−76, ICV: Inhibits spontaneous feeding	[18]

Pathophysiological Implications

Clinical Implications

In an obese 10-year-old boy, where the onset of obesity was at the age of 2 years, a G-to-C transversion at nucleotide 729 of the CART gene was identified, resulting in a leucine to phenylalanine substitution at codon 34 (L34F) [20]. The leucine at this position is well-conserved among species.

The L34F mutation segregated with severe obesity over three generations in this family, and was not identified in the control population. The affected subjects demonstrated reduced resting energy expenditures. Further, sequence variability in the CART gene has been identified as being related to obesity in several hundred French subjects.

References

1. Douglass J, McKinzie AA, Couceyro P. PCR differential display identifies a rat brain mRNA that is transcriptionally regulated by cocaine and amphetamine. *J Neurosci*. 1995;15:2471–2481.
2. Thim L, Kristensen P, Nielsen PF, et al. Tissue-specific processing of cocaine- and amphetamine-regulated transcript peptides in the rat. *Proc Natl Acad Sci USA*. 1999;96:2722–2727.
3. Dylag T, Kotlinska J, Rafalski P, et al. CART (85–102)-inhibition of psychostimulant-induced hyperlocomotion: importance of cyclization. *Peptides*. 2006;27:1926–1933.
4. Douglass J, Daoud S. Characterization of the human cDNA and genomic DNA encoding CART: a cocaine- and amphetamine-regulated transcript. *Gene*. 1996;169:241–245.
5. Murphy KG, Abbott CR, Mahmoudi M, et al. Quantification and synthesis of cocaine- and amphetamine-regulated transcript peptide (79–102)-like immunoreactivity and mRNA in rat tissues. *J Endocrinol*. 2000;166:659–668.
6. Bech P, Winstanley V, Murphy KG, et al. Elevated cocaine- and amphetamine-regulated transcript immunoreactivity in the circulation of patients with neuroendocrine malignancy. *J Clin Endocrinol Metab*. 2008;93:1246–1253.
7. Yermolaieva O, Chen J, Couceyro PR, et al. Cocaine- and amphetamine-regulated transcript peptide modulation of voltage-gated Ca^{2+} signaling in hippocampal neurons. *J Neurosci*. 2001;21:7474–7480.
8. Sen A, Lv L, Bello N, et al. Cocaine- and amphetamine-regulated transcript accelerates termination of follicle-stimulating hormone-induced extracellularly regulated kinase 1/2 and Akt activation by regulating the expression and degradation of specific mitogen-activated protein kinase phosphatases in bovine granulosa cells. *Mol Endocrinol*. 2008;22:2655–2676.
9. Volkoff H, Peter RE. Characterization of two forms of cocaine- and amphetamine-regulated transcript (CART) peptide precursors in goldfish: molecular cloning and distribution, modulation of expression by nutritional status, and interactions with leptin. *Endocrinol*. 2001;142:5076–5088.
10. Kristensen P, Judge ME, Thim L, et al. Hypothalamic CART is a new anorectic peptide regulated by leptin. *Nature*. 1998;393:72–76.
11. Thim L, Nielsen PF, Judge ME, et al. Purification and characterisation of a new hypothalamic satiety peptide, cocaine and amphetamine regulated transcript (CART), produced in yeast. *FEBS Lett*. 1998;428:263–268.
12. Kimmel HL, Gong W, Vechia SD, et al. Intra-ventral tegmental area injection of rat cocaine and amphetamine-regulated transcript peptide 55–102 induces locomotor activity and promotes conditioned place preference. *J Pharmacol Exp Ther*. 2000;294:784–792.
13. Jaworski JN, Kozel MA, Philpot KB, et al. Intra-accumbal injection of CART (cocaine-amphetamine regulated transcript) peptide reduces cocaine-induced locomotor activity. *J Pharmacol Exp Ther*. 2003;307:1038–1044.
14. Bannon AW, Seda J, Carmouche M, et al. Multiple behavioral effects of cocaine- and amphetamine-regulated transcript (CART) peptides in mice: CART 42–89 and CART 49–89 differ in potency and activity. *J Pharmacol Exp Ther*. 2001;299:1021–1026.
15. Chaki S, Kawashima N, Suzuki Y, et al. Cocaine- and amphetamine-regulated transcript peptide produces anxiety-like behavior in rodents. *Eur J Pharmacol*. 2003;464:49–54.
16. Wierup N, Björkqvist M, Kuhar MJ, et al. CART regulates islet hormone secretion and is expressed in the beta-cells of type 2 diabetic rats. *Diabetes*. 2006;55:305–311.
17. Kask A, Schiöth HB, Mutulis F, et al. Anorexigenic cocaine- and amphetamine-regulated transcript peptide intensifies fear reactions in rats. *Brain Res*. 2000;857:283–285.
18. Lambert PD, Couceyro PR, McGirr KM, et al. CART peptides in the central control of feeding and interactions with neuropeptide Y. *Synapse*. 1998;29:293–298.
19. Asnicar MA, Smith DP, Yang DD, et al. Absence of cocaine- and amphetamine-regulated transcript results in obesity in mice fed a high caloric diet. *Endocrinology*. 2001;142:4394–4400.
20. del Giudice EM, Santoro N, Cirillo G, et al. Mutational screening of the CART gene in obese children: identifying a mutation (Leu34Phe) associated with reduced resting energy expenditure and cosegregating with obesity phenotype in a large family. *Diabetes*. 2001;50:2157–2160.

Supplemental Information Available on Companion Website

- Accession numbers of genes, mRNA, and precursor protein for CART s/E-Table 10C.1
- Primary structure of pro-CART sequences in various vertebrates/E-Figure 10C.1

CHAPTER 11

Urotensin II

Norifumi Konno

Abbreviation: UII, U-II, UTS2

A neuropeptide isolated from the urophysis of teleost fish with a potent vasoconstrictor activity, urotensin II exerts behavioral effects and regulates cardiovascular, renal, and immune functions.

Discovery

Urotensin II (UII) was initially isolated from the urophysis of the goby fish (*Gillichthys mirabilis*) in 1980, on the basis of its ability to contract smooth muscle [1]. A paralog, called UII-related peptide (URP), was isolated from the rat brain in 2003 [2].

Structure

Structural Features

Human prepro-UII consists of 124 aa residues with bioactive mature UII at the C-terminus [3]. Human UII consists of 11 aa residues with a cyclic structure of 6 aa residues flanked by two cystein residues, which is fully conserved among UIIs (Figure 11.1) [3]. URP is a cyclic octapeptide whose sequence differs by only 1 aa residue from that of the C-terminal region of human UII (Figure 11.1). UII has been identified in all vertebrate species, from lampreys to mammals. URP exists in tetrapods and teleosts, but has not been reported in other vertebrate classes. URP1 and URP2, a teleost counterpart of mammalian URP, has been characterized in teleosts, and URP1 is absent in tetrapods [4].

Primary Structure

The C-terminal portion of UII contains a highly conserved cyclic hexapeptide sequence which is responsible for the majority of its biological activities, whereas the N-terminal portion shows low sequence identity between species.

Properties

Human UII: Mr 1,388.6, pI 4.37. Human URP, Mr 1,017.2, pI 8.09. Freely soluble in water, ethanol, and 20% acetonitrile/water.

Synthesis and Release

Gene, mRNA, and Precursor

The human UII gene is located at chromosome 1 (1p36) and contains five exons (E-Figure 11.1). Human UII mRNA has 551 bp of nucleotides [3]. The human URP gene is found at 3q28 and contains five exons (E-Figure 11.2). Human URP mRNA has 439 bp. The UII and URP genes originate from the same ancestral gene as do the somatostatin and cortistatin genes.

Distribution of mRNA

UII mRNA is detected most abundantly in the brain and spinal cord of fishes, frogs, and mammals. In the frog brain, transcripts are found in the trochlear nucleus, facial motor nucleus, abducens nucleus, glossopharyngeal nucleus, and hypoglossal nucleus [3]. In human peripheral tissues, UII mRNA has been detected in the kidney, spleen, pancreas, small intestine, thymus, prostate, pituitary gland, and adrenal gland [3]. In teleosts, UII and URP mRNAs are abundantly expressed in the caudal neurosecretory system (CNSS) and medulla oblongata [5].

Tissue and Plasma Concentrations

Tissue: Goby urophysis, $7-10$ μg/mg protein; sucker urophysis, $0.8-1.5$ μg/mg protein; flounder (FW), 0.253 ± 0.133 nmoles/gland ($n=7$) [6]; flounder (SW), 0.371 ± 0.178 nmoles/gland ($n=7$) [6].

Plasma: Sucker, 55 pg/ml; flounder (FW), 11 ± 1.5 fmol/ml ($n=7$) [6]; flounder (SW), 38 ± 7.7 fmol/ml ($n=7$) [6]; human (normal value), 4.4 fmol/ml, hemodialysis patients, 13.1 fmol/ml. There are no significant differences in plasma UII levels between males and females.

Regulation of Synthesis and Release

UII is released depending on intracellular Ca^{2+} from the motoneuron terminal. Although plasma UII concentrations rise in diseases such as heart failure, essential hypertension, renal disease, diabetes, and liver cirrhosis, the regulatory factor for UII synthesis and release is unknown in mammals. In flounder, plasma UII levels are higher in saltwater (SW)-acclimated fish than in freshwater (FW) ones, and decrease after transfer of SW-acclimated fish to FW [7].

Receptors

Structure and Subtype

UII binds the orphan GPR14 receptor with high affinity, and the receptor was renamed as a UII receptor (UTR). The UTR belongs to the class A, rhodopsin-like GPCR family, and shares high sequence identity with somatostatin receptors. Structural characteristics of the receptor include two potential N-glycosylation sites in the N-terminal domain and two cysteine residues in the first and second extracellular loops (E-Figures 11.6, 11.7). Human UTR has 389 aa residues with an Mr of 42,130. UII and URP both activate UTR with similar

Y. Takei, H. Ando, & K. Tsutsui (Eds): Handbook of Hormones. DOI: http://dx.doi.org/10.1016/B978-0-12-801028-0.00011-8

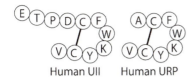

Figure 11.1 Primary structure of human UII and URP.

potency. UTR is expressed in various human tissues, notably in the cardiovascular system, including the left atrium, ventricle, coronary artery, and aorta, and in the spleen, kidney, urinary bladder, and skeletal muscle.

Signal Transduction Pathway

The UTR is coupled to the $G\alpha_{q/11}$ signal transduction pathway, the activation of which leads to an increase in inositol triphosphate and mobilization of intracellular Ca^{2+}. Vasoconstriction and smooth muscle contraction by UII involve small GTPase RhoA and its downstream effector Rho-kinase, phospholipase C, protein kinase C and tyrosine kinase, and PKC-independent phosphorylation of myosin light chain (MLC-2), as well as the Ca^{2+}-calmodulin/MLC kinase system, extracellular signal-regulated kinase (ERK) and p38 mitogen-activated protein kinase [8]. UII-induced mitogenic/hypertrophic response is mediated via the ERK activation pathway and the RhoA/Rho kinase pathway.

Agonists

The octapeptide humanUII (4–11) has been demonstrated as the minimum active fragment necessary for high affinity binding to the UTR. [Orn⁸]UII acts as a full agonist in calcium mobilization assay with a maximal effect similar to that of UII.

Antagonists

Urantide, [Pen5, D-Trp7, Orn8] human UII (4–11), is one of the most potent peptide antagonists to UTR. SB-611812 is a non-peptide receptor antagonist. Palosran (ACT-058362) is another potent non-peptide UTR antagonist with therapeutic potential. The somatostatin antagonist SB-710411 is also a rat UTR antagonist.

Biological Functions

Target Cells/Tissues and Functions

UII induces vasoconstriction, migration, and proliferation of vascular smooth muscle cells [8] (Table 11.1 [9–19]). UII also induces NO synthesis in the intact endothelium, resulting in vasodilation. In the human heart, UII exhibits cardiac stimulant effects *in vitro*. In fish, UII exerts osmoregulatory functions, in particular epithelial ion transport across the isolated skin, intestine, urinary bladder, and gill (Table 11.1). In mammalian kidney, UII has vasodilator and natriuretic effects. UII may exert indirect actions on hydromineral homeostasis via stimulation of prolactin and cortisol secretion. UII inhibits glucose-induced insulin release from the rat pancreas. In the central nervous system, UII exerts anxiogenic effects, hyperlocomotion, orexigenic and dipsigenic effects, and increased REM sleep duration.

Phenotype in Gene-Modified Animals

There are no reports on UII knockout animals to date. Loss of the mouse UTR does not show an overt change in cardiovascular phenotype *in vivo* [20]. Mouse UTR gene deletion leads to a reduction in body weight gain, elevated systolic and pulse pressure, decreased LDL uptake in the liver, hypertriglyceridemia, reduced hepatic steatosis, and reduced liver mass [20].

Table 11.1 Functions of UII

Species	Function	Reference
Human	Contraction of vascular smooth muscle	[9]
Human/rat	Acceleration of migration of monocytes	[10]
Rat	Decrease in corticosterone secretion in adrenocortical cells	[11]
Rat	Increase in REM sleep duration	[12]
Rat	Inhibition of glucose-induced insulin release from pancreas	[13]
Mouse	Exertion of hyperlocomotion, anxiogenic and depressant effects	[14]
Bull frog	Contraction of vascular smooth muscle and fall in cardiac output	[15]
Tilapia	Stimulation of prolactin release from pituitary	[16]
Trout	Increase in arterial blood pressure and decrease in heart rate	[17]
Goby	Stimulation of NaCl absorption in posterior intestine	[18]
Flounder (SW)	Stimulation of cortisol secretion of interrenal/head kidney	[19]

Pathophysiological Implications

Clinical Implications

Plasma UII concentration is elevated in renal failure, congestive heart failure, diabetes mellitus, systemic hypertension, and portal hypertention caused by liver cirrhosis [21]. In addition to the diseases mentioned above, UII is thought to play a central role in the pathogenesis of the metabolic syndrome, including obesity, hyperlipidemia, hypertension, hyperglycemia, and insulin resistance, leading to the development of type 2 diabetes, cardiovascular disease, non-alcoholic fatty liver disease, and renal impairment [21]. UTR is expressed in rat and human peripheral blood mononuclear cells, and UII acts as a chemoattractant for monocytes. UII activates migration and differentiation of adventitial fibroblasts from rat aorta. Thus, the UII system may be involved in the pathogenesis of atherosclerosis.

Use for Diagnosis and Treatment

Altered plasma UII concentration in diseases such as heart failure, essential hypertension, renal disease, diabetes, and liver cirrhosis has potential as a useful biomarker in detecting disease onset or progression [21]. In heart failure, UII plasma concentrations are not only increased but also correlated with big endothelin-1 and brain natriuretic peptide levels. UII is also a useful biomarker for diabetic retinopathy and carotid atherosclerosis.

References

1. Pearson D, Shively JE, Clark BR, et al. Urotensin II: a somatostatin-like peptide in the caudal neurosecretory system of fishes. *Proc Natl Acad Sci USA*. 1980;77:5021–5024.
2. Sugo T, Murakami Y, Shimomura Y, et al. Identification of urotensin II-related peptide as the urotensin II-immunoreactive molecule in the rat brain. *Biochem Biophys Res Commun*. 2003;310:860–869.
3. Coulouarn Y, Lihrmann I, Jegou S, et al. Cloning of the cDNA encoding the urotensin II precursor in frog and human reveals intense expression of the urotensin II gene in motoneurons of the spinal cord. *Proc Natl Acad Sci USA*. 1998;95:15803–15808.
4. Parmentier C, Hameury E, Dubessy C, et al. Occurrence of two distinct urotensin II-related peptides in zebrafish provides new insight into the evolutionary history of the urotensin II gene family. *Endocrinology*. 2011;152:2330–2341.
5. Nobata S, Donald JA, Balment RJ, et al. Potent cardiovascular effects of homologous urotensin II (UII)-related peptide and UII in unanesthetized eels after peripheral and central injections. *Am J Physiol*. 2011;300:R437–R446.

6. Winter MJ, Hubbard PC, McCrohan CR, et al. A homologous radioimmunoassay for the measurement of urotensin II in the euryhaline flounder, *Platichthys flesus*. *Gen Comp Endocrinol*. 1999;114:249−256.

7. Lu W, Greenwood M, Dow L, et al. Molecular characterization and expression of urotensin II and its receptor in the flounder (*Platichthys flesus*): a hormone system supporting body fluid homeostasis in euryhaline fish. *Endocrinology*. 2006;147:3692−3708.

8. Ross B, McKendy K, Giaid A. Role of urotensin II in health and disease. *Am J Physiol Regul Integr Comp Physiol*. 2010;298: R1156−R1172.

9. Ames RS, Sarau HM, Chambers JK, et al. Human urotensin-II is a potent vasoconstrictor and agonist for the orphan receptor GPR14. *Nature*. 1999;401:282−286.

10. Segain JP, Rolli-Derkinderen M, Gervois N, et al. Urotensin II is a new chemotactic factor for UT receptor-expressing monocytes. *J Immunol*. 2007;179:901−909.

11. Albertin G, Casale V, Ziokowska A, et al. Urotensin-II and UII-receptor expression and function in the rat adrenal cortex. *Int J Mol Med*. 2006;17:1111−1115.

12. de Lecca L, Bourgin P. Neuropeptide interactions and REM sleep: a role for Urotensin II? *Peptides*. 2008;29:845−851.

13. Silvestre RA, Rodríguez-Gallardo J, Eqido EM, et al. Inhibition of insulin release by urotensin II − a study on the perfused rat pancreas. *Horm Metab Rev*. 2001;33:379−381.

14. Do-Rego JC, Chatenet D, Orta MH, et al. Behavioral effects of urotensin-II centrally administered in mice. *Psychopharmacology*. 2005;183:103−117.

15. Yano K, Hicks JW, Vaudry H, et al. Cardiovascular actions of frog urotensin II in the frog, *Rana catesbeiana*. *Gen Comp Endocrinol*. 1995;97:103−110.

16. Grau EG, Nishioka RS, Bern HA. Effects of somatostatin and urotensin II on tilapia pituitary prolactin release and interactions between somatostatin, osmotic pressure Ca^{++}, and adenosine 3′,5′-monophosphate in prolactin release *in vitro*. *Endocrinology*. 1982;110:910−915.

17. Le Mevel JC, Olson KR, Conklin D, et al. Cardiovascular actions of trout urotensin II in the conscious trout, *Oncorhynchus mykiss*. *Am J Physiol*. 1996;271:R1335−R1343.

18. Loretz CA, Howard ME, Siegel AJ. Ion transport in goby intestine: cellular mechanism of urotensin II stimulation. *Am J Physiol*. 1985;249:G284−G293.

19. Kelsall CJ, Balment RJ. Native urotensins influence cortisol secretion and plasma cortisol concentration in the euryhaline flounder, *Platichthys flesus*. *Gen Comp Endocrinol*. 1998;112:210−219.

20. Behm DJ, Harrison SM, Ao Z, et al. Deletion of the UT receptor gene results in the selective loss of urotensin-II contractile activity in aortae isolated from UT receptor knockout mice. *Br J Pharmacol*. 2003;139:464−472.

21. Barrette PO, Schwertani AG. A closer look at the role of urotensin II in the metabolic syndrome. *Front Endocrinol*. 2012;3:165.

Supplemental Information Available on Companion Website

- Gene, mRNA, and precursor structures of the human UII and URP/E-Figures 11.1, 11.2
- Precursor and mature hormone sequences of various animals/E-Figures 11.3, 11.4
- Gene, mRNA, and precursor structure of the human UII receptor (UTR)/E-Figure 11.5
- Primary structure of the human UTR (389 aa residues)/E-Figure 11.6
- Primary structure of UTR of various animals/E-Figure 11.7
- Accession numbers of genes and cDNAs for UII and URP and UII receptor/E-Tables 11.1−11.3

CHAPTER 12

Neurotensin

Norifumi Konno

Abbreviations: NT, NTS

A hypotensive peptide first isolated from the bovine hypothalamus, neurotensin acts as a neurotransmitter and neurotransmodulator in the central nervous system, and also as a local hormone in the small intestine.

Discovery

Neurotensin (NT) is a tridecapeptide that was originally isolated from extracts of bovine hypothalamus in 1973, based on its ability to cause a visible vasodilation in the exposed cutaneous regions of anesthetized rats [1].

Structure

Structural Features

Neurotensin and neuromedin N (NMN) are synthesized by a common precursor (pro-NT/NMN). Human pro-NT/NMN consists of a conserved polypeptide of 170 aa residues starting with a signal peptide of 23 aa residues (Figure 12.1). NT is a tridecapeptide with a highly conserved C-terminal portion (8–13) which is responsible for its biological activities (E-Figure 12.2). NT peptides have been identified in various vertebrates except for cyclostomes and elasmobranchs. NMN and the frog skin peptide, xenopsin, exhibit strong similarity with the C-terminal portion of NT.

Primary Structure

NT and NMN sequences are located in tandem near the C-terminal portion of pro-NT/NMN and are separated from it by a Lysine–Arginine sequence (Figure 12.1). Four Lysine–Arginine cleavage sites are located in the C-terminal position of pro-NT/NMN [2].

Properties

Human, rat, and mouse: Mr 1,671.9; pI 8.6. Soluble in water and 5% acetic acid.

Synthesis and Release

Gene and mRNA

The human NT/NMN gene is located on chromosome 12q21 and its mRNA has 1,264 bp. The gene encompasses 8.7 kb and is divided into four exons by three introns (E-Figure 12.1). The fourth exon encodes both NT and NMN.

Distribution of mRNA

NT/NMN mRNA or protein is detected in the central nervous system (hypothalamus, pituitary, forebrain) and gastrointestinal tract (predominantly in the small intestinal mucosa) of a variety of tetrapods. NT is also detected in the adrenal medulla of mammals. Two distinct mRNAs, 1.0 and 1.5 kb in size, are present in the central nervous system and intestine of the rat [2]. The smaller mRNA is predominantly expressed in the intestine and anterior pituitary, while both mRNAs are expressed equally in the hypothalamus, brainstem, and cortex.

Tissue and Plasma Concentrations

Tissue: In rats, the anterior lobe of the pituitary, the median eminence, and the lateral preoptic area contain 3–5 ng/mg protein [3], the median preoptic area, paraventricular, supraoptic, and ventromedian nuclei contain 1.5–2 ng/mg protein [4], and the suprachiasmatic and accumbens nucleus contain 1 ng/mg protein [4], while the intestine contains 130 pmol/g wet tissue [3].

Plasma: Rat, 50 fmol/ml; calf, 15–25 fmol/ml; fasting adult human, 29 fmol/ml; fed adult human, 53 fmol/ml [5].

Regulation of Synthesis and Release

Regulation of NT/NMN gene transcription depends on specific *cis*-regulatory elements located in the proximal 5′ flanking region of the gene. The *cis*-regulatory element of the human NT/NMN gene contains a CRE/AP-1-like element that binds both AP-1 and CREB/ATF proteins, the glucocorticoid response element (GRE), and the AP-1 site [6]. NT/NMN gene expression and NT peptide synthesis are also induced in response to combined treatment with nerve growth factor, dexamethasone, lithium, and the adenylate cyclase activator forskolin. In the hypothalamus and anterior pituitary, depolarization by high K^+ concentration causes release of NT and NMN from synaptic vesicles or secretory granules in a Ca^{2+}-dependent manner. In the gastrointestinal tract, NT is released in response to increased intraluminal fats.

Receptors

Structure and Subtype

Three NT receptors, termed NTR1 (E-Figures 12.3, 12.6), NTR2 (E-Figures 12.4, 12.7), and NTR3 (E-Figures 12.5, 12.8), have been identified [7]. Both NTR1 and NTR2 are seven-transmembrane-spanning, G-protein coupled receptors. Rat and human NTR1s consist of 424 and 418 aa residues, respectively. The human NTR1 gene is located at the long arm (20q13) of chromosome 20 and three introns in the coding

Y. Takei, H. Ando, & K. Tsutsui (Eds): Handbook of Hormones. DOI: http://dx.doi.org/10.1016/B978-0-12-801028-0.00012-X

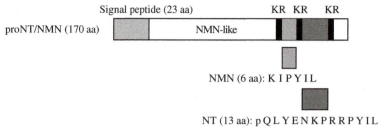

Figure 12.1 Structure of human NT/NMN precursor.

regions (E-Figure 12.3). Rodent and human NTR2s (416 and 410 aa residues, respectively) share only around 40% amino acid identities with NTR1s. NTR2 has a shorter N-terminal extracellular tail and a longer third intracytoplasmic loop than NTR1. NTR3, also known as sortilin, belongs to a new receptor family sharing an N-terminal luminal domain related to the yeast sorting receptor Vps10p, which includes a single transmembrane domain. The human NTR3 gene encodes a protein of 833 aa residues.

Signal Transduction Pathway

NTR1 has high affinity for NT and activates PKC, stimulating IP3 production through the $G_{q/11}$-coupled pathway. NTR1 is also linked to the G_s-mediated pathway, which increases intracellular cAMP levels. Affinity of NT for NTR2 is lower than that for NTR1. Human NTR2 is coupled to the $G_{q/11}$-dependent phospholipase C signal pathway, but not to G_s. Human NTR2 also interacts with $G_{i/o}$ and $G_{12/13}$. NTR3 activates the IP3−PKC signaling pathway in HT29 cells.

Agonists

Cyclic C-terminal NT (8−13) analog binds human NTR1 and NTR2 with high affinity. Four NT analogs (Eisai compound, NT66L, NT67L, and NT69) and the highly selective NTR1 agonist PD149163 have antipsychotic efficacy.

Antagonists

The non-peptide NT antagonist SR 48692 has higher affinity for NTR1 ($IC_{50} = 5.6$ nM) than for NTR2 ($IC_{50} = 300$ nM) and inhibits several of the central and peripheral effects of NT. SR 142948A has high affinity for NTR1 and NTR2. SR 142948A is able to antagonize a more diverse array of NT-induced effects compared to SR 48692.

Biological Functions

Target Cells/Tissues and Functions

NT was first isolated based on hypotensive action and peripheral vasodilation [1]. In the central nervous system, NT exerts various CNS effects, including hypothermia, analgesia, modulation of dopamine neurotransmission, stimulation of anterior pituitary hormone (ACTH, CRH, DA, GnRH, and SS) release, central control of blood pressure, inhibition of food intake, and sleep−wake regulation. In the gastrointestinal tract, NT exerts the effects of pancreatic endocrine secretion and colonic motility, and decrease in gastric acid secretion (Table 12.1 [8−17]).

Phenotype in Gene-Modified Animals

NT acts as a neurotransmitter or a neuromodulator affecting dopaminergic, serotonergic, GABAergic, glutamatergic, and cholinergic systems implicated in schizophrenia. NT peptide knockout mice ($NT^{-/-}$) exhibit reduced prepulse inhibition (PPI) and are not sensitive to the PPI-disrupting effects of amphetamine. NT plays a role in regulating feeding behavior.

Table 12.1 Functions of NT

Species	Function	Reference
Rat	Elicitation of hypothermia	[8]
Rat	Exertion of a potent analgesia	[9]
Rat	Increase in DAergic neuron firing rate and DA release	[10]
Rat	Stimulation of ACTH secretion in pituitary	[11]
Rat	Stimulation of synthesis and secretion of GnRH, SS, and CRH	[12]
Rat	Decrease in blood pressure	[13]
Rat	Inhibition of food intake	[14]
NTR1-KO mice	Decrease in a percentage of sleep time spent in REM sleep in dark phase	[15]
Dog	Simulation of exocrine pancreatic secretion	[16]
Rat	Decrease in gastric acid secretion	[17]

NTR1 knockout mice cause hyperphagia and abnormal weight gain, and display decreased sensitivity to the anorectic action of leptin. They also display altered REM sleep regulation and also show increased anxious and despair behaviors. $NTR2^{-/-}$ mice have reduced freezing response in a fear conditioning test. NT is involved in modulation of analgesia and pain response. NT and NTR2 knockout mice display defects in nociceptive responses [18].

Pathophysiological Implications

Clinical Implications

NT is related to the pathophysiology of a series of disorders, such as schizophrenia, drug abuse, Parkinson's disease (PD), feeding disorders, cancer, cerebral stroke, and other neurodegenerative diseases [19]. Furthermore, NT is involved in the physiology of pain induction, central control of blood pressure, and inflammation [19]. The levels of endogenous NT and NTR expression are decreased in patients with the symptoms of schizophrenia. The number of NT binding sites and expression levels of NTR1 mRNA were decreased in the substantia nigra of patients with PD. NT and its analog NT69L reduce body weight and food intake in healthy and obese rats. The expression of NT is increased in the hypothalamus of anorexic animals. NT has potential as a therapeutic drug in feeding disorders. NT shows possible implications in cancer; the size of colon or lung tumors increase in the presence of NT and become smaller in the presence of NTR1 antagonists. In cystic fibrosis patients with pancreatic insufficiency, the plasma concentration of NT is significantly increased compared with healthy controls.

Use for Diagnosis and Treatment

High expression of NT receptors has potential as a useful marker for diagnosis and assessing the progression of pancreatic cancers [20], although plasma NT levels do not differ between healthy controls and pancreatic cancer patients. Many EIA kits for measurement of plasma NT are sold by various companies (e.g., Peninsula Laboratories, Phoenix Pharmaceuticals, and Bachem).

References

1. Carraway R, Leeman SE. The isolation of a new hypotensive peptide, neurotensin, from bovine hypothalami. *J Biol Chem*. 1973;248:6854−6861.
2. Kislauskis E, Bullock B, McNeil S, et al. The rat gene encoding neurotensin and neuromedin N. Structure, tissue-specific expression, and evolution of exon sequences. *J Biol Chem*. 1988;263:4963−4968.
3. Leeman SE, Aronin N, Ferris CF. Substance P and neurotensin. *Recent Prog Horm Res*. 1982;38:93−132.
4. Beck B, Burtlet A, Nicolas JP, et al. Neurotensin in microdissected brain nuclei and in the pituitary of the lean and obese Zucker rats. *Neuropeptides*. 1989;13:1−7.
5. Shulkes A, Chick P, Wong H, et al. A radioimmunoassay for neurotensin in human plasma. *Clinica Chimica Acta*. 1982;125:49−58.
6. Wang X, Gulhati P, Li J, et al. Characterization of promoter elements regulating the expression of the human neurotensin/neuromedin N gene. *J Biol Chem*. 2011;286:542−554.
7. Vincent JP, Mazella J, Kitabgi P. Neurotensin and neurotensin receptors. *Trends Pharmacol Sci*. 1999;20:302−309.
8. Bissette CB, Loosen PT, Prange Jr. AJ, et al. Hypothermia and intolerance to cold induced by intracisternal administration of the hypothalamic peptide neurotensin. *Nature*. 1976;262:607−609.
9. Clinschmidt BV, McGuffin JC. Neurotensin administered intracisternally inhibits responsiveness of mice to noxious stimuli. *Eur J Pharmacol*. 1977;46:395−396.
10. Okuma Y, Fukuda Y, Osumi Y. Neurotensin potentiates the potassium-induced release of endogenous dopamine from rat striatal slices. *Eur J Pharmacol*. 1983;93:27−33.
11. Fuxe K, Aqnati LF, Andersson K, et al. Studies on neurotensin catecholamine interactions in the hypothalamus and in the forebrain of the male rat. *Neurochem Int*. 1984;6:737−750.
12. Rowe W, Viau V, Meaney MJ, et al. Stimulation of CRH-mediated ACTH secretion by central administration of neurotensin: evidence for the participation of the paraventricular nucleus. *J Neuroendocrinol*. 1995;7:109−117.
13. Rioux F, Quirion R, St-Pierre S, et al. The hypotensive effect of centrally administered neurotensin in rats. *Eur J Pharmacol*. 1981;69:241−247.
14. Cador M, Kelley AE, Le Moal M, et al. Ventral tegmental area infusion of substance P, neurotensin and enkephalin: differential effects on feeding behavior. *Neuroscience*. 1986;18:659−669.
15. Fitzpatrick K, Winrow CJ, Gotter AL, et al. Altered sleep and affect in the neurotensin receptor 1 knockout mouse. *Sleep*. 2012;35:949−956.
16. Nustede R, Schmidt WE, Köhler H, et al. Role of neurotensin in the regulation of exocrine pancreatic secretion in dogs. *Regul Pept*. 1993;44:25−32.
17. Osumi Y, Nagasaka Y, Wang, et al. Inhibition of gastric acid secretion and mucosal blood flow induced by intraventricularly applied neurotensin in rats. *Life Sci*. 1978;23:2275−2280.
18. Dobner PR. Neurotensin and pain modulation. *Peptides*. 2006;27:2405−2414.
19. St-Gelais F, Jomphe C, Trudeau LE. The role of neurotensin in central nervous system pathophysiology: what is the evidence? *J Psychiatry Neurosci*. 2006;31:229−245.
20. Reubi JC, Waser B, Friess H, et al. Neurotensin receptors: a new marker for human ductal pancreatic adenocarcinoma. *Gut*. 1998;42:546−550.
21. Kyte J, Doolittle RF. A simple method for displaying the hydropathic character of a protein. *J Mol Biol*. 1982;157:105−132.

Supplemental Information Available on Companion Website

- Gene, mRNA, and precursor structures of the human NT/E-Figure 12.1
- Precursor and mature hormone sequences of various animals/E-Figure 12.2
- Gene, mRNA, and precursor structure of the human NT receptors (NTR)/E-Figures 12.3−12.5
- Primary structure of the human NTRs/E-Figures 12.6−12.8
- Primary structure of NTR of various animals/E-Figure 12.9 [21]
- Accession numbers of genes and cDNAs for NT and NT receptor/E-Tables 12.1, 12.2

CHAPTER 13

Neuromedin U/S

Kenji Mori and Mikiya Miyazato

Abbreviations: NMU/NMS

NMU is an anorexigenic neuropeptide involved in the regulation of feeding behavior; NMS is a neuropeptide that is structurally related to NMU. Both peptides activate the same receptors.

Discovery

The neuromedin family of neuropeptides comprises seven members: neuromedin B and C (bombesin-like), K and L (tachykinin-like), N (neurotensin-like), and U and S (NMU-like). Most of these neuropeptides were isolated based on their contractile activity on smooth muscle preparations. NMU was originally isolated from porcine spinal cord by using the rat uterus contraction assay in 1985 [1]. NMS was identified and purified from rat brain as an endogenous ligand for orphan GPCRs in 2005 [2].

Structure

Structural Features

Human NMU consists of 25 aa residues with C-terminal amidation (Table 13.1). Human NMS comprises 33 aa residues and is C-terminally amidated (Table 13.2). These C-terminal amidated structures are essential for the biological activities of these peptides [1,3,4].

Primary Structure

NMU has been identified in mammals and other species (Table 13.1, E-Table 13.1) [3,4]. Its C-terminal seven amino acids and amidated structure are highly conserved among all species, although sequence identity is low in the N-terminal portion. NMS, however, has only been identified in humans, mice, rats, and toads (Table 13.2, E-Table 13.2). Like NMU, the C-terminal structures of NMS, but not its N-terminal, are highly conserved. In mammals, the C-terminal heptapeptide sequences with amidation (FLFRPRN-NH$_2$) are identical between NMU and NMS, and this highly conserved region is important for their biological activities [1–4].

Properties

Human NMU, Mr 3,080.4; human NMS, Mr 3,791.3. Both peptides are highly soluble in aqueous solution.

Synthesis and Release

Gene, mRNA, and Precursor

The human NMU gene is located on chromosome 4 (4q12) (E-Figure 13.1) [2]. An mRNA encoding human NMU produces a preproNMU of 174 aa residues (E-Figure 13.2). The amino acid sequence of NMU corresponds to positions 142–166 of the preproNMU, and a glycine residue at position 167 serves as an amide donor for the C-terminal residue of the mature peptide. These structures are flanked by cleavage sites for subtilisin-like proprotein convertases (SPCs) such as PC1/3 and PC2. The human NMS gene is located on chromosome 2 (2q11.2) (E-Figure 13.1) [2]. Human NMS mRNA encodes a preproNMS of 153 aa residues (E-Figure 13.2). The amino acid sequence of NMS (positions 109–141) and a glycine residue at position 142 (amino donor for C-terminal amidation) are flanked by cleavage sites for SPCs.

Distribution of mRNA

NMU mRNA is highly expressed in the human central nervous system (CNS), pituitary, stomach, intestine, and bone marrow, and to a lesser extent in various other tissues, including spleen, lymphocytes, adipose tissue, prostate, and placenta [3,4]. In the human CNS, prominent expression of NMU mRNA is observed in the hypothalamus, cingulate gyrus, locus ceruleus, medial frontal gyrus, medulla oblongata, nucleus accumbens, parahippocampal gyrus, substantia nigra, superior frontal gyrus, and thalamus [3,4]. NMS mRNA is abundant in the rat brain, spleen, and testis [2]. In the rat brain, NMS mRNA is expressed predominantly in the suprachiasmatic nucleus (SCN) of the hypothalamus, although low-level expression of NMS mRNA is also found in other hypothalamic nuclei, such as the arcuate nucleus, paraventricular nucleus, and supraoptic nucleus [2].

Tissue Content

RIAs have been used to determine the levels of NMU and NMS in various tissues. In the rat, the highest levels of NMU are found in the pituitary and gastrointestinal tract, with substantial levels in the brain, spinal cord, and genito-urinary tract [3]. NMU-like immunoreactivities are also detected in the human cerebral cortex and ileum [3]. In the rat brain, high levels of NMS are found in the hypothalamus, midbrain, and pons–medulla oblongata, although abundant expression of NMS mRNA is detected only in the hypothalamus [5].

Regulation of Synthesis

The levels of NMU mRNA in the brain are influenced by fasting. The expression of NMU mRNA is reduced in the

Y. Takei, H. Ando, & K. Tsutsui (Eds): Handbook of Hormones. DOI: http://dx.doi.org/10.1016/B978-0-12-801028-0.00013-1

Table 13.1 Structures of NMU*

Human	FRVDEEFQSPFASQSRGYFLFRPRN-NH$_2$
Porcine	FKVDEEFQGPIVSQNRRYFLFRPRN-NH$_2$
Rabbit	FPVDEEFQSPFGSRSRGYFLFRPRN-NH$_2$
Canine	FRLDEEFQGPIASQVRRQFLFRPRN-NH$_2$
Guinea pig	GYFLFRPRN-NH$_2$
Mouse	FKA-EYQSPSVGQSKGYFLFRPRN-NH$_2$
Rat	YKV-NEYQGP-VAPSGGFLFRPRN-NH$_2$
Chicken	YKVDEDLQGAGGIQSRGYFFFRPRN-NH$_2$
Frog (Rana temporaria)	LKPDEELQGPGGVLSRGYFVFRPRN-NH$_2$
Tree frog (Litoria caerulea)	SDEEVQVPGGVISNGYFLFRPRN-NH$_2$
Goldfish	MKLNDDLQGPGRIQSRGFFLYRPRN-NH$_2$

*Conserved residues are shaded.

Table 13.2 Structures of NMS*

Human	ILQRGSGTAAVDFTKKDHTATWGRPFFLFRPRN-NH$_2$
Mouse	LPRLLRLDSRMATVDFPKKDPTTSLGRPFFLFRPRN-NH$_2$
Rat	LPRLLHTDSRMATIDFPKKDPTTSLGRPFFLFRPRN-NH$_2$
Toad (Bombina orientalis)	DSSGIVGRPFFLFRPRN-NH$_2$

*Conserved residues are shaded.

Table 13.3 Effects of NMU Revealed by Intracerebroventricular Administration in Rodents

Biological Function	Reference
Inhibit food intake	[9]
Reduce body weight	[10]
Increase energy expenditure	[10]
Induce stress response	[11]
Increase blood pressure and heart rate	[12]
Induce phase shift in circadian rhythms	[13]
Stimulate nociceptive reflexes	[14]
Reduce urine volume	[15]
Elevate plasma levels of luteinizing hormone	[16]
Increase milk secretion	[17]
Induce anxiolytic action	[18]
Induce antidepressant-like effect	[19]

Table 13.4 Phenotypes in NMU Knockout Mice

Phenotype	Reference
Hyperphagic obesity	[21]
Reduced nociceptive reflexes	[14]
Reduced complete Freund's adjuvant-induced inflammation	[22]
Resistance to septic shock	[23]
Reduced airway eosinophilia	[24]
Early onset of virginal opening	[25]
High bone mass	[26]
Diarrhea	[27]
Less severe arthritis than in control mice	[28]

ventromedial hypothalamus of fasted rats [3]. In contrast, NMU mRNA expression is elevated in the dorsomedial hypothalamic nucleus of fasted mice [3]. In the rat, both NMU and NMS mRNAs are expressed in the SCN, which is the master circadian pacemaker in mammals [2,4]. Pronounced rhythmic expression of their mRNAs is observed in the SCN of rats maintained under 12-hour light/dark cycles. Under constant darkness, the expression of NMU mRNA, but not NMS mRNA, shows a circadian rhythm in the SCN, suggesting that the expression of NMU is under the control of clock-gene families. In addition, hypothalamic levels of both NMU and NMS mRNAs fluctuate during the estrous cycle in female rats [4,6].

Receptors

Structure and Subtype

There are two NMU receptors, namely NMU receptor types 1 (NMUR1) and 2 (NMUR2), which belong to the class A family of GPCRs (E-Figures 13.3, 13.4; E-Tables 13.3, 13.4) [3,4]. NMU and NMS are physiological ligands for these receptors, and share similar potency and efficacy for both recombinant NMUR1 and NMUR2 [2]. Human NMUR1 and NMUR2 consist of 403 and 412 aa residues, respectively (E-Figure 13.4). In humans and rats, NMUR1 mRNA expression is more abundant in peripheral tissues, whereas NMUR2 mRNA is expressed predominantly in the CNS [3,4].

Signal Transduction Pathway

In recombinant cell lines, ligand-induced activation of these receptors results in an increase in intracellular calcium ion concentrations and suppression of forskolin-induced cAMP production [3,4].

Agonists

There are selective hexapeptide agonists to NMUR1 and NMUR2 [7]. NMUR2 agonists also include the natural products EUK2010, EUK2011, EUK2012, and icariin [4]. In addition, two non-peptidic small molecules have been identified as non-selective agonists [4].

Antagonist

The non-peptidic small molecule (R)-5'-(phenylaminocarbonylamino)spiro [1-azabicyclo[2.2.2]octane-3,2' (3'H)-furo[2,3-b]pyridine] is a highly potent and selective NMUR2 antagonist [8].

Biological Functions

Target Cells/Tissues and Functions

Intracerebroventricular (ICV) administration of NMU induces several interesting responses (Table 13.3 [9–19]). In particular, ICV-administered NMU inhibits food intake in rats, mice, Japanese quail, chicks, and goldfish [3,4]. Peripheral administration of NMU in rodents also reduces food intake and body weight [20]. ICV administration of NMS produces similar effects to those of NMU; however, the activity of NMS is approximately 10 times greater than that of NMU in the brain [2,4].

Phenotypes in Gene-Modified Animals

NMU knockout mice exhibit hyperphagic obesity and several interesting phenotypes (Table 13.4 [21–28]). Transgenic mice overexpressing NMU become hypophagic and lean [4]. NMS knockout mice exhibit a decrease in heart rate, and prolongations of the QRS and RR interval on electrocardiograms. No expression of mRNA for receptors is observed in the mouse heart. In addition, ICV-administered NMS-induced increase in heart rate is inhibited by pretreatment with a sympathetic nerve blocker, timolol, in mice. These data suggest that endogenous NMS may regulate cardiovascular function through the sympathetic nervous system [29].

Pathophysiological Implications

Clinical Implications

Human NMU gene variants are associated with excess weight and obesity. In middle-aged Caucasians, the Ala19Glu polymorphism correlates with an overweight or obese phenotype. The rare Arg165Trp variant was associated with childhood obesity in a Czech family [4]. Changes in NMU mRNA expression have been reported in several forms of cancer. The expression of NMU mRNA is increased in bladder carcinoma, ovarian carcinoma, non-small cell lung cancers, and acute myeloid leukemia. In contrast, decreased expression is found in oral tumors and esophageal squamous cell carcinoma [4].

Use for Diagnosis and Treatment

To date, there are no known uses for NMU/NMS in diagnosis or treatment.

References

1. Minamino N, Kangawa K, Matsuo H. Neuromedin U-8 and U-25: novel uterus stimulating and hypertensive peptides identified in porcine spinal cord. *Biochem Biophys Res Commun.* 1985;130:1078−1085.
2. Mori K, Miyazato M, Ida T, et al. Identification of neuromedin S and its possible role in the mammalian circadian oscillator system. *EMBO J.* 2005;24:325−335.
3. Brighton PJ, Szekeres PG, Willars GB. Neuromedin U and its receptors: structure, function, and physiological roles. *Pharmacol Rev.* 2004;56:231−248.
4. Mitchell JD, Maguire JJ, Davenport AP. Emerging pharmacology and physiology of neuromedin U and the structurally related peptide neuromedin S. *Br J Pharmacol.* 2009;158:87−103.
5. Mori M, Mori K, Ida T, et al. Different distribution of neuromedin S and its mRNA in the rat brain: NMS peptide is present not only in the hypothalamus as the mRNA, but also in the brainstem. *Front Endocrinol.* 2012;3:152.
6. Vigo E, Roa J, López M, et al. Neuromedin S as novel putative regulator of luteinizing hormone secretion. *Endocrinology.* 2007;148:813−823.
7. Takayama K, Mori K, Taketa K, et al. Development of selective hexapeptide agonists to human neuromedin U receptors types 1 and 2. *J Med Chem.* 2014;57:6583−6593.
8. Liu JJ, Payza K, Huang J, et al. Discovery and pharmacological characterization of a small-molecule antagonist at neuromedin U receptor NMUR2. *J Pharmacol Exp Ther.* 2009;330:268−275.
9. Howard AD, Wang R, Pong SS, et al. Identification of receptors for neuromedin U and its role in feeding. *Nature.* 2000;406:70−74.
10. Nakazato M, Hanada R, Murakami N, et al. Central effects of neuromedin U in the regulation of energy homeostasis. *Biochem Biophys Res Commun.* 2000;277:191−194.
11. Hanada R, Nakazato M, Murakami N, et al. A role for neuromedin U in stress response. *Biochem Biophys Res Commun.* 2001;289:225−228.
12. Chu C, Jin Q, Kunitake T, et al. Cardiovascular actions of central neuromedin U in conscious rats. *Regul Pept.* 2002;105:29−34.
13. Nakahara K, Hanada R, Murakami N, et al. The gut-brain peptide neuromedin U is involved in the mammalian circadian oscillator system. *Biochem Biophys Res Commun.* 2004;318:156−161.
14. Nakahara K, Kojima M, Hanada R, et al. Neuromedin U is involved in nociceptive reflexes and adaptation to environmental stimuli in mice. *Biochem Biophys Res Commun.* 2004;323:615−620.
15. Sakamoto T, Mori K, Nakahara K, et al. Neuromedin S exerts an antidiuretic action in rats. *Biochem Biophys Res Commun.* 2007;361:457−461.
16. Vigo E, Roa J, Pineda R, et al. Novel role of the anorexigenic peptide neuromedin U in the control of LH secretion and its regulation by gonadal hormones and photoperiod. *Am J Physiol Endocrinol Metab.* 2007;293:E1265−E1273.
17. Sakamoto T, Mori K, Miyazato M, et al. Involvement of neuromedin S in the oxytocin release response to suckling stimulus. *Biochem Biophys Res Commun.* 2008;375:49−53.
18. Telegdy G, Adamik A. Anxiolytic action of neuromedin-U and neurotransmitters involved in mice. *Regul Pept.* 2013;186:137−140.
19. Tanaka M, Telegdy G. Neurotransmissions of antidepressant-like effects of neuromedin U-23 in mice. *Behav Brain Res.* 2014;259:196−199.
20. Peier AM, Desai K, Hubert J, et al. Effects of peripherally administered neuromedin U on energy and glucose homeostasis. *Endocrinology.* 2011;152:2644−2654.
21. Hanada R, Teranishi H, Pearson JT, et al. Neuromedin U has a novel anorexigenic effect independent of the leptin signaling pathway. *Nat Med.* 2004;10:1067−1073.
22. Moriyama M, Sato T, Inoue H, et al. The neuropeptide neuromedin U promotes inflammation by direct activation of mast cells. *J Exp Med.* 2005;202:217−224.
23. Moriyama M, Matsukawa A, Kudoh S, et al. The neuropeptide neuromedin U promotes IL-6 production from macrophages and endotoxin shock. *Biochem Biophys Res Commun.* 2006;341:1149−1154.
24. Moriyama M, Fukuyama S, Inoue H, et al. The neuropeptide neuromedin U activates eosinophils and is involved in allergen-induced eosinophilia. *Am J Physiol Lung Cell Mol Physiol.* 2006;290:L971−L977.
25. Fukue Y, Sato T, Teranishi H, et al. Regulation of gonadotropin secretion and puberty onset by neuromedin U. *FEBS Lett.* 2006;580:3485−3488.
26. Sato S, Hanada R, Kimura A, et al. Central control of bone remodeling by neuromedin U. *Nat Med.* 2007;13:1234−1240.
27. Nakashima Y, Ida T, Sato T, et al. Neuromedin U is necessary for normal gastrointestinal motility and is regulated by serotonin. *Ann N Y Acad Sci.* 2010;1200:104−111.
28. Rao SM, Auger JL, Gaillard P, et al. The neuropeptide neuromedin U promotes autoantibody-mediated arthritis. *Arthritis Res Ther.* 2012;14:R29.
29. Sakamoto T, Nakahara K, Maruyama K, et al. Neuromedin S regulates cardiovascular function through the sympathetic nervous system in mice. *Peptides.* 2011;32:1020−1026.

Supplemental Information Available on Companion Website

- Gene, mRNA, and precursor structures of human NMU and mouse NMS/E-Figure 13.1
- Precursor and mature peptide sequences/E-Figure 13.2
- Gene, mRNA, and precursor structures of human NMUR/E-Figure 13.3
- Primary structures of human NMUR1 and NMUR2/E-Figure 13.4
- Accession numbers of NMU, NMS, NMUR1, and NMUR2/E-Tables 13.1−13.4

Adenohypophysial Hormones

CHAPTER 14

Glycoprotein Hormone Family

Hironori Ando

History

The pituitary has long been known to contain gonad- and thyroid-stimulating materials. Gonad-stimulating substance was also found in human pregnancy urine, and later purified as placental-origin human chorionic gonadotropin (hCG). Since the 1930s, isolation of the gonadotropic and thyrotropic substances from the pituitary has been the focus of many researchers, and in the 1950s, as protein purification methods improved, the preparations of follicle-stimulating hormone (FSH), luteinizing hormone (LH), and thyroid-stimulating hormone (TSH) of high potency were purified from mammalian species such as bovines, pigs, horses, rats, and humans. In the 1970−1980s, partial and full amino acid sequences of α- and β-subunits of these hormones were determined using HPLC and peptide sequencing, and later, in the 1980−1990s, their putative primary structures were determined by molecular cloning of cDNA and genes. In lower vertebrates, the presence of these hormones and functional similarities and differences between mammalian hormones have been examined since the 1970s, and the evolution of theses hormones in vertebrates was proposed by Fontain and Burzawa-Gerard in 1977 [1]. In fish, the presence of two different gonadotropins (GTHs) has long been controversial, and two GTH preparations, later designated FSH and LH, were identified in salmon in 1988 [2]. Recently, another heterodimeric glycoprotein hormone (GPH), thyrostimulin, has been identified in humans [3]. Thyrostimulin is expressed not only in the pituitary but also in various peripheral tissues. Homologous genes for thyrostimulin are widely distributed in vertebrates and invertebrates, and thyrostimulin is considered to be an ancestral molecule of the vertebrate pituitary GPHs.

Structure

Structural Features

FSH, LH, TSH, CG, and thyrostimulin belong to the GPH family. Placental CG is present only in primates and equids. The pituitary GPHs and CG are heterodimers consisting of a common GPH α-subunit non-covalently bound to a hormone-specific β-subunit (Figure 14.1) [4]. Thyrostimulin consists of glycoprotein α-subunit 2 (GPA2) and glycoprotein β-subunit 5 (GPB5), which are considered to be ancestral forms of the pituitary GPH α- and β-subunits, respectively. These α- and β-subunits contain a cysteine-knot motif, found in growth factors such as TGF-β family members. They associate in a head-to-tail arrangement to form the dimer [5]. They also contain N-linked oligosaccharide chains that are important for intracellular folding, secretion, metabolic clearance, and biological activity of the hormone [6].

Molecular Evolution of Family Members

The GPH α- and β-subunits are considered to have evolved from a single ancestor gene through multiple gene duplications (Figure 14.2, E-Figure 14.1) [7, 8]. The presence of GPA2 and GPB5 genes in amphioxus and the presence of a single set of GPH α- and β-subunits and thyrostimulin in agnathans suggest that GPA2 and GPB5 are ancestral forms of the GPH α- and β-subunits, respectively, in vertebrates, and that the hormone-specific GPH β-subunits diverged in the vertebrate lineage (Figure 14.2, E-Figures 14.2, 14.3). The GPH β-subunit gave rise to the LH β-subunit, TSH β-subunit, and FSH β-subunit genes with their pituitary-specific expression in vertebrates. In equids, the single LH β-subunit gene is expressed not only in the pituitary but also in the placenta, while in primates a new CG β-subunit gene evolved that was expressed specifically in the placenta.

Receptors

Structure and Subtypes

The receptors for GPHs belong to the membrane-bound GPCR superfamily, representing a subclass of the rhodopsin/β-adrenergic subfamily. The GPH receptors have a characteristic structure of a large N-terminal extracellular domain which contains 14 sequence repeats.

Signal Transduction Pathway

GPH receptors are preferentially coupled to G_s protein, which activates AC to enhance the synthesis of cAMP.

Biological Functions

Target Cells/Tissues and Functions

TSH and GTHs are critical for the development and function of the thyroid gland and gonads, respectively. Targeted disruption of the common GPH α-subunit resulted in hypothyroidism and hypogonadism [9]. In homozygous mutant mice, development of the thyroid gland was arrested in late gestation, while the fetal and neonatal gonadal development were normal but led to a block in folliculogenesis and spermatogenesis.

Y. Takei, H. Ando, & K. Tsutsui (Eds): Handbook of Hormones. DOI: http://dx.doi.org/10.1016/B978-0-12-801028-0.00014-3

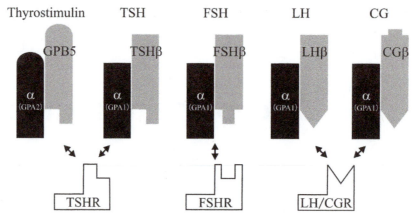

Figure 14.1 Schematic representation of the structure of five GPHs and their receptors.

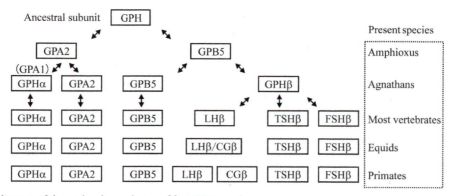

Figure 14.2 Putative diagram of the molecular evolution of five GPH members.

References

1. Fontain YA, Burzawa-Gerard E. Evolution of gonadotropic and thyrotropic hormones in vertebrates. *Gen Comp Endocrinol.* 1977;32:341−347.
2. Suzuki K, Kawauchi H, Nagahama Y. Isolation and characterization of two distinct gonadotropins from chum salmon pituitary glands. *Gen Comp Endocrinol.* 1988;71:292−301.
3. Nakabayashi K, Matsui H, Bhalla H, et al. Thyrostimulin, a heterodimer of two new human glycoprotein hormone subunits, activates the thyroid-stimulating hormone receptor. *J Clin Invest.* 2002;109:1445−1452.
4. Pierce JC, Parsons TF. Glycoprotein hormones: structure and function. *Annu Rev Biochem.* 1981;50:465−495.
5. Fox KM, Dias JA, Roey PV. Three-dimensional structure of human follicle-stimulating hormone. *Mol Endocrinol.* 2001;15:378−389.
6. Ulloa-Aquirre A, Maldonado A, Damian-Matsumura P, et al. Endocrine regulation of gonadotropin glycosylation. *Arch Med Res.* 2001;32:520−532.
7. Li MD, Ford JJ. A comprehensive evolutionary analysis based on nucleotide and amino acid sequences of the alpha- and beta-subunits of glycoprotein hormone gene family. *J Endocrinol.* 1998;156:529−542.
8. Kubokawa K, Tando Y, Roy S. Evolution of the reproductive endocrine system in chordates. *Integr Comp Biol.* 2010;50:53−62.
9. Kendall SK, Samuelson LC, Saunders TL, et al. Targeted disruption of the pituitary glycoprotein hormone alpha-subunit produces hypogonadal and hypothyroid mice. *Genes Dev.* 1995;9:2007−2019.
10. Jones DT, Taylor WR, Thornton JM. The rapid generation of mutation data matrices from protein sequences. *Comput Appl Biosci.* 1992;8:275−282.
11. Tamura K, Stecher G, Peterson D, et al. MEGA6: molecular evolutionary genetics analysis version 6.0. *Mol Biol Evol.* 2013;30:2725−2729.

Supplemental Information Available on Companion Website

- Primary structure of four GPH subunits in human/E-Figure 14.1
- Phylogenetic tree of the GPH family members/E-Figures 14.2, 14.3

Follicle-Stimulating Hormone

Hironori Ando

Abbreviation: FSH
Additional name: Follitropin

A gonadotropic glycoprotein hormone secreted from the anterior pituitary, FSH stimulates follicular growth and estrogen production in the ovary and promotes spermatogenesis in the testis.

Discovery

In the 1960s, purification and isolation of FSH was reported from ovine, porcine, bovine, and human pituitaries. Human FSH was first fully purified in 1968 [1].

Structure

Structural Features

FSH is a glycoprotein consisting of non-covalently linked glycoprotein hormone (GPH) α- and FSH β-subunits. The GPH α-subunit is common to luteinizing hormone (LH), thyroid-stimulating hormone (TSH), and chorionic gonadotropin (CG). Both GPH α- and FSH β-subunits contain a cysteine-knot motif, which is critical for the heterodimer assembly and biological activity of FSH [2,3]. The *N*- and *O*-linked oligosaccharide chains are important for the intracellular folding, secretion, metabolic clearance, and biological activity of FSH.

Primary Structure

The primary structures of the human GPH α-subunit (92 aa residues) and the human FSH β-subunit (110 aa residues) are shown in Figure 14A.1. The human GPH α-subunit contains 10 cysteine residues, and 6 of them are conserved in the human FSH β-subunit that contains 12 cysteine residues (see E-Figure 14.1 in Chapter 14).

Properties

Mr 25,000−41,000, pI 3.0−5.0. Multiple isoforms exist due to the microheterogeneity of oligosaccharide chains. Soluble in water, physiological saline solution, 50% alcohol and 50% acetone; insoluble in alcohol, acetone, benzene, chloroform, and ether. Stable in 6-M urea; dissociated into two subunits in 1-M propionic acid.

Synthesis and Release

Gene, mRNA, and Precursor

The human GPH α-subunit gene, *CGA*, location 6q12−q21, consists of four exons (E-Figure 14A.1). Human GPH α-subunit mRNA has 397 b that encode a signal peptide of 24 aa residues and a mature protein of 92 aa residues (E-Figure 14A.3). The human FSH β-subunit gene, *FSHB*, location 11p13, consists of three exons (E-Figure 14A.2). Human FSH β-subunit mRNA has 506 b that encode a signal peptide of 19 aa residues and a mature protein of 110 aa residues (E-Figure 14A.4).

Distribution of mRNA

The GPH α- and FSH β-subunit genes are expressed in the basophilic gonadotropes in the anterior pituitary. In tetrapods, FSH and LH are co-expressed in gonadotropes, whereas they are produced in different cells in teleosts [4]. The GPH α-subunit gene is also expressed in thyrotropes in the anterior pituitary and the placental trophoblast.

Tissue and Plasma Concentrations

Tissue: Rat, female 400−600 (ng NIH-FSH-S1/pituitary).
Plasma: Rat, female in proestrus stage 20 (ng NIH-FSH-S1/ml); chicken, mature female 50, ovariectomized female 950, immature male 5−10, mature male 225, castrated male 900; quail, mature male 300, castrated male 1500.

Regulation of Synthesis and Release

The synthesis and release of FSH are regulated by gonadotropin-releasing hormone (GnRH); gonadal proteins such as activin, inihibin, and follistatin; and the feedback effects of gonadal steroids [5]. In tetrapods, GnRH acts on a single cell type of gonadotrope and differentially regulates FSH and LH secretion through changes in the pattern of GnRH pulse secretion. Activin stimulates FSH synthesis and release through binding to activin receptor at the cell membrane of the gonadotropes, which is antagonized by binding of inhibin and follistatin. FSH secretion is also regulated by gonadal steroids such as estradiol and progesterone. Gonadal steroids exert their effects at the level of the hypothalamus by changing GnRH secretion; and directly at the level of the gonadotropes, where they exert different effects, depending on the species and reproductive condition of the animals [6].

Receptors

Structure and Subtype

The receptor of FSH (FSHR) is a glycoprotein that belongs to a subclass of the rhodopsin/β-adrenergic subfamily of the membrane-bound GPCR superfamily [7]. The FSHR consists of 650−700 aa residues that contain a large N-terminal extracellular domain (∼350 aa residues), 7 transmembrane domains, and a C-terminal intracellular domain (E-Figures 14A.6, 14A.7).

Y. Takei, H. Ando, & K. Tsutsui (Eds): Handbook of Hormones. DOI: http://dx.doi.org/10.1016/B978-0-12-801028-0.00129-X

```
Human GPH α-subunit
        10        20        30        40        50        60
APDVQDCPECTLQENPFFSQPGAPILQCMGCCFSRAYPTPLRSKKTMLVQKNVTSESTCC
        70        80        90
VAKSYNRVTVMGGFKVENHTACHCSTCYYHKS
```

```
Human FSH β-subunit
        10        20        30        40        50        60
SCELTNITIAIEKEECRFCISINTTWCAGYCYTRDLVYKDPARPKIQKTCTFKELVYETV
        70        80        90        100       110
RVPGCAHHADSLYTYPVATQCHCGKCDSDSTDCTVRGLGPSYCSFGEMKE
```

Figure 14A.1 Primary structure of the human GPH α- and FSH β-subunits. *N*-linked glycosylation sites are underlined and Cys residues are shaded.

Table 14A.1 Biological Functions of FSH in Various Vertebrate Species

Species	Function	Reference
Rat	Stimulation of estrogen production by granulosa cells	
	Stimulation of inhibin production by granulosa and Sertoli cells	[8]
Chicken	Stimulation of progesterone secretion by granulosa cells	[9]
	Stimulation of estrogen secretion by theca cells	[10]
	Stimulaton of DNA synthesis in granulosa cells	[11]
Green sea turtle	Stimulation of testosterone secretion by testis	[12]
Salmon	Stimulation of estrogen production by ovarian follicles	[13]
Carp	Stimulation of estrogen production by ovarian follicles and testosterone poduction by testis	[14]

Signal Transduction Pathway

The receptor mainly couples to G_s protein, and FSH stimulates production of cAMP in target cells.

Agonists

Purified and recombinant human FSH, thiazolidinone compounds, cyclic and acyclic α- and β-aminocarboxamide derivatives, and biaryl derivatives are agonists.

Antagonists

Antibodies to FSHR, aminoalkylamide derivatives, tetrahydroquinoline derivatives, 7-{4-[bis-(2-carbamoyl-ethyl)-amino]-6-chloro-(1,3,5)-triazin-2-ylamino)-4-hydroxy-3-(4-methoxyphenylazo)-naphthalene}-2-sulfonic acid, and suramin (a nonselective FSHR-antagonist) are antagonists.

Biological Functions

Target Cells/Tissues and Functions

In vertebrates, FSH binds to FSHR located on the membrane of granulosa cells and stimulates estrogen production and maturation of developing follicles in the ovary. In the testis, FSH binds specifically to FSHR located on the membrane of Sertoli cells and promotes spermatogenesis. Biological functions of FSH in various vertebrate species are shown in Table 14A.1 [8—14]

Phenotype in Gene-Modified Animals

In females, both FSH knockout mice [15] and FSHR knockout mice [16] are infertile due to a block in folliculogenesis before antral follicle formation. In males, both FSH knockout mice and FSHR knockout mice are fertile but display small testes and partial spermatogenic failure.

Pathophysiological Implications

Clinical Implications

Hypergonadotropic ovarian dysgenesis is a disease characterized by a normal karyotype, highly elevated gonadotropins, and streaky gonads associated with primary amenorrhea. A mutation (Ala 189 Val) in the extracellular domain of FSHR is considered a probable cause of the disease [17]. Functional studies revealed a lack of cAMP production by the mutated receptor upon FSH stimulation. FSH levels are useful in the investigation of menstrual irregularities, and in the diagnosis of pituitary disorders or diseases involving the ovaries or testes. Conditions with high FSH levels include premature menopause (i.e., premature ovarian failure), poor ovarian reserve, gonadal dysgenesis, Turner syndrome, Swyer syndrome, and Klinefelter syndrome. Conditions with low FSH levels include hypopituitarism, polycystic ovarian syndrome, Kallmann syndrome, Sheehan syndrome, and Chiari-Frommel syndrome.

Use for Diagnosis and Treatment

Recombinant human FSH is used for treatment of infertility, to stimulate follicular development.

References

1. Roos P. Human follicle-stimulating hormone. Its isolation from the pituitary gland and from postmenopausal urine and a study of some chemical, physical immunological, and biological properties of the hormone from these two sources. *Acta Endocrinol Suppl.* 1968;131:3—93.
2. Hiro'oka T, Maassen D, Berger P, et al. Disulfide bond mutations in follicle-stimulating hormone result in uncoupling of biological activity from intracellular behavior. *Endocrinology.* 2000;141:4751—4756.
3. Fox KM, Dias JA, Van Roey P. Three-dimensional structure of human follicle-stimulating hormone. *Mol Endocrinol.* 2001;15:37889.
4. Yaron Z, Gur G, Melamed P, et al. Regulation of fish gonadotropins. *Int Rev Cytol.* 2003;225:131—185.
5. Melamed P. Hormonal signaling to follicle stimulating hormone β-subunit gene expression. *Mol Cell Endocriol.* 2010;314:204—221.
6. Ando H, Urano A. Molecular regulation of gonadotropin secretion by gonadotropin-releasing hormone in salmonid fishes. *Zool Sci.* 2005;22:379—389.
7. Simoni M, Gromoll J, Nieschlag E. The follicle-stimulating hormone receptor: biochemistry, molecular biology, physiology, and pathophysiology. *Endocr Rev.* 1997;18:739—773.
8. Bicsak TA, Tucker EM, Cappel S, et al. Hormonal regulation of granulosa cell inhibin biosynthesis. *Endocrinology.* 1986;119:2711—2719.
9. Krishnan KA, Proudman JA, Bahr JM. Purification and characterization of chicken follicle-stimulating hormone. *Comp Biochem Physiol B.* 1992;102:67—75.
10. Onagbesan OM, Peddie MJ. Calcium-dependent stimulation of estrogen secretion by FSH from theca cells of the domestic hen (*Gallus domesticus*). *Gen Comp Endocrinol.* 1989;75:177—186.
11. McElroy AP, Caldwell DJ, Proudman JA, et al. Modulation of *in vitro* DNA synthesis in the chicken ovarian granulosa cell follicular hierarchy by follicle-stimulating hormone and luteinizing hormone. *Poult Sci.* 2004;83:500—506.
12. Licht P, Papkoff H. Reevaluation of the relative activities of the pituitary glycoprotein hormones (follicle-stimulating hormone, luteinizing hormone, and thyrotrophin) from the green sea turtle, *Chelonia mydas. Gen Comp Endocrinol.* 1985;58:443—451.
13. Suzuki K, Nagahama Y, Kawauchi H. Steroidogenic activities of two distinct salmon gonadotropins. *Gen Comp Endocrinol.* 1988;71:452—458.
14. Van der Kraak G, Suzuki K, Peter RE, et al. Properties of common carp gonadotropin I and gonadotropin II. *Gen Comp Endocrinol.* 1992;85:217—229.

15. Kumar TR, Wang Y, Lu N, et al. Follicle stimulating hormone is required for ovarian follicle maturation but not male fertility. *Nat Genet*. 1997;15:201–204.

16. Dierich A, Sairam MR, Monaco L, et al. Impairing follicle-stimulating hormone (FSH) signaling *in vivo*: targeted disruption of the FSH receptor leads to aberrant gametogenesis and hormonal imbalance. *Proc Natl Acad Sci USA*. 1998;95:13612–13617.

17. Aittomäki K, Dieguez Lucena JL, et al. Mutation in the follicle-stimulating hormone receptor gene causes hereditary hypergonadotropic ovarian failure. *Cell*. 1995;82:959–968.

Supplemental Information Available on Companion Website

- Gene, mRNA. and precursor strucures of the human GRH α- and FSH β-subunits/E-Figures 14A.1, 14A.2
- Precursor and mature hormone sequences of various animals/E-Figures 14A.3, 14A.4
- Gene, mRNA and precursor structures of the human FSHR/E-Figure 14A.5
- Primary structure of the human FSHR/E-Figure 14A.6
- Primary structure of FSHR of various animals/E-Figure 14A.7
- Accession numbers of genes and cDNAs for GRH α- and FSH β-subunits and FSHR/E-Tables 14A.1–14A.3.

Luteinizing Hormone

Hironori Ando

Abbreviation: LH
Additional names: lutropin, interstitial cell-stimulating hormone (ICSH)

Luteinizing hormone is a gonadotropic glycoprotein hormone secreted by the anterior pituitary. In females, an LH surge triggers ovulation and stimulates the development of the corpus luteum. In males, LH stimulates androgen production and spermatogenesis.

Discovery

Gonadotropic fractions with the properties of LH were purified from ovines in the late 1950s, and subsequently LH has been isolated from many other species. Human LH was first fully purified in 1964 [1]. The structure and full nucleotide sequence of the human LH β-subunit gene was determined in 1984 [2].

Structure

Structural Features

LH is a glycoprotein consisting of non-covalently linked glycoprotein hormone (GPH) α- and LH β-subunits. The GPH α-subunit is common to follicle-stimulating hormone (FSH), thyroid-stimulating hormone (TSH), and chorionic gonadotropin (CG) (see Subchapter 14A). The LH β-subunit contains a cysteine-knot motif, which is critical for the heterodimer assembly and biological activity of the hormone [3]. The N-linked oligosaccharide chain is important for the intracellular folding, secretion, metabolic clearance, and biological activity of the hormone [4].

Primary Structure

The primary structure of the human LH β-subunit (121 aa residues) is shown in Figure 14B.1.

Properties

Mr 26,000−48,000. pI: human LH, 7.2−9.2; rat LH, 8.6−9.3; pig LH, 7.2−9.2; horse LH, 4.5−7.5. Multiple isoforms exist due to the microheterogeneity of oligosaccharide chains. Soluble in water; insoluble in alcohol and acetone. Partially (50%) dissociated to two subunits at pH 1.9. Inactivated by oxidation (hydrogen peroxide, periodic acid), reduction (cysteine, ketone), and treatment with trypsin, chymotrypsin, and pepsin. Picrolonic, flavianic, picric, and trichloroacetic acids precipitate LH with retention of its activity.

Synthesis and Release

Gene, mRNA, and Precursor

The human LH β-subunit gene, *LHB*, location 2p21, consists of three exons (E-Figure 14B.1). The human LH β-subunit mRNA has 523 b that encode a signal peptide of 20 aa residues and a mature protein of 121 aa residues (E-Figure 14B.2).

Distribution of mRNA

The LH β-subunit gene is expressed in the basophilic gonadotropes in the anterior pituitary. In tetrapods, FSH and LH are co-expressed in gonadotropes, whereas they are produced in different cells in teleosts.

Tissue and Plasma Concentrations

Tissue: Rat, male 10 (μg NIH-LH-S1/pituitary), female 5; quail, male under short-day condition 0.25, male under long-day condition 1.

Plasma: Human, adult male ∼3 (mIU/ml), adult female in ovulatory phase ∼14, adult female in follicular phase ∼4, adult female in luteal phase ∼14, adult female in menopausal phase ∼18; rat, female in proestrus stage 20−25 (ng NIH-LH-S1/ml), female in other stages 1−2; chicken, immature female 3−4 (ng/ml), mature female 1−2, female at peak of ovulation 3−6, gonadotropin-releasing hormone (GnRH)-treated immature female 30, GnRH-treated mature female 5, castrated male 15; chum salmon, seawater female in spawning season 15 (ng/ml), freshwater female in spawning season 50, seawater male in spawning season 4, freshwater male in spawning season 10.

Regulation of Synthesis and Release

Synthesis and release of LH are regulated by gonadotropin-releasing hormone (GnRH) and the feedback effects of gonadal steroids. In tetrapods, GnRH acts on a single cell type of gonadotrope and differentially regulates FSH and LH secretion through changes in the pattern of GnRH pulse secretion. LH secretion is also regulated by gonadal steroids such as estradiol and testosterone. Gonadal steroids exert their effects at the level of the hypothalamus by changing GnRH secretion, and directly at the level of the gonadotropes, where they exert different effects depending on the species and reproductive condition of animals [5].

Y. Takei, H. Ando, & K. Tsutsui (Eds): Handbook of Hormones. DOI: http://dx.doi.org/10.1016/B978-0-12-801028-0.00130-6

```
Human LH β-subunit
      10        20        30        40        50        60
SREPLRPWCHPINAILAVEKEGCPVCITVNTTICAGYCPTMMRVLQAVLPPLPQVVCTYR
      70        80        90       100       110       120
DVRFESIRLPGCPRGVDPVVSFPVALSCRCGPCRRSTSDCGGPKDHPLTCDHPQLSGLLFL
```

Figure 14B.1 The primary structure of the human LH β-subunit. *N*-linked glycosylation sites are underlined and Cys residues are shaded.

Receptors

Structure and Subtype

The receptor of LH (LHR) is a glycoprotein that belongs to a subclass of the rhodopsin/β-adrenergic subfamily of the membrane-bound GPCR superfamily [6]. The LHR consists of around 700 aa residues and contains a large N-terminal extracellular domain ($\sim$360 aa residues), seven transmembrane domains, and a C-terminal intracellular domain (E-Figure 14B.3).

Signal Transduction Pathway

The receptor mainly couples to G_s protein, and LH stimulates production of cAMP in target cells.

Agonists

hCG, and purified and recombinant human LH, are agonists.

Antagonists

Deglycosylated hCG and deglycosylated LH are antagonists.

Biological Functions

Target Cells/Tissues and Functions

LH binds to the LHR located on the membrane of granulosa cells in the ovary and stimulates estrogen production and follicular maturation in cooperation with FSH. LH also promotes ovulation and development of the corpus luteum. In the testis, LH binds to the LHR located on the membrane of Leydig cells. LH stimulates androgen production and promotes spermatogenesis.

Phenotype in Gene-Modified Animals

The LH β-subunit knockout mice show hypogonadism, defects in gonadal steroidogenesis, and infertility in both sexes [7]. In mutant males, spermatogenesis is blocked at the round spermatid stage and Leydig cell hypoplasia is prominent, resulting in a decrease in testicular size and testosterone levels. Mutant females show defects in folliculogenesis, including degenerating antral follicle and the absence of corpora lutea, although normal theca cells, resulting in a decrease in estradiol and progesterone levels. A similar phenotype of hypogonadism was observed in LHR knockout mice of both sexes [8,9].

Pathophysiological Implications

Clinical Implications

A surge of LH is tested to predict ovulation using urinary ovulation predictor kits. A single mutation (Gly 578 Asp) in the sixth transmembrane domain of LHR resulting in the constitutive activation of the LHR causes familial male precocious puberty [10]. A missense mutation (Ala 593 Pro) in the sixth transmembrane domain of LHR causes Leydig cell hypoplasia. Conditions with high LH levels include premature menopause, gonadal dysgenesis, Turner syndrome, castration, Swyer syndrome, polycystic ovary syndrome, certain forms of congenital adrenal hyperplasia, testicular failure, and pregnancy. Conditions with low FSH levels include Kallmann syndrome, hypothalamic suppression, hypopituitarism, eating disorder, female athlete triad, hyperprolactinemia, and hypogonadism.

Use for Diagnosis and Treatment

Recombinant human LH is used for the treatment of female infertility. Menotropins (human menopausal gonadotropin, hMG), a mixture of FSH and LH, are used to treat infertility in women. hCG derived from the urine of pregnant women is used as an LH substitute.

References

1. Reichert J, Parlow AF. Further studies on the purification of human pituitary luteinizing hormone. *Endocrinology*. 1964;75:815–817.
2. Talmadge K, Vamvakopoulos NC, Fiddes JC. Evolution of the genes for the beta subunits of human chorionic gonadotropin and luteinizing hormone. *Nature*. 1984;307:37–40.
3. Lapthorn AJ, Harris DC, Littlejohn A, et al. Crystal structure of human chorionic gonadotropin. *Nature*. 1994;369:455–461.
4. Ulloa-Aquirre A, Maldonado A, Damian-Matsumura P, et al. Endocrine regulation of gonadotropin glycosylation. *Arch Med Res*. 2001;32:520–532.
5. Yaron Z, Gur G, Melamed P, et al. Regulation of fish gonadotropins. *Int Rev Cytol*. 2003;225:131–185.
6. Dufau ML. The luteinizing hormone receptor. *Annu Rev Physiol*. 1998;60:461–496.
7. Ma X, Dong Y, Matzuk MM, et al. Targeted disruption of luteinizing hormone β-subunit leads to hypogonadism, defects in gonadal steroidogenesis, and infertility. *Proc Natl Acad Sci USA*. 2004;101:117294–117299.
8. Zhang F-P, Poutanen M, Wilbertz J, et al. Normal prenatal but arrested postnatal sexual development of luteinizing hormone receptor knockout (LuRKO) mice. *Mol Endocrinol*. 2001;15:172–183.
9. Lei ZM, Mishra S, Zou B, et al. Targeted disruption of luteinizing hormone/human chorionic gonadotropin receptor gene. *Mol Endocrinol*. 2001;15:184–200.
10. Shenker A, Laue L, Kosugi S, et al. A constitutively activating mutation of the luteinizing hormone receptor in familial male precocious puberty. *Nature*. 1993;365:652–654.

Supplemental Information Available on Companion Website

- Gene, mRNA, and precursor strucures of the human LH β-subunit/E-Figure 14B.1
- Precursor and mature hormone sequences of various animals/E-Figure 14B.2
- Primary structure of LHR of various animals/E-Figure 14B.3
- Accession numbers of genes and cDNAs for LH β-subunit and LHR/E-Tables 14B.1, 14B.2

Thyroid-Stimulating Hormone

Hironori Ando

Abbreviation: TSH
Additional names: thyrotropin, thyrotropic hormone, thyreotrophic hormone

A thyrotropic glycoprotein hormone secreted from the anterior pituitary, TSH stimulates thyroid growth and function and the synthesis and release of thyroid hormones.

Discovery

Bovine TSH was the first purified TSH, and its primary structure was determined in 1971 [1]. The structure and full nucleotide sequence of the human TSH β-subunit gene was determined in 1985 [2].

Structure

Structural Features

TSH is a glycoprotein consisting of non-covalently linked glycoprotein hormone (GPH) α- and TSH β-subunits. The GPH α-subunit is common to follicle-stimulating hormone (FSH), luteinizing hormone (LH), and chorionic gonadotropin (CG) (see Subchapter 14A). The TSH β-subunit contains a cysteine-knot motif, which is critical for the heterodimer assembly and its biological activity [3]. The *N*-linked oligosaccharide chain is important for the intracellular folding, secretion, metabolic clearance, and biological activity of the hormone.

Primary Structure

The primary structure of the human TSH β-subunit (119 aa residues) is shown in Figure 14C.1.

Properties

Mr 25,000–30,000, pI 6.8–8.5. Multiple isoforms exist due to the microheterogeneity of oligosaccharide chains. Soluble in water. Inactivated by heating; treatment with trypsin, chymotrypsin, and pepsin; and oxidation (potassium permanganate, elemental iodine).

Synthesis and Release

Gene, mRNA, and Precursor

The human TSH β-subunit gene, *TSHB*, location 1p13, consists of three exons (E-Figure 14C.1). Human TSH β-subunit mRNA has 867 b that encode a signal peptide of 19 aa residues and a mature protein of 119 aa residues (E-Figure 14C.2).

Distribution of mRNA

The TSH β-subunit mRNA is present in the basophilic gonadotropes.

Tissue and Plasma Concentrations

Tissue: Rat, female 182.9 (mU/pituitary).
  Plasma: Human, normal 0.3–5.0 (μU/ml); rat, normal male 146, cold-exposed male 351, normal female 17.9, pregnant female 30.3, thyrotropin-releasing hormone (TRH)-treated female 152; sheep, thyroidectomized female, 5 days after treatment 0.026 (μg NIH-TSH-S6/ml), 42 days after treatment 0.117; cow, normal 14.7 (μg NIH-TSH-S3/ml), TRH-treated 32.4.

Regulation of Synthesis and Release

The synthesis and release of TSH are directly stimulated by thyrotropin-releasing hormone (TRH). The administration of estrogen and insulin, and cold exposure, have stimulatory effects on TSH release. Thyroid hormones (T_3 and T_4) inhibit the synthesis and release of TSH through binding to thyroid receptor. Somatostatin and melanin-concentrating hormone (MCH) also inhibit TSH release.

Receptors

Structure and Subtype

The receptor of TSH (TSHR) is a glycoprotein that belongs to a subclass of the rhodopsin/β-adrenergic subfamily of the membrane-bound GPCR superfamily [4]. The TSHR consists of around 770 aa residues containing a large N-terminal extracellular domain (~410 aa residues), 7 transmembrane domains, and a C-terminal intracellular domain (E-Figure 14C.3). TSHR also acts as a receptor for thyrostimulin (GPA2 + GPB5, see Chapter 14).

Signal Transduction Pathway

The receptor is mainly coupled to G_s protein, and TSH stimulates production of cAMP.

Agonists

Recombinant human TSH and small molecule TSHR ligands [5] are agonists

Antagonists

Deglycosylated TSH [6] and small molecule TSHR antagonist [5] are antagonists.

Y. Takei, H. Ando, & K. Tsutsui (Eds): Handbook of Hormones. DOI: http://dx.doi.org/10.1016/B978-0-12-801028-0.00131-8

Human TSH β-subunit
```
        10        20        30        40        50        60
SFCIPTEYTMHIERRECAYCLTINTTICAGYCMTRDINGKLFLPKYALSQDVCTYRDFIY
        70        80        90       100       110       120
RTVEIPGCPLHVAPYFSYPVALSCKCGKCNTDYSDCIHEAIKTNYCTKPQKSYLVGFSV
```

Figure 14C.1 The primary structure of the human TSH β-subunit. *N*-linked glycosylation sites are underlined and Cys residues are shaded.

Biological Functions

Target Cells/Tissues and Functions

TSH binds to the TSHR located on the membrane of thyroid follicular cells. TSH stimulates thyroid hormone production, iodine uptake and organification, and thyroid growth. TSHR is also expressed in a variety of extrathyroidal tissues, including fat, fibroblasts, bone, and cardiomyocytes.

Phenotype in Gene-Modified Animals

TSHR knockout mice exibit developmental and growth delays, and show severe hypothyroidism with no detectable thyroid hormone and elevated TSH [7]. TSHR knockout mice produce un-iodinated thyroglobulin, but the ability to organify iodide can be restored with culturing in the presence of forskolin. TSHR is required for expression of sodium-iodide symporter.

Pathophysiological Implications

Clinical Implications

Familial TSH deficiency with point mutations in *TSHB* and numerous mutations in the TSHR gene have been identified and associated with thyroid diseases [3]. Graves' disease is an autoimmune thyroid disorder in which the body produces antibodies to TSHR (TSAbs) that mimic the effects of TSH [8]. TSAbs induce increased metabolic rate, and the associated symptoms result in thyroid eye disease, exophthalmos (protuberance of one or both eyes), fatigue, weight loss with increased appetite, and other symptoms of hyperthyroidism.

Use for Diagnosis and Treatment

TSH levels are measured as part of a thyroid function test in patients suspected of hyperthyroidism and hypothyroidism. Recombinant human TSH is used in the follow-up of patients with thyroid cancer to ablate the thyroid remnant following thyroid cancer surgery, and to obtain reliable diagnostic test results for the recurrence of differentiated thyroid cancer.

References

1. Shome B, Liao TH, Howard SM, et al. The primary structure of bovine thyrotropin. I. Isolation and partial sequences of cyanogen bromide and tryptic peptides. *J Biol Chem.* 1971;246:833−849.
2. Hayashizaki Y, Miyai K, Kato K, et al. Molecular cloning of the human thyrotropin-beta subunit gene. *FEBS Lett.* 1985; 188:394−400.
3. Szkudlinski MW, Fremont V, Ronin C, et al. Thyroid-stimulating hormone and thyroid-stimulating hormone receptor structure−function relationships. *Physiol Rev.* 2001;82:473−502.
4. Kleinau G, Krause G. Thyrotropin and homologous glycoprotein hormone receptors: structural and functional aspects of extracellular signaling mechanisms. *Endocr Rev.* 2009;30:133−151.
5. Gershengorn MC, Neumann S. Update in TSH receptor agonists and antagonists. *J Clin Endocrinol Metab.* 2012;97:4287−4292.
6. Fares FA, Levi F, Reznick AZ, et al. Engineering a potential antagonist of human thyrotropin and thyroid-stimulating antibody. *J Biol Chem.* 2001;276:4543−4548.
7. Marians RC, Ng L, Blair HC, et al. Defining thyrotropin-dependent and -independent steps of thyroid hormone synthesis by using thyrotropin receptor-null mice. *Proc Natl Acad Sci USA.* 2002;99:15776−15781.
8. Michalek K, Morshed SA, Latif R, et al. TSH receptor autoantibodies. *Autoimmun Rev.* 2009;9:113−116.

Supplemental Information Available on Companion Website

- Gene, mRNA, and precursor strucures of the human TSH β-subunit/E-Figure 14C.1
- Precursor and mature hormone sequences of various animals/E-Figure 14C.2
- Primary structure of TSHR of various animals/E-Figure 14C.3
- Accession numbers of genes and cDNAs for TSH β-subunit and TSHR/E-Tables 14C.1, 14C.2

CHAPTER 15

Growth Hormone/Prolactin Family

Tatsuya Sakamoto

History

Growth hormone (GH) was first isolated in 1944, and the primary structures of bovine and ovine GHs were determined in early 1970s [1]. Prolactin (PRL) was discovered in rabbits and pigeons as the milk-secreting factor around 1930, and later confirmed in humans in 1970 [2]. Somatolactin (SL) was first isolated and characterized in the early 1990s [3]. So far, SL has been found only in fish [4]. In addition, placental lactogen (PL), a mammalian polypeptide placental hormone related to GH and PRL, was identified in humans in 1963.

Structure

Structural Features

GH, PRL, SL, and PL form a family of polypeptide hormones, despite the low sequence identity between PL and the other three hormones of only approximately 25% [5]. These hormones contain highly conserved intramolecular disulfide bonds that are important for biological activity.

Molecular Evolution of Family Members

The chromosomal location of the genes and phylogenetic analysis suggest that, in vertebrate evolution, the genes encoding the family members have gone through a complex expansion involving several duplication events, including local and whole genome duplications (Figures 15.1, 15.2).

The GH and PRL genes are present in both ray-finned and lobe-finned fishes and tetrapods, while the third member, SL, is found only in ray-finned fishes and lungfishes [6]. For receptors, it has been shown that the GH receptor became duplicated in the proposed teleost-specific tetraploidization (3R), giving rise to the SL receptor, but that the emergence of SL predated this event [7]. Also, a second PRL gene in assembled fish genomes has been identified and its origin has been suggested to be through the duplication of a large chromosome block, possibly through a whole genome duplication event; however, its time point is still unclear [4].

Receptors

Structure and Subtypes

The receptors for these family members belong to the large superfamily of class 1 cytokine receptors [8]. They consist of an extracellular region that binds hormones, a transmembrane region, and a cytoplasmic region.

Signal Transduction Pathway

The hormones produce their biological effects through dimerizing with specific single transmembrane-domain receptors. These family members bind their receptors and activate the JAK/STAT cascade as a major signaling pathway.

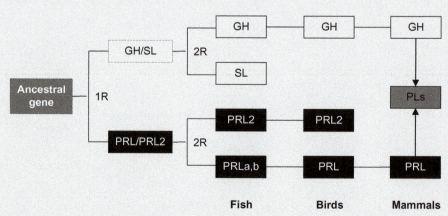

Figure 15.1 History of diversification of the GH/PRL family in vertebrates [4].

Y. Takei, H. Ando, & K. Tsutsui (Eds): Handbook of Hormones. DOI: http://dx.doi.org/10.1016/B978-0-12-801028-0.00015-5

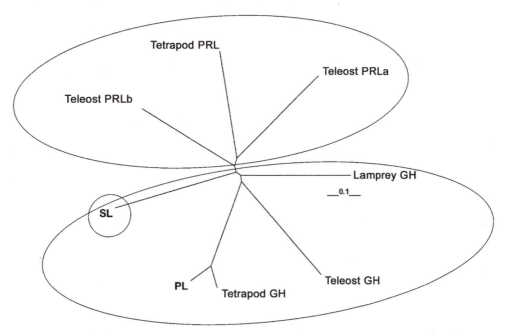

Figure 15.2 The unrooted phylogenetic tree of GH/PRL family members constructed using sequences from human and medaka[4].

Biological Functions

Target Cells/Tissues and Functions

The principal biological role of GH, the control of postnatal growth, has remained quite consistent throughout vertebrate evolution and is largely mediated by insulin-like growth factors (IGFs). PRL has many and diverse roles, including control of water and salt balance. SL has a wide range of biological activities, including body-color regulation. PL modifies the metabolic state of the mother during pregnancy to facilitate the energy supply of the fetus. Correlation between the molecular evolution and functional evolution of these hormones is unlikely. The functional evolution of these hormones may involve changes in regulatory elements of their gene promoters, such as altering tissue of origin and posttranscriptional processing, rather than changes in the protein structure.

References

1. Kopchick JJ. History and future of growth hormone research. *Horm Res.* 2003;60(Suppl 3):103−112.
2. Bole-Feysot C, Goffin V, Edery M, et al. Prolactin (PRL) and its receptor: actions, signal transduction pathways and phenotypes observed in PRL receptor knockout mice. *Endocr Rev.* 1998;19:225−268.
3. Ono M, Takayama Y, Rand-Weaver M, et al. cDNA cloning of somatolactin, a pituitary protein related to growth hormone and prolactin. *Proc Natl Acad Sci USA.* 1990;87:4330−4334.
4. Ocampo Daza D, Sundstrom G, Larsson TA, et al. Evolution of the growth hormone-prolactin-somatolactin system in relation to vertebrate tetraploidizations. *Ann N Y Acad Sci.* 2009;1163:491−493.
5. Forsyth IA, Wallis M. Growth hormone and prolactin − molecular and functional evolution. *J Mammary Gland Biol Neoplasia.* 2002;7:291−312.
6. Amemiya Y, Sogabe Y, Nozaki M, et al. Somatolactin in the white sturgeon and African lungfish and its evolutionary significance. *Gen Comp Endocrinol.* 1999;114:181−190.
7. Fukamachi S, Meyer A. Evolution of receptors for growth hormone and somatolactin in fish and land vertebrates: lessons from the lungfish and sturgeon orthologues. *J Mol Evol.* 2007;65:359−372.
8. Huising MO, Kruiswijk CP, Flik G. Phylogeny and evolution of class-I helical cytokines. *J Endocrinol.* 2006;189:1−25.

Growth Hormone

Shunsuke Moriyama

Abbreviation: GH
Additional names: somatotropic hormone, somatotropin

Secreted by the anterior pituitary in all classes of vertebrates, growth hormone exhibits a wide range of biological activities, including somatic growth, energy metabolism, sexual maturation, and immune functions. In euryhaline teleosts, GH is involved in seawater adaptability.

Discovery

Pituitary gland-derived growth promoting activity was discovered in 1921, and the first GH was isolated in 1944. The primary structure of human GH was first proposed in 1969 [1]. Subsequently, several corrections have been added to this sequence, and its primary structure was determined in 1971. During that same period, the structure of bovine and ovine GH was also determined. In 2002 the structure of GH was determined in the sea lamprey, an extant representative of the Agnatha, a group of the most ancient vertebrates, but it is not yet determined in hagfish [2].

Structure

Structural Features

GH is a single-chain polypeptide. Human GH contains two highly conserved intramolecular disulfide bonds that are important for biological activity.

Primary Structure

GH consists of 178−191 aa residues with two disulfide bonds (Figure 15A.1). In ostariophysan fish, GH contains an additional unpaired cysteine residue.

Properties

Mr 20,000−22,000, pI 4.9−6.8. Soluble in weak acidic or alkaline buffers; insoluble in water, alcohol, acetone, benzene, and chloroform. Human GH is stable in 10-M urea for 24 hours at room temperature, and stable at 100°C for 10 minutes.

Synthesis and Release

Gene, mRNA, and Precursor

In humans, the q22−24 region of chromosome 17 contains two GH genes (*GH1*, *GH2*) interspersed with three placental lactogen (PL) genes. *GH1* has a Pit-1/GHF-1 regulatory element in the promoter region. In fish, there are two types of exon−intron structures: a five-exon type in the Cypriniformes, Siluriformes, Chondrichthyes, and Agnatha, and a six-exon type in the Salmoniformes, Perciformes, and Tetradontiformes. The six-exon type has an additional intron inserted in Exon 5 of the six-exon type (E-Figure 15A.1) [3,4]. Human GH mRNA has 804 bp that encode a signal peptide of 26 aa residues and a mature protein of 191 aa residues.

Tissue Distribution of mRNA

GH gene is expressed in the acidophilic somatotropes in the pituitary. In sea lamprey, GH cells are localized in the dorsal half of the proximal pars distalis of the pituitary.

Tissue and Plasma Concentrations

Tissue: Human, 60 mg/g pituitary; rainbow trout (freshwater), 20 mg/g pituitary.
   Plasma: Human, 5 ng/ml; rainbow trout (freshwater), 5 ng/ml.

Regulation of Synthesis and Release

Synthesis and release of GH is stimulated by growth hormone-releasing hormone and ghrelin, and is inhibited by somatostatin [5]. Gonadotropin-releasing hormone, thyroid hormone, cortisol, insulin-like growth factor (IGF)-I, and activin are also involved in the regulation of GH synthesis and release.

Receptors

Structure and Subtypes

The GH receptor (GHR) belongs to the type I cytokine receptor family [6]. Human GHR consists of 638 aa residues that contain single extracellular, transmembrane, and intracellular domains. The extracellular domain of GHR has conserved cysteine residues and two antiparallel β-sheets holding the GH binding site in between, while the intracellular domain has two conserved Pro-rich sequences, Box 1 and Box 2 (E-Figure 15A.6).

Signal Transduction Pathway

The initial step in GH signaling is the dimerization of two GHRs [6], which leads to the activation of JAK2 by tyrosine phosphorylation. The activated JAK2 then interacts with and directly activates other signaling molecules, such as STAT, SHC, FAK, and PLC. GH also activates the MAPK pathway.

Agonists

Purified and recombinant human GH are agonists.

Y. Takei, H. Ando, & K. Tsutsui (Eds): Handbook of Hormones. DOI: http://dx.doi.org/10.1016/B978-0-12-801028-0.00132-X

```
        10        20        30        40        50        60
FPTIPLSRLFDNAMLRAHRLHQLAFDTYQEFEEAYIPKEQKYSFLQNPQTSLCFSESIPT
        70        80        90       100       110       120
PSNREETQQKSNLELLRISLLLIQSWLEPVQFLRSVFANSLVYGASDSNVYDLLKDLEEG
       130       140       150       160       170       180
IQTLMGRLEDGSPRTGQIFKQTYSKFDTNSHNDDALLKNYGLLYCFRKDMDKVETFLRIV
       190
QCRSVEGSCGF
```

Figure 15A.1 The primary structure of human GH (191 aa residues). Cysteine residues are shaded.

Antagonists

Antibodies to GHR, and pegvisomant (GHR antagonist), are antagonists.

Biological Functions

Target Cells/Tissues and Functions

In various vertebrate species, GH binds to GHR located in the liver, muscle, adipose tissue, mammary gland, bone, kidney, and embryonic stem cells. GHR is also localized in the immune system [6]. Many of the actions of GH on somatic growth are mediated by IGF. GH stimulates absorption of amino acids and protein synthesis, and is an important regulator of blood glucose and amino acid utilization. GH also contributes to sexual maturation, gametogenesis, and gonadal steroidogenesis. In euryhaline fish, GH is essential for seawater adaptation.

Phenotype in Gene-Modified Animals

GH knockout mice show normal fetal growth and normal birth weight, but growth is impaired after birth [7]. GH transgenic animals, including frog and fish, achieve accelerated growth rates.

Pathophysiological Implications

Clinical Implications

GH deficiency is a medical condition caused by problems arising in the pituitary that prevent the production of GH, commonly in children manifested as growth failure and short stature. Whereas excess production is rare in childhood but causes gigantism, it is common after epiphyseal plate closure and results in acromegaly with increased size of digits.

Use for Diagnosis and Treatment

GH deficiency is treated with purified and recombinant human GH [8]. Acromegaly is treated with somatostatin analogs, dopamine antagonists, and GHR antagonists.

References

1. Li CH, Dixon JS, Liu WK. Human pituitary growth hormone. XIX. The primary structure of the hormone. *Arch Biochem Biophys*. 1969;133:70–91.
2. Kawauchi H, Suzuki K, Yamazaki T, et al. Identification of growth hormone in the sea lamprey, an extant representative of a group of the most ancient vertebrates. *Endocrinology*. 2002;143: 4916–4921.
3. Kawauchi H, Sower SA. The dawn and evolution of hormones in the adenohypophysis. *Gen Comp Endocrinol*. 2006;148:3–14.
4. Moriyama S, Oda M, Takahashi A, et al. Genomic structure of the sea lamprey growth hormone-encoding gene. *Gen Comp Endocrinol*. 2006;148:33–40.
5. Pombo M, Pombo CM, Garcia A, et al. Hormonal control of growth hormone secretion. *Horm Res*. 2001;55:11–16.
6. Kopchick JJ, Andry JM. Growth hormone (GH), GH receptor, and signal transduction. *Mol Genet Metab*. 2000;71:293–314.
7. Bartke A, Chandrashekar V, Turyn D, et al. Effects of growth hormone overexpression and growth hormone resistance on neuroendocrine and reproductive functions in transgenic and knock-out mice. *Proc Soc Exp Biol Med*. 1999;222:113–123.
8. Molitch ME, Clemmons DR, Malozowski S, et al. Evaluation and treatment of adult growth hormone deficiency: an Endocrine Society clinical practice guideline. *J Clin Endocrinol Metab*. 2011;96:1587–1609.

Supplemental Information Available on Companion Website

- Gene, mRNA, and precursor structure of human GH/E-Figure 15A.1
- GH gene structure of the various animals/E-Figure 15A.2
- Precursor and mature hormone sequences of various animals/E-Figure 15A.3
- Gene, mRNA, and precursor structure of human GHR/E-Figure 15A.4
- Primary structure of human GHR/E-Figure 15A.5
- Primary structure of GHR of various animals/E-Figure 15A.6
- Accession numbers of genes and cDNAs for GH and GHR/E-Tables 15A.1, 15A.2

Prolactin

Tatsuya Sakamoto

Tatsuya Sakamoto

Abbreviation: PRL
Additional names: luteotropic hormone, luteotropin, mammotropin, lactogen, lactogenic hormone, galactin

Prolactin is best known for its role in enabling female mammals to produce milk, and in a variety of biological functions in vertebrates, including an essential role in the control of water and salt balance, metabolism, and regulation of the immune system [1].

Discovery

PRL was first discovered in rabbits and pigeons as the milk-secreting factor around 1930, and confirmed in humans in 1970.

Structure

Structural Features

PRL consists of about 200 aa residues in most species. In tetrapods, sturgeon, and lungfish, PRL has three disulfide bonds, the first forming a small loop near the N-terminus, the second linking distant parts of the polypeptide chain, and the third forming a loop close to the C-terminus (Figure 15B.1). In Halecomorphi (teleosts and the Amiiformes), PRL lacks the first disulfide bond. Molecular heterogeneity in regard to glycosylation, phosphorylation, and sulfation has been described. The non-glycosylated form of PRL is the dominant form of PRL that is secreted from the pituitary.

Primary Structure

The evolution of PRL in mammals shows a similar pattern to that for GH, with a fairly slow basal evolutionary rate but accelerated evolution in lineages leading to humans, ruminants, and rodents (E-Figure 15B.1). In fishes, including teleosts and sturgeon, PRL shows a great deal of structural variation as seen for GH [2].

Properties

Mr ~23,000; pI 6.5 (human), 5.7 (sheep), 5.9 (pig), 5.1 (rat), 6.0 (salmon). There is a tendency to form dimers. Unstable in solution, even at low temperatures.

Synthesis and Release

Gene, mRNA, and Precursor

The human PRL gene, *PRL*, location 6p22.3, consists of five exons (E-Figure 15B.1). Alternative splicing gives rise to variant mRNAs and therefore proteins.

Distribution of mRNA

PRL mRNA is detected most abundantly in the acidophilic lactotropes in the anterior pituitary. Transcripts are also found in the decidua, myometrium, breast, lymphocytes, leukocytes, intestine, and prostate [3].

Tissue and Plasma Concentrations

Tissue: Human, male 1.5 (mg/g pituitary), female 1.3.
   Plasma: In humans, the upper threshold of normal PRL is 25 ng/ml for women and 20 ng/ml for men [4]. PRL deficiency (hypoprolactinemia) is defined as PRL levels below 3 ng/ml for women and 5 ng/ml for men. PRL levels vary with age, sex, menstrual cycle stage, and pregnancy. In rats, the levels increase during proestrus from 10 to 60 ng/ml.

Regulation of Synthesis and Release

Pit-1 is a transcription factor that binds to several sites in the promoter of the PRL gene to allow for synthesis of PRL in the anterior pituitary. Most extrapituitary production of PRL is controlled by a superdistal promoter [3]. Neurosecretory dopamine neurons inhibit PRL secretion via D2 receptors. Estrogens are also key regulators of PRL production by enhancing PRL cell growth as well as by stimulating PRL production directly. Thyrotropin-releasing hormone and PRL-releasing peptide-2 have a stimulatory effect on PRL release. Vasoactive intestinal peptide regulates prolactin secretion in humans.

Receptors

Structure and Subtype

PRL receptor belongs to the type I cytokine receptor family, and consists of an extracellular region that binds PRL, a transmembrane region, and a cytoplasmic region. There is a variation of PRL receptor isoforms among different tissues due to expression from multiple promoters and alternative splicing [5].

Signal Transduction Pathway

When PRL binds to its receptor, it causes dimerization of the receptor. This results in the activation of JAK2, a tyrosine

Y. Takei, H. Ando, & K. Tsutsui (Eds): Handbook of Hormones. DOI: http://dx.doi.org/10.1016/B978-0-12-801028-0.00133-1

```
        10         20         30         40         50         60
LPICPGGAARCQVTLRDLFDRAVVLSHYIHNLSSEMFSEFDKRYTHGRGFITKAINSCHT
        70         80         90        100        110        120
SSLATPEDKEQAQQMNQKDFLSLIVSILRSWNEPLYHLVTEVRGMQEAPEAILSKAVEIE
       130        140        150        160        170        180
EOTKRLLEGMELIVSQVHPETKENEIYPVWSGLPSLQMADEESRLSAYYNLLHCLRRDSH
       190        200
KIDNYLKLLKCRIIHNNNC
```

Figure 15B.1 Primary structure of human PRL (199 aa residues).

kinase that initiates the JAK/STAT pathway. Activation of the PRL receptor also results in the activation of MAPK and Src kinase [4].

Antagonists

Various candidates have been generated by different laboratories, but none has yet entered clinical trials. One of them, namely Δ1–9-G129R-hPRL, appears to be promising, as it is the only one displaying both PRL receptor specificity and pure antagonism [6].

Biological Functions

Target Cells/Tissues and Functions

PRL receptors are present in various tissues, including the mammary glands, ovaries, pituitary, heart, lung, thymus, spleen, liver, pancreas, kidney, adrenal gland, uterus, skeletal muscle, skin, gill, and central nervous system [4]. Although often associated with milk production, PRL plays a wide range of physiological roles in humans and other vertebrates. Control of water and salt balance is an important function, especially in teleost fishes. PRL controls the levels of sex steroids as a gonadotropic hormone. PRL has important cell cycle-related functions, acting as growth, differentiating, and anti-apoptotic factors, especially in hematopoiesis, angiogenesis, and the immune system [1]. Prolactin also stimulates proliferation of oligodendrocyte precursor cells that differentiate into oligodendrocytes responsible for the formation of myelin coating in the central nervous system, and may control behavior [7].

Phenotype in Gene-Modified Animals

PRL-related knockout models have mainly highlighted its irreplaceable role in lactation and reproduction, suggesting that most of its other target tissues are presumably modulated by PRL rather than strictly dependent on it [8].

Pathophysiological Implications

Clinical Implications

Hyperprolactinemia or excess serum PRL is associated with hypoestrogenism, anovulatory infertility, oligomenorrhoea, amenorrhoea, unexpected lactation, and loss of libido in women, and erectile dysfunction and loss of libido in men. Hypoprolactinemia is associated with ovarian dysfunction in women, and metabolic syndrome, anxiety, arteriogenic erectile dysfunction, premature ejaculation, oligozoospermia, asthenospermia, hypofunction of seminal vesicles, and hypoandrogenism in men. In hypoprolactinemic men, normal sperm characteristics were restored when PRL levels were brought up to normal values.

Use for Diagnosis and Treatment

PRL levels are checked as part of a sex hormone work-up, because elevated PRL secretion can suppress the secretion of follicle-simulating hormone (FSH) and gonadotropin-releasing hormone (GnRH), leading to hypogonadism and erectile dysfunction in men. Serum PRL levels are of some use in distinguishing epileptic seizures from psychogenic non-epileptic seizures, as the levels rise following an epileptic seizure.

References

1. Bole-Feysot C, Goffin V, Edery M, et al. Prolactin (PRL) and its receptor: actions, signal transduction pathways and phenotypes observed in PRL receptor knockout mice. *Endocr Rev.* 1998;19:225–268.
2. Forsyth IA, Wallis M. Growth hormone and prolactin – molecular and functional evolution. *J Mammary Gland Biol Neoplasia.* 2002;7:291–312.
3. Gerlo S, Davis JR, Mager DL, et al. Prolactin in man: a tale of two promoters. *Bioessays.* 2006;28:1051–1055.
4. Mancini T, Casanueva FF, Giustina A. Hyperprolactinemia and prolactinomas. *Endocrinol Metab Clin North Am.* 2008;37:67–99.
5. Binart N, Bachelot A, Bouilly J. Impact of prolactin receptor isoforms on reproduction. *Trends Endocrinol Metab.* 2010;21:362–368.
6. Tallet E, Rouet V, Jomain JB, et al. Rational design of competitive prolactin/growth hormone receptor antagonists. *J Mammary Gland Biol Neoplasia.* 2008;13:105–117.
7. Gregg C, Shikar V, Larsen P, et al. White matter plasticity and enhanced remyelination in the maternal CNS. *J Neurosci.* 2007;27:1812–1823.
8. Goffin V, Binart N, Touraine P, et al. Prolactin: the new biology of an old hormone. *Annu Rev Physiol.* 2002;64:47–67.

Supplemental Information Available on Companion Website

- Gene structure of human PRL/E-Figure 15B.1
- Primary structure of PRL of various animals/E-Figure 15B.2
- Primary structure of the rat PRL receptor/E-Figure 15B.3
- Primary structure of the non-mammalian PRL receptor/E-Figure 15B.4
- Accession numbers of gene and cDNAs for PRL and PRL receptor/E-Table 15B.1

Somatolactin

Sho Kakizawa

Abbreviation: SL

Secreted by the pars intermedia of fish pituitary, somatolactin is involved in a wide range of physiological functions, including sexual maturation, plasma ion metabolism, acid—base regulation. and body-color regulation.

Discovery

SL was first isolated and characterized from cod in the early 1990s [1], during the course of characterizing growth hormone. In the first decade of this century, two distinct forms coded by separate genes were reported from several teleosts [2]. So far, SL has only been found in fish (teleosts, sturgeon, and lungfish), and there are no reports in other vertebrates.

Structure

Structural Features

SL belongs to the class I helical cytokine family. Two types of SLs, SLα and SLβ, are coded by separate genes. SLα is widely present in fish, whereas SLβ has been found in a limited number of fish species. Cod SL (SLα) contains eight cysteine residues in mature hormone, six of which form disulfide bonds (Figure 15C.1). Asparagine at position 147 is N-glycosylated. The mature zebrafish SLβ has six cysteine residues.

Primary Structure

The amino acid sequences of cod SLα (235 aa residues), zebrafish SLα (233 aa residues), and zebrafish SLβ (230 aa residues) are shown in Figure 15C.1. The primary structure of SLα and SLβ is conserved in fish (E-Figures 15C.3, 15C.4).

Properties

Mr 23,000—26,000; pI 5.7—5.9 (cod SL) [1], 5.2 (zebrafish SLβ) [2]. Soluble in alkaline buffers.

Synthesis and Release

Gene, mRNA, and precursor

The zebrafish SLα gene, *smtla*, location 18, consists of five exons (E-Figure 15C.1). Zebrafish SLα mRNA has 778 bp that encode a signal peptide of 26 aa residues and a mature protein of 209 aa residues. The zebrafish SLβ gene, *smtlb*, location 10, consists of five exons (E-Figure 15C.2). Zebrafish SLβ mRNA (854 bp) encodes a signal peptide of 24 aa residues and a mature protein of 206 aa residues.

Tissue Distribution of mRNA

The SL gene is expressed in the pars intermedia in the pituitary. In zebrafish, SLα and SLβ mRNA are expressed in different cells in the pars intermedia.

Tissue and Plasma Concentrations

Tissue: Cod, 1.0 (mg/g pituitary); salmon, 0.4.
  Plasma: Salmon, mature female ~40 ng/ml.

Regulation of Synthesis and Release

SL synthesis and release are affected by dopamine, epinephrine, serotonin, corticotropin-releasing hormone, gonadotropin-releasing hormone, melanin-concentrating hormone, adenylate cyclase-activating polypeptide, leptin, neuropeptide Y, and SL itself. Synthesis and plasma concentration of SL are also affected by sexual maturation, low environmental calcium, salinity and pH, acute stress, seasonal rhythms, background color, and plasma acidosis and bicarbonate concentration [3—7].

Receptors

Structure and Subtypes

A single type of SL receptor (SLR), belonging to the type I cytokine receptor family, is reported in teleosts, but not in sturgeon and lungfish [8]. In some teleost species, SLR was originally identified as a type 1 GH receptor with binding activity to GH but at a lower level than to SL [8,9]. Salmon SLR consists 657 aa residues that contain a signal peptide of 20 aa residues, an FGEFS motif, seven cysteine residues, a single transmembrane region, and Box 1 and 2 regions in the intracellular domain [9].

Signal Transduction Pathway

Carp SLα and SLβ activate the JAK2/STAT5, PI3K/Akt, and MAPK signaling pathways.

Biological Functions

Target Cells/Tissues and Functions

In various fish species, the expression of the SLR gene is observed in the liver, visceral fat, muscle, skin, pituitary, brain, heart, spleen, testis, ovary, gill, stomach, gall bladder, intestine, and kidney [9,10]. SL is involved in sexual maturation, plasma ion and pH regulation, lipid metabolism, and body-color regulation (Table 15C.1 [7,11—14]). Although changes in plasma concentration and expression level of SL are reported in a number of studies, the effects of administration and/or deficiency of SL have been demonstrated in few.

Y. Takei, H. Ando, & K. Tsutsui (Eds): Handbook of Hormones. DOI: http://dx.doi.org/10.1016/B978-0-12-801028-0.00134-3

```
           10        20        30        40        50        60
cSLα  MHTLAAVVVLQVCWAAVLWPCPPTHSSPVDCREEQAGSSQCPTISQEKLLDRVIQHTELI
zSLα  MNTVKVLQVCVCVLILGQFLVSGAVPLDCKDDAG  SRCASISQEKLLDRVIQHAELI
zSLβ  MKKATALQVCV VVLVCLKKAVIGSPVECPDQEIGG TSCTISAEKLLDRAVQHAELI
           70        80        90       100       110       120
cSLα  YRVSEESCSMFEDMFVPFPVRLQRNQAGNTCITKDFPIPTSKNELQQISDTWLLHSVLML
zSLα  YRVSEECCTLFEDMFVPYPLHVLINQAGNTCHSKHIPIPTSKSEIQQISDSWLLHSVLFL
zSLβ  YRFSEEAKLLFDELLDLFGGMNQYIPGGAVCAPKTVPVPLSKSEIQQISDRWLLHSVLIL
          130       140       150       160       170       180
cSLα  VQSWIEPLVYLQTTLDRYDDVPDVLLNKTKWMSEKLISLEQGVVVLIRKMLDGAILNSS
zSLα  VQSWMEPLLYLQTTLDRYDDAPNALLSKTKWVSDKLLSLEQGVVVLIRKMLDEGIMASST
zSLβ  VQFWIEPLVNLQKSLENYKSAPIGLLSRNQWIASKLSSLEEGLLVLIRQILGEGGLVL
          190       200       210       220       230
cSLα  YNEYSAVQLDVQ PEVLESILRDYNVLCCFKKDAHKIETILKLLKCRQIDKYNCALY
zSLα  IFDHTQSPYDGQFPEVLESVIRDYHLLTCFKKDTHKMETFLKLLKCRQSNKLSCLPQ
zSLβ  GPPEDVSDNTLS VDAFETVRRDYSVIYCFRKDAHKIQTFLKLLKCRQVDRENCSLF
```

Figure 15C.1 The amino acid sequences of cod SLα (cSLα), zebrafish SLα (zSLα), and zebrafish SLβ (zSLβ). The signal peptide is in gray lettering, the N-linked glycosylation site is underlined, and cysteine residues are shaded.

Phenotype in Gene-Modified Animals

The mutant medaka "*color interfere*," having a mutated SL gene with 11-bp deletion, shows gray skin (instead of brown skin in wild type), increased lipids (triglycerides and cholesterols) in the liver/muscle, and decreased plasma cortisol levels [11].

References

1. Rand-Weaver M, Noso T, Muramoto K, et al. Isolation and characterization of somatolactin, a new protein related to growth hormone and prolactin from Atlantic cod (*Gadus morhua*) pituitary glands. *Biochemistry*. 1991;30:1509−1515.
2. Zhu Y, Stiller JW, Shaner MP, et al. Cloning of somatolactin alpha and beta cDNAs in zebrafish and phylogenetic analysis of two distinct somatolactin subtypes in fish. *J Endocrinol*. 2004;182:509−518.
3. Kawauchi H, Sower SA. The dawn and evolution of hormones in the adenohypophysis. *Gen Comp Endocrinol*. 2006;148:3−14.
4. Kaneko T. Cell biology of somatolactin. *Int Rev Cytol*. 1996;169:1−24.
5. Azuma M, Suzuki T, Mochida H, et al. Pituitary adenylate cyclase-activating polypeptide (PACAP) stimulates release of somatolactin (SL)-α and SL-β from cultured goldfish pituitary cells via the PAC_1 receptor-signaling pathway, and affects the expression of SL-α and SL-β mRNAs. *Peptides*. 2013;43:40−47.
6. Kakizawa S, Kaneko T, Hirano T. Effects of hypothalamic factors on somatolactin secretion from the organ-cultured pituitary of rainbow trout. *Gen Comp Endocrinol*. 1997;105:71−78.
7. Kakizawa S, Ishimatsu A, Takeda T, et al. Possible involvement of somatolactin in the regulation of plasma bicarbonate for the compensation of acidosis in rainbow trout. *J Exp Biol*. 1997;200:2675−2683.

Table 15C.1 Physiological Functions of SL

Animal	Function	Reference
Medaka	Body-color regulation (gene mutation)	[11]
Rainbow trout	Acid−base regulation (immunoneutralization)	[7]
Rainbow trout	Lipid metabolism (fewer SL-producing cells)	[12]
Winter flounder	Phosphate metabolism (administration *in vitro*)	[13]
Coho salmon	Steroidogenesis (administration *in vitro*)	[14]

8. Fukamachi S, Meyer A. Evolution of receptors for growth hormone and somatolactin in fish and land vertebrates: lessons from the lungfish and sturgeon orthologues. *J Mol Evol*. 2007;65:359−372.
9. Fukuda H, Ozaki Y, Pierce AL, et al. Identification of the salmon somatolactin receptor, a new member of the cytokine receptor family. *Endocrinolgy*. 2005;146:2354−2361.
10. Pierce AL, Fox BK, Davis LK, et al. Prolactin receptor, growth hormone receptor, and putative somatolactin receptor in Mozambique tilapia: tissue specific expression and differential regulation by salinity and fasting. *Gen Comp Endocrinol*. 2007;154:31−40.
11. Fukamachi S, Sugimoto M, Mitani H, et al. Somatolactin selectively regulates proliferation and morphogenesis of neural-crest derived pigment cells in medaka. *Proc Natl Acad Sci USA*. 2004;101:10661−10666.
12. Kaneko T, Kakizawa S, Yada T. Pituitary of "cobalt" variant of the rainbow trout separated from the hypothalamus lacks most pars intermedial and neurohypophysial tissue. *Gen Comp Endocrinol*. 1993;92:31−40.
13. Lu M, Swanson P, Renfro JL. Effect of somatolactin and related hormones on phosphate transport by flounder renal tubule primary cultures. *Am J Physiol*. 1995;268:R577−R582.
14. Planas JV, Swanson P, Rand-Weaver MP, et al. Somatolactin stimulates in vitro gonadal steroidogenesis in coho salmon, *Oncorhynchus kisutch*. *Gen Comp Endocrinol*. 1992;87:1−5.

Supplemental Information Available on Companion Website

CHAPTER 16

Proopiomelanocortin Family

Akiyoshi Takahashi

History

Proopiomelanocortin (POMC) is the common precursor of adrenocorticotropic hormone (ACTH), α-melanocyte-stimulating hormone (α-MSH), and β-endorphin (β-END). Elucidation of the structure of POMC as a precursor of multiple peptide hormones has occurred over three major periods [1]. First, two groups of peptides were identified during the 1950s to 1970s. One group consists of ACTH, α-MSH, and corticotropin-like intermediate lobe peptide (CLIP). The amino acid sequences of the latter two peptides are identical to the N-terminal half and C-terminal half of ACTH, respectively. These interrelationships suggest that ACTH is a precursor of the small peptides. The second group consists of β-lipotropin (β-LPH), β-MSH, and β-END. The latter two peptides are identical to the middle and C-terminal portions of β-LPH, respectively, indicating that β-LPH is a precursor for β-MSH and β-END. During the second period, in the latter half of the 1970s, a single protein that contained ACTH and β-LPH was discovered. In these studies, the gene products of AtT20/D$_{16v}$ cells (a mouse pituitary cell line derived from ACTH cells) were subjected to double immunoprecipitation using antisera against ACTH and β-LPH or β-END. As a result, a protein that reacted to both antisera was identified. The precursor, named proopiocortin, was purified from the camel pituitary. During the third period, the complete cDNA nucleotide sequence for the common precursor of ACTH and β-LPH was obtained from the bovine pituitary pars intermedia [2]. That study showed that POMC consists of three segments: N-POMC, ACTH, and β-LPH. Characteristically, each part contains one MSH sequence, γ-MSH in N-POMC, α-MSH in ACTH, and β-MSH in β-LPH. The presence of γ-MSH was originally demonstrated by a cDNA cloning study. MSH and ACTH both contain the core sequence His−Phe−Arg−Trp, and are collectively called melanocortin. Thus, the name "proopiomelanocortin" for the common precursor was proposed [3].

Structure

Structural Features

POMC is composed of three segments: N-POMC, which is located at the N-terminus; ACTH, which is located in the middle; and β-LPH, which is located at the C-terminus (Figure 16.1). Each of these segments contains one MSH sequence: γ-MSH in N-POMC, α-MSH in ACTH, and β-MSH in β-LPH. β-LPH also contains an opioid peptide, β-END. The segments are flanked by dibasic amino acid sequences. POMC is synthesized in the pars distalis and pars intermedia of the pituitary. However, because of posttranslational processing that occurs in a tissue-specific manner, a number of final products can be produced from POMC (E-Figure 16.1). The major products in the pars distalis are ACTH and β-END, whereas the end-products in the pars intermedia are α-MSH, which corresponds to N-acetyl ACTH(1−13)-amide, and N-terminally acetylated and C-terminally truncated β-END. ACTH is the major end-product in the human pituitary, which lacks the pars intermedia in adulthood. ACTH and MSH are collectively called melanocortins (MCs) because of the presence of the common sequence His−Phe−Arg−Trp. In humans, POMC is located on chromosome 2 (2p23.3).

Molecular Evolution of Family Members

Although the tetrapod POMC contains three copies of MSH and one copy of β-END, in fish the number of copies of MSH in POMC varies depending on the taxonomic group (Figure 16.2). POMCs of lobe-finned fish and primitive ray-finned fish have three copies of MSH, namely α-MSH, β-MSH, and γ-MSH, while the more derived ray-finned fish POMC lacks γ-MSH, and the cartilaginous fish POMC has four copies of MSH. Moreover, lampreys have two different forms of POMC that contain either one or two copies of MSH. Thus, POMC is thought to have diverged by the insertion and deletion of an MSH segment during the early evolution of vertebrates.

Receptors

Structure and Subtype

The receptors for MC peptides belong to the GPCR superfamily, and are classified into five subtypes: MC1R, MC2R, MC3R, MC4R, and MC5R. MC1R is a classical α-MSH receptor, and MC2R is a classical ACTH receptor. In human chromosomes, MC1R is located at 16q24.3, MC2R at 18p11.2, MC3R at 20q13.2−q13.3, MC4R at 18q22, and MC5R at 18q11.2 [4].

Signal Transduction Pathway

The MCRs are the smallest known GPCRs and have short N- and C-terminal ends. They are all positively coupled to AC via G proteins, and mediate their effects primarily by activating the cAMP-dependent signaling pathway. MCRs display many features that are similar to other GPCRs, including potential N-glycosylation sites in their N-terminal domains, recognition sites for PKA and PKC phosphorylation, and conserved cysteine residues in their C-terminal ends [4]. See Chapter 7 for β-END.

Y. Takei, H. Ando, & K. Tsutsui (Eds): Handbook of Hormones. DOI: http://dx.doi.org/10.1016/B978-0-12-801028-0.00016-7

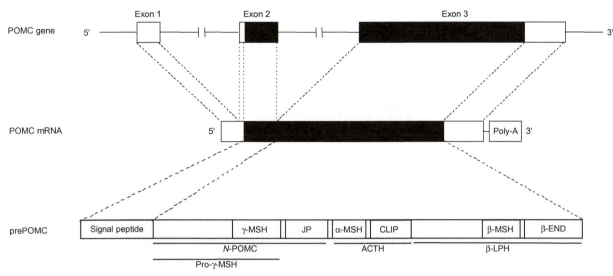

Figure 16.1 Schematic diagram for POMC gene, mRNA, and POMC in mammals. JP, joining peptide.

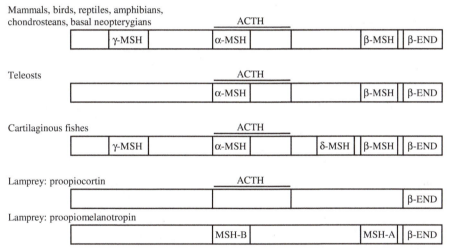

Figure 16.2 Outline of POMC structure.

Biological Functions

Target Cells/Tissues and Functions

The MCs are involved in a diverse number of physiological functions, including pigmentation, steroidogenesis, energy homeostasis, exocrine secretion, sexual function, analgesia, and inflammation [5]. MC1R is expressed on cutaneous melanocytes, where it has a key role in determining skin and hair pigmentation. MC1R is also expressed in leukocytes, where it may mediate anti-inflammation. MC2R is expressed in the adrenal cortex in the zona reticularis and zona fasciculata, where it mediates the effects of ACTH on steroid synthesis and secretion. MC3R is expressed in many areas of the central nervous system and peripheral tissues, and is involved in energy homeostasis. MC4R is expressed predominantly in the central nervous system, and regulates both food intake and sexual function. MC5R is expressed in numerous human peripheral tissues, and is mainly involved in exocrine function, particularly sebaceous gland secretion. β-END exhibits physiological functions via the μ-opioid receptor. See Chapter 7 for a detailed outline.

References

1. Takahashi A, Kawauchi H. Diverse structures and functions of melanocortin, endorphin and melanin-concentrating hormone in fish. In: Reinecke M, Zaccone G, Kapoor BG, eds. *Fish Endocrinology.* Enfield, NH, USA: Science Publishers; 2006:325−392.
2. Nakanishi S, Inoue A, Kita T, et al. Nucleotide sequence of cloned cDNA for bovine corticotropin-β-lipotropin precursor. *Nature.* 1979;278:423−427.
3. Chrétien M, Benjannet S, Gossard F, et al. From β-lipotropin to β-endorphin and 'pro-opio-melanocortin'. *Can J Biochem.* 1979;57:1111−1121.
4. Cooray SN, Clark AJL. Melanocortin receptors and their accessory proteins. *Mol Cell Endocrinol.* 2011;331:215−221.
5. Yang Y. Structure, function and regulation of the melanocortin receptors. *Eur J Pharmacol.* 2011;660:125−130.

Supplemental Information Available on Companion Website

- Post-translational processing of POMC/E-Figure 16.1
- Phylogenetic tree of POMC/E-Figure 16.2
- Accession numbers of genes and cDNAs for POMC/E-Table 16.1

Adrenocorticotropic Hormone

Akiyoshi Takahashi

Abbreviation: ACTH
Additional names: adrenocorticotrophin, corticotropin

A pituitary hormone secreted from the pars distalis, ACTH was one of the first adenohypophysial hormones demonstrated to be present in vertebrates from jawless fish to mammals, together with melanocyte-stimulating hormone (MSH).

Discovery

The fact that ACTH is derived from a precursor protein called proopiomelanocortin (POMC) was demonstrated in 1979 using the pars intermedia of the bovine pituitary.

Structure

Structural Features

ACTH consists of 39 aa residues and is composed of two regions corresponding to α-MSH and corticotropin-like intermediate lobe peptide (CLIP). These regions are connected by basic amino acid sequences. ACTH is generated from the common precursor POMC in the pars distalis of the pituitary. α-MSH and CLIP are generated in the pars intermedia of the pituitary in many vertebrates except for some mammals, such as humans, which lack the lobe in adults. ACTH is a type of melanocortin (MC), which has the common sequence His-Phe-Arg-Trp in the α-MSH segment. The first 24-aa sequence of ACTH, i.e., ACTH(1−24), is highly conserved in gnathostomes ranging from cartilaginous fishes to humans. Furthermore, mammalian ACTH(1−24) is equivalent in potency to full-length ACTH, ACTH(1−39). In the ACTH sequence the His−Phe−Arg−Trp domain is referred to as the "message" sequence, and the Lys−Lys−Arg−Arg domain is referred to as the "address" signal. ACTH(1−24) has full activity, yet ACTH(1−16) has minimal activity, and α-MSH has no activity in various adrenal cortex bioassay systems. ACTH (11−24) can function as an antagonist of ACTH(1−24) activation of adrenal cortex cells [1,2].

Primary Structure

Human ACTH is composed of 39 aa residues (Mr 4,541) (Figure 16A.1). The amino acid sequences of various vertebrate ACTHs are shown in E-Figure 16A.1.

Synthesis and Release

Gene, mRNA, and Precursor

The human POMC gene, *POMC*, location 2p23.3, contains three exons. *POMC* transcription requires cooperation between Tpit and Pitx 1, which bind to contiguous sites within the same regulatory element [3].

Distribution of mRNA

Although *POMC* mRNA is expressed in the pituitary gland and other tissues, ACTH is generated from POMC by post-translational processing in corticotropes in the pars distalis of the pituitary in normal subjects.

Tissue Content

The average level of plasma ACTH in normal subjects is 72.0 pg/ml. The plasma ACTH concentration undergoes circadian variation, with highest levels in the early morning [4].

Regulation of Synthesis and Release

The release of ACTH from corticotropes is primarily stimulated by corticotropin-releasing hormone (CRH), which is produced in the hypothalamus. ACTH release is also positively stimulated by arginine vasopressin (AVP). CRH and AVP co-localize in hypothalamic neurons. ACTH release is also stimulated by neurotransmitters such as norepinephrine, epinephrine, serotonin, and acetylcholine, and neuropeptides such as oxytocin, vasoactive intestinal peptide, peptide histidine isoleucine, and angiotensin-II. ACTH release is suppressed by negative feedback effects on corticotropes and CRH neurons by glucocorticoids from the adrenal cortex [5].

Receptors

Structure

ACTH interacts with all fives subtypes of melanocortin receptors (MC1R−MC5R), which are members of the GPCR family. Among them, MC2R specifically binds ACTH; thus, MC2R is a classical ACTH receptor. MC2R is retained at the endoplasmic reticulum and is unable to traffic to the cell surface in the absence of MC receptor accessory protein (MRAP). MRAP assists MC2R trafficking to the cell surface. Once at the cell surface, MRAP enhances MC2R signaling [6].

Y. Takei, H. Ando, & K. Tsutsui (Eds): Handbook of Hormones. DOI: http://dx.doi.org/10.1016/B978-0-12-801028-0.00135-5

SYSMEHFRWG KPVGKKRRPV KVYPNGAEDE SAEAFPLEF
α-MSH   CLIP

Figure 16A.1 Amino acid sequence of human ACTH. α-MSH and CLIP are shaded. Spaces are inserted every ten aa.

Signal Transduction

ACTH activates the AC/PKA pathway via G proteins. See also Chapter 16.

Agonists

Cosyntropin and corticotropin zinc hydroxide are agonists.

Antagonists

ACTH(11−24) and ASIP [90−132(L89Y)] (inverse agonist) are antagonists.

Biological Functions

ACTH stimulates the release of glucocorticoids, which alter protein and carbohydrate metabolism, from the adrenal cortex where MC2R is expressed. ACTH also exhibits a variety of extra-adrenal physiological functions, such as lipolysis in fat tissues, immune modulation, and exocrine gland function via MC1R and MC3R−MC5R.

Phenotype in Gene-Deficiency

A congenital lack of POMC causes obesity and glucocorticoid deficiency. *Pomc*-deficient mice are hypersensitive to the adverse metabolic effects of glucocorticoids [7]. Mutation of *MC2R* is partly related to familial glucocorticoid deficiency (FGD; hereditary unresponsiveness to ACTH), which is an autosomal recessive disorder resulting from resistance to the action of ACTH on the adrenal cortex [8]. Manifestations of FGD include hyperpigmentation, hypoglycemia, failure to thrive, and recurrent infections. Mutations of MRAP also cause FGD, indicating that the mutated MRAP does not interact with MC2R.

Pathophysiological Implications

Clinical Implications

Cushing's syndrome refers to the manifestations of chronic glucocorticoid excess and may arise from various causes, including non-pituitary ACTH oversecretion. In Cushing's disease, pituitary ACTH oversecretion induces bilateral adrenocortical hyperplasia and excess production of corticosteroids such as cortisol, adrenal androgens, and 11-deoxycorticosterone. In this disease, the CRH receptor is overexpressed in corticotropes in comparison to normal pituitaries. Cells containing the glucocorticoid receptor are abundantly present. *MC2R* is predominantly expressed in the adrenal cortex. In Addison's disease, or primary adrenal insufficiency, steroid hormones such as cortisol and dehydroepiandrosterone concentrations are decreased due to adrenal destruction.

Use for Diagnosis and Treatment

Cosyntropin is used to diagnose adrenal insufficiency. Corticotropin zinc hydroxide is a drug for ulcerative colitis and other colonic disorders, and treatment of some collagen diseases. ACTH is used to treat the symptoms related to allergic disorders.

References

1. Dores RM, Baron AJ. Evolution of POMC: origin, phylogeny, post-translational processing, and the melanocortins. *Ann NY Acad Sci.* 2011;1220:34−48.
2. Dores RM, Liang L. Analyzing the activation of the melanocortin-2 receptor of tetrapods. *Gen Comp Endocrinol.* 2014;203:3−9.
3. Budry L, Couture C, Balsalobre A, et al. The Ets factor Etv1 interacts with Tpit protein for pituitary pro-opiomelanocortin (POMC) gene transcription. *J Biol Chem.* 2011;286:25387−25396.
4. Aronin N, Krieger DT. Measurements of ACTH and lipotropins. In: van Wimersma Greidanus TB, ed. *Frontiers of Hormone Research.* Vol 8. Basel: Karger; 1981:62−79.
5. Papadimitriou A, Priftis KN. Regulation of the hypothalamic−pituitary−adrenal axis. *Neuroimmunomodulation.* 2009;16:265−271.
6. Cooray SN, Clark AJL. Melanocortin receptors and their accessory proteins. *Mol Cell Endocrinol.* 2011;331:215−221.
7. Coll AP, Challis BG, López M, et al. Proopiomelanocortin-deficient mice are hypersensitive to the adverse metabolic effects of glucocorticoids. *Diabetes.* 2005;54:2269−2276.
8. Metherell LA, Chapple JP, Cooray S, et al. Mutations in MRAP, encoding a new interacting partner of the ACTH receptor, cause familial glucocorticoid deficiency type 2. *Nat Genetics.* 2005;37:166−170.

Supplemental Information Available on Companion Website

- Amino acid sequence of vertebrate ACTH/E-Figure 16A.1
- Phylogenetic tree of MC2R/E-Figure 16A.2
- Accession numbers of genes and cDNAs for MC2R/E-Table 16A.1

Melanocyte-Stimulating Hormone

Akiyoshi Takahashi

Abbreviation: MSH
Additional names: melanotropin, melanophore-stimulating hormone

A pituitary hormone secreted from the pars intermedia, MSH was one of the first adenohypophysial hormones demonstrated to be present in vertebrates, from jawless fish to mammals, together with adrenocorticotropic hormone (ACTH).

Discovery

The fact that MSH is derived from a precursor protein called proopiomelanocortin (POMC) was demonstrated in 1979, using the pars intermedia from the bovine pituitary.

Structure

Structural Features

Three types of MSH molecules, with different amino acid sequences, are contained in the common precursor POMC in mammals. α-Melanocyte-stimulating hormone (α-MSH) is composed of 13 aa residues. This peptide is generated from the N-terminal region of adrenocorticotropic hormone (ACTH), and corresponds to acetyl ACTH(1−13)-amide. In MSH, the N-terminal serine is free, or monoacetylated at the N position, or diacetylated at the N and O positions, whereas the carboxyl terminal is consistently in the amide form. These variations of MSH are called desacetyl α-MSH, α-MSH, and diacetyl α-MSH, respectively. Of these peptides, α-MSH is a classical α-MSH. β-MSH, which is generated from POMC via β-lipotropin (β-LPH), is composed of 18 aa residues. Unlike α-MSH, in β-MSH both termini are free. γ-MSH is produced from POMC via pro-γ-MSH or *N*-POMC, which consists of γ-MSH together with a joining peptide. γ-MSH (also known as γ_1-MSH) is composed of 12 aa residues in which the N-terminus and the C-terminus are free and amide, respectively. γ_3-MSH is composed of 25 aa residues in which the N-terminal region corresponds to γ_{11}-MSH. Each MSH segment is flanked by basic amino acid residues. Cartilaginous fish such as shark, ray, and ratfish possess δ-MSH in addition to the three other MSH peptides. Comparison with the amino acid sequence and topology of POMC suggests that δ-MSH might have evolved from β-MSH. Accordingly, α-MSH and γ-MSH are suggested to share an antecedent. Teleost POMC lacks γ-MSH [1].

Primary Structure

The amino acid sequences of MSH peptides are shown in Figure 16B.1. See Figure 16.2 of Chapter 16 for locations of these peptides on the precursor protein.

Synthesis and Release

Gene, mRNA, and Precursor

See Subchapter 16A.

Distribution of mRNA

POMC is mainly expressed in the pituitary. A major cell type generating MSH peptides is the melanotrope in the pars intermedia of the pituitary; however, the pars intermedia is absent in the adult human pituitary. The presence of α-MSH and γ-MSH is still controversial. Human β-MSH was originally reported to consist of 22 aa residues, although this might be an artifact produced during the isolation process. Nevertheless, MSH peptides of 18 aa residues corresponding to human β-MSH(5−22) are also produced in human extrapituitary tissues [2].

Tissue Content

The pars intermedia in the pituitary is the main source of α-MSH. The tissue content in the rat pars intermediate is 1.1 (μg/mg) and in the pars distalis is 4.5 (ng/mg). Other tissues also contain α-MSH, but the concentration is remarkably lower (less than 1%) than in the pars intermedia. Acid−ethanol extracts of various regions of the rat brain showed that the highest concentration of α-MSH, excluding the pituitary, was 163 (pg/mg) in the pineal gland, and the contents in other brain regions were hypothalamus 143 (pg/mg), thalamus 25, brainstem 11, cerebrum 5.7, and cerebellum 2.3 [3]. Diurnal changes are observed in the rat plasma levels. Over a 24-hour period, two peaks were observed: one at 4 a.m. (142 pg/ml) and the other at 4 p.m. (139 pg/ml) [4]. Plasma levels of α-MSH change in response to changes in background color in fish and amphibians.

Regulation of Synthesis and Release

In rodents, α-MSH release is under strong inhibitory control by direct innervation from hypothalamic neurons. Dopamine plays a role as a physiological melanotropin release-inhibiting hormone (MRIH). In contrast to ACTH release from corticotropes in the pars distalis of the pituitary, there is no apparent negative feedback control on α-MSH release from

Y. Takei, H. Ando, & K. Tsutsui (Eds): Handbook of Hormones. DOI: http://dx.doi.org/10.1016/B978-0-12-801028-0.00136-7

α–MSH (mammals)	:	+SYSMEHFRWGKPVa	[Mr 1,665]
β–MSH (human)	:	AEKKDEGPYRMEHFRWGSPPKD	[Mr 2,204]
β–MSH (bovine)	:	DSGPYKMEHFRWGSPPKD	[Mr 2,134]
γ₁–MSH (human)	:	KYVMGHFRWDRFa	[Mr 1,641]
δ–MSH (dogfish)	:	DGKIYKMTHFRWa	[Mr 1,582]

Figure 16B.1 Amino acid sequence of MSH peptides. +, acetyl group; a, amide group.

Table 16B.1 Pharmacological Properties of MCR [6]

Receptors	Potency of Ligands
MC1R	α-MSH = ACTH > β-MSH > γ-MSH
MC2R	ACTH
MC3R	α-MSH = β-MSH = γ-MSH = ACTH
MC4R	α-MSH = ACTH > β-MSH > γ-MSH
MC5R	α-MSH > ACTH > β-MSH > γ-MSH

melanotropes. Similar mechanisms of regulation have also been demonstrated in amphibians [5]. In humans, the pars intermedia is functional in the fetus and neonate, whereas adults lack the pars intermedia.

Receptors

Structure

α-MSH and other MSH peptides interact with four of the five subtypes of melanocortin receptors (MC1R, MC3R, MC4R, and MC5R, excluding the ACTH-specific receptor MC2R), which are members of the GPCR family. Among them, MC1R is a classical α-MSH receptor. ACTH interacts with all five MC receptors. The pharmacological properties of the mammalian MC receptors are shown in Table 16B.1 [6].

Signal Transduction

α-MSH activates the AC/PKA pathway via G proteins. See also Chapter 16.

Agonists

NDP-MSH and MT-II are agonists for human MC1R, MC3R, MC4R, and MC5R.

Antagonist

HS024 is an antagonist for human MC1R, MC3R, MC4R, and MC5R.

Biological Functions

α-MSH and other MSH peptides are associated with a wide spectrum of biological functions through MC receptors that are distributed in many tissues. MC1R is expressed in melanocytes, keratinocytes, macrophages, leukocytes, and adipose tissue; MC3R is expressed in the central nervous system (CNS), kidney, testis, ovary, skeletal muscle, placenta, and mammary gland; MC4R is expressed in the CNS; and MC5R is expressed in exocrine glands, muscle, and the CNS. Representative physiological functions of MSH peptides mediated by MC receptors are stimulation of melanocytes in the skin to synthesize melanin, including regulation of the eumelanin−pheomelanin switch via MC1R; energy homeostasis and natriuresis via MC3R; energy homeostasis and erectile function via MC4R; and synthesis and secretion of exocrine gland products via MC5R [7]. β-MSH is associated with aldosterone release from the adrenal cortex, and γ-MSH is related to sodium metabolism and blood pressure regulation as cardiovascular effects. α-MSH is also associated with physiological body-color change, which occurs when fish and amphibians disperse pigments in chromatophores. In mammals, α-MSH has been established to be a representative anorexigenic neuropeptide. β-LPH, which is a precursor for β-MSH and β-END, exhibits lipolytic activity, but its potency is weaker than that of the MSH peptides and ACTH.

Phenotype in Gene-Deficiency

Individuals carrying MC1R receptor variants, especially those associated with red hair color, fair skin, and poor tanning ability, are more prone to melanoma. Deletion of MC3R produces a moderate obesity syndrome, with increased weight observed in females but increased adipose mass observed in both sexes. MC4R-deficient mice are obese, hyperphagic, and hyperinsulinemic. They exhibit the same endocrine profile as that seen in the obese agouti strain, which results from the absence of either leptin or the leptin receptor. MC5R knock-out mice absorbed more water in their coats when performing a forced swim, and were missing a particular class of sebaceous lipids [7,8]. See also Subchapter 16A.

Pathophysiological Implications

Clinical Implications

The anti-inflammatory activity of α-MSH includes immuno-modulatory effects on several resident skin cells and antifibrogenic effects mediated via MC1R that is expressed by dermal fibroblasts [9]. In human mast cells, α-MSH appears to be proinflammatory due to histamine release. α-MSH exhibits cytoprotective activity against ultraviolet B-induced apoptosis and DNA damage, which is associated with the increased risk of cutaneous melanoma in individuals with loss of function *MC1R* mutation. Congenital deficiency of POMC results in a syndrome of hypoadrenalism, severe obesity, and altered skin and hair pigmentation. In one case from a Turkish family, a child who was homozygous for a frameshift mutation in the N-terminal region of POMC, and was thus predicted to have a loss of all POMC-derived peptides, showed typical symptoms of POMC deficiency. However, this child did not have red hair, unlike cases of Northern European origin [10].

Use for Diagnosis and Treatment

Measurement of the blood concentration of MSH has not been validated for routine clinical use. MSH analogs have recently been developed as anti-obesity medication. The potential use of radiolabeled α-MSH peptides in melanoma imaging and treatment of disseminated disease has also been reported.

References

1. Takahashi A, Kawauchi H. Diverse structures and functions of melanocortin, endorphin and melanin-concentrating hormone in fish. In: Reinecke M, Zaccone G, Kapoor BG, eds. *Fish Endocrinology.* Enfield, NH, USA: Science Publishers; 2006:325−392.
2. Bertagna X, Lenne F, Comar D, et al. Human β-melanocyte-stimulating hormone revisited. *Proc Natl Acad Sci USA.* 1986;83:9719−9723.
3. Oliver C, Porter JC. Distribution and characterization of α-melanocyte-stimulating hormone in the rat brain. *Endocrinology.* 1978;102:697−705.
4. Usategui R, Oliver C, Vaudry H, et al. Immunoreactive α-MSH and ACTH levels in rat plasma and pituitary. *Endocrinology.* 1976;98:189−196.

5. Vazquez-Martinez R, Castaño JP, Tonon MC, et al. Melanotrope secretory cycle is regulated by physiological inputs via the hypothalamus. *Am J Physiol Endocrinol Metab*. 2003;285: E1039—E1046.

6. Gantz I, Fong TN. The melanocortin system. *Am J Physiol Endocrinol Metab*. 2003;284:E468—E474.

7. Cone RD. Studies on the physiological functions of the melanocortin system. *Endocr Rev*. 2006;27:736—749.

8. Cao J, Wan L, Hacker E, et al. MC1R is a potent regulator of PTEN after UV exposure in melanocytes. *Mol Cell*. 2013;51: 409—422.

9. Böhm M, Luger TA, Tobin DJ, et al. Melanocortin receptor ligands: new horizons for skin biology and clinical dermatology. *J Inv Dermatol*. 2006;126:1966—1975.

10. Farooqi IS, Drop S, Clements A, et al. Heterozygosity for a POMC-null mutation and increased obesity risk in humans. *Diabetes*. 2006;55:2549—2553.

Supplemental Information Available on Companion Website

- Amino acid sequence of α-MSH, β-MSH, γ-MSH, and δ-MSH/E-Figures 16B.1—16B.4
- Phylogenetic trees of MC1R, MC3R, MC4R, and MC5R/E-Figures 16B.5—16B.8
- Accession numbers of genes and cDNAs for MC1R, MC3R, MC4R, and MC5R/E-Tables 16B.1—16B.4

Endorphin

Akiyoshi Takahashi

Abbreviation: END

Endorphin is an endogenous morphine secreted from the pituitary. Sources are both corticotropes in the pars distalis and melanotropes in the pars intermedia.

Discovery

The fact that END is derived from a precursor protein called proopiomelanocortin (POMC) was demonstrated in 1979 using the pars intermedia in the bovine pituitary [1].

Structure

Structural Features

β-END is derived from the C-terminal region of the common precursor POMC. It is composed of 31 aa residues in mammals (Figure 16C.1). The N-terminal 5-aa sequence, Tyr−Gly−Gly−Phe−Met, is identical to Met-enkephalin, which is derived from proenkephalin, and is essential for analgesic activity. Acetylation occurs at the N-terminal tyrosine in a substantial amount of β-END in the pars intermedia. Non-acetylated β-END is the active form, and acetylated β-END is the inactive form. α-END and γ-END correspond to β-END(1−16) and β-END(1−17), respectively [2]. In the frog *Xenopus laevis*, N-acetyl β-END(1−8) has been shown to be the terminal product of β-END processing. The N-terminal 5-aa sequence sometimes undergoes mutation in species having two or more POMC genes. In the barfin flounder, in which three POMC sequences have been reported, Tyr−Gly−Gly−Phe−Met is changed to Ser−Gly−Arg−Phe−Met [3]. In lampreys, one of the two POMCs (arbitrarily termed proopiomelanocortin, POM), which is synthesized in the pars intermedia, could be a source of met-enkephalin. The met-enkephalin sequence is segregated with basic amino acids from the rest of the β-END sequence [4].

Primary Structure

The N-terminal pentapeptide sequence is identical to met-enkephalin.

Synthesis and Release

Gene, mRNA, and Precursor

See Subchapter 16A.

Distribution of mRNA

Major cell types expressing the POMC gene, from which β-END is generated, are the corticotropes and melanotropes

of the pars distalis and pars intermedia in the pituitary, respectively; however, the pars intermedia is absent in the adult human pituitary.

Tissue Content

In many animals, the pars intermedia in the pituitary is the major source of β-END. β-END is also present in the pars distalis of the pituitary and hypothalamus. In the rat, the contents in the pars distalis are: β-END 729 (pmol/g of wet weight), N-acetyl-β-END 39.2, β-END(1−27) 3,168, and N-acetyl-β-END(1−27) 1,868. Those in the neurointermediate lobe are: β-END 12,790, N-acetyl-β-END 24,910, β-END(1−27) 186,400, and N-acetyl-β-END(1−27) 103,600. Those in the hypothalamus are: β-END 131.5, N-acetyl-β-END 21.9, β-END(1−27) 53.2, and N-acetyl-β-END(1−27) 29.1 [5].

Regulation of Synthesis and Release

See Subchapter 16A for the release from the pars distalis of the pituitary, and Subchapter 16B for the release from the pars intermedia of the pituitary.

Receptors

Structure

β-END and a related peptide such as β-END(1−27) are agonists for the μ-opioid receptor (MOR), although these peptides also bind to the δ-opioid receptor with slightly weaker affinities [6]. MOR is a subtype of opioid receptor belonging to the GPCR superfamily. The location of *MOR* in human chromosomes is 6q24−q25.

Signal Transduction

See Subchapter 7A.

Agonists

DAMGO, sufentanil, hydromorphone, fentanyl, and nalbuphine are agonists.

Antagonists

Naloxone, naltrexone, nalmefene, alvimopan, and levallorpan are antagonists.

Biological Functions

β-END exhibits opiate-like analgesic activity that is 18 to 33 times more potent than that of morphine, and its actions are blocked by the specific opiate antagonist naloxone hydrochloride [7]. It also affects feeding, sexual behavior, and learning, and modulates neuroendocrine function when

Y. Takei, H. Ando, & K. Tsutsui (Eds): Handbook of Hormones. DOI: http://dx.doi.org/10.1016/B978-0-12-801028-0.00137-9

YGGFMTSEKS QTPLVTLFKN AIIKNAYKKG E [Mr 3, 465]

Figure 16C.1 Amino acid sequence of the human β-END. Spaces are inserted every 10 aa.

administered into the cerebral ventricle or the brain. In peripheral tissues β-END is associated with the cardiovascular and immune systems, although the significance of peripherally circulating β-END remains to be fully elucidated. N-acetylated forms of β-END do not bind to opioid receptors. See also Subchapter 7A.

Phenotype of Gene-Deficiency

Analgesia, considered the main therapeutic activity, was absent in all MOR mutant mice, at doses that produce maximal analgesia in wild-type mice. Reward, the other main biological action of morphine, was also ablated in mutant mice. Morphine was even found to be aversive in self-administration experiments. The investigation of other acute morphine effects showed no respiratory depression, constipation, inhibition of vas deferens contractions, or increased production of stress hormones in those mice. Other morphine effects were also abolished, including naloxone-precipitated withdrawal and downregulation of dynamin in the brain, or immunosuppression [8]. See also Subchapter 7A.

Pathophysiological Implications

Clinical Implications

There is a correlation between endogenous opioid peptides, especially β-END, and alcohol abuse. The consumption of alcohol activates the endogenous opioid system. Consumption of alcohol results in an increase in β-END levels in the brain regions that are associated with reward. However, it has also been observed that habitual alcohol consumption leads to β-END deficiency. People with a genetic deficit in β-END are particularly susceptible to alcoholism.

The plasma levels of β-END in subjects genetically at high risk of excessive alcohol consumption show lower basal activity of β-END [9].

References

1. Nakanishi S, Inoue A, Kita T, et al. Nucleotide sequence of cloned cDNA for bovine corticotropin-β-lipotropin precursor. *Nature.* 1979;278:423–427.
2. Kosanam H, Ramagiri S, Dass C. Quantification of endogenous α- and γ-endorphins in rat brain by liquid chromatography-tandem mass spectrometry. *Anal Biochem.* 2009;392:83–89.
3. Takahashi A, Amano M, Itoh T, et al. Nucleotide sequence and expression of three subtypes of proopiomelanocortin mRNA in barfin flounder. *Gen Comp Endocrinol.* 2005;141:291–303.
4. Takahashi A, Mizusawa K. Posttranslational modifications of proopiomelanocortin in vertebrates and their biological significance. *Front Endocrinol.* 2013;4. Available from: http://dx.doi.org/10.3389/fendo.2013.00143.
5. Zakarian S, Smyth D. Distribution of active and inactive forms of endorphins in rat pituitary and brain. *Proc Natl Acad Sci USA.* 1979;76:5972–5976.
6. Mansour A, Hoversten MT, Taylor LP, et al. The cloned μ, δ and κ receptors and their endogenous ligands: evidence for two opioid peptide recognition cores. *Brain Res.* 1995;700:89–98.
7. Loh HH, Tseng LF, Wei E, et al. β-Endorphin is a potent analgesic agent. *Proc Natl Acad Sci USA.* 1976;73:2895–2898.
8. Kieffer BL, Gavériaux-Ruff C. Exploring the opioid system by gene knockout. *Prog Neurobiol.* 2002;66:285–306.
9. Zalewska-Kaszubska J, Czarnecka E. Deficit in beta-endorphin peptide and tendency to alcohol abuse. *Peptides.* 2005;26:701–705.

Supplemental Information Available on Companion Website

- Amino acid sequence of β-END of a variety of vertebrates/E-Figure 16C.1
- Phylogenetic tree of MOR/E-Figure 16C.2
- Accession numbers of genes and cDNAs for MOR/E-Table 16C.1

SECTION I.3

Gastrointestinal Hormones

CHAPTER 17

Glucagon Family

Min Kyun Park

History

The glucagon family of peptides includes glucagon, the glucagon-like peptides (GLP-1 and GLP-2), and gastric inhibitory peptide (GIP) (Figure 17.1). Glucagon was first isolated from a side fraction of purified insulin [1] as a hyperglycemic pancreatic factor [2]. Following sequencing of the anglerfish proglucagon peptide in the early 1890s, mammalian proglucagon sequences were determined from cattle, hamster, human, and rat in the 1980s. In these proglucagons, two additional glucagon-like sequences (GLP-1 and GLP-2) were identified with high similarity in sequences and properties. GIP, however, was isolated from porcine intestinal extracts in the early 1970s. After the incretin effect of GIP was discovered, an alternative definition of the GIP acronym was introduced: glucose-dependent insulinotropic polypeptide. Incretins are gut hormones that are secreted from enteroendocrine cells and stimulate insulin secretion in a glucose-dependent manner. GLP-1 was also identified as a mammalian incretin [3], and only two physiological incretins have been identified so far.

Structure

Structural Features

In mammalian species, a single proglucagon gene expresses a single mRNA encoding a single precursor protein that contains three distinct peptides: glucagon, GLP-1, and GLP-2 [4] (E-Figure 17.1). Non-mammalian vertebrates use more complex mechanisms for proglucagon gene expression [5]. GIP is also a structurally related peptide, and its insulinotropic function is shared by GLP-1. DPP-4 is an important enzyme in regulation of circulating levels of endogenous peptides of the glucagon family, and prefers substrates with an amino-terminal proline or alanine at position 2. Therefore, GIP, GLP-1, and GLP-2 are excellent substrates for this enzyme. DPP-4 may also cleave substrates with non-preferred amino acids at position 2 [6], and glucagon has also been shown to be a substrate for DPP-4

in vitro. However, whether DPP-4 regulates physiological levels of endogenous glucagon remains unclear.

Molecular Evolution of Family Members

It has been suggested that ancestral glucagon and GIP genes were divided from an ancestral gene by whole genome duplication more than a billion years ago. Exon duplication of the ancestral glucagon then presumably created a GLP that subsequently diverged into GLP-1 and GLP-2 by another exon duplication ~700 million years ago. For the glucagon receptor family, a different evolutionary scheme has been proposed. When considering phylogenetic evolution, the GLP-2 receptor is thought to be of the most ancient origin. This result indicates that evolution of the peptides and receptors was independent, and ligand-receptor specificities evolved much later [7].

Receptors

Structure and Subtypes

All the receptors for glucagon family peptides belong to the superfamily of G protein-coupled receptors, exhibit considerable amino acid sequence identity, and share similar structural properties with respect to ligand binding and signal transduction (Figure 17.2). With these characteristics, these receptors are grouped as the glucagon receptor family within the secretin (Class B) superfamily, like their ligand peptides. Each receptor is named for its principal and only physiologically relevant ligand [8] (E-Figures 17.2, 17.3).

Signal Transduction Pathway

Both the peptides and their receptors exhibit considerable amino acid sequence identity and share similar structural properties. However, there is no significant biologically meaningful cross-reactivity among related peptides and receptors [9]. They signal through G_s leading to the activation of AC and increased levels of cAMP [8].

```
hGlucagon: HSQGTFTSDY SKYLDSRRAQ DFVQWLMNT
     hGIP: YAE..I... .IAM.KIHQ. ..N..LAQK GKKNDWKHNI TQ
   hGLP-1: .AE.....V .S..EGQA.K E.IA..VKGR G
   hGLP-2: .AD.S.SDEM NTI..NLA.R ..IN..IQ.K ITD
```

Figure 17.1 Comparison of amino acid sequences of the glucagon family in humans. The amino acids identical to those of human glucagon are indicated by dots, and the completely conserved aa residues are shaded gray.

Y. Takei, H. Ando, & K. Tsutsui (Eds): Handbook of Hormones. DOI: http://dx.doi.org/10.1016/B978-0-12-801028-0.00017-9

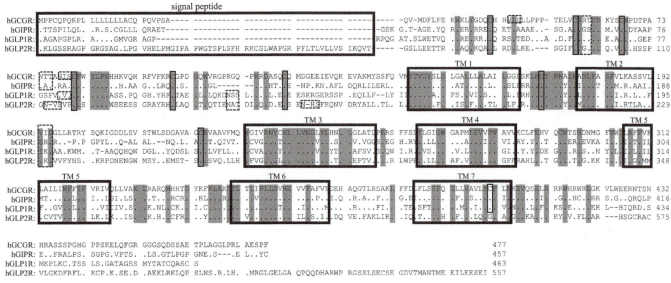

Figure 17.2   Comparison of amino acid sequences of the glucagon receptor family in humans. Bases identical to those of human glucagon receptor are indicated by dots. Gaps are indicated by hyphens. Conserved cysteine residues and potential N-glycosylation sites are boxed by solid and dashed lines, respectively. Signal peptide and putative seven-transmembrane (TM) domains are boxed within bold lines. Accession numbers: hGCGR, NM_000160; hGIPR, NM_000164; hGLP1R, NM002062; hGLP2R, NM_004246.

Biological Functions

Target Cells/Tissues and Functions

Glucagon family peptides and their receptors are expressed in the gastrointestinal tract, pancreas, and/or brain, and exert multiple biological actions on the regulation of somatic growth, energy intake, nutrient absorption and disposal, and cell proliferation and apoptosis. GIP and GLP-1 share glucose-dependent insulinotropic function [10], and are considered as major incretin hormones in response to ingested food.

References

1. Kenny AJ. Extractable glucagon of the human pancreas. *J Clin Endocrinol Metab*. 1955;15:1089−1105.
2. Kimball C, Murlin J. Aqueous extracts of pancreas. *J Biol Chem*. 1923;58:337−348.
3. Lund PK. The discovery of glucagon-like peptide 1. *Regul Pept*. 2005;128:93−96.
4. Irwin DM. Molecular evolution of proglucagon. *Regul Pept*. 2001;98:1−12.
5. Richards MP, McMurtry JP. The avian proglucagon system. *Gen Comp Endocrinol*. 2009;163:39−46.
6. Bongers J, Lambros T, Ahmad M, et al. Kinetics of dipeptidyl peptidase IV proteolysis of growth hormone-releasing factor and analogs. *Biochim Biophys Acta*. 1992;1122:147−153.
7. Ng SYL, Lee LT, Chow B. Insights into evolution of proglucagon-derived peptides and receptors in fish and amphibians. *Ann NY Acad Sci*. 2010;1200:15−32.
8. Mayo KE, Miller LJ, Bataille D, et al. International union of pharmacology. XXXV. The glucagon receptor family. *Pharmacol Rev*. 2003;55:167−194.
9. Drucker DJ. Minireview: the glucagon-like peptides. *Endocrinology*. 2001;142:521−527.
10. Unger RH, Eisentraut AM. Entero-insular axis. *Arch Intern Med*. 1969;123:261−266.

Supplemental Information Available on Companion Website

- The structures and processing of proglucagon and proGIP/E-Figure 17.1
- Phylogenetic tree of the glucagon family members/E-Figure 17.2
- Phylogenetic tree of the glucagon receptor family members/E-Figure 17.3

Glucagon

Min Kyun Park

Abbreviation: GCG

Glucagon is the principal hyperglycemic hormone, and acts as a counterbalancing hormone to insulin. Glucagon generally elevates the level of blood glucose by promoting gluconeogenesis and glycogenolysis.

Discovery

Glucagon was first described in 1923 as an additional substance with hyperglycemic properties from pancreatic extracts. The aa sequence of glucagon was confirmed in the 1950s. After establishment of a specific radioimmunoassay in the 1970s, the role of glucagon in physiology and disease was rapidly elucidated. The entire coding sequence of preproglucagon was initially characterized from anglerfish by using the recombinant DNA technique [1].

Structure

Structural Features

Glucagon is a single-chain polypeptide that contains 29 aa residues in all vertebrates except the paddlefish, where it is 31 aa long (Figure 17A.1). The primary structure of glucagon is identical in most mammals, including humans, though some differences have been noted in the guinea pig or in non-mammals. *In vivo*, the plasma half-life is approximately 5 minutes due to DPP-4 action. The N-terminal two aa residues are cleaved off by the DPP-4 and the truncated glucagon fragment, glucagon(3–29), behaves as a partial agonist of the glucagon receptor but shows no glycemic effect *in vivo* [2].

Properties

Human glucagon: Mr 3,483, theoretical pI 6.75. Crystalline glucagon is a white to off-white powder. It is relatively insoluble in water, methanol, ethanol, and ether, but soluble at a pH <3 or >9.

Synthesis and Release

Gene, mRNA, and Precursor

In all tetrapods a single proglucagon gene has been determined. Moreover, a single mRNA transcript has been reported from mammals, whereas two different splicing patterns were reported in non-mammalian tetrapods. Bony fish species have duplicated genes, one of which has two transcripts. Jawless fish species also have two genes, but only a single transcript has been characterized for each gene. While in amniotic species proglucagon genes all encode the three glucagon-like sequences (glucagon, GLP-1, and GLP-2), some non-amniotic vertebrates have different numbers of glucagon-like sequences that range from two to five. The human proglucagon gene (*GCG*), located on chromosome 2 (2q36−q37), spans approximately 9.4 kb. The preproglucagon encoded by this gene is 180 amino acids in length and contains glucagon and two glucagon-like peptides (GLP-1 and GLP-2) (E-Figure 17A.1).

Distribution of mRNA

Expression of proglucagon has been reported in pancreatic α cells, intestinal L cells, and the brain.

Regulation of Synthesis and Release

Secretion of bioactive glucagon is regulated by cell-specific expression of prohormone convertase (PC) enzymes (Table 17A.1), and the major bioactive hormone is glucagon in pancreatic α cells. The existence of several glucagon sequence-containing peptides, such as oxyntomodulin (glucagon plus a carboxyl-terminal extension), glicentin (glucagon plus both amino- and carboxyl-terminal extension), and miniglucagon (the carboxyl-terminal [19−29] glucagon sequence), should be noted (E-Figure 17A.2); however, their signaling mechanism and physiological role remain unclear.

Plasma Concentration

Serum glucagon levels increase within the first 48 hours (from 126 to 189 pg/ml) and continue to rise throughout the first several days of fasting in normal humans. Glucagon is an important test to diagnose glucagonomas, where in almost all patients the plasma glucagon level is elevated (>150 pg/ml).

Receptors

Structure and Subtype

The receptor of glucagon is a glycoprotein that belongs to the glucagon receptor family within the secretin receptor superfamily of G protein-coupled receptors [3]. The first cloning of glucagon receptor cDNA was achieved from the rat. The rat glucagon receptor is a 485-aa residue with a predicted Mr of 54,963 (for the human glucagon receptor, see E-Figure 17A.3). The first non-mammalian glucagon receptor was characterized from *Rana tigrina rugulosa* (for alignment, see E-Figure 17A.4) [4]. The glucagon receptor is mainly expressed in the liver and kidney, with lesser amounts found in the heart, adipose tissue, spleen, thymus, adrenal glands, pancreas, cerebral cortex, and gastrointestinal tract [5,6].

Y. Takei, H. Ando, & K. Tsutsui (Eds): Handbook of Hormones. DOI: http://dx.doi.org/10.1016/B978-0-12-801028-0.00138-0

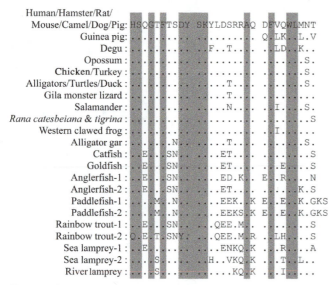

Figure 17A.1 Comparison of aa sequences of the vertebrate glucagon. Amino acids identical to those of human glucagon are indicated by dots. The completely conserved aa residues are shaded gray.

Table 17A.1 Regulating Factors for Glucagon Secretion

Stimulating Factors	Inhibiting Factors
Hypoglycemia	Glucose
Epinephrine (via β_2, α_2, and α_1 adrenergic receptors)	Free fatty acids, ketone bodies
Amino acids (especially arginine, alanine)	Somatostatin (via SST-2 receptor)
Acetylcholine	Insulin (via GABA)
cAMP	Somatostatin
GIP	GLP-1
Bombesin	Secretin
Gastrin	PPARγ/retinoid X receptor heterodimer
Cholecystokinin	Increased urea production
Neurotensin	

Signal Transduction Pathway

Numerous studies reveal that at least $G_{s\alpha}$ and G_q are involved in the signal transduction of the glucagon receptor. The activation of $G_{s\alpha}$ leads to activation of AC, an increase in the intracellular level of cAMP, and subsequent activation of PKA. Activation of G_q leads to the activation of PLC, production of IP3, and subsequent release of Ca^{2+} from endoplasmic reticulum [7,8].

Agonists

Synthetic and recombinant human glucagon are agonists.

Antagonists

Bay 27-9955, glucagon receptor antagonist, des-His[1]-[Nle[9]-Ala[11]-Ala[16]] glucagon, des-His[1]-des-[Phe[6]-Glu[9]] glucagon-NH2, and NNC 92-1687 — a non-peptidic competitive receptor antagonist — are antagonists.

Biological Functions

Target Cells/Tissue and Functions

Glucagon plays a key role in glucose metabolism and homeostasis, and regulates the concentration of blood glucose by increasing gluconeogenesis and decreasing glycolysis (see E-Table 17A.3 for details).

Phenotype in Gene-Modified Animals

Glucagon receptor knockout mice exhibit a significant increase of total pancreatic weight, marked islet α-cell hyperplasia, extremely elevated circulating levels of glucagon and GLP-1, and mild reproductive abnormalities [9]. The importance of the glucagon receptor for hepatocyte cell survival was also demonstrated by using glucagon receptor knockout mice [10].

Pathophysiological Implications

Clinical Implications

Hyperglucagonemia is usually caused by excessive production of glucagon by a tumor of the α cells of the pancreatic islets (glucagonoma). However, hyperglucagonemia can also occur in diverse conditions such as acute pancreatitis, severe stress, acromegaly, Cushing's syndrome, hepatocirrhosis, chronic hepatitis, chronic renal insufficiency, diabetic ketoacidosis, prolonged starvation, hypercorticism, septicemia, and familial hyperglucagonemia. Manifestations of hyperglucagonemia include dermatitis, diabetes, diarrhea, weight loss, abdominal pain, anemia, and thromboembolic disease. Hypoglucagonemia is rare, and can occur in conditions such as chronic pancreatitis, anterior pituitary failure, Addison's disease, and glucagon deficiency.

Use for Diagnosis and Treatment

The glucagon test is performed to identify glucagonoma. It is also performed clinically to measure glucose control in patients who have developed type 2 diabetes, or in patients who may be insulin resistant. Glucagon administration is used in treatment of hypoglycemia, in anaphylactic reaction in patients on beta-blocker therapy, in beta-blocker- or calcium channel blocker-induced myocardial depression unresponsive to standard measures, and in temporary inhibition of gastrointestinal movement during radiologic examinations.

References

1. Lund PK, Goodman RH, Dee PC, et al. Pancreatic preproglucagon cDNA contains two glucagon-related coding sequences arranged in tandem. *Proc Natl Acad Sci USA*. 1982;79:345–349.
2. Hinke SA, Pospisilik JA, Demuth HU, et al. Dipeptidyl peptidase IV (DPIV/CD26) degradation of glucagon. Characterization of glucagon degradation products and DPIV-resistant analogs. *J Biol Chem*. 2000;275:3827–3834.
3. Mayo KE, Miller LJ, Bataille D, et al. International union of pharmacology. XXXV. The glucagon receptor family. *Pharmacol Rev*. 2003;55:167–194.
4. Ngan ESW, Chow LSN, Tse DLY, et al. Functional studies of a glucagon receptor isolated from frog *Rana tigrina rugulosa*: implications on the molecular evolution of glucagon receptors in vertebrates. *FEBS Lett*. 1999;457:499–504.
5. Svoboda M, Tastenoy M, Vertongen P, et al. Relative quantitative analysis of glucagon receptor mRNA in rat tissues. *Mol Cell Endocrinol*. 1994;105:131–137.
6. Dunphy JL, Taylor RG, Fuller PJ. Tissue distribution of rat glucagon receptor and GLP-1 receptor gene expression. *Mol Cell Endocrinol*. 1998;141:179–186.
7. Burcelin R, Katz EB, Charron MJ. Molecular and cellular aspects of the glucagon receptor: role in diabetes and metabolism. *Diabetes Metab*. 1996;22:373–396.
8. Christophe J. Glucagon receptors: from genetic structure and expression to effector coupling and biological responses. *Biochim Biophys Acta*. 1995;1241:45–57.
9. Gelling RW, Du XQ, Dichmann DS, et al. Lower blood glucose, hyperglucagonemia, and pancreatic alpha cell hyperplasia in glucagon receptor knockout mice. *Proc Natl Acad Sci USA*. 2003;100:1438–1443.

10. Sinclair EM, Yusta B, Streutker C, et al. Glucagon receptor signaling is essential for control of murine hepatocyte survival. *Gastroenterology*. 2008;135:2096−2106.
11. Cavanaugh ES, Nielsen PF, Conlon JM. Isolation and structural characterization of proglucagon-derived peptides, pancreatic polypeptide, and somatostatin from the urodele *Amphiuma tridactylum*. *Gen Comp Endocrinol*. 1996;101:12−20.
12. Irwin DM, Sivarajah P. Proglucagon cDNAs from the leopard frog, *Rana pipiens*, encode two GLP-1-like peptides. *Mol Cell Endocrinol*. 2000;162:17−24.

Supplemental Information Available on Companion Website

- Comparison of amino acid sequences of the glucagon receptor of various vertebrates/E-Figure 17A.1
- The structures and differential posttranslational processing of proglucagon in the pancreas, and in the gut and brain/E-Figure 17A.2
- Gene, mRNA, and precursor structures of the human glucagon receptor/E-Figure 17A.3
- Comparison of amino acid sequences of the glucagon receptor of various vertebrates/E-Figure 17A.4
- Accession numbers of genes, cDNAs, and peptides for proglucagon and glucagon receptor/E-Tables 17A.1 [11,12], 17A.2
- Biological activities of glucagon/E-Table 17A.3

Gastric Inhibitory Peptide

Min Kyun Park

Abbreviation: GIP
Additional names: gastric inhibitory polypeptide, gastrointestinal inhibitory peptide, glucose-dependent insulinotropic peptide

GIP was originally isolated as a gastric inhibitory polypeptide. After the discovery of its glucose-dependent insulinotropic activity, known as the incretin effect, GIP was renamed as glucose-dependent insulinotropic peptide.

Discovery

GIP was originally isolated from porcine intestinal extract on the basis of its acid inhibitory activity in dogs (gastric inhibitory polypeptide) in the early 1970s, and subsequently renamed glucose-dependent insulinotropic peptide after finding its major physiologically important role as a potentiator of glucose-stimulated insulin secretion [1].

Structure

Structural Features

Human GIP is a single 42-aa peptide. The structure of vertebrate GIP is well conserved and both the N-terminal and central regions are important for biological activity, because truncated forms of GIP, GIP_{1-39}, and GIP_{1-30} showed a high degree of biological activity (Figure 17B.1) [2]. The N-terminal two aa residues are cleaved off by dipeptidyl-peptidase 4 (DPP-4) in the circulation to form GIP_{3-42}, which has no insulinotropic activity.

Properties

Human GIP: Mr 4,983.6, theoretical pI 6.92. Soluble in water, but insoluble in ethanol.

Synthesis and Release

Gene, mRNA, and Precursor

Human gene coding for GIP, located on chromosome 17 (17q21.3−q22), spans approximately 10 kb. This gene consists of six exons and has potential binding sites for a number of transcriptional factors, including Sp1, AP-1, and AP-2. Human cDNA clones have a 459-bp open reading frame that encodes the 153-aa preproGIP. The GIP sequence in proGIP is flanked by single arginine residues, sites for cleavage by prohormone convertase (PC) enzymes. GIP and PC1/3 are co-expressed in K cells, and PC1/3 is sufficient to produce bioactive GIP [3].

Distribution of mRNA

GIP is secreted from enteroendocrine K cells, which are located in the mucosa of the duodenum and the jejunum of the gastrointestinal tract. The majority of intestinal K cells are located in the proximal duodenum. In the rodent gut, GIP distribution extends through to the ileum. Expression of GIP is also reported in the submandibular salivary gland, stomach, and brain.

Plasma Concentration

Plasma GIP levels range from non-detectable to 420 pg/ml (mean range, 203 pg/ml) in overnight fasting humans. This level may increase fivefold or more (mean level, 1573 pg/ml) after a mixed meal [4].

Regulation of Synthesis and Release

Expression of the GIP is regulated by nutrients. Administration of glucose and lipid into the rat gastrointestinal tract increases GIP mRNA levels. Circulating GIP levels are low in the fasted state and increase within minutes of food ingestion. The postprandial level of circulating GIP is dependent on meal size, and the degree to which nutrients regulate GIP secretion is species dependent. Fat is a more potent stimulator than carbohydrates in humans, whereas in the rodent and pig carbohydrates are more potent than fat [5]. Once released, GIP is rapidly deactivated by DPP-4.

Receptors

Structure and Subtype

Both the relatively long extracellular amino-terminal domain and the first transmembrane domain are important for ligand binding and receptor activation. The carboxyl-terminal cytoplasmic domain of the receptor is important for receptor desensitization and internalization. Like their peptide ligands, GIP receptor and GLP-1 receptor exhibit high degrees of amino acid sequence identity, with similar molecular structures and signaling processes. However, GIP does not bind to GLP-1 receptor, and vice versa.

Signal Transduction Pathway

Ligand binding to GIP receptor primarily activates adenylyl cyclase and increases intracellular cAMP [6]. Activation of the MAP kinase pathway phospholipase A2, as well as the phosphatidylinositol 3-kinase/protein kinase B pathway, have also been reported.

Y. Takei, H. Ando, & K. Tsutsui (Eds): Handbook of Hormones. DOI: http://dx.doi.org/10.1016/B978-0-12-801028-0.00139-2

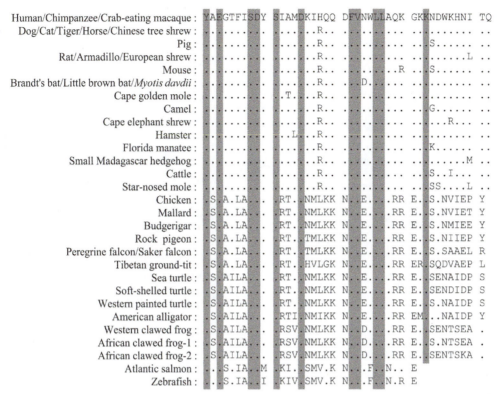

Figure 17B.1 Comparison of aa sequences of vertebrate GIP. Amino acids identical to those of human GIP are indicated by dots. The completely conserved aa residues are shaded gray.

Agonist

[D-Ala2]-GIP is an agonist.

Antagonists

GIP(6−30), ANTGIP (GIP-(7−30)-NH$_2$) (a truncated GIP peptide antagonist), and [Pro3]-GIP; Exendin(9−39)amide are antagonists.

Biological Functions

Target Cells/Tissues and Functions

The mRNA of GIP receptor is expressed in both α and β cells in the pancreatic islets, gastrointestinal tract, adipose tissues, adrenal cortex, pituitary, heart, testis, endothelium of major blood vessels, bone, trachea, spleen, thymus, lung, kidney, thyroid, and several regions of the brain. The primary physiological action of GIP is stimulation of glucose-dependent insulin secretion (Table 17B.1). GIP also exerts a number of additional actions in various tissues and organs, but some of these additional effects are achieved only at pharmacological levels, and their physiological significance is unclear. GIP and GLP-1 share common properties as incretins, but they also possess different biological characteristics.

Phenotype in Gene-Modified Animals

GIP receptor knockout mice exhibit reduced adipocyte mass, and are resistant to the development of diet-induced obesity. Double knockout mice of both GIP and GLP-1 receptors show comparatively greater glucose tolerance compared with single incretin-receptor knockout mice. This result suggests that each incretin hormone plays an additive role in the regulation of glucose homeostasis. These double knockout mice also show only modest impairment in glucose homeostasis,

Table 17B.1 Major Physiological Roles of GIP

Organ/Tissue	Functions
Pancreas	Stimulates β-cell proliferation and survival
	Stimulates insulin secretion
	Stimulates glucagon secretion
Adipose tissue	Stimulates lipoprotein lipase activity
	Stimulates adipokine secretion
Bone	Stimulates bone formation
Central nervous system	Stimulates memory

suggesting the presence of a compensating mechanism for the absence of the action of GIP and GLP-1 [7].

Pathophysiological Implications

Clinical Implications

GIP and GLP have implications in the following diseases and conditions.

- *Type 2 diabetes.* In type-2 diabetic patients, GLP-1 retained much of its insulinotropic activity, but the maximum effect of GIP was significantly lower than in normal subjects. It has been suggested that this decreased response is a result of a decreased receptor expression in the pancreas. A reduced insulinotropic effect of GIP was also reported in first-degree relatives of patients with type 2 diabetes [8].
- *Food-dependent Cushing's syndrome.* In food- or GIP-dependent Cushing's syndrome, ectopic adrenal expression of the GIP receptor has been identified. The secretion of cortisol from the adrenal gland was stimulated only briefly, but repeatedly after each food ingestion [9].
- *Celiac disease.* The serum level of GIP was significantly lower in patients with celiac disease than in healthy people. The increase of glucose in serum from patients was significantly smaller than that occurring in healthy people only

during the first hour after the meal, but the absolute number of K cells was not significantly reduced. These results suggest that the release of GIP is influenced by the rate of absorption of nutrients in patients with celiac disease.

- *Acromegaly*. Acromegaly is often associated with fasting and postprandial hyperinsulinemia. In patients with acromegaly, fasting and postprandial GIP levels are abnormally high, suggesting GIP hypersecretion might play a role in the pathogenesis of the hyperinsulinemia that characterizes acromegaly.
- *Obesity*. GIP also links over-nutrition to obesity by acting on adipocytes. GIP-reduced mice demonstrate that partial reduction of GIP alleviates obesity and lessens the degree of insulin resistance under high-fat diet conditions, suggesting a potential therapeutic value [10].

Use for Diagnosis and Treatment

GIP possessing incretin activity enhances glucose-stimulated insulin release. GIP agonists are potentially useful for the treatment of diabetes. Moreover, DPP-4 inhibitors are approved for use in diabetes patients because GIP is rapidly deactivated by DPP-4.

References

1. Brown JC, Dryburgh JR, Ross SA, et al. Identification and actions of gastric inhibitory polypeptide. *Recent Prog Horm Res*. 1975;31:487−532.
2. Wheeler MB, Gelling RW, McIntosh CH, et al. Functional expression of the rat pancreatic islet glucose-dependent insulinotropic polypeptide receptor: ligand binding and intracellular signaling properties. *Endocrinology*. 1995;136:4629−4639.
3. Fehmann HC, Göke R, Göke B. Cell and molecular biology of the incretin hormones glucagon-like peptide-I and glucose-dependent insulin releasing polypeptide. *Endocr Rev*. 1995;16:390−410.
4. Morgan LM, Morris BA, Marks V. Radioimmunoassay of gastric inhibitory polypeptide. *Ann Clin Biochem*. 1978;15:172−177.
5. Baggio LL, Drucker DJ. Biology of incretins: GLP-1 and GIP. *Gastroenterology*. 2007;132:2131−2157.
6. Lu M, Wheeler MB, Leng XH, et al. The role of the free cytoplasmic calcium level in β-cell signal transduction by gastric inhibitory polypeptide and glucagon-like peptide 1(7−37). *Endocrinology*. 1993;132:94−100.
7. Hansotia T, Baggio LL, Delmeire D, et al. Double incretin receptor knockout (DIRKO) mice reveal an essential role for the enteroinsular axis in transducing the glucoregulatory actions of DPP-IV inhibitors. *Diabetes*. 2004;53:1326−1335.
8. Meier JJ, Hücking K, Holst JJ, et al. Reduced insulinotropic effect of gastric inhibitory polypeptide in first-degree relatives of patients with type 2 diabetes. *Diabetes*. 2001;50:2497−2504.
9. Lacroix A, Ndiaye N, Tremblay J, et al. Ectopic and abnormal hormone receptors in adrenal Cushing's syndrome. *Endocr Rev*. 2001;22:75−110.
10. Nasteska D, Harada N, Suzuki K, et al. Chronic reduction of GIP secretion alleviates obesity and insulin resistance under high fat diet condition. *Diabetes*. 2014;63:2332−2343.

Supplemental Information Available on Companion Website

- Gene, mRNA, and precursor structures of human GIP/E-Figure 17B.1
- The structures of preproGIP and mature GIP/E-Figure 17B.2
- Gene, mRNA, and precursor structures of the human GIP receptor/E-Figure 17B.3
- Comparison of amino acid sequences of the GIP receptors of various vertebrates/E-Figure 17B.4
- Comparison of the physiological roles of GIP and GLP-1/E-Table 17B.1
- Accession numbers of genes, cDNAs, and peptides for GIP/E-Table 17B.2
- Accession numbers of genes, cDNAs, and peptides for GIP receptor/E-Table 17B.3

Glucagon-Like Peptide-1

Min Kyun Park

Abbreviation: GLP-1
Additional name: glucagon-like peptide-I (GLP-I)

The main physiological function of GLP-1 is stimulation of pancreatic β cells to secrete insulin as an incretin hormone. GLP-1 also has additional non-incretin actions, including suppression of glucagon secretion, inhibition of gastric motility, and promotion of satiety.

Discovery

GLP-1 was first identified in the early 1980s, following the cloning of genes for proglucagon. The biological activity of GLP-1 was investigated initially by using the full length amino-terminal extended form of GLP-1 (1−37 and 1−36 amide). However, these GLP-1 molecules were devoid of biological activity. In the late 1980s, naturally occurring GLP-1 was identified by sequencing of the peptides purified from gut extraction, and subsequent researches revealed that GLP-1(7−37) and GLP-1(7−36)-amide are the natural bioactives in GLP-1. In current literature, the unqualified designation of GLP-1 covers only the truncated peptides.

Structure

Structural Features

GLP-1 is a 31-aa peptide hormone and is secreted from intestinal endocrine L cells in two major molecular forms, as GLP-1(7−37) and its amide (Figure 17C.1). Amidation is not always important for its biological activity, and these two molecular forms have similar biological activities. The majority of circulating GLP-1 is found in the GLP-1(7−36) amide, with lesser amounts of the bioactive GLP-1(7−37) form also detectable. The structure of GLP-1, including GIP and GLP-2, reveals a highly conserved alanine at position 9, rendering these peptides ideal substrates for dipeptidyl peptidase 4 (DPP-4).

Primary Structure

The primary structure of GLP-1 is highly conserved in most mammals, although platypus GLP-1 has a large number of changes (11 substitutions). Excluding the platypus sequence, only one site replacement of the carboxyl-terminal glycine with glutamate has been reported, in squirrel GLP-1.

Properties

Human GLP-1(7−37): Mr 3,355.7, theoretical pI 5.53. Soluble in water.

Synthesis and Release

Gene, mRNA, and Precursor

The human proglucagon gene (*GCG*), located on chromosome 2 (2q36−q37), spans approximately 9.4 kb and comprises six exons and five introns. It encodes a preproglucagon of 180 amino acids that contains 3 peptide hormones (glucagon, GLP-1, and GLP-2).

Distribution of mRNA

The glucagon gene is mainly expressed in the pancreatic α cells and the intestinal L cells.

Plasma Concentration

In fasting plasma, the mean concentrations of GLP-1 (7−36)-amide and GLP-1 (7−37) were 7 ± 1 and 6 ± 1 pM, respectively, and the concentration of GLP-1 (7−36)-amide significantly increased to 41 ± 5 pM 90 minutes after ingestion of a meal, whereas the concentration of GLP-1 (7−37) only increased slightly to a maximum of 10 ± 1 pM [1].

Regulation of Synthesis and Release

Proglucagon contains not only glucagon but also two glucagon-like peptides (GLP-1 and GLP-2), and is posttranslationally processed in pancreatic α cells and intestinal L cells in a tissue-specific manner. Prohormone convertase (PC) enzymes are responsible for the tissue-specific proteolytic processing of many precursor proteins, including proglucagon. Glucagon is liberated from pancreatic α cells by PC2, whereas GLP-1 and GLP-2 are liberated as bioactive hormones from the intestinal L cells by PC1/3. As a result, GLP-1 and GLP-2 are co-released in a 1:1 ratio following nutrient ingestion, primarily by meals rich in carbohydrates and lipids. *In vivo*, the biological half-life of GLP-1 is only 1−2 minutes due to DPP-4 action. The GLP-1 secretagogs are numerous, and are secreted by nutrients such as lipids and carbohydrates. In addition, several hormones, such as cholecystokinin (CCK), GIP, somatostatin, and numerous neuromediators, also regulate GLP-1 secretion [2]. Insulin has been reported to inhibit GLP-1 release, indicating that a feedback loop mechanism regulates GLP-1 secretion.

Receptors

Structure and Subtype

GLP-1 receptors were first identified by radioligand binding experiments and measurements of cyclic AMP accumulation using the rat insulinoma cell line [3]. The human GLP-1

Y. Takei, H. Ando, & K. Tsutsui (Eds): Handbook of Hormones. DOI: http://dx.doi.org/10.1016/B978-0-12-801028-0.00140-9

135

```
                      Human/Hamster/Rat/Mouse/Camel/
                      Dog/Pig/Guinea pig/Degu : HAEGTFTSDV SSYLEGQAAK EFIAWLVKGR G
                                     Squirrel : .......... .......... .......... E
                                     Platypus : .S....N... TRL..EK.TV ......L..L E
          Chicken/Turkey/Turtles/American alligator : ...Y...I T.......... ......N... .
                             Chinese alligator : ...Y...I T.......... ......N... R
                            Gila monster lizard : .D.Y...I .......... ......N... .
                                   Salamander : ..D.L...I ..F..K..T. .......S... .
                               Rana catesbeiana : ..D......M ....EK... .VD..I... P
                                Rana tigrina-2 : .........M T...EK... .VD..I... P
                                Rana tigrina-1 : ...Y.N... TQF..EK... ..D..I..K P
                                  Alligator gar : .D.Y.... QD.... K.VT..KQ.Q D
                                       Catfish : .D.Y.... ...QD.... D..T..KS.Q P
                                   Anglerfish-1 : .D..... KD..I. D.VDR.KA.Q V
                                   Anglerfish-2 : .D.Y.... QD.... D.VS..KA.. .
                                    Paddlefish : .D.Y..A ..F.QE...R D..S..K..Q
                                 Rainbow trout-1 : .D.Y.... .T.QK... D.VS..KS.. A
                                 Rainbow trout-2 : .D.Y.... .T.QD.... D.VS..KS.P A
                                   Sea lamprey : .D....N.M T...DAK..R D.VS..ARSD KS
```

Figure 17C.1   Comparison of amino acid sequences of vertebrate GLP-1. Bases identical to those of human GLP-1 are indicated by dots. Completely conserved amino acid residues are shaded gray (for accession numbers, see E-Table 17C.2).

receptor is located on chromosome 6 (6p21). The GLP-1 receptor sequence contains a large hydrophilic extracellular domain and seven hydrophobic transmembrane domains. The GLP-1 receptor protein has three potential N-linked glycosylation sites, and glycosylation may modulate receptor function [4].

Signal Transduction Pathway

The GLP-1 receptor is functionally coupled to AC [3] via G_s. Through activation of AC, cAMP is formed and activates PKA. Ligand stimulation of the receptor also increases the cytoplasmic concentration of Ca^{2+}; this is thought to be executed both through Na^+-dependent uptake of extracellular Ca^{2+} and through release of Ca^{2+} from intracellular Ca^{2+} stores. The increased cytosolic Ca^{2+} in conjunction with the activated PKA stimulates the translocation and exocytosis of insulin-containing secretory granules [5].

Agonists

Exendin-4, subsequently renamed exenatide, was the first FDA approved "incretin mimetic"; liraglutide, a long-acting DPP-4-resistant GLP-1 receptor agonist, was the second. Other agonists are CJC-1131, a GLP-1 analog engineered for covalent coupling to albumin; albiglutide (originally referred to as albugon), a recombinant human albumin-GLP-1 protein; ZP10, an exendin-4 derivative; BIM51077 (subsequently renamed taspoglutide); LY315902, a DPP-4-resistant GLP-1 analog; LY2428757, a pegylated GLP-1; and LY2199265 (dulaglutide), an Fc immunoglobulin fusion protein.

Antagonists

Exendin (9−39) and T-0632, a small non-peptide ligand, are antagonists.

Biological Functions

Target Cells/Tissues and Functions

GLP-1 receptor does not exhibit cross-reactivity with structurally related members of the glucagon family such as GIP, GLP-2, and glucagon. Tissue distribution of GLP-1 receptor mRNA is fairly wide, and has been determined in the endocrine pancreas, central and peripheral nervous systems, gastrointestinal tract, cardiovascular system, kidney, and lung [6]. The main action of GLP-1 is to stimulate insulin secretion as an incretin hormone. GLP-1 also has other important roles in

Table 17C.1   Pancreatic and Extra-Pancreatic GLP-1 Receptor Agonist-Dependent Actions

Organ/Tissues	Functions
Pancreas	Stimulates insulin secretion
	Inhibits glucagon secretion
	Stimulates β-cell proliferation and neogenesis
	Stimulates somatostatin secretion
	Stimulates expression of genes that modify β-cell function
	Stimulates β-cell survival
Central nervous system	Decreases food intake
	Induces satiety
	Stimulates neuronal cell proliferation and neogenesis
	Stimulates neuronal cell survival
	Stimulates learning and memory
Stomach and intestine	Inhibits gastric emptying
	Inhibits bowel motility
Liver, fat, and muscle	Stimulates glycogen synthesis
	Stimulates lipogenesis
Heart	Cardioprotection

glycemic control, such as inhibition of gastric motility, suppression of glucagon secretion, and stimulation of satiety [2] (Table 17C.1).

Pathophysiological Implications

Clinical Implications

GIP and GLP have implications in the following diseases and conditions.

- *Diabetes.* Type 2 diabetes is characterized by a severely reduced or absent incretin effect. The incretin defect is due to almost complete loss of the insulinotropic effect of GIP, whereas secretion of GIP is normal. In contrast to GIP, GLP-1 was found to retain its insulinotropic effect in patients, and this fact provides part of the background for the clinical use of GLP-1 receptor agonists in diabetes treatment [7].
- *Obesity.* A role for GLP-1 in the development of obesity was suggested partly because of the apparent physiological effects of the hormones on appetite and food intake [8], and reduced GLP-1 secretion in obesity.

- *Dumping syndrome.* If food reaches the distal intestine very rapidly — for example, after gastrectomy — exaggerated plasma concentration of GLP-1 may lead to reactive hyperinsulinemia and hypoglycemia [9].

Use for Diagnosis and Treatment

GLP-1, possessing incretin activity, enhances glucose-stimulated insulin release. GLP-1 agonists are resistant to degradation by DPP-4; thus these agonists are used clinically for treatment of diabetes. DPP-4 inhibitors are also acceptable for use in diabetes patients because GLP-1 is rapidly deactivated by DPP-4.

References

1. Orskov C, Rabenhøj L, Wettergren A, et al. Tissue and plasma concentrations of amidated and glycin-extended glucagon-like peptide I in humans. *Diabetes*. 1994;43:535–539.
2. Holst JJ. The physiology of glucagon-like peptide 1. *Physiol Rev*. 2007;87:1409–1439.
3. Drucker DJ, Phillippe J, Mojsov S, et al. Glucagon-like peptide I stimulates insulin gene expression and increases cyclic AMP levels in a rat islet cell line. *Proc Natl Acad Sci USA*. 1987;84:3434–3438.
4. Chen Q, Miller LJ, Dong M. Role of N-linked glycosylation in biosynthesis, trafficking, and function of the human glucagon-like peptide 1 receptor. *Am J Physiol Endocrinol Metab*. 2010;299: E62–E68.
5. Ahren B. Glucagon-like peptide-1 (GLP-1): a gut hormone of potential interest in the treatment of diabetes. *BioEssays*. 1998;20:642–651.
6. Bullock BP, Heller RS, Habener JF. Tissue distribution of messenger ribonucleic acid encoding the rat glucagon-like peptide-1 receptor. *Endocrinology*. 1996;137:2968–2978.
7. Kjems LL, Holst JJ, Vølund A, et al. The influence of GLP-1 on glucose-stimulated insulin secretion: effects on beta-cell sensitivity in type 2 and nondiabetic subjects. *Diabetes*. 2003;52:380–386.
8. Verdich C, Flint A, Gutzwiller JP, et al. A meta-analysis of the effect of glucagon-like peptide-1 (7–36) amide on ad libitum energy intake in humans. *J Clin Endocrinol Metab*. 2001;86:4382–4389.
9. Yamamoto H, Mori T, Tsuchibashi H, et al. A possible role of GLP-1 in the pathophysiology of early dumping syndrome. *Dig Dis Sci*. 2005;50:2263–2267.

Supplemental Information Available on Companion Website

- Gene, mRNA, and precursor structures of human GLP-1 and other derived peptides/E-Figure 17C.1
- The structures and differential posttranslational processing of proglucagon in the pancreas, and in the gut and brain/ E-Figure 17C.2
- Gene, mRNA, and precursor structures of the human GLP-1 receptor/E-Figure 17C.3
- Comparison of amino acid sequences of the GLP-1 receptors of various vertebrates/E-Figure 17C.4
- Comparison of clearly demonstrated physiological roles of GLP-1 and GIP/E-Table 17C.1
- Accession numbers of genes, cDNAs, and peptides for proglucagon and GLP-1 receptor/E-Tables 17C.2, 17C.3

Glucagon-Like Peptide-2

Min Kyun Park

Abbreviation: GLP-2
Additional name: glucagon-like peptide-II (GLP-II)

GLP-2 is co-secreted with GLP-1 in response to nutrient ingestion. The principal role of GLP-2 appears to be the maintenance of growth and absorptive function of the intestinal mucosal villus epithelium.

Discovery

GLP-2 was first identified as a novel peptide following the cloning of genes for proglucagon in the early 1980s, and subsequently the biosynthesis and release of GLP-2 were confirmed by isolation and characterization from porcine and human small intestine [1]. The biological action of GLP-2 was first demonstrated in 1996 by Drucker and colleagues [2].

Structure

Structural Features

GLP-2 is a 33-aa peptide, and has high sequence homology as a member of glucagon family within the secretin superfamily (Figure 17D.1). The N-terminal two aa residues are cleaved off by DPP-4, and the amino-terminally truncated GLP-2 fragment, GLP-2(3–33), acts as a competitive antagonist of the GLP-2 receptor, inhibiting nutrient- and GLP-2-induced mucosal growth in rodents. All placental mammalian GLP-2 sequences differ from reptilian or avian sequences at 14–18 residues. Phylogenetic analyses suggest that GLP-2 sequences have evolved most rapidly among the members of the glucagon family [3,4].

Properties

Human GLP-2: Mr 3,766.1, theoretical pI 4.17.

Synthesis and Release

Gene, mRNA, and Precursor

The human proglucagon gene (*GCG*), located on chromosome 2 (2q36–q37), spans approximately 9.4 kb. Preproglucagon encoded, it is 180 amino acids in length and contains glucagon and two glucagon-like peptides (GLP-1 and GLP-2). Expression of proglucagon has been detected in the pancreatic α cells, intestinal L cells, and brain.

Plasma Concentration

The plasma level of GLP-2 was 16 ± 3 pmol/l in overnight fasted humans, and increased to 73 ± 10 pmol/l at 90 minutes after mixed meal ingestion [5].

Regulation of Synthesis and Release

Proglucagon contains not only GLP-2 but also glucagon and GLP-1, and is posttranslationally processed in a tissue-specific manner in pancreatic α cells and intestinal L cells. In pancreatic α cells the major bioactive hormone is glucagon cleaved by PC2, whereas in the intestinal L cells PC1/3 liberates GLP-1 and GLP-2 as bioactive hormones. As a result, GLP-1 and GLP-2 are co-released from intestinal L cells in a 1:1 ratio following nutrient ingestion, primarily of meals rich in carbohydrates and lipids. The hormone secretion is also regulated by GIP and somatostatin, gastrin-releasing peptide, and neural stimuli in a species-specific manner. The biological half-life of circulating GLP-2 is very short (approximately 7 minutes in the rat and human), owing to inactivation via amino-terminal cleavage by DPP-4.

Receptors

Structure and Subtypes

The human GLP-2 receptor gene (*GLP2R*) is located on chromosome 17 (17p13.3), and the human and rat GLP-2 receptor cDNAs were cloned from intestinal and hypothalamic cDNA libraries in 1999 [6]. The rat GLP-2 cDNA comprises 2,357 bp and encodes an open reading frame of a 550-aa precursor. The GLP-2 receptor is highly specific to GLP-2, with increased cAMP production ($EC_{50} = 0.58$ nM), but not to related members of the glucagon peptide family (no significant cAMP production at 10 nM) [7]. In the rat, GLP-2 receptor is highly restricted and tissue-specific, being most abundant in the jejunum, followed by the duodenum, ileum, colon and stomach, but found only at very low levels in several other tissues, including the central nervous system.

Signal Transduction Pathway

GLP-2 activates cAMP production in rodent and human cells transfected with rat or human GLP-2 receptor [6]. GLP-2 receptor is not localized to the known target cells of GLP-2 tropic action, but to scattered enteroendocrine cells, enteric neurons, and subepithelial myofibroblasts. These results suggest that paracrine and/or neural pathways may mediate the intestinal tropic actions of GLP-2. It has been reported that GLP-2 acts through a neural pathway to affect intestinal crypt cell *c-fos* expression, and through a nitric oxide-dependent mechanism to affect intestinal blood flow. For GLP-2-induced epithelial growth, KGF and

Y. Takei, H. Ando, & K. Tsutsui (Eds): Handbook of Hormones. DOI: http://dx.doi.org/10.1016/B978-0-12-801028-0.00141-0

```
              Human : HADGSFSDEM NTILDNLAAR DFINWLIQTK ITD
              Mouse : .......... S......... .......... ...
Hamster/Rat/Guinea pig : .......... .....S..T. .......... ...
              Camel : .......... ........TQ ......L... ...
                Dog : .......... ..V..T..T. ......L... ...
               Degu : .......... ..V..H..TK .......... ...
                Pig : ......K... ..V.....T. ......LH.. ...SL
Chicken/Turkey/Alligators/Painted turtle : ...T.TSDI .K...DM..K E.LK...N.. V.Q
   Soft-shelled turtle : ...T.TSDF .K...DM..K E.LK...N.. V.Q
     Gila monster lizard : ...T.TSDY .QL..DI.TQ E.LK...NQ. V.Q
   Rana catesbeiana & tigrina : .....TSDF .KA..IK..Q E.LD.I.N.P VKE
    Western clawed frog : .....TNDI .KV..II..Q E.LD.V.N.Q V.E
           Salamander : .....TSDI .KV..TI..K E.L....S.. V.E
   Rainbow trout-1 and -2 : V....TSDV .KV..S...K EYLL.VMTS. TSG
        Sea lamprey-1 : ..EDVMALIL R.MAKTDFEN WEKQNSNTQT D
        Sea lamprey-2 : .S....TND. .VM..RMS.K N.LE..K.QG RG
```

Figure 17D.1 Comparison of aa sequences of vertebrate GLP-2. Amino acids identical to those of human GLP-2 are indicated by dots. The completely conserved amino acid residues are shaded gray.

IGF-1, secreted from subepithelial myofibroblasts, act as essential mediators in responses to GLP-2 [6].

Agonists

[Glycine2]-GLP-2 is a long-acting agonist; teduglutide is a protease-resistant analog of GLP-2.

Antagonists

There are no high affinity specific GLP-2 receptor antagonists.

Biological Functions

Target Cells/Tissues and Functions

Compared with GLP-1 and glucagon receptors, GLP-2 receptor expression is more restricted, occurring predominantly in the gastrointestinal tract, brain, and lung. The gastrointestinal tract, from the stomach to the colon, is the principal target for GLP-2 action. It has been suggested that GLP-2 may exert diverse actions involved mainly in the control of gastrointestinal growth and function (for example, epithelial integrity, motility, and secretion; local blood flow; and nutrient uptake and utilization). A stimulatory effect of GLP-2 on glucagon secretion from the human pancreas has also been reported [8]. GLP-2 receptor mRNA expression has also been found in normal human cervix, but its functional relevance is not yet known.

Phenotype in Gene-Modified Animals

In the GLP-2 receptor knockout mice cell lineage allocation, growth and development of the small and large bowel was normal; however, the intestinal response to GLP-2 was completely extinguished. Mucosal adaptation was also defective, and exogenous EGF could restore re-feeding induced mucosal adaptation in the knockout mice [9].

Pathophysiological Implications

Clinical Implications

The pharmacological application of GLP-2 has been recognized and assessed in preclinical and clinical investigations to prevent or treat a number of intestinal diseases, including short bowel syndrome, Crohn's disease, inflammatory bowel disease, chemotherapy-induced intestinal mucositis, colon carcinogenesis, and small bowel enteritis [10,11].

Use for Diagnosis and Treatment

The US Food and Drug Administration (FDA) has accepted teduglutide, an analog of GLP-2, for use in adult patients with short bowel syndrome.

References

1. Buhl T, Thim L, Kofod H, et al. Naturally occurring products of proglucagon 111−160 in the porcine and human small intestine. *J Biol Chem.* 1988;263:8621−8624.
2. Drucker DJ, Erlich P, Asa SL, et al. Induction of intestinal epithelial proliferation by glucagon-like peptide 2. *Proc Natl Acad Sci.* 1996;93:7911−7916.
3. Irwin DM. Molecular evolution of mammalian incretin hormone genes. *Regl Pept.* 2009;155:121−130.
4. Estall JL, Drucker DJ. Glucagon-like peptide-2. *Annu Rev Nutr.* 2006;26:391−411.
5. Hartmann B, Harr MB, Jeffersen PB, et al. *In vivo* and *in vitro* degradation of glucagon-like peptide-2 in humans. *J Clin Endocrinol Metab.* 2000;85:2884−2888.
6. Dube PE, Forse CL, Bahrami J, et al. The essential role of insulin-like growth factor-1 in the intestinal tropic effects of glucagon-like peptide-2 in mice. *Gastroenterology.* 2006;131:589−605.
7. Munroe DG, Gupta AK, Kooshesh F, et al. Prototypic G protein-coupled receptor for the intestinotrophic factor glucagon-like peptide 2. *Proc Natl Acad Sci USA.* 1999;96:1569−1573.
8. Meier JJ, Nauck MA, Pott A, et al. Glucagon-like peptide 2 stimulates glucagon secretion, enhances lipid absorption, and inhibits gastric acid secretion in humans. *Gastroenterology.* 2006;130:44−54.
9. Bahrami JI, Yusta B, Drucker DJ. ErbB activity links the glucagon-like peptide-2 receptor to refeeding-induced adaptation in the murine small bowel. *Gastroenterology.* 2010;138:2447−2456.
10. Yazbeck R, Abbott CA, Howarth GS. The use of GLP-2 and related growth factors in intestinal diseases. *Curr Opin Investig Drugs.* 2010;11:440−446.
11. Kannen V, Garcia SB, Stopper H, et al. Glucagon-like peptide 2 in colon carcinogenesis: possible target for anti-cancer therapy? *Pharmacol Ther.* 2013;139:87−94.
12. Cavanaugh ES, Nielsen PF, Conlon JM. Isolation and structural characterization of proglucagon-derived peptides, pancreatic polypeptide, and somatostatin from the urodele, *Amphiuma tridactylum. Gen Comp Endocrinol.* 1996;101:12−20.
13. Irwin DM, Sivarajah P. Proglucagon cDNAs from the leopard frog, *Rana pipiens*, encode two GLP-1-like peptides. *Mol Cell Endocrinol.* 2000;162:17−24.
14. Nguyen TM, Mommsen TP, Mims SM, Conlon JM. Characterization of insulins and proglucagon-derived peptides from a phylogenetically ancient fish, the paddlefish (*Polyodon spathula*). *Biochem J.* 1994;300(Pt 2):339−345.

Supplemental Information Available on Companion Website

- Gene, mRNA, and precursor structures of the human GLP-2 and other derived peptides/E-Figure 17D.1
- The structures and differential posttranslational processing of proglucagon in the pancreas, and in the gut and brain/E-Figure 17D.2
- Gene, mRNA, and precursor structures of the human GLP-2 receptor/E-Figure 17D.3
- Comparison of amino acid sequences of the GLP-2 receptor of various vertebrates/E-Figure 17D.4
- Schematic representation of interactions between GLP-2 and growth factor signaling systems in the regulation of intestinal growth/E-Figure 17D.5
- Accession numbers of genes, cDNAs, and peptides for proglucagon and GLP-2 receptor/E-Tables 17D.1 [12,13], 17D.2

CHAPTER 18

Secretin (Pituitary Adenylate Cyclase-Activating Polypeptide) Family

Yoshio Takei

History

The secretin-glucagon superfamily or pituitary adenylate cyclase-activating peptide (PACAP)/glucagon superfamily can be divided into two families, the secretin (PACAP) family and the glucagon family. The secretin family was initially so named because secretin was the first to be named as "hormone" in 1902. However, secretin is absent in fishes, and the ancient gene of the family seems to code PACAP. The secretin (PACAP) family consists of PACAP, PACAP-related peptide (PRP), vasoactive intestinal peptide (VIP), peptide histidine methionine/isoleucine (PHM/PHI), secretin, and growth hormone-releasing hormone (GHRH). VIP was first identified in the porcine small intestine in 1970, PACAP in the ovine hypothalamus in 1989, and GHRH in a human pancreatic tumor in 1982.

Structure

Structural Features

Human PACAP consists of 27 or 38 aa residues with amidated C-terminus (Figure 18.1) [1]. PACAP is the most conserved peptide among the secretin-glucagon superfamily, and human PACAP-27 is identical to those of some teleosts [2]. PRP, consisting of 29 aa residues in humans, locates in the N-terminal region of proPACAP, and is the most versatile among the family members (E-Figure 18.1) [3]. VIP consists of 28 aa residues with an amidated C-terminus, and is the second most conserved peptide in the family. PHM/PHI consists of 27 aa residues and exists in the N-terminal region of proVIP (Figure 18.1) [2]. Two PRP/PACAP and PHI/VIP genes exist in some teleosts because of the additional whole genome duplication. Secretin is a 27-aa peptide with C-terminal amidation. GHRH is a 44-aa peptide in mammals, and is highly variable among species.

Molecular Evolution of Family Members

Two PACAPs and two PRPs are found in a tunicate (*Chelyosoma productum*) [2]. Both peptides consist of 27 aa residues and are well conserved, only 1 aa different from human PACAP, suggesting that ancestral molecule of the family is PACAP-like (Figure 18.2). PACAP has also been identified in protostomes such as crab, cockroach, planarian, and hydra [4]. It is amazing that hydra PACAP-27 is identical to the human counterpart. It is apparent that PRP/PACAP coexists in the same gene already in the protochordates by exon duplication (Figure 18.2). The PRP/PACAP gene may then have duplicated by genome duplication sometime before the divergence of ray-finned fishes and lobe-finned fishes, from one of which GHRH may have been produced from PRP because of high sequence similarity (Figure 18.1). Another whole genome duplication may have occurred for the PRP/PACAP gene, which resulted in the production of the PHI/VIP gene. The order of phylogenetic appearance of the GHRH and PHI/VIP genes needs further investigation; cartilaginous fishes possess the PHI/VIP gene but the GHRH gene has not yet been identifiable in this group [2,4]. Secretin is found in birds and mammals, but not in teleosts. It is not yet known when the secretin gene appeared during evolution (Figure 18.2).

Receptors

Structure and Subtypes

At least seven types of receptors have been identified in the family: PAC1 for PACAP, VPAC1 and VPAC2 for both PACAP and VIP, GHRH receptor in teleosts and mammals, secretin receptor only in mammals, PRP receptor in birds and teleosts, and PHI/PHM receptor in teleosts [4]. PAC1 and VPAC1/2 have been duplicated in teleosts, as is the case for their ligands, but only a single PRP and PHI/PHM receptor has been reported. All these receptors belong to the class B or II GPCRs [5]. This class of GPCR includes receptors for the glucagon family, the GPCR family, and others.

Signal Transduction Pathway

All secretin family receptors are coupled with G_s protein and primarily activate adenylyl cyclase to increase the levels of intracellular cAMP. PAC1 is able to stimulate the phospholipase C and IP3 pathway in mammals and teleosts [4]. Class B GPCRs often interact with the receptor activity-modifying proteins (RAMPs), and the secretin receptor is found to be coupled with RAMP 3 and VPAC1 with RAMP1/2/3 [6]. However, how RAMPs regulate the receptor activity (insertion of the intracellular receptor proteins in vesicles into cell membrane, modification of signal transduction, etc.) has not yet been fully elucidated.

```
VIP       HSDAVFTDNYTRLRKQMAVKKYLNSILN–NH2
PACAP     HSDGIFTDSYSRYRKQMAVKKYLAAVLGKRYKQRVKNK–NH2
PRP       DVAHGILNEAYRKVLDQLSAGKHLQSLVARGVGGSLGGGAGDDAEPLS
PHM       HADGVFTSDFSKLLGQLSAKKYLESLM
Secretin  HSDGTFTSELSRLREGARLQRLLQGLV–NH2
GHRH      YADAIFTNSYRKVLGQLSARKLLQDIMSRQQGESNQERGARARL
```

Figure 18.1 Comparison of amino acid sequences of secretin family peptides in human.

Y. Takei, H. Ando, & K. Tsutsui (Eds): Handbook of Hormones. DOI: http://dx.doi.org/10.1016/B978-0-12-801028-0.00018-0

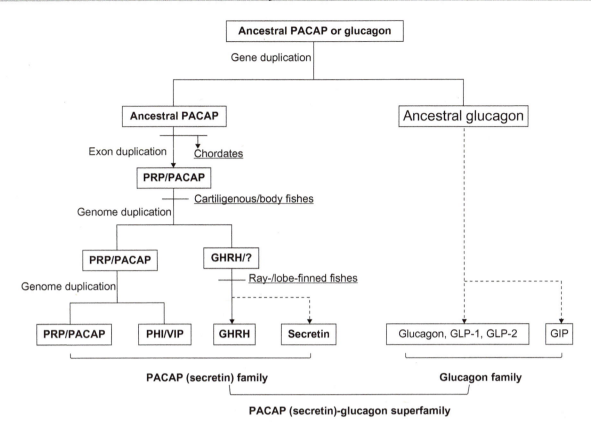

Figure 18.2 Evolutionary history of the secretin—glucagon family in chordates.

Biological Functions

Target Cells/Tissues and Functions

As the biological functions of respective family members are explained in detail in the following chapters, the main action of each member will be introduced briefly here, and then an attempt will be made to extract a common and essential function of the family. Members of the secretin (PACAP) family appear to be divided into two groups based on diversity of actions; PACAP and VIP exert pleiotropic actions, while secretin, GHRH, PRP, and PHI/PHM have rather specific functions. In fact, the functions of PRP and PHI/PHM are largely unknown as yet even in mammals. It is of interest to note that PACAP and VIP, which exhibit diverse actions, have more conserved aa sequences across species than other members with a specific action, suggesting some are essential for life [1,2].

PACAP is the most pleiotropic hormone, exhibiting a spectrum of actions not only in the CNS but also in the peripheral tissues. In the CNS, it is involved in the basic functions for life, such as nerve growth and differentiation, and neuroprotection [2]. Thus, PACAP gene knockout in mice often leads to early death after birth. In the periphery, PACAP acts as a smooth muscle relaxant and secretagog of pituitary, pancreatic, and other hormones. VIP has similar actions to PACAP, but vasorelaxant actions and regulation of ion transport in the intestine and other tissues are more prominent. Probably due to the weaker CNS effect, VIP knockout is not lethal. The biological functions of PRP and PHI/PHM are not evident in mammals, probably related to their unidentifiable receptors. PRP stimulates GH release in goldfish, which is the cause of its initial name ("GHRH-like peptide") for teleost PRP. PHI stimulates prolactin, insulin, and glucagon secretion in rats, but its potency and efficacy are much lower than those of VIP and PACAP. GHRH is the secretagog of GH in mammals, as its name indicates, but not in non-mammalian species (birds and teleosts). Secretin is an important regulator of gastrointestinal endocrine/exocrine secretion, but its actions are much more diverse than initially thought, as indicated by the wide distribution of its receptor. Collectively, the main action of the secretin (PACAP) family is to coordinate the central and peripheral network for homeostatic regulation of various systems by governing the development and activity of neural and endocrine tissues.

References

1. Sherwood NM, Krueckl SL, McRory JE. The origin and function of the pituitary adenylate cyclase-activating polypeptide (PACAP)/glucagon superfamily. *Endocr Rev.* 2000;21:619—670.
2. Vaudry D, Falluel-Morel A, Bourgault S, et al. Pituitary adenylate cyclase-activating polypeptide and its receptors: 20 years after the discovery. *Pharmacol Rev.* 2009;61:283—357.
3. Tam JK, Lee LT, Chow BK. PACAP-related peptide (PRP) — Molecular evolution and potential functions. *Peptides.* 2007;28:1920—1929.
4. Cardoso JCR, Vieira FA, Gomes AS, et al. PACAP, VIP and their receptors in the metazoan: insights about the origin and evolution of the ligand—receptor pair. *Peptides.* 2007;28:1902—1919.
5. Hollenstein K, de Graaf C, Bortolato A, et al. Insights into the structure of class B GPCRs. *Trends Pharm Sci.* 2014;35:12—22.
6. Archbold JK, Flanagan JU, Watkins HA, et al. Structural insights into RAMP modification of secretin family G protein-coupled receptors: implications for drug development. *Trends Pharm Sci.* 2011;32:591—600.

Supplemental Information Available on Companion Website

- Gene structures of secretin family peptides in human/E-Figure 18.1
- Phylogenetic tree of secretin family peptides/E-Figure 18.2
- Phylogenetic tree of secretin family receptors/E-Figure 18.3

Secretin

Tomoya Nakamachi

Abbreviation: SCT
Additional names: oxykrinin, secretina

Secretin was the first hormone identified. It is a secretagog of pancreatic digestive juices from the porcine duodenum, used to aid in the diagnosis of gastrinoma.

Discovery

Secretin was discovered by Bayliss and Starling in 1902 [1], as a secretagog of pancreatic juices from the porcine duodenum. It was the first hormone identified.

Structure
Structural Features

In 1970, the aa sequence of secretin was determined by Mutt to be a polypeptide comprising 27 aa residues. Secretin belongs to secretin/glucagon family, which shares significant structural and conformational homology [2]. The family includes vasoactive intestinal peptide (VIP), pituitary adenylate cyclase activating peptide (PACAP), growth hormone-releasing hormone (GHRH), peptide histidine isoleucine (PHI) or peptide histidine methionine (PHM), glucagon, glucagon-like peptide 1 (GLP-1), glucagon-like peptide 2, and gastric inhibitory peptide (GIP). Secretin is highly conserved among mammalian species, but the sequence identity in non-mammalian species up to birds is lower than in mammals (Figure 18A.1) [3]. Secretin-like sequences have not been found in teleost and lungfish genomes.

Properties

Porcine secretin, Mr 3,055; human secretin, 3,039; chicken secretin, 3,073. Soluble in water, physiological saline solution, and aqueous organic solvents. Stable in dilute hydrochloric acid at $-20°C$, but unstable in aqueous solution.

Human	HSDGT	FTSEL	SRLRE GARLQ	RLLQG LV-
Pig	HSDGT	FTSEL	SRLRD SARLQ	RLLQG LV-
Dog	HSDGT	FTSEL	SRLRE SARLQ	RLLQG LV-
Guinea pig	HSDGT	FTSEL	SRLRD SARLQ	RLLQG LV-
Rat	HSDGT	FTSEL	SRLQD SARLQ	RLLQG LV-
Mouse	HSDGM	FTSEL	SRLQD SARLQ	RLLQG LV-
Chiken	HSDGL	FTSEY	SKMRG NAQVQ	KFIQN LM-
X. laevis	HVDGR	FTSEF	SRARG SAAIR	KIINS ALA
X. tropicalis	HVDGM	FTSEF	SRARG SAAIR	KIINS ALA

Figure 18A.1 Comparison of Secretin Sequences in Representative Species (Conserved Sequence is Shaded).

Synthesis and Release
Gene, mRNA, and Precursor

Human secretin is derived from a 121-aa residue secretin precursor consisting of a signal peptide, short N-terminal peptide, secretin, and C-terminal peptide. The human secretin precursor gene, SCT, is located on chromosome 11p15.5. SCT consist of four exons, and the secretin coding region is exon 2 (Figure 18A.2). The chicken SCT gene consists of seven exons. Exons 1 and 2 are non-coding, exon 4 encodes the secretin-like peptide, and exon 5 encodes the secretin peptide (Figure 18A.2). The accession numbers of the SCT genes and cDNAs are listed in E-Table 18A.1.

Distribution of mRNA

Secretin mRNA has been mainly detected in the duodenum, jejunum, and ileum, and slightly less so in the colon, by northern blotting analysis [4]. By PCR analysis, secretin mRNA has been detected in hypothalamus, brain stem, brain cortex, heart, lung, kidney, liver, and testis, but secretin peptide is not detected in the organs.

Tissue and Plasma Concentrations

Tissue: In humans, secretin tissue content is 73 ± 7 pmol/g in the duodenum, 32 ± 4 pmol/g in the jejunum, and 5 ± 0.5 pmol/g in the ileum.

Plasma: Human (fasting) plasma concentration of secretin is 500−10,000 fmol/ml.

Regulation of Synthesis and Release

Secretin is synthesized and secreted by S cells in the small intestine, and neurons in the brain. Secretin release is mainly stimulated by gastric acid delivered into the duodenal lumen. In addition, secretin is released by digested products of fat and protein. In canine duodenal mucosal explants, somatostatin did not alter the basal secretion of secretin but inhibited secretin secretion stimulated by pH 4.5 [5]. GABA stimulated both basal and pH 4.5-stimulated secretin secretion. In the brain, estrogen-related receptor α upregulates SCT promoter and gene expression.

Receptors
Structure and Subtype

The receptor of secretin (SCTR) is a seven-transmembrane GPCR that belongs to a subclass of the family B (E-Figure 18A.1). The human SCTR consists of 440 aa residues that contain a large N-terminal extracellular domain. Cys residue and

Y. Takei, H. Ando, & K. Tsutsui (Eds): Handbook of Hormones. DOI: http://dx.doi.org/10.1016/B978-0-12-801028-0.00142-2

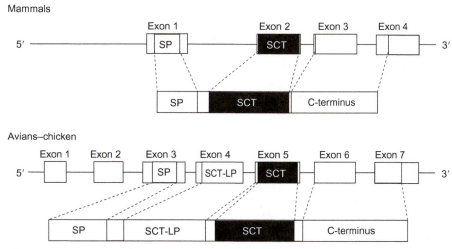

Figure 18A.2   Gene and mRNA organization of mammal and avian secretin. Exons are shown as boxes and introns as lines. SP, signal peptide; SCT-LP, secretin-like peptide. *Figure adapted from Tam et al. 2014 [3].*

disulfide bonds in the extracellular domain have been suggested to play a key role in agonist binding. The accession numbers of SCTR genes and cDNA are listed in E-Table 18A.2.

Signal Transduction Pathway

The pathway is mainly coupled to G_s protein. Secretin stimulates production of cAMP, with the exception of teleost secretin receptors.

Biological Functions

Target Cells/Tissues and Functions

Using northern blotting, human SCTR mRNA was detected in the pancreas, intestine, colon, kidney, lung, and liver [6]. Secretin stimulates the secretion of bicarbonate, water, and electrolytes from the ductal epithelium. In the stomach, secretin acts as an enterogastrone that inhibits gastric acid release and gastric emptying. In the kidney, secretin regulates urine output and activates adenylyl cyclase in rats. In the brain, SCTR is widely distributed in the hippocampus, central amygdala, thalamus, hypothalamus, posterior pituitary, cerebellum, and medulla oblongata [7]. Secretin regulates social interaction, water and food intake, motor coordination, and spatial and motor learning behaviors [7]. Zebrafish secretin receptor has been shown to be non-bioactive.

Phenotype in Gene-Modified Animals

SCT knockout mice demonstrate impairment in synaptic plasticity in the CA1 area of the hippocampus. In addition, SCT knockout mice exhibit reduced survival of neural progenitor cells in the subgranular zone of the dentate gyrus, reduced long-term potentiation, and impaired spatial learning ability in adults [7]. SCTR knockout mice show a mild polydipsia and polyuria. SCTR knockout mice show altered glomerular and tubular morphology and reduction of AQP2 and AQP4 renal expression, suggesting possible disturbances in the filtration and/or water reabsorption process [8].

Pathophysiological Implications

Clinical Implications

The stimulatory effect of secretin on pancreatic secretion has been used in a range of clinical applications. Zollinger-Ellison syndrome is caused by a gastrin-secreting tumor (gastrinoma),

and the secretin test is used for diagnosis of the syndrome [9]. In the secretin test, the gastrin level is raised after intravenous secretin administration in Zollinger-Ellison syndrome patients, but not in healthy individuals. Secretin is elevated in Zollinger-Ellison patients and in patients with duodenal ulcers. Secretin levels are low in patients with pernicious anemia and achlorhydria. A clinical trial of secretin in autism has been carried out, but the clinical efficacy remains unreliable [10].

Use for Diagnosis and Treatment

Recombinant human secretin (SecreFlo, ChiRhoStim) is approved for use in stimulating gastrin secretion to aid in the diagnosis of gastrinoma.

References

1. Chey WY, Chang TM. Secretin, 100 years later. *J Gastroenterol.* 2003;38:1025–1035.
2. Ng SS, Yung WH, Chow BK. Secretin as a neuropeptide. *Mol Neurobiol.* 2002;26:97–107.
3. Tam JK, Lee LT, Jin J, et al. Molecular evolution of GPCRs: secretin/secretin receptors. *J Mol Endocrinol.* 2014;52:T1–14.
4. Kopin AS, Wheeler MB, Leiter AB. Secretin: structure of the precursor and tissue distribution of the mRNA. *Proc Natl Acad Sci USA.* 1990;87:2299–2303.
5. Murthy SN, Lavy A, Morgantini DS, et al. Neurohormonal regulation of secretin secretion in canine duodenal mucosa *in vitro. Peptides.* 1986;7:229–236.
6. Chow BK. Molecular cloning and functional characterization of a human secretin receptor. *Biochem Biophys Res Commun.* 1995;212:204–211.
7. Zhang L, Chow BK. The central mechanisms of secretin in regulating multiple behaviors. *Front Endocrinol (Lausanne).* 2014;5:77.
8. Chu JY, Chung SC, Lam AK, et al. Phenotypes developed in secretin receptor-null mice indicated a role for secretin in regulating renal water reabsorption. *Mol Cell Biol.* 2007;27:2499–2511.
9. McGuigan JE, Wolfe MM. Secretin injection test in the diagnosis of gastrinoma. *Gastroenterology.* 1980;79:1324–1331.
10. Williams K, Wray JA, Wheeler DM. Intravenous secretin for autism spectrum disorders (ASD). *Cochrane Database Syst Rev.* 2012;4:CD003495.

Supplemental Information Available on Companion Website

Growth Hormone-Releasing Hormone

Zhifang Xu

Abbreviation: GHRH
Additional names: growth-hormone-releasing factor (GHRF, GRF), somatoliberin, somatocrinin

Expressed and secreted from the hypothalamic neurons of the arcuate nucleus, GHRH stimulates the release of growth hormone (GH) in the anterior pituitary.

Discovery

In 1982, three isoforms of GHRH (1−37, 1−40, 1−44 aa residues) were initially isolated from human pancreatic tumors that caused acromegaly, and the latter two were found in human hypothalamus. The sequence of GHRH was also identified in various vertebrates, from rodents to fish, including a protochordate [1,2].

Structure

Structural Features

GHRH$_{1-29}$ is the bioactive core of human GHRH (hGHRH, Figure 18B.1). The N-terminal tyrosine residue with selected aromatic rings is important for the high bioactivity in human and non-rodent mammalian GHRH.

Primary Structure

The amino acid sequence of hypothalamic GHRH shows higher identities between human, porcine, bovine, and caprine species, but the rat and mouse are exceptions. The sequence of the C-terminus is highly variable among species, while the identity of the N-terminal is more conserved [3] (E-Figure 18B.2).

Properties

Mr 12,447, pI 10.3. Soluble in acidic aqueous solution (e.g., 1% acetic acid). Lyophilized GHRH is stable at room temperature for 2 months, and recommended storage is below −18° with desiccation.

Synthesis and Release

Gene, mRNA, and Precursor

The hGHRH gene, located on chromosome 20 (20q11.2), consists of five exons. The mRNA has 459 bases that encode a signal peptide of 24 aa residues, a mature protein of 44 aa residues, and a C-peptide of 31 aa residues with unknown function (E-Figure 18B.1). In non-mammalian vertebrates and protochordates, GHRH and pituitary adenylate cyclase-activating polypeptide (PACAP) are believed to be encoded by the same gene and processed from the same transcript and prepropolypeptide.

Distribution of mRNA

The hGHRH gene is mainly expressed in the brain, especially in the ARC of the hypothalamus, and has also been detected in the medulla nephrica, ovary, testis, and placenta [2].

Tissue and Plasma Concentrations

Tissue: Content in the brain is greater than in other tissues, at 2.7 ± 0.24 ng/mg in the hypothalamus of male rats.
  Plasma: Human, 10−60 pg/ml.

Regulation of Synthesis and Release

The synthesis and release of GHRH are regulated by transcriptional factors, sex hormones, aging, the negative feedback effect of GH, and diverse pathological conditions. Gsh-1 has been considered as a transcriptional factor of Ghrh gene expression in rat hypothalamus. GHRH synthesis is inhibited by somatostatin, whose receptor sst2A on GHRH neurons is higher in female than in male mice. Production of hypothalamic GHRH was also observed to be decreased by aging. It is also negatively regulated by GH feedback, whereas ghrelin stimulates GHRH release.

Receptors

Structure and Subtype

GHRH-R belongs to the GPCR B II subclass, highly selective for this cognate ligand [4]. The GHRH-R of most mammals consists of 423 aa residues. The N-terminal extracellular domain contains a site for N-glycosylation as well as six cysteine residues and an aspartate residue that are conserved in this receptor family, while the third intracellular loop and C-intracellular domain contain several potential phosphorylation sites, which may regulate signaling and receptor internalization. It is mainly expressed in the pituitary [5] (E-Figures 18B.4 and 18B.5).

Signal Transduction Pathway

Transduction primarily involves cAMP and Ca^{2+}-dependent pathways, but also IP3/DAG and other minor pathways.

Agonists

Tesamorelin and sermorelin (GHRH (1−29)-NH$_2$) are agonists.

Y. Takei, H. Ando, & K. Tsutsui (Eds): Handbook of Hormones. DOI: http://dx.doi.org/10.1016/B978-0-12-801028-0.00143-4

	1	10	20	30	40
hGHRH			YADAIFTNSYRKVLGQLSARKLLQDIMSRQQGESNQE RGARARL-NH$_2$		

Figure 18B.1 The amino acid sequence of hGHRH.

Table 18B.1 The Functions of GHRH

Species	Function	Reference
Human	Stimulation of GH release	[6]
Rat	Stimulation of somatotroph proliferation	[7]
Mouse	Pit-1-dependent expression of GHRH-R mediates pituitary cell growth	[8]
Rat	Regulation of sleep−wake cycle and locomotion	[9]
Dog	Control of jejunal motility in fed but not in fasted condition	[10]
Mouse	Human GHRH transgenic mouse as a model of modest obesity	[11]

Antagonists

Antagonists comprise the antibodies or peptides to GHRH-R: JV-1-10, JV-1-36, JV-1-37, JV-1-38, JV-1-39, JV-1-40, JV-1-41, JV-1-42, JV-1-43, JV-1-62, JV-1-63, MZ-4-71, MZ-4-169, MZ-4-181, MZ-4-243, MZ-5-78, MZ-5-156, MZ-5-192, MZ-6-55, [Ac-Tyr1, D-Arg2]GHRH(1−29)-NH$_2$.

Biological Functions

Target Cells/Tissues and Functions

The primary function of GHRH is to stimulate GH synthesis and secretion from anterior pituitary somatotrophs. GHRH activates cell proliferation, cell differentiation, and growth of somatotrophs, and is also involved in the modulation of appetite and feeding behavior, regulation of sleeping, control of jejunal motility, and increase in leptin levels in modest obesity (Table 18B.1 [5−11]).

Phenotype in Gene-Modified Animals

Mutation of the mouse metallothionein I/human GHRH fusion gene induces somatotroph hyperplasia. Gsh-1, PC-1 deletion rodent models show GHRH deficiency, and Hpt mutant mice present low GH release and development of dwarfism. Pit-1 and prop-1 knockout mice show Ames dwarfism due to a transcriptional defect of GHRH-R [4].

Pathophysiological Implications

Clinical Implications

Mutations in the GHRH gene have never been described. A single base change in the GHRH-R gene in human somatotropinomas confers hypersensitivity to GHRH binding. Pit-1 mutation inducing the low gene expression of GHRH-R can lead to development of dwarfism [12].

Use for Diagnosis and Treatment

Sermorelin, a functional peptide fragment of GHRH1−29, has been used in the diagnosis and treatment of children with idiopathic growth hormone deficiency [13]. Tesamorelin, a stabilized synthetic peptide analog of GHRH$_{1-44}$, received US Food and Drug Administration approval in 2010 for the treatment of lipodystrophy in HIV patients under highly active antiretroviral therapy, and was investigated for effects on certain cognitive functions in adults with cognitive impairment and healthy older adults [4].

References

1. Müller EE, Locatelli V, Cocchi D. Neuroendocrine control of growth hormone secretion. *Physiol Rev.* 1999;79:511−607.
2. Lee LT, Siu FK, Tam JK, et al. Discovery of growth hormone-releasing hormones and receptors in nonmammalian vertebrates. *Proc Natl Acad Sci USA.* 2007;104:2133−2138.
3. Ling N, Baird A, Wehrenberg WB, et al. Synthesis and *in vitro* bioactivity of human growth hormone-releasing factor analogs substituted at position-1. *Biochem Biophys Res Commun.* 1984;122:304−310.
4. Spooner LM, Olin JL. Tesamorelin: a growth hormone-releasing factor analogue for HIV-associated lipodystrophy. *Ann Pharmacother.* 2012;46:240−247.
5. Gaylinn BD. Molecular and cell biology of the growth hormone-releasing hormone receptor. *Growth Horm IGF Res.* 1999;9:37−44.
6. Bloch B, Brazeau P, Ling N, et al. Immunohistochemical detection of growth hormone-releasing factor in brain. *Nature.* 1983;301:607−608.
7. Billestrup N, Swanson LW, Vale W. Growth hormone-releasing factor stimulates proliferation of somatotrophs *in vitro. Proc Natl Acad Sci USA.* 1986;83:6854−6857.
8. Lin C, Lin SC, Chang CP, et al. Pit-1-dependent expression of the receptor for growth hormone releasing factor mediates pituitary cell growth. *Nature.* 1992;360:765−768.
9. Ehlers CL, Reed TK, Henriksen SJ. Effects of corticotropin-releasing factor and growth hormone-releasing factor on sleep and activity in rats. *Neuroendocrinology.* 1986;42:467−474.
10. Bueno L, Fioramonti J, Primi MP. Central effects of growth hormone-releasing factor (GRF) on intestinal motility in dogs: involvement of dopaminergic receptors. *Peptides.* 1985;6:403−407.
11. Cai A, Hyde JF. The human growth hormone-releasing hormone transgenic mouse as a model of modest obesity: differential changes in leptin receptor (OBR) gene expression in the anterior pituitary and hypothalamus after fasting and OBR localization in somatotrophs. *Endocrinology.* 1999;140:3609−3614.
12. Martari M, Salvatori R. Diseases associated with growth hormone-releasing hormone receptor (GHRHR) mutations. *Prog Mol Biol Transl Sci.* 2009;88:57−84.
13. Prakash A, Goa KL. Sermorelin: a review of its use in the diagnosis and treatment of children with idiopathic growth hormone deficiency. *BioDrugs.* 1999;12:139−157.

Supplemental Information Available on Companion Website

- Gene, mRNA and precursor structures of human GHRH /E-Figure 18B.1
- Precursor and mature hormone sequences of various animals/E-Figure 18B.2
- Gene, mRNA and precursor structures of the human GHRH receptor/E-Figure 18B.3
- Primary structure of the human GHRH-R/E-Figure 18B.4
- Primary structure of GHRH receptor of various animals/E-Figure 18B.5
- Accession numbers of genes, cDNAs, and proteins for GHRH and GHRH-R/E-Tables 18B.1, 18B.2

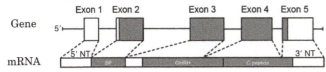

E-Figure 18B.1 Gene, mRNA and precursor structures: human GHRH, location 20q11.2.

Pituitary Adenylate Cyclase-Activating Polypeptide

Tomoya Nakamachi

Abbreviation: PACAP

PACAP consists of 27 or 38 aa residues and belongs to the secretin/glucagon family. PACAP has a pleiotropic functions, acting as a neurotransmitter, neurotrophic factor in the central nervous system, and vasodilator, insulin secretagog, smooth muscle relaxant, and immunosuppressor in peripheral tissue.

Discovery

PACAP was first isolated in 1989 from extract of ovine hypothalamus on the basis of its ability to elevate cAMP levels in rat anterior pituitary cells [1].

Structure

Structural Features

PACAP38, consisting of 38 aa residues, was the first isolated, followed by PACAP27, consisting of 27 aa residues and lacking 11 aa residues from the C-terminal of PACAP38. The C-terminals of both PACAPs was amidated. PACAP belongs to the secretin/glucagon superfamily, and its closest analog is vasoactive intestinal polypeptide (VIP) which shows 68% aa residue homology (Figure 18C.1). Human PACAPs are derived from a 176-aa residue precursor, located in the C-terminal, downstream of PACAP-related peptide (PRP) (E-Figure 18C.1). The PACAP precursor is cleaved by various prohormone convertases to generate PACAP27 or PACAP38. From the evolutionary aspect of the superfamily, the ancestral peptide was created 700 million years ago, and PACAP was established by gene duplication, exon duplication, and exon deletion [2]. The PACAP sequence is well conserved in vertebrates, and is perfectly matched in mammals (Figure 18C.2). Stingray PACAP has 44 aa residues with 2 expected processing sites, suggesting that the processing site of preproPACAP shows diversity in species.

Properties

Human PACAP27, Mr 3,147.6; human PACAP38, Mr 4.534.3. Freely soluble in water and ethanol. PACAP solution in water is unstable at 4°C, but is stable for a year at −80°C at 10^{-4}M concentration.

Synthesis and Release

Gene, mRNA, and Precursor

The human preproPACAP gene (*ADCYAP1*), located on chromosome 18 (18p11), consists of five exons (E-Figure 18C.1). PACAP mRNA of about 3.0 kb has been detected in the human and rat by northern blotting. The preproPACAP gene first appeared in tunicates. The gene structure and its mRNA size are well conserved among teleosts, amphibians, and mammals.

Distribution of mRNA

PACAP is widely distributed in the peripheral and central nervous tissues, being detected most abundantly in the central nervous tissues [3]. A high concentration of PACAP was detected in the hypothalamus, but it is also present in hippocampus, cortex, amygdala, brainstem, posterior pituitary gland, and retina. In the peripheral organs, abundant PACAP is found in the testis, and lower levels are found in the adrenal gland, intestine, and pancreas.

Tissue Concentration

PACAP38 concentrations (ng/g) in the organs of male rats are: hypothalamus, 540 ± 19.9; hippocampus, 36.5 ± 7.8; cerebral cortex, 24.2 ± 3.9; posterior pituitary gland, 40.2 ± 4.2; testis, 54.6 ± 3.9; adrenal gland, 14.4 ± 0.7; jejunum, 6.1 ± 1.3; duodenum, 8.8 ± 1.3; and pancreas, 0.6 ± 0.1. PACAP27 contents are almost 10% those of PACAP38 in mammals.

Regulation of Synthesis and Release

PACAP synthesis and release are stimulated by NGF and cAMP inducer in nervous tissues. PACAP stimulates biosynthesis of itself, indicating that PACAP regulates its expression level via the autocrine/paracrine system.

Receptors

Structure and Subtype

PACAP and VIP share three types of G protein-coupled receptors with seven transmembrane domains: PAC1 receptor (PAC1R), and VPAC1 and VPAC2 receptors (VPAC1R, VPAC2R). PACAP binds to VPAC1-R and VPAC2-R with a similar affinity to that of VIP, while PACAP interacts with PAC1R with 1,000 times higher affinity than does VIP [4]. PAC1R has various splicing forms, which differ in the presence or absence of two cassettes, termed HIP and HOP, in the region of the gene encoding the third cytoplasmic loop

Y. Takei, H. Ando, & K. Tsutsui (Eds): Handbook of Hormones. DOI: http://dx.doi.org/10.1016/B978-0-12-801028-0.00144-6

```
hPACAP38    HSDGI FTDSY SRYRK QMAVK KYLAA VLGKR YKQRV KNK
hPACAP27    ----- ----- ----- ----- ----- --
hVIP        ---AV ---N- T-L-- ----- ---NS I-N
hGlucagon   --Q-T --SD- -K-LD SRRAQ DFVQW LMNT
hSecretin   ----T --SEL --L-E GARLQ RL-QG LV
hGHRH       YA-A- --N-- -KVLG -LSAR -L-QD IMSRQ QGESN QERGA RARL
hPRP        DVAH- -LNEA -RKVL D-LSA G-H-Q SLVA
hPHI        -A--V --SDF -RLLG -LSA- ---ES LI
```

Figure 18C.1 Comparison of Sequence of Human PACAP and Secretin/Glucagon Family.

```
Human/Sheep/Rat/Mouse HSDGI FTDSY SRYRK QMAVK KYLAA VLGKR YKQRV KNKa
Chicken               -I--- ----- ----- ----- ----- --|-| ----- ---a
Frog                  ----- ----- ----- ----- ----- --|-| ----I ---a
Salmon                ----- ----- ----- ----- ----- --|-| -R--Y RS-?
Catfish               ----- ----- ----- ----- ----- ---R-| -R--F R--?
Stargazer             ----- ----- ----- ----Q ----- ---R-| -R--- R--a
Stingray              ----- ----- ----- ----- ----- --|-| ----- --SGR RVFYa
Tunicate-I            ----- ----- ----N ----- ----- --?
Tunicate-II           ----- ----- ----N ----- ---N- L-?
```

Figure 18C.2 Comparison of PACAP sequences in representative species of each vertebrate class. Outlined aa residues indicate predicted sites of processing.

(e.g., PAC1R-short, PAC1R-hop1, PAC1R-hop2, PAC1R-hip) [5]. These cassettes contribute to regulation of the signal transduction pathway.

Signal Transduction Pathway

The G_s pathway is a major pathway of PAC1R, to produce cAMP and stimulate phosphorylation of PKA and/or CREB. The G_q pathway is another important signaling pathway, activating PKC. PACAP also activates mitogen-activating protein kinase (MAPK; ERK) via PAC1R.

Antagonist and Agonist

PACAP6−38, missing 1 to 5 aa residues from the N-terminal of PACAP38, has potential as an antagonist against PAC1R and VPAC2-R. VIP-6-28, missing 1 to 5 aa residues from the N-terminal of VIP, has potential as an antagonist against VPAC1-R and VPAC2-R. Maxadilan is a specific agonist of PAC1R isolated from sand fly saliva, and the modification peptide M65 works as a PAC1R-specific antagonist.

Biological Functions

Target Cells/Tissues and Functions

PACAP has a pleiotropic function in the central nervous system and peripheral tissues. PACAP act as a neurotransmitter (neuromodulator), neuroprotectant, neurite outgrowth factor, and inducer of neural stem cell differentiation into astrocytes in nervous tissues [6,7]. In peripheral tissues, PACAP act as a bronchodilator, vasodilator, smooth muscle relaxant, adrenergic secretagog in the adrenal gland, insulin secretagog in the pancreas, inducer of spermatogenesis in testis, and immunosuppressor [8]. PACAP also regulates the synthesis and secretion of hormones from the pituitary gland.

Phenotype in Gene-Modified Animals

Most PACAP gene knockout mice of C57Bl/6 background suffered sudden death 1−2 weeks after birth because of a low respiratory response against hypoxia in the brainstem [9]. The sudden death was partially prevented by keeping them at a higher temperature of around 26°C. The PACAP knockout mouse indicates many other phenotypes, with weakened neurons in the brain, spinal cord, and retina, and delayed development of the cerebellum. The PACAP knockout mouse shows abnormal behavior, abnormalities of circadian rhythm, impairment of memory and learning, and schizophrenia-related behavior [10]. It also shows metabolic abnormality, and low prolactin and sex steroid levels.

References

1. Miyata A, Arimura A, Dahl RR, et al. Isolation of a novel 38 residue-hypothalamic polypeptide which stimulates adenylate cyclase in pituitary cells. *Biochem Biophys Res Commun*. 1989;164:567−574.
2. Sherwood NM, Krueckl SL, McRory JE. The origin and function of the pituitary adenylate cyclase-activating polypeptide (PACAP)/glucagon superfamily. *Endocr Rev*. 2000;21:619−670.
3. Arimura A, Somogyvari-Vigh A, Miyata A, et al. Tissue distribution of PACAP as determined by RIA: highly abundant in the rat brain and testes. *Endocrinology*. 1991;129:2787−2789.
4. Harmar AJ, Fahrenkrug J, Gozes I, et al. Pharmacology and functions of receptors for vasoactive intestinal peptide and pituitary adenylate cyclase-activating polypeptide: IUPHAR review 1. *Br J Pharmacol*. 2012;166:4−17.
5. Dickson L, Finlayson K. VPAC and PAC receptors: from ligands to function. *Pharmacol Ther*. 2009;121:294−316.
6. Arimura A. Perspectives on pituitary adenylate cyclase activating polypeptide (PACAP) in the neuroendocrine, endocrine, and nervous systems. *Jpn J Physiol*. 1998;48:301−331.
7. Nakamachi T, Farkas J, Watanabe J, et al. Role of PACAP in neural stem/progenitor cell and astrocyte − from neural development to neural repair. *Curr Pharm Des*. 2011;17:973−984.
8. Vaudry D, Falluel-Morel A, Bourgault S, et al. Pituitary adenylate cyclase-activating polypeptide and its receptors: 20 years after the discovery. *Pharmacol Rev*. 2009;61:283−357.
9. Arata S, Nakamachi T, Onimaru H, et al. Impaired response to hypoxia in the respiratory center is a major cause of neonatal death of the PACAP-knockout mouse. *Eur J Neurosci*. 2013;37:407−416.
10. Hashimoto H, Shintani N, Baba A. New insights into the central PACAPergic system from the phenotypes in PACAP- and PACAP receptor-knockout mice. *Ann NY Acad Sci*. 2006;1070:75−89.

Supplemental Information Available on Companion Website

- Gene, mRNA, and peptide organization of human PACAP/E-Figure 18C.1
- Accession numbers of genes and cDNAs for PACAP and PAC1R/E-Tables 18C.1, 18C.2.

Pituitary Adenylate Cyclase-Activating Polypeptide-Related Peptide

Tomoya Nakamachi

Abbreviation: PRP
Additional name: GHRH-like peptide (GHRH-LP)

Pituitary adenylate cyclase-activating polypeptide-related peptide (PRP) is identified as a structurally related peptide of PACAP on the PACAP precursor. Some of the PRP in lower vertebrates has the potential to stimulate GH release from pituitary cells, but not in mammalian species.

Discovery

PRP was identified as a structurally related peptide of PACAP on the PACAP precursor [1].

Structure

Structural Features

Human PRP is a 29-aa residue peptide encoded in Exon 4, at a region immediately upstream of PACAP on the 175-aa residue PRP/PACAP precursor. The amino acid sequence of mammalian PRP is relatively conserved (Figure 18D.1). Mammalian PRP peptides comprise a 29-aa residue flanked by paired basic amino acids recognized by the cleaving enzyme [2]. Two PRP-PACAP transcripts, 46 and 43 aa residues in length, have been cloned from the chicken *Gallus gallus*. Avian and fish PRPs are coded on Exons 4 and 5 in the PRP/PACAP precursor. Fish PRP is mainly divided into two types, carp-like PRP and catfish-like PRP. Two tunicate PRP/PACAP precursor genes encode PRP-like peptides consisting of 27 aa residues. As PRP-like peptide is encoded by PACAP in tunicates, it is likely that this is a result of exon duplication from an ancestral gene.

Synthesis and Release

Gene, mRNA, and Precursor

The human preproPRP/PACAP gene (*ADCYAP1*), located on chromosome 18 (18p11), consists of five exons (E-Figure 18D.1). PRP/PACAP mRNA of about 3.0 kb has been detected in the human and rat by northern blotting. The preproPRP/PACAP gene first appeared in tunicates.

Distribution of mRNA

PRP/PACAP precursor is detected most abundantly in the central nervous tissues, especially in the hypothalamus [3]. PRP-immunopositive nerve fibers were observed in the rat median eminence. In the chicken, PRP/PACAP precursor mRNA was detected in the brain and gonads, but not in the pituitary, heart, liver, kidney, intestine, eye, or muscle. Tunicate PRP-like peptide/PACAP precursor I was localized in the neural ganglion but not in the neural gland, gonad, gonad/digestive gland, intestine, or heart, while PRP-like peptide/PACAP precursor II was detected in the neural ganglion, dorsal strand, and intestine.

Receptors

Structure and Subtype

PRP receptor, formerly referred to as GHRH-receptor, was first isolated and characterized in goldfish [4]. The receptors are G protein-coupled receptors with seven transmembrane helical domains. PRP receptor mRNA has been detected in the brain, pituitary, gill, testis, ovary, lower gut, upper gut, heart, liver, skeletal muscle, and spleen, but not in the gallbladder or kidney. By assay using Chinese hamster ovary cells, both carp PRP and goldfish PRP salmon-like peptide could increase intracellular cAMP in a dose-dependent manner, while goldfish PRP catfish-like peptide could not [5]. To date, no specific receptor has been found to be stimulated by PRP catfish-like peptide. Therefore, this catfish-like peptide could be a duplicated product of PRP via teleost-specific duplication, but its function is lost subsequently. In chicken, PRP could activate cGHRH-LPR (cPRP-R), and it also activated cGHRHR1 and cGHRHR2. Those receptors are G protein-coupled receptors functionally coupled to the intracellular PKA signaling pathway. Expression of cGHRHR1, cGHRHR2, and cGHRH-LPR is restricted mainly to the pituitary and/or brain [6]. Although a receptor highly specific for PRP has been characterized in non-mammals, a specific receptor for PRP has not been isolated in any mammalian species. The presence of a specific receptor for PRP as shown in goldfish and chicken suggests that this peptide may play a role in the physiology of these animals, while these functions of PRP might have been lost in mammals due to the loss of its receptor [7].

Y. Takei, H. Ando, & K. Tsutsui (Eds): Handbook of Hormones. DOI: http://dx.doi.org/10.1016/B978-0-12-801028-0.00145-8

	DVAHG	ILNEA	YRKVL	DQLSA	GKHLQ	SLVAR	GVGGS	LGGGA	GDDA*	EPLS
human	DVAHG	ILNEA	YRKVL	DQLSA	GKHLQ	SLVAR	GVGGS	LGGGA	GDDA*	EPLS
Sheep	-----	--DK-	-----	-----	RRY--	T-M-K	-L--Y	P----	D--S*	----
rat	----E	-----	-----	-----	R-Y--	-M---	-M-EN	-AAA-	V--R*	A--T
mouse	----E	-----	-----	-----	R-Y--	-V---	-A-DE	PRRH-	V--P*	A--T
chicken I	*H-D-	-FSK-	---L-	G----	R-Y-H	--M-K	R---*	*ASSG	LG-EA	----
chicken II	*H-D-	-FSK-	---L-	G----	R-Y-H	--M-K	R--**	***SG	LG-EA	----
xenopus	*H-DE	L--KV	--N--	GH---	R-Y-H	T-M-Q	RL-T*	*VSSS	LE-ES	----
carp	*H-D-	MF-K-	---A-	G----	R-Y-H	T-M-K	R--**	*--SM	IE-DN	----
catfish	*H-D-	L-DR-	L-DI-	V----	R-Y-H	--T-V	R--**	*EEEE	DEEDS	----
goldfish (carp-like)	*H-D-	MF-K-	---A-	G----	R-Y-H	T-M-K	R--**	*--ST	IE-DN	----
goldfish (catfish-like)	*H-D-	L-DR-	L-DI-	V----	R-Y-H	--M-V	R--**	*--SS	EE-ES	----
salmon	*H-D-	MF-K-	---A-	G----	R-Y-H	--M-K	R--**	*--ST	ME-DX	----
zebrafish	*H-D-	L-DR-	L-DI-	V----	R-Y-H	--M-V	R--**	*--SS	EE-ES	----
tunicate I	*HSD-	-FTKD	---Y-	G--R-	Q-F--	W-M**	*****	*****	*****	****
tunicate II	*HSD-	-FTSD	--RY-	G----	Q-F--	W-M**	*****	*****	*****	****

Figure 18D.1 Comparison of Amino Acid Sequence Alignment of Vertebrate PRPs and Protochordate PRP-like Peptides*. *Extended forms of mammalian PRPs are shown. Consensus residues with human PRP are marked by dashes. Asterisks (*) represent gaps introduced for maximizing identity.

Biological Functions

Target Cells/Tissues and Functions

The function of PRP is well summarized in Tam's review [8]. The majority of past studies of PRP focused on GH-releasing activity in fish. Carp PRP could stimulate GH release from cultured goldfish pituitary glands, elevating serum GH levels 30 minutes after injection (0.1 mg/g) in goldfish. However, contrasting results regarding the GH-releasing capability of PRP have been indicated in more recent studies. In the presence of somatostatin in turbot pituitary culture, PACAP was capable of counteracting the inhibitory effects of somatostatin, resulting in a significant GH release, while both PRP salmon-like and PRP catfish-like proteins were unable to produce the same effects. Another study revealed that synthetic sockeye salmon PRP was unable to release GH significantly in coho salmon pituitary cell cultures at 100-nM concentration. Chicken PRP had poor activity in stimulating GH release from somatotrophs *in vitro* and only a slight *in vivo* effect at very high doses. Moreover, synthetic PRP(1–33) and PRP (1–44) had minimal effects on 4- to 6-week-old chickens. In mammals, PRP is shortened (29 aa residues) and structural modifications, together with the absence of PRP receptors in mammalian genomes, may contribute to the loss of function of PRP in [8].

References

1. Kimura C, Ohkubo S, Ogi K, et al. A novel peptide which stimulates adenylate cyclase: molecular cloning and characterization of the ovine and human cDNAs. *Biochem Biophys Res Commun.* 1990;166:81–89.
2. Okazaki K, Itoh Y, Ogi K, et al. Characterization of murine PACAP mRNA. *Peptides.* 1995;16:1295–1299.
3. Mikkelsen JD, Hannibal J, Fahrenkrug J, et al. Pituitary adenylate cyclase activating peptide-38 (PACAP-38), PACAP-27, and PACAP related peptide (PRP) in the rat median eminence and pituitary. *J Neuroendocrinol.* 1995;7:47–55.
4. Chan KW, Yu KL, Rivier J, et al. Identification and characterization of a receptor from goldfish specific for a teleost growth hormone-releasing hormone-like peptide. *Neuroendocrinology.* 1998;68:44–56.
5. Kee F, Ng SS, Vaudry H, et al. Aspartic acid scanning mutation analysis of a goldfish growth hormone-releasing hormone (GHRH) receptor specific to the GHRHsalmon-like peptide. *Gen Comp Endocrinol.* 2005;140:41–51.
6. Wang Y, Li J, Wang CY, et al. Characterization of the receptors for chicken GHRH and GHRH-related peptides: identification of a novel receptor for GHRH and the receptor for GHRH-LP (PRP). *Domest Anim Endocrinol.* 2010;38:13–31.
7. Lee LT, Siu FK, Tam JK, et al. Discovery of growth hormone-releasing hormones and receptors in nonmammalian vertebrates. *Proc Natl Acad Sci USA.* 2007;104:2133–2138.
8. Tam JK, Lee LT, Chow BK. PACAP-related peptide (PRP) — molecular evolution and potential functions. *Peptides.* 2007;28:1920–1929.

Supplemental Information Available on Companion Website

- Gene, mRNA, and peptide organization of human PRP/E-Figure 18D.1
- Accession numbers of genes and cDNAs for PRP and PRP (GHRH) receptor/E-Tables 18D.1, 18D.2

Vasoactive Intestinal Peptide

Jun Watanabe

Abbreviation: VIP
Additional name: Vasoactive intestinal polypeptide

Vasoactive intestinal peptide (VIP) is a neuropeptide consisting of 28 aa residues, with wide distribution in the central and peripheral nervous systems. VIP has a broad spectrum of biologic actions, and acts as a neurotransmitter or neuromodulator but sometimes also as a bloodborne hormone.

Discovery

First discovered as a smooth muscle-relaxant vasodilator peptide in the lung, VIP was isolated from porcine intestine in 1970 [1]. Its original label as a "candidate gastrointestinal hormone" was soon replaced by its new and apparently true identity as a neuropeptide with neurotransmitter and neuromodulator properties.

Structure

Structural Features

VIP possesses two segments of secondary structures: a random coil structure in the N-terminal region containing approximately 10 amino acid residues between positions 1 and 9, and a long α-helical structure in the C-terminal region stretching from position 10 to its C-terminus (Table 18E.1) [2].

Primary Structure

Sequence conservation is shown by the complete identity of the amino acid sequence for human/monkey/pig/cattle/sheep/goat/dog/rabbit/rat/mouse VIP peptides. VIP has also been identified from the guinea pig/opossum/chicken/turkey/alligator/frog/cod/trout/bowfin/dogfish, and has four or five differences from human VIP [3].

Table 18E.1 Primary Structure of VIP in Various Vertebrates

Human/monkey/pig/cattle/sheep/goat/dog/rabbit/rat/mouse	HSDAVFTDNY	TRLRKQMAVK	DYLNSILN
Guinea pig	HSDALFTDTY	TRLRKQMAMK	KYLNSVLN
Opossum	HSDAVFTDSY	TRLIKQMAMR	KYLDSILN
Chicken/turkey/alligator/frog	HSDAVFTDNY	SRFRKQMAVK	KYLNSVLT
Cod	HSDAVFTDNY	SRFRKQMAAK	KYLNSVLA
Trout/bowfin	HSDAIFTDNY	SRFRKQMAVK	KYLNSVLT
Dogfish	HSDAVFTDNY	SRIRKQMAVK	KYINSILA

Properties

Mr 3,325.80, pI >11. Soluble in water to 20 mg/ml.

Synthesis and Release

Gene, mRNA, and Precursor

The human VIP gene, located in the human chromosomal region 6q24, contains seven exons and six introns. Each exon encodes a distinct functional domain of the VIP precursor or its mRNA [1]. The VIP precursor polypeptide (preproVIP) contains sequences encoding several additional biologically active peptides, including peptide histidine isoleucine (PHI, found in non-human mammals), peptide histidine methionine (PHM, the human equivalent of PHI), and peptide histidine valine (PHV, a C-terminally extended form of PHI and PHM) (Figure 18E.1) [4].

Distribution of mRNA

VIP shows widespread tissue−organ distribution. VIP has been shown to be present in the cerebral cortex, suprachiasmatic nuclei of the hypothalamus, hippocampus, pituitary body, adrenal glands, nerve endings of the respiratory system, gastrointestinal tract, and reproductive system, and in immune system cells [5].

Tissue and Plasma Concentrations

Tissue: VIP tissue content is described in E-Tables 18E.1 and 18E.2 [6].
   Plasma: In 110 healthy fasting volunteers, plasma VIP concentrations were estimated to lie between 0.5 and 21 fmol/ml (median 1.7). No significant change was seen after ingestion of a standard test meal [7].

Regulation of Synthesis and Release

VIP-PHM mRNA synthesis is stimulated by dibutyryl cyclic (c) AMP and by increased protein kinase-C activity induced by tumor-promoting phorbol esters. When acting together, cAMP and phorbol esters synergistically stimulate VIP gene transcription, apparently via different sites on the gene [1].

Receptors

Structure and Subtype

VIP and PACAP share a wide spectrum of biological activity, and common receptors belonging to the so-called class II of G protein-coupled receptors (GPCR). Three VIP or PACAP

Y. Takei, H. Ando, & K. Tsutsui (Eds): Handbook of Hormones. DOI: http://dx.doi.org/10.1016/B978-0-12-801028-0.00146-X

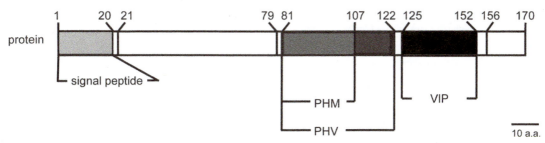

Figure 18E.1 Structure of the human preproVIP.

receptor types deriving from independent genes are officially recognized, named VPAC1, VPAC2, and PAC1 by the IUPHAR (International Union of Pharmacology). Two of these receptors, VPAC1 and VPAC2, share a common high affinity in the nanomolar range for VIP and PACAP. A third receptor type, PAC1, has been characterized for its high affinity for PACAP, but low affinity for VIP [8].

Signal Transduction Pathway

By binding to ligand, VPAC acts on the G_s protein and activates PKA via elevation of adenylate cyclase activity and production of cAMP (cAMP-dependent pathway). PKA either inhibits phosphorylation of the downstream MAP/ERK kinase or promotes phosphorylation of cAMP response element binding protein (CREB), which finally leads to inhibition of NF-κB. Meanwhile, studies have also identified a pathway that inhibits nuclear entry of NF-κB through VPAC signaling, which inhibits IκB phosphorylation (cAMP-independent pathway) [8].

Agonists

Agonists: PAC1, maxadilan; VPAC1, $[K^{15}R^{16}L^{27}]VIP(1-7)GRF$ $(8-27)NH_2$ $[Ala^{2,8,9,11,19,24,25,27,28}]VIP$; VPAC2, Ac-His^1[Glu^8, Lys^{12}, Nle^{17}, Ala^{19}, Asp^{25}, Leu^{26}, Lys^{27,28}, Gly^{29,30}, Thr^{31}]VIP (cyclo21−25), and Ac-His^1 [Glu^8, OCH^3-Tyr^{10}, Lys^{12}, Nle^{17}, Ala^{19}, Asp^{25}, Leu^{26}, Lys^{27,28}] VIP(cyclo 21−25).

Antagonists

Antagonists: PAC1, PACAP(6−38); VPAC1, [Ac-His^1, D-Phe^2, Lys^{15}, Arg^{16}] VIP(3−7)GRF(8−27)NH$_2$ antagonists.

Biological Functions

Target Cells/Tissues and Functions

VIP is extensively distributed in central and peripheral tissues, where it acts as a neurotransmitter and neuromodulator [8]. The main activity of VIP is vasodilatation. VIP induces dilation of vessels, which results in increased blood flow, decreased peripheral vascular resistance, and hypotension. While VIP has a suppressive effect on intestinal smooth muscle, it exerts relaxant effects on the lower esophageal sphincter, the sphincter of Oddi, anal sphincter, and bronchial smooth muscle. In the small intestine, VIP facilitates the secretion of electrolytes and aqueous liquid. In the stomach, it inhibits gastric secretion. In the pancreas, VIP facilitates external secretion of bicarbonic acid and aqueous liquid from pancreatic epithelial cells, and facilitates enzyme secretion from acinar cells. VIP promotes the secretion of both insulin and glucagon in a glucose-dependent manner. VIP is broadly distributed in the central nervous system, especially in the hypothalamus and pituitary anterior lobe, and is associated with the regulation of pituitary hormones. VIP is produced by lymphoid tissue and exerts a wide spectrum of immunological functions, controlling the homeostasis of the immune system through

different receptors expressed in various immunocompetent cells. VIP acts as a growth factor. It functions as a proliferative factor in normal tissue cells as well as in cancer cells. VIP might inhibit apoptosis by stimulating the expression of the apoptosis-inhibiting gene Bcl2 or by inhibiting the activity of caspase 3.

Phenotype in Gene-Modified Animals

It has been reported that VIP knockout mice have macroscopic and microscopic intestinal abnormalities [9]. VIP knockout mouse bowel presents increased weight but decreased length compared to control mice, associated with thickening of the smooth muscle layers, increased villus length, and higher abundance of goblet cells, which were also deficient in mucus secretion. The differences were more pronounced in the small intestine, but less apparent or absent in the colon with the exception of mucus secretion defects. Physiologically, intestinal transit was impaired in VIP knockout mice.

Pathophysiological Implications

Clinical Implications

Various diseases have been reported to involve VIP activity, including bronchial asthma, transmission of pain, cluster headache, Alzheimer's disease, Parkinson's disease, and brain injury [8].

Use for Diagnosis and Treatment

Several clinical trials have been reported regarding the use of VIP or its analog for asthma and sarcoidosis [8]. In an open clincial Phase II study, patients with histologically proven sarcoidosis and active disease were treated with nebulized VIP. VIP inhalation was safe and well-tolerated, and significantly reduced the production of tumor necrosis factor-α by cells isolated from bronchoalveolar lavage fluids of these patients.

References

1. Said SI. Vasoactive intestinal polypeptide: biologic role in health and disease. *Trends Endocrinol Metab.* 1991;2:107−112.
2. Onoue S, Misaka S, Yamada S. Structure−activity relationship of vasoactive intestinal peptide (VIP): potent agonists and potential clinical applications. *Naunyn-Schmiedeberg's Arch Pharmacol.* 2008;377:579−590.
3. Sherwood NM, Krueckl SL, McRory JE. The origin and function of the pituitary adenylate cyclase-activating polypeptide (PACAP)/glucagon superfamily. *Endocr Rev.* 2000;21:619−670.
4. Harmar AJ, Fahrenkrug J, Gozes I. Pharmacology and functions of receptors for vasoactive intestinal peptide and pituitary adenylate cyclase-activating polypeptide: IUPHAR review 1. *Br J Pharmacol.* 2012;166:4−17.

5. Dejda A, Sokołowska P, Nowak JZ. Neuroprotective potential of three neuropeptides PACAP, VIP and PHI. *Pharmacol Rep.* 2005;57:307—320.

6. Fahrenkrug J. Vasoactive intestinal polypeptide: measurement, distribution and putative neurotransmitter function. *Digestion.* 1979;19:149—169.

7. Mitchell SJ, Bloom SR. Measurement of fasting and postprandial plasma VIP in man. *Gut.* 1978;19:1043—1048.

8. Igarashi H, Fujimori N, Ito T, et al. Vasoactive intestinal peptide (VIP) and VIP receptors — elucidation of structure and function for therapeutic applications. *Int J Clin Med.* 2011;2:500—508.

9. Abad C, Gomariz R, Waschek J, et al. VIP in inflammatory bowel disease: state of the art. *Endocr Metab Immune Disord Drug Targets.* 2012;12:316—322.

Supplemental Information Available on Companion Website

- Tissue distribution and concentration of VIP in human/E-Tables 18E.1, 18E.2
- Gene and mRNA structures of preproVIP, VPAC1-R, and VPAC2-R/E-Figures 18E.1—18E.3
- Alignment of amino acid sequences of VIP and its receptors in vertebrates/E-Figures 18E.4—18E.7
- Accession numbers of genes, mRNA and protein for VIP and its receptors in various animals /E-Tables 18E.3—18E.6

Peptide Histidine Isoleucine/Methionine

Jun Watanabe

Abbreviation: PHI/PHM

PHI/M (peptide histidine isoleucine, or its human counterpart peptide histidine methionine) is a 27-aa peptide (Chr 6q26–q27) with a molecular weight of 20 kDa. PHI/M belongs to the glucagon/secretin family and causes vasodilation. PHI/M is expressed in gastrointestinal tissues and neural tissues, possibly acting as a neurotransmitter.

Discovery

PHI was discovered and isolated from porcine upper intestinal tissue in 1980, by using a chemical method for finding peptide hormones and other active peptides [1]. The human form of PHI (PHM) was found from vasoactive intestinal peptide (VIP) precursor protein, which contains not only VIP but also a peptide of 27 amino acids, different by only 2 amino acids from PHI [2].

Structure

Structural Features

PHI/M shows considerable sequence homology with secretin and VIP (Table 18F.1) [3].

Primary Structure

PHI seems to be a conservative peptide (Table 18F.2). The different forms of PHI identified so far vary from one another in no more than 6 aa residues. The rat PHI differs from bovine, porcine, human, and goldfish forms by 1, 2, 4, and 6 aa residues, respectively [3,4].

Properties

Mr 2,985.41, pI 7.7. Soluble in water to 1 mg/ml.

Synthesis and Release

Gene, mRNA, and Precursor

The human PHM gene is located in the human chromosomal region 6q26–q27. Each exon encodes a distinct functional domain of the PHM precursor or its mRNA. The PHM precursor contains sequences encoding several additional biologically active peptides, including PHI/M, VIP, and peptide histidine valine (PHV, a C-terminally extended form of PHI and PHM) [5].

Distribution of mRNA

PHI, VIP, and PACAP show widespread and similar tissue—organ distribution. They have been shown to be present in the cerebral cortex, pituitary body, adrenal glands, nerve endings of the respiratory system, gastrointestinal tract, and reproductive system. PHI has been identified in the temporal lobes, striatum, and medulla oblongata oblongata. VIP and PHI have been detected in the suprachiasmatic nuclei of the hypothalamus and hippocampus [4].

Tissue and Plasma Concentrations

Tissue: PHM tissue content is described in E-Table 18F.1 [6].

Plasma: The mean immunoreactive PHM concentrations in plasma were 3.5 ± 0.7 ($\pm$ SD) fmol/ml in the six normal women, 3.3 ± 0.7 fmol/ml in six normal men, and 3.6 ± 1.1 fmol/ml in five women with prolactinomas. There were no significant differences among the groups [6].

Regulation of Synthesis and Release

VIP-PHM mRNA synthesis is stimulated by dibutyryl cAMP and by increased PKC activity induced by tumor-promoting phorbol esters. When acting together, cAMP and phorbol esters synergistically stimulate VIP gene transcription, apparently via different sites on the gene [7].

Receptors

Structure and Subtype

Evidence from binding studies suggests that the PHM/PHI molecule may have a specific receptor, but the receptor in mamals has not been isolated or characterized. The VPAC1 and VPAC2 receptors bind PHM/PHI, but to a lesser extent than they bind VIP or PACAP [3]. Some groups reported that the PHI/PHV specific receptors were identified in goldfish and zebrafish [8].

Signal Transduction Pathway, Agonist, and Antagonist

See Subchapter 18E for the signal transduction pathway, agonists, and antagonists of VPAC1 and VPAC2 receptors.

Y. Takei, H. Ando, & K. Tsutsui (Eds): Handbook of Hormones. DOI: http://dx.doi.org/10.1016/B978-0-12-801028-0.00147-1

Table 18F.1 Primary Structure of Rat PHI/VIP/Secretin

PHI	HADGVFTSDY SRLLGQISAK KYLESLI
VIP	HSDAVFTDNY TRLRKQMAVK KYLNSILN
Secretin	HSDGTFTSEL SRLQDSARLQ RLLQGLV

Table 18F.2 Comparison of PHI Primary Structures in Different Species

Rat/mouse	HADGVFTSDY SRLLGQISAK KYLESLI
Guinea pig	HADGVFTSDY SRLLGQLSAR KYLESLI
Cow/sheep	HADGVFTSDY SRLLGQLSAK KYLESLI
Pig/rabbit	HADGVFTSDF SRLLGQLSAK KYLESLI
Human	HADGVFTSDF SKLLGQLSAK KYLESLM
Goldfish	HADGLFTSGY SKLLGQLSAK EYLESLL
Chicken	HADGIFTSVY SHLLAKLAVK RYLHSLI
Turkey	HADGIFTTVY SHLLAKLAVK RYLHSLI

Biological Functions

Target Cells/Tissues and Functions

PHM and VIP generally exert similar biological effects, but PHM is less potent. Like VIP and PACAP, PHM has been reported to increase the release of prolactin, insulin, and glucagon [3].

Phenotype in Gene-Modified Animals

PHI and VIP double knockout mice showed exhibited diurnal rhythms in activity which were largely indistinguishable from wild-type controls in a light–dark cycle. In constant darkness, the VIP/PHI-deficient mice exhibited pronounced abnormalities in their circadian system [9].

Pathophysiological Implications

Clinical Implications

PACAP, VIP, PHI, ADNF, and ADNP, as well as some of their synthetic analogs, may become useful therapeutic agents in many neurological disorders characterized by neuronal degeneration, such as Alzheimer's and Parkinson's diseases, amyotrophic lateral sclerosis, or stroke [4].

Use for Diagnosis and Treatment

The effect of PHM on ACTH and cortisol secretion was examined in female patients with Cushing's disease. PHM was given as an IV bolus in a dose of 100 μg, and plasma levels of ACTH and cortisol were measured. Of the patients with Cushing's disease, five (42%) patients were responsive to VIP and three (25%) to PHM, showing significant increases in both ACTH and cortisol [10].

References

1. Tatemoto K, Mutt V. Isolation of two novel candidate hormones using a chemical method for finding naturally occurring polypeptides. *Nature.* 1980;285:417–418.
2. Itoh N, Obata K, Yanaihara N, Okamoto H. Human preprovasoactive intestinal polypeptide contains a novel PHI-27-like peptide, PHM-27. *Nature.* 1983;304:547–549.
3. Sherwood NM, Krueckl SL, McRory JE. The origin and function of the pituitary adenylate cyclase-activating polypeptide (PACAP)/glucagon superfamily. *Endocr Rev.* 2000;21:619–670.
4. Dejda A, Sokołowska P, Nowak JZ. Neuroprotective potential of three neuropeptides PACAP, VIP and PHI. *Pharmacol Rep.* 2005;57:307–320.
5. Harmar AJ, Fahrenkrug J, Gozes I. Pharmacology and functions of receptors for vasoactive intestinal peptide and pituitary adenylate cyclase-activating polypeptide: IUPHAR review. *Br J Pharmacol.* 2012;166:4–17.
6. Sasaki A, Sato S, Go MG, et al. Distribution, plasma concentration, and *in vivo* prolactin-releasing activity of peptide histidine methionine in humans. *J Clin Endocrinol Metab.* 1987;65:683–688.
7. Said SI. Vasoactive intestinal polypeptide: biologic role in health and disease. *Trends Endocrinol Metab.* 1991;2:107–112.
8. Wu S, Roch GJ, Cervini LA, Rivier JE, Sherwood NM. Newly-identified receptors for peptide histidine-isoleucine and GHRH-like peptide in zebrafish help to elucidate the mammalian secretin superfamily. *J Mol Endocrinol.* 2008;41:343–366.
9. Colwell CS, Michel S, Itri J, et al. Disrupted circadian rhythms in VIP- and PHI-deficient mice. *Am J Physiol Regul Integr Comp Physiol.* 2003;285:R939–R949.
10. Watanobe H, Tamura T. Stimulation by peptide histidine methionine (PHM) of adrenocorticotropin secretion in patients with Cushing's disease: a comparison with the effect of vasoactive intestinal peptide (VIP) and a study on the effect of combined administration of corticotropin-releasing hormone with PHM or VIP. *J Clin Endocrinol Metab.* 1994;78:1372–1377.

Supplemental Information Available on Companion Website

- Tissue distribution and concentration of PHM in human/E-Table 18F.1
- The alignment of amino acid sequence of the PHI/M in various animals/E-Figure 18F.1

CHAPTER 19

Insulin Family

Tadashi Andoh

History

The insulin family consists of insulin, insulin-like growth factor (IGF)-I and -II, relaxin-1, -2, -3, and insulin-like peptide (INSL)3, INSL4, INSL5, INSL6. Insulin was first isolated in 1921, when its biological functions were elucidated. Its primary structure was determined in 1953. Relaxin was extracted for the first time in 1926, and its primary structure was determined in 1977. After the 1980s, relaxin was redesignated relaxin H2, and was revealed to constitute a family including insulin, IGFs and INSLs. IGFs were discovered in the 1960s, and their primary structures were elucidated in the late 1970s. Members of the insulin family have been characterized in coelenterates, nematodes, platyhelminths, mollusks, arthropods, echinoderms, protochordates, and vertebrates (see Chapter 45).

Structure

Structural Features

Prepropeptides of this family consist of a signal peptide, B-chain, C-peptide, A-chain, D domain, and E domain. After covalent linking via three disulfide bonds, some of the members receive processing at the signal peptide, C-peptide, D domain, and E domain. Accordingly, mature peptides are the heterodimers comprising the A- and B-chains. The locations of the six cysteine residues that form the three disulfide bonds are highly conserved. In particular, the cysteine motif (Cys−Cys−3X−Cys−8X−Cys) in the A-chain has been termed the insulin signature (Figure 19.1).

Molecular Evolution of Family Members

The insulin family is one of the most phylogenetically ancient groups of hormones in animals. Insulin and IGF-II genes are composed of a gene cluster together with tyrosine hydroxylase, and this cluster is termed the "tyrosine hydroxylase−insulin cluster." This gene cluster is well conserved across the animal kingdom [1]. Ten members have been found in the human, including insulin, IGF-I and -II, and seven peptides related to relaxin (E-Figure 19.1). Relaxin-related peptides have been diversified especially in teleosts and primates [2,3]. A bioinformatics approach using genome sequences from 46 organisms identified putatively 45 genes in humans, 19 family members in the squirrel, and 58 family members in *Xenopus* [4], suggesting amplification of the gene in each animal species.

Receptors

Structure and Subtypes

Receptors for insulin and IGF-I are transmembrane proteins with a tyrosine kinase domain at the intracellular region and two α- and two β-subunits. The IGF-II receptor is a monomer with a large extracellular domain and a relatively short cytoplasmic tail, and is also called the cation-independent mannose-6-phosphate receptor (Figure 19.2). Relaxin receptors are a family of GPCRs containing a seven-transmembrane domain. The receptors for the insulin family in vertebrates are different structurally among hormones, and are apparently polyphyletic, suggesting that there are different lineages between ligands and receptors in the insulin family. Additionally, an orphan insulin receptor-related receptor has been described in vertebrates except for fish [5]. Therefore, the insulin family is a good model for understanding the mechanisms of gene amplification in the genome, co-evolution of ligands and receptors [3], and the acquisition process of hormone functions.

Biological Functions

Target Cells/Tissues and Functions

Target tissues and major functions of the insulin family members are widely divergent among the ligands [6]. Cross-binding with heterogeneous receptors, or sharing receptors and post-transcriptional modification of both ligands and receptors, produces functional divergence in this family [6,7]. Insulin mainly regulates anabolism and nutrient metabolism. IGF-I acts as a mediator of growth hormone. Relaxins play roles mainly in female and male reproduction. Modulation of the neural network and activation of colon motility are also functions of this family. Intermediate metabolite and connecting peptides between the A- and B-chains of family members also have functions. For example, the precursor (proinsulin) is a potential neuroprotective factor in mammals, and the C-peptide ameliorates some diabetic symptoms [1,8]. These multifunctional properties may account for the evolutionary divergence of the insulin family genes. Moreover, specific binding proteins in blood (insulin-like growth factor binding protein, IGFBP) are involved in the regulation of IGFs. There is no binding protein at high affinity in other members of the insulin family in vertebrates. The origin of the interrelationship of IGFBP and IGFs was established after the divergence of vertebrates and protochordates in the chordate lineage [9]. Meanwhile, crustacean IGFBP, which is structurally related to

Y. Takei, H. Ando, & K. Tsutsui (Eds): Handbook of Hormones. DOI: http://dx.doi.org/10.1016/B978-0-12-801028-0.00019-2

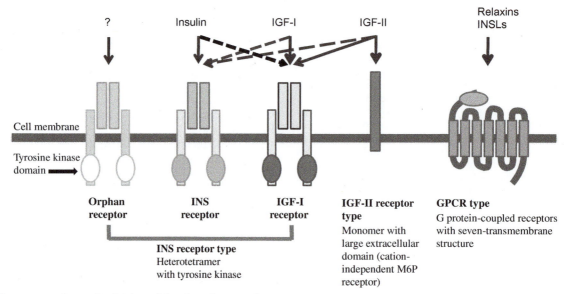

	B-chains		A-chains
INS	FVNQHLCGSHLVEALYLVCGERGFFYTPKT		GIVEQCCTSICSLYQLENYCN
IGF-I	GPETLCGAELVDALQFVCGDRGFYFNKPT		GIVDECCFRSCDLRRLEMYCA
IGF-II	AYRPSETLCGGELVDTLQFVCGDRGFYF		GIVEECCFRSCDLALLETYCA
RLN1	VAAKWKDDVIKLCGRELVRAQIAICGMSTWS		PYVALFEKCCLIGCTKRSLAKYC
RLN2	DSWMEEVIKLCGRELVRAQIAICGMSTWS		QLYSALANKCCHVGCTKRSLARFC
RLN3	RAAPYGVRLCGREFIRAVIFTCGGSRW		DVLAGLSSSCCKWGCSKSEISSLC
INSL3	LGPAPTPEMREKLCGHHFVRALVRVCGGPRWSTEA		AAATNPARYCCLSGCTQQDLLTLCPY
INSL4	AELRGCGPRFGKHLLSYCPMPEKTFTTTPGGWL		SGRHRFDPFCCEVICDDGTSVKLCT
INSL5	KESVRLCGLEYIRTVIYICASSRW		QDLQTLCCTDGCSMTDLSALC
INSL6	RELSDISSARKLCGRYLVKEIEKLCGHANWSQF		GYSEKCCLTGCTKEELSIACLPYIDF

Insulin signature

Figure 19.1 Alignment of the amino acid sequences of the A- and B-chains of the insulin family in human. Conserved residues are indicated in shaded areas and cysteine residues are indicated in gray. INS, insulin; IGF, insulin-like growth factor; RLN, relaxin; INSL, insulin-like peptide.

Figure 19.2 Receptors for insulin family and their ligands in vertebrates.

vertebrate IGFBP, has been identified in crayfish [10]. This may be explained as parallel evolution from an ancestral protein of IGFBP.

References

1. de la Rosa EJ, de Pablo F. Proinsulin: from hormonal precursor to neuroprotective factor. *Front Mol Neurosci.* 2011;4:20.
2. Arroyo JI, Hoffmann FG, Opazo JC. Evolution of the relaxin/insulin-like gene family in anthropoid primates. *Genome Biol Evol.* 2014;6:491–499.
3. Good S, Yegorov S, Joran Martijn J, Franck J, Bogerd J. New insights into ligand-receptor pairing and coevolution of relaxin family peptides and their receptors in teleosts. *Int J Evol Biol.* 2012;2012:310278.
4. Pinero-Gonzalez J, Gonzalez-Perez A. The ubiquity of the insulin superfamily across the eukaryotes detected using a bioinformatics approach. *OMICS J Integr Biol.* 2011;15:439–447.
5. Hernández-Sánchez C, Mansilla A, de Pablo F, Zardoya R. Evolution of the insulin receptor family and receptor isoform expression in vertebrates. *Mol Biol Evol.* 2008;25:1043–1053.
6. Bathgate RAD, Halls ML, van der Westhuizen ET, Callander GE, Kocan M, Summers RJ. Relaxin family peptides and their receptors. *Physiol Rev.* 2013;93:405–480.
7. Panda AC, Grammatikakis I, Yoon JH, Abdelmohsen K. Posttranscriptional regulation of insulin family ligands and receptors. *Int J Mol Sci.* 2013;14:19202–19229.
8. Wang S, Wei W, Zheng Y, et al. The role of insulin C-peptide in the coevolution analyses of the insulin signaling pathway: a hint for its functions. *PLoS One.* 2012;7:e52847.
9. Zhou J, Xiang J, Zhang S, Duan C. Structural and functional analysis of the amphioxus IGFBP gene uncovers ancient origin of IGF-independent functions. *Endocrinology.* 2013;154:3753–3763.
10. Rosen O, Weil S, Manor R, Ziv Roth Z, Khalaila I, Sagi A. A crayfish insulin-like-binding protein: another piece in the androgenic gland insulin-like hormone puzzle is revealed. *J Biol Chem.* 2013;288:22289–22298
11. Mita M, Yoshikuni M, Ohno K, et al. A relaxin-like peptide purified from radial nerves induces oocyte maturation and ovulation in the starfish, *Asterina pectinifera. Proc Natl Accad Sci USA.* 2009;106:9507–9512.

Supplemental Information Available on Companion Website

- Phylogenetic tree of the insulin family in vertebrates and protochordates/E-Figure 19.1

Insulin

Tadashi Andoh

Abbreviation: INS
Additional names: none

Insulin is the only hormone lowering blood sugar levels in vertebrates. However, the importance of insulin (INS) in carbohydrate metabolism may be different between mammals and other vertebrates.

Discovery

Frederick G. Banting and Charles H. Best discovered a substance that lowered blood sugar levels in the dog pancreas in 1921, and it was immediately applied to diabetes care. The primary structure of INS has been reported in over 100 species, including agnathans, fish, and tetrapods.

Structure

Structural Features

Human preproINS consists of 110 aa residues, of which 24 are in the signal peptide, 30 in the B-chain, 31 in the C-peptide, and 21 in the A-chain (Figure 19A.1). The signal and C-peptides are excised from the preproINS to produce the mature peptide, consisting of heterodimeric A- and B-chains (Figure 19A.2).

Primary Structure

Six cysteine residues are completely conserved through evolution from agnathans to mammals (E-Figure 19A.1). The N-terminal seven aa residues of the A-chain and the receptor binding region of the B-chain (Gly—Phe—Phe—Tyr) are highly conserved.

Properties

Mr 5,808, pI 5.3. Almost insoluble at neutral pH and soluble under acidic and alkaline conditions. Can be stored for up to 6 months at 4°C in 1-M acetic acid. Long storage under an alkaline pH increases the rate of deamidation and aggregation. One IU is equivalent to 0.0347 mg human INS.

Synthesis and Release

Gene, mRNA, and Precursor

The human preproINS gene, *INS*, location 11p15.5, consists of three exons. There are many regulatory elements, such as the A box, GG box, cAMP response element, in the promoter region. The coding region of human and tilapia preproINS mRNAs comprises 330 and 339 bp, respectively (E-Figure 19A.2).

Distribution of mRNA

The primary source of INS is the β cells of the islets of Langerhans in the pancreas, and the mRNA has been detected in the islet tissue and the brain of mammals and trout [1]. Most teleost species have anatomically discrete islet organs called Brockmann bodies (BB).

Tissue and Plasma Concentrations

Tissue: Rat, 15 µg/g pancreas; flounder, 1 mg/g BB.
   Plasma: The main effecter of plasma INS concentration is nutritional condition. The levels are summarized in Table 19A.1. The half-life of INS is estimated to be 4.8 minutes in the human, and is much longer in fish.

Regulation of Synthesis

INS gene transcription is stimulated by glucose in mammals. The glucose treatment results in a two- to five-fold elevation within 60—90 minutes. Long-term exposure to physiological concentrations of estradiol-17β increases the β-cell content, INS gene expression, and INS release in mice [2]. Fasting reduces the expression levels of two INS genes in trout.

Regulation of Release

Glucose-stimulated release is the principal mechanism for INS release in mammals. Glucose, incorporated by the β cell via glucose transporter 2, is phosphorylated by glucose kinase and activated in mitochondria, resulting in a rise in the cytoplasmic ATP/ADP ratio that leads to closure of ATP-sensitive K^+ channels, depolarization of the plasma membrane, and opening of voltage-sensitive L-type Ca^{2+} channels. The entry of Ca^{2+} then triggers exocytosis of INS granules [3]. The C-peptide, which ameliorates common complications of diabetes [4], is also released in equimolar concentrations with INS. In contrast, the islet tissues of chicken, reptiles, fish, and fetus of mammals are not sensitive to glucose for INS release, but are sensitive to amino acids (E-Table 19A.3) [3]. It is considered that the regulatory role of INS in blood sugar levels is a derivative function in vertebrate evolution [5]. INS release shows a biphasic pattern caused by two pools (a readily releasable pool and a reserve pool) in the β cells of mammals [6]. The first phase of INS release is lost in patients with type 2 diabetes. Most catabolic hormones, such as adrenaline, cortisol and thyroxine, inhibit INS release. In birds and lizards, α cells are more abundant in islet tissue than β cells, suggesting that glucagon plays a more important role in blood sugar regulation in these animals.

Y. Takei, H. Ando, & K. Tsutsui (Eds): Handbook of Hormones. DOI: http://dx.doi.org/10.1016/B978-0-12-801028-0.00148-3

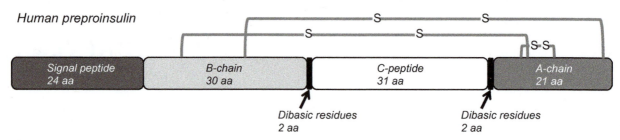

Figure 19A.1 Structure of human preproINS.

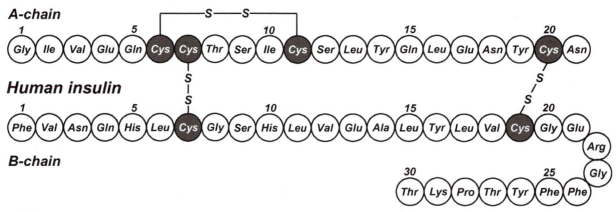

Figure 19A.2 Primary structure of human INS.

Table 19A.1 INS Concentrations in Blood Samples of Various Animals

Animal	Concentration (ng/ml)
Human	0.1–6.1
Rat	0.1–5.4
Chicken	0.5–2.3
Flounder	0.5–67.6
Salmon	0.9–14.3

Receptors

Structure and Subtypes

The INS receptor (INSR) is a heterotetrameric glycoprotein. Many aa residues are phosphorylated or glycosylated. INS binds to the two extracellular α-subunits linked by a disulfide bond (Figure 19A.3). The two β-subunits are connected to α-subunits by a disulfide bond and contain a tyrosine kinase domain in their intracellular region [7,8]. The human INSR is encoded by a single gene, *INSR*, comprising 22 exons (E-Figure 19A.4) [9] and located on chromosome 19 (19p13.3–p13.2). Human INSR consists of 1,382 aa residues containing a signal peptide, and α-and β-subunits. *INSR* generates two alternative splicing variants (INSR-A and -B) that differ at the carboxyl terminus of α-subunits. INSR-A lacks the terminus encoded by Exon 11, and INSR-B is not detected in chicken or *Xenopus*. There are three types of hybrid receptors among these INSRs and the IGF-I receptor (IGF-IR) (Figure 19A.3). The expression ratio of these hybrid receptors influences INS action at target organs [7,10]. Guinea pig INSR shows exceptionally high affinity ($K_d = 8.3 \times 10^{-11}$). Several fish species have plural INSR genes, which show different expression patterns among tissues [1].

Signal Transduction Pathway

After binding, the β subunit tyrosine kinase phosphorylates different substrate adaptors (INSR substrate [IRS] proteins), followed by activation of numerous signaling partners. PI3K is one of the signaling partners, and plays a role by activating the Akt and PKC ζ cascades. Activated Akt induces synthesis of glycogen and protein, and also mediates translocation of the glucose transporter to and fusion with the plasma membrane followed by influx of glucose [7,10]. Moreover, aa transport systems are also responsive to INS [11].

Agonists

IGFs bind to INSR, but INS binds IGF-IR with low affinity. Several synthetic peptides (S519 and RB539) show agonistic effects with high affinity for INSR. Fish INSs bind mammalian INSR with high affinity, and vice versa.

Antagonists

RB537, S661, and S961 are INSR antagonists with high affinity. S961 shows an agonistic effect only on ^{3}H-thymidine incorporation.

Biological Functions

Target Cells/Tissues and Functions

INS is a key regulator of glucose homeostasis in mammals, because the peptide is the only hormone that lowers blood sugar. Moreover, many functions relating to anabolism have been elucidated for INS, and are summarized in Table 19A.2 [1,11]. INS actions are mediated via two receptors (INSR-A and -B) in mammals. INSR-A mainly enhances the effects of IGF-II during embryogenesis and fetal development, whereas INSR-B is predominantly expressed in the adult and enhances metabolic effects of INS [7].

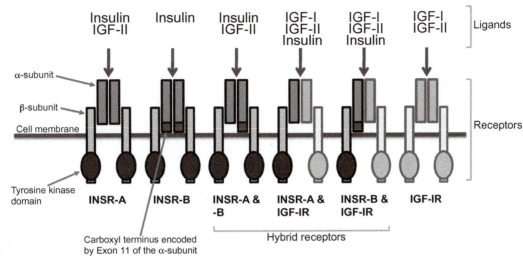

Figure 19A.3 Cross-binding relationships for receptors among INS and IGFs in mammals.

Table 19A.2 Actions of Insulin in Vertebrates

	Liver	Adipose Tissue	Muscle	Brain
Glucose uptake	⇧	⇧	⇧	
Glycogenesis	⇧		⇧	⇧
Glycogenolysis	⬇		⬇	
Glycolysis	⇧	⇧	⇧	
Gluconeogenesis	⬇			
Fatty acid synthesis	⇧			
Ketogenesis	⬇			
Lipolysis		⬇		
Amino acid uptake			⇧	
Protein synthesis			⇧	
Proteolysis			⬇	

Phenotype in Gene-Modified Animals

Mice lacking INSRs are born with normal features but develop early postnatal diabetes and die of ketoacidosis. Several mutants have been reported in humans and mice.

Pathophysiological Implications

Clinical Implications

Diabetes is classified into two types. Type 1 diabetes, or "insulin-dependent diabetes mellitus," results from autoimmune destruction of β cells. Type 2 diabetes is "non-insulin-dependent diabetes mellitus," which is a metabolic disorder characterized by hyperglycemia due to an absolute or relative lack of INS or cellular resistance to INS. β cells are destroyed selectively by administration of alloxan or streptozotocin, producing a type 2 diabetes model in mammals and fish.

Use for Diagnosis and Treatment

INS levels are routinely measured using commercial kits in mammals and chicken. Several studies have developed specific immunoassays for fish [5]. The insulin-induced hypoglycemia test is used clinically to estimate the hypothalamic−pituitary response to stress in patients with suspected ACTH deficiency or growth hormone deficiency. Many analogs, which show different times of action, have been developed to treat diabetes.

References

1. Caruso MA, Sheridan MA. New insights into the signaling system and function of insulin in fish. *Gen Comp Endocrinol.* 2011;173:227−247.
2. Alonso-Magdalena P, Ropero AB, Carrera MP, et al. Pancreatic insulin content regulation by the estrogen receptor ER alpha. *PLoS One.* 2008;3:e2069.
3. Polakof S, Mommsen TP, Soengas JL. Glucosensing and glucose homeostasis: from fish to mammals. *Comp Biochem Physiol B.* 2011;160:123−149.
4. Wahren J, Kallas Å, Sima AAF. The clinical potential of C-peptide replacement in type 1 Diabetes. *Diabetes.* 2012;61:761−772.
5. Andoh T. Amino acids are more important insulinotropins than glucose in a teleost fish, barfin flounder (*Verasper moseri*). *Gen Comp Endocrinol.* 2007;151:308−317.
6. Seino S, Shibasaki T, Minami K. Dynamics of insulin secretion and the clinical implications for obesity and diabetes. *J Clin Invest.* 2011;121:2118−2125.
7. Belfiore A, Frasca F, Pandini G, et al. Insulin receptor isoforms and insulin receptor/insulin-like growth factor receptor hybrids in physiology and disease. *Endocr Rev.* 2009;30:586−623.
8. Menting JG, Whittaker J, Margetts MB, et al. How insulin engages its primary binding site on the insulin receptor. *Nature.* 2013;493:241−248.
9. Seino S, Seino M, Nishi S, Bell GI. Structure of the human insulin receptor gene and characterization of its promoter. *Proc Natl Acad Sci USA.* 1989;86:114−118.
10. Hale LJ, Coward RJM. Insulin signalling to the kidney in health and disease. Clin. Sci. (Lond) 2013;124:351−370.
11. Flakoll PJ, Jensen MD, Cherrington AD. Physiological action of insulin. In: LeRoith D, Taylor SI, Olefasky JM, eds. *Diabetes Mellitus: A Fundamental and Clinical Text.* 3rd ed. Philadelphia, PA: Lippincott, Williams & Wilkins; 2004:165−181.
12. Fajans SS, Floyd Jr JC, Knopf RF, Conn FW. Effect of amino acids and proteins on insulin secretion in man. *Recent Prog Horm Res.* 1967;23:617−662.
13. Newsholme P, Bender K, Kiely A, Brennan L. Amino acid metabolism, insulin secretion and diabetes. *Biochem Soc Trans.* 2007;35:1180−1186.
14. Jun T, Catalano M. Nitric oxide is involved in the insulin release in rats by L-arginine. *Int J Angiol.* 1997;6:187−189.
15. Milner RD, Ashworth MA, Barson AJ. Insulin release from human foetal pancreas in response to glucose, leucine and arginine. *J Endocrinol.* 1972;52:497−505.
16. Giroix MH, Portha B, Kergoat M, et al. Glucose insensitivity and amino-acid hypersensitivity of insulin release in rats with non-insulin-dependent diabetes. A study with the perfused pancreas. *Diabetes.* 1983;32:445−451.
17. King DL, Hazelwood RL. Regulation of avian insulin secretion by isolated perfused chicken pancreas. *Am J Physiol.* 1976;231:1830−1839.

18. Colca JR, Hazelwood RL. Amino acids as *in vitro* secretogogues of avian pancreatic polypeptide (APP) and insulin from the chicken pancreas. *Gen Comp Endocrinol.* 1982;47:104−110.

19. Rhoten WB. Perifusion of saurian pancreatic islets and biphasic insulin release following glucose stimulation. *Comp Biochem Physiol A Comp Physiol.* 1973;45:1001−1007.

20. Ince BW. Insulin secretion from the in situ perfused pancreas of the European silver eel, *Anguilla anguilla* L. *Gen Comp Endocrinol.* 1979;37:533−540.

21. Ince BW. Amino acid stimulation of insulin secretion from the *in situ* perfused eel pancreas; modification by somatostatin, adrenaline, and theophylline. *Gen Comp Endocrinol.* 1980;40:275−282.

Supplemental Information Available on Companion Website

- Alignment of amino acid sequence of INSs in vertebrates/ E-Figure 19A.1
- Structures of preproINS gene in human and tilapia/ E-Figure 19A.2
- Phylogenetic tree of INSs from various vertebrates/ E-Figure 19A.3
- Structure of human INSR gene/E-Figure 19A.4
- Accession numbers of genes, cDNAs, and peptides for INSs/E-Table 19A.1
- Accession numbers of genes and cDNAs for INSRs/E-Table 19A.2
- Relative insulinotropic effects of amino acids and glucose/ E-Table 19A.3 [5,12−21]

Insulin-Like Growth Factor-I

Munetaka Shimizu

Abbreviation: IGF-I
Additional names: somatomedin C, non-suppressible insulin-like activity soluble in acid/ethanol (NSILA-S)

IGF-I is a multifunctional polypeptide structurally related to proinsulin, IGF-I promotes cell proliferation, differentiation, growth, migration, and survival through autocrine/paracrine and endocrine pathways. It mediates part of the growth hormone actions and is essential for normal prenatal and postnatal growth.

Discovery

In 1957 a "sulfation factor" that mediates the action of growth hormone (GH) on the incorporation of [^{35}S] sulfate into cartilage segment was discovered in rats, and named as somatomedin, now known as IGF-I [1]. IGF-I was also identified in 1963 as non-suppressible insulin-like activity soluble in acid/ethanol (NSILA-S). In 1978, IGF-I was isolated from the Cohn fraction of plasma proteins together with IGF-II.

Structure

Structural Features

IGF-I is a single chain polypeptide sharing high structural homology with proinsulin (about 50%) and IGF-II (about 70%) (Figure 19B.1) [1,2]. Three disulfide bonds that are involved in structural maintenance of the insulin family peptides are well conserved. PreproIGF-I is composed of a signal peptide and five domains; B, C, A, D, and E domains (Figure 19B.2). The E domain is proteolytically cleaved before secretion.

Primary Structure

The aa sequence of IGF-I is highly conserved in vertebrates (~80% identical; E-Figure 19B.2).

Properties

Mr ~7,500, pI 8.5. Lyophilized peptide should be reconstituted in 10-mM HCl. Lyophilized peptide can be stored at 2–4°C for at least 2 years.

Synthesis and Release

Gene, mRNA, and Precursor

The human IGF-I gene, *IGF1*, location 12q22–q23, consists of six exons [2,3]. There are multiple transcript variants that differ in promoter use, transcription start site, splicing, and/or polyadenylation (E-Figure 19B.1) [2,3]. These IGF-I mRNAs are classified based on the alternative splicing of exons encoding the E domain. In this regard, human has two IGF-I mRNAs, IGF-IA and IGF-IB. Both types encode the signal peptide of 48 or 33 aa residues depending on the transcription start site on exon 1 (class 1) or 2 (class 2), the same mature protein of 70 aa residues, and different E domains (35 aa residues for IGF-IA and 77 aa residues for IGF-IB) [2,3]. Salmon have four transcript variants alternatively spliced at the E domain. Two non-allelic genes for IGF-I have been identified in *Xenopus*, zebrafish, and salmon [4]. In zebrafish and tilapia, a gonad-specific *igf-3* (or *igf-1b*) has been identified.

Distribution of mRNA

In postnatal animals, the liver is the major site of expression. The IGF-I gene is also expressed widely by connective tissue cell types such as stromal cells.

Tissue and Plasma Concentrations

Tissue: Tissue levels of IGF-I are much lower than serum levels, being 6–9% equivalent in rats [1].

Plasma: The characteristics, tissue distribution, and plasma concentration of IGF-I in vertebrates can be seen in Table 19B.1. The levels are approximately 50 ng/ml (childhood), 400 ng/ml (puberty), and 100–200 ng/ml (>25 years old) in humans. It is essential to extract IGFs before assay to avoid interference by IGF-binding proteins (IGFBPs) [5]. Size-exclusion chromatography under acidic conditions is the gold standard; however, acid/ethanol extraction is most commonly used for IGF extraction due to its simplicity. After extraction, an excess amount of IGF-II is often added to saturate residual IGFBPs.

Regulation of Synthesis and Release

Mammalian IGF-I genes have two promoters (P1 and P2) that lack TATA and CAAT elements [2,3]. P1 is the potent, major promoter, and is conserved widely in vertebrates. The proximate promoter region of the IGF-I gene contains binding sites for liver-enriched transcription factors such as HNF-1α, C/EBPα, and C/EBPβ. GH is the primary hormone regulating the synthesis and release of IGF-I in the liver after birth. The action of GH is mediated chiefly by the JAK2/Stat5b pathway. Several GH-inducible Stat5b binding sites have been found in introns and distal regions of the *Igf1* loci. IGF-I is also regulated at the transcription level by other hormones, such as insulin, cortisol, and sex steroids, and by developmental stage independently of GH action. Nutritional status regulates IGF-I

Y. Takei, H. Ando, & K. Tsutsui (Eds): Handbook of Hormones. DOI: http://dx.doi.org/10.1016/B978-0-12-801028-0.00149-5

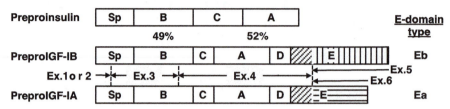

```
                B domain                        C domain                      A domain      D domain
ProIns  FVNQHLCGSHLVEALYLVCGERGFFYTPKTRREAEDLQVGQVELGGGPGAGSLQPLALEGSLQKRGIVEQCCTSICSLYQLENYCN
IGF-I   GPETLCGAELVDALQFVCGDRGFYFNKPT-------------------GYGSSSRR-------APQTGIVDECCFRSCDLRRLEMYCAPLKPAKSA
```

Figure 19B.1 Comparison of amino acid sequence of human mature IGF-I with that of proinsulin (ProIns). Consensus amino acid residues are shaded.

Figure 19B.2 Structure of the human preproinsulin and two preproIGF-Is. Ex., exon.

Table 19B.1 Characteristics, Tissue Distribution, and Plasma Concentration of IGF-I in Vertebrates

Species	# of aa	Mr	Gene Location	# of Exon	Tissue (Prenatal)	Tissue (Postnatal)	Blood (ng/ml) (Prenatal)	Blood (ng/ml) (Postnatal)
Human	70	7,649	12	6	brain, stomach, placenta	liver, muscle, bone, cartilage	10−60	150−400
Rat	70	7,687	7	6	lung, stomach	liver, kidney, lung, testis	160−240	400−1,000
Mouse	70	7,677	10	6	liver, lung, brain, kidney	liver, kidney, spleen, lung, pancreas, testis	low	400−900
Chicken	70	7,738	1	4	pancreas, brain, eye, muscle	liver, muscle	5−20	30−50
Xenopus (I")	70	7,796	?	?	heart	liver, lung, heart, kidney, perioneal fat	?	?
Zebrafish (Ia)	70	7,747	LG4	5	anterior part of embryo	liver, brain, eye	?	?
Salmon	70	7,705	?	5	+ (whole embyro)	liver, brain, kideny, gill, gonad, intestine, spleen	?	5−100

mRNA at the post-transcriptional level by affecting mRNA processing and stability.

Receptors

Structure and Subtype

The receptor of IGF-I (type I IGF receptor, IGF-IR) belongs to a family of the receptor tyrosine kinase (RTK) containing a single transmembrane domain, and shares high sequence homology with the insulin receptor (60%). The human IGF-IR gene is located on chromosome 15q25−26, and consists of 21 exons encoding an extracellular α-subunit (706 aa residues), which contains a ligand binding domain, and a transmembrane β-subunit (627 aa residues), which contains tyrosine kinase activity (E-Figures 19B.3, 19B.4) [6]. The α- and β-subunits are synthesized as a single chain prepropeptide and cleaved after translation, bridged by a disulfide bond to form the IGF-I half-receptor (αβ). Two half-receptors dimerize to form a functional IGF-IR (α2β2). The IGF-I half-receptor can also form a hybrid receptor with the insulin half-receptor to bind mainly IGF-I. Teleosts have two paralogs of IGF-IR.

Signal Transduction Pathway

Human IGF-IR binds IGF-I and IGF-II with a high affinity (0.1−1 nM) but with a 100-fold lower affinity for insulin. Ligand binding to the α-subunit induces autophosphorylation of the intracellular domain of the β-subunit, which in turn activates the RTK. The activated RTK phosphorylates several signaling elements such as insulin receptor substrates (IRSs), which provide docking sites for signaling effectors mainly through the SH domain, and those effectors branch into downstream signal pathways. The two main pathways are the PI3K/Akt pathway, which induces many metabolic responses, and the RAS/RAF/MAPK pathway, which generally controls cell growth and proliferation. IGF-IR is also capable of translocating into the nucleus and acts as a transcription factor.

Agonists

IGF-II, insulin, Des [1−3] IGF-I, Long R3 IGF-I (reduced binding to IGFBPs), and LL-37 are agonists.

Antagonists

IGFBPs, JB1 and JB3 (12-aa synthetic peptides), and M1557 (D domain analog) are antagonists.

Biological Functions

Target Cells/Tissues and Functions

IGF-I acts on most tissues, but liver may not be a major target. IGF-I is involved in growth and metabolism at the organismal level, and in cell proliferation, migration, differentiation, and survival at the cellular level. IGF-I inhibits apoptosis. An important role of circulating IGF-I is to regulate GH synthesis/secretion at the pituitary and hypothalamus through a negative feedback loop [7]. A single IGF-I allele is responsible for small size in dogs. In euryhaline fish, IGF-I is involved in the development of hypo-osmoregulatory ability by acting on osmoregulatory organs such as gills.

Phenotype in Gene-Modified Animals

The growth of *Igf1*-null mice is severely retarded (30% of wild adult size) and they become infertile. Null mutants for the *Igf-1r* die shortly after birth. Studies using conditional *Igf1* knockout mice, in which the hepatic expression of *Igf1* was inactivated, showed that liver-derived IGF-I is the principal source of endocrine IGF-I (about 75%) but is not required for normal postnatal growth [7,8]. An important contribution of endocrine IGF-I to postnatal longitudinal bone growth (about 30%) has been revealed by knocking-in an *Igf1* cDNA into *Igf-1* null mice [8].

Pathological Implications

Clinical Implications

IGF-I deficiency is related to Laron syndrome (short stature due to GH resistance or insensitivity), liver cirrhosis, and age-related cardiovascular and neurological diseases [9]. Epidemiologic studies suggest relationships between IGF-I and cancer risks such as prostate, colon, and breast cancers.

Use of Diagnosis and Treatment

IGF-I levels are routinely used for diagnosis in patients with suspected acromegaly or GH/IGF-I deficiency. The US Food and Drug Administration has approved recombinant human IGF-I for treatment of IGF-I deficiency patients. Due to its anti-apoptotic and proliferative actions, inhibition of the IGF-I signaling using antagonists or antibodies is a potential therapy for cancers and atherosclerosis [10].

References

1. Daughaday WH, Rotwein P. Insulin-like growth factors I and II. Peptide, messenger ribonucleic acid and gene structures, serum, and tissue concentrations. *Endocrine Rev*. 1989;10:68−91.
2. Potwein P. Molecular biology of IGF-I and IGF-II. In: Rosenfeld RG, Roberts CT, eds. *The IGF System: Molecular Biology, Physiology, and Clinical Applications*. Totowa NJ: Humana Press; 1999:19−35.
3. Oberbauer AM. The regulation of IGF-1 gene transcription and splicing during development and aging. *Front Endocrinol*. 2013;4:39.
4. Wood AW, Duan C, Bern HA. Insulin-like growth factor signaling in fish. *Int Rev Cytol*. 2005;243:215−285.
5. Rikke H, Frystyk J. Determination of IGFs and their binding proteins. *Best Practice Res Clin Endocrinol Metab*. 2013;27:771−781.
6. Chitnis MM, Yuen JSP, Protheroe AS, et al. The type 1 insulin-like growth factor receptor pathway. *Clin Cancer Res*. 2008;14:6364−6370.
7. LeRoith D, Bondy C, Yakar S, et al. The somatomedin hypothesis: 2001. *Endocrine Rev*. 2001;22:53−74.
8. Ohlsson C, Mohan S, Sjgren K, et al. The role of liver-derived insulin-like growth factor-I. *Endocrine Rev*. 2009;30:494−535.
9. Puche JE, Castilla-Cortázar I. Human conditions of insulin-like growth factor-I (IGF-I) deficiency. *J Trans Med*. 2012;10:224.
10. Clemmons DR. Modifying IGF1 activity: an approach to treat endocrine disorders, atherosclerosis and cancer. *Nat Rev Drug Discov*. 2007;6:821−833.

Supplemental Information Available on Companion Website

- Gene and precursor structures of IGF-I/E-Figure 19B.1
- Mature hormone sequences of various animals/E-Figure 19B.2
- Gene, mRNA, and precursor structures of the human IGF-IR/E-Figure 19B.3
- Primary structure of the human IGF-IR/E-Figure 19B.4
- Accession numbers of genes, cDNAs, and proteins for IGF-I and IGF-IR/E-Table 19B.1

Insulin-Like Growth Factor-II

Munetaka Shimizu

Abbreviation: IGF-II
Additional names: somatomedin A, multiplication-stimulating activity (MSA)

IGF-II is a polypeptide having multifunctions similar to those of IGF-I in vitro *but mainly regulating embryonic/fetal growth and placental function* in vivo. *It is encoded by an imprinted gene expressed paternally to regulate fetal growth.*

Discovery

IGF-II was discovered as a component of somatomedins in 1957, and in 1972 was found to have multiplication-stimulating activity. In 1978, IGF-II was co-purified with IGF-I from Cohn factions of plasma protein as a somatomedin that mediates growth hormone (GH) action on the cartilage segment [1,2]. A basic peptide was named IGF-I whereas the other peptide, which was slightly acidic, was named IGF-II.

Structure

Structural Features

IGF-II is a single polypeptide of 67 aa residues in humans, sharing high structural homology with IGF-I and proinsulin [2,3]. PreproIGF-II is composed of a signal peptide and five domains: B, C, A, D, and E domains (Figure 19C.1). The E domain is processed before secretion. Three disulfide bonds that are involved in the structural integrity of the peptide are conserved. Unlike IGF-I, there are no variable E domains, but isoforms with different O-glycosylation status exist. There are also "big" variants of IGF-IIs (1−104 and 1−87) that are differently cleaved by proteases [4,5].

Primary Structure

Although the total number of aa residues varies slightly among species, mature IGF-II sequence is well conserved (Figure 19C.2, E-Figure 19C.2).

Properties

Mr ~7,500, pI 6.4. Should be reconstituted in 10-mM HCl. Lyophilized peptide can be stored at 2−4°C for at least 2 years.

Synthesis and Release

Gene, mRNA, and Precursor

The human IGF-II gene, *IGF2*, location 11p15.5, consists of 10 exons, but only the latter 3 exons (8−10) encode the mature peptide (E-Figure 19C.1) [3,4]. There are four transcription start sites creating five transcripts alternatively spliced at 5′-UTR, and two polyadenylation sites in exon 10 [3]. Two non-allelic genes for IGF-II have been identified in zebrafish but not in salmon [6].

Distribution of mRNA

The liver is the main site of expression after birth.

Tissue and Plasma Concentrations

Tissue Content

Rat: liver 17 (ng/g tissue), bone 1,750, anterior pituitary 150−190.

Plasma Concentration

In humans, plasma IGF-II levels are relatively low during development, increase after birth, and remain high throughout life. There is no IGF-II peak at puberty. In contrast, plasma IGF-II levels in rodents are high during development and become very low after birth. See Table 19C.1 for the characteristics, tissue distribution, and plasma concentration of IGF-II in vertebrates

Regulation of Synthesis and Release

Four promoters, P1−P4, are present upstream of the human IGF-II gene. P3 and P4 are the main promoters regulating IGF-II synthesis during development, but P1 becomes dominant after birth in humans. Rodents lack P1, which may be a reason for the limited expression and very low circulating IGF-II after birth. Unlike IGF-I, GH has little effect on inducing IGF-II gene expression, although the human P2 and P4 promoters are responsive to GH. In teleosts, GH stimulates hepatic expression of IGF-II. Prolonged fasting reduces IGF-II levels.

Receptors

Structure and Subtype

The biological actions of IGF-II are mainly mediated by IGF-I receptor (IGF-IR) and insulin receptor isoform A (INSR-A), an alternatively spliced isoform lacking exon 11. In addition, a type I transmembrane glycoprotein receptor that is homologous to the cation-independent mannose-6-phosphate receptor (MPR) exhibits the highest affinity to IGF-II, and is called type II IGF-R (IGF-IIR/MPR) (Figure 19C.3) [7]. The human IGF-IIR/MPR gene, *IGF2R*, is located on chromosome 6q25−q37 and consists of 48 exons (E-Figure 19C.3). The mature receptor is 270 kDa, consisting of 15 repeat sequences in the extracellular domain (2,269 aa) followed by transmembrane (23-aa) and intracellular (163-aa) domains

Y. Takei, H. Ando, & K. Tsutsui (Eds): Handbook of Hormones. DOI: http://dx.doi.org/10.1016/B978-0-12-801028-0.00150-1

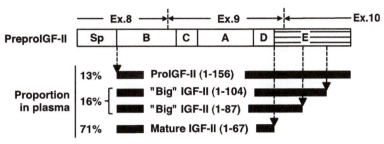

Figure 19C.1 Structure of the human prepro-, pro-, "big"- and mature IGF-II. Ex., exon.

```
                                B domain              C domain        A domain       D domain
Human IGF-II       AYRPSETLCGGELVDTLQFVCGDRGFYFSRPASRVSRR----SRGIVEECCFRSCDLALLETYCATPAKSE
Chicken IGF-II     YGTAETLCGGELVDTLQFVCGDRGFYFSRPVGRNNRRI---NRGIVEECCFRSCDLALLETYCAKSVKSE
Frog IGF-II        AYRATETLCGGELVDTLQFVCGDRGFYFSTNNGRSNRRP--NRGIVDVCCFKSCDLELLETYCAKPTKNE
Zebrafish IGF-IIa    SAETLCGGELVDALQFVCEDRGFYFSRPTSRSNSRRSQNRGIVEECCFSSCNLALLEQYCAKPAKSER
Zebrafish IGF-IIb    GETLCGGELVDTLQFVCGEDGFYISRPNRSNSRRP--QRGIVEECCFRSCELHLLQQYCAKPVKSERD
Salmon IGF-II      EVASAETLCGGELVDALQFVCEDRGFYFSRPTSRSNSRRSQNRGIVEECCFRSCDLNLLEQYCAKPAKSE
```

Figure 19C.2 Comparison of amino acid sequence of mature IGF-II in vertebrates. Asterisks indicate the conserved cysteine residues. Consensus/major amino acid residues are shaded.

Table 19C.1 Characteristics, Tissue Distribution, and Plasma Concentration of IGF-II in Vertebrates

Species	# of aa	Mr	Gene Location	# of Exon	Tissue (Prenatal)	Tissue (Postnatal)	Blood (ng/ml) (Prenatal)	Blood (ng/ml) (Postnatal)
Human	67	7,469	11	9	liver, adrenal, muscle	liver, skin, heart, bone, kidney, pituitary, muscle, reproductive organs, choroid plexus, leptomeninges	5–500	400–700
Rat	68	7,515	1	6	liver, brain, kidney, lung, muscle	pituitary, kidney, choroid plexus, leptomeninges	1,800–4,400	10–20
Mouse	68	7,386	7	6	liver	choroid plexus, leptomeninges	?	low
Chicken	67	7,512	5	3	eye, heart, limb bud	liver?, muscle?	75–120	35–55
Xenopus (IIA)	68	7,642	?	?	head region, dorsal mid line	?	?	?
Zebrafish (2a)	68	7,610	LG25	4	ubiquitous	liver, brain, eye, gut, gonad	?	?
Salmon	70	7,866	?	4	+ (whole embryo)	liver, kidney, gill, spleen, pyloric ceca	?	10–50

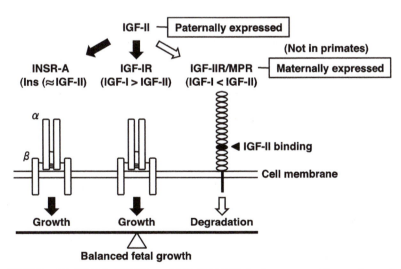

Figure 19C.3 Interaction of IGF-II with INSR-A, IGF-IR, and IGF-IIR/MPR during mammalian embryonic development. Relative binding preference is shown in parentheses.

(E-Figure 19C.4). A truncated form of the extracellular domain of the receptor circulates in the blood. The IGF-IIR/MPR has no intracellular signal pathway but, rather, disposes of IGF-II from the circulation by transporting it to lysosomes. The MPRs of non-mammalian species (chickens, lizards, and teleosts) were thought to have little ability to bind IGF-II [8], but some studies have revealed that they do indeed possess IGF-II binding ability, with high affinity [9].

Signal Transduction Pathway

The signal transduction pathway through the IGF-IR is similar to that of IGF-I. During fetal development, signaling through INSR-A is of particular importance. The IGF-IIR/MPR has no known intrinsic kinase activity, and is thought to be a scavenger receptor involved in controlling the extracellular IGF-II level.

Agonists

IGF-I, insulin, [Arg6] IGF-II, and Des [1−6] IGF-II (reduced binding to IGF-binding proteins, IGFBPs) are agonists.

Antagonists

IGFBPs and IGF-IIR 11^{E1554K}-Fc (IgG Fc domain) are antagonists.

Biological Functions

Target Cells/Tissues and Functions

During development in mammals, IGF-II regulates balanced fetal growth through being expressed only from the paternal allele (genomic imprinting) (Figure 19C.3) [4]. On the other hand, IGF-IIR/MPR is maternally expressed; however, this genomic imprinting is restricted to marsupials, rodents, and artiodactyls, and is not seen in primates and monotremes.

Phenotype in Gene-Modified Animals

Inactivation of IGF-II results in growth retardation of the fetus (60% of wild-type), while inactivation of IGF-IIR causes severe overgrowth.

Pathological Implications

Clinical Implications

Loss of imprinting (LOI) results in abnormal expression of IGF-II, which often causes growth disorders, cancer (breast, colon, lung), cardiovascular disease, Beckwith-Wiedemann syndrome (BWS, fetal overgrowth) with increased risk of Wilms' tumor (childhood cancer of the kidney), or Silver-Russell syndrome (severe intrauterine and postnatal growth retardation) [10]. The IGF-IIR/MPR gene functions as a tumor suppressor gene, and its single-nucleotide polymorphisms cause an increased risk of cancer. IGF-II accelerates tumor metastasis. Under non-islet cell tumor hypoglycemia (NICTH), circulating levels of "big" IGF-IIs are increased.

Use of Diagnosis and Treatment

Trapping circulating IGF-II using soluble IGF-IIR/MPR or its synthetic IGF-II binding domain may be useful in inhibiting tumor growth. Gene therapy using anti-cancer adenovirus that targets LOI of IGF-II to selectively kill cancer cells is being investigated.

References

1. Pierson RW, Temin HM. The partial purification from calf serum of a fraction with multiplication-stimulating activity for chicken fibroblasts in cell culture and with non-suppressible insulin-like activity. *J Cell Physiol.* 1972;79:319−330.
2. Daughaday WH, Rotwein P. Insulin-like growth factors I and II. Peptide, messenger ribonucleic acid and gene structures, serum, and tissue concentrations. *Endocrine Rev.* 1989;10:68−91.
3. Potwein P. Molecular biology of IGF-I and IGF-II. In: Rosenfeld RG, Roberts CT, eds. *The IGF System: Molecular Biology, Physiology, and Clinical Applications.* Totowa NJ: Humana Press; 1999:19−35.
4. Bergman D, Halje M, Nordin M, et al. Insulin-like growth factor 2 in development and disease: a mini-review. *Gerontology.* 2013;59:240−249.
5. Marks AG, Carroll JM, Purnell JQ, et al. Plasma distribution and signaling activities of IGF-II precursors. *Endocrinology.* 2011;152:922−930.
6. Zou S, Kamei H, Modi Z, Duan C. Zebrafish IGF genes: gene duplication, conservation and divergence, and novel roles in midline and notochord development. *PLoS One.* 2009;4:e7026.
7. Nadimpalli SK, Amancha PK. Evolution of mannose 6-phosphate receptors (MPR300 and 46): lysosomal enzyme sorting proteins. *Curr Protein Peptide Sci.* 2010;11:68−90.
8. McMurtry JP, Francis GL, Upton Z. Insulin-like growth factors in poultry. *Domest Anim Endocrinol.* 1997;14:199−229.
9. Mendez E, Planas JV, Castillo J, et al. Identification of a type II insulin-like growth factor receptor in fish embryos. *Endocrinology.* 2001;142:1090−1097.
10. Chao W, D'Amore PA. IGF2: epigenetic regulation and role in development and disease. *Cytokine Growth Factor Rev.* 2008;19:111−120.

Supplemental Information Available on Companion Website

- Gene and precursor structures of IGF-II/E-Figure 19C.1
- Mature hormone sequences of various animals/E-Figure 19C.2
- Gene, mRNA, and precursor structures of mouse IGF-IIR/E-Figure 19C.3
- Primary structure of the human IGF-IIR/E-Figure 19C.4
- Accession numbers of genes, cDNAs, and proteins for IGF-II and IGF-IIR/E-Table 19C.1

Insulin-Like Growth Factor Binding Proteins

Munetaka Shimizu

Abbreviation: IGFBPs
Additional names: none

IGFBPs are a family of cysteine-rich proteins that specifically bind insulin-like growth factor (IGF)-I and IGF-II with high affinity. This family consists of IGFBP-1 to -6. IGFBPs prolong the half-life of IGFs in the circulation, target them to certain tissues, and either potentiate or inhibit the availability of IGFs to the receptors. Some IGFBPs have IGF-independent actions on cell growth by translocating into the nucleus or interacting with cell surface receptors.

Discovery

In 1975, the presence of a serum protein that specifically binds IGF-I was reported in humans [1]. IGFBP-1 was first purified from human amniotic fluid in 1984.

Structure

Structural Features

IGFBPs are single chain polypeptides consisting of three domains: conserved N- and C-terminal domains, and a variable L (linker) domain [2,3]. The structure of IGFBPs is not related to that of IGF receptors. Twelve and six cysteine residues in the N- and C-terminal domains, respectively, are important for the high affinity binding to IGFs, and are conserved among IGFBP-1 to -5. IGFBP-4 has two additional cysteine residues in the L domain and IGFBP-6 lacks two cysteine residues in the N-terminal domain. The L domain contains modification sites for N-glycosylation, phosphorylation, and proteolytic cleavage, and nuclear localization signal motif and/or other binding sites that are specific to particular IGFBP types (Figure 19D.1). The N-terminal domain shares high sequence homology with IGFBP-related proteins (IGFBP-rPs). There are about 10 IGFBP-rPs, but they have lower affinities for IGFs due to the lack of the C-terminal domain. IGFBP and IGFBP-rPs could arise from exon shuffling of the N-terminal domain.

Primary Structure

The N- and C-terminal domains are well conserved, while the central L domain is highly variable (E-Figure 19D.2).

Properties

Mr 22,500−31,500, pI 5.1−8.8.

Synthesis and Release

Gene, mRNA, and Precursor

Six human IGFBP genes consist of four exons and three introns (E-Figure 19D.1). There are no particular transcript variants. Pairs of IGFBP-1 and -3, and IGFBP-2 and -5, are on the same chromosomes, respectively, aligned in a tail-to-tail fashion (Table 19D.1) [4]. During vertebrate evolution, the six IGFBPs likely arose from a local gene duplication followed by the two rounds of whole genome duplication and loss of two genes [4]. Teleosts could have 12 IGFBPs due to an extra genome duplication.

Distribution of mRNA

IGFBP genes are expressed in virtually all cell types, but the types and proportion of IGFBPs differ among cells. In the liver, hepatocytes express IGFBP-1, -2, -4, -5, and -6, and Kupffer and endothelial cells express IGFBP-3. Bone produces mainly IGFBP-4 and -5.

Tissue and Plasma Concentrations

See Table 19D.1.

Regulation of Synthesis and Release

Nutritional status/calorie intake, hormones, illness/injury, and others influence the synthesis and release of IGFBPs. See Table 19D.1 and below.

Receptors

Structure, Subtype, and Signal Transduction Pathway

Although IGFBPs are not categorized as hormones, there are some receptors or binding partners that mediate signal transduction pathways. IGFBP-1 and -2 have integrin-binding motifs (Arg−Gly−Asp, RGD), which bind $\alpha5\beta1$ integrin (fibronectin receptor) and activate the downstream signals [2,3]. IGFBP-3 can associate with the type V transforming growth factor (TGF)-β receptor on the cell membrane [5]. IGFBP-3 also associates with the nuclear retinoid X receptor-α [5].

Biological Functions

IGFBPs prolong the half-life of circulating IGFs by protecting them from degradation and glomerular filtration [6]. Most circulating IGFs are bound to one of six IGFBPs in humans.

Y. Takei, H. Ando, & K. Tsutsui (Eds): Handbook of Hormones. DOI: http://dx.doi.org/10.1016/B978-0-12-801028-0.00151-3

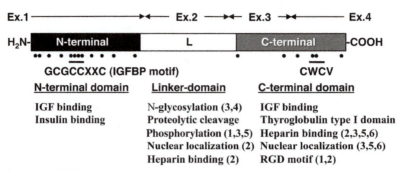

Figure 19D.1 **Structure of human IGFBP.** Ex., exon. Dots indicate cysteine residues conserved among IGFBP-1 to -5. Numbers in parentheses indicate characters applicable to particular IGFBP(s). RGD, Arg—Gly—Asp.

Table 19D.1 Characters, Tissue Distribution, and Plasma Concentration of Six Human IGFBPs

Type	# of aa	Core Mr	On gel (kDa)	Gene Location	Structural Feature	Affinity	Tissue/Fluid	Blood (ng/ml)	Regular (*in vivo*)
IGFBP-1	234	25,271	30	7p12—13	P, RGD	I = II	liver, kidney, amniotic fluid, serum	1.3—95	insulin(−), cortisol
IGFBP-2	289	31,355	34	2q33—34	RGD, NLS, H	I < II	liver, CSF, serum, milk	100—700	
IGFBP-3	264	28,717	43, 45	7p-12—13	N, P, NLS, ALS	I = II	liver, serum milk	2,000—6,000	GH, IGF-I
IGFBP-4	237	25,957	24, 28	17q	N	I = II	liver, bone, placenta, serum	50—500	vitamin D, PTH
IGFBP-5	252	28,553	30, 31	2q33	O, P, H, NLS, ALS	I < II	liver, bone, muscle, ovary, serum	220—660	IGF-I
IGFBP-6	216	22,847	28—30	12q13	O, H, NLS	I << II	liver, lung, CNS, serum	70—300	

P: Phosphorylation; RGD: Arg—Gly—Asp; NLS: nuclear localization signal; H: heparin binding; N: N-glycosylation; O: O-glycosylation; ALS: ALS binding; CSF: cerebrospinal fluid; CNS: central nervous system; PTH: parathyroid hormone.

A ternary complex (150 kDa) of one each molecule of IGF, IGFBP-3, and acid-labile subunit (ALS) carries about 75—80% of IGFs in the circulation and extends the IGF's half-life to 12—16 hours since this complex is too large to leave the capillary barrier. Binary complexes (40—50 kDa) of IGF and IGFBP carry about 20% of circulating IGFs with a half-life of 1—2 hours, and less than 1% of circulating IGFs exists as the free form (7.5 kDa), which is active but has a much shorter half-life (5—10 minutes). IGFBPs target IGFs to specific tissues/cells through association with an extracellular matrix such as heparin. As the affinity for IGFs of IGFBP is equal to or higher than that of the IGF receptors, IGFBPs' actions are generally inhibitory by preventing IGFs from binding to their receptors. However, cell association or proteolytic cleavage reduces the affinity of IGFBPs, making IGFs available to the proximate receptors in the peripheral tissues. As well as the IGF-modulating actions described above, some IGFBPs act directly on cell growth and gene expression without IGFs, through cell membrane receptors and nuclear transport.

IGFBPs vary in function:

IGFBP-1 [7] was the first member of the family to be identified by cDNA cloning. It is not abundant in the circulation, but shows dynamic diurnal changes in response to meals and fasting. Negatively regulated by insulin, it may be involved in glucoregulation. IGFBP-1 is largely unsaturated and may sequester free IGFs from the circulation. Phosphorylated forms have greater affinity for IGFs and act as inhibitors of the IGF actions. IGFBP-1 is abundant in amniotic fluid, at a 1,000-fold higher level than in serum.

IGFBP-2 [7] is the second most abundant circulating form. It is increased by long-term fasting and severe chronic stress, and may be a marker of insulin sensitivity. IGFBP-2 is unsaturated, and may sequester free IGFs from the circulation.

IGFBP-3 [5,6] is the most abundant form, carrying 75—80% of circulating IGFs by forming a ternary complex with IGF and ALS. IGFBP-3 is produced mainly by the Kupffer and endothelial cells in the liver in response to growth hormone (GH), whereas hepatocytes produce ALS and IGFs presumably to avoid formation of the 150-kDa complex. N-glycosylated proteolyzed fragments of IGFBP-3 with reduced affinity for IGFs are abundant in pregnant mammals. IGFBP-3 also has IGF-independent actions through translocation into the nucleus, while it interacts with cell surface receptors such as the type V TGF-β receptor.

ALS [2,6] is a component of a ternary complex with IGF and IGFBP-3 (or -5) which itself has no ability to bind IGFs. ALS expression is induced by GH. ALS is a ∼85-kDa leucine-rich, N-glycosylated protein conserved among a wide range of vertebrates. However, there is no evidence of the formation of a ternary complex in birds, amphibians, teleosts, or agnathans.

IGFBP-4 [8] is the smallest form, having two additional cysteine residues, and is the only form known as a consistent inhibitor of IGF actions. Present as N-glycosylated (28 kDa) and non-glycosylated (24 kDa) forms, IGFBP-4 is involved in the regulation of bone and pregnancy. Pregnancy-associated plasma protein A was the first enzyme to be identified that cleaves IGFBP.

IGFBP-5 [9] is the most conserved across species. IGFBP-5 has the ability to bind extracellular matrix proteins/glycosaminoglycans and hydroxylapatite, and is important for bone growth and mammary gland involution. It forms a ternary complex with IGF and ALS in the circulation despite being a minority fraction. It also has IGF-independent actions through translocation into the nucleus.

IGFBP-6 [10] exhibits much higher binding preference to IGF-II, probably due to the absence of two cysteine residues in the N-terminus. IGFBP-6's primary structure is the most divergent compared with other family members. It may have IGF-independent action(s).

Target Cells/Tissues and Functions

Virtually all cell/tissue types are affected. IGFBPs modulate IGF actions or affect cell function independent of IGFs.

Pathological Implications

Clinical Implications

IGFBPs are involved in many types of cancer, Alzheimer's disease, diabetes, cardiovascular disease, and other diseases through IGF-dependent and/or IGF-independent manners [5–10].

Use for Diagnosis and Treatment

Circulating IGFBP-1 may be a marker of insulin status and cardiovascular disease risk. IGFBP-2 is a prognostic marker of cancers such as prostate, breast, and gliomas. Plasma IGFBP-3 levels are generally associated with a reduced risk of cancers, although the relationship is less consistent compared to IGF-I.

References

1. Zapf J. Physiological role of the insulin-like growth factor binding proteins. *Eur J Endocrinol.* 1995;132:645–654.
2. Firth SM, Baxter RC. Cellular actions of the insulin-like growth factor binding proteins. *Endocr Rev.* 2002;23:824–854.
3. Forbes BE, McCarthy P, Norton RS. Insulin-like growth factor binding proteins: a structural perspective. *Front Endocrinol.* 2012;3:38.
4. Daza DO, Sundstrom G, Bergqvist CA, et al. Evolution of the insulin-like growth factor binding protein (IGFBP) family. *Endocrinology.* 2011;152:2278–2289.
5. Jogie-Brahim S, Feldman D, Oh Y. Unraveling insulin-like growth factor binding protein-3 actions in human disease. *Endocr Rev.* 2009;30:417–437.
6. Rajaram S, Baylink DJ, Mohan S. Insulin-like growth factor-binding proteins in serum and other biological fluids: regulation and functions.. *Endocr Rev.* 1997;18:801–831.
7. Wheatcroft SB, Kearney MT. IGF-dependent and IGF-independent actions of IGF-binding protein-1 and -2: implications for metabolic homeostasis. *Trends Endocrinol Metab.* 2009;20:153–162.
8. Mazerbourg S, Callebaut I, Zapf J, et al. Update on IGFBP-4: regulation of IGFBP-4 levels and functions, *in vitro* and *in vivo.* *Growth Horm IGF Res.* 2004;14:71–84.
9. Schneider MR, Wolf E, Hoeflich A, et al. IGF-binding protein-5: flexible player in the IGF system and effector on its own. *J Endocrinol.* 2002;172:423–440.
10. Bach LA, Fu P, Yang Z. Insulin-like growth factor-binding protein-6 and cancer. *Clin Sci.* 2013;124:215–229.

Supplemental Information Available on Companion Website

- Gene, mRNA, and precursor structures of IGFBP/E-Figure 19D.1
- Primary structure of the human IGFBPs/E-Figure 19D.2
- Accession numbers of genes and cDNAs for IGFBP and ALS/E-Table 19D.1

Relaxins

Masatoshi Mita

Abbreviation: RLN
Additional names: relaxin-like factor (RLF), insulin-like peptide (INSL)

Relaxin has been viewed as a hormone of pregnancy, and facilitates parturition in placental animals.

Discovery

Relaxin (RLN) was discovered by Hisaw in 1926 as a substance influencing the reproductive tract [1]. In recent years, seven relaxin family peptides have been identified [2]. RLN family peptides have been found in the majority of vertebrates [3]. Among invertebrates, starfish gonad-stimulating substance (GSS) was also identified as an RLN-like peptide (see Chapter 45E). [4].

Structure

Structural Features

RLN belongs to the insulin/IGF/RLN superfamily. There are seven RLN family peptides: relaxin -1, -2, and -3, and insulin-like peptide (INSL)3, INSL4, INSL5, and INSL6. All peptides exhibit a two-chain structure which is stabilized by a single intrachain and two interchain disulfide bonds (Figure 19E.1). Relaxin-1, -2 and -3, but not INSLs, display the relaxin-binding motif (Arg−X−X−X−Arg−X−X−Ile/Val) [2].

Primary Structure

Each RLN family pepetide shares high sequence homology of up to 60% in vertebrates. Six cysteine residues that confer the three disulfide bonds are completely conserved in vertebrates (E-Figures 19E.1−19E.5).

Properties

Mr 5,000−6,000, pI ∼7.0. Soluble in water and in acidic conditions.

Synthesis and Release

Gene, mRNA, and Precursor

The human relaxin-2 gene, *RLN2*, location 9p24.1, consists of two exons (E-Figure 19E.6). The relaxin-1 gene arose as a result of an ancestral duplication of the relaxin-2 gene [5]. All RLN family genes evolved from a relaxin-3 like ancestral gene (Figure 19E.2) [2]. RLN family peptides are synthesized as pre-prohormones composed of a signal sequence and B, C, and A domains. Cleavage of the C-peptide produces a heterodimeric peptide of A- and B-chains [2].

Distribution of mRNA

In mammals, RLN is produced mainly by the corpus luteum and placenta. RLNs are also produced in the prostate and seminal vesicles. Relaxin-3 is highly expressed in the brain [6]. INSL3 is produced by testicular Leydig cells. INSL5 is highly expressed in the gastrointestinal tract.

Tissue and Plasma Concentrations

In polytocous species such as the rat, peripheral RLN levels start to increase 10 or more days after estrus, coincident with the establishment of the corpus luteum of pregnancy. RLN concentration continues to rise throughout pregnancy, and declines precipitously within the hours or a day preceding birth. In humans, an increase in RLN is detectable at day 8 after the LH surge and reaches a maximum at 10−14 weeks of gestation, followed by a decline to a steady level of about 0.5 ng/ml throughout pregnancy, essentially mimicking the pattern of hCG secretion.

Regulation of Synthesis and Release

Regulation of synthesis and release of RLF is not yet well understood.

Receptors

Structure and Subtype

Receptors for the RLN family peptides belong to the GPCR family [2]. The receptor for relaxin-1 and -2 in humans is a leucine-rich repeat containing GPCR (LGR), termed LGR7, and that of INSL3 is closely related (LGR8). Relaxin-3 and INSL5 interact with type I GPCRs with a short N-terminal domain, GPCR135 and GPCR142, respectively. These receptors form the RLN family peptide (RXFP) receptor family, which includes LGR7, LGR8, GPCR135, and GPCR142, being renamed RXFP1, RXFP2, RXFP3, and RXFP4, respectively (Table 19E.1) [7]. The receptors for INSL4 and INSL6 are currently unknown.

Signal Transduction Pathway

RXFP1 activates AC, PKA, PKC, PI3K, and Erk1/2, and also interacts with NO signaling. The activation of adenylate cyclase by RXFP1 is complex and involves interaction with several G proteins, resulting in a biphasic pattern of cAMP accumulation [6]. RXFP2 activates AC *in vitro*, but some physiological responses are sensitive to pertussis toxin [6].

Y. Takei, H. Ando, & K. Tsutsui (Eds): Handbook of Hormones. DOI: http://dx.doi.org/10.1016/B978-0-12-801028-0.00152-5

Relaxin-1
(human)
RPYVALFEKCCLIGCTKRSLAKYC
KWKDDVIKLCGRELVRAQIAICGMSTWS

Relaxin-2
(human)
QLYSALANKCCHVGCTKRSLARFC
SWMEEVIKLCGRELVRAQIAICGMSTWS

Relaxin-3
(human)
DVLVAGLSSSCCKWGCSKSEISSLC
RAAPYGVRLCGREFIRAVIFTCGGSRW

INSL3
(human)
AAATNPARYCCLSGCTQQDLLTLCPY
APTPEMREKLCGHHFVRALVRVCGGPRWSTEA

INSL4
(human)
SGRHRFDPFCCEVICDDGTSVKLCT
AELRGCGPRFGKHLLSYCPMPEKTFTTTPGGWL

INSL5
(human)
QDLQTLCCTDGCSMTDLSALC
KESVRLCGLEYIRTVIYICASSRW

INSL6
(human)
GYSEKCCLTGCTKEELSIACLPYIDF
RELSDISSARKLCGRYLVKEIEKLCGHANWSQF

Figure 19E.1 Primary structures of human relaxin and INSL.

Table 19E.1 Relaxin Family Peptide (RXFP) Receptors and their Ligands

Receptor	Alternative Name	Ligand
RXFP1	LGR7	Relaxin-1, relaxin-2, relaxin-3
RXFP2	LGR8	Relaxin-1, relaxin-2, INSL3
RXFP3	GPCR135, SALPR, RLN3R1	Relaxin-3
RXFP4	GPCR142, GPR100, RLN3R2	INSL5, relaxin-3

The signaling pathways activated by RXFP3 or RXFP4 result in the inhibition of AC and a decrease in cAMP accumulation.

Agonists and Antagonists

There are no known relaxin agonists or antagonists.

Biological Functions

Target Cells/Tissues and Functions

In placental animals, RLN widens the pubic bones and facilitates labor. It induces cervical ripening and relaxes the uterine musculature [1]. RLN inhibits collagen synthesis [2]. It also enhances angiogenesis and is a potent renal vasodilator [8].

Phenotype in Gene-Modified Animals

In RLN knockout mice, cervical development is severely compromised, and nipples with dense collagen fiber bundles are undeveloped and small.

Pathophysiological Implications

Clinical Implications

Plasma levels of RLN rise during the first trimester of pregnancy, and thereafter decline to low levels. RLN may act indirectly to promote myometrial relaxation by stimulating myometrial prostacyclin production [9]. It also mediates hemodynamic changes during pregnancy.

Use for Diagnosis and Treatment

RLN concentration in serum and urine may be expected as a marker of pregnancy. Serelaxin (RLX030) is a recombinant

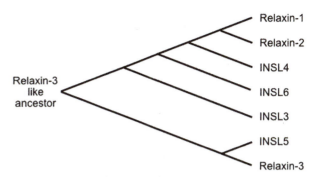

Figure 19E.2 Phylogenetic tree of the relaxin peptide family.

form of human relaxin-2 and is used as an investigational drug for the treatment of acute heart failure, targeting the RLN receptor [10].

References

1. Hisaw F. Experimental relaxation of the pubic ligament of the guinea pig. *Proc Soc Exp Biol Med.* 1926;23:661−663.
2. Bathgate RA, Samuel CS, Brazin TCD, et al. Human relaxin gene 3 (H3) and the equivalent mouse relaxin (M3) gene. Novel members of the relaxin peptide family. *J Biol Chem.* 2002;277:1148−1157.
3. Yegorov S, Good S. Using paleogenomics to study the evolution of gene families: origin and duplication history of the relaxin family hormones and their receptors. *PLoS One.* 2012;7:e32923.
4. Mita M, Yoshikuni M, Ohno K, et al. A relaxin-like peptide purified from radial nerves induces oocyte maturation and ovulation in the starfish, *Asterina pectinifera. Proc Natl Acad Sci USA.* 2009;106:9507−9512.
5. Kong RC, Shilling PJ, Lobb DK, et al. Membrane receptors: structure and function of the relaxin family peptide receptors. *Mol Cell Endocrinol.* 2010;320:1−15.
6. Bathgate RAD, Halls ML, van der Westhuizen ET, et al. Relaxin family peptides and their receptors. *Physiol Rev.* 2013;93: 405−480.
7. Bathgate RAD, Ivell R, Sanborn BM, et al. International Union of Pharmacology LVII: Recommendations for the nomenclature of receptors for relaxin family peptides. *Pharmacol Rev.* 2006;58:7−31.
8. Sherwood OD. Relaxin's physiological roles and other diverse actions. *Endocr Rev.* 2004;25:205−234.
9. MacLennan AH, Nicolson R, Green RC. Serum relaxin and pelvic pain of pregnancy. *Lancet.* 1986;2:241−243.
10. Teerlink JR, Cotter G, Davison BA. Serelaxin, recombinant human relaxin-2, for treatment of acute heart failure (RELAX-AHF): a randomised, placebo-controlled trial. *Lancet.* 2013;381:29−39.

Supplemental Information Available on Companion Website

- Precursor and mature hormone sequences of various animals/E-Figures 19E.1−19E.5
- Gene, mRNA and precursor strucures of the human relaxin-2/E-Figure 19E.6
- Primary structure of the human RXFP receptors/E-Figures 19E.7−19E.10
- Primary structure of RXFP receptors of various animals/E-Figures 19E.11−19E.14
- Accession numbers of genes and cDNAs for RLNs, INSLs, and RXFP receptors/E-Tables 19E.1−19E.3

CHAPTER 20

Gastrin Family

Toshio Sekiguchi

History

Gastrin was first described in 1905. In 1928, cholecystokinin (CCK) was first identified from pig intestine as a substrate that stimulates gallbladder contraction. In 1964, gastrin was isolated from porcine stomach and sequenced. Caerulein was first discovered in the skin of the Australian frog *Hyla caerulea* in 1967. A 33-aa residue sequence of porcine CCK was determined in 1968.

Structure

Structural Features

Gastrin, CCK, and caerulein share a four amino-acid C-terminal consensus sequence (Trp−Met−Asp−Phe−NH$_2$) (Figure 20.1), which is required for ligand activity. Furthermore, they possess a sulfated tyrosine residue. In mammals, gastrin has an O-sulfated tyrosine at position 6 from the C-terminus, while CCK has a sulfated tyrosine at position 7 from the C-terminus (Figure 20.1). However, non-mammalian CCK and gastrin have a sulfated tyrosine at position 7 from the C-terminus. The sulfated tyrosine residue of caerulein is located at position 7 from the C-terminus (Figure 20.1).

Molecular Evolution of Gastrin Family Members

In vertebrates, gastrin and CCK share common structural features (Figure 20.2). Phylogenetic and synteny analysis of vertebrate gastrins and CCKs demonstrate that they share a common ancestor [1]. Evolutionary studies of the gastrin family in gnathostomes indicate that gastrin and CCK may have diverged during the evolution of cartilaginous fish. Furthermore, the gastrin family peptide cionin has been identified in the protochordate *Ciona intestinalis* [1,2]. Cionin shares the C-terminal common tetrapepide, and possesses two sulfated tyrosines located at positions 6 and 7 from the C-terminus, suggesting that it possesses a hybrid hallmark of gastrin and CCK. Thus, gastrin family peptides are conserved in chordates. In addition, gastrin family-like sulfated peptides have been reported in protostomes − for example, sulfakinin and neuropeptide-like protein-12 (NLP-12) have been identified in arthropods and nematodes, respectively. However, the C-terminal tetrapeptide sequences of sulfakinin (His−Met−Arg−Phe−NH$_2$) and NLP-12 (Pro−Leu−Gln−Phe−NH$_2$) differ from that of vertebrate gastrin family peptides (Trp−Met−Asp−Phe−NH$_2$). It is suggested that vertebrate gastrin family peptides and these protostome sulfated peptides might have evolved in independent lineages [2].

Caerulein has only been detected in frogs, and is believed to share a common ancestor of gastrin family peptides. However, phylogenetic analysis using transcriptome and genome data demonstrates that the *Caerulein* genes in *Xenopus tropicalis* and *Litoria splendida* convergently evolved from *CCK* and *Gastrin* ancestor genes, respectively [3].

Receptors

Structure and Subtypes

In vertebrates, two GPCRs, CCK receptor type 1 (CCK1R) and type 2 (CCK2R), have been identified. These two proteins have seven transmembrane domains and show high amino acid sequence homology with each other. Phylogenetic analysis of vertebrate CCKRs demonstrates that CCK1r and CCK2R share a common ancestor (E-Figure 20.2) [4]. CCK1R is a CCK-specific receptor that recognizes sulfated CCK (Figure 20.2). CCK and gastrin are CCK2R ligands with similar affinity and potency (E-Figure 20.2).

Signal Transduction Pathways

CCK binds to CCK1R and CCK2R, which are coupled with G$_q$ protein, inducing intracellular calcium mobilization. Gastrin and CCK induce the accumulation of intracellular cAMP via CCK2R binding to G$_s$ protein.

Biological Functions

Target Cells/Tissues and Functions

Gastrin secretion induced by nutrients results in the release of gastric acid from the oxyntic mucosa and affects growth of the gastric mucosa. CCK acts as a brain−gut hormone. In the gut, CCK stimulates gallbladder contraction and pancreatic enzyme secretion. In addition, CCK has various neural functions, affecting food intake, anxiogenesis, satiety, nociception, memory, and learning. Caerulein is expected to be involved

Mammalian Gastrin (human)	−Glu−Ala−Tyr(SO$_3$H)−Gly−	Trp−Met−Asp−Phe−NH$_2$
Non-mammalian Gastrin-8 (Bullfrog)	Asp−Tyr(SO$_3$H)−Met−Gly−	Trp−Met−Asp−Phe−NH$_2$
CCK-8 (human)	Asp−Tyr(SO$_3$H)−Met−Gly−	Trp−Met−Asp−Phe−NH$_2$
Caerulein (Australian frog)	−Asp−Tyr(SO$_3$H)−Thr−Gly−	Trp−Met−Asp−Phe−NH$_2$

Figure 20.1 Amino acid sequence comparison of gastrin family peptides. The conserved C-terminal four amino-acid sequence is indicated by the gray box.

Y. Takei, H. Ando, & K. Tsutsui (Eds): Handbook of Hormones. DOI: http://dx.doi.org/10.1016/B978-0-12-801028-0.00020-9

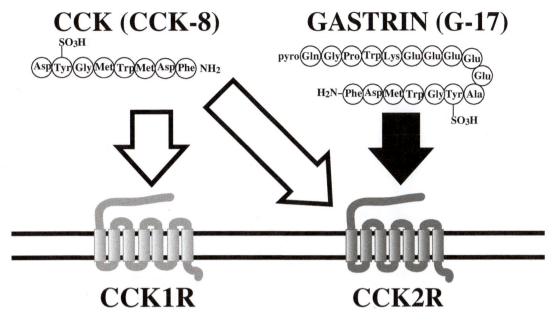

Figure 20.2 Ligand–receptor relationships in the gastrin family.

in the defense system because preproCaerulein contains an antimicrobial peptide. For details, see Subchapters 20A–20C.

References

1. Johnsen AH. Phylogeny of the cholecystokinin/gastrin family. *Front Neuroendocrinol.* 1998;19:73–99.
2. Sekiguchi T, Ogasawara M, Satake H. Molecular and functional characterization of cionin receptors in the ascidian, *Ciona intestinalis*: the evolutionary origin of the vertebrate cholecystokinin/gastrin family. *J Endocrinol.* 2012;213:99–106.
3. Roelants K, Fry BG, Norman JA, et al. Identical skin toxins by convergent molecular adaptation in frogs. *Curr Biol.* 2010;20:125–130.
4. Staljanssens D, Azari EK, Christiaens O, et al. The CCK(-like) receptor in the animal kingdom: functions, evolution and structures. *Peptides.* 2011;32:607–619.

Supplemental Information Available on Companion Website

- Primary structure of gastrin, cholecystokinin in human, and caerulein in frog/E-Figure 20.1
- Phylogenetic tree of the cholecystokinin receptors/E-Figure 20.2

Gastrin

Toshio Sekiguchi

Abbreviations: G14, G17, G34
Additional names: small gastrin, little gastrin, big gastrin

Gastrin is released from G cells in the antral mucosa in response to nutrients, and stimulates acid secretion in the gastric mucosa. In addition, gastrin is associated with the growth and differentiation of the gastric mucosa.

Discovery

Gastrin was first purified from pig antral mucosa, and identified as two types of 17 aa peptides, in 1964 [1]. One peptide is O-sulfated on the tyrosine located at the sixth residue from the C-terminus (G17-II). The other peptide is not sulfated (G17-I) [2].

Structure

Structural Features

Human preprogastrin is composed of 101 aa residues. Three mature peptides with different N-terminal extensions, big gastrin (G34), little gastrin (G17), and small gastrin (G14), are formed by posttranslational processing (Figure 20A.1). Furthermore, non-amidated glycine-extended gastrin is found in blood and tissues. Human G17 contains a pyroglutamic acid residue in the N-terminus, and is amidated at the C-terminus. In mammals, all gastrins possess a sulfated Tyr located as the sixth residue from the C-terminus.

Primary Structure

Among vertebrates, preprogastrin has low sequence identity, but mature gastrin is conserved. The C-terminal four amino-acid sequence (Trp−Met−Asp−Phe−NH$_2$) that is responsible for receptor activation is conserved in vertebrates. The position of the sulfated tyrosine differs between mammalian and non-mammalian gastrin, being located either sixth or seventh from the C-terminus, respectively (Figure 20A.2).

Properties

Human G14-I, Mr 1,647; human G17-I, Mr 2,096; human G34-I, Mr 3,839. Half-lives (in blood) of human G14, G17, and G34 are 1.8, 3.2, and 15.8 minues, respectively. Proteins are soluble in DMSO and low alkaline solutions (e.g., 0.1-M ammonium hydrogen carbonate and 0.01-M sodium hydroxide) and insoluble in acidic solutions. Solutions of gastrin should be stored below −20°C.

Synthesis and Release

Gene, mRNA, and Precursor

The human gastrin gene (*GAS*), located on chromosome 17 (17q21), consists of three exons. The gene structure of gastrin is conserved in mammals, except for that of porcine *Gastrin*, which has five exons. In mammals, the latter two exons are the coding region of the gastrin gene. *GAS* mRNA is a 475-bp transcript that encodes the 101-aa preprogastrin protein.

Distribution of mRNA

Gastrin mRNA is mainly expressed in the G cells of the antral mucosa. The gastrin genes are also transcribed in the duodenum, small intestine, pancreas, colon, pituitary, and testis [3].

Tissue and Plasma Concentrations

Tissue: Human pyloric antrum, 2,342 ± 144 pmol/g; duodenum, 1,397 ± 192 pmol/g; jejunum, 190 ± 24 pmol/g; gastric body, 23.5 ± 12 pmol/g [4].
   Plasma: In humans, fasting gastrin levels range from 5 to 290 pg/ml [5].

Regulation of Synthesis and Release

Nutrients, including amino acids (particularly aromatic amino acids such as L-phenylalanine and L-tryptophan), amines, and calcium, are sensed by the luminal side of antral G cells, and stimulate gastrin secretion. Gastrin-releasing peptide (GRP) that is secreted from the gastric nerve induces the release of gastrin from G cells. The negative regulator of gastrin secretion is somatostatin. Increases in the H$^+$ concentration of the stomach stimulates somatostatin release from antral D cells, inhibiting gastrin secretion from G cells via somatostatin receptor type 2.

Receptors

Structure and Subtype

The gastrin receptor, known as CCK receptor 2 (CCK2R), is a seven transmembrane GPCR. Human CCK2R is 447 aa residues in length. CCK, non-sulfated CCK, gastrin, and non-sulfated gastrin bind to CCK2R with similar affinity and potency [6].

Signal Transduction Pathways

CCK2R is coupled with G$_q$ protein. Gastrin stimulates intracellular calcium mobilization and protein kinase C via the PLC−IP3 pathway and the PLC−GAG pathway. In addition,

Y. Takei, H. Ando, & K. Tsutsui (Eds): Handbook of Hormones. DOI: http://dx.doi.org/10.1016/B978-0-12-801028-0.00153-7

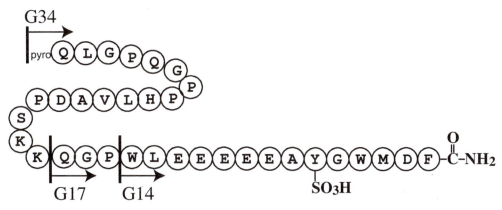

Figure 20A.1 Primary structures of human GASTRIN. G34, G17, and G14 indicate big GASTRIN, little GASTRIN, and small GASTRIN, respectively.

Human G34	QLGPQGPPHLVADPSKKQGPWLEEEEEAYGWMDF-NH₂
Pig G34	QLGLQGPPHLVADLAKKQGPWMEEEEEAYGWMDF-NH₂
Chicken G36	FLPHVFAELSDRKGFVQGNGAVEALHDHFYPDWMDF-NH₂
Bullfrog G47	DLLASLTHEQKQLIMSQLLPELLSELSNAEDHLHPMRDRDYAGWMDF-NH₂
Turtle G47	LSQDQKLLMAKFLPHIYAELANREGNWHEDAALRPLHDHDYPGWMDF-NH₂
Dogfish G49	AAPLRAEELISKLLPQIQEAGLLNQADRYQLRDVLHQMHDRDYTGWMDF-NH₂

Figure 20A.2 Amino acid sequence comparison of gastrins in vertebrates. Y indicates sulfated tyrosine. The conserved C-terminal four amino-acid sequence is indicated by the gray box.

CCK2R activates the MAPK pathway and PI3K pathway. Moreover, gastrin induces the processing of HB-EGF precursor in gastric epithelial cells.

Agonists

CCK2R agonists include gastrin, non-sulfated gastrin, CCK, non-sulfated CCK, RB-400, and PBC-264.

Antagonists

CCK2R antagonists include YF-476, GV150013, L-740093, YM-022, JNJ-26070109, L-365260, PR73870, and LY262691.

Biological Functions

Target Cells/Tissues and Functions

Gastrin is released into the bloodstream from antrum G cells, and induces the synthesis and release of histamine from enterochromaffin-like cells (ECL cells) through CCK2R. Histamine binds to histamine-2 receptors and stimulates acid secretion in parietal cells. Furthermore, gastrin directly activates acid secretion in parietal cells via CCK2R. Studies of the mouse model of hypergastrinemia and genetically modified mice have demonstrated that gastrin is responsible for normal growth and differentiation of the gastric mucosa.

Phenotype in Gene-Modified Animals

Gastrin gene knockout mice exhibit achlorhydria and atrophy of the gastric mucosa because of reduction of the parietal cell mass and non-activation of ECL cells. Bacterial overgrowth is observed in the stomachs of gastrin-deficient mice. *Cck2r*-deficient mice show 10-fold higher plasma concentrations of gastrin compared with wild-type mice. In addition, chlorhydria and hypoplasia of the oxyntic mucosa is observed in *Cck2r* null mice. Investigations using electron microscopy to study *Cck2r*-deficient mice report abnormal ECL cell differentiation [7].

Pathophysiological Implications

Clinical Implications

Zollinger-Ellison syndrome (ZES) is a clinical syndrome due to an ectopic Gastrin-secreting tumor in the duodenum or pancreas. Patients with ZES show gastric acid hypersecretion, which induces refractory peptic ulcers, severe gastroesophageal reflux disease, diarrhea, and even death [8]. Gastrin with a C-terminal glycin is produced by some tumors, including colorectal carcinoma [9].

Use for Diagnosis and Treatment

The plasma concentration of gastrin is measured by RIA or ELISA [10], and can be used to diagnose gastrin-secreting tumors, atrophic gastritis, gastric ulcers, and pernicious anemia. The gastrin test is performed to estimate acid secretory capacity and assess the extent of surgical vagotomy. It is also used to stimulate bioactive neuropeptide secretion in patients with neuroendocrine tumors such as vasoactive intestinal polypeptide secreting tumors.

References

1. Gregory RA, Tracy HJ. The constitution and properties of two gastrins extracted from hog antral mucosa. *Gut*. 1964;5:103−114.
2. Gregory H, Hardy PM, Jones DS, et al. The antral hormone gastrin. structure of gastrin. *Nature*. 1964;204:931−933.
3. Friis-Hansen L. Characteristics of gastrin controlled ECL cell specific gene expression. *Reg Peptides*. 2007;139:5−22.
4. Bloom SR, Polak JM. Gut hormone overview. In: Bloom SR, ed. *Gut Hormones*. New York: Churchill Livingstone; 1978:3−19.
5. Hansky J, Cain MD. Radioimmunoassay of gastrin in human serum. *Lancet*. 1969;294:1388−1390.

6. Dufresne M, Seva C, Fourmy D. Cholecystokinin and gastrin receptors. *Physiol Rev*. 2006;86:805−847.

7. Chen D, Zhao CM, Hakanson R, et al. Gastric phenotypic abnormality in cholecystokinin 2 receptor null mice. *Pharmacol Toxicol*. 2002;91:375−381.

8. Ito T, Cadiot G, Jensen RT. Diagnosis of Zollinger-Ellison syndrome: increasingly difficult. *World J Gastroenterol*. 2012;18:5495−5503.

9. Watson SA, Grabowska AM, El-Zaatari M, et al. Gastrin − active participant or bystander in gastric carcinogenesis? *Nat Rev Cancer*. 2006;6:936−946.

10. Rehfeld JF, Bardram L, Hilsted L, et al. Pitfalls in diagnostic gastrin measurements. *Clin Chem*. 2012;58:831−836.

Supplemental Information Available on Companion Website

- Gene structure of human gastrin/E-Figure 20A.1
- Preprogastrins in various animals/E-Figure 20A.2
- Gene structure of the human CCK2 receptor (*CCK2R*)/E-Figure 20A.3
- Primary structure of CCK2 receptor of various animals/E-Figure 20A.4
- Accession numbers of genes, cDNAs, and proteins for *Gastrin* and *Cck2r*/E-Tables 20A.1, 20A.2

Cholecystokinin

Toshio Sekiguchi

Abbreviation: CCK
Additional name: Pancreozymin

Cholecystokinin (CCK) is released from I cells in the small intestine. CCK stimulates gallbladder contraction and pancreatic enzyme secretion. Actions of CCK in the central nervous system include inhibition of food intake, anxiogenesis, satiety, nociception, memory, and learning.

Discovery

In 1928, cholecystokinin (CCK) was first described as a substance derived from the small intestine of dogs and cats, and that induced gallbladder contractions [1]. Pig intestine extracts were known to contain a substance called pancreozymin, which stimulated the secretion of pancreatic enzymes. In 1962, a 33-aa peptide isolated from pig intestines was reported to activate both gallbladder contractions (CCK action) and pancreatic enzyme secretion (pancreozymin action). The sequence of porcine CCK/pancreozymin was determined in 1968 [2].

Structure

Structural Features

Human CCK preproprotein contains 115 aa residues. As shown in Figure 20B.1, bioactive peptides of varying lengths with different N-terminal extensions (CCK-58, CCK-33, CCK-22, CCK-12, CCK-8, and CCK-4) are synthesized by enzymatic processing. CCK possesses a sulfated tyrosine at the seventh amino acid residue from the C-terminus. The C-terminal phenylalanine residue is amidated.

Primary Structure

In vertebrates, preproCCK shows low sequence identity; however, mature CCK is conserved. The C-terminal four amino-acid sequence (Trp−Met−Asp−Phe−NH$_2$) is responsible for receptor activation, and is conserved among vertebrates (Table 20B.1). In addition, the location of the sulfated tyrosine is conserved in vertebrates and is required for activation of one of the CCK receptor subtypes (Table 20B.1).

Properties

Human CCK-33, Mr 3,945.4; human CCK-8, Mr 1,143.3. Soluble in water, acidic solutions, and methanol.

Synthesis and Release

Gene, mRNA, and Precursor

The human CCK gene (*CCK*) is located on chromosome 3 (3p22.1), and contains five exons. *CCK* mRNA has a length of 1511 bp and encodes a precursor of 115 amino-acid residues. The preproCCK encodes a signal peptide and a mature CCK peptide.

Distribution of mRNA

CCK is expressed in I cells of the duodenum and small intestine. In rats, the *CCK* gene is expressed in the brain, heart, lung, kidney, and small intestine [3].

Tissue and Plasma Concentrations

Tissue: Human duodenum, 62.5 ± 8 pmol/g; upper jejunum, 26 ± 5 pmol/g [4].

Plasma: In healthy humans, fasting concentration is 1.13 ± 0.10 pmol/l ($n = 26$). Concentration rises to 4.92 ± 0.34 pmol/l after a mixed meal [5].

Regulation of Synthesis and Release

In intestinal I cells, dietary lipids and proteins are sensed by GPR40 and calcium-sensing receptor, respectively, resulting in stimulation of CCK release.

Receptors

Structure and Subtypes

CCK receptor 1 (CCK1R) and CCK receptor 2 (CCK2R) are GPCRs that have seven transmembrane domains. Human CCK1R is 428 aa long. CCK binds and responds to CCK1R with 500 to 1,000 times the affinity and potency of gastrin and non-sulfated CCK [7]. Sulfation of CCK is responsible for receptor binding and response. A more detailed description of CCK2R is presented in Subchapter 20A.

Signal Transduction Pathway

CCK1R and CCK2R are coupled with G$_q$ protein. The binding of CCK to CCK1R induces mobilization of intracellular calcium and activation of PKC via the PLC−IP3 and DAG pathways, respectively. In addition, CCK1R activates the mitogen-activated protein MAPK and PI3K pathway. CCK1R is also coupled with G$_s$ protein and induces cAMP accumulation. The signal transduction pathway of CCK2R is described in Subchapter 20A.

Y. Takei, H. Ando, & K. Tsutsui (Eds): Handbook of Hormones. DOI: http://dx.doi.org/10.1016/B978-0-12-801028-0.00154-9

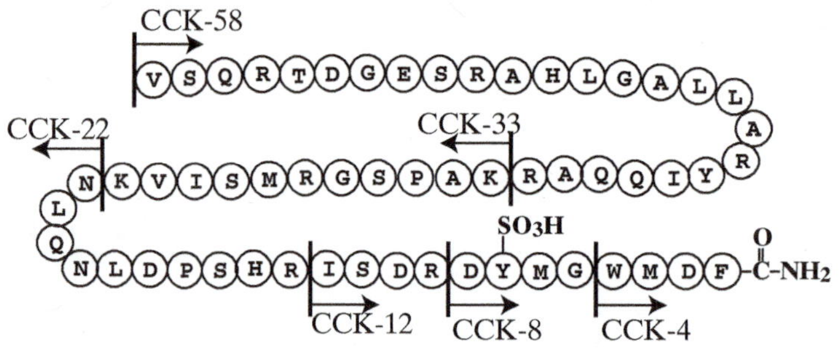

Figure 20B.1 Amino acid structure of human CCK subtypes. The N-termini of CCK subtypes are indicated by lines and arrows.

Table 20B.1 Amino Acid Alignment of Vertebrate CCKs

Human CCK-8	DYMGWMDF-NH₂
Pig Cck-8	DYMGWMDF-NH₂
Chick Cck-8	DYMGWMDF-NH₂
Bullfrog Cck-8	DYMGWMDF-NH₂
Dogfish Cck-8	DYMGWMDF-NH₂
Goldfish CcK-8	DYLGWMDF-NH₂

Y indicates sulfated tyrosine; the conserved four amino-acid sequence is boxed in gray.

Agonists

CCK1R agonists include CCK, A-71623, and GW-5823.

Antagonists

CCK1R antagonists include devazepide, lintitript, lorglumide, PD-140548, and T-0632.

Biological Functions

Target Cells/Tissues and Functions

CCK is released in the bloodstream, and induces contraction of the gallbladder and secretion of pancreatic enzymes via CCK1R. CCK affects the vagal afferent neurons in a paracrine and hormonal manner via CCK1R, and inhibits gastric emptying and food intake through these neurons [6]. Insulin secretion is induced by CCK stimulation. In the central nervous system, CCK is implicated in anxiogenesis, satiety, nociception, memory, and learning [7].

Phenotype in Gene-Modified Animals

Although *Cck* null mice are viable, several dysfunctions are observed, such as deficiency of gallbladder motility, increased gallbladder size, and cholesterol crystallization [8]. High amylase secretion and low insulin secretion are detected in the pancreas of *Cck*-deficient mice. *Cck1r* null mice have larger gallbladder volumes, decreased small intestinal transit times, and a higher propensity for developing gallstones. *Cck2r* knockout mice show defects of memory, suggesting that CCK is involved in memory. Otsuka Long-Evans Tokushima Fatty (OLETF) rats have a mutation in the *Cck1r* locus that exhibits the phenotype of little or no *CCK1R* mRNA expression. OLETF rats develop hyperphagia due to lack of satiety [9].

Pathophysiological Implications

Clinical Implications

CCK is associated with pancreatitis and pancreatic cancer growth [10].

Use for Diagnosis and Treatment

Antagonists of CCK1R have been developed as therapeutic agents for the treatment of diseases associated with CCK [7].

References

1. Ivy AC, Oldberg E. Hormone mechanism for gallbladder contraction and evacuation. *Am J Physiol.* 1928;86:599−613.
2. Mutt V, Jorpes JE. Structure of porcine cholecystokinin-pancreozymin. 1. Cleavage with thrombin and with trypsin. *Eur J Biochem.* 1968;6:156−162.
3. Funakoshi A, Tanaka A, Kawanami T, et al. Expression of the cholecystokinin precursor gene in rat tissues. *J Gastroenterol.* 1994;29:125−128.
4. Bloom SR, Polak JM. Gut hormone overview. In: Bloom SR, ed. *Gut Hormones.* London: Churchill Livingstone; 1978:3−18.
5. Rehfeld JF. Accurate measurement of cholecystokinin in plasma. *Clin Chem.* 1998;44:991−1001.
6. Dockray GJ. Cholecystokinin. *Curr Opin Endocrinol Diabetes Obes.* 2012;19:8−12.
7. Herranz R. Cholecystokinin antagonists: pharmacological and therapeutic potential. *Med Res Rev.* 2003;23:559−605.
8. Miyasaka K, Kanai S, Ohta M, et al. Lack of satiety effect of cholecystokinin (CCK) in a new rat model not expressing the CCK-A receptor gene. *Neurosci Lett.* 1994;180:143−146.
9. Wang HH, Portincasa P, Liu M, et al. Effect of gallbladder hypomotility on cholesterol crystallization and growth in CCK-deficient mice. *Biochim Biophys Acta.* 2010;1801:138−146.
10. Smith JP, Solomon TE. Cholecystokinin and pancreatic cancer: the chicken or the egg? *Am J Physiol Gastrointest Liver Physiol.* 2014;306:G91−G101.

Supplemental Information Available on Companion Website

- Gene structure of human CCK/E-Figure 20B.1
- Precursor sequences of various animals/E-Figure 20B.2
- Gene structures of the human CCK1 receptor/E-Figure 20B.3
- Primary structure of the Cck1 receptor of various animals/E-Figure 20B.4
- Accession numbers of genes and cDNAs for Cck and Cck1 receptors/E-Tables 20B.1, 20B.2

Caerulein

Toshio Sekiguchi

Additional name: ceruletide

Caerulein is a peptide secreted from the skin cells of frogs. Caerulein shares the common amino acid sequence that is responsible for receptor activity with vertebrate gastrin and cholecystokinin (CCK), and could function as an agonist of vertebrate gastrin and CCK.

Discovery

Caerulein was first described in a number of Australian amphibians as a polypeptide that elicits a decrease in blood pressure, external secretion, and extravascular smooth muscle contraction in mammals [1]. Caerulein was first purified from the Australian tree frog *Hyla caerulea* in 1967 [2].

Structure

Structural Features

Caerulein is a decapeptide that contains a 4-aa sequence (Gly−Trp−Met−Asp−Phe−NH$_2$) that is conserved in vertebrate gastrin and CCK. A pyroglutamate residue is present in the N-terminus, and a C-terminal phenylalanine residue is amidated. Caerulein possesses a sulfated tyrosine at the seventh residue from the C-terminus (Figure 20C.1).

Primary Structure

Two preprocaeruleins, caerulein precursor fragment (CPF)-st6 and CPF-st7, have been reported in the western clawed frog *Silurana tropicalis* (Figure 20C.2) [3]. The precursor of caerulein contains a signal peptide, an antimicrobial peptide called caerulein precursor fragment, and mature caerulein (Figure 20C.2). Sauvage's leaf frog, *Phyllomedusa sauvagei*, possesses a caerulein-like nonapeptide called phyllocaerulein. Caerulein has been identified in various frog species, including *Xenopus laevis*, *Litoria splendida*, and *Hylambates maculatus* [4]. Although caeruleins of several frog species share the C-terminal 4-aa sequence, exceptions also exist. For example, one variant (caerulein 1.2; Gln−Gln−Asn−Tyr (SO$_3$H)−Thr−Gly−Trp−Phe−Asn−Phe−NH$_2$) found in the

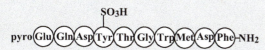

Figure 20C.1 Primary structure of caerulein peptide in *Hyla caerulea*. pyroGlu, pyroglutamate.

magnificent tree frog (*Litoria splendida*) does not have the consensus 4-aa sequence [5].

Properties

H. caerulea caerulein, Mr 1,352.4. Soluble in DMSO, but insoluble in acetone and diethyl ether.

Synthesis and Release

Gene, mRNA, and Precursor

In *S. tropicalis*, two caerulein genes have been identified in a four-exon structure [3]. The lengths of the caerulein mRNA are 428 and 418 bp, encoding peptides of 98 and 91 aa residues, respectively [3].

Distribution of mRNA

The caerulein gene is expressed in the skin.

Tissue Content

In adult *H. caerulea*, the content of caerulein is approximately 300−1,000 µg/g of fresh skin [6]. Caerulein content in the dorsal region of the skin is higher compared with that in the ventral region [6]. In contrast, the caerulein level is below 1 µg/g of wet weight in the dorsal region of larval skin [7].

Regulation of Synthesis and Release

Dockray and Hopkins reported that a caerulein-like substance is released by *X. laevis* in response to treatment with adrenaline [8]. This substance stimulates contraction of the guinea pig gallbladder and pancreatic secretions in rats [8]. Amino acid analysis of the secreted substance has a similar aa composition to that of caerulein [8]. Seasonal changes in caerulein synthesis have been observed [5]. For example, *L. splendida* synthesizes caerulein that stimulates smooth muscle contraction during the reproductive summer season. In the winter, synthesis of a less active, desulfated form of caerulein increases, and another caerulein peptide subtype (caerulein 1.2), which has relatively low activity, is released [5].

Receptors

No endogenous receptor has been identified in frogs. However, caerulein responds to mammalian CCK1R and CCK2R as agonists.

Y. Takei, H. Ando, & K. Tsutsui (Eds): Handbook of Hormones. DOI: http://dx.doi.org/10.1016/B978-0-12-801028-0.00155-0

```
Cpf-st6          MFKGLFLCVLFAVLSAQSMAQPTASADEEANANERVARKLGFENFLVKALKTVMHVPTSP
Cpf-st7          MFKGLFLCVLLAVLSAQSMAQPKASADEEE-MNERVAR-----NLLGSLLKTGLKVGSN-
Clustal Consensus **********:**********.******    ******    *:* . *** ::* :.

Cpf-st6          LLGRREANDRRFADGPNAVGQTEYEGWMDFGRRSAEEE
Cpf-st7          LLGRREANDRRFADGPNAVGQTEYEGWMDFGRRSAEEE
Clustal Consensus *************************************
```

Figure 20C.2 **Amino acid sequence comparison of preprocaeruleins in *Silurana tropicalis*.** Cpf and caerulein are indicated by gray and black boxes, respectively.

Biological Functions

Target Cells/Tissues and Functions

Treatment of the outer side of frog skin with caerulein results in an influx of sodium ions, while treatment of the inner side of frog skin represses sodium ion influx. These observations suggest that caerulein is associated with dermal functions within the frog itself. In addition, preprocaerulein contains an antimicrobial peptide, and the mature caerulein peptide affects endogenous signaling of gastrin and CCK in other animals. These observations suggest that preprocaerulein functions as a defensive peptide against microbes and predators. The function of the caerulein peptide in the frog has not been entirely characterized.

Phenotype in Gene-Modified Animals

No gene-modified animals have been reported.

Pathophysiological Implications

Clinical Implications

Caerulein is used in clinical research as an agonist of CCK1R and CCK2R due to its structural similarity to gastrin and CCK. Treatment with high-dose caerulein induces the secretion of pancreatic juice and results in acute pancreatitis. Therefore, caerulein is used to generate rodent models of pancreatitis treatment [9].

References

1. Erspamer V, Roseghini M, Endean R, et al. Biogenic amines and active polypeptides in the skin of Australian amphibians. *Nature.* 1966;212:204.

2. Anastasi A, Erspamer V, Endean R. Isolation and structure of caerulein, an active decapeptide from the skin of *Hyla caerulea*. *Experientia.* 1967;23:699–700.

3. Roelants Fry BG, Ye L, Stijlemans B, et al. Origin and functional diversification of an amphibian defense peptide arsenal. *PLoS Genet.* 2013;9:e1003662.

4. Bowie JH, Tyler MJ. Host defense peptides from Australian amphibians: Caerulein and other neuropeptides. In: Kastin AJ, ed. *Handbook of Biologically Active Peptides.* San Diego, CA: Academic Press; 2006:283–289.

5. Sherman PJ, Jackway RJ, Nicholson E, et al. Activities of seasonally variable caerulein and rothein skin peptides from the tree frogs *Litoria splendida* and *Litoria rothii*. *Toxicon.* 2009;54:828–835.

6. De Caro G, Endean R, Erspamer V, et al. Occurrence of caerulein in extracts of the skin of *Hyla caerulea* and other Australian hylids. *Br J Pharmacol Chemother.* 1968;33:48–58.

7. Seki T, Kikuyama S, Yanaihara N. Development of *Xenopus laevis* skin glands producing 5-hydroxytryptamine and caerulein. *Cell Tissue Res.* 1989;258:483–489.

8. Dockray GJ, Hopkins CR. Caerulein secretion by dermal glands in *Xenopus laevis*. *J Cell Biol.* 1975;64:723–733.

9. Lerch MM, Gorelick FS. Models of acute and chronic pancreatitis. *Gastroenterology.* 2013;144:1180–1193.

Supplemental Information Available on Companion Website

- Gene structure of the caerulein/E-Figure 20C.1
- Precursor of caerulein in various frogs/E-Figure 20C.2
- Accession numbers of genes, cDNAs, and proteins for caerulein/E-Table 20C.1

CHAPTER 21

Ghrelin—Motilin Family

Hiroyuki Kaiya

History

The definition of ghrelin (GHRL) and motilin (MLN) as a family originates from a common ancestral gene. Although this definition is not widely accepted, there are several articles to support their being a GHRL—MLN family owing to the similarity of these hormones and their receptors [1,2]. Historically, MLN was discovered earlier than GHRL. MLN is mainly produced in the duodenum, and the complete aa sequence of porcine MLN was determined in 1973 [3]. GHRL was isolated from rat and human stomach extracts in 1999 [4]. Another group reported identification of the MLN-related peptide (MTLRP) in mice [5], and this sequence was identical to that of GHRL {1—18} without acyl modification. In non-mammalian vertebrates, GHRL was identified in species higher than elasmobranchs [6]. Strictly, GHRLs identified in elasmobranchs such as sharks and stingray have been called "GHRL-like peptides" (GHRL-LPs) because of their low sequence homology with other GHRLs. However, GHRL-LP has acyl modification and an ability to bind to rat GHRL receptor. Recently, MLN-like peptide has been identified in fish, but there are some structural differences between tetrapod and fish MLNs. This casts some doubt on the family relationship between MLN and GHRL.

Structure

Structural Features

Human GHRL and MLN are straight-chain peptides consisting of 22 and 28 aa residues, respectively [2—4] (Figure 21.1). In common, the mature peptides are encoded in two exons, and the boundary between the two is at the fourteenth glutamine residue. Eight aa residues are identical at the mature portion [1], but overall identicality between precursors is not high (24.8%).

Receptors

Structure and Subtypes

There are two isoforms for each MLN and GHRL receptor, which are generated by alternative splicing of the mRNA: GPR38A and 38B for MLN [7], and growth hormone secretagog receptor (GHS-R) 1a and 1b for GHRL [8]. Type A receptors (GPR38A and GHS-R1a) have seven transmembrane domains (TMDs) and are involved in biological function (Figure 21.2), whereas type B receptors lack TMDs 6 and 7. In humans, identity of the overall aa sequence between GPR38A and GHS-R1a is 52%, and much higher (87%) sequence identicality was shown in the TMDs [1,2] (Figure 21.2). GHRL and MLN both seem to cross-react with the other receptor at high concentrations and activate signal transduction, although the potency is low [9].

Signal Transduction Pathway

Both receptors belong to class A, rhodopsin-like GPCR, and basically a $G_{\alpha q}$-mediated signal transduction pathway induces the production of inositol triphosphate that releases Ca^{2+} from intracellular calcium stores and diacylglycerol that activates protein kinase C [6,10]. GRLN also induces G_s-coupled Ca^{2+} increase via the cAMP-protein kinase A signaling pathway in NPY-containing neurons [7], whereas MLN is known to activate $G_{\alpha 13}$ as well [10].

Biological Functions

Target Cells/Tissues and Functions

Both MLN and GHRL are involved in gastrointestinal (GI) contraction [1,2], but this action seems to be region-specific in animals with both peptides. In chicken and quail, GHRL dominantly acts on the esophagus, crop sac, and colon, while MLN mainly affects contraction of the middle intestine such as

```
MLN:   MVSRKAVAAL LVVHVAAMLA SQTEAFVPIF TYGELQR-MQ EKERNKGQKK SLSVWQRSGE
       *  *   * * *       **    .    *   * ***  *  ** *   *  *    *
GHRL:  MPSPGTVCSL L---LLGMLW LDLAMAGSSF LSPEHQRVQQ RKESKKPPAK -LQPRALAGW


MLN:   EGPVDPAEPI REEENEMIKL TAPLEIGMRM NSRQLEKYPA TL-----EGL LSEMLPQHAA K
        *  *          *          ** *        *            *      *   *  *
GHRL:  LRPEDGGQAE GAEDELEVRF NAPFDVGIKL SGVQYQQHSQ ALGKFLQDIL WEEAKEAPAD K
```

Figure 21.1 Comparison of amino acid sequence of human GHRL and MLN precursors. Red letters indicate the portion of mature peptide, and asterisks show the amino acids identical between precursors.

Y. Takei, H. Ando, & K. Tsutsui (Eds): Handbook of Hormones. DOI: http://dx.doi.org/10.1016/B978-0-12-801028-0.00021-0

```
MLNR1A:   M--GSPWNGS  DGPEGAREPP  WPALPPCD--  -ERRCSPFPL  GALVPVTAVC  LCLFVVGVSG  NVVTVMLIGR  YRDMRTTTNL  YLGSMAVSDL
          *      *          *    **          **       * * ***  *     * ****  *       * ******  ** *** ***
GHS-R1a:  MWNATPSEEP  GFNLTLADLD  WDASPGNDSL  GDELLQLFPA  PLLAGVTATC  VALFVVGIAG  NLLTMLVVSR  FRELRTTTNL  YLSSMAFSDL


MLNR1A:   LILLGLPFDL  YRLWRSRPWV  FGPLLCRLSL  YVGEGCTYAT  LLHMTALSVE  RYLAICRPLR  ARVLVTRRRV  RALIAVLWAV  ALLSAGPFLF
          ** *   * **  ***  ***   ** ***  *     * * *****  *    ******  ** *** ***  * *  **   **   *  *  *** *  ****
GHS-R1a:  LIFLCMPLDL  VRLWQYRPWN  FGDLLCKLFQ  FVSESCTYAT  VLTITALSVE  RYFAICFPLR  AKVVVTKGRV  KLVIFVIWAV  AFCSAGPIFV


MLNR1A:   LVGVEQDPGI  SVVPGLNGTA  RIASSPLASS  PPLWLSRAPP  PSPPSGPETA  EAAALFSREC  RP--SPAQLG  ALRVMLWVTT  AYFFLPFLCL
          *****          ***                                                    ** **       *  * ** **    ****  *
GHS-R1a:  LVGVE-----  ----HENGT-  ----------  ----------  ----------  --DPWDTNEC  RPTEFAVRSG  LLTVMVWVSS  IFFFLPVFCL


MLNR1A:   SILYGLIGRE  LWSSRRPLRG  PAASGRERGH  RQTVRVLLVV  VLAFIICWLP  FHVGR---II  YINTEDSRMM  YFSQYFNIVA  LQLFYLSASI
          **  ****  ** ** **         ** *      * ***  * **  **  *** **** *****               *** * *       ****** *
GHS-R1a:  TVLYSLIGRK  LWRRRRGDAV  VGASLRDQNH  KQTVKMLAVV  VFAFILCWLP  FHVGRYLFSK  SFEPGSLEIA  QISQYCNLVS  FVLFYLSAAI


MLNR1A:   NPILYNLISK  KYRAAAFKLL  LARKSRPRGF  HRSRDTAGEV  AGDTGGDTVG  YTETSANVKT  MG
          ******  ** *** * **       **                  *                ** * *
GHS-R1a:  NPILYNIMSK  KYRVAVFRLL  --------GF  EPFSQRKLST  LKD--ESSRA  WTESSIN---  -T
```

Figure 21.2 Alignment of GHRL and MLN receptors in humans. Pink letters indicate transmembrane domains. Asterisks indicate the amino acids identical between receptors.

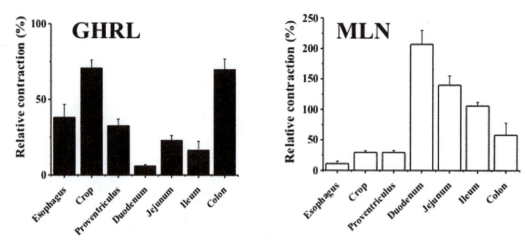

Figure 21.3 Gastrointestinal region-dependent difference in amplitude of contraction induced by chicken GHRL and chicken MLN. The effect of 1-μM ligand on smooth muscle strips from GI regions is shown. The values are expressed as a percentage of high-K^+ (50-mM) induced contraction.

the duodenum, jejunum, and ileum [7] (Figure 21.3). Rats and mice are natural MLN knockout animals [2], and GHRL compensates for MLN actions in all GI regions.

References

1. Peeters TL. Ghrelin: a new player in the control of gastrointestinal functions. *Gut.* 2005;54:1638−1649.
2. Sakata I, Sakai T. The gut peptide hormone family, motilin and ghrelin. In: Aimaretti G, Marzullo P, Prodam F, eds. *Update on Mechanisms of Hormone Action − Focus on Metabolism, Growth and Reproduction.* Rijeka, Croatia: InTech; 2011.
3. Brown JC, Cook MA, Dryburgh JR. Motilin, a gastric motor activity stimulating polypeptide: the complete amino acid sequence. *Can J Biochem.* 1973;51:533−537.
4. Kojima M, Hosoda H, Date Y, et al. Ghrelin is a growth-hormone-releasing acylated peptide from stomach. *Nature.* 1999;402:656−660.
5. Tomasetto C, Karam SM, Ribieras S, et al. Identification and characterization of a novel gastric peptide hormone: the motilin-related peptide. *Gatroenterology.* 2000;119:395−405.
6. Kaiya H, Miyazato M, Kangawa K. Recent advances in the phylogenetic study of ghrelin. *Peptides.* 2011;32:2155−2174.
7. Feighner SD, Tan CP, McKee KK, et al. Receptor for motilin identified in the human gastrointestinal system. *Science.* 1999;284:2184−2188.
8. Howard AD, Feighner SD, Cully DF, et al. A receptor in pituitary and hypothalamus that functions in growth hormone release. *Science.* 1996;273:974−977.
9. Nunoi H, Matsuura B, Utsunomiya S, et al. A relationship between motilin and growth hormone secretagogue receptors. *Regul Pept.* 2012;176:28−35.
10. Huang J, Zhou H, Mahavadi S, et al. Signaling pathways mediating gastrointestinal smooth muscle contraction and MLC20 phosphorylation by motilin receptors. *Am J Physiol.* 2005;288:G23−G31.
11. Liu Y, Li S, Huang X, et al. Identification and characterization of a motilin-like peptide and its receptor in teleost. *Gen Comp Endocrinol.* 2013;186:85−93.

Supplemental Information Available on Companion Website

- Molecular phylogenetic tree of hormones/E-Figure 21.1 [6,11]
- Molecular phylogenetic tree of receptors/E-Figure 21.2

Ghrelin

Hiroyuki Kaiya

Abbreviations: GHRL, GLRN
Additional names: appetite-regulating hormone, motilin-related peptide (MTLRP)

A unique fatty acid modification mainly with n-octanoic acid at the third serine residue of ghrelin is essential for eliciting ghrelin's activity. Ghrelin was the first peripheral hormone found to stimulate feeding behavior.

Discovery

In 1999, Kojima *et al.* [1] isolated GHRL rat stomach extracts by the orphan receptor strategy using the growth hormone secretagog receptor type-1a (GHS-R1a). The name originates from its growth hormone-releasing activity, with reference to the Proto-Indo-European word "*ghre,*" meaning "*grow,*" and G*row*th H*or*mone R*el*ease + *in* (suffix of *inducing*) to give "ghrelin."

Structure

Structural Features

The serine residue at position 3 (Ser-3) is modified by middle-chain fatty acids, and GHRL usually refers to the octanoylated form (acylated ghrelin) (Figure 21A.1). While *n*-decanoic acid contributes to the modification, species difference was seen [2]. In the genus *Rana* (frogs), acylated aa has substituted for threonine [3]. The first seven-aa sequence (GSSFLSP) is highly conserved across animal species. Teleost GHRLs have an amide structure at the C-terminus [3,4]. Stingray GHRL has *O*-glycosylation at Ser-10 and/or Thr-11 in addition to the acylation at Ser-3 [3].

Primary Structure

Typical aa sequences of GHRL are shown in Figure 21A.2. Mammalian GHRLs consist of 28 aa in general, but a 27-aa molecule lacking the fourteenth glutamine residue by an alternative splicing, namely des-Gln14 ghrelin, has also been identified (E-Figures 21A.1, 21A.6). Amino acid sequences have been determined in non-mammalian vertebrates, including cartilaginous fishes (E-Figure 21A.2), bony fishes (E-Figure 21A.3), amphibians (E-Figure 21A.4), and reptiles and birds (E-Figure 21A.5). The number of amino acids ranges from 16 (stingray) to 28 residues (frogs) depending on species [3].

Properties

Human GHRL: Mr 3,371.9. Lyophilized material is stable at $-30°C$, but $-80°C$ is more desirable. Soluble in water. Because of high adsorptivity, it is desirable to maintain storage concentration at $>10^{-4}$ M, or to dilute with saline containing 5% mannitol or 0.2% BSA for treatment.

Synthesis and Release

Gene, mRNA, and Precursor

The mRNA from the GHRL gene (chromosomal location: 3p26−p25 for humans, 6 E3|6 52.84 cM for mice, and 4q42 for rats) codes for 117-aa preproGHRL, consisting of five exons (E-Figure 21A.1). Mammalian GHRL mRNA is approximately 480−510 bp in length. The prohormone convertase 1/3 (PC1/3) cleaves a 28-aa unacylated ghrelin, and C-peptide, in which obestatin is presumed to be cleaved from proGHRL (Figure 21A.3) [5].

Distribution of mRNA

The GHRL gene predominantly expresses in the stomach in almost all vertebrates, followed by the intestine, pancreas, and gonads. GHRL immunopositive cells are detected in the arcuate nucleus of the hypothalamus, but gene expression is very low [1].

Tissue and Plasma Concentrations

Total (acylated and unacylated) and acylated form of ghrelin in male Sprague-Dawley rats (mean ± SE [$n = 5$], total/acylated form, nmol/g tissue) are: $1.8 ± 0.3/ < 0.05$, hypothalamus; $8.5 ± 3.1/ < 0.05$, pituitary; $2.6 ± 0.6/0.2 ± 0.1$, pancreas; $1779.8 ± 533.9/377.3 ± 55.8$, stomach; $106.7 ± 7.3/20.6 ± 0.7$, duodenum; $60.2 ± 17.2/10.7 ± 5.4$, jejunum; $20.5 ± 5.1/0.2 ± 0.1$, ileum; $2.8 ± 0.2/ < 0.05$, testis; and $219.6 ± 71.8/4.0 ± 1.9$ fmol/ml in plasma [6].

Regulation of Synthesis

Gastric GHRL expression increases by fasting, possibly via decreased plasma carbohydrates. In the rat stomach or in thyroid removal, insulin or leptin stimulates, and growth hormone (GH) treatment inhibits, gene expression. Expression in the rat stomach appears from the first day after birth and increases with age, reaching a constant level around 40 days after birth. Ghrelin-*O*-acyltransferase (GOAT) contributes acylation [7].

Regulation of Release

GHRL has a pulsatile secretion pattern, and the frequency and quantity of secretion rise at the time of fasting. Inhibition

Y. Takei, H. Ando, & K. Tsutsui (Eds): Handbook of Hormones. DOI: http://dx.doi.org/10.1016/B978-0-12-801028-0.00156-2

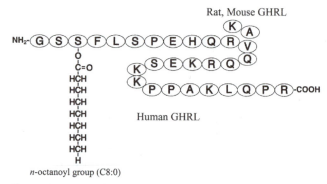

Rat, Mouse GHRL

Human GHRL

n-octanoyl group (C8:0)

Figure 21A.1 Primary structures of ghrelin in humans and rats.

Hammerhead shark	GVSF-HPRLKEKDDNSSGNSRKSKNP
Zebrafish	GTSFLSP-TQKPQ---GRR-PPRVa
Rainbow Trout	GSSFLSP-SQKPQVRQGKGKPPRVa
Bullfrog	GLTFLSPADMQKIAERQSQNKLRHGNMN
Red-eared slider	GSSFLSPEYQNTQQRKDPKKHT-KLN
Chicken	GSSFLSPTYKNIQQQKDTRKPTARLH
Human	GSSFLSPEHQRVQQRKESKKPPAKLQPR

Figure 21A.2 Amino acid sequences of mature ghrelin in vertebrates of various animal classes.

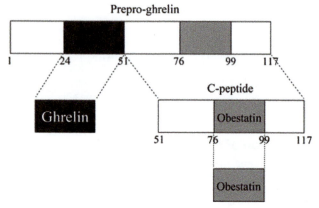

Figure 21A.3 Schematic drawing of prepro-ghrelin.

of GHRL release after re-feeding is not controlled by the physical gastric extension, but by carbohydrate intake. The adrenergic (sympathetic) nerve accelerates, and the cholinergic (parasympathetic) nerve inhibits, release. Insulin, leptin, and noradrenaline promote, and melatonin reduces, the release. Gastrin does not influence the release.

Receptors

Structure and Subtypes

Ghrelin is a member of the Class A GPCR family. The GHRL (growth hormone secretagog, GHS) receptor gene comprises two exons and one intron (E-Figure 21A.7). There are two isoforms, GHS-R1a and -R1b, in the same gene. GHS-R1a is a functional GHRL receptor [8], and GHS-R1b is a mRNA splicing variant. Human GHS-R1a and 1b have 366 aa and 289 aa, respectively (E-Figure 21A.8). GHS-R1b lacks the structure after TMD 5 of GHS-R1a, and 24 amino acids that originate from the intron after the 265th leucine in TMD 6 elongate at the C-terminus.

Signal Transduction Pathway

GHS-R1a activates $G_{\alpha q/11}$, which induces IP3 production and releases Ca^{2+} from intracellular Ca stores, while DAG activates PKC. In addition, GHS-R1a is constitutively active and produces high basal IP3 under the absence of agonists, which causes PLC—PKC-dependent Ca^{2+} mobilization by the L-type voltage-gated calcium channel.

Agonists

In addition to GHRL and its fragments containing N-terminal four amino acids minimum with acylation, peptidyl agonists, [His^1, Lys^6]-GHRP-6, hexarelin, D-Ala-β-[2-Nal]-D-Ala-Trp-D-Phe-Lys-amide (KP-102), Aib-His-D-2-Nal-D-Phe-Lys-amide, and capromorelin, or non-peptidyl agonists L-692,585 and tabimorelin hemifunarate, are commercially available.

Antagonists

[D-Lys3]-GHRP-6 is the most frequently used antagonist, and Cortistatin-8 [D-Arg^1, D-Phe^5, D-Trp^7, ^{9}, Leu^{11}]-Substance P, YIL781 are commercially available. [D-Arg^1, D-Phe^5, D-Trp^{7,9}, Leu^{11}]- Substance P is also known to act as an antagonist.

Biological Functions

Target Cells/Tissues and Functions

GHRL receptors are expressed in the hypothalamic neuropeptide Y (NPY), somatostatin (SST), growth hormone-releasing hormone (GHRH), proopiomelanocortin (POMC), orexin and Agouti-related peptide (AgRP) neurons, in the pituitary, lung, heart, stomach, intestine, pancreas, thymus, gonads, thyroid, kidney, and adipose tissue [1,3,4]. GHRL's functions include stimulation of GH release, feeding, anxiety, learning, and memory, gastrointestinal contraction, fat accumulation, depression, gastric acid secretion, sleep duration, and inhibition of drinking [3,4]. Species differences are seen — for example, feeding is stimulated in goldfish and frogs, but inhibited in chickens and rainbow trout [3,4].

Phenotype in Gene-Modified Animals

Lack of the *Ghrl* or *Ghsr* gene does not influence GH secretion, and the animal grows up normally. The quantity of meal and the weight and growth rate do not change even if the *Ghrl* and/or *Ghsr* gene is lost. GH secretion through GHRL does not occur at the time of fasting during nutritional deficiency, and the mechanism to maintain blood-sugar level fails when the *Ghrl* gene is lost. Abnormalities of blood pressure and temperature, gastrointestinal hypokinesia, and circadian rhythm, regulated by autonomic nerves, are seen. Loss of both *Ghrl* and *Ghsr* genes is accompanied by a decrease in body weight and length, and increases in energy consumption and activity.

Pathophysiological Implications

Clinical Implications

In humans, an association between variations of the *GHRL* gene and obesity has been suggested. *GHRL* gene polymorphisms (604G/A, Arg51Gln, Leu72Met, and Leu90Gln) seem to be associated with cardiovascular disease, carcinoma, type 2 diabetes, and Alzheimer's disease. There is a possibility that the Brachman-de Lange syndrome with poor growth occurs due to overlap or translocation of the chromosome on which the GHSR gene is located [9].

Use for Diagnosis and Treatment

Application of GHRL for treatments is still in at the clinical trial stage [10]. The expectation is that it might be used as an appetite inducer, for treatment of GH deficiency and improvement of cardiac function, to stimulate breathing in chronic obstructive pulmonary disease (COPD), to stimulate gastrointestinal motility, and in the treatment of obesity and type 2 diabetes.

References

1. Kojima M, Hosoda H, Date Y, et al. Ghrelin is a growth-hormone-releasing acylated peptide from stomach. *Nature*. 1999;402:656−660.
2. Kojima M, Ida T, Sato T. Structure of mammalian and nonmammalian ghrelins. *Vitam Horm*. 2008;77:31−46.
3. Kaiya H, Miyazato M, Kangawa K. Recent advances in the phylogenetic study of ghrelin. *Peptides*. 2011;32:2155−2174.
4. Kaiya H, Miyazato M, Kangawa K, et al. Ghrelin: a multifunctional hormone in non-mammalian vertebrates. *Comp Biochem Physiol A*. 2008;149:109−128.
5. Zhu X, Cao Y, Voogd K, et al. On the processing of proghrelin to ghrelin. *J Biol Chem*. 2006;281:38867−38870.
6. Hosoda H, Kojima M, Matsuo H, et al. Ghrelin and des-acyl ghrelin: two major forms of rat ghrelin peptide in gastrointestinal tissue. *Biochem Biophys Res Commun*. 2000;279:909−913.
7. Mohan H, Unniappan S. Discovery of ghrelin O-acytransferase. In: Benso A, Casanueva FF, Ghigo E, Granata R, eds. *The Ghrelin System*. vol. 25. Basel: Karger;2013:16−24.
8. Howard AD, Feighner SD, Cully DF, et al. A receptor in pituitary and hypothalamus that functions in growth hormone release. *Science*. 1996;273:974−977.
9. Gueorguiev M, Korbonits M. Genetics of the ghrelin system. In: Benso A, Casanueva FF, Ghigo E, Granata R, eds. *The Ghrelin System*. vol. 25. Basel: Karger; 2013:25−40.
10. Allas S, Abribat T. Clinical perspectives for ghrelin-derived therapeutic products. In: Benso A, Casanueva FF, Ghigo E, Granata R., eds. The Ghrelin System. vol. 25. Basel: Karger; 2013:157−166.
11. He C, Tsend-Ayush E, Myers MA, et al. Changes in the ghrelin hormone pathway may be part of an unusual gastric system in monotremes. *Gen Comp Endocrinol*. 2013;191:74−82.

Supplemental Information Available on Companion Website

- Precursor and mature ghrelin sequences in mammals/E-Figure 21A.1
- Precursor and mature ghrelin sequences in cartilaginous fishes/E-Figure 21A.2
- Precursor and mature ghrelin sequences in bony fishes/E-Figure 21A.3
- Precursor and mature ghrelin sequences in amphibians/E-Figure 21A.4
- Precursor and mature ghrelin sequences in reptiles and birds/E-Figure 21A.5
- Gene, mRNA and precursor structures of ghrelin/E-Figure 21A.6
- Gene and isoforms of the human ghrelin receptor/E-Figure 21A.7
- Primary structure of the ghrelin receptors/E-Figure 21A.8
- Synteny of the ghrelin gene/E-Figure 21A.9 [11]
- Synteny of the ghrelin receptor (GHS-R)/E-Figure 21A.10

Motilin

Takafumi Sakai

Abbreviations: MTL
Additional name: housekeeper of the gut

Motilin was first isolated from the duodenojejunal mucosa and found to control motor activity of the digestive tract. It is a potential therapeutic drug target for improving digestive dysmotility.

Discovery

Motilin was first isolated from a side fraction produced during the purification of secretin in 1971 [1], and was found to stimulate contractility in the fundus of the stomach. Complete porcine and human motilins were purified and sequenced in 1973 [2] and 1983 [3], respectively.

Structure

Structural Features

Motilin is highly conserved across species, and is synthesized as part of a larger inactive prohormone. Structure–activity studies with analogs and fragments of porcine motilin have shown an NH_2-terminal region, which is considered a physiological and biological active site, and a COOH-terminal α-helical domain [4,5]. Human motilin is synthesized as a preprohormone composed of 133 aa residues, each consisting of a 25-aa signal peptide followed by a 22-aa (motilin itself) and a C-terminal motilin-associated peptide (MAP).

Primary Structure

In the N-terminal region, identical aa sequences exist in human and porcine motilin but differ from canine motilin at residues 7, 8, 12, 13, and 14 [6,7] (Table 21B.1). Chicken motilin also differs from human and porcine residues by six residues (4, 7–10, and 12) [6,7], and the binding affinity and pharmacological potency against the chicken motilin receptor differ from those for mammalian motilin.

Properties

Human motilin: Mr 2698. Water soluble.

Synthesis and Release

Gene, mRNA, and Precursor

The human motilin gene has been mapped to the p21.3 region of chromosome 6. It is a single copy gene composed of five exons, spanning approximately 9 kb [4]. The 22-aa motilin peptide gene has a structure similar to that of ghrelin, and is encoded by Exon 3 and part of Exon 2 (E-Figure 21B.1). The aa sequences of prepromotilin from pig, rabbit, human, sheep, and monkey have been deduced from the cDNA sequence. The C-terminal MAP is largely encoded by part of Exons 3 and 4, with the last two amino acids of the motilin precursor and 3′ untranslated region encoded by Exon 5.

Distribution of mRNA

Motilin has been identified throughout the gastrointestinal (GI) tract of numerous species. Motilin is found predominantly in the endocrine cells of the duodenal mucosa but also in the myenteric plexus, where it is co-localized with neurons immunoreactive for neuronal nitric oxide synthase [8]. Motilin-producing cells decrease distally in the small intestine. Motilin is also present in the thyroid and the brain, where the highest concentration is found in the hypothalamus [8].

Tissue and Plasma Concentrations

Tissue: Motilin is most abundant in the duodenal mucosa, and is located in endocrine M cells. In humans, motilin has been observed in the duodenum (165.7 ± 15.9 fmol/mg) more commonly than in the jejunum (37.5 ± 2.8 fmol/mg), with negligible amounts in the remainder of the gut [9].
   Plasma: Increased plasma motilin concentration has been found with cyclical interdigestive contractions of the stomach in humans and dogs. The mean plasma motilin concentration during fasting in humans is reported to be approximately 5–300 fmol/ml [9].

Regulation of Synthesis and Release

It is well accepted that plasma motilin is released at approximately 100-minute intervals in the interdigestive state, and ingestion of food during this period prevents the secretion of motilin [6]. Reports have shown that the cholinergic pathway is an important regulator of the release of motilin. Also, muscarinic3 (M3) receptors have been found responsible for motilin release from canine motilin cells in the perifusion system [6]. Moreover, it has been demonstrated that exogenous motilin treatment stimulates endogenous motilin release through muscarinic receptors on motilin-producing cells via preganglionic pathways involving 5-hydroxytryptamine 3 receptors [10]. However, the detailed mechanism of motilin release remains unknown.

Receptors

Structure and Subtypes

Feighner and colleagues (1999) [5] first identified the orphan GPCR, GPR38, as the human motilin receptor (MTLR), for

Y. Takei, H. Ando, & K. Tsutsui (Eds): Handbook of Hormones. DOI: http://dx.doi.org/10.1016/B978-0-12-801028-0.00157-4

Table 21B.1 Amino Acid Sequence of Motilin in Various Species of Mammals and Chicken

	1–5	6–10	11–15	16–20	21–22
Human	FVPIF	TYGEL	QRMQE	KERNK	GQ
Pig	FVPIF	TYGEL	QRMQE	KERNK	GQ
Dog	FVPIF	THSEL	QKIRE	KERNK	GQ
Suncus	FMPIF	TYGEL	QKMQE	KEQNK	GQ
Cat	FVPIF	THSEL	QRIRE	KERNK	GQ
Cow	FVPIF	TYGEV	RRMQE	KERYK	GQ
Rabbit	FVPIF	TYSEL	QRMQE	RERNR	GH
Chicken	FVPIF	TQSDI	QKMQE	KERNK	GQ
	*:***	*:.*:	:::*	:*:::	*:

Table 21B.2 Biological Actions of Motilin

Regions	Human	Dog
LES	LES Pressure↑	LES Pressure↑
Gallbladder	Emptying↑	—
Stomach	Fatty cream emptying↓	Liquids emptying↑
	Glucose Solution emptying↑	Phase III MMC↑
	Phase III MMC↑	
Small Intestine	Phase III MMC↑	Phase III MMC↑
Large Intestine	Transit↑/no effect	Transit↑

LES: lower esophageal sphincters; MMC: migrating motor complex.

which two alternatively spliced forms exist (E-Figure 21B.2). An mRNA, GPR38-A (splice variant 1a), encodes a polypeptide of 412 aa with seven predicted α-helical transmembrane domains, the hallmark feature of GPCRs, and is an active form of the receptor, whereas GPR38-B (variant 1b) mRNA encodes a 386-aa polypeptide with five predicted transmembrane domains [5].

Signal Transduction Pathway

The signal transduction pathway of MTLR is unknown. However, the molecular and cellular mechanisms involved in motilin-induced MTLR desensitization have been observed. After motilin-induced stimulation, the MTLR becomes phosphorylated, probably via GPCR kinases. This leads to recruitment of β-arrestin-2, which targets the receptor to clathrin-coated pits. Upon internalization, the β-arrestin dissociates from the receptor, and the motilin–receptor complex is subsequently sorted to the recycling endosomes that transport the MTLR back to the plasma membrane [8].

Agonists

Erythromycin has been extensively used as an effective agent to accelerate gastric emptying of food in patients with diabetic gastroparesis through MTLR. Several pharmaceutical companies have generated motilin-like macrolides (motilides) that are erythromycin derivatives devoid of antibiotic activity but with strong affinity to MTLR. However, the first drugs, EM-523 and its successor EM-574, failed because of their chemical instability and low bioavailability [8]. Another compound, ABT-229, was also unsuccessful for treating functional dyspepsia and diabetic gastroparesis, possibly because of its strong desensitizing properties [8]. To overcome these limitations, second-generation compounds were developed. Based on the N-terminal region of motilin and its biological activity, a novel synthetic human motilin analog, atilmotin, which lacks the C-terminal end, was developed to accelerate gastric emptying in healthy subjects. However, its effect was poor, and only detected during the first 30 minutes [8]. An acid-resistant non-peptide motilin agonist, mitemcinal, was developed, which is orally active and could be beneficial for the treatment of delayed gastric emptying and transit [8]. However, symptom relief occurred only in a subset of patients.

Antagonists

The macrocyclic peptidomimetic TZP-201 is a potent semisynthetic non-peptide motilin antagonist that has been used for the treatment of various forms of moderate to severe diarrhea associated with irritable bowel syndrome, cancer, and infectious diseases [8]. Preclinical data showed that TZP-201 was efficacious in a dog model of chemotherapy-induced diarrhea [8]. Similarly, an orally active motilin receptor antagonist, MA-2029, inhibits motilin-induced gastrointestinal motility without affecting basal gastrointestinal tone or gastric emptying rate.

Biological Functions

Target Cells/Tissues and Functions

MTLRs are mainly found in the GI tract, but their exact localization is species dependent. In humans, through binding experiments with iodinated porcine [Leu13] motilin, MTLR density was found to be highest in the gastroduodenal region, and decreased distally in the small intestine towards the colon [8]. MTLR immunoreactivity is present also in muscle cells and the myenteric plexus, but not in mucosal or submucosal cells, in humans. In rabbits, the highest MTLR density is found in the colon. MTLRs have been found outside the GI tract in the hypothalamus, nodose ganglion, central nervous system, thyroid, and bone marrow [5]. MTLR gene expression has also been found in the lung and heart in *Suncus murinus*, suggesting that motilin could have an unknown function in the respiratory and cardiovascular systems. As summarized in Table 21B.2, the main biological functions of MTLRs' activation are to increase lower esophageal sphincter (LES) pressure and induce the interdigestive motor complex (IMC) to remove debris and cleanse the GI tract.

Phenotype in Gene-Modified Animals

Progress in the development of a motilin gene-modified animal has been delayed because no one has successfully isolated motilin cDNA from rats or mice. Therefore, research related to motilin has been mainly limited to large animals such as dogs. Recent studies with rat and mouse genomes showed that motilin and its receptor only exist as pseudogenes in these species [8], and they are considered natural motilin knockouts. A recent report demonstrated the existence of motilin and a functional receptor in *Suncus murinus* [7], and the genome sequencing of *Suncus* is under progress, which could potentially lead to the development of therapeutic applications by establishing genetically engineered *Suncus*.

Pathophysiological Implications

Clinical Implications

Plasma motilin concentration is enhanced significantly in patients with diabetic gastroparesis who maintain a normal migrating motor complex (MMC), even without antral phase III activity. Similarly, a higher plasma motilin concentration is reported in hypergastrinemic chronic atrophic gastritis and chronic renal failure, whereas decreased motilin release has been observed in patients with functional bowel disorders such as chronic idiopathic constipation or idiopathic megacolon. Abnormal fluctuations of motilin also occur with severe pancreatic insufficiency. Furthermore, fasting and postprandial levels of motilin are significantly raised in patients with infectious diarrhea. Hypermotilinemia is often associated with Crohn's disease, ulcerative colitis, and tropical malabsorption [5,6].

Use for Diagnosis and Treatment

To date, almost all macrolides have lacked effectiveness as motilin agonists. However, pharmacies have been trying to develop new types of macrolides, such as PF-04548043 (formerly known as KOS-2187), for the treatment of GI motility disorders like gastroparesis and gastroesophageal reflux disease [8]. Moreover, the new MTLR agonist RQ-00201894 is a promising drug for the treatment of gastroparesis, postoperative ileus, and functional dyspepsia.

References

1. Brown JC, Mutt V, Dryburgh JR. The further purification of motilin, a gastric motor activity stimulating polypeptide from the mucosa of the small intestine of hogs. *Can J Physiol Pharmacol.* 1971;49:399–405.
2. Brown JC, Cook MA, Dryburgh JR. Motilin, a gastric motor activity stimulating polypeptide: the complete amino acid sequence. *Can J Biochem.* 1973;51:533–537.
3. Poitras P, Reeve Jr. JR, Hunkapiller MW, Hood LE, Walsh JH. Purification and characterization of canine intestinal motilin. *Regul Pept.* 1983;5:197–208.
4. Huang Z, Depoortere I, De Clercq P, Peeters T. Sequence and characterization of cDNA encoding the motilin precursor from chicken, dog, cow and horse. Evidence of mosaic evolution in prepromotilin. *Gene.* 1999;240:217–226.
5. Feighner SD, Tan CP, McKee KK, et al. Receptor for motilin identified in the human gastrointestinal system. *Science.* 1999;284:2184–2188.
6. Itoh Z. Motilin and clinical application. *Peptides.* 1997;18: 593–608.
7. Tsutsui C, Kajihara K, Yanaka T, et al. House musk shrew (*Suncus murinus*, order: Insectivora) as a new model animal for motilin study. *Peptides.* 2009;30:318–329.
8. De Smet B, Mitselos A, Depoortere I. Motilin and ghrelin as prokinetic drug targets. *Pharmacol Ther.* 2009;123:207–223.
9. Bloom SR, Mitznegg P, Bryant MG. Measurement of human plasma motilin. *Scandinavian J Gastroenterol Suppl.* 1976;39: 47–52.
10. Mochiki E, Satoh M, Tamura T, et al. Exogenous motilin stimulates endogenous release of motilin through cholinergic muscarinic pathways in the dog. *Gastroenterology.* 1996;111:1456–1464.
11. Seino Y, Tanaka K, Takeda J, et al. Sequence of an intestinal cDNA encoding human motilin precursor. *FEBS Lett.* 1987;223:74–76.
12. Yano H, Seino Y, Fujita J, et al. Exon-intron organization, expression, and chromosomal localization of the human motilin gene. *FEBS Lett.* 1989;249:248–252.
13. Daikh DI, Douglass JO, Adelman JP. Structure and expression of the human motilin gene. *DNA.* 1989;8:615–621.

Supplemental Information Available on Companion Website

- Gene, mRNA, and precursor sequences of human MTL/E-Figure 21B.1
- Gene, mRNA, and precursor of human MTLR/E-Figure 21B.2
- Accession numbers of MTL and MTLR in vertebrates/E-Table 21B.1

CHAPTER 22

Bombesin-Like Peptide Family

Hiroko Ohki-Hamazaki

History

Bombesin (BN) was first isolated in pure form from the skin of two European amphibians, *Bombina bombina* and *B. variegata*, in 1971 [1]. At the same time, alytesin, a peptide having a structure very similar to that of BN, was isolated from the skin of another European amphibian of the same family, *Alytes obstetricans* [1]. Many other structurally related peptides were discovered from amphibian skins and divided into three groups according to the structure: the BN family, which includes BN and alytesin; the ranatensin family, which includes ranatensin and litorin; and the phyllolitorin family. Two structurally related mammalian peptides were then identified. Gastrin-releasing peptide (GRP) was the first to be isolated, in 1979, from porcine gastric tissue; it belongs to the BN family and was subsequently identified in other mammals and non-mammals, such as chicken, alligator, toad, trout, and dogfish. The second was neuromedin B (NMB), isolated from porcine spinal cord, which is structurally related to ranatensin [2–4].

Structure

Structural Features

The BN family peptides have His—Leu—Met at their C-terminus, but the ranatensin family peptides have His—Phe—Met (Table 22.1). Multiple forms of BN exist in all frogs. In *B. orientalis*, [Leu^3] BN, [Phe^{13}] BN and [Ser^3, Arg^9, Phe^{13}] BN (SAP-BN) are present. The structural characteristic

Table 22.1 Comparison of Amino Acid Sequences of the Bombesin-Like Peptide Family in Mammals and Amphibians

Bombesin Family	
Bombesin (Leu^{13})	pEQRLGNQWAVGHLM-NH$_2$
Bombesin (Phe^{13})	pEQRLGNQWAVGHFM-NH$_2$
SAP-Bombesin	pEQSLGNQWEVGHFM-NH$_2$
Alytesin	pEGRLGTQWAVGHLM-NH$_2$
Human GRP	VPLP .. AGGGTVLTKMYPRGNHWAVGHLM-NH$_2$
pig GRP	APVS .. VGGGTVLAKMYPRGNHWAVGHLM-NH$_2$
Toad GRP	SPTSQQHNDAASLSKIYPRGSHWAVGHLM-NH$_2$
Ranatensin Family	
Ranatensin	pEVPQWAVGHFM-NH$_2$
Litorin	pEQWAVGHFM-NH$_2$
Human NMB-32	APLSWDLPEPRSRASK IRVHSRGNLWATGHFM-NH$_2$
Pig NMB-32	APLSWDLPEPRSRAGKIRVHPRGNLWATGHFM-NH$_2$
Pig NMB	GNLWATGHFM-NH$_2$
Phyllolitor in Family	
Phyllolitor in (Leu8)	pELWAVGSLM-NH$_2$
Phyllolitor in (Ph8)	pELWAVGSFM-NH$_2$

pE, pyroglutamate.

of the phyllolitorin family peptides resides in the third residue (Ser) from the C-terminus (Table 22.1). The N-terminal residue of the amphibian BN-like peptides is pyroGlu except for GRP. The Met residue at the C-terminus of all family members is amidated [3,4].

Receptors

Structure and Subtypes

BN-like peptide receptors are G protein-coupled and have seven membrane-spanning domains. The GRP-preferring receptor (GRPR or BB2) was first cloned from Swiss 3T3 cells (murine fibroblast cell line) and characterized in 1990–1991. A high-affinity receptor for NMB (NMBR or BB1) was cloned from rat esophagus in 1991. Human GRPR and NMBR were identified in 1991, mouse GRPR in 1997, and chick GRPR and NMBR in 2003 and 2006, respectively (Figure 22.1) [2–5]. The third receptor showing structural homology to GRPR and NMBR, named BRS3 (BN receptor subtype 3 or BB3), was cloned from the guinea pig in 1992, human in 1993 and mouse in 1997, but it has low affinity for both GRP and NMB [2–4]. A natural ligand for BRS3 is not known to date. In 1995, frog GRPR and NMBR were cloned from *B. orientalis*, and at the same time a fourth BN receptor, BB4, which has high affinity for [Phe^{13}]BN, was also identified [6]. In chicken, BRS3.5, which has moderate affinity for BN and low affinity for both GRP and NMB, was also identified in 2003 [5].

Signal Transduction Pathway

BN-like peptide receptors are coupled with G$_q$ protein and, upon agonist binding, activate phospholipase C$_\beta$, which catalyzes the hydrolysis of PIP2. This leads to the production of second messengers, IP3 and DAG, and results in the mobilization of intracellular Ca^{2+} and activation of PKC [4].

Biological Functions

Target Cells/Tissues and Functions

Most of the function of BN-like peptides has been studied in mammals or with cell lines [2,4]. All peptides of this family identified so far can stimulate contraction of urogenital and gastrointestinal smooth muscle. *In vivo*, BN affects exocrine secretions from the digestive organs, endocrine secretions, gastrointestinal motility, blood pressure, heart rate, body temperature, plasma glucose, and oxygen consumption. BN exerts its effect on physiology and behavior, including food intake, scratching/grooming, perception of pain, locomotor activity, sleep, and memory, but its effect is highly complex

Y. Takei, H. Ando, & K. Tsutsui (Eds): Handbook of Hormones. DOI: http://dx.doi.org/10.1016/B978-0-12-801028-0.00022-2

```
                                                                    I                                    II
                                                             _____                          _____
human NMBR   MP-----SKS LSNLSVTTGA NESGSVPEGW ERDFLPASDG TTTELVIRCV IPSLYLLIIT VGLLGNIMLV KIFITNSAMR SVPNIFISNL AAGDLLLLLT
human GRPR   MA-----LND CFLLNLEVDH FMHCNISS-- HSADLPVND- DWSHPGILYV IPAVYGVIIL IGLIGNITLI KIFCTVKSMR NVPNLFISSL ALGDLLLLIT
frog  BB4    MPEGFQSLNQ TLPSAISSIA HLES-LNDSF ILGAKQSED- VSPGLEILAL ISVTYAVIIS VGILGNTILI KVFFKIKSMQ TVPNIFITSL AFGDLLLLLT
chick BRS3.5 MSQVYLHPSN QTLCAATNGT ELKS-ILDNE TTNEKWTED- SFPGLEILCT IYVTYAVIIS VGLLGNAILI KVFFKIKSMQ TVPNIFITSL AFGDLLLLLT
human BRS3   MAQRQPHSPN QTLISITNDT ESSSSVVSND NTNKGWSGD- NSPGIEALCA IYITYAVIIS VGILGNAILI KVFFKTKSMQ TVPNIFITSL AFGDLLLLLT

                 III                                           IV
             _____                               _____
human NMBR   CVPVDASRYF FDEWMFGKVG CKLIPVIQLT SVGVSVFTLT ALSADRYRAI VNPMDMQTSG ALLRTCVKAM GIWVVSVLLA VPEAVFSEVA RISSLD-NSS
human GRPR   CAPVDASRYL ADRWLFGRIG CKLIPFIQLT SVGVSVFTLT ALSADRYKAI VRPMDIQASH ALMKICLKAA FIWIISMLLA IPEAVFSDLH PFHEESTNQT
frog  BB4    CVPVDASRYI VDTWMFGRAG CKIISFIQLT SVGVSVFTLT VLSADRYRAI VKPLQLQTSD AVLKTCGKAV CVWIISMLLA APEAVFSDLY EFGSSEKNTT
chick BRS3.5 CVPVDATRYI VDTWIFGRIG CKLLSFIQLT SVGVSVFTLT VLSADRYRAI VKPLELQTSD ALLKTCCKAG CVWIVSMVFA IPEAVFSDLY SFSNPEKNVT
human BRS3   CVPVDATHYL AEGWLFGRIG CKVLSFIRLT SVGVSVFTLT ILSADRYKAV VKPLERQPSN AILKTCVKAG CVWIVSMIFA LPEAIFSNVY TFRDPNKNMT

                         V                                               VI
                     _____                                 _____
human NMBR   FTACIPYPQT DELHPKIHSV LIFLVYFLIP LAIISIYYYH IAKTLIKSAH NLPGEYNEHT KKQMETRKRL AKIVLVFVGC FIFCWFPNHI LYMYRSFNYN
human GRPR   FISCAPYPHS NELHPKIHSM ASFLVFYVIP LSIISVYYYF IAKNLIQSAY NLPVEGNIHV KKQIESRKRL AKTVLVFVGL FAFCWLPNHV IYLYRSYHYS
frog  BB4    FEACAPYPVS EKILQETHSL ICFLVFYIVP LSIISAYYFL IAKTLYKSTF NMPAEEHTHA RKQIESRKRV AKTVLVLVAL FAVCWLPNHM LYLYRSFTYH
chick BRS3.5 FEACAPYPVS EKILQEVHSL VCFLVFYIVP LAVISVYYFL IARTLYKSTF NMPAEEHGHA RKQIESRKRV AKTVLVLVAL FAFCWLPNHI LYLYRSFTYH
human BRS3   FESCTSYPVS KKLLQEIHSL LCFLVFYIIP LSIISVYYSL IARTLYKSTL NIPTEEQSHA RKQIESRKRI ARTVLVLVAL FALCWLPNHL LYLYHSFTSQ

                             VII
                         _____
human NMBR   -EIDPSLGHM IVTLVARVLS FGNSCVNPFA LYLLSESFRR HFNSQLCCGR KSYQERGTSY LLSSSAVRMT SLKSN----AK NMVTNSVLLN GHSMKQEMAL
human GRPR   -EVDTSMLHF VTSICARLLA FTNSCVNPFA LYLLSKSFRK QFNTQLLCCQ PGLIIR--SH STGRSTTCMT SLKST----NP SVATFS-LIN GNICHERYV-
frog  BB4    SAVNSSAFHL SATIFARVLA FSNSCVNPFA LYWLSRSFRQ HFKKQVYCCK TEPPASQQSP THSSTITGIT AVKGN--IQM SEISITLLSA YDVKKE-----
chick BRS3.5 TSVDASTFHL IVTIFSRALA FSNSCVNPFA LYWLSRSFRQ HFKKQVSCCK AKLCTKPPSA PHSNSPPRAL SVTGS--THG SEISVTLLTD YSITKEEESV
human BRS3   TYVDPSAMHF IFTIFSRVLA FSNSCVNPFA LYWLSKSFQK HFKAQLFCCK AERPEPPVAD TSLTTLAVMG TVPGTGSIQM SEISVTSFTG CSVKQAEDRF
```

Figure 22.1 Comparison of amino acid sequences of the bombesin-like peptide receptors. Predicted transmembrane domains (I–VII) are shown.

because opposite or unrelated effects were observed after administration into the brain or spinal cord, or via systemic injection. This is due to the dual effect of BN on peripheral organs and on the central nervous system. Moreover, the opposite effect can be observed depending on the species of animal tested. The function of GRP and NMB largely overlaps, but the specific function through each receptor can be distinguished by using specific agonists/antagonists to GRPR and NMBR, or analyzing the GRPR and NMBR knockout mice. BRS3 knockout mice became obese, and analysis revealed that BRS3 is implicated in the regulation of food intake, lipid metabolism, and energy homeostasis [7].

References

1. Anastasi A, Erspamer V, Bucci M. Isolation and structure of bombesin and alytesin, 2 analogous active peptides from the skin of the European amphibians *Bombina* and *Alytes*. *Experientia*. 1971;27:166–167.
2. Ohki-Hamazaki H. Neuromedin B. *Prog Neurobiol*. 2000;62: 297–312.
3. Ohki-Hamazaki H, Iwabuchi M, Maekawa F. Development and function of bombesin-like peptides and their receptors. *Int J Dev Biol*. 2005;49:293–300.
4. Jensen RT, Battey JF, Spindel ER, et al. International Union of Pharmacology. LXVIII. Mammalian bombesin receptors: nomenclature, distribution, pharmacology, signaling, and functions in normal and disease states. *Pharmacol Rev*. 2008;60:1–42.
5. Iwabuchi M, Ui-Tei K, Yamada K, et al. Molecular cloning and characterization of avian bombesin-like peptide receptors: new tools for investigating molecular basis for ligand selectivity. *Br J Pharmacol*. 2003;139:555–566.
6. Nagalla SR, Barry BJ, Creswick KC, et al. Cloning of a receptor for amphibian [Phe13]bombesin distinct from the receptor for gastrin-releasing peptide: identification of a fourth bombesin receptor subtype (BB4). *Proc Natl Acad Sci USA*. 1995;92: 6205–6209.
7. Ohki-Hamazaki H, Watase K, Yamamoto K, et al. Mice lacking bombesin receptor subtype-3 develop metabolic defects and obesity. *Nature*. 1997;390:165–169.

Supplemental Information Available on Companion Website

- Phylogenetic tree of BN-like peptide family members/E-Figure 22.1
- Phylogenetic tree of the BN-like peptide receptors/E-Figure 22.2

Gastrin-Releasing Peptide

Hirotaka Sakamoto

Abbreviation: GRP
Additional names: bombesin-like peptide (BLP)

A brain—gut peptide widely conserved in vertebrates, GRP is a neuropeptide that modulates the autonomic system. It is an anorexigenic factor in the brain, regulates male sexual function at the spinal cord level, conveys the itch sensation, and is involved in fear memory consolidation. GRP is also a specific tumor marker for small cell lung cancer. It has the GXHWAVGHLM amide (X = N or S) sequence at its C-terminus.

Discovery

First isolated from the porcine stomach and named gastrin-releasing peptide (GRP) [1], GRP was originally believed to be a mammalian counterpart of the amphibian peptide, bombesin. Its structure has been determined in many mammals, including pigs, rats, mice, guinea pigs, dogs, and humans. In addition, GRP_{18-27}, a possible fragment of mature GRP, was later isolated from porcine spinal cord and originally called neuromedin C [2], although a more appropriate name is either GRP-10 or GRP_{18-27}. GRP orthologs have been identified in birds, reptiles, amphibians, and teleost fishes (Table 22A.1).

Structure

Structural Features

Human GRP precursor consists of 148 aa residues (rat, 147 aa). A predicted preproGRP translation product consists of a signal peptide, GRP_{1-27}, and a carboxyl-terminal extension peptide termed pro-GRP_{31-125} [3].

Primary Structure

Most GRP peptides have a GXHWAVGHLM amide (X = N or S) motif at their C-terminus (Figure 22A.1). Although the N-terminal sequences are less conserved across vertebrate species, many identified GRP peptides in mammals consist of 27 amino acids, excluding rats and mice (29 aa). In non-mammals, amphibian GRP peptide consists of 29 amino acids, reptile peptide of 28 aa, and avian peptide of 27 aa, while teleost peptide appears to be shorter (24–25 aa).

Properties

Human GRP: Mr 2,859. Normally white and crystalline. Freely soluble in water.

Synthesis and Release

Gene, mRNA, and Precursor

The human GRP precursor gene is located on chromosome 18, consisting of three exons (850 bp) (E-Figure 22A.1). Its mRNA produces a 125-aa preproGRP as a precursor [3]. In chickens, the GRP precursor gene is located on chromosome Z. In zebrafish, the GRP precursor gene is predicted to be located on chromosome 21, from the information of the draft genome. The chromosomal localization has not been reported in other non-mammalian GRP genes.

Distribution of mRNA

GRP mRNA is produced in the central nervous system, stomach, intestine, pancreas, and lung in humans as well as other mammals. No information is currently available for non-mammals. It has also been reported that GRP mRNA is overexpressed in patients suffering from small cell lung cancer [4].

Regulation of Synthesis and Release

In the central nervous system, bombesin- or GRP-immunoreactivity was observed in approximately 5% of dorsal root ganglion neurons [5] and approximately 10% of trigeminal ganglion neurons [6]. In the brain, abundant GRP-immunoreactivity was observed in the cerebral cortex, hypothalamus, and medulla oblongata. Expression of GRP mRNA was detected in the limbic system, including the amygdala as well as the hippocampus [7]. In the gastrointestinal tract, abundant GRP-immunoreactive fibers were observed in the stomach (suggesting gastrin release), small intestine, colon, and pancreas, hence the so-called *brain—gut peptide*. In addition, overexpression of GRP has been demonstrated at both the mRNA and protein levels in various types of tumors, including lung, prostate, breast, stomach, pancreas, and colon [4].

Receptors

Structure and Subtype

GRPR is a seven-transmembrane-domain receptor of the GPCR superfamily. GRPR is directly coupled to the G_q protein, and GRPR activation leads to an increase in cellular $[Ca^{2+}]$ and stimulation of the PLC/PKC and ERK/MAPK pathways. In mammals, bombesin-like peptides act on a family of at least three GPCRs: GRP-preferring receptor (GRPR), NMB-preferring receptor (NMBR), and bombesin receptor subtype-3, which is considered an orphan receptor. Human GRP binds GRPR (BB_2 receptor) with higher affinity than that of NMBR (BB_1 receptor).

Y. Takei, H. Ando, & K. Tsutsui (Eds): Handbook of Hormones. DOI: http://dx.doi.org/10.1016/B978-0-12-801028-0.00158-6

Table 22A.1 GRP Orthologs Identified in Birds, Reptiles, Amphibians, and Teleost Fishes

Bombesin	EQRL	GNQWAVGHLM-NH$_2$
GRP-10		GNHWAVGHLM-NH$_2$
Human	VPLP--AGG GTVLTKMYPR	GNHWAVGHLM-NH$_2$
Bovine	APVT---AGR GGALAKMYTR	GNHWAVGHLM-NH$_2$
Porcine	APVS--VGG GTVLAKMYPR	GNHWAVGHLM-NH$_2$
Dog	APVP--GGQ GTVLDKMYPR	GNHWAVGHLM-NH$_2$
Rat	APVSTGAGG GTVLAKMYPR	GNHWAVGHLM-NH$_2$
Chick	APLQ--PGG GPALTKIYPR	GNHWAVGHLM-NH$_2$
Alligator	APAP-SGGG SAPLAKIYPR	GNHWAVGHLM-NH$_2$
Dogfish	APVE----N QGSFPKMFPR	GNHWAVGHLM-NH$_2$
Trout	SENT GAIGKVFYPR	GNHWAVGHLM-NH$_2$
Toad	SPTSQQHND AASLSKIYPR	GNHWAVGHLM-NH$_2$

Bombesin		EQRL	GNQWAVGHLMa
GRP-10			GNHWAVGHLMa
Human	VPLP	AGGGTVLTKMYPR	GNHWAVGHLMa
Bovine	APVT	AGRGGALAKMYTR	GNHWAVGHLMa
Porcine	APVS	VGGGTVLAKMYPR	GNHWAVGHLMa
Dog	APVP	GGQGTVLDKMYPR	GNHWAVGHLMa
Rat		APVSTGAGGGTVLAKMYPR	GSHWAVGHLMa
Chick	APLQ	PGGSPALTKIYPR	GSHWAVGHLMa
Alligator	APAP	SGGGSAPLAKIYPR	GSHWAVGHLMa
Dogfish	APVE	NQGSFPKMFPR	GSHWAVGHLMa
Trout		SENTGAIGKVFYPR	GNHWAVGHLMa
Toad		SPTSQQHNDAASLSKIYPR	GSHWAVGHLMa

Figure 22A.1 Primary structure of GRP peptides in vertebrates.

Agonist

A fragment of the C-terminal of human GRP, GRP$_{18-27}$ (GRP-10), is reported to be a selective and potent agonist for GRPR [8]. Central administration of GRP-10 significantly induced many physiological responses, similar to those with mature GRP.

Antagonist

RC3095 has been reported as a selective and potent antagonist of GRPR [8]. Central administration of RC-3095 significantly attenuated many physiological responses, as mentioned above.

Biological Functions

Target Cells and Function

GRP functions via GRPR. GRPR is highly expressed in the pancreas, and is also expressed in the stomach, adrenal cortex, and brain. Considerable evidence indicates that GRP plays a role in many physiological processes, including food intake [9], male sexual functions [8], circadian rhythms [10], and fear-memory consolidation [7] in the mammalian central nervous system. In addition, the itch sensation is a major biological function of GRP in mammals [5]. In chicks, GRP also functions as an anorexigenic factor in the brain. Biological functions of GRP in the other non-mammalian species have not yet been reported.

Phenotype in Gene-Modified Animals

Sun and Chen [5] reported, using GRPR null mutants, that activation of GRPR in the spinal cord is an important pathway in mediating pruritic signals, which bears profound clinical implications. The itch sensation is conveyed by a spinal cord system distinct from that mediating pain, and this system relies on GRP/GRPR signaling in the dorsal horn of the spinal cord [5].

Pathophysiological Implications

Clinical Implications

GRP and GRPR are frequently expressed, and their expression levels are increased by cancers of the gastrointestinal tract, lung, and prostate [4]. GRP as well as bombesin also acts to increase tumor cell proliferation [4].

Use for Diagnosis and Treatment

It has been reported that GRP mRNA is overexpressed in patients suffering from small cell lung cancer. Clinically, therefore, a higher level of circulating pro-GRP$_{31-125}$ is a specific tumor marker for the small cell lung cancer [4]. No GRP and compounds have yet been used clinically as therapeutic agents.

References

1. McDonald TJ, Jornvall H, Nilsson G, et al. Characterization of a gastrin releasing peptide from porcine non-antral gastric tissue. *Biochem Biophys Res Commun.* 1979;90:227–233.
2. Minamino N, Kangawa K, Matsuo H. Neuromedin C: a bombesin-like peptide identified in porcine spinal cord. *Biochem Biophys Res Commun.* 1984;119:14–20.
3. Sausville EA, Lebacq-Verheyden AM, Spindel ER, et al. Expression of the gastrin-releasing peptide gene in human small cell lung cancer. Evidence for alternative processing resulting in three distinct mRNAs. *J Biol Chem.* 1986;261:2451–2457.
4. Patel O, Shulkes A, Baldwin GS. Gastrin-releasing peptide and cancer. *Biochim Biophys Acta.* 2006;1766:23–41.
5. Sun YG, Chen ZF. A gastrin-releasing peptide receptor mediates the itch sensation in the spinal cord. *Nature.* 2007;448:700–703.
6. Takanami K, Sakamoto H, Matsuda KI, et al. Distribution of gastrin-releasing peptide in the rat trigeminal and spinal somatosensory systems. *J Comp Neurol.* 2014;522:1858–1873.
7. Shumyatsky GP, Tsvetkov E, Malleret G, et al. Identification of a signaling network in lateral nucleus of amygdala important for inhibiting memory specifically related to learned fear. *Cell.* 2002;111:905–918.
8. Sakamoto H, Matsuda K-I, Zuloaga DG, et al. Sexually dimorphic gastrin releasing peptide system in the spinal cord controls male reproductive functions. *Nat Neurosci.* 2008;11:634–636.
9. Ladenheim EE, Taylor JE, Coy DH, et al. Hindbrain GRP receptor blockade antagonizes feeding suppression by peripherally administered GRP. *Am J Physiol.* 1996;271:R180–R184.
10. Karatsoreos IN, Romeo RD, McEwen BS, et al. Diurnal regulation of the gastrin-releasing peptide receptor in the mouse circadian clock. *Eur J Neurosci.* 2006;23:1047–1053.

Supplemental Information Available on Companion Website

- Mature hormone sequences of GRP in various animals/E-Table 22A.1
- Gene, mRNA, and precursor structures of human GRPR/E-Figure 22A.1
- Primary structure of GRPR of various animals/E-Figure 22A.2
- Accession numbers of cDNAs for vertebrate GRP thus far identified/E-Table 22A.2

Neuromedin B

Hiroko Ohki-Hamazaki

Abbreviation: NMB

NMB is distributed in the central nervous system as well as the gastrointestinal tract, and has many regulatory functions in physiology — exocrine and endocrine secretions, smooth muscle contraction, feeding, blood pressure, blood glucose, body temperature, nociception, emotion, and cell growth.

Discovery

NMB was purified from porcine spinal cord in 1983, as one of the peptides having a stimulant effect on rat uterus contraction [1]. NMB is structurally related to ranatensin, an amphibian peptide belonging to the bombesin-like peptide family, which was purified in 1970 by Nakajima. NMB cDNA was first isolated in the human and then in the rat [2,3].

Structure

Structural Features

NMB was first identified as a decapeptide, but subsequently 30 (NMB-30) or 32 (NMB-32) forms were also identified. The decapeptide (NMB) is also noted as NMB_{23-32}. The C-terminal residue of the peptide is amidated.

Primary Structure

Amino acid sequences for the pig, human, and rat NMB-32 are highly conserved and only differ in 4 aa. Their sequences of 11 aa from the C-terminus are completely identical (Table 22B.1) [4].

Synthesis and Release

Gene, mRNA, and Precursor

Human *NMB* is located on chromosome 15 (15q22—qter) and consists of three exons. Human NMB is encoded in a 76-aa precursor (preproNMB), which consists of a 24-aa signal peptide, a 32-aa mature peptide (NMB-32), and a 17-aa C-terminal extension peptide. The C-terminus of NMB-32 is flanked by a Gly—Lys—Lys sequence, which is a target for proteolytic processing [2]. *Nmb* in the rat consists of three exons, and the intron—exon border is well conserved in the human and rat. Rat preproNMB is 117 aa long, and its structure is similar to that of human preproNMB; the difference exists in the length of the extension peptide, which is 52 aa long instead of 17, and the existence of a cleavage recognition site at the C-terminus of the extension peptide [3].

Distribution of mRNA

In humans, NMB mRNA is expressed in the central nervous system, stomach, and colon [2]; in rat brain, it is expressed prominently in the olfactory bulb and dentate gyrus, and moderately in the wider brain, such as the amygdala, basal ganglia, and brainstem. Expression is also found in the peripheral nervous system, and the trigeminal and dorsal root ganglia [3].

Receptors

Structure and Subtypes

NMB binds NMB-preferring receptor (NMBR, also called BB1) with high affinity ($K_i = 4$ nM for rat NMBR-transfected BALB3T3 cells), and also binds gastrin-releasing peptide-preferring receptor with much lower affinity ($K_i = 174$ nM for mouse GRPR-transfected BALB3T3 cells). NMBR is a GPCR with seven transmembrane domains. The TM5 segment is shown to be critical for high affinity and selectivity for agonist binding. NMBR has been cloned in the human, rat, mouse, chick, and frog. Both human and rat NMBR cDNA encode a protein 390 aa residues in length, and calculated Mr is 43 kDa in the rat. The sequence similarity of NMBR genes in mouse, rat, and human is 90–97%. In the rat, NMBR mRNA is detected notably in the olfactory regions, hippocampal formations, amygdala, thalamus, and central core, and moderate expression is found in many other brain regions. In the peripheral organs, strong expression is found in rat and mouse esophagus, and significant levels of expression have been found, by RT-PCR, in the intestines, testis, and uterus in mouse [4,5].

Signal Transduction Pathway

NMBR is coupled to $G_{\alpha q}$ protein. Binding of NMB to NMBR activates PLCβ, which catalyzes the hydrolysis of PIP2 in the cell membrane, yielding second messengers IP3 and DAG. IP3 stimulates Ca^{2+} release and results in increased cytosol Ca^{2+}, and DAG activates PKC [4,5].

Agonist

NMB is an agonist.

Table 22B.1 Comparison of Amino Acid Sequences of Neuromedin B in Mammals

Rat NMB-32	TPFSWDLPEPRSRASKIRVHPRGNLWATGHFM-NH$_2$
Pig NMB-32	APLSWDLPEPRSRAGKIRVHPRGNLWATGHFM-NH$_2$
Human NMB-32	APLSWDLPEPRSRASKIRVHSRGNLWATGHFM-NH$_2$
Pig NMB-30	LSWDLPEPRSRAGKIRVHPRGNLWATGHFM-NH$_2$
Pig NMB	GNLWATGHFM-NH$_2$

Y. Takei, H. Ando, & K. Tsutsui (Eds): Handbook of Hormones. DOI: http://dx.doi.org/10.1016/B978-0-12-801028-0.00159-8

Table 22B.2 Pharmacological Effects of NMB

Smooth muscle contraction	Esophagus (rat↑)	Gall bladder (guinea-pig↑)	Intestine (rabbit↑)	Stomach (cat↑, guinea pig↑, rat↑)	Urinary bladder (guinea pig↑, rat↑)	Uterus (rat↑)
Exocrine & endocrine secretions	Amylase (rat↑)	CCK (rat↑)	Enteroglucagon (rat↑)	GIP (rat↑)	Gastrin (dog↑ rat↑)	Insulin (dog ↑, rat↑) Thyrotropin (rat↓)
Physiology & Behavior	Food intake (rat↓)	Grooming & scratching (rat↑)	Hyperglycemia (rat↑)	Hypothermia (rat↑)	Locomotion (rat↓)	Nociception (mouse↑)

Antagonists

D-Nal-Cys-Tyr-D-Trp-Lys-Val-Cys-Nal-NH$_2$ (with disulfide bridge between Cys^2 and Cys^7) and PD168368 [5] are antagonists.

Biological Functions

Target Cells/Tissues and Functions

NMB regulates smooth muscle contraction, exocrine and endocrine secretions, blood glucose, body temperature, food intake, energy homeostasis, grooming and scratching, emotion, nociception, and cell growth (Table 22B.2) [4–6]. The contractile effect of NMB is found to be more potent than that of GRP in rat urinary bladder and in esophagus.

Phenotype in Gene-Modified Animals

Analysis of mice lacking NMBR shows that the NMBR has an essential role in regulation of body temperature, but not in smooth muscle contraction of the stomach or for satiety [7]. NMBR-deficient mice show an altered stress response along with a modulated serotonergic system [8]. When NMBR is overexpressed in chicken embryonic brain, the brain size is enlarged [9].

Pathophysiological Implications

Clinical Implications

NMBR is expressed in various human cancers including small cell lung carcinoma, non-small cell lung carcinoma, colon cancer, and various carcinoid tumors, and NMB may have a growth effect. Since NMB has a regulatory function for thyrotropin release [10], NMB could be implicated in human thyroid disorders.

References

1. Minamino N, Kangawa K, Matsuo H. Neuromedin B: a novel bombesin-like peptide identified in porcine spinal cord. *Biochem Biophys Res Commun*. 1983;114:541–548.
2. Krane IM, Naylor SL, Helin-Davis D, et al. Molecular cloning of cDNAs encoding the human bombesin-like peptide neuromedin B. Chromosomal localization and comparison to cDNAs encoding its amphibian homolog ranatensin. *J Biol Chem*. 1988;263: 13317–13323.
3. Wada E, Way J, Lebacq-Verheyden AM, et al. Neuromedin B and gastrin-releasing peptide mRNAs are differentially distributed in the rat nervous system. *J Neurosci*. 1990;10:2917–2930.
4. Ohki-Hamazaki H. Neuromedin B. *Prog Neurobiol*. 2000;62: 297–312.
5. Jensen RT, Battey JF, Spindel ER, et al. International Union of Pharmacology. LXVIII. Mammalian bombesin receptors: nomenclature, distribution, pharmacology, signaling, and functions in normal and disease states. *Pharmacol Rev*. 2008;60:1–42.
6. Mishra SK, Holzman S, Hoon MA. A nociceptive signaling role for neuromedin B. *J Neurosci*. 2012;32:8686–8695.
7. Ohki-Hamazaki H, Sakai Y, Kamata K, et al. Functional properties of two bombesin-like peptide receptors revealed by the analysis of mice lacking neuromedin B receptor. *J Neurosci*. 1999;19:948–954.
8. Yamano M, Ogura H, Okuyama S, et al. Modulation of 5-HT system in mice with a targeted disruption of neuromedin B receptor. *J Neurosci Res*. 2002;68:59–64.
9. Iwabuchi M, Maekawa F, Tanaka K, et al. Overexpression of gastrin-releasing peptide receptor induced layer disorganization in brain. *Neuroscience*. 2006;138:109–122.
10. Oliveira KJ, Ortiga-Carvalho TM, Cabanelas A, et al. Disruption of neuromedin B receptor gene results in dysregulation of the pituitary–thyroid axis. *J Mol Endocrinol*. 2006;36:73–80.

Supplemental Information Available on Companion Website

- Gene, mRNA, and precursor structure of rat NMB/E-Figure 22B.1
- Precursor and mature hormone sequences of various animals/E-Figure 22B.2
- Gene, mRNA and structure of the rat NMBR/E-Figure 22B.3
- Amino acids sequences of the NMBR of various animals/E-Figure 22B.4
- Accession numbers of genes and cDNAs for NMB and NMBR/E-Tables 22B.1, 22B.2

CHAPTER 23

Guanylin Family

Shinya Yuge

History

Guanylin (GN) and uroguanylin (UGN) were first discovered in the rat in 1992 and the opossum in 1993, respectively, as ligands of guanylyl/guanylate cyclase C receptor (GC-C) [1,2]. The discovery was based on anticipation of the existence of endogenous ligands for GC-C that are structurally similar to heat-stable enterotoxin (ST) produced by enterotoxigenic *E. coli* [1,3]. Subsequent studies have revealed that (1) GN and UGN are possible endocrine factors secreted from the intestine and inducing natriuresis in the kidney after oral salt intake, and (2) UGN is an intestinal diarrheal inducer secreted from the enterochromaffin cells. Both concepts were proposed in 1970s, despite the presence of GN and UGN being unknown [4,5]. Currently, GN and UGN as entero-renal natriuretic factors are being re-evaluated [6]. Additionally, *lymphoguanylin (lgn)* and *renoguanylin (rgn)* were identified in the opossum in 1999 and in the eel in 2003, respectively [1,7]. Nowadays, *GN* and/or *UGN* are found in the nucleotide databases of various vertebrate species; however, *lgn* is believed to be unique to marsupials and *rgn* to eels. In invertebrates, a *ugn*-like sequence has been observed in the genome database of a chordate, *Ciona intestinalis*.

Structure

Structural Features

Guanylin family peptides consist of 15−17 aa residues, although the mature sequence has not been determined in non-mammals [1,2,7]. GN and UGN are distinguished by the ninth aa residue: an aromatic aa residue (Tyr/Phe) in GN, and a non-aromatic one (Asn/His) in UGN (Figure 23.1) [1,2,7]. This feature can classify Lgn and Rgn into the UGN and GN groups, respectively. Some UGNs and Rgn possess an extra 1−2 aa residues at the C-terminus. GN, UGN, and Rgn have two disulfide rings derived from four Cys residues, which allows the peptide to form two interchangeable stereoisomers, though only one isomer is biologically active [1,7]. Only Lgn has a single disulfide ring due to the lack of a Cys residue [1]. In terms of the ninth aa residue and the C-terminal extension, UGN is more similar to ST than the other peptides.

Molecular Evolution of Family Members

GN and *UGN* are localized closely to each other on the same chromosome in mammals and fishes, based on the genome databases, suggesting that the two genes were duplicated in tandem from a single ancestral gene [1]. *lgn* could have emerged from *ugn*, based on the location in the opossum genome database. Eel *rgn* appears to be evolutionarily related to *gn* in phylogenetic analysis. Also intriguing is the co-evolution of the *GN* family and *ST*, namely of endogenous and exogenous ligands, which might be an example of interchange of genes between host and parasite (*E. coli*) [1].

Receptors

Structure and Subtypes

GC-C, a membrane protein consisting of an extracellular ligand-binding domain, a membrane-spanning domain, a kinase-like domain, and a catalytic GC domain, is the primary receptor for both GN and UGN [1–3,7]. In cell lines expressing GC-C, binding affinity and cGMP production potency of UGN for GC-C exhibit $EC_{50} = 10^{-8}-10^{-7}$ M, which is 3- to 10-fold and >100-fold greater in terms of efficacy and potency than GN and Lgn, respectively. However, even UGN is >10 times less efficacious than ST [1,2]. In contrast to the single *GC-C* in mammals, at least two isoforms of *gc-c* exist in teleost fishes such as eel and medaka [7,8]. In the eel, both Gc-c1 and Gc-c2 expressed in COS cells produce cGMP in response to all three eel Gns; Ugn > Gn ≥ Rgn (>ST) for Gc-c1, and Gn ≥ Rgn > Ugn (>ST) for Gc-c2 [6]. At least one *gc-c*-like sequence is found in the nucleotide databases of birds, reptiles, and amphibians. There are further three interesting reports/postulations: (1) a shorter *GC-C*, a splice variant, is expressed in some colorectal cancer cells in mammals [1]; (2) another type of GC, GC-D, on the murine olfactory epithelia, is stimulated by GN and UGN [9,10]; (3) the presence of novel GN/UGN receptor(s) has been predicted [1].

Signal Transduction Pathway

In the representative mammalian GC-C signaling cascade, binding of GN and UGN to the extracellular domain triggers catalytic activity of GC in the intracellular domain [1,2]. The cyclase catalyzes conversion of GTP to cGMP, and increased levels of cGMP stimulate PKGII. PKGII activates the cystic fibrosis conductance regulator (CFTR) Cl^- channel, whereas cGMP directly activates a CFTR-independent HCO_3^- transport pathway and directly inhibits sodium hydrogen exchanger 3 (NHE3) [1,2]. These functions are also affected by the cAMP and PKAII pathway activated by cGMP [1,2]. Basics of Lgn/Gc-c signaling would be similar. The cGMP/CFTR pathway appear to be present also in eel Gn/Ugn/Rgn signaling [7]. cGMP is also produced by GC-D stimulated by GN/UGN in murine olfaction [9,10], but this cascade needs to be studied in detail.

Y. Takei, H. Ando, & K. Tsutsui (Eds): Handbook of Hormones. DOI: http://dx.doi.org/10.1016/B978-0-12-801028-0.00023-4

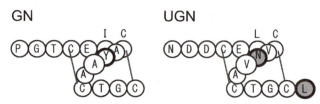

Figure 23.1 Structures of human guanylin (GN) and uroguanylin (UGN).

Biological Functions

Target Cells/Tissues and Functions

The primary target sites of GN/UGN are the intestine and kidney, where their receptor GC-C expression is plentiful in mammals and teleost fishes. The liver, gall bladder, bile duct, testis, trachea, salivary glands, and nasal mucosa are other tissues expressing *GC-C*, and thereby the target sites [1,2]. In eel and medaka, mRNA expression of the two *gc-c* is also abundant in the intestine and kidney, but in other tissues the two expression patterns are different [7,8]. Where *GN/UGN* and *GC-C* expressions overlap they may act locally, as exemplified by the intestine. GN/UGN produced in intestinal epithelial cells is secreted into the lumen and/or into the circulation. Luminocrine GN/UGN binds to apically located GC-C, resulting in activation of Cl^- and HCO_3^- secretion via CFTR and a non-CFTR type transport pathway, and in inhibition of Na^+ absorption via NHE3 [1,2]. The ion regulation eventually drives water secretion into the intestinal lumen. The basic Rgn action may be similar in the intestine [7]. In the human and rodents, GN/UGN is capable of inhibiting proliferation of intestinal epithelial cells via GC-C/cGMP signaling [1]. Circulating and locally-produced GN/UGN also acts on renal tubules to induce natriuresis, kariuresis, and diuresis, presumably by affecting the NHE3 and K^+ channel [1,2]. Regarding enteric and renal effects, UGN is more effective than GN in mammals, while three Gn are similarly effective in the eel intestine [1,2,7]. Actions of Lgn and Rgn are described in the UGN and GN subchapters, respectively. Other novel functions have recently been suggested by a correlation between reduced intestinal GN/UGN expression and obesity, and by stimulation of GC-D/cGMP signaling in olfaction.

References

1. Forte Jr. LR. Uroguanylin and guanylin peptides: pharmacology and experimental therapeutics. *Pharmacol Ther.* 2004;104:137−162.
2. Sindic A, Schlatter E. Cellular effects of guanylin and uroguanylin. *J Am Soc Nephrol.* 2006;17:607−616.
3. Schulz S, Green CK, Yuen PS, Garbers DL. Guanylyl cyclase is a heat-stable enterotoxin receptor. *Cell.* 1990;63:941−948.
4. Lennane RJ, Carey RM, Goodwin TJ, Peart WS. A comparison of natriuresis after oral and intravenous sodium loading in sodium-depleted man: evidence for a gastrointestinal or portal monitor of sodium intake. *Clin Sci Mol Med.* 1975;49:437−440.
5. Fujita T, Kobayashi S. Structure and function of gut endocrine cells. *Int Rev Cytol Suppl.* 1977;:187−233.
6. Preston RA, Afshartous D, Forte LR, et al. Sodium challenge does not support an acute gastrointestinal−renal natriuretic signaling axis in humans. *Kidney Int.* 2012;82:1313−1320.
7. Takei Y, Yuge S. The intestinal guanylin system and seawater adaptation in eels. *Gen Comp Endocrinol.* 2007;152:339−351.
8. Nakauchi M, Suzuki N. Enterotoxin/guanylin receptor type guanylyl cyclases in non-mammalian vertebrates. *Zoolog Sci.* 2005;22:501−509.
9. Leinders-Zufall T, Cockerham RE, Michalakis S, et al. Contribution of the receptor guanylyl cyclase GC-D to chemosensory function in the olfactory epithelium. *Proc Natl Acad Sci USA.* 2007;104:14507−14512.
10. Basu N, Visweswariah SS. Defying the stereotype: non-canonical roles of the peptide hormones guanylin and uroguanylin. *Front Endocrinol (Lausanne).* 2011;2:14.

Supplemental Information Available on Companion Website

- Molecular phylogenetic tree of guanylin family peptides/E-Figure 23.1
- Molecular phylogenetic tree of guanylin receptors (GC-C)/E-Figure 23.2

Guanylin

Shinya Yuge

Abbreviation: GN
Additional name: Guanylyl cyclase activator 2A (GUCA2A)

Guanylin is a peptide hormone that consists of 15 aa residues and binds to guanylyl/guanylate cyclase C receptor (GC-C) to produce cGMP for regulation of ion and water transport in the intestine and kidney. Eel renoguanylin (RGN), more similar to guanylin (GN) than to uroguanylin (UGN), is also described here.

Discovery

GN was first isolated from rat jejunum, and its gene was identified in human and rat in 1992 [1,2]. In non-mammalian vertebrates, *gn* and *rgn* were first cloned from the Japanese eel in 2003 [3]. In nucleotide databases, *GN*-like sequences are found in other vertebrates from fishes to mammals, whereas so far *rgn* has been found only in the eel.

Structure

Structural Features

GN consists of 15 aa residues that are cleaved from the C-terminus of proGN [2,4,5] (Figure 23A.1a, E-Figure 23A.1). The fourth and twelfth Cys and the seventh and fifteenth Cys form disulfide bonds, respectively (Figure 23A.1b), and the two rings give rise to formation of two stereoisomers, of which only one is active [4]. The aromatic aa residue (Tyr/Phe) at the ninth position is targeted by chymotrypsin-like proteases in the renal tubule [2,5] (Figure 23A.1c). These features are shared among all GN found and eel Rgn, although the mature peptide has not yet been verified by isolation in non-mammalian Gn.

Primary Structure

Most of the 15 aa residues are conserved across vertebrate species, whereas eel Rgn possesses an extra Leu residue at the C-terminus [3] (Figure 23A.1). The prosegment is variable with the exception of a conserved Leu−Lys-rich region located upstream of proGN (E-Figure 23A.2).

Properties

Human GN: Mr 1,458.7, pI 4.56. GN is soluble in water, alcohol, and water-containing organic solvents. It is stable at neutral pH, but at acidic pH it is more easily converted to the inactive stereoisomer. GN in water can be stored at −20°C for more than 1 year. Empirically, eel Rgn has been solubilized and stored like GN.

Synthesis and Release

Gene, mRNA, and Precursor

GN is located near the *UGN* on chromosome 1 in the human, and similar co-localization is found in other vertebrate genome databases. Three exons divided by two introns are transcribed and translated into preproGN of 115 aa residues containing the signal peptide and the mature peptide [2,4,5] (Figure 23A.1a, E-Figure 23A.1). ProGN is released from cells and processed into GN during circulation or on the intestinal and renal tubular lumen [2,4,5]. The mature region is conserved among vertebrate GN and in eel RGN (E-Figure 23A.2), although the precise processing mechanism needs to be studied in non-mammalian Gn and Rgn. So far, no significant splice variant is known.

Distribution of mRNA

GN transcripts are most abundant in alimentary tracts other than the esophagus [2,4,5]. The large intestine contains more *GN* transcripts than the small intestine in human and rodents, but a converse tendency is seen in the opossum. The liver, pancreas, kidney, ovary, testis, adrenal gland, trachea, salivary glands, and nasal mucosa are the sites of detectable *GN* mRNA expression in mammals [2,4,5]. In eels, plentiful *GN* mRNA expression is detected only in the intestine [3]. Eel *rgn* transcripts are also ample in the intestine but are most plentiful in the kidney; this is a characteristic difference between GN and RGN [3].

Tissue and Plasma Concentration

Tissue: GN is stored as proGN in human and rodent tissues: duodenum and jejunum, ∼9 fmol/mg; ileum, ∼33 fmol/mg; ascending and transverse colon, ∼28 fmol/mg; descending colon, ∼14 fmol/mg in humans [2,4]. No data are available regarding tissue Gn and Rgn levels in other species.

Plasma: Both GN and proGN are present in plasma [2,4,5]. Plasma GN levels are 30–42 fmol/ml in humans and ∼60 fmol/ml in the rat [2,4]. The level is increased by more than 20- to 100-fold in patients with chronic renal failure [2,4]. No data are available regarding other species.

Regulation of Synthesis

The GN promoter sequence in the 5′-flanking region lacks typical TATA and CAAT boxes, but possesses a putative TATA-alternative motif, TTTAAAA, and several potential binding/responsive sites for transcription factors/elements such as AP-1, AP-2, SP1, GRE, GCF, and HNF-1 in the human and mouse [4]. HNF-1 could play a role in restricting GN

Y. Takei, H. Ando, & K. Tsutsui (Eds): Handbook of Hormones. DOI: http://dx.doi.org/10.1016/B978-0-12-801028-0.00160-4

(a) preproGN

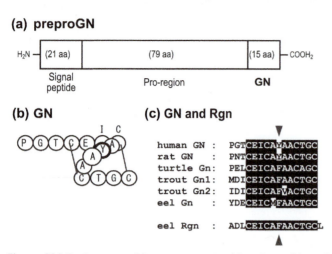

(b) GN   (c) GN and Rgn

```
human GN  : PGT CEICAMAACTGC
rat GN    : PNT CEICAMAACTGC
turtle Gn : PEL CEICAFAACAGC
trout Gn1 : MDI CEICAFAACTGC
trout Gn2 : IDI CEICAFVACTGC
eel Gn    : YDE CEICMFAACTGC

eel Rgn   : ADL CEICAFAACTGCL
```

Figure 23A.1 Structure of human preproGN (a) and GN (b), and comparison of vertebrate GNs and eel Rgn (C). (B) The schematic view is based on previous reports. (C) Two Gn-like peptides are predicted in the salmon gene databases. Arrowheads indicate the ninth aa residue characteristic of GN.

expression to the intestine. This information is unknown in other species. In rat and eel intestine, Gn is synthesized in the goblet cells [2−5].

Regulation of Release

GN release and/or mRNA expression in the intestine is increased or decreased by a high- or low-salt diet, respectively, in rats [2,4,5]. In eels, mRNA expression of *gn*, but not *rgn*, is enhanced in the intestine after transfer to seawater [3]. In a desert rodent, *Notomys alexis*, intestinal *Gn* mRNA expression is promoted after water deprivation [6]. At least in mammals and eels, intestinal epithelia sense changes in luminal salt levels and regulate the production and release of GN. In mammals, *GN* mRNA expression is downregulated in some colonic tumors. In the rat, intestinal *Gn* mRNA expression is higher during the night than during the day. In the rat, intestinal GN secretion into the lumen is promoted by vascular perfusion of VIP, bethanechol, bombesin, PGE_2, interleukin-1β, forskolin, and SNP. In the perfused rat kidney, natriuresis is enhanced synergistically by combinatorial GN and ANP.

Receptors

Structure and Subtypes

GC-C (GUCY2C), a membrane protein consisting of an extracellular ligand-binding domain, a membrane-spanning domain, a kinase-like domain, and a cyclase catalytic domain, is the primary receptor for both GN and UGN (E-Figure 23A.3) [2,4,5]. A shorter GC-C, a splice variant, is expressed in some colorectal cancer cells. In contrast to mammals, at least two isoforms exist in teleost fish (E-Figure 23A.4). For cGMP production in COS cells expressing the eel isoforms, the potency ranking of Gc-c2 and Gc-c1 is Gn ≥ Rgn > Ugn and Rgn ≤ Gn < Ugn, respectively [7]. In birds, reptiles, and amphibians, one *Gc-c* is present in the genome databases. On the murine olfactory epithelia, another type of membrane GC, GC-D, is stimulated by GN and UGN, though this mechanism remains to be studied [8]. Moreover, the presence of novel GN/UGN receptor(s) has been postulated [5].

Signal Transduction Pathway

Binding of GN triggers the catalytic activity of GC in the intracellular domain, the activated cyclase catalyzes conversion of

GTP to cGMP, and increased levels of cGMP stimulate PKGII [2,5]. PKGII activates the cystic fibrosis conductance regulator (CFTR) Cl^- channel, whereas cGMP directly activates CFTR-independent HCO_3^- transport and inhibits sodium hydrogen exchanger 3 (NHE3) [2,5]. cGMP also activates PKAII, and PKAII is involved in those functions [2,5]. In mammals, GN is lower than UGN regarding affinity with and cGMP-producing potency for GC-C, whereas the converse applies in the eel regarding GMP-producing potency for Gc-c2. cGMP is also utilized in GN-induced GC-D signaling in murine olfaction, but the precise mechanism is not yet clear [8].

Agonists

ST produced by enterotoxigenic *E. coli* is an exogenous ligand of GC-C and exerts stronger effects than GN and UGN [2]. In teleost fishes, substantial effects of ST are found on one medaka Gc-c (Olgc9), but not on medaka Olgc6 and eel Gc-c and 2. A synthetic peptide modified from GN/UGN, linaclotide, is another agonist, and is used for regulation of visceral pain [9]. Plecanatide and SP-333 are other known agonists for GC-C.

Antagonists

No antagonists are yet known.

Biological Functions

Target Cells/Tissues and Functions

The primary site of GN action is the intestine, where GC-C expression is most plentiful [2,4,5]. The kidney, liver, gall bladder, bile duct, testis, trachea, salivary glands, and nasal mucosa are also possible target sites. In eels and medaka, the major sites of *gc-c* expression are the intestine and kidney [7]. In the intestine of rodents [2,4] and eels [3], GN is produced in goblet cells and secreted into the lumen ("luminocrine" fashion) to activate Cl^- and HCO_3^- secretion via CFTR and a non-CFTR type HCO_3^- transport pathway, and inhibit Na^+ absorption via NHE3. These ion movements eventually drive water secretion into the intestinal lumen. GN is more active at pH 8.0 than at pH 5.0. In humans and rodents, GN is capable of inhibiting the proliferation of intestinal epithelial cells [2,5], which might contribute to preventing some colorectal cancers. In rodents, the recent finding of a relationship between reduced *Gn* expression and obesity is notable. In the rodent renal tubule, GN secreted into the circulation induces natriuresis, kariuresis, and diuresis, presumably by affecting the NHE3 and K^+ channel, though this GN's effect may be limited due to its quick degradation by proteases on the brush border of renal tubules [5]. In these functions, GN is less effective than UGN in mammals. In the eel intestine, however, Gn is as effective as or slightly more effective than Ugn and Rgn [10]. As intestinal and renal *rgn* mRNA expression do not significantly change after seawater transfer, RGN may have a role(s) different from those of GN and UGN [3]. In a heterologous system, eel Rgn inhibits NHE3, as does rat UGN, but is less effective than rat UGN in the rat perfused renal tubule. As mentioned above, olfactory GC-D and its signaling in mice may be linked to a novel GN function [8].

Phenotype in Gene-Modified Animals

Gn-null mice exhibit normal growth, normal fertility, and no abnormality in intestinal and renal absorption, but increased epithelial proliferation with reduced levels of epithelia-produced GMP in colon [5].

Pathophysiological Implications

Clinical Implications

The fact that *GN* mRNA expression is decreased in some colorectal cancers and that epithelial proliferation in the colon is promoted with the loss of *GN* allows the presumption that changes in *GN* activity are related to or result from the development of colorectal tumors [5]. A GN level in plasma might become a marker for some kidney diseases [5]. A function in intestinal fluid secretion has led scientists to apply GN and UGN for anti-constipation treatment [8].

Use for Diagnosis and Treatment

An GN agonist, linaclotide, was approved in the US and EU in 2012 for treatment of irritable bowel syndrome with constipation [9].

References

1. Currie MG, Fok KF, Kato J, et al. Guanylin: an endogenous activator of intestinal guanylate cyclase. *Proc Natl Acad Sci U S A.* 1992;89:947–951.
2. Forte LR, London RM, Krause WJ, Freeman RH. Mechanisms of guanylin action via cyclic GMP in the kidney. *Annu Rev Physiol.* 2000;62:673–695.
3. Yuge S, Inoue K, Hyodo S, Takei Y. A novel guanylin family (guanylin, uroguanylin, and renoguanylin) in eels: possible osmoregulatory hormones in intestine and kidney. *J Biol Chem.* 2003;278:22726–22733.
4. Nakazato M. Guanylin family: new intestinal peptides regulating electrolyte and water homeostasis. *J Gastroenterol.* 2001;36:219–225.
5. Sindic A, Schlatter E. Cellular effects of guanylin and uroguanylin. *J Am Soc Nephrol.* 2006;17:607–616.
6. Donald JA, Bartolo RC. Cloning and mRNA expression of guanylin, uroguanylin, and guanylyl cyclase C in the Spinifex hopping mouse. *Notomys Alexis. Gen Comp Endocrinol.* 2003;132:171–179.
7. Yuge S, Yamagami S, Inoue K, Suzuki N, Takei Y. Identification of two functional guanylin receptors in eel: multiple hormone-receptor system for osmoregulation in fish intestine and kidney. *Gen Comp Endocrinol.* 2006;149:10–20.
8. Basu N, Visweswariah SS. Defying the stereotype: non-canonical roles of the peptide hormones guanylin and uroguanylin. *Front Endocrinol (Lausanne).* 2011;2:14.
9. Hannig G, Tchernychev B, Kurtz CB, et al. Guanylate cyclase-C/cGMP: an emerging pathway in the regulation of visceral pain. *Front Mol Neurosci.* 2014;7:31.
10. Yuge S, Takei Y. Regulation of ion transport in eel intestine by the homologous guanylin family of peptides. *Zool Sci.* 2007;24:1222–1230.
11. Fan X, Wang Y, London RM, et al. Signaling pathways for guanylin and uroguanylin in the digestive, renal, central nervous, reproductive, and lymphoid systems. *Endocrinology.* 1997;138:4636–4648.
12. London RM, Eber SL, Visweswariah SS, et al. Structure and activity of OK-GC: a kidney receptor guanylate cyclase activated by guanylin peptides. *Am J Physiol.* 1999;276:F882–F891.

Supplemental Information Available on Companion Website

- Structure of GUANYLIN (GN) precursor sequence /E-Figure 23A.1
- Comparison of GUANYLIN (GN) precursor sequences among various animal species/E-Figure 23A.2
- Structure of GUANYLYL CYCLASE C (GC-C) sequence/E-Figure 23A.3
- Comparison of GUANYLYL CYCLASE C (GC-C) sequences among various animal species/E-Figure 23A.4
- Accession numbers of *GUANYLIN* and *GUANYLYL CYCLASE C (GC-C)* in vertebrates/E-Table 23A.1 [11,12]

Uroguanylin

Shinya Yuge

Abbreviation: UGN
Additional name: Guanylyl cyclase activator 2B (GUCA2B)

Uroguanylin is a peptide hormone that consists of 15−17 aa residues and binds to guanylyl/guanylate cyclase C receptor (GC-C) as well as guanylin (GN). It exerts regulation of ion and water transport in the intestine and kidney. Opossum lymphoguanylin (LGN) is similar to UGN and is also described here.

Discovery

UGN was first isolated from opossum urine in 1993 [1], and its gene was identified in humans in 1994 [2]. In non-mammals, *ugn* was first cloned from eel in 2001 [3,4]. *UGN*-like sequences are found in the nucleotide databases of other vertebrates, from fishes to mammals, and even in the ascidian *Ciona intestinalis*. *lgn* was cloned from opossum in 1999 [5].

Structure

Structural Features

UGN consists of 15−17 aa residues that are released from the C-terminus of proUGN [2,6,7] (Figure 23B.1a, E-Figure 23B.1). The fourth and twelfth Cys and the seventh and fifteenth Cys form disulfide bonds, respectively (Figure 23B.1b), and the two rings give rise to the formation of two stereoisomers, of which only one is active [6]. Some UGNs have a C-terminal extension (Figure 23B.1), which may contribute to stabilizing the active isomer. The ninth residue is non-aromatic and thus resistant to chymotrypsin-like proteases in the renal tubule (Figure 23B.1c) [2,7]. These features are shared among all UGNs found. Lgn possesses a non-aromatic aa residue at the same position, but lacks a fifteenth Cys residue to form a second disulfide ring (Figure 23B.1) [5].

Primary Structure

Most of the 15−17 aa residues are conserved across vertebrate species, though some are different in Lgn (Figure 23B.1). In some species, 1−2 extra aa residues are present at the C-terminus. The prosegment is variable except for a conserved Leu−Lys-rich region located upstream of proUGN (E-Figure 23B.2).

Properties

Human UGN: Mr 1,667.9, pI 3.19. UGN is soluble in water, alcohol, and water-containing organic solvent. UGN in solution is stable at neutral pH but relatively unstable at acidic pH. UGN in water can be stored at −20°C for more than 1 year.

Synthesis and Release

Gene, mRNA, and Precursor

In the human, *UGN* consisting of three exons is located on chromosome 1 with *GN* as in other vertebrate species. The translated preproUGN of 112 aa residues contains the signal peptide and the mature peptide (Figure 23B.1, E-Figure 23B.1) [2,6,7]. ProUGN is released from cells and processed into UGN during circulation, or in the intestinal and renal tubular lumen [2,6,7]. The mature region is conserved among vertebrate UGN (E-Figure 23B.2), although the precise processing mechanism needs to be investigated in non-mammalian Ugn. A splice variant is suggested in the European eel [3].

Distribution of mRNA

UGN mRNA expression is most abundant in the alimentary tract, more in the small intestine than in the large intestine in humans and rodents, and is also notable in the kidney [2,6,7]. Some or most of the tissues among the liver, pancreas, heart, spleen, ovary, testis, adrenal gland, trachea, salivary glands, and nasal mucosa are also sites of detectable *UGN* mRNA expression in mammals [2,6–8]. In eels, *ugn* mRNA expression is abundant in the intestine and moderate in the kidney, liver, and esophagus [4]. Opossum *lgn* transcripts are present mainly in the heart, spleen, thymus, lymph node, white blood cells, kidney, and testis [5].

Tissue and Plasma Concentrations

Tissue: UGN is stored as proUGN in human and rodent tissues: stomach, ~2.4 and 0.2 fmol/mg; duodenum, ~2.5 and 1.5 fmol/mg; jejunum, ~2.8 and 5.2 fmol/mg; ileum, ~1.2 and 1.8 fmol/mg; colon, ~0.2 and 0.15 fmol/mg; kidney, ~1.5 and 2.0 fmol/mg [2,6]. No data are available in other species, or regarding Lgn.

Plasma: Both UGN and proUGN are present in plasma, and UGN is present in urine [1,2,6,7]. Plasma UGN is ~5 fmol/ml in humans [2,6]. The level is increased by >6-to 8-fold in patients with chronic renal failure [2,6]. No data are available in other species or regarding Lgn.

Regulation of Synthesis

The human *UGN* promoter sequence possesses TATA and CAAT boxes, and several potential binding/responsive sites for AP-1, AP-2, SP1, CRE, and GCF [6]. Similar features are

Y. Takei, H. Ando, & K. Tsutsui (Eds): Handbook of Hormones. DOI: http://dx.doi.org/10.1016/B978-0-12-801028-0.00161-6

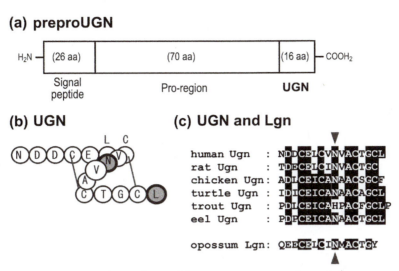

(a) preproUGN

(b) UGN

(c) UGN and Lgn

human Ugn :	NDDCELCVNVACTGCL
rat Ugn :	TDECELCINVACTGC
chicken Ugn:	ADLCEICANAACSGCF
turtle Ugn :	IDICEICANAACAGCL
trout Ugn :	PDLCEICAHPACFGCLP
eel Ugn :	PDPCEICANAACTGCL
opossum Lgn:	QEECELCINMACTGY

Figure 23B.1 Structure of human preproUGN (A) and UGN (B), and comparison of vertebrate UGN and opossum Lgn (C). (B) The schematic view is based on previous reports. (C) Arrowheads indicate the ninth aa residue characteristic of UGN.

present in the mouse and rat genes. In the rodent intestine, UGN is synthesized in enterochromaffin cells [2].

Regulation of Release

Release and/or mRNA expression of UGN in the intestine is increased and decreased by high- and low-salt diets, respectively, in rats [2,6,7]. High salt intake also promotes renal *Ugn* mRNA expression in rats. In eels, intestinal and renal *ugn* mRNA expression is enhanced after seawater transfer [3,4]. In a desert rodent, *Notomys alexis*, intestinal *Ugn* mRNA expression is enhanced after water deprivation [8]. At least in mammals and eels, the intestine senses changes in luminal salt levels, the kidney responds to fluctuations in body salt levels, and both organs regulate the production and release of UGN. In mammals, *UGN* mRNA expression is downregulated in some colonic tumors. In rats, intestinal *Ugn* mRNA expression is higher (more markedly than *Gn*) during the night than during the daytime. In perfused rat kidney, natriuresis caused by UGN is enhanced synergistically by ANP.

Receptors

Structure and Subtypes

GC-C (GUCY2C), a membrane protein consisting of an extracellular ligand-binding domain, a membrane-spanning domain, a kinase-like domain, and a cyclase catalytic domain, is the primary receptor for both UGN and GN (E-Figure 23B.1) [2,6,7]. A shorter GC-C, a splice variant, is expressed in some colorectal cancer cells. In contrast to mammals, at least two *gc-c* exist in teleost fish (E-Figure 23A.4). In eels, at the cGMP production, Gc-c1 is stimulated primarily by Ugn, while Gc-c2 responds to Ugn less effectively than Gn and Rgn [9]. One *gc-c* is detectable in the databases of birds, reptiles, and amphibians. On the murine olfactory epithelia, another type of GC, GC-D, is stimulated by UGN for cGMP activation. The presence of novel UGN/GN receptor(s) has been suggested [7].

Signal Transduction Pathway

Binding of UGN to the extracellular domain triggers catalytic activity of GC in the intracellular domain, resulting in increased levels of cGMP and subsequent activation of PKGII [2,7]. PKGII activates the cystic fibrosis conductance regulator (CFTR) Cl^- channel, whereas cGMP also directly activates a CFTR-independent HCO_3^- transport pathway and inhibits sodium hydrogen exchanger 3 (NHE3). These functions also involve the cAMP/PKAII pathway activated by cGMP [2,7]. UGN exhibits higher affinity with and stronger cGMP-producing potency for GC-C than GN. cGMP is also produced in UGN-induced GC-D signaling in murine olfaction, although this cascade remains to be studied.

Agonists

ST produced by enterotoxigenic *E. coli* is an agonist of GC-C and acts more strongly than UGN. However, ST appears to be a weak agonist or ineffective in teleost fishes [9]. A synthetic peptide modified from UGN/GN, "linaclotide," is another agonist, and is used for regulation of visceral pain. Plecanatide and SP-333 are other known agonists of GC-C.

Antagonists

No antagonists are yet known.

Biological Functions

Target Cells/Tissues and Functions

The primary sites of UGN action are the intestine and kidney, where its receptor *GC-C* mRNA expression is plentiful. The liver, gall bladder, bile duct, testis, trachea, salivary glands, and nasal mucosa also express *GC-C* [2,7]. In eels and medaka, the major sites of the two *gc-c* mRNA expression are also the intestine and kidney, while expression patterns are different in other tissues [9]. Where *UGN* and *GC-C* mRNA expression is found, local UGN action is likely. In the rodent intestine, UGN is produced in enterochromaffin cells, secreted into the lumen, and acts on apically-located GC-C in a luminocrine fashion, resulting in activation of Cl^- and HCO_3^- secretion via CFTR and a non-CFTR type HCO_3^- transport pathway, and in depression of Na^+ absorption via NHE3 [2,7]. These ion movements eventually drive water secretion in the intestine. In these actions, UGN is stronger than GN. UGN is more active at pH = 5.0 than at pH = 8.0. In humans and rodents, UGN is capable of inhibiting the proliferation of intestinal epithelial cells with higher potency than GN [2,7]. In rodents, a relationship between reduction of intestinal *UGN* expression and obesity has recently been reported. Circulating UGN also reaches the renal tubule and induces natriuresis, kariuresis, and diuresis, presumably by affecting

NHE3 and the K^+ channel. Regarding renal function, UGN is more effective than GN because of its resistance to proteases in renal tubule [2,7]. In the eel intestine, Ugn is also important for ion transport, but its effectiveness is similar to that of Gn and Rgn [10]. Lgn also exhibits similar actions via Gc-c, but is less effective at cGMP production and renal actions [2,5]. As mentioned above, olfactory GC-D and its signaling in mice may be linked to a novel UGN function.

Phenotype in Gene-Modified Animals

Ugn-null mice exhibit reduction of electrogenic anion secretion in the intestine, decreases in renal natriuresis, and increases in blood pressure.

Pathophysiological Implications

Clinical Implications

There is no known disease related to UGN. Its function in intestinal fluid secretion has led scientists to apply UGN and GN as an anti-constipation medication. The decreased UGN expression in some colorectal cancers suggests its clinical application for prevention of enlargement of colorectal tumors.

Use for Diagnosis and Treatment

An GN agonist, Linaclotide, was approved in the US and EU in 2012 for treatment of irritable bowel syndrome with constipation.

References

1. Hamra FK, Forte LR, Eber SL, et al. Uroguanylin: structure and activity of a second endogenous peptide that stimulates intestinal guanylate cyclase. *Proc Natl Acad Sci USA.* 1993;90:10464−10468.
2. Forte LR, London RM, Krause WJ, Freeman RH. Mechanisms of guanylin action via cyclic GMP in the kidney. *Annu Rev Physiol.* 2000;62:673−695.
3. Comrie MM, Cutler CP, Cramb G. Cloning and expression of guanylin from the European eel (*Anguilla anguilla*). *Biochem Biophys Res Commun.* 2001;281:1078−1085.
4. Yuge S, Inoue K, Hyodo S, Takei Y. A novel guanylin family (guanylin, uroguanylin, and renoguanylin) in eels: possible osmoregulatory hormones in intestine and kidney. *J Biol Chem.* 2003;278:22726−22733.
5. Forte LR, Eber SL, Fan X, et al. Lymphoguanylin: cloning and characterization of a unique member of the guanylin peptide family. *Endocrinology.* 1999;140:1800−1806.
6. Nakazato M. Guanylin family: new intestinal peptides regulating electrolyte and water homeostasis. *J Gastroenterol.* 2001;36:219−225.
7. Sindic A, Schlatter E. Cellular effects of guanylin and uroguanylin. *J Am Soc Nephrol.* 2006;17:607−616.
8. Donald JA, Bartolo RC. Cloning and mRNA expression of guanylin, uroguanylin, and guanylyl cyclase C in the Spinifex hopping mouse *Notomys Alexis*. *Gen Comp Endocrinol.* 2003;132:171−179.
9. Yuge S, Yamagami S, Inoue K, et al. Identification of two functional guanylin receptors in eel: multiple hormone-receptor system for osmoregulation in fish intestine and kidney. *Gen Comp Endocrinol.* 2006;149:10−20.
10. Yuge S, Takei Y. Regulation of ion transport in eel intestine by the homologous guanylin family of peptides. *Zool Sci.* 2007;24:1222−1230.

Supplemental Information Available on Companion Website

- Structure of UROGUANYLIN (UGN) precursor sequence/E-Figure 23B.1
- Comparison of UROGUANYLIN (UGN) precursor sequences among various animal species/E-Figure 23B.2
- Structure of GUANYLYL CYCLASE C (GC-C) sequence/E-Figure 23A.3
- Comparison of GUANYLYL CYCLASE C (GC-C) sequences among various animal species /E-Figure 23A.4
- Accession numbers of *UROGUANYLIN (UGN)* in vertebrates/E-Table 23B.1
- Accession numbers of *GUANYLYL CYCLASE C (GC-C)* in vertebrates/Table 23A.1

CHAPTER 24

Galanin Peptide Family

Nobuhiro Wada

History

Galanin was first discovered 30 years ago by Tatemoto and colleagues [1]. The discovery of galanin was followed by identification and characterization of secondary peptides originating from a paralogous gene: galanin-like peptide (GALP) [2], and alarin (Figure 24.1) [3]. These peptides can bind galanin receptors 1−3 (GalR1−3). Galanin was first isolated from the pig intestine in 1983, followed by several vertebrate species. The N-terminal end of galanin is crucial for its biological activity, and the first 15 aa residues are conserved in all species excluding the tuna fish. GALP was initially isolated from the pig hypothalamus in 2002 by Ohtaki and colleagues [2]. In mammals, GALP has been identified in various species such as the human, rhesus monkey, pig, rat, and mouse, but it has not been identified thus far in non-mammalian species. The newest member of the galanin family is alarin, which was first identified in the gangliocytes of human neuroblastic tumors. Alarin is an alternative splicing variant from the GALP gene. Alarin has been reported only in mammals such as the human, monkey, and mouse, and its aa sequence is conserved among these animals.

Structure

Structural Features

Galanin has 29 aa residues with an amidated C-terminus. The N-terminal aa residues 1−15 are highly conserved among species, although human galanin is unique in having 30 aa residues with no amidation of the C-terminus. GALP has 60 aa residues, and appears to be cleaved from a prohormone of 114−120 aa residues, depending on the species. This peptide shows a high degree of sequence identity in the mouse, rat, pig, rhesus monkey, and human. Alarin has 25 aa residues, and is formed by an alternative splicing of the GALP gene.

Molecular Evolution of Family Members

These hormones are produced mainly in the brain of mammals. GALP and alarin have a highly conserved sequence of 29 aa residues from the N-terminal in humans, but galanin has a low degree of homology in comparison with GALP and alarin. Thus there is no homology between galanin and GALP.

Receptors

Structure and Subtypes

Galanin receptors have three subtypes, referred to as galanin receptor types 1 to 3 (GalR1−3). The human GalR1 gene contains three exons, and is translated into a 349-aa protein. The homology between species is rather high, as 93% of the aa residues are identical between rat and human GalR1. The expression of GalR1 is regulated by cAMP through the transcription factor of the cAMP response element binding protein (CREB). GalR2 protein also has high sequence identity of 92% between human and rat GalR2. However, there is a notable difference in the 15-aa extension of the C-terminal end in human GalR2. GalR2 is expressed in a wider pattern, compared to GalR1, as it is found in several peripheral tissues, including the pituitary gland, gastrointestinal tract, skeletal muscle, heart, kidney, uterus, ovary, and testis, as well as in regions of the CNS. GalR3 was first isolated from the rat hypothalamic cDNA library. The 368-aa human GalR3 shares 36% aa identity with human GalR1, 58% with human GalR2, and approximately 90% with rat GalR3. Specific receptors for GALP and alarin have not yet been demonstrated, but GALP and alarin can bind GalRs. GALP shows affinity constants K_i of 77, 28, and 10 nM with GalR1, GalR2, and GalR3, respectively, and alarin shows K_i values of $>10^3$, $>10^3$ and $>10^7$ nM with these receptors, respectively [4].

Signal Transduction Pathway

The different galanin receptors are coupled to a variety of signal transduction pathways, with both GalR1 and GalR3 acting mainly on the inhibitory G_i proteins, while GalR2 can influence either stimulatory G_q proteins or inhibitory G_i proteins [4]. GalR1 can influence inwardly rectifying K^+ channels (GIRK). Activation of GalR1 can also stimulate mitogen-associated protein kinase (MAPK) activity through a PKC-independent mechanism.

GalR2 is able to activate the stimulatory pathway of $G_{\alpha q/11}$ class of G proteins. GalR2 is also able to activate MAPK through a typical protein kinase C (PKC) and $G_{\alpha o}$ class of G protein-dependent mechanism. Both GalR1 and GalR2 activation can inhibit CREB. Signaling properties of GalR3 are still poorly defined. Activation of GalR3 receptors results in inward K^+ currents that are characteristic of $G_{\alpha i}$ coupled receptors.

Biological Functions

Target Cells/Tissues and Functions

GalR1 is prominently widely distributed in the hypothalamus, amygdala, hippocampus, thalamus, brainstem, and spinal cord, dorsal root ganglion, gut, heart, lung, kidney, skeletal muscle, adipocytes, and testis of rat. GalR2 mainly exists in the cerebral cortex, hypothalamus, hippocampus, amygdala, cerebellum, and dorsal root ganglion (DRG) in the brain, and

Y. Takei, H. Ando, & K. Tsutsui (Eds): Handbook of Hormones. DOI: http://dx.doi.org/10.1016/B978-0-12-801028-0.00024-6

```
GALP      1: MAPPSVPLVL LLVLLLSLAE TPASAPAHRG RGGWTLNSAG YLLGPVLH-- LPQMGDQDGK RETALEILDL 70
Alarin    1: MAPPSVPLVL LLVLLLSLAE TPASAPAHRS S--------- ---------- ---------- ---------- 31
Galanin   1: MARGSALLLA SLLAAALSA  SAGLWSPAKE KRGWTLNSAG YLLGPHAVGN HRSFSDKNGL TSKRELRPED 70


GALP     71: WKAIDGLPYS HPPQPSKRNV METFAKPEIG DLGMLSMKIP KEEDVLKS-- --- 118
Alarin   32: ---------- -------TF  PKWVTKTERG RQPLRS---- ---------- --- 49
Galanin  71: DMKPGSFDRS IPENNIMRTI IEFLSFLHLK EAGALDRLLD LPAAASSEDI ERS 123
```

Figure 24.1 Alignment of the amino acid sequences of the human galanin peptide family. Conserved amino acid residues are indicated in green. Accession numbers: galanin, NP_057057.2; GALP, NP_001139018.1; alarin, NP_149097.

in the heart, liver, lung, kidney, intestine, uterus, ovary, stomach, pancreas, and testis in the periphery. GALR3 is found in the hypothalamus, dorsal raphe nucleus (DRN), LC, and amygdala [5]. Galanin is thought to regulate numerous physiological systems in the adult mammalian CNS and periphery, including arousal/sleep regulation, energy and osmotic homeostasis, reproduction, nociception, and cognition.

References

1. Tatemoto K, Rökaeus A, Jörnvall H, et al. Galanin — a novel biologically active peptide from porcine intestine. *FEBS Lett.* 1983;164:124–128.
2. Ohtaki T, Kumano S, Ishibashi Y, et al. Isolation and cDNA cloning of a novel galanin-like peptide (GALP) from procine hypothalamus. *J Biol Chem.* 1999;274:37041–37045.
3. Santic R, Fenninger K, Graf K, et al. Gangliocytes in neuroblastic tumors express alarin, a novel peptide derived by differential splicing of the galanin-like peptide gene. *J Mol Neurosci.* 2006;29:145–152.
4. Webling KEB, Runesson J, Bartfai T, et al. Galanin receptors and ligands. *Front Endocrinol.* 2012;3:146.
5. Hawes JJ, Picciotto MR. Characterization of GalR1, GalR2, GalR3 immunoreactivity in catecholaminergic nuclei of the mouse brain. *J Comp Neurol.* 2004;479:410–423.

Supplemental Information Available on Companion Website

- Comparison of galanin family sequences in humanGene structures of the family in human/E-Figure 24.1
- Phylogenetic tree of the family galanin peptides family members/E-Figure 24.2
- Phylogenetic tree of the family galanin peptides family receptors/E-Figure 24.3

Galanin

Satoshi Hirako

Abbreviation: GAL

Galanin is expressed in the brain and peripheral organs and has diverse physiological actions, including energy homeostasis, reproduction, nociception, and cognition.

Discovery

Galanin was isolated in 1983 from porcine intestine by Tatemoto and colleagues [1]. Galanin peptide has been purified from chicken, alligator, and fish species, including trout, tuna fish, bowfin, dogfish, and sturgeon.

Structure

Structural Features

Galanin is a 29-aa C-terminally amidated peptide, apart from human galanin, which has 20 amino acids and no amidated N-terminus [2]. The N-terminal 1−15 amino acids are highly conserved.

Primary Structure

Galanin has been identified in different classes of vertebrates; the human preprogalanin sequence is shown in Figure 24A.1.

Properties

Human progalanin: Mr 13,302.1, pI 6.84. Soluble in water and physiological saline solution.

Synthesis and Release

Gene, mRNA, and Precursor

The galanin gene is located on human chromosome 11q13.3, rat chromosome 1q42, and mouse chromosome 19A. The peptide precursor of galanin is encoded by a single-copy gene consisting of six small exons spanning about 6 kb of genomic DNA. Regarding its structure, a sequence of galanin is followed by the signal peptide, and the remainder is called galanin message-associated peptide (GMAP) [3].

Distribution of mRNA

Galanin is widely expressed in the central and peripheral nervous systems in many mammalian species [4]. In the brain, galanin is synthesized in the dorsal raphe nucleus, locus coeruleus, rostral ventrolateral medulla, central nucleus of the amygdala, paraventricular nucleus, and supraoptic nucleus. Galanin is also found in the spinal cord and gut.

Tissue Concentration

In mice, galanin content in terms of picomoles per gram of tissue weight was found in the following descending ranking order: ileum, duodenum, hypothalamus, stomach, and cortex.

Regulation of Synthesis

Galanin co-localizes in vasopressin neurons and increases during salt loading, suggesting its influence on osmotically stimulated vasopressin release [5]. Galanin gene expression in magnocellular neurons is increased by estrogen.

Receptors

Structure and Subtypes

Galanin receptors have three subtypes, referred to as galanin receptor types 1, 2, and 3 (GalR1, GalR2, and GalR3) [6]. The human GalR1 gene contains three exons and is translated into a 349-aa protein. Homology between species is 93% for rat and human GalR1. Expression of GalR1 is regulated by cAMP through the transcription factor CREB. Human GalR2 has 92% sequence identity to rat GalR2, although there is a 15-aa extension of the C-terminal end in human GalR2. The GalR2 gene is expressed more ubiquitously compared with that of GalR1, as it is found in several peripheral tissues, including the pituitary gland, gastrointestinal tract, skeletal muscle, heart, kidney, uterus, ovary, and testis in addition to the CNS. The GalR3 transcript was first isolated from rat hypothalamic cDNA libraries. Human GalR3 consists of 368 amino acids and shares 36% identity with human GalR1 and 58% with human GalR2, and approximately 90% with rat GalR3.

Signal Transduction Pathway

The different galanin receptors are coupled to a variety of signal transduction pathways, GalR1 and GalR3 mainly with G_i proteins, and GalR2 with G_q or G_i proteins [7]. GalR1 can influence inwardly rectifying K^+ channels (GIRK). Activation of GalR1 can also stimulate MAPK activity through a PKC-independent mechanism. GalR2 is able to activate the stimulatory pathway of $G_{\alpha q/11}$ class of G proteins. GalR2 is also able to activate MAPK through a PKC- and $G_{\alpha o}$-dependent mechanism. Both GalR1 and GalR2 activation can inhibit CREB. The signaling properties of GalR3 are not yet fully defined. Activation of GalR3 receptors results in an inward K^+ current that is characteristic of $G_{\alpha i}$-coupled receptors.

Agonists

Agonists are the GalR1 agonist M617, and GalR2 agonists M1153 and M1145.

Y. Takei, H. Ando, & K. Tsutsui (Eds): Handbook of Hormones. DOI: http://dx.doi.org/10.1016/B978-0-12-801028-0.00162-8

```
            10        20        30        40        50        60
    MARGSALLLA SLLLAAALSA SAGLWSPAKE KRGWTLNSAG YLLGPHAVGN HRSFSDKNGL
            70        80        90       100       110       120
    TSKRELRPED DMKPGSFDRS IPENNIMRTI IEFLSFLHLK EAGALDRLLD LPAAASSEDI

    ERS
```

Figure 24A.1 Human preprogalanin sequence.

Antagonists

Non-selective galanin receptor antagonists are M35 and galantide. Specific antagonists are GalR1 antagonist RWJ-57408, GalR2 antagonist M871, and GalR3 antagonists SNAP37889 and SNAP398299.

Biological Functions

Target Cells/Tissues and Functions

GalR1 is prominently distributed in the hypothalamus, amygdala, hippocampus, thalamus, brainstem, and spinal cord, dorsal root ganglion (DRG), gut, heart, lung, kidney, muscle, adipocytes, and testis of rat. GalR2 mainly exists in the cortex, hypothalamus, hippocampus, amygdala, cerebellum, and DRG, heart, liver, lung, kidney, intestine, uterus, ovary, stomach, pancreas, and testis. GalR3 is found in the hypothalamus, dorsal raphe nucleus, locus coeruleus, and amygdala. Galanin is thought to regulate numerous physiological processes in the adult mammalian nervous system, including sleep/wake regulation, energy and osmotic homeostasis, reproduction, nociception, and cognition [8].

Phenotype in Gene-Modified Animals

In galanin knockout mice, lactation and developmental disorders of the mammary gland are reduced in females. In addition, galanin knockout mice displayed a reduced pain response and autotomy in chronic pain models [9].

Pathophysiological Implications

Clinical Implications

Galanin increases food intake and body weight via GalR1. Recent studies have found that intranasal administration is an effective route for delivery of galanin-related agents into the brain by use of various cyclodextrins. Thus, intranasal administration of a GalR1 antagonist offers an attractive approach to combat obesity [10].

References

1. Tatemoto K, Rökaeus A, Jörnvall H, et al. Galanin — a novel biologically active peptide from porcine intestine. *FEBS Lett.* 1983;164:124−128.
2. Evans HF, Shine J. Human galanin: molecular cloning reveals a unique structure. *Endocrinology.* 1991;129:1682−1684.
3. Lang R, Gundlach AL, Kofler B. The galanin peptide family: receptor pharmacology, pleiotropic biological actions, and implications in health and disease. *Pharmacol Ther.* 2007;115:177−207.
4. Lundström L, Elmquist A, Bartfai T, et al. Galanin and its receptors in neurological disorders. *Neuromol Med.* 2005;7:157−180.
5. Gundlach AL, Burazin TC, Larm JA. Distribution, regulation and role of hypothalamic galanin systems: renewed interest in a pleiotropic peptide family. *Clin Exp Pharmacol Physiol.* 2001;28:100−105.
6. Gundlach AL. Galanin/GALP and galanin receptors: role in central control of feeding body weight/obesity and reproduction? *Eur J Pharmacol.* 2002;440:255−268.
7. Kristin EB, Webling KEB, Runesson J, et al. Galanin receptors and ligands. *Front Endocrinol (Lausanne).* 2012;3:146.
8. Lang R, Gundlach AL, Kofler B. The galanin peptide family: receptor pharmacology, pleiotropic biological actions, and implications in health and disease. *Pharmacol Ther.* 2007;115:177−207.
9. Kerr BJ, Cafferty WB, Gupta YK, et al. Galanin knockout mice reveal nociceptive deficits following peripheral nerve injury. *Eur J Neurosci.* 2000;12:793−802.
10. Fang P, Yu M, Guo L, et al. Galanin and its receptors: a novel strategy for appetite control and obesity therapy. *Peptides.* 2012;36:331−339.

Supplemental Information Available on Companion Website

- Gene, mRNA, and precursor structures of human galanin/E-Figure 24A.1
- Human galanin sequences of various animals/E-Figure 24A.2
- Gene, mRNA and precursor structures of human GalR1 GalR2 and GalR3/E-Figures 24A.3−24A.5
- Primary structure of GalR1, GalR2 and GalR3 of various animals/E-Figures 24A.6−24A.8
- Accession numbers of genes and cDNAs for galanin and GalR1, GalR2 and GalR3/E-Tables 24A.1−24A.4

Galanin-Like Peptide

Fumiko Takenoya

Additional name/Abbreviation: GALP

Galanin-like peptide (GALP) was originally isolated from porcine hypothalamic extracts and is well known as a neuropeptide regulating feeding behavior and energy metabolism.

Discovery

In 1999, GALP was isolated from porcine hypothalamus on the basis of its ability to bind to and activate galanin receptors *in vitro* [1].

Structure

Structural Features

GALP appears to be cleaved from a prohormone of 115–120 amino acids, depending on the species. The GALP peptide shows a high degree of sequence identity in mice, rats, macaques, pigs, and humans.

Primary Structure

Human GALP consists of 60 aa residues. GALP and galanin share a 13-aa residue identity in their peptide sequence. The aa residues at positions 9–21 of GALP are identical to those at positions 1–13 of galanin (Figure 24B.1) [2].

Properties

Human GALP: Mr 6,500.3; pI is 6.28.

Synthesis and Release

Gene, mRNA, and Precursor

In humans, the GALP gene is located on chromosome 19 (19q12.13) while the galanin gene is on chromosome 11 (11q13.3). Although in rats both peptides' genes are located on chromosome 1, GALP and galanin are still encoded by two separate genes (1q12 and 1q42, respectively). In humans, the GALP and galanin genes both comprise six exons, and share other structural similarities.

Distribution of mRNA

In rodents, GALP mRNA is distributed in the periventricular regions of the arcuate nucleus, median eminence, and pituitary gland [3]. Immunohistochemical studies have shown that GALP-immunoreactive neuronal cell bodies are observed in the ARC and posterior pituitary gland. Furthermore, immunoreactive fibers are distributed in the ARC, paraventricular nucleus, bed nucleus of the stria terminalis, medial preoptic area, and lateral septal nucleus.

Plasma Concentration

The plasma concentration of immunoreactive GALP is 200–500 pg/ml in rats [4].

Regulation of Synthesis

Fasting reduces the expression of GALP mRNA in the ARC [5]. GALP mRNA levels are reduced in leptin-deficient *ob/ob* mice, in leptin receptor-deficient *db/db* mice, and in Zucker obese rats, compared to their wild-type counterparts. Moreover, GALP mRNA expression in *ob/ob* mice can be restored by leptin administration. It is suggested that GALP-expressing neurons are a directly regulated target of leptin. In addition, insulin regulates GALP gene expression [6].

Receptors

Structure and Subtypes

Receptor binding experiments suggest that the receptor for GALP is actually the galanin receptor (GalR). *In vitro* studies have shown that GalR3 has the highest affinity for GALP, followed by GalR2 and then GalR1 [7]. However, central administration of a GalR2/3 agonist to rats had no effect on GALP. The GALP receptor may be GalR3, but the true identity of GALP receptors has not yet been established.

Biological Functions

Target Cells/Tissues and Functions

GALP has an orexigenic action in rats, although the effect is short term. Twenty-four hours after intracerebroventricular administration of GALP, food intake and body weight significantly decrease. In the mouse, GALP administration decreases food intake within 2 hours, and suppressed both food intake and body weight for approximately 24 hours. In addition, GALP has a thermogenesis effect via the action of prostaglandin in the brain and increased expression of uncoupling protein-1 in peripheral tissues. GALP is also involved in reproduction. GALP stimulates luteinizing hormone (LH) secretion and evokes GnRH release [8].

Phenotype in Gene-Modified Animals

GALP knockout mice were not affected regarding sexual development, body weight, food and water consumption, and motor behaviors. However, male GALP knockout mice

Y. Takei, H. Ando, & K. Tsutsui (Eds): Handbook of Hormones. DOI: http://dx.doi.org/10.1016/B978-0-12-801028-0.00163-X

Human GALP

```
        10         20         30         40         50         60
APAHRGRGGW TLNSAGYLLG PVLHLPQMGD QDGKRETALE ILDLWKAIDG LPYSHPPQPS
```

Figure 24B.1 Amino acid sequence of human galanin-like peptide.

consumed less food during re-feeding after a fast than did wild-type controls [9].

Pathophysiological Implications

Clinical Implications

GALP has an anti-obesity effect owing to the enhanced energy metabolism and feeding behavior, and may therefore have potential as an anti-obesity drug. Recently, Nonaka *et al.* reported that uptake by the whole brain, olfactory bulb, and cerebrospinal fluid after intranasal administration is greater than that after intravenous injection [10]. The intranasal delivery method may thus potentially be useful in treating lifestyle-related diseases and obesity in humans.

References

1. Ohtaki T, Kumano S, Ishibashi Y, et al. Isolation and cDNA cloning of a novel galanin-like peptide (GALP) from porcine hypothalamus. *J Biol Chem*. 1999;274:37041–37045.
2. Cunningham MJ. Galanin-like peptide as a link between metabolism and reproduction. *J Neuroendocrinol*. 2004;16:717–723.
3. Kageyama H, Takenoya F, Kita T, et al. Galanin-like peptide in the brain: effects on feeding, energy metabolism and reproduction. *Regul Pept*. 2005;126:21–26.
4. Kastin AJ, Akerstrom V, Hackler L. Food deprivation decreases blood galanin-like peptide and its rapid entry into the brain. *Neuroendocrinology*. 2001;74:423–432.
5. Man PS, Lawrence CB. Galanin-like peptide: a role in the homeostatic regulation of energy balance? *Neuropharmacology*. 2008;55:1–7.
6. Lawrence C, Fraley GS. Galanin-like peptide (GALP) is a hypothalamic regulator of energy homeostasis and reproduction. *Front Neuroendocrinol*. 2011;32:1–9.
7. Shiba K, Kageyama H, Takenoya F, et al. Galanin-like peptide and the regulation of feeding behavior and energy metabolism. *FEBS J*. 2010;277:5006–5013.
8. Cunningham MJ. Galanin-like peptide as a link betwenn metabolism and reproduction. *J Neuroendocrinol*. 2004;16:717–723.
9. Dungan Lemko HM, Clifton DK, Steiner RA, et al. Altered response to metabolic challenges in mice with genetically targeted deletions of galanin-like peptide. *Am J Physiol*. 2008;295:E605–12.
10. Nonaka N, Farr SA, Kageyama H, et al. Delivery of galanin-like peptide to the brain: targeting with intranasal delivery and cyclodextrins. *J Pharmacol Exp Ther*. 2008;325:513–519.

Supplemental Information Available on Companion Website

- Gene, mRNA, and precursor structures of human GALP/E-Figure 24B.1
- GALP sequences of various animals/E-Figure 24B.2
- Accession numbers of genes and cDNAs for GALP /E-Table 24B.1

Alarin

Nobuhiro Wada

Alarin is a member of the galanin family of neuropeptides that includes galanin and galanin-like peptide. It has potent effects on increases in food intake and body weight.

Discovery

Alarin was first identified in 2006 as a peptide originating from a splice variant of the GALP gene [1].

Structure
Structural Features

Alarin is a 25-aa peptide located at the C-terminus of preproalarin, which is translated from the alarin mRNA transcribed from the galanin-like peptide (GALP) gene (Figure 24C.1). Alarin was so named after the N-terminal residue (alanine) and the C-terminal residue (serine).

Properties

Human alarin: Mr, 5,321.3; pI is 11.55.

Synthesis and Release
Gene, mRNA, and Precursor

The GALP gene that includes the alarin sequence is localized on chromosome 19q13.43 in humans, chromosome 1q12 in rats, and chromosome 7A1 in mice, spanning 11 kb of genomic DNA [2]. The alarin mRNA is a splice variant of the GALP mRNA from the GALP gene, and lacks exon 3 (Figure 24C.2). The amino acid sequence identity of alarin is 88% between murine and rat, 96% between human and macaque 96%, but only ca. 60% between primates and rodents [3].

Distribution of mRNA

Alarin mRNA was first detected in ganglionic cells of neuroblastomic tumors, and it has a much wider distribution than that of GALP in the central nervous system of humans. Alarin mRNA has been detected in the mouse brain, skin, and thymus [4]. Alarin can also be found locally around blood vessels

```
        10            20
APAHRSSTFP KWVTKTERGR QPLRS
```

Figure 24C.1 Amino acid sequence of human alarin.

in the skin. In the dermal vascular system, alarin-like immunoreactivity (alarin-LI) has been observed in pericytes of microvascular arterioles and venules, as well as in layers of smooth muscle cells in larger vessels. Alarin-LI has been observed in different areas of the murine brain. A high intensity of alarin-LI was detected in the accessory olfactory bulb, medial preoptic area, and amygdala, in different nuclei of the hypothalamus such as the arcuate nucleus and ventromedial hypothalamic nucleus, and in the trigeminal complex, locus coeruleus, ventral cochlear nucleus, facial nucleus, and epithelial layer of the plexus choroideus [5].

Tissue and Plasma Concentrations

Tissue and plasma concentrations have not been determined yet.

Regulation of Synthesis and Release

Regulation of the synthesis and release of alarin is as yet unclear.

Receptors
Structure and Subtypes

An alarin-specific receptor has not yet been identified. The effects of alarin seem to be mediated by specific alarin receptors, because alarin lacks homology to galanin and is not able to compete with galanin for galanin receptors (GalRs) [4]. Alarin has low affinity for GalR1, GalR2, and GalR3 ($K_i > 1,000$ for GalR1, $K_i > 1,000$ for GalR2, and $K_i > 10^6$ for GalR3).

Biological Functions
Target Cells/Tissues and Functions

Alarin stimulates feeding behavior, increases body weight, and influences luteinizing hormone (LH) secretion [6]. Male sexual behavior was not altered by alarin treatment in intact or castrated rats, but luteinizing hormone secretion was significantly increased by alarin injection in male rats. Furthermore, alarin is able to stimulate the release of gonadotropin-releasing hormone (GnRH) in both hypothalamic explants and a hypothalamic cell line [7]. When hypothalamic explants were treated with alarin, neuropeptide Y release from the arcuate nucleus was significantly increased, suggesting that alarin may yield its effects through activation of NPY neurons [7].

Y. Takei, H. Ando, & K. Tsutsui (Eds): Handbook of Hormones. DOI: http://dx.doi.org/10.1016/B978-0-12-801028-0.00164-1

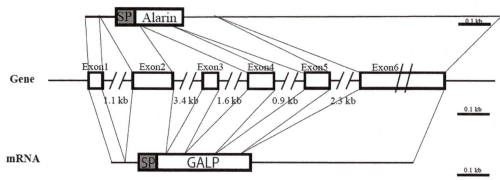

Figure 24C.2 Translation of alarin from the GALP gene in the human.

Phenotype in Gene-Modified Animals

Mice transgenic for alarin have not yet been created.

Pathophysiological Implications

Clinical Implications

Alanine is a recently discovered hormone, and its receptor and the functions or mechanisms of action are not fully elucidated. Therefore, further studies are necessary regarding clinical application.

References

1. Santic R, Fenninger K, Graf K, et al. Gangliocytes in neuroblastic tumors express alarin, a novel peptide derived by differential splicing of the galanin-like peptide gene. *J Mol Neurosci.* 2006;29:145–152.
2. Cunningham MJ, Scarlett JM, Steiner RA. Cloning and distribution of galanin-like peptide mRNA in the hypothalamus and pituitary of the macaque. *Endocrinology.* 2002;143:755–763.
3. Fraley GS, Leathley E, Nickols A, et al. Alarin 6-25Cys antagonizes alarin-specific effects on food intake and luteinizing hormone secretion. *Neuropeptides.* 2013;47:37–41.
4. Santic R, Schmidhuber SM, Lang R, et al. Alarin is a vasoactive peptide. *Proc Natl Acad Sci USA.* 2007;104:10217–10222.
5. Eberhard N, Mayer C, Santic R, et al. Distribution of alarin immunoreactivity in the mouse brain. *J Mol Neurosci.* 2012;46:18–32.
6. Fraley GS, Leathley E, Lundy N, et al. Effects of alarin on food intake, body weight and luteinizing hormone secretion in male mice. *Neuropeptides.* 2012;46:99–104.
7. Boughton CK, Patterson M, Bewick GA, et al. Alarin stimulates food intake and gonadotrophin release in male rats. *Br J Pharmacol.* 2010;161:601–613.

Supplemental Information Available on Companion Website

- Gene, mRNA, and precursor structures of alarin in various vertebrates/E-Figure 24C.1
- Comparison of amino acid sequences of the vertebrate alarin /E-Figure 24C.2
- Accession number of alarin/E-Table 24C.1

CHAPTER 25

Neuropeptide Y Family

Yoshio Takei

History

The neuropeptide Y family consists of neuropeptide Y (NPY), peptide YY (PYY), and pancreatic polypeptide (PP). NPY was first isolated from the porcine brain in 1982 using a chemical assay that permits detection of peptides with C-terminal amidation [1,2]. PYY was isolated from the porcine intestine, using the same methodology, in 1982. PP was discovered earlier, in 1968, as a byproduct of insulin isolation from the chicken pancreas, and its sequence was determined in 1975. Since then, NPY family members have been identified in various vertebrate species as a peptide or a cDNA. NPY and PYY are found from cyclostomes (lampreys) to mammals, but PP is identifiable only in tetrapods, including amphibians, reptiles, birds, and mammals [2].

Structure

Structural Features

All members of the NPY family consist of 36 aa residues with an amidated C-terminus. NPY and PYY have tyrosine (Y) residues at both termini, which is the origin of their names. NPY has the highest sequence identity (only 1 aa different between human and chicken), followed by PYY, and PP is the most variable peptide even in mammals. Fish PYYs are equally similar to mammalian NPY and PYY, but they are distinguishable by proline at position 14, where NPY has alanine in all species (Figure 25.1). The common structural feature is the presence of a "PP-fold," characterized by the helix structure with three intermittent proline residues and that with two tyrosine residues facing each other following an abrupt turn [1]. The N-terminal dipeptide of PYY is readily cleaved by dipeptidyl peptidase 4 (EC 3.4.14.5) to produce PYY_{3-36}, which is a major circulating form in the blood [3].

Molecular Evolution of Family Members

It appears that NPY and PYY existed in ancestral vertebrates, as both have been identified in the most primitive extant jawless vertebrates, cyclostomes. The NPY and PYY genes in human (*NPY* and *PYY*) and *npy* and *pyy* in zebrafish are linked to a hox gene, indicating that the two genes are the products of whole genome duplication. Another *PYY*, named *PYY2*, has been found in ungulates and primates, but only bovine seminal plasmin seems to be a functional peptide produced from the PYY2 gene. The additional whole genome duplication that occurred only in the teleost lineage may have produced two NPY genes (*npya* and *npyb*) and two PYY genes (*pyya* and *pyyb*). Judging from the CNS localization and the highest sequence conservation, NPY may retain characteristics of the ancestral molecule. It is not known whether the NPY family genes originated from chordates or exist in protostomes. The PP gene may be generated by tandem duplication of the PYY gene after divergence of ray-finned fishes and lobe-finned fishes, as both genes exist on the same chromosome and PP was found only in tetrapods. It is not yet known whether the PP gene exists in lobe-finned fishes.

Receptors

Structure and Subtypes

At least five types of receptors, named Y receptors, have been identified for NPY family peptides in mammals: Y1, Y2, Y4, Y5, and Y6 [4,5]. No biological function was assigned to Y6, and it is a pseudogene in human and other tetrapods, so Y6 is often distinguished from other Y receptors by a lower-case letter (y6). The presence of Y3 is suggested only by pharmacological evidence. Additional Y receptors, Yb and Y7, were found in zebrafish and other teleosts, and frogs seem to retain Y7. Y1 and Y5, important receptors of NPY for appetite arousal, are lost in teleosts. The ancestral Y receptors may have existed as three subfamilies (Y1, Y2, and Y5), and they are generated by tandem duplication on the same chromosome in primitive vertebrates, as at least Y1 and Y2 already existed in lampreys. Because of their ancient origin, sequence identity of the three subfamily receptors is low ($\sim$30%) compared with those of other GPCRs that share the same ligand. It seems that some Y receptor genes have disappeared in different lineages after whole genome duplications. Y4, Y6, and Yb are classified as the Y1 subfamily, and Y2 and Y7 are grouped as the Y2 subfamily. NPY and PYY exhibit similar high affinities for Y1, Y2, and Y5 subfamily receptors, while PP preferentially binds to Y4 of the Y1 subfamily, and, interestingly, NPY_{3-36} and PYY_{3-36} are preferred ligands for the Y2 subfamily receptors. In the brain, the Y1 and Y2 genes are ubiquitously expressed in various regions, but the expression of Y4 and Y5 genes is restricted to specific loci involved in the regulation of appetite, circadian rhythm, and anxiety.

Signal Transduction Pathway

The Y receptors belong to the GPCRs that couple to pertussis toxin-sensitive $G_{i/o}$ protein [6]. Thus, the NPY family peptides basically exhibit their function through inhibition of the cAMP dependent kinase (A-kinase).

Y. Takei, H. Ando, & K. Tsutsui (Eds): Handbook of Hormones. DOI: http://dx.doi.org/10.1016/B978-0-12-801028-0.00025-8

```
Neuropeptide Y
Human           YPSKPDNPGEDAPAEDMARYYSALRHYINLITRQRY-NH2
Duck            YPSKPDSPGEDAPAEDMARYYSALRHYINLITRQRY-NH2
Clawed frog     YPSKPDNPGEDAPAEDMAKYYSALRHYINLITRQRY-NH2
Eel             YPSKPDNPGEDAPAEELAKYYSALRHYINLITRQRY-NH2
Elephant shark  YPSKPDNPGEGAPAEDLAKYYSALRHYINLITRQRY-NH2

Peptide YY
Human           YPIKPEAPGEDASPEELNRYYASLRHYLNLVTRQRY-NH2
Duck            YPPKPESPGEDASPEELARYFSALRHYINLVTRQRY-NH2
Clawed frog     YPTKPENPGNDASPEEMAKYLTALRHYINLVTRQRY-NH2
Eel             YPPKPENPGEDASPEEQAKYYTALRHYINLITRQRY-NH2
Elephant shark  YPPKPENPGEDAPPEELAKYYSALRHYINLITRQRY-NH2
Sea lamprey     FPPKPDNPGDNASPEQMARYKAAVRHYINLITRQRY-NH2

Pancreatic polypeptide
Human           APLEPVYPGDNATPEQMAQYAADLRRYINMLTRPRY-NH2
Duck            GPSQPTYPGDDAPVEDLVRFYNDLQQYLNVVTRHRY-NH2
Clawed frog     APSEPMHPGDQASPEQLAKYYDDWWQYITFITRPRY-NH2
```

Figure 25.1 Comparison of amino acid sequences of the NPY family in different vertebrate species. Pancreatic polypeptide is absent in fishes. For abbreviations, see text.

Biological Functions

Target Cells/Tissues and Functions

The NPY family constitutes important members of the brain—gut hormones, which play pivotal roles in neural and humoral communication between the digestive tract and the CNS (gut—brain axis) [4,7]. NPY is basically a neuropeptide synthesized in the CNS, enteric neurons, and primary afferent neurons, and acts locally in the brain to regulate food intake, energy metabolism, anxiety, cognition, stress resilience, and nociception, and in the gastrointestinal tract to regulate its function via Y1, Y2, and Y5 subfamily receptors [4]. By contrast, PYY and PP are gastrointestinal hormones, and their synthesis in the brain is minimal. PYY is secreted from the intestinal L cells and PP from the pancreatic F cells; this is stimulated following a meal. NPY and PYY act in a paracrine fashion in the gastrointestinal tract to inhibit their motility and electrolyte secretion from crypt cells, which is in contrast to the stimulatory actions by various hormones, including vasoactive intestinal peptide, natriuretic peptides, and guanylin [8]. PYY co-localizes with proglucagon-derived peptides, suggesting a synergistic action on insulin secretion. The main action of PP is to inhibit exocrine secretions from the pancreas and gall bladder. The most prominent action of NPY is an enhancement of food intake (see Chapter 10), but NPY gene knockout did not resulted in reduced food intake. Interestingly, NPY induces appetite but PYY and PP inhibit it, although both NPY and PYY bind to Y1 and Y5 with high affinities. It seems that PYY is cleaved to PYY_{3-36} immediately and acts on Y2. The expression of the PP gene is scarce in the brain, but its receptor (Y4) knockout resulted in reduced anxiety and depression [9].

References

1. Conlon MJ. The origin and evolution of peptide YY (PYY) and pancreatic polypeptide (PP). *Peptides.* 2002;23:269—278.
2. Larhammer D. Evolution of neuropeptide Y, peptide YY and pancreatic polypeptide. *Reg Pept.* 1996;62:1—11.
3. Medeiros MD, Turner AJ. Processing and metabolism of peptide-YY: pivotal roles of dipeptidylpeptidase-IV, aminopeptidase-P, and endopeptidase-24.11. *Endocrinology.* 1994;134:2088—2094.
4. Holzer P, Reichmann F, Farzi A, Neuropeptide Y. peptide YY and pancreatic polypeptide in the gut—brain axis. *Neuropeptides.* 2012;46:261—274.
5. Larhammer D, Salaneck E. Molecular evolution of NPY peptide subtypes. *Neuropeptides.* 2004;38:141—151.
6. Alexander SPH, Mathie A, Peters JA. G protein-coupled receptors.. *Br J Pharmacol.* 2011;164:S5—S113.
7. Field BCT, Chaudhri OB, Bloom SR. Bowels control brain: gut hormones and obesity. *Nat Rev Endocrinol.* 2010;6:444—453.
8. Cox HM. Neuropeptide Y receptors; antisecretory control of intestinal epithelial function. *Auton Neurosci.* 2007;133:76—85.
9. Painsipp E, Wultsch T, Edelsbrunner ME, et al. Reduced anxiety-like and depression-related behavior in neuropeptide Y Y4 receptor knockout mice. *Genes Brain Behav.* 2008;7:532—542.

Supplemental Information Available on Companion Website

- Gene structures of the NPY family members in human/ E-Figure 25.1
- Phylogenetic tree of the NPY family members/E-Figure 25.2
- Phylogenetic tree of the NPY family receptors/E-Figure 25.3

Pancreatic Polypeptide

Goro Katsuura and Akio Inui

Abbreviation: PP

Pancreatic polypeptide is a peptide hormone secreted by the pancreas to inhibit pancreatic exocrine secretion, gallbladder contraction, gastric emptying, and gut motility, and to modulate anxiolytic and depressive behaviors.

Discovery

Pancreatic polypeptide (PP) was isolated initially in 1975 as a by-product of insulin purification from the pancreas [1].

Structure

Structural Features

Mature PP is a linear peptide consisting of 36 aa residues in mammals, including humans. PP is a member of the neuropeptide Y (NPY) family, and has about 50% homology with NPY and 70% homology with peptide YY (PYY) in pigs.

Primary Structure

Amino acid sequences of PP are conserved in mammals and other non-mammalian species (Table 25A.1).

Properties

Human PP: Mr 4,182. Soluble in water, acid solutions, and methanol.

Synthesis and Release

Gene and mRNA

The human PP gene is localized with the NPY gene on chromosome 17 (p11.1) and consists of four exons. In humans, preproPP has 95 aa residues, comprising a 29-aa signal peptide, a 36-aa mature PP, and a 30-aa C-terminal peptide.

Distribution of mRNA

PP gene expression is restricted to the pancreas in the fourth cell type, named F cells. The PP-producing F cells often take up a peripheral position in the islets, and are distinct from the three other major islet cell populations, such as insulin-producing β cells, glucagon-producing α cells, and somatostatin-producing S cells.

Tissue and Plasma Concentrations

Tissue: F cells are often located in the peripheral region of the islets of Langerhans in normal pancreas, and their population is 6% of the total islet cells [2]. The percentage distribution of F cells increases significantly after the onset of diabetes [2].

Plasma: Plasma PP levels after fasting is 20 fmol/ml in normal subjects.

Regulation of Synthesis and Release

A potential AP-1 binding site and several potential AP-2 binding sequences, which are activated by cAMP and phorbol ester, are located upstream of the transcriptional initiation site of the human PP gene. PP is secreted from the F cells of the pancreatic islets [3]. The secretion of PP is increased by protein meal ingestion, fasting, and exercise.

Receptors

Structure and Subtype

PP binds preferentially to the Y4 receptor of NPY receptor subtypes, but not to other subtypes [4].

Signal Transduction Pathway

The Y4 receptor is a seven-transmembrane-spanning GPCR. The receptor is associated with a member of the G_i and G_o protein family, and thus ligand binding results in inhibition of adenylyl cyclase and second messenger cAMP production [4].

Agonists and Antagonists

Agonists and antagonists for the Y4 receptor are discussed in Subchapter 25B.

Biological Functions

Target Cells/Tissues and Functions

PP secreted from the pancreas stimulates gastric juice secretion. Peripheral administration of PP decreases food intake in rodents, while central administration of PP increases food intake [5]. PP inhibits pancreatic exocrine secretion, gallbladder contraction, gastric emptying, and gut motility [6–8]. Peripheral administration of PP causes Y4 receptor-dependent c-Fos expression in the brainstem, hypothalamus, and amygdala [9]. Chronic peripheral administration of PP reduces anxiety [10].

Phenotype in Gene-Modified Animals

Overexpression of the PP gene shows reduction of body weight and food intake, and protection against diet-induced and genetic obesity. Conversely, deficiency of the Y4 receptor in mice results in reduced food intake and body weight. Mice with overexpression of the PP gene show an anxiogenic

Y. Takei, H. Ando, & K. Tsutsui (Eds): Handbook of Hormones. DOI: http://dx.doi.org/10.1016/B978-0-12-801028-0.00165-3

Table 25A.1 Amino Acid Sequence of PP in Different Vertebrate Species*

Species				
Human	APLEPVYPGD	NATPEQMAQY	AADLRRYINM	LTRPRY
Monkey	APLEPVYPGD	NATPEQMAQY	AADLRRYINM	LTRPRY
Cow	APLEPQYPGD	DATPEQMAQY	AAELRAYINM	LTRPRY
Buffalo	SPLEPVYPGD	NATPEEMAQY	AAELRRYINM	LTRPRY
Pig	APLEPVYPGD	DATPQQMAQY	AAEMRRYINM	LTRPRY
Zebra	APMEPVYPGD	NATPEQMAQY	AAELRRYINM	LTRPRY
Dog	APLEPVYPGD	DATPEQMAQY	AADLRRYINM	LTRPRY
Rabbit	APPEPVYPGD	DATPEQMAEY	VADLRRYINM	LTRPRY
Tapir	APLEPVYPGD	NATPEQMAQY	AAELRRYINM	LTRPRY
Chinchilla	APLEPVYPGD	NATPEQMAQY	AAEMRRYINM	LTRPRY
Rat	APLEPMYPGD	YATHEQRAQY	ETQLRRYINT	LTRPRY
Hedgehog	VPLEPVYPGD	NATPEQMAHY	AAELRRYINM	LTRPRY
Ostrich	GPAQPTYPGD	DAPVEDLVRF	YDNLQQYLNV	VTRHRY
Chicken	GPSQPTYPGD	DAPVEDLIRF	YNDLQQYLNV	VTRHRY
Duck	GPSQPTYPGD	DAPVEDLVRF	YNDLQQYLNV	VTRHRY
Goose	GPSQPTYPGN	DAPVEDLXRF	YDNLQQYRLN	VFRHRY
Magpie	APAQPAYPGD	DAPVEDLLRF	YNDLQQYLNV	VTRPRY
Gull	GPVQPTYPGD	DAPVEDLVRF	YNDLQQYLNV	VTRHRY
Crocodile	TPLQPKYPGD	GAPVEDLIQF	YNDLQQYLNV	VTRPRF
Bullfrog	APSEPHHPGD	QATPDQLAQY	YSDLYQYITF	ITRPRF
Rana ridibunda	YPPKPENPGE	DASPEEMTKY	LTALRHYINL	VTRQRY
Bombina bombina	APSEPHHPGD	QATQDQLAQY	YSDLYQYITF	VTRPRF
Toad	TPSEPQHPGD	QASPEQLAQY	YSDLWQYITF	VTRPRF
Xenopus	APSEPMHPGD	QASPEQLAKY	YDDWWQYITF	ITRPRF
Salmon	YPPKPENPGE	DAPPEELAKY	YTALRHYINL	ITRQRY

*Shaded amino acids are identical to human amino acids.

phenotype, which is consistent with the anxiolytic behavior of Y4 receptor knockout mice.

Pathophysiological Implications

Clinical Implications

Plasma PP levels are reduced following increased food intake, and elevated in anorexia nervosa. PP administration reduces food intake in lean humans as well as in obese patients with Prader-Willi syndrome. No PP-related peptides and compounds have been used clinically as therapeutic agents.

Use for Diagnosis and Treatment

PP secretion from the pancreas in response to vagal nerve stimulation has been used as a diagnostic test of vagal nerve function.

References

1. Kimmel JR, Hayden LJ, Pollock HG. Isolation and characterization of a new pancreatic polypeptide hormone. *J Biol Chem*. 1975;250:9369–9376.
2. Lin TM, Evans DC, Chance RE, et al. Bovine pancreatic peptide: action on gastric and pancreatic secretion in dogs. *Am J Physiol*. 1977;232:E311–E315.
3. Hazelwood RL. The pancreatic polypeptide (PP-fold) family: gastrointestinal, vascular, and feeding behavioral implications. *Proc Soc Exp Biol Med*. 1993;202:44–63.
4. Adrian TE, Mitchenere P, Sagor G, et al. Effect of pancreatic polypeptide on gallbladder pressure and hepatic bile secretion. *Am J Physiol*. 1982;243:G204–G207.
5. Asakawa A, Inui A, Yuzuriha H, et al. Characterization of the effects of pancreatic polypeptide in the regulation of energy balance. *Gastroenterology*. 2003;124:1325–1336.
6. Ueno N, Inui A, Iwamoto M, et al. Decreased food intake and body weight in pancreatic polypeptide-overexpressing mice. *Gastroenterology*. 1999;117:1427–1432.
7. Boey D, Lin S, Enriquez RF, et al. PYY transgenic mice are protected against diet-induced and genetic obesity. *Neuropeptides*. 2008;42:19–30.
8. McGowan BMC, Bloom SR. Peptide YY and appetite control. *Curr Opin Pharmacol*. 2004;4:583–588.
9. Ekblad E, Sundler F. Distribution of pancreatic polypeptide and peptide YY. *Peptides*. 2002;23:251–261.
10. Tasan RO, Lin S, Hetzenauer A, et al. Increased novelty-induced motor activity and reduced depression-like behavior in neuropeptide Y (NPY)-Y4 receptor knockout mice. *Neuroscience*. 2009;158:1717–1730.

Supplemental Information Available on Companion Website

- Primary structure of the human PP gene/E-Figure 25A.1
- Amino acid sequences of human preproPP/E-Figure 25A.2
- Accession numbers of PP in various vertebrates/E-Table 25A.1

Neuropeptide Y

Goro Katsuura and Akio Inui

Abbreviation: NPY

Neuropeptide Y is a neuropeptide released by neurons in the central nervous system and sympathetic nerve system to induce several actions in the brain.

Discovery

In 1982, the isolation and identification of NPY from porcine brain were reported [1].

Structure

Structural Features

Mature NPY is a linear peptide consisting of 36 aa in mammals, including humans. Both C- and N-terminal residues are Tyr (Y), which is responsible for its name. C-terminal Tyr is amidated. The amino acid sequence of NPY is homologous to those of peptide YY (PYY: 70%) and pancreatic polypeptide (PP: 50%).

Primary Structure

Amino acid sequences of NPY are conserved in mammals and other non-mammalian species (Table 25B.1).

Properties

Human NPY: Mr 4,272. Solubility is 1.5 mg/ml in water and 1 mg/ml in 5% acetic acid.

Synthesis and Release

Gene, mRNA, and Precursor

The human NPY gene locus is on chromosome 7 (7p15.1), the rat NPY gene locus is on chromosome 4 (4q24), and the mouse NPY gene locus is on chromosome 6 (6 24.04 cM). The NPY RNA transcript contains 4 exons and is 551 bp in length. In the rat, the preproNPY has 98 aa residues, comprising a 29-aa signal peptide, a 36-aa mature NPY, and a 33-aa C-terminal peptide.

Distribution of mRNA

NPY mRNA is predominantly expressed in the arcuate nucleus of the hypothalamus [2]. There is high expression of NPY mRNA in the nucleus accumbens and the ventral region of the caudate. NPY is abundantly present in the central nervous system, especially in the cerebral cortex, paraventricular nucleus, dorsomedial nucleus, medial preoptic area, and hippocampus. In the sympathetic nervous system, NPY is expressed with norepinephrine. NPY is also present in the non-neuronal organs, including the liver, heart, gastrointestinal tract, salivary gland, thyroid gland, adrenal gland, and pancreas.

Tissue and Plasma Concentrations

Tissue: The distribution of NPY in the human brain is shown in Table 25B.2 [3].
Plasma: The plasma concentration of NPY in humans is 2.4 pmol/l [4].

Regulation of Synthesis and Release

A potential AP-1 binding site and several potential AP-2 binding sequences, which are activated by cAMP and phorbol ester, are located upstream from the transcriptional initiation site. NPY in the neuron is released in a Ca^{2+}-dependent manner. NPY levels in the hypothalamus of the brain in particular are regulated by energy status and leptin. NPY in the sympathetic nervous system is co-released with norepinephrine.

Receptors

Structure and Subtype

Five distinct NPY receptors have been cloned, namely Y1, Y2, Y4, Y5, and y6, as seven transmembrane-spanning G protein-coupled receptors. They are associated with a member of the G_i and G_0 family (Table 25B.3) [5], and thus ligand binding results in inhibition of adenylyl cyclase and second messenger cAMP production. Among the cloned receptors, the Y1, Y2, Y4, and Y5 receptors represent fully defined subtypes, while no functional protein of y6 is expressed in primates due to a truncation in the sixth transmembrane domain. The Y3 receptor has not been identified because the evidence is not sufficient to grant its presence.

Signal Transduction Pathway

Y1, Y2, Y4, and Y5 receptor subtypes are coupled with $G_{i/0}$, resulting in inhibition of adenylate cyclase. Additional signaling responses include inhibition of Ca^{2+} channels in neurons and modulation of K^+ channels, for example in cardiomyocytes and vascular smooth muscle cells.

Agonists

The subtypes Y1, Y2, and Y5 preferentially bind NPY and peptide YY, whereas subtype Y4 preferentially binds pancreatic polypeptide (Table 25B.3).

Y. Takei, H. Ando, & K. Tsutsui (Eds): Handbook of Hormones. DOI: http://dx.doi.org/10.1016/B978-0-12-801028-0.00166-5

Table 25B.1 Amino Acid Sequence of NPY in Different Vertebrate Species

Human	YPSKPDNPGE	DAPAEDMARY	YSALRHYINL	ITRQRY
Monkey	YPSKPDNPGE	DAPAEDMARY	YSALRHYINL	ITRQRY
Cow	YPSKPDNPGE	DAPAEDLARY	YSALRHYINL	ITRQRY
Bullfrog	YPSKPDNPGE	DAPAEDMAKY	YSALRHYINL	ITRQRY
Porcine	YPSKPDNPGE	DAPAEDLARY	YSALRHYINL	ITRQRY
Sheep	YPSKPDNPGD	DAPAEDLARY	YSALRHYINL	ITRQRY
Rat	YPSKPDNPGE	DAPAEDMARY	YSALRHYINL	ITRQRY
Mouse	YPSKPDNPGE	DAPAEDMARY	YSALRHYINL	ITRQRY
Chicken	YPSKPDSPGE	DAPAEDMARY	YSALRHYINL	ITRQRY
Duck	YPSKPDSPGE	DAPAEDMARY	YSALRHYINL	ITRQRY
Alligator	YPSKPDNPGE	DAPAEDMARY	YSALRHYINL	ITRQRY
Toad	YPSKPDNPGE	DAPAEDMAKY	YSALRHYINL	ITRQRY
Salmon	YPVKPETPGE	DAPAEELAKY	YSALRHYINL	ITRQRY
Rainbow trout	YPVKPENPGE	DAPTEELAKY	YTALRHYINL	ITRQRY
Cod	YPIKPENPGE	DAPADELAKY	YSALRHYINL	ITRQRY
Torpedo	YPSKPDNPGE	GAPAEDLAKY	YSALRHYINL	ITRQRY
Eel	FPNKPDSPGE	DAPAEDLARY	LSAVRHYINL	ITRQRY

Shaded amino acids are identical to human amino acids.

Table 25B.2 Distribution of NPY in the Human Brain (Mean ± SEM)

Area	pmol/g wet wt	n
Brodmann area 4	23.2 ± 3.4	4
Brodmann area 10	38.1 ± 3.4	20
Brodmann area 11	21.0 ± 3.6	5
Brodmann area 38	24.6 ± 1.7	8
Hippocampus	21.6 ± 1.0	6
Hypothalamus	18.4 ± 2.8	4
Substantia nigra	1.4 ± 0.4	5
Amygdala	27.0 ± 5.8	5
Lateral globus pallidus	20.8 ± 0.9	8
Nucleus accumbens	15.5 ± 1.6	5
Anterior thalamic nucleus	1.4 ± 0.3	4
Caudate nucleus	20.9 ± 3.4	8

Table 25B.3 Characterization of NPY Receptor Subtypes

	Y1	Y2	Y4	Y5	y6
Agonist order of potency	NPY ≥ PYY ≫ PP	NPY, PYY ≫ PP	PP > NPY, PYY	NPY ≥ PYY ≥ PP	NPY, PYY > PP
Selective agonists	[Leu31, Pro34]NPY [Pro34]NPY [Leu31, Pro334]PYY [Pro34]PYY	NPY 13−36 NPY 3−36 PYY 13−36 PYY 3−36	PP	[D-Trp34]NPY	
Selective antagonists	BIBO 3304 BIBP 3226 BMS 193885 PD 160170	BIIE 0246 CYM 9484 JNJ 5207787 SF 11	NPY ≥ PYY ≫ PP	CGP 71683 L-152804 LU AA33810 NPY 5RA972 NTNCB hydrochloride S 25585	
Signal transduction	$G_{i/o}$, adenylyl cyclase inhibition	$G_{i/o}$, adenylyl cyclase inhibition	$G_{i/o}$, adenylyl cyclase inhibition	$G_{i/o}$, adenylyl cyclase inhibition	adenylyl cyclase inhibition

Antagonists

Non-peptide low molecule weight antagonists that are selective for NPY receptor subunits are synthesized (Table 25B.3).

Biological Functions

Target Cells/Tissues and Functions

In the central nervous system, NPY induces food intake and decreases energy expenditure. NPY induces potent vasoconstriction. It also acts as a chemical mediator that controls the light−dark cycle entrainment of circadian rhythms. NPY modulates memory processes, depressive state, anxiety, stress, and seizure [6], and in the sympathetic nerve it stimulates catecholamine secretion.

Phenotype in Gene-Modified Animals

NPY-deficient or transgenic mice exhibit no effect regarding food intake. However, fasting-induced re-feeding is increased in NPY-deficient mice [7]. Y1- or Y5-deficient mice show late-onset obesity [7]. Mild seizures occur in NPY-deficient mice [8].

Pathophysiological Implications

Clinical Implications

Numerous findings suggest that selective antagonists for NPY receptor subtypes could be useful for the treatment of metabolic syndromes and obesity. However, no compound has been successful in clinical trials to date.

References

1. Tatemoto K. Neuropeptide Y: complete amino acid sequence of the brain peptide. *Proc Natl Acad Sci USA*. 1982;79:5485−5489.
2. Morris BJ. Neuronal localisation of neuropeptide Y gene expression in rat brain. *J Comp Neurol*. 1989;290:358−368.
3. Dawbarn D, Hunt SP, Emson PC. Neuropeptide Y: regional distribution chromatographic characterization and immunohistochemical demonstration in post-mortem human brain. *Brain Res*. 1984;296:168−173.
4. Grouzmann E, Comoy E, Bohuon C. Plasma neuropeptide Y concentrations in patients with neuroendocrine tumors. *J Clin Endocrinol Metab*. 1989;64:808−813.
5. Michel MC, Beck-Sickinger A, Cox H, et al. XVI International Union of Pharmacology recommendations for the nomenclature of neuropeptide Y, peptide YY, and pancreatic polypeptide receptors. *Pharmacol Rev*. 1998;50:143−150.
6. Pedrazzini T, Pralong F, Grouzmann E. Neuropeptide Y: the universal soldier. *Cell Mol Life Sci*. 2003;60:350−377.
7. Lin S, Boey D, Herzog H. NPY and Y receptors: lessons from transgenic and knockout models. *Neuropeptides*. 2004;38:189−200.
8. Erickson JC, Clegg KE, Palmiter RD. Sensitivity to leptin and susceptibility to seizures of mice lacking neuropeptide Y. *Nature*. 1996;381:415−418.

Supplemental Information Available on Companion Website

- Primary structure of the rat NPY gene/E-Figure 25B.1
- Amino acid sequences of the rat preproNPY/E-Figure 25B.2
- Accession numbers of NPY in various vertebrates/E-Table 25B.1

Peptide YY

Goro Katsuura and Akio Inui

Abbreviation: PYY
Additional names: peptide tyrosine tyrosine

Peptide YY is released postprandially from intestinal L cells as a satiety signal.

Discovery

In 1980, PYY was isolated and identified from the porcine intestine by means of unique chemical assay identifying the C-terminal amide structure [1,2].

Structure

Structural Features

Mature PYY is a linear peptide consisting of 36 aa residues in mammals and other non- mammalian vertebrates, and both C- and N-terminal residues are tyrosine (Y), which is responsible for its name (Table 25C.1). C-terminal Y is amidated. PYY is the member of neuropeptide Y (NPY) family, and its amino acid sequence similarity is ~70% with NPY, and 50% with pancreatic polypeptide (PP).

Primary Structure

The amino acid sequences of PYY are conserved in mammals and in non-mammalian species (Table 25C.1). Two major molecular forms of PYY circulating in blood are PYY1−36 and PYY3−36.

Properties

Porcine PYY: Mr 4,240.7. Readily soluble in water, acid solutions, and methanol.

Synthesis and Release

Gene, mRNA, and Precursor

The human PYY gene is localized on chromosome 17 (q21.1). The PYY gene transcript contains three exons. In the human, preproPYY has 97 aa residues, comprising a 28-aa signal peptide, a 36-aa mature PYY, and a 33-aa C-terminal peptide.

Distribution of mRNA

PYY mRNA is expressed in epithelial cells of the jejunum, cecum, and colon in rats [3]. In the rat brain, PYY mRNA is distributed in the rostral brainstem [4].

Tissue and Plasma Concentrations

Tissue: PYY is expressed in the early embryonic stages of mouse intestine, colon, and pancreas to regulate gastrointestinal and pancreatic functions during development.

Plasma: The main circulating form of PYY is PYY3−36. In the human, the basal plasma PYY level is 77 fmol/ml, and it is increased to 128 fmol/ml 1 hour after a meal.

Regulation of Synthesis and Release

A potential AP-1 binding site and several potential AP-2 binding sequences, which are activated by cAMP and phorbol ester, are located upstream of the transcriptional initiation site. PYY is synthesized in enteroendocrine L cells in the distal ileal, colonic, and rectal mucosae. In the pancreas, PYY is co-produced with glucagon and PP. In the lower intestine, PYY is also synthesized in the same cells with glucagon. Dietary fat strongly stimulates PYY secretion into the bloodstream [5], and plasma PYY levels increase significantly within 15−30 minutes after nutrient ingestion [6]. Gastric bypass surgery elevates plasma PYY levels [7]. Postprandial levels of PYY are lower in obese subjects compared with lean subjects [8].

Receptors

Structure and Subtype

Five distinct NPY receptor subtypes have been cloned, namely Y1, Y2, Y4, Y5, and y6. Among these receptors, PYY preferentially binds to the Y2 receptor and, to a lesser extent, to the Y1 and Y5 receptors.

Signal Transduction Pathway

The Y1, Y2, and Y5 receptor subtypes are seven transmembrane-spanning GPCRs (for details, see Chapter 25). They are associated with a member of the G_i and G_o family, and thus ligand binding results in inhibition of adenylyl cyclase and second messenger cAMP production.

Agonists and Antagonists

Agonists and antagonists for Y1, Y2, and Y5 are given in Subchapter 25B.

Biological Functions

Target Cells/Tissues and Functions

PYY inhibits stomach acid secretion, stomach emptying, and pancreatic exocrine secretion. PYY administration reduces appetite and weight gain in rodents and obese humans [9,10]. In contrast, administration of PYY into the central nervous

Y. Takei, H. Ando, & K. Tsutsui (Eds): Handbook of Hormones. DOI: http://dx.doi.org/10.1016/B978-0-12-801028-0.00167-7

Table 25C.1 Amino Acid Sequence of PYY in Different Vertebrate Species

Human	YPIKPEAPGE	DASPEELNRY	YASLRHYLNL	VTRQRY
Porcine	YPAKPEAPGE	DASPEELSRY	YASLRHYLNL	VTRQRY
Rat	YPAKPEAPGE	DASPEELSRY	YASLRHYLNL	VTRQRY
Mouse	YPAKPEAPGE	DASPEELSRY	YASLRHYLNL	VTRQRY
Cow	YPAKPQAPGE	HASPDELNRY	YTSLRHYLNL	VTRGRF
Pig	YPAKPEAPGE	DASPEELSRY	YASLRHYLNL	VTRGRY
Rabbit	YPSKPDNPGE	DAPAEDMARY	YSALRHYINL	ITRGRY
Toad	YPSKPDNPGE	DAPAEDMAKY	YSALRHYINL	ITRGRY
Xenopus	YPTKPENPGN	DASPEEMAKY	LTALRHYINL	VTRGRY
Salmon	YPPKPENPGE	DAPPEELAKY	YTALRHYINL	ITRGRY

Shaded amino acids are identical to human amino acids.

system has a potent stimulatory effect on food intake, just like NPY [8].

Phenotype in Gene-Modified Animals

PYY knockout mice have increased body weight and fat mass. Moreover, PYY knockout increases depression-like behavior but does not alter anxiety. Gene deletion of PYY receptor (Y2) results in increased food intake, body weight, and fat deposition in mice. The re-feeding response after starvation is strongly elevated in Y2 knockout mice.

Pathophysiological Implications

Clinical Implications

Intravenous administration of PYY at physiological levels induces anorectic effects in both normal and obese subjects [8].

References

1. Tatemoto K, Mutt V. Isolation of two novel candidate hormones using a chemical method for finding naturally occurring polypeptides. *Nature*. 1980;285:417−418.
2. Tatemoto K. Isolation and characterization of peptide YY (PYY), a candidate gut hormone that inhibits pancreatic exocrine secretion. *Proc Natl Acad Sci USA*. 1982;79:2514−2518.
3. Zhou J, Hegsted M, McCutcheon KL, et al. Peptide YY and proglucagon mRNA expression patterns and regulation in the gut. *Obesity*. 2006;14:683−689.
4. Pieribone VA, Brodin L, Friberg K, et al. Differential expression of mRNAs for neuropeptide Y-related peptides in rat nervous tissues: possible evolutionary conservation. *J Neurosci*. 1992;12:3361−3371.
5. Adrian TE, Ferri GL, Bacarese-Hamilton AJ, et al. Human distribution and release of a putative new gut hormone, peptide YY. *Gastroenterology*. 1985;89:1070−1077.
6. Greeley GH, Jemg YJ, Gomez G, et al. Evidence for regulation of peptide-YY release by the proximal gut. *Endocrinology*. 1989;124:1438−1443.
7. Moringo R, Moize V, Musri M, et al. Glucagon-like peptide-1, peptide YY, hunger, and satiety after gastric bypass surgery in morbidly obese subjects. *J Clin Endocrinol Metab*. 2006;91:1735−1740.
8. McGowan BMC, Bloom SR. Peptide YY and appetite control. *Curr Opin Pharmacol*. 2004;4:583−588.
9. Batterham RL, Cowley MA, Small CJ, et al. Gut hormone PYY(3-36) physiologically inhibits food intake. *Nature*. 2002;418:650−654.
10. Parkinson JR, Dhillo WS, Small CJ, et al. PYY3-36 injection in mice produces an acute anorexigenic effect followed by a delayed orexigenic effect not observed with other anorexigenic gut hormones. *Am J Physiol*. 2008;294:E698−E708.

Supplemental Information Available on Companion Website

- Primary structure of the human PYY gene/E-Figure 25C.1
- Amino acid sequences of human preproPYY/E-Figure 25C.2
- Accession numbers of PYY in various vertebrates/E-Table 25C.1

Parathyroid Gland, Ultimobranchial Gland, and Stannius Corpuscle Hormones

CHAPTER 26

Parathyroid Hormone Family

Nobuo Suzuki

History

Both parathyroid hormone (PTH) and parathyroid hormone-related protein (PTHrP) were discovered as hypercalcemic factors [1,2]. Removal of parathyroid glands in cats and dogs as well as rodents is associated with death from tetany. Because the infusion of calcium or oral administration of large doses of calcium can prevent tetany, the parathyroid gland was believed to have hypercalcemic factor(s). In 1925, it was reported that hot acid extracts of parathyroid gland completely relieved the tetany caused by experimental parathyroidectomy, indicating the presence of PTH in the parathyroid gland. PTH was then isolated from bovine parathyroid glands in 1970. Hypercalcemia in patients with cancer was first described in the 1920s. In the early 1980s, the close similarities between the biochemical features of humoral hypercalcemia of malignancy (HHM) syndrome and those of primary hyperparathyroidism were recognized. Using a sensitive biological assay that measured the cyclic AMP response in parathyroid hormone-responsive osteogenic sarcoma cells, PTHrP was isolated from a human lung cancer cell line in 1987.

Structure

Structural Features

PTH is a single-chain non-glycosylated peptide. In the human, PTH is synthesized as a 115-aa precursor polypeptide. The signal peptide is cleaved and the mature peptide is packed in secretory vesicles and then secreted as an 84-aa peptide. It was believed that PTH is absent in fish because amphibians were phylogenetically the first animals in which the parathyroid glands were discovered. However, cloning of the corresponding cDNAs established that two PTH genes are expressed in both the zebrafish and pufferfish [3]. Human PTHrP encodes a signal peptide of 36 aa residues and a mature protein of 141 aa residues. In teleosts, another PTHrPB gene which shares 67% homology with PTHrPA in pufferfish, *T. rubripes*, and a further gene PTH-like protein, PTH-L, have been identified (E-Figure 26.1) [3].

Molecular Evolution of Family Members

The parathyroid hormone family comprises 3 principal members, namely PTH, PTHrP, and the recently identified PTH-L (Figure 26.1). In teleosts, there are five genes that encode two PTHrPs, two PTHs, and a PTH-L, whereas tetrapods have three genes (PTHrP, PTH, and PTH-L), although placental mammals lack a PTH-L gene [4,5]. In cartilaginous fish (elephant shark), PTH1, PTH2, and PTHrp have recently been

determined [5], although questions remain unanswered regarding the classification of these hormones. In addition, immunohistochemical studies with heterologous antisera suggest that PTH-like peptide exists in invertebrates such as the snail, cockroach, and amphioxus [5]. Furthermore, a similar size of immunoreactive substance to rat PTH was detected in the ganglia of the snail. Thus, the vertebrate PTH/PTHrP family may have occurred prior to or early in the origin of vertebrates [5].

Receptors

Structure and Subtypes

Two receptors that belong to a class B G protein-coupled receptor have been identified. PTHrP and PTH share some sequence identities between amino acids 1 and 34 on the N-termini, which explains their ability to bind with equal affinity to a single common PTH/PTHrP receptor, PTH receptor type 1 (PTH1R) [3]. In contrast, type 2 PTH receptor (PTH2R) is not activated by PTHrP but by PTH and a tuberoinfundibular peptide of 39 amino acids (TIP39) [3,5]. TIP39 is known as a ligand of PTH2R [3,5], which exhibits only limited amino acid sequence homology with PTH (E-Figure 26.2). PTH2R is expressed primarily in the hypothalamus, but little is known about the possible functions of the PTH−PTH2R or TIP39−PTH2R systems [3]. In addition to PTH1R and PTH2R, zebrafish and other teleosts possess a third receptor, PTH3R, which may be a result of gene/genome duplication of PTH1R [5]. In some insects, PTHR-like sequences were reported [5], suggesting the existence of a similar PTH−PTHR system in invertebrates.

Signal Transduction Pathway

PTH binds to the PTH1R, which is abundantly expressed in kidneys and bones. The signaling pathways for this receptor are related to the G_s-linked cAMP production and $G_{q/11}$-dependent activation of phospholipase C and, subsequently, the calcium transients and protein kinase C activation [6]. It was initially thought that PTH and PTHrP bind and activate PTHR1 equally. However, biochemical and biophysical data have recently demonstrated that PTH and PTHrP bind to PTHR1 differently, and generate a different temporal pattern of downstream signals [7]. Portions of both peptides from amino acids 14 to 30 form an amphipathic helix, which binds to a specific cleft in the extracellular domain of the receptor. However, the α-helix formation in PTH is slightly longer than that in PTHrP. Thus, PTH binds more tightly to PTH1 receptor than PTHrP.

Y. Takei, H. Ando, & K. Tsutsui (Eds): Handbook of Hormones. DOI: http://dx.doi.org/10.1016/B978-0-12-801028-0.00026-X

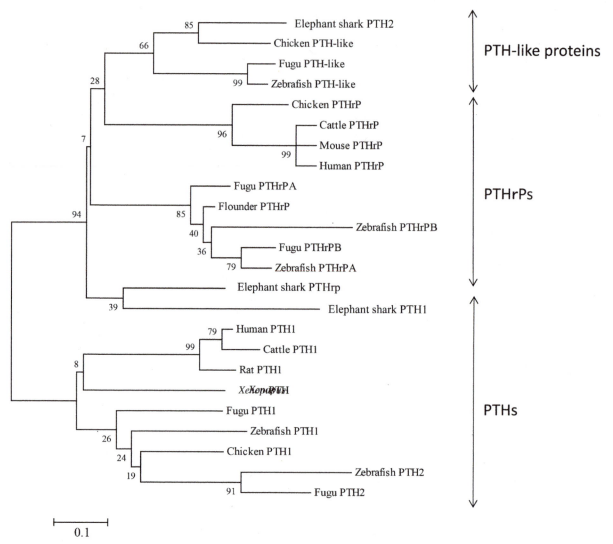

Figure 26.1 A phylogenetic tree showing the divergence of PTH, PTHrP, and PTH-like protein. A neighbor-joining tree with 100 bootstrap replicates. Bootstrap values are shown. The scale bar indicates an evolutionary distance of 0.1 amino acid substitutions per protein.

Biological Functions

Target Cells/Tissues and Functions

PTH is considered to be a mixed skeletal anabolic and catabolic agent. To induce bone formation, the preferable bone resorption performed by PTH may be needed. Therefore, PTH promotes bone resorption and then induces bone formation [8]. On the other hand, PTHrP is known to be a critical regulator of cellular and organ growth, development, migration, differentiation, survival, and epithelial calcium ion transport. In mammals, the placenta is one of the main sources of PTHrP during gestation [2]. Moreover, maternal milk contains high concentrations of PTHrP [2]; therefore, PTHrP has a greater importance for the fetus and newborn baby [2,9]. In addition, PTHrP, PTH, and the PTH1R receptor play important roles in endochondral bone development [9]. Mutations in PTH1R have been identified in Jansen metaphyseal chondrodysplasia, Blomstrand's lethal chondrodysplasia, and enchondromatosis. The signaling of PTHR1 has a critical role in both proliferation and differentiation of chondrocytes. PTH appears to play a paracrine role in fish, and an endocrine role in tetrapods [10]. This is suggested by the fact that parathyroid glands have not been found in fish. The high concentration of PTHrP in fish blood suggests that PTHrP has a classical endocrine function in fish; however, to date no endocrine gland producing PTHrP has been identified [3].

References

1. Potts Jr. JT. A short history of parathyroid hormone, its biological role, and pathophysiology of hormone excess. *J Clin Densitom.* 2013;16:4−7.
2. Martin TJ, Moseley JM, Gillespie MT. Parathyroid hormone-related protein: biochemistry and molecular biology. *Crit Rev Biochem Mol Biol.* 1991;26:377−395.
3. Guerreiro PM, Renfro JL, Power DM, et al. The parathyroid hormone family of peptides: structure, tissue distribution, regulation, and potential functional roles in calcium and phosphate balance in fish. *Am J Physiol.* 2007;292:R679−R696.
4. Pinheiro PLC, Cardoso JC, Gomes AS, et al. Gene structure, transcripts and calciotropic effects of the PTH family of peptides in *Xenopus* and chicken. *BMC Evol Biol.* 2010;10:373.
5. Pinheiro PLC, Cardoso JC, Power DM, et al. Functional characterization and evolution of PTH/PTHrP receptors: insights from the chicken. *BMC Evol Biol.* 2012;12:110.

6. Esbrit P, Alcaraz MJ. Current perspectives on parathyroid hormone (PTH) and PTH-related protein (PTHrP) as bone anabolic therapies. *Biochem Pharmacol*. 2013;85:1417−1423.

7. Wysolmerski JJ. Parathyroid hormone-related protein: an update. *J Clin Endocrinol Metab*. 2012;97:2947−2956.

8. Toulis KA, Anastasilakis AD, Polyzos SA, et al. Targeting the osteoblast: approved and experimental anabolic agents for the treatment of osteoporosis. *Hormone*. 2011;10:174−195.

9. Mannstadt M, Jüppner H, Gardella TJ. Receptors for PTH and PTHrP: their biological importance and functional properties. *Am J Physiol*. 1999;277:F665−F675.

10. McCauley LK, Martin TJ. Twenty-five years of PTHrP progress: From cancer hormone to multifunctional cytokine. *J Bone Miner Res*. 2012;27:1231−1239.

Supplemental Information Available on Companion Website

- Alignment of the amino acid sequences of the fugu PTH family/E-Figure 26.1
- Amino acid sequences of human TIP39/E-Figure 26.2
- Accession numbers of the PTH family peptides/E-Table 26.1

Parathyroid Hormone

Nobuo Suzuki

Abbreviation: PTH
Additional names: parathormone, parathyrin

The first hormone isolated from the parathyroid gland to have hypercalcemic action, PTH is a drug target for hypoparathyroidism and osteoporosis.

Discovery

The presence of PTH in the parathyroid gland was reported in 1925. It was isolated in 1970 from bovine parathyroid glands.

Structure

Structural Features

PTH is a single-chain non-glycosylated peptide. In humans, PTH is synthesized as a 115-aa precursor polypeptide. The signal peptide is cleaved, and the mature peptide is packed in secretory vesicles and then secreted as an 84-aa peptide (Figure 26A.1). Chicken PTH consists of 88 aa residues, with a striking similarity to the mammalian hormones in the N-terminus. In teleost zebrafish and pufferfish, two PTH genes and one PTH-like protein gene have been discovered [1]. In addition, elephant shark has PTH1 and PTH2 [2], although questions are left unanswered about the classification of these hormones in the PTH/PTHrP family.

Primary Structure

The mature PTH in mammals comprises an 84-aa peptide, the sequence of which varies in non-mammalian vertebrates (E-Figure 26A.1). Among teleost fish, humans, and chicken, the amino acid identity of mature PTHs is around 20—25% and the sequence similarity is close to 40% [1]. However, high sequence conservation is observed among the first 34 N-terminal amino acid residues of all PTHs (E-Figure 26A.2).

Properties

Mammalian PTHs: Mr 9,500. Soluble in water and stable in solution at pH 4.5.

Synthesis and Release

Gene and mRNA

The human *PTH* gene is located on chromosome 11 (11p5.3—p15.1). The human *PTH* and chicken *pth* genes comprise two introns that divide the gene into three exons, encoding the 5′-untranslated region (UTR), the signal peptide, and the mature peptide and the 3′-UTR (E-Figure 26A.3) [1]. The *pth* gene structures of the recently identified pufferfish are similar to those of tetrapods, which include three exons divided by two introns (E-Figure 26A.3 [1]).

Distribution of mRNA

PTH mRNA is detected most abundantly in the parathyroid gland of terrestrial vertebrates. In zebrafish, PTH1 and PTH2 mRNAs were detected in the neuromasts of the lateral line and central nervous system during embryogenesis [1]. In the elephant shark, PTH1 is expressed in the eye, heart, ovary, and uterus, while PTH2 is expressed in the spleen and brain [2].

Plasma Concentration

The half-life of endogenous PTH is extremely short (2—4 minutes), and therefore plasma PTH levels vary according to the peptide moiety selected for measurement. Two monoclonal antibodies against the first six N-terminal residues and against the C-terminal region (39—84) have been developed [3]. By measuring PTH levels using these assays, the normal values in healthy human subjects were 7—36 pg/ml [3]. In non-mammalian vertebrates, mean concentrations of PTH in the chickens given a low-calcium diet (12.85 pg/ml) were more than twice as high as in the control chickens (5.95 pg/ml) [4].

Regulation of Synthesis and Release

Extracellular calcium is the primary regulator of PTH secretion from the parathyroid gland. In mammals and birds, the circulating levels of PTH are greatly altered by minute changes in the blood calcium concentration [1]. PTH secretion is modulated by the action of the calcium-sensing receptor (CaSR), which is located in the parathyroid gland [1]. Under hypercalcemic conditions, Ca^{2+} binding to CaSR inhibits PTH secretion [1]; conversely, under hypocalcemic conditions the receptor is not occupied by extracellular Ca^{2+}, resulting in PTH secretion [1]. In addition, $1,25(OH)_2D_3$ acts directly on the parathyroid gland and suppresses the transcription rate of the PTH gene [5].

Receptors

Structure and Subtype

The organization of the PTH1 receptor (PTH1R) gene is highly homologous in three mammalian species, namely rat, human, and mouse. This gene extends over 22 kb and contains at least 15 exons and 14 introns. The human gene was mapped to chromosome 3p22—p21.1 Both PTH and parathyroid hormone-related protein (PTHrP) bind to the

Y. Takei, H. Ando, & K. Tsutsui (Eds): Handbook of Hormones. DOI: http://dx.doi.org/10.1016/B978-0-12-801028-0.00168-9

```
         10          20          30
MIPAKDMAKV MIVMLAICFL TKSDGKSVKK
         40          50          60
RSVSEIQLMH NLGKHLNSME RVEWLRKKLQ
         70          80          90
DVHNFVALGA PLAPRDAGSQ RPRKKEDNVL
        100         110
VESHEKSLGE ADKADVNVLT KAKSQ
```

Figure 26.1 Sequences of the precursor polypeptides in the human PTH. Mature aa residues (84) are underlined.

PTH1R [1]. As both peptides signal through the same receptor, it is also called the PTH/PTHrP receptor. In contrast, type 2 PTH receptor (PTH2R) is not activated by PTHrP but by PTH and the tuberoinfundibular peptide of 39 amino acids (TIP39) (E-Figure 26A.4) [1]. TIP39 is known as a ligand of PTH2R, although similarity of sequence between TIP39 and PTH is low. In addition to PTH1R and PTH2R, zebrafish and other teleosts possess a third receptor, PTH3R, which may be derived by the duplication of PTH1R [1]. PTHRs share the same basic structure (seven transmembrane domains) with that of the class B G protein-coupled receptor (GPCR) super-family. A defining feature of the class B receptors is the relatively long extracellular N-terminal domain, which comprises approximately 150 amino acids and is important for ligand binding [6]. The presence of six cysteine residues that are strictly conserved within the N-terminal domain of all class B GPCRs suggests that three disulfide linkages are present and are of critical importance [6].

Signal Transduction Pathway

Binding of PTH to PTH1R triggers the signaling pathways related to G_s-linked cAMP production and $G_{q/11}$-dependent activation of phospholipase C (PLC) and, subsequently, the calcium transients and protein kinase C activation [7].

Agonists

The complete molecule is not essential to exert a biological effect. The N-terminal fragments of PTH and PTHrP, comprising amino acids 1−34, have biological activities similar to those of the complete molecules [6]. PTH (1−14) and PTH (1−21) analogs show high-affinity binding and efficient activation of PTH1R [6].

Antagonists

[Leu[11], D-Trp[12]] PTHrP (7−34) and [Ile[5], Trp[23]] PTHrP (5−36) bind with high affinity but are devoid of signaling activity [6]. The PTH2R agonist, TIP39, binds efficiently to the PTH1R but does not activate it, and thus functions as a weak antagonist at this receptor [6].

Biological Functions

Target Cells/Tissues and Functions

PTH1R is expressed at high levels in bones and kidneys, where it mediates the classical effects of PTH and PTHrP on Ca^{2+} and P_i homeostasis. In mammals, PTH1R is located in cells of the osteoblast lineage, where it increases osteoclast formation and activity *via* the receptor activator of the NF-κB ligand (RANKL) and its receptor RANK pathways [7]. In a goldfish scale *in vitro* assay system, the fugu PTH1 (1−34) acts on osteoblasts and increases osteoclastic activity [8], although to date no PTH-producing gland has been identified in fish [1]. In this assay system, both RANK and RANKL mRNA expression increased significantly in osteoclasts that were

treated with fugu PTH1 [8]. Furthermore, removal of the parathyroid gland from various amphibians and reptiles, except salamanders and turtles, causes a decline in blood calcium levels and causes tetanic convulsions. In contrast, PTH2R is expressed primarily in the hypothalamus, but little is known about the possible function of the PTH−PTH2R system. It has been suggested that it is involved with the regulation of growth hormone secretion, release of arginine vasopressin, and cardiovascular and renal hemodynamics [1].

Phenotype in Gene-Modified Animals

Analysis of PTH-null (PTH[−/−]) mice shows that the absence of PTH causes mean erythrocyte volume and reticulocyte counts to increase, while decreasing erythrocyte counts, hemoglobin, hematocrit, and the mean corpuscular hemoglobin concentration [9]. These changes are accompanied by increases in erythrocyte cation content, a denser cell population, and increased K^+ permeability [9]. In addition, erythrocyte osmotic fragility is enhanced in PTH[−/−] mice compared with that of wild-type mice [9]. Several changes in the erythrocyte parameters of the PTH[−/−] mice were rescued by the deletion of the CaSR gene in the background of PTH[−/−] mice [9]. This suggests that CaSR and PTH are functionally coupled to maintain erythrocyte homeostasis. Activating mutations in the PTH1R receptor result in Jansen-type metaphyseal chondrodysplasia [6]. In the growth plate of patients, constitutive PTH1R signaling, which mimics the actions of PTHrP, leads to chondrodysplasia with abnormal growth plate elongation and some endochondral remnants in the diaphyses. Inactivating mutations in the PTH1R result in a perinatal lethal disorder called Blomstrand-type lethal chondrodysplasia [6]. This disorder is characterized by short limbs, advanced dentition, and osteosclerosis — i.e., phenotypic changes that are similar to those seen in mice with homozygous ablation of either PTHrP or PTH1R.

Pathophysiological Implications

Clinical Implications

Plasma PTH measurement is integral to the diagnosis and management of hypoparathyroidism and hyperparathyroidism. However, PTH is relatively unstable and hydrolyzes easily in storage. It is recommended that blood samples for PTH measurement should be taken using tubes containing EDTA, and plasma should be separated within 24 hours [10].

Use for Diagnosis and Treatment

Plasma PTH concentration is routinely measured to diagnose a variety of calcium metabolic disorders, such as hyperparathyroidism, hypoparathyroidism, hypercalcemia of malignancy, and chronic renal failure. The skeletal actions of PTH are dependent on the pattern of its administration [7]. Intermittent exposure to PTH leads to a net increase in bone formation, whereas its continuous administration produces bone loss. Thus, the bone anabolic capability of PTH strictly depends upon its administration producing a transient peak level in plasma [7]. PTH(1−34) is used for osteoporosis therapy because of its ability to increase osteoblastogenesis and osteoblast survival [7]. Furthermore, PTH is involved in proliferation and differentiation of chondrocytes, because mutations in PTH1R have been identified in Jansen-type metaphyseal chondrodysplasia and Blomstrand-type lethal chondrodysplasia.

References

1. Guerreiro PM, Renfro JL, Power DM, et al. The parathyroid hormone family of peptides: structure, tissue distribution, regulation, and potential functional roles in calcium and phosphate balance in fish. *Am J Physiol.* 2007;292:R679−R696.
2. Liu Y, Ibrahim AS, Tay BH, et al. Parathyroid hormone gene family in a cartilaginous fish, the elephant shark (*Callorhinchus milii*). *J Bone Miner Res.* 2010;25:2613−2623.
3. Gannagé-Yared MH, Farès C, Ibrahim T, et al. Comparison between a second and a third generation parathyroid hormone assay in hemodialysis patients. *Metabolism.* 2013;62:1416−1422.
4. Singh R, Joyner CJ, Peddie MJ, et al. Changes in the concentrations of parathyroid hormone and ionic calcium in the plasma of laying hens during the egg cycle in relation to dietary deficiencies of calcium and vitamin D. *Gen Comp Endocrinol.* 1986;61:20−28.
5. Kumar R, Thompson JR. The regulation of parathyroid hormone secretion and synthesis. *J Am Soc Nephrol.* 2011;22:216−224.
6. Gensure RC, Gardella TJ, Jüppner H. Parathyroid hormone and parathyroid hormone-related peptide, and their receptors. *Biochem Biophys Res Commun.* 2005;328:666−678.
7. Esbrit P, Alcaraz MJ. Current perspectives on parathyroid hormone (PTH) and PTH-related protein (PTHrP) as bone anabolic therapies. *Biochem Pharmacol.* 2013;85:1417−1423.
8. Suzuki N, Danks JA, Maruyama Y, et al. Parathyroid hormone 1 (1−34) acts on the scales and involves calcium metabolism in goldfish. *Bone.* 2011;48:1186−1193.
9. Romero JR, Youte R, Brown RM, et al. Parathyroid hormone ablation alters erythrocyte parameters that are rescued by calcium-sensing receptor gene deletion. *Eur J Haematol.* 2013;91:37−45.
10. Hanon EA, Sturgeon CM, Lamb EJ, et al. Sampling and storage conditions influencing the measurement of parathyroid hormone in blood samples: a systematic review. *Clin Chem Lab Med.* 2013;51:1925−1941.

Supplemental Information Available on Companion Website

- Alignment of the amino acid sequences of PTHs/E-Figure 26A.1
- Comparison of amino acid sequences among N-terminus 1−34 PTHs/E-Figure 26A.2
- Gene structures of PTH in fugu, chicken, and human/E-Figure 26A.3
- Amino acid sequences of human TIP39/E-Figure 26A.4
- Accession numbers of vertebrate PTH and PTH-like protein/E-Table 26A.1
- Accession numbers of vertebrate PTHRs/E-Table 26A.2

Parathyroid Hormone-Related Protein

Nobuo Suzuki

Additional names/abbreviation: parathyroid hormone-related peptide, adenylate cyclase-stimulating protein /PTHrP

The first hormone isolated as a hypercalcemic factor involved in humoral hypercalcemia of malignancy, PTHrP is a drug target for osteoporosis.

Discovery

The existence of a PTH-like factor in cancers has been reported. It was isolated in 1987 from a human lung cancer cell line.

Structure

Structural Features

Human PTHrP encodes a 36-aa signal peptide and a 141-aa mature protein (Figure 26B.1). In teleosts, another PTHrPB gene that shares 67% homology with PTHrPA in the pufferfish *Takifugu rubripes* has been found (E-Figure 26B.1) [1]. The mature PTHrPAs of the pufferfish, the sea bream, and the European flounder comprise 126, 125, and 129 aa, respectively, and share approximately 90% sequence similarity [1]. Although the total amino acid identity among teleost and mammalian PTHrPs is low ($\sim$36%), it is higher ($\sim$60%) in the N-terminal region [1].

Primary Structure

Eight of the first 13 residues of the human PTHrP are identical to those in PTH (E-Figure 26B.2). In PTHrP as well as PTH, the first 34 amino acids have the structural requirements for full PTHrP biological activity [2]. Comparison of the C-terminal region of PTHrPs reveals it to be much shorter in fish than in tetrapods (E-Figure 26B.1) [1].

Properties

Mammalian PTHrPs: Mr $\sim$17 kDa. Soluble in water, and stable in solution at pH 4.8.

Synthesis and Release

Gene and mRNA

The human PTHrP gene is located on chromosome 12 (12p12.1−p11.2). The human PTHrP gene has eight exons and seven introns, and produces several splice variants encoding mature proteins with 139, 141, and 173 amino acids (E-Figure 26B.3) [3]. In chicken, two different mRNA isoforms encode mature proteins with 139 and 141 amino acids (E-Figure 26B.3), while the mouse and rat genes encode only a single isoform [3]. On the other hand, there is a high similarity of the organization of pufferfish pthrp gene with human PTH, chicken pth, and pufferfish pth genes, having three exons spanning a 2.25-kb region (E-Figure 26B.3) [1,3].

Distribution of mRNA

After PTHrP was identified in tumor tissues, the first normal tissue shown to contain PTHrP protein and mRNA was the lactating mammary gland. In addition, the following tissues express PTHrP: cardiovascular system (endothelial, smooth, and cardiac muscle), connective tissues (fetal cartilarge and bone), urogenital system (kidney and urinary bladder), reproductive tissues (testes and ovary), uteroplacental tissues (amnion, chorion, and placenta), central nervous system (brain and pituitary gland), immune system (lymphocytes and T lymphocytes), and keratinocytes [3]. Also, in teleosts, PTHrP is widely expressed in various tissues [1]. In particular, mRNA expression levels in the muscle of sea bream are approximately 20-fold higher than those in the skin, brain, kidney, gill, or duodenum, and approximately eight-, four-, and two-fold higher than those in the pituitary, heart, and hindgut, respectively [1].

Tissue and Plasma Concentrations

Circulating levels of PTHrP in healthy humans are very low, although PTHrP levels can easily be detected in 80−100% of patients with humoral hypercalcemia of malignancy (HHM, around 2.8 to 51.2 pmol/l). Possible functions of PTHrP in the fetus and early development are supported by the tissue and plasma concentrations shown in Table 26B.1 [1,4−8].

In mammals, the placenta is the one of the main sources of PTHrP during gestation because placenta, amnion over placenta, and choriodecidua released PTHrP (3.7 ± 0.5, 139.3 ± 43.1, and 65.7 ± 20.8 pmol/g protein, respectively) using late pregnant human gestational tissues in an *in vitro* explant system [6]. In addition, maternal milk has been shown to contain high concentrations of PTHrP [5]. Meanwhile, the high concentrations of PTHrP in fish plasma suggest that PTHrP has a classical endocrine function in fish [1].

Y. Takei, H. Ando, & K. Tsutsui (Eds): Handbook of Hormones. DOI: http://dx.doi.org/10.1016/B978-0-12-801028-0.00169-0

```
           10         20         30         40         50
        MQRRLVQQWS VAVFLLSYAV PSCGRSVEGL SRRLKRAVSE HQLLHDKGKS
           60         70         80         90        100
        IQDLRRRFFL HHLIAEIHTA EIRATSEVSP NSKPSPNTKN HPVRFGSDDE
          110        120        130        140        150
        GRYLTQETNK VETYKEQPLK TPGKKKKGKP GKRKEQEKKK RRTRSAWLDS
          160        170
        GVTGSGLEGD HLSDTSTTSL ELDSRRH
```

Figure 26B.1 Sequence of precursor polypeptides in human PTHrP. Mature amino acid residues (141) are underlined.

Table 26B.1 PTHrP Concentration in Plasma and Tissues

Species	Plasma or Tissues	Concentration	Reference
Human	Patients with humoral hypercalcemia	Plasma, 2.8–51.2 pmol/l	[4]
Human	Milk	100 ng/ml	[5]
Human	Maternal plasma	2.7 ± 0.4 pmol/l	[6]
	Umbilical artery	6.6 ± 2.0 pmol/l	
	Umbilical vein	3.2 ± 0.6 pmol/l	
	Amniotic fluid	41.8 ± 3.8 pmol/l	
Cow	Periparturient period	Plasma, not detectable	[7]
		Milk, 14.9–41.2 nmol/l	
Flounder	Plasma	0.4–1.2 nmol/l	[1]
Sea bream	Pituitary	37.7 ± 6.1 ng/g	[8]
	Esophagus	2.3 ± 0.7 ng/g	

Regulation of Synthesis and Release

Increasing Ca^{2+} concentrations stimulate PTHrP secretion via CaSR in a number of normal and cancerous cell types [1]. CaSR is important in the modulation of PTHrP secretion for maintenance of placental–fetal Ca^{2+} transportation [1]. In addition, the PTHrP gene contains consensus regulatory motifs for cAMP, 1,25-dihydroxy vitamin D_3 $(1,25(OH)_2D_3)$, and glucocorticoids [5]. Dexamethasone, vitamin D, cortisol, estrogens, and androgens have been shown to inhibit PTHrP production *in vitro* [1].

Receptors

Structure and Subtype

PTH1R, but not PTH2R, is activated by PTHrP. The organization of the PTH1R gene is conserved in three mammalian species: rat, human, and mouse. As described in the PTH section, PTH1R interacts with the first 34 amino acids of the N-termini of both PTH and PTHrP [1]. Zebrafish and other teleosts possess a third receptor, PTH3R, which may be a result of duplication of PTH1R [1].

Signal Transduction Pathway

PTHrP and PTH bind mammalian PTH1R with an indistinguishable affinity and stimulate cAMP and inositol 1,4,5-trisphosphate accumulation with equivalent efficacy. However, it is now known that PTHrP and PTH generate different temporal patterns of downstream signals [9]. The crystal structures of PTHrP (1–36) and PTH (1–34) that are bound to the extracellular domain of PTHR1 showed that PTH binds more tightly to the receptor than PTHrP, perhaps explaining why PTH can maintain the receptor in a more active conformation for longer periods of time [9].

Agonists

PTHrP (1–36) as well as PTH (1–34) is thought to interact with the PTH1R through distinct binding and activating domains. PTH (1–14) and PTH (1–21) analogs show high-affinity binding and efficient activation of the PTH1R [10].

Antagonists

[Leu^{11}, D-Trp^{12}] PTHrP (7–34) and [Ile^5, Trp^{23}] PTHrP (5–36) bind with high affinity but are devoid of signaling activity [10]. Substitution or modification of the second valine in PTH (1–34) or PTHrP (1–36) with a bulky amino acid or an amino acid derivative (Trp, Arg, or L-p-benzoylphenylalanine) also results in the generation of potent antagonists [10].

Biological Functions

Target Cells/Tissues and Functions

PTHrP is known to be a critical regulator of cellular and organ growth, development, migration, differentiation, survival, and epithelial calcium ion transport [11]. In mammals, judging from the distribution and expression of mRNA, PTHrP has a greater importance in early development, which determines the rate of chondrogenesis and structure mineralization [3,5,6]. In teleosts, on the other hand, possible involvement of PTHrP in the process of expansion and relaxation of the organs has been suggested [1]. In terrestrial vertebrates, PTHrP acts as a paracrine or autocrine regulator. While high concentrations (0.4–1.2 nM) in the blood of fish predict a classical endocrine function, to date no PTHrP-producing gland has been clearly identified in fish [1].

Phenotype in Gene-Modified Animals

The critical role of PTHrP in daily life has been demonstrated using PTHrP- or PTH1R-knockout mice, which die at birth or *in utero*. These experiments have suggested that PTHrP has a PTH-like role in bone formation [12]. In preclinical models, PTHrP$^{-/-}$ mice died postnatally, probably from asphyxia, and had widespread abnormalities in endochondral bone development. PTHrP$^{+/-}$ mice exhibit osteopenia characterized by altered trabecular architecture within 3 months of age. However, daily administration of PTH (1–34) to PTHrP$^{+/-}$ mice improved all the parameters of skeletal microarchitecture.

Pathophysiological Implications

Clinical Implications

Plasma PTHrP levels have been used for the differential diagnosis of humoral HHM based on a secreted humoral bone resorption factor and a judgment of its curative effect from various tumors [9,11].

Use for Diagnosis and Treatment

Plasma PTHrP concentration is routinely measured to diagnose malignant tumors (including lung, esophageal, oral, head and neck, uterine, breast, pancreatic, ovarian, hepatocellular, kidney, and bladder cancers). It has been noted that specimens for PTHrP testing are collected according to the requirements given by the referral laboratory (collection into chilled EDTA tubes with or without protease inhibitors with subsequent freezing of the plasma) [13].

References

1. Guerreiro PM, Renfro JL, Power DM, et al. The parathyroid hormone family of peptides: structure, tissue distribution, regulation, and potential functional roles in calcium and phosphate balance in fish. *Am J Physiol.* 2007;292:R679–R696.
2. McCauley LK, Martin TJ. Twenty-five years of PTHrP progress: from cancer hormone to multifunctional cytokine. *J Bone Miner Res.* 2012;27:1231–1239.
3. Ingleton PM, Danks JA. Distribution and functions of parathyroid hormone-related protein in vertebrate cells. *Int Rev Cytol.* 1996;166:231–280.
4. Grill V, Ho P, Body JJ, et al. Parathyroid hormone-related protein: elevated levels in both humoral hypercalcemia of malignancy and hypercalcemia complicating metastatic breast cancer. *J Clin Endocrinol Metab.* 1991;73:1309–1315.
5. Martin TJ, Moseley JM, Gillespie MT. Parathyroid hormone-related protein: biochemistry and molecular biology. *Crit Rev Biochem Mol Biol.* 1991;26:377–395.
6. Farrugia W, Ho PW, Rice GE, et al. Parathyroid hormone-related protein (1–34) in gestational fluids and release from human gestational tissues. *J Endocrinol.* 2000;165:657–662.
7. Onda K, Sato A, Yamaguchi M, et al. Parathyroid hormone-related protein (PTHrP) and Ca levels in the milk of lactating cows. *J Vet Med Sci.* 2006;68:709–713.
8. Rotllant J, Worthington GP, Fuentes J, et al. Determination of tissue and plasma concentrations of PTHrP in fish: development and validation of a radioimmunoassay using a teleost 1–34 N-terminal peptide. *Gen Comp Endocrinol.* 2003;133:146–153.
9. Wysolmerski JJ. Parathyroid hormone-related protein: an update. *J Clin Endocrinol Metab.* 2012;97:2947–2956.
10. Gensure RC, Gardella TJ, Jüppner H. Parathyroid hormone and parathyroid hormone-related peptide, and their receptors. *Biochem Biophys Res Commun.* 2005;328:666–678.
11. Fiaschi-Taesch NM, Stewart AF. Parathyroid hormone-related protein as an intracrine factor – trafficking mechanisms and functional consequences. *Endocrinology.* 2003;144:407–411.
12. Toulis KA, Anastasilakis AD, Polyzos SA, et al. Targeting the osteoblast: approved and experimental anabolic agents for the treatment of osteoporosis. *Hormones.* 2011;10:174–195.
13. Fritchie K, Zedek D, Grenache DG. The clinical utility of parathyroid hormone-related peptide in the assessment of hypercalcemia. *Clin Chim Acta.* 2009;402:146–149.
14. Pinheiro PLC, Cardoso JC, Power DM, et al. Functional characterization and evolution of PTH/PTHrP receptors: insights from the chicken. *BMC Evol Biol.* 2012;12:110.

Supplemental Information Available on Companion Website

- Alignment of PTHrP amino acid sequences from the fugu PTHrPA, zebrafish PTHrPA, flounder PTHrP, fugu PTHrPB, zebrafish PTHrPB, and human PTHrP/E-Figure 26B.1
- Comparison of amino acid sequences between human PTHrP and human PTH1/E-Figure 26B.2
- Gene structures of PTHrP and PTH in fugu, chicken, and human/E-Figure 26B.3
- Accession numbers of vertebrate PTHrP/E-Table 26B.1
- Accession numbers of vertebrate PTH1R/E-Table 26B.2 [14]

CHAPTER 27

Calcitonin/Calcitonin Gene-Related Peptide Family

Nobuo Suzuki

History

The calcitonin (CT)/calcitonin gene-related peptide (CGRP) family belongs to a multigenic family that includes amylin (AMY) and adrenomedullin (AM). The details of these hormones are described in Subchapters 27F and 27D, respectively. CT, a 32-aa peptide, was isolated from porcine thyroid gland in 1968, and CGRP, a 37-aa neuropeptide, was discovered by molecular cloning of the CT gene in 1983. The tissue specificity of this transcript is regulated by alternative splicing of the CT gene. CT precursor mRNA is synthesized in thyroidal C cells; CGRP precursor mRNA is synthesized in neural tissues. AMY, a 37-aa peptide, was first isolated from an insulinoma in 1987 and was thus initially named "insulinoma amyloid peptide," later "diabetes-associated peptide," and finally "islet amyloid polypeptide" or "AMY." AM was discovered in 1993 in human pheochromocytoma tissue, and later found to be a circulating hormone. Thereafter, two novel AM-related peptides, named AM2 (or intermedin) and AM5, were identified during genomic screening in mammals as well as in fish.

Structure

Structural Features

A common structural feature of the family members is a conserved intramolecular ring consisting of six or seven aa residues and an amidated C-terminus (Figure 27.1).

Molecular Evolution of Family Members and Peptides

The presence of CT-like peptides in invertebrates such as mollusks and arthropods has recently been reported [1]. Therefore, the ancestral molecule of the CT/CGRP peptide family may have appeared before the divergence of protostomes and deuterostomes [2]. In animal groups, including lophotrochozoa (mollusks) and ecdysozoa (arthropods), CT-like molecules function in ionic regulation [2]. Furthermore, CT/CGRP-like receptors identified in *Drosophila melanogaster* possess a signaling pathway similar to that of the mammalian CGRP receptor (CT-like receptor, CLR) [2]. The mature CT sequences of non-mammalian vertebrates such as teleosts, urodeles, reptiles, and birds are considerably conserved. In mammals, although the primary structure is changed, a C-terminal amidated proline and an N-terminal circular structure formed by a disulfide bridge between Cys 1 and Cys 7 are

conserved [3]. CTs of salmons and porcines share 41% identity at the amino acid level. In contrast, CGRP is conserved among vertebrates. The aa sequence of flounder CGRP is more conserved than that of CT among vertebrates, and share 78%, 78%, 78%, 81%, and $73 \sim 78\%$ identity with salmon, cod, frog, chicken, and mammalian CGRP, respectively. Similarly to CGRP, the aa sequence of AMY is also conserved. Goldfish AMY shares 71%, 78%, 89%, and 91% identity with human, rat, salmon, and pufferfish AMY, respectively. In fish, AM has diversified, and five AM paralogs (AM 1–5) form an independent subfamily [4]. Based on this discovery in fishes, AM2 and AM5 have also been found in mammals [4].

Receptors

Structure and Subtypes

The CT peptide family comprises CT, AMY, CGRP, AM1, AM2, and AM5. The receptors are divided into two subtypes: calcitonin receptor (CTR), which binds CT, and CTR-like receptor (CLR), for which the endogenous ligand is unknown. Receptor activity modifying proteins (RAMPs) are a family of single transmembrane proteins that form heterodimers with G protein-coupled receptors, which are divided into three subtypes: RAMP1, RAMP2, and RAMP3. RAMPs are involved in the regulation of glycosylation and trafficking of receptors, and may drive the pharmacological studies of CLR (E-Figure 27.1) [5]. CGRP is able to activate CLR when this receptor is co-expressed with RAMP1. In contrast, when CLR interacts with RAMP2 or RAMP3, the stable complex formed is not activated by CGRP but by another related peptide, AM. Interaction of CLR with RAMP2 creates an AM_1 receptor phenotype, whereas its interaction with RAMP3 generates an AM_2 receptor phenotype. cAMP production in the AM_1 receptor is higher than that in the AM_2 receptor, suggesting that the AM_1 receptor is the primary receptor for AM [4]. In the absence of RAMPs, CTR binds to CT. CTR exhibits three different subtypes of AMY receptors when it is co-expressed with the three RAMPs (E-Figure 27.1) [5]. The interplay between two receptor subtypes, three RAMPs, and various peptide ligands generates various pharmacological receptor patterns.

Signal Transduction Pathway

The most important pathway for the CT family of peptides is cAMP-mediated signal transduction. CTR can also activate the phospholipase C (PLC) enzyme pathway [5]. The PLC

Y. Takei, H. Ando, & K. Tsutsui (Eds): Handbook of Hormones. DOI: http://dx.doi.org/10.1016/B978-0-12-801028-0.00027-1

```
Human CT                            CGNLSTCMLGTYTQDFNKFHTFPQTAIGVGAP-HN2
Human CGRP                          ACDTATCVTHRLAGLLSRSGGVVKNNFVPTNVGSKAF-NH2
Human AMY                           KCNTATCATQRLANFLVHSSNNFGAILSSTNVGSNTY-NH2
Human AM1         YRQSMNNFQGLRSFGCRFGTCTVQKLAHQIYQFTDKDKDNVAPRSKISPQGY-NH2
Human AM2            TQAQLLRVGCVLGTCQVQNLSHRLWQLMGPAGRQDSAPVDPSSPHSY-NH2
Porcine AM5         HQVSLKSGRLCSLGTCQTHRLPEIIYWLRSASTKELSGKAGRKPQDPYSY-NH2
```

Figure 27.1 **Amino acid sequences of CT, CGRP, AMY, AM1, AM2, and AM5.** The rectangles represent cysteine residues. Identical amino acids among three kinds of sequences are indicated by shading.

pathway, along with the cAMP pathway, can be initiated by the coupling of receptors to multiple G proteins. Calmodulin (CaM) is a ubiquitous calcium-binding protein that regulates many intracellular proteins, such as cytoskeletal elements, ion channels, kinases, or phosphatases, and various enzymes involved in G protein-coupled receptor signaling [5]. An interaction between CaM and CTR was observed by the analysis of a glutathione S-transferase (GST) pulldown assay with GST-CaM [5]. The CLR for CGRP and AM requires an intracellular peripheral membrane protein named CGRP-receptor component protein (RCP) for signal transduction. RCP is required for CGRP and AM receptor signaling [6].

Biological Functions

Target Cells/Tissues and Functions

CT is derived from the C cells of the thyroid gland in mammals or the ultimobranchial gland in non-mammalian vertebrates. CT reduces bone resorption by inhibiting multinucleated osteoclast (active type of osteoclasts) formation and increases renal calcium excretion. Therefore, it is used as a drug in the management of postmenopausal osteoporosis, Paget's disease of bone, Sudeck's atrophy, and malignancy-associated hypercalcemia [7]. CGRP and its receptors are widely distributed in the body, and it is a major regulator of vascular tone and cardiovascular development [8]. AMY is primarily expressed in pancreatic islet β cells and plays a physiological role in glucose regulation [9]. In certain species, AMY is linked to β-cell death and type 2 diabetes [10]. AM as well as CGRP have several physiological effects, including vasodilation through the CLR [8]. CT, CGRP, AMY, AM1, AM2, and AM5 have overlapping biological effects because of their similar structures and cross-reactivity among receptors.

References

1. Rowe ML, Achhala S, Elphick MR. Neuropeptides and polypeptide hormones in echinoderms: new insights from analysis of the transcriptome of the sea cucumber *Apostichopus japonicus*. *Gen Comp Endocrinol*. 2014;197:43–55.
2. Lafont AG, Dufour S, Fouchereau-Peron M. Evolution of the CT/CGRP family: comparative study with new data from models of teleosts, the eel, and cephalopod molluscs, the cuttlefish and the nautilus. *Gen Comp Endocrinol*. 2007;153:155–169.
3. Sasayama Y. Hormonal control of Ca homeostasis in lower vertebrates: considering the evolution. *Zool Sci*. 1999;16:857–869.
4. Takei Y, Ogoshi M, Nobata S. Exploring new CGRP family peptides and their receptors in vertebrates. *Curr Prot Pept Sci*. 2013;14:282–293.
5. Couvineau A, Laburthe M. The family B1 GPCR: structural aspects and interaction with accessory proteins. *Curr Drug Targets*. 2012;13:103–105.
6. Dickerson IM. Role of CGRP-receptor component protein (RCP) in CLR/RAMP function. *Curr Protein Pept Sci*. 2013;14:407–415.
7. Wimalawansa SJ. Amylin, calcitonin gene-related peptide, calcitonin, and adrenomedullin: a peptide superfamily. *Crit Rev Neurobiol*. 1997;11:167–239.
8. Chang CL, Hsu SYT. Roles of CLR/RAMP receptor signaling in reproduction and development. *Curr Protein Pept Sci*. 2013;14:393–406.
9. Westermark P, Andersson A, Westermark GT. Islet amyloid polypeptide, islet amyloid, and diabetes mellitus. *Physiol Rev*. 2011;91:795–826.
10. Wu C, Shea JE. Structural similarities and differences between amyloidogenic and non-amyloidogenic islet amyloid polypeptide (IAPP) sequences and implications for the dual physiological and pathological activities of these peptides. *PLoS Comput Biol*. 2013;9:e1003211.

Supplemental Information Available on Companion Website

- CT family peptides binding to CTR and CLR, and interaction with RAMPs/E-Figure 27.1

Calcitonin

Nobuo Suzuki

Abbreviation: CT
Additional names: thyrocalcitonin, ultimobranchial-calcitonin

The first hormone reported to have potent hypocalcemic activity resulting from the suppression of osteoclastic activity, CT is used as a drug to treat Paget's disease (osteitis deformans), pain associated with osteoporosis, Sudeck's atrophy, and hypercalcemia.

Discovery

In 1962, Copp discovered the presence of a hypocalcemic factor released by the thyroid—parathyroid complex. In 1964, Hirsch and colleagues showed that a hypocalcemic substance is produced and secreted by the thyroid gland and not the parathyroid gland; thus they named the substance thyrocalcitonin. In 1966, Pearse demonstrated that calcitonin (CT) is secreted from the parafollicular cells or C cells of the thyroid gland. Thereafter, Pearse and Carvalheira also showed that in non-mammalian vertebrates these cells are located in the ultimobranchial gland, which has a common origin with that of the thyroid gland in the ultimobranchial cells of the primitive pharynx. In 1968, CT was isolated from the porcine thyroid gland.

Structure

Structural Features

In vertebrates, a mature CT is composed of 32 aa residues. These CTs share a basic feature, i.e, a disulfide bridge between the cysteine residues at positions 1 and 7, which results in a ring of 7 aa residues at the N-terminus. They also share a proline amide group at the C-terminus [1]. An invertebrate CT sequence was recently determined from the protochordate *Ciona intestinalis* (AB485672). The protochordate CT sequence was composed of 30 aa residues but was shown to display high similarity to that of vertebrate CTs. They share the N-terminal circular region and C-terminal amidated proline. Human CT has an α-helical structure between the ninth and sixteenth residues, and shows both hydrophilic and hydrophobic properties (Figure 27A.1).

Primary Structure

The mature CT sequences of non-mammalian vertebrates, including teleosts, urodeles, reptiles, and birds, are considerably conserved, although those of cartilaginous fishes and anurans are slightly variant from those of other non-mammals [2,3]. In mammals, the primary structure is quite different

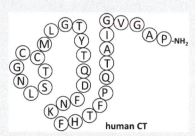

Figure 27A.1 Primary structure of mature human CT.

from that of non-mammals. Mammalian CTs share a conserved C-terminal amidated proline and an N-terminal circular structure, which is formed by a disulfide bridge between Cys^1 and Cys^7 [2,3]. The sequence identities shared by one of the three salmon CTs (CTI) and CTs of other vertebrates are shown in Table 27A.1 (see also E-Tables 27A.1, 27A.2)

Properties

Human CT, Mr 3,418, pI 8.7; salmon CTI, Mr 3,432, pI 10.4; for other CTs, Mr is around 3,500. CT is soluble in water but is insoluble in ethanol. It unstably oxidizes in neutral solution and easily in basic solution. It is stable in a solution of pH 4.5.

Synthesis and Release

Gene Structure and mRNA

The human *CT* gene, located on chromosome 11p15.2—p15.1, consists of six exons and five introns. CT mRNA is co-encoded with calcitonin gene-related peptide (CGRP) mRNA in a single gene [4]. In mammals, synthesis of the mRNAs encoding these two hormones is controlled by tissue-specific alternative splicing. The CT precursor mRNA is synthesized in thyroidal C cells, whereas the CGRP precursor mRNA is synthesized in neural tissues (E-Figure 27A.1) [4]. The pufferfish *ct* gene consists of four coding exons. Splicing of exons 1, 2, and 3 yields CT, whereas that of exons 1, 2, and 4 yields CGRP (E-Figure 27A.2) [2].

Distribution of mRNA

CT mRNA is detected most abundantly in the thyroidal C cells of mammals and the ultimobranchial gland of non-mammalian vertebrates. Some transcripts are found in the nervous tissue, digestive tract, pituitary gland, and gonads in mammals as well as other animals [3].

Y. Takei, H. Ando, & K. Tsutsui (Eds): Handbook of Hormones. DOI: http://dx.doi.org/10.1016/B978-0-12-801028-0.00170-7

Table 27A.1 Sequence Identity of Salmon to Other CTs

	Stingray CT	Lungfish CT	Eel CT	Salamander CT	Frog CT	Turtle CT	Chicken CT	Human CT	Porcine CT
Salmon	66%	88%	91%	91%	69%	84%	84%	50%	41%

Note: There are three kinds of salmon CT, but shown here is identity with salmon CTI.

Tissue and Plasma Concentrations

Tissue: Two months after a thyroidectomy in monkeys, immunoreactive CT at approximately 0.2 to 1 ng/g wet weight is detected in the liver, lung, digestive tract, kidney, hypothalamus, and other tissues [1]. In the lung and brain of lizards, immunoreactive CT is also detected at 0.1−0.4 ng/g and 0.06−0.1 ng/g wet weight, respectively [1].

Plasma: Under normal conditions, the plasma level of human CT is less than 100 pg/ml. A large quantity (ng/ml) of CT can be detected in the blood of non-mammalian vertebrates using RIA. However, from the analysis of plasma CT levels by a specific sandwich ELISA technique, the plasma CT concentration in eels is similar to that in humans [5].

Regulation of Synthesis and Release

CT synthesis in thyroid C cells and its release are stimulated principally by increased blood calcium levels. In fish as well as mammals, a calcium-sensing receptor has been cloned [2]. In fasted eels, plasma CT levels were not detectable by the specific sandwich ELISA technique (detection limit: 30 pg/ml) [5]. In eels that were fed a high amount of calcium−consommé solution (Ca^{2+}: 1.25 M; 1 ml/100 g body weight), plasma CT concentration increased rapidly (CT: below detection level at 0 h to 1,118.2 pg/ml at 3 h) corresponding to increased plasma calcium levels [5]. In fish as well as mammals, the trigger for CT secretion appears to be primarily a change in blood calcium levels.

Receptors

Structure and Subtype

Calcitonin receptor (CTR) is a member of a subfamily of the seven-transmembrane domain G protein-coupled receptor superfamily that includes several peptide receptors. Porcine CTR cDNA was obtained for the first time in 1991. Subsequent cloning of the gene demonstrated that it is approximately 70 kb in length and contains at least 14 exons, 12 of which encode porcine CTR [6]. Different isoforms of CTR resulting from alternative splicing of the gene have been described in various animal species with differential tissue expression of the transcripts [6]. The human *CTR* gene has been mapped to chromosome 7q21.3. CTRs from vertebrates were listed (E-Table 27A.3). Invertebrate CTR (AB485673) has also been sequenced from the protochordate *Ciona intestinalis*.

Signal Transduction Pathway

CTR is divided into three functional domains: an N-terminal extracellular domain, a transmembrane domain, and a C-terminal receptor domain [7]. The intracellular loops in the TM and receptor C-terminus interact with intracellular proteins such as G proteins and β-arrestin [7]. The principal mechanism of action of CT is the ability of its receptor to couple with at least two signal transduction pathways: the cAMP pathway and the phospholipase C pathway [6].

Agonists

The relative potencies of agonistic effects are salmon CT = eel CT > human CT > porcine CT [1]. CT family peptide hormones such as CGRP and amylin also have agonist activity.

Antagonists

Salmon CT-(8−32)(VLGKLSQELHKLQTYPRTNTGSGTP) and AC512 (Lys^{11}-Bolton Hunter, $R^{18}N^{30}Y^{32}$-salmon CT-(9−32)) have an antagonist activity [8].

Biological Functions

Target Cells/Tissues and Functions

Osteoclasts are the major target for the action of CT in mammals. As CT acts on the central nervous system, it prevents painful osteoporosis in humans. CT has specific binding sites within the central nervous system in mammals. The intracerebral injection of CT suppresses food and water intake in rats [4]. In addition, CT functions in various tissues and organs, including the gastrointestinal tract, kidney, breast, and hypothalamo-pituitary axis [4].

Phenotype in Gene-Modified Animals

αCGRP and CT are the products of the same gene. αCGRP is predominantly produced in the nervous system through tissue-specific alternative splicing of the initial gene transcript. The human *CT/αCGRP* gene has six exons. The mRNA encoding the CT precursor consists of exons 1, 2, 3, and 4, and is predominantly expressed in thyroid C cells. The mRNA encoding the αCGRP precursor in the nervous system contains exons 1, 2, 3, 5, and 6 (E-Figure 27A.1). CT/αCGRP-knockout mice experience increased bone mass and bone formation rate [9]. Increased bone formation has also been observed in heterozygous $CTR^{+/−}$ mice. Homozygous CTR null mice cannot be studied because the embryos die before skeletogenesis [9].

Pathophysiological Implications

Clinical Implications

CT reduces bone resorption by inhibiting mature active osteoclast formation, and increases renal calcium excretion. As described above, the activity of fish CTs is stronger than that of mammalian CTs [1]. Therefore, fish CTs are used in the management of postmenopausal osteoporosis, Paget's disease of bone, Sudeck's atrophy, and malignancy-associated hypercalcemia [1,8]. In addition to its effect on osteoclasts and renal tubules, CT has an analgesic effect, possibly mediated through β-endorphins and the central modulation of pain perception [4]. CT has been shown to relieve pain in patients with a number of conditions [4].

Use for Diagnosis and Treatment

Plasma CT concentration is usually measured using commercially available ELISA kits. The plasma CT level is a specific and sensitive marker of medullary thyroid carcinoma (MTC) [1]. Tumor size is significantly correlated with plasma CT levels, which can rise to more than 4,000 pg/ml in MTC patients. ProCT is the most established biomarker used to diagnose and treat septic patients, and is available in clinical practice [10]. ELISA kits for proCT and CT such as for humans, chickens, eels, and salmons are sold by various companies.

References

1. Azria M. *The Calcitonins: Physiology and Pharmacology.* Basel: Karger; 1989.
2. Power DM, Ingleton PM, Clark MS. Application of comparative genomics in fish endocrinology. *Int Rev Cytol.* 2002;221:149–190.
3. Sasayama Y. Hormonal control of Ca homeostasis in lower vertebrates: considering the evolution. *Zool Sci.* 1999;16:857–869.
4. Wimalawansa SJ. Amylin, calcitonin gene-related peptide, calcitonin, and adrenomedullin: a peptide superfamily. *Crit Rev Neurobiol.* 1997;11:167–239.
5. Suzuki N, Suzuki D, Sasayama Y, et al. Plasma calcium and calcitonin levels in eels fed a high calcium solution or transferred to seawater. *Gen Comp Endocrinol.* 1999;114:324–329.
6. Masi L, Brandi ML. Calcitonin and calcitonin receptors. *Clin Cases Miner Bone Metab.* 2007;4:117–122.
7. Barwell J, Gingell JJ, Watkins HA, et al. Calcitonin and calcitonin receptor-like receptors: common themes with family B GPCRs? *Br J Pharmacol.* 2012;166:51–65.
8. Hilton JM, Dowton M, Houssami S, et al. Identification of key components in the irreversibility of salmon calcitonin binding to calcitonin receptors. *J Endocrinol.* 2000;166:213–226.
9. Muff R, Born W, Lutz TA, et al. Biological importance of the peptides of the calcitonin family as revealed by disruption and transfer of corresponding genes. *Peptides.* 2004;25:2027–2038.
10. Walley KR. Biomarkers in sepsis. *Curr Infect Dis Rep.* 2013;15:413–420.
11. Martins R, Vieira FA, Power DM. Calcitonin receptor family evolution and fishing for function using *in silico* promoter analysis. *Gen Comp Endocrinol.* 2014;209:61–73.

Supplemental Information Available on Companion Website

- Primary structure of CT in vertebrates/E-Table 27A.1
- A schematic representation of the structural organization of human CT/α CGRP gene/E-Figure 27A.1
- Comparison of structural organization of human and fugu CT/CGRP genes/E-Figure 27A.2
- Accession numbers of CT in vertebrates/E-Table 27A.2
- Accession numbers of CTR in vertebrates/E-Table 27A.3

Calcitonin Gene-Related Peptide

Maho Ogoshi

Abbreviation: CGRP
Additional names: calcitonin-related polypeptide, calcitonin gene-related polypeptide

CGRP is a vasodilative neuropeptide synthesized mainly in the central nervous system. It is a transmitter of nociceptive signals in relation to migraine.

Discovery

Presence of CGRP-α mRNA formed by alternative splicing from the calcitonin gene (*CALCA*) was reported in 1982, and it was isolated in 1984 from human medullary thyroid calcinoma [1]. CGRP-β was identified by human genome analysis in 1985 [2].

Structure

Structural Features

Two types of CGRP (CGRP-α and CGRP-β) are conserved among several mammalian species in Euarchontglires. Human proCGRP-α and mature CGRP-α consists of 128 and 37 aa residues, respectively. Human proCGRP-β and mature CGRP-β consists of 127 and 37 aa residues, respectively. Sequence identity between CGRP-α and CGRP-β is 92% in the human and 97% in the rat. The sequence of CGRP-α is highly conserved among vertebrates from fish to mammals (Figure 27B.1). Glycine[14] and leucine[15] sequences are conserved in mammalian CGRP-α peptides, whereas aspartic acid and phenylalanine are conserved in the region of non-mammalian CGRPs.

Primary Structure

CGRP-α and CGRP-β both contain a ring structure formed by the single disulfide bond between cysteine[2] and cysteine[7] residues. The carboxyl end is amidated.

Properties

CGRP-α, Mr 3,789.4; CGRP-β, Mr 3,793.4. pI 5.3. Soluble in water and physiological saline solution.

Synthesis and Release

Gene, mRNA, and Precursor

The human CGRP-α gene (*CALCA*), located on chromosome 11 (11p15.2), consists of six exons. CGRP-α mRNA is synthesized by alternative splicing of exons 1, 2, 3, 5, and 6 of the *CALCA* gene. Exon 4 codes for calcitonin (CT). The human CGRP-β gene (*CALCB*), located in the vicinity of *CALCA* on chromosome 11 (11p15.2), consists of five exons. *CALCB* codes only for CGRP-β and does not contain a CT-like sequence. The sequence and structure of *CALCA* are well conserved among all vertebrate species. *CALCB* is preserved in primates and in rodents. In fishes, medaka also possesses two types of CGRP gene (*cgrp1* and *cgrp2*), but medaka *cgrp2* is not the ortholog of mammalian *CALCB*. Mammalian *CALCA* and *CALCB* are considered to have been generated by the tandem duplication which occurred later than the separation of the Euarchontoglires lineages, for they are located in the same chromosome. In contrast, medaka *cgrp* genes are located in separate chromosomes, indicating that they were produced by the teleost-specific whole genome duplication which occurred after the separation of the teleost lineage [3].

Distribution of mRNA

Messenger RNA of mammalian CGRP-α has been found both in the central and peripheral nervous system, but is predominantly expressed in the central nervous system. CGRP-α mRNA is detected in the sensory ganglia, primary spinal afferent fibers, and motor neurons. CGRP-β is predominantly expressed in the enteric nervous system and in the human pituitary.

Plasma Concentration

CGRP-α: human, 40–50 pg/ml; rat, 80–100 pg/ml.

Regulation of Synthesis and Release

CGRP is synthesized in a variety of central and peripheral neurons, and is packaged into vesicles which are transported to nerve terminals. CGRP release from sensory nerve terminals is stimulated by vascular wall tension, steroids such as estrogens and androgens, bradykinin/prostaglandins, and endothelin. During migraine, activation of trigeminal nerves stimulated by inflammatory substances from the cerebrovasculature leads to the release of CGRP [4]. In the gastrointestinal system, the synthesis and release of CGRP are triggered by activation of capsaicin receptor TRPV1 in the sensory nerves [5]. However, the exact mechanism of CGRP release is largely unknown.

Receptors

Structure and Subtype

The functional CGRP receptor is derived from calcitonin receptor-like receptor (CLR), whose phenotype is determined by co-expression with receptor activity-modifying protein (RAMP). Co-expression of CLR with RAMP1 results in the

Y. Takei, H. Ando, & K. Tsutsui (Eds): Handbook of Hormones. DOI: http://dx.doi.org/10.1016/B978-0-12-801028-0.00171-9

```
human CGRP-α   ACDTATCVTHRLAGLLSRSGGVVKNNFVPTNVGSKAF-NH₂
human CGRP-β   ACNTATCVTHRLAGLLSRSGGMVKSNFVPTNVGSKAF-NH₂
               ** ****************** ** ***********

human          ACDTATCVTHRLAGLLSRSGGVVKNNFVPTNVGSKAF
rat            SCNTATCVTHRLAGLLSRSGGVVKDNFVPTNVGSEAF
mouse          SCNTATCVTHRLAGLLSRSGGVVKDNFVPTNVGSEAF
pig            SCNTATCVTHRLAGLLSRSGGMVKSNFVPTDVGSEAF
chicken        ACNTATCVTHRLADFLSRSGGVGKNNFVPTNVGSKAF
Xenopus        ACNTATCVTHRLADFLSRSGGMGKNNFVPTNVGSKAF
Takifugu       ACKTATCVTHRLADFLSRSGGLGYSNFVPTNVGAQAF
medaka         ACNTATCVTHRLADFLSRSGGLGHSNFVPTNVGAQAF
               .*.********** ****** . *****.**. **
```

Figure 27B.1 Mature sequences of human CGRPs and alignment of CGRP-α sequences among vertebrates. Asterisks show identical residues and dots indicate residues identical in more than five sequences. The disulfide bridge is bracketed. Aspartic acid[14] and phenylalanine[15] in non-mammalian species are boxed.

CGRP receptor, whereas co-expression with RAMP2 or RAMP3 produces the receptor for adrenomedullin, another member of the CGRP family. CLR is a seven-transmembrane-domain G protein-coupled receptor consisting of 474–548 aa residues in mammals, which shares 55% sequence identity with the CT receptor. RAMP is a single-transmembrane accessory protein which regulates the activities of several G protein-coupled receptors. Besides contributing to receptor specificity, RAMPs are required for transportation of CLRs from the endoplasmic reticulum to the plasma membrane. Three types of RAMPs consisting of 148–175 aa residues exist in mammals, and five types are identified in teleost fishes. The functional CGRP receptor requires another accessory protein, the receptor component protein (RCP).

Signal Transduction Pathway

CGRP action is mainly mediated by adenylate cyclase coupled with the cAMP pathway. RCP protein is regarded to enable signaling by binding directly to the receptor. CGRP also acts via the nitric oxide (NO)-dependent cGMP pathway in rat aorta.

Agonists

Recombinant human CGRP-α and amylin are agonists.

Antagonists

CGRP$_{8-37}$ fragment, BIBN4094BS, MK3207, BI44370A, SB-273779, and WO98/11128 are antagonists.

Biological Functions

Target Cells/Tissues and Functions

CGRP exerts vasodilation in the microvasculature in the brain, heart, gastrointestinal system, limbs, and skin. CGRP, released from the terminal of sensory nerves, binds its receptor in the vascular smooth muscle cells and causes vasodilation. CGRP also has a cardioprotective role against ischemia/reperfusion injury. CGRP is likely to be involved in various biological actions; its plasma level increases during pregnancy in rats, it inhibits gastric acid secretion in rats and dogs, and gene transfer of CGRP restores erectile function in aged rats [6].

Phenotype in Gene-Modified Animals

CGRP-α knockout mice with normal CT expression are viable, fertile, and do not display obvious abnormalities. CT/

CGRP-α double knockout mice also have no obvious defects, but the number of viable fetuses of homozygous mothers is reduced. CT/CGRP-α double knockout mice show high systolic blood pressure and elevated mean arterial pressure [7]. CGRP-α deficient mice reduce morphine-induced analgesia, and exhibit reduced edema formation and nociception associated with inflammation [8].

Pathophysiological Implications

Clinical Implications

CGRP is actively studied in relation to migraine. The jugular level of CGRP is increased during migraine attacks, and intravenous CGRP administration induces migraine-like headache in humans [9]. CGRP receptor antagonists are considered to be effective for treatment of migraine. Increased plasma CGRP levels are also observed in myocardial ischemia and sepsis. CGRP may be implicated in Raynaud's disease, the syndrome characterized by severe peripheral vasospasm. Administration of CGRP leads to peripheral vasodilation and promotes the healing of ulcers in patients with this disease [10].

Use for Diagnosis and Treatment

Undergoing clinical studies.

References

1. Morris HR, Panico M, Etienne T, et al. Isolation and characterization of human calcitonin gene-related peptide. *Nature*. 1984;308:746–748.
2. Hoppener JWM, Steenbergh PH, Zandberg J, et al. The second human calcitonin/CGRP gene is located on chromosome 11. *Hum Genet*. 1985;70:259–263.
3. Ogoshi M, Inoue K, Naruse K, et al. Evolutionary history of the calcitonin gene-related peptide family in vertebrates revealed by comparative genomic analyses. *Peptides*. 2006;27:3154–3164.
4. Durham PL, Russo AF. Stimulation of the calcitonin gene-related peptide enhancer by mitogen-activated protein kinases and repression by an antimigraine drug in trigeminal ganglia neurons. *J Neurosci*. 2003;23:807–815.
5. Luo XJ, Liu B, Dai Z, et al. Stimulation of calcitonin gene-related peptide release through targeting capsaicin receptor: a potential strategy for gastric mucosal protection. *Dig Dis Sci*. 2013;58:320–325.
6. Ma H. Calcitonin gene-related peptide (CGRP). *Nat Sci*. 2004;2:41–47.

7. Zhang L, Hoff AO, Wimalawansa SJ, et al. Arthritic calcitonin/ alpha calcitonin gene-related peptide knockout mice have reduced nociceptive hypersensitivity. *Pain*. 2001;89:265−273.

8. Muff R, Born W, Lutz TA, et al. Biological importance of the peptides of the calcitonin family as revealed by disruption and transfer of corresponding genes. *Peptides*. 2004;25:2027−2038.

9. Bigal ME, Walter S, Rapoport AM. Calcitonin gene-related peptide (CGRP) and migraine current understanding and state of development. *Headache*. 2013;53:1230−1244.

10. Brain SD, Grant AD. Vascular actions of calcitonin gene-related peptide and adrenomedullin. *Physiol Rev*. 2004;84:903−934.

Supplemental Information Available on Companion Website

- Primary structure of CGRP sequences in various vertebrates/E-Figure 27B.1
- Gene structure of mammalian CALCA and CALCB/E-Figure 27B.2
- Primary structure of CLR in various vertebrates/E-Figure 27B.3
- Primary structure of RAMP in various vertebrates/E-Figure 27B.4
- Accession numbers of cDNAs for CGRP/E-Table 27B.1

Calcitonin Receptor-Stimulating Peptide

Maho Ogoshi

Abbreviation: CRSP

A peptide hormone synthesized in the central nervous system and in the thyroid gland, CRSP is a potent stimulator of calcitonin receptors. It is frequently misidentified as calcitonin gene-related peptide (CGRP) in mammalian species other than primates and rodents.

Discovery

CRSP (CRSP-1) was isolated in 2003 from porcine brain by monitoring cAMP production in the porcine kidney cell line [1]. Other members in the pig were identified by cDNA cloning based on the CRSP-1 sequence.

Structure

Structural Features

Three types of *crsp* gene are identified in the pig. Porcine proCRSP-1 and CRSP-1 consist of 126 and 38 aa residues respectively. Porcine proCRSP-2 and proCRSP-3 consist of 117 and 125 aa residues respectively. The putative mature CRSP-2 and -3 both consist of 37 aa residues, but the *in vivo* structure is unknown since there is no prohormone convertase sequence in the N-terminal region [2]. In addition to pigs, CRSP is conserved in dogs, cow, goats, sheep, and horses. The sequence identities among the species are 57−97%. The structure and sequence of CRSP are highly similar to those of calcitonin gene-related peptide (CGRP) and often confused with CGRP-β.

Primary Structure

CRSPs contain a ring structure formed by the single disulfide bond between cysteine[2] and cysteine[7] residues and an amidated carboxyl end (Figure 27C.1).

Properties

CRSP-1: Mr (pig) 4,130.6. Soluble in water and physiological saline solution.

Synthesis and Release

Gene, mRNA, and Precursor

Porcine *crsp* genes are located on chromosome 2. *crsp-1* and *crsp-2* genes consist of five exons. The porcine *crsp-3* gene contains an additional exon encoding a calcitonin (CT)-like sequence, but this exon could be a pseudo-exon since its mRNA is undetectable in porcine tissues. Three types of *crsp* genes exist in several species of Laurasiatheria, such as pigs, cattle, dogs, goats, sheep, and horses, but have not been identified in human, mice and rats. On account of the sequence similarity and location of the genes, CRSPs are frequently misidentified and registered as CGRPs in these species. The absence of a CT-like sequence would discriminate between CRSP and CGRP, but there are several *crsp* genes that contain the sequences. Therefore, the peptides whose stimulative effects on the calcitonin receptor (CTR) are undetermined are regarded as CT/CGRP/CRSP family members. *Crsp* genes are likely to have been generated by tandem duplication of *calca* (gene coding CGRP) later than the separation of Euarchontoglires and Laurasiatheria, since *crsp* genes are not possessed by Euarchontoglires (Figure 27C.2).

Distribution of mRNA

Messenger RNA of porcine *crsp-1* is detected in the hypothalamus, midbrain, pituitary, and thyroid gland [3].

Tissue and Plasma Concentrations

Tissue and plasma concentrations of porcine CRSP-1 are shown in Table 27C.1. Plasma concentrations of CRSPs have not been determined.

Regulation of Synthesis and Release

Neither the regulation mechanism nor the stimuli of the synthesis and release of CRSP is yet known.

Receptors

Structure and Subtype

Porcine CRSP-1 binds to CTR, the seven-transmembrane-domain G protein-coupled receptor, either accompanied by or without any of the receptor-activity modifying proteins (RAMPs). Mammalian CTR consists of 474−548 aa residues. Porcine CRSP-2 and -3 do not stimulate CTR, and their receptors are unknown.

Signal Transduction Pathway

CTR is mainly coupled to G_s protein, and CRSP-1 is considered to stimulate cAMP production in target cells.

Y. Takei, H. Ando, & K. Tsutsui (Eds): Handbook of Hormones. DOI: http://dx.doi.org/10.1016/B978-0-12-801028-0.00172-0

```
pig CRSP-1     SCNTATCMTHRLVGLLSRSGSMVRSNLLPTKMGFKVFG-NH2
pig CRSP-2     SCNTASCVTHKMTGWLSRSGSVAKNNFMPTNVDSKIL-NH2
pig CRSP-3     SCNTAICVTHKMAGWLSRSGSVVKNNFMPINMGSKVL-NH2
pig CGRP       SCNTATCVTHRLAGLLSRSGGMVKSNFVPTDVGSEAF-NH2
human CGRP-α   ACDTATCVTHRLAGLLSRSGGVVKNNFVPTNVGSKAF-NH2
               .*.**.*.**...*.*****.....*. *...... .
```

Figure 27C.1   Comparison of CRSP and CGRP sequences in human and pig. Asterisks show residues identical in all sequences, and dots indicate residues identical in more than three sequences. The disulfide bridge is bracketed.

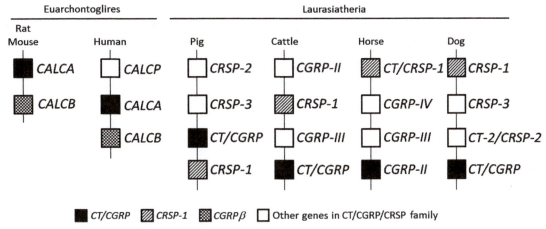

Figure 27C.2 Schematic representation of CT/CGRP/CRSP family genes on the orthologous chromosome.

Table 27C.1 Concentrations of Immunoreactive CRSP-1 in Porcine Tissues [1]

Tissue	CRSP (pmol/g wet tissue)
Midbrain	7.5 ± 0.9
Thalamus	3.5 ± 0.4
Hypothalamus	9.9 ± 1.2
Anterior pituitary	14 ± 2
Posterior pituitary	96 ± 15
Thyroid gland	68 ± 39

Agonists

Salmon calcitonin, amylin, PHM 27, and SUN-B 8155 are agonists.

Antagonist

Salmon calcitonin$_{8-32}$ is an antagonist.

Biological Functions

Target Cells/Tissues and Functions

Porcine CRSP-1 administration in the central nervous system reduces food intake, increases body temperature, and decreases locomotion activity in rats [4]. Intravenously administered CRSP-1 reduces the plasma CT level less potently than calcitonin in rats and sheep, but has no effect on blood pressure. CRSP-1 also inhibits plasma calcium release and bone resorption in mice [5].

Phenotype in Gene-Modified Animals

There are no reports of *crsp* gene-modified animals thus far.

Pathophysiological Implications

No study of CRSP related to pathophysiological condition have been reported thus far, as CRSP is not conserved in humans and rodents.

References

1. Katafuchi T, Kikumoto K, Hamano K, et al. Calcitonin receptor-stimulating peptide, a new member of the calcitonin gene-related peptide family. Its isolation from porcine brain, structure, tissue distribution, and biological activity. *J Biol Chem.* 2003;278:12046−12054.
2. Katafuchi T, Minamino N. Structure and biological properties of three calcitonin receptor-stimulating peptides, novel members of the calcitonin gene-related peptide family. *Peptides.* 2004;25:2039−2045.
3. Katafuchi T, Yasue H, Osaki T, et al. Calcitonin receptor-stimulating peptide: its evolutionary and functional relationship with calcitonin/calcitonin gene-related peptide based on gene structure. *Peptides.* 2009;30:1753−1762.
4. Sawada H, Yamaguchi H, Shimbara T, et al. Central effects of calcitonin receptor-stimulating peptide-1 on energy homeostasis in rats. *Endocrinology.* 2006;147:2043−2050.
5. Notoya M, Arai R, Katafuchi T, et al. A novel member of the calcitonin gene-related peptide family, calcitonin receptor-stimulating peptide, inhibits the formation and activity of osteoclasts. *Eur J Pharmacol.* 2007;560:234−239.

Supplemental Information Available on Companion Website

- Primary structure of CRSP sequences in various vertebrates/E-Figure 27C.1
- Schematic representation of CT/CGRP/CRSP family genes/E-Figure 27C.2
- Gene structure of mammalian CRSP/E-Figure 27C.3
- Molecular phylogenetic analysis of CRSPs in vertebrates/E-Figure 27C.4
- Primary structure of CTR in various vertebrates/E-Figure 27C.5
- Accession numbers of cDNAs for CRSP/E-Table 27C.1

Adrenomedullin

Maho Ogoshi

Abbreviation: AM, ADM
Additional name: adrenomedullin 1 (AM1)

AM is a potent hypotensive peptide synthesized in vasculature and various tissues. It is a promising drug target for hypertension and cardiac diseases.

Discovery

AM was isolated and identified from human pheochromocytoma tissue in 1993, by monitoring the activity elevating intracellular cAMP in rat platelets [1].

Structure

Structural Features

Human preproAM consists of 185 aa residues and contains a 21-aa residue signal peptide. The gene structure and synthetic pathway of human AM are summarized in Figure 27D.1. AM shares structural homology of an intramolecular ring structure and C-terminal amidation with calcitonin gene-related peptide (CGRP) and amylin; therefore, AM is considered to be a member of the CGRP family. The sequence identities between AM and other members are not high: 20% in the human.

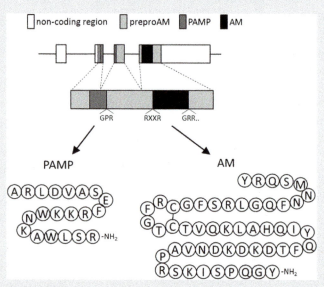

Figure 27D.1 The exon–intron structure of the human AM gene and the biosynthetic process of AM and PAMP.

ProAM includes a unique sequence of 20 aa residues followed by a typical amidation signal, Gly–Lys–Arg. This peptide with an amidated carboxyl terminus is termed "proadrenomedullin N-terminal 20 peptide" (PAMP). AM is conserved among various vertebrate species, including teleost fishes, though the sequence identities are low among species. The sequence of the disulfide ring region and following C-terminal region is highly conserved in tetrapods, but it varies in teleost fishes (Figure 27D.2). Teleost fishes possess two types of palarogous *am* genes; *am1* and *am4*.

Primary Structure

Mature human AM consists of 52 aa residues and contains a ring structure formed by the intramolecular disulfide bond between cysteine[16] and cysteine[21]. The tyrosine residue of the carboxyl end is amidated according to the subsequent amidation signal.

Properties

Human: Mr 6,028.7, pI 5.1. Soluble in water and physiological saline solution.

Synthesis and Release

Gene, mRNA, and Precursor

The human AM gene (*ADM*), located on chromosome 11 (11p15.4), consists of four exons. Human AM mRNA has 1,449 bp. Human *ADM* has TATA, CAAT, and GC boxes in the 5′-flanking region. The promoter region of human *ADM* contains binding sites for several transcription factors, including nuclear factor for interleukin-6 expression (NF-IL6), hypoxia-inducible factor-1 (HIF-1), and activator protein-1 (AP-1) and -2 (AP-2) [2].

Distribution of mRNA

AM mRNA is widely distributed throughout the body, especially in the vasculature, including the vascular endothelial cells and smooth muscle cells, in endocrine tissues, and in the central nervous system.

Plasma Concentration

Human 10–40 pg/ml; rat 90–120 pg/ml.

Regulation of Synthesis and Release

AM synthesis and release are stimulated by interleukin-1 (IL-1β), tumor necrosis factor-α (TNF-α), and lipopolysaccharide (LPS). AM production is increased by inflammatory cytokines and NO, possibly mediated by NF-IL6. The expression of *ADM* is also induced by hypoxia, ischemia, and

Y. Takei, H. Ando, & K. Tsutsui (Eds): Handbook of Hormones. DOI: http://dx.doi.org/10.1016/B978-0-12-801028-0.00173-2

```
Human      YRQSMNNFQGLRSF--GCRFGTCTVQKLAHQIYQFTDKDKDNVAPRSKISPQGY-NH2
rat        YRQSMN--QGSRST--GCRFGTCTMQKLAHQIYQFTDKDKDGMAPRNKISPQGY-NH2
mouse      YRQSMN--QGSRSN--GCRFGTCTFQKLAHQIYQLTDKDKDGMAPRNKISPQGY-NH2
chicken    YRQSVNSFPHLPTFRMGCRFGTCTVQKLAHQLYQLTDKVKDGAAPVNKISPQGY-NH2
frog       YRHT---FHHLQSVRVGCRFGTCTVQNLAHQIFQYTDKDKDSTAPVNKISSQGY-NH2
Takifugu   -----SKNLVNQSRKNGCSLGTCTVHDLAFRLHQLGFQYKIDIAPVDKISPQGY-NH2
medaka     -----SKISNSQSRRQGCSLGTCTVHDLAHRLHELN--LRIGSAPADKISPKGY-NH2
eel        -----SKNSVSSARRPGCSLGTCTVHDLAHRIHQLNNKLKVGSAPMDKISPLGY-NH2
             ..        .     **..****.. **... ... ... ** ***..**
```

Figure 27D.2 Alignment of mature AM sequences among vertebrates. Asterisks show residues identical in all sequences and dots indicate residues identical in more than five sequences. The disulfide bridge is bracketed.

oxidative stress. Several conflicting results have been reported regarding the effect of mechanical stimuli on *ADM* expression, but the existence of consensus sequences for shear stress responsive element (SSRE) in *ADM* suggest that AM synthesis is influenced by these stimuli. AM secretion from vascular endothelial cells and smooth muscle cells is stimulated by various substances, such as angiotensin II, thyroid hormone, and steroid hormones, including aldosterone and glucocorticoids.

Receptors

Structure and Subtype

Functional AM receptor is derived from calcitonin receptor-like receptor (CLR), whose phenotype is determined by co-expression with receptor activity-modifying protein (RAMP). Co-expression of CLR with RAMP2 or RAMP3 produces receptor for AM, whereas co-expression with RAMP1 results in CGRP receptor [3]. RAMP2 and RAMP3 are indistinguishable in terms of AM binding, and the differential roles of the CLR-RAMP2 and CLR-RAMP3 receptors have not been fully clarified. CLR is a seven-transmembrane-domain G protein-coupled receptor consisting of 474–548 aa residues in mammals, and shares 55% sequence identity with the CT receptor. RAMP is a single-transmembrane accessory protein which regulates the activities of several G protein-coupled receptors. Besides contributing to receptor specificity, RAMPs are required for the transportation of CLRs from the endoplasmic reticulum to the plasma membrane. Three types of RAMPs consisting of 148–175 aa residues exist in mammals, and five types have been identified in teleost fishes. A functional AM receptor requires another accessory protein, the receptor component protein (RCP).

Signal Transduction Pathway

AM is likely to act via two pathways. Cyclic AMP was the first-reported second messenger after AM was discovered by monitoring cAMP elevation in rat platelets. Besides the cAMP-mediated pathway, AM elicits phosphatidylinositol 3-kinase activation, which leads to stimulation of NO synthase, suggesting its action is mediated by the NO–cGMP pathway.

Agonists

AM_{13-52}, AM_{15-52}, and recombinant human amylin are agonists.

Antagonist

AM_{22-52} fragment is an antagonist.

Biological Functions

Target Cells/Tissues and Functions

AM has a wide range of biological functions. Receptor expression is detected in the vascular smooth muscle cells, vascular endothelial cells, microvessels, heart, lung, spleen, fat, and kidney. The main functions of AM are vasodilation, hypotension, angiogenesis, and regulation of fluid and electrolyte homeostasis [4]. AM is also synthesized in the hypothalamus and induces oxytocin release. In non-mammalian species, AM decreases blood pressure by intravenous injection in eels [5].

Phenotype in Gene-Modified Animals

ADM null knockout mice are embryonic lethal with abnormalities of vascular development. Heterozygotes of *ADM* knockout ($ADM^{+/-}$) mice show elevated blood pressure and reduced neovascularization. Conditional knockout mice lacking AM in the CNS have impaired motor coordination and show lower survival under stress conditions [6].

Pathophysiological Implications

Clinical Implications

AM is linked to considerable numbers of diseases, such as hypertention, congestive heart failure, ischemic heart injury, pulmonary hypertension, sepsis, cancers, renal impairment, and diabetes. Elevated plasma levels of AM are useful in assessing the progression of these diseases [7]. AM has cardio-protective and vasoprotective roles in pathophysiological conditions; therefore, therapeutic use of AM is considered to be promising in both acute-phase disorders and chronic diseases. Circulating AM is also increased after tissue transplantation, suggesting its protective role against oxidative damage. AM is abundantly expressed in tumor cells and is considered to be involved in carcinogenesis, promotion of tumor proliferation, angiogenesis, and the inhibition of apoptosis [8].

Use for Diagnosis and Treatment

AM is currently undergoing clinical studies.

References

1. Kitamura K, Kangawa K, Kawamoto M, et al. Adrenomedullin: a novel hypotensive peptide isolated from human pheochromocytoma. *Biochem Biophys Res Commun.* 1993;192:553–560.
2. Ishimitsu T, Ono H, Minami J, et al. Pathophysiologic and therapeutic implications of adrenomedullin in cardiovascular disorders. *Pharmacol Ther.* 2006;111:909–927.
3. McLatchie LM, Fraser NJ, Main MJ, et al. RAMPs regulate the transport and ligand specificity of the calcitonin-receptor-like receptor. *Nature.* 1998;393:333–339.

4. Eto T. A review of the biological properties and clinical implications of adrenomedullin and proadrenomedullin N-terminal 20 peptide (PAMP), hypotensive and vasodilating peptides. *Peptides.* 2001;22:1693–1711.

5. Takei Y, Ogoshi M, Nobata S. Exploring new CGRP family peptides and their receptors in vertebrates. *Curr Protein Pept Sci.* 2013;14:282–293.

6. Fernandez AP, Serrano J, Tessarollo L, et al. Lack of adrenomedullin in the mouse brain results in behavioral changes, anxiety, and lower survival under stress conditions. *Proc Natl Acad Sci USA.* 2008;105:12581–12586.

7. Bunton DC, Petrie MC, Hillier C, et al. The clinical relevance of adrenomedullin: a promising profile? *Pharmacol Ther.* 2004;103:179–201.

8. Nikitenko LL, Fox SB, Kehoe S, et al. Adrenomedullin and tumour angiogenesis. *Br J Cancer.* 2006;94:1–7.

Supplemental Information Available on Companion Website

- Gene, mRNA, and precursor sequences of human AM and PAMP/E-Figure 27D.1
- Primary structure of AM in various vertebrates/E-Figure 27D.2
- Primary structure of PAMP in various vertebrates/E-Figure 27D.3
- Primary structure of CLR in various vertebrates/E-Figure 27D.4
- Primary structure of RAMP in various vertebrates/E-Figure 27D.5
- Accession numbers of cDNAs for AM and PAMP/E-Table 27D.1

Adrenomedullin 2 and 5

Maho Ogoshi

Abbreviations: AM2, IMD, and AM5
Additional names: adrenomedullin 2/intermedin

AM2 is a potent systemic and pulmonary vasodilator synthesized in various tissues; AM5 is a novel member of the CGRP family predominantly synthesized in mammalian spleen.

Discovery

AM2 was identified by two independent groups in 2004, using genomic analyses [1,2]. AM5 was first identified in teleost fishes, by searching for the human AM(1)-like sequence, in 2003 [3], and its mammalian ortholog was determined in 2006 [4].

Structure

Structural Features

Human preproAM2 consists of 148 aa residues. AM5 is mutated in humans, but is conserved in various species such as pigs, dogs, and horses. PreproAM5 consists of 165 aa residues in pigs. Both AM2 and AM5 are conserved among various vertebrate species, including teleost fishes (Figure 27E.1). Teleost fishes possess two types of paralogous *am2* genes: *am2* and *am3*. AM2 and AM5 share structural homology of an intramolecular ring structure and C-terminal amidation with AM, calcitonin gene-related peptide (CGRP), and amylin; therefore, they are considered to be the members of the CGRP family. The sequence identities between these peptides are not high. ProAM2 and proAM5 do not contain PAMP-like sequences.

Primary Structure

AM2 and AM5 both contain a ring structure formed by a single disulfide bond between two cysteine residues (Figure 27E.1). The carboxyl end is amidated. The C-terminal glycine of proAM2 is followed by a stop codon, whereas that of proAM5 is followed by arginine residue(s) as observed in proAM.

Properties

Human AM2: Mr 5,100.8. Soluble in water and physiological saline solution.

Synthesis and Release

Gene, mRNA, and Precursor

The human AM2 gene (*ADM2*), located on chromosome 22 (22q13.33), consists of two exons. Human AM2 mRNA has 4,045 bp. The synteny analysis of fish *am2* and *am3* between chromosomes clarified that both of them are in an orthologous relationship with mammalian *ADM2*, and these genes are considered to be produced by teleost-specific whole genome duplication. The porcine AM5 gene (*ADM5*) is located on chromosome 18. Putative porcine mRNA consists of 1,840 bp. The putative human AM5 gene exists in chromosome 19 (19q13.33), but the mature sequence is disrupted by the deletion of 2 bp. Deletion at the same region on AM5 mRNA is also observed in chimpanzees but not in gorillas, suggesting that this frameshift occurred later than the separation of Hominini. In rodents the AM5 gene seems to be deleted, since no AM5-like sequence is detected in the expected location of their chromosomes. Teleost *am5* is likely to have duplicated in teleost-specific whole genome duplication, but only one type of AM5 gene has been detected thus far, suggesting that the duplicated counterpart has been deleted during evolution.

Distribution of mRNA

AM2 mRNA is detected at particularly high levels in the kidney, gastrointestinal tract, brain (especially the hypothalamus and intermediate and anterior lobes of the pituitary), skin, submaxillary gland, pancreas, lung, spleen, thymus, and ovary [5]. Expression of AM2 is also detected in human fetoplacental tissues [6]. AM2 is expressed in neonatal cardiomyocytes, but is absent or sparse in adult myocardia in rodents. Porcine AM5 mRNA is detected in the thymus and spleen.

Plasma Concentration

AM2: human, 7−120 pg/ml; rat, 70−200 pg/ml. Plasma AM5 levels are undetermined.

Regulation of Synthesis and Release

ADM2 expression is upregulated in the myocardium and aorta in spontaneously hypertensive rats. In human tissue, the expression levels of *ADM2* vary between individuals. The factors that influence the AM2 level in plasma and tissues are not currently known. Regulation of *ADM5* gene expression and release are also yet to be investigated.

Receptors

Structure and Subtype

AM2 binds to calcitonin receptor-like receptor (CLR) in association with any of the receptor activity-modifying proteins (RAMP)-1, -2 or -3. Although it appears that AM2 shows slight selectivity for CLR-RAMP3, the affinity is less potent compared to major ligands for the receptor, CGRP and AM. The receptor specific to AM2 is yet unknown. CLR is a seven-transmembrane-domain G protein-coupled receptor consisting of 474−548 aa residues in mammals, which shares 55%

Y. Takei, H. Ando, & K. Tsutsui (Eds): Handbook of Hormones. DOI: http://dx.doi.org/10.1016/B978-0-12-801028-0.00174-4

```
human AM2    -----------TQAQLLRVGCVLGTCQVQNLSHRLWQLMGPAGRQDSAPVDPSSPHSY-NH2
rat AM2      -----------PHAQLLRVGCVLGTCQVQNLSHRLWQLVRPSGRRDSAPVDPSSPHSY-NH2
mouse AM2    -----------PHAQLLRVGCVLGTCQVQNLSHRLWQLVRPAGRRDSAPVDPSSPHSY-NH2
frog AM2     ----------SHGPRLMRVGCSLGTCQVQILNHRLWQLMGQSGKED-SPIELSNPHSY-NH2
medaka AM2   -----HANGSNGRGHMMRVGCVLGTCQVQNLSHRLYQLIGQSGKEDSPPINPRSPHSY-NH2
medaka AM3   QTHPRGVHQYPHNAQLMRVGCFLGTCQVQNLSHRLYQLVGQKGREESSPFNPKSPHSY-NH2
eel AM2      HAHGRGRGHHHHHPQLMRVGCVLGTCQVQNLSHRLYQLIGQSGREDTSPMNPKSPHSY-NH2
eel AM3      HAHHGARG-HHHHPQLMRVGCVLGTCQVQNLSHRLYQLIGQSGREDSSPINPHSPHSY-NH2
                            ...****.*******.*.***.**...*....*..*..****


pig AM5      HQVSLKSGRLCSLGTCQTHRLPEIIYWLRSASTKELSGKAGRKPQDPYSY-NH2
frog AM5     -YISPISMRGCHLGTCQIQNLASMLYRLGNNGYKDGSNRDTK---DPLGY-NH2
medaka AM5   -----ALQRGCQLGTCQLHNLANTLYHINKTNGKEESTK-AH---DPQGY-NH2
eel AM5      -----APSRGCQLGTCQLHNLANTLYRIGQTNGKDESKK-AN---DPQGY-NH2
               *.* *****..*..* * * . ** .*
```

Figure 27E.1 Alignment of putative mature AM2/AM3 and AM5 sequences among vertebrates. Asterisks show residues identical in all sequences, and dots indicate residues identical in more than half of the species. The disulfide bridge is bracketed.

sequence identity with the CT receptor. RAMP is a single-transmembrane accessory protein that regulates the activities of several G protein-coupled receptors. Besides contributing to receptor specificity, RAMPs are required for the transportation of CLRs from the endoplasmic reticulum to the plasma membrane. Three types of RAMPs, consisting of 148−175 aa residues, exist in mammals, and five types have been identified in teleost fishes. Functional CGRP/AM receptor requires another accessory protein, the receptor component protein (RCP); thus, AM2 action via CLR-RAMPs is also likely to require RCP. AM5 shows some affinity to CLR-RAMPs, but its specific receptor is undetermined, as is the case with AM2.

Signal Transduction Pathway

The signal transduction pathway is possibly mediated by adenylate cyclase coupled with cAMP.

Agonists

Functional agonists for CLR are AM_{13-52}, AM_{15-52}, and recombinant human amylin.

Antagonists

$AM2_{23-30}$, $AM2_{31-35}$, and $AM2_{39-47}$ fragments are antagonists [7].

Biological Functions

Target Cells/Tissues and Functions

AM2 causes hypotension by peripheral administration, but when administered into the CNS it increases blood pressure and activates sympathetic nerves. AM2 also increases oxytocin and prolactin release, causes antidiuretic and natriuretic effects, and reduces food intake. AM5 again causes hypotension by intravenous injection, and elevates blood pressure and plasma oxytocin levels by intracerebroventricular injection [8]. In non-mammalian species, AM2 and AM5 decrease blood pressure more potently than AM1 in eels [9].

Phenotype in Gene-Modified Animals

Knockout animals of *ADM2* or *ADM5* have not yet been generated.

Pathophysiological Implications

Clinical Implications

Increased AM2 expression is detected in drug-induced hypertrophied myocardium in rats. AM2 also attenuates

ischemia−reperfusion injury when administered during the reperfusion period in rat heart. The cardioprotective effects of AM2 are similar to those of CGRP and AM, and are possibly mediated by cAMP [5].

Use for Diagnosis and Treatment

No clinical studies have been reported thus far.

References

1. Takei Y, Inoue K, Ogoshi M, et al. Identification of novel adrenomedullin in mammals: a potent cardiovascular and renal regulator. *FEBS Lett.* 2004;556:53−58.
2. Roh J, Chang CL, Bhalla A, et al. Intermedin is a calcitonin/calcitonin gene-related peptide family peptide acting through the calcitonin receptor-like receptor/receptor activity-modifying protein receptor complexes. *J Biol Chem.* 2004;279:7264−7274.
3. Ogoshi M, Inoue K, Takei Y. Identification of a novel adrenomedullin gene family in teleost fish. *Biochem Biophys Res Commun.* 2003;311:1072−1077.
4. Ogoshi M, Inoue K, Naruse K, et al. Evolutionary history of the calcitonin gene-related peptide family in vertebrates revealed by comparative genomic analyses. *Peptides.* 2006;27:3154−3164.
5. Bell D, McDermott BJ. Intermedin (adrenomedullin-2): a novel counter-regulatory peptide in the cardiovascular and renal systems. *Br J Pharmacol.* 2008;153:S247−S262.
6. Chauhan M, Yallampalli U, Dong YL, et al. Expression of adrenomedullin 2 (ADM2)/intermedin (IMD) in human placenta: role in trophoblast invasion and migration. *Biol Reprod.* 2009;81:777−783.
7. Hong Y, Hay DL, Quirion R, et al. The pharmacology of adrenomedullin 2/intermedin. *Br J Pharmacol.* 2012;166:110−120.
8. Takei Y, Hashimoto H, Inoue K, et al. Central and peripheral cardiovascular actions of adrenomedullin 5, a novel member of the calcitonin gene-related peptide family, in mammals. *J Endocrinol.* 2008;197:391−400.
9. Nobata S, Ogoshi M, Takei Y. Potent cardiovascular actions of homologous adrenomedullins in eels. *Am J Physiol Regul Integr Comp Physiol.* 2008;294:R1544−R1553.

Supplemental Information Available on Companion Website

- Primary structure of AM2−AM5 sequences in various vertebrates/E-Figure 27E.1
- Molecular phylogenetic analysis of AM2/AM5 precursors in vertebrates/E-Figure 27E.2
- Model of evolutionary history of vertebrate AM genes/E-Figure 27E.3
- Accession numbers of cDNAs for AM2−AM5/E-Table 27E.1

Amylin

Maho Ogoshi

Abbreviation: AMY
Additional names: islet amyloid polypeptide (IAPP), insulinoma amyloid peptide, diabetes-associated peptide

AMY is a peptide hormone predominantly co-secreted with insulin from pancreatic β cells. It aggregates to form islet amyloid in type 2 diabetes.

Discovery

Deposition of amyloid in the islets of Langerhans in type 2 diabetes has been observed, and was described as hyalinization in 1901. The genuine nature of human pancreatic islet amyloid was described as AMY or IAPP by two independent groups in 1987 [1,2].

Structure

Structural Features

Human AMY is derived after a 67-aa residue proAMY. The short C- and N-terminal flanking peptides are cleaved by the prohormone convertases PC2 and PC1/3 to form the 37-aa residues of mature AMY. Human AMY_{20-29} is considered to be the responsive region that forms amyloid fibrils in type 2 diabetes, as synthesized 20−29 residues are extremely fibrillogenic. However, rat and mouse models of diabetes lack islet amyloid. The AMY_{20-29} regions vary among humans and rodents, and rat/mouse AMY has three proline residues, known as β-sheet breakers, in this region. Since amyloid is the aggregated protein in which molecules in a β-sheet structure are bound to each other, the lack of islet amyloid in rodents appears to be due to the presence of proline residues in the AMY_{20-29} region and therefore the peptides in these species are saved from fibrillogenic conformation [3]. The sequence of mature AMY is highly conserved across vertebrate lineages (60−98% identity, Figure 27F.1), but the sequence of 20−29 regions is variable, which also supports the theory that islet amyloid is observed only in humans and cats.

Primary Structure

Mature AMY contains a disulfide bridge between cystein[2] and cysteine[7], and an amidated C-terminus − structures similar to those in calcitonin gene-related peptide (CGRP). AMY is normally soluble and is natively unfolded in its monomeric state, but aggregates to form islet amyloid in human type 2 diabetes. The amyloid fibril form is proposed to be a double β-hairpin with three β-strands consisting of residues 12−37 (Figure 27F.2). It is suggested that the 12−17, 22−27, and 31−37 regions of human AMY form antiparallel β sheets, with the 18−21 and 28−30 regions forming the β turns [4].

Properties

Human: Mr 3,903.4, pI 10.2. Soluble in water and physiological saline solution.

Synthesis and Release

Gene, mRNA, and Precursor

The human AMY gene (*IAPP*), located on chromosome 12 (12p12.1), consists of three exons and is regulated by a transcription factor, PDX-1 [5]. Human AMY mRNA has 1,992 bp. The gene structure and its mRNA size are well conserved among vertebrates.

Distribution of mRNA

Mammalian *IAPP* is expressed in islet β cells, in δ cells, in the gastrointestinal tract, and in sensory neurons. In teleosts, *amy* transcripts are detected and are found in the optic tectum, hypothalamus, posterior brain, and testis of goldfish [6].

Plasma Concentration

Human, 2−20 pmol/l; cat, 97 pmol/l.

Regulation of Synthesis and Release

ProAMY is processed in the Golgi complex and stored in the insulin secretory granules of pancreatic β cells. AMY is released together with insulin in response to glucose elevation in a 20:1 molar ratio of insulin to AMY. The slightly acidic intragranular pH (5−6), close to the isoelectric point of insulin, is favorable for the solution of AMY molecules. This low pH likely contributes to inhibit aggregation of AMY in the normal state, since the rate of AMY amyloid formation is reported to be strongly pH dependent and is slower at intragranular pH [7].

Receptors

Structure and Subtype

The functional receptors for AMY are generated from the calcitonin receptor (CTR) in a complex with one of three receptor activity-modifying proteins, RAMP-1, -2, or -3. Pairing of CTR and RAMP3 is considered to be the dominant AMY receptor, judged by the binding affinity. AMY also binds to CTR without RAMPs, but the affinity is low. CTR is a seven-transmembrane-domain G protein-coupled receptor that is highly conserved among vertebrates and existed before the

Y. Takei, H. Ando, & K. Tsutsui (Eds): Handbook of Hormones. DOI: http://dx.doi.org/10.1016/B978-0-12-801028-0.00175-6

```
human     KCNTATCATQRLANFLVHSSNNFGAILSSTNVGSNTY-NH2
rat       KCNTATCATQRLANFLVRSSNNLGPVLPPTNVGSNTY-NH2
cat       KCNTATCATQRLANFLIRSSNNLGAILSPTNVGSNTY-NH2
chicken   KCNTATCVTQRLADFLVRSSSNIGAIYSPTNVGSNTY-NH2
turtle    KCNTATCVTQRLADFLVRSSNNIGAIYSPTDVGSNTY-NH2
Takifugu  KCNTATCVTQRLADFLVRSSNTIGTVYAPTNVGSTTY-NH2
          ****** ***** **..**.. *.. ..*.***.**
```

Figure 27F.1 Alignment of mature AMY sequences among vertebrates. Asterisks show identical residues. The disulfide bridge is bracketed, and dots indicate residues identical in more than four sequences. The human AMY$_{20-29}$ region, which is considered to determine the ability of forming amyloid fibrils, is in bold.

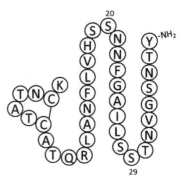

Figure 27F.2 Primary structure of mature human AMY.

separation of this lineage, for it is identified in invertebrate *Ciona intestinalis*. RAMP is a single-transmembrane accessory protein that regulates the activities of several G protein-coupled receptors. Three types of RAMPs consisting of 148–175 aa residues exist in mammals, and five types are identified in teleost fishes.

Signal Transduction Pathway

CTR is mainly coupled to G$_s$ protein and AMY stimulates production of cAMP in target cells.

Agonist

Salmon calcitonin is an agonist.

Antagonists

Salmon calcitonin$_{8-32}$ and AMY$_{8-37}$ are antagonists.

Biological Functions

Target Cells/Tissues and Functions

AMY reduces blood glucose levels. AMY is reported to suppress glucagon release from pancreatic β cells and is therefore considered to play a role in glucose homeostasis. There have been contradictory reports regarding *in vitro* effects of AMY on insulin secretion. AMY may have dual effects on insulin release, which stimulates basal insulin secretion and suppresses it when insulin secretion is augmented [3]. A number of studies have been carried out on the autocrine/paracrine functions of pancreatic AMY, but the mechanisms are still largely unknown. AMY is believed to inhibit food intake and gastric emptying in relation to satiety center stimulation. AMY has also been reported to inhibit insulin-stimulated glucose uptake and the synthesis of glycogen in isolated rat skeletal muscle [7].

Phenotype in Gene-Modified Animals

Male *IAPP* knockout mice show increased insulin responses paralleled with more rapid blood glucose elimination compared to wild-type. *IAPP* knockout mice also show high systolic blood pressure, elevated mean arterial pressure, and reduced nociceptive hypersensitivity [3].

Pathophysiological Implications

Clinical Implications

AMY aggregation forms islet amyloid in the β cells found in type 2 diabetes. Aggregation occurs in a stepwise manner, with soluble monomeric AMY forming oligomeric structures, protofibrils, and eventually amyloid fibrils, which are toxic and lead to cell death of pancreatic β cells. The proposed mechanisms of AMY-induced toxicity during amyloid formation start with cell membrane disruption; then endoplasmic reticulum

stress causes unfolded protein release and mitochondrial dysfunction, which eventually leads to oxidative stress and apoptosis. The human AMY$_{20-29}$ sequence is considered to determine its ability to form amyloid fibrils, since AMY in other species, such as rodents, which have variations within this region, does not form islet amyloids [8]. AMY is co-secreted with insulin, and thus is not produced in type 1 diabetes.

Use for Diagnosis and Treatment

Pramlintide, whose sequence is based on rat AMY to avoid amyloid formation, is used for treatment of type 1 and type 2 diabetes to reduce blood glucose levels.

References

1. Westermark P, Wernstedt C, Wilander E, et al. Amyloid fibrils in human insulinoma and islets of Langerhans of the diabetic cat are derived from a neuropeptide-like protein also present in normal islet cells. *Proc Natl Acad Sci USA*. 1987;84:3881–3885.
2. Cooper GJ, Willis AC, Clark A, et al. Purification and characterization of a peptide from amyloid-rich pancreases of type 2 diabetic patients. *Proc Natl Acad Sci USA*. 1987;127:414–417.
3. Westermark P, Andersson A, Westermark GT. Islet amyloid polypeptide, islet amyloid, and diabetes mellitus. *Physiol Rev*. 2011;91:795–826.
4. Pillay K, Govender P. Amylin uncovered: a review on the polypeptide responsible for type II diabetes. *Biomed Res Int*. 2013;:826706.
5. Watada H, Kajimoto Y, Kaneto H, et al. Involvement of the homeodomain-containing transcription factor PDX-1 in islet amyloid polypeptide gene transcription. *Biochem Biophys Res Commun*. 1996;229:746–751.
6. Martinez RM, Volkoff H, Munoz Cueto JA, et al. Molecular characterization of calcitonin gene-related peptide (CGRP) related peptides (CGRP, amylin, adrenomedullin and adrenomedullin-2/intermedin) in goldfish (*Carassius auratus*): cloning and distribution. *Peptides*. 2008;29:1534–1543.
7. Cao P, Marek P, Noor H, et al. Islet amyloid: from fundamental biophysics to mechanisms of cytotoxicity. *FEBS Lett*. 2013;587:1106–1118.
8. Westermark P. Amyloid in the islets of Langerhans: thoughts and some historical aspects. *Ups J Med Sci*. 2011;116:81–89.

Supplemental Information Available on Companion Website

- Alignment of AMY sequences in various vertebrates/E-Figure 27F.1
- Primary structure of human AMY/E-Figure 27F.2
- Primary structure of CTR in various vertebrates/E-Figure 27F.3
- Primary structure of RAMP in various vertebrates/E-Figure 27F.4
- Accession numbers of cDNAs for AMY/E-Table 27F.1

CHAPTER 28

Stanniocalcin

Nobuo Suzuki

Abbreviation: STC
Additional names: hypocalcin, teleocalcin, parathyrin

The first hormone isolated from the corpuscles of Stannius of teleosts, STC shows potent regulation of Ca^{2+}/P_i homeostasis.

Discovery

Presence of a hypocalcemic factor in the corpuscles of Stannius (CS) of a teleost was reported in 1964, and stanniocalcin (STC) was isolated in 1988 from salmon CS [1]. Thereafter, mammalian forms of STC were identified in 1995 and 1996 in two independent laboratories. An STC 1 paralog, STC2, was identified from the expressed sequence tag databases.

Structure

Structural Features

The preprotein of STC in coho salmon is composed of 256 aa residues. The first 33 residues comprise signal peptide, and the remaining 223 residues comprise the mature form of the hormone. It has been reported that the hormone from CS cross-reacts with an antiserum against parathyroid hormone. However, STC has no sequence homology with parathyroid hormone [2]. The most prominent structural features of STCs are the 11 cysteine residues and an N-linked glycosylation consensus sequence (Asn–X–Thr/Ser), which are completely conserved in fish and mammals [3–6]. The eleventh cysteine is important for the formation of a disulfide-linked homodimer [4–6]. A feature of mammalian STC2 is the presence of histidine residues compared with those of STC1 [4]. These histidine residues are able to interact with transition metals [4]. The sequences of STCs are aligned in E-Figure 28.1. STC was comparatively recently determined in tunicate *Ciona intestinalis* [7]. The eleventh cysteine, which is used to form a homodimer in vertebrates, is not conserved in tunicate STC, suggesting that it exists as a monomer (E-Figure 28.2).

Primary Structure

The sequence of coho salmon STC1, which consists of 256 aa residues (GenBank accession AAB26419.1) is shown in Figure 28.1.

Properties

STC is a glycosylated 50-kDa disulfide-linked homodimeric protein hormone. It is soluble in water and physiological saline solution.

Synthesis and Release

Gene and mRNA

The gene encoding human STC1 is located on the short arm of chromosome 8 (8p21.2). Human and mouse genes contain four exons spanning 13 kb of DNA. Two transcripts (2 and 4 kb) were detected in mammals using northern blotting, corresponding to STC1 (4 kb) and STC2 (2 kb) [2,5]. STC2 shares 34% overall homology with STC1. The human *STC2* gene is located on chromosome 5q35.1, and contains four exons. In mammals, the exon–intron boundaries are fully conserved between STC1 and STC2 (Figure 28.2) [4]. Meanwhile, sockeye salmon and fugu *stc1* genes are composed of five exons, while fugu *stc2*, and zebrafish *stc1* and *stc2* genes, contain four exons.

Distribution of mRNA

STC1 mRNA has been detected in extracorpuscular tissues such as the kidney, gonad, brain, heart, gills, muscle, and intestine of salmon [1]. Flounder STC2 mRNA is also widely expressed in the pituitary, brain, heart, kidney, gills, stomach, spleen, skin, dorsal fin, skeletal muscle, liver, intestine, ovary, and testis, in addition to the corpuscles of Stannius [8]. In mammals as well as fish, STC1 is expressed in diverse tissues such as the heart, lung, liver, adrenal gland, kidney, prostate, and ovary [1]. Among these tissues, STC1 expression levels are remarkably high in the ovary. In mammals, the most abundant expression of STC2 is found in the pancreas, spleen, kidney, and skeletal muscles [1]. STC mRNA is also expressed in a variety of tissues in tunicates, most prominently in the neural complex, heart, and gonads [7]. Locally produced tunicate STC may function as a paracrine factor in these tissues.

Tissue and Plasma Concentrations

Tissue: STC-like immunoreactivity can be detected in the human kidney using western blotting and immunocytochemistry [9]. Human kidney extracts have STC-like bioactivity when injected into fish [9].

   Plasma: STC was detected at concentrations of 2–200 ng/ml in the serum of sockeye salmon, using a double-antibody salmon STC radio-immunoassay [9]. In general, STC is absent from mammalian blood [1,4].

Regulation of Synthesis and Release

In teleosts, a rise in serum Ca^{2+} levels is known to be the primary stimulus for promoting STC release from CS. In the primary culture of trout CS, STC expression is activated by extracelluar Ca^{2+} and mediated via a calcium-sensing

Y. Takei, H. Ando, & K. Tsutsui (Eds): Handbook of Hormones. DOI: http://dx.doi.org/10.1016/B978-0-12-801028-0.00028-3

```
        10         20         30         40         50         60
    MLAKFGLCAV FLVLGTAATF DTDPEEASPR RARFSSNSPS DVARCLNGAL AVGCGTFACL
        70         80         90        100        110        120
    ENSTCDTDGM HDICQLFFHT AATFNTQGKT FVKESLRCIA NGVTSKVFQT IRRCGVFQRM
       130        140        150        160        170        180
    ISEVQEECYS RLDICGVARS NPEAIGEVVQ VPAHFPNRYY STLLQSLLAC DEETVAVVRA
       190        200        210        220        230        240
    GLVARLGPDM ETLFQLLQNK HCPQGSNQGP NSAPAGWRWP MGSPPSFKIQ PSMRGRDPTH
       250
    LFARKRSVEA LERVME
```

Figure 28.1 Amino acid sequence of coho salmon STC1. The N-linked glycosylation sites are underlined, and the 11 cysteine residues in the mature form are shaded.

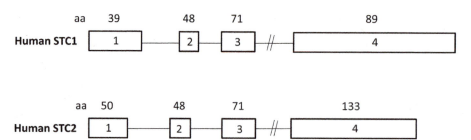

Figure 28.2 Gene structures of human STC1 and STC2. Exons are represented by boxes, introns by the intervening lines between exons. Number of amino acid residues (aa) is shown for each exon.

receptor [1]. In addition to its Ca^{2+}/P_i regulatory function, studies using various cultured cell models have shown that STC gene expression is upregulated by growth factors such as VEGF, bFGF, and HGF; the following immune and environmental stimuli also cause the upregulation of STC gene expression: cytokines such as TNF-α and IL-1, acidic pH, hypertonicity, toxic chemicals, and bacterial and viral infections [5]. *In vivo* studies have shown that STC gene expression is elevated at sites of inflammation, angiogenesis, and ischemia, and in various forms of malignancy [5]. Positive correlation of STC1 and STC2 expressions with cancers and estrogen receptor expression has been noted (see "Pathophysiological Implications," below, for details).

Receptors

No receptor has been identified for STC. From receptor binding analyses performed using an STC-alkaline phosphatase fusion protein, the majority of STC binding sites have been localized to the inner mitochondrial membranes in kidney and liver [5]. This may imply a role for STC in cellular metabolism [4]. G protein-coupled signaling has been implicated in the activity of STC because the activation of cAMP—PKA signaling has been detected in STC-treated flounder proximal tubule cells [1].

Biological Functions

Target Cells/Tissues and Functions

The main target organ for STC is not the bone but the entry routes of calcium, particularly the intestine, kidney, and gills in fish [2]. In adult fish, STC is known to regulate serum Ca^{2+} homeostasis via inhibition of branchial/intestinal Ca^{2+} uptake and renal P_i excretion. The Ca^{2+}/P_i regulatory function of STC1 is evolutionarily conserved in mammals, as well as in fish [1]. In mammals, the contribution of STC1 to renal inorganic phosphate excretion in rats and intestinal calcium

absorption in swine and rats has been demonstrated [1]. These notions are supported by localization of STC1 signals in the renal tubule cells of the mouse, rat, and human [1]. In addition, upregulation of STC in angiogeneic endothelial cells and immune and inflammatory responses has been reported [1,5].

Phenotype in Gene-Modified Animals

Serum calcium and phosphate levels are normal in STC1$^{(-/-)}$ and STC2$^{(-/-)}$ mice [10], suggesting that mammalian STCs are unlikely to be major regulators of mineral metabolism. Growth inhibition has been reported in STC2$^{(-/-)}$ mice [10]. However, in most transgenic and knockout studies, the effects of STC1 and STC2 have been negative regarding muscle and bone development [1].

Pathophysiological Implications

Relationship with Cancers

Elevated expression levels of STC1 and STC2 have been observed in breast and ovarian cancers [5,6]. In breast cancers, the expression levels of both STC1 and STC2 show a positive correlation with the levels of estrogen receptor [1]. In addition, an increase in STC1 expression has been detected in tumor samples from human colorectal cancers/hypatocelluar carcinomas, non-small-cell lung cancers, breast carcinomas, and leukemia [1]. In human gastric cancers, neuroblastomas, castration-resistant prostate cancers, breast cancer, colorectal cancer, esophageal squamous cell cancer, and renal cell carcinomas, the STC2 expression levels are elevated compared with those in the respective normal tissues [1]. The human *STC1* gene is present on chromosome 8p, which is linked with tumor progression and metastasis [1]. The human *STC2* gene has been mapped to 5q35.1, which is also linked with tumor progression [1]. Genetic and epigenetic analysis may be useful to elucidate the relationship of STC1 and STC2 to cancers.

References

1. Yeung BHY, Law AYS, Wong CKC. Evolution and roles of stanniocalcin. *Mol Cell Endocrinol*. 2012;349:272–280.
2. Wendelaar Bonga SE, Pang PKT. Control of calcium regulating hormones in the vertebrates: parathyroid hormone, calcitonin, prolactin, and stanniocalcin. *Int Rev Cytol*. 1991;128:139–213.
3. Wagner GF, DiMattia GE. The stanniocalcin family of proteins. *J Exp Zool*. 2006;305A:769–780.
4. Chang AC-M, Jellinek DA, Reddel RR. Mammalian stanniocalcins and cancer. *Endocr Relat Cancer*. 2003;10:359–373.
5. Gerritsen ME, Wagner GF. Stanniocalcin: no longer just a fish tale. *Vitam Horm*. 2005;70:105–135.
6. Ishibashi K, Imai M. Prospect of a stanniocalcin endocrine/paracrine system in mammals. *Am J Physiol Renal Physiol*. 2002;282: F367–F375.
7. Roch GJ, Sherwood NM. Genomics reveal ancient forms of stanniocalcin in amphioxus and tunicate. *Integr Comp Biol*. 2010;50:86–97.
8. Shin J, Sohn YC. cDNA cloning of Japanese flounder stanniocalcin 2 and its mRNA expression in a variety of tissues. *Comp Biochem Physiol*. 2009;153:24–29.
9. Wagner GF, Guiraudon CC, Milliken C, et al. Immunological and biological evidence for a stanniocalcin-like hormone in human kidney. *Proc Natl Acad Sci USA*. 1995;92:1871–1875.
10. Chang AC-M, Hook J, Lemckert FA, et al. The murine stanniocalcin 2 gene is a negative regulator of postnatal growth. *Endocrinology*. 2008;149:2403–2410.

Supplemental Information Available on Companion Website

- Alignment of the amino acid sequences of STCs in fishes and mammals/E-Figure 28.1
- Amino acid sequences of tunicate STC and teleost STC/ E-Figure 28.2
- Accession numbers of STCs/E-Table 28.1

SECTION I.5

Other Peripheral Hormones

CHAPTER 29

Renin-Angiotensin System

Marty K.S. Wong

History

In 1898, the term *renin* was coined based on the observation that injection of saline extracts from fresh rabbit kidneys increased arterial blood pressure. In 1958, the product of the renin-angiotensin system (RAS), *angiotensin*, which increases blood vessel (angio-) tension (-tensin), was identified and so named [1]. Renin was cloned from the submaxillary gland of mice, and its substrate, angiotensinogen, was cloned from the mouse liver in 1983. In addition to the endocrine role of the RAS, the local RAS and the intracellular RAS are now established [2].

Structure

RAS Cascade

Angiotensinogen (AGT) is the sole precursor protein that harbors angiotensin I (Ang I) at the N-terminus, which is specifically cleaved by renin (Figure 29.1). Ang I is cleaved by angiotensin converting enzyme (ACE) to form Ang II. Ang III and IV are formed by sequential removal of N-terminal aa by aminopeptidase A and N (APA/N). Another active component, Ang (1−7), is cleaved by ACE2, but can also be produced through different pathways (Figure 29.1).

The RAS cascade in non-mammals is similar to that of mammals, and homologous renin, ATG, ACE1, ACE2, angiotensin type 1 receptor (AT1), angiotensin type-2 receptor (AT2), and angiotensin type-4 receptor/isulin-regulated aminopeptidase (AT4/IRAP) are present according to genome data, but Mas receptor (Mas1) is not present in any teleost genomes (Table 29.1). AGT, ACE1, and AT1 were identified in the sea lamprey genome, but the existence of other components is currently unknown.

RAS Components

AGT was originated from the serine-protease inhibitor (SERPIN) family and belongs to SERPIN clade A, member 8 [3]. Although structurally resembling a SERPIN, AGTs are non-inhibitory SERPIN except for lamprey AGT [4]. AGT sequences are highly variable, but the sequences of the functional domain, Ang I, are relatively conserved. Molecular evolution of the RAS components was studied using available genome data [5]. The phylogenetic analysis suggested monophyletic relationships within AT1 and AT2 in non-mammalian vertebrates [6,7].

Receptors

Structure and Subtypes

Renin receptor binds to both prorenin and renin to increase the affinity of the enzyme. At least four angiotensin receptors were discovered in mammals: AT1, AT2, Mas1, and AT4/IRAP (Figure 29.1). AT1, AT2, and Mas1 are GPCRs with seven transmembrane domains and are composed of 359, 363, and 325 aa residues respectively in human. AT4/IRAP is an insulin-regulated aminopeptidase that is composed of 916 aa residues in human.

Signal Transduction Pathway

AT1 is the most frequently studied angiotensin receptor and it mediates most of the known classical angiotensin effects such as vasoconstriction and drinking stimulation [1]. AT1 signals through the Ca/IP3 pathway and activates acute responses, such as vascular smooth muscle contraction, MAPK activities, and EGF receptor in the plasma membrane. In mammals, AT2 signals through MAPK-1, SHP-1, and PP2A, which modulate the cGMP/NO pathway [8]. Mas1 is activated by Ang (1−7) but the intracellular signaling pathway is not well understood. The AT4/IRAP receptor activates intracellular signals including an increase in intracellular Ca concentration, modulation of MAPKs, activation of NF-κB signaling, and production of cGMP.

Biological Functions

Target Cells/Tissues and Functions

Ang II acts on the brain to control thirst, blood pressure, and ventilation rate, and also participates in the memory process (Figure 29.2). The Ang II/AT1 axis mediates vasoconstriction, thirst, release of vasopressin and aldosterone, renal sodium reabsorption, fibrosis, inflammation, angiogenesis, vascular aging, and atherosclerosis. AT2 is mostly embryonic and its expression decreases in adults. AT2 is confined in certain tissues such as kidney, and the expression and function are usually associated with pathophysiological conditions induced by AT1 signaling. The Ang (1−7)/Mas1 axis is also known to antagonize the effects of the AT1 axis, including anti-hypertrophic action, anti-thrombotic and anti-fibrotic effects, and vasodilation via stimulation of nitric oxide (NO) synthesis in endothelium and potentiation of the bradykinin effect. AT4/IRAP has a broad tissue distribution in the kidney, aorta, heart, liver, lung, uterus, adrenal gland, and brain, especially in neurons associated with memory function [9].

Y. Takei, H. Ando, & K. Tsutsui (Eds): Handbook of Hormones. DOI: http://dx.doi.org/10.1016/B978-0-12-801028-0.00029-5

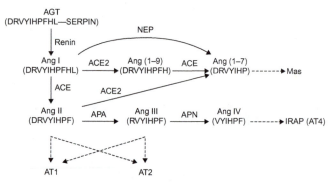

Figure 29.1 Cascade pathways and major receptors of RAS. Classical RAS pathway includes cleavage of angiotensinogen (AGT) by renin to produce Ang I, which is subsequently converted into Ang II by angiotensin converting enzyme (ACE). Ang II is subsequently cleaved to form Ang III and Ang IV. Ang II and Ang III act on AT1 and AT2 receptors while Ang IV acts on the AT4/IRAP receptor. ACE2 provides the alternative pathways to form Ang (1−7) that act through the Mas receptor in mammals.

Table 29.1 Presence of RAS Components in Vertebrates

RAS components	Lamprey	Teleosts	Tetrapods
Renin	?	+	+
AGT	+	+	+
ACE1	+	+	+
ACE2	?	+	+
AT1	+	+	+
AT2	?	+	+
AT4/IRAP	?	+	+
Mas1	?	−	+

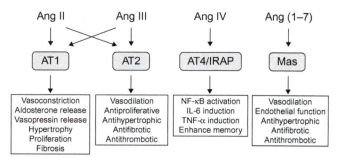

Figure 29.2 Functional summary of RAS via various receptors. Major actions of Ang II take place via AT1 receptors that promote water and salt retention and cell growth. Actions via AT2 and Mas receptors are often antagonistic to those of AT1. AT4/IRAP is involved in the memory process.

References

1. Kobayashi H, Takei Y. Renin-angiotensin system: comparative aspects. In: Bradshaw SD, Burggren W, Heller HC, et al., eds. *Zoophysiology.* vol 35. NY: Academic Press; 1996:1−245.

2. Kumar R, Thomas CM, Yong QC, et al. The intracrine renin-angiotensin system. *Clin Sci.* 2012;123:273−284.

3. Wong MKS, Takei Y. Characterization of a native angiotensin from an anciently diverged serine protease inhibitor in lamprey. *J Endocrinol.* 2011;209:127−137.

4. Wang Y, Ragg H. An unexpected link between angiotensinogen and thrombin. *FEBS Lett.* 2011;585:2395−2399.

5. Fournier D, Luft FC, Bader M, et al. Emergence and evolution of the renin-angiotensin-aldosterone system. *J Mol Med.* 2012;90:495−508.

6. Evans AN, Henning T, Gelsleichter J, et al. Molecular classification of an elasmobranch angiotensin receptor: quantification of angiotensin receptor and natriuretic peptide receptor mRNAs in saltwater and freshwater populations of the Atlantic stingray. *Comp Biochem Physiol B.* 2010;157:423−431.

7. Wong MKS, Takei Y. Angiotensin AT2 receptor activates the cyclic-AMP signaling pathway in eel. *Mol Cell Endocrinol.* 2013;365:292−302.

8. Gallinat S, Busche S, Raizada MK, et al. The angiotensin II type 2 receptor: an enigma with multiple variations. *Am J Physiol.* 2000;278:E357−E374.

9. Chai SY, Fernando R, Peck G, et al. The angiotensin IV/AT4 receptor. *Cell Mol Life Sci.* 2004;61:2728−2737.

Supplemental Information Available on Companion Website

- Gene, mRNA, and domain structure of the human angiotensinogen/E-Figure 29.1
- Phylogenetic tree of the angiotensinogen in vertebrates/E-Figure 29.2
- Accession numbers of vertebrate angiotensinogen genes (*agt*)/ E-Table 29.1

Renin

Marty K.S. Wong

Abbreviation: Ren
Additional names: Angiotensinogenase, angiotensin forming enzyme

Rate-limiting enzyme for the initiation of the cascade of the renin-angiotensin system. Plasma renin activity is a marker for hypertension. Drug target for hypertension and renal disease.

Discovery

In the late 18th century, renin was discovered based on the observation that injection of saline extracts from fresh rabbit kidneys into other rabbits increased arterial blood pressure.

Structure

Structural Features

Renin is a highly specific endopeptidase, whose only known function is to generate angiotensin I (Ang I) from angiotensinogen (AGT), initiating a cascade of reactions that produces an elevation of blood pressure and increases sodium retention by the kidney.

Primary Structure

Ren1 is a protein coding gene with 406 aa residues in humans and it forms an aspartyl protease that has high specificity on AGT (Figure 29A.1). Renin cleaves Ang I from AGT (Figure 29A.2) and the cleavage specificity is determined by the [HPF]-domain on Ang I, the enzyme hydrolyzing the peptide bond two amino acids after this signature sequence [1]. The [HPF]-domain is highly conserved among vertebrates (see Subchapter 29B, Angiotensin II), indicating a similar functional constraint could be present in the other vertebrate renins.

Synthesis and Release

Gene and mRNA

Ren1 is located at chromosome 1q32.1 in humans. Renin is synthesized as an inactive form called prorenin but recent studies have indicated that prorenin can bind to renin receptors and activate the cleavage site reversibly by changing the conformation of the prosegment. Renin secretion by kidney is stimulated by β-adrenergic response and cAMP augmentation. The release of active renin from kidney is inhibited by increased arterial pressure or Ang II, and an increased intracellular Ca concentration. This is in contrast to the other secretory cells, in which an increase in intracellular calcium

```
        10         20         30         40         50         60
MDGWRRMPRWGLLLLLWGSCTFGLPTDTTTFKRIFLKRMPSIRESLKERGVDMARLGPEW
        70         80         90        100        110        120
SQPMKRLTLGNTTSSVILTNYMDTQYYGEIGIGTPPQTFKVVFDTGSSNVWVPSSKCSRL
       130        140        150        160        170        180
YTACVYHKLFDASDSSSYKHNGTELTLRYSTGTVSGFLSQDIITVGGITVTQMFGEVTEM
       190        200        210        220        230        240
PALPFMLAEFDGVVGMGFIEQAIGRVTPIFDNIISQGVLKEDVFSFYYNRDSENSQSLGG
       250        260        270        280        290        300
QIVLGGSDPQHYEGNFHYINLIKTGVWQIQMKGVSVGSSTLLCEDGCLALVDTGASYISG
       310        320        330        340        350        360
STSSIEKLMEALGAKKRLFDYVVKCNEGPTLPDISFHLGGKEYTLTSADYVFQESYSSKK
       370        380        390        400
LCTLAIHAMDIPPPTGPTWALGATFIRKFYTEFDRRNNRIGFALAR
```

Figure 29A.1   Amino acid sequence and structural features of human renin. Underlined region indicates the A1-propeptide that covers the active site of renin. The shaded sequence indicates the aspartyl protease domain.

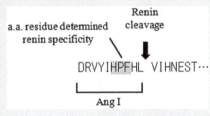

Figure 29A.2   Renin action on angiotensinogen. The [HPF] residue that determines the renin specificity is shaded.

usually enhances the depletion of secretory granules. Thus, this is often called the calcium paradox. The regulatory elements of the promoter of the renin gene were termed "renin enhancer," and were later found to be a compound of regulatory elements with several stimulatory and inhibitory activities. In the renin enhancer, cAMP response elements have been identified. Besides transcriptional regulation, a posttranslational regulation at the untranslated region (a 200 bp segment beyond the coding region) was also known, for which the mRNA stability can be enhanced by cAMP stimulation [2].

Distribution of mRNA

Renin (*Ren1*) is synthesized mainly by kidney but mRNA is also found in many other tissues including heart, brain, and adrenal glands.

Y. Takei, H. Ando, & K. Tsutsui (Eds): Handbook of Hormones. DOI: http://dx.doi.org/10.1016/B978-0-12-801028-0.00176-8

Tissue and Plasma Concentrations

The term plasma renin concentration (PRC) is often confused with the term plasma renin activity (PRA). PRC is determined by an immunoradiometric assay. This assay uses two monoclonal antibodies and measures the number of active renin molecules independently of their enzymatic activity (expressed in pg/ml, IU/L, or pmol/L). PRA describes and quantifies the enzyme's catalytic activity and is determined by the amount of Ang I generated per unit time during incubation of plasma samples *in vitro* (expressed either in pmol or in ng Ang I/ml plasma per hour incubation). Normal human adult PRA is about 1.8 ng/ml/hour. Under physiological conditions, PRA is proportional to the concentration of active renin as long as the AGT concentration is constant.

Receptors

Structure and Subtype

The renin receptor (*ATP6AP2*) binds to both prorenin and renin. The renin receptor gene is located on chromosome Xp11.4 in humans. The major known function is that the binding increases the catalytic activity of renin on AGT by 5- to 10-fold: K_m for AGT (in the absence of renin receptor) is 1 μM and the K_m of membrane-bound renin receptor complex is 0.15 μM. Renin receptors are abundantly expressed in heart, brain, and placenta, and lower expression is found in kidney and liver [3]. In the brain, the renin receptor is expressed in the subfornical organ, paraventricular nucleus, supraoptic nucleus, tractus solitarius nucleus, and rostral ventrolateral medulla, which are likely involved in the regulation of cardiovascular and volume regulation [4]. Genome data also suggest that orthologous renin receptor is present in other vertebrates including teleosts and tetrapods.

Signal Transduction Pathway

Renin binding to the renin receptor leads to activation of mitogen-active protein kinase p44/42 and TGF-β, which in turn increases contractility, hypertrophy, and fibrosis.

Agonist

No available agonist.

Antagonist

Aliskiren is the first in a new class of orally active, non-peptide, low molecular weight renin inhibitors [5]. It is a transition-state mimetic, with favorable physico-chemical properties including high aqueous solubility (350 mg/ml at pH 7.4) and high hydrophilicity (log Poct/water = 2.45 at pH 7.4), which are are important properties for oral bioavailability. Aliskiren is designed through a combination of molecular modeling techniques and crystal structure elucidation, and it effectively reduces the blood pressure as a mono therapy as well as in combination therapy.

Biological Functions

Target Cells/Tissues and Functions

Renin is the rate-limiting enzyme that initiates the cascade generating the angiotensin peptides that regulate blood pressure, cell growth, apoptosis and electrolyte balance. Renin is a hormone that ultimately integrates cardiovascular and renal function in the control of blood pressure as well as salt and volume homeostasis.

Prorenin is activated by proteolytic enzymes such as trypsin and kallikrein. Acid treatment (optimal pH at 3.3) fully activates prorenin and low temperature, partially ($\sim$15%). Binding of prorenin to renin receptor changes the conformation of the pro-segment to expose the active site to activate the enzyme (Figure 29A.3).

Phenotype in Gene-Modified Animals

Ren1-knockout mice have a decreased heart rate, an increased susceptibility to bacterial infection, an increase in plasma creatinine, and a lower mean arterial pressure.

Pathophysiological Implications

Clinical Implications

High renin activities eventually stimulate the RAS cascade, which leads to vasoconstriction and retention of sodium and water, resulting in hypertension.

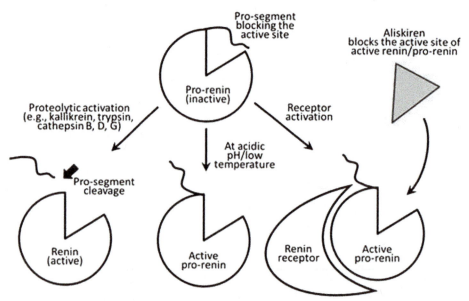

Figure 29A.3 Mode of activation of prorenin. Proteolytic activation (irreversible) via cleavage of the pro-segment. Physical activation (reversible) via low temperature and low pH. Protein−protein interaction activation (reversible) occurs via receptor binding.

Use for Diagnosis and Treatment

Renin inhibitors can be used for the treatment of hypertension. In current medical practice, PRA is the indicator of RAS overactivity, this being more commonly treated using either ACE blockers and/or AT1 blockers rather than a direct oral renin inhibitor. ACE inhibitors and/or AT1 blockers are also part of the standard treatment after a heart attack. The diagnosis of kidney cancer includes juxtaglomerular cell tumor (reninoma), Wilms' tumor, and renal cell carcinoma, all of which may produce renin, leading to hypertension in patients.

References

1. Nakagawa T, Akaki J, Satou R, et al. The His-Pro-Phe motif of angiotensinogen is a crucial determinant of the substrate specificity of renin. *Biol Chem.* 2007;388:237–246.
2. Persson PB. Renin: origin, secretion and synthesis. *J Physiol.* 2003;552:667–671.
3. Jan Danser AH, Batenburg WW, van Esch JH. Prorenin and the (pro)renin receptor — an update. *Nephrol Dial Transplant.* 2007;22:1288–1292.
4. Nabi AHMN, Biswas KB, Ebihara A, et al. Renin angiotensin system in the context of renin, prorenin, and the (pro)renin receptor. *Rev Agric Sci.* 2013;1:43–60.
5. Allikmets K. Aliskiren — an orally active renin inhibitor. Review of pharmacology, pharmacodynamics, kinetics, and clinical potential in the treatment of hypertension. *Vasc Health Risk Manag.* 2007;3:809–815.

Supplemental Information Available on Companion Website

- Gene, mRNA and domain structure of human renin /E-Figure 29A.1
- Phylogenetic tree of renin in vertebrates /E-Figure 29A.2
- Accession numbers of vertebrate renin genes /E-Table 29A.1

Angiotensin II

Marty K.S. Wong

Abbreviation: Ang II, ANG II, AII, AngII, ANGII, Ang(1−8)
Additional names: hypertensin, angiotonin

Active peptide hormone of the renin-angiotensin system with broad functions including vasopressor effect and thirst stimulation. Drug target for hypertension and renal diseases.

Discovery

In 1940, two groups of researchers discovered separately a potent vasoconstrictor by incubating plasma with renin, and named this substance "angiotonin" and "hypertensin" respectively. In 1958, they agreed to name this substance "angiotensin" (angio = blood vessel; tensin = tension) [1].

Structure

Structural Features

Ang II is a linear peptide with no known secondary modification.

Primary Structure

Ang II is produced by subsequent cleavages of angiotensinogen (AGT) by renin and angiotensin I (Ang I) by ACE (see Chapter 29, Renin-Angiotensin System).

Synthesis and Release

Gene and mRNA

AGT is located on chromosome 1q42.2 in humans. AGT belongs to the SERPIN family. AGT protein sequences are highly variable but the functional domains (Ang I) are relatively conserved among vertebrates (Table 29B.1). Ang I sequences were identified by the incubation of a renin source (kidney extract) and an angiotensinogen source (plasma). Lamprey Ang I with sequence NRVYVHPFTL was first sequenced using peptide purification. However, lamprey genome data suggested a different Ang I sequence, and it was later confirmed as the native AGT [2]. The purified teleost-type Ang I that is produced in the buccal gland was suggested to be involved in an endocrine mimicry to reduce host immunorejection [3].

Distribution of mRNA

AGT is widely expressed in various tissues including brain and kidney, but the main site is the liver. In addition, fish *AGT* is also expressed in the digestive tract and the shark rectal gland. Lamprey *AGT* is expressed in most tissues studied, including heart, gill, muscle, and gonad [4].

Tissue and Plasma Concentrations

Circulating Ang II was measured by RIAs in different vertebrate species (Table 29B.2). Although the *AGT* of teleosts produces [Asn[1]]-AGT, [Asp[1]]-Ang II is also present in their plasma, possibly due to the conversion of asparagine to aspartic acid by asparinginase in the liver and other tissues. A higher plasma concentration is usually associated with hypertension as a result of increased salt loading. A relatively high plasma Ang II is associated with seawater acclimation in fishes.

Regulation of Synthesis and Release

AGT production in liver is constitutive and the regulation of the synthesis is regulated negatively by renin and positively by Ang II. Ang II formation is determined by renin activity. Conversion of Ang I to Ang II by ACE is not rate-limiting under normal conditions.

Receptors

Structure and Subtype

Ang II activates AT1 (AGTR1) and AT2 (AGTR2). Two major methods were used to identify the angiotensin receptor subtypes: (1) binding affinities/antagonist specificities; and (2) molecular cloning. Early studies have used mammalian receptor antagonists to identify the receptor subtypes in non-mammalian vertebrates. Extrapolation of the effects of receptor antagonists as a definition of receptor subtypes (e.g., the losartan-sensitive receptor in fish) has been proven invalid [2].

Pharmacological and physiological studies of avian angiotensin receptors suggested three types of receptors: (1) a vascular endothelial receptor that mediates vasorelaxation via cGMP and nitric oxide production; (2) a vascular smooth muscle receptor with high affinity binding sites but unknown function; and (3) an adrenergic nerve ending receptor that mediates norepinephrine release. Homologous chicken and turkey AT1 receptors were cloned and expressed, and they exhibit affinities to Ang II and peptide analogs, but not to non-peptide AT1 and AT2 receptor antagonists such as losartan [4]. Trout glomeruli exhibited losartan binding but losartan lacks a physiological effect to inhibit the RAS in fish [5].

Signal Transduction Pathway

AT1 is the most thoroughly studied Ang II receptor, mediating most of the known classical angiotensin effects, such as vasoconstriction and drinking stimulation [1]. In mammals, AT1 signals through the Ca/IP3 pathway and activates acute

Y. Takei, H. Ando, & K. Tsutsui (Eds): Handbook of Hormones. DOI: http://dx.doi.org/10.1016/B978-0-12-801028-0.00177-X

Table 29B.1 Ang I in Vertebrates

Human	-----DRVYIHPFHL-
Rat	-----DRVYIHPFHL-
Bovine	-----DRVYVHPFHL-
Microbat	-----DRLYIHPFHML-
Kangaroo rat	-----DRIYVHPFHFL
Opossum	-----DRVYVHPFHL-
Tasmanian devil	-----DRVVYHPFYL-
Platypus	-----DRVYVHPFHFL-
Chicken	-----DRVYVHPFSL-
Anole lizard	-----DRVYVHPFYL-
Western clawed frog	-----NRVYVHPFKL-
Spotted gar	-----NRVYVHPFKL-
Coelacanth	-----NRVYVHPFNL-
Seabream	-----NRVYIHPFHL-
Eel	-----NRVYVHPFGL-
Stingray	-----DRPYIHPFFL-
Little skate	-----YRPYIHPFSL-
Dogfish	-----NRPYIHPFQL-
Elephantfish	-----NRPHIHPFLL-
Sea lamprey	EEDYDDRPYMQPFHL-

The shaded residues indicate difference in amino acid compared to
   human.

Table 29B.2 Circulating Levels of Ang II in Vertebrates

Species		Plasma Ang II (fmol/ml)
Human		8.6 ± 2.9
Rat		99.3 ± 28.4
Duck		102.7 ± 646.9
Frog		12.5 ± 2.3
Toad		84.6 ± 11.8
Eel	FW	48.3 ± 9.0
	SW	85.2 ± 14.1
	DSW	383.1 ± 151.9
Seabream	FW	2097.6 ± 478.5
	SW	1409.9 ± 295.1
	DSW	5946.1 ± 614.6
Dogfish	SW	109.9 ± 15.3
Lamprey	FW	157.4 ± 35.2

FW; freshwater, SW; seawater, DSW; double-seawater.

responses, such as vascular smooth muscle contraction, MAPK, and the EGF receptor. The Ang II-induced Ca/IP3 pathway in tissues was demonstrated in fishes, amphibians, reptiles, and mammals, but not in birds [4]. In chicken, Ang II elicits a transient rise in cGMP to induce relaxation. *Xenopus* AT1 receptor was cloned and the expressed receptor triggers a Ca/IP3 signal but has no binding affinity to losartan. Ang II binding sites are present in snake heart but the receptors have low affinity to losartan and PD compounds. Ang II constricts vascular smooth muscle in turtle and snake, depending on adrenergic stimulation. In fish, Ang II elicits a typical Ca/IP3 response [5]. In toadfish, Ang II increased blood pressure but the vasopressor effect was not selectively reduced by AT1 and AT2 antagonists. In spiny dogfish, Ang II increased blood pressure and plasma catecholamines, and α-adrenergic blockers abolished the vasopressor effect but not the catecholamine-releasing effect. Although AT1 receptors were cloned in elasmobranchs and teleosts, no expression study has demonstrated a Ca/IP3 signal in the recombinant receptors, which leads to the question whether the same receptor is responsible for the observed AT1-like signaling in isolated tissues or cells [5]. At least three specific phosphatases were stimulated upon AT2 activation: MAPK-1, SH2 domain-containing phosphatase 1 (SHP-1), and protein phosphatase 2A (PP2A). The AT2 also modulated the cGMP/NO pathway via both soluble and particulate guanylyl cyclases (GC), but the increase or decrease in cGMP was highly dependent on the cell types used in the experiments [6]. Recombinant AT2

of eel induced a cAMP signaling pathway, indicating AT2 receptors may have undergone significant evolutionary changes in intracellular signaling [7].

Agonist

A non-peptide ligand, L-162,313, was shown to bind to AT1A, AT1B, and AT2 receptors and act as an agonist on AT1A and AT1B receptors in mice (34.9 and 23.3% of maximum response of Ang II, respectively). Early studies used CGP42112A as agonist for the AT2 receptor but Compound 21, with little affinity for AT1 receptor, was recently developed as a specific agonist of the AT2 receptor in humans [8].

Antagonist

Saralasin ([Sar^1], [Ala^8]-Ang II) is a partial agonist of angiotensin II receptors, though it is commonly mistaken as a competitive antagonist. Antagonists including losartan, irbesartan, olmesartan, candesartan, and valsartan are commonly used clinically as AT1 blockers. Two AT2 receptor antagonists, PD123177 (discontinued) and PD123319 (available), were developed to inhibit AT2 activities.

Biological Functions

Target Cells/Tissues and Functions

Ang II/AT1 axis mediates vasoconstriction, thirst, release of vasopressin and aldosterone, renal sodium reabsorption, fibrosis, inflammation, angiogenesis, vascular aging, and atherosclerosis. Ang II-induced effects included blood pressure control, increased drinking, adrenergic stimulation, modulation of ion pump and transporter activities in the gill, kidney, and intestine in fish, control of filtering nephron population in fish, and regulation of ventral skin absorption in amphibians [1]. Injection of Ang II significantly increases ventral skin drinking in the frog. Lamprey Ang II is a vasodepressor instead of a vasopressor when injected intra-arterially [2]. Intracerebroventricular (ICV) injection of Ang II into trout increases systemic blood pressure, heart rate, and ventilation rate. ICV injection of Ang II elicits tachycardia in contrast to bradycardia when injected peripherally. Central Ang II injection also inhibits the vagal-mediated baro-reflex, indicating brain RAS is involved in heart-rate control [9]. The AT2 receptor is mostly embryonic and expression is decreased in adults and is confined in certain tissues such as kidney. The effects of AT2 are often antagonistic to AT1, and activation of AT2 receptors usually indicates a pathophysiological condition of AT1-mediated action with potential harmful consequences. AT2 is abundantly expressed in the spleen of adult eel, which suggested an immune-related function [7].

Phenotype in Gene-Modified Animals

AGT-knockout mice displayed marked hypotension, abnormal kidney morphology, reduced survival rates of newborns, and impaired blood–brain barrier function after cold injury. AT1A-knockout mice exhibit decreased cardiac fibrosis, attenuation of diet-induced body weight gain and adiposity, low blood pressure, increase in water intake under high salt diet, disturbed circadian rhythm, and enhanced sodium sensitivity [10].

Pathophysiological Implications

Clinical Implications

High levels of Ang II are often related to hypertension, renal failure, and cardiac fibrosis.

Use for Diagnosis and Treatment

Renin antagonists, ACE inhibitors, and AT1 receptor blockers have been used, in singular or multiple blockade, to treat hypertension and other RAS-related diseases.

References

1. Kobayashi H, Takei Y. Renin-angiotensin system: comparative aspects. In: Bradshaw SD, Burggren W, Heller HC, et al., eds. *Zoophysiology*. vol 35. NY: Academic Press; 1996:1−245.
2. Wong MKS, Takei Y. Characterization of a native angiotensin from an anciently diverged serine protease inhibitor in lamprey. *J Endocrinol*. 2011;209:127−137.
3. Wong MKS, Sower SA, Takei Y. The presence of teleost-type angiotensin components in lamprey buccal gland suggests a role in endocrine mimicry. *Biochimie*. 2012;94:637−648.
4. Nishimura H. Angiotensin receptors − evolutionary overview and perspectives. *Comp Biochem Physiol A Mol Integr Physiol*. 2001;128:11−30.
5. Russell MJ, Klemmer AM, Olson KR. Angiotensin signaling and receptor types in teleost fish. *Comp Biochem Physiol A*. 2001;128:41−51.
6. Gallinat S, Busche S, Raizada MK, et al. The angiotensin II type 2 receptor: an enigma with multiple variations. *Am J Physiol Endocrinol Metab*. 2000;278:E357−E374.
7. Wong MKS, Takei Y. Angiotensin AT2 receptor activates the cyclic-AMP signaling pathway in eel. *Mol Cell Endocrinol*. 2013;365:292−302.
8. Unger T, Dahlöf B. Compound 21, the first orally active, selective agonist of the angiotensin type 2 receptor (AT2): implications for AT2 receptor research and therapeutic potential. *J Renin Angiotensin Aldosterone Syst*. 2010;11:75−77.
9. Le Mével JC, Lancien F, Mimassi N. Central cardiovascular actions of angiotensin II in trout. *Gen Comp Endocrinol*. 2008;157:27−34.
10. Zhuo JL, Ferrao FM, Zheng Y, et al. New frontiers in the intrarenal Renin-Angiotensin system: a critical review of classical and new paradigms. *Front Endocrinol (Lausanne)*. 2013;4:166.

Supplemental Information Available on Companion Website

- Gene, mRNA and domain structure of the human angiotensin type-1 receptor /E-Figure 29B.1
- Phylogenetic tree of the angiotensin type-1 receptors in vertebrates /E-Figure 29B.2
- Gene, mRNA and domain structure of the human angiotensin type-2 receptor /E-Figure 29B.3
- Phylogenetic tree of the angiotensin type-2 receptors in vertebrates /E-Figure 29B.4
- Accession numbers of vertebrate angiotensin type-1 and type-2 receptors (AT1 and AT2) /E-Table 29B.1

Other Angiotensins

Marty K.S. Wong

Additional names/abbreviations: Angiotensin III or angiotensin (2−8) (Ang III, AIII, Ang(2−8)); angiotensin IV or angiotensin (3−8) (Ang IV, AIV, Ang(3−8)); angiotensin (1−7) (Ang(1−7))

Originally thought to be inactive metabolites of the renin-angiotensin system, but these peptides were recently shown to possess different receptors and the functions are often antagonistic to those of Ang II. Drug targets for hypertension and Ang II-induced cardiovascular and renal diseases.

Discovery

Originally thought to be biological inactive, these angiotensin peptide subtypes were found to have physiological roles and some possess specific receptors and signaling pathways [1].

Structure

Structural Features

Ang III, Ang IV, and Ang(1−7) are linear peptides with no known secondary modification (Table 29C.1).

Primary Structure

Ang III is produced by subsequent cleavage of Ang II by aminopeptidase A. Ang IV is produced by cleavage of Ang III by aminopeptidase N. Ang(1−7) is produced by various pathways involving ACE2 (See Chapter 29, Renin-Angiotensin System).

Synthesis and Release

Gene and mRNA

See Chapter 29B, Angiotensin II.

Distribution of mRNA

See Chapter 29B, Angiotensin II.

Tissue and Plasma Concentrations

Ang III and Ang IV are short-lived peptides, and their concentrations in human plasma are kept at undetectable levels. Plasma Ang(1−7) baseline concentration in humans is 4.7 ± 0.9 fmol/ml. In eels, plasma Ang III and Ang IV are present in plasma but their levels are low compared to Ang II (Table 29C.2) [2]. In trout brain, Ang III is not detectable but Ang IV is present [3].

Regulation of Synthesis and Release

The regulation of synthesis and release of Ang III and Ang IV is not clear. Ang(1−7) synthesis and release are regulated by the activity of ACE2 (see Chapter 29, Renin-Angiotensin System).

Receptors

Structure and Subtypes

Ang III binds to AT1 and AT2. Ang IV binds to the angiotensin type-4 receptor (AT4), also known as insulin-regulated aminopeptidase (IRAP). Ang(1−7) binds to the Mas receptor (*Mas1*).

Signal Transduction Pathway

Ang III stimulates AT1 and AT2 receptors and its signaling pathway is similar to that of Ang II. Ang III has preferential binding on the AT2 receptor. The Mas receptor is activated by Ang(1−7) but the intracellular signaling pathway is not well understood. The receptor function is usually associated with Ang II-dependent effects and is known to counter the AT1-dependent signaling. Ang(1−7)/Mas activation inhibits the AT1-dependent activation of MAPK kinase in the epithelial cells of proximal tubules. In cardiovascular epithelium of mammals, Mas receptor activation stimulates phosphorylation of AKT and increases endothelial NO synthesis, leading to vasorelaxation that counters the vasoconstriction effects of Ang II. A combination of angiotensin signaling has been recently noticed, which states that the physiological effect depends not only on a single pathway, but is a result of a specific ratio among various angiotensins, acting through their own receptor pathways. Renal mesangial cell proliferation was stimulated by Ang II and Ang(1−7) independently via AT1 and Mas receptors respectively [4]. However, a combination of stimulatory concentration of Ang II and Ang(1−7) counters the stimulation, indicating the complex interaction within RAS signaling. AT4/IRAP activates intracellular signals including an increase in intracellular Ca concentration, modulation of MAPK kinases, activation of NF-κB signaling, and production of cGMP. However, the effects are largely dependent on the cell types and, in some cases, no classical signaling could be demonstrated despite the presence of AT4 binding sites.

Agonist

AVE 0991 is a non-peptide Ang(1−7) agonist and it stimulates the Mas receptor to produce cardiovascular protective effects to counter the pathophysiological effects of AT1 [5].

Y. Takei, H. Ando, & K. Tsutsui (Eds): Handbook of Hormones. DOI: http://dx.doi.org/10.1016/B978-0-12-801028-0.00178-1

Table 29C.1 Peptide Sequences of Ang III, Ang IV, and Ang(1−7) in Human

Ang III	RVYIHPF
Ang IV	VYIHPF
Ang (1−7)	DRVYIHP

Table 29C.2 Plasma Concentrations of Ang III and Ang IV in Eel

	Ang III	Ang IV
FW	28.3 ± 3.8	26.5 ± 3.6 fmol/ml
SW	56.3 ± 11.9	60.1 ± 13.3 fmol/ml

[Nle^1]-angiotensin IV has been suggested as a specific Ang IV agonist that constitutively activates AT4/IRAP.

Antagonist

A779 is a non-peptide Ang(1−7) agonist and it inhibits the signaling of Mas receptors and attenuates the cardiovascular effects of Ang(1−7) [6].

Biological Functions

Target Cells/Tissues and Functions

Ang III has preferential stimulation to release aldosterone in adrenal cortex, which is partially via AT2 but not AT1 [7]. Intra-arterial injection of Ang III or Ang IV increases blood pressure in teleosts but Ang III has a higher potency than Ang IV. Intracerebroventricular (ICV) injection of Ang III increases the heart rate without affecting the blood pressure and ventilation rate in trout, in contrast to the effect of Ang II, which increases all three parameters [3]. ICV injection of Ang IV does not affect blood pressure, heart rate, and ventilation rate even though it is detected in the brain by immunoassays. This suggested that a specific receptor for Ang III (and Ang II) could be present to elicit the preferential effect of heart rate control in trout [8]. The Mas receptor expresses in brain, testis, ovary, and endothelial cells of blood vessels. Besides AT2, the Ang(1−7)/Mas receptor axis is also known to antagonize the effects of the AT1 axis. These antagonistic effects include anti-hypertrophic action, anti-thrombotic and anti-fibrotic effects, and vasodilation via stimulation of NO synthesis in endothelium and potentiation of the bradykinin effect. The localization and physiological effects of the Mas receptor are not clear in non-mammalian vertebrates. AT4/IRAP is broadly distributed in kidney, aorta, heart, liver, lung, uterus, adrenal gland, and brain, especially in neurons associated with memory function [9]. The Ang IV/AT4 axis is involved in facilitation of memory and can reverse memory deficits caused by alcohol abuse and ischemia. AT4 antagonist decreases renal blood flow and increases urinary sodium excretion, and these effects are independent of the AT1 pathway. The large variation in signaling and function of AT4 poses difficulties for researchers. Ang IV was detected in considerable amounts in the brain of trout, indicating a possible role of memory function as in the case of mammals. There is so far no information on AT4/IRAP in non-mammalian vertebrates.

Phenotype in Gene-Modified Animals

ACE-deleted human patients exhibit hypertension and this is related to a low plasma Ang(1−7) concentration. Baroreflex bradycardia was lowered but vascular responsiveness to Ang II was enhanced in Mas-knockout mice. Mas-knockout also impaired post-ischemic neovascularization and endothelial NO formation.

Pathophysiological Implications

Clinical Implications

Ang(1−7) reduces mechanical stretch-induced cardiac hypertrophy through downregulation of AT1. ACE2 was found to function as a receptor for the coronavirus that caused the infamous severe acute respiratory syndrome (SARS) in 2002−2003. The SARS virus attaches to ACE2 and diminishes the expression and thus the production of Ang(1−7), leading to an intensified activation of AT1. Injection of recombinant ACE2 into mice protected the lung from sepsis and thus ACE can be a target treatment in lung injury associated with SARS.

Use for Diagnosis and Treatment

A low Ang(1−7) is associated with hypertension and cardiac hypertrophy. Agonists for the Mas receptor have been targets to control cardiovascular diseases caused by hyperactive AT1. AngIV/AT4 is involved in memory and is a treatment target of Alzheimer's and Parkinson's diseases.

References

1. Fyhrquist F, Saijonmaa O. Renin-angiotensin system revisited. *J Intern Med.* 2008;264:224−236.
2. Wong MKS, Takei Y. Changes in plasma angiotensin subtypes in Japanese eel acclimated to various salinities from deionized water to double-strength seawater. *Gen Comp Endocrinol.* 2012;178:250−258.
3. Lancien F, Wong MKS, Arab AA, et al. Central ventilatory and cardiovascular actions of angiotensin peptides in trout. *Am J Physiol.* 2012;303:R311−R320.
4. Xue H, Zhou L, Yuan P, et al. Counteraction between angiotensin II and angiotensin-(1−7) via activating angiotensin type I and Mas receptor on rat renal mesangial cells. *Regul Pept.* 2012;177:12−20.
5. Santos RA, Ferreira AJ. Pharmacological effects of AVE 0991, a non-peptide angiotensin-(1−7) receptor agonist. *Cardiovasc Drug Rev.* 2006;24:239−246.
6. Peiró C, Vallejo S, Gembardt F, et al. Complete blockade of the vasorelaxant effects of angiotensin-(1-7) and bradykinin in murine microvessels by antagonists of the receptor Mas. *J Physiol.* 2013;591:2275−2285.
7. Yatabe J, Yoneda M, Yatabe MS, et al. Angiotensin III stimulates aldosterone secretion from adrenal gland partially via angiotensin II type 2 receptor but not angiotensin II type 1 receptor. *Endocrinology.* 2011;152:1582−1588.
8. Mimassi N, Lancien F, Le Mével JC. Central and peripheral cardiovascular effects of angiotensin III in trout. *Ann N Y Acad Sci.* 2009;1163:469−471.
9. Chai SY, Fernando R, Peck G, et al. The angiotensin IV/AT4 receptor. *Cell Mol Life Sci.* 2004;61:2728−2737.

Supplemental Information Available on Companion Website

- Gene, mRNA and domain structure of the human mas-1 receptor/E-Figure 29C.1
- Phylogenetic tree of the mas-1 oncogene receptors in vertebrates/E-Figure 29C.2
- Gene, mRNA and domain structure of the human AT4 receptor/IRAP/E-Figure 29C.3
- Phylogenetic tree of the angiotensin type-4 receptors in vertebrates/E-Figure 29C.4
- Accession numbers of vertebrate Mas-1 oncogene (MAS1) for angiotensin (1-7)/E-Table 29C.1
- Accession numbers of vertebrate AT4 receptor/IRAP gene (*Inpep*) for angiotensin IV/E-Table 29C.2

Angiotensin Converting Enzymes

Marty K.S. Wong

Additional names/abbreviations: Angiotensin converting enzyme, dipeptidyl carboxypeptidase I, peptidase P, kininase II, angiotensin I-converting enzyme/ACE; angiotensin converting enzyme 2, peptidyl-dipeptidase A, peptidyl-dipeptidase A/ACE2

ACE possesses dual actions to convert Ang I to Ang II and degrade bradykinin. The development of ACE inhibitor was the first effective drug for hypertension caused by high renin activity. ACE2 was identified as the receptor for SARS (severe acute respiratory syndrome) coronavirus, which caused the outbreak of an epidemic in 2002–2003.

Discovery

ACE was discovered in the mid-1950s by the observation that dialysis of plasma and kidney extract with water and saline before incubation had produced two separate pressor substances, Ang I and Ang II respectively [1]. It was discovered for a second time in 1966 during the characterization of bradykinin (BK) degrading enzyme from kidney and this enzyme was named kininase II; it later was found to be the same enzyme as ACE. ACE2 was discovered in 2000 when two independent research groups cloned homologous ACE that could convert Ang I to Ang(1−9) and yet also is captopril-insensitive [2,3].

Structure

Structural Features

Two isozymes of ACE are present in mammals: somatic ACE and testis ACE. Somatic ACE possesses two catalytic domains (N- and C-domains) and a C-terminal transmembrane segment (stalk) (Figure 29D.1). Somatic and testis ACEs in humans contain 1,306 and 665 aa residues, respectively. Testis ACE only possesses one catalytic domain. Both catalytic domains are zinc-metallopeptidase with the active motif HEMGH where the two histidine residues coordinate the zinc ion. The stalk anchors the enzyme on the membrane and is susceptible to be cleaved by shedding enzymes, resulting in plasma ACE activity (Figure 29D.1). ACE2 is a chimaera protein with a single catalytic domain of ACE, and a C-terminal highly resembling collectrin, which may act as a chaperone protein to deliver other proteins to the brush border membrane.

Synthesis and Release

Gene and mRNA

ACE and *ACE2* genes are located at chromosome 17q23 and Xp22 in humans, respectively. Testis *ACE* is transcribed from the same gene with an alternative transcription starting site on the 13th intron of the *ACE* gene, resulting in only C-domain and stalk segment with a unique additional 67 aa N-terminal sequence in humans. The two catalytic domains are the result of gene/domain duplication and the duplication occurred multiple times in evolution as the cnidarians, crustaceans, insects, and vertebrates possess ACE-like enzymes with one or two catalytic domains. No expression studies so far have been performed for non-mammalian *ACE* and *ACE2*.

Distribution of mRNA

Somatic *ACE* is expressed in various tissues including blood vessels, kidney, intestine, adrenal gland, liver, and uterus, and is especially abundant in highly vascular organs such as retina and lung. Testis *ACE* is expressed by postmeiotic male germ cells and high-level expression is found in round and elongated spermatids. *ACE2* is expressed in lung, liver, intestine, brain, testis, heart, and kidney.

Tissue and Plasma Concentrations

Lung possesses the highest amount of ACE and contributes to 0.1% of total protein. Serum ACE levels in humans ranged from $299.3 \pm 49\,\mu g/l$ (DD) to $494.1 \pm 88.3\,\mu g/l$ (II) with heterozygous individuals $392.6 \pm 66.8\,\mu g/l$ [4]. (ID: see the section "Pathophysiological Implications" for the genotype definition.) Several enzymatic assays have been developed for the measurement of ACE activity in plasma and tissues and usually involve artificial substrates such as hippuryl-His-Leu or N-[3-(2-furyl)acryloyl]L-phenylalanyl-glycyl-glycine (FAPGG), in combination with captopril inhibition. These methods were developed in mammals but were also extended to other vertebrates including birds, amphibians, and fishes [5]. However, these enzymatic methods may be erroneous because the enzyme specificity on the artificial substrates could be different. Lamprey ACE activities in different tissues were measured but captopril failed to decrease the ACE activities, indicating a possible nonspecific enzyme measurement. In amphibian, high captopril-sensitive ACE activities were found in gonad, intestine, kidney, and lung, moderate activities were presented in liver, heart, and skin, and low or negligible activities were observed in plasma, muscle, and erythrocytes.

Y. Takei, H. Ando, & K. Tsutsui (Eds): Handbook of Hormones. DOI: http://dx.doi.org/10.1016/B978-0-12-801028-0.00254-3

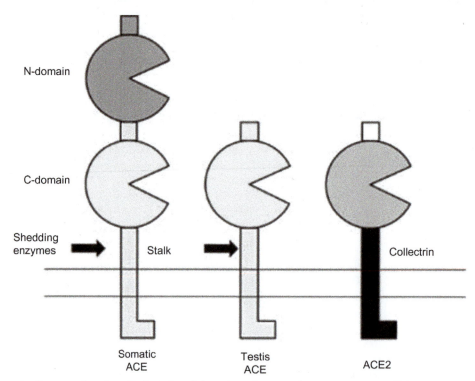

Figure 29D.1 Schematic diagram showing the functional domains of ACE and ACE2. Two extracellular enzymatic domains are present in ACE. An alternative transcript of ACE form testicular ACE which only possess a single enzymatic domain. The stalk anchors the enzyme on the membrane and is suspectible to be cleaved by shedding enzymes, resulting in plasma ACE activity. ACE2 is a chimaera protein with a single catalytic domain of ACE, and a C-terminal highly resemble collectrin, which may act as a chaperone protein to deliver other proteins to the brush border membrane.

Regulation of Synthesis and Release

Expression of *ACE* is affected by steroids and thyroid hormone, but the details of the regulation are not clear. *ACE* is under promoter regulation by hypoxia-inducing factor 1α (HIF-1α), which upregulates the *ACE* expression under hypoxic conditions, resulting in an increase in Ang II concentration. Under hypoxia, *ACE2* will be downregulated but it was shown that it is indirectly controlled by Ang II, but not HIF-1α [6]. Testis *ACE* expression control is highly specific and regulated by a tissue-specific promoter located immediately −59 bp of the transcription start site, which is frequently used in testis-specific overexpression studies. Hypoxia induced by high temperature decreased gill ACE activity but had no effect on kidney in carp. Promoters of *ACE2* from mammals, amphibians, and teleosts drive specific expression in the heart. *Cis*-element search results discovered WGATAR motifs in all putative *ACE2* promoters from different vertebrates, suggesting a possible role of GATA family transcriptional factors in *ACE2* expression regulation.

Receptors

None.

Inhibitors

The first ACE inhibitor was a peptide antagonist called SQ 20,881 (GWPRPEIPP) discovered from snake venom but it was not orally active. The snake venom peptides were further studied to produce the first orally active form, captopril, that lowers the blood pressure of essential hypertensive patients [7]. The most common side effects of captopril are cough, skin rash, and loss of taste, and therefore derivatives such as enalapril, lisinopril, and ramipril were developed with fewer

side effects. After the discovery of N- and C-domains of ACE, specific domain inhibitors were developed to increase specificity. Ang I is mainly hydrolyzed by the C-domain *in vivo* but BK is hydrolyzed by both domains. By developing a C-domain selective inhibitor (RXPA380) some degradation of BK by the N-domain would be permitted and this degradation could be enough to prevent accumulation of excess BK causing angioedema [8].

Biological Functions

Target Cells/Tissues and Functions

The well-known function of ACE is the conversion of Ang I to Ang II and degradation of BK, which all play an important role in controlling blood pressure. ACE also acts on other natural substrates including encephalin, neurotensin, and substance P. Besides being involved in blood pressure control, ACE possesses widespread functions including renal development, male fertility, hematopoiesis, erythropoiesis, myelopoiesis, and immune responses [1]. ACE2 can convert Ang II to Ang(1−7), thereby reducing the concentration of Ang II and increasing that of Ang(1−7). ACE2 can also convert Ang I to Ang(1−9), which is subsequently converted into Ang(1−7) by ACE. The high expression of ACE2 favors the balance of Ang(1−7) over Ang II, which accounts for the cardioprotective role of ACE2 via the Ang(1−7)/Mas signaling pathway [9].

Phenotype in Gene-Modified Animals

ACE-knockout mice display normal blood pressure under normal conditions, but are sensitive to changes in blood pressure such as exercise. ACE-knockout also affects renal function, renal development, serum and urine electrolyte composition, haematocrit, and male reproductive capacity

[10]. Deficiency in testis ACE affects male fertility but its exact role is still not clear. Although mice with testis ACE deficiency mate normally and their sperm quantity and motility are no different from those of wild-type mice, the survival of sperm in the oviduct and fertilization rate are highly reduced [1]. Overexpression of *ACE2* in hypertensive models, but not in normotensive animals, reduced blood pressure. ACE2-knockout mice displayed progressive cardiac dysfunction resembling that of long-term hypoxia after coronary artery disease or bypass surgery in human, which could be reversed by concurrent ACE-knockout. It was suggested that the cardioprotective function of ACE2 is to counterbalance the effects of ACE.

Pathophysiological Implications

Clinical Implications

Inclusion (II) or deletion (DD) of 287 bp Alu repeats in the 16th intron affects the human plasma ACE levels and the DD genotype was more frequently found in patients with myocardial infarction but no convincing evidence was available on the association of the DD genotype with hypertension [4]. ACE2 was identified as the receptor for SARS (severe acute respiratory syndrome) coronavirus. SARS virus binding downregulates the cellular expression of *ACE2*, and the binding induces clathrin-dependent internalization of virus/receptor (SARS/ACE2) complex. Not only has ACE2 facilitated the invasion of SARS virus for rapid replication, but also ACE2 is depleted from the cell membrane and therefore the damaging effects of Ang II are enhanced, resulting in acute deterioration of lung tissues.

Use for Diagnosis and Treatment

ACE has been the target of hypertension control since the 1970s. ACE inhibitors are prescribed as the sole or combinational treatment of high blood pressure, for the dual effects of lowering Ang II and slowing down BK degradation. In human hypertensive patients, ACE2 levels are lower in both kidney and heart compared to normotensive volunteers.

References

1. Bernstein KE, Ong FS, Blackwell WL, et al. A modern understanding of the traditional and nontraditional biological functions of angiotensin-converting enzyme. *Pharmacol Rev.* 2012;65:1−46.
2. Tipnis SR, Hooper NM, Hyde R, et al. A human homolog of angiotensin-converting enzyme. Cloning and functional expression as a captopril-insensitive carboxypeptidase. *J Biol Chem.* 2000;275(43):33238−33243.
3. Donoghue M, Hsieh F, Baronas E, et al. A novel angiotensin-converting enzyme-related carboxypeptidase (ACE2) converts angiotensin I to angiotensin 1-9. *Circ Res.* 2000;87:E1−E9.
4. Soubrier F, Wei L, Hubert C, et al. Molecular biology of the angiotensin I converting enzyme: II. Structure-function. Gene polymorphism and clinical implications. *J Hypertens.* 1993;11:599−604.
5. Chou CF, Loh CB, Foo YK, et al. ACE2 orthologues in non-mammalian vertebrates (*Danio, Gallus, Fugu, Tetraodon* and *Xenopus*). *Gene.* 2006;377:46−55.
6. Zhang R, Wu Y, Zhao M, et al. Role of HIF-1alpha in the regulation ACE and ACE2 expression in hypoxic human pulmonary artery smooth muscle cells. *Am J Physiol Lung Cell Mol Physiol.* 2009;297:L631−L640.
7. Erdös EG. The ACE and I: how ACE inhibitors came to be. *FASEB J.* 2006;20:1034−1038.
8. Georgiadis D, Cuniasse P, Cotton J, et al. Structural determinants of RXPA380, a potent and highly selective inhibitor of the angiotensin-converting enzyme C-domain. *Biochemistry.* 2004;43:8048−8054.
9. Clarke NE, Turner AJ. Angiotensin-converting enzyme 2: the first decade. *Int J Hypertens.* 2012;2012:307315.
10. Cole J, Ertoy D, Bernstein KE. Insights derived from ACE knockout mice. *J Renin Angiotensin Aldosterone Syst.* 2000;1:137−141.

Supplemental Information Available on Companion Website

- Protein sequences and structural features of human ACE and ACE2/E-Figure 29D.1
- Schematic diagram showing the functional domains of ACE and ACE2/E-Figure 29D.2
- Protein sequences and structural features of ACE and ACE2 of human/E-Figure 29D.3. E-Figure 29D.4
- Accession number of ACEs/E-Table 29D.1

CHAPTER 30

Kallikrein-Kinin System

Marty K.S. Wong

History

The observation in the early 19th century that injection of human urine into dog blood led to hypotension is considered to be the first evidence for the presence of the kallikrein-kinin system (KKS). Kininogen was found in the serum when incubation with kallikrein produced a factor that contracts isolated smooth muscles. Incubation of venom from *Bothrops jararaca* with dog plasma also produced a powerful hypotensive and smooth muscle relaxant peptide, which was coined as bradykinin (BK) later. Using protein purification, the bradykinin sequences were determined in a number of vertebrate representatives, including teleosts, reptiles, and birds. However, the homologous KKS in amphibians has been lost and an elaborate skin kinin system has evolved. The plasma KKS system has either been lost or not evolved in teleosts [1].

Structure

KKS Cascade

Two cascades, a plasma KKS and a tissue KKS, are major pathways for the formation of kinins in mammals (Figure 30.1). In the plasma KKS, high molecular weight (HMW) kininogen (KNG) is cleaved by plasma kallikrein (KLKB1) to form a nonapeptide known as BK. In the tissue KKS, low molecular weight (LMW) KNG is cleaved by tissue kallikreins (KLKs) to form a decapeptide [Lys^0]-BK or kallidin.

Tissue kallikreins are serine proteases that were known as glandular kallikreins, but the names and symbols of this protease family were unified recently [2]. Tissue KKS may be lost in amphibians and plasma KKS is lost in teleosts [3]. In elasmobranchs, only BK was purified and it is not clear whether the KKS exists or not [4]. There is no evidence for a functional KKS in lamprey as only the putative *KNG1* was cloned [5] (Table 30.1).

BK is a short-lived peptide and the main degradation enzyme is kininase II, which is also known as angiotensin converting enzyme (ACE). Besides ACE, BK is also degraded by a large number of enzymes including kininase I, aminopeptidase P, and cathepsin K. (Figure 30.2) [6]. Among different species, the ratio of these peptidases in plasma is very different, indicating that the regulation of BK metabolism is highly diverse [7].

The KKS cascade is closely related to that of the renin-angiotensin system (RAS). KLKB1 is known to convert prorenin to renin. ACE and kininase II are same enzyme that converts inactive Ang I to Ang II and inactivates BK. Kininase II possesses a higher affinity to BKs than Ang I, and is thought to originate from the KKS and predate its role in the RAS. Inhibition of ACE not only decreases Ang II, but also delays BK degradation. Besides active peptides, the KKS and RAS also cross-talk at the receptor level by heterologous dimerization among the angiotensin type-1 receptor (AT1), angiotensin type-2 receptor (AT2), and bradykinin B2 receptor, which modulates receptor signaling pathways [8].

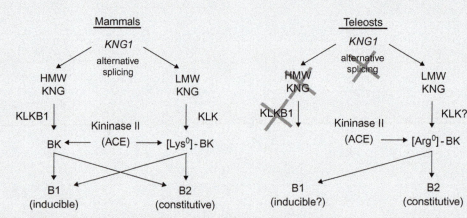

Figure 30.1   Cascade pathways of KKS of mammals vs. teleosts. Two cascades, a plasma KKS and a tissue KKS, are major pathways for the formation of kinins in mammals. In the plasma KKS, high molecular weight (HMW) kininogen (KNG) is cleaved by plasma kallikrein (KLKB1) to form a nonapeptide known as bradykinin (BK). In the tissue KKS, low molecular weight (LMW) KNG is cleaved by tissue kallikreins (KLKs) to form a decapeptide [Lys^0]-BK or kallidin. Plasma KKS cascade has not been identified in teleosts.

Y. Takei, H. Ando, & K. Tsutsui (Eds): Handbook of Hormones. DOI: http://dx.doi.org/10.1016/B978-0-12-801028-0.00030-1

Table 30.1 Presence of KKS Components in Vertebrates

KKS Components		Lamprey	Cartilaginous Fish	Teleosts	Coelacanths	Amphibians	Raptiles/Birds/Mammals
Kininogen (KNGI)	HWM	?	?	−	+	pseudogene	+
	LMW	+	?	+	?	−	+
Plasma kallikrein (KLKBI)		?	?	−	+	+	+
Tissue kallikrein (KLK)		?	?	?	?	?	+
Bradykinin (BK)		?	+	+	+	skin-BK	+
B1 receptor (BDKRB1)		?	?	+	?	?	+
B2 receptor (BDKRB2)		?	+	+	+	+	+

The KKS components were identified through genome database and published data. The components of KKS in amphibians, fishes, and cyclostomes have not yet been fully determined.

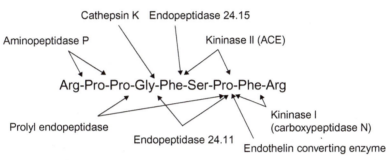

Figure 30.2 Metabolism of bradykinin by various enzymes. BK is a short-lived peptide and the main degradation enzyme is kininase II, which is also known as angiotensin converting enzyme (ACE). In addition to ACE, BK is also degraded by a large number of enzymes including kininase I, aminopeptidase P, and cathepsin K.

Receptors

Structure and Subtypes

Two receptors are known for KKS: inducible B1 and constitutive B2 receptors. Both receptors are GPCRs. The B1 receptor is induced during inflammation, especially in wound areas. B2 is present in most tissues [1]. BK receptors in non-mammalian vertebrates such as fishes were characterized using pharmacological antagonists. However, the affinities and effects of the blockers were not validated in those studies, and, in addition, many were ineffective in blocking the effects of teleost BKs [9]. Therefore, it is important to identify the source of definition of BK receptors in non-mammalian vertebrates. Overexpressed zebrafish B2 has demonstrated a similar Ca signaling pathway to that in mammals [10].

Signal Transduction Pathway

Both B1 and B2 are coupled with G_q protein and utilize Ca signaling, and involve endothelium-dependent relaxing factors such as NO production and prostaglandin synthesis.

Biological Functions

Target Cells/Tissues and Functions

Due to the rapid degradation of kinins, KKS actions are considered to be paracrine rather than endocrine. BKs target various tissues and are involved in the regulation of blood pressure, inflammation, membrane permeability, pain perception, drinking control, etc. Many functions are also antagonistic to those of the RAS including vasodilation, anti-

hypertrophy, anti-apoptosis, anti-fibrosis and natriuresis/diuresis at the kidney.

References

1. Conlon JM. The kallikrein-kinin system: evolution of form and function. *Ann NY Acad Sci.* 1998;839:1−8.
2. Diamandis EP, Yousef GM, Clements J, et al. New nomenclature for the human tissue kallikrein gene family. *Clin Chem.* 2000;46:1855−1858.
3. Wong MKS, Takei Y. Lack of plasma kallikrein-kinin system cascade in teleosts. *PLoS One.* 2013;8:e81057.
4. Anderson WG, Leprince J, Conlon JM. Lack of plasma kallikrein-kinin system cascade in teleosts. *Peptides.* 2008;29:1280−1286.
5. Zhou L, Liu X, Jin P, et al. Cloning of the kininogen gene from *Lampetra japonica* provides insights into its phylogeny in vertebrates. *J Genet Genomics.* 2009;36:109−115.
6. Campell DJ. Towards understanding the kallikrein-kinin system: insights from measurement of kinin peptides. *Braz J Med Biol Res.* 2000;33:665−677.
7. Décarie A, Raymond P, Gervais N, et al. Serum interspecies differences in metabolic pathways of bradykinin and [des-Arg9] BK: influence of enalaprilat. *Am J Physiol.* 1996;271:H1340−H1347.
8. Schmaier AH. The kallikrein-kinin and the renin-angiotensin systems have a multilayered interaction. *Am J Physiol Regul Integr Comp Physiol.* 2003;285:R1−13.
9. Jensen J, Conlon JM. Effects of trout bradykinin on the motility of the trout stomach and intestine: evidence for a receptor distinct from mammalian B1 and B2 subtypes. *Br J Pharmacol.* 1997;121:526−530.
10. Bromée T, Conlon JM, Larhammar D. Bradykinin receptors in the zebrafish (*Danio rerio*). *Ann NY Acad Sci.* 2005;1040:246−248.

Kininogen

Marty K.S. Wong

Abbreviation: KNG

Additional names/abbreviation: Alpha-2-thiol proteinase inhibitor, Fitzgerald factor, kinin precursor

The gene structure of kininogen in vertebrates increased its complexity during evolution, and the existence of high molecular weight (HMW) and low molecular weight (LMW) kininogen is evolutionarily advanced. Besides serving as the precursor proteins of the kallikrein-kinin system (KKS), novel functions of kininogens have been discovered continually.

Discovery

The discovery of kininogen (KNG) dates back to the observation in 1937 that addition of serum to kallikrein was necessary for the contraction of isolated smooth muscule tissues.

Structure

Structural Features

KNG is the precursor protein of the KKS. HMW and LMW KNGs are formed by alternative splicing of *KNG1* in mammals. The *KNG1* in human locates at chromosome 3q27. Human HMW and LMW KNGs contain 644 and 427 aa residues, respectively (Figure 30A.1).

In HMW KNG, the bradykinin (BK) domain (D4) is preceded by two to three cystatin domains (D1−D3, depending on species), and followed by a histidine-rich domain (D5) and a kallikrein-interacting domain (D6). The D5 is positively charged, and interacts with negatively charged surfaces. The D6 is known to interact with the apple-domains of plasma kallikrein (KLKB1). The cystatin domains contribute to the protease inhibitory function. In LMW KNG, D5 and D6 are not present. From reptiles to mammals, the domain structures of KNG are relatively conserved. However, in amphibians, the KKS underwent a lineage-specific modification where the BK domain on the homologous KNG was lost (Figure 30A.2). Whether amphibians possess the homologous KKS cascade has not yet been defined but they were known to produce "skin kinins" as poisons for self-protection. In teleosts, a single KNG is present and it is similar to LMW KNG of mammals, which possesses two cystatin domains and a BK domain [1]. The D6 domain in teleosts is not clear as they also lackKLKB1 and therefore a homologous protein−protein interaction is not assumed. In lampreys, KNG possesses only one cystatin-like domain and a BK domain, but whether the lamprey BK is genuine or not remains undetermined as the suggested cleavage site is not compatible with kallikrein [2]. HMW KNG is a glycosylated globular protein and possesses a disulfide bond connecting D1 and D6. D1−D3 is referred to as the light chain and D5−D6 as the heavy chain. No disulfide bond is present in LMW KNG. The BK domain is buried within the globular protein. Mr values of HMW KNG and LMW KNG in humans are 120 kDa and 68 kDa, respectively. N- and O-glycosylation are present: percentage glycosylation of human HMW KNG and LMW KNG are 40% and 30%, respectively. Phosphorylation sites are present. The Mr of teleost KNGs ranges from 51.0 to 60.7 kDa with 24.0−29.8% glycosylation.

Synthesis and Release

Gene and mRNA

The gene that produces HMW and LMW KNG is *KNG1*, which is located at chromosome 3q21 in humans. The gene transcript alternatively splices at the 3′-region and forms two different transcripts, resulting in HMW and LMW KNGs. The frog skin-kinin gene mostly contains only a signal peptide and a kinin domain and it is expressed in the skin glands (Figure 30A.3) [3].

Distribution of mRNA

HMW *KNG1* transcripts are mainly found in the liver. LMW *KNG1* transcripts are widely distributed and expressed in most tissues. In teleosts, KNG is ubiquitously expressed in most tissues and a higher expression is found in the liver.

Tissue and Plasma Concentrations

The plasma concentrations of HMW KNG and LMW KNG in humans are 100 and 90 μg/ml, respectively. HMW KNG is mostly present in the plasma but LMW KNG is present in both plasma and tissues as a local precursor.

Regulation of Synthesis and Release

HMW KNG is synthesized and released from the liver and the secretion is constitutive. Human *KNG1* was upregulated by agonists of the farnesoid X receptor (FXR), a nuclear receptor for bile acids, which act on the conserved FXR responsive element at the promoter region of *KNG1* [4]. Lipopolysaccharide treatment induced LMW *KNG1* expression in cultured vascular smooth muscle cells [5].

Receptors

Structure and Subtype

None.

Y. Takei, H. Ando, & K. Tsutsui (Eds): Handbook of Hormones. DOI: http://dx.doi.org/10.1016/B978-0-12-801028-0.00180-X

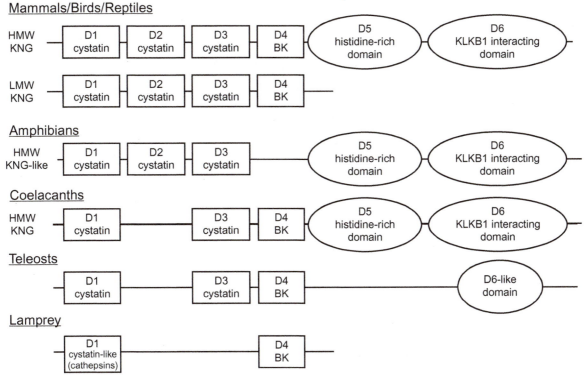

```
          10        20        30        40        50        60                    10        20        30        40        50        60
MKLITILFLCSRLLLSLTQESQSEEIDCNDKDLFKAVDAALKKYNSQNQSNNGFVLYRIT          MKLITILFLCSRLLLSLTQESQSEEIDCNDKDLFKAVDAALKKYNSQNQSNNGFVLYRIT
          70        80        90       100       110       120                    70        80        90       100       110       120
EATKTVGSDTFYSFKYEIKEGDCPVQSGKTWQDCEYKDAAKAATGECTATVGKRSSTKFS          EATKTVGSDTFYSFKYEIKEGDCPVQSGKTWQDCEYKDAAKAATGECTATVGKRSSTKFS
         130       140       150       160       170       180                   130       140       150       160       170       180
VATQTCQITPAEGPVVTAQYDCLGCVHPISTQSPDLEPILRHGIGYFNNNTQHSSLFMLN          VATQTCQITPAEGPVVTAQYDCLGCVHPISTQSPDLEPILRHGIQYFNNNTQHSSLFMLN
         190       200       210       220       230       240                   190       200       210       220       230       240
EVKRAGRQVVAGLNFRITYSIVQTNCSKENFLFLTPDCKSLWNGDTGECTDNAYIDIQLR          EVKRAGRQVVAGLNFRITYSIVQTNCSKENFLFLTPDCKSLWNGDTGECTDNAYIDIQLR
         250       260       270       280       290       300                   250       260       270       280       290       300
IASFSQNCDIYPGKDFVQPPTKICVGCPRDIPTNSPELEETLTHTITKLNAENNATFYFK          IASFSQNCDIYPGKDFVQPPTKICVGCPRDIPTNSPELEETLTHTITKLNAENNATFYFK
         310       320       330       340       350       360                   310       320       330       340       350       360
IDNVKKARVQVVAGKKYFIDFVARETTCSKESNEELTESCETKKLGQSLDCNAEVYVVPW          IDNVKKARVQVVAGKKYFIDFVARETTCSKESNEELTESCETKKLGQSLDCNAEVYVVPW
         370       380       390       400       410       420                   370       380       390       400       410       420
EKKIYPTVNCQPLGMISLMKRPPGFSPFRSSRIGEIKEETTVSPPHTSMAPAQDEERDSG          EKKIYPTVNCQPLGMISLMKRPPGFSPFRSSRIGEIKEETTSHLRSCEYKGRPPKAGAEP
         430       440       450       460       470       480
KEQGHTRRHDWGHEKQRKHNLGHGHKHERDQGHGHQRGHGLGHGHEQQHGLGHGHKFKLD          ASEREVS
         490       500       510       520       530       540
DDLEHQGGHVLDHGHKHKHGHGHGKHKNKGKKNGKHNGWKTEHLASSSEDSTTPSAQTQE
         550       560       570       580       590       600
KTEGPTPIPSLAKPGVTVTFSDFQDSDLIATMMPPISPAPIGSDDDWIPDIQIDPNGLSF
         610       620       630       640
NPISDFPDTTSPKCPGRPWKSVSEINPTTQMKESYYFDLTDGLS
```

Figure 30A.1 Protein sequences and structural features of HMW and LMW KNGs of human. Three cystatin domains are shaded. BK sequence is boxed. Histidine-rich domain is underlined. N- and O-glycosylation sites are wave-underlined and bolded respectively.

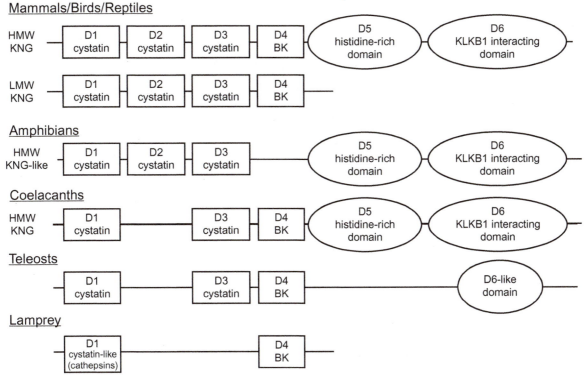

Figure 30A.2 Domain composition of KNGs in various vertebrate lineages. Mammalian *kng1* can be alternatively spliced to form HMW and LMW KNGs but no evidence of alternative splicing is present in non-mammalian vertebrates. The histidine-rich domain is absent in teleosts but a D6-like (glycosylation-rich domain) is present.

```
          10        20        30        40        50        60
MFTMKKSLLLFFFLGIVSLSLCKEERDADEEENEETGGEVKVEEVKRQKTYNRRPPGWSP
LRIAPTNV
```

Figure 30A.3 Skin BK precursor of frog *Pelophylax* kl. *esculentus.* The bradykinin sequence (shaded) is identical to those of teleosts. Signal peptide is underlined.

Biological Functions

Target Cells/Tissues and Functions

KNGs are inhibitors of thiol proteases. HMW KNG interacts with factor XII, factor XI and prekallikrein to initiate the contact activation pathway (also called the intrinsic pathway) of coagulation. HMW KNG optimizes the interacting positions of prekallikrein and factor XI next to factor XII. HMW KNG inhibits the thrombin- and plasmin-induced aggregation of thrombocytes. BK released from KNG increases smooth muscle contraction, induces hypotension, elicits natriuresis and diuresis, and decreases blood glucose levels. BK also mediates inflammation, increases vascular permeability, stimulates nociceptors to cause pain, and induces prostaglandin and endothelium-derived relaxing factor release to dilate blood vessels. The D5 of HMW KNG releases anti-microbial and anti-fungal peptides that contribute to innate immune functions [6]. The D5 domain also downregulates endothelial cell proliferation and migration, inhibits angiogenesis, and reduces apoptosis [7]. LMW KNG inhibits the aggregation of thrombocytes but it is not involved in blood clotting. The cystatin domains on KNG are identical to those of alpha-cysteine proteinase inhibitors and are acting as natural inhibitors for cysteine cathepsins.

Phenotype in Gene-Modified Animals

Patients with HWM KNG deficiency do not have a hemorrhagic tendency, but they exhibit abnormal surface-mediated activation of fibrinolysis. *Kng1*-knockout mice were protected from thrombosis after artificial vessel wall injury, had reduced BK-induced inflammation and blood—brain barrier damage [8].

Pathophysiological Implications

Clinical Implications

Higher KNG levels in plasma and tissues are associated with diabetes, myocardial infarction, and injury and inflammation.

Use for Diagnosis and Treatment

KNG appears to be instrumental in pathologic thrombus formation and inflammation but dispensable for hemostasis. KNG inhibition may offer a selective and safe strategy for combating stroke and other thromboembolic diseases [9].

References

1. Wong MKS, Takei Y. Lack of plasma kallikrein-kinin system cascade in teleosts. *PLoS One*. 2013;8:e81057.
2. Zhou L, Liu X, Jin P, et al. Cloning of the kininogen gene from *Lampetra japonica* provides insights into its phylogeny in vertebrates. *J Genet Genomics*. 2009;36:109—115.
3. Chen X, Wang L, Wang H, et al. A fish bradykinin (Arg0, Trp5, Leu8-bradykinin) from the defensive skin secretion of the European edible frog, Pelophylax kl. esculentus: structural characterization; molecular cloning of skin kininogen cDNA and pharmacological effects on mammalian smooth muscle. *Peptides*. 2011;32:26—30.
4. Zhao A, Lew JL, Huang L, et al. Human kininogen gene is transactivated by the farnesoid X receptor. *J Biol Chem*. 2003;278:28765—28770.
5. Yayama K. Expression of low-molecular-weight kininogen in mouse vascular smooth muscle cells. *Biol Pharm Bull*. 1998;21:772—774.
6. Langhauser F, Göb E, Kraft P, et al. Kininogen deficiency protects from ischemic neurodegeneration in mice by reducing thrombosis, blood-brain barrier damage, and inflammation. *Blood*. 2012;120:4082—4092.
7. Lalmanach G, Naudin C, Lecaille F, et al. Kininogens: More than cysteine protease inhibitors and kinin precursors. *Biochimie*. 2010;92:1568—1579.
8. Sonesson A, Nordahl EA, Malmsten M, et al. Antifungal activities of peptides derived from domain 5 of high-molecular-weight kininogen. *Int J Pept*. 2011;2011:761037.
9. Colman RW, Jameson BA, Lin Y, et al. Domain 5 of high molecular weight kininogen (kininostatin) down-regulates endothelial cell proliferation and migration and inhibits angiogenesis. *Blood*. 2000;95:543—550.

Supplemental Information Available on Companion Website

- Gene, mRNA and domain structure of the human kininogen/E-Figure 30A.1
- Phylogenetic tree of the kininogens in vertebrates/E-Figure 30A.2
- Accession numbers of vertebrate kininogen genes (kng1 and kng2)/E-Table 30A.1

Kallikrein

Marty K.S. Wong

Abbreviation: KLK
Additional names: Plasma kallikrein (PK), plasma prekallikrein (PPK), kallikrein B (KLKB1), kininogenin, Fletcher factor; tissue kallikrein (TK), kidney/pancreas/salivary gland kallikrein, glandular kallikrein (GK)

Plasma kallikrein possesses a unique protein structure with four apple domains and a trypsin domain, which evolved before coagulation factor XI. Tissue kallikreins are trypsin-based enzymes and some members are highly correlated with prostate cancer.

Discovery

The evidence that human urine induces hypotension when injected intravenously into anesthetized dogs was first found in 1909. Two major kallikreins, plasma kallikrein (KLKB1) and tissue (glandular) kallikrein (KLK), have been found in mammals, which are transcribed by different genes. Glandular KLK is an old name and has been replaced by tissue KLK in modern nomenclature.

Structure

Structural Features

Plasma kallikrein possesses a unique structure in vertebrates and is composed of four apple domains followed by a trypsin domain (Figure 30B.1). The apple domain is a conserved protein folding with three disulfide bridges, and is often found on diverse proteins for protein–protein or protein–carbohydrate interactions. The apple domains on KLKB1 interact with the D6 domain of high molecular weight (HMW) kininogen (KNG), which is highly glycosylated. In teleost genomes, the KLKB1-like gene is absent and only lectins with four apple domains but no trypsin domain were found [1]. These lectins were not found in other vertebrates, suggesting the two genes could be sharing the same origin but the trypsin domain was lost in the teleost lineage during evolution. Tissue kallikreins possess a single trypsin domain and are closely related to other serine proteases [2]. It is not clear what enzyme is involved in the cleavage of KNG to form bradykinin (BK) in fishes and lamprey given that KLKB1 and KLK are not identified.

Synthesis and Release

Gene and mRNA

Plasma kallikrein (EC 3.4.21.34) is transcribed by *KLKB1* located on chromosome 4q35 in human. In mammals, *KLKB1*

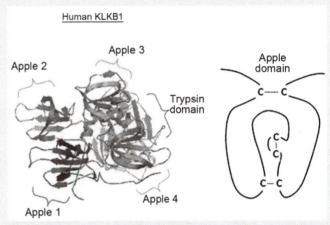

Figure 30B.1 3D model of human plasma kallikrein (left) and schematic structure of an apple domain (right). Plasma kallikrein possesses a unique structure in vertebrates and is composed of four apple domains followed by a trypsin domain.

is duplicated to form factor XI (F11) by tandem duplication [1]. KLKB1 in plasma is mostly in an inactive form known as prekallikrein and it is activated by contact activation involving HMW KNG and factor XII. Tissue kallikreins (EC 3.4.21.35) are transcribed by *KLK* genes located on chromosome 19q13 in human. In tetrapods, the tissue kallikreins were tandemly duplicated into a large family (*KLK1–KLK15* on human chromosome 19) [2]. Human KLKB1 possesses four apple domains followed by a trypsin domain and KLK possesses only a trypsin domain (Figure 30B.2). The sequences of the trypsin domain are highly variable but the catalytic domains (trypsin triads) are similar [2]. KLK-like sequences were identified in teleosts but the catalytic domains are mutated, indicating a possible loss of function of these enzymes and the genes became pseudogenes [1]. Therefore, other trypsin-like enzymes may have taken the role of these KLK-like enzymes in teleosts to produce [Arg^0]-BK from KNG.

Distribution of mRNA

KLKB1 is mostly synthesized in the liver and the enzyme is secreted into the circulation but minute expression is also found in the brain, kidney, pancreas, testes, ovary, carotid gland, esophagus, skin, respiratory tract, prostate, and breast. KLKs are widely distributed including kidney, blood vessels, central nervous system, pancreas, gut, salivary and sweat glands, spleen, adrenal, plasma, and neutrophils. The wide

Y. Takei, H. Ando, & K. Tsutsui (Eds): Handbook of Hormones. DOI: http://dx.doi.org/10.1016/B978-0-12-801028-0.00181-1

```
        10        20        30        40        50        60
MILFKQATYFISLFATVSCGCLTQLYENAFFRGGDVASMYTPNAQYCQMRCTFHPRCLLF
        70        80        90       100       110       120
SFLPASSINDMEKRFGCFLKDSVTGTLPKVHRTGAVSGHSLKQCGHQISACHRDIYKGVD
       130       140       150       160       170       180
MRGVNFNVSKVSSVEECQKRCTSNIRCQFFSYATQTFHKAEYRNNCLLKYSPGGTPTAIK
       190       200       210       220       230       240
VLSNVESGFSLKPCALSEIGCHMNIFQHLAFSDVDVARVLTPDAFVCRTICTYHPNCLFF
       250       260       270       280       290       300
TFYTNVWKIESQRNVCLLKTSESGTPSSSTPQENTISGYSLLTCKRTLPEPCHSKIYPGV
       310       320       330       340       350       360
DFGGEELNVTFVKGVNVCQETCTKMIRCQFFTYSLLPEDCKEEKCKCFLRLSMDGSPTRI
       370       380       390       400       410       420
AYGTQGSSGYSLRLCNTGDNSVCTTKTSTRIVGGTNSSWGEWPWQVSLQVKLTAQRHLCG
       430       440       450       460       470       480
GSLIGHQWVLTAAHCFDGLPLQDVWRIYSGILNLSDITKDTPFSQIKEIIIHQNYKVSEG
       490       500       510       520       530       540
NHDIALIKLQAPLNYTEFQKPICLPSKGDTSTIYTNCWVTGWGFSKEKGEIQNILQKVNI
       550       560       570       580       590       600
PLVTNEECQKRYQDYKITQRMVCAGYKEGGKDACKGDSGGPLVCKHNGMWRLVGITSWGE
       610       620       630
GCARREQPGVYTKVAEYMDWILEKTQSSDGKAQMQSPA
```

```
        10        20        30        40        50        60
MWFLVLCLALSLGGTGAAPPIQSRIVGGWECEQHSQPWQAALYHFSTFQCGGILVHRQWV
        70        80        90       100       110       120
LTAAHCISDNYQLWLGRHNLFDDENTAQFVHVSESFPHPGFNMSLLENHTRQADEDYSHD
       130       140       150       160       170       180
LMLLRLTEPADTITDAVKVVELPTEEPEVGSTCLASGWGSIEPENFSFPDDLQCVDLKIL
       190       200       210       220       230       240
PNDECKKAHVQKVTDFMLCVGHLEGGKDTCVGDSGGPLMCDGVLQGVTSWGYVPCGTPNK
       250       260
PSVAVRVLSYVKWIEDTIAENS
```

Figure 30B.2 Protein sequences and structural features of plasma and tissue kallikreins of human. Four apple domains of plasma kallikrein are shaded. Trypsin domains are underlined.

distribution of KLKs suggests a paracrine function, acting principally in the local tissues or organs.

Tissue and Plasma Concentrations

The circulating concentration of KLKB1 is 450 nM, which is 15-fold higher than that of F11 (30 nM). Moreover, the free Zn concentration required for KLKB1 to bind to HMW KNG is 0.3 μM, which is lower than that of F11 (7 μM). Therefore, the KLKB1/HMW KNG binding predominates in the circulation unless a high Zn concentration is triggered by cell lysis such as injury. The teleost KLKB1-like lectin is widely distributed in the skin, gill, esophagus, and liver [3]. A recent study indicated that the KLK concentration in patients with coronary artery disease (346 ± 0.87 μg/ml) is significantly higher than that in controls (256 ± 0.087 μg/ml) [4].

Regulation of Synthesis and Release

Although KLKB1 and F11 are tandem duplicated orthologs, their regulation is different. Hepatic nuclear factor 4α (HNF4α) deletion decreased F11 but not KLKB1. Estrogen or thyroid hormone treatment increased F11 expression but not KLKB1, while a high-fat diet increased both F11 and KLKB1 expression [5]. For KLK, the regulation of synthesis and release was most studied in KLK3 and multiple androgen-responsive elements were identified upstream of KLK3 [6].

Receptors

Structure and Subtype

None.

Biological Functions

Target Cells/Tissues and Functions

KLKB1 selectively cleaves Arg/Xaa and Lys/Xaa bonds, including Lys/Arg and Arg/Ser bonds in human KNG to release BK. It also digests plasminogen, forming plasmin and participates in surface-dependent activation of blood coagulation, fibrinolysis, and inflammation. It converts prorenin to renin to activate the renin-angiotensin system (RAS). Teleost KLKB1-like lectin facilitates hemagglutination by binding to

the pathogen-like glycoproteins and could be involved in immune function [2].

KLK is highly selective to release [Lys^0]-BK from both HMW and LMW KNGs, which involves hydrolysis of Met/Xaa or Leu/Xaa. Besides acting as an enzyme for KKS, KLK is involved in proteolytic cascades for semen liquefaction through hydrolysis of seminogelin and desquamation of the skin by cleavage of cellular adhesion proteins [7].

Phenotype in Gene-Modified Animals

Trangenic mice with overexpressed *Klk1* possess 10- to 40-fold higher concentration than that of normal and they experienced chronic hypotension that can be rescued by aprotonin and B2 antagonist treatments [8]. KLKB1 deficiency patients exhibit prolonged activated partial thromboplastin time, which is a performance indicator measuring the efficacy of coagulation pathways.

Pathophysiological Implications

Clinical Implications

Kallikreins are drug targets for the control of hypertension, inflammation, and blood coagulation diseases. They are also possible biomarkers for cancer.

Use for Diagnosis and Treatment

KLK2 and KLK3 [prostate-specific antigen (PSA)] are used as serum biomarkers for prostate cancer [9].

References

1. Wong MKS, Takei Y. Lack of plasma kallikrein-kinin system cascade in teleosts. *PLoS One*. 2013;8:e81057.
2. Pavlopoulou A, Pampalakis G, Michalopoulos I, et al. Evolutionary history of tissue kallikreins. *PLoS One*. 2010;5:e13781.
3. Tsutsui S, Okamoto M, Ono M, et al. A new type of lectin discovered in a fish, flathead (*Platycephalus indicus*), suggests an alternative functional role for mammalian plasma kallikrein. *Glycobiology*. 2011;21:1580−1587.
4. Zhang Q, Ran X, Wang DW. Relation of plasma tissue kallikrein levels to presence and severity of coronary artery disease in a Chinese population. *PLoS One*. 2014;9:e91780.

5. Safdar H, Cleuren AC, Cheung KL, et al. Regulation of the F11, Klkb1, Cyp4v3 gene cluster in livers of metabolically challenged mice. *PLoS One*. 2013;8:e74637.

6. Paliouras M, Diamandis EP. The kallikrein world: an update on the human tissue kallikreins. *J Biol Chem*. 2006;387: 643−652.

7. Pampalakis G, Sotiropoulou G. Tissue kallikrein proteolytic cascade pathways in normal physiology and cancer. *Biochim Biophys Acta*. 2007;1776:22−31.

8. Chao J, Chao L. Functional analysis of human tissue kallikrein in transgenic mouse models. *Hypertension*. 1996;27: 491−494.

9. Lawrence MG, Lai J, Clements JA. Kallikreins on steroids: structure, function, and hormonal regulation of prostate-specific antigen and the extended kallikrein locus. *Endocr Rev*. 2010;31: 407−446.

Supplemental Information Available on Companion Website

- Gene, mRNA and domain structure of the human plasma kallikrein /E-Figure 30B.1
- Phylogenetic tree of the plasma kallikrein in vertebrates /E-Figure 30B.2
- Gene, mRNA and domain structure of the human tissue kallikreins /E-Figure 30B.3
- Phylogenetic tree of the tissue kallikreins in vertebrates /E-Figure 30B.4
- Accession numbers of vertebrate plasma kallikrein genes (*klkb1*) /E-Table 30B.1
- Accession numbers of vertebrate tissue kallikrein genes (klk1-klk15)/E-Table 30B.2

Bradykinin

Marty K.S. Wong

Abbreviation: BK
Additional names: [Lys^0]-bradykinin ([Lys^0]-BK), Lys-bradykinin (KBK), kallidin (KD)

Bradykinin is a short-lived, potent vasodilating peptide hormone which is highly involved in inflammation and edema formation. It is the main active hormone in the kallikrein-kinin system. Inhibition of angiotensin converting enzyme (ACE) or kininogen (KNG) II partly delays the degradation rate; therefore it is a target for drug development against hypertension and inflammatory diseases.

Discovery

BK was discovered through the observation that incubation of plasma with a snake venom from *Bothrops jararaca* produced a powerful hypotensive and smooth muscle stimulating factor. The BK sequences are relatively conserved among different vertebrate lineages (Table 30C.1). The skate BK sequence is highly different from other BKs and it was discovered by biochemical purification while molecular evidence of a functional KNG in cartilaginous fishes is lacking. On the other hand, lamprey BK was discovered via genomic evidence but whether such a BK sequence actually exists is uncertain because the prediction of trypsin processing of lamprey KNG does not produce lamprey BK.

Structure

Structural Features

Both BK and [Lys^0]-BK are linear peptides with no known secondary structures. The N-terminal dibasic amino acid residues are resistant to trypsin digestion because of the presence of proline residues that follow them (Table 30C.1).

Synthesis and Release

Gene, mRNA and mRNA

See Chapter 30A, Kininogen.

Distribution of mRNA

See Chapter 30A, Kininogen.

Tissue and Plasma Concentrations

BKs are notorious for their rapid degradation in the body and are considered to be more paracrine than endocrine. Concentrations of BKs in tissues may well exceed blood levels and could contribute to the anti-hypertensive and vasodilator effects in humans. Circulating BK concentrations are very low, usually lower than 3 fmol/ml, but tissue BK levels can reach 100–300 fmol/g tissue (Table 30C.2) [1]. The plasma concentrations need to be increased to 1,000 fmol/ml to acutely lower the blood pressure. Although blood BK levels may increase to such an extent in some conditions, it is rarely high enough to explain the changes in blood pressure under physiological conditions. It is a common misconception that ACE inhibitors dramatically increase circulating BKs by lowering their degradation rate. However, plasma BKs are unchanged or moderately increase after ACE inhibitor treatment, never reaching a sufficient level to induce hypotension. Nevertheless, the ACE inhibition may increase local BK levels significantly to regulate vascular resistance in a paracrine fashion [2]. An increase in renal BK level after ACE inhibition contributes to the increase of Na$^+$ and water excretion. Various BK subtypes were measured by a combination of HPLC purification and various specific immunoassays in mammals and teleosts [1,3].

Regulation of Synthesis and Release

BK is released by the action of plasma kallikrein on high molecular weight (HMW) KNG. [Lys^0]-BK is released by the action of tissue kallikrein on both HMW and LMW KNGs.

Receptors

Structure and Subtypes

BKs signal via two GPCRs known as B1 (BDKRB1) and B2 (BKDRB2) receptors. These BK receptor genes were the result of tandem duplication that predated the teleost tetraploidization while no extra copy was found in teleosts [4]. B2 is constitutively expressed and predominant in different tissues while B1 expression is induced by inflammation. B1 expression in inflammatory tissues such as wound areas and vascular injury contributes to the inflammatory edema by its vasodilatory effect on local blood vessels. B2 is ubiquitiously expressed in different tissues. In the aorta and large muscular arteries including carotid and mesenteric arteries, B2 is localized predominantly in the endothelium. However, in small arterioles towards urinary bladder, myometrium, breast, etc., B2 locates predominantly in the smooth muscle rather than endothelium [5]. Overexpressed zebrafish *bdkrb2* induced IP3 accumulation by BK with an EC$_{50}$ of 6.6 nM.

ignal Transduction Pathway

Both B1 and B2 trigger a typical Ca signaling of GPCRs. Receptor activation leads to PLC activation, intracellular Ca mobilization, release of endothelium-dependent relaxation

Y. Takei, H. Ando, & K. Tsutsui (Eds): Handbook of Hormones. DOI: http://dx.doi.org/10.1016/B978-0-12-801028-0.00182-3

Table 30C.1 Bradykinin Sequences in Vertebrates

Human	KRPPGFSPFR
Tasmanian devil	KRPPGFSPFR
Chicken	RRPPGFTPLR
Lizard	RRPPGFTPFR
Turtle	RRPPGFTPFR
Lungfish	-YGPGFSAPFR
Coelacanths	RLPAGMSPFR
Medaka	RRPPGWSPLR
Zebrafish	KRPPGWSPLR
Fugu	RRPPGWTPLR
Tilapia	RRPPGWSPLR
Stickleback	RRPAGWSPLR
Gar	RRPPGWSPFR
Sturgeon	-MPPGMSPFR
Skate	-GITSWLPF-
Lamprey	-RRPGFMHIA

The shaded residues indicate difference in amino acids compared to human.

Table 30C.2 Bradykinin (BK) Contents in Rat

Tissue	BK concentration (fmol/ml or fmol/g)
Blood	1.9 ± 0.2
Heart	18 ± 3
Aorta	342 ± 38
Brown adipose Tissue	239 ± 58
Adrenal	106 ± 26
Lung	107 ± 23
Brain	28 ± 8

Bradykinin levels were measured by radioimmunoassay after HPLC purification in different tissues.

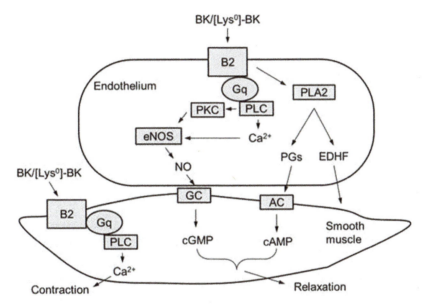

Figure 30C.1   B2 receptor signaling in endothelial and smooth muscle cells. B2 activation leads to phospholipase C (PLC) activation, intracellular Ca mobilization, release of endothelium-dependent relaxation factor (EDRF), and activation of PKC and cytosolic PLA2 that results in the release of prostaglandins (PGs). Endothelial nitric oxide synthase (eNOS) is also stimulated by B2 activation, leading to an increase in NO, stimulation of guanylyl cyclase, and an increase in cGMP. In smooth muscle cells, B2 activation activates PLC directly, leading to a transient increase in Ca flux to induce muscle contraction.

factor (EDRF), and activation of PKC and cytosolic PLA2 that results in the release of prostaglandins (PGs) [6]. The endothelial nitric oxide synthase (eNOS) is also stimulated by B2 activation, leading to an increase in NO, stimulation of guanylyl cyclase, and an increase in cGMP. Most of these factors are EDRFs and contribute to the hypotensive effect of BK. However, in smooth muscle cells, B2 activation activates PLC directly, leading to a transient increase in Ca flux to induce muscle contraction. Although BK mostly induces vasorelaxation via endothelium-dependent signalings, in small arterioles where B2 is predominantly localized in the smooth muscle cells, BK induces vasoconstriction (Figure 30C.1). In the light of this dual function, [Arg⁰]-BK injected in teleosts was a vasopressor and this could be related to a direct stimulation of B2 receptors on smooth muscles as the endothelium-dependent relaxing system is not prominent in teleosts [7].

Agonist

[Lys⁰, des-Arg⁹]-BK is used as a B1 receptor agonist. Cereport (RMP-7) is a selective B2 agonist and has been shown to transiently increase permeability of the blood—brain barrier [8].

Antagonists

B1 antagonist BI-113823 possesses an anti-inflammatory function. Icatibant (HOE 140 or JE 049) is a potent, specific and selective peptidomimetic B2 antagonist.

Biological Functions

Target Cells/Tissues and Functions

BKs are potent endothelium-dependent vasodilators that contribute to vasodilation and hypotension in the systemic circulation. BKs induce thirst in mammals. BKs in addition cause contraction of non-vascular smooth muscle in the bronchus and gut, and increased vascular permeability, and are involved in pain perception. In the kidney, BKs induce natriuresis and diuresis, which leads to a decrease in blood pressure. BKs induce inflammatory responses via B1 and B2 in injured tissues. Via B2, BKs also induce anti-proliferation, anti-apoptosis, anti-fibrosis, anti-hypertrophy, and angiogenesis [6]. In teleosts, [Arg⁰]-BK or BK injection induced vasopressor or triphasic vasoactive responses and are anti-dipsogenic [9]. In skate, BK injection elicited transient hypertension and induced vasoconstriction in isolated branchial and mesenteric arteries.

Phenotype in Gene-Modified Animals

Controversial findings on knockout models were found among different research groups. Blood pressure and cardiovascular function are normal in B1-knockout or B2-knockout mice. B2-knockout mice developed hypertension under a high-salt diet or upon aging. On the other hand, B2-knockout mice experience hypertension, and develop marked dilatation of the left ventricle, cardiac hypertrophy and fibrosis. B2-knockout mice experienced a diminished cardioprotective response to ACE inhibition and an increased urinary concentration in response to vasopressin.

Pathophysiological Implications

Clinical Implications

The activation of KKS is important in blood pressure regulation and inflammatory responses. The effects of BKs are mostly antagonistic to those of the renin-angiotensin system; thus, KKS is being targeted in anti-hypertensive studies clinically.

Use for Diagnosis and Treatment

The ACE has long been a target for hypertension treatment and the inhibition of ACE not only decreases the production of vasoconstrictive Ang II, but also delays the degradation of vasodilatory BK. The use of ACE inhibitors for treating hypertension has been a hallmark and specific agonists for KKS will continue to be a popular topic in hypertension research. Besides blood pressure related disease, kinins are also potential markers for respiratory allergic reactions, septic shock, heart diseases, pancreatitis, hereditary and acquired angioedema, Alzheimer's disease, and liver cirrhosis [10].

References

1. Campell DJ. Towards understanding the kallikrein-kinin system: insights from measurement of kinin peptides. *Braz J Med Biol Res*. 2000;33:665−677.
2. Rhaleb NE, Yang XP, Carretero OA. The kallikrein-kinin system as a regulator of cardiovascular and renal function. *Compr Physiol*. 2011;1:971−993.
3. Wong MKS, Takei Y. Lack of plasma kallikrein-kinin system cascade in teleosts. *PLoS One*. 2013;8:e81057.
4. Bromée T, Venkatesh B, Brenner S, et al. Uneven evolutionary rates of bradykinin B1 and B2 receptors in vertebrate lineages. *Gene*. 2006;373:100−108.
5. Figueroa CD, Marchant A, Novoa U, et al. Differential distribution of bradykinin B(2) receptors in the rat and human cardiovascular system. *Hypertension*. 2001;37:110−120.
6. Blaes N, Girolami JP. Targeting the 'Janus face' of the B2-bradykinin receptor. *Expert Opin Ther Targets*. 2013;17:1145−1166.
7. Conlon JM. The kallikrein-kinin system: evolution of form and function. *Ann N Y Acad Sci*. 1998;839:1−8.
8. Borlongan CV, Emerich DF. Facilitation of drug entry into the CNS via transient permeation of blood brain barrier: laboratory and preliminary clinical evidence from bradykinin receptor agonist, Cereport. *Brain Res Bull*. 2003;60:297−306.
9. Takei Y, Tsuchida T, Li Z, et al. Antidipsogenic effects of eel bradykinins in the eel *Anguilla japonica*. *Am J Physiol Regul Integr Comp Physiol*. 2001;281:R1090−R1096.
10. Golias Ch, Charalabopoulos A, Stagikas D, et al. The kinin system − bradykinin: biological effects and clinical implications. Multiple role of the kinin system − bradykinin. *Hippokratia*. 2007;11:124−128.

Supplemental Information Available on Companion Website

- Gene, mRNA and domain structure of the human bradykinin B1 receptor/E-Figure 30C.1
- Phylogenetic tree of the bradykinin B1 receptor in vertebrates/E-Figure 30C.2
- Gene, mRNA and domain structure of the human bradykinin B2 receptor/E-Figure 30C.3
- Phylogenetic tree of the bradykinin B2 receptor in vertebrates/E-Figure 30C.4
- Accession numbers of vertebrate bradykinin B1 and B2 receptors (*bdkrb1/bdkrb2*)/E-Table 30C.1

CHAPTER 31

Apelin

Sho Kakizawa

Secreted from various tissues in the cardiovascular, digestive, urinary, and central nervous systems. Regulates a wide range of physiological/pathophysiological functions, including cardiovascular function, blood pressure, and angiogenesis and drinking behavior.

Discovery

In 1998, apelin was isolated and characterized from bovine stomach extracts as an endogenous ligand for a G protein-coupled receptor, APJ (putative receptor protein related to the angiotensin receptor, AT1), and the peptide sequence of bovine apelin was used to obtain human and bovine cDNA encoding preproapelin [1].

Structure

Structural Features

Human apelin gene encodes a pre-proprotein of 77 aa residues, containing a signal peptide of 22 aa residues [2]. After cleavage of the signal peptide, the proprotein of 55 aa residues generates several active fragments, including apelin 36 (aa 42–77), apelin 17 (aa 61–77) and apelin 13 (aa 65–77). Apelin 13 is highly active and responsible for the APJ binding and biological activities of mature apelin [1].

Primary Structure

Apelin has been identified in mammals, birds, reptiles, amphibians, and teleosts. The human pre-proapelin sequence is shown in Figure 31.1. Comparison of the mature apelin sequences among different vertebrate classes is shown in Figure 31.2.

Properties

Mr 8,569 (human pre-proapelin), 4,195.87 (human apelin 36), 1,550.84 (human apelin 13). The mature apelin, apelin 13, contains no cysteine and N-glycosylation site. Soluble in water and physiological saline solution. Apelin is a specific substrate of angiotensin converting enzyme 2 (ACE2) [3].

Synthesis and Release

Gene, mRNA, and Precursor

Human apelin gene *APLN*, location Xq25–26.1, consists of three exons [2]. Human apelin mRNA has 3,238 nucleotides that encode a pre-proprotein of 77 aa. Cleavage of the proprotein (by subtilisin/kexin 3 (PCSK3 or furin) probably) produces apelin 13 (aa 65–77), the mature apelin peptide with biological activities.

Tissue Distribution of mRNA

In mammals, expression of apelin mRNA is seen in the heart, lung, kidney, liver, adipose tissue, intestine, vas deferens, mammary gland, pituitary, adrenal gland, endothelium, and brain [2,3].

Tissue Content

Not determined yet.

Plasma Concentration

The concentration of the sum of apelin 36, 17, and 13 in the plasma of normal adult humans (male and female) is 0.85–5.13 ng/ml, and that in the plasma of adult humans with chronic heart failure is 0.53–2.04 ng/ml [4].

Regulation of Synthesis and Release

Hypoxia, diabetic retinopathy, angiopoeitin, and FGF2 induce apelin expression [5–6]. In human adipocytes, apelin expression and secretion are strongly induced under hypoxic conditions, whereas these are decreased by aldosterone.

Receptors

Structure and Subtype

The apelin receptor, APJ, is the 7-transmembrane G protein-coupled receptor. Human APJ consists of 380 aa residues [7]. Although human APJ shows high homology with human angiotensin receptor 1 (AT 1; 30% in total aa and 54% in transmembrane regions), angiotensin II does not bind APJ.

```
         10        20        30        40        50        60        70
MNLRLCVQALLLLWLSLTAVCGGSLMPLPDGNGLEDGNVRHLVQPRGSRNGPGPWQGGRRKFRRQRPRLSHKGPMPF
```

Figure 31.1 Human apelin pre-proprotein (77 aa residues). The aa sequence of the mature apelin, apelin 13, is shaded.

Y. Takei, H. Ando, & K. Tsutsui (Eds): Handbook of Hormones. DOI: http://dx.doi.org/10.1016/B978-0-12-801028-0.00031-3

```
                  10        20        30        40        50        60
Rat       MNLSFCVQALLLLWLSLTAVCGVPLMLPPDGKGL-EEGNMRYLVKPRTSRTGPGAWQGGR
Mouse     MNLRLCVQALLLLWLSLTAVCGVPLMLPPDGTGL-EEGSMRYLVKPRTSRTGPGAWQGGR
Human     MNLRLCVQALLLLWLSLTAVCGGSLMPLPDGNGL-EDGNVRHLVQPRGRSNGPGPWQGGR
Bovine    MNLRRCVQALLLLWLCLSAVCGGPLLQTSDGKEM-EEGTIRYLVQPRGPRSGPGPWQGGR
Frog      MNLRLWALALLLFILTLTSAFGAPLAEGSDRND--EEQNIRTLVNPKMVRNPAPQRQANW
Turtle    MSVRRWLLALLLLWLALSAGSGAPLSEVPDGRDL-EKGNIRNLVHPKGMRNGAGHRQNGW
Chicken   ----------------------MDP--------EDGLIRTLVRPRGARRGNVRRPGGW
Zebrafish MNVKILTLVIVLVVSLLCSASAGPMASTEHSKEIEEVGSMRTPLRQNPARAGRSQRPAGW
Alligator MRRR------------EETAGPLAEAPDGKD-PEEGNIRNLVYPRGARNGAGHRQNGW
```

```
                  70
Rat       RKFRRQRPRLSHKGPMPF
Mouse     RKFRRQRPRLSHKGPMPF
Human     RKFRRQRPRLSHKGPMPF
Bovine    RKFRRQRPRLSHKGPMPF
Frog      RKIRRQRPRLSHKGPMPF
Turtle    RKSRRPRPRLSHKGPMPF
Chicken   RRLRRPRPRLSHKGPMPF
Zebrafish RR-RRPRPRLSHKGPMPF
Alligator RKYRRPRPRLSHKGPMPF
```

Figure 31.2 Amino acid sequences of apelin in vertebrates. The aa sequence of the mature apelin, apelin 13, is shaded.

Signal Transduction Pathway

Coupled to G_i/G_o, protein and apelin inhibit production of cAMP in target cells [8]. The activation of APJ is also mediated by Akt/mTOR/p70S6 pathway.

Agonists

Apelin proprotein (55 aa residues), apelin 36, and apelin 13 (mature apelin).

Antagonist

(Ala^{13})-apelin 13 (functional antagonist).

Biological Functions

Target Cells/Tissues and Functions

Apelin binds to the apelin receptor, APJ, located on the cell membrane of various tissues. Apelin induces a very wide range of biological effects, such as hypotensive effect through nitric oxide (NO) release, angiogenesis, and stimulation of cardiac contractility, water intake, and diuretic effect (Table 31.1). Apelin inhibits HIV infection by blocking the HIV co-receptor APJ.

Phenotype of Gene-Modified Animals

Apelin-knockout mice show impaired retinal vascularization and ocular development [9]. The apelin-induced hypotensive effect is abolished in APJ-deficient mice [10]. APJ-deficient mice also show increased vasopressor response to the vaso-constrictor angiotensin II. Base line blood pressure of double mutant mice homozygous for the deletion of both APJ and angiotensin type 1a receptor is significantly higher than that in mice deficient of angiotensin type 1a receptor alone.

Pathophysiological Implications

Clinical Implications

Downregulation of the apelin receptor APJ is suggested to be a possible cause for the development of heart failure. Apelin is indicated to present a future drug target for the treatment of hypertension and heart failure. APJ agonists are expected to be blockers of HIV infection [3].

Use for Diagnosis and Treatment

None.

Table 31.1 Physiological Functions of Apelin in Vertebrates

Animal	Function	Ref.
Rat	Activation of endothelial nitric oxide synthase	11
Mouse	Induces hypotension	12
Rat	Inhibition of vasopressin release	13
Frog	Angiogenesis	14
Zebrafish	Angiogenesis	15
Rat	Neuroprotection	16
Rat	Inotropic effect	17

References

1. Takemoto K, Hosoya M, Habata Y, et al. Isolation and characterization of a novel endogenous peptide ligand for the human APJ receptor. *Biochem Biophys Res Commun.* 1998;251:471–476.
2. Lee DK, Cheng R, Nguyen T, et al. Characterization of apelin, the ligand for the APJ receptor. *J Neurochem.* 2000;74:34–41.
3. Kleinz MJ, Davenport AP. Emerging roles of apelin in biology and medicine. *Pharmacol Ther.* 2005;107:198–211.
4. Chong KS, Gardner RS, Morton JJ. Plasma concentrations of the novel peptide apelin are decreased in patients with chronic heart failure. *Eur J Heart Fail.* 2006;8:355–360.
5. Eyries M, Siegfried G, Ciumas M, et al. Hypoxia-induced apelin expression regulates endothelial cell proliferation and regenerative angiogenesis. *Circ Res.* 2008;103:432–440.
6. Kasai A, Ishimaru Y, Kinjo T, et al. Apelin is a crucial factor for hypoxia-induced retinal angiogenesis. *Arterioscler Thromb Vasc Biol.* 2010;30:2182–2187.
7. O'Dowd BF, Heiber M, Chan A, et al. A human gene that shows identity with the gene encoding the angiotensin receptor is located on chromosome 11. *Gene.* 1993;136:355–360.
8. Masri B, Lahlou H, Mazarguil H, et al. Apelin (65–77) activates extracellular signal-regulated kinases via a PTX-sensitive G protein. *Biochem Biophys Res Commun.* 2002;290:539–545.
9. Kasai A, Shintani N, Kato H, et al. Retardation of retinal vascular development in apelin-deficient mice. *Arterioscler Thromb Vasc Biol.* 2008;28:1717–1722.
10. Ishida J, Hashimoto T, Hashimoto Y, et al. Regulatory roles for APJ, a seven-transmembrane receptor related to angiotensin-type 1 receptor in blood pressure in vivo. *J Biol Chem.* 2004;279:26274–26279.
11. Takemoto K, Takayama K, Zou MX, et al. The novel peptide apelin lowers blood pressure via a nitric oxide-dependent mechanism. *Regul Pept.* 2001;99:87–92.
12. Ishida J, Hashimoto T, Hashimoto Y, et al. Regulatory roles for APJ, a seven-transmembrane receptor related to angiotensin-type 1 receptor in blood pressure in vivo. *J Biol Chem.* 2004;279:26274–26279.
13. DeMota N, Reaux-Le Goazigo A, El Messari S, et al. Apelin, a potent diuretic neuropeptide counteracting vasopressin actions through inhibition of vasopressin neuron activity and vasopressin release. *Proc Natl Acad Sci USA.* 2004;101:10464–10469.
14. Cox CM, D'Agostino SL, Miller MK, et al. Apelin, the ligand for the endothelial G-protein-coupled receptor, APJ, is a potent angiogenic factor required for normal vascular development of the frog embryo. *Dev Biol.* 2006;296:177–189.
15. Zeng XX, Wilm TP, Sepich DS, et al. Apelin and its receptor control heart field formation during zebrafish gastrulation. *Dev Cell.* 2007;12:391–402.
16. O'Donnell LA, Agrawal A, Sabnekar P, et al. Apelin, an endogenous neuronal peptide, protects hippocampal neurons against excitotoxic injury. *J Neurochem.* 2007;102:1905–1917.
17. Szokodi I, Tavi P, Földes G, et al. Apelin, the novel endogenous ligand of the orphan receptor APJ, regulates cardiac contractility. *Circ Res.* 2002;91:434–440.

Supplemental Information Available on Companion Website

- Gene of human apelin/E-Figure 31.1
- Gene of human apelin receptor/E-Figure 31.2
- Apelin receptor sequences of various animals/E-Figure 31.3
- Accession numbers of gene and cDNA for apelin and apelin receptor/E-Table 31.1 and E-Table 31.2

CHAPTER 32

Natriuretic Peptide Family

Yoshio Takei

History

Atrial natriuretic peptide (ANP) was first isolated from rat heart in 1983, followed by brain natriuretic peptide (BNP) in 1988 and C-type natriuretic peptide (CNP, ortholog of teleost CNP4) in 1990 from pig brain [1,2]. In birds, BNP was isolated from chicken heart in 1988 and CNP3 from chicken brain in 1991. In amphibians, ANP was isolated in 1988 and BNP in 1996 from frog heart, and CNP4 in 1990 and CNP3 in 1994 from frog brain. In fishes, ANP was isolated from eel atria 1989, ventricular natriuretic peptide (VNP) from eel ventricle in 1991, CNP1 from killifish and eel brain in 1990, CNP3 from shark heart and brain in 1991, and CNP4 from lamprey heart and brain in 2006 and CNP4-like NP from hagfish heart and brain in 2003 [3]. Four CNPs (CNP1-4) were first identified in the pufferfish genome database in 2003 [4]. CNP1 and VNP-like NP were also detected in the chicken genome database.

Structure

Structural Features

A common structural feature of the family members is a conserved intramolecular ring structure consisting of 17 aa residues and N- and C-terminal extensions. CNPs lack the C-terminal extension. The C-terminus of teleost ANP and hagfish NP is amidated [3]. Mature sequences of human ANP, BNP, and CNP and of eel ANP, BNP, VNP, and CNP1-4 are shown in Figure 32.1.

Molecular Evolution of Family Members

Figure 32.2 shows the evolutionary history of the NP family. The ancestral molecule of the NP family may be CNP4; then CNP3 was produced in the gnathostomes and linked to the enolase gene [4]. CNP1 and CNP2 were generated by block duplication of CNP3 together with the enolase gene, and three cardiac peptides (ANP, BNP, and VNP) by tandem duplication of the CNP3 gene on the same chromosome. Three cardiac NPs and four CNPs already existed when lobe-finned fish and ray-finned fish diverged, and some became silenced in respective vertebrate classes. For instance, VNP and CNP1-3 were lost in mammals, and ANP and VNP disappeared in medaka. BNP is the only cardiac NP that exists in all species thus far examined [5]. It is of interest to note that sequences of elasmobranch CNP3, teleost CNP1, amphibian CNP3, avian CNP1, and mammalian CNP4 are highly conserved (>90% identity), showing convergent evolution.

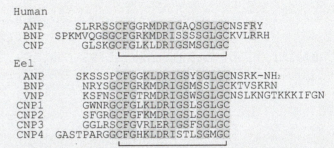

Figure 32.1 Comparison of amino acid sequences of the natriuretic peptide family in human and eel. Disulfide bond is connected by bracket.

Receptors

Structure and Subtypes

In mammals, two single-chain receptors with a cytoplasmic guanylyl cyclase (GC) domain (NPR-A/GC-A and NPR-B/GC-B) and a GC-deficient receptor with a short intracellular domain (NPR-C) have been identified [6] (Table 32.1). The GC domain and another cytoplasmic domain, kinase-like domain, highly conserved high, showing their important function for signal transduction. In teleost fishes, NP receptors were also diversified and two NPR-A types and a NPR-C type named NPR-D have been identified in addition to NPR-B and NPR-C. Ligand selectivity to each NP receptor is variable among NPs.

Signal Transduction Pathway

NPR-A and NPR-B are membrane GC that use cGMP and G-kinase cascade as an intracellular signal transduction (Table 32.1). As other membrane GCs are receptors for external physical and chemical signals, the NP receptors are the first truly endocrine GC-coupled receptors. NPR-C may be a clearance receptor that is internalized immediately after binding for ligand metabolism, but inhibition of cAMP production has been suggested recently. A study on NPR-D may help elucidate the function of GC-deficient NP receptors.

Biological Functions

Target Cells/Tissues and Functions

The NP family can be divided into two groups in terms of function and site of production: endocrine hormones (ANP, BNP, and VNP) secreted from the heart [7] and paracrine factors (CNP1−4) produced in the brain, osteocytes, and

Y. Takei, H. Ando, & K. Tsutsui (Eds): Handbook of Hormones. DOI: http://dx.doi.org/10.1016/B978-0-12-801028-0.00032-5

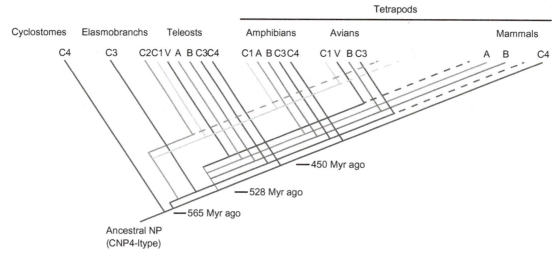

Figure 32.2 History of diversification of the NP family in vertebrates.

Table 32.1 The NP Receptors Identified Thus Far and Their Characteristics

Receptors	NPR-A	NPR-B	NRP-C	NPR-D (eel)
Additional Name	GC-A, NPR1	GC-B, NPR2	NPR3	
Ligands	ANP, BNP, VNP, CNP3	CNP1, 2, 4	ANP, BNP, CNPs, VNP	ANP, CNPs, VNP (BNP: ND)
Structure	homotetramer	homotetramer	homodimer	homotetramer
Guanylyl cyclase	present	present	absent	absent
Second messenger	cGMP	cGMP	cAMP?	ND
Internalization	+	+	+	ND
Antagonist	HS-142-1	HS-142-1	C-ANF	C-ANF, HS-142-1
Tissue expression	adrenal, brain, vascular, smooth muscle, lung, kidney, adipose, heart, intestine	chondrocytes, brain, lung, vascular smooth muscle, uterus, gill (eel), rectal gland (shark)	most tissues	brain, gills (eel)
Major action	natriuresis, diuresis, vasodilation, drinking inhibition, decrease in intestinal NaCl uptake	endochondral ossification, vasodilation	clearance	ND
Mouse KO phenotype	hypertension, ventricular fibrosis, cardiac hypertrophy	dwarfism, seizures, female sterility, decreased adiposity	hypotension, giantism	

ND: Not determined.

endothelial cells [8]. Cardiac NPs act on various tissues and cells that are involved in body fluid and cardiovascular regulation, and they have been used for the treatment of cardiac failure except for VNP [9]. The knockout of the ANP and BNP gene or their receptor (NPR-A) gene increased blood pressure and heart weight, and the high blood pressure was exaggerated further by high salt intake [10] (Table 32.1). CNPs are local hormones whose functions are more diversified than cardiac NPs (for details see the description of each hormone). The knockout of the CNP or NPR-B gene impaired normal skeletal growth, resulting in shorter naso-anal length in mice [10] (Table 32.1). A cohort study also showed that human population with SNP in the CNP gene has shorter height. Judging from the modest expression of CNP and NPR-B genes in cartilaginous tissues, the main function of CNP may be skeletal growth, although their expression is highest in the brain.

References

1. Takei Y. Structural and functional evolution of the natriuretic peptide system in vertebrates. *Int Rev Cytol*. 2000;194:1–66.
2. Potter LR, Abbey-Hosch S, Dickey DM. Natriuretic peptides, their receptors, and cyclic guanosine monophosphate-dependent signaling functions. *Endocrine Rev*. 2006;27:47–72.
3. Toop T, Donald JA. Comparative aspects of natriuretic peptide physiology in non-mammalian vertebrates: a review. *J Comp Physiol B*. 2004;174:189–204.
4. Inoue K, Naruse K, Yamagami S, et al. Four functionally distinct C-type natriuretic peptides found in fish reveal evolutionary history of the natriuretic peptide system. *Proc Natl Acad Sci USA*. 2003;100:10079–10084.
5. Takei Y, Inoue K, Trajanovska S, et al. B-type natriuretic peptide (BNP), not ANP, is the principal cardiac natriuretic peptide in vertebrates as revealed by comparative studies. *Gen Comp Endocrinol*. 2012;171:258–266.
6. Hirose S, Hagiwara H, Takei Y. Comparative molecular biology of natriuretic peptide receptors. *Can J Physiol Pharmacol*. 2001;79:665–672.
7. Brenner BM, Ballerman BJ, Gunning ME, et al. Diverse biological actions of atrial natriuretic peptide. *Physiol Rev*. 1990;70:665–699.
8. Teixeina CC, Agoston H, Beier F. Nitric oxide, C-type natriuretic peptide and cGMP as regulators of endochondral ossification. *Dev Biol*. 2008;319:171–178.
9. Saito Y. Roles of atrial natriuretic peptide and its therapeutic use. *J Cardiol*. 2010;56:262–270.
10. Kishimoto I, Tokudome T, Nakao K, et al. Natriuretic peptide system: an overview of studies using genetically engineered animal models. *FEBS J*. 2011;278:1830–1841.

Supplemental Information Available on Companion Website

- Gene structures of NPs and NPRs in human/E-Figure 32.1
- Phylogenetic tree of the NP family members/E-Figure 32.2
- Phylogenetic tree of the NP receptors/E-Figure 32.3

Atrial Natriuretic Peptide

Yoshio Takei

Abbreviation: ANP
Additional names: A-type natriuretic peptide, atrial natri-uretic factor (ANF), atriopeptin, auriculin, cardiodilatin, natriodilatin

First cardiac hormone isolated from atria with potent hypo-tensive and natriuretic/diuretic actions. Drug target for hyper-tension and cardiac/renal failure.

Discovery

Presence of a natriuretic factor in the rat heart was first reported in 1981, which was isolated in 1983 from rat and human atria [1].

Structure

Structural Features

Human proANP consists of 126 aa residues with bioactive mature ANP at the C-terminus. Human ANP, or ANP-(99–126), consists of 28 aa residues with an intramolecular ring structure of 17 aa residues, as with other NPs [1] (Figure 32A.1). Amphibians and bony fishes also possess ANP, but birds, reptiles (except for turtles), cartilaginous fishes, and cyclostomes do not [2]. N-terminal truncated forms exist in the brain and an N-terminal elongated form named urodilatin is present in the kidney (Figure 32A.2).

Primary Structure

Sequence identity is low in the pro-segment. Mature ANP sequence is conserved (only one aa difference) in mammals, but is variable across different classes (Figure 32A.2). Most ANPs of teleosts have amidated C-terminus [3].

Properties

Mr of human ANP is 3082 and isoelectric point is ca. 10.7. It is freely soluble in water, ethanol and 70% acetone, and insolu-ble in acetone, benzene, chloroform and ether. ANP solution in water at $>10^{-4}$ M is stable for more than a year at $-20°C$.

Synthesis and Release

Gene and mRNA

Human ANP gene *(NPPA)*, located on chromosome 1 (1p36), consists of three exons (E-Figure 32A.1) and has AP-1, GRE, and other regulatory elements in the promoter region [4]. Human ANP mRNA has 858 bp. The ANP gene first appeared in early bony fishes by tandem duplication of the CNP3 gene. The gene structure and its mRNA size are well conserved among teleosts, amphibians, and mammals.

Distribution of mRNA

ANP mRNA is detected most abundantly in the atrium [4]. Some transcripts are found in the cardiac ventricle, brain, kid-ney, adrenal, lung, gonads, and lymphoid tissues in mammals. In the brain, immunoreactive ANP is most abundant in the olfactory bulb, followed by the hypothalamus and midbrain in the pig. In non-mammals, the heart (atrium) is the main tissue of ANP synthesis followed by the brain, kidney, and interrenal in frogs and teleost fishes [2].

Tissue and Plasma Concentrations

These concentrations are shown in Table 32A.1 [5–10]. The right atrium has greater concentrations than found in the left atrium in mammals. Eels in fresh water and seawater have similar concentrations.

Regulation of Synthesis and Release

Atrial ANP synthesis and release are stimulated principally by atrial stretch (increased blood volume) in mammals [5]. Corticosteroids, α-adrenergic stimulation, and hypoxia also stimulate the gene expression as inferred by the presence of such responsive elements in the human gene. Increased ANP secretion in patients with congestive and ischemic heart fail-ure is related to the atrial stretch and hypoxia, respectively. In eels, ANP release is increased more profoundly by osmotic stimulus than volemic stimulus, but the latter is a major stimu-lus in trout as in mammals. Regulation of ANP gene expres-sion and release has not been examined in non-cardiac tissues yet.

Receptors

Structure and Subtype

ANP binds A-type natriuretic peptide receptor (NPR-A or GC-A) with high affinity ($K_d = 2–3$ nM) [11]. NPR-A is a single chain receptor with an extracellular ligand-binding domain, membrane-spanning domain, and intracellular guanylyl-cyclase (GC) and kinase-like domains. The human NPR-A has 1061 aa residues with an Mr of 118,923. NPR-A appears to exist as a tetramer although ANP binds a monomeric receptor. NPR-A has been cloned in the bullfrog and eel, and a second NPR-A type has been found in medaka and eels [12]. ANP also binds NPR-C that has only a short intracellular domain. The human NPR-C consists of 540 aa residues with an Mr

Y. Takei, H. Ando, & K. Tsutsui (Eds): Handbook of Hormones. DOI: http://dx.doi.org/10.1016/B978-0-12-801028-0.00183-5

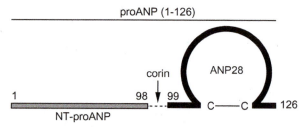

Figure 32A.1 Primary structure of mammalian ANP.

```
     Human ANP   SLRRSS-CFGGRMDRIGAQSGLGCNSFRY
  Bullfrog ANP   SMRRSSDCFGSRIDRIGAQSGMGC-GRRF
       Eel ANP   SKSSSP-CFGGKLDRIGSYSGLGCNSRK-NH₂

Urodilatin  TAPRSLRRSS-CFGGRMDRIGAQSGLGCNSFRY
```

Figure 32A.2 Alignment of mature ANP sequences in representative species of each vertebrate class. ANP has not been identified in other classes. Disulfide bond is connected by bracket. In human kidney, urodilatin is formed by a different post-translational processing of proANP.

Table 32A.1 Tissue and Plasma Concentrations of ANP in Human, Rat, and Eel

Species	Tissues	Concentration	Ref.
human	atrium	13.1 ± 2.3 nmol/g	[5]
	ventricle	0.006 ± 0.02 nmol/g	
	plasma	6.4 ± 0.9 fmol/ml	[6]
rat	atrium	16–20 nmol/g	[7]
	ventricle	0.002–0.006 nmol/g	
	plasma	55 ± 5.5 fmol/ml	[8]
eel	atrium (FW)	11 ± 2 nmol/g	[9]
	atrium (SW)	56 ± 10 nmol/g	
	ventricle (FW)	0.056 ± 0.008 nmol/g	
	ventricle (SW)	0.041 ± 0.013 nmol/g	
	plasma (FW)	68.1 ± 2.9 fmol/ml	[10]
	plasma (SW)	60.7 ± 3.6 fmol/ml	

The values show average concentration of right/left atrium or ventricle in human and rat. FW, freshwater-adapted eel; SW, seawater-adapted eel.

Table 32A.2 Biological Actions of ANP

Animal	Brain/Skin (Ingestion)		Intestine (Absorption)		Kidney (Excretion)		Adrenal Corticosteroid (Secretion)	Pituitary ADH (Secretion)	Blood Vessel (Constriction)
	Water	NaCl	Water	NaCl	Water	NaCl			
Rat	↓	↓	↓	↓	↑↑	↑↑	↓↓	↓	↓↓
Frog	↓↓	↓↓	n.d.	n.d.	n.d.	n.d.	↓	n.d.	↓↓
Eel	↓↓	↓↓	↓↓	↓↓	↓	↑	↓↓	n.d.	↓↓

of 59,768. NPR-C appears to be generated by exon shuffling of the GC-coupled receptor. In addition, NPR-D, a second GC-deficient receptor, has been cloned in the eel. While NPR-C is a dimeric receptor, NPR-D exists as a tetramer as does NPR-A [12].

Signal Transduction Pathway

After ANP binding, the GC domain of NPR-A is activated to catalyze the production of cGMP, which serves as a second messenger for biological actions. NPR-C is thought to be a clearance receptor to regulate local ANP concentration as it exists ubiquitously in various tissues, but inhibition of adenylyl cyclase was suggested recently.

Agonists

Other cardiac NPs such as BNP and VNP bind NPR-A with high affinities, while all NPs readily bind NPR-C and NPR-D. Some NPs from snake venom such as DNP and synthetic chimeric NPs are used clinically as agonists.

Antagonist

HS-142-1 isolated from a bacterium serves as a sole antagonist for ANP binding to NPR-A. C-ANF, an ANP analog with modified intra-ring sequences, is an antagonist for NPR-C but not for NPR-A. Thus, C-ANF administration increases plasma ANP and enhances its biological effects.

Biological Functions

Target Cells/Tissues and Functions

As expected from the secretory stimulus, ANP acts to restore increased blood volume to normal by decreasing levels of sodium and water in mammals [13]. Detailed actions are shown in Table 32A.2. In addition to the vascular effect, ANP augments cardiac performance. In eels, ANP effects on the brain and intestine are much more potent and efficacious but the effect on the kidney is less efficacious than in mammals [3]. ANP induces weak antidiuresis in eels but brisk diuresis in trout. Comparative studies in eels suggest that the essential action of ANP is on sodium extrusion but not water [3].

Phenotype in Gene Modified Animals

ANP gene knockout mice have no difference in water balance and blood pressure compared with normal littermates, but increased salt intake induced severer hypertension in the knockout animals. BNP that shares NPR-A may have compensated the ANP functions [9]. On the other hand, transgenic mice overexpressing the ANP gene exhibited lower arterial pressure compared with their littermates. In medaka where the ANP gene is lost, CNP3, from which ANP is generated by tandem duplication, took over the role of ANP, and its gene knockdown in embryo resulted in abnormal development of the atrium. A cohort study in humans with ANP gene mutations detected a significant correlation between plasma ANP concentration and hypertension.

NPR-A knockout mice, in which both ANP and BNP are unable to exhibit their functions, have elevated blood volume, cardiac hypertrophy, and hypertension [14]. Targeted deletion of the NPR-A gene in endothelial cells resulted in sustained vasorelaxant response to ANP but increased blood pressure, suggesting that hypertension is due to the failure of normal salt excretion.

Pathophysiological Implications

Clinical Implications

Plasma ANP concentration is enhanced in proportion to the severity of heart failure in the New York Heart Association (NYHA) functional classification, which explains the use of plasma ANP measurement for diagnosis of heart failure [15]. There are innumerable studies on the role of ANP in cardiac failure. In relation to hypertension, significant inverse correlation was detected between plasma ANP concentration and arterial pressure in humans, and administration of ANP to hypertension patients decreased arterial pressure to a normal range. In addition, plasma ANP concentration increases in patients with renal failure and infectious diseases.

Use for Diagnosis and Treatment

Plasma ANP concentration is routinely measured to diagnose cardiac failure. The N-terminal fragment of proANP (NT-proANP) is also measured for diagnosis because of its longer half life in plasma. Synthetic human ANP (Carperitide1® or Hanp®) is used for treatment of acute heart failure. Many kits for measurement of plasma ANP are sold by various companies. Recently, ANP has been gradually replaced by BNP for diagnosis and treatment of cardiac failure because of its immediate secretion in the earlier phase.

References

1. de Bold AJ. Thirty years of research on atrial natriuretic factor: historical background and emerging concepts. *Can J Physiol Pharmacol.* 2011;89:527−531.
2. Takei Y, Inoue K, Trajanovska S, et al. B-type natriuretic peptide (BNP), not ANP, is the principal cardiac natriuretic peptide in vertebrates as revealed by comparative studies. *Gen Comp Endocrinol.* 2011;171:258−266.
3. Takei Y, Hirose S. The natriuretic peptide system in eels: a key endocrine system for euryhalinity? *Am J Physiol.* 2002;282: R940−R951.
4. Kuwahara K, Nakao K. Regulation and significance of atrial and brain natriuretic peptides as cardiac hormones. *Endocrine J.* 2010;57:555−565.
5. Saito Y, Nakao K, Arai H, et al. Augmented expression of atrial natriuretic polypeptide gene in ventricle of human failing heart. *J Clin Invest.* 1989;83:298−305.
6. Potter LR, Abbey-Hosch S, Dickey DM, et al. Natriuretic peptides, their receptors, and cyclic guanosine monophosphate-dependent signaling functions. *Endocr Rev.* 2006;27:47−72.
7. Arai H, Nakao K, Saito Y, et al. Simultaneous measurement of atrial natriuretic polypeptide (ANP) messenger RNA and ANP in rat heart − evidence for a preferentially increased synthesis and secretion of ANP in left atrium of spontaneously hypertensive rats (SHR). *Biochem Biophys Res Commun.* 1987;148:239−245.
8. Morii N, Nakao K, Kihara M, et al. Decreased content in left atrium and increased plasma concentration of atrial natriuretic polypeptide in spontaneously hypertensive rats (SHR) and SHR stroke-prone. *Biochem Biophys Res Commun.* 1986;26:74−81.
9. Takei Y, Balment RJ. Biochemistry and physiology of a family of natriuretic peptide in eels. *Fish Physiol Biochem.* 1993;11: 183−188.
10. Kaiya H, Takei Y. Atrial and ventricular natriuretic peptide concentrations in plasma of freshwater- and seawater-adapted eels. *Gen Comp Endocrinol.* 1996;102:183−190.
11. Gerbers DL. Guanylyl cyclase receptors and their endocrine, paracrine, and autocrine ligands. *Cell.* 1992;71:1−4.
12. Hirose S, Hagiwara H, Takei Y. Comparative molecular biology of natriuretic peptide receptors. *Can J Physiol Pharmacol.* 2001;79:665−672.
13. Brenner BM, Ballerman BJ, Gunning ME, et al. Diverse biological actions of atrial natriuretic peptide. *Physiol Rev.* 1990;70:665−699.
14. Kishimoto I, Tokudome T, Nakao K, et al. Natriuretic peptide system: an overview of studies using genetically engineered animal models. *FEBS J.* 2011;278:1830−1841.
15. Saito Y. Roles of atrial natriuretic peptide and its therapeutic use. *J Cardiol.* 2010;56:262−270.

Supplemental Information Available on Companion Website

- Gene, mRNA and precursor sequences of human ANP/E-Figure 32A.1
- Primary structure of ANP sequences in various vertebrates/E-Figure 32A.2
- Molecular phylogenetic analysis of ANP precursors in vertebrates/E-Figure 32A.3
- Gene, mRNA and precursor of human NPR-A/E-Figure 32A.4
- Primary structure of NPR-A in various vertebrates/E-Figure 32A.5
- Molecular phylogenetic analysis of NPR-As in vertebrates/E-Figure 32A.6
- Accession numbers of cDNA for ANP/E-Table 32A.1
- Accession numbers of cDNA for NPR-A/E-Table 32A.2

B-type Natriuretic Peptide

Takehiro Tsukada

Abbreviation: BNP
Additional names: Brain natriuretic peptide (BNP)

The only cardiac natriuretic peptide common to all vertebrate species thus far examined. Plasma BNP level is used as a robust and reliable biomarker for diagnosis and prognosis of heart failure.

Discovery

BNP was first isolated in 1988 from porcine brain extracts and thus given the name "Brain" natriuretic peptide [1]. Soon thereafter, BNP was found to be expressed abundantly in the cardiac ventricle and scarcely in the brain of human and rats. Currently, BNP has been recognized as a principal cardiac hormone and referred to as "B-type" natriuretic peptide.

Structure

Structural Features

Human proBNP consists of 108 aa residues with bioactive mature BNP (32 aa residues) at the C-terminus (Figure 32B.1). Like other natriuretic peptides (NPs), BNP has an intramolecular ring consisting of 17 aa residues and N- and C-terminal extensions of varying length. All vertebrate species (tetrapods and fishes) except for chondrichthyes and cyclostomes possess BNP [2]. Thus, the BNP gene is considered to have occurred before divergence of ray-finned fishes and lobe-finned fishes.

Primary Structure

Sequence identity of mature BNP is quite variable in mammals (<50% identity between human and mouse BNP), whereas the identity is relatively high among non-mammalian vertebrates (avian, reptilian, amphibian, and teleost species; Figure 32B.2) [2].

Properties

Mr of mature human BNP is 3,466 and pI is ca. 10. Both pro- and mature BNP are freely soluble in water, acid, and 67% acetone, but insoluble in 99% acetone. BNP solution in water at $>10^{-4}$ M is stable for more than a year at $-20°C$.

Synthesis and Release

Gene, mRNA, and Precursor

The BNP and atrial natriuretic peptide (ANP) genes are thought to be generated by tandem duplication of the CNP3 gene [2]. The human BNP and ANP genes (*NPPA* and *NPPB*) are located on chromosome 1 (1q36) [3]. The mouse BNP and ANP genes (*Nppb* and *Nppa*) are on chromosome 4. The BNP gene is composed of three exons and two introns. One of the characteristics of BNP mRNA is a repetitive AUUUA motif in the 3′-untranslated region [4]. The motif is considered to destabilize mRNA and is not present in ANP mRNA, suggesting that the BNP gene is regulated differently from the ANP gene. Human proBNP is *O*-glycosylated posttranslationally (Figure 32B.1) [4]. *O*-glycosylated proBNP is further processed by specific convertases (probably furin and corin) to give rise to the bioactive, mature form of BNP and inactive N-terminal (NT)-proBNP. It is suggested that cleavage of proBNP is regulated by *O*-glycosylation at threonine-71 [4].

Distribution of mRNA and Peptide

BNP mRNA is highly expressed in the heart. BNP mRNA is also found in the porcine brain, but it is below detectable levels in human and rat brain [4]. Thus, there is a species difference in BNP mRNA distribution. Concentration of BNP is higher in the cardiac atrium than in the ventricle, but compared with ANP, the total amount of BNP is greater in cardiac ventricle [4]. Thus, BNP is thought to be a ventricular hormone in humans. In teleosts, relative expression of the BNP gene is also greater than that of the ANP gene, but the ventricular natriuretic peptide (VNP) is apparently a ventricular hormone in teleost species that have VNP [2]. Plasma BNP concentration is ca. 1 fmol/ml in humans [3]. Half-life of plasma BNP is longer than that of ANP (ca. 20 min vs. ca. 2 min) [3].

Regulation of Synthesis and Release

BNP in the cardiac ventricle is secreted via a constitutive pathway as opposed to ANP secretion that occurs via a regulatory pathway (except BNP in the atrium, where BNP is stored and secreted with ANP) [3]. BNP production is regulated transcriptionally and various regulatory elements are located in the 5′-flanking region of the BNP gene, including GATA, M-CAT, AP-1/CRE-like elements, NRSE, shear stress-responsive elements, thyroid hormone-responsive element, and the nuclear factor of activated T-cells (NFAT) binding site [4]. Putative transcription factors are GATA4, YY1, and KLF13 [5]. BNP expression in cardiomyocytes is enhanced by mechanical stretch, TGF-β, and ET-1 [6].

Receptors

Structure and Subtype

BNP shares two receptors with ANP and VNP (cardiac NP found in ray-finned fishes). A-type NP receptor (NPR-A or GC-A) is a biological receptor and has a guanylyl cyclase domain that produces second messenger cGMP. The order of potency for cGMP production via NPR-A is ANP ≥ BNP >>

Y. Takei, H. Ando, & K. Tsutsui (Eds): Handbook of Hormones. DOI: http://dx.doi.org/10.1016/B978-0-12-801028-0.00184-7

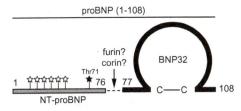

Figure 32B.1 Schematic of BNP structure and processing. *O*-glycosylation sites are shown with stars. The seventh *O*-glycosylation site (threonine-71: Thr71) is thought to be critical for cleavage of proBNP.

human	SPKMVQGSGCFGRKMDRISSSSGLGCKVLRRH
mouse	NSKVTHISSCFGHKIDRIGSVSRLGCNALKLL
chicken	MMRDSGCFGRRIDRIGSLSGMGCNGSRKN
crocodile	MMRDSGCFGRRIDRIGSLSGMGCNGSRKN
Xenopus	MMRGSGCFGRRIDRIDSLSGMGCNGSRRY
pufferfish	RRSSSCFGRRMDRIGSMSSLGCNTVGKYNPK

Figure 32B.2 An alignment of mature BNP sequences in representative species of each vertebrate class. Two cysteine residues form a disulfide bond bridge (asterisks).

CNP in humans [3]. Another receptor is C-type NP receptor (NPR-C). NPR-C lacks a guanylyl cyclase domain and acts as a clearance receptor. Affinities of NPR-C to NPs are ANP > CNP > BNP in humans [3]. The longer half-life of plasma BNP is attributable mainly to the lower susceptibility to NPR-C. In mammals, NPR-A is highly expressed in the adrenal, brain, kidney, adipose, aortic, and lung tissues, whereas NPR-C is found ubiquitously in most tissues [3]. For more details on NPR-A and NPR-C receptors, signaling transduction pathway, and agonist/antagonist, see Chapter 32A, "Atrial Natriuretic Peptide."

Biological Functions

Target Cells/Tissues and Functions

Biological actions of BNP are overlapped with those of ANP, because they share the same biological receptor, NPR-A (see Chapter 32A, "Atrial Natriuretic Peptide"). Briefly, BNP induces natriuresis and diuresis by increasing glomerular filtration rate and decreasing water and sodium reabsorption by the kidney [2]. In addition, BNP decreases aldosterone secretion from the adrenal and vasopressin secretion from the posterior pituitary, which also enhances natriuresis and diuresis. BNP also decreases systemic vascular resistance by its direct vasorelaxant action. Thus, the net effect of BNP is to decrease blood volume (preload) and systemic blood pressure (afterload), thereby protecting the heart. In eels, again similarly to ANP actions, BNP decreases water intake, plasma sodium concentration, and aortic pressure, but the effect is weaker than that of ANP and VNP in the case of eels that have VNP [7].

Phenotype in Gene-Modified Animals

BNP null mice displayed ventricular fibrosis, suggesting that BNP acts as an antifibrotic factor [8]. Since this phenotype was not seen in ANP null mice, the antifibrotic effect is probably BNP-specific. No ventricular hypertrophy and hypertension was observed in BNP null mice [8], although mice overexpressing BNP had a lower blood pressure than non-transgenic littermates [8].

Pathophysiological Implications

Clinical Implications

Elevation of plasma BNP concentration is seen in patients with congestive heart failure, myocardial infarction, hypertension, left ventricular hypertrophy, and chronic renal failure [3,4]. Although the plasma ANP level is also elevated by such pathological conditions, the increment of BNP level is quicker and greater than that of ANP. In congestive heart failure, for instance, plasma BNP and ANP concentrations increase 200−300 fold and 10−30 fold, respectively, compared with those in normal humans. The greater increase in BNP level is probably due to its longer half-life in plasma (see above). Thus, plasma BNP level is used as a more reliable biomarker of both ischemic and congestive heart failure. NT-proBNP secreted with BNP has a longer half-life (ca. 120 min), and thus is used as a more reliable marker for diagnosis of heart failure [9]. Recent studies showed that a high level of *O*-glycosylated proBNP circulates in patients with severe heart failure [4].

Use for Diagnosis and Treatment

Severity of congestive heart failure assessed by the New York Heart Association (NYHA) functional classification is correlated positively with the elevation of plasma BNP and NT-proBNP levels [4]. Therefore, these peptides are used for diagnosis and management of patients with congestive heart failure. Usage of recombinant BNP, nesiritide, for treatment has been approved by the U.S. Food and Drug Administration. It has a beneficial effects in patients with acute, decompensated heart failure; however, it is reported to increase the risk of renal dysfunction and mortality [3]. Further evaluation is required for clinical usage of nesiritide.

References

1. Sudoh T, Kangawa K, Minamino N, et al. A new natriuretic peptide in porcine brain. *Nature*. 1988;332:78−81.
2. Takei Y, Inoue K, Trajanovska S, et al. B-type natriuretic peptide (BNP), not ANP, is the principal cardiac natriuretic peptide in vertebrates as revealed by comparative studies. *Gen Comp Endocrinol*. 2011;171:258−266.
3. Potter LR, Abbey-Hosch S, Dickey DM. Natriuretic peptides, their receptors, and cyclic guanosine monophosphate-dependent signaling functions. *Endocr Rev*. 2006;27:47−72.
4. Nishikimi T, Kuwahara K, Nakao K. Current biochemistry, molecular biology, and clinical relevance of natriuretic peptides. *J Cardiol*. 2011;57:131−140.
5. Hayek S, Nemer M. Cardiac natriuretic peptides: from basic discovery to clinical practice. *Cardiovasc Ther*. 2011;29:362−376.
6. Sergeeva IA, Christoffels VM. Regulation of expression of atrial and brain natriuretic peptide, biomarkers for heart development and disease. *Biochim Biophys Acta*. 2013;1832:2403−2413.
7. Miyanishi H, Nobata S, Takei Y. Relative antidipsogenic potencies of six homologous natriuretic peptides in eels. *Zool Sci*. 2011;28:719−726.
8. Kishimoto I, Tokudome T, Nakao K, et al. Natriuretic peptide system: an overview of studies using genetically engineered animal models. *FEBS J*. 2011;278:1830−1841.
9. Vanderheyden M, Bartunek J, Goethals M. Brain and other natriuretic peptides: molecular aspects. *Eur J Heart Fail*. 2004;6:261−268.

Supplemental Information Available on Companion Website

- Alignment of amino acid sequence of BNP precursor in vertebrates/E-Figure 32B.1
- Molecular phylogenetic analyses of BNP precursors in vertebrates/E-Figure 32B.2
- Molecular phylogenetic analyses of NPR-C and NPR-D in vertebrates/E-Figure 32B.3
- Accession numbers of gene and cDNA for BNP/E-Table 32B.1
- Accession numbers of cDNA for NPR-C and NPR-D/E-Table 32B.2

C-type Natriuretic Peptides

Takehiro Tsukada

Abbreviation: CNPs

CNPs are ancient members of the natriuretic peptide family, which is characterized by the absence of C-terminal tail extensions from the intramolecular ring. In mammals, CNP is highly expressed in the brain, vascular endothelial cells, and chondrocytes, and acts locally as a paracrine/autocrine factor.

Discovery

CNP was first isolated in 1990 from killifish brain and shortly thereafter from pig brain [1]. Four distinct CNPs (CNP1−4) were then discovered in 2003 from teleost genome databases [2]. Later, comparative genomic analyses inferred that only CNP4 is retained in mammals, CNP1 and CNP3 in birds, CNP3 and CNP4 in amphibians, CNP1−4 in most teleosts, CNP3 in elasmobranchs, and CNP4 or CNP4-like NP in cyclostomes (Figure 32C.1) [3].

Structure

Structural Features

Human proCNP consists of 103 aa residues with bioactive mature CNP at the C-terminus (Figure 32C.2) [4]. All four types of CNPs consist of 22 aa residues except for some teleost CNP4 [3]. Like other NPs, CNPs have a 17 aa intramolecular ring, but lack a C-terminal tail sequence.

Primary Structure

CNPs are the most conserved NPs although they contain different groups of peptides [3]. Sequence identity of mature CNP4 is >95% in mammals, except for unique CNP in the little brown bat (*Myotis lucifugus*). CNP1 and CNP3 are also well-conserved (both >80% identity) in each vertebrate class that possesses both peptides. In contrast, CNP2 in teleosts is not so conserved. Although major CNPs are of different types among vertebrate classes, mature sequences are highly conserved among these CNPs (Figure 32C.3), suggesting that CNPs have undergone convergent evolution [3].

Properties

Mr of mature human CNP is 2199 and pI is ca. 9. It is soluble in water, acid, and 67% acetone, and partially soluble in 99% acetone. However, CNPs are not as soluble in water as other NPs because of higher hydrophobicity. CNP solution in water at $>10^{-4}$ M is stable for more than a year at $-20°C$.

Synthesis and Release

Gene, mRNA, and Precursor

The human CNP gene (*NPPC*) is located on chromosome 2 (2q24), which is different from the loci of atrial natriuretic peptide (ANP) and B-type natriuretic peptide (BNP) genes (*NPPA* and *NPPB*, chromosome 1) [5]. The mouse CNP gene (*Nppc*) is on chromosome 1. As well as mammalian CNP, all four types of CNP genes in teleosts are composed of three exons and two introns, although the exon−intron arrangement between CNP1/2 and CNP3/4 is different [2]. In elasmobranchs and cyclostomes, only one intron have been found in their CNP genes. Human CNP mRNA is 701 bp in length. Human proCNP (1−103) is cleaved into CNP-53 (51−103) and N-terminal (NT)-proCNP (1−50) by intracellular prohormone convertase furin (Figure 32C.2) [4,5]. CNP-53 is further processed to give rise to the bioactive mature form CNP-22 and inactive NT-CNP-53.

Distribution of mRNA

CNPs are highly expressed in the brain of tetrapods [6,7,8]. In mammals, CNP transcripts are also identified in the pituitary, cardiovascular system, endochondral bone, and reproductive systems [6]. In teleosts, CNP1, CNP2, and CNP4 are predominantly expressed in the brain, whereas CNP3 is expressed in a variety of tissues including the heart, pituitary, kidney, spleen, ovary, spinal cord, and intestine [2]. In elasmobranchs and cyclostomes, where only a single CNP but no cardiac ANP/BNP/Ventricular natriuretic peptide (VNP) is found, the CNP is strongly expressed in both the brain and the heart [7,8]. It is noteworthy that the CNP acts both as an endocrine hormone and as a paracrine/autocrine factor in these primitive vertebrates.

Tissue and Plasma Concentrations

Although CNP-22 and CNP-53 have similar biological functions and potencies, their expression pattern differs among tissues. CNP-22 is identified in the plasma and cerebral spinal fluid, whereas CNP-53 is found in the brain, endothelial cells, and heart [5]. In the shark, CNP-22 is the major form stored in the brain, but proCNP (1−115) is predominantly found in the heart. Half-life of plasma CNP-22 is ca. 2.6 min in humans [5]. Tissue and plasma CNP concentrations are summarized in Table 32C.1 [5,9−11].

Regulation of Synthesis and Release

The human and mouse CNP genes have a Y-box, a CRE-like sequence, and a GC box in the proximal 5′-flanking region (NF-κB recognition site is present only in the mouse

Y. Takei, H. Ando, & K. Tsutsui (Eds): Handbook of Hormones. DOI: http://dx.doi.org/10.1016/B978-0-12-801028-0.00185-9

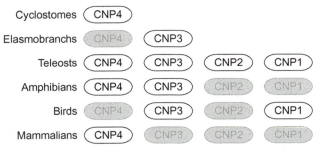

Figure 32C.1 Distribution of the CNP genes in different classes of vertebrates. Shaded and non-shaded CNP genes represent absence and presence of the CNP genes, respectively. Presence of the CNP3 gene in cyclostomes has not been determined.

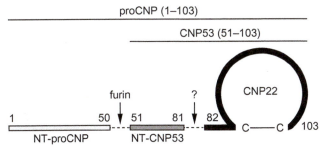

Figure 32C.2 Schematic of CNP structure and processing.

human	CNP4	G*SKGCFGLKLDRIGSMSGLGC
chicken	CNP1	AVPRGCFGLKMDRIGAFSGLGC
Xenopus	CNP3	GFSRGCFGMKLDRIGS*SGLGC
pufferfish	CNP1	GWNRGCFGLKLDRIGSMSGLGC
dogfish	CNP3	GPSRGCFGVKLDRIGAMSGLGC
		* *

Figure 32C.3 An alignment of major CNPs in different classes of vertebrates. Two cysteine residues form a disulfide bond bridge (asterisks).

Table 32C.1 Tissue and Plasma Concentration of CNPs in Vertebrates

Species	Tissues	Concentration	Ref.
human	brain	ca. 1 fmol/mg (wet weight)	[9]
	heart	ca. 5 fmol/mg (wet weight)	
	plasma	1.4 ± 0.6 fmol/ml	[5]
eel	brain (SW/FW)	20−40 fmol/mg (wet weight)	[10]
	heart (SW/FW)	4−7 fmol/mg (wet weight)	
	plasma (FW)	ca. 120 fmol/ml	
	plasma (SW)	ca. 20 fmol/ml	
dogfish (elasmobranch)	brain	16.2 ± 3.0 pmol/mg (protein)	[11]
	atrium	110.3 ± 13.2 pmol/mg (protein)	
	cardiac ventricle	87.1 ± 23.2 pmol/mg (protein)	
	plasma	1.92 ± 0.38 pmol/ml	

SW: seawater-adapted eel, FW: freshwater-adapted eel.

CNP gene) [6]. Putative transcription factors are Sp-1, CREB, TSC22D1, and STK16 [6]. Regulation of CNP release has been extensively studied using cultured endothelial cells. CNP secretion is enhanced by TGF-β, TNF-α, IL-1, ANP, and sheer stress, and suppressed by insulin and VEGF [6]. Unlike ANP, CNPs are not stored in large secretory granules.

Receptors

Structure and Subtype

The biological receptor for CNPs is B-type NP receptor (NPR-B or GC-B). Similarly to an receptor for ANP/BNP (NPR-A or GC-A), NPR-B is coupled with guanylyl cyclase which is involved in cGMP mediated intracellular signaling cascade and forms homotetramer [5]. In medaka, CNP1, 2, 4 bind to OlGC1 (a NPR-B homolog), while CNP3 prefers OlGC2 (one of two NPR-A homologs) [2]. The human NPR-B gene (*NPR1*) is located on chromosome 9 (9p21−12) and its protein contains 1047 aa residues with an Mr of 117,022 [5]. The NPR-B gene is widely expressed throughout the body, including the brain, chondrocytes, lung, vascular smooth muscle, adrenal, kidney, ovary, and uterus [5]. In shark, NPR-B is expressed in the rectal gland, a distinct salt-secreting organ [7,8]. In the rainbow trout, expression of NPR-B decreased upon transfer to seawater [7]. CNPs also bind to the guanylyl cyclase-deficient receptor, C-type NP receptor (NPR-C), with high affinity.

Antagonist

HS-142-1 inhibits CNP binding to NPR-B, but not NPR-C. C-ANF (an NP peptide with a truncated ring sequence) inhibits binding to NPR-C and thus enhances the paracrine and endocrine CNP actions.

Biological Functions

Target Cells/Tissues and Functions

CNPs do not induce natriuresis at physiological concentrations. In mammals, the most recognized function of CNP is regulation of long bone growth [5]. CNP targets chondrocytes to promote endochondral ossification. CNP also has a vasodepressor effect in mammals, teleosts (eel and trout), and elasmobranchs, presumably through its action on the vascular smooth muscle [5,7,8]. In medaka that lack the ANP gene, CNP3 appears to be important for normal atrial development as CNP3 knockdown resulted in hypertrophy of atria [12]. Thus, CNP3 appears to take over the roles of ANP in this teleost. In shark, CNP3 stimulates release of vasoactive intestinal peptide (VIP) from the nerve terminals within the rectal gland [7,8].

Phenotype in Gene-Modified Animals

CNP or NPR-B knockout mice showed severe dwarfism due to impaired endochondral ossification [5,6,13]. However, these mutants did not show growth abnormality until birth, suggesting that the CNP/NPR-B system may be involved in postnatal bone growth. The opposite phenotype, namely, skeletal overgrowth, was observed in mice overexpressing CNP or in NPR-C knockout mice with super-physiological levels of local CNP resulting from reduced clearance [13]. Genome-wide association study using northern European population-based cohorts found single nucleotide polymorphisms (SNPs) near the CNP gene and the NPR-C gene to be associated with the height of each individual [6].

Pathophysiological Implications

Clinical Implications

Some reports have shown that plasma CNP levels increase in cardiac diseases [6]. However, the increment is not as large as that for cardiac NPs.

Use for Diagnosis and Treatment

CNP has not been used as a diagnostic or therapeutic tool. Several studies suggested potential uses of CNP, e.g. for the treatment for growth disorders and as a biomarker of growth plate activity.

References

1. Sudoh T, Minamino N, Kangawa K, et al. C-type natriuretic peptide (CNP): a new member of natriuretic peptide family identified in porcine brain. *Biochem Biophys Res Commun.* 1990;168:863−870.

2. Inoue K, Naruse K, Yamagami S, et al. Four functionally distinct C-type natriuretic peptides found in fish reveal evolutionary history of the natriuretic peptide system. *Proc Natl Acad Sci USA.* 2003;100:10079−10084.

3. Takei Y, Inoue K, Trajanovska S, et al. B-type natriuretic peptide (BNP), not ANP, is the principal cardiac natriuretic peptide in vertebrates as revealed by comparative studies. *Gen Comp Endocrinol.* 2011;171:258−266.

4. Volpe M, Rubattu S, Burnett J. Natriuretic peptides in cardiovascular diseases: current use and perspectives. *Eur Heart J.* 2014;35:419−425.

5. Potter LR, Abbey-Hosch S, Dickey DM. Natriuretic peptides, their receptors, and cyclic guanosine monophosphate-dependent signaling functions. *Endocr Rev.* 2006;27:47−72.

6. Sellitti DF, Koles N, Mendonça MC. Regulation of C-type natriuretic peptide expression. *Peptides.* 2011;32:1964−1971.

7. Johnson KR, Olson KR. Comparative physiology of the piscine natriuretic peptide system. *Gen Comp Endocrinol.* 2008;157:21−26.

8. Toop T, Donald JA. Comparative aspects of natriuretic peptide physiology in non-mammalian vertebrates: a review. *J Comp Physiol B.* 2004;174:189−204.

9. Minamino N, Makino Y, Tateyama H, et al. Characterization of immunoreactive human C-type natriuretic peptide in brain and heart. *Biochem Biophys Res Commun.* 1991;179:535−542.

10. Takei Y, Inoue K, Ando K, et al. Enhanced expression and release of C-type natriuretic peptide in freshwater eels. *Am J Physiol Regul Integr.* 2001;280:R1727-R1735

11. Suzuki R, Togashi K, Ando K, et al. Distribution and molecular forms of C-type natriuretic peptide in plasma and tissue of a dogfish, *Triakis scyllia'. Gen Comp Endocrinol.* 1994;96:378−384.

12. Miyanishi H, Okubo K, Nobata S, et al. Natriuretic peptides in developing medaka embryos: implications in cardiac development by loss-of-function studies. *Endocrinology.* 2013;154:410−420.

13. Kishimoto I, Tokudome T, Nakao K, et al. Natriuretic peptide system: an overview of studies using genetically engineered animal models. *FEBS J.* 2011;278:1830−1841.

14. Trajanovska S, Inoue K, Takei Y, et al. Genomic analyses and cloning of novel chicken natriuretic peptide genes reveal new insights into natriuretic peptide evolution. *Peptides.* 2007;28:2155−2163.

Supplemental Information Available on Companion Website

- Alignment of amino acid sequence of CNP4 precursor in vertebrates/E-Figure 32C.1
- Molecular phylogenetic analyses of CNP precursors in vertebrates/E-Figure 32C.2 [14]
- Molecular phylogenetic analyses of NPR-B in vertebrates/E-Figure 32C.3
- Accession numbers of gene and cDNA for CNP1/E-Table 32C.1
- Accession numbers of gene and cDNA for CNP2/E-Table 32C.2
- Accession numbers of gene and cDNA for CNP3/E-Table 32C.3
- Accession numbers of gene and cDNA for CNP4 and CNP4-like NP/E-Table 32C.4
- Accession numbers of cDNA for NPR-C and NPR-D/E-Table 32C.5

Ventricular Natriuretic Peptide

Takehiro Tsukada

Abbreviation: VNP

Cardiac natriuretic peptide presents abundantly in the ventricle of early ray-finned fishes such as bichir and sturgeon and of primitive teleosts such as eel and salmonids.

Discovery

VNP was first purified in 1991 from the eel ventricle [1].

Structure

Structural Features

Deduced eel proVNP consists of 128 aa residues (Figure 32D.1). Bioactive mature eel VNP (36 aa residues) is located at the C-terminus. Eel VNP has 17 aa residues of an intramolecular ring, from which a long C-terminal tail sequence (14 aa residues) extends [1]. The C-terminally truncated form, VNP (1−25), is also present in eel plasma [2]. An NP gene abundantly expressed in the chicken kidney, named renal NP (RNP), may be an ortholog of VNP [3].

Primary Structure

Sequence identity is >75% in the mature sequences of eel, salmon, sturgeon, and bichir VNP (Figure 32D.2) [3]. The VNP gene is absent in the genome database of several advanced teleost species (e.g., medaka and pufferfish) and other advanced classes of vertebrates [3], except in birds (chicken).

Properties

Mr of mature eel VNP is 3940 and pI is ca. 10. It is freely soluble in water, acid, and 67% acetone, but insoluble in 99% acetone. VNP solution in water at $>10^{-4}$ M is stable for more than a year at $-20°C$.

Synthesis and Release

Gene and mRNA

By linkage mapping, the rainbow trout VNP gene was found to be localized in tandem with the atrial natriuretic peptide (ANP) and B-type natriuretic peptide (BNP) genes on the same chromosome [3]. More recently, the VNP gene has been placed on the same scaffold as those of ANP and BNP in the eel genome database [4]. However, precise chromosomal location of the VNP gene has not been determined yet in the two species. The size of eel VNP mRNA is 1024 bp. Unlike BNP mRNA, there is no repetitive AUUUA motif in the 3′-untranslated region of VNP mRNA.

Distribution of mRNA

VNP mRNA is exclusively expressed in the heart (both atrium and ventricle) of eel, trout, bichir, and sturgeon. In the bichir, the transcripts are also found in the gill to a lesser extent.

Tissue and Plasma Concentrations

Tissue and plasma concentrations of eel VNP are shown in Table 32D.1[5,6]. VNP is stored in the heart as proVNP. The concentration in the atrium is somewhat higher than that in the ventricle, but the total amount of VNP is greater in the ventricle because of its larger size. The major circulating form of eel VNP is VNP (1−36), although proVNP and VNP (1−25) also circulate in eel plasma [2]; the amount of plasma proVNP is approximately one quarter that of mature VNP. Plasma concentration of eel VNP (1−36) is ca. 60 fmol/ml, which is comparable to that of eel ANP. There is no difference in basal plasma VNP levels between seawater- and freshwater-adapted eels.

Regulation of Synthesis and Release

In eels, a major stimulus for VNP release is osmotic stimulus, particularly an acute increase in plasma osmolality. Plasma eel VNP level is transiently and rapidly increased after seawater transfer, or hypertonic NaCl or mannitol injections [6,7]. Increased blood volume (volume stimulus) also induces VNP release, although it is less potent. Increased VNP is cleared from the circulation quickly due to the high metabolic clearance rate [8]. Similarly to ANP in mammals, chronic volume load, but not salt load, is a major stimulus for VNP secretion in trout [9]. The promoter region and potential transcription factors have not been identified yet for the VNP gene.

Receptors

Structure and Subtype

Eel VNP binds eel A-type NP receptor (NPR-A: $K_d = 0.1$ nM), C-type NP receptor (NPR-C: $K_d = 0.15$ nM), and D-type NP receptor (NPR-D: $KK_a = 1$ nM) which are transiently expressed in COS-7 cells [10]. Although binding affinity to the eel B-type NP receptor (NPR-B) has not been determined yet, eel VNP stimulates guanylyl cyclase activity of eel NPR-B expressed in COS-7 cells at 10 nM. Therefore, it is assumed that VNP is a ligand not only for NPR-A but also for NPR-B when its secretion is enhanced. Eel NPR-A, -B, and -C are widely distributed, while eel NPR-D is specifically expressed in the brain and gills [10]. A VNP-specific receptor has not yet been found.

Y. Takei, H. Ando, & K. Tsutsui (Eds): Handbook of Hormones. DOI: http://dx.doi.org/10.1016/B978-0-12-801028-0.00186-0

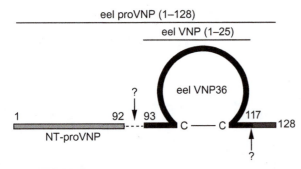

Figure 32D.1 Schematic of eel VNP structure and processing.

```
   eel   KSFNSCFGTRMDRIGSWSGLGCNSLK*GTKKKIFGN
 trout   KSFNSCFGNRIERIGSWSGLGCNNVKTG*KKRIFGN
sturgeon RSM*GCFGNRIERIGSWSSLGCNNSRFGSKKRIF
 bichir  KAFNGCFGNRIERIGSWSALGCNSPKLGAKKRIFK
              *                 *
```

Figure 32D.2 An alignment of VNP mature peptide sequences. Two cysteine residues form a disulfide bond bridge (asterisks).

Table 32D.1 Tissue and Plasma Concentrations of Eel VNP

Tissue (concentration)	SW	FW	Ref.
Atrium (fmol/mg)	3344 ± 502	254 ± 310	[5]
Ventricle (fmol/mg)	1265 ± 94	1397 ± 83	
Plasma (fmol/ml)	55.0 ± 3.7	51.6 ± 3.0	[6]

SW: seawater-adapted eel, FW: fresh-water-adapted eel.

Antagonists

HS-142-1 blocks binding of VNP to eel NPR-A, -B and -D, but not NPR-C. C-ANF is a blocker for binding to both NPR-C and -D.

Biological Functions

Target Cells/Tissues and Functions

VNP is as potent as ANP and more potent than BNP for cardiovascular effects in eels and trout. Systemic injection of eel VNP at doses of 0.1−1 nmoles/kg decreased blood pressure and increased hematocrit [8,11]. The hypotensive action of eel VNP lasts longer than that of eel ANP [8]. In seawater-adapted eels, VNP decreased plasma Na^+ concentration by inhibiting drinking rate and subsequent intestinal NaCl absorption [12]. Thus, VNP is thought to be a hormone important for seawater acclimation in eels. Eel VNP potentiates steroidogenic action of ACTH. In rats, natriuretic and hypotensive effects of eel VNP are observed at 1−10 nmoles/kg [2].

Phenotype in Gene-Modified Animals

A VNP gene knockout animal has not been generated.

Pathophysiological Implications

Clinical Implications

None.

Use for Diagnosis and Treatment

None.

References

1. Takei Y, Takahashi A, Watanabe TX, et al. A novel natriuretic peptide isolated from eel cardiac ventricles. *FEBS Lett.* 1991;282:317−320.
2. Takei Y, Takahashi A, Watanabe TX, et al. Eel ventricular natriuretic peptide: isolation of a low molecular size form and characterization of plasma form by homologous radioimmunoassay. *J Endocrinol.* 1994;141:81−89.
3. Takei Y, Inoue K, Trajanovska S, et al. B-type natriuretic peptide (BNP), not ANP, is the principal cardiac natriuretic peptide in vertebrates as revealed by comparative studies. *Gen Comp Endocrinol.* 2011;171:258−266.
4. ZF-Genomics, Eel Genome website, 2015, available at <http://www.zfgenomics.org/sub/eel>.
5. Takei Y, Balment R. Biochemistry and physiology of a family of eel natriuretic peptides. *Fish Physio Biochem.* 1993;11:183−188.
6. Kaiya H, Takei Y. Changes in plasma atrial and ventricular natriuretic peptide concentrations after transfer of eels from freshwater to seawater or vice versa. *Gen Comp Endocrinol.* 1996;104:337−345.
7. Kaiya H, Takei Y. Osmotic and volaemic regulation of atrial and ventricular natriuretic peptide secretion in conscious eels. *J Endocrinol.* 1996;149:441−447.
8. Nobata S, Ventura A, Kaiya H, et al. Diversified cardiovascular actions of six homologous natriuretic peptides (ANP, BNP, VNP, CNP1, CNP3, and CNP4) in conscious eels. *Am J Physiol.* 2010;298:R1549−R1559.
9. Johnson KR, Olson KR. Comparative physiology of the piscine natriuretic peptide system. *Gen Comp Endocrinol.* 2008;157:21−26.
10. Takei Y, Hirose S. The natriuretic peptide system in eels: a key endocrine system for euryhalinity? *Am J Physiol.* 2002;282: R940−R951.
11. Miyanishi H, Nobata S, Takei Y. Relative antidipsogenic potencies of six homologous natriuretic peptides in eels. *Zool Sci.* 2011;28:719−726.
12. Tsukada T, Takei Y. Integrative approach to osmoregulatory action of atrial natriuretic peptide in seawater eels. *Gen Comp Endocrinol.* 2006;147:31−38.

Supplemental Information Available on Companion Website

- Alignment of amino acid sequence of VNP and RNP precursor in vertebrates/E-Figure 32D.1
- Accession numbers of gene and cDNA for VNP and other NPs/E-Table 32D.1

CHAPTER 33

Gonadal Hormones

Hironori Ando

History

Gonadal functions are under the control of the hypothalamic-pituitary-gonadal (HPG) axis that governs sexual development and reproduction. Gonadotropin-releasing hormone (GnRH) secreted from the hypothalamus regulates the secretion of follicle-stimulating hormone (FSH) and luteinizing hormone (LH) from the pituitary, and then FSH and LH act on the gonads to stimulate gonadal development and maturation. Besides producing eggs and sperm, the gonads also are endocrine organs that secrete sex steroid hormones and protein hormones to exert feedback effects. The sex steroid hormones are described in Chapter 94. This chapter, therefore, focuses on the gonadal protein hormones including inhibin, activin, follistatin, and anti-Müllerian hormone (AMH). Among them, inhibin, activin, and AMH are glycoproteins belonging to the TGF-β superfamily. Inhibin was originally identified as an FSH-inhibiting factor in 1985 [1,2], followed by identification of activin [3] and follistatin [4], both of which also are modulators of FSH secretion. Since then, activin and follistatin have been shown to be present in many tissues, and diverse local actions of activin have been the focus of considerable attention with a particular emphasis on its roles in cell differentiation and embryonic development [5]. It is now evident that activin acts largely as an autocrine and/or paracrine growth factor during development and in reproductive physiology [6]. AMH is a glycoprotein that causes regression of Müllerian ducts during male differentiation and was first purified from the bovine testis in 1984 [7].

Structures and Biological Functions

Inhibin and activin share α-, β_A-, and β_B-subunits. Inhibin is a disulfide-linked heterodimer consisting of an α-subunit and either a β_A-subunit (inhibin A) or a β_B-subunit (inhibin B). Hetero- or homodimers of β_A- and β_B-subunits give rise to three forms of activins (activin A, activin B, and activin AB; see Figure 33A.1 in Subchapter 33A). Follistatin is structurally unrelated to inhibin and activin, but binds to activin and neutralizes activin actions (Figure 33.1). Inhibins also antagonize activin actions by forming complexes with activin receptors with the aid of betaglycan. Activin signaling is mediated through specific type I and type II serine/threonine kinase receptors (Figure 33.1), whereas no specific receptor has been identified for inhibin and follistatin. Activin binds to the type II receptor that leads to the recruitment, phosphorylation, and activation of the type I receptor. The activated type I receptor then phosphorylates Smad2 and Smad3, resulting in the activation of the Smad signaling pathway leading to activation or repression of target genes. Activins exert pluripotent actions throughout the endocrine,

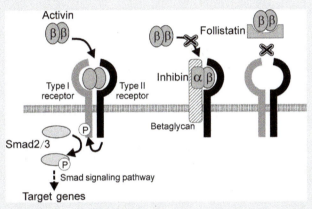

Figure 33.1 Activin signaling pathway and antagonism of inhibin and follistatin.

reproductive, immune and hematopoietic systems. AMH is a dimeric glycoprotein that belongs to the TGF-β superfamily, and its actions are mediated through specific type I and type II serine/threonine kinase receptors like activin is. After birth, AMH is produced in testicular Sertoli and Leydig cells and ovarian granulosa cells and is involved in the regulation of Leydig cell differentiation and ovarian follicle development, respectively.

References

1. Miyamoto K, Hasegawa Y, Fukuda M, et al. Isolation of porcine follicular fluid inhibin of 32K daltons. *Biochem Biophys Res Commun*. 1985;129:396−403.
2. Robertson DM, Foulds LM, Leversha L, et al. Isolation of inhibin from bovine follicular fluid. *Biochem Biophys Res Commun*. 1985;126:220−226.
3. Ling N, Ying SY, Ueno N, et al. Pituitary FSH is released by a heterodimer of the beta-subunits from the two forms of inhibin. *Nature*. 1986;321:779−782.
4. Ueno N, Ling N, Ying S-Y, et al. Isolation and partial characterization of follistatin: a single-chain Mr 35,000 monomeric protein that inhibits the release of follicle-stimulating hormone. *Proc Natl Acad Sci USA*. 1987;84:8282−8286.
5. Asashima M, Nakano H, Shimada K, et al. Mesodermal induction in early amphibian embryos by activin A (erythroid differentiation factor). *Roux's Arch Dev Biol*. 1990;198:330−335.
6. Welt C, Sidis Y, Keutmann H, et al. Activins, inhibins, and follistatins: from endocrinology to signaling. A paradigm for the new millennium. *Exp Biol Med (Maywood)*. 2002;227:724−752.
7. Picard J-Y, Josso N. Purification of testicular anti-Müllerian hormone allowing direct visualization of the pure glycoprotein and determination of yield and purification factor. *Mol Cell Endocrinol*. 1984;34:23−29.

Y. Takei, H. Ando, & K. Tsutsui (Eds): Handbook of Hormones. DOI: http://dx.doi.org/10.1016/B978-0-12-801028-0.00033-7

Inhibin

Hiroyuki Kaneko

Abbreviation: None
Additional names: FSH-inhibiting factor

A glycoprotein, a member of transforming growth factor (TGF)-β superfamily, and secreted mainly from the gonads. Inhibits follicle-stimulating hormone (FSH) secretion from the anterior pituitary and in turn regulates gonadal function and development.

Discovery

Presence of FSH-inhibiting activity had been reported in the gonads from the 1970s, and the FSH-inhibiting factor "inhibin" was first isolated in 1985 from porcine [1] and bovine [2] follicular fluid.

Structure

Structural Features

Inhibins and activins are structurally related. Inhibins are disulfide-linked heterodimers composed of an α-subunit and either a β_A-subunit (inhibin A) or a β_B-subunit (inhibin B), whereas activins are dimers made up of β-subunits (Figure 33A.1). The monomeric α-subunit, devoid of FSH-suppressing activity, has been also identified. Approximate 80% identity is seen in the sequences of the human, porcine, bovine, and rat α-subunits (E-Figure 33A.1). The mature β_A-subunit shares the same sequence among the above species (E-Figure 33A.2). The β_B-subunit shows an approximate 90% identity (E-Figure 33A.3). The α-subunit has *N*-linked glycosylation sites and their degree of glycosylation modifies biological activity [3].

Primary Structure

The amino acid sequence of human mature inhibin [4] is shown in Figure 33A.2.

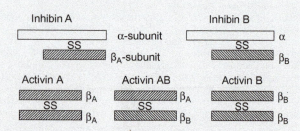

Figure 33A.1 Mature forms of inhibins and activins.

Properties

Mr approximately 32,000 (mature inhibin). Additional Mr forms (55,000–105,000) exist due to the processing of the largest precursor form (Mr 105,000) [5]. pI 6.9–7.3. Stable in 8 M urea. Dissociated to two subunits in 2% (v/v) 2-mercaptoethanol.

Synthesis and Release

Gene, mRNA, and Precursor

The inhibin α- and β-subunits are encoded by separate genes. The human inhibin α-subunit gene, *INHA*, location 2q33–qtr, consists of two exons, and has AP-1, GATA, and SF-1 binding elements and CRE in the promoter region. The α-subunit mRNA has 1098 bp that encodes a precursor protein (proαNαC) of 348 aa residues. The human β_A-subunit gene, *INHBA*, location 7p15–p14, consists of two exons. The mRNA has 1,278 bp that encode a precursor protein (pro$\beta_A\beta_A$) of 398 aa residues. The human β_B-subunit gene, *INHBB*, location 2cen–q13, consists of two exons. The mRNA has 1,221 bp encoding a precursor protein (pro$\beta_B\beta_B$) of 379 aa residues.

Tissue Distribution of mRNA

Coexpression of inhibin α- and β-subunit mRNAs suggests production of inhibin molecules. In the ovaries of mammals, birds, and fish, mRNAs for α-, β_A- and β_B-subunits are expressed in granulosa cells. Human and primate luteal cells express α- and β_A-subunits. In the placenta of mouse and human, α- and β_A-subunit mRNAs are expressed. In males, α- and β-subunit mRNAs are detected in the Sertoli cells and Leydig cells. In the adrenal cortex, α-, β_A-, and β_B-subunit mRNAs are detected.

Tissue Content

Adult rat, ovary, inhibin A, 2–30 (pg/mg tissue), inhibin B, 200–1200; adult rat, testis, inhibin A, 10 (pg/testis), inhibin B, 2000–3000.

Plasma Concentrations

Plasma concentrations of inhibins in various mammals are shown in Table 33A.1 [6–13].

Regulation of Synthesis and Release

FSH, cAMP, and forskolin stimulate the secretion of inhibins from granulosa cells and Sertoli cells in various mammals. Inhibin production from granulosa cells is suppressed by EGF. Inhibin production in primate and human luteal cells is promoted by luteinizing hormone (LH) and human chorionic

Y. Takei, H. Ando, & K. Tsutsui (Eds): Handbook of Hormones. DOI: http://dx.doi.org/10.1016/B978-0-12-801028-0.00187-2

Inhibin α-subunit (134 a.a. residues)
```
      10        20        30        40        50        60        70        80
STPLMSWPWSPSALRLLQRPPEEPAAHANCHRVALNISFQELGWERWIVYPPSFIFHYCHGGCGLHIPPNLSLPVPGAPP
      90       100       110       120       130
TPAQPYSLLPGAQPCCAALPGTMRPLHVRTTSDGGYSFKYETVPNLLTQHCACI
```

Inhibin β_A-subunit (116 a.a. residues), upper, and β_B-subunit (115 a.a. residues), lower
```
      10        20        30        40        50        60        70        80
GLECDGKVNICCKKQFFVSFKDIGWNDWIIAPSGYHANYCEGECPSHIAGTSGSSLSFHSTVINHYRMRGHSPFANLKSC
GLECDGRTNLCCRQQFFIDFRLIGWNDWIIAPTGYYGNYCEGSCPAYLAGVPGSASSFHTAVVNQYRMRGLNPGTVNSCC
      90       100       110
CVPTKLRPMSMLYYDDGQNIIKKDIQNMIVEECGCS
IPTKLSTMSMLYFDDEYNIVKRDVPNMIVEECGCA
```

Figure 33A.2 Sequence of human mature inhibin. The *N*-linked glycosylation sites are underlined.

Table 33A.1 Circulating Inhibin Concentrations in Various Mammals

Species	Status	Isoform	Concentration	Ref
human ♀	menstrual cycle	A	5–60 pg/ml	[6]
		B	5–150 pg/ml	
♂	adult	B	~200 pg/ml	[7]
rat ♀	estrous cycle	A	50–300 pg/ml	[8]
		B	50–550 pg/ml	
♂	adult	A	<5 pg/ml	[9]
		B	~200 pg/ml	
sheep ♀	estrous cycle	A	200–500 pg/ml	[10]
♂	adult	A	~200 pg/ml	[11]
cattle ♀	estrous cycle	A	100–200 pg/ml	[12]
horse ♀	estrous cycle	A	150–450 pg/ml	[13]

Table 33A.2 Effects of Inhibin-Related Proteins on Various Cells or Tissues

Cells or Tissues	Species	Inhibin	Effects	Ref
granulosa cell	cattle	A	↓estrogen production	[16]
theca cell	human	A	↑basal or LH-induced androgen production	[17]
follicle	frog	A	↓pituitary homogenate stimulated progesterone production	[18]
oocyte	zebrafish	A	↑germinal vesicle breakdown	[19]
cumulus-oocyte	mouse	A	↓germinal vesicle breakdown	[20]
complex	cattle	α-subunit	↓early embryonic development	[21]
Leydig cell	rat	A	↑LH-induced androgen production	[22]
spermatogonia	rat	A	↓DNA synthesis	[23]
adrenocortex cell	mouse	A, B	antagonizing activin-reduced steroid synthesis	[24]

gonadotropin (hCG). Adrenocorticotropic hormone (ACTH) stimulates inhibin secretion from adrenal cortex cells.

Receptors

Structure and Subtype

A specific receptor for inhibin has not been identified. It is now accepted that inhibin actions result from antagonism of activin signaling in the presence of betaglycan. Human betaglycan (type III TGF-β receptor) has 851 aa residues that are expressed on the surface of pituitary cells, granulosa cells, theca cells, Sertoli cells, Leydig cells, and adrenal cortex cells. It consists of an extracellular domain, a transmembrane region, and a short intracellular domain but lacks a signaling domain [14]. Inhibins bind to activin type II receptors via their β-subunits, and to betaglycan via the α-subunit to form a stable complex. The complex occupies activin type II receptors and prevents activin from activation of the type I receptors, resulting in blockade of the Smad 2/3 signaling pathway [15].

Agonist

Recombinant and purified inhibin.

Antagonist

Activin.

Biological Functions

Target Cells/Tissues and Functions

Inhibins have an endocrine role in various animals, being released into the circulation from the gonads. Inhibins suppress the expression of the FSH β-subunit in the pituitary, and thereby regulate gonadal functions and development. Inhibins also have paracrine and autocrine effects (Table 33A.2) [16–24].

Phenotype of Gene-Modified Animals

Inhibin α-subunit gene knockout mice develop Sertoli cell tumors in males and granulosa cell tumors in females, resulting in death until 17 weeks of age [25]. When the knockout mice were gonadoectomized, life expectancy increased; however, these mice developed adrenal tumors around 30 weeks of age. Inhibin β_A-subunit gene knockout mice die perinatally and have defects in tooth, palate, and retinal formation [26]. Inhibin β_B gene knockout mice are viable but have defects in mammary gland function [27].

Pathophysiological Implications

Clinical Implications

From the findings that inhibin α-subunit knockout mice develop gonadal and adrenal tumors, the α-subunit gene is expected to act as a tumor suppressor gene. However, no consistent gene mutations have been identified in cancer patients. Correlated with tumor growth, inhibin production is enhanced in several types of adrenal and gonadal tumors. Women affected with premature ovarian failure show low serum levels of inhibin A and inhibin B. Decrease in testicular inhibin B production is noted in men with testicular dysfunction. Pregnancies affected with Down's syndrome accompany high circulating concentrations of inhibin A. Hyperplasia of the adrenal cortex (Cushing's syndrome) often raises inhibin A secretion from the adrenal gland.

Use for Diagnosis and Treatment

Inhibin levels in the circulation are a reliable marker for granulosa cell tumors and serous and mucinous epithelial carcinomas. This may be true for granulosa cell tumors in mares. Evaluation of plasma inhibin A levels until the second trimester of pregnancy is useful for screening for Down's syndrome. The circulating inhibin B is a good predictor for the conditions of spermatogenesis. High levels of inhibin are noted in humans, rats, and dogs with Leydig cell tumors, whereas a low level clinically suggests a premature ovarian failure.

References

1. Miyamoto K, Hasegawa Y, Fukuda M, et al. Isolation of porcine follicular fluid inhibin of 32 K daltons. *Biochem Biophys Res Commun.* 1985;129:396–403.
2. Robertson DM, Foulds LM, Leversha L, et al. Isolation of inhibin from bovine follicular fluid. *Biochem Biophys Res Commun.* 1985;126:220–226.
3. Makanji Y, Harrison CA, Stanton PG, et al. Inhibin A and B in vitro bioactivities are modified by their degree of glycosylation and their affinities to betaglycan. *Endocrinology.* 2007;148:2309–2316.
4. Mason AJ, Niall HD, Seeburg PH. Structure of two human ovarian inhibins. *Biochem Biophys Res Commun.* 1986;135:957–964.
5. Sugino K, Nakamura T, Takio K, et al. Purification and characterization of high molecular weight forms of inhibin from bovine follicular fluid. *Endocrinology.* 1992;130:789–796.
6. Groome NP, Illingworth PJ, O'Brien M, et al. Measurement of dimeric inhibin B throughout the human menstrual cycle. *J Clin Endocrinol Metab.* 1996;81:1401–1405.
7. Anawalt BD, Bebb RA, Matsumoto AM, et al. Serum inhibin B levels reflect Sertoli cell function in normal men and men with testicular dysfunction. *Clin Endocrinol Metab.* 1996;81:3341–3345.
8. Kenny HA, Woodruff TK. Follicle size class contributes to distinct secretion patterns of inhibin isoforms during the rat estrous cycle. *Endocrinology.* 2006;147:51–60.
9. Buzzard JJ, Loveland KL, O'Bryan MK, et al. Changes in circulating and testicular levels of inhibin A and B and activin A during postnatal development in the rat. *Endocrinology.* 2004;145:3532–3541.
10. Knight PG, Feist SA, Tanetta DS, et al. Measurement of inhibin-A (alpha beta A dimer) during the oestrous cycle, after manipulation of ovarian activity and during pregnancy in ewes. *J Reprod Fertil.* 1998;113:159–166.
11. McNeilly AS, Souza CJ, Baird DT, et al. Production of inhibin A not B in rams: changes in plasma inhibin A during testis growth, and expression of inhibin/activin subunit mRNA and protein in adult testis. *Reproduction.* 2002;123:827–835.
12. Kaneko H, Noguchi J, Kikuchi K, et al. Alterations in peripheral concentrations of inhibin A in cattle studied using a time-resolved immunofluorometric assay: relationship with estradiol and follicle-stimulating hormone in various reproductive conditions. *Biol Reprod.* 2002;67:38–45.
13. Medan MS, Nambo Y, Nagamine N, et al. Plasma concentrations of ir-inhibin, inhibin A, inhibin pro-alphaC, FSH, and estradiol-17beta during estrous cycle in mares and their relationship with follicular growth. *Endocrine.* 2004;25:7–14.
14. López-Casillas F, Cheifetz S, Doody J, et al. Structure and expression of the membrane proteoglycan betaglycan, a component of the TGF-beta receptor system. *Cell.* 1991;67:785–795.
15. Lewis KA, Gray PC, Blount AL, et al. Betaglycan binds inhibin and can mediate functional antagonism of activin signalling. *Nature.* 2000;404:411–414.
16. Jimenez-Krassel F, Winn ME, Burns D, Ireland JL, Ireland JJ. Evidence for a negative intrafollicular role for inhibin in regulation of estradiol production by granulosa cells. *Endocrinology.* 2003;144:1876–1886.
17. Hillier SG, Yong EL, Illingworth PJ, Baird DT, Schwall RH, Mason AJ. Effect of recombinant inhibin on androgen synthesis in cultured human thecal cells. *Mol Cell Endocrinol.* 1991;75:R1–R6.
18. Lin YW, Petrino T, Landin AM, et al. Inhibitory action of the gonadopeptide inhibin on amphibian (*Rana pipiens*) steroidogenesis and oocyte maturation. *J Exp Zool.* 1999;284:232–240.
19. Wu T, Patel H, Mukai S, et al. Activin, inhibin, and follistatin in zebrafish ovary: expression and role in oocyte maturation. *Biol Reprod.* 2000;62:1585–1592.
20. O WS, Robertson DM, de Kretser DM. Inhibin as an oocyte meiotic inhibitor. *Mol Cell Endocrinol.* 1989;62:307–311.
21. Silva CC, Groome NP, Knight PG. Demonstration of a suppressive effect of inhibin alpha-subunit on the developmental competence of in vitro matured bovine oocytes. *J Reprod Fertil.* 1999;115:381–388.
22. Hsueh AJ, Dahl KD, Vaughan J, et al. Heterodimers and homodimers of inhibin subunits have different paracrine action in the modulation of luteinizing hormone-stimulated androgen biosynthesis. *Proc Natl Acad Sci USA.* 1987;84:5082–5086.
23. Hakovirta H, Kaipia A, Söder O, et al. Effects of activin-A, inhibin-A, and transforming growth factor-beta 1 on stage-specific deoxyribonucleic acid synthesis during rat seminiferous epithelial cycle. *Endocrinology.* 1993;133:1664–1668.
24. Farnworth PG, Stanton PG, Wang Y, et al. Inhibins differentially antagonize activin and bone morphogenetic protein action in a mouse adrenocortical cell line. *Endocrinology.* 2006;147:3462–3471.
25. Matzuk MM, Finegold MJ, Su JG, et al. Alpha-inhibin is a tumour-suppressor gene with gonadal specificity in mice. *Nature.* 1992;360:313–319.
26. Matzuk MM, Kumar TR, Bradley A. Different phenotypes for mice deficient in either activins or activin receptor type II. *Nature.* 1995;374:356–360.
27. Vassalli A, Matzuk MM, Gardner HA, et al. Activin/inhibin beta B subunit gene disruption leads to defects in eyelid development and female reproduction. *Genes Dev.* 1994;8:414–427.

Supplemental Information Available on Companion Website

- Primary structure of the inhibin α-subunit precursor of various animals/E-Figure 33A.1
- Primary structure of the inhibin β_A-subunit precursor of various animals/E-Figure 33A.2
- Primary structure of the inhibin β_B-subunit precursor of various animals/E-Figure 33A.3
- Accession numbers of genes and cDNAs for inhibin α-, β_A- and β_B-subunits/E-Table 33A.1

Activin

Hiroyuki Kaneko

Abbreviation: none
Additional names: erythroid differentiation factor (EDF)

A member of transforming growth factor (TGF)-β superfamily, produced in a wide range of tissues. Exerts pluripotent effects on tissue growth and function in an autocrine/paracrine manner.

Discovery

Activin was initially isolated from follicular fluid as a stimulator for follicle-stimulating hormone (FSH) secretion in 1986 [1]. Also independently isolated from a human monocytic cell line (THP-1 cell) as a differentiation factor for erythroleukemic cells in 1987 [2].

Structure

Structural Features

Activins are disulfide-linked dimers composed of β_A- and β_A-subunits (activin A), β_B- and β_B-subunits (activin B), or β_A- and β_B-subunits (activin AB) (see Figure 33A.1 in Subchapter 33A). Mature β_A- and β_B-subunits have nine cysteine residues that are required for inter- and intramolecular disulfide bonds. Unlike the inhibin α-subunit, β-subunits lack glycosylation sites. Amino acid sequences of the mature β_A- and β_B-subunits are highly conserved among species (see E-Figures 33A.2 and 33A.3 in Subchapter 33A). Additionally, genes encoding the β_C-subunit [3] and the β_E-subunit [4] have been identified in mammals, and the β_D-subunit gene [5] in *Xenopus laevis*, potentially giving rise to more activin/inhibin isoforms.

Primary Structure

Amino acid sequences of the human β-subunits are shown in Figure 33A.2 in Subchapter 33A.

Properties

Mr 25,000–28,000 (mature activin). Additional Mr forms (70,000–110,000) exist in follicular fluid due to the processing of the largest precursor form (Mr 110,000). These larger molecular species have no biological activity. pI 4.0–5.0. Stable in 8 M urea. Dissociated to two subunits in 2% (v/v) 2-mercaptoethanol.

Synthesis and Release

Gene, mRNA, and Precursor

Each β-subunit is independently encoded. Gene location and the mRNAs encoding precursor proteins are described in subchapter 33A. In the promoter region, the bovine β_A-subunit gene has an AP-2 binding site and the rat β_B-subunit gene has SP1 and AP-2 binding sites.

Tissue Distribution of mRNA

β_A- and β_B-subunit genes are widely expressed (Table 33B.1). The β_C-subunit gene is expressed in the liver, lung, epididymis, testis, ovary, uterus, pituitary, adrenal gland, and prostate. The β_E-subunit gene is highly expressed in the liver.

Tissue Content

Adult rat, testis, activin A, 150 (pg/testis).

Plasma Concentrations

Human menstrual cycle 300–500 (pg/ml), pregnancy 400–4000.

Regulation of Synthesis and Release

FSH and cAMP enhance the expression of the β_B-subunit gene in rat Sertoli cells and the β_A-subunit gene in rat granulosa cells, less effectively as compared to the inhibin α-subunit gene. Bacterial lipopolysaccharide (LPS) or interleukin (IL)-1 promotes β_A-subunit gene expression and activin A production in rat Sertoli cells and sheep monocytes.

Table 33B.1 Expression of β-subunit mRNAs

Cells or Tissues	mRNA	Species
hypothalamus	β_A and β_B	rat
pituitary cell	β_A and β_B	various mammals
granulosa cell	β_A and β_B	various mammals,
follicle cell	β_A and β_B	zebrafish, goldfish
luteal cell	β_A	human, primate
endometrium	β_A and β_B	human, primate, sheep
early embryo	β_A	cattle
	β_A, β_B, and β_D	frog
Sertoli cell	β_A and β_B	various mammals
Leydig cell	β_A and/or β_B	various mammals
male germ cell	β_A and β_B	various mammals
prostate	β_A and β_B	human
adenocortex cell	β_A and β_B	human, sheep
hepatocyte	β_A and β_B	human
pancreas	β_B	human, rat
kidney	β_A and β_B	human
bone marrow	β_A	various mammals
macrophage, T cell	β_A	human

Y. Takei, H. Ando, & K. Tsutsui (Eds): Handbook of Hormones. DOI: http://dx.doi.org/10.1016/B978-0-12-801028-0.00188-4

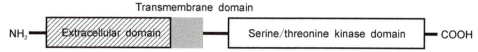

Figure 33B.1 Structure of activin receptor.

The expression of the β_A-subunit gene in the adrenocortical cells is stimulated by adrenocorticotropic hormone (ACTH).

Receptors

Structure and Subtype

Two types of activin receptors are identified: type I (ActRIB, also known as activin receptor-like kinase 4 (ALK4) [6] and ActRIC (ALK7) [7]) and type II (ActRIIA [8] and ActRIIB [9]). Both types of activin receptors have approximately 500 aa residues, consisting of a cysteine rich extracellular domain (~ 110 aa residues) with ligand binding activity, a transmembrane domain (~ 26 aa residues), and an intracellular domain (~ 360 aa residues) that contains serine/threonine kinase activity (Figure 33B.1 and E-Figure 33B.1). The type II receptors bind activin with high affinity, but the type I receptor (ALK4) is unable to bind activin in the absence of the type II receptor.

Signal Transduction Pathway

Activin binds primarily the type II receptor and cooperatively recruits the type I receptor, which forms a ternary receptor complex. In the receptor complex, the type II receptor phosphorylates and activates the type I receptor. The activated type I receptor then phosphorylates the cytoplasmic transcription factors, Smad2 and Smad3, resulting in the formation of a complex with Smad4. The Smad complex then moves to the nucleus, leading to activation or repression of target genes.

Agonists

Recombinant and purified activins.

Antagonists

Antibodies to ActRII, follistatin, follistatin-related protein-3 (FRP-3), inhibin.

Biological Functions

Target Cells/Tissues and Functions

Activin A and B have paracrine and autocrine actions throughout the endocrine, reproductive, immune, and hematopoietic systems in adult animals (Table 33B.2) [10−30]. Activin is an inducer of dorsal mesodermal and neural tissues in *Xenopus* embryos. β_C- and β_E-subunit genes are expressed in several tissues; however, their biological functions have not been fully clarified.

Phenotype of Gene-Modified Animals

β_A-subunit gene null mice die perinatally. The lacking of activin type II receptor expression lowers the growth of testis and ovary, most likely through disruption of FSH production [31]. Overexpression of the β_A-subunit gene results in degeneration of seminiferous tubules.

Pathophysiological Implications

Clinical Implications

Activin A expression is increased in several inflammatory diseases. Elevation in the concentrations of serum activin is noted in patients with septicemia. Experimental exposure to

Table 33B.2 Effects of Activin on Various Cells or Tissues

Cells or Tissues	Species	Effects	Ref
GnRH neuron	cell line	↑GnRH production	[10]
gonadotroph	rat	↑FSH production	[11]
somatotroph	rat	↓basal or GRF stimulated GH production,	[12]
lactotrophs	rat	↓TRH-stimulated PRL production	[13]
granulosa cell			
premature	rat	↑FSH-induced aromatase activity	[14]
	rat	↑FSH-induced progesterone production	[15]
	rat	↑inhibin production	[15]
	rat	↑FSH receptor expression	[16]
	rat	↑proliferation	[17]
mature	rat	↓basal or LH-stimulated progesterone production	[18]
luteinized	human	anti-luteinization (FSH receptor↑, LH receptor↓)	[19]
theca cell	rat	↓LH-induced androgen production	[20]
oocyte	zebrafish	↑germinal vesicle breakdown (GVBD)	[21]
	rat	↑GVBD	[22]
embryo	frog	mesodermal and neural tissue induction	[23]
	cattle	↑early embryonic development	[24]
Sertoli cell	rat	↑proliferation	[25]
Leydig cell	rat	↓LH-induced androgen production	[20]
spermatogonia	rat	↑DNA synthesis	[26]
bone marrow	human	↑colony formation of erythroid and multipotential progenitor cells	[27]
endothelial cell	human	↓proliferation	[28]
pancreatic cell	human	↑insulin production, ↑β-cell differentiation	[29]
hepatocyte	rat	↓EGF induced DNA synthesis	[30]

LPS induces a rapid increase in the circulating activin A, far earlier than with IL-6 but concurrently with tumor necrosis factor α (TNF-α). The activin response, occurring through the Toll-like receptor 4 signaling pathway, promotes an inflammatory reaction by stimulation of monocytes and macrophages to produce IL-6, TNF-α, and prostaglandin E2.

Use for Diagnosis and Treatment

None.

References

1. Ling N, Ying SY, Ueno N, et al. Pituitary FSH is released by a heterodimer of the beta-subunits from the two forms of inhibin. *Nature*. 1986;321:779−782.
2. Eto Y, Tsuji T, Takezawa M, et al. Purification and characterization of erythroid differentiation factor (EDF) isolated from human leukemia cell line THP-1. *Biochem Biophys Res Commun*. 1987;142:1095−1103.
3. Lau AL, Nishimori K, Matzuk MM. Structural analysis of the mouse activin beta C gene. *Biochem Biophys Acta*. 1996;1307:145−148.
4. Fang J, Yin W, Smiley E, et al. Molecular cloning of the mouse activin beta E subunit gene. *Biochem Biophys Res Commun*. 1996;228:669−674.
5. Oda S, Nishimatsu S, Murakami K, et al. Molecular cloning and functional analysis of a new activin beta subunit: a dorsal mesoderm-inducing activity in *Xenopus*. *Biochem Biophys Res Commun*. 1995;210:581−588.
6. Attisano L, Cárcamo J, Ventura F, et al. Identification of human activin and TGF beta type I receptors that form heteromeric kinase complexes with type II receptors. *Cell*. 1993;75:671−680.
7. Tsuchida K, Sawchenko PE, Nishikawa S, et al. Molecular cloning of a novel type I receptor serine/threonine kinase for the TGF beta superfamily from rat brain. *Mol Cell Neurosci*. 1996;7:467−478.

8. Mathews LS, Vale WW. Expression cloning of an activin receptor, a predicted transmembrane serine kinase. *Cell*. 1991;65: 973−982.

9. Attisano L, Wrana JL, Cheifetz S, et al. Novel activin receptors: distinct genes and alternative mRNA splicing generate a repertoire of serine/threonine kinase receptors. *Cell*. 1992;68:97−108.

10. González-Manchón C, Bilezikjian LM, Corrigan AZ, et al. Activin-A modulates gonadotropin-releasing hormone secretion from a gonadotropin-releasing hormone-secreting neuronal cell line. *Neuroendocrinology*. 1991;54:373−377.

11. Vale W, Rivier J, Vaughan J, et al. Purification and characterization of an FSH releasing protein from porcine ovarian follicular fluid. *Nature*. 1986;321:776−779.

12. Billestrup N, González-Manchón C, Potter E, et al. Inhibition of somatotroph growth and growth hormone biosynthesis by activin in vitro. *Mol Endocrinol*. 1990;4:356−362.

13. Kitaoka M, Kojima I, Ogata E. Activin-A: a modulator of multiple types of anterior pituitary cells. *Biochem Biophys Res Commun*. 1988;157:48−54.

14. Hutchinson LA, Findlay JK, de Vos FL, et al. Effects of bovine inhibin, transforming growth factor-beta and bovine Activin-A on granulosa cell differentiation. *Biochem Biophys Res Commun*. 1987;146:1405−1412.

15. Sugino H, Nakamura T, Hasegawa Y, et al. Erythroid differentiation factor can modulate follicular granulosa cell functions. *Biochem Biophys Res Commun*. 1988;153:281−288.

16. Hasegawa Y, Miyamoto K, Abe Y, et al. Induction of follicle stimulating hormone receptor by erythroid differentiation factor on rat granulosa cell. *Biochem Biophys Res Commun*. 1988;156: 668−674.

17. Kaipia A, Toppari J, Huhtaniemi I, et al. Sex difference in the action of activin-A on cell proliferation of differentiating rat gonad. *Endocrinology*. 1994;134:2165−2170.

18. Miró F, Smyth CD, Hillier SG. Development-related effects of recombinant activin on steroid synthesis in rat granulosa cells. *Endocrinology*. 1991;129:3388−3394.

19. Myers M, van den Driesche S, McNeilly AS, et al. Activin A reduces luteinisation of human luteinised granulosa cells and has opposing effects to human chorionic gonadotropin in vitro. *J Endocrinol*. 2008;199:201−212.

20. Hsueh AJ, Dahl KD, Vaughan J, et al. Heterodimers and homodimers of inhibin subunits have different paracrine action in the modulation of luteinizing hormone-stimulated androgen biosynthesis. *Proc Natl Acad Sci USA*. 1987;84:5082−5086.

21. Wu T, Patel H, Mukai S, et al. Activin, inhibin, and follistatin in zebrafish ovary: expression and role in oocyte maturation. *Biol Reprod*. 2000;62:1585−1592.

22. Itoh M, Igarashi M, Yamada K, et al. Activin A stimulates meiotic maturation of the rat oocyte in vitro. *Biochem Biophys Res Commun*. 1990;166:1479−1484.

23. Thomsen G, Woolf T, Whitman M, et al. Activins are expressed early in *Xenopus* embryogenesis and can induce axial mesoderm and anterior structures. *Cell*. 1990;63:485−493.

24. Yoshioka K, Kamomae H. Recombinant human activin A stimulates development of bovine one-cell embryos matured and fertilized in vitro. *Mol Reprod Dev*. 1996;45:151−156.

25. Boitani C, Stefanini M, Fragale A, et al. Activin stimulates Sertoli cell proliferation in a defined period of rat testis development. *Endocrinology*. 1995;136:5438−5444.

26. Hakovirta H, Kaipia A, Söder O, et al. Effects of activin-A, inhibin-A, and transforming growth factor-beta 1 on stage-specific deoxyribonucleic acid synthesis during rat seminiferous epithelial cycle. *Endocrinology*. 1993;133:1664−1668.

27. Broxmeyer HE, Lu L, Cooper S, et al. Selective and indirect modulation of human multipotential and erythroid hematopoietic progenitor cell proliferation by recombinant human activin and inhibin. *Proc Natl Acad Sci USA*. 1988;85:9052−9056.

28. McCarthy SA, Bicknell R. Inhibition of vascular endothelial cell growth by activin-A. *J Biol Chem*. 1993;268:23066−23071.

29. Demeterco C, Beattie GM, Dib SA, et al. A role for activin A and betacellulin in human fetal pancreatic cell differentiation and growth. *J Clin Endocrinol Metab*. 2000;85:3892−3897.

30. Yasuda H, Mine T, Shibata H, et al. Activin A: an autocrine inhibitor of initiation of DNA synthesis in rat hepatocytes. *J Clin Invest*. 1993;92:1491−1496.

31. Matzuk MM, Kumar TR, Bradley A. Different phenotypes for mice deficient in either activins or activin receptor type II. *Nature*. 1996;374:356−360.

Supplemental Information Available on Companion Website

- Primary structure of the human type I activin receptor/E-Figure 33B.1
- Accession numbers of vertebrate angiotensinogen genes (*agt*)/E-Table 33B.1

Follistatin

Hiroyuki Kaneko

Abbreviation: none
Additional names: activin binding protein, follicle-stimulating hormone-suppressing protein (FSP).

A single chain glycoprotein structurally unrelated to the inhibin and activin proteins. Binds activin and neutralizes activin action.

Discovery

Follistatin was originally isolated from porcine [1] and bovine [2] follicular fluid for its ability to suppress follicle-stimulating hormone (FSH) secretion from the rat pituitary. Subsequently, it was discovered that follistatin binds activin with high affinity.

Structure

Structural Features

Resulting from alternative splicing of the common precursor gene, two mature forms of follistatin exist: follistatin 315 and follistatin 288, which lacks the carboxyl-terminal glutamic acid region (27 aa residues) of follistatin 315 (Figure 33C.1) [3]. A third isoform is follistatin 303, produced by proteolysis of follistatin 315 [4]. Follistatin has two *N*-linked glycosylation sites, producing additional size variants. Additionally, follistatin-like 3 (follistatin-related protein) has a domain architecture similar to that of follistatin [5]. Two follistatin molecules bind one activin molecule via the activin β-subunits.

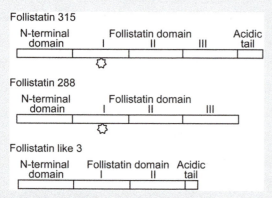

Figure 33C.1 Structure of follistatin family. Stars: heparin/heparan sulfate binding sites.

Primary Structure

The amino acid sequence of the human mature follistatins is shown in Figure 33C.2. Follistatins 288 and 315 contain three cysteine-rich follistatin domains of 73—77 aa residues (Figure 33C.1) and they have a similar binding affinity for activin. Both follistatins have a biding sequence for heparin and heparan sulfate, a major component of proteoglycans on the cell surface, in the first follistatin domain (Figure 33C.2). Follistatin 288 shows high affinity for heparan sulfate, while follistatin 315 has low affinity for cell-surface proteoglycans and its glutamic acidic tail is thought to be involved in the weak affinity for proteoglycan [4]. Follistatin-like 3 consists of two follistatin domains, but lacks the heparin binding site. The amino acid sequence of mature follistatin is highly conserved between species (E-Figure 33C.1).

Properties

Mr 35,000 (follistatin 315), 32,000 (follistatin 288). pI 5.0—6.0 (follistatin 315), 8.0—9.0 (follistatin 288). Soluble in water.

Synthesis and Release

Gene, mRNA, and Precursor

Human follistatin gene, *FST*, location 5q11.2, consisting of six exons (E-Figure 33C.2). Human follistatin-like 3 gene, *FSTL3*, location 19p13.3, consisting of five exons. *Fst* (mouse) has SP1 and AP-2 binding sites and CRE in the promoter region.

Tissue Distribution of mRNA

The follistatin gene is expressed in most tissues where activin mRNAs are detected: hypothalamus, pituitary, ovary, testis, liver, bone marrow, embryo, uterus, placenta, skin, blood vessel. The follistatin like-3 gene is expressed in the ovary, testis, adrenal gland and heart.

Tissue Content

Adult mouse, testis, 30 (pg/testis).

Plasma Concentration

Human, menstrual cycle ∼0.6 (ng/ml), pregnancy: ∼0.8 (first trimester) to ∼2.8 (third trimester), normal male ∼0.45.

Regulation of Synthesis and Release

FSH and forskolin stimulate follistatin synthesis in rat granulosa cells. Follistatin synthesis in the rat anterior pituitary cells is enhanced by gonadotropin-releasing hormone via cAMP signaling and by activin via Smad proteins. Dexamethasone upregulates follistatin gene expression in osteoblasts [6].

Y. Takei, H. Ando, & K. Tsutsui (Eds): Handbook of Hormones. DOI: http://dx.doi.org/10.1016/B978-0-12-801028-0.00189-6

```
          10        20        30        40        50        60        70        80
GNCWLRQAKNGRCQVLYKTELSKEECCSTGRLSTSWTEEDVNDNTLFKWMIFNGGAPNCIPCKETCENVDCGPGKKCRMN
          90       100       110       120       130       140       150       160
KKNKPRCVCAPDCSNITWKGPVCGLDGKTYRNECALLKARCKEQPELEVQYQGRCKKTCRDVFCPGSSTCVVDQTNNAYC
         170       180       190       200       210       220       230       240
VTCNRICPEPASSEQYLCGNDGVTYSSACHLRKATCLLGRSIGLAYEGKCIKAKSCEDIQCTGGKKCLWDFKVGRGRCSL
         250       260       270       280       290       300       310
CDELCPDSKSDEPVCASDNATYASECAMKEAACSSGVLLEVKHSGSCNSISEDTEEEEEDEDQDYSFPISSILEW
```

Figure 33C.2 Amino acid sequence of the human follistatin 288 (underlined) and follistatin 315. Heparin and heparan sulfate binding sequence is shaded.

Peroxisome proliferator-activated receptor (PPAR) γ downregulates follistatin gene expression in intestinal epithelial cells through SP1.

Receptors

Structure and Subtype

A specific receptor for follistatin has not been identified. Proteoglycan on the cell surface binds follistatin 288 with high affinity.

Signal Transduction Pathway

A specific pathway for follistatin has not been identified.

Agonist

Recombinant and purified follistatin.

Antagonist

Activin.

Biological Functions

Target Cells/Tissues and Functions

Follistatin 288 is associated with proteoglycan on the cell surface because of its high affinity for heparan sulfate, and antagonizes the effects of activins by blocking the binding of activin to their receptors. Complexes of follistatin and activins associated with proteoglycan are then endocytosed and degraded by lysosomal enzymes. Follistatin 315 with low affinity for heparan sulfate is the predominant form in the human circulation but its exact role is unclear. Follistatin-like 3 that lacks heparan sulfate binding sites is considered to circulate but its exact function is unclear. Follistatin can neutralize the responses induced by exogenous or endogenous activins in various cells and tissues (Table 33C.1 [7–18]; also see Table 33B.2 in subchapter 33B). In addition, follistatin binds to several members of the transforming growth factor (TGF) β family and blocks the interaction of these cytokines with their cognate receptors.

Phenotype of Gene-Modified Animals

Mice in which the follistatin gene was inactivated do not survive long after birth due to a variety of skeletal and cutaneous abnormalities [19]. Overexpression of the follistatin gene resulted in the degeneration of seminiferous tubules in male mice and defects in follicular development in female mice, when the mice show the highest levels of follistatin expression [20]. Granulosa cell-specific inactivation of the follistatin gene results in reduced numbers of ovarian follicles and ovulation and elevated levels of FSH in mice [21]. Follistatin-like 3 knockout mice exhibited increased pancreatic islet size, enhanced circulating insulin levels, and improved glucose tolerance [22].

Pathophysiological Implications

Clinical Implications

Patients with septicemia show high serum concentrations of follistatin and activin. The exposure of bacterial lipopolysaccharide induces an elevation in serum follistatin concentration in response to a rapid increase in the circulating activin A. Activin A promotes release of inflammatory cytokines such as TNF and IL-1, whereas follistatin is able to suppress these cytokines' release. Serum levels of follistatin increased in patients with acute liver failure.

Use for Diagnosis and Treatment

None.

Table 33C.1 Effects of Follistatin on Activin Effects in Various Cells or Tissues

Cells or Tissues	Species	Responses Induced by Activin	Follistatin Effects, (Solely Induced)	Ref
gonadotroph	rat	↑FSH production	neutralizing the activin effect (↓basal FSH production)	[7]
granulosa cell	rat	↑FSH-induced progesterone production	neutralizing the activin effect	[8]
	rat	↑inhibin production	neutralizing the activin effect	[8]
	rat	↑FSH/LH receptor expression	neutralizing the activin effect	[8]
luteinized	human	↓basal progesterone production	neutralizing the activin effect	[9]
oocyte	zebrafish	↑germinal vesicle breakdown (GVBD)	neutralizing the activin effect	[10]
	rat	↑GVBD	neutralizing the activin effect (↓GVBD)	[11]
embryo	frog	↑mesoderm induction	neutralizing activin effect	[12]
	cattle	↑early embryonic development	neutralizing activin effect (↓early embryo development)	[13]
gonocyte	rat	↑proliferation	neutralizing the activin effect	[14]
adenocortex cell	cattle	↓ACTH-induced cortisol and androgen production	partially neutralizing the activin effect	[15]
bone marrow	mouse	↑colony formation of erythroid progenitor cell	neutralizing the activin effect (↓colony formation)	[16]
pancreatic cell	rat	↑glucose stimulated insulin release from β cells	neutralizing the activin effect	[17]
hepatocyte	rat	↑apoptosis	neutralizing the activin effect	[18]

References

1. Ueno N, Ling N, Ying S-Y, et al. Isolation and partial characterization of follistatin: a single-chain Mr 35,000 monomeric protein that inhibits the release of follicle-stimulating hormone. *Proc Natl Acad Sci USA.* 1987;84:8282−8286.
2. Robertson DM, Klein R, de Vos FL, et al. The isolation of polypeptides with FSH suppressing activity from bovine follicular fluid which are structurally different to inhibin. *Biochem Biophys Res Commun.* 1987;149:744−749.
3. Shimasaki S, Koga M, Esch F, et al. Primary structure of the human follistatin precursor and its genomic organization. *Proc Natl Acad Sci USA.* 1988;85:4218−4222.
4. Sugino K, Kurosawa N, Nakamura T, et al. Molecular heterogeneity of follistatin, an activin-binding protein. Higher affinity of the carboxyl-terminal truncated forms for heparan sulfate proteoglycans on the ovarian granulosa cell. *J Biol Chem.* 1993;268:15579−15587.
5. Schneyer A, Tortoriello D, Sidis Y, et al. Follistatin-related protein (FSRP): a new member of the follistatin gene family. *Mol Cell Endocrinol.* 2001;180:33−38.
6. Hayashi K, Yamaguchi T, Yano S, et al. BMP/Wnt antagonists are upregulated by dexamethasone in osteoblasts and reversed by alendronate and PTH: potential therapeutic targets for glucocorticoid-induced osteoporosis. *Biochem Biophys Res Commun.* 2009;379:261−266.
7. Kogawa K, Nakamura T, Sugino K, et al. Activin-binding protein is present in pituitary. *Endocrinology.* 1991;128:1434−1440.
8. Nakamura T, Hasegawa Y, Sugino K, et al. Follistatin inhibits activin-induced differentiation of rat follicular granulosa cells in vitro. *Biochim Biophys Acta.* 1992;1135:103−109.
9. Cataldo NA, Rabinovici J, Fujimoto VY, et al. Follistatin antagonizes the effects of activin-A on steroidogenesis in human luteinizing granulosa cells. *J Clin Endocrinol Metab.* 1994;79:272−277.
10. Wu T, Patel H, Mukai S, et al. Activin, inhibin, and follistatin in zebrafish ovary: expression and role in oocyte maturation. *Biol Reprod.* 2000;62:1585−1592.
11. Sadatsuki M, Tsutsumi O, Yamada R, et al. Local regulatory effects of activin A and follistatin on meiotic maturation of rat oocytes. *Biochem Biophys Res Commun.* 1993;196:388−395.
12. Asashima M, Nakano H, Uchiyama H, et al. Presence of activin (erythroid differentiation factor) in unfertilized eggs and blastulae of *Xenopus laevis. Proc Natl Acad Sci USA.* 1991;88:6511−6514.
13. Yoshioka K, Suzuki C, Iwamura S. Activin A and follistatin regulate developmental competence of in vitro-produced bovine embryos. *Biol Reprod.* 1998;59:1017−1022.
14. Meehan T, Schlatt S, O'Bryan MK, et al. Regulation of germ cell and Sertoli cell development by activin, follistatin, and FSH. *Dev Biol.* 2000;220:225−237.
15. Nishi Y, Haji M, Tanaka S, et al. Human recombinant activin-A modulates the steroidogenesis of cultured bovine adrenocortical cells. *J Endocrinol.* 1992;132:R1−R4.
16. Shiozaki M, Sakai R, Tabuchi M, et al. Evidence for the participation of endogenous activin A/erythroid differentiation factor in the regulation of erythropoiesis. *Proc Natl Acad Sci USA.* 1992;89:1553−1556.
17. Brown ML, Kimura F, Bonomi LM, et al. Differential synthesis and action of TGFβ superfamily ligands in mouse and rat islets. *Islets.* 2011;3:367−375.
18. Schwall RH, Robbins K, Jardieu P, et al. Activin induces cell death in hepatocytes in vivo and in vitro. *Hepatology.* 1993;18:347−356.
19. Matzuk MM, Lu N, Vogel H, et al. Multiple defects and perinatal death in mice deficient in follistatin. *Nature.* 1995;374:360−363.
20. Guo Q, Kumar TR, Woodruff T, et al. Overexpression of mouse follistatin causes reproductive defects in transgenic mice. *Molecular Endocrinol.* 1998;12:96−106.
21. Jorgez CJ, Klysik M, Jamin SP, et al. Granulosa cell-specific inactivation of follistatin causes female fertility defects. *Mol Endocrinol.* 2004;18:953−967.
22. Mukherjee A, Sidis Y, Mahan A, et al. FSTL3 deletion reveals roles for TGF-beta family ligands in glucose and fat homeostasis in adults. *Proc Natl Acad Sci USA.* 2007;104:1348−1353.

Supplemental Information Available on Companion Website

- Primary structure of the follistatin precursor of various animals/E-Figure 33C.1
- Gene, mRNA and precursor structures of the human follistatin/E-Figure 33C.2
- Accession numbers of genes and cDNAs for follistatin/E-Table 33C.1

Anti-Müllerian Hormone

Atsushi P. Kimura

Abbreviation: AMH
Additional names: Müllerian inhibiting (inhibitory) substance (MIS), Müllerian inhibiting (inhibitory) factor (MIF)

Glycoprotein hormone secreted by testicular Sertoli and ovarian granulosa cells. Responsible for regression of Müllerian ducts during male differentiation.

Discovery

The presence of an anti-Müllerian substance was predicted in the 1940s. The hormone was first purified from bovine fetal testes in 1984 [1], and cDNAs for bovine and human AMH genes were cloned in 1986 [2,3].

Structure

Structural Features

Mature human AMH is a glycoprotein consisting of 535 aa residues and forming a homodimer by disulfide bonds. Proteolytic cleavage produces a 426-aa N-terminal domain (AMH-N) and a 109-aa C-terminal domain (AMH-C). AMH-N and AMH-C remain associated in a non-covalent complex. Only AMH-C is biologically active and AMH-N is required for AMH to be fully active. An approximately 100-aa C-terminal sequence shows a similarity with transforming growth factor (TGF)-β family proteins, forming a cysteine knot via seven canonical cysteine residues, and therefore AMH belongs to the TGF-β family.

Primary Structure

Refer to (Figure 33D.1).

Properties

By cleavage of a 140-kDa homodimer, 110-kDa AMH-N and 25-kDa AMH-C are produced. pI is approximately 7.0. AMH solution in 4 mM hydrochloric acid with 0.1% bovine serum albumin is stable for 3 months at −20°C to −70°C.

Synthesis and Release

Gene and mRNA

Human AMH gene *AMH*, location 19p13.3, consists of five exons encompassing a 2.75-kb genome sequence (E-Figure 33D.1). Human *AMH* mRNA has 2,065 b and contains a 1,683-bp open reading frame. The AMH gene is evolutionarily conserved among vertebrates such as black porgy, *Xenopus*, alligator, and chicken (E-Figure 33D.2).

Distribution of mRNA

AMH mRNA is expressed in testicular Sertoli and ovarian granulosa cells. During rat development, *Amh* mRNA is detected at embryonic day 14 in the testis during gestation and the expression is retained until birth. After birth, the testicular *Amh* expression decreases and becomes almost undetectable after postnatal day 5. Conversely, the ovary increasingly expresses *Amh* after birth. In the ovary, *Amh* expression is observed at maximum levels in granulosa cells of preantral or small antral follicles.

Plasma Concentrations

Human, 0−25-year-old female, 1−5 (ng/ml); 0−10-year-old male, 10−70; both in female and male, serum AMH levels gradually decrease thereafter. Mouse, 0−6-day-old male, 160−200 (ng/ml); 7−13-day-old male, 60−160; 15−60-day-old male, approximately 20; 4−8-month-old female, approximately 28; 10−12-month-old female, approximately 21; 14−18-month-old female, approximately 6. Bovine, 0−1-year-old female, approximately 90 (ng/ml); 4−9-year-old female, 0.025−0.359; 0−5-month-old males, 700−1,000; in males, the serum AMH level gradually decreases thereafter and reaches around 200 ng/ml at 10 months. Ewe lamb, 0−0.59 (ng/ml). Mare, 0.22−2.94 (ng/ml).

Regulation of Synthesis and Release

In humans, AMH is secreted by testicular Sertoli cells from 6 weeks of gestation. This secretion is stimulated by follicle-stimulating hormone (FSH), and the AMH level remains high until birth. The testicular AMH secretion temporarily declines at birth but is kept high until puberty when Sertoli cells stop proliferation. In adults, AMH is secreted at a low level. Androgen, secreted by Leydig cells in response to LH, negatively regulates AMH production. In mice, a 180-bp *Amh* promoter is sufficient for appropriate activation of transcription [4], and SF-1, SOX-9, and GATA bind to the promoter and play important roles in gene activation.

Receptors

Structure and Subtype

AMH binds to a heterodimer consisting of a type I and a type II AMH receptor. Activin-like kinase (ALK)2, ALK3, and ALK6, which function in other TGF-β signaling pathways, are considered to be type I AMH receptors, and the type II AMH receptor (AMHR2) mediates ligand specificity. Human AMHR2 consists of 573 aa residues, containing a 21-aa transmembrane domain and a 309-aa serine/threonine kinase

Y. Takei, H. Ando, & K. Tsutsui (Eds): Handbook of Hormones. DOI: http://dx.doi.org/10.1016/B978-0-12-801028-0.00190-2

```
         10        20        30        40        50        60
RAEEPAVGTSGLIFREDLDWPPGSPQEPLCLVALGGDSNGSSSPLRVVGALSAYEQAFLG
         70        80        90       100       110       120
AVQRARWGPRDLATFGVCNTGDRQAALPSLRRLGAWLRDPGGQRLVVLHLEEVTWEPTPS
        130       140       150       160       170       180
LRFQEPPPGGGAGPPELALLVLYPGPGPEVTVTRAGLPGAQSLCPSRDTRYLVLAVDRPAG
        190       200       210       220       230       240
AWRGSGLALTLQPRGEDSRLSTARLQALLFGDDHRCFTRMTPALLLLPRSEPAPLPAHGQ
        250       260       270       280       290       300
LDTVPFPPPRPSAELEESPPSADPFLETLTRLVRALRVPPARASAPRLALDPDALAGFPQ
        310       320       330       340       350       360
GLVNLSDPAALERLLDGEEPLLLLLRPTAATTGDPAPLHDPTSAPWATALARRVAAELQA
        370       380       390       400       410       420
AAAELRSLPGLPPATAPLLARLLALCPGGPGGLGDPLRALLLLKALQGLRVEWRGRDPRG
        430       440       450       460       470       480
PGRAQRSAGATAADGPCALRELSVDLRAERSVLIPETYQANNCQGVCGWPQSDRNPRYGN
        490       500       510       520       530
HVVLLLKMQARGAALARPPCCVPTAYAGKLLISLSEERISAHHVPNMVATECGCR
```

Figure 33D.1 Mature human AMH (535 aa residues). Cysteine residues are shaded, and putative N-glycosylation sites are boxed. The aa sequence corresponding to AMH-C is underlined.

domain (E-Figures 33D.3 and 33D.4). *AMHR2* cDNA has been cloned in various vertebrate species (E-Table 33D.2).

Signal Transduction Pathway

By binding to AMHR2, AMH forms a heterodimer of type I and type II receptors, and an intracellular serine/threonine kinase domain of AMHR2 phosphorylates the type I receptor to be activated. Then, the type I receptor binds to Smad proteins and activates the Smad signaling pathway.

Agonist

Recombinant human AMH.

Biological Functions

Target Cells/Tissues and Functions

AMHR2 is localized in Müllerian duct mesenchyme, testicular Sertoli and Leydig cells, and ovarian granulosa cells. Müllerian ducts regress in response to AMH during male differentiation. After birth, AMH blocks the Leydig cell differentiation in the testis. In the ovary, AMH inhibits the entry of primordial follicles into growing stages and reduces follicle sensitivity to FSH during the menstrual cycle.

Phenotype of Gene-Modified Animals

Both *Amh* knockout mice and *Amhr2* knockout mice showed persistent Müllerian ducts in male, resulting in retention of female reproductive organs after birth, as well as Leydig cell hyperplasia [5,6]. The same phenotype was observed in double knockout mice of *Amh* and *Amhr2* genes [6]. Mutation of a SOX9-binding site in the mouse *Amh* promoter resulted in pseudohermaphrodites [7]. Although *Amh* knockout female mice were fertile, the number of primordial follicles reduced at the age of 13 months [8]. In medaka fish, *hotei* mutants, which contained a mutation in the *amhr2* gene, showed excessive germ cell proliferation and male-to-female sex reversal [9]. When AMH was chronically expressed, female mice lost uterus and oviducts [10].

Physiological Implications

Clinical Implications

Persistent Müllerian duct syndrome (PMDS) is a disease characterized by persistence of the uterus and Fallopian tubes in males. In most cases, PMDS is due to mutation of *AMH* or *AMHR2* genes that cause the lack of AMH secretion or the receptor insensitivity. Many kinds of mutations such as nonsense and missense mutation, insertion, and deletion were reported throughout the coding sequences of both genes.

Use for Diagnosis and Treatment

AMH is thought to be a good marker of ovarian reserve, because its serum level is correlated with the number of small antral follicles. However, high levels of serum AMH are often associated with polycystic ovary syndrome (PCOS). For *in vitro* fertilization (IVF), the AMH level may predict the response to ovarian stimulation.

References

1. Picard J-Y, Josso N. Purification of testicular anti-Müllerian hormone allowing direct visualization of the pure glycoprotein and determination of yield and purification factor. *Mol Cell Endocrinol.* 1984;34:23−29.
2. Cate RL, Mattaliano RJ, Hession C, et al. Isolation of the bovine and human genes for Müllerian inhibiting substance and expression of the human gene in animal cells. *Cell.* 1986;45:685−698.
3. Picard J-Y, Benarous R, Guerrier D, et al. Cloning and expression of cDNA for anti-Müllerian hormone. *Proc Natl Acad Sci USA.* 1986;83:5464−5468.
4. Giuili G, Shen W-H, Ingraham HA. The nuclear receptor SF-1 mediates sexually dimorphic expression of Mullerian Inhibiting Substance, in vivo. *Development.* 1997;124:1799−1807.
5. Behringer RR, Finegold MJ, Cate RL. Müllerian-inhibiting substance function during mammalian sexual development. *Cell.* 1994;79:415−425.
6. Mishina Y, Rey R, Finegold MJ, et al. Genetic analysis of the Müllerian-inhibiting substance signal transduction pathway in mammalian sexual differentiation. *Genes Dev.* 1996;10:2577−2587.

7. Arango NA, Lovell-Badge R, Behringer RR. Targeted mutagenesis of the endogenous mouse Mis gene promoter: in vivo definition of genetic pathways of vertebrate sexual development. *Cell.* 1999;99:409–419.

8. Durlinger ALL, Kramer P, Karels B, et al. Control of primordial follicle recruitment by anti-Müllerian hormone in the mouse ovary. *Endocrinology.* 1999;140:5789–5796.

9. Morinaga C, Saito D, Nakamura S, et al. The hotei mutation of medaka in the anti-Mullerian hormone receptor causes the dysregulation of germ cell and sexual development. *Proc Natl Acad Sci USA.* 2007;104:9691–9696.

10. Behringer RR, Cate RL, Froelick GJ, et al. Abnormal sexual development in transgenic mice chronically expressing Müllerian inhibiting substance. *Nature.* 1990;345:167–170.

Supplemental Information Available on Companion Website

- Gene, mRNA and precursor structures of the human AMH/E-Figure 33D.1
- C-terminal AMH sequences of various animals/E-Figure 33D.2
- Gene, mRNA and precursor structures of the human AMHR2/E-Figure 33D.3
- Primary structure of the human AMHR2/E-Figure 33D.4
- Accession numbers of genes and cDNAs for AMH and AMHR2/E-Tables 33D.1 and 33D.2

CHAPTER 34

Adipocyte Hormones

Satoshi Hirako

History

Adipose tissues are energy storage tissues mainly composed of adipocytes, and are also known as endocrine tissues that secret several kinds of peptide hormones, called adipocyte hormones. The adipocyte hormones include leptin, adiponectin, acylation stimulating protein (ASP), and resistin. Of these, ASP was cloned first. ASP was initially discovered in 1989 deriving from human plasma [1] and identified as being mainly produced in white adipose tissues. ASP was identified in mouse, rat, pig, and cattle. Leptin was isolated from mouse white adipose tissues in 1994 by Friedman and co-workers. [2]; it is the most famous adipocyte hormone in non-mammal species. Leptin was cloned primarily in chicken liver and white adipose tissues. Adiponectin was first cloned in mouse 3T3-L1 and 3T3-F442A cells. Human adiponectin was first cloned from adipose tissue by Matsuzawa, who called it *adipose most abundant gene transcript 1* (apM1) [3]. Since then, it has been identified in various animals such as chicken, zebrafish, rainbow trout, and lizard. Resistin was originally discovered in research on mice in 2001, as a factor that is resistant to insulin [4]. Since then, it has been identified in many mammalian species such as mouse, rat, pig, and rabbit.

Structure

Structural Features

Leptin is a 16-kDa polypeptide consisting of 167 amino acids (aa) in its precursor form. The amino acid sequence homology of leptin precursors between mouse and rat is 96%, and between mouse and human, is 83%. Human adiponectin is a 244 aa protein of approximately 28 kDa, which belongs to the complement factor C1q family of proteins. In plasma, adiponectin forms numerous multimers, including trimer and hexamer, collectively described as low molecular weight (LMW) oligomers, and high molecular weight (HMW) multimers. ASP is identical to C3adesArg, which is a 76-aa protein (8.9 kDa); it is produced by the interaction of the proteins of the alternative complement pathway, complement proteins C3, factor B, and adipsin (complement factor D). Human resistin is a 12.5-kDa cysteine-rich peptide with 108 aa residues. Mouse resistin consists of 114 aa and the sequence homology between human and mouse is 53%. In mouse plasma, resistin appears to exist as a trimer or hexamer. See Figure 34.1.

Molecular Evolution of Family Members

These hormones are secreted from adipose tissues, but there is no homology between such hormones.

Receptors

Structure and Subtypes

The leptin receptor (Ob-R) was cloned in 1995 by Tartaglia et al. [5]. It has five isoforms by alternative splicing (Ob-Ra, Ob-Rb, Ob-Rc, Ob-Rd, and Ob-Rf). Among them, the intracellular region of Ob-Rb is the longest, containing intracellular motifs that activate the Janus-kinase (JAK)-Signal transducer activators of transcription (STAT) signal transduction pathway. Adiponectin receptors process two subtypes of adiponectin receptor 1 and 2 (AdipoR1 and AdipoR2). They have seven transmembrane domains, and an intracellular N-terminus and an extracellular C-terminus. C5L2 is considered a functional receptor of ASP and was first cloned in 2000 as a 337-aa protein, belonging to the G protein-coupled receptor as well. C5L2 shares its closest sequence homology with C5aR (38%) and is conserved across mammalian species (human, rat, and mouse). The receptor of resistin remains unclear.

Signal Transduction Pathway

The signal pathways of leptin mainly include JAK-STAT, IRS-PI3K, and AMPK pathways. Adiponectin binds to AdipoR and induces the degradation of fat by phosphorylation of AMPK and activation of PPARα. ASP signaling involves sequential activation of PI3K, with downstream activation of protein kinase C, Akt, and MAPK/ERK. Resistin induces hepatic insulin resistance via inhibiting the phosphorylation of Akt and GSK3.

Biological Functions

Target Cells/Tissues and Functions

Ob-R is mainly expressed in hypothalamus regions. Leptin regulates feeding by affecting both the NPY (orexigenic) and POMC (anorexigenic) neurons in the hypothalamus [6]. AdipoR1 is abundantly expressed in skeletal muscle, while AdipoR2 is expressed mainly in the liver. Adiponectin enhances energy metabolism and fatty acid oxidation by activating PPARα [7]. C5L2 is highly expressed in adipose tissue, heart, liver, brain, placenta, testis, and kidney. ASP stimulates triacylglycerol synthesis and glucose transport in adipocytes. Resistin induces insulin resistance and inhibits glucose uptake by insulin in skeletal muscle.

Y. Takei, H. Ando, & K. Tsutsui (Eds): Handbook of Hormones. DOI: http://dx.doi.org/10.1016/B978-0-12-801028-0.00034-9

```
ASP           1: ---------- ---------- ----------  -SVQLTEKRM DKVGKYPKEL RKCCEDGMRE  29
Resistin      1: ---------- ---------- ----------  --MKALCLLL LPVLGLLVSS KTLCSMEEAI  28
Leptin        1: ---------- ---------- ----------  MHWGTLCGFL WLWPYLFYVQ AVPIQKVQDD  30
Adiponectin   1: MLLLGAVLLL LALPGHDQET TTQGPGVLLP  LPKGACTGWM AGIPGHPGHN GAPGRDGRDG  60

ASP          30: NP-------- --------MR FSC------- --------QR RTRFISLGEA CKKVFLDCCN  58
Resistin     29: NERIQEVAGS LIFRAISSIG LEC------- --------QS VTSRGDLATC PRGFAVTGCT  73
Leptin       31: TKTLIKTIVT RINDISHTQS VSS------- --------KQ KVTGLDFIPG LHPILTLSKM  75
Adiponectin  61: TPGEKGEKGD PGLIGPKGDI GETGVPGAEG PRGFPGIQGR KGEPGEGAYV YRSAFSVGLE 120

ASP          59: YITELRRQHA RAS------- ---------- ---------- --HLGLA--- ----------  76
Resistin     74: CGSACGSWDV RAET------ ---------- ---------- TCHCQCAG-- --MDWTG--- 101
Leptin       76: DQTLAVYQQI LTSMPSRNVI QISNDLENLR DLLHVLAFSK SCHLPWASGL ETLDSLGGVL 135
Adiponectin 121: TYVTIPNMPI RFTKIFYNQQ NHYDGSTGKF HCNIPGLYYF AYHITVYMKD VKVSLFKKDK 180

ASP          76: ---------- ---------- ---------- ---------- ---------- ----------  76
Resistin    102: ---------A RCCRVQP--- ---------- ---------- ---------- ---------- 108
Leptin      136: EASGYSTEVV ALSRLQGS-- ------LQDM LWQLDLSPGC ---------- ---------- 167
Adiponectin 181: AMLFTYDQYQ ENNVDQASGS VLLHLEVGDQ VWLQVYGEGE RNGLYADNDN DSTFTGFLLY 240

ASP          76: ----  76
Resistin    110: ---- 108
Leptin      170: ---- 167
Adiponectin 241: HDTN 244
```

Figure 34.1   The alignment of amino acid sequences of ASP, resistin, leptin and adiponectin in humans. Conserved amino acid residues are indicated in green, and Cys residues are indicated in yellow. The O-linked glycosylation sites are indicated in blue.

References

1. Cianflone KM, Sniderman AD, Walsh MJ, et al. Purification and characterization of acylation stimulating protein. *J Biol Chem.* 1989;264:426−430.
2. Zhang Y, Proenca R, Maffel M, et al. Positional cloning of the mouse obese gene and its human homologue. *Nature.* 1994;372:425−432.
3. Maeda K, Okubo K, Shimomura I, et al. Paradoxical decrease of an adipose-specific protein, adiponectin, in obesity. *Biochem Biophys Res Commun.* 1996;221:286−289.
4. Steppan CM, Bailey ST, Bhat S, et al. The hormone resistin links obesity to diabetes. *Nature.* 2001;409:307−312.
5. Tartaglia LA, Dembski M, Weng X, et al. Identification and expression cloning of a leptin receptor, OB-R. *Cell.* 1995;29:1263−1271.
6. Jéquier E. Leptin signaling, adiposity, and energy balance. *Ann N Y Acad Sci.* 2002;967:379−388.
7. Kadowaki T, Yamauchi T. Adiponectin and adiponectin receptors. *Endocr Rev.* 2005;26:439−451.

Supplemental Information Available on Companion Website

- The alignment of amino acid sequence of the human adipocytokine family
- The unrooted phylogenetic tree of the adipocyte hormone family members/E-Figure 34.1
- The unrooted phylogenetic tree of the adipocyte hormone receptor family members/E-Figure 34.2

Leptin

Nobuhiro Wada

Abbreviation: LEP, OB, OBS

Leptin is an adipocyte hormone secreted from white adipose tissues which regulates energy expenditure and thermogenesis by acting on receptors located in the hypothalamus.

Discovery

During searching for the genetic mutation responsible for obesity (*ob*), Friedman and his colleagues cloned the mouse *ob* gene, and named the hormone coded by the *ob* gene leptin [1].

Structure

Structural Features

Leptin is produced as a 167-aa residue preprohormone including an N-terminal 21-aa signal peptide and is released as a 146-aa peptide after enzymatic cleavage.

Primary Structure

Human preproleptin is shown in Figure 34A.1.

Properties

Mol. Wt. is 16,000. Isoelectric point is 5.8 in human leptin. Soluble in water.

Synthesis and Release

Gene, mRNA and Precursor

The human leptin gene locates on chromosome 7 (7q31.3), and consists of three exons. Human leptin mRNA has a 3.5 kb length that encodes a signal peptide of 21 aa residues and a mature protein of 146 aa residues [2].

Tissue Distribution of mRNA

The leptin gene is strongly expressed in white adipose tissues (WAT). The gene also expresses in human mammary epithelial cells and bone marrow. In mice, leptin mRNA is expressed in WAT, liver, and pituitary gland. In rat, the mRNA is distributed in stomach, muscle, and pituitary gland [2].

Tissue Content

Obese human adipose tissues (subcutaneous) contain 17.6 ± 2.4 ng/ml of leptin [3].

Plasma Concentration

In humans, the plasma leptin concentration in non-obese subjects is ∼ 20 ng/ml, obese subjects ∼ 100 ng/ml, and subjects with leptin gene abnormalities, 300−700 ng/ml [4].

Regulation of Synthesis and Release

Synthesis and release of leptin are regulated by environmental factors. Obesity increases synthesis and plasma concentration of leptin, as found in human and rodent studies. Exercise decreases leptin concentrations in rat blood. Exposure to low temperature reduces circulating leptin level concentrations and leptin gene expression in WAT of rodents. While fasting attenuates leptin gene expression, restoration of feeding restores levels of gene expression [2].

Receptors

Structure and Subtypes

Leptin receptor (Ob-R) is a member of the class I cytokine receptor family and is found in three classes with six isoforms. The soluble form (Ob-Re) is produced by proteolytic cleavage of membrane binding-type Ob-R. The long form (Ob-Rb) has an extracellular domain with leptin binding site, transmembrane domain, and long intracellular domain which enables activation of intracellular signaling. The short form (Ob-Ra, Ob-Rc, Ob-Rd, and Ob-Rf) has similar structure to Ob-Rb except for the intracellular domain. The short form has a short intracellular domain, for which the role is still unclear. Therefore, the long form Ob-Rb is considered to be involved in signaling transduction [5].

Signal Transduction Pathway

Binding of leptin to Ob-Rb activates Janus kinase 2 (JAK2) and signal transducers and activator of transcription (STAT3) cascade. The dimerized STATs translocates into the nucleus and increases transcription of a gene contributing to feeding suppression. Ob-Rb also activates ERK1/2 mitogen-activated protein kinase (MAPK), PI3/Akt, insulin receptor substrate 1 (IRS-1) pathways [5].

Agonist

A synthetic Ob-Rb agonist is [D-LEU-4]-OB3 (Ser-Cys-Ser-Leu-Pro-Gln-Thr) which consists of a mouse leptin sequence between aa residues 116-122 [6].

Antagonist

No synthetic Ob-R antagonist has been reported.

Y. Takei, H. Ando, & K. Tsutsui (Eds): Handbook of Hormones. DOI: http://dx.doi.org/10.1016/B978-0-12-801028-0.00191-4

```
        10        20        30        40        50        60
MHWGTLCGFLWLWPYLFYVQAVPIQKVQDDTKTLIKTIVTRINDISHTQSVSSKQKVTGL
        70        80        90       100       110       120
DFIPGLHPILTLSKMDQTLAVYQQILTSMPSRNVIQISNDLENLRDLLHVLAFSKSCHLP
       130       140       150       160
WASGLETLDSLGGVLEASGYSTEVVALSRLQGSLQDMLWQLDLSPGC
```

Figure 34A.1 Amino acid sequence of human preproleptin. Mature leptin is shaded.

Biological Functions

Target Cell/Tissues and Functions

Ob-Rb is distributed throughout the entire brain, including arcuate nucleus, ventromedial hypothalamus, dorsomedial hypothalamus, paraventricular nucleus, and laternal hypothalamus [7]. Ob-Rb is observed also in peripheral tissues and organs including islet β cells of the pancreas, epithelial cells of the intestine, vascular endothelial cells, placenta, ovary, and adrenal cortex in human and rodents [2]. Leptin promotes thermogenesis, and glucose and lipid metabolism, and decreases food intake. Leptin regulates neuroendocrine hormones such as lutenizing hormone (LH), follicle stimulation hormone (FSH), prolactin, and growth hormone (GH). Blood leptin levels have been correlated with the bone mass of non-obese women [8].

Phenotype of Gene-Modified Animals

Leptin encoding gene (*ob*) mutation mice (*ob/ob* mice) express moderate obesity, infertility, and hyperglycemia. The receptor of leptin encoding gene (db) mutation mice (*db/db* mice) increases blood levels of insulin, leptin, and glucose, and causes moderate obesity and infertility. Leptin gene over-expressing mice exhibit hyperleptinemia, increased glucose metabolism, and insulin sensitivity [9].

Pathophysiological Implications

Clinical Implications

Mutation of the leptin gene has been reported in humans; such patients have hyperphagia, modest obesity, and increased leptin resistance in childhood. However, the mutation caused neither a decrease of energy expenditure nor hypercorticosterone [10].

Use for Diagnosis and Treatment

Human leptin analog (Metreleptin®) is used for treatment of lipodystrophy in clinical settings.

References

1. Zhang Y, Proenca R, Maffei M, et al. Positional cloning of the mouse obese gene and its human homologue. *Nature*. 1994;372:425–432.
2. Margetic S, Gazzola C, Pegg GG, et al. Leptin: a review of its peripheral actions and interactions. *Int J Obesity*. 2002;26:1407–1433.
3. Russell CD, Ricci MR, Brolin RE, et al. Regulation of the leptin content of obese human adipose tissue. *Am J Physiol*. 2001;280: E399–E404.
4. Klein S, Coppack SW, Mohamed-Ali V, et al. Adipose tissue leptin production and plasma leptin kinetics in humans. *Metabolism*. 1996;45:984–987.
5. Zhou Y, Rui L. Leptin signaling and leptin resistance. *Front Med*. 2013;7:207–222.
6. Rozhavskaya AM, Lee DW, Leinung MC, et al. Design of a synthetic leptin agonist: effects on energy balance, glucose homeostasis, and thermoregulation. *Endocrinology*. 2000;141: 2501–2507.
7. Funahashi H, Yada T, Suzuki R, et al. Distribution, function, and properties of leptin receptors in the brain. *Int Rev Cytol*. 2003;224:1–27.
8. Pasco JA, Henry MJ, Kotowicz MA, et al. Serum leptin levels are associated with bone mass in nonobese women. *J Clin Endocr Metab*. 2001;86:1884–1887.
9. Masuzaki H, Ogawa Y, Aizawa-Abe M, et al. Glucose metabolism and insulin sensitivity in transgenic mice overexpressing leptin with lethal yellow agouti mutation: usefulness of leptin for the treatment of obesity-associated diabetes. *Diabetes*. 1999;48:1615–1622.
10. Montague CT, Farooqi IS, Whitehead JP, et al. Congenital leptin deficiency is associated with severe early-onset obesity in humans. *Nature*. 1997;387:903–908.

Supplemental Information Available on Companion Website

- Gene, mRNA and precursor structures of the human leptin/E-Figure 34A.1
- Precursor and mature hormone sequences of various animals/E-Figure 34A.2
- Gene, mRNA and precursor structures of the human long form leptin receptor/E-Figure 34A.3
- Primary structure of the human leptin receptor/E-Figure 34A.4
- Primary structure of leptin receptor of various animals/E-Figure 34A.5
- Accession numbers of gene and cDNA for leptin and leptin receptors/E-Tables 34A.1 and 34A.2

Adiponectin

Satoshi Hirako

Additional names: Most abundant gene transcript 1 (apM1), adipocyte complement-related protein of 30 kDa (ACRP30), adipoQ, gelatin binding protein of 28 kDa (GBP28)

Adiponectin is a secretory protein produced by adipocytes. It increases insulin sensitivity and affects obesity and energy homeostasis.

Discovery

Mouse cDNA for adiponectin was cloned by mouse 3T3-L1 and 3T3-F442A cells and called adipocyte complement-related protein of 30 kDa (Acrp30). In 1996, human adiponectin was first cloned from adipose tissues by Matsuzawa, who called it adipose most abundant gene transcript 1 (apM1) [1] (Figure 34B.1). In 1996, adiponectin protein was identified by Nakano *et al*. from human plasma.

Structure

Structural Features

Adiponectin belongs to the structure of the complement 1q (C1q) family and it exists in a variety of multimer complexes in plasma, which includes trimer, hexamer, and high molecular weight (HMW) multimer (12-, 18-mers and possibly larger) [2,3].

Primary Structure

Human adiponectin consists of 244 amino acids. Adiponectin is composed of an N-terminal collagen-like sequence (collagen domain) and a C-terminal globular region (globular domain). The amino terminus is a varied region important for multimer formation and is conserved among species (Figure 34B.1).

Properties

Mr of human adiponectin is 26,414 and pI is 5.42. Soluble in water and physiological saline solution.

Synthesis and Release

Gene, mRNA, and Precursor

The human adiponectin gene (*ADIPOQ*) locates on chromosome region 3p27, spans 16 kbp, and contains three exons. The adiponectin mRNA in humans has a relatively high homology with chimpanzee (99%), cattle (87%), pig (87%), dog (87%) mouse (84%), and rat (83%).

Tissue Distribution of mRNA

In mammals, adiponectin mRNA is mainly expressed in the white and brown adipose tissues. In fish, however, it is expressed in liver, adipose tissue, muscle, and brain [4].

Plasma Concentration

Plasma adiponectin levels are 5–10 μg/ml in humans; it is very high compared to commonly occurring hormones such as leptin and insulin.

Regulation of Synthesis

Plasma adiponectin levels decrease in obesity and diabetes. Adiponectin gene expression is increased by peroxisome proliferator-activated receptor (PPAR) γ, C/EBPs, nuclear factor Y, sterol-regulatory-element-binding protein (SREBP)-1c, SIRT1, and Foxo1 [5,6]. In contrast, the gene expression is inhibited by tumor necrosis factor (TNF)-α, IL-6 and NFATc4 [7].

Regulation of Release

Insulin and PPARγ stimulates adiponectin secretion via the PI3K pathway.

```
         10         20         30         40         50         60
MLLLGAVLLL LALPGHDQET TTQGPGVLLP LPKGACTGWM AGIPGHPGHN GAPGRDGRDG
         70         80         90        100        110        120
TPGEKGEKGD PGLIGPKGDI GETGVPGAEG PRGFPGIQGR KGEPGEGAYV YRSAFSVGLE
        130        140        150        160        170        180
TYVTIPNMPI RFTKIFYNQQ NHYDGSTGKF HCNIPGLYYF AYHITVYMKD VKVSLFKKDK
        190        200        210        220        230        240
AMLFTYDQYQ ENNVDQASGS VLLHLEVGDQ VWLQVYGEGE RNGLYADNDN DSTFTGFLLY

HDTN
```

Figure 34B.1   Amino acid sequence of human adiponectin. Cysteine residues are shaded. Accession No. NP_004788.1.

Y. Takei, H. Ando, & K. Tsutsui (Eds): Handbook of Hormones. DOI: http://dx.doi.org/10.1016/B978-0-12-801028-0.00192-6

Receptors

Structure and Subtypes

Adiponectin receptors are G protein-coupled receptors (GPCRs) and are of two receptor subtypes, (AdipoR) 1 (375 aa, 42.4 kDa) and AdipoR2 (311 aa, 35.4 kDa). AdipoR 1 was cloned from a cDNA library of human skeletal muscle for globular adiponectin binding. AdipoR2 cloned only in yeast and is 67% amino acid homologous with AdipoR1 [8]. In both AdipoR1 and AdipoR2, the N-terminal domain exists in intracellular space and the C-terminal domain presents in the external region of cells. Usually, GPCR has an N-terminal extracellular domain. Therefore, AdipoRs are a new receptor group.

Signal Transduction Pathway

Adiponectin activates AMP-activated protein kinase (AMPK) by an adaptor protein containing a pleckstrin homology domain and a phosphotyrosine binding (PTB) domain; also, the leucine zipper motif (APPL) 1 binds to the intracellular domain of AdipoR. Further, Rab5 (a small GTPase) is attached to the PH domain of APPL1, and this enhances GLUT4 membrane translocation. In addition, adiponectin stimulates PPARα activity in the cell by an increase in PPARα ligand and enhancement of the transcription of PPARα itself.

Agonists

Osmotin, AdipoRon.

Biological Functions

Target Cells/Tissues and Functions

AdipoRs is expressed in many tissues and organs such as liver, skeletal muscle, pancreas, and blood vessels. AdipoR1 is mainly expressed in skeletal muscle and AdipoR2 mainly in liver. Adiponectin enhances fatty acid biosynthesis and inhibition of gluconeogenesis in the liver. In addition, it enhances fatty acid oxidation and glucose uptake in skeletal muscle. These effects relate to the activation of AMPK by adiponectin. Adiponectin improves insulin resistance by reducing the amount of intracellular fat through increasing oxidation of fatty acid via activation of PPARα and enhancement of the IRS signaling in skeletal muscle and liver. Furthermore, adiponectin has antioxidant, anti-inflammatory, and anti-atherosclerotic effect.

Phenotype of Gene-Modified Animals

Adiponectin-gene deficient mice have the property of insulin resistance and impaired glucose tolerance in high-fat diet feeding [9].

Adiponectin-overexpressing mice improve insulin resistance and glucose tolerance. This is consistent with the results of administration of adiponectin. These findings indicate that adiponectin is implicated in the development of type 2 diabetes [10].

Pathophysiological Implications

Clinical Implications

Adiponectin improves several dysfunctions such as insulin resistance, hyperlipidemia, hypertension, arteriosclerosis and NASH. In addition, the plasma adiponectin concentration was decreased in patients with lifestyle-related diseases and metabolic syndrome associated with obesity. Thus adiponectin has a crucial role in diabetes and other metabolic syndromes.

Use for Diagnosis and Treatment

Adiponectin has not yet been used for clinical diagnosis and treatment.

References

1. Maeda K, Okubo K, Shimomura I, et al. cDNA cloning and expression of a novel adipose specific collagen-like factor, apM1 (AdiPose Most abundant Gene transcript 1). *Biochem Biophys Res Commun.* 1996;221:286−289.
2. Shapiro L, Scherer PE. The crystal stucture of a complement-1q family protein suggests an evolutionary link to tumor necrosis factor. *Curr Biol.* 1998;8:335−338.
3. Yokota T, Oritani K, Takahashi I, et al. Adiponectin, a new member of the family of soluble defense collagens, negatively regulates the growth of myelomonocytic progenitors and the functions of macrophages. *Blood.* 2000;96:1723−1732.
4. Nishio S, Gibert Y, Bernard L, et al. Adiponectin and adiponectin receptor genes are coexpressed during zebrafish embryogenesis and regulated by food deprivation. *Dev Dyn.* 2008;237:1682−1690.
5. Yamauchi T, Kamon J, Waki H, et al. The fat-derived hormone adiponectin reverses insulin resistance associated with both lipoatrophy and obesity. *Nat Med.* 2001;7:941−946.
6. Seo JB, Moon HM, Noh MJW, et al. Adipocyte determination-and differentiation-dependent factor 1/sterol regulatory element-binding protein 1c regulates mouse adiponectin expression. *J Biol Chem.* 2004;279:22108−22117.
7. Kim HB, Kong M, Kim TM, et al. NFATc4 and ATF3 negatively regulate adiponectin gene expression in 3T3-L1 adipocytes. *Diabetes.* 2006;55:1342−1352.
8. Karpichev IV, Cornivelli L, Small GM. Multiple regulatory roles of a novel *Saccharomyces cerevisiae* protein, encoded by YOL002c, in lipid and phosphate metabolism. *J Biol Chem.* 2002;277:19609−19617.
9. Nawrocki AR, Rajala MW, Tomas E, et al. Mice lacking adiponectin show decreased hepatic insulin sensitivity and reduced responsiveness to peroxisome proliferator-activated receptor gamma agonists. *J Biol Chem.* 2006;281:2654−2660.
10. Bauche IB, El Mkadem SA, Pottier AM, et al. Overexpression of adiponectin targeted to adipose tissue in transgenic mice: impaired adipocyte differentiation. *Endocrinology.* 2007;148:1539−1549.

Supplemental Information Available on Companion Website

- Gene, mRNA and precursor strucures of the human adiponectin/E-Figure 34B.1
- Human adiponectin sequences of various animals/E-Figure 34B.2
- Gene, mRNA and precursor structures of the human adipoR1 and R2/E-Figures 34B.3 and 34B.4
- Primary structure of adipoR1 and R2 of various animals/E-Figures 34B.5 and 34B.6
- Accession numbers of gene and cDNA for adiponectin and adipoR1 and R2/E-Tables 34B.1, 34B.2, and 34B.3

Acylation Stimulating Protein

Eiji Ota

Abbreviation: ASP
Additional names: C3a desArg

ASP is an adipokine and was isolated as a potent fat storage factor. ASP binds to C5L2, which is a functional receptor of ASP and promotes triglyceride synthesis in adipose tissues.

Discovery

ASP was first purified in human plasma as a factor that enhanced lipid synthesis in cultured normal human skin fibroblasts [1,2].

Structure

Structural Features

ASP is a cleavage peptide from complement C3 and has 76 (human, bovine, and porcine) or 77 (mouse and rat) aa residues in mammals (Figure 34C.1). While complement C3 is glycoprotein, no glycosylation site exists in ASP. ASP has six cysteine residues, which make three disulfide bonds. Recently, C3a proteins have been purified in teleost fish, rainbow trout, and carp.

Properties

ASP Mr is 8,700–9,000, with a pH 7.5–9.5 of pI; it dissolves in water and phosphate buffer.

Synthesis and Release

Gene, mRNA, and Precursor

Human complement C3 gene (*CPAMD1*), located on chromosome 19 (19p13.3–p13.2). Human C3 consists of 41 exons and the mRNA has 5,101 bp. Human C3 involves 22 aa signaling peptides and 1,641 aa residues.

Distribution of mRNA

Although ASP is mainly produced in adipose tissues, the precursor protein, complement C3, gene expression is recognized in the brain, liver, adipocytes, macrophages, and monocytes.

Plasma Concentration

Plasma or serum levels of ASP has been reported in adult human plasma (27 fmol/ml), porcine serum (1.54 fmol/ml), female rat serum (1.71 fmol/ml as C3a, and 1.44 fmol/ml as ASP). Human plasma ASP levels are 10,000–58,000 fmol/ml in non-obese people and increases 25 to 400% in people with obesity, cardiovascular disease, and diabetes [3].

Regulation of Synthesis

ASP is mainly produced in adipose tissues from the precursor protein complement C3 by posttranscriptional enzymatic cleavage. Ezymatic cleavage of complement C3 by factor B and D (adipsin) forms C3a. Thereafter, the C3a is rapidly digested by carboxypeptidase B into ASP (C3a desArg).

Receptors

Structure and Signal Transduction Pathway

Complement component 5a receptor 2 (C5L2, also known as GPR77) is considered to be a functional receptor of ASP. ASP and its carboxy-terminal peptides do not bind to or activate the C3a receptor. C5L2 is highly expressed in adipose tissues, heart, liver, lung, spleen, bone marrow, and leukocytes. C5L2 consists of approximately 340 aa residues and contains an N-terminal extracellular domain, seven transmembrane domains and a C-terminal intracellular domain. C5L2 shows high homology with the C5a receptor (C5aR). However, C5L2 does not have the aspartic acid–arginine–phenylalanine (Asp–Arg–Phe) (DRF) motif due to the substitution of arginine with leucine. Therefore, C5L2 is considered to be a decoy receptor of C5aR [4]. However, C5L2 is considered to be a functional receptor, and the transduction pathway is not understood in detail because of this. An idea has emerged that C5L2 mediates a G protein-independent signaling pathway, such as β-arrestin signaling. After stimulation with C5a, C5a desArg, C3a, or C3a desArg, β-arrestin 1-GFP fusion protein was translocated to human C5L2 in transfected HEK293 cells [5].

Biological Functions

ASP increases uptake of glucose and fatty acids, and stimulates triglyceride synthesis and storage in fat cells. This is independent of and additive to the lipogenic effect of insulin. ASP activates diacylglycerol acyltransferase, which is a rate-limiting enzyme of triglyceride synthesis, and activates lipoprotein lipase indirectly by relieving fatty acid inhibition. ASP suppresses lipolysis of triglyceride in fat tissues. ASP inhibits hormone-sensitive lipase and decreases fatty acid release from adipocytes [6]. Furthermore, ASP induces cytokine synthesis by human peripheral blood mononuclear cells (PBMC and tonsil-derived B cells) as well as inhibiting cytotoxicity of natural killer cells [7]. Even in the absence of leptin, ASP contributes to alterations in food intake and energy expenditure [8].

Y. Takei, H. Ando, & K. Tsutsui (Eds): Handbook of Hormones. DOI: http://dx.doi.org/10.1016/B978-0-12-801028-0.00193-8

```
            10          20          30          40          50          60
   S V Q L T E K R M D K V G K Y P K E L R K C C E D G M R E N P M R F S C Q R R T R F I  S L G E A C K K V F L D C C N Y I

            70
   T E L R R Q H A R A S H L G L A
```

Figure 34C.1 Amino acid sequence of human ASP.

Phenotype in Gene-Modified Animals

ASP gene deficient (knockout) mice show a significant increase in food intake although they are leaner. They are counterbalanced by an increase of energy expenditure and the energy expenditure is mediated at the least through elevation of muscle fatty acid uptake and oxidation. ASP knockout mice on a high-fat diet utilize fat preferentially as an energy substrate. Mice lacking C5L2 showed a similar phenotype to ASP knockout mice, such as delayed postprandial triglyceride clearance and increase of dietary food intake and muscle fatty acid oxidation [8].

Pathophysiological Implications

Clinical Implications

ASP is associated with glucose and lipid metabolism in healthy subjects. However, the effect of ASP seems to become dysregulated during metabolic disturbances in diabetes [9]. Chronic recombinant ASP administration in mice enhanced the high-fat diet induced inflammatory response, leading to an insulin-resistant state. Injection of recombinant ASP in DIO mice failed to accelerate fat clearance. Furthermore, this injection of ASP increased basal levels of plasma ASP and decreased C5L2 expression in adipose tissues [10]. ASP might express different faces during pathological conditions such as obesity and cardiovascular disease, and could be a progressive factor [7].

Use for Diagnosis and Treatment

Potential use of ASP for diagnosis of some disorders, including diabetes, obesity, and acute coronary disease, is being considered. ASP has not yet been applied for clinical treatment.

References

1. Cianflone K, Kwiterovich PO, Sniderman AD, et al. Stimulation of fatty acid uptake and triglyceride synthesis in human cultured skin fibroblasts and adipocytes by a serum protein. *Biochem Biophys Res Commun*. 1987;144:94−100.
2. Cianflone MK, Sniderman AD, Rodriguez MA, et al. Purification and characterization of acylation stimulating protein. *J Biol Chem*. 1989;264:426−430.
3. Cianflone K, Xia Z, Chen LY, et al. Critical review of acylation-stimulating protein physiology in humans and rodents. *Biochim Biophys Acta*. 2003;1609:127−143.
4. Boshra H, Li J, Sunyer JO, et al. Cloning, expression, cellular distribution, and role in chemotaxis of a C5a receptor in rainbow trout: the first identification of a C5a receptor in a nonmammalian species. *J Immunol*. 2004;172:4381−4390.
5. Klos A, Wende E, Monk PN, et al. International Union of Basic and Clinical Pharmacology. [corrected]. LXXXVII. Complement peptide C5a, C4a, and C3a receptors. *Pharmacol Rev*. 2013;65:500−543.
6. Saleh J, Al-Wardy N, Cianflone K, et al. Acylation stimulating protein: a female lipogenic factor? *Obes Rev*. 2011;12:440−448.
7. Munkonda MN, Lapointe M, Cianflone K, et al. Recombinant acylation stimulating protein administration to C3 −/− mice increases insulin resistance via adipocyte inflammatory mechanisms. *PLoS One*. 2012;7:e46883.
8. Roy C, Roy MC, Cianflone K, et al. Acute injection of ASP in the third ventricle inhibits food intake and locomotor activity in rats. *Am J Physiol Endcorinol Metabo*. 2011;301:E232−E241.
9. Koistinen H, Vidal H, Ebeling P, et al. Plasma acylation stimulating protein concentration and subcutaneous adipose tissue C3 mRNA expression in nondiabetic and type 2 diabetic men. *Arterioscler Thromb Vasc Biol*. 2001;21:1034−1039.
10. Fisette A, Lapointe M, Cianflone K, et al. Obesity-inducing diet promotes acylation stimulating protein resistance. *Biochem Biophys Res Commun*. 2013;437:403−407.

Supplemental Information Available on Companion Website

- Gene, mRNA and precursor structures of the human ASP and C5L2 /E-Figure 34C.1
- Primary structure of ASP of various animals /E-Figure 34C.2
- Primary structure of C5L2 of various animals /E-Figure 34C.3
- Accession numbers of genes and cDNAs for ASP and C5L2/E-Table 34C.1

Resistin

Haruaki Kageyama

Additional names: Adipose tissue-specific secretory factor (ADSF), C/EBP-epsilon-dependent promyelocyte-specific gene (XCP1). Found in inflammatory zone 3 (FIZZ3).

Resistin is a cysteine-rich hormone secreted from white adipocytes. Resistin is involved in insulin resistance.

Discovery

The resistin gene was discovered in 2001 using a screening assay for genes downregulated in mature adipocytes treated with peroxisome proliferator-activated receptor γ (PPARγ) agonist [1].

Structure
Structural Features

Resistin has a unique motif of cysteines (X_{11}-C-X_8-C-X-C-X_3-C-X_{10}-C-X-C-X-C-X_9-C-C-X_{3-6}-END) [1]. The crystal structure of resistin reveals an unusual hexameric structure. Each protomer comprises a carboxy-terminal disulfide-rich β-sandwich "head" domain and an amino-terminal α-helical "tail" segment. The α-helical segments associate to form three-stranded coils, and surface-exposed interchain disulfide linkages mediate the formation of tail-to-tail hexamers. Resistin forms hexamers and trimers in circulation [2].

Primary Structure

The amino terminal region is hydrophobic. The unique pattern of cysteines (C-X_{28}-C-X_{11}-C-X_8-C-X-C-X_3-C-X_{10}-C-X-C-X-C-X_9-C-C-X_{3-6}-END) is conserved in mammals (Figure 34D.1).

Properties

Mr 15,060.71 (human). pI 7.5947 (human).

Synthesis and Release
Gene, mRNA, and Precursor

The human resistin gene is located on chromosome 19 (19p13.2) and consists of four exons (E-Figure 34D.1). A splicing variant with excluded Exon 3 exists. Human resistin mRNA has 327 bp that encodes a signal peptide of 16 aa residues and a mature protein of 92 aa residues.

Tissue Distribution of mRNA

Resistin mRNA is expressed in monocytes, macrophages, and epithelial cells in primates, pigs, and dogs. In rodents, it is expressed in the white adipose tissue.

Tissue Content

Concentrations of resistin in different white adipose tissues are shown in Table 34D.1 [3].

Plasma Concentration

Serum concentrations of resistin in 48-hour fasted and fed humans are 15.2 ± 5.2 ng/ml and 29.2 ± 1.4 pg/ml, respectively [4].

Concentrations of resistin in lean subjects and obese subjects were reported to be 21.5 ± 3.2 ng/ml and 28.8 ± 5.8, respectively [5].

Regulation of Synthesis

Analysis of the human resistin promoter region revealed that sterol regulatory element binding protein (SREBP)-1c and CCAAT enhancer binding protein (C/EBP)-α increased resistin mRNA expression in adipocytes [6].

Regulation of Release

Hyperglycemia, treatment of dexamethasone and thiazolidinedione increases the gene expression of resistin in 3T3-L1 adipocytes and white adipose tissues of mouse. Insulin and thyroid hormone suppress the expression of resistin mRNA in mice. Proinflammatory cytokine such as TNF-α, IL-6 and IL-1β, and bacterial endotoxins such as lipopolysaccharide (LPS) increase the gene expression in human primary blood monocytes differentiated into macrophages and in healthy human participants [7].

Receptors
Structure and Subtypes

A proteolytic cleavage product of decorin that truncated the glycanation site is a candidate receptor for resistin. Mature decorin is a secreted proteoglycan that mediates the assembly of collagen-I into functional fibers. It belongs to the small leucine-rich proteoglycan family [8].

Biological Functions
Target Cells/Tissues and Functions

Resistin suppresses the uptake of glucose and insulin sensitivity in mice. Various proinflammatory stimuli promote the expression and secretion of resistin from human macrophages, suggesting that resistin plays a crucial role in inflammation [9].

Y. Takei, H. Ando, & K. Tsutsui (Eds): Handbook of Hormones. DOI: http://dx.doi.org/10.1016/B978-0-12-801028-0.00194-X

```
        -20        -10          1        10        20        30        40
Mouse   MKNLSFPLLFLFFLVPELLGSSMPLCPIDEAIDKKIKQDFNSLFPNAIKNIGLNCWTVSS

Human      MKALCLLLLPVLGLLVSSKTLCSMEEAINERIQEVAGSLIFRAISSIGLECQSVTS


         50        60        70        80        90
Mouse   RGKLASCPEGTAVLSCSCGSACGSWDIREEKVCHCQCARIDWTAARCCKLQVAS

Human   RGDLATCPRGFAVTGCTCGSACGSWDVRAETTCHCQCAGMDWTGARCCRVQP
```

Figure 34D.1 Primary structure of mouse and human resistin. Cysteine residues are shaded. Putative signal peptide sequence is underlined.

Table 34D.1 Concentrations of Resistin in Different White Tissues and Plasma

Animal	Sex	White Adipose Tissue (ng/g wet Tissue)			Plasma (ng/ml)
		Perirenal	Abdominal	Omental	
Mouse (6 wks)	Male	990	500	1020	7.7
	Female	940	~1090	3770	23.5

Phenotype of Gene-Modified Animals

Resistin knockout mice exhibit low blood glucose levels after fasting, due to reduced hepatic glucose production. This is partly mediated by activation of adenosine monophosphate-activated protein kinase (AMPK) and decreased expression of gluconeogenic enzymes in the liver [7]. Resistin-overexpressing rats infected by adenovirus encoding mouse resistin displayed glucose intolerance and hyperinsulinemia [10]. Resistin-overexpressing transgenic mice showed glucose intolerance and promotion of phosphoenolpyruvate carboxy-kinase (PEPCK) in the liver [7].

Pathophysiological Implications

Clinical Implications

Epidemiological studies suggest that resistin is related to insulin resistance and patients with type 2 diabetes mellitus. Several single-nucleotide polymorphisms (SNPs) have been demonstrated to be associated with elevated resistin levels. Polymorphisms in the promoter region of the human resistin gene were associated with resistin levels in Japanese obese individuals. SNPs in the human resistin gene were associated with an increased risk of coronary disease in a Chinese population. However, this variant in Europeans and Caucasians was found not to correlate with carotid and coronary atherosclerosis. Increased resistin levels have been observed in patients with rheumatoid arthritis and inflammatory bowel disease and shown to be associated with disease activity [7].

Use for Diagnosis and Treatment

None.

References

1. Steppan CM, Bailey ST, Bhat S, et al. The hormone resistin links obesity to diabetes. *Nature*. 2001;409:307−312.
2. Patel SD, Rajala MW, Rossetti L, et al. Disulfide-dependent multimeric assembly of resistin family hormones. *Science*. 2004;304:1154−1158.
3. Fujinami A, Ohta K, Matsui H, et al. Resistin concentrations in murine adipose tissue and serum measured by a new enzyme immunoassay. *Obesity (Silver Spring)*. 2006;14:199−205.
4. Rajala MW, Qi Y, Patel HR, et al. Regulation of resistin expression and circulating levels in obesity, diabetes, and fasting. *Diabetes*. 2004;53:1671−1679.
5. Silha JV, Krsek M, Skrha JV, et al. Plasma resistin, adiponectin and leptin levels in lean and obese subjects: correlations with insulin resistance. *Eur J Endocrinol*. 2003;149:331−335.
6. Seo JB, Noh MJ, Yoo EJ, et al. Functional characterization of the human resistin promoter with adipocyte determination- and differentiation-dependent factor 1/sterol regulatory element binding protein 1c and CCAAT enhancer binding protein-alpha. *Mol Endocrinol*. 2003;17:1522−1533.
7. Park HK, Ahima RS. Resistin in rodents and humans. *Diabetes Metab J*. 2013;37:404−414.
8. Daquinag AC, Zhang Y, Amaya-Manzanares F, et al. An isoform of decorin is a resistin receptor on the surface of adipose progenitor cells. *Cell Stem Cell*. 2011;9:74−86.
9. Devanoorkar A, Kathariya R, Guttiganur N, et al. Resistin: a potential biomarker for periodontitis influenced diabetes mellitus and diabetes induced periodontitis. *Dis Markers*. 2014;2014:930206.
10. Satoh H, Nguyen MT, Miles PD, et al. Adenovirus-mediated chronic "hyper-resistinemia" leads to in vivo insulin resistance in normal rats. *J Clin Invest*. 2004;114:224−231.

Supplemental Information Available on Companion Website

- Primary structure of mammalian resistin /E-Figure 34D.1
- Comparison of resistin sequences in representative species of each mammalian/E-Figure 34D.2
- Accession numbers of gene and cDNA for resistin/E-Table 34D.1

CHAPTER 35

Hematopoietic Growth Factors

Takashi Kato

History

The first hematopoietic humoral regulator, termed *hemopoietine* for the production of red blood cells, was identified in 1906 [1], and was later named *erythropoietin* (EPO). However, it was not until 1977 that purification of human EPO was achieved. A hierarchical view of proliferation and differentiation of hematopoietic progenitors derived from multipotential hematopoietic stem cells was developed, and the abilities of hematopoietic growth factors, including colony stimulating factors (CSFs) were confirmed by *in vitro* bioassay of hematopoietic colony formations in semi-solid culture [2,3]. The molecular identifications of most hematopoietic regulators and their receptors [4–6] became reality after the 1980s with the modern development of molecular approaches (Table 35.1). Subsequently, clinical applications of recombinant factors and receptor antagonists/agonists have been well-established [1,4,5].

Structure

Structural Features

Most hematopoietic growth factors are glycosylated, consist of 150–350 aa residues, and display four-α-helical bundle structures (Figure 35.1) [6,7], as was first found in growth hormone. The tertiary structures of EPO, G-CSF, and TPO (E-Figure 35.1) are classified as long-chain helical bundles, whereas GM-CSF, M-CSF, IL-5, and IL-3 are classified as short-chain cytokines [6].

The Molecular Evolution of Family Members

Cytokines existed before the fish–tetrapod divergence that occurred 450 million years ago [7,8]; thereafter, hematopoietic lineage-specific cytokines such as EPO, G-CSF, and TPO (Figure 35.2 and E-Figure 35.2) could emerge to generate blood cells [9,10].

Receptors

Structure and Subtypes

Most hematopoietic receptors, for example those for EPO, TPO, GM-CSF, IL-3, and IL-5, belong to the type I cytokine receptor superfamily [4–6]. They do not have tyrosine kinase activity in their intracellular portion, and they share common properties of four conserved cysteine residues, fibronectin type III domains, a WSXWS motif in their extracellular domain, and box1 and box2 motifs that recruit JAK in the membrane proximal intracellular region. EPOR, c-Mpl, and G-CSFR consist of two identical subunits, whereas GM-CSFR, IL-3R, and IL-5R form heterodimeric complexes consisting of specific α chains and common β chains (Table 35.1). Soluble receptors comprising the extracellular domain of the receptors, produced by proteolysis/shedding of membrane proteins or alternative mRNA splicing, are detected in various body fluids. However, their physiological roles are not fully elucidated.

Signal Transduction Pathway

The ligand-receptor binding triggers changes of the intracellular conformation of the receptor complex into juxtaposition to initiate cross-phosphorylate tyrosine residues of JAK2 [4–6]. Then the activated JAKs phosphorylate secondary signaling cascades, including the major JAK-STAT3/5 pathway, the Ras/MAPK pathway, and the PI3K/Akt pathway. In some cases, the combination of factors exerts synergistic or additive effects, probably due to the crosstalk of cellular signaling.

Biological Functions

Target Cells/Tissues and Functions

The number of peripheral blood cells is spontaneously maintained, since hematopoietic growth factors act on hematopoietic progenitors derived from pluripotent stem cells, residing in hematopoietic organs, e.g., human fetal liver and adult bone marrow, to promote the survival, proliferation, and differentiation of immature hematopoietic progenitor cells. Their intrinsic cellular signaling cascades bring about various events in the target organs/cells (Table 35.1). In general, they consist of two groups. One is a group of intermediate-acting lineage-nonspecific factors (e.g., GM-CSF and IL-3) that support the proliferation of immature, multipotential progenitors, but only after they exit from G_0 [3]. Another group (e.g., EPO, G-CSF, and TPO) consists of late-acting lineage-specific factors that promote committed progenitors toward their terminal differentiation. EPO and TPO are produced predominantly in the kidney and the liver, and circulate into the bone marrow; therefore, both molecules act in an endocrine manner. These discoveries were made after those of other hormones, due to considerably low blood levels in body fluids. The other factors listed in Table 35.1 are secreted by various cells, but the blood levels vary in response to the pathophysiological state.

Y. Takei, H. Ando, & K. Tsutsui (Eds): Handbook of Hormones. DOI: http://dx.doi.org/10.1016/B978-0-12-801028-0.00035-0

Table 35.1 Major Hematopoietic Growth Factors and Their Receptors

Ligand Receptor [CD Number]	Receptor Complex	Year of Gene Cloning (*)	Major Target Lineage (Major Target Cells)	Predominant Sites of Production	Primary Functions
EPO EPOR	Homodimer	1985 (h) 1989 (m)	Erythroids (BFU-E, CFU-E)	Kidney, liver	Production of red blood cells in response to renal hypoxia
G-CSF (CSF3, CSF-β) G-CSFR [CD114]	Homodimer	1986 (h) 1990 (m)	Granulocytes, neutrophils (CFU-G, CFU-Mix)	Endothelial cells, fibroblasts, macrophages	Homeostatic and on-demand neutrophil production
TPO (c-Mpl ligand) c-Mpl (TPOR) [CD110]	Homodimer	1994 (h, m) 1990 (m)	Megakaryocytes, platelets (CFU-Mk, HSC)	Liver, kidney, bone marrow	Platelet production and maintenance of hematopoietic stem cells
GM-CSF (CSF2, CSF-α) GM-CSFR (CSF2R) [CD116 + CD131]	Heterodimer (α/β)	1985 (h, m) 1986 (h)	Granulocytes, monocyte (CFU-GM, CFU-Mix)	T cells, macrophages, mast cells	Macrophage and granulocyte production in acute inflammation
M-CSF (CSF1) c-Fms (M-CSFRc, CSF1R) [CD115]	Homodimer	1985 (h) 1983 (h)	Monocytes, macrophages, osteoclasts (CFU-M)	Fibroblasts, endothelial cells, macrophages	Macrophage and osteoclast production
IL-5 IL-5R [CD125 + CD131]	Heterodimer (α/β)	1986 (h) 1990 (m)	Eosinophils, B cells, [human] basophils	T cells, mast cells, macrophages	Eosinophil production
IL-3 (Multi-CSF) IL-3R [CD123 + CD131]	Heterodimer (α/β)	1984 (m) 1984 (m)	Multilineage progenitors, basophils (CFU-Mix)	T cells, macrophages	Proliferation and differentiation of immature myeloid progenitors

*h, human; m, mouse.

The year of gene cloning is that of the earliest identification of cDNA or genomic DNA. Other factors, including c-kit ligand (stem cell factor), Flt-3 ligand, oncostatin M, leukemia inhibitory factor, IL-6, IL-7, IL-9, IL-11, and IL-17, have been identified as factors acting on hematopoiesis.

CD, cluster of differentiation; EPO, erythropoietin; G-CSF, granulocyte CSF; TPO, thrombopoietin; M-CSF, macrophage CSF; IL, interleukin; BFU-e, burst forming unit-erythroid; CFU, colony forming unit; CFU-E, CFU-erythroid; CFU-G, CFU-granulocyte; CFU-Mk, CFU-megakaryocyte; HSC, hematopoietic stem cells; CFU-GM, CFU-granulocyte-macrophage; CFU-M, CFU-macrophage; α, α chain of specific cytokine receptor; β, β chain of common cytokine receptor [CD131].

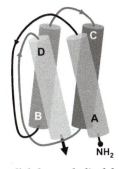

Figure 35.1 Antiparallel four-α-helical bundle structure. The long-chain helical bundles consist of four helices (A, B, C, and D) with two long loops (A to B and C to D). The whole structure becomes tightly packed by forming the pair of the antiparallel A and D helices crossovers another pair of the antiparallel B and C helices.

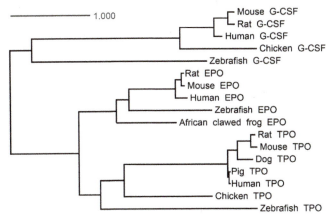

Figure 35.2 Molecular evolution of EPO, G-CSF, and TPO. The tree was generated by the use of Robust Phylogenetic Analysis for the Non-Specialist [10].

References

1. Jelkmann W. Erythropoietin after a century of research: younger than ever. *Eur J Haematol.* 2007;78:183–205.
2. Metcalf D. Hematopoietic cytokines. *Blood.* 2008;111:485–491.
3. Ogawa M. Differentiation and proliferation of hematopoietic stem cells. *Blood.* 1993;81:2844–2853.
4. Sieff CA. Hematopoietic growth factors. *J Clin Invest.* 1987;79:1549–1557.
5. Kaushansky K. Lineage-specific hematopoietic growth factors. *New Engl J Med.* 2006;354:2034–2045.
6. Thomas D, Lopez AF. *Haematopoietic growth factors. Encyclopedia of Life Sciences (eLS)*, 1–12. Chichester, UK: John Wiley & Sons, Ltd; Mar 2009.
7. Huising MO, Kruiswijk CP, Flik G. Phylogeny and evolution of class-I helical cytokines. *J Endocrinol.* 2006;189:1–25.
8. Liongue C, Ward AC. Evolution of Class I cytokine receptors. *BMC Evol Biol.* 2007;7:120–128.
9. Wen D, Boissel JP, Tracy TE, et al. Erythropoietin structure-function relationships: high degree of sequence homology among mammals. *Blood.* 1993;82:1507–1516.
10. Dereeper A, Guignon V, Blanc G, et al. Phylogeny.fr: robust phylogenetic analysis for the non-specialist. *Nucleic Acids Res.* 2008 Jul 1;36(Web Server issue):W465–W469.
11. Berman HM, Westbrook J, Feng Z, et al. The protein data bank. *Nucleic Acids Res.* 2000;28:235–242.
12. Syed RS, Reid SW, Li C, et al. Efficiency of signalling through cytokine receptors depends critically on receptor orientation. *Nature.* 1998;395:511–516.
13. Tamada T, Honjo E, Maeda Y, et al. Homodimeric cross-over structure of the human granulocyte colony-stimulating factor (GCSF) receptor signaling complex. *Proc Natl Acad Sci USA.* 2006;103:3135–3140.
14. Feese MD, Tamada T, Kato Y, et al. Structure of the receptor-binding domain of human thrombopoietin determined by complexation with a neutralizing antibody fragment. *Proc Natl Acad Sci USA.* 2006;101:1816–1821.
15. Moreland JL, Gramada A, Buzko OV, Zhang Q, Bourne PE. The Molecular Biology Toolkit (MBT): a modular platform for developing molecular visualization applications. *BMC Bioinformatics.* 2005;6:21:1–7

Supplemental Information Available on Companion Website

- Tertiary structure of human EPO, G-CSF and TPO /E-Figure 35.1
- Molecular evolution of factors with four-α-helical bundles /E-Figure 35.2

Erythropoietin

Takashi Kato

Abbreviation: EPO, Epo
Additional names: Epoetin is used as a recombinant non-proprietary drug name.

Primary regulator of erythropoiesis. Produced predominantly in the fetal liver and the adult kidney to promote survival, proliferation, and differentiation of erythroid progenitor cells. Other non-hematopoietic roles include neuro-, cardio-, and reno-protection and wound healing.

Discovery

The relationship between anemia and hypoxia/atmospheric pressure was perceived around the 16th century, predicted experimentally by Carnot and Deflandre [13], and then the tentative name *erythropoietin* (EPO) was proposed independently by Komiya [14] and by Bonsdorff and Jalavisto [15]. Reissmann [16] and Erslev [17] confirmed the humoral activity in the blood. Jacobson and colleagues found that the kidney primarily produced EPO. The human EPO protein was finally purified directly from approximately 2,550 liters of human urine of patients with aplastic anemia [1]. The subsequent molecular cloning of cDNA and genomic DNA of human EPO was accomplished concurrently by two research groups in 1985 [2,3].

Structure

Structural Features

Mature EPOs are heavily glycosylated (Figure 35A.1) to reach 30–40% (w/w) of the whole molecule [1,4]. The terminal sialic acids provide the stability in the circulation, and are essential for *in vivo* activity [1,4]. Tertiary structure displays four-α-helical bundles shared among typical cytokines [5].

Primary Structure

The mature human EPO consists of 165 aa residues after post-translational cleavage of an Arg[166] at the C-terminus. A diagram of the structure of human, rat, mouse, African clawed frog, and zebrafish EPOs is shown in Figure 35A.1.

Properties

Mr of human EPO polypeptide backbone is 18.2 kDa, 34 kDa on SDS-PAGE (glycosylated native form), or more than 40 kDa (fully glycosylated recombinant EPO expressed by CHO cells). Isoelectric point (pH 3–5) varies depending on glycosylation. The level of biological activity is reasonable against short-term heat treatment, 6 M urea, 6 M GuHCl or ordinary surfactants, when renaturation is performed properly.

Synthesis and Release

Gene, mRNA, and Precursor

The human EPO gene locates in the q11 to q22 region of chromosome 7 [NC_000007.14 (100720800..100723700)] (Figure 35A.2). Human EPO mRNA encodes a 193-aa polypeptide. Heterogeneity in the size of EPO mRNA has been reported; for example, EPO mRNA of larger size is found in the brain.

Tissue Distribution of mRNA

The production sites vary depending on either species or ontogenic stages. In mammals, the predominant sites of EPO production shifts from the fetal liver to the adult kidney along with the transition of haematopoietic tissues [8]. In mice, hepatocytes in the fetus and adult renal cells in the peritubular interstitium produce EPO [9]. In non-mammalian vertebrates, EPO is generated predominantly in the lung and the liver of the African clawed toad. In teleost fishes, the heart is one of the major production sites of EPO. EPO for murine primitive erythropoiesis is produced in neuroepithelial and neural crest cells [10].

Tissue and Blood Concentration

The international unit (IU) has been defined by WHO based on *in vivo* activity in model animals. Currently recombinant human EPO produced by CHO cells (1650 IU in approximately 11 µg of epoetin-α) is distributed in ampoules as the third WHO International Standard for the calibration of recombinant DNA-derived EPO activity by bioassay [11].

In normal human serum, the EPO levels are 8 to 36 mIU/ml by RIA or ELISA. The circulating EPO concentrations generally increase more than several hundred-fold in association with the severity of hypoxia or anemia [6–8]. The blood levels of EPO decrease in patients with kidney diseases due to impaired production of renal EPO. Conversely, the levels increase in patients with bone marrow failure, e.g., aplastic anemia, by negative feedback regulation induced by insufficient oxygen supply to the kidney.

Regulation of Synthesis and Release

Human EPO regulates erythropoiesis in a hypoxia-inducible manner [6–8]. Levels of HIF-1α in EPO-producing cells increase exponentially as O_2 concentration declines, since ubiquitination and proteasomal degradation of HIF-1α

Y. Takei, H. Ando, & K. Tsutsui (Eds): Handbook of Hormones. DOI: http://dx.doi.org/10.1016/B978-0-12-801028-0.00195-1

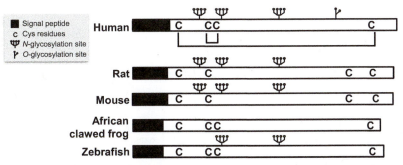

Figure 35A.1 Schematic structure of human, rat, mouse African clawed frog and zebrafish EPOs. Human mature EPO protein is a glycoprotein consisting of two disulfide bridges, three *N*-glycans and one *O*-glycan. *N*-glycosylation is completely absent in EPO in African clawed frog [18].

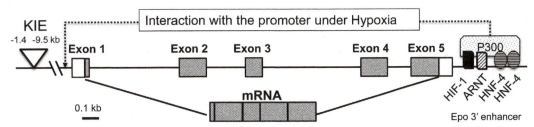

Figure 35A.2 Human EPO gene consists of five exons shown as gray rectangles. An open triangle indicates a kidney-inducible element (KIE). A critical 3′ enhancer region to which HNF-4, ARNT and HIF-1α binds forms a complex with P300 to interact with the 5′ promoter under hypoxia [6,7].

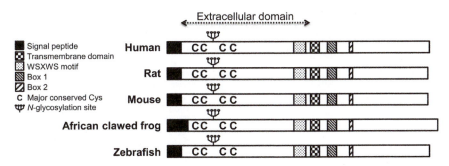

Figure 35A.3 The essential structures of EPORs are evolutionarily conserved.

decrease. Then EPO expression is directly upregulated by transcriptional activation via interaction of a 3′ enhancer complex with the 5′ promoter as shown in Figure 35A.2. The number of EPO-producing cells in the human kidney correlates positively with circulating levels of EPO, but the range of EPO expression per cell has not been determined. Ninety percent of circulating EPO originates from the kidneys and the rest from various organs, including the liver, brain, spleen, lung and testis.

Receptors

Structure and Subtype

The EPO receptor (EPOR) is a glycoprotein that belongs to the type I superfamily of single-transmembrane cytokine receptors [12]. A comparison among species and the structure of the human EPOR gene are respectively shown in Figures 35A.3 and 35A.4. A soluble form of EPOR lacking a transmembrane region generated by alternative splicing of EPOR mRNA is found in human blood. The tertiary structure of human EPO and the homodimerized EPOR complex has been determined [5].

Signal Transduction Pathway

Intracellular EPO-EPOR signaling is triggered by the binding of EPO, and EPORs homodimerize to activate a cascade of JAK2 and STAT3/5, PI3K, and/or RAS/ MAPK.

Agonists

Small mimetic peptides such as EMP1 [5] and their derivatives, agonistic antibodies to EPOR that mimic the conformation of EPO-EPOR binding (Ab12.6, also known as ABT007), and EPO fused with hybrid immunoglobulin Fc (EPO-hyFc), have been reported.

Antagonists

Other than specific antibodies to EPO or EPOR, soluble EPOR inhibits the EPO/EPOR dependent cell proliferation in glioma cells.

Biological Functions

Target Cells/Tissues and Functions

As a primary target of EPO for erythropoiesis, EPOR is expressed in the organs of hematopoiesis including the fetal

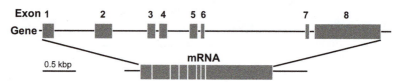

Figure 35A.4 The human EPOR gene locates in the p13.3–p13.2 region of chromosome 19 and consists of eight exons.

liver, and the adult bone marrow and spleen. EPO-EPOR signaling regulates proliferation/differentiation and survival of the erythroid progenitors, providing an important stage-specific function of erythroid differentiation. The numbers of EPOR expressed in various cells are relatively low and range between 100 to 3000 per cell with the binding affinity (ED_{50}) to EPO ranging from 0.1 to 3 nM. A wide distribution of EPOR expression is confirmed on renal cells, endothelial cells, cardiomyocytes, brain, and peripheral nervous system where EPO may exert pleiotropic or anti-apoptotic effects.

The Phenotype of Gene-Modified Animals

Mice lacking a functional EPO or EPOR gene are embryonic lethal. EPO-deficient adult animals display severe anemia, which has been found to be curable by the administration of EPO. Overexpression of EPO in mice results in extreme polycythemia with increased numbers of erythroid precursors in hematopoietic tissue, and increased circulating EPO levels.

Pathophysiological Implications

Clinical Implications

EPO maintains the number of circulating red blood cells, i.e., hemoglobin levels, by stimulating proliferation and differentiation of erythrocyte progenitors in the bone marrow. Therefore, EPO is mainly administered as hormonal replacement therapy to patients with renal failure who have experienced compromised production of EPO.

Use for Diagnosis and Treatment

Recombinant human EPO has been among the most successful therapeutic biologics. Originally, epoetin-α and -β produced by CHO cells were developed and launched for treatment of chronic renal anemia to achieve optimal hemoglobin levels and improve the QOL. Its application has extended to cancer-related anemia involving chemotherapy/radiation, inflammatory bowel disease (Crohn's disease and ulcer colitis), and others. The circulating levels of human EPO have been determined by RIA or ELISA for the diagnosis of different types of anemia. The second generation of erythropoiesis-stimulating agents (ESAs), such as a long-acting analog darbepoetin with two additional glycans and its various generics, is following the first generation. Doping with EPO has become a serious issue in athletics competitions.

References

1. Miyake T, Kung CK, Goldwasser E. Purification of human erythropoietin. *J Biol Chem*. 1977;252:5558–5564.
2. Lin FK, Suggs S, Lin CH, et al. Cloning and expression of the human erythropoietin gene. *Proc Natl Acad Sci USA*. 1985;82:7580–7584.
3. Jacobs K, Shoemaker C, Rudersdorf R, et al. Isolation and characterization of genomic and cDNA clones of human erythropoietin. *Nature*. 1985;313:806–810.
4. Takeuchi M, Takasaki S, Miyazaki H, et al. Comparative study of the asparagine-linked sugar chains of human erythropoietins purified from urine and the culture medium of recombinant Chinese hamster ovary cells. *J Biol Chem*. 1988;263:3657–3663.
5. Syed RS, Reid SW, Li C, et al. Efficiency of signalling through cytokine receptors depends critically on receptor orientation. *Nature*. 1998;395:511–516.
6. Ebert BL, Bunn HF. Regulation of the erythropoietin gene. *Blood*. 1999;94:1864–1877.
7. Semenza GL. Involvement of oxygen-sensing pathways in physiologic and pathologic erythropoiesis. *Blood*. 2009;114:2015–2019.
8. Palis J. Primitive and definitive erythropoiesis in mammals. *Front Physiol*. 2014;5:3.
9. Obara N, Suzuki N, Kim K, et al. Repression via the GATA box is essential for tissue-specific erythropoietin gene expression. *Blood*. 2008;111:5223–5232.
10. Suzuki N, Hirano I, Pan X, et al. Erythropoietin production in neuroepithelial and neural crest cells during primitive erythropoiesis. *Nat Comm*. 2013;4:2902.
11. Storring PL, Gaines Das RE. The International Standard for Recombinant DNA-derived Erythropoietin: collaborative study of four recombinant DNA-derived erythropoietins and two highly purified human urinary erythropoietins. *J Endocrinol*. 1992;134:459–484.
12. D'Andrea AD, Zon LI. Erythropoietin receptor. Subunit structure and activation. *J Clin Invest*. 1990;86:681–687.
13. Carnot P, Deflandre C. Sur l'activité hémopoiétique du sérum au cours de la régénération du sang. *C R Acad Sci Paris*. 1906;143:384–386.
14. Komiya E. Nomenclature of molecules that promote hematopoiesis [in Japanese]. *J Kumamoto Med Soc (Kumamoto Igakkai Zasshi)*. 1936;12:2355.
15. Bonsdorff E, Jalavisto E. A humoral mechanism in anoxic erythrocytosis. *Acta Physiol Scand*. 1948;16:150–170.
16. Reissmann KR. Studies on the mechanism of erythropoietic stimulation in parabiotic rats during hypoxia. *Blood*. 1950;5:372–380.
17. Erslev A. Humoral regulation of red cell production. *Blood*. 1953;8:349–357.
18. Nogawa-Kosaka N, Hirose T, Kosaka N, et al. Structural and biological properties of erythropoietin in Xenopus laevis. *Exp Hematol*. 2010;38(5):363–372.

Supplemental Information Available on Companion Website

- Multiple alignment of amino acid sequences of EPO /E-Figure 35A.1
- Multiple alignment of amino acid sequences of EPOR/ E-Figure 35A.2
- Accession numbers of gene and cDNA for EPO, selected/ E-Table 35A.1
- Accession numbers of gene and cDNA for EPOR, selected E-Table 35A.2

Thrombopoietin

Takashi Kato

Abbreviation: TPO, THPO, Tpo
Additional names: c-Mpl ligand/Mpl ligand (ML), megakaryocyte growth and development factor (MGDF) termed as a truncated recombinant pharmaceutical agent.

A glycoprotein that primarily regulates thrombopoiesis: i.e., platelet production. Produced constitutively in the liver and inductively in the bone marrow, to promote proliferation and differentiation of megakaryocytes and their progenitors, and maintain hematopoietic stem cells.

Discovery

After the nature of platelets as described by Donné, Addison, Osler, and Bizzozero in the 1800s, Wright [30] reported that the origin of human platelets (thrombocytes) was the bone marrow [1]. The tentative name *thrombopoietin* (TPO) was proposed for a humoral factor of platelet production, independently by Komiya [31] and by Kelemen [32]. It was not apparent whether TPO activity was identical to *in vitro* activity defined as a megakaryocyte colony stimulating factor. TPO molecules were finally identified simultaneously by independent research groups [1–4] (Table 35B.1) [5–13]. Concurrently, c-Mpl (Mpl) expressing megakaryocytes [14,15,26] was confirmed as the TPO receptor.

Structure

Structural Features

Schematic structures of TPO proteins are shown in Figure 35B.1.

Primary Structure

The mature human TPO consists of 332 aa residues.

Properties: MW of Human TPO

Native TPOs originally isolated were truncated N-terminal domains (17 to 36 kDa) [3,6,11–13]. Size distribution of circulating human TPO in patients with various hematologic disorders does not differ markedly, indicating that the truncation of TPO is not related to disease pathophysiology.

Synthesis and Release

Gene, mRNA, and Precursor

A human TPO gene (termed *thpo*) is located in chromosome 3q26.3–27.

Tissue Distribution of mRNA

The expression of TPO mRNA is detected predominantly in the hepatocytes, and also in several tissues, including kidney, brain, spleen, skeletal muscle, intestine, and bone marrow. In murine embryogenesis, TPO mRNA was detected by *in situ* hybridization from 12.5 days postcoitus in hepatocytes of fetal liver under hematopoiesis. Northern blot analysis showed that murine adult liver transcribed the same size TPO mRNA as in the fetus; whereas adult rat TPO mRNA displays 1.8 kb in the liver, kidney, and skeletal muscle, but it is 6.5 kb in the brain [29].

Tissue and Blood Concentration

TPO levels were inversely correlated to the peripheral platelet count in general [1,2]. The values were expressed by weight concentration (pg/ml), or by the molar concentration (1 fmol/ml equivalent to approximately 100 pg of the full length TPO), because of the possible existence of truncated native forms [16]. TPO levels [16] are approx. 0.4 to 1.2 fmol/ml in normal human serum, and increase to $5\sim25$ fmol/ml in patients with aplastic anemia (AA) and essential thrombocythemia (ET), but remain at a normal level in patients with hepatic cirrhosis due to the deteriorated production of hepatic TPO. The concentrations of free TPO in the circulating blood are regulated by the binding capacity of TPO to total Mpl mass of platelets plus megakaryocytes in the body [1,2]. According to this "sponge" theory, which was proposed by Kuter [33], TPO levels are elevated in immune thrombocytopenic purpura (ITP) patients with consumptive platelet degradation, but moderately above normal owing to the systemic total Mpl mass [1].

Regulation of Synthesis and Release

TPO is constitutively produced in the liver, and inductively in the bone marrow microenvironment. Specific induction of TPO mRNA expression has not been clarified yet. Hepatic and renal TPO expressions in animal models with severe thrombocytopenia are retained at normal levels, while circulating TPO levels increase, since the total mass of Mpl in the body decreases.

Receptors

Structure and Subtype

Mpl is a glycoprotein (~80 kDa) that belongs to a member of the cytokine receptor superfamily [24,25,27,28] (Figure 35B.2). The sequence contains a retroviral oncogene (v-*mpl*) that induces multi-lineage myeloproliferative leukemia in mice. A

Y. Takei, H. Ando, & K. Tsutsui (Eds): Handbook of Hormones. DOI: http://dx.doi.org/10.1016/B978-0-12-801028-0.00196-3

Table 35B.1 The Simultaneous Discovery of TPO/Mpl Ligand in 1994

Research Group	Source	Molecular Identification	Reference
Kirin	Irradiated rat plasma	Direct purification with 11 steps from rat plasma	[5]
		→ Internal AA sequencing	[6]
		→ Screen rat TPO producing tissues	
		→ Rat TPO cDNA cloning	
		→ Human TPO cDNA cloning	
Genetech & Mayo Clinic	Irradiated porcine plasma	Mpl-IgG Fc affinity purification of porcine irradiated plasma	[7]
		→ N-terminal AA sequencing	[8]
		→ Porcine TPO genomic DNA cloning	
		→ Human TPO genomic DNA cloning	
		→ Human TPO cDNA cloning	
ZymoGenetics & Univ. of Washington	Chemically induced mutations of murine cells	Expression cloning of murine TPO based on acquiring auto-proliferation by EMS-mutation of Mpl expressing murine Ba/F3 cells	[9]
			[10]
		→ Murine TPO cDNA cloning	
		→ Human TPO cDNA cloning	
Amgen	Irradiated canine plasma	Mpl affinity purification of canine plasma	[11]
		→ N-terminal AA sequencing	[12]
		→ Canine cDNA cloning	
		→ Human cDNA cloning	
Massachusetts Inst. of Tech. & Harvard Med. School	Busulfan-treated sheep plasma	Direct purification with 11 steps from sheep plasma	[13]
		→ Internal AA sequencing	
		→ Gene cloning unfinished	

Since Mpl is specifically expressed in megakaryocyte-linage cells [15], the addition of *c-mpl* synthetic antisense oligodeoxy-nucleotides specifically decreased *in vitro* colony formation of megakaryocyte progenitors (CFU-MK) in human CD34$^+$ cells [26]. The ligand-hunting of Mpl was then triggered. Meanwhile, rat and sheep TPOs were directly purified based on the biological activity to promote megakaryocyte development.

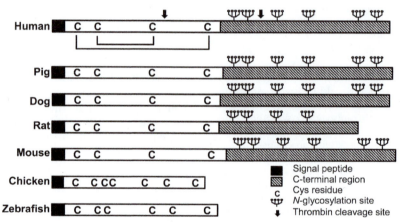

Figure 35B.1 **The N-terminal half domains are essential for the binding to Mpl, and display similarity of erythropoietin (EPO)** [1−4]. The C-terminal half domains are rich in *N*-glycans. Chicken and fish TPO lack the C-terminal domain.

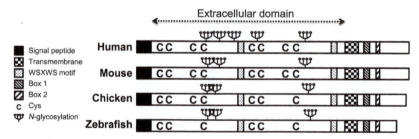

Figure 35B.2 Schematic structure of human, murine, chicken, and zebrafish Mpls.

human Mpl gene [14] locates in the chromosome 1p34.2. Splicing variants of soluble (S-form) and truncated (K-form) types are identified [14]. Positions of cysteine are mostly conserved, but the disulfide bonds are not fully confirmed. The motifs of WSXWS, and box 1 and box 2 are conserved among species.

Signal Transduction Pathway

The dimeric Mpl binding to TPO activates a number of secondary messengers that promote cell proliferation,

differentiation, and survival in megakaryocytes/platelets, and leukemic cell lines, through the JAK2-STAT3/5 signaling cascade activating PI3K and/or RAS/MAPK.

Agonists

Various agonists have been developed. After the development [1,2,19] of pegylated recombinant human megakaryocyte growth and development factor (PEG-rhMGDF) produced by *E. coli* and recombinant human TPO (rhTPO) produced by CHO cells, an artificial mimetic peptide of romiplostim (AMG-531) and an orally available small molecule of eltrombopag (SB-497115) were approved as therapeutic agents. Other non-peptide agonists (AKR-501 (E5501), LGA-4665, NIP-004, NIP-022, butyzamide, ALXN4100TPO) and several agonist antibodies or related derivatives such as MA01G4344U and VB22B were also reported.

Antagonist

An antagonistic compound specific to TPO-Mpl intracellular signaling has not been reported yet.

Biological Functions

Target Cells/Tissues and Functions

TPO binds to Mpl on the surface of megakaryocyte/platelet progenitors to stimulate their proliferation and differentiation. One mature megakaryocyte with a higher ploidy class are capable of releasing thousands of platelets. TPO is also produced by osteoblastic/stromal cells in the bone marrow microenvironment, and maintains hematopoietic stem cells that express Mpl. Cultured megakaryocytes derived from either peripheral or cord blood express a single class of high-affinity Mpl (2,000 sites with $K_d \cong 90$ pM and 180 sites per cell with $K_d \cong 125$ pM), respectively, whereas single peripheral platelets display 20–30 sites ($K_d = 20$–60 pM) [34].

The Phenotype of Gene-Modified Animals

In both Mpl $(-/-)$ [17,18] and TPO $(-/-)$ [19] mice, the platelet count became almost null, and bone marrow megakaryocytes decreased to less than 15% of normal; however, neither type of mouse developed major bleeding. TPO $(+/-)$ mice displayed thrombocytopenia, while the platelet counts in Mpl $(+/-)$ deficient mice were normal. Both TPO $(+/-)$ and Mpl $(+/-)$ mice decreased the number of hematopoietic progenitors in the other lineages as well as megakaryocyte progenitors. The number of hematopoietic progenitors decreased and the definitive hematopoiesis was delayed in the Mpl deficient mice. Overexpression of TPO in mice exhibits extreme thrombocytosis.

Pathophysiological Implications

Clinical Implications

TPO acts primarily on megakaryocytes as a late-acting differentiation factor to platelet production, as well as the maintenance of hematopoietic stem cells. Stimulating TPO–Mpl cellular signaling is effective for the treatment of thrombocytopenia caused by various diseases and therapies including chemotherapy and irradiation. Likewise, TPO is expected to be applicable to hematopoietic stem cell expansion [1–4]. Based on the physiology of the TPO–Mpl system, however, TPO does not induce the immediate release of platelets from megakaryocytes.

Use for Diagnosis and Treatment

The clinical trials [1,2] of a first generation of recombinant TPO molecules (i.e., PEG-rhMGDF and rhTPO) started in 1995, and proved the increase of platelet counts in humans for the first time, but were discontinued due to the development of a neutralizing antibody to endogenous TPO in 13 human subjects in the PEG-MGDF study. The second generation of synthetic TPO-Mpl agonists (romiplostim and eltrombopag) followed, and were approved throughout the world for the treatment of immune thrombocytopenia (ITP) [1]. Other indications are further considered.

References

1. Kuter DJ. Milestones in understanding platelet production: a historical overview. *Br J Haematol.* 2014;165:248–258.
2. Kaushansky K. Thrombopoietin. *N Engl J Med.* 1998;339: 746–754.
3. Kato T, Matsumoto A, Ogami K, et al. Native thrombopoietin: structure and function. *Stem Cells.* 1998;16:322–328.
4. Wendling F. Thrombopoietin: its role from early hematopoiesis to platelet production. *Haematologica.* 1999;84:158–166.
5. Sohma Y, Akahori H, Seki N, et al. Molecular cloning and chromosomal localization of the human thrombopoietin gene. *FEBS Lett.* 1994;353:57–61.
6. Kato T, Ogami K, Shimada Y, et al. Purification and characterization of thrombopoietin. *J Biochem.* 1995;118:229–236.
7. de Sauvage FJ, Hass PE, Spencer SD, et al. Stimulation of megakaryocytopoiesis and thrombopoiesis by the c-Mpl ligand. *Nature.* 1994;369:533–538.
8. Gurney AL, Xie MH, Malloy BE, et al. Genomic structure, chromosomal localization, and conserved alternative splice forms of thrombopoietin. *Blood.* 1995;85:981–988.
9. Lok S, Kaushansky K, Holly RD, et al. Cloning and expression of murine thrombopoietin cDNA and stimulation of platelet production in vivo. *Nature.* 1994;369:565–568.
10. Kaushansky K, Lok S, Holly RD, et al. Promotion of megakaryocyte progenitor expansion and differentiation by the c-Mpl ligand thrombopoietin. *Nature.* 1994;369:568–571.
11. Hunt P, Li YS, Nichol JL, et al. Purification and biologic characterization of plasma-derived megakaryocyte growth and development factor. *Blood.* 1995;86:540–547.
12. Bartley TD, Bogenberger J, Hunt P, et al. Identification and cloning of a megakaryocyte growth and development factor that is a ligand for the cytokine receptor Mpl. *Cell.* 1994;77:1117–1124.
13. Kuter DJ, Beeler DL, Rosenberg RD. The purification of megapoietin: a physiological regulator of megakaryocyte growth and platelet production. *Proc Natl Acad Sci USA.* 1994;91:11104–11108.
14. Vigon I, Mornon JP, Cocault L, et al. Molecular cloning and characterization of MPL, the human homolog of the v-mpl oncogene: identification of a member of the hematopoietic growth factor receptor superfamily. *Proc Natl Acad Sci USA.* 1992;89:5640–5644.
15. Debili N, Wendling F, Cosman D, et al. The Mpl receptor is expressed in the megakaryocytic lineage from late progenitors to platelets. *Blood.* 1995;85:391–401.
16. Tahara T, Usuki K, Sato H, et al. A sensitive sandwich ELISA for measuring thrombopoietin in human serum: serum thrombopoietin levels in healthy volunteers and in patients with haemopoietic disorders. *Br J Haematol.* 1996;93:783–788.
17. Gurney AL, Carver-Moore K, de Sauvage FJ, Moore MW. Thrombocytopenia in c-mpl-deficient mice. *Science.* 1994;265: 1445–1447.
18. Alexander WS, Roberts AW, Nicola NA, et al. Deficiencies in progenitor cells of multiple hematopoietic lineages and defective megakaryocytopoiesis in mice lacking the thrombopoietic receptor c-Mpl. *Blood.* 1996;87:2162–2170.
19. de Sauvage, Carver-Moore K, Luoh SM, et al. Physiological regulation of early and late stages of megakaryocytopoiesis by thrombopoietin. *J Exp Med.* 1996;183:651–656.
20. Kato T, Oda A, Inagaki Y, et al. Thrombin cleaves recombinant human thrombopoietin: one of the proteolytic events that generates truncated forms of thrombopoietin. *Proc Natl Acad Sci USA.* 1997;94:4669–4674.
21. Feese MD, Tamada T, Kato Y, et al. Structure of the receptor-binding domain of human thrombopoietin determined by complexation with a neutralizing antibody fragment. *Proc Natl Acad Sci USA.* 2004;101:1816–1821.
22. Matsumoto A, Tahara T, Morita H, et al. Characterization of native human thrombopoietin in the blood of normal individuals and of patients with haematologic disorders. *Thromb Haemost.* 1999;82:24–29.
23. Wendling F, Varlet P, Charon M. A retrovirus complex inducing an acute myeloproliferative leukemia disorder in mice. *Virology.* 1986;149:242–246.
24. Souyri M, Vigon I, Penciolelli JF, et al. A putative truncated cytokine receptor gene transduced by the myeloproliferative leukemia virus immortalizes hematopoietic progenitors. *Cell.* 1990;63:1137–1147.

25. Vigon I, Mornon JP, Cocault L, et al. Characterization of the murine Mpl proto-oncogene, a member of the hematopoietic cytokine receptor family: molecular cloning, chromosomal location and evidence for a function in cell growth. *Oncogene*. 1993;8:2607−2615.

26. Methia N, Louache F, Vainchenker W, et al. Oligodeoxynucleotides antisense to the proto-oncogene c-mpl specifically inhibit megakaryocytopoiesis. *Blood*. 1993;82:1395−1401.

27. Skoda RC, Seldon DC, Chiang MK, et al. Murine c-mpl: a member of the hematopoietic growth factor receptor superfamily that transduces a proliferative signal. *EMBO J*. 1993;12:2645−2653.

28. Vigon I, Dreyfus F, Melle J, et al. Expression of the c-mpl proto-oncogene in human hematologic malignancies. *Blood*. 1993;82:877−883.

29. Shimada Y, Kato T, Ogami K, et al. Production of thrombopoietin (TPO) by rat hepatocytes and hepatoma cell lines. *Exp Hematol*. 1995;23:1388−1396.

30. Wright JH. The histogenesis of the blood platelets. *J Exp Medic*. 1910;21:263−278.

31. Komiya E. Hematopoietic regulation in abdominal organs [in Japanese]. *Japan J Gastro-enterol. (Nihon Shokakibyo Gakkai Zasshi)*. 1938;37:443−456.

32. Kelemen E, Cserhati I, Tanos B. Demonstration and some properties of human thrombopoietin in thrombocythaemic sera. *Acta Haematologica*. 1958;20:350−355.

33. Kuter DJ, Rosenberg RD. The reciprocal relationship of thrombopoietin (c-Mpl ligand) to changes in the platelet mass during busulfan-induced thrombocytopenia in the rabbit. *Blood*. 1995;85:2720−2730.

34. Kuwaki T, Hagiwara T, Yuki C, et al. Quantitative analysis of thrombopoietin receptors on human megakaryocytes. *FEBS Lett*. 1998;427:46−50.

Supplemental Information Available on Companion Website

- Multiple alignment of amino acid sequences of TPO/ E-Figure 35B.1
- Gene, mRNA and precursor structures of the human TPO/ E-Figure 35B.2
- Gene, mRNA and precursor structures of the human Mpl/ E-Figure 35B.3
- Multiple alignment of amino acid sequences of Mpl/ E-Figure 35B.4
- Accession numbers of gene and cDNA for TPO, selected/ E-Table 35B.1
- Accession numbers of gene and cDNA for Mpl, selected/ E-Table 35B.2

Granulocyte Colony-Stimulating Factor

Takashi Kato

Abbreviation: G-CSF

Additional names: colony stimulating factor 3 (CSF3), pluripotent CSF, granulopoietin; filgrastim, lenograstim (as recombinant nonproprietary drug names)

The primary regulator of proliferation and differentiation, maturation, survival and functions of neutrophils/granulocytes to exert the biological defense mechanism via neutrophil progenitors in the bone marrow.

Discovery

The bone marrow colony forming activity of maturing granulocytes was recognized in various cell and tissue cultures [1,2]. Murine G-CSF was first purified from lung conditioned medium from mice injected with bacterial endotoxin [3]. The human G-CSF was purified from conditioned media of tumor cell lines, and the cDNA was cloned in 1986, independently by two groups (Table 35C.1) [4−7].

Structure

Structural Features

Gross structures of G-CSFs are shown in Figure 35C.1. The human G-CSF and its receptor (G-CSFR) form a 2:2 complex with a crossover interaction between the Ig-like domains of the G-CSFR and GCSF.

Primary Structure

The predominant form of mature human G-CSF consists of 174 aa reduces, internally two disulfide bridges (C^{36}−C^{42} and C^{64}−C^{74}), and one O-glycan at T^{133}. In the splicing variant, consisting of 177 aa residues with insertion of V-S-E between L^{35} and C^{36}, the biological activity decreases 10-fold due to the modification of the ligand−receptor conformation. The reason for the difference in biological role(s) of the two forms has not yet been elucidated.

Properties

Mr of human G-CSF is 18,671 (calculated polypeptide backbone), 18−19 kDa on SDS-PAGE (O-glycosylated form secreted from tumor cell lines). The pI varies from 5 to 6 depending on O-glycosylation. The non-glycosylated form produced by *E. coli* retains biological activity.

Synthesis and Release

Gene, mRNA, and Precursor

The structure of human G-CSF gene is shown in Figure 35C.2.

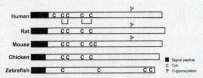

Figure 35C.1 Schematic structures of human, rat, murine, chicken, and zebrafish G-CSF. The tertiary structure of human G-CSF displays four helix bundles with β-sheets. Human G-CSF displays cross-reactivity with murine G-CSF, sharing moderate similarity (72%) in the deduced amino acid sequence. In spite of the sequence similarity to human interleukin 6 (45%), they are not cross-reactive. The similarity among oncostatin M, leukemia-inhibitory factor, interleukin 6, and G-CSF suggests that they evolved from a common ancestor, sharing functional activities to induce differentiation of a variety of cell types [8].

Table 35C.1 *Simultaneous* Identification of Human G-CSF

Research Group	Source	Biological Assay	Molecular Identification	Ref.
Memorial Sloan-Kettering Cancer Center, Amgen, Univ. of Erlangen, Germany & USA	0.2% Fetal calf serum-containing conditioned media of human bladder carcinoma cell line 5637	Differentiation induction of murine WEHI-3B (D +) human HL-60 leukemic cells, and bone marrow colony formation assay	Direct Purification with 4 steps → 18 kDa on SDS-PAGE (14 μg) pI 5.5 → NH₂-terminal aa sequencing → Human cDNA cloning	[4,6]
Chugai Pharma, Tokai Univ., Univ. of Tokyo, Japan	Serum-free conditioned medium of human oral cavity squamous tumor cell line CHU-2	Colony forming assay of human and murine bone marrow cells	Direct Purification with 4 steps → 19 kDa on SDS-PAGE pI 5.5, 5.8, 6.1, O-glycosylated → NH₂-terminal aa sequencing → Human cDNA cloning	[5,7]

The discoveries in Table 35C.1 published in the literature: Refs [4−7].

Y. Takei, H. Ando, & K. Tsutsui (Eds): Handbook of Hormones. DOI: http://dx.doi.org/10.1016/B978-0-12-801028-0.00197-5

Figure 35C.2 The structure of human G-CSF gene. The human G-CSF (CSF3) gene locates in chromosome 17q11.2–q12, and consists of five exons and four introns. Seventy-three organisms have orthologs with human G-CSF. An alternative use of 5′ splice donor sequence in intron 2 is responsible for the production of two different mRNAs encoding G-CSF splicing variants consisting of 174 or 177 aa residues [10].

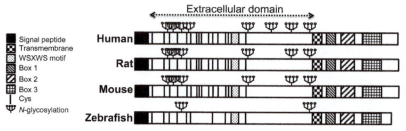

Figure 35C.3 Structure of G-CSFR. Mature G-CSFR consists of four highly conserved cysteine residues and a tryptophan-serine repeat (WSXWS) in the extracellular domain. Compositions of an extracellular region consisting of an Ig-like domain, a cytokine receptor homologous domain, three fibronectin type III-like domains, a transmembrane region, and a cytoplasmic domain are conserved. Murine and human G-CSFRs comprise 812 and 813 aa residues, respectively, sharing 62.5% homology. The extracellular domain of the human G-CSFR contains 603 aa residues.

Figure 35C.4 Gene, mRNA, and precursor structures of human G-CSFR. The human G-CSFR gene maps to chromosome 1p35-p34.3, and consists of 17 exons and 16 introns.

Tissue Distribution of mRNA

G-CSF is produced by multiple types of cells including monocytes, fibroblasts, and endothelial cells, and tissues including lung, skin, liver, placenta, cervix, intestine, uterus, muscle, eye, brain, ovary, and bone marrow.

Tissue and Blood Concentration

G-CSF levels in human serum of healthy individuals are low (33 to 163 pg/ml), and elevate significantly up to 3,200 pg/ml by infection [9]. Higher G-CSF blood levels were positively correlated with peripheral neutrophil counts in patients with various disorders, including aplastic anemia, myelodysplastic syndrome, and leukemia. In addition, G-CSF blood levels rise in patients with leukemia/solid cancers after chemotherapy or hematopoietic stem cell transplantation.

Regulation of Synthesis and Release

The expression of G-CSF gene is regulated by pathogen-mediated transcriptional and posttranscriptional pathways. Inflammatory factors such as bacterial lipopolysaccharide, interleukin-1β, tumor necrosis factor α, IL-1, and IL-17 from Th17 cells induce G-CSF expression via intracellular signaling though NF-κB, C/EBPα, and C/EBPβ. The increase in the number of circulating neutrophils reduces the production of G-CSF in the bone marrow.

Receptors

Structure and Subtype

The structures of human and other G-CSFR genes [11] are shown in Figures 35C.3 and 35C.4.

Signal Transduction Pathway

The binding of GCSF to the extracellular Ig-like and cytokine receptor homologous domain triggers homodimerization to engage JAK 1/2 and STAT, Ras/Raf/Ras/ MAPK, and PKB/Akt pathways.

Agonists

Filgrastim produced by *E. coli* attached an additional N-terminal methionine, and lenograstim produced by CHO cells attached a single *O*-glycan; these were developed as therapeutic biologicals. Other molecules, including pegylated filgrastim, follow as next-generation therapeutic reagents. The fusion proteins of G-CSF with IgG$_1$-Fc and IgG$_4$-Fc are reported.

Antagonist

An antagonistic compound specific to the G-CSF-G-CSFR signaling has not been reported yet.

Biological Functions

Target Cells/Tissues and Functions

G-CSF stimulates proliferation, differentiation, and survival of neutrophil precursors in the bone marrow to promote their maturation process. G-CSF exerts minimal direct effects on the production of hematopoietic cell types other than the neutrophil lineage, as obtained in white blood cell differentials during clinical trials. The G-CSF-G-CSFR signaling in mature neutrophils activates multiple effector functions in response to bacterial infections, such as superoxide anion generation, release of arachidonic acid, and production of leukocyte alkaline phosphatase and myeloperoxidase.

Neurons of the CNS express both G-CSF and G-CSF-R, suggesting an autocrine neuroprotection system, as a non-hematopoietic function.

The Phenotype of Gene-Modified Animals

Both G-CSF-null [12] and G-CSFR null mice [13] were viable, fertile, and superficially healthy but neutrophil progenitors reduced and became a chronic neutropenia, exhibiting 15−30% of blood neutrophil levels compared to the wild-type littermates.

Pathophysiological Implications

Clinical Implications

G-CSF participates in biological defense mechanisms by stimulating proliferation and differentiation of neutrophil progenitors in the bone marrow, to maintain the number of mature and functional neutrophils.

Use for Diagnosis and Treatment

Recombinant G-CSF therapies by filgrastim and lenograstim have been established in several indications [14,15]. Primarily, G-CSF is administered to patients with severe congenital or chronic neutropenia caused by a myeloid maturation arrest in the bone marrow [14]. G-CSF is also applicable to therapy-induced neutropenia developed in cancer patients receiving myelosuppressive chemotherapy and bone marrow transplant and in patients with acute myeloid leukemia receiving induction or consolidation chemotherapy. In addition, G-CSF induces the release of hematopoietic stem and progenitor cells from the bone marrow into the peripheral blood. Therefore, G-CSF is used in transplantation therapy for the mobilization and isolation of peripheral hematopoietic stem cells. The stem cell mobilization by G-CSF is supported by multiple mechanisms including proteolytic enzyme release, modulation of adhesion molecules, and activation of CXCR4 chemokine receptors.

References

1. Asano S, Urabe A, Okabe T, et al. Demonstration of granulopoietic factor(s) in the plasma of nude mice transplanted with a human lung cancer and in the tumor tissue. *Blood.* 1977;49:845−852.
2. Metcalf D. The granulocyte-macrophage colony-stimulating factors. *Science.* 1985;229:16−22.
3. Nicola NA, Metcalf D, Matsumoto M, et al. Purification of a factor inducing differentiation in murine myelomonocytic leukemia cells. Identification as granulocyte colony-stimulating factor. *J Biol Chem.* 1983;258:9017−9023.
4. Welte K, Platzer E, Lu L, et al. Purification and biochemical characterization of human pluripotent hematopoietic colony-stimulating factor. *Proc Natl Acad Sci USA.* 1985;82:1526−1530.
5. Nomura H, Imazeki I, Oheda M, et al. Purification and characterization of human granulocyte colony-stimulating factor (G-CSF). *EMBO J.* 1986;5:871−876.
6. Souza LM, Boone TC, Gabrilove J, et al. Recombinant human granulocyte colony-stimulating factor: effects on normal and leukemic myeloid cells. *Science.* 1986;232:61−65.
7. Nagata S, Tsuchiya M, Asano S, et al. The chromosomal gene structure and two mRNAs for human granulocyte colony-stimulating factor. *Nature.* 1986;319:415−418.
8. Rose TM, Bruce AG. Oncostatin M is a member of a cytokine family that includes leukemia-inhibitory factor, granulocyte colony-stimulating factor, and interleukin 6. *Proc Natl Acad Sci USA.* 1991;88:8641−8645.
9. Watari K, Asano S, Shirafuji N, et al. Serum granulocyte colony-stimulating factor levels in healthy volunteers and patients with various disorders as established by enzyme immunoassay. *Blood.* 1989;73:117−122.
10. Nagata S, Tsuchiya M, Asano S, et al. The chromosomal gene structure and two mRNAs for human granulocyte colony-stimulating factor. *EMBO J.* 1986;5:575−581.
11. Tweardy DJ, Anderson K, Cannizzaro LA, et al. Molecular cloning of cDNAs for the human granulocyte colony-stimulating factor receptor from HL-60 and mapping of the gene to chromosome region 1p32−34. *Blood.* 1992;79:1148−1154.
12. Lieschke GJ, Grail D, Hodgson G, et al. Mice lacking granulocyte colony-stimulating factor have chronic neutropenia, granulocyte and macrophage progenitor cell deficiency, and impaired neutrophil mobilization. *Blood.* 1994;84:1737−1746.
13. Liu F, Wu HY, Wesselschmidt R, et al. Impaired production and increased apoptosis of neutrophils in granulocyte colony-stimulating factor receptor-deficient mice. *Immunity.* 1996;5:491−501.
14. Welte K, Gabrilove J, Bronchud MH, et al. Filgrastim (r-metHuG-CSF): the first 10 years. *Blood.* 1996;88:1907−1929.
15. Berliner N. Lessons from congenital neutropenia: 50 years of progress in understanding myelopoiesis. *Blood.* 2008;111:5427−5432.
16. Aritomi M, Kunishima N, Okamoto T, et al. Atomic structure of the GCSF-receptor complex showing a new cytokine-receptor recognition scheme. *Nature.* 1999;401:713−717.
17. Tamada T, Honjo E, Maeda Y, et al. Homodimeric cross-over structure of the human granulocyte colony-stimulating factor (GCSF) receptor signaling complex. *Proc Natl Acad Sci USA.* 2006;103:3135−3140.

Supplemental Information Available on Companion Website

- Multiple alignment of amino acid sequences of G-CSF/E-Figure 35C.1
- Multiple alignment of amino acid sequences of G-CSFR/E-Figure 35C.2
- Accession numbers of gene and cDNA for G-CSF, selected/E-Table 35C.1
- Accession numbers of gene and cDNA for G-CSFR, selected/E-Table 35C.2

CHAPTER 36

Endothelins

Hirokazu Ohtaki

Abbreviation: ET, EDN

Endothelin (ET), also called ET-1, was first isolated from culture supernatant of porcine endothelial cells. The hormone shows a potent vasoconstrictive activity. There are three isopeptides, ET-1, -2 and -3.

Discovery

Presence of a potent vasoconstructor in the supernatant of porcine endothelial cells was reported and isolated as ET-1 in 1988 [1].

Structure

Structural Features

ET-1 consists of 21 aa residues with free amino- and carboxyl-termini including a hydrophobic C-terminus and four cysteine residues which form two disulfide bonds (Figure 36.1). There are two isopeptides, ET-2, and ET-3, differing by the 2nd and 6th amino acids, respectively. ET-1 is the most abundant isoform.

Primary Structure

Mature ET-1 is conserved (only one aa difference) in fish and the higher organisms.

Properties

Mr of porcine ET-1 is 2,492 and predicted pI is 9.5. Soluble in water. ET-1 solution in water at 1 mg/ml is stable for more than a year at $-20°C$.

Synthesis and Release

Gene and mRNA

In the human, the ET-1 (*EDN1*), ET-2 (*EDN2*), and ET-3 (*EDN3*) genes are located on chromosome-6 (6p24.1), -1 (1p34), and -20 (20q13.2-q13.3). Human ET-1, ET-2 and ET-3 mRNA have 2109, 1258, and 2438 bp respectively. Genetic fine mapping studies of human *EDN1* localized it to the telomeric region of chromosome 6p, close to the gene encoding the α-subunit of clotting factor XIII. The human *EDN1* contains five exons, four introns, and 5′- and 3′-flanking regions and spans approximately 6.8 kb of DNA [2]. The ET-1 mRNA is 2.3 kb in length and directs translation of the precursor prepro-ET-1 peptide, including the sequence of pro-ET-1 (big ET) and the ET-like peptide (ETLP). The ET-1 promoter contains two functional transcription start sites including TATAA and CAAT boxes, nuclear factor -1 (NF-1), GATA-2, AP-1/jun, and acute phase reaction regulatory elements (APRE).

Precursors

All three ET precursors are processed by two proteases to digest the mature active forms [3]. For ET-1, the 212-residue preproendothelin, including 19 residues of signal sequence which is characteristic of a secretary signal sequence, are cleaved at the dibasic site by fulin-like endopoptidase to produce big endothelins (big ETs). The big ETs consist of 37 to 41 aa residues and are an inactive intermediate form. Mature and active ET-1 is formed by a family of membrane-bound zinc metalloproteinases, from the neprilysin superfamily, termed endothelin-converting enzyme (ECEs) [4]. Other enzymes such as non-ECE metalloproteinase, cathepsin A, and chymase contribute to the degradation.

Tissue Distribution of mRNA

ET-1 is most abundantly localized in vascular endothelial cells, including umbilical vein, mesenteric artery, glomerulus, corpus cavernosum, aorta, and brain microvessels [1,2]. However, ET isopeptides are expressed in a variety of tissues and cell types and the isopeptides express a tissue-specific pattern. ET-1 mRNA expression is observed in breast epithelium, keratinocytes, endometrial stromal and glandular epithelial cells, macrophages, bone marrow mast cells, astrocytes, mesangial cells, neurons of the spinal cord and dorsal root ganglia, avascular human amnion, cardiomyocytes, and some tumor-derived cells. Few data are available for ET-2 and ET-3 in specific cell types, although a human renal adenocarcinoma cell line was shown to express ET-2. At the tissue level of rat, ET-1 expresses in vascular endothelium, lung, brain, uterus, stomach, heart, adrenal gland, and kidney [5]. ET-3 mRNA expresses in eyeball, submandibular gland, brain, kidney, jejunum, stomach, and spleen with northern blotting. Although ET-2 mRNA expresses in large and small intestine, skeletal muscle, heart, and stomach, large and small intestine expression is greatest.

Tissue and Plasma Concentrations

Plasma level of ET-1 in healthy human is ~2.0 pg/ml by specific radioimmunoassay (RIA). Other studies of ET-1 levels with RIA demonstrated in rat that the inner medulla of the kidney had the highest concentration ($8.7 ± 2.2$ pg/mg of wet weight).

Y. Takei, H. Ando, & K. Tsutsui (Eds): Handbook of Hormones. DOI: http://dx.doi.org/10.1016/B978-0-12-801028-0.00036-2

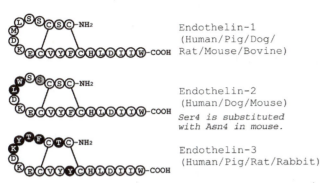

Figure 36.1 Primary structure of mammal endothelins. Ser⁴ is
substituted with Asn⁴ in mouse.

Regulation of Synthesis and Release

ET-1 is transcriptionally regulated by stimuli such as shear
stress, hypoxia, cytokines, lipopolysaccharide, and growth
factors. These enhance ET-1 mRNA transcription and protein
secretion; ET-1 shows autocrine activity. Some of them may
contribute to the stability of preproET-1 mRNA.

Receptors
Structure and Subtype

Two G protein-coupled 7-transmembrane receptors (ET_A and
ET_B) have been identified in humans. Both receptor types
contain 7-transmembrane domain 22−26 hydrophobic aa in
their ~400-aa sequences. The ET_A exhibits subnanomolar
affinities for ET-1 and ET-2 and 100-fold lower affinity for ET-
3 [3]. The ET_B receptor has equal subnanomolar affinities for
all ETs. The ET_A is considered the primary vasoconstrictor
and growth-promoting receptor, and the ET_B inhibits cell
growth and vasoconstriction in the vascular system. The ET_B
also functions as a "clearance receptor" which metabolites
ET-1. This clearance mechanism is particularly important in
the lung with metabolites being about 80% of ET-1 in the
circulation.

Signal Transduction Pathway

After ET-1 binding to ET_A on smooth muscle cells, PLC is activated via G-coupled protein and produces IP3 and DG. IP3
induces the release of Ca^{2+} in endoplasmic reticulum.
Activation of ET_A also induces the influx of extracellular Ca^{2+}
via the voltage-dependent Ca^{2+} channel, receptor-operated
Ca^{2+} channel, and nonselective cation channel. The Ca^{2+}
influx increases $[Ca^{2+}]i$ level and finally induces the phosphorylation of the myosin light chain and leads to contraction
of the smooth muscle cells.

Agonist

No synthetic agonist is shown for ET_A. Agonists for ET_B are
{Ala^{1,3,11,15}} ET-1, BQ3020, IRL-1620, and S6c.

Antagonist

Peptide and non-peptide ET receptor selective or non-
selective antagonists have been developed by pharmaceutical
companies. Ro47-0203 (Bosentan) is a non-selective ET
receptor antagonist used in the treatment of pulmonary artery
hypertension (PAH). Representative selective ET_A receptor
antagonists are BQ-123, BMS182874, LU135252 (darusentan)
and PD156707. Representative selective ET_B receptor antagonists are BQ-788 and RES701-1. Representative non-selective
ET_A/ET_B receptor antagonists are TAK-044 and Ro47-0203
(bosentan).

Biological Functions
Target Cells/Tissues and Functions

ET-1 causes strong and lasting vasoconstriction and raises
blood pressure. Therefore, ET-1 was expected to contribute
to the functioning of the cardiovascular system. Moreover, in
addition to vasoconstriction, ET-1 acts on various cardiovascular cells directly or indirectly causing cell proliferation and
production of diverse active substances, including extracellular matrix. ET-1 is involved in renal diseases, cancer, diabetes
and insulin resistance, allograft rejection, and renal diseases,
as well as cardiovascular diseases.

Phenotype in Gene-Modified Animals

A series of gene-deficient mice of ETs and the receptors are
reported from the Yanagisawa group. The ligand or receptor
gene-deficient mice showed several developmental abnormalities. Homozygous ET-1 knockout mice have craniofacial and
cardiac malformations that lead to neonatal death [6]. ET-2
knockout mice exhibit growth retardation, hypothermia, hyp-
oxemic hypoxia, hypercapnia, emphysema, and premature
death [7]. ET-3 knockout mice exhibit aganglionic megacolon
with white spotting of the hair coat due to impaired expansion and differentiation of epidermal melanoblasts. Mutants
die around weaning with impacted colons [8]. Homozygous
inactivation of ET_A causes perinatal lethality and craniofacial
deformities including middle ear defects. The knockout mice
also show severe cardiac outflow and great vessel anomalies,
respiratory and PNS defects, thymus and tongue hypoplasia,
and absent salivary glands [9]. ET_B knockout mice show
pigmentation limited to small patches on the head and rump,
exhibit abnormal neural epithelium of the inner ear, and die
from megacolon resulting from impaired neuronal migration
and aganglia [10].

Pathophysiological Implications
Clinical Implications

Plasma ET1 level was measured in diverse abnormal states of
humans with RIA and was reported to increase in abnormal
states such as pulmonary hypertension, acute myocardial
infarction, essential hypertension, subarachnoid hemorrhage,
and diabetes mellitus. The plasma level of ET-1 increases with
the severity of congestive heart failure (CHF) and, in particular, increases in patients nominated as Class IV of the New
York Heart Association (NYHA) functional classification. The
plasma ET-1 and big-ET-1 levels are inversely correlated with
the left ventricle ejection fraction (LVEF) and are independent
predictors of survival in patients with severe CHF.

Use for Diagnosis and Treatment

ET-1 concentration is not measured to diagnosis in clinical
settings. A non-selective ET_A/ET_B antagonist (bosentan;
Tracleer®) is used for treatment of pulmonary hypertension.
A selective ET_A antagonist (ambrisentan; Volibris®) has also
been used for the disease since 2010.

References

1. Yanagisawa M, Kurihara H, Kimura S, et al. A novel potent vasoconstrictor peptide produced by vascular endothelial cells. *Nature*.
   1988;332:411−415.
2. Rubanyi GM, Polokoff MA. Endothelins: molecular biology, biochemistry, pharmacology, physiology, and pathophysiology.
   Pharmacol Rev. 1994;46:325−415.
3. Kedzierski RM, Yanagisawa M. Endothelin system: the double-
   edged sword in health and disease. *Annu Rev Pharmacol Toxicol*.
   2001;41:851−876.

4. Barton M, Yanagisawa M. Endothelin: 20 years from discovery to therapy. *Can J Physiol Pharmacol.* 2008;86:485–498.
5. Sakurai T, Masaki T. Endothelin receptors—an overview. *Tanpakushitsu Kakusan Koso.* 1991;36:2381–2388.
6. Kurihara Y, Kurihara H, Suzuki H, et al. Elevated blood pressure and craniofacial abnormalities in mice deficient in endothelin-1. *Nature.* 1994;368:703–710.
7. Chang I, Bramall AN, Baynash AG, et al. Endothelin-2 deficiency causes growth retardation, hypothermia, and emphysema in mice. *J Clin Invest.* 2013;123:2643–2653.
8. Baynash AG, Hosoda K, Giaid A, et al. Interaction of endothelin-3 with endothelin-B receptor is essential for development of epidermal melanocytes and enteric neurons. *Cell.* 1994;79:1277–1285.
9. Clouthier DE, Hosoda K, Richardson JA, et al. Cranial and cardiac neural crest defects in endothelin-A receptor-deficient mice. *Development.* 1998;125:813–824.
10. Hosoda K, Hammer RE, Richardson JA, et al. Targeted and natural (piebald-lethal) mutations of endothelin-B receptor gene produce megacolon associated with spotted coat color in mice. *Cell.* 1994;79:1267–1276.

Supplemental Information Available on Companion Website

- Human endothelin-1 (EDN1) gene located on chromosome 6 (6p24.1)./E-Figure 36.1
- The alignment of amino acid sequence of the EDN1 in vertebrates./E-Figure 36.2
- Human endothelin-1 receptor type A (EDNRA) gene located on chromosome 4 (4q31.22)./E-Figure 36.3
- Human endothelin-1 receptor type B (EDNRB) gene located on chromosome 13 (13q22)./E-Figure 36.4
- The alignment of amino acid sequence of the EDNRA in vertebrates./E-Figure 36.5
- The alignment of amino acid sequence of the EDNRB in vertebrates./E-Figure 36.6
- Lists of agonists and antagonists for ETRs/E-Table 36.1
- Accession numbers of EDN1/E-Table 36.2
- Accession numbers of EDNRA/E-Table 36.3
- Accession numbers of EDNRB/E-Table 36.4

CHAPTER 37

Irisin

Hirokazu Ohtaki

Additional names: fibronectin type III domain containing 5 (FNDC5), FRCP2

Irisin is a recently discovered hormone, isolated from mouse skeletal muscle in 2012. Irisin is secreted from muscles in response to exercise and may mediate some beneficial effects of exercise in humans, such as weight loss and thermoregulation.

Discovery

Irisin was discovered from mouse skeletal muscle as a secreted peptide in 2012 [1].

Structure

Structural Features

Irisin consists of 112 aa residues (Figure 37.1). The irisin molecule may be glycosylated since multiple bands were detected in Western blot analyses, which is a general feature of secreted and glycosylated protein [1]. However, the exact nature of glycosylation has not yet been clarified. Irisin is a cleavage protein of fibronectin type III domain 5 (FNDC5), which consists of 212 aa residues (Mr 23,321 Da). Therefore, most of the information for irisin, including gene, homology, and/or properties, is reported as FNDC5 as well.

Primary Structure

Mouse FNDC5 consists of three parts: a 29-aa signal peptide, a 94-amino acid single FNIII fibronectin domain, and a C-terminal. Transmembrane FNDC5, which has an Mr of about 32 kDa, is bigger than cellular FNDC5 (23 kDa). Irisin has been well preserved in mammals through the evolutionary process, except for *Macaca mulatta* and *Canis lupus familiaris*. The structure of irisin in animals has a 50-aa N-terminal insertion for human irisin. Mouse, rat, and human irisin are 100% identical (the rates of similarity are 85% for insulin, 90% for glucagon, and 83% for leptin) [2].

Properties

The Mr of irisin is 12,587 Da. No information is shown on the isoelectric point of irisin. However, there is some information about FNDC5, the precursor of irisin. Human FNDC5 consists of 212 aa residues and has an Mr of 23,321 Da. Part of FNDC5 is cleaved to irisin, so FNDC5 is a precursor of irisin. Predicted pI of FNDC5 is 6.88.

Synthesis and Release

Gene, mRNA, and Precursor

The gene symbol for irisin is FNDC5, the precursor of irisin. Although irisin and FNDC5 are often recognized as having the same expression characteristics at the protein level, this is not strictly correct. Here all information related to gene, mRNA and posttranscriptional synthesis are for FNDC5. In humans, the FNDC5 gene is located on chromosome 1 (1p35.1). Human FNDC5 mRNA, has 2,099 bp. The human FNDC5 gene is estimated to have six exons and five introns. The FNDC5 gene is conserved with a high percentage of identification in rhesus monkey, dog, mouse, rat, chicken, and zebrafish.

Distribution of mRNA

Irisin was discovered from mouse skeletal muscle; muscle tissue expresses a large amount of FNDC5 mRNA. FNDC5 mRNA was shown to be present in the following tissues: muscle, rectum, pericardium, intracranial artery, heart, tongue, optic nerve, uvula, brain, ovary, oviduct, pituitary, seminal vesicles, adrenal gland, esophagus, vena cava, kidney, penis, retina, testis, urethra, urinary bladder, spinal cord, liver, small intestine, tonsil, thyroid, and vagina [3]. Komolka *et al.* recently showed that FNDC5 mRNA was abundant in bovine skeletal muscle and was detected at lower levels in adipose tissue and liver [4].

Tissue and Plasma Concentrations

Boström *et al.* were the first to report that irisin can be detected in human and mouse plasma and increased approximately 1.5-fold compared with control after exercise, using Western blotting [1]. The study is supported by many other studies using human subjects [5]. The plasma or serum level of irisin measured by the enzyme-linked immunosorbent assay (ELISA) kit for human reveals that irisin exists on the order of several dozens to hundreds of nanograms per milliliter [3]. For instance, a young healthy volunteer showed 20.7 ± 6.3 ng/ml and a disease-free centenarian was 35.3 ± 5.5 ng/ml [6]. Another study reported 257 ± 24 ng/ml in control subjects [7]. Many recent studies have compared irisin levels in the circulation of human subjects with diabetes, obesity, cardiac failure, and pregnancy with those of healthy volunteers. Lower circulation of irisin is associated with type 2 diabetes mellitus and these other diseases [3].

Regulation of Synthesis and Release

Irisin is a cleavage product of FNDC5. The gene expression of FNDC5 is mediated by the transcriptional co-activator peroxisome proliferator-activated receptor-γ (PPAR γ) and the

Y. Takei, H. Ando, & K. Tsutsui (Eds): Handbook of Hormones. DOI: http://dx.doi.org/10.1016/B978-0-12-801028-0.00037-4

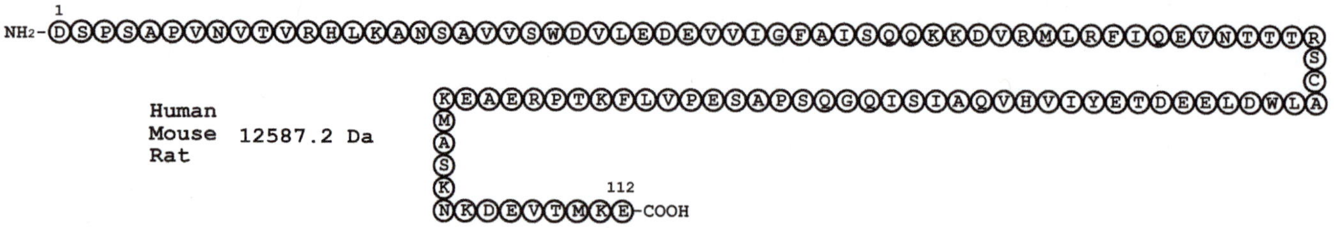

Figure 37.1   Primary structure of mammalian irisin. The figure is modified from Aydin, 2014 [2].

coactivator-1 α (PGC1α). The cleavage and release of irisin is similar to that for transmembrane polypeptides such as epidermal growth factor (EGF) and transforming growth factor-α (TGF-α). After the N-terminal signal peptide is removed, the peptide is cleaved proteolytically from the C-terminal moiety, glycosylated, and released as a hormone of 112 aa (in human, amino acids 32−143 of the full-length protein; in mouse and rat, amino acids 29−140) that comprises most of the FNIII repeat region [2]. However, the protease has not been identified in detail.

Receptors

Although it is suggested that irisin is mediated by a receptor, the irisin receptor has not been identified yet.

Biological Functions

Target Cells/Tissues and Functions

Irisin is synthesized in many other tissues as well as skeletal muscle. Although the main site of irisin synthesis was originally claimed to be skeletal muscle tissue, later studies showed that irisin was synthesized primarily in the heart muscle. The main physiological role of irisin is to travel into the white adipose tissue through the bloodstream, and to cause fat destruction via converting the white adipose tissue to brown adipose tissue. What happens during the fat destruction in the brown adipose tissue is not ATP synthesis, but heat generation. This biochemical event that does not include ATP synthesis results from the elevation of UCP1 amounts in the brown adipose tissue by irisin. UCP1 is an uncoupling agent. The heat generated in this pathway probably serves to regulate body temperature in adults, just as it does in babies. In addition, irisin reduces insulin levels and resistance, while increasing mRNA and oxygen consumption. It also leads to a rise in the expression of Elov13, Cox7a, and Otop1 genes. All these genes, the expressions of which are enhanced, are molecules mediating energy expenditure. Increased irisin expression in mice was reported to lead to weight loss and improved glucose tolerance. The adipose tissue of obese rats produces more irisin than that of control groups. Irisin is a thermogenic agent that serves anti-obesity and anti-diabetic functions and acts through a cell surface receptor.

Phenotype in Gene-Modified Animals

No gene-modified animals have been reported for irisin.

Pathophysiological Implications

Clinical Implications

The recently discovered hormone irisin has gained much attention as a potential therapeutic agent for the treatment of obesity and its related conditions. FNDC5 induces expression and converts to irisin after exercise. The increased irisin induces the browning of human and mouse white adipose tissue. However, it is still unclear whether the synthesized irisin or FNDC5 including recombinant peptides has similar effects to endogenous peptides. It is also still unclear regarding the nature of its receptor. Serum or plasma levels of irisin are decreased in many patients with metabolic syndromes and related diseases. Therefore, measurement of irisin could be useful for diagnosis. However, levels of irisin measured with ELISA show diverse values between studies and assay kit, dependent on the suppliers. Stable and conclusive antibodies that can detect irisin are required for development of the diagnosis.

References

1. Boström P, Wu J, Jedrychowski MP. A PGC1-α-dependent myokine that drives brown-fat-like development of white fat and thermogenesis. *Nature*. 2012;481:463−468.
2. Aydin S. Three new players in energy regulation: preptin, adropin and irisin. *Peptides*. 2014;56:94−110.
3. Huh JY, Panagiotou G, Mougios V. FNDC5 and irisin in humans: I. Predictors of circulating concentrations in serum and plasma and II. mRNA expression and circulating concentrations in response to weight loss and exercise. *Metabolism*. 2012;61:1725−1738.
4. Komolka K, Albrecht E, Schering L, et al. Locus characterization and gene expression of bovine FNDC5: is the myokine irisin relevant in cattle? *PLoS One*. 2014;9:e88060.
5. Irving BA, Still CD, Argyropoulos G. Does IRISIN have a BRITE future as a therapeutic agent in humans? *Curr Obes Rep*. 2014;3:235−241.
6. Emanuele E, Minoretti P, Pareja-Galeano H, et al., Serum irisin levels, precocious myocardial infarction, and healthy exceptional longevity. *Am J Med*. 127(9):888−890.
7. Liu JJ, Wong MD, Toy WC, et al. Lower circulating irisin is associated with type 2 diabetes mellitus. *J Diabetes Complications*. 2013;27:365−369.

Supplemental Information Available on Companion Website

- Human FNDC5 gene that is precursor of irisin located on chromosome 1 (1p35.1)/E-Figure 37.1
- The alignment of amino acid sequence of FNDC5 and irisin in vertebrates/E-Figure 37.2
- Accession numbers of irisin/E-Table 37.1

PART II

Peptides and Proteins in Invertebrates

SECTION II.1

Neuropeptides Related to Vertebrate Hormones

CHAPTER 38

Gonadotropin-Releasing Hormone-Like Peptide Family

Honoo Satake

History

In 1996, tunicate gonadotropin-releasing hormones (t-GnRH-1 and t-GnRH-2) were isolated from the central nervous system of an ascidian *Chelyosoma productum*, as the first invertebrate GnRHs [1]. Thereafter, t-GnRH-3 to -9 plus t-GnRH-X were characterized in the ascidians *Ciona intestinalis* and *Ciona savignyi* [2,3]. Another ascidian, *Halocynthia roretzi*, was found to possess t-GnRH-10 and -11 [4]. In amphioxus *Branchiostoma floridae*, a GnRH-like peptide, Bf-GnRH-like peptide gene, was characterized [5]. A GnRH-like peptide sequence, sp-GnRH, was also identified in the genome of the sea urchin *Strongylocentrotus purpuratus* [6]. The first protostome GnRH-like peptide, oct-GnRH, was isolated from the central nervous system of the octopus *Octopus vulgaris* [7], followed by characterization of the homologs of other mollusks and annelids [8].

Structure

Structural Features and Primary Structure

All ascidian GnRHs, like vertebrate GnRHs, consist of 10 aa residues and share the vertebrate GnRH consensus sequence pyroGlu−His−Trp−Ser and Pro−Gly−NH$_2$ (E-Table 38.1). Ci-GnRH-X is composed of 16 aa residues, and conserves the N-terminal sequence and C-terminal Gly−NH$_2$, while it lacks the penultimate proline. Sequence alignment optimization implies that the Phe^8−Thr^9 and the C-terminal Trp in Bf-GnRH-like peptide may correspond to Trp/Phe^3−Ser/Thr^4 and Trp/Tyr^7 in ascidian or vertebrate GnRHs. All non-chordate GnRH-like peptide sequences are composed of 12 aa residues, and conserve the GnRH-like C-terminal Pro−Gly−NH$_2$ and N-terminal consensus inserted by two aa residues at positions 2 and 3 (E-Table 38.1).

Properties

Mr 900−1800. Soluble in water, physiological saline solution, aqueous acetonitrile, and methanol.

Synthesis and Release

Gene, mRNA, and Precursor

All non-chordate GnRH and Bf-GnRH-like peptide precursors encode a single GnRH-like peptide sequence [8]. In contrast, *Ciona* GnRH encodes three different GnRH peptide sequences [2]. Two GnRHs are encoded in the single precursor in *H. roretzi* [4].

Distribution of mRNA

Ciona and *Octopus* GnRH-like peptide mRNAs have been detected in neurons of the central nervous system [3,8].

Receptors

Structure and Subtype

Invertebrate GnRH-like peptide receptors (GnRHRs) have been characterized from *C. intestinalis* (four receptor subtypes) [9], *O. vulgaris* (one receptor) [10], and *B. floridae* (four subtypes) [5], and these receptors all belong to the class A (rhodopsin-like) GPCR family [8]. In addition, *in silico* gene predictions suggest the presence of putative GnRHRs in other invertebrates, namely three sea urchin GnRHRs, a hemichordate GnRHR, a marine worm GnRHR, a limpet GnRHR, and a sea hare GnRHR. Ligand−receptor interactions have yet to be investigated [8].

Signal Transduction Pathway

C. intestinalis GnRHRs, Ci-GnRHR-1 to -4, expressed in HEK293 cells or cultured cells, trigger the intracellular calcium mobilization or cAMP production in response to the cognate t-GnRHs with prominent ligand selectivity, and multiple receptor heterodimerization pairs modulate Ci-GnRHR functions [8]. Oct-GnRHR expressed in *Xenopus* oocytes also elicits the intracellular calcium mobilization [8]. No cognate ligands for amphioxus GnRHR-1, -2, or -4 have so far been identified [5], whereas amphioxus GnRHR-3 responded not only to GnRHII but also to oct-GnRH and adipokinetic hormone (AKH) [6,8].

Biological Functions

Target Cells/Tissues and Functions

Ci-GnRHRs are expressed in the neural complex, heart, intestine, endostyle, and branchial sac, and in the ovary [8,9]. In particular, Ci-GnRHR-1, -2, and -4 are largely co-localized in test cells of vitellogenic follicles [8,10]. Oct-GnRHR mRNA was detected in digestive tissues, aorta, heart, salivary glands, branchiae, radula retractor muscle, egg, and genital organs [8]. t-GnRH-3 to -9 increase water flow and then induce the release of eggs and sperm by injection into the gonaducts,

Y. Takei, H. Ando, & K. Tsutsui (Eds): Handbook of Hormones. DOI: http://dx.doi.org/10.1016/B978-0-12-801028-0.00038-6

ovary, stomach, or posterior body cavity of *C. intestinalis* [2]. In *O. vulgaris*, oct-GnRH induced contraction of the oviduct and the radula retractor muscle [7,8]. Moreover, treatment of the follicle and spermatozoa with oct-GnRH resulted in release of sex steroids, including testosterone, progesterone, and 17β-estradiol [8]. Sea hare GnRH (ap-GnRH) stimulates the parapodial opening, inhibition of feeding, and promotion of substrate attachment, whereas no reproductive action by ap-GnRH has been observed [8].

References

1. Powell JF, Reska-Skinner SM, Prakash MO, et al. Two new forms of gonadotropin-releasing hormone in a protochordate and the evolutionary implications. *Proc Natl Acad Sci USA.* 1996;93:10461−10464.
2. Adams BA, Tello JA, Erchegyi J, et al. Six novel gonadotropin-releasing hormones are encoded as triplets on each of two genes in the protochordate, *Ciona intestinalis. Endocrinology.* 2003;149:4346−4356.
3. Kawada T, Aoyama M, Okada I, et al. A novel inhibitory gonadotropin-releasing hormone-related neuropeptide in the ascidian, *Ciona intestinalis. Peptides.* 2009;30:2200−2205.
4. Hasunuma I, Terakado K. Two novel gonadotropin-releasing hormones (GnRHs) from the urochordate ascidian, *Halocynthia roretzi*: implications for the origin of vertebrate GnRH isoforms. *Zool Sci.* 2013;30:311−318.
5. Roch GJ, Tello JA, Sherwood NM. At the transition from invertebrates to vertebrates, a novel GnRH-like peptide emerges in amphioxus. *Mol Biol Evol.* 2014;31:765−778.
6. Rowe ML, Elphick MR. The neuropeptide transcriptome of a model echinoderm, the sea urchin *Strongylocentrotus purpuratus. Gen Comp. Endocrinol.* 2012;179:331−344.
7. Iwakoshi E, Takuwa-Kuroda K, Fujisawa Y, et al. Isolation and characterization of a GnRH-like peptide from *Octopus vulgaris. Biochem Biophys Res Commun.* 2002;291:1187−1193.
8. Kawada T, Aoyama M, Sakai T, et al. Structure, function, and evolutionary aspects of invertebrate GnRHs and their receptors. In: Sills ES, ed. *Gonadotropin-Releasing Hormone (GnRH): Production, Structure and Functions.* Hauppauge, NY: Nova Science Publishers, Inc.; 2013:1−16.
9. Tello JA, Rivier JE, Sherwood NM. Tunicate gonadotropin-releasing hormone (GnRH) peptides selectively activate *Ciona intestinalis* GnRH receptors and the green monkey type II GnRH receptor. *Endocrinology.* 2005;146:4061−4073.
10. Sakai T, Aoyama M, Kusakabe T, et al. Functional diversity of signaling pathways through G protein-coupled receptor heterodimerization with a species-specific orphan receptor subtype. *Mol Biol Evol.* 2010;27:1097−1106.

Supplemental Information Available on Companion Website

- Amino acid sequences of invertebrate GnRH-like peptides/E-Table 38.1
- Accession numbers for cDNAs encoding GnRH-like peptides and their receptors/E-Table 38.2

Protochordata Gonadotropin-Releasing Hormone

Honoo Satake

Abbreviation: protochordata GnRH
Additional names: t-GnRH, Ci-GnRH, Bf-GnRH-like peptide

Ascidians and amphioxus also conserve peptides structurally related to vertebrate GnRHs, although they possess no tissues corresponding to either the hypothalamus or the pituitary.

Discovery

In 1996, tunicate gonadotropin-releasing hormones (t-GnRH-1 and t-GnRH-2), as the first invertebrate GnRHs, were isolated from an ascidian, *Chelyosoma productum* [1]. Thereafter, t-GnRH-3 to -9 and Ci-GnRH-X were identified in two cosmopolitan species, *Ciona intestinalis* and *Ciona savignyi* [2–4]. In 2013, t-GnRH-10 and -11 were characterized as GnRHs in another ascidian, *Halocynthia roretzi* [5]. Quite recently, their structurally related peptide was found in an amphioxus *Branchiostoma floridae* [6]. Identification of chicken GnRHII or mammalian GnRH in *C. intestinalis* or *B. floridae* was also reported. Unfortunately, these "ascidian and amphioxus GnRHs" have widely been recognized as "contaminants" of reference peptides during experimental manipulation, given that no sequences of chicken GnRHII or mammalian GnRH are present on the *C. intestinalis* or *B. floridae* genome (Table 38A.1) [2,4].

Structure

Structural Features and Primary Structure

All t-GnRHs, like vertebrate orthologs, consist of 10 aa residues, and contain the consensus sequences of pyroGlu−His−Trp−Ser and Pro−Gly−NH$_2$. Ci−GnRH−X is composed of 16 aa residues, and conserves the N-terminal sequence and C-terminal Gly-amide, while it lacks the penultimate Pro. *B. floridae* GnRH-like peptide (Bf-GnRH-like peptide) shares neither the N-terminal nor C-terminal consensus, except for the N-terminal pyroGlu (Table 38A.1). Nevertheless, sequence alignment optimization implies that the Phe[8]−Thr[9] and the C-terminal Trp in Bf-GnRH-like peptide may correspond to Trp/Phe[3]−Ser/Thr[4] and Trp/Tyr[7] in ascidian or vertebrate GnRHs [7]. Identification of the Bf-GnRH-like peptide as an amphioxus authentic GnRH ortholog awaits further investigation.

Properties

Mr 1,000−1,800. Soluble in water, physiological saline solution, aqueous acetonitrile, methanol.

Synthesis and Release

Gene, mRNA, and Precursor

As shown in E-Figure 38A.1, *Ci-gnrh-1* and *-2* each encode three different GnRH peptide sequences, t-GnRH-3, -5, -6, and t-GnRH-4, -7, -8, respectively [2]. Similarly, *Cs-gnrh-1* and *-2*: *Cs-gnrh-1* encodes two t-GnRH-5 sequences and one t-GnRH-6, and *Cs-gnrh-2* encodes one sequence each of t-GnRH-7, -8, -9 (E-Figure 38A.1) [2]. t-GnRH-10 and t-GnRH-11 are encoded in the single precursor (E-Figure 38A.1) [5]. Contrary to these multiple GnRH sequences in a precursor in t-GnRHs, Ci-GnRH-X and Bf-GnRH-like peptide, like vertebrate and protostome GnRH, are encoded as a single copy in the respective precursor (E-Figure 38A.1) [2–4].

Distribution of mRNA

Ci-gnrh-1 and *Ci-gnrh-X* are expressed in numerous neurons residing in the cerebral ganglion of *C. intestinalis*, which is compatible with the presence of mature forms of t-GnRH-3, -5, -6 and Ci-GnRH-X in this tissue [2–4]. Additionally, *Ci-gnrh-X* mRNA is largely co-localized with *Ci-gnrh-1* mRNA [3]. Expression of *Ci-gnrh-1* and *-2* is observed in the cerebral ganglion of *C. intestinalis* larva [6]. *Hr-gnrh*, encoding t-GnRH-10 and -11, is expressed in neurons in the cerebral ganglion and dorsal cord [5]. The gene expression of Bf-GnRH-like peptide is localized to the central canal of the anterior nerve cord [7].

Receptors

Structure and Subtype

To date, four *C. intestinalis* GnRH receptors (GnRHRs), Ci-GnRHR-1 to -4 [4,8], and four Bf-GnRHR-like receptors (amphioxus GnRHR) have been identified [7]. GnRHRs belong to the class A (rhodopsin-like) GPCR family. Ci-GnRHR-1, -2, and -3 sequences were found to harbor a long C-terminal tail, whereas a short tail is present in the C-terminus of Ci-GnRHR-4 [9]. Amphioxus GnRHRs are categorized into two paralogous phylogenetic groups: amphioxus GnRHR-1 and -2, which are closer in sequence to vertebrate GnRHRs than other non-chordate GnRHRs; and amphioxus

Y. Takei, H. Ando, & K. Tsutsui (Eds): Handbook of Hormones. DOI: http://dx.doi.org/10.1016/B978-0-12-801028-0.00198-7

Table 38A.1 Amino Acid Sequences of Protochodate GnRHs

Ascidian	*Chelyosoma productum*	t-GnRH-1	pEHWSYGLRPG-NH₂
Amphioxus	*Ciona intestinalis*	t-GnRH-2	pEHWSLCHAPG-NH₂
	Ciona savignyi	t-GnRH-3	pEHWSYEFMPG-NH₂
	Halocynthia roretzi	t-GnRH-4	pEHWSNQLTPG-NH₂
	Branchiostoma floridae	t-GnRH-5	pEHWSYEYMPG-NH₂
		t-GnRH-6	pEHWSKGYSPG-NH₂
		t-GnRH-7	pEHWSYALSPG-NH₂
		t-GnRH-8	pEHWSLALSPG-NH₂
		Ci-GnRH-X	pEHWSNWWIPGAPGYNG-NH₂
		t-GnRH-5	pEHWSYEYMPG-NH₂
		t-GnRH-6	pEHWSKGYSPG-NH₂
		t-GnRH-7	pEHWSYALSPG-NH₂
		t-GnRH-8	pEHWSLALSPG-NH₂
		t-GnRH-9	pEHWSNKLAPG-NH₂
		t-GnRH-10	pEHWSYGFSPG-NH₂
		t-GnRH-11	pEHWSYGFLPG-NH₂
		Bf-GnRH-like peptide	pEILCARAFTYTHTW-NH₂

Table 38A.2 Ligand Selective Signaling via Ci-GnRHRs

Receptor	Ligand	Second Messenger	Effect by Ci-GnRH-X
Ci-GnRHR-1	t-GnRH-6	Ca^{2+}, cAMP	moderate inhibition
Ci-GnRHR-2	t-GnRH-7,-8,-6	cAMP	no effect
Ci-GnRHR-3	t-GnRH-3,-5	cAMP	moderate inhibition
Ci-GnRHR-4	no ligand	none	none

GnRHR-3 and -4, which are highly homologous to protostome GnRHRs [7]. Both Ci-GnRHRs and amphioxus GnRHRs are included in independent clades of the GnRHR phylogenetic tree (E-Figure 38A.2), indicating that they were generated via gene duplication in a species-specific lineage [7,8].

Signal Transduction Pathway

t-GnRH-6 induces the mobilization of intracellular calcium ion and cAMP production via Ci-GnRHR-1 (Table 38A.2) [8–10]. Additionally, Ci-GnRHR-1 expressed in HEK293 cells triggers ERK phosphorylation via activation of PKC-α and -ξ [9]. Ci-GnRHR-2 stimulates cAMP production in response to t-GnRH-7, -8, -6, in this order of potency [8–10]. t-GnRH-3 and -5 evoke cAMP production via Ci-GnRHR-3 to a similar extent [6,8]. Ci-GnRHR-4 exhibits neither elevation of intracellular calcium nor cAMP production [8–10]. Ci-GnRH-X moderately (50%) inhibits the elevation of intracellular calcium and cAMP production by t-GnRH-6 at Ci-GnRHR-1, and cAMP production by t-GnRH-3, and t-GnRH-5 via Ci-GnRHR-3 is also inhibited by Ci-GnRH-X [3]. Ci-GnRHR-4 heterodimerizes with Ci-GnRHR-1 and -2 [9,10]. The Ci-GnRHR-1 to -4 heteromer potentiates the elevation of intracellular calcium via both calcium-dependent and -independent PKC subtypes, and ERK phosphorylation, in a ligand-selective fashion [9]. The Ci-GnRHR-2 to -4 heterodimer decreases cAMP production by 50% in a non-ligand selective manner by shifting of activation from G_s protein to G_i protein by Ci-GnRHR-2, compared with the Ci-GnRHR-2 monomer/homodimer [10]. Collectively, Ci-GnRHR-4 has been concluded to regulate differential GnRH signaling cascades via heterodimerization with Ci-GnRHR-1 and -2 as an endogenous allosteric modulator. In *B. floridae*, amphioxus GnRHR-3 induces the elevation of intracellular calcium in response to *B. floridae* GnRH-like peptide [6]. No cognate ligands for amphioxus GnRHR-1, -2, or -4 have so far been identified [9,10]. On the other hand, amphioxus GnRHR-1 and -2 stimulate IP3 accumulation in response to both mammalian GnRH and chicken GnRHII, and amphioxus GnRHR-2 exhibits high selectivity for mammalian GnRH. In addition, amphioxus GnRHR-3 responded not only to chicken GnRHII but also to oct-GnRH (octopus GnRH) and adipokinetic hormone (AKH) [4].

Biological Functions

Target Cells/Tissues and Functions

Ci-GnRHRs are expressed in the cerebral ganglion and ovary. In particular, Ci-GnRHR-1, -2, and -4 are largely co-localized in test cells of vitellogenic follicles [9,10]. t-GnRH-3 to -9 increase water flow and then induce the release of eggs and sperm on injection into the gonaduct, ovary, stomach, or posterior body cavity of *C.intestinalis* [2].

References

1. Powell JF, Reska-Skinner SM, Prakash MO, et al. Two new forms of gonadotropin-releasing hormone in a protochordate and the evolutionary implications. *Proc Natl Acad Sci USA*. 1996;93:10461–10464.
2. Adams BA, Tello JA, Erchegyi J, et al. Six novel gonadotropin-releasing hormones are encoded as triplets on each of two genes in the protochordate, *Ciona intestinalis*. *Endocrinology*. 2003;149:4346–4356.
3. Kawada T, Aoyama M, Okada I, et al. A novel inhibitory gonadotropin-releasing hormone-related neuropeptide in the ascidian, *Ciona intestinalis*. *Peptides*. 2009;30:2200–2205.
4. Kawada T, Aoyama M, Sakai T, et al. Structure, function, and evolutionary aspects of invertebrate GnRHs and their receptors. In: Sills ES, ed. *Gonadotropin-Releasing Hormone (GnRH): Production, Structure and Functions*. Hauppauge, NY: Nova Science Publishers, Inc.; 2013:1–16.
5. Hasunuma I, Terakado K. Two novel gonadotropin-releasing hormones (GnRHs) from the urochordate ascidian, *Halocynthia roretzi*: implications for the origin of vertebrate GnRH isoforms. *Zool Sci*. 2013;30:311–318.
6. Kusakabe TG, Sakai T, Aoyama M, et al. A conserved non-reproductive GnRH system in chordates. *PLoS One*. 2012;7: e41955.
7. Roch GJ, Tello JA, Sherwood NM. At the transition from invertebrates to vertebrates, a novel GnRH-like peptide emerges in amphioxus. *Mol Biol Evol*. 2014;31:765–778.
8. Tello J.A., Rivier J.E., Sherwood N.M. Tunicate gonadotropin-releasing hormone (GnRH) peptides selectively activate *Ciona intestinalis* GnRH receptors and the green monkey type II GnRH receptor.
9. Sakai T, Aoyama M, Kusakabe T, et al. Functional diversity of signaling pathways through G protein-coupled receptor heterodimerization with a species-specific orphan receptor subtype. *Mol Biol Evol*. 2010;27:1097–1106.
10. Sakai T, Aoyama M, Kawada T, et al. Evidence for differential regulation of GnRH signaling via heterodimerization among GnRH receptor paralogs in the protochordate, *Ciona intestinalis*. *Endocrinology*. 2012;153:1841–1849.

Supplemental Information Available on Companion Website

- Primary structure of Protochordate GnRH precursors/E-Figure 38A.1
- Phylogenetic tree of GnRHRs/E-Figure 38A.2

Octopus Gonadotropin-Releasing Hormone

Hiroyuki Minakata

Abbreviation: oct-GnRH

Oct-GnRH induces the gonadal maturation and oviposition of the octopus Octopus vulgaris by regulating sex steroidogenesis and a series of egg-laying behaviors. Oct-GnRH is suggested to act as a multifunctional modulatory factor in higher brain functions.

Discovery

Oct-GnRH has been isolated from the brain of the common octopus *Octopus vulgaris*, as a stimulating factor of the systemic heart of the octopus [1].

Structure

Structural Features

Oct-GnRH is a dodecapeptide having structural features similar to those of vertebrate GnRHs. Oct-GnRH has a pyroGlu residue at the N-terminus and a Pro—Gly—NH$_2$ sequence at the C-terminus. Oct-GnRH conserves a His—aromatic amino acid residue—Ser sequence and a Gly residue at the corresponding positions of GnRH, except for the insertion of two amino acid residues after the pyroGlu residue [1].

Structure and Subtype

Oct-GnRH has also been characterized from the central nervous systems of the common cuttlefish *Sepia officinalis* and the swordchip squid *Uroteuthis edulis* (Table 38B.1) [2].

Table 38B.1 Amino Acid Sequence of Oct-GnRH*

Origin		Name		Sequence
Mollusca	Cephalopoda	*Octopus vulgaris*	oct-GnRH	pENYHFSNGWH PG-NH$_2$
		*Sepia officinalis**		-QNYHFSNGWH PGGKR-
		*Uroteuthis edulis**		-QNYHFSNGWH PGGKR-

*Deduced amino acid sequences in the precursor cDNA or from *in silico* analysis of the gene are shown. pE, pyroglutamic acid.

Synthesis and Release

Gene, mRNA, and Precursor

The total length of mRNA (770 bp) encoding oct-GnRH produces 87 aa residues of the precursor protein, which is composed of a signal peptide, oct-GnRH, a Gly—Lys—Arg sequence, and a GnRH-associated protein (GAP), of which the function is not known in the octopus [1]. cDNA having a similar sequence to oct-GnRH precursor has been partially cloned from *S. officinalis* [2]. The open reading frame of the cDNA cloned from the brain of *U. edulis* is 80.5% similar to that of oct-GnRH in its nucleotide sequence (E-Figure 38B.1) [2].

Distribution of mRNA

Oct-GnRH mRNA is expressed in the optic gland, which regulates the proliferation of reproductive cells, the maturation of gonads, and the synthesis of yolk protein of the octopus [1]. Oct-GnRH mRNA-expressing cell bodies are located in the lobes of the supra-esophageal and sub-esophageal part of the brain and in the optic lobe as follows: the superior buccal lobe, which controls feeding behavior; the inferior frontal lobe and the optic lobes, which contain the touch and visual memory systems; the brachial lobes, which control arm movement; and the palliovisceral lobe and the vasomotor lobe, which are centers of cardiac control [3].

Receptors

Structure and Subtype

The receptor for oct-GnRH (oct-GnRHR) has been identified as a class A GPCR from the brain of octopus (E-Figure 38B.2) [4]. Subtypes of oct-GnRHR have not been reported [4].

Signal Transduction Pathway

Oct-GnRH activated the oct-GnRHR expressed in *Xenopus* oocytes with an induction of membrane Cl$^-$ currents coupled to the IP/Ca^{2+} pathway [4]. Oct-GnRH mRNA expression is enhanced by N-methyl-D-aspartic acid (NMDA) and nitric oxide (NO) via a Glu/NMDA/NO signal transduction pathway in the olfactory lobe [5].

Y. Takei, H. Ando, & K. Tsutsui (Eds): Handbook of Hormones. DOI: http://dx.doi.org/10.1016/B978-0-12-801028-0.00199-9

Biological Functions

Target Cells and Function

Oct-GnRH immunoreactive cell bodies and the immunoreactive nerve fibers have been detected in the subpedunculate lobe and the posterior olfactory lobe, which together control the activities of the optic gland [3]. Some of the immunoreactive fibers may reach the optic gland, which expresses oct-GnRHR [4]. Oct-GnRH modulates contractions of the oviduct [3]. Oct-GnRH induces steroidogenesis of testosterone, progesterone, and estradiol-17β in the octopus ovary and testis, where oct-GnRHR is abundantly expressed [4]. These results suggest that oct-GnRH is a key peptide in the subpedunculate lobe and/or the posterior olfactory lobe—optic gland—gonadal axis, an octopus analog of the hypothalamo-hypophysial—gonadal axis [3,4]. In addition to reproductive functions, oct-GnRH also acts as a multifunctional modulatory factor in controlling higher brain functions such as feeding, memory, cardiac control, arm movement, postural regulation, and sensory and autonomic functions [2,4].

References

1. Iwakoshi E, Takuwa-Kuroda K, Minakata H, et al. Isolation and characterization of a GnRH-like peptide from *Octopus vulgaris*. *Biochem Biophys Res Commun*. 2002;291:1187—1193.

2. Minakata H. Oxytocin/vasopressin and gonadotropin-releasing hormone from cephalopods to vertebrates. *Ann NY Acad Sci*. 2010;1200:33—42.

3. Iwakoshi-Ukena E, Ukena K, Minakata H, et al. Expression and distribution of octopus gonadotropin-releasing hormone in the central nervous system and peripheral organs of the octopus (*Octopus vulgaris*) by *in situ* hybridization and immunohistochemistry. *J Comp Neurol*. 2004;477:310—323.

4. Kanda A, Takahashi T, Minakata H, et al. Molecular and functional characterization of a novel gonadotropin-releasing-hormone receptor isolated from the common octopus (*Octopus vulgaris*). *Biochem J*. 2006;395:125—135.

5. Di Cristo C, De Lisa E, Di Cosmo A. Control of GnRH expression in the olfactory lobe of *Octopus vulgaris*. *Peptides*. 2009;30:538—544.

Supplemental Information Available on Companion Website

- cDNA and deduced amino acid sequence alignment of the cephalopod oct-GnRH precursor/E-Figure 38B.1
- cDNA and deduced amino acid sequence alignment of oct-GnRH receptor and GnRH receptors of other animals/E-Figure 38B.2
- Exon—intron structure of oct-GnRH receptor/E-Figure 38B.3
- Accession numbers of genes and cDNAs for oct-GnRH and oct-GnRH receptor/E-Table 38B.1

CHAPTER 39

Corticotropin-Releasing Factor-Like Peptide

Shinji Nagata

Abbreviation: CRF
Additional names: corticotropin-releasing factor-like diuretic hormone (CRF/DH), DH, DH$_{44}$

Corticotropin-releasing factor (CRF)-like peptide in insects is a peptide hormone capable of regulating fluid excretion from the renal organ (Malpighian tubules). Therefore, this peptidyl factor is a diuretic hormone in insect species.

Discovery

CRF/DH was first identified from the crude head extract of the tobacco hornworm *Manduca sexta*. CRF/DH has been purified by evaluation of the fluid (meconium: pupal waste fluid) secretion immediately after eclosion of the cabbage white butterfly *Pieris rapae* [1].

Structure

Structural Features

M. sexta CRF/DH is a linear peptide composed of 44 amino acids with C-terminal amidation (Table 39.1). Although the length of CRF/DH ranges from 30 to 47 aa residues according to species, amino acid sequences exhibit low similarity among insect species. The amino acid sequence of CRF/DH also shows low similarity with those of mammalian CRFs.

Structure and Subtype

Several subtypes of CRF/DH are observed in one species, although the number of subtypes differs according to species. In *M. sexta*, another peptidyl factor having similar biological activity has been identified in addition to CRF/DH. That peptide was composed of a 30-aa designated diuretic peptide II (DP II). In the mealworm *Tenebrio molitor*, two CRF/DHs, DH$_{47}$ and DH$_{37}$ (named after the length of their amino acid sequences), have been identified. Although most CRF/DHs are C-terminally amidated, *T. molitor* CRF/DHs are,

exceptionally, not amidated. In addition to CRF/DH, the other peptidyl factor that shows similarity with calcitonin has been identified; however, this calcitonin-like diuretic hormone (CT/DH) is thought to belong to different class of diuretic hormones from CRF/DHs.

Synthesis and Release

Gene, mRNA, and Precursor

A cDNA of CRF/DH encodes a preproprotein composed of a signal sequence and predomain and a mature peptide [2].

Distribution of mRNA

The CRF/DH gene is expressed in the nervous system, especially in the brain and terminal ganglion. In some species, CRF/DH expression is confirmed not only in nervous cells but also in intestinal endocrinal cells. In *M. sexta* larvae, the expression of CRF/DH is confirmed in the brain, CNS, and midgut [2].

Receptors

Structure and Subtype

The CRF/DH receptor is a class of G protein-coupled receptor (G protein-coupled receptor (GPCR). *M. sexta* CRF/DH receptor is the first receptor that has been cloned using COS-7 expression cloning techniques from invertebrates [3]. As a paralogous receptor is present in mammalian species, most invertebrate species have two subtypes of GPCRs for CRF/DH. However, exceptionally only a single copy of GPCR for CRF/DH has been identified in some species, including *M. sexta*.

Signal Transduction Pathway

CRF/DH activates adenylate cyclase to produce intracellular cAMP via GPCR. The consequent intracellular cAMP affects fluid secretion by the Malpighian tubules [4].

Table 39.1 Alignment of Amino Acid Sequences of CRF/DH and CRF*

Manduca sexta CRF/DH	RMPSLSIDL PMSVLRQKLS LEKERKVHAL RAAANRNFLN DI-NH$_2$
DP II	SFSV-N PAVDILQHRY MEKV------ -AQNNRNFLN RV-NH$_2$
Drosophila melanogaster	PSLSIVNPLD VLRQRLLLEI ARRQMKENSR QVELNRAILK NV-NH$_2$
homo sapiens	SEEPPISLDL TFHLLREVL- -EMARAELAQ QAHSNRKLME II-NH$_2$

*Conserved amino acid residues are shaded.

Y. Takei, H. Ando, & K. Tsutsui (Eds): Handbook of Hormones. DOI: http://dx.doi.org/10.1016/B978-0-12-801028-0.00039-8

Biological Functions

Target Cells and Functions

CRF/DH modulates diuresis by controlling fluid secretion from the Malpighian tubules [4]. Additionally, CRF/DH modulates feeding behavioral duration via unknown mechanisms by which the release of CRF/DH can trigger the endogenous signal for the end of feeding [5,6].

References

1. Kataoka H, Troetschler RG, Li JP, et al. Isolation and identification of a diuretic hormone from the tobacco hornworm, *Manduca sexta*. *Proc Natl Acad Sci USA*. 1989;86:2976−2980.
2. Digan ML, Roberts DN, Enderlin FE, et al. Characterization of the precursor for *Manduca sexta* diuretic hormone *Mas*-DH. *Proc Natl Acad Sci USA*. 1991;89:11074−11078.
3. Reagan JD. Expression cloning of an insect diuretic hormone receptor. A member of the calcitonin/secretin receptor family. *J Biol Chem*. 1993;269:9−12.
4. Coast GM. Neuropeptides implicated in the control of diuresis in insects. *Peptides*. 1996;17:327−336.
5. Goldsworhty GJ, Owusu M, Ross KTA, et al. The synthesis of an analogue of the locust CRF-like diuretic peptide, and the biological activities of this and some C-terminal fragments. *Peptides*. 2003;24:1607−1613.
6. Audsley N, Goldsworthy GJ, Coast GM. Circulating levels of *Locusta* diuretic hormone: the effects of feeding. *Peptides*. 1997;18:59−65.

Supplemental Information Available on Companion Website

- Accession numbers of CRF/DH preproproteins and CRF/DH receptors/E-Table 39.1

CHAPTER 40

Oxytocin/Vasopressin Superfamily

Tsuyoshi Kawada

History

Locusta migratoria diuretic hormone (Lom-DH) was the first oxytocin (OXT)/vasopressin (VP) superfamily peptide isolated from invertebrate species, in 1987 [1]. Subsequently, Lys-conopressin was identified from a snail [2] and a leech [3]. Moreover, annetocin was found from an earthworm in 1994 [4], and two octopus peptides, cephalotocin and octopressin, were isolated in 1992 and 2003, respectively [2]. In deuterostome invertebrates, Ci-VP and *Styela* oxytocin-related peptide (SOP) were identified from different ascidians, *Ciona intestinalis* and *Styela plicata*, respectively in 2008 [5], while genes encoding [Ile^4]-vasotocin or echinotocin were detected on genome databases of the amphioxus or sea urchin in 2009 [5]. Likewise, genome database searches revealed the presence of OXT/VP superfamily peptides in nematodes and insects. In 2008, a gene encoding inotocin was found in the genome of a red flour beetle, and its amino acid sequence is identical to that of Lom-DH [6]. In 2012, a gene encoding nematocin was identified in the genome of the nematode *Caenorhabditis elegans* [7]. In addition, a gene encoding sepiatocin was cloned from the cDNA library of cuttlefish *Sepia officinalis* [8]. However, no gene encoding an OXT/VP superfamily peptide was detected in genome sequences of several insects; a mosquito, a silkworm, a honeybee, and fruit flies.

Structure

Structural Features

In vertebrates, all OXT/VP superfamily peptides are composed of nine aa residues and bear a circular structure formed by an intramolecular disulfide bridge between Cys^1 and Cys^6 (Table 40.1). The consensus amino acid residues (Pro^7, Gly^9, and C-terminal amide) are also present in invertebrate OXT/VP superfamily peptides, with some exceptions: Ci-VP, SOP, and nematocin. These peptides are elongated at the C-termini, resulting in the composition of 11−14 aa residues (Table 40.1). Moreover, the C-terminal glycine has been lost in Ci-VP, SOP, and nematocin, and Ci-VP is free at the C-terminus (Table 40.1).

Molecular Evolution of Family Members

Only a VP family peptide is present in the cyclostomes, which are an ancient species of vertebrates, suggesting that the ancestral OXT/VP superfamily peptide belonged to VP family peptides. Indeed, only a VP-like peptide, [Ile^4]vasotocin, was identified from an amphioxus, which is one of the closest species to vertebrates [5]. Only one type of OXT/VP superfamily peptide has been identified in a single species of most invertebrates. Exceptionally, two types of OXT/VP superfamily peptides, cephalotocin and octopressin, are present in a single species of octopus, *Octopus vulgaris* [2]. The precursors of cephalotocin and octopressin are highly homologous with each other, showing that the duplication of the OXT/VP superfamily peptide gene occurred during the evolution of mollusks [2].

Receptor

Structure and Subtypes

OXT/VP superfamily peptide receptors are included in the Class A GPCR superfamily. Most invertebrate species have one type of OXT/VP superfamily peptide receptor, whereas the pond snail and octopus possess two and three types, respectively (E-Figure 40.1) [2,9]. Likewise, two OXT/VP superfamily peptide receptor genes were found in the genome of a nematode, *C. elegans* [7].

Signal Transduction Pathway

An increase of intracellular Ca^{2+} is evoked by the activation of receptors for Ci-VP [3], annetocin [10], Lys-conopressin [9], cephalotocin [2], octopressin [2], inotocin [6], and nematocin [7]. Furthermore, nematocin increases cAMP levels via its receptor [7].

Biological Functions

Target Cells/Tissues and Functions

Injection of annetocin into the earthworm results in the induction of egg-laying behavior [5]. Furthermore, Lys-conopressin induces contractile activity of the vas deferens muscle of a pond snail *Lymnaea stagnalis* [2]. In *C. elegans*, males lacking nematocin or its receptors perform poorly regarding reproductive behaviors [7]. These findings suggest that several invertebrate OXT/VP superfamily peptides are involved in reproductive functions. On the other hand, osmoregulatory functions have been induced by inotocin and SOP [5]. Inotocin indirectly stimulates the Malpighian tubules, while SOP evokes contractions with increased tonus in the siphon of the ascidian [5]. Nematocin is involved in short-term gustatory associative learning in the nematode [7]. Moreover, Bardou and colleagues suggest that cephalotocin and octopressin induce long-term memory of passive avoidance in the cuttlefish *S. officinalis* [11].

Y. Takei, H. Ando, & K. Tsutsui (Eds): Handbook of Hormones. DOI: http://dx.doi.org/10.1016/B978-0-12-801028-0.00040-4

Table 40.1 Amino acid Sequences of OXT/VP Superfamily Peptides*

Peptide	Sequence	Organism
oxytocin	C*YIQNC*PLG—NH₂	mammals
vasopressin	C*YFQNC*PRG—NH₂	mammals
mesotocin	C*YIQNC*PIG—NH₂	marsupial, non-mammalian tetrapods
isotocin	C*YISNC*PIG—NH₂	lungfish
vastocin	C*YIQNC*PRG—NH₂	non-mammalian vertebrates
Lys-conopressin	C*FIRNC*PKG—NH₂	pond snail, leech, sea hare
cephalotocin	C*YFRNC*PIG—NH₂	octopus
octopressin	C*FWTSC*PIG—NH₂	octopus
sepiatocin	C*FWTTC*PIG—NH₂	cuttlefish
annetocin	C*FVRNC*PTG—NH₂	earthworm
inotocin	C*LITNC*PRG—NH₂	insects
nematocin	C*FLNSC*PYRRY—NH₂	nematodes
Ci-VP	C*FFRDC*SNMDWYR	ascidian
SOP	C*YISDC*PNSRFWST—NH₂	ascidian
[11e⁴]-vasotocin	C*YIINC*PRG—NH₂	amphioxus
echinoptocin	C*FISNC*PKG—NH₂	sea urchin

*Cysteine residues denoted by asterisks form an intramolecular disulfide bond. Moreover, consensus amino acids in OXT/VP superfamily peptides are shown by shading.

References

1. Proux JP, Miller CA, Li JP, et al. Identification of an arginine vasopressin-like diuretic hormone from *Locusta migratoria*. *Biochem Biophys Res Commun*. 1987;149:180—186.
2. Minakata H. Oxytocin/vasopressin and gonadotropin-releasing hormone from cephalopods to vertebrates. *Ann NY Acad Sci*. 2010;1200:33—42.
3. Salzet M, Bulet P, Van Dorsselaer A, et al. Isolation, structural characterization and biological function of a lysine-conopressin in the central nervous system of the pharyngobdellid leech *Erpobdella octoculata*. *Eur J Biochem*. 1993;217:897—903.
4. Oumi T, Ukena K, Matsushima O, et al. Annetocin: an oxytocin-related peptide isolated from the earthworm, *Eisenia foetida*. *Biochem Biophys Res Commun*. 1994;198:393—399.
5. Kawada T, Sekiguchi T, Sakai T, et al. Neuropeptides, hormone peptides, and their receptors in *Ciona intestinalis*: an update. *Zoolog Sci*. 2010;27:134—153.
6. Stafflinger E, Hansen KK, Hauser F, et al. Cloning and identification of an oxytocin/vasopressin-like receptor and its ligand from insects. *Proc Natl Acad Sci USA*. 2008;105:3262—3267.
7. Beets I, Temmerman L, Jansen T, et al. Ancient neuromodulation by vasopressin/oxytocin-related peptides. *Worm*. 2013;2:e24246.
8. Henry J, Cornet V, Bernay B, et al. Identification and expression of two oxytocin/vasopressin-related peptides in the cuttlefish *Sepia officinalis*. *Peptides*. 2013;46:159—166.
9. van Kesteren RE, Tensen CP, Smit AB, et al. Co-evolution of ligand-receptor pairs in the vasopressin/oxytocin superfamily of bioactive peptides. *J Biol Chem*. 1996;271:3619—3626.
10. Kawada T, Kanda A, Minakata H, et al. Identification of a novel receptor for an invertebrate oxytocin/vasopressin superfamily peptide: molecular and functional evolution of the oxytocin/vasopressin superfamily. *Biochem J*. 2004;382:231—271.
11. Bardou I, Leprince J, Chichery R, Vaudry H, Agin V. Vasopressin/oxytocin-related peptides influence long-term memory of a passive avoidance task in the cuttlefish, *Sepia officinalis*. *Neurobiol Learn Mem*. 2010;93:240—247.

Supplemental Information Available on Companion Website

- Phylogenic analysis of OXT/VP superfamily peptide receptors/E-Figure 40.1
- Accession numbers of genes encoding OXT/VP superfamily peptide/E-Table 40.1

Lys-Conopressin

Tsuyoshi Kawada

An oxytocin/vasopressin (OXT/VP) superfamily peptide isolated from the pond snail *Lymnaea stagnalis*, the sea hare *Aplysia kurodai*, and the leech *Erpobdella octoculata*, Lys-conopressin potently induces contractions in the vas deferens of pond snail.

Discovery

In 1992, van Kesteren and colleagues isolated a mollusk OXT/VP superfamily peptide named Lys-conopressin from the pond snail *Lymnaea stagnalis* [1]. In 1993, Lys-conopressin was also discovered in the leech *Erpobdella octoculata* [2].

Structure

Structure and Subtypes

Lys-conopressin is composed of nine aa residues containing consensus amino acids of the OXT/VP superfamily peptides: Cys^1, Asn^5, Cys^6, Pro^7, Gly^9; and C-terminal amidation (Table 40A.1).

Table 40A.1 Amino Acid Sequence of Lys-Conopressin

Lys-conopressin	CFIRNCPKG-NH$_2$

Lys-conopressin possesses a positively charged amino acid, lysine, at position 8, like VP (Table 40A.1).

Synthesis and Release

Gene, mRNA, and Precursor

The Lys-conopressin precursor is composed of typical structural units of the OXT/VP superfamily peptide precursors: a signal peptide domain, an OXT/VP superfamily peptide domain, and a neurophysin domain (Figure 40A.1). The *lys-conopressin* gene in the pond snail genome is composed of three exons and two introns at exactly the same locations as in the OXT/VP genes of vertebrates (Figure 40A.1) [3]. The *lys-conopressin* gene is a single copy gene in the pond snail genome, and no related gene has been identified [3].

Distribution of mRNA

In situ hybridization analysis of the pond snail showed that Lys-conopressin mRNA is present in neurons of the anterior lobes of cerebral ganglia [3]. These neurons were shown to project into the penis nerve to innervate the penis complex and the vas deferens [3].

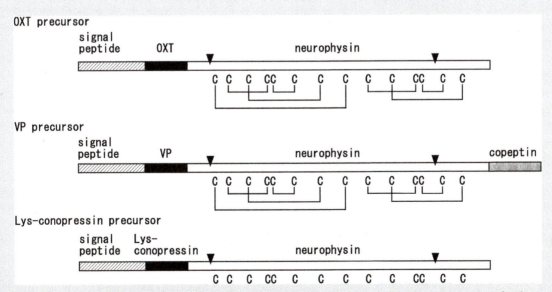

Figure 40A.1 Schematic representations of precursors encoding OXT, VP, and Lys-conopressin. OXT/VP superfamily peptides are indicated by black boxes, while signal peptides and neurophysin regions are represented by slashed and white boxes, respectively. Arrowheads show intron-inserting positions, and "C" indicates cysteines in each neurophysin region. In the OXT precursor and VP precursor, each disulfide bridge is connected by a line. A copeptin-encoding domain is indicated by a dotted box.

Y. Takei, H. Ando, & K. Tsutsui (Eds): Handbook of Hormones. DOI: http://dx.doi.org/10.1016/B978-0-12-801028-0.00200-2

Receptors

Structure and Subtypes

Two Lys-conopressin receptors (LSCPR1 and LSCPR2) have been identified from the pond snail *L. stagnalis*, and the sequences of LSCPRs show high homologies to those of OXT/VP superfamily peptide receptors [4,5]. Both *lscpr* genes are expressed in the brain and reproductive tissues. LSCPR1 mRNA is distributed in the vas deferens, while the sperm oviduct contains LSCPR2 mRNA [5]. In the brain, the *lscpr1* gene is expressed in neurons in the anterior lobe of the right cerebral ganglion, pedal Ib cluster, and neuroendocrine light green cells [4], while LSCPR2 mRNA is distributed in neurons of the visceral ganglion and right parietal ganglion [5].

Signal Transduction Pathway

Lys-conopressin evokes Ca^{2+}-dependent Cl^- currents in *Xenopus* oocytes injected with LSCPR cRNAs. EC_{50} values of Lys-conopressin for LSCPR1 or LSCPR2 were shown to be approximately 22 nM and 86 nM, respectively [4,5]. A synthetic analog, Ile-conopressin, also induced responses of LSCPR1 or LSCPR2 with EC_{50} values of 180 nM and 92 nM, respectively [4,5], suggesting that the Lys^8 residue is not crucial for the ligand selectivity of LSCPRs.

Biological Functions

Target Tissues and Function

Lys-conopressin potently induces contractions in the vas deferens [3], indicating that Lys-conopressin is involved in the ejaculation of semen during intromission in the pond snail; this corresponds to the role of oxytocin in mammals. In addition, Lys-conopressin induces the depolarization of *Lymnaea* muscle cells from the anterior vas deferens, neurons in the anterior lobe of the right cerebral ganglion, and neuronal light green cells [4].

References

1. van Kesteren RE, Smit AB, Dirks RW, et al. Evolution of the vasopressin/oxytocin superfamily: characterization of a cDNA encoding a vasopressin-related precursor, preproconopressin, from the mollusc *Lymnaea stagnalis*. *Proc Natl Acad Sci USA*. 1992;89:4593—4597.
2. Salzet M, Bulet P, Van Dorsselaer A, et al. Isolation, structural characterization and biological function of a lysine-conopressin in the central nervous system of the pharyngobdellid leech *Erpobdella octoculata*. *Eur J Biochem*. 1993;217:897—903.
3. van Kesteren RE, Smit AB, De Lange RP, et al. Structural and functional evolution of the vasopressin/oxytocin superfamily: vasopressin-related conopressin is the only member present in *Lymnaea*, and is involved in the control of sexual behavior. *J Neurosci*. 1995;15:5989—5998.
4. van Kesteren RE, Tensen CP, Smit AB, et al. A novel G protein-coupled receptor mediating both vasopressin- and oxytocin-like functions of Lys-conopressin in *Lymnaea stagnalis*. *Neuron*. 1995;15:897—908.
5. van Kesteren RE, Tensen CP, Smit AB, et al. Co-evolution of ligand—receptor pairs in the vasopressin/oxytocin superfamily of bioactive peptides. *J Biol Chem*. 1996;271:3619-26.1

Cephalotocin/Octopressin

Tsuyoshi Kawada

Abbreviation: CT/OP

CT and OP are oxytocin (OXT)/vasopressin (VP) superfamily peptides identified from an octopus, Octopus vulgaris. OP evokes contractions of the smooth muscles such as the rectum, oviduct, efferent branchial vessel, and anterior aorta.

Discovery

In 1992, an OXT/VP superfamily peptide was found from an octopus, *Octopus vulgaris*, and was named cephalotocin (*CT*) [1]. In 2003, another octopus OXT/VP superfamily peptide, octopressin (OP), was identified from the same octopus species [2].

Structure

Structure and Subtypes

CT and OP are each composed of nine aa residues containing consensus amino acids of the OXT/VP superfamily peptides: Cys^1, Cys^6, Pro^7, Gly^9; and C-terminal amidation (Table 40B.1). CT conserves Asn^5 like OXT/VP, whereas the fifth aa residue in OP is serine (Table 40B.1). Both CT and OP contain a neutral amino acid, isoleucine, at position 8, unlike Lys-conopressin, which locates a basic amino acid at position 8 (see Chapter 40). Recently, an OXT/VP superfamily peptide named as sepiatocin was found from another cephalopod, the cuttlefish *Sepia officinalis*. The amino acid sequence of sepiatocin is almost identical to that of OP except for Thr^5.

Synthesis and Release

Gene, mRNA, and Precursor

The CT and OP precursors consist of three domains: a signal peptide domain, a hormone domain, and a neurophysin domain, like OXT/VP precursors. The *CT* and *OP* genes consist of a single exon [3], which is unlike the structure of the Lys-conopressin gene in the pond snail *Lymnaea stagnalis*, suggesting that the octopus has lost introns in the *OXT/VP* family peptide genes during the evolutionary process in mollusks. Southern blot analysis revealed that the octopus genome encodes a single copy of the *CT* or *OP* gene [3].

Table 40B.1 Amino Acid Sequences of Octopus OXT/VP Superfamily Peptides

cephalotocin	$CYFRNCPIG-NH_2$
octopressin	$CFWTSCPIG-NH_2$

Distribution of mRNA

OP mRNA is distributed in the supraesophageal and subesophageal brains, and in the buccal and gastric ganglia [2]. The *CT* gene, however, is mostly expressed in the subesophageal brain [2]. *In situ* hybridization in the brain showed that OP mRNA was localized in many lobes, and CT mRNA was almost limited to the ventral median vasomotor lobe [2].

Receptors

Structure and Subtypes

Three genes encoding the OXT/VP superfamily peptide receptors were cloned from the brain of an octopus *O. vulgaris*. [4,5]. Two of them are subtypes of the CT receptors (CTR-1 and CTR-2), while one is the OP receptor (OPR). CTR-1 mRNA was strongly detected in the central and peripheral nervous systems, and slightly detected in peripheral tissues, the pancreas, oviduct, and ovary [4]. CTR-2 mRNA is mainly distributed in peripheral tissues, abundantly in the branchia and vas deferens [5]. OPR transcripts are widely detected in both the nervous system and peripheral tissues [5].

Signal Transduction Pathway

As stated above, CT exerts its activity via CTR-1 and CTR-2 but not OPR, whereas OP interacts with only OPR [5]. Administration of CT to *Xenopus* oocytes expressing CTR-1 or CTR-2 induced Ca^{2+} signaling at EC_{50} values of 14 nM and 67 nM, respectively [4,5]. The *Xenopus* oocyte assay showed an EC_{50} value of OP with OPR of 68 nM [5].

Biological Functions

Target Tissues and Function

OP evokes contractions of the smooth muscles of such as the rectum, oviduct, efferent branchial vessel, and anterior aorta [2]. In contrast, CT has no effect on these tissues. In addition, long-term memory formation of the cuttlefish *Sepia officinalis* can be enhanced after *in vivo* administration of CT or OP [6].

References

1. Reich G. A new peptide of the oxytocin/vasopressin family isolated from nerves of the cephalopod *Octopus vulgaris*. *Neurosci Lett*. 1992;134:191–194.
2. Takuwa-Kuroda K, Iwakoshi-Ukena E, Kanda A, et al. Octopus, which owns the most advanced brain in invertebrates, has two members of vasopressin/oxytocin superfamily as in vertebrates. *Regul Pept*. 2003;115:139–149.

Y. Takei, H. Ando, & K. Tsutsui (Eds): Handbook of Hormones. DOI: http://dx.doi.org/10.1016/B978-0-12-801028-0.00201-4

3. Kanda A, Takuwa-Kuroda K, Iwakoshi-Ukena E, et al. Single exon structures of the oxytocin/vasopressin superfamily peptides of octopus. *Biochem Biophys Res Commun*. 2003;309:743–748.

4. Kanda A, Takuwa-Kuroda K, Iwakoshi-Ukena E, et al. Cloning of octopus cephalotocin receptor, a member of the oxytocin/vasopressin superfamily. *J Endocrinol*. 2003;179:281–291.

5. Kanda A, Satake H, Kawada T, et al. Novel evolutionary lineages of the invertebrate oxytocin/vasopressin superfamily peptides and their receptors in the common octopus (*Octopus vulgaris*). *Biochem J*. 2005;387:85–91.

6. Bardou I, Leprince J, Chichery R, et al. Vasopressin/oxytocin-related peptides influence long-term memory of a passive avoidance task in the cuttlefish, *Sepia officinalis*. *Neurobiol Learn Mem*. 2010;93:240–247.

Annetocin

Tsuyoshi Kawada

Annetocin is an oxytocin (OXT)/vasopressin (VP) superfamily peptide identified from the earthworm *Eisenia foetida*. Injection of annetocin into the earthworm resulted in induction of stereotypical egg-laying behavior.

Discovery

In 1994, an OXT/VP superfamily peptide was isolated from an earthworm *Eisenia foetida*, and named annetocin [1].

Structure

Structure and Subtypes

Annetocin is composed of nine aa residues containing consensus amino acids of the vertebrate OXT/VP family peptides: Cys^1, Asn^5, Cys^6, Pro^7, Gly^9; and C-terminal amidation (Table 40C.1; see also Chapter 40).

Synthesis and Release

Gene, mRNA, and

The annetocin precursor is composed of three domains: a signal peptide domain, a hormone domain, and a neurophysin domain, like OXT/VP precursors.

Distribution of mRNA

The annetocin gene is expressed in the anterior region of the earthworm, and annetocin mRNA was detected in neurons symmetrically positioned in two separate regions of the subesophageal ganglia [2].

Receptors

Structure and Subtypes

Genomic analysis of the annectocin receptor (AnR) gene revealed that the intron-inserted position is conserved between the *AnR* gene and mammalian OXT/VP genes [3]. *In situ* hybridization using an earthworm showed that the *AnR* gene was specifically expressed in the nephridia

located in the clitellum region [3]. Another annelid receptor homolog for the OXT/VP suprafamily peptide was cloned from the leech *Theromyzon tessulatum* [4], although no endogenous ligand has been found in the leech. The receptor homolog gene is expressed in the genital tract, ovary, and brain of the leech. Furthermore, leech AnR mRNA gradually increases during the period of sexual maturation, and disappears after egg-laying [4].

Signal Transduction Pathway

Administration of annetocin to *Xenopus* oocytes expressing AnR induces Ca^{2+}-dependent signal transduction. The EC_{50} value of annetocin for AnR was calculated to be approximately 40 nM by dose−response curves for current shifts [3].

Biological Functions

Target Tissues and Function

Injection of annetocin into the earthworm resulted in induction of stereotypical egg-laying behavior. Likewise, injection of annetocin into a leech *Whitmania pigra*, leads to egg-laying behavior [5]. Furthermore, annetocin evokes contractions of the nephridia and gut of the earthworm, and injection of annetocin into the leech induces the a reduction of body weight [5].

Table 40C.1 Amino Acid Sequence of Annetocin

annetocin	CFVRNCPTG-NH$_2$

References

1. Oumi T, Ukena K, Matsushima O, et al. Annetocin: an oxytocin-related peptide isolated from the earthworm, *Eisenia foetida*. *Biochem Biophys Res Commun.* 1994;198:393−399.
2. Satake H, Takuwa K, Minakata H, et al. Evidence for conservation of the vasopressin/oxytocin superfamily in Annelida. *J Biol Chem.* 1999;274:5605−5611.
3. Kawada T, Kanda A, Minakata H, et al. Identification of a novel receptor for an invertebrate oxytocin/vasopressin superfamily peptide: molecular and functional evolution of the oxytocin/vasopressin superfamily. *Biochem J.* 2004;382:231−237.
4. Levoye A, Mouillac B, Riviere G, et al. Cloning, expression and pharmacological characterization of a vasopressin-related receptor in an annelid, the leech *Theromyzon tessulatum*. *J Endocrinol.* 2005;184:277−289.
5. Oumi T, Ukena K, Matsushima O, et al. Annetocin, an annelid oxytocin-related peptide, induces egg-laying behavior in the earthworm, *Eisenia foetida*. *J Exp Zool.* 1996;276:151−156.

Y. Takei, H. Ando, & K. Tsutsui (Eds): Handbook of Hormones. DOI: http://dx.doi.org/10.1016/B978-0-12-801028-0.00202-6

Inotocin

Tsuyoshi Kawada

An insect oxytocin (OXT)/vasopressin (VP) superfamily peptide, inotocin, indirectly stimulates the Malpighian tubules through the CNS of a red flour beetle, leading to induction of diuretic activity.

Discovery

Locusta migratoria diuretic hormone (Lom-DH) was the first invertebrate OXT/VP superfamily peptide isolated from a locust, *Locusta migratoria,* in 1987 [1]. Genome sequencing for various invertebrate species has led to the identification of *inotocin* genes encoding the OXT/VP superfamily peptide in several insect species, such as a red flour beetle (*Tribolium castaneum*), a parasitic wasp (*Nasonia vitripennis*), and ants (*Atta cephalotes, Camponotus floridanus,* and *Harpegnathos saltator*) [2,3]. In contrast, no *OXT/VP* superfamily peptide gene was detected in genome sequences of a mosquito (*Anopheles gambiae*), a silkworm (*Bombyx mori*), a honeybee (*Apis mellifera*), or fruitflies (12 *Drosophila* species), indicating that several insects have lost the OXT/VP superfamily peptide gene during evolution. Moreover, the amino acid sequence of Lom-DH also corresponds to inotocin.

Structure

Structure and Subtypes

Inotocin is composed of nine aa residues, which contain the consensus amino acids of vertebrate OXT/VP family peptides: Cys^1, Asn^5, Cys^6, Pro^7, Gly^9; and C-terminal amidation (Table 40D.1; see also Chapter 40).

Synthesis and Release

Distribution of mRNA

The *Tribolium inotocin* gene is expressed only in the head of the adult, and its mRNA has been detected in cells on the ventral surface of the subesophageal ganglion [2]. In the developmental process, *Tribolium inotocin* gene expression is upregulated in the early larval stage but downregulated in the late larval stage [2]. Thereafter, the inotocin mRNA increases again in pupae [2].

Table 40D.1 Amino Acid Sequence of Inotocin

inotocin	CLITNCPRG-NH$_2$

Receptors

Structure and Subtypes

In silico gene predictions have suggested the presence of OXT/VP superfamily peptide receptors in arthropods: the red flour beetle, parasitic wasp, and water flea [2]. *Tribolium* inotocin receptor mRNA is mainly distributed in the head of the adult, and is slightly present in the hindgut and Malpighian tubes [2]. In the developmental process, *Tribolium* inotocin receptor mRNA was detected in the early larval stage, whereas no inotocin receptor mRNA was detected in the late larval stage [2].

Signal Transduction Pathway

Administration of inotocin to CHO/G16 cells expressing the *Tribolium* inotocin receptor induced Ca^{2+} signaling with an EC_{50} value of 5 nM [2].

Biological Functions

Function

The osmoregulatory function was revealed by administration of inotocin to *T. castaneum*. Inotocin indirectly stimulates the Malpighian tubules through the CNS, including the endocrine organs corpora cardiaca and corpora allata, leading to induction of diuretic activity [4].

References

1. Proux JP, Miller CA, Li JP, et al. Identification of an arginine vasopressin-like diuretic hormone from *Locusta migratoria*. *Biochem Biophys Res Commun.* 1987;149:180−186.
2. Stafflinger E, Hansen KK, Hauser F, et al. Cloning and identification of an oxytocin/vasopressin-like receptor and its ligand from insects. *Proc Natl Acad Sci USA.* 2008;105:3262−3267.
3. Gruber CW, Muttenthaler M. Discovery of defense- and neuropeptides in social ants by genome-mining. *PLoS One.* 2012;7: e32559.
4. Aikins MJ, Schooley DA, Begum K, et al. Vasopressin-like peptide and its receptor function in an indirect diuretic signaling pathway in the red flour beetle. *Insect Biochem Mol Biol.* 2008;38: 740−748.

Y. Takei, H. Ando, & K. Tsutsui (Eds): Handbook of Hormones. DOI: http://dx.doi.org/10.1016/B978-0-12-801028-0.00203-8

Nematocin

Tsuyoshi Kawada

Abbreviation: NTC

Nematocin is a nematode oxytocin (OXT)/vasopressin (VP) superfamily peptide. Null mutant assay suggests that nematocin is involved in mating behaviors and gustatory associative learning.

Discovery

In 2012, Garrison and colleagues identified a gene encoding an OXT/VP superfamily peptide, nematocin (NTC) by *in silico* cloning of the *Caenorhabditis elegans* genome [1], and mass spectrum analysis identified NTC in the nematode *C. elegans*. Likewise, *NTC* genes have also been identified in the genome or EST data of various nematodes; *C. briggsae*, *C. remanei*, *C. brenneri*, *C. japonica*, *Pristionchus pacificus*, *Necator americamus*, *Ancylostoma caninum*, *Strongyloides stercoralis*, and *Bursaphelenchus xylophilus* [2].

Structure

Structure and Subtypes

NTC is composed of 11 aa residues, and has an amidated tyrosine located at the C-terminus (Table 40E.1). NTC conserves the OXT/VP consensus amino acids Cys^1, Cys^6, Pro^7, and C-terminal amidation. By contrast, nematocin is elongated at the C-terminus when compared with mammalian OXT and VP, like ascidian OXT/VP superfamily peptides (i.e., Ci-VP and *Styela* oxytocin-related peptide; see also Table 40.1 in Chapter 40).

Synthesis and Release

Gene, mRNA, and Precursor

The NTC precursor possesses typical structural units of the OXT/VP superfamily peptide precursors: a signal peptide domain, an OXT/VP superfamily peptide domain, and a neurophysin domain.

Gene Expression

Expression of the *NTC* gene, *ntc-1*, was detected using GFP reporters. Garrison *et al.* constructed a GFP gene regulated by *ntc-1* promoter, and observed the GFP fluorescence via *ntc-1* promoter activation in transgenic nematodes [1]. Beets and colleagues observed the GFP-tagged NTC precursor protein expressed by *ntc-1* promoter activation in transgenic nematodes [2]. These results suggested that *ntc-1* is expressed in thermosensory neurons, mechanosensory neurons, neurosecretory cells, interneurons, and pharyngeal neurons in both sexes, while *ntc-1* is also expressed in motor neurons that control turning behavior during mating in males [1,2].

Receptors

Structure and Subtypes

NTC receptor (NTR)-1 and -2 encoded in the *C. elegans* genome share approximately 30% amino acid identity with each other, and 50% identity with mammalian OXT and VP receptors. *ntr-1* and *ntr-2* are expressed in part-overlapping sets of head and tail neurons in both sexes [1]. In addition, the *NTR* genes are expressed in male-specific neurons and muscles, suggesting their roles in mating. *ntr-1* is expressed in the hook and tail sensory neurons and spicule protractor muscles, while *ntr-2* promoter activation has been detected in the male-specific sensorimotor tail neuron [1]. Beet *et al.* showed that *ntr-1* was expressed in the gustatory neurons, chemosensory neuron pairs, ray neurons, and most likely tail neurons [2]. In particular, several *ntr-1*-expressing neurons are involved in important functions in the chemotaxis of *C. elegans*.

Signal Transduction Pathway

NTC evokes a Ca^{2+} response to NTR-1-expressing CHO cells at an EC_{50} value of 32.2 nM [2]. By contrast, NTR-2 does not respond to NTC or affect the Ca^{2+} response by activation of NTR-1 [2]. Activation of NTR-1 upregulates the intracellular cAMP level in the CHO cells [2], whereas no cAMP response was induced by activation of only NTR-1 or NTR-2 in HEK293T cells [1]. NTC suppressed forskolin-stimulated cAMP production in HEK 293T cells transfected with three genes, *ntr-1*, *ntr-2*, and *Ga15*, indicating that NTR-1 and NTR-2 can function as heterodimers [1].

Biological Functions

Phenotype in Gene-Modified Animals

Null mutants for *ntc-1*, *ntr-1*, and *ntr-2* are viable and fertile as hermaphrodites with normal locomotion speed, egg-laying behavior, and numbers of progeny [1]. The male null mutants are also viable, although they have reduced mating success [1]. Selective inactivation of *ntc-1* in mechanosensory neurons leads to suppression of mating behaviors in initial contact and vulva location [1]. By contrast, each null mutant for *ntc-1* or *ntr-1* shows normal salt chemotaxis and osmotic avoidance [2]. Furthermore, the gustatory plasticity assay using each null

Y. Takei, H. Ando, & K. Tsutsui (Eds): Handbook of Hormones. DOI: http://dx.doi.org/10.1016/B978-0-12-801028-0.00204-X

Table 40E.1 Amino Acid Sequence of Nematocin

nematocin	CFLNSCPYRRY-NH$_2$

mutant for *ntc-1* or *ntr-1* suggested that NTC facilitated the experience-driven modulation of salt chemotaxis that is a type of gustatory associative learning [2]. Cell-specific expression of *ntr-1* in the gustatory neuron and of *ntc-1* in interneurons rescues the plasticity defect of *ntr-1* and of *ntc-1* mutants, respectively [2].

References

1. Garrison JL, Macosko EZ, Bernstein S, et al. Oxytocin/vasopressin-related peptides have an ancient role in reproductive behavior. *Science*. 2012;338:540–543.
2. Beets I, Janssen T, Meelkop E, et al. Vasopressin/oxytocin-related signaling regulates gustatory associative learning in *C. elegans*. *Science*. 2012;338:543–545.

Ci-Vasopressin

Tsuyoshi Kawada

Abbrevation: Ci-VP
Additional name: *Ciona* vasopressin

An oxytocin (OXT)/vasopressin (VP) superfamily peptide identified from an ascidian, Ciona intestinalis, *Ci-VP is elongated at the C-terminus, and its C-terminus is not amidated.*

Discovery

In 2008, an OXT/VP superfamily peptide was identified from an ascidian *Ciona intestinalis* [1]. This ascidian peptide was the first deuterostome invertebrate OXT/VP superfamily peptide, and was named Ci-VP.

Structure

Structure and Subtypes

Although most OXT/VP superfamily peptides consist of nine aa residues and conserve the consensus sequences Cys^1, Asn^5, Cys^6, Pro^7, Gly^9, and C-terminal amidation, Ci-VP consists of 13 aa residues with an elongated C-terminus (Table 40F.1). Moreover, the C-terminus of Ci-VP is non-amidated, unlike in typical OXT/VP superfamily peptides. Another ascidian OXT/VP superfamily peptide, *Styela* oxytocin-related peptide (SOP), characterized from *Styela plicata*, is also elongated at the C-terminus, although its C-terminus is amidated [2]. Ile^4-vasotocin is another notochord OXT/VP superfamily peptide identified from an amphioxus *Branchiostoma floridae* [3], while an echinoderm OXT/VP superfamily peptide, echinotocin, was found from a sea urchin *Strongylocentrotus purpuratus* [4]. Ile^4-vasotocin and echinotocin are nonapeptides that completely share the OXT/VP consensus amino acids. Therefore, the ascidian OXT/VP superfamily peptides have been diversified during ascidian-specific evolution.

Synthesis and Release

Gene, mRNA, and Precursor

All OXT/VP superfamily peptide precursors encode the hormone peptide and the associated protein, neurophysin. *ci-vp* gene encodes a neurophysin domain that harbors only 10 cysteines, although typical neurophysins contain 14 cysteines [1]. The 14 consensus cysteines in the neurophysin are completely conserved in the SOP precursor [2], indicating the intraphyletic molecular diversity of neurophysin domains within urochordate species. Precursors of Ile^4-vasotocin

and echinotocin locate a neurophysin domain containing 14 cysteines [3,4].

Distribution of mRNA

Ci-VP mRNA is specifically distributed in the neural complex. prominent *ci-vp* gene expression has been detected in several neurons in both the cortex and medulla regions of the adult brain ganglion, but not in the neural gland [1]. Likewise, SOP mRNA has been detected only in the ganglion of the neural complex. SOP mRNA-containing cells are clustered in the area surrounding the cerebral ganglion [2]. No expressional analysis has been performed for the genes encoding Ile^4-vasotocin and echinotocin.

Receptors

Structure and Subtypes

Ci-VP receptor (Ci-VP-R) was characterized from *C. intestinalis*. Two putative amphioxus OXT/VP superfamily peptide receptors were identified in the *B. floridae* genome. An ortholog of the OXT/VP receptor was also found in the genome of a sea urchin *S. purpuratus*. No OXT/VP superfamily peptide receptor has been identified in other deuterostome invertebrates.

Signal Transduction Pathway

Application of Ci-VP to the Ci-VP-R-expressing *Xenopus* oocytes evokes a typical Ca^{2+}-dependent inward Cl^- current. A maximal response was observed at more than 10 nM, and the EC_{50} was calculated to be approximately 52.4 nM [1]. *ci-vp-r* gene was expressed predominantly in the neural complex, digestive tract, gonad, and heart to a similar degree, but considerably less in the endostyle [1].

Biological Functions

Target Tissues and Function

No function has been clarified for Ci-VP. However, the OXT/VP superfamily peptides are likely to participate in osmoregulation in ascidians. The amount of SOP mRNA in hypotonic sea water is about two-fold greater than in isotonic and hypertonic sea water [2]. Furthermore, SOP evokes contractions with increased tonus in the siphon of *S. plicata* at 10 nM [2]. These results suggest the functional correlation of SOP with osmoregulation. Echinotocin induces contraction of the tube foot and esophagus preparations from another sea urchin *Echimus esculentus* [4], suggesting that echinotocin is involved in locomotion and feeding behavior.

Y. Takei, H. Ando, & K. Tsutsui (Eds): Handbook of Hormones. DOI: http://dx.doi.org/10.1016/B978-0-12-801028-0.00255-5

Table 40F.1 Amino Acid Sequences of Deuterostome Invertebrate OXT/VP Superfamily Peptides

Ci-VP	CFFRDCSNMDWYR
SOP	CYISDCPNSRFWST-NH$_2$
Ile^4-vasotocin	CYIINCPRG-NH$_2$
echinotocin	CFISNCPKG-NH$_2$

References

1. Kawada T, Sekiguchi T, Itoh Y, et al. Characterization of a novel vasopressin/oxytocin superfamily peptide and its receptor from an ascidian, *Ciona intestinalis. Peptides.* 2008;29:1672−1678.

2. Ukena K, Iwakoshi-Ukena E, Hikosaka A. Unique form and osmoregulatory function of a neurohypophysial hormone in a urochordate. *Endocrinology.* 2008;149:5254−5261.

3. Gwee PC, Tay BH, Brenner S, et al. Characterization of the neurohypophysial hormone gene loci in elephant shark and the Japanese lamprey: origin of the vertebrate neurohypophysial hormone genes. *BMC Evol Biol.* 2009;9:47.

4. Elphick MR, Rowe ML. NGFFFamide and echinotocin: structurally unrelated myoactive neuropeptides derived from neurophysin-containing precursors in sea urchins. *J Exp Biol.* 2009;212:1067−1077.

CHAPTER 41

Neuropeptide F

Shinji Nagata

Abbreviation: NPF
Additional names: long NPF

NPF functions in behavioral and motivational modulation in insects, including feeding behavior and reproductive behavior. NPF is a possible invertebrate counterpart peptide of the mammalian neuropeptide Y (NPY), regulating feeding and its related behaviors.

Discovery

The first invertebrate NPF was identified as a homologous peptide to mammalian NPY from the tapeworm *Moniezia expansa*, using an antibody against RY-NH$_2$ [1]. A similar method using an antibody against RY-NH$_2$ led to identification of another shorter RF-NH$_2$-possessing peptide, designated short NPF (sNPF). The original NPF is thus also called long NPF, to discriminate between these peptides.

Structure

Structural Features

NPFs are composed of 28—45 aa residues with the conserved C-terminal tetrapeptide, R(or K)XRF-NH$_2$ [2,3]. Only a few aa residues within NPFs are identical with those of mammalian NPY and pancreatic polypeptide Y(PPY) (Table 41.1). NPF is also identified in molluskan species such as the snail and *Aplysia*.

Structure and Subtype

Transcriptional and genomic information has revealed the presence of one to three NPF subtypes in one species. In *D. rosophila melanogaster* only one NPF is present in the genomic sequence [4], whereas three NPF-like peptides NPF1a, NPF1b, and NPF2 are present in the silkworm *Bombyx mori* [5]. An N-terminally truncated form is present as the subtype in some species, including *Locusta migratoria*.

Synthesis and Release

Gene, mRNA, and Precursor

An mRNA encoding NPF produces a deduced precursor peptide of approximately 60 aa residues. The precursor peptide is composed of a signal sequence, a biologically active NPF portion, and a predomain. NPF is produced by cleavages of signal sequence and a dibasic cleavage site.

Distribution of mRNA

NPF is expressed in the lateral protocerebrum neurons in the brain, subesophageal ganglion, and other ganglia throughout the central nervous system in *D. melanogaster*. In *D. melanogaster*, the number of NPF-expressing neurons differs according to sex: there are 26 neurons in male flies and 20 in females [5]. In *B. mori*, NPF1 and NPF2 are also expressed, not only in the ganglia but in the frontal ganglion, which modulates the pattern of feeding [6]. NPF is also expressed in the clock neurons. Also, most species show NPF expression by intestinal endocrinal cells such as in the midgut and hindgut [3,7].

Receptors

Structure and Subtype

The receptor for NPF (NPFR) is a GPCR belonging to the clade of receptors for peptides including C-terminal RF-NH$_2$. To date, no subtypes of NPFR have been reported.

Signal Transduction Pathway

NPF signaling is involved in the insulin signaling pathway in *D. melangaster*.

Biological Functions

Target Cells and Function

To date, a number of investigations regarding the physiological contributions of NPF have been performed. Therefore, NPF is a pleiotropic factor in insects [3,8]. Among the functions of NPF, predominant biological activities include the following.

1. Modulation of feeding behavior and motivation, including the response to starvation. The function of NPF in feeding effects is involved in the insulin-signaling pathway.
2. As clock-related neurons are co-expressed with NPF, clock-related functions have been demonstrated.
3. NPF contributes to locomotor activity for reproduction.
4. NPF's involvement in aggressive behaviors such as territorial competition and male-specific aggression has also been demonstrated.
5. Using genetic manipulation of NPF in *D. melanogaster*, learning and memory effects are also confirmed. Behavioral modulation, including disturbance of social behavior, is also observed in the nematode *Caenorhabditis elegans*.

Y. Takei, H. Ando, & K. Tsutsui (Eds): Handbook of Hormones. DOI: http://dx.doi.org/10.1016/B978-0-12-801028-0.00041-6

Table 41.1 Alignment of Sequences of NPFs and NPY*

Moniezia expansa	NPF	PDKDFIVNPS	DLVLDNKAAL	R-DYLRQINE	YFAIIGRPRF-NH$_2$
Drosophila melanogaster	NPF	SNSRPPR	KNDVNTMADA	YK-FLQDLDT	YYGDRARVRF-NH$_2$
Bombyx mori	NPF1a	REE	GPN--NVAEA	-LRILQLLDN	YYTQAARPRF-NH$_2$
	NPF1b	DVDAAGDRV	DPELLDRAVR	-LLWLEKLDR	IYSYHTRPRF-NH$_2$
	NPF2	QYPRPR	RPERFDTAEQ	ISNYLKELQE	YYSVHGRGRY-NH$_2$
Homo sapiens	NPY	YPSKPD	NPGEDAPAED	MARYYSALRH	YINLITRQRY-NH$_2$

*Conserved amino acid residues are shaded.

6. Release of other hormone sand biologically active compounds, including dopamine, octopamine, and GABA, is observed in NPF-deficient flies.
7. It has been demonstrated that sensitivity to ethanol is controlled via NPFergic neurons in the mushroom body.

Basically, NPF appears to contribute to the biological events concerning emotional behaviors and locomotor activities.

Phenotype in Gene-Modified Animals

D. melanogaster and *C. elegans* have been extensively investigated using transgenic animals in which NPF neurons are genetically deficient. The predominant phenotypes illuminate the perturbation of behaviors, including social behavior, as described above. These behavioral abuse are caused by dysfunction of normal locomotor and other activities. In particular, the relevance of NPF and NPFR to feeding behavior and food consumption has been demonstrated using RNAi experiments in several insect species, including *D. melanogaster*, *L. migratoria*, and *C. elegans*.

References

1. Maule AG, Shaw C, Halton DW, et al. Neuropeptide F: a novel parasitic flatworm regulatory peptide from *Moniezia expansa* (Cestoda: Cyclophillidea). *Parasitology*. 1991;102:309−316.
2. Mair GR, Halton DW, Shaw C, et al. The neuropeptide F (NPF) encoding gene from the cestode, *Moniezia expansa*. *Parasitology*. 2005;120:71−77.
3. Nässel DR, Wegener C. A comparative review of short and long neuropeptide F signaling in invertebrates: any similarities to vertebrate neuropeptide Y signaling? *Peptides*. 2011;32:1335−1355.
4. Brown MR, Crim JW, Arata RC, et al. Identification of a *Drosophila* brain-gut peptide related to the neuropeptide Y family. *Peptides*. 1999;20:1035−1042.
5. Lee G, Bahn JH, Allada R. Sex- and clock-controlled expression of the neuropeptide F gene in *Drosophila*. *Proc Natl Acad Sci USA*. 2006;103:12580−12585.
6. Roller L, Yamanaka N, Watanabe K, et al. The unique evolution of neuropeptide genes in the silkworm *Bombyx mori*. *Insect Biochem Mol Biol*. 2008;38:1147−1157.
7. Veenstra JA. Peptidergic paracrine and endocrine cells in the midgut of the fruit fly maggot. *Cell Tissue Res*. 2009;336:309−323.
8. Nässel DR, Winther AM. *Drosophila* neuropeptides in regulation of physiology and behavior. *Prog Neurobiol*. 2010;92:42−104.

Supplemental Information Available on Companion Website

- Accession numbers of NPF and NPF receptor/E-Table 41.1

CHAPTER 42

Short Neuropeptide F

Shinji Nagata

Abbreviation: sNPF

sNPF is a neuropeptide modulating extensive biological events involving feeding behavior and reproductive behavior.

Discovery

sNPF was identified from the brain extract of the Colorado potato beetle *Leptinotarsa decemlineata* as a peptide recognized by an antiserum against RYamide. The C-terminal aa sequence of this peptide shares that of another longer neuropeptide, neuropeptide F (NPF), and it is called sNPF because it is shorter than NPF [1].

Structure

Structural Features

sNPF is composed of 6−14 aa residues with the conserved C-terminal sequence RXRF-NH$_2$.

Structure and Subtype

Generally, there are several subtypes within in the precursor protein [2]. In the fruit fly *Drosophila melanogaster*, four subtypes are present in the precursor (Table 42.1) [3]. Exceptionally, only one sNPF is encoded in the preproprotein of the mosquito *Aedes aegypti*.

Synthesis and Release

Gene, mRNA, and Precursor

Genomic and transcriptomic data show the presence of a single mRNA encoding a precursor peptide of sNPF.

Distribution of mRNA

Almost all ganglia contain sNPF-expressing cells. The number of sNPF-expressing cells differs according to the species and developmental stages. In the *D. melanogaster* larva, about 70 cells in the brain, more than 1,000 Kenyon cells in the mushroom body, and about 120 cells in the ventral nerve cord express sNPF, while in adult flies the cell numbers increase to 280 and 4,000 in the brain and mushroom body, respectively. sNPF is expressed in various types of neuronal cells, such as interneurons, lateral neurosecretory cells, and lobular optic lobe neurons. sNPF is also expressed in intestinal endocrine cells [2,4].

Regulation of Synthesis and Release

In *Bombyx mori*, starvation induces sNPF secretion from the brain or CNS, consistent with the suggested function of sNPFs in feeding behavior [5].

Receptors

Structure and Subtype

The receptor for sNPF (sNPFR) is identified as a GPCR in several insect species [6]. All annotated sNPFRs exhibit 60−90% similarity to one another. Two sNPFR subtypes are present in *B. mori*, whereas only one sNPFR is present in *D. melanogaster* and *Anopheles gambiae*. The sequence of sNPFR is similar to that of vertebrate NPY receptor type 2. *sNPFR* is predominantly expressed in the CNS and brain. It is also expressed in the midgut, hindgut, malpighian tubules, fat body, ovary, and testis [7].

Biological Functions

Target Cells and Functions

The small size of sNPF and the expression sites of sNPF and sNPFR indicate that sNPF functions as possibly as a hormone and possibly as a neurotransmitter. sNPF contributes to a wide range biological functions, including the response to osmotic and metabolic stress, locomotor activity, and sensing olfactory modification [2]. In particular, sNPF plays an essential role in feeding behavior and growth. The fact that sNPF expresses in insulin-producing cells and in the fan-shaped tachykinin-producing cells in the brain supports the biological effects of sNPF on these mechanisms, because insulin and tachykinin are strongly related to metabolic and locomotor activities.

Phenotype in Gene-Modified Animals

In *D. melanogaster*, knockdown strains targeting sNPF and sNPFR have been reported. Phenotypes of those strains support the functions of sNPF, as mentioned above. Significant disturbance of locomotor behavior and reduction of feeding behavior are observed in sNPF-disrupted flies. Also, sensing of olfactory stimulation is modified in sNPF knockdown flies. Similar results have been obtained using RNAi-treated *Locusta migratoria* [8].

Table 42.1 Alignment of *D. melanogaster* sNPFs*

sNPF-1	AQRSPSLRLRF−NH$_2$
sNPF-2	SPSLRLRF−NH$_2$
sNPF-3	PQRLRW−NH$_2$
sNPF-4	PMRLRW−NH$_2$

*Conserved amino acid residues are shaded.

Y. Takei, H. Ando, & K. Tsutsui (Eds): Handbook of Hormones. DOI: http://dx.doi.org/10.1016/B978-0-12-801028-0.00042-8

References

1. Spittaels K, Verhaert P, Shaw C, et al. Insect neuropeptide F (NPF)-related peptides: isolation from Colorado potato beetle (*Leptinotarsa decemlineata*) brain. *Insect Biochem Mol Biol.* 1996;26:375−382.
2. Nässel DR, Wegener C. A comparative review of short and long neuropeptide F signaling in invertebrates: any similarities to vertebrate neuropeptide Y signaling? *Peptides.* 2011;32:1335−1355.
3. Vanden Broeck J. Neuropeptides and their precursors in the fruitfly, *Drosophila melanogaster. Peptides.* 2001;22:241−254.
4. Johard HA, Enell LE, Gustafsson E. Intrinsic neurons of *Drosophila* mushroom bodies express short neuropeptide F: relations to extrinsic neurons expressing different neurotransmitters. *J Comp Neurol.* 2008;507:1479−1496.
5. Nagata S, Matsumoto S, Nakane T, et al. Effects of starvation on brain short neuropeptide F-1, -2, and -3 levels and short neuropeptide F receptor expression levels of the silkworm, *Bombyx mori. Front Endocrinol.* 2012;3:3.
6. Mertens I, Meeusen T. Characterization of the short neuropeptide F receptor from *Drosophila melanogaster. Biochem Biophys Res Commun.* 2002;297:1140−1148.
7. Garczynski SF, Brown MR, Shen P, et al. Characterization of a functional neuropeptide F receptor from *Drosophila melanogaster. Peptides.* 2002;23:773−780.
8. Wegener C, Gorbashov A. Molecular evolution of neuropeptides in the genus *Drosophila. Genome Biol.* 2009;9:R131.

Supplemental Information Available on Companion Website

- Accession numbers of *D. melanogaster* sNPF and sNPFR/ E-Table 42.1

CHAPTER 43

RYamide

Shinji Nagata

Abbreviation: RYa

Deorphanization of Drosophila melanogaster *GPCRs revealed a novel peptide with C-terminal RY-NH$_2$. RYa regulates feeding behavior in insects, similarly to mammalian neuropeptide Y (NPY).*

Discovery

RYa was identified as a ligand for orphan GPCR (CG13968) of *D. melanogaster*, from the extract of flies [1]. Also, *Nasonia* pyro-sequencing supported the presence of RYa [2].

Structure

Structural Features

D. melanogaster RYa is composed of 9 aa residues. The C-terminal RY-NH$_2$ is highly conserved among RYas. In most species, another RYa subtype, RYa-2, composed of 10 aa residues, is present in the preproprotein (Table 43.1).

Structure and Subtype

Transcriptomic and genomic researches revealed that RYa preproproteins have two subtypes, RYa-1 and RYa-2, sharing a C-terminal RY-NH$_2$. The length and aa sequences of RYa-1 vary over species, while RYa-2 shares a C-terminal GXRY-NH$_2$.

Synthesis and Release

Gene, mRNA, and Precursor

A preproprotein mRNA encodes two RYa subtypes between dibasic cleavage sites.

Distribution of mRNA

The RYa gene is expressed in the anterior midgut and CNS in *Bombyx mori* larvae. Immunostaining using an antiserum against any peptides with C-terminal RY-NH$_2$ suggests the presence of RYa in *D. melanogaster* CNS and gut.

Receptors

Structure and Subtype

cDNAs encoding the RYa receptor are conserved among insect species [2,3]. This receptor is a GPCR belonging to the clade of receptors for peptides including RY-NH$_2$ and RF-NH$_2$.

Biological Functions

Target Cells and Function

RYa is thought to function in the regulation of feeding behavior, like mammalian RY-NH$_2$ peptides including NPY. In *D. melanogaster*, RYa injection decreases the proboscis extension accompanied by reduction of sucrose intake. Thus, this peptide is an additional counterpart of mammalian NPY, like neuropeptide F (NPF) and short NPF (sNPF).

References

1. Ida T, Takahashi T, Tominaga H, et al. Identification of the novel bioactive peptides dRYamide-1 and dRYamide-2, ligands for a neuropeptide Y-like receptor in *Drosophila*. *Biochem Biophys Res Commun*. 2011;410:872−877.
2. Hauser F, Neupert S, Williamson M, et al. Genomics and peptidomics of neuropeptides and protein hormones present in the parasitic wasp *Nasonia vitripennis*. *J Proteome Res*. 2010;9:5296−5310.
3. Collin C, Hauser F, Krogh-Meyer P, et al. Identification of the *Drosophila* and *Tribolium* receptors for the recently discovered insect RYamide neuropeptides. *Biochem Biophys Res Commun*. 2011;412:578−583.

Supplemental Information Available on Companion Website

- Accession numbers of RYa and RYa receptor/E-Table 43.1

Table 43.1 Alignment of RYa*

Drosophila melanogaster	RYa-1	PVFFVASRY–NH$_2$
Tribolium castaneum	RYa-1	VQNLATFKTMMRY–NH$_2$
Drosophila melanogaster	RYa-2	NEHFFLGSRY–NH$_2$
Tribolium castaneum	RYa-2	ADAFFLGPRY–NH$_2$

*Conserved amino acid residues are shaded.

Y. Takei, H. Ando, & K. Tsutsui (Eds): Handbook of Hormones. DOI: http://dx.doi.org/10.1016/B978-0-12-801028-0.00043-X

CHAPTER 44

Tachykinin-Like Peptide Family

Honoo Satake

History

In 1962, the first tachykinin (TK), eledoisin, was isolated from the posterior salivary gland of a Mediterranean octopus, *Eledone moschata* [1]. Eledoisin-like salivary TKs have been isolated from the salivary gland of another octopus, *Octopus vulgaris*, and a mosquito *Aedes aegypti* [2−4]. In 1990, Lom-TKs were isolated as the first invertebrate TK-like neuropeptides from the central nervous system of the locust *Locusta migratoria* [5], followed by identification of their structurally related peptides in a wide range of protostomes, including insects, annelids, mollusks, and sea anemone [2−4]. Such protostome TK-related peptides have been designated "TKRPs" or "TRPs" [2−4]. Moreover, the authentic TK in invertebrates, Ci-TK, was detected in the brain and gut of the protochordate *Ciona intestinalis* in 2004 [6]. In 2013, another type of TKRP, natalisin, was identified in holometabolous insect species, including *Drosophila melanogaster*, *Tribolium castaneum*, and *Bombyx mori* [7]. Collectively, "invertebrate TK-like peptides" have thus far been categorized into three groups: salivary TKs in several protostomes, TKRPs including natalisins in various protostomes, and authentic TKs in protochordates (Ci-TKs).

Structure

Structural Features and Primary Structure

All salivary TKs and Ci-TKs share the vertebrate TK consensus sequence Phe−Xaa−(Gly/Ala)−Leu−Met−NH_2. TKRP bears the consensus sequences of Phe−Xaa^1−(Gly/Ala)−Xaa^2−Arg−NH_2. The natalisin consensus sequence is Phe−Xaa^1−Xaa^2−Xaa^3−Arg−NH_2.

Properties

Mr 800−1,200. Soluble in water, physiological saline solution, aqueous acetonitrile, and methanol.

Synthesis and Release

Gene, mRNA, and Precursor

All salivary TK precursors encode a single TK sequence. Ci-TK precursor, like vertebrate substance P (SP) precursors, encodes two TK sequences (E-Figure 44.1). TKRP precursors encode multiple (5−16) TKRP sequences. In all TK sequences, TKs are flanked by a C-terminal amidation signal and mono- or dibasic endoproteolytic sites [2−6]. The number of TKRP sequences in the precursors differs among species [2−6].

Distribution of mRNA

Salivary TKs are expressed exclusively in the salivary gland. Ci-TK expression is detected in neurons of the brain, unidentified cells in the intestine, and putative endocrine cells in the endostyle. All TKRP mRNAs are expressed in neurons of the central nervous system [2−4]. In several insects, the TKRP mRNA is detected in digestive tissues [2−4].

Receptors

Structure and Subtype

Ci-TK receptor (Ci-TK-R) and TKRP receptors (TKRPRs) belong to the class A (rhodopsin-like) GPCR family. Salivary TKs do not respond to endogenous receptors, but activate several vertebrate TK-Rs.

Signal Transduction Pathway

All TKRPRs and Ci-TK-R, expressed in *Xenopus* oocytes or cultured cells, trigger intracellular calcium mobilization in response to the cognate TKRPs and Ci-TK, respectively [2−5,8].

Biological Functions

Target Cells/Tissues and Functions

Ci-TK-R is expressed in the brain, intestine, endostyle, and gonad. TKRPR is distributed to the central nervous system, digestive tract, smooth muscle, and reproductive organs [2−4,6,8,9]. Ci-TK enhances vitellogenic follicle growth [8,9], and elicits contraction of the guinea pig ileum comparable to vertebrate SP [6]. Depolarization or hyperpolarization of various neurons is induced by TKRPs in insect neural ganglia [2−4]. Most TKRPs, like vertebrate TKs, stimulate spontaneous contraction of the insect foregut, midgut, and/or hindgut, probably via secretion of an insect neuropeptide, proctolin [2−5,10]. Several TKRPs also induce contraction of the oviduct [2,3,10]. RNAi against mRNA encoding *Drosophila* TKRP, DTK, causes almost 100% lethality of the cognate embryos [10]. Similarly, DTK-RNAi disturbs sensitivity of the *Drosophila* larva to odorants [10]. Lom-TKs are involved in the release of other peptide hormones such as adipokinetic hormones [2,3]. Natalisin-RNAi reduces mating behavior of male and female *D. melanogaster* [7]. Fecundity is also suppressed by natalisin knockdown in *T. castaneum* [7].

Y. Takei, H. Ando, & K. Tsutsui (Eds): Handbook of Hormones. DOI: http://dx.doi.org/10.1016/B978-0-12-801028-0.00044-1

References

1. Erspamer V, Falconieri Elspamer G. Pharmacological actions of eledoisin on extravascular smooth muscle. *Br J Pharmacol.* 1962;19:337–354.
2. Satake H, Kawada T, Nomoto K, et al. Insight into tachykinin-related peptides, their receptors, and invertebrate tachykinins: a review. *Zool Sci.* 2003;20:533–549.
3. Satake H, Kawada T. Overview of the primary structure, tissue-distribution, and functions of tachykinins and their receptors. *Curr Drug Target.* 2006;7:963–974.
4. Satake H, Aoyama M, Sekiguchi T, et al. Insight into molecular and functional diversity of tachykinins and their receptors. *Prot Pept Lett.* 2013;20:615–627.
5. Schoofs L, Holman GM, Hayes TK, et al. Locustatachykinin I and II, two novel insect neuropeptides with homology to peptides of the vertebrate tachykinin family. *FEBS Lett.* 1990;261:397–401.
6. Satake H, Ogasawara M, Kawada T, et al. Tachykinin and tachykinin receptor of an ascidian, *Ciona intestinalis*: evolutionary origin of the vertebrate tachykinin family. *J Biol Chem.* 2004;279:53798–53805.
7. Jiang H, Lkhagva A, Daubnerová I, et al. Natalisin, a tachykinin-like signaling system, regulates sexual activity and fecundity in insects. *Proc Natl Acad Sci USA.* 2013;110:E3526–E3534.
8. Aoyama M, Kawada T, Fujie M, et al. A novel biological role of tachykinins as an up-regulator of oocyte growth: identification of an evolutionary origin of tachykininergic functions in the ovary of the ascidian, *Ciona intestinalis. Endocrinology.* 2008;149: 4346–4356.
9. Aoyama M, Kawada T, Satake H. Localization and enzymatic activity profiles of the proteases responsible for tachykinin-directed oocyte growth in the protochordate, *Ciona intestinalis. Peptides.* 2012;34:186–192.
10. Van Loy T, Vandersmissen HP, Poels J, et al. Tachykinin-related peptides and their receptors in invertebrates: a current view. *Peptides.* 2010;31:520–524.

Supplemental Information Available on Companion Website

- Primary sequences of typical TKs and TK-like peptides /E-Table 44.1
- Accession numbers for cDNAs encoding representative invertebrate TKs and TKRPs/E-Table 44.2
- Schematic structure of representative TK precursors/E-Figure 44.1

Protochordate Tachykinin

Honoo Satake

Abbreviation: Ci-TK

A tachykinin identified in an ascidian, Ciona intestinalis, *Ci-TK is the first authentic tachykinin identified in invertebrates.*

Discovery

In 2004, a combination of sequence homology search and mass spectrometric analyses detected Ci-TK-I and -II as the first authentic TKs in invertebrates in the central nervous system of an ascidian, *Ciona intestinalis* [1]. An N-terminally extended form of Ci-TK-II was also detected by a peptidomic analysis of the central nervous system in 2011 [2].

Structure

Structural Features and Primary Structure

Ci-TK-I and -II consist of nine and seven aa residues, respectively, and share the chordate TK consensus sequences of Phe−Xaa−Gly−Leu−Met−NH_2 at the C-terminus [1,3,4]. N-terminally extended Ci-TK-II contains 19 aa residues. Primary sequences of Ci-TKs and mammalian TKs are shown in Table 44.1.

Properties

Mr 725 (Ci-TK-II), 1,159 (Ci-TK-I), and 2,085 (N-terminally extended Ci-TK-II). Soluble in water, physiological saline solution, aqueous acetonitrile, methanol.

Synthesis and Release

Gene, mRNA, and Precursor

A Ci-TK precursor encodes a single sequence of Ci-TK-I and -II flanked by a glycine C-terminal amidation signal (E-Figure 44A.1) [1]. The Ci-TK-I sequence possesses dibasic endoproteolytic sites at both termini, whereas mono Arg and the endoproteolytic site are located at the N- and C-termini of the Ci-TK-II sequence, respectively [1]. Unlike vertebrate TAC1 precursors [3−5], the Ci-TK gene yields no splicing variants [1].

Table 44A.1 Primary Sequences of Ci-TK-I, Ci-TK-II, and N-terminally extended Ci-TK-II*

Ci-TK-I	HVRHFYGLM-NH_2
Ci-TK-II	ASFTGLM-NH_2
Ci-TK-II (N-term. extended)	SIGDQPSIFNERASFTGLM-NH_2

*TK consensus sequence is shaded.

Distribution of mRNA

Ci-TK mRNA is localized in neurons of the brain ganglion, unidentified small cells in the intestine, and endocrine cell-like cells in the endostyle [1].

Receptors

Structure and Subtype

The Ci-TK receptor, Ci-TK-R, was cloned from the CNS as the sole TK receptor in *C. intestinalis* [1] and, like vertebrate TK receptors [3−5], belongs to the class A GPCR family [1,4,5].

Signal Transduction Pathway

Ci-TK-R, expressed in *Xenopus* oocytes or cultured cells, triggers intracellular calcium mobilization in response to Ci-TKs and several mammalian TKs such as substance P and neurokinin A, but not to neurokinin B or invertebrate TK-related peptides [1,3−5]. Moreover, Ci-TK-R is also activated by an NK1 (mammalian receptor selective to substance P)-specific agonist [Sar^9, Met(O$_2$)^{11}]SP, whereas an NK2 (mammalian receptor selective to neurokinin B)-specific antagonist GR94800 blocks activation of Ci-TK-R by Ci-TK-I [6].

Biological Functions

Target Cells/Tissues and Functions

Ci-TK-R is distributed in the brain ganglion, intestine, endostyle, and ovary. In the ovary, Ci-TK-R is expressed exclusively in test cells (putative counterparts to vertebrate ovarian granulosa cells) of vitellogenic follicles [6,7]. Ci-TK-I specifically enhances growth of *C. intestinalis* vitellogenic follicles to post-vitellogenic ones via upregulation of gene expression and enzymatic activities of cathepsin D, carboxypeptidase B1, and chymotrypsin, which participate in proteolytic processing of follicular component proteins [6,7]. [Sar^9, Met(O$_2$)^{11}] SP also elicits follicle growth activity, while GR94800 antagonizes it [6]. Moreover, cathepsin D is directly upregulated by Ci-TK-I in test cells, whereas upregulation of the other proteases is a secondary action in inner follicular cells [7]. In addition, Ci-TK-I-triggered follicle growth is arrested by a *C. intestinalis* neurotensin-like peptide, Ci-NTLP6, via downregulation of gene expression of cathepsin D, carboxypeptidase B1, and chymotrypsin [6]. Ci-TK-I also elicits TK-typical contraction of the guinea pig ileum comparable to vertebrate substance P [1].

Y. Takei, H. Ando, & K. Tsutsui (Eds): Handbook of Hormones. DOI: http://dx.doi.org/10.1016/B978-0-12-801028-0.00205-1

References

1. Satake H, Ogasawara M, Kawada T, et al. Tachykinin and tachykinin receptor of an ascidian, *Ciona intestinalis*: evolutionary origin of the vertebrate tachykinin family. *J Biol Chem.* 2004;279:53798−53805.
2. Kawada T, Ogasawara M, Sekiguchi T, et al. Peptidomic analysis of the central nervous system of the protochordate, *Ciona intestinalis*: homologs and prototypes of vertebrate peptides and novel peptides. *Endocrinology.* 2011;152:2416−2427.
3. Satake H, Kawada T, Nomoto K, et al. Insight into tachykinin-related peptides, their receptors, and invertebrate tachykinins: a review. *Zool Sci.* 2003;20:533−549.
4. Satake H, Kawada T. Overview of the primary structure, tissue-distribution, and functions of tachykinins and their receptors. *Curr Drug Target.* 2006;7:963−974.
5. Satake H, Aoyama M, Sekiguchi T, et al. Insight into molecular and functional diversity of tachykinins and their receptors. *Prot Pept Lett.* 2013;20:615−627.
6. Aoyama M, Kawada T, Fujie M, et al. A novel biological role of tachykinins as an up-regulator of oocyte growth: identification of an evolutionary origin of tachykininergic functions in the ovary of the ascidian, *Ciona intestinalis. Endocrinology.* 2008;149:4346−4356.
7. Aoyama M, Kawada T, Satake H. Localization and enzymatic activity profiles of the proteases responsible for tachykinin-directed oocyte growth in the protochordate, *Ciona intestinalis. Peptides.* 2012;34:186−192.

Supplemental Information Available on Companion Website

- Primary structure of a Ci-TK precursor/E-Figure 44A.1

Tachykinin-Related Peptides

Honoo Satake

Abbreviations: TKRP, TRP
Additional names: insect tachykinin

Neuropeptides structurally related to chordate tachykinins (TKs) in protostomes, TKRPs possess diverse primary sequences and structural organization of the precursors among species.

Discovery

In 1990, Lom-TKs were isolated as the first invertebrate tachykinin (TK)-like peptides from the central nervous system of the locust *Locusta migratoria*, and were shown to elicit contractile activity of the gut tissues [1]. Successively, their structurally related peptides have been identified in a wide range of protostomes, including insects, annelids, mollusks, and sea anemone [2–4]. These protostome TK-like peptides have been designated TKRPs or TRPs.

Structure

Structural Features and Primary Structure

All TKRPs consist of 7–12 aa residues, and contain the consensus sequences of Phe–Xaa^1–(Gly/Ala)–Xaa^2–Arg–NH$_2$ at the C-terminus (Table 44B.1) [2–4]. Novel TKRP-like peptides, natalisins, identified in holometabolous insect species including *Drosophila melanogaster*, *Tribolium castaneum*, and *Bombyx mori*, bear the analogous sequence Phe–Xaa^1–Xaa^2–Xaa^3–Arg–NH$_2$ (Table 44B.1) [5].

Properties

Mr 800–1,200. Soluble in water, physiological saline solution, aqueous acetonitrile, methanol.

Synthesis and Release

Gene, mRNA, and Precursor

The first cDNA encoding TKRPs was cloned from the echiuroid worm *Urechis unicinctus* [2–6]. In general, TKRP precursors encode multiple (5–16) TKRP sequences flanked by a C-terminal amidation signal at the C-terminus and mono- or dibasic endoproteolytic sites (E-Figure 44B.1) [2–6]. The number of TKRP sequences in the precursors and the exon–intron organization differ among species [2–6].

Distribution of mRNA

All TKRP mRNAs are expressed in neurons of the CNS [2–4]. In several insects, TKRP mRNA is detected in the digestive tissue [2–4].

Receptors

Structure and Subtype

TKRP receptors (TKRPRs) have been identified in an echiuroid worm, octopus, and insects [2–5,7,8]. All TKRPRs belong to the class A (rhodopsin-like) GPCR family. In contrast to multiple TKRPs, only a single TKRPR is present in each organism [2–5,7,8]. The first TKRPR was cloned from *D. melanogaster*, and the cognate TKRP and receptor interaction was revealed in 2003 [2–4,7,8].

Signal Transduction Pathway

All TKRPRs, expressed in *Xenopus* oocytes or cultured cells, trigger intracellular calcium mobilization in response to the cognate TKRPs [2–5,8]. *U. unicinctus* TKRPR, UTKRPR, and *D. melanogaster* TKRPR, DTKR, exhibit redundant responses to multiple cognate TKRPs, whereas octopus TKRPR, oct-TKRPR, and natalasin receptors, including NKD (also known as tachykinin receptor 86C in *D. melanogaster*), possess modest and prominent ligand preferences, respectively [2–5,8–10]. Vertebrate TKs fail to activate any TKRPRs, and *vice versa* [2–4,8–10]. Nonetheless, simple replacement of C-terminal Met and Arg with one another in TKs and TKRPs results in acquisition of opposite ligand selectivity [2–5,8–10].

Biological Functions

Target Cells/Tissues and Functions

TKRPR is distributed in the CNS, digestive tracts, smooth muscle, and reproductive organs [2–4,8–10]. Depolarization or hyperpolarization of various neurons is induced by TKRPs in insect neural ganglia [2–4]. Most TKRPs, like vertebrate TKs, stimulate spontaneous contraction of the insect foregut, midgut, and/or hindgut, probably via secretion of an insect neuropeptide, proctolin [2–4,10]. Several TKRPs also induce contraction of the oviduct [2,3,10]. RNAi against mRNA encoding *Drosophila* TKRP, DTK, causes almost 100% lethality of the cognate embryos [10]. Similarly, DTK-RNAi disturbs sensitivity of the *Drosophila* larva to odorants [10]. Lom-TKs are involved in the release of other peptide hormones, such as adipokinetic hormones [2,3]. Natalisin-RNAi reduces mating behavior of male and female *D. melanogaster* [5]. Fecundity is also suppressed by natalisin knockdown in *T. castaneum* [5].

Y. Takei, H. Ando, & K. Tsutsui (Eds): Handbook of Hormones. DOI: http://dx.doi.org/10.1016/B978-0-12-801028-0.00206-3

Table 44B.1 Primary Sequences of Typical Tachykinins and Tachykinin-Like Peptides

Species	Peptide	Sequence
Chordate tachykinins		
Mammals	Substance P	RPKPQQFFGLM-NH$_2$
	Neurokinin A	HKTDSFVGLM-NH$_2$
	Neurokinin B	DMHDFVGLM-NH$_2$
Rat and mouse	Hemokinin-1	SRTRQFYGLM-NH$_2$
Ascidian (*Ciona intestinalis*)	Ci-TK-I	HVRHFYGLM-NH$_2$
	Ci-TK-II	SIGDQPSIFNERASFTHVRHFYGLM-NH$_2$
Protostome tachykinin-related peptides		
Locust (*Locusta migratoria*)	Lom-TK-I	GPSGFYGVR-NH$_2$
	Lom-TK-II	APLSGFYGVR-NH$_2$
Cockroach (*Leucophaea maderae*)	Lem-TRP-1	APSGFLGYR-NH$_2$
	Lem-TRP-3	NGERAPGSKKAPSGFLGTR-NH$_2$
Fruit fly (*Drosophila melanogaster*)	DTK-1	APTSSFIGMR-NH$_2$
	DTK-2	APLAFVGLR-NH$_2$
Echiuroid worm (*Urechis unicinctus*)	Uru-TK I	LRQSQFVGAR-NH$_2$
	Uru-TK II	AAGMGFFGAR-NH$_2$
Octopus (*Octopus vulgaris*)	Oct-TKRP I	VNPYSFQGTR-NH2
Shrimp (*Penaeus vannamei*)	Pev-tachykinin	APSGFLGMR-NH$_2$
Fruit fly (*Drosophila melanogaster*)	Natalisin-1	EKLFDGYQFGEDMSKENDPFIPPR-NH$_2$
Cnidarian tachykinin-related peptides		
Sea anemone (*Nematostella vetensis*)	Nv-TK I	YQVIFEGVR-NH$_2$
	Nv-TK II	TLQVGRR-NH$_2$

pE, pyro-Glu amidation; a, C-terminal amidation.

References

1. Schoofs L, Holman GM, Hayes TK, et al. Locustatachykinin I and II, two novel insect neuropeptides with homology to peptides of the vertebrate tachykinin family. *FEBS Lett*. 1990;261:397−401.
2. Satake H, Kawada T, Nomoto K, et al. Insight into tachykinin-related peptides, their receptors, and invertebrate tachykinins: a review. *Zool Sci*. 2003;20:533−549.
3. Satake H, Kawada T. Overview of the primary structure, tissue-distribution, and functions of tachykinins and their receptors. *Curr Drug Target*. 2006;7:963−974.
4. Satake H, Aoyama M, Sekiguchi T, et al. Insight into molecular and functional diversity of tachykinins and their receptors. *Prot Pept Lett*. 2013;20:615−627.
5. Jiang H, Lkhagva A, Daubnerová I, et al. Natalisin, a tachykinin-like signaling system, regulates sexual activity and fecundity in insects. *Proc Natl Acad Sci USA*. 2013;110:E3526−E3534.
6. Kawada T, Satake H, Minakata H, et al. Characterization of a novel cDNA sequence encoding invertebrate tachykinin-related peptides isolated from the echiuroid worm, *Urechis unicinctus*. *Biochem Biophys Res Commun*. 1999;261:848−852.
7. Li XJ, Wolfgang W, Wu YN, et al. Cloning, heterologous expression and developmental regulation of a *Drosophila* receptor for tachykinin-like peptides. *EMBO J*. 1991;10:3221−3229.
8. Kawada T, Furukawa Y, Shimizu Y, et al. A novel tachykinin-related peptide receptor. Sequence, genomic organization, and functional analysis. *Eur J Biochem*. 2002;269:4238−4246.
9. Kanda A, Takuwa-Kuroda K, Aoyama M, et al. A novel tachykinin-related peptide receptor of *Octopus vulgaris* — evolutionary aspects of invertebrate tachykinin and tachykinin-related peptide. *FEBS J*. 2007;274:2229−2239.
10. Van Loy T, Vandersmissen HP, Poels J, et al. Tachykinin-related peptides and their receptors in invertebrates: a current view. *Peptides*. 2010;31:520−524.

Supplemental Information Available on Companion Website

• Primary structure of TKRP precursors/E-Figure 44B.1

CHAPTER 45

Insulin Superfamily

Naoki Okamoto

History

The existence of hypoglycemic hormones in invertebrates was first demonstrated in the honeybee in 1964, and in several other species in the 1970s and 1980s. The first insulin-like peptide identified in invertebrates was bombyxin (initially called 4K-prothoracicotropic hormone), which was purified from the silkmoth *Bombyx mori* in 1984 [1]. In 1988, the cDNA coding for molluskan insulin-related peptide (MIP) was cloned and the sequence of MIP precursor determined in the pond snail *Lymnaea stagnalis* [2]. In 1999, androgenic gland hormone (AGH), which is essential for male sexual differentiation in malacostracan crustaceans, was isolated and characterized from the androgenic glands of the male isopod *Armadillidium vulgare* [3]. The precursor of AGH is structurally related to that of insulin superfamily members. In 2009, the first insulin-like growth factor (IGF)-like peptide and relaxin-like peptide in invertebrates were isolated from the hemolymph of *B. mori* pupae and from the radial nerve of the starfish *Asterina pectinifera*, respectively [1,4]. The latter is called a gonad-stimulating substance (GSS). Recent application of genomic and EST sequencing analyses has identified genes for the insulin superfamily members in an increasingly wide variety of invertebrate phyla, including insects (E-Figure 45.1, E-Table 45.1) [5].

Structure

Structural Features

The common structural feature of invertebrate insulin superfamily peptides is a conserved domain organization of their precursor consisting of a signal peptide, B-chain, C-peptide, and A-chain, as in the vertebrate insulin superfamily members. After cleavage of a signal peptide, the C-peptide is removed to generate a mature heterodimeric peptide consisting of an A- and a B-chain (Figure 45.1), except that IGF-like peptides retain the C-peptide [1]. The B- and A-chains typically contain two and four conserved Cys residues, respectively, which are linked by two inter- and one intra-chain disulfide bonds [6]. However, atypical domain organization and disulfide bond patterns have been identified in some peptides from mollusks, crustaceans, and the nematode *Caenorhabditis elegans* [2,7].

Molecular Evolution of Family Members

Most invertebrate genomes encode multiple insulin superfamily peptides, ranging from only one or two in the Hymenoptera genome to more than 40 in *C. elegans* and *B. mori* [1,5,8]. It is difficult to achieve a phylogenetic analysis

of the invertebrate insulin superfamily based on sequence homology because of a short length of the peptides and of highly diverged amino acid sequences except for some critical residues necessary for appropriate processing and tertiary structure formation [5]. In the course of evolution, ancestral molecules of the insulin superfamily may have been duplicated and diverged within each species.

Receptors

Structure and Subtypes

In vertebrates, insulin and IGFs activate insulin receptor and IGF type-I receptor, respectively, both of which are RTKs. By contrast, another class of insulin superfamily peptides, relaxins or relaxin-like peptides, activate leucine-rich repeat-containing GPCRs. An insulin-like receptor (InR) has been identified in several invertebrate species, including insects (E-Table 45.2). All such InRs show high similarity with mammalian insulin and IGF type-I receptors. Although multiple insulin superfamily peptides exist in each invertebrate species, its genome typically encodes only one InR [5]. Using the fruit fly *Drosophila melanogaster* and *C. elegans*, genetic interactions between insulin/IGF-like peptides and InR have been extensively demonstrated, although there is little evidence that invertebrate insulin/IGF-like peptides directly bind to InR. A relaxin-like peptide (also known as GSS) in *A. pectinifera* has been suggested to interact with a GPCR [4].

Signal Transduction Pathway

The insulin/IGF signaling (IIS) pathway is well conserved between invertebrates and mammals. IIS pathway components have been identified and characterized in *D. melanogaster* and *C. elegans* (E-Figure 45.2) [8,9].

Biological Functions

Target Cells/Tissues and Functions

In invertebrates, insulin superfamily peptides are involved in multiple biological functions, including growth, metabolism, reproduction (including sexual differentiation), immunity, behavior, stress resistance, diapause, and lifespan [5,6]. Physiological studies in moths and genetic studies in *D. melanogaster* have revealed their roles in the regulation of growth and metabolism (see Subchapters 42A, 42B) [1,5,9,10]. In blood-feeding mosquitoes, the mechanisms by which insulin-like peptides regulate egg production via ovarian steroidogenesis are well demonstrated [5]. Their functions in reproduction are also reported in crustaceans (AGH) and in

Y. Takei, H. Ando, & K. Tsutsui (Eds): Handbook of Hormones. DOI: http://dx.doi.org/10.1016/B978-0-12-801028-0.00045-3

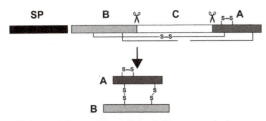

Figure 45.1 A schematic model of the proteolytic conversion of insulin superfamily peptides. Black oblong, signal peptide (SP); light gray oblong, B-chain (B); white oblong, C-peptide (C); dark gray oblong, A-chain (A). Disulfide bonds (S–S) are indicated by lines.

the starfish (GSS) [3,4]. Major progress on the functions of the IIS pathway in regulating lifespan has been made through numerous studies on long-lived mutants of *C. elegans* during the past 20 years. In *C. elegans*, genetic manipulation of the IIS pathway can more than double the lifespan of adults [8]. Extension of lifespan by manipulation of the IIS pathway is observed in *D. melanogaster* and also the mouse.

References

1. Mizoguchi A, Okamoto N. Insulin-like and IGF-like peptides in the silkmoth *Bombyx mori*: discovery, structure, secretion, and function. *Front Physiol*. 2013;4:217.
2. Smit AB, Van Kesteren RE, Li KW, et al. Towards understanding the role of insulin in the brain: lessons from insulin-related signaling systems in the invertebrate brain. *Prog Neurobiol*. 1998;54:35–54.
3. Okuno A, Hasegawa Y, Ohira T, et al. Characterization and cDNA cloning of androgenic gland hormone of the terrestrial isopod *Armadillidium vulgare*. *Biochem Biophys Res Comm*. 1999;264:419–423.
4. Mita M. Relaxin-like gonad-stimulating substance in an echinoderm, the starfish: a novel relaxin system in reproduction of invertebrates. *Gen Comp Endocrinol*. 2013;181:241–245.
5. Antonova Y, Arik AJ, Moore W, et al. Insulin-like peptides: Structure, signaling, and function. In: Gilbert LI, ed. *Insect Endocrinology*. NY: Elsevier/Academic Press; 2012:63–92.
6. Claeys I, Simonet G, Poels J, et al. Insulin-related peptides and their conserved signal transduction pathway. *Peptides*. 2002;23:807–816.
7. Pierce SB, Costa M, Wisotzkey R, et al. Regulation of DAF-2 receptor signaling by human insulin and *ins-1*, a member of the unusually large and diverse *C. elegans* insulin gene family. *Genes Dev*. 2001;15:672–686.
8. Murphy C.T., Hu P.J. Insulin/insulin-like growth factor signaling in *C. elegans*. In: The *C. elegans* Research Community, ed. *WormBook*. 2013:1–43.
9. Teleman AA. Molecular mechanisms of metabolic regulation by insulin in *Drosophila. Biochem J*. 2010;425:13–26.
10. Géminard C, Arquier N, Layalle S, et al. Control of metabolism and growth through insulin-like peptides in *Drosophila. Diabetes*. 2006;55(Suppl. 2):S5–S8.
11. Brogiolo W, Stocker H, Ikeya T, et al. An evolutionarily conserved function of the *Drosophila* insulin receptor and insulin-like peptides in growth control. *Curr Biol*. 2001;11:213–221.
12. Grönke S, Clarke D-F, Broughton S, et al. Molecular evolution and functional characterization of *Drosophila* insulin-like peptides. *PLoS Genet*. 2010;6:e1000857.
13. Li B, Predel R, Neupert S, et al. Genomics, transcriptomics, and peptidomics of neuropeptides and protein hormones in the red flour beetle *Tribolium castaneum. Genome Res*. 2008;8: 113–122.
14. Grönke S, Partridge L. The functions of insulin-like peptides in insects. In: Clemmons DR, Robinson CAF, Christen Y, eds. *IGFs: Local Repair and Survival Factors Throughout Life Span*. Heidelberg: Springer; 2010:105–124.
15. Garelli A, Gontijo AM, Miguela V, et al. Imaginal discs secrete insulin-like peptide 8 to mediate plasticity of growth and maturation. *Science*. 2012;336:579–582.
16. Colombani J, Anderson DS, Léopold P. Secreted peptide Dilp8 coordinates *Drosophila* tissue growth with developmental timing. *Science*. 2012;336:582–585.
17. Veenstra JA. Neurohormones and neuropeptides encoded by the genome of *Lottia gigantea*, with reference to other mollusks and insects. *Gen Comp Endocrinol*. 2010;167:86–103.

Supplemental Information Available on Companion Website

- Insulin superfamily in insects/Text & E-Figure 45.1
- Insulin/IGF signaling (IIS) pathway in invertebrates/Text & E-Figure 45.2
- Accession numbers of genes and cDNAs for the insulin superfamily and insulin-like receptors in insects/E-Tables 45.1, 45.2

Insect Insulin-Like Peptides

Naoki Okamoto

Abbreviation: insect ILP
Additional names: insect insulin-related peptide (IRP)

A brain-derived insulin family peptides in insects, insect ILP is released in response to nutritional signals and controls cellular and organismal growth, metabolism, reproduction, and lifespan.

Discovery

The existence of hypoglycemic hormones and of insulin-immunoreactive materials in insects was demonstrated in the 1960s–1980s. The first insulin-like peptide (ILP) identified in insects (and in invertebrates) was bombyxin (Figure 45A.1), which was isolated from 678,000 adult heads of the silkmoth *Bombyx mori* and partially sequenced in 1984 [1], followed by determination of the complete amino acid sequence in 1986 [1].

Structure

Structural Features

Bombyxin is a heterodimeric peptide consisting of an A- and a B-chain, like vertebrate insulin. Two disulfide bonds between the two chains and one disulfide bond within the A-chain are formed in the same fashion as in insulin (Figure 45A.1). The structure of the mature forms of most other insect ILPs, which are all deduced from cDNA, is predicted to be the same as that of bombyxin and insulin, because they have Cys residues at conserved positions in the predicted A- and B-chains and because both chains are flanked by dibasic cleavage sites.

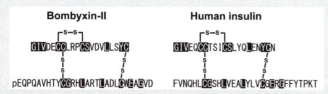

Figure 45A.1 Amino acid sequences of bombyxin II and human insulin. The identical residues between the two peptides are boxed. Disulfide bonds (S−S) are indicated by lines and pyroglutamic acid is indicated by pE.

Primary Structure

The amino acid sequences of insect ILPs are highly diverged between insect orders, except for some critical residues (such as cysteine) required for tertiary structure formation.

Synthesis and Release

Gene, mRNA, and Precursor

Most insect genomes encode multiple ILPs, ranging from only one or two in the Hymenoptera genome to more than 40 in the silkmoth *B. mori* [2]. ILP mRNAs encode preproILPs consisting of the signal peptide and three domains corresponding to the B-chain, C-peptide, and A-chain of insulin precursor (proinsulin).

Distribution of mRNA

In insects, *ILP* genes are mainly expressed in several pairs of medial neurosecretory cells (MNCs) in the brain (Figure 45A.2) [2]. In *B. mori*, bombyxin genes are all expressed in four pairs of MNCs in the brain [1]. In the fruit fly *Drosophila melanogaster*, *Drosophila* ILP (DILP)2, -3, and -5 genes are expressed in seven pairs of MNCs in the brain [3,4]. Some *ILP* transcripts are also found in cells other than the brain MNCs.

Plasma Concentrations

In *B. mori*, the bombyxin titer is of the order of 100 fmol/ml [1].

Regulation of Synthesis and Release

In *B. mori*, bombyxins are axonally transported to and released from the corpora allata (CA), a pair of neurohemal organs associated with the brain [1]. Similarly to insulin release in mammals, glucose stimulates the secretion of bombyxin into hemolymph (Figure 45A.2) [1]. In *D. melanogaster*, although *DILP2*, -3, and -5 are co-expressed in the MNCs, transcription of each *DILP* is differentially regulated [4]. *DILP3* and -5 are expressed in a nutrient-dependent manner, but *DILP2* is not [4]. MNC-derived DILPs are axonally transported to the corpora cardiaca, a pair of neurohemal organs located on the ring gland. The *Drosophila* MNCs also extend processes to the dorsal vessel (insect heart), allowing for the direct release of DILPs into the bloodstream [5]. DILP release from the MNCs depends on the nutritional conditions. The availability of nutrients is remotely sensed by the fat body, which in turn regulates DILP secretion through a humoral signal(s) (Figure 45A.2) [6]. In *D. melanogaster*, the main nutritional signal in the diet is amino acids [6].

Y. Takei, H. Ando, & K. Tsutsui (Eds): Handbook of Hormones. DOI: http://dx.doi.org/10.1016/B978-0-12-801028-0.00207-5

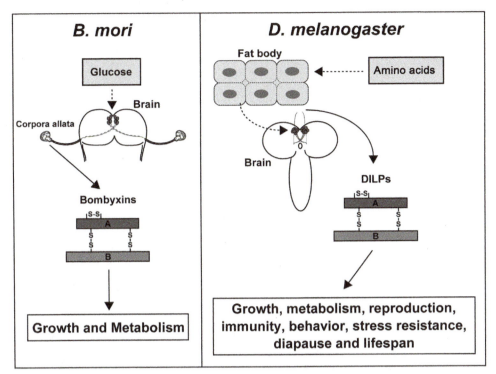

Figure 45A.2 Schematic model of the structural feature, sites of production, secretion control, and actions of (left) bombyxin in *B. mori* and (right) of DILPs in *D. melanogaster*.

Receptors

Structure and Subtype

Insulin-like receptors (InRs) have been identified in several insect species [2]. *Drosophila* InR has features characteristic of the insulin receptor and IGF type-I receptor, possessing a Cys-rich region, a potential dibasic cleavage site, a transmembrane domain, and a TK domain, although it contains a 400 additional amino acid sequence at its C-terminus [2]. *Drosophila* InR can be activated by human insulin [2]. However, there is little evidence that insect ILPs bind to InR, although it was reported that in the yellow fever mosquito *Aedes aegypti*, ILP3 directly binds to *A. aegypti* InR with high affinity [2]. Recently, a secreted decoy of InR (SDR) that is structurally similar to the extracellular domain of InR was identified in *D. melanogaster* [7]. SDR can directly interact with several DILPs.

Signal Transduction Pathway

Insect ILPs mainly activate the PI3K/AKT signaling pathway — called the insulin/IGF signaling (IIS) pathway (see Chapter 45). Recent analyses show that IIS is associated with target of rapamycin (TOR) signaling [2].

Biological Functions

Target Cells/Tissues and Functions

Pleiotropic functions of ILPs have been reported in several insect species [2]. In *D. melanogaster*, DILPs and downstream IIS components are involved in multiple biological processes, including growth, metabolism, reproduction, immunity, behavior, stress resistance, diapause, and lifespan (Figure 45A.2) [2]. In Lepidoptera, bombyxin regulates imaginal disc growth and carbohydrate metabolism (Figure 45A.2) [1]. In the beetle *T. castaneum*, ILP2 is responsible for carbohydrate metabolism, stress resistance, and lifespan [8]. In the mosquito

A. aegypti, ILPs stimulate ecdysteroid biosyn thesis in the ovaries and regulate carbohydrate and lipid metabolism [2]. In another mosquito, *Culex pipiens*, ILPs are involved in the regulation of diapause induction [2].

Phenotype in Gene-Modified Animals

Genetic approaches to analyze ILP functions have been used extensively in *D. melanogaster*. When the DILP-producing MNCs in the brain are genetically ablated, larval growth is severely retarded and adult size is significantly reduced [5]. Similar phenotypes are also observed in the mutation combinations of *DILP1−5*, in which all MNC-derived *DILPs* are mutated [9]. In contrast, when individual *DILPs* are overexpressed during larval development, a proportionate increase in body size is observed in adult flies, with *DILP2* showing the highest potency for growth promotion [4]. The *DILP*-overexpressing flies exhibit larger wing size due to an increase in both cell size and cell number. Genetic ablation of MNCs or combined *DILP* gene knockout also affects carbohydrate and lipid metabolism, fecundity, stress resistance, and lifespan [9,10].

References

1. Mizoguchi A, Okamoto N. Insulin-like and IGF-like peptides in the silkmoth *Bombyx mori*: discovery, structure, secretion, and function. *Front Physiol*. 2013;4:217.
2. Antonova Y, Arik AJ, Moore W, et al. Insulin-like peptides: structure, signaling, and function. In: Gilbert LI, ed. *Insect Endocrinology*. NY: Elsevier/Academic Press; 2012:63−92.
3. Brogiolo W, Stocker H, Ikeya T, et al. An evolutionarily conserved function of the *Drosophila* insulin receptor and insulin-like peptides in growth control. *Curr Biol*. 2001;11:213−221.
4. Ikeya T, Galic M, Belawat P, et al. Nutrient-dependent expression of insulin-like peptides from neuroendocrine cells in the CNS contributes to growth regulation in *Drosophila*. *Curr Biol*. 2002;12:1293−1300.

5. Rulifson EJ, Kim SK, Nusse R. Ablation of insulin-producing neurons in flies: growth and diabetic phenotypes. *Science*. 2002;296:1118–1120.

6. Géminard C, Rulifson EJ, Léopold P. Remote control of insulin secretion by fat cells in *Drosophila*. *Cell Metab*. 2009;10: 199–207.

7. Okamoto N, Nakamori R, Murai T, et al. A secreted decoy of InR antagonizes insulin/IGF signaling to restrict body growth in *Drosophila*. *Genes Dev*. 2013;27:87–97.

8. Xu J, Sheng Z, Palli SR. Juvenile hormone and insulin regulate trehalose homeostasis in the red flour beetle, *Tribolium castaneum*. *PLoS Genet*. 2013;9:e1003535.

9. Grönke S, Clarke D-F, Broughton S, et al. Molecular evolution and functional characterization of *Drosophila* insulin-like peptides. *PLoS Genet*. 2010;6:e1000857.

10. Broughton SJ, Piper MWD, Ikeya T, et al. Longer lifespan, altered metabolism, and stress resistance in *Drosophila* from ablation of cells making insulin-like ligands. *Proc Natl Acad Sci USA*. 2005;22:3105–3110.

Supplemental Information Available on Companion Website

- See Chapter 45.

Insect Insulin-Like Growth Factor-Like Peptides

Naoki Okamoto

Abbreviation: insect IGFLP
Additional names: insect IGF-like peptide

IGFLPs are fat body-derived single-chain insulin family peptides in insects that act as a growth factor during the post-feeding metamorphic period.

Discovery

The presence of remarkably high concentrations of bombyxin-immunoreactive substance in the pupal hemolymph of the silkmoth *Bombyx mori* was reported in 1992 [1]. This substance (peptide) was isolated, sequenced, and characterized in 2009, revealing that this peptide is a member of the insulin family of peptides and is more similar to insulin-like growth factor (IGF) than to insulin in structural and functional features [2]. This IGF-like peptide is called *Bombyx* IGFLP or BIGFLP.

Structure

Structural Features

BIGFLP is a single-chain peptide consisting of the B, C, and A domains (Figure 45B.1), like proinsulin or IGFs, although the C domain (C-peptide) is flanked by dibasic amino acid motifs. One of the structural features of BIGFLP, as compared with bombyxins (insulin-like peptides [ILPs] in *B. mori*), is a relatively short C-peptide, the feature characteristic of vertebrate IGFs. Such insulin family peptides are found ubiquitously in insect genomes. Among them are DILP6 in the fruit fly *Drosophila melanogaster*, AmILP1 in the honey bee *Apis mellifera*, TcILP3 in the red flour beetle *Tribolium castaneum*, and AaegILP6 in the yellow fever mosquito *Aedes aegypti* [2,3]. The C-peptide of these ILPs lacks a dibasic cleavage site (s) at the boundary with the B domain and/or A domain, which is also a structural feature of vertebrate IGFs. Thus,

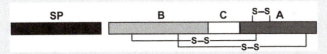

Figure 45B.1 Structure of insect IGF-like peptide precursor. Oblongs indicate the predicted domains: black, signal peptide (SP); light gray, B-chain (B); white, C-peptide (C); dark gray, A-chain (A). Predicted disulfide bonds (S—S) are indicated by lines.

these peptides are potential candidates for IGFLP in these insects. Based on the structure, production site, and physiological functions, as described below, DILP6 is considered to be *Drosophila* IGFLP.

Primary Structure

The amino acid sequences of IGF-like peptides are highly diverged between insect orders, except for some critical residues (such as cysteine) that are necessary for tertiary structure formation.

Synthesis and Release

Gene and mRNA

The *BIGFLP* gene is located on chromosome 1, consisting of a single exon [3]. *DILP6* is located on chromosome X (FlyBase), consisting of four exons. *BIGFLP* mRNA has 433 nucleotides (nt) [3]. *DILP6* mRNA has four isoforms — isoforms A (1,908 nt), B (634 nt), C (941 nt), and D (702 nt) — but the resulting peptide sequences are identical to each other (FlyBase).

Distribution of mRNA

In both *B. mori* and *D. melanogaster*, the genes encoding IGF-like peptides (*BIGFLP* and *DILP6*) are predominantly expressed in the fat body (Figure 45B.2), the functional equivalent of vertebrate liver, which secretes IGFs [2–5]. In *A. aegypti*, the candidate IGF-like peptide *AaegILP6* is highly expressed in the thoracic/abdominal regions including the fat body [6]. In *B. mori*, *BIGFLP* transcripts are also found in the brain medial neurosecretory cells (identical to the cells that produce bombyxins) and gonads (ovariole sheath and testis sheath) (Figure 45B.2) [3]. In the ovariole sheath, *BIGFLP* expression is especially high in the area where follicles are immature [3]. In *D. melanogaster*, *DILP6* transcript is found in a subset of glial cells in the larval central nervous system as well (Figure 45B.2) [7–9].

Plasma Concentration

In *B. mori*, the BIGFLP titer is very low during the larval stage but increases slightly at the pharate pupal stage (100 pmol/ml in females and 50 pmol/ml in males) and steeply after pupal ecdysis (800 pmol/ml in females and 200 pmol/ml in males) [2]. The concentration of BIGFLP in the hemolymph is remarkably high compared with that of bombyxin, which is of the order of 100 fmol/ml [3].

Y. Takei, H. Ando, & K. Tsutsui (Eds): Handbook of Hormones. DOI: http://dx.doi.org/10.1016/B978-0-12-801028-0.00208-7

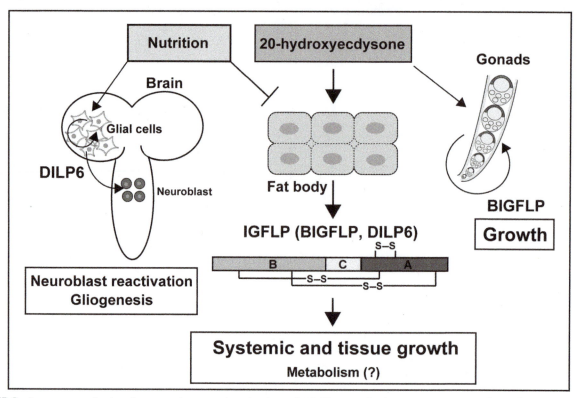

Figure 45B.2 Structure, production sites, secretion control, and actions of IGF-like peptides (BIGFLP in *B. mori* and DILP6 in *D. melanogaster*).

Regulation of Synthesis and Release

Expression of *BIGFLP* and *DILP6* in the fat body is stimulated by the steroid hormone 20-hydroxyecdysone (20E) (Figure 45B.2), producing a large peak during metamorphosis [2–5]. In *B. mori*, *BIGFLP* expression in the gonads is also stimulated by 20E (Figure 45B.2) [3]. However, *BIGFLP* expression in the brain is not induced by 20E [3]. In *D. melanogaster*, *DILP6* expression is also regulated by nutritional conditions (Figure 45B.2). *DILP6* expression in the fat body is upregulated by nutrient restriction through regulation of the transcription factor FoxO [5,10]. In contrast, glial *DILP6* expression is positively regulated by nutrition [7,8]. These observations indicate that the regulatory mechanisms of *IGFLP* expression differ among tissues.

Receptors

Structure and Subtype

IGF-like peptides are presumed to share the same insulin/IGF-like receptor (InR) with insulin-like peptides, because the genomes of most insects, including *B. mori* and *D. melanogaster*, contain only a single InR gene. However, direct binding of IGF-like peptides to InR has yet to be demonstrated.

Signal Transduction Pathway

IGF-like peptides are presumed to share the same insulin/IGF-signaling pathway with insulin-like peptides.

Biological Functions

Target Cells/Tissues and Functions

IGFLP functions as a systemic hormone to regulate tissue growth during the post-feeding metamorphic period

(Figure 45B.2) [2–5]. BIGFLP promotes the growth of adult tissues (genital discs, sperm ducts, flight muscle anlagen, and wing discs) but not of larval tissues (fat body, midgut, and epidermis, all of which are degenerated or reconstructed during metamorphosis) *in vitro*, suggesting that BIGFLP functions as a growth factor to regulate adult tissue development during metamorphosis [2]. In addition to the systemic function, IGFLP acts locally. The glia-derived *DILP6* regulates neuroblast reactivation and gliogenesis in a paracrine/autocrine manner (Figure 45B.2) [7–9]. The ovariole sheath-derived BIGFLP may regulate early follicular growth in a paracrine manner (Figure 45B.2) [3].

Phenotype in Gene-Modified Animals

DILP6 mutants show approximately 10% reduced final adult body size and weight, with a normal cell size but a reduced cell number [4,5]. This phenotype is rescued by the fat body-specific expression of *DILP6* during the post-feeding period [4]. Specific knockdown of *DILP6* in the fat body also causes a systemic growth defect similar to that observed in *DILP6* mutants, indicating that the most important tissue for DILP6 production is the fat body [5]. In contrast, when *DILP6* is overexpressed during the post-feeding period, the body weight of adult flies is increased [4,5]. Overexpression of *DILP6* specifically in cortex glia is sufficient to induce neuroblast reactivation during nutrient restriction [7,8]. *DILP6* mutants show significantly reduced gliogenesis [9]. In adults, overexpression of *DILP6* in the fat body systemically reduces insulin/IGF signaling through the non-autonomous repression of the secretion of DILP2 from the brain insulin-producing cells [10]. In *DILP6*-overexpressing flies, lifespan is extended and carbohydrate and fat storage are elevated.

References

1. Saegusa H, Mizoguchi A, Kitahora H, et al. Changes in the titer of bombyxin-immunoreactive material in hemolymph during the postembryonic development of the silkmoth *Bombyx mori*. *Dev Growth Differ*. 1992;34:595–605.
2. Okamoto N, Yamanaka N, Satake H, et al. An ecdysteroid-inducible insulin-like growth factor-like peptide regulates adult development of the silkmoth *Bombyx mori*. *FEBS J*. 2009;276:1221–1232.
3. Mizoguchi A, Okamoto N. Insulin-like and IGF-like peptides in the silkmoth *Bombyx mori*: discovery, structure, secretion, and function. *Front Physiol*. 2013;4:217.
4. Okamoto N, Yamanaka N, Yagi Y, et al. Fat body derived IGF-like peptide regulates post-feeding growth in *Drosophila*. *Dev Cell*. 2009;17:885–891.
5. Slaidina M, Delanoue R, Grönke S, et al. A *Drosophila* insulin-like peptide promotes growth during nonfeeding states. *Dev Cell*. 2009;17:874–884.
6. Riehle MA, Fan Y, Cao C, et al. Molecular characterization of insulin-like peptides in the yellow fever mosquito, *Aedes aegypti*: expression, cellular localization, and phylogeny. *Peptide*. 2006;27:2547–2560.
7. Chell JM, Brand AH. Nutrition-responsive glia control exit of neural stem cells from quiescence. *Cell*. 2010;143:1161–1173.
8. Sousa-Nunes R, Yee LL, Gould AP. Fat cells reactivate quiescent neuroblasts via TOR and glial insulin relays in *Drosophila*. *Nature*. 2011;471:508–512.
9. Avet-Rochex A, Kaul AK, Gatt AP, et al. Concerted control of gliogenesis by InR/TOR and FGF signalling in the *Drosophila* post-embryonic brain. *Development*. 2012;139:2763–2772.
10. Bai H, Kang P, Tatar M. *Drosophila* insulin-like peptide-6 (dilp6) expression from fat body extends lifespan and represses secretion of *Drosophila* insulin-like peptide-2 from the brain. *Aging Cell*. 2012;11:978–985.

Supplemental Information Available on Companion Website

- See Chapter 45.

Molluscan Insulin-Related Peptides

Naoki Okamoto

Abbreviation: MIPs

MIPs are neuroendocrine cell-derived insulin family peptides in mollusks that control growth, metabolism, reproduction, and memory consolidation.

Discovery

In the pond snail *Lymnaea stagnalis*, clusters of neuroendocrine cells called light green cells (LGCs) in the cerebral ganglia have been suggested to be involved in the control of a wide variety of physiological processes associated with animal growth. In 1988, the primary structure of molluskan insulin-related peptides (MIPs) was deduced from the nucleotide sequence of LGC-specific cDNA clones of *L. stagnalis*, and these peptides were suggested to be responsible for the growth-regulating roles of the LGCs [1].

Structure

Structural Features

MIPs are synthesized as prepropeptides consisting of a signal peptide, B-chain, C-peptide, and A-chain (Figure 45C.1) [1]. In some MIPs, a unique D-peptide exists in the C-terminal part of the propeptide [2,3]. It is presumed that a mature MIP peptide consisting of the A- and B-chains is formed in the same way as vertebrate insulin (Figure 45C.1) [4]. However, each of the two chains of MIPs has an additional Cys residue (Figure 45C.2), suggesting the existence of an extra disulfide bond between the A- and B-chains (Figure 45C.1). Five MIPs (MIP-I, II, III, V, and VII) have been identified and characterized in *L. stagnalis*, while only one MIP has been found in the California sea hare *Aplysia californica* and the Pacific oyster *Crassostrea gigas* (Figure 45C.2, E-Table 45C.1) [2,4,5]. More recently, using genomic or EST sequencing analyses, multiple MIPs (including some atypical MIPs) have been identified in several other species: four in the owl limpet *Lottia gigantea*, six in the polychaete worm *Capitella teleta*, and four in the Californian leech *Helobdella robusta* [3,6].

Primary Structure

The amino acid sequences of MIPs are highly diverged among orders of mollusks, except for some critical residues (such as Cys) that are necessary for tertiary structure formation.

Properties

Mr 6,500–7,000 (mature MIPs in *L. stagnalis*), 9,146 (mature MIP in *A. californica*), as determined by MALDI-TOF MS analysis [2,4].

Synthesis and Release

Gene and mRNA

In *L. stagnalis*, seven MIP genes (*MIP-I to -VII*) have been identified [4]. *MIP-I*, *-II*, *-III*, *-V*, and *-VII* are functionally transcribed, but *MIP-IV* and *-VI* are pseudogenes. All MIP genes consist of three exons. The number and the positions of introns are identical to those in the vertebrate insulin gene.

Distribution of mRNA

In *L. stagnalis*, all five functional *MIP* genes are expressed in the neuroendocrine LGCs in the central nervous system [7,8]. Subsets of LGCs (types A and B) in the cerebral ganglia, and ectopic LGCs, canopy cells (CCs), in the lateral lobes show differential expression pattern of the MIP genes [7]. Type A cells and CCs express all *MIPs*, while type B cells do not express *MIP-1*. *MIP-III* is highly expressed in type B cells, while *MIP-II* and *-V* are relatively weakly expressed in all groups of LGCs. In addition to the LGCs, *MIP-VII* is also expressed in neurons of the buccal ganglia, which are thought to be involved in the regulation of feeding behavior [8]. In *A. californica*, *MIP* is expressed in the clusters of neurons (F and C clusters) in the central region of the cerebral ganglia [2]. In *C. gigas*, *MIP* mRNAs are only detected in the visceral ganglia [5].

Regulation of Synthesis and Release

In *L. stagnalis*, expression of *MIP* is regulated in a nutrient-dependent manner. In the LGCs, *MIP-II* and *-V* are downregulated by starvation, whereas the expression of the other genes is unaffected [4]. Re-feeding after starvation restores the *MIP* expression to normal levels. Therefore, the expression of each *MIP* is regulated independently, even if they are expressed in the same cell. In *A. californica*, *MIP* expression is also significantly downregulated by starvation [2]. The LGC-produced MIPs are transported to the neurohemal axon terminals in the periphery of the ipsilateral median lip nerves, where they are stored and released into hemolymph [4]. MIPs are also suggested to be released within the central nervous system and act in a paracrine manner. It was demonstrated

Y. Takei, H. Ando, & K. Tsutsui (Eds): Handbook of Hormones. DOI: http://dx.doi.org/10.1016/B978-0-12-801028-0.00209-9

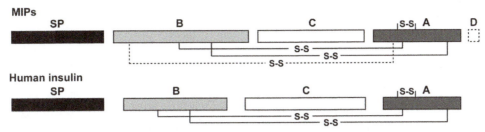

Figure 45C.1 Schematic representation of the precursors of MIP and human insulin. Oblongs indicate the predicted domains in the precursor peptides: black, signal peptide (SP); light gray, B-chain (B); white, C-peptide (C); dark gray, A-chain (A); dotted square, D-peptide (in some MIPs). Disulfide bonds (S—S) are indicated by lines. The putative extra disulfide bond in MIP is indicated by a dotted line.

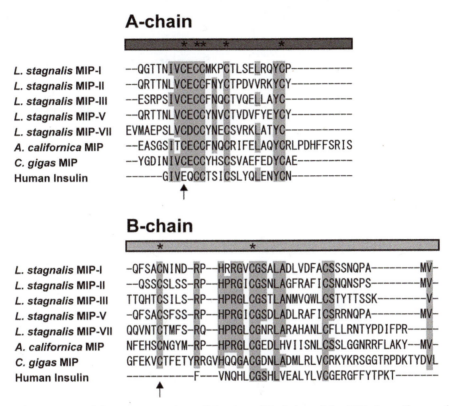

Figure 45C.2 Amino acid sequences of the representatives of the A- and B-chains of the MIPs from the pond snail *L. stagnalis*, the California sea hare *A. californica*, and the Pacific oyster *C. gigas* aligned with that of human insulin. Highlighted amino acid residues indicate highly conserved residues. Asterisks in the square bars denote Cys residues. The extra Cys residues in MIP are indicated by arrows.

in vitro that dissociated LGCs release MIPs in response to various factors, such as dopamine and glucose [4,9]. Moreover, *in vitro* studies suggested that an increase in the intracellular levels of cAMP and Ca^{2+} is necessary for the release of MIPs from the LGCs [9]. However, the physiological mechanisms that stimulate MIP release *in vivo* are still unclear.

Receptors

Structure and Subtype

A cDNA encoding a putative *L. stagnalis* MIP receptor (MIPR) was cloned by using a PCR strategy with degenerate primers based on conserved cytoplasmic tyrosine kinase (TK) domains of the human insulin receptor and the fruit fly insulin-like receptor (E-Table 45C.2) [4]. *L. stagnalis* MIPR has typical features of insulin or IGF type I receptor, including a Cys-rich region, a potential dibasic cleavage site, a transmembrane domain, and a TK domain. The TK domain shows 70% amino acid sequence identity with the human IGF type-I

receptor, whereas the C-terminal domain of the MIPR is very long and shows low sequence conservation with other IRs. Genomic southern blot analysis revealed that there is only one MIPR gene in *L. stagnalis*, suggesting that the different MIPs bind to the same MIPR [4]. The direct binding of MIPs to MIPR has yet to be demonstrated.

Signal Transduction Pathway

MIP stimulation of HEK293 cells co-expressing *L. stagnalis* MIPR and human insulin receptor substrate 1 (IRS-1) resulted in phosphorylation of IRS-I, suggesting that the signal transduction pathway downstream of MIPR is common to that of insulin receptor [4].

Biological Functions

Target Cells/Tissues and Functions

The LGC-derived MIPs are suggested to be involved in regulation of the growth of both the soft body parts and the shell, and of protein and carbohydrate metabolism. The buccal ganglion neuron-derived MIP-VII may be involved in feeding behavior as a neurotransmitter/neuromodulator [4]. Recently, the involvement of *L. stagnalis* MIPs in long-term synaptic plasticity and long-term memory was reported [10].

References

1. Smit AB, Vreugdenhil E, Ebberink RHM, et al. Growth-controlling molluscan neurons produce the precursor of an insulin-related peptide. *Nature*. 1988;331:335–338.
2. Floyd PD, Li L, Rubakhin SS, et al. Insulin prohormone processing, distribution, and relation to metabolism in *Aplysia californica*. *J Neurosci*. 1999;15:7732–7741.
3. Veenstra JA. Neuropeptide evolution: neurohormones and neuropeptides predicted from the genomes of *Capitella teleta* and *Helobdella robusta*. *Gen Comp Endocrinol*. 2011;171:160–175.
4. Smit AB, van Kesteren RE, Li KW, et al. Towards understanding the role of insulin in the brain: Lessons from insulin-related signaling systems in the invertebrate brain. *Prog Neurobiol*. 1998;54:35–54.
5. Hamano K, Awaji M, Ustuki H. cDNA structure of an insulin-related peptide in the Pacific oyster and seasonal changes in the gene expression. *J Endocrinol*. 2005;187:55–67.
6. Veenstra JA. Neurohormones and neuropeptides encoded by the genome of *Lottia gigantea*, with reference to other mollusks and insects. *Gen Comp Endocrinol*. 2010;167:86–103.
7. Meester I, Ramkema MD, van Minnen J, et al. Differential expression of four genes encoding molluscan insulin-related peptides in the central nervous system of the pond snail *Lymnaea stagnalis*. *Cell Tissue Res*. 1992;269:183–188.
8. Smit AB, Spijker S, van Minnen J, et al. Expression and characterization of molluscan insulin-related peptide VII from the mollusc *Lymnaea stagnalis*. *Neuroscience*. 1996;70:589–596.
9. Geraerts WPM, Smit AB, Li KW, et al. The light green cells of *Lymnaea*: a neuroendocrine model for stimulus-induced expression of multiple peptide genes in a single cell type. *Experientia*. 1992;48:464–473.
10. Murakami J, Okada R, Sadamoto H, et al. Involvement of insulin-like peptide in long-term synaptic plasticity and long-term memory of the pond snail *Lymnaea stagnalis*. *J Neurosci*. 2013;33: 371–383.

Supplemental Information Available on Companion Website

- Accession numbers of genes and cDNAs for molluskan insulin-related peptides and MIP receptors/E-Tables 45C.1, 45C.2

Androgenic Gland Hormone

Hidekazu Katayama

Abbreviation: AGH
Additional name: androgenic hormone

An insulin-like peptide involved in sex differentiation of crustaceans, AGH acts on young animals to differentiate them into a functional male. Thus, AGH is the sex determination factor in crustaceans.

Discovery

AGH was discovered as the hormone responsible for the differentiation of sex-specific characteristics in thte amphipod crustacean *Orchestia gammarella* [1]. AGH was first isolated from the terrestrial isopod *Armadillidium vulgare* as a peptide that is capable of masculinizing young female animals [2].

Structure

Structural Features

The chemical structure of AGH has been determined only in the terrestrial isopod *A. vulgare*, and the AGH structure of other animals is merely presumed from its cDNA sequence. AGH is an insulin-like heterodimeric peptide that belongs to the insulin superfamily (Figure 45D.1). The A- and B-chains consist of 29−31 aa residues and 44−46 aa residues, respectively. In spite of the sequence similarity to the mammalian insulin, the disulfide bond arrangement in AGH is different from that in insulin [3]. In addition, AGH has an additional disulfide bond in the B-chain, and a total of four disulfide bonds exist in the AGH molecule. Isopod AGH has an asparagine-linked glycan in the A-chain, which is significant for AGH biological function [3]. These structural characteristics are completely conserved in isopod AGHs [4]. In decapod crustaceans, an insulin-like molecule termed insulin-like androgenic gland factor (IAG) exists in the androgenic glands (AGs), male-specific organs located beside the testes producing AGH, and it is presumed that IAG is the AGH in decapods [5,6]. The native IAG protein has not yet been isolated from any decapod species, and the structure of IAG is presumed from its mRNA sequence to be an insulin-like heterodimeric peptide. The length of the A- and B-chains varies widely (30−50 aa residues for the A-chain and 30−45 aa residues for the B-chain). Putative N-glycosylation sites are found in many IAG sequences from various decapod species, although the glycosylation positions are not conserved among decapod IAGs. Six cysteine residues found in insulin superfamily members are completely conserved in IAGs, but some IAGs have two additional cysteine residues, one in the A-chain and the other in the B-chain, suggesting the possibility that an additional interchain disulfide bond is formed in these IAGs [7].

Synthesis and Release

Gene, mRNA, and Precursor

Similarly to mammalian insulin, the mRNA encoding AGH produces a single-chain 140- to 150-aa precursor polypeptide consisting of a signal peptide, B-chain, C-peptide, and A-chain. For activation of AGH, the C-peptide portion is enzymatically cleaved from the precursor protein at two dibasic cleavage sites, and the mature heterodimeric structure is formed. The length of decapod IAG precursor proteins is more variable than those in isopod AGHs, but the mechanism of precursor processing is presumed to be similar.

Distribution of mRNA

In isopods, the AGH gene is expressed in the AGs, and no expression is found in other male tissues or in females. The distribution of IAG mRNA in decapods is essentially the same as that of AGH mRNA in isopods, except that a low level of *IAG* expression is found in the hepatopancreas [6].

Regulation of Synthesis

Regulation of *AGH* expression in isopods has not yet been investigated, and information regarding transcriptional regulation is still quite limited. In decapod crustaceans, eyestalk ablation increases IAG mRNA levels in the AGs, and it is presumed that *IAG* expression is negatively regulated by substances present in the eyestalk ganglia [6]. Detailed molecular mechanisms of the eyestalk−AG axis have not yet been clarified, and information about transcriptional regulation of IAG is also limited.

Biological Functions

Target Tissues and Function

No specific target organ of AGH has been determined. AGH promotes sex differentiation in young animals. Injection of 38 pg of AGH masculinizes approximately 50% of young female *A. vulgare*, and maximal activity is found at a dose of 76 pg/animal [2]. An N-linked glycan at the A-chain and the correct disulfide bond arrangement are significant for AGH activiy [3]. Information regarding the function of decapod IAG peptide is, however, quite limited.

Y. Takei, H. Ando, & K. Tsutsui (Eds): Handbook of Hormones. DOI: http://dx.doi.org/10.1016/B978-0-12-801028-0.00210-5

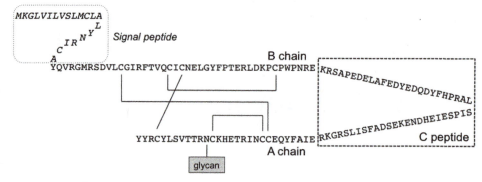

Figure 45D.1 **Primary structure of *Armadillidium vulgare* AGH precursor polypeptide.** Solid lines represent disulfide bonds.

Phenotype in Gene-modified Animals

In decapod crustaceans, gene silencing of IAG by RNAi in intersex young animals induces the arrest of testicular spermatogenesis and spermatophore development in the sperm duct, and also induces the expression of female-specific genes [8]. IAG gene silencing in males reduces male secondary sex characteristics and feminizes male-related phenotypes, such as the shape of the internal branch of the swimming legs [9].

References

1. Charniaux-Cotton H. Discovery in an amphipod crustacean (*Orchestia gammarella*) of an endocrine gland responsible for the differentiation of primary and secondary male sex characteristics. *C R Acad Sci Paris.* 1954;239:780−782.
2. Okuno A, Hasegawa Y, Nagasawa H. Purification andproperties of androgenic gland hormone from the terrestrial isopod *Armadillidium vulgare. Zoolog Sci.* 1997;14:837−842.
3. Katayama H, Hojo H, Ohira T, et al. Correct disulfide pairing is required for the biological activity of crustacean androgenic gland hormone (AGH): Synthetic studies of AGH. *Biochemistry.* 2010;49:1798−1807.
4. Cerveau N, Bouchon D, Grève P, et al. Molecular evolution of the androgenic hormone in terrestrial isopods. *Gene.* 2014;540:71−77.
5. Manor R, Weil S, Sagi A, et al. Insulin and gender: an insulin-like gene expressed exclusively in the androgenic gland of the male crayfish. *Gen Comp Endocrinol.* 2007;150:326−336.
6. Chung JS, Manor R, Sagi A. Cloning of an insulin-like androgenic gland factor (IAG) from the blue crab, *Callinectes sapidus*: implications for eyestalk regulation of IAG expression. *Gen Comp Endocrinol.* 2011;173:4−10.
7. Banzai K, Izumi S, Ohira T. Molecular cloning and expression analysis of cDNAs encoding an insulin-like androgenic gland factor from three Palaemonid species, *Macrobrachium lar, Palaemon paucidens* and *P. pacificus. Jpn Agr Res Q.* 2012;46:105−114.
8. Rosen O, Manor R, Sagi A, et al. A sexual shift induced by silencing of a single insulin-like gene in crayfish: ovarian upregulation and testicular degeneration. *PLoS One.* 2010;5:e15281.
9. Ventura T, Manor R, Sagi A, et al. Temporal silencing of an androgenic gland-specific insulin-like gene affecting phenotypical gender differences and spermatogenesis. *Endocrinology.* 2009;150: 1278−1286.

Supplemental Information Available on Companion Website

- Accession numbers of cDNAs for AGH and IAG/E-Table 45D.1

Gonad-Stimulating Substance

Masatoshi Mita

Abbreviation: GSS
Additional names: gamete-shedding substance (GSS), relaxin-like gonad-stimulating peptide (RGP)

GSS is a neuropeptide hormone in starfish secreted from the radial nerves responsible for final gamete maturation, rendering it functionally analogous to gonadotropins in vertebrates.

Discovery

Chaet and McConnaughy first reported in 1959 that an extract of radial nerves induces spawning when injected into a starfish *Asterias forbesi* [1]. The active substance contained in the nerve extract was first named "gamete-shedding substance," and later "gonad-stimulating substance." GSS purified from the radial nerves of the starfish *Asterina pectinifera* was identified as a relaxin-like peptide [2].

Structure

Structural Features

GSS in *A. pectinifera* is a heterodimer composed of two different peptides, A- and B-chains with disulfide cross-linkages. The A-chain is composed of 24 aa residues and the B-chain of 19 aa residues (Figure 45E.1) [2]. The A-chain of GSS harbors a cysteine motif identical to the signature pattern [Cys−Cys−X3−Cys−X8−Cys] of the insulin/insulin-like growth factor (IGF)/relaxin superfamily [3]. Phylogenetic analysis shows that GSS is a relaxin-like peptide (E-Figure 45E.1) [2]. GSS is present in all starfish species. However, a few amino acid residues constituting GSS are different among species.

Primary Structure

The cDNA sequence of *A. pectinifera* GSS (AB496611) consists of an open reading frame (ORF) encoding a peptide of 116 aa residues, including a signal peptide of 29 aa residues at the N-terminus (E-Figure 45E.2) [2,3]. The sequence of the GSS B-chain occurs after the signal peptide, and the GSS-A chain is located in the C-terminal end of the ORF. There is an intermediate sequence (C-peptide) of 44 aa residues between the GSS-B and GSS-A chains, which is flanked by typical proteolytic sites, Lys and Arg (KR), at both ends [2]. Thus, GSS is probably synthesized in the cells as a preproGSS initially translated from its mRNA [3]. After the formation of three disulfide cross-linkages between the A- and B-chains and within the A-chain, the C-peptide is eliminated by a trypsin-like proteolytic enzyme to generate GSS.

Properties

Mr 4,737 (*A. pectinifera* GSS) [2]; soluble in water.

Synthesis and Release

Distribution of mRNA

GSS is distributed mainly in the radial nerves and nerve rings in starfish [4]. *In situ* hybridization reveals the presence of its mRNA in the peripheral area of radial nerves proximal to the tube feet but not at the side of the hemal sinus [2].

Tissue Concentration

GSS content is high in radial nerves and the nerve ring (Figure 45E.2) [4]. Trace amounts of GSS are observed in cardiac stomachs and in the tube feet. However, GSS is not detectable in the pyloric ceca, ovaries, and testes. The concentration of GSS in the radial nerves is found to be identical in both sexes and to remain constant throughout the year regardless of the breeding/non-breeding season. However, GSS activity in coelomic fluid can be observed only during the breeding season [4].

Regulation of Synthesis and Release

Little is known about factors that regulate GSS production and secretion. However, GSS secretion from radial nerves is induced by ionomycin, a Ca^{2+} ionophre [5]. It is thus considered that stimulation of GSS secretion is mediated by an increase in intracellular Ca^{2+} concentration.

Receptors

Structure and Subtype

In the gonads, GSS specifically binds to its receptors located on the surface of ovarian follicle cells and testicular interstitial cells [6]. The K_d value estimated from Scatchard plots is about 0.6 nM in the membranes of follicle cells. However, the GSS receptor has not yet been identified. It is considered that GPCRs are involved in mediating the action of GSS [7], though GSS does not possess the relaxin-specific receptor-binding motif (Arg−X3−Arg−X2−(Ile/Val)) typical of vertebrate relaxins [2,3].

Signal Transduction Pathway

Isolated follicle cells cultured with GSS show a dose-related increase in cAMP production [2,7]. Two types of G proteins have been identified in the membranes of follicle cells: stimulatory (G_s) and inhibitory (G_i) [7]. Thus, the action of GSS is

Y. Takei, H. Ando, & K. Tsutsui (Eds): Handbook of Hormones. DOI: http://dx.doi.org/10.1016/B978-0-12-801028-0.00211-7

(A-chain) SEYSGIASYCCLHGCTPSELSVVC

(B-chain) EKYCDDDFHMAVFRTCAVS

Figure 45E.1 Primary structure of GSS in *A. pectinifera*. The cysteine bridges are shown in lines.

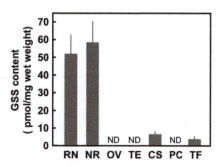

Figure 45E.2 Distribution of GSS in various organs of the starfish *A. pectinifera*. RN, radial nerve; NR, nerve rings; OV, ovaries; TE, testes; CS, cardiac stomachs; PC, pyloric ceca; TF, tube feet. Each column and vertical line represents the mean value for three independent samples and standard error of mean, respectively. ND, not detectable.

thought to be mediated through the activation of its receptor, G proteins, and AC in follicle cells.

Agonists

Concanavalin A (Con A) mimics the action of GSS [8]. α-Methyl-D-mannoside, a competitive inhibitor of Con A, completely inhibits the effect of Con A on follicle cells. However, α-methyl-D-mannoside has no effect on GSS action.

Biological Functions

Target Cells/Tissues and Functions

GSS initiates spawning behavior and shedding of mature gametes in both male and female starfish [2]. Although GSS is the primary mediator of oocyte maturation in starfish, its effect is indirect. GSS mediates oocyte maturation in starfish by acting on the ovary to produce a maturation-inducing hormone (MIH), 1-methyladenine (1-MeAde), which further induces maturation of the oocytes (E-Figure 45E.3) [9]. In this sense, GSS is functionally identical to the vertebrate luteinizing

hormone (LH), especially piscine and amphibian LHs, acting on the ovarian follicle cells to produce an MIH to induce the final maturation or meiotic resumption of the oocyte. Testicular somatic interstitial cells are responsible for 1-MeAde production under the influence of GSS [10]. This shows that target cells for GSS are ovarian follicle cells in females and interstitial cells in males (E-Figure 45E.3). Although the role of 1-MeAde in females and the mechanism of its action in ovaries is well known, its role in males remains unknown.

References

1. Chaet AB, McConnaughy RA. Physiologic activity of nerve extracts. *Biol Bull*. 1959;117:407—408.
2. Mita M, Yoshikuni M, Ohno K, et al. A relaxin-like peptide purified from radial nerves induces oocyte maturation and ovulation in the starfish, *Asterina pectinifera*. *Proc Nat Acad Sci USA*. 2009;106:9507—9512.
3. Mita M. Relaxin-like gonad-stimulating substance in an echinoderm, the starfish: a novel relaxin system in reproduction of invertebrates. *Gen Comp Endocrinol*. 2013;181:241—245.
4. Mita M, Ito C, Kubota E, et al. Expression and distribution of gonad-stimulating substance in various organs of the starfish *Asterina pectinifera*. *Ann NY Acad Sci*. 2009;1163:472—474.
5. Mita M. Release of relaxin-like gonad-stimulating substance from starfish radial nerves by ionomycin. *Zool Sci*. 2013;30:602—606.
6. Mita M, Yamamoto K, Yoshikuni M, et al. Preliminary study on the receptor of gonad-stimulating substance (GSS) as a gonadotropin of starfish. *Gen Comp Endocrinol*. 2007;153:299—301.
7. Mita M, Nagahama Y. Involvement of G-proteins and adenylate cyclase in the action of gonad-stimulating substance on starfish ovarian follicle cells. *Dev Biol*. 1991;144:262—268.
8. Kubota J, Kanatani H, Concanavalin A. Its action in inducing oocyte maturation-inducing substance in starfish follicle cells. *Science*. 1975;187:654—655.
9. Kanatani H, Shirai H, Nakanishi K, et al. Isolation and identification of meiosis-inducing substance in starfish, *Asterias amurensis*. *Nature*. 1969;221:273—274.
10. Kubota J, Nakao K, Shirai H, et al. 1-Methyladenine-producing cell in starfish testis. *Exp Cell Res*. 1977;106:63—70.

Supplemental Information Available on Companion Website

- Phylogenetic tree of GSS and the insulin/IGF/relaxin superfamily/E-Figure 45E.1
- Coding DNA sequence and precursor strucures of GSS in the starfish/E-Figure 45E.2
- Signal transduction pathway/E-Figure 45E.3

CHAPTER 46

Sulfakinin

Shinji Nagata

Abbreviation: SK
Additional names: cholecystokinin-like peptide, leucosulfakinin

SK exerts myoinhibitory effects, causing decreasing in feeding behavior in arthropods. It is thought to be a structural and functional counterpart of vertebrate cholecystokinin (CCK) and gastrin.

Discovery

SK was first isolated from head extracts of the Madeira cockroach *Leucophaea maderae*, and showed myotropic activity on the isolated cockroach hindgut. It was thus designated leucosulfakinin [1].

Structure

Structural Features

Most SKs possess a sulfated tyrosine residue (underlined) in their characteristic C-terminal consensus sequence -D(or E)YGHMRF-NH$_2$ (Table 46.1). A number of transcriptomic analyses have revealed that SKs are conserved over insect species and composed of 10–17 aa residues [2,3]. Most species have two different homologous peptides in a precursor protein.

Structure and Subtype

Two different SK subtypes (true SK and SK-related peptides, SKRPs) are present in a precursor peptide from one species. Although the possible sulfated tyrosine residues are conserved between true SK and SKRPs, the sulfation is not confirmed by experimental analyses.

Synthesis and Release

Gene, mRNA, and Precursor

SK precursor peptide mRNA in most insect species encodes a SK and two SKRPs. In *D. melanogaster*, the precursor peptide of SK encodes three possible peptides: true SK (DSK-I), SKRP (DSK-II), and non-related peptide (DSK-0) (E-Figure 46.1).

Distribution of mRNA

Most SKs are confirmed as being expressed in the brain and central nervous system (CNS). In *D. melanogaster*, only four pairs of endocrine cells are identified within the larval CNS. In addition to expression in the nervous system, SK is expressed in the intestinal endocrine cells.

Regulation of Synthesis and Release

It has been suggested that SK expression and release are regulated by nutritional or feeding states, because gene disruption of SKs or the SK receptor (SKR) activates feeding behavior.

Receptors

Structure and Subtype

SKR has been identified as a GPCR, which belongs to the clade of receptors for the peptides with C-terminal RF-NH$_2$. Most species have two different subtypes of SKRs. In *D. melanogaster*, SKs activate two SKR subtypes, one of which is similar to the CCK receptor [4]. As two SKR subtypes are present in the red flour beetle *Tribolium castaneum*, and as observed in *D. melanogaster*, most species are thought to have two GPCRs for SKs [3].

Signal Transduction Pathway

A second messenger of SK signaling is cAMP mediated by the G$_s$ signaling pathway. This is similar to CCK signaling in vertebrates. A recent study revealed that intracellular Ca^{2+} is also a second messenger for SK [5].

Biological Functions

Target Cells and Function

It has been demonstrated that SKs inhibit contraction of the foregut, heart, ejaculatory duct, and oviduct. The myoinhibitory effects of SKs are also observed on visceral muscle contractions and the gut in *D. melanogaster*. Eventually, such inhibitory effects decrease locomotor and feeding activity. Physiological and pharmacological studies have revealed that the targets of SKs are any somatic tissues related to muscular construction.

Phenotype in Gene-Modified Animals

In several insect species, it has been demonstrated that knockdown of the SK gene or SK receptor gene increases food consumption, possibly because of increased locomotor and feeding behavioral activities, sometimes involved in digestive enzyme release [5–8]. In *D. melanogaster*, the nematodes, and crustacean species, more detailed observations have demonstrated that SKs inhibit larval and adult locomotion, odor preference, and synaptic growth [6–8].

Y. Takei, H. Ando, & K. Tsutsui (Eds): Handbook of Hormones. DOI: http://dx.doi.org/10.1016/B978-0-12-801028-0.00046-5

Table 46.1 Alignment of SKs and CCK*

Drosophila melanogaster	DSK-0	NQKTMSF-NH₂
	DSK-1	FDDY*GHMRF-NH₂
	DSK-2	GGDDQFDDY*GHMRF-NH₂
Leucophaea maderae	LSK-I	EQFEDY*GHMRF-NH₂
	LSK-II	pESDDY*GHMRF-NH₂
Homo sapiens	CCK-8	DY*MGWMDF-NH₂

*Asterisks indicate sulfated tyrosine residues; conserved amino acid residues are shaded.

References

1. Nachman RJ, Holman GM, Haddon WF, et al. Leucosulfakinin, a sulfated insect neuropeptide with homology to gastrin and cholecystokinin. *Science*. 1986;234:71−73.
2. Nichols R, Schneuwly S, Dixon J. Identification and characterization of a *Drosophila* homologue to the vertebrate neuropeptide cholecystokinin. *J Biol Chem*. 1988;263:12167−12170.
3. Yu N, Nachman RJ, Smagghe G. Characterization of sulfakinin and sulfakinin receptor and their roles in food intake in the red flour beetle *Tribolium castaneum*. *Gen Comp Endocrinol*. 2013;188:196−203.
4. Kubiak TM, Larsen MJ, Burton KJ, et al. Cloning and functional expression of the first *Drosophila melanogaster* sulfakinin receptor DSK-R1. *Biochem Biophys Res Commun*. 2002;291:313−320.
5. Downer KE, Haselton A, Nachman RJ, et al. Insect satiety: sulfakinin localization and the effect of drosulfakinin on protein and carbohydrate ingestion in the blow fly, *Phormia regina* (Diptera: Calliphoridae). *J Insect Physiol*. 2007;53:106−112.
6. Meyering-Vos M, Müller A. RNA interference suggests sulfakinins as satiety effectors in the cricket *Gryllus bimaculatus*. *J Insect Physiol*. 2007;53:840−848.
7. Maestro JL, Aguilar R, Pascual N, et al. Screening of antifeedant activity in brain extracts led to the identification of sulfakinin as a satiety promoter in the German cockroach. Are arthropod sulfakinins homologous to vertebrate gastrins-cholecystokinins? *Eur J Biochem*. 2001;268:5824−5830.
8. Janssen T, Meelkop E, Lindemans M, et al. Discovery of a cholecystokinin-gastrin-like signaling system in nematodes. *Endocrinology*. 2008;149:2826−2839.

Supplemental Information Available on Companion Website

- *D. melanogaster* SK preproprotein/E-Figure 46.1
- Accession numbers of SK preproproteins and SK receptors/E-Table 46.1

CHAPTER 47

Cionin

Honoo Satake

Additional names: Ci-cholecystokinin (Ci-CCK)

An ascidian ortholog of vertebrate CCK, cionin is the only authentic CCK in invertebrates.

Discovery

Cionin was isolated from the neural complex of an ascidian, *Ciona intestinalis*, in 1990 [1].

Structure

Structural Features and Primary Structure

Cionin consists of eight aa residues. The C-terminal amidated tetrapeptide is conserved with vertebrate CCK/gastrin. Two tyrosines at positions 2 and 3are sulfated (Table 47.1).

Properties

Mr 1,254; soluble in water, saline solution, aqueous acetonitrile, and methanol.

Synthesis and Release

Gene, mRNA, and Precursor

Cionin is encoded as a single copy in its precursor [2].

Distribution of mRNA

Cionin mRNA is distributed in numerous neurons in the central nervous system and intestine of *C. intestinalis* [2,3].

Receptors

Structure and Subtype

Two cionin receptors (CioRs), CioR1 and CioR2, have been identified [3,4]. CioRs belong to the the class A (rhodopsin-like) GPCR family, and display 58–62% sequence identity to human CCK receptors 1 and 2 [3].

Signal Transduction Pathway

Cionin induces the mobilization of intracellular Ca^{2+} via CioRs expressed in HEK293 MSR cells, whereas the invertebrate CCK analog, sulfakinin, fails to activate CioRs [3]. Monosulfated cionin derivatives exhibit activity that is much less potent and efficient. The non-sulfated cionin derivative is devoid of any activation of CioRs [3].

Biological Functions

Target Cells/Tissues and Functions

*coir*1 and *coir*2 are expressed mainly in the neural complex, digestive organs, ovary, oral siphon, and atrial siphon sperma of *C. intestinalis* [3]. *cior1* is highly expressed in the stomach and middle intestine, whereas expression of *coir2* is low in these tissues. In contrast, expression of *cior2* is higher than that of *cior1* [3]. No biological role for cionin in *C. intestinalis* has been elucidated. Nevertheless, CCK- or gastrin-like activity has been shown in the vertebrate organ, the contraction of rainbow trout gallbladders [5], and the release of histamine and gastric acid from the rat stomach [6].

References

1. Johnsen AH, Rehfeld JF. Cionin: a disulfotyrosyl hybrid of cholecystokinin and gastrin from the neural ganglion of the protochordate *Ciona intestinalis*. *J Biol Chem*. 1990;265:3054–3058.
2. Monstein HJ, Thorup JU, Folkesson R, et al. cDNA deduced procionin. Structure and expression in protochordates resemble that of procholecystokinin in mammals. *FEBS Lett*. 1993;331:60–64.
3. Sekiguchi T, Ogasawara M, Satake H. Molecular and functional characterization of cionin receptors in the ascidian, *Ciona intestinalis*: the evolutionary origin of the vertebrate cholecystokinin/gastrin family. *J Endocrinol*. 2012;213:99–106.
4. Nilsson IB, Svensson SP, Monstein HJ. Molecular cloning of a putative *Ciona intestinalis* cionin receptor, a new member of the CCK/gastrin receptor family. *Gene*. 2003;323:79–88.
5. Schjoldager B, Jorgensen JC, Johnsen AH. Stimulation of rainbow trout gallbladder contraction by cionin, an ancestral member of the CCK/gastrin family. *Gen Comp Endocrinol*. 1995;98:269–278.
6. Marvik R, Johnsen AH, Rehfeld JF, et al. Effect of cionin on histamine and acid secretion by the perfused rat stomach. *Scand J Gastroenterol*. 1994;29:591–594.

Table 47.1 Primary Sequences of Cionin, Cholecystokinin, and Gastrin*

Cionin	NYYGWMNDF-NH₂
Cholecystokinin	DYMGWMDF-NH₂
Gastrin	pEGPWLEEEEEAYGWMDH-NH₂

*Y is a sulfated tyrosin. Conserved amino acid residues are shaded.

Y. Takei, H. Ando, & K. Tsutsui (Eds): Handbook of Hormones. DOI: http://dx.doi.org/10.1016/B978-0-12-801028-0.00047-7

CHAPTER 48

Ci-Galanin-Like Peptide

Honoo Satake

Abbreviation: Ci-GALP

A Ciona intestinalis peptide structurally related to vertebrate galanin and GALP, Ci-GALP is the only GALP identified in invertebrates.

Discovery

In 2011, a peptidomic analysis of the *C. intestinalis* central nervous system detected Ci-GALP as the first invertebrate peptide structurally related to vertebrate galanin and GALP [1].

Structure

Structural Features and Primary Structure

Ci-GALP consists of 26 aa residues, and shares the vertebrate galanin/GALP consensus sequences of Gly—Trp—Thr—Leu—Asn—Ser—Ala—Gly—Tyr—Leu—Leu—Gly—Pro at the moiety (Table 48.1) [1,2]. Ci-GALP contains a longer sequence than vertebrate GALP, but harbors an N-terminally extended sequence from the consensus sequence [1,2]. Moreover, Ci-GALP is C-terminally free, as seen in GALP and non-mammalian galanin [1,2].

Table 48.1 Primary Sequences of Ci-GALP and Mammalian Galanin and GALP*

Ci-GALP	PFRGQGGWTLNSVGYNAGLGALRKLFE
Galanin (human)	GWTLNSAGYLLGPHAVGNHRSFSDKNGLTS-NH$_2$
GALP (human)	PAHRGRGGWTLNSAGYLLGPVLHLPQMGDQD-GKRETALEILDLWKAIDGLPYSHPPQPS

*Conserved amino acid residues are shaded.

Properties

Mr 2,909; soluble in water, physiological saline solution, aqueous acetonitrile, and methanol.

Synthesis and Release

Gene, mRNA, and Precursor

A Ci-GALP precursor encodes a single sequence of Ci-GALP, as seen in vertebrate galanin or GALP precursors.

Distribution of mRNA

Ci-GALP mRNA is localized in neurons of the brain ganglion of *C. intestinalis* [1].

Receptors and Biological Functions

Neither the cognate receptor nor the biological function of Ci-GALP has been reported.

References

1. Kawada T, Ogasawara M, Sekiguchi T, et al. Peptidomic analysis of the central nervous system of the protochordate, *Ciona intestinalis*: homologs and prototypes of vertebrate peptides and novel peptides. *Endocrinology*. 2011;152:2416—2427.
2. Lang R, Gundlach AL, Kofler B. The galanin peptide family: receptor pharmacology, pleiotropic biological actions, and implications in health and disease. *Pharmacol Ther*. 2007;115:177—207.

Y. Takei, H. Ando, & K. Tsutsui (Eds): Handbook of Hormones. DOI: http://dx.doi.org/10.1016/B978-0-12-801028-0.00048-9

CHAPTER 49

Allatostatin-C

Akira Mizoguchi

Abbreviations: AST-C, PISCF-AST
Additional names: *C-type allatostatin (C-type AST),*
Manduca sexta *allatostatin (Manse-AS)*

One of three types of allatostatins, AST-C inhibits juvenile hormone (JH) biosynthesis by the corpora allata (CA) mainly in lepidopteran insects and exerts myoinhibitory effects in a wide variety of insects. It is likely an arthropod homolog of vertebrate somatostatin.

Discovery

AST-C was isolated in 1991 from the brain of the tobacco hornworm *Manduca sexta* as the peptide that inhibits juvenile hormone biosynthesis by the corpora allata [1,2].

Structure

Structural Features

There are two subtypes of AST-C peptides. The members of the first type, which include *M. sexta* AST (Manse-AS), are non-amidated, N-terminally blocked, 15-aa peptides containing a disulfide bridge between the cysteine residues at positions 7 and 14. The second type of peptide lacks the N-terminal pyroglutamate (or glutamine) and possesses the amidated C-terminus, although two cysteines and some other residues are common with the first type [2,3].

Primary Structure

Peptides identical to Manse-AS have been identified in all the lepidopteran insects examined. Similar peptides (the first type AST-Cs) have been found in other orders of insects, and even in crustaceans [2,4,5]. The sequence is remarkably conserved and is represented as pEXRXRQCYFNPISCF, with only a small number of aa substitutions. The members of the second type of AST-C are found in some insects (honeybee, aphid, louse, locust), ticks, and crustaceans. The structure is again highly conserved and is represented as SYWKQCAFNAVSCF-NH$_2$, with a small number of aa substitutions (Table 49.1) [2,4].

Synthesis and Release

Gene, mRNA, and Precursor

In the armyworm *Pseudaletia unipuncta*, the precursor protein consists of 125 aa residues, containing a Manse-AS sequence at its C-terminus. The expression of the gene in the brain is low in the final instar larva, prepupae, and early pupa, but is relatively high in the late pupa and early adult [6]. In other species,

the developmental changes in gene expression are somewhat different. Until recently, there was considered to be only a single AST-C in any animal species. However, recent studies in both insects and crustaceans have revealed, by different approaches, that most species have two AST-C peptides. In crustaceans, mass-based studies have demonstrated that each animal has the two known subtypes of AST-C [5]. In insects, genome-based studies have disclosed the presence of an additional AST-C-like peptide in each species, which differs from typical AST-C in the N-terminal sequence and, more importantly, in the structural features of the precursor protein [4]. These newly discovered peptides are called AST-CC (double C). The genes for AST-C and AST-CC are closely located in the same or opposite directions [4].

Distribution of Peptide

Manse-AS-producing cells have been immunochemically identified in several insects [1]. In larvae of the tomato moth *Lacanobia oleracea*, two groups of Manse-AS-immunoreactive neurosecretory cells are localized in the pars lateralis of the brain. The axons from these cells exit the brain, pass through the corpora cardiaca (CC), and innervate the CA and the mandibular (salivary) glands. Two more perikarya situated between the two cell groups in the pars lateralis also show strong Manse-AS immunoreactivity. The axons from these cells give rise to numerous arborizations at the center of the brain before exiting the brain [7]. A pair of large neurons in the frontal ganglion, which are Manse-AS immunoreactive, send their axons to the muscles of the foregut [8]. A study using ELISA showed that Manse-AS is also present in the midgut and Malpighian tubules [9]. In *D. melanogaster*, its expression in the midgut endocrine cells has been reported [4].

Tissue Content

Tissue contents (fmol) of Manse-AS in the sixth instar larvae of *L. oleracea* as determined by ELISA are 73.1 in the brain, 48.6 in the CC/CA, 31.6 in the subesophageal ganglion, 34.1 in thoracic ganglia, and 41.7 in the abdominal ganglia [9]. The titer in hemolymph is 6.92 fmol/μl (4.54 in plasma and 2.38 in hemocytes) [9].

Receptors

Structure

Two AST-C receptor genes (*CG7285* and *CG13702*) have been identified in *D. melanogaster* by using *Xenopus* oocytes-based functional assays [10]. These genes encode for GPCRs of 467 and 489 aa residues, respectively. Both receptors (called Drostar 1 and 2) show high sequence identity

Y. Takei, H. Ando, & K. Tsutsui (Eds): Handbook of Hormones. DOI: http://dx.doi.org/10.1016/B978-0-12-801028-0.00049-0

Table 49.1 Comparison of the aa Sequences of AST-Cs in Representative Species from Various Orders of Arthropod*

Manduca sexta (moth)	pEVRFRQCYFNPISCF
Drosophila melanogaster (fruit fly)	pEVRYRQCYFNPISCF
Tribolium castaneum (beetle)	pESRYRQCYFNPISCF
Litopenaeus vannamei (shrimp)	pEIRYHQCYFNPISCF
Apis mellifera (honeybee)	SYWKQCAFNAVSCF—NH_2
Pediculus humanus (louse)	SYWKQCAFNAVSCF—NH_2
Locusta migratoria (locust)	SYWKQCAFNAVSCF—NH_2
Daphnia pulex (water flea)	SYWKQCAFNAVSCF—NH_2
Ixodes scapularis (tick)	SGWKQCSFNAVSCF—NH_2

*Identical residues across all AST peptides are shaded. pE, pyroglutamate.

(60% overall identity and 76% identity at the transmembrane regions), suggesting that they are paralogs. The two receptors appear to be homologs of the mammalian somatostatin receptors, sharing 42% aa identity in the transmembrane regions [1,10]. They are not activated by known mammalian agonists. Site-directed mutagenesis of a residue in transmembrane region 3 and the loop between transmembrane regions 6 and 7 affect ligand binding, as with somatostatin receptors [10]. The two receptor genes are expressed in various tissues, including the brain and midgut, with similar tissue specificity [4]. Only a single homolog of these receptors has been identified in the silkmoth *Bombyx mori*, the beetle *Tribolium castaneum*, and the honeybee *Apis mellifera* [4]. In *B. mori*, it is highly expressed in the CC/CA. The structural similarity between AST-C receptors and vertebrate somatostatin receptors, together with the presence of a disulfide bridge within the AST-C peptides and general inhibitory actions of these peptides, suggests that AST-C is an arthropod somatostatin homolog [4].

Biological Functions

Target Cells/Tissues and Functions

Manse-AS inhibits JH biosynthesis *in vitro* by the CA from lepidopteran insects and mosquitoes [2,3]. In *M. sexta*, its effect on the CA is dose-dependent over the range of 0.5–10 nM [3]. In some other moths, however, higher doses (up to 10 µM) are required to inhibit JH biosynthesis, and the maximal inhibition is less than 100% (50–77%, depending on the species, sex, and developmental stage) [3]. This peptide is ineffective on most of non-lepidopteran insects, as far as its action on the CA is concerned. Manse-AS also regulates muscle contraction [3,8]; it inhibits spontaneous contractile activity of the foregut and stomodeal valve in *L. oleracea*, with 50% inhibition at 10^{-12} M and complete inhibition at 10^{-7} M. An injection of Manse-AS into *L. oleracea* larvae severely inhibits feeding and growth, due probably to its myoinhibitory action [3]. The myoinhibitory actions of AST-C have also been reported in non-lepidopteran insects in which AST-C does not inhibit JH biosynthesis. In *D. melanogaster*, AST-C decreases spontaneous muscle contractions in the heart and crop [2]. As AST-C peptides show common activity in inhibiting muscle contraction in most insects, myoregulation is probably the primary function of these peptides. Although the physiological roles of AST-C in crustaceans remain to be determined, the distribution of AST-C in the stomatogastric ganglion and posterior midgut cecum suggests its role in the modulation of feeding-related behavior [5].

Structure—Activity Relationship

Structure-modification studies on Manse-AS have demonstrated the importance of the disulfide bridge between Cys^7 and Cys^{14} and aromatic residues Tyr^8, Phe^9, and Phe^{15} for its biologic activity [3]. Amidation of the C-terminus does not affect its activity [3]. Deletion of N-terminal several residues severely reduces the activity [3].

References

1. Stay B, Tobe SS. The role of allatostatins in juvenile hormone synthesis in insects and crustaceans. *Annu Rev Entomol.* 2007;52:277–299.
2. Weaver RJ, Audsley N. Neuropeptide regulators of juvenile hormone synthesis. *Ann NY Acad Sci.* 2009;1163:316–329.
3. Audsley N, Matthews HJ, Weaver RJ. Allatoregulatory peptides in Lepidoptera, structures, distribution and functions. *J Insect Physiol.* 2008;54:969–980.
4. Veenstra JA. Allatostatin C and its paralog allatostatin double C: The arthropod somatostatins. *Insect Biochem Mol Biol.* 2009;39:161–170.
5. Stemmler EA, Bruns EA, Cashman CR, et al. Molecular and mass spectral identification of the broadly conserved decapod crustacean neuropeptide pQIRYHQCYFNPISCF: The first PISCF-allatostatin (*Manduca sexta*- or C-type allatostatin) from a non-insect. *Gen Comp Endocrnol.* 2010;165:1–10.
6. Jansons IS, Cusson M, McNeil JN, et al. Molecular characterization of a cDNA from *Pseudaletia unipuncta* encoding the *Manduca sexta* allatostatin peptide (Mas-AST). *Insect Biochem Mol Biol.* 1996;26:767–773.
7. Audsley N, Duve H, Thorpe A, et al. Morphological and physiological comparisons of two types of allatostatin in the brain and retrocerebral complex of the tomato moth, *Lacanobia oleracea* (Lepidoptera: Noctuidae). *J Comp Neurol.* 2000;424:37–46.
8. Duve H, Audsley N, Weaver RJ, et al. Triple co-localisation of two types of allatostatin and an allatotropin in the frontal ganglion of the lepidopteran *Lacanobia oleracea* (Noctuidae): innervation and action on the foregut. *Cell Tissue Res.* 2000;300:153–163.
9. Audsley N, Weaver RJ, Edwards JP. Enzyme linked immunosorbent assay for *Manduca sexta* allatostatin (Mas-AS), isolation and measurement of Mas-AS immunoreactive peptide in *Lacanobia oleracea. Insect Biochem Mol Biol.* 1998;28:775–784.
10. Kreienkamp HJ, Larusson HJ, Witte I, et al. Functional annotation of two orphan G-protein-coupled receptors, Drostar1 and -2, from *Drosophila melanogaster* and their ligands by reverse pharmacology. *J Biol Chem.* 2002;277:39937–39943.

Supplemental Information Available on Companion Website

- Accession numbers of the cDNAs for allatostatin-C and allatostatin-C receptor/E-Tables 49.1, 49.2

CHAPTER 50

Calcitonin-Related Peptide

Shinji Nagata

Abbreviation: CT/DH, CTLP
Additional names: diuretic hormone 31 (DH31)

CT/DH has been identified from the cockroach Diploptera punctata as a diuretic factor. This peptide contributes to homeostasis by controlling diuresis in arthropods.

Discovery

CT/DH was isolated from the cockroach *Diploptera punctata*, on the basis of its fluid excretion activity, in 2000 [1].

Structure

Structural Features

The amino acid sequence of CT/DH appears to be weakly similar to that of mammalian CT. Most CT/DHs are composed of 31 aa residues with C-terminal amidation, while recent transcriptomic and genomic analyses revealed that the length of the aa sequence of CT/DH ranges from 30 to 34 aa residues. The presence of CT/DH is highly conserved in arthropods (Table 50.1) [2].

Synthesis and Release

Gene, mRNA, and Precursor

CT/DH mRNA encodes a 100- to 150-aa preprotein. After cleavage of a signal sequence and dibasic sites and C-terminal amidation, CT/DH is produced [3]. In some species, several subtypes of CT/DH preproproteins occurring by alternative splicing have been reported. In the kissing bug *Rhodnius prolixus*, three subtypes are present (Figure 50.1) [3]. However, the mature domain is not substituted with any different sequences.

Distribution of mRNA

In *D. melanogaster*, CT/DH is widely expressed in the CNS. Endocrine cells within the midgut are other expression sites of CT/DH [3]. By contrast, CT/DH is not expressed in the midgut of *R. prolixus* [4]. Therefore, intestinal expression of CT/DH is dependent on species.

Receptors

Structure and Subtype

An ortholog of mammalian CT receptor L-type has been characterized and identified as a receptor for CT/DH [5]. The CT/DH receptor belongs to the class B GPCRs, and is highly conserved in arthropods. No subtypes of CT/DH receptor have been reported.

Signal Transduction Pathway

Like other diuretic hormones in insects, such as corticotropin-releasing factor-like diuretic hormone (CRF/DH), CT/DH activates an intracellular AC to produce cAMP. cAMP elevation affects fluid secretion from the Malpighian tubules. In almost all the species reported so far, a similar signaling pathway via cAMP after CT/DH stimulation is utilized in the Malpighian tubules for diuresis.

Table 50.1 Alignment of Amino Acid Sequences of CT/DHs and CT

Diploptera punctata	CT/DH	GLDLGLSRG FSGSQAAKHL MGLAAANYAG GP—NH$_2$
Rhodnius prolixus	CT/DH	GLDLGLSRG FSGSQAAKHL MGLAAANYAG GP—NH$_2$
Drosophila melanogaster	CT/DH	TVDFGLARG YSGTQEAKHR MGLAAANYAG GP—NH$_2$
Homo sapiens	CT	CGNLSTOMLG TYTQDFNKFH TFPQTAIGVG AP—NH$_2$

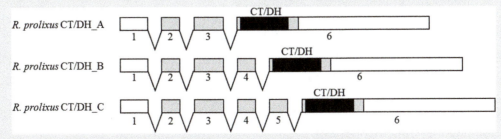

Figure 50.1 Schematic diagram of *R. prolixus* CT/DH preproprotein splicing variants.
Gray and black boxes indicate open reading frame and mature CT/DH, respectively.

Y. Takei, H. Ando, & K. Tsutsui (Eds): Handbook of Hormones. DOI: http://dx.doi.org/10.1016/B978-0-12-801028-0.00050-7

Biological Functions

Target Cells and Function

The predominant function of CT/DH is to activate diuresis mediated by the malpighian tubules. Recently, some reports have shown that CT/DH inhibits foregut and hindgut contraction *in vitro*, possibly modulating midgut peristalsis in *D. melanogaster* larvae [6]. This gut peristalsis function is linked to the CT/DH-expressing intestinal endocrine cells in some species, possibly innervating the central stomatogastric nervous system. It is thus considered that CT/DH is involved in food intake and consumption.

References

1. Furuya K, Milchak RH, Shegg KM, et al. Cockroach diuretic hormones: characterization of a calcitonin-like peptide in insects. *Proc Natl Aca Sci USA*. 2000;97:6469–6474.
2. Ianowski JP, Paluzzi JP, Te Brugge VA, et al. The antidiuretic neurohormone RhoprCAPA-2 downregulates fluid transport across the anterior midgut in the blood-feeding insect *Rhodnius prolixus*. *Am J Physiol Reg Integr Comp Physiol*. 2010;298:R548–R557.
3. Zandawala M. Calcitonin-like diuretic hormones in insects. *J Insect Physiol*. 2012;42:816–825.
4. Te Brugge VA, Orchard I. Amino acid sequence and biological activity of a calcitonin-like diuretic hormone (DH31) from *Rhodnius prolixus*. *J Exp Biol*. 2008;211:382–390.
5. Hewes RS, Taghert PH. Neuropeptides and neuropeptide receptors in the *Drosophila melanogaster* genome. *Genome Res*. 2000;11:1126–1142.
6. Lajeunesse DR, Johnson B, Presnell JS, et al. Peristalsis in the junction region of the *Drosophila* larval midgut is modulated by DH31 expressing enteroendocrine cells. *BMC Physiol*. 2010;10:14.

Supplemental Information Available on Companion Website

- Accession numbers of CT/DH preproproteins and CT/DH receptor/E-Table 50.1

CHAPTER 51

Ci-Calcitonin

Honoo Satake

Abbreviation: Ci-CT

A peptide structurally and functionally related to vertebrate calcitonin, Ci-CT was the first calcitonin-like peptide identified in invertebrates.

Discovery

Ci-CT was identified in 2009 by a homology search on the draft genome database of *C. intestinalis* using salmon calcitonin (sCT) [1].

Structure

Primary Structure

C. intestinalis calcitonin (Ci-CT) is composed of 30 amino acids and conserves a C-terminal amidated proline and an N-terminal CT consensus circular region formed by a disulfide bond between Cys^1 and Cys^7 (Table 51.1), which is requisite for the biological activities of vertebrate CTs. The primary sequence of Ci-CT displays 26.7—40% identity with vertebrate CTs [1].

Synthesis and Release

Gene, mRNA, and Precursor

The *ci-ct* gene is organized with five exons, and encodes a single copy of the Ci-CT sequence [1]. Unlike vertebrate CT genes generating CT-gene related peptide (CGRP) via alternative splicing [2,3], *ci-ct* yields no splicing variants [1].

Table 51.1 Primary Structure of Ci-CT and Human CT/CGRP Family Peptides*

Peptide	Sequence
Ci-CT	CDGVSTCWLHELGNSVHATAGGKQNVGFG--P-NH$_2$
Human CT	CGNLSTCMLGTYTQDFNKFHTFPQTAIGVGAP-NH$_2$
Human CGRP	ACDTATCVTHRLAGLL SRSGGVVKNNFVPTNVGSKAF-NH$_2$
Human adrenomedullin	YRQSMNNFQGLRSFGCRFGTCTVQKLAHQIYQFTDKDKD-NVAPRSKISPQGY-NH$_2$
Human amylin	KCNTATCATQRLANFLVHSSNNFGAILSSTNVGSNTY-NH$_2$
Pig CRSP	SCNTATCMTHRLVGLLSRSGSMVRSNLLPTKMGFKVFG-NH$_2$

*Circular regions are underlined.

Distribution of mRNA

The *ci-ct* gene is expressed in the neural gland (a non-neuronal ovoid body), stigmata cells of gills, the gastrointestinal tract, blood cells, and endostyle [1]. No expression of *ci-ct* has been detected in any neural tissues [1].

Receptors

Structure and Subtype; Signal Transduction Pathway

The Ci-CT receptor (Ci-CTR) was cloned from the *C. intestinalis* central nervous system [1]; neither *Xenopus* oocytes nor mammalian culture cells express functional Ci-CTR [1]. Nevertheless, Ci-CT elicits weak but significant cAMP production via human CTR expressed in COS-7 cells [1]. No gene encoding a receptor activity-modifying protein (RAMP), which is a prerequisite for ligand selectivity and endoplasmic reticulum—plasma membrane translocation of some receptors in vertebrates, has been detected in the *C. intestinalis* genome [1].

Biological Functions

Target Cells/Tissues and Functions

The biological functions of Ci-CT in *C. intestinalis* have yet to be elucidated [1]. However, Ci-CT, like vertebrate CTs, suppresses osteoclast activity in goldfish scales via inhibition of the activity of tartrate-resistant acid phosphatase [1,4].

References

1 Sekiguchi T, Suzuki N, Fujiwara N, et al. Calcitonin in a protochordate, *Ciona intestinalis* − the prototype of the vertebrate calcitonin/calcitonin gene-related peptide superfamily. *FEBS J.* 2009;276:4437−4447.

2 Wimalawansa SJ. Amylin, calcitonin gene-related peptide, calcitonin, and adrenomedullin: a peptide superfamily. *Crit Rev Neurobiol.* 1997;11:167−239.

3 Ogoshi M, Inoue K, Naruse K, et al. Evolutionary history of the calcitonin gene-related peptide family in vertebrates revealed by comparative genomic analyses. *Peptides.* 2006;27:3154−3164.

4 Suzuki N, Suzuki T, Kurokawa T. Suppression of osteoclastic activities by calcitonin in the scales of goldfish (freshwater teleost) and nibbler fish (seawater teleost). *Peptides.* 2000;21:115−124.

Y. Takei, H. Ando, & K. Tsutsui (Eds): Handbook of Hormones. DOI: http://dx.doi.org/10.1016/B978-0-12-801028-0.00051-9

SECTION II.2

Invertebrate-Unique Peptides

SUBSECTION II.2.1

Regulation of Development and Metabolism

CHAPTER 52

FXPRLamide Peptide Family

Toshinobu Yaginuma and Teruyuki Niimi

History

The FXPRLamide peptide family includes pyrokinin (PK)/ myotropin (MT), pheromone biosynthesis activating neuropeptide (PBAN)/melanization and reddish coloration hormone (MRCH), and diapause hormone (DH), as they contain a common C-terminal sequence of FXPRL-NH$_2$ [1]. In 1986, leucopyrokinin (LPK, an 8-aa peptide) was first isolated as a myostimulatory peptide from the cockroach *Leucophaea maderae* [1,2]. In 1989, PBAN (a 33-aa peptide) was isolated as a peptide regulating sex pheromone production in female pheromone glands from the armyworm *Helicoverpa zea* [1,2]. Thereafter, Lom-PK-I and -II, Lom-MT-I to -IV, Pea-PK-5 to -6, and two isoforms of PBANs were respectively purified from the migratory locust *Locusta migratoria*, the American cockroach *Periplaneta americana*, and the silkworm *Bombyx mori* [3–6]. MRCH was identified from *B. mori* in 1990, disclosing that it is structurally identical to PBAN. In 1991, DH (a 24-aa peptide) was purified from *B. mori*. As deduced from *B. mori* DH-PBAN cDNA, five peptides, including DH and PBAN/MRCH, are processed from a precursor protein as described in Subchapter 52A [7]. In 1985, neuromedin U (NMU) was first isolated from porcine tissue as a peptide to contract uterine smooth muscle [8,9]. NMU is highly conserved among vertebrates, and has a common C-terminal sequence of FXPRNamide [8,9]. Therefore, based on such C-terminal similarity, insect members of FXPRLamide peptide family are also called NMU-like peptides.

Structure

Structural Features

The common structural feature of family members is a conserved C-terminal sequence FXPRLamide, which is similar to the FXPRNamide shared by the members of the NMU family (Table 52.1, Figure 52.1) [1–9].

Table 52.1 Amino Acid Sequences of Insect FXPRLamide Peptides and Mammalian Neuromedin U (FXPRNamide Peptides) *

L. maderae	PK	pETSFTPRL-NH$_2$
D. melanogaster	PK2	SVPFKPRL-NH$_2$
D. melanogaster	CAPA-PK	TGPSASSGLWFGPRL-NH$_2$
B. mori	DH	TDMKDESDRGAHSERGALWFGPRL-NH$_2$
B. mori	αSGNP	IIFTPKL-NH$_2$
B. mori	βSGNP	SVAKPQTHESLEFIPRL-NH$_2$
B. mori	PBAN/MCRH	RLSEDMPATPADQEMYQPDPEEMESRTR-YFSPRL-NH$_2$
B. mori	γSGNP	TMSFSPRL-NH$_2$
B. mori	CAPA-PK	NEPHDDLLGLHLDDPGMWFGPRL-NH$_2$
Homo sapiens	NMU25	FRVDEEFQSPFASQSRGYFLFRPRN-NH$_2$
Rattus norvegicus	NMU23	YKVNEYQGPVAPSGGFFLFRPRN-NH$_2$
D. melanogaster	hugγ	QLQSNGEPAYRVRTPRL-NH$_2$

*Letters or W with gray background indicate aa residues conserved in a C-terminal FXPRLamide sequence or DH signature (WFXPRLamide). pE, pyroglutamate.

Figure 52.1 Precursor proteins for *Drosophila* and *Bombyx* CAPA peptides. Boxes indicate CAPA peptides or other potentially amidated peptides. Letters in bold indicate aa residues characteristic of PXPRLamide and FPRVamide (PVK) peptide families. See chapters on individual hormones chapter for the precursor proteins of other FXPRLamide peptide family members.

Y. Takei, H. Ando, & K. Tsutsui (Eds): Handbook of Hormones. DOI: http://dx.doi.org/10.1016/B978-0-12-801028-0.00052-0

Table 52.2 Main Functions of Members of the FXPRLamide Peptide Family (Subchapters 52A–C)

Species Name	Hormone	Function
Cockroach	PK/MT	Stimulation of the hindgut contraction
Armyworm	MRCH	Stimulation of the larval cuticle melanization
Lepidoptera	PBAN	Activation of the biosynthesis of sex pheromone
B. mori	DH	Induction of embryonic diapause
Heliothis/ Helicoverpa	DH	Termination of pupal diapause

Receptors

Structure and Subtypes

The receptors are divided into two groups: PK1/DH receptors (higher affinity for WFXPRLa, DH signature), and PK2/PBAN receptors [1,2,10]. Both groups of receptors are typical GPCRs with seven transmembrane domains.

Signal Transduction Pathway

Although each signal transduction pathway is dependent on each function, commonly Ca^{2+} and PKC are involved [2,4–6,10]. For details, see Subchapters 52A–C.

Biological Functions

Target Cells/Tissues and Functions

As described above, members of the FXPRLamide peptide family play various roles. Therefore, the target organs include various cells/tissues [1–7]. Recently, DH has been found to function to break the diapause of corn earworm pupae [1]. The main function of each class of FXPRLamide peptides is shown in Table 52.2 (see also Subchapters 52A–C).

References

1. Altstein M, Hariton A, Nachman RJ. The FXPRLamide (pyrokinin/PBAN) peptide family. In: Kastin AJ, ed. *Handbook of Biologically Active Peptides*, 2nd ed. San Diego: Elsevier Press; 2013:255–266.
2. Jurenka R, Nusawardani T. The pyrokinin/ pheromone biosynthesis-activating neuropeptide (PBAN) family of peptides and their receptors in Insecta: evolutionary trace indicates potential receptor ligand-binding domains. *Insect Mol Biol.* 2011;20:323–334.
3. Predel R, Wegener C. Biology of the CAPA peptides in insects. *Cell Mol Life Sci.* 2006;63:2477–2490.
4. Rafaeli A. Diapause hormone of the silkworm, *Bombyx mori*: structure, gene expression and function. *Gen Comp Endocrinol.* 2009;162:69–78.
5. Matsumoto S, Ohnishi A, Lee JM, et al. Unraveling the pheromone biosynthesis activating neuropeptide (PBAN) signal transduction cascade that regulates sex pheromone production in moths. *Vitam Horm.* 2010;83:425–445.
6. Jurenka R, Rafaeli A. Regulatory role of PBAN in sex pheromone biosynthesis of heliothine moths. *Front Endocrinol.* 2011;2:1–8.
7. Yamashita O. Diapause hormone of the silkworm, *Bombyx mori*: structure, gene expression and function. *J Insect Physiol.* 1996;42:669–679.
8. Melcher C, Bader R, Walther S, et al. Neuromedin U and its putative *Drosophila* homolog hugin. *PloS Biol.* 2006;4:e68.
9. Brighton PJ, Szekeres PG, Willars GB. Neuromedin U and its receptors: structure, function, and physiological roles. *Pharmacol Rev.* 2004;56:231–248.
10. Homma T, Watanabe K, Tsurumaru S, et al. G protein-coupled receptor for diapause hormone, an inducer of *Bombyx* embryonic diapause. *Biochem Biophys Res Commun.* 2006;344:386–393.

Supplemental Information Available on Companion Website

- Accession numbers of cDNAs and precursor proteins for CAPA/E-Table 52.1

Diapause Hormone

Toshinobu Yaginuma and Teruyuki Niimi

Abbreviation: DH (note that diuretic hormone is also abbreviated in the same way)
Additional name: FXPRLamide peptide hormone

DH is synthesized by the subesophageal ganglion (SG), released into the hemolymph via the corpus cardiacum—corpus allatum (CC—CA) complex, and targets the developing ovaries during pupal—adult development to induce embryonic diapause of the silkworm Bombyx mori.

Discovery

DH was first discovered as a hormone that is synthesized by SG and released into the hemolymph to induce embryonic diapause of *B. mori* [1–3]. It was isolated from *B. mori* SG, and its structure was determined in 1991 [4]. Recently, DH has been shown to induce the termination of pupal diapause in the corn earworm, *Heliothis/Helicoverpa* species [5,6].

Structure

Structural Features

DH is a C-terminally amidated peptide consisting of 24 aa residues having a C-terminal sequence Phe—Gly—Pro—Arg—Leu—NH$_2$ [4]. Therefore, DH is a member of the FXPRLamide peptide family. DH is widely conserved among lepidopteran insects (Table 52A.1).

Synthesis and Release

Gene, mRNA, and Precursor

DH mRNA produces an approximately 190-aa precursor peptide (Figure 52A.1) [4]. After cleavage of a signal sequence and processing, five FXP(R/K)Lamide peptides are produced, including DH, αSGNP (subesophageal ganglion neuropeptide), βSGNP, PBAN (pheromone biosynthesis activating

neuropeptide; see Subchapter 52B), and γSGNP. Therefore, the DH gene is also called the DH-PBAN gene, PBAN gene, or pyrokinin-PBAN gene.

Distribution of mRNA

DH mRNA is detected in the neurosecretory cells (seven paired cells) of *B. mori* SG [4,7].

Regulation of Synthesis and Release

B. mori DH is biosynthesized in the neurosecretory cells in the SG, transported, via nervus corporis cardiaci-III, maxillary nervus, nervi corporis cardiaci-ventralis, and nervi corporis cardiaci-nervus recurrens, to the CC—CA complex, and from there released into the hemolymph [7]. Expression of the *B. mori* DH-PBAN gene is switched on by the interaction of a *cis*-element (−1,117 to −1,088 nt) on the 5′-upstream with a transcription factor (*B. mori* Pitx1) [7].

Receptors

Structure and Subtype

DH receptor (DHR) was first identified in *B. mori* as a class of GPCR with seven transmembrane domains [8], followed by identification of orthologous proteins in four other lepidopteran insects (*Orgyia thyellina*, *Ostrina nubilalis*, *Danaus plexippus*, and *Helicoverpa zea*) [8,9]. Although the subtypes of DHR have been cloned from *B. mori* and *H. zea*, the function of the variants remains to be understood.

Distribution of mRNA

In *B. mori*, DHR mRNA is detected in the developing ovaries, the pheromone glands, and the prothoracic glands [8,10].

Biological Functions

Target Cells and Function

B. mori DH acts on the developing ovaries during the middle pupal stage, and transfers the DH signal into the oocyte via DHR on the oolemma. Two to three days after egg-laying the eggs reach the diapause stage, with the enlargement of both extremities of the embryo and the completion of mesoderm segregation. Therefore, there is a time lag of approximately 1 week between the action of DH and the initiation of diapause. The signal transduction pathway between these events still remains to be understood. In the developing ovaries, DH is known to enhance trehalase activity on the oolemma, which leads to accelerated incorporation of hemolymph trehalose by the oocyte and, finally, to higher levels of glycogen

Table 52A.1 Amino Acid Sequence Alignment of *B. mori* Diapause Hormone with DHs from Other Lepidopteran Insects*

B. mori	DH	TDMK DESDRGAHSE RGALWFGPRL-NH$_2$
Manduca sexta	DH	NDIK DEGDRGAHSD RGALWFGPRL-NH$_2$
Helicoverpa zea	DH	NDVK DGAASGAHAD RLGLWFGPRL-NH$_2$
Helicoverpa armigera	DH	NDVK DGAASGHSDD RLGLWFGPRL-NH$_2$
Heliothis virescens	DH	NDDK DGAASGAHSD RLGLWFGPRL-NH$_2$
Spodoptera littoralis	DH	EIKD GGSDRGAHSD RAGLWFGPRL-NH$_2$
Plutella xylostella	DH	DDLK DEDIQRDARD RASMWFGPRL-NH$_2$

*Letters with gray background indicate aa residues that show more than 50% identity among seven DHs.

Y. Takei, H. Ando, & K. Tsutsui (Eds): Handbook of Hormones. DOI: http://dx.doi.org/10.1016/B978-0-12-801028-0.00212-9

```
         10          20          30          40          50          60          70          80
MYKTNIVFNVLALALFSIFFASC[TDMKDESDRGAHSERGALWFGPRL]GKRSMKPSTEDNRQTFLRLLEAADALFYYDQLPYERQADEPE
                       DH
         90         100         110         120         130         140         150         160         170
TKVTKK[IIFTPKL]GR[SVAKPQTHESLEFIPRL]GR[RLSEDMPATPADQEMYQPDPEEMESRTRYFSPRL]GR[TMSFSPRL]GRELSYDYPT
       αSGNP         βSGNP                         PBAN                                      γSGNP
        180         190
KYRVARSVNKTMDN
```

Figure 52A.1 *B. mori* **DH-PBAN precursor protein** [4].
Boxes indicate DH, αSGNP, βSGNP, PBAN, and γSGNP with common C-terminals FXP(R/K)Lamide.

deposition compared to the oocyte without the DH signal. This glycogen is utilized for production of sorbitol and glycerol along with the diapause. DH promotes trehalase activity via activation of the trehalase gene, during which Ca^{2+} and protein kinase C are involved in signal transduction [4]. When DH induces the termination of pupal diapause, DH-DHR is thought to enhance the ecdysteroidogenesis in the prothoracic glands [5,6,9,10].

Interaction of DHR with FXPRLamide Peptides

In *B. mori* DHR expressed in *Xenopus laevis* oocytes, concentrations of ligands for 50% of maximal activity (EC_{50}) are 70, 1,500, >10,000, 3,000, and 1,700 nM for DH, PBAN, αSGNP, βSGNP, and γSGNP, respectively [8]. Therefore, *B. mori* DHR has the highest affinity for DH among these FXPRLamide peptides [8]. In *H. zea* DHR expressed in Chinese hamster ovary cells, EC_{50} values for *H. zea* DH and truncated GLWFGPRLa peptide are 41 and 49 nM, respectively [9]. *H. zea* DHR is thought to be cross-reactive with the endogenous PBAN [9].

References

1. Fukuda S. The production of the diapause eggs by transplanting the suboesophageal ganglion in the silkworm. *Proc Jpn Acad.* 1951;27:672—677.
2. Hasegawa K. Studies on the voltinism in the silkworm, *Bombyx mori* L., with special reference to the organs concerning determination of voltinism (a preliminary note). *Proc Jpn Acad.* 1951;27:667—671.
3. Hasegawa K. The diapause hormone of the silkworm, *Bombyx mori. Nature.* 1957;178:1300—1301.
4. Yamashita O. Diapause hormone of the silkworm, *Bombyx mori*: structure, gene expression and function. *J Insect Physiol.* 1996;42:669—679.
5. Xu WH, Denlinger DL. Molecular characterization of prothoracicotropic hormone and diapause hormone in *Heliothis virescens* during diapause, and a new role for diapause hormone. *Insect Mol Biol.* 2003;12:509—516.
6. Zhang Q, Nachman RJ, Kaczmarek K, et al. Disruption of insect diapause using agonists and an antagonist of diapause hormone. *Proc Natl Acad Sci USA.* 2011;108:16922—16926.
7. Shiomi K, Fujiwara Y, Yasukochi Y, et al. The *Pitx* homeobox gene in *Bombyx mori*: regulation of *DH-PBAN* neuropeptide hormone gene expression. *Mol Cell Neurosci.* 2007;34:209—218.
8. Homma T, Watanabe K, Tsurumaru S, et al. G protein-coupled receptor for diapause hormone, an inducer of *Bombyx* embryonic diapause. *Biochem Biophys Res Commun.* 2006;344:386—393.
9. Jiang H, Wei Z, Nachman RJ, et al. Molecular cloning and functional characterization of the diapause hormone receptor in the corn earworm *Helicoverpa zea. Peptides.* 2014;53:243—249.
10. Watanabe K, Hull JJ, Niimi T, et al. FXPRL-amide peptides induce ecdysteroidogenesis through a G-protein coupled receptor expressed in the prothoracic gland of *Bombyx mori. Mol Cell Endocrinol.* 2007;273:51—58.

Supplemental Information Available on Companion Website

- Accession numbers of cDNAs for DH and DHR/E-Table 52A.1

Pheromone Biosynthesis Activating Neuropeptide

Toshinobu Yaginuma and Teruyuki Niimi

Abbreviations: PBAN, PT, MRCH
Additional names: pheromonotropin (PT), FXPRLamide peptide hormone, melanization and reddish coloration hormone (MRCH), melanotropic hormone

A peptide hormone that activates pheromone biosynthesis in lepidopteran sex pheromone glands, MRCH also activates cuticle melanization of armyworms. MRCH is molecularly identical to βSGNP-like peptide/PBAN. PBAN/MRCH is a member of the FXPRLamide peptide family and is processed from a precursor peptide encoded by the DH-PBAN gene.

Discovery

PBAN was first discovered as the peptide hormone that is produced by the brain—subesophageal ganglion (Br—SG) complex and activates the biosynthesis of sex pheromone in the pheromone glands of the corn earworm, *Heliothis/Helicoverpa zea* [1]. This hormone was isolated from *H. zea* Br—SG complexes. Two variants have been isolated in the silkworm *Bombyx mori* [2,3]. MRCH was first discovered as the hormone controlling melanization of the larval cuticle of the armyworm *Mythimna* (*Leucania, Pseudaletia*) *separata* [3]. The peptide has been isolated from *M. separata* and *B. mori*.

Structure

Structural Features

PBAN/MRCH is a C-terminally amidated peptide consisting of 33—34 aa residues having a C-terminal sequence Phe—(Ser/Thr)—Pro—Arg—Leu—NH_2 [2,3]. Therefore, PBAN/MCRH is a member of the FXPRLamide peptide family. PBAN is widely conserved among insect species (Table 52B.1).

Synthesis and Release

Gene, Precursor, and mRNA

PBAN/MRCH is encoded by the same gene and mRNA as for diapause hormone, and thus arises from the same precursor protein as diapause hormone (DH) (see Chapter 52A for details). DH-PBAN mRNAs have been identified from at least 19 insect species, including many moths, the red fire ant *Solenopsis invicta*, and the blood-sucking bug *Rhodnius prolixus*.

Distribution of mRNA

Distribution of mRNA is the same as for diapause hormone (see Chapter 52A).

Regulation of Synthesis and Release

In *B. mori*, two paired mandibular neuromere cells and three paired maxillary neuromere cells of the SG are involved in PBAN secretion. Regulation of synthesis and the route of secretion are the same as for diapause hormone (see Chapter 52A).

Receptors

Structure and Subtype

The receptor for PBAN (PBANR) belongs to a class of GPCRs with seven transmembrane domains. The PBANR-coding genes have been cloned from at least 11 insect species [4,5], including many moths and the yellow fever mosquito *Aedes aegypti*. Recently, PBANR splice variants (As, A, B, and C subtypes) have been cloned from *B. mori*, *Helicoverpa armigera*, and *M. separata* [6]. PBAN-C is thought to be the principal receptor molecule involved in PBAN signaling. The phylogenetic relationship indicates that the PBANR group is clearly distinguishable from the DH receptor group [7].

Distribution of mRNA

In *B. mori*, PBANR mRNA is detected mainly in the pheromone glands (PGs) immediately after adult emergence, and in the developing ovaries. Lower levels of mRNA are detected in the midgut and the brain of the 6-day-old fifth instar larvae [8].

Signal Transduction Pathway

In *B. mori*, the binding of PBAN with PBANR on the PG cell surface triggers an influx of Ca^{2+} via a canonical signal transduction cascade that culminates in the activation of stromal interaction molecule 1 and calcium release-activated calcium channel protein 1. The concomitant increase in intracellular Ca^{2+} accelerates lipolysis and fatty-acyl reduction through a calmodulin/calcineurin-mediated phosphorylation/dephosphorylation cascade culminating in bombykol production [9]. In the heliothine moth PG, the increase in intracellular Ca^{2+} activates calcium—calmodulin sensitive AC that promotes the production of cyclic AMP, which in turn activates kinase and/or phosphatase, leading to the stimulation of fatty acid biosynthesis in the pheromone biosynthesis pathway [7].

Y. Takei, H. Ando, & K. Tsutsui (Eds): Handbook of Hormones. DOI: http://dx.doi.org/10.1016/B978-0-12-801028-0.00213-0

Table 52B.1 Amino Acid Sequence Alignment of PBAN/MRCHs*

H. zea	PBAN/MRCH	LSD DMPATPADQE MYRQDPEQID SRTKYFSPRL-NH₂
B. mori	PBAN/MRCH-I	LSE DMPATPADQE MYQPDPEEME SRTRYFSPRL-NH₂
B. mori	PBAN/MRCH-II	RLSE DMPATPADQE MYQPDPEEME SRTRYFSPRL-NH₂
Lymantria dispar	PBAN	LAD DMPATMADQE VYRPEPEQID SRNKYFSPRL-NH₂
M. separata	MRCH (βSGNP like)	KLSYDDVK FENVEFTPRL-NH₂

*Letters with gray background indicate aa residues that show more than 50% identity among five PBAN/MRCHs; bold letters indicate aa residues common to the FXPRLamide family members.

Biological Functions

Target Cells and Function

The target of PBAN in adult female moths is the PG cells, which are found as intersegmental tissues located between the eighth and ninth abdominal segments in the heliothines and *B. mori* [7,9]. Injection or administration of PBAN into intact female moths, isolated abdomens, or *in vitro* cultured PG stimulates pheromone production by the PG [7,9]. In *H. armigera*, PBAN is known to regulate the production of hair-pencil pheromone in males [7]. When larvae of some kinds of armyworms grow under crowded conditions, their cuticle is heavily melanized. MRCH is responsible for melanin synthesis in this process, but the mode of action of this hormone has not been understood in depth [6,9].

Enzymes Activated by PBAN

Before *B. mori* adult eclosion, the bombykol precursor generated from fatty acid synthesis-derived palmitic acid is stored as a triacylglycerol within lipid droplets in the PG cells. Immediately after eclosion, PBAN accelerates the triacylglycerol lipase and fatty-acyl reductase, which in turn leads to production of bombykol [9]. In the heliothine moth, PBAN regulates the acetyl CoA carboxylase that catalyzes the rate-limiting enzyme of fatty acid biosynthesis, prior to the action of fatty acid synthetase [7].

Phenotype in Gene-Modified Animals

When double-stranded RNAs (1−10 μg) for mRNAs encoding PG fatty acyl reductase, PG Z11/Δ10,12 desaturase, PG acyl-CoA-binding protein, and PBANR were injected into 1-day-old female pupae of *B. mori*, all of the RNAi decreased the bombykol production after adult eclosion [9]. In *H. armigera*, RNAi for PBANR reduced male pheromone production [7].

Interaction of PBANR with FXPRLamide Peptides

The responses of PBANRs of *B. mori*, *H. zea*, and *S. littoralis* to PBAN, DH, and some other FXPRLamide peptides are similar, as determined by functional assays using PBANR-expressing cells [4,8,10]. However, *H. peltigera* PBANR does not respond to *S. littoralis* DH [10].

References

1. Raina AK, Klun JA. Brain factor control of sex pheromone production in the corn earworm moth. *Science*. 1984;225:531−533.
2. Raina AK, Jaffe H, Kempe TG, et al. Identification of a neuropeptide hormone that regulates sex pheromone production in female moths. *Science*. 1989;244:796−798.
3. Kitamura A, Nagasawa H, Kataoka H, et al. Amino acid sequence of pheromone-biosynthesis-activating neuropeptide (PBAN) of the silkworm, *Bombyx mori*. *Biochem Biophys Res Commun*. 1989;163:520−526.
4. Choi MY, Fuerst EJ, Rafaeli A, et al. Identification of a G-protein coupled receptor for pheromone biosynthesis activating neuropeptide from pheromone gland of the moth *Helicoverpa zea*. *Proc Natl Acad Sci USA*. 2003;100:9721−9726.
5. Hull JJ, Ohnishi A, Moto K, et al. Cloning and characterization of the pheromone biosynthesis activating neuropeptide receptor from the silkworm, *Bombyx mori*. Significance of the carboxyl terminus in receptor internalization. *J Biol Chem*. 2004;279:51500−51507.
6. Lee JM, Hull JJ, Kawai T, et al. Re-evaluation of the PBAN receptor molecule: characterization of PBANR variants expressed in the pheromone glands of moths. *Front Endocrinol*. 2012;3:1−12.
7. Jurenka R, Rafaeli A. Regulatory role of PBAN in sex pheromone biosynthesis of heliothine moths. *Front Endocrinol*. 2011;2:1−8.
8. Watanabe K, Hull JJ, Niimi T, et al. FXPRL-amide peptides induce ecdysteroidogenesis through a G-protein coupled receptor expressed in the prothoracic gland of *Bombyx mori*. *Mol Cell Endocrinol*. 2007;273:51−58.
9. Matsumoto S, Ohnishi A, Lee JM, et al. Unraveling the pheromone biosynthesis activating neuropeptide (PABN) signal transduction cascade that regulates sex pheromone production in moths. *Vitam Horm*. 2010;83:425−445.
10. Hariton-Shalev A, Shalev M, Adir N, et al. Structural and functional differences between pheromonotropic and melanotropic PK/PBAN receptors. *Biochim Biophys Acta*. 2013;1830:5036−5048.

Supplemental Information Available on Companion Website

- Accession numbers of cDNAs for PBAN and PBANR/E-Table 52B.1

Pyrokinin

Toshinobu Yaginuma and Teruyuki Niimi

Abbreviation: PK, MT, hugγ
Additional names: myotropin (MT), *hugin* peptide (hugγ and PK2)

PK/MT controls the muscle contraction of the heart, gut, and malpighian tubules. This class of peptides was isolated from cerebral and abdominal neurohemal organs of many insects. PK/MT peptides share the C-terminal FXPRL-NH₂, but their length and N-terminal sequences are variable.

Discovery

A small peptide having a C-terminal FXPRLamide was first isolated from the brain extract of the cockroach *Leucophaea maderae* as a peptide hormone that controls hindgut contraction [1]. Since the N-terminal end of the peptide is pyroglutamate (pE), it was named pyrokinin (PK) [1]. Subsequently, FXPRLamide peptides with myotropic activity but without N-terminal pE were isolated from the locust *Locusta migratoria*, and called myotropin (MT). Now, they are grouped as pyrokinin/myotropin (Table 52C.1) [2–4]. In the fruit fly *Drosophila melanogaster*, MT-II (PK2, βSGNP-like) and hugγ peptides arise from a precursor peptide (191 aa residues) encoded by the *hugin* gene (Figure 52C.1) [5,6]. Although *D. melanogaster* MT-II/PK2 shows moderate myostimulatory activity, hugγ peptide is structurally and functionally more similar to ecdysis-triggering hormone (ETH-1, DDSSPGFFLKITKNVPRLamide) [5,6]. Further, the *capability* (*capa*) gene encoding CAPA-PK (diapause hormone-like) has been cloned from insects, including *D. melanogaster* [7,8].

Table 52C.1 Amino Acid Sequences of Insect Pyrokinin/Myotropins and hugγ Peptide*

Leucophaea maderae	PK	pETSFTPRL-NH₂
Locusta migratoria	PK-I	pEDSDG WPQQPFVPRL-NH₂
L. migratoria	PK-II	pESVPTFTPRL-NH₂
L. migratoria	MT-I	GA VPAAQFSPRL-NH₂
L. migratoria	MT-II	EGDFTPRL-NH₂
L. migratoria	MT-III	RQQPFVPRL-NH₂
L. migratoria	MT-IV	RLH QNGMPFSPRL-NH₂
D. melanogaster	MT-I (CAPA-PK, DH-like)	TGPSA SSGLWFGPRL-NH₂
D. melanogaster	MT-II (PK2, βSGNP-like)	SVPFKPRL-NH₂
D. melanogaster	hugγ	QLQSNGE PAYRVRTPRL-NH₂

*Letters with gray background indicate a.a. residues conserved among the FXPRLamide family peptides; pE, pyroglutamate.

Structure

Structural Features

Most PK/MT peptides share a common C-terminal sequence FXPRL-NH₂. Therefore, these peptides are members of the FXPRLamide peptide family. PK/MT is widely conserved among insects (Table 52C.1).

Synthesis and Release

Gene, mRNA, and Precursor

In *D. melanogaster*, the *hugin* gene encoding MT-II (PK2) and hugγ peptides and the *capa* gene encoding MT-I (PK1) have been cloned. In the silkworm *Bombyx mori* the *capa* gene codes for CAPA-PK1, but its function is as yet unclear [4–9]. Unfortunately, the genes encoding PK/MTs of *L. maderae* and *L. migratoria* have not yet been reported [9].

Distribution of mRNA

The *hugin* mRNA in *D. melanogaster* is expressed in the subesophageal ganglion (SG) [4–9]. See also Subchapters 52A, 52B.

Tissue Content

L. maderae PK is found in the brain (1.4 pmol) [1]. *L. migratoria* PK and MT are detected in the brain, the corpus cardiacum–corpus allatum (CC–CA) complex, and the SG [4]. CAPA peptides are detected in the CNS, including the SG in *D. melanogaster*, the cockroach *Periplaneta americana*, and the moth *Manduca sexta* [8].

Receptors

Structure and Subtype

PK1/DH receptor (CG9918) and PK2/PBAN receptors (CG8784, CG8795) have been identified in *D. melanogaster*. The PK1/DH receptor expressed in CHO/G-16 cells shows higher affinity for MT-I/CAPA-PK1 (DH-like) than for MT-II/PK2, hugγ, and *L. maderae* PK, while the PK2/PBAN receptors show higher affinity for MT-II/PK2, hugγ, and *L. maderae* PK than for MT-I/CAPA-PK1 (DH-like) [9,10].

Biological Functions

Target Cells and Function

See Subchapters 52A and 52B for the distribution of DH and PBAN receptors, respectively. *L. maderae* PK shows moderate myostimulatory activity regarding hindgut contraction, but

Y. Takei, H. Ando, & K. Tsutsui (Eds): Handbook of Hormones. DOI: http://dx.doi.org/10.1016/B978-0-12-801028-0.00214-2

```
         10        20        30        40        50        60        70        80
MCGPSYCTLLLIAASCYILVCSHAKSLQGTSKLDLGNHISAGSARGSLSPASPALSEARQKRAMGDYKELTDIIDELEENSLAQKASA
         90        100       110       120       130       140       150       160       170
TMQVAAMPPQGQEFDLDTMPPLTYYLLLQKLRQLQSNGEPAYRVRTPRLGRSIDSWRLLDAEGATGMAGGEEAIGGQFMQRMVKKSV
                                   hugγ
         180       190
PFKPRLGKRAQVCGGD
PK2 (βSGNP-like)
```

Figure 52C.1 *D. melanogaster hugin* precursor protein. Boxes indicate hugγ and PK2 with a common C-terminal PRL.

strong activity regarding foregut and oviduct contraction [1]. *L. migratoria* PK and MT have a myostimulatory effect on cockroach hindgut contraction [2]. On the other hand, hugγ is thought to function as a molting controlling factor and to be involved in feeding behavior [5,6].

Phenotype in Gene-Modified Animals

Experiments to modify gene expression have been performed in *D. melanogaster* for the *hugin* gene and PK receptor gene. Ubiquitous ectopic *hugin* expression leads to larval death immediately after ecdysis to the third instar by interfering with the regulation of ecdysis [5]. Further, overexpression of *hugin* results in a strong reduction in growth, with no larvae surviving to pupal stage. Blocking synaptic transmission of hugγ neurons alters food intake behavior [6]. RNAi silencing of PK2 receptor causes embryonic lethality, while silencing of another PK2 receptor reduces embryonic and first instar larval survival, suggesting that PK2 is essential for *D. melanogaster* embryonic development [9,10].

References

1. Holman GM, Cook BJ, Nachman RJ. Isolation, primary structure and synthesis of a blocked neuropeptide isolated from the cockroach *Leucophaea maderae*. *Comp Biochem Physiol*. 1986;85C:219−224.
2. Schoofs L, Broeck JV, de Loof A. The myotropic peptides of *Locusta migratoria*: Structures, distribution, function and receptors. *Insect Biochem Mol Biol*. 1993;23:859−881.
3. Gade G, Hoffmann KH, Spring JH. Hormonal regulation in insects: facts, gaps, and future direction. *Physiol Rev*. 1997;77:963−1032.
4. Predel R, Kellner R, Nachman RJ, et al. Differential distribution of pyrokinin-isoforms in cerebral and abdominal neurohemal organs of the American cockroach. *Insect Biochem Mol Biol*. 1999;29:139−144.
5. Meng X, Wahlstrom G, Immonen T, et al. The *Drosophila* gene codes for mystimulatory and ecdysis-modifyng neuropeptides. *Mech Dev*. 2002;17:5−13.
6. Melcher C, Pankratz MJ. Candidate gustatory interneurons modulating feeding behavior in the *Drosophila* brain. *PloS Biol*. 2005;3: e305.
7. Roller L, Yamanaka N, Watanabe K, et al. The unique evolution of neuropeptide genes in the silkworm *Bombyx mori*. *Insect Biochem Mol Biol*. 2008;38:1147−1157.
8. Predel R, Wegener C. Biology of the CAPA peptides in insects. *Cell Mol Life Sci*. 2006;63:2477−2490.
9. Altstein M, Hariton A, Nachman RJ. The FXPRLamide (Pyrokinin/PBAN) peptide family. In: Kastin AJ, ed. *Handbook of Biologically Active Peptides*, 2nd ed. San Diego: Elsevier Press; 2013:255−266.
10. Cazzamali G, Torp M, Hauser F, et al. The *Drosophila* gene CG9918 codes for a pyrokinin-1 receptor. *Biochem Biophys Res Commun*. 2005;335:14−19.

Supplemental Information Available on Companion Website

- Accession numbers of cDNAs for PK/MT and receptors/ E-Table 52.1

CHAPTER 53

Crustacean Hyperglycemic Hormone

Tsuyoshi Ohira

Abbreviation: CHH

Additional names: family members include molt-inhibiting hormone (MIH), vitellogenesis-inhibiting hormone (VIH), and mandibular organ-inhibiting hormone (MOIH)

Pleiotropic hormones referred to as eyestalk hormones are produced by the most important crustacean endocrine organ, the X-organ sinus gland complex, and are members of the CHH family peptides.

Discovery

A CHH regulating hemolymph glucose levels as a hyperglycemic factor was first isolated and sequenced from the shore crab *Carcinus maenas* in 1989 [1]. Subsequently, a MIH and a VIH were isolated and characterized from *C. maenas* [2] and from the American lobster *Homarus americanus* [3], respectively. MIH controls molting by suppressing synthesis and/or secretion of ecdysteroids by the Y-organ. VIH regulates the onset of vitellogenesis by suppression of vitellogenin synthesis. Later, MOIH, inhibiting methyl farnesoate synthesis and/or secretion by the mandibular organ, was purified and structurally determined from the crab *Cancer pagurus* [4]. In *C. maenus*, it has been shown that CHH is produced in non-eyestalk tissues such as the pericardial organ and the gut [5,6].

Structure

Structural Features

CHH, MIH, VIH, and MOIH are members of the CHH family peptides. These peptides are mostly 72−78 aa residues in length, having 6 conserved cysteine residues that form three intramolecular disulfide bonds. CHH family peptides are divided into two subtypes based on the absence (type I) or presence (type II) of a glycine residue at position 12 in the mature peptide (Figure 53.1).

Primary Structure

Most eyestalk CHH belongs to type I and consists of 72 aa residues with a C-terminal amide. The C-terminal amide is important for its hyperglycemic activity [5]. In *H. americanus* and several crayfish species, stereoisomer isoforms (D-phenylalanine) have been found [5]. Long-type CHH isoforms (non-eyestalk type) produced by alternative splicing have been found in *C. maenas* and the giant freshwater prawn *Macrobrachium rosenbergii* [5]. N-terminal aa sequences of eyestalk CHH and non-eyestalk CHH are common (positions 1−40), but the remaining C-terminal sequences differ considerably. MIH, VIH, and MOIH are grouped as type II, because they have a characteristic glycine residue at position 12 [6]. MIH, VIH, and MOIH are longer than CHH. All of the peptides have a free N-terminus, and most of them are not amidated at the C-terminus. The molecular weights of CHHs range from 8,200 to 8,700, and those of MIHs, VIHs, and MOIHs range from 9,000 to 9,500.

Synthesis and Release

Gene, mRNA, and Precursor

CHH genes occur in multiple isoforms of two-exon (*Penaeus monodon* CHH1) [7], three-exon (*Metapenaeus ensis* CHHA and CHHB) [6], or four-exon genes (*C. maenas* and *M. rosenbergii* CHHs) [5]. Four-exon CHH genes express multiple mRNAs, eyestalk CHH mRNA (Exons 1, 2, and 4), and non-eyestalk CHH mRNA (Exons 1−4) by alternative splicing. The CHH precursor consists of a signal peptide, CHH-precursor related peptide, dibasic cleavage site, mature CHH, and an amidation signal (eyestalk CHH only). MIH and MOIH genes consist of three-exon genes only. There have been no data for VIH genes. Type II peptide (MIH, VIH, and MOIH) precursors consist of a signal peptide and mature peptide. The organization of a type II peptide precursor is apparently different from that of type I (CHH).

Distribution of mRNA

Eyestalk CHH mRNA is detected most abundantly in the X-organ, but also in non-eyestalk tissues such as the ventral nerve cord and gut. Long-type CHH isoforms (non-eyestalk type) are expressed in the pericardial organ in *C. maenus*, and in the heart, gills, antennal glands, and thoracic ganglion in *M. rosenbergii* [5]. Most type II peptide genes are expressed only in eyestalk neurons.

Regulation of Synthesis and Release

CHH is secreted to hemolymph in response to stressful conditions such as emersion, hypoxia, drastic thermal change, bacterial endotoxins, parasitism by dinoflagellates, and heavy metal pollutants [5]. Exercise also stimulates CHH release from the sinus gland. Serotonin (5-hydroxy-tryptamine; 5HT) is a CHH-releasing agent, while Leu-enkephalin and glucose are inhibitory chemicals.

Y. Takei, H. Ando, & K. Tsutsui (Eds): Handbook of Hormones. DOI: http://dx.doi.org/10.1016/B978-0-12-801028-0.00053-2

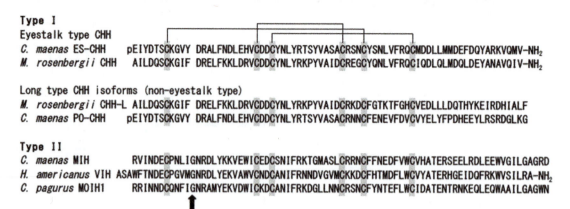

Figure 53.1 Alignment of CHH, MIH, VIH, and MOIH. N-terminal "pE" indicates pyroglutamic acid residue. Conserved cysteine residues are gray-boxed. Disulfide bridges are indicated by lines. Characteristic glycine residues in type II are indicated by an arrow. ES, eyestalk; PO, pericardial organ.

Receptors

CHH-, MIH-, VIH-, and MOIH-receptors have not yet been identified. Membrane-bound guanylate cyclase (GC) is presumed to be a candidate as a CHH receptor, because cGMP is used as a second messenger of CHH signaling [5]. Regarding second messengers involved in MIH signaling, both cAMP and cGMP are implicated in the regulation of ecdysteroidogenesis in various crustacean species [5]. Therefore, several reports suggest that the MIH receptor is a GPCR but not a membrane-bound GC.

Biological Functions

Crustacean Hyperglycemic Hormone

Eyestalk CHH controls crustacean hemolymph glucose levels as a hyperglycemic factor. Hyperglycemia is a result of mobilization of glycogen in target tissues such as the hepatopancreas and abdominal muscles. Steady CHH hemolymph levels of *Carcinus maenas* are 25 fmol/ml in the winter and 50–55 fmol/ml in the summer [8]. Recent studies have revealed that eyestalk CHHs in many species have pleiotropic functions, such as MIH, MOIH, and VIH activities [5,8,9]. Eyestalk CHH also possesses a role in osmo/ionoregulation at the gill in euryhaline crustacean species. CHH expressed in the gut, the aa sequence of which is identical with that of eyestalk CHH, contributes as an another osmoregulator in, for example, ion transport at the gills and water uptake for the purpose of enlarging body size after molting. A few studies have shown that eyestalk CHH is probably involved with inhibitory regulation of the androgenic gland (AG). The biological functions of long-type CHH isoforms (non-eyestalk type) have not yet been elucidated.

Molt-Inhibiting Hormone

MIH inhibits molting by suppressing synthesis and/or secretion of the molting hormone (ecdysteroid) by the Y-organ, except at the premolt stage when ecdysteroids are released to trigger ecdysis. The MIH hemolymph level of the American crayfish *Procambarus clarkii* is around 5–6 fmol/ml at the intermolt stage and decreases to 1.28 fmol/ml at the early premolt stage [10]. American lobster, *H. americanus*, CHH also showed MIH activity [5].

Vitellogenesis-Inhibiting Hormone

VIH inhibits vitellogenin synthesis by the ovary and/or hepatopancreas as well as vitellin accumulation in the oocytes, thus negatively controlling crustacean vitellogenesis. In some penaeid shrimp species, CHHs decrease vitellogenin gene expression in the cultured ovary *in vitro* [8].

Mandibular Organ-Inhibiting Hormone

The mandibular organ is the site of synthesis of methyl farnesoate (MF), the unepoxidated precursor of insect juvenile hormone III. The biological functions of MF in crustaceans have not yet been well defined. MOIH inhibits MF synthesis and/or secretion by the mandibular organ. In the spider crab *Libinia emarginata*, CHH showed both hyperglycemic activity and MF synthesis inhibition activity [5].

References

1. Kegel G, Reichwein B, Weese S, et al. Amino acid sequence of the crustacean hyperglycemic hormone (CHH) from the shore crab, *Carcinus maenas*. *FEBS Lett*. 1989;255:10–14.
2. Webster SG. Amino acid sequence of putative moult-inhibiting hormone from the crab *Carcinus maenas*. *Proc R Soc Lond B*. 1991;244:247–252.
3. Soyez D, Le Caer JP, Noel PY, Rossier J. Primary structure of two isoforms of the vitellogenesis inhibiting hormone from the lobster *Homarus americanus*. *Neuropeptides*. 1991;20:25–32.
4. Wainwright G, Webster SG, Wilkinson MC, Chung JS, Rees HH. Structure and significance of mandibular organ-inhibiting hormone in the crab, *Cancer pagurus*. *J Biol Chem*. 1996;271:12749–12754.
5. Webster SG, Keller R, Dircksen H. The CHH-superfamily of multifunctional peptide hormones controlling crustacean metabolism, osmoregulation, moulting, and reproduction. *Gen Comp Endocrinol*. 2012;175:217–233.
6. Wiwegweaw A, Udomkit A, Panyim S. Molecular structure and organization of crustacean hyperglycemic hormone genes of *Penaeus monodon*. *J Biochem Mol Biol*. 2004;37:177–184.
7. Chan SM, Gu PL, Chu KH, Tobe SS. Crustacean neuropeptide genes of the CHH/MIH/GIH family: implications from molecular studies. *Gen Comp Endocrinol*. 2003;134:214–219.
8. Chung JS, Zmora N, Katayama H, Tsutsui N. Crustacean hyperglycemic hormone (CHH) neuropeptides family: functions, titer, and binding to target tissues. *Gen Comp Endocrinol*. 2010;166:447–454.
9. Katayama H, Ohira T, Nagasawa H. Crustacean peptide hormones: structure, gene expression and function. *Aqua-BioSci Monogr*. 2013;6:49–90.
10. Nakatsuji T, Sonobe H. Regulation of ecdysteroid secretion from the Y-organ by molt-inhibiting hormone in the American crayfish, *Procambarus clarkii*. *Gen Comp Endocrinol*. 2004;135:358–364.

Supplemental Information Available on Companion Website

- Phylogenetic tree of the CHH family members/E-Figure 53.1
- Accession numbers of genes and cDNAs for precursors of CHH, MIH, VIH, and MOIH/E-Table 53.1

CHAPTER 54

Ion Transport Peptide

Naoaki Tsutsui

Abbreviation: ITP

An antidiuretic hormone that stimulates chloride-dependent fluid reabsorption in the ileum, ITP is a member of the crustacean hyperglycemic hormone family found in insects.

Discovery

ITP was originally purified and identified based on the stimulatory effect on chloride ion transport measured as changes in short circuit current through the epithelium of the ileum [1]. Its complete primary structure was determined by RT-PCR-based cDNA cloning in the desert locust *Schistocerca gregaria* [2]. The cDNA cloning also revealed the existence of a longer isoform of ITP called ITP-like (ITPL). Genes and mRNAs encoding ITP/ITPL and their homologs have been found in approximately 20 arthropod species.

Structure

Structural Features

Both ITP and ITPL share structural characteristics with the crustacean hyperglycemic hormone (CHH); six cysteine residues which form three intramolecular disulfide bonds are located in corresponding positions, and the bonding patterns are inferred to be the same as those of other peptides belonging to the CHH family.

Primary Structure

Most ITPs consist of 72 aa, with the exception of dipteran ITP which is composed of 73 aa. The length of ITPL ranges from 79 to 94 aa. Amino acid sequences of the N-terminal half are common in ITP and ITPL, while the C-terminal half is distinctive in length and aa sequence (Figure 54.1). ITP possesses an amidated C-terminus, which is necessary to confer full biological activity [3], except in the case of the red flour beetle *Tribolium castaneum* [4]. The Mr values of ITPs range from 8,360 to 8,910, and those of ITPLs range from 9,130 to 10,820. The predicted isoelectric point ranges from 4.5 to 6.9.

Synthesis and Release

Gene, mRNA, and Precursor

The ITP gene of the tobacco hornworm *Manduca sexta* consists of three exons, from which mRNAs for ITPL (Exons 1 and 3) and ITP (Exons 1, 2, and 3) are produced by alternative splicing [5]. MasITP mRNA has 909 bases that encode a putative signal peptide of 24 aa, an ITP precursor-related

peptide (IPRP) of 11 aa, a dibasic processing site, a mature hormone of 72 aa, and an amidation/cleavage site of 3 aa. The MasITPL precursor encoded by 1,030 bases of mRNA has similar structure to the MasITP precursor — a signal peptide, an IPRP, and a dibasic processing site — but the amidation/cleavage site is absent.

In *Drosophila melanogaster*, the ITP gene consists of five exons, and is thought to generate three mRNA splice forms: ITPL1 from Exons 1–5; ITPL2 from Exons 1, 2, 4, and 5; and ITP from Exons 1, 2, and 5 [6].

Distribution of mRNA

ITP mRNA is expressed in several pairs of brain cells, some of which have projections to the corpora cardiaca. ITPL mRNA is detected in the neurosecretory cells of the thoracic ganglia and abdominal ganglia as well as the brain [5,6].

Regulation of Synthesis and Release

In adult *S. gregaria*, ITP and ITPL are secreted into the hemolymph in response to the feeding stimulus, probably for postprandial osmoregulation, and to exercise stress, probably for metabolic regulation; concentrations of ScgITP and ScgITPL are below the limit of detection and 9.6 nM in the starved condition, and 0.5 nM and 41.5 nM in the fed condition, respectively [7]. ScgITP/ITPL becomes detectable in the hemolymph of the last instar nymph before the final molt to the adult [7], and MasITP/ITPL detected as CHH-related immunoreactive material is released into the hemolymph just before ecdysis [8]. For ScgITP, there are suggestions of a certain role for ITP/ITPL in the control of insect ecdysis. The amount of ITP immunoreactivity in reversed-phase HPLC separated fractions that coelute with synthetic ITP is about 0.5 nM, whereas the immunoreactivity to ITP has also been observed in more hydrophobic fractions: 0.2 nM (starved), and 14 nM (fed). The entity of this ITP immunoreactive factor has yet to be characterized.

Receptors

ITP receptors remain unidentified in any insect species. GPCR is presumed to be the candidate, since ion transport-stimulating effects are mediated by cAMP. cGMP, as well as cAMP, is reported to act as a second messenger in the signal transduction process of ITP in the desert locust, suggesting the existence of two distinct receptors — probably membrane-bound guanylate cyclase, and GPCR [9].

Biological Functions

ITP facilitates chloride ion transport from the lumen to the hemolymph and inhibits acid secretion into the lumen in the

Y. Takei, H. Ando, & K. Tsutsui (Eds): Handbook of Hormones. DOI: http://dx.doi.org/10.1016/B978-0-12-801028-0.00054-4

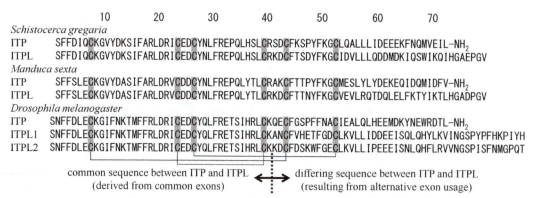

Figure 54.1 **Comparison of amino acid sequences of ITP/ITPL in selected insect species.** Positions of disulfide bridges are indicated by solid lines.

ileum, leading to generation of the electrochemical gradient driving water resorption [1]. As ITPL inhibits the stimulatory effect on ion transport exerted by ITP, it is proposed that ITPL released from peripheral neurosecretory cells may act as a competitive inhibitor of ITP [3,6]. Moreover, ITP and ITPL are thought to be involved in various biological processes. In *T. castaneum*, knockdown of ITP/ITPL-1/ITPL-2 using double-stranded RNAs in the larval stages causes an increase of mortality in subsequent developmental stages, finally attaining almost 100% around the time of eclosion. Knockdown during the pupal stage results in a significant reduction in the number of eggs produced and in the survival rate of offspring, accompanied by developmental defects of the ovary [4]. ITP and/or ITPL possibly contribute to the control of ecdysis [8]. Additionally, ITP and ITPL are each supposed to function as a neurotransmitter/neuromodulator as well as a hormone circulating in the hemolymph. For instance, it is suggested that ITP is one of the neurotransmitters released from the clock neurons of *Drosophila* brain [10].

References

1. Audsley N, McIntosh C, Phillips JE. Isolation of a neuropeptide from locust corpus cardiacum which influences ileal transport. *J Exp Biol*. 1992;173:261–274.
2. Meredith J, Ring M, Macins A, et al. Locust ion transport peptide (ITP): primary structure, cDNA and expression in a baculovirus system. *J Exp Biol*. 1996;199:1053–1061.
3. Wang Y-J, Zhao Y, Meredith J, et al. Mutational analysis of the C-terminus in ion transport peptide (ITP) expressed in *Drosophila* Kc1 cells. *Arch Insect Biochem Physiol*. 2000;45:129–138.
4. Begum K, Li B, Beeman RW, et al. Functions of ion transport peptide and ion transport peptide-like in the red flour beetle *Tribolium castaneum*. *Insect Biochem Mol Biol*. 2009;39:717–725.
5. Dai L, Zitnan D, Adams ME. Strategic expression of ion transport peptide gene products in central and peripheral neurons of insects. *J Comp Neurol*. 2007;500:353–367.
6. Dircksen H, Tesfai LK, Albus C, et al. Ion transport peptide splice forms in central and peripheral neurons throughout postembryogenesis of *Drosophila melanogaster*. *J Comp Neurol*. 2008;509: 23–41.
7. Audsley N, Meredith J, Phillips JE. Haemolymph levels of *Schistocerca gregaria* ion transport peptide and ion transport-like peptide. *Physiol Entomol*. 2006;31:154–163.
8. Drexler AL, Harris CC, Dela Pena MG, et al. Molecular characterization and cell-specific expression of an ion transport peptide in the tobacco hornworm, *Manduca sexta*. *Cell Tissue Res*. 2007;329: 391–408.
9. Audsley N, Jensen D, Schooley DA. Signal transduction for *Schistocerca gregaria* ion transport peptide is mediated via both cyclic AMP and cyclic GMP. *Peptides*. 2013;41:74–80.
10. Johard HA, Yoishii T, Dircksen H, et al. Peptidergic clock neurons in *Drosophila*: ion transport peptide and short neuropeptide F in subsets of dorsal and ventral lateral neurons. *J Comp Neurol*. 2009;516:59–73.

Supplemental Information Available on Companion Website

- Gene and mRNA structures/E-Figure 54.1
- Accession numbers of genes and cDNAs for ITP and ITPL/E-Table 54.1

CHAPTER 55

Prothoracicotropic Hormone

Tadafumi Konogami, Kazuki Saito and Hiroshi Kataoka

Abbreviation: PTTH
Additional name: prothoracicotropin

One of the earliest hormones ever discovered from insects, PTTH regulates insect molting and metamorphosis by promoting biosynthesis and release of a steroidal molting hormone, ecdysone, from the prothoracic glands.

Discovery

The prothoracicotropic hormone (PTTH) was discovered in 1922 as a hormonal factor causing insect pupation [1], and the partial structure of PTTH of the silkworm *Bombyx mori* was first identified in 1987 [2].

Structure

Structural Features of Silkworm PTTH

Mature silkworm PTTH is composed of two identical 109-aa residue peptide chains, each of which possesses 7 cysteine residues (Figure 55.1) [3]. The hormone forms a homodimer linked by an interchain disulfide bond, $Cys^{15}-Cys^{15'}$, while each chain may take a cysteine knot structural motif, characterized by three intrachain disulfide bonds: $Cys^{17}-Cys^{54}$, $Cys^{40}-Cys^{96}$, and $Cys^{48}-Cys^{98}$ [3]. Silkworm PTTH has been reported to have a carbohydrate chain at asparagine[41] [4], although its recombinant protein lacking the sugar chain also exhibits activity to induce adult development of the silkworms [3].

Structures of Other Insect PTTHs

Although positions of the seven cysteine residues are conserved among PTTHs of any insects, their primary sequences vary with insect species (Figure 55.2). For example, the sequence of the silkworm PTTH shares 40–60% identity with those of other lepidopteran PTTHs, while it shows much lower identity (~20%) with that of a dipteran fruit fly, *Drosophila melanogaster*. In addition, most of the lepidopteran hormones have a potential N-glycosylation site(s), while that of *D. melanogaster* does not have the site.

Synthesis and Release

Gene, mRNA, and Precursor

The silkworm PTTH gene, composed of five exons and four introns [5], is located on chromosome 22. The gene, ~3,300 bp in length, is transcribed and translated into a precursor preproPTTH protein of ~200 aa residues, and converted into the active mature peptide by sequential removals of the signal peptide and N-terminal prepro sequences.

Distribution of mRNA

An mRNA coding PTTH is expressed in two pairs of dorsolateral neurosecretory cells in the brain.

Regulation of Synthesis and Release

The silkworm PTTH produced in the brain cells is transported through axons and stored in the corpus allata, while that of *D. melanogaster* remains in the hormonogenic neuronal cells. After the stored PTTH is released in response to extrinsic stimuli, it acts on the prothoracic glands (PGs) to promote the synthesis and release of a steroidal molting hormone, ecdysone. In the case of silkworms, PTTH is released from the corpus allata to the hemolymph, and transient increases of PTTH concentration in the hemolymph cause activation of ecdysone biosynthesis in PGs [6]. In contrast, fruit fly PTTH is secreted directly near to the PG cells in the ring gland, from termini of neurons projecting to the gland. Although various extrinsic environmental stimuli can be triggers for PTTH release from the storage organs, a change in light–dark cycle hours is a major factor, at least for the tobacco hornworm *Manduca sexta* and silkworm *B. mori* [7], in altering the PTTH titers in their hemolymph. Since PTTH release is induced by carbachol and serotonin *in vitro*, the titers may be regulated through neuronal activation in response to extrinsic environmental stimuli.

Receptor

Structure and Subtype

A receptor for PTTH, named Torso, was identified as one of the receptor tyrosine kinases (RTKs) (E-Figure 55.1) [8]. The intracellular kinase domain of the receptor is classified into the type 3 subfamily of RTK that has an additional sequence of 50–100 residues in the domain, and phosphorylation of tyrosine residues in the sequence may be involved in engagement with the downstream signal transduction. However, due to poor homology of its peptide sequence with those of other RTKs, the domain structure of the extracellular region has not been predicted for Torso.

Signal Transduction Pathway

In *D. melanogaster*, ligand activation of Torso causes phosphorylation of downstream enzymes such as Ras, Raf, and

Y. Takei, H. Ando, & K. Tsutsui (Eds): Handbook of Hormones. DOI: http://dx.doi.org/10.1016/B978-0-12-801028-0.00055-6

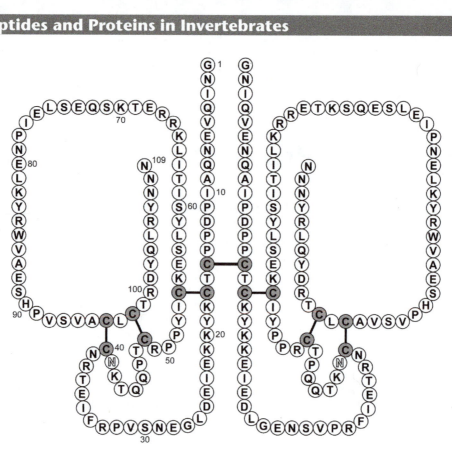

Figure 55.1 Primary structure of silkworm PTTH.

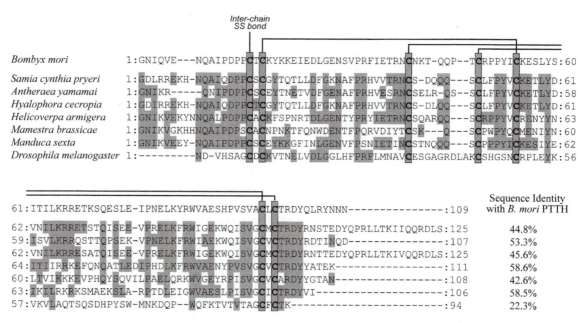

Figure 55.2 Sequence alignment of PTTHs from different insect species.

ERK [8]. Torso may be one of the portals for the ERK signaling pathway, because ERK phosphorylation has been observed in culture cells expressing Torso, as well as in the silkworm PGs, when stimulated by PTTH (Figure 55.3) [8]. On the other hand, in the tobacco hornworm *M. sexta* and silkworm *B. mori*, PTTH-dependent increases of intracellular calcium concentration and subsequently cAMP were observed in their PG cells.

Biological Functions

Target Cells/Tissues and Function

PTTH acts on PGs and promotes the ecdysone biosynthesis therein. PTTH sensitivity of the glands varies with the growth of insects. For example, in fifth-instar silkworm larvae the sensitivity is lower in the early days but becomes higher as they grow. PTTH has also been found to be involved in the

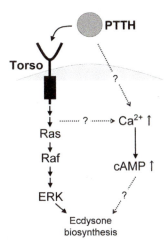

Figure 55.3 Intracellular signaling pathways in insect steroidogenic cells activated by PTTH stimulation.

		Tested debrained pupa *of*		
		Bombyx mori	*Samia cynthia ricini*	*Manduca sexta*
Injected PTTH *from*	*Bombyx mori*	+	−	−
	Samia cynthia ricini	−	+	−
	Manduca sexta	−	−	+

+ ; induced the eclosion at ~0.3 ng/pupa
− ; not induced the eclosion

Figure 55.4 Cross-activities of various PTTHs between different insects.

regulation of diapause. In the cabbage army moth *Mamestra brassicae*, the hemolymph PTTH titer is reduced during the diapause period, as compared to those before the introduction of diapause [9], and recovers to a higher original level after the moth breaks away from the diapause state.

Specificity of Biological Activity

In the debrained pupal assay, silkworm PTTH induces adult development of silkworms themselves but it works on neither the Eri-silkmoth *Samia cynthia ricini* nor the tobacco hornworm *M. sexta*. The converse is also true (Figure 55.4). PTTH may be much less effective in insects other than an original source; that is, the species specificity of PTTH is quite strict [10].

Bioassay

Using dissected PG samples, PTTH activity can be estimated by measuring ecdysone released from the isolated prothoracic glands to the culture medium. Amounts of the released ecdysteroids are usually determined by radioimmunoassay or LC-MS/MS. Alternatively, using debrained insect pupae, PTTH activity can be measured as titers for induction of adult development, after injection of PTTH.

References

1. Kopeć S. Studies on the necessity of the brain for the inception of insect metamorphosis. *Biol Bull*. 1922;42:323–342.
2. Kataoka H, Nagasawa H, Isogai A, et al. Isolation and partial characterization of a prothoracicotropic hormone of the silkworm, *Bombyx mori. Agric Biol Chem*. 1987;51:1067–1076.
3. Ishibashi J, Kataoka H, Isogai A, et al. Assignment of disulfide bond location in prothoracicotropic hormone of the silkworm,

Bombyx mori: A homodimeric peptide. *Biochemistry*. 1994;33:5912–5919.
4. Nagata S, Kobayashi J, Kataoka H, et al. Structural determination of an *N*-glycan moiety attached to the prothoracicotropic hormone from the silkmoth *Bombyx mori. Biosci Biotech Biochem*. 2014;78:1381–1383.
5. Adachi-Yamada T, Iwami M, Kataoka H, et al. Structure and expression of the gene for the prothoracicotropic hormone of the silkmoth *Bombyx mori. Eur J Biochem*. 1994;220:633–643.
6. Mizoguchi A, Ohashi Y, Hosoda K, et al. Developmental profile of the changes in the prothoracicotropic hormone titer in hemolymph of the silkworm *Bombyx mori*: correlation with ecdysteroid secretion. *Insect Biochem Mol Biol*. 2001;31:349–358.
7. Mizoguchi A, Dedos SG, Fugo H, et al. Basic pattern of fluctuation in hemolymph PTTH titers during larval-pupal and pupal-adult development of the silkworm, *Bombyx mori. Gen Comp Endocrinol*. 2002;127:181–189.
8. Rewitz KF, Yamanaka N, Gilberty LI, et al. The insect neuropeptide PTTH activates receptor tyrosine kinase Torso to initiate metamorphosis. *Science*. 2009;326:1403–1405.
9. Mizoguchi A, Ohsumi S, Kobayashi K, et al. Prothoracicotropic hormone acts as a neuroendocrine switch between pupal diapause and adult development. *PLoS One*. 2013;8:e60824.
10. Nagata S, Kataoka H, Suzuki A. Silk moth neuropeptide hormones: prothoracicotropic hormone and others. *Ann NY Acad Sci*. 2005;1040:38–52.

Supplemental Information Available on Companion Website

- Primary structures of the PTTH receptor Torso/E-Figure 55.1
- Accession numbers of genes and cDNAs for PTTH and Torso/E-Tables 55.1, 55.2

CHAPTER 56

Bursicon

Dušan Žitňan and Ivana Daubnerová

Abbreviation: burs (burs-α), pburs (burs-β)

Bursicon is a heterodimer glycoprotein hormone that is involved in the regulation of ecdysis and post-ecdysis processes associated with expansion, tanning, and sclerotization of the cuticle in insects and probably other arthropods.

Discovery

Initially bursicon was isolated as a 30- to 40-kDa protein in several insects, based on cuticle tanning assays. The partial aa sequences of a protein with bursicon-like bioactivity were identified in the cockroach central nervous system (CNS) [1]. A BLAST search with these sequences was used to identify a gene encoding bursicon (burs; CG13419) in the *Drosophila melanogaster* (fruit fly) genome. Later, a partner of burs (pburs; CG15284) along with the bursicon receptor (rickets) was identified and characterized. Descriptions of both bursicon subunits, named either burs and pburs or burs-α and burs-β, and its receptor were independently published the same year [1–3]. Bursicon genes are highly conserved in many arthropods.

Structure

Structural Features

Bursicon is a cysteine-knot heterodimer glycoprotein composed of two subunits, burs (burs-α) and pburs (burs-β) (Figure 56.1). However, in several neurons it is also produced as a monomer or homodimer [1,2].

Primary Structure

The homologs of *burs* and *pburs* have been identified or deduced in many insects, crustaceans, a tick, and a sea urchin. The position of cysteines is conserved in all animals [1,2].

Properties

burs, Mr 15,678; pburs, Mr 13,059.

Synthesis and Release

Gene, mRNA, and Precursor

In the moth *Bombyx mori*, mRNA of burs (burs-α) has 483 nt that encode a 160-aa precursor containing a 20-aa signal peptide and a mature protein of 140 aa residues; pburs (burs-β) mRNA has 611 nt and encodes 137-aa precursor that consists of a signal peptide of 24 aa residues and a mature peptide of 113 aa residues.

Distribution of mRNA

Bursicon is expressed in the CNS in a specific segmental network of neurons named 27/704. In the larval CNS of the moth *Manduca sexta*, bursicon expression is restricted to the subesophageal ganglion (SG), thoracic ganglia, and first abdominal ganglion. Pharate pupae and pharate adults show expression of this heterodimer in all ganglia [4]. In *Drosophila* larvae, expression of a bursicon heterodimer is confined to abdominal neuromeres 1–4; the additional neurons in the ventral nerve cord express only burs. In pharate adults, bursicon is produced by 2 neurons in the SG and 14 abdominal neurons [5,6].

Hemolymph and Tissue Concentration

Tissue: Crab *Callinectes sapidus*, 36 pmol per animal [7].

Hemolymph: *C. sapidus*, ~20 pM during the molt cycle and ~90 pM at ecdysis [7].

Regulation of Synthesis and Release

In larvae of *D. melanogaster*, bursicon appears to be released in two temporal waves, the first prior to the onset of ecdysis and the second immediately after ecdysis onset. In adults, bursicon is also released in two consecutive waves. Initial bursicon secretion from two neurons in the SG subsequently elicits its secondary release from a cluster of abdominal neurons, which controls expansion, hardening, and pigmentation of the cuticle and wings [5–7].

Receptors

Structure and Subtype

The *Drosophila* leucine-rich G protein-coupled receptor (*DLGR2*) encoded by *rickets* is the bursicon receptor. It belongs to the leucine-rich subfamily of GPCRs. The *DLGR2* gene contains 15 exons and 5,407 bp mRNA codes for a 1,360-aa precursor protein [1–3].

Signal Transduction Pathway

In *D. melanogaster*, a cAMP and protein kinase A signaling pathway is involved [8].

Biological Functions

Target Cells/Tissues and Functions

In *D. melanogaster*, bursicon controls cuticle sclerotization and melanization [5,6,8]. In *D. melanogaster*, *B. mori*, and the beetle *Tribolium castaneum* it is necessary for expansion and hardening of the wings [1,2]. Expansion of wings requires bursicon-induced extensibility of the wing cuticle, as

Y. Takei, H. Ando, & K. Tsutsui (Eds): Handbook of Hormones. DOI: http://dx.doi.org/10.1016/B978-0-12-801028-0.00056-8

burs (burs-α)

```
        10        20        30        40        50        60        70
FPVTGHEVQLPPGTKFFCQECQMTAVIHVLKHRGCKPKAIPSFACIGKCTSYVQVSGSKIWQMERTCNCC
        80        90       100       110       120       130       140
QESGEREATVVLFCPDAQNEEKRFRKVSTKAPLQCMCRPCGSIEESSIIPQEVAGYSEEGPLYNHFRKSL
```

pburs (burs-β)

```
        10        20        30        40        50        60        70
EENCETVASEVHVTKEEYDEMGRLLRSCSGEVSVNKCEGMCNSQVHPSISSPTGFQKECFCCREKFLRER
        80        90       100       120
LVTLTHCYDPDGIRFEDEENALMEVRLREPDECECYKCGDFSR
```

Figure 56.1 Primary structures of burs and pburs in *D. melanogaster*.

demonstrated in the moth *M. sexta* [4]. In *D. melanogaster*, pburs is an important signal for controlling ecdysis behavior [9]. In the crab *C. sapidus*, bursicon controls deposition and thickening of new cuticle, as well as granulation of hemocytes [7].

Phenotype in Gene-Modified Animals

Drosophila mutants lacking bursicon or its receptor (rickets) display apparent defects in melanization of the cuticle and wing expansion. Targeted inhibition of excitability of bursicon neurons in the SG supresses the release of bursicon from abdominal neurons and consequent hardening and tanning of cuticle and wing expansion. It also suppresses programmed cell death of the abdominal bursicon neurons after eclosion of adults [5,6,8]. Genetically null mutants of *pburs* alone show severe defects at ecdysis, while surviving adults display defects in cuticle pigmentation/hardening and wing expansion [9].

References

1. Honegger HW, Dewey EM, Ewer J. Bursicon, the tanning hormone of insects: recent advances following the discovery of its molecular identity. *J Comp Physiol*. 2008;194:989–1005.

2. Van Loy T, Van Hiel MB, Vandersmissen HP, et al. Evolutionary conservation of bursicon in the animal kingdom. *Gen Comp Endocrinol*. 2007;153:59–63.

3. Luo CW, Dewey EM, Sudo S, et al. Bursicon, the insect cuticle-hardening hormone, is a heterodimeric cystine knot protein that activates G protein-coupled receptor LGR2. *Proc Natl Acad Sci USA*. 2005;102:2820–2825.

4. Dai L, Dewey EM, Zitnan D, et al. Identification, developmental expression, and functions of bursicon in the tobacco hawkmoth, *Manduca sexta*. *J Comp Neurol*. 2008;506:759–774.

5. Luan H, Lemon WC, Peabody NC, et al. Functional dissection of a neuronal network required for cuticle tanning and wing expansion in *Drosophila*. *J Neurosci*. 2006;26:573–584.

6. Peabody NC, Diao F, Luan H, et al. Bursicon functions within the *Drosophila* CNS to modulate wing expansion behavior, hormone secretion, and cell death. *J Neurosci*. 2008;28:14379–14391.

7. Chung JS, Katayama H, Dircksen H. New functions of arthropod bursicon: inducing deposition and thickening of new cuticle and hemocyte granulation in the blue crab, *Callinectes sapidus*. *PLoS One*. 2012;7(9):e46299.

8. Loveall BJ, Deitcher DL. The essential role of bursicon during *Drosophila* development. *BMC Dev Biol*. 2010;10(92):8.

9. Lahr EC, Dean D, Ewer J. Genetic analysis of ecdysis behavior in *Drosophila* reveals partially overlapping functions of two unrelated neuropeptides. *J Neurosci*. 2012;32:6819–6829.

Supplemental Information Available on Companion Website

- Accession numbers of genes for bursicon and its receptor (rickets) in insects and crustaceans/E-Tables 56.1 and 56.2.
- Comparison of bursicon precursors in the silkmoth *Bombyx mori* (*B.m.*), the fruitfly *Drosophila melanogaster* (*D.m.*), the green crab *Carcinus maenas* (*C.m.*), and the deer tick *Ixodes scapularis* (*I.s.*)/E-Figure 56.1.

CHAPTER 57

Allatotropin

Akira Mizoguchi

Abbreviation: AT
Additional names: accessory gland myotropin (locust), myoactive peptide (mollusks)

Allatotropin is a multifunctional peptide found in a wide variety of invertebrates, showing myoregulatory activity in common. The name derives from its activity in regulating juvenile hormone biosynthesis in insects.

Discovery

This peptide was originally identified in a lepidopteran insect, *Manduca sexta*, in 1989 as a neuropeptide that stimulates the biosynthesis of juvenile hormone (JH) by the corpora allata *in vitro* [1]; however, subsequent studies have revealed multiple biological activities of this peptide. Similar peptides have been identified in other families of insects. Some of the myotropic factors identified in other phyla including mollusks, annelids, and hydras show high similarity to insect allatotropins, constituting a large family of peptides [2].

Structure

Structural Features

M. sexta allatotropin (Manse-AT) is a tridecapeptide with its C-terminus amidated. All the members of this family of peptides, either isolated or predicted from cDNA sequences, have a similar sequence length (13−16 aa residues) and an amidated C-terminus (Table 57.1).

Primary Structure

The presence of peptides identical to Manse-AT has been confirmed by sequence analysis of isolated peptides, or predicted from cDNA sequences of several lepidopteran insects. Similar peptides have been isolated or predicted in other orders of insects, including Orthoptera, Coleoptera, and Diptera [2]. Several aa sequences at the N- and/or C-terminus are highly conserved among insect ATs. Molluskan and annelid ATs also share the N-terminal GF(R/K) and C-terminal GFamide sequences with insect ATs. The predicted precursor proteins of Manse-AT contain AT-related peptides, Manse-ATL (allatotropin-like)-I, -II, and −III [3].

Properties

Extractable by acid alcohol. Heat stable [1].

Synthesis and Release

Gene, mRNA, and Precursor

Three alternatively spliced mRNAs for Manse-AT have been characterized [3]. Each mRNA encodes a precursor protein, from which one copy of Manse-AT is derived. The two longer mRNAs contain the insertion of one or two alternative exons, each of which codes for one or two additional peptides structurally related to Manse-AT (Manse-ATL-I, -II, -III). A cDNA coding for a precursor protein containing Manse-AT and a peptide similar to Manse-ATL-III have been reported in other Lepidoptera, suggesting the presence of AT-like peptide in addition to AT is not limited to *M. sexta*.

Distribution of mRNA and Peptide

In *M. sexta*, the Manse-AT transcripts are alternatively spliced in a tissue-specific manner [3]: the brain and frontal ganglion predominantly express the mRNA isoform I, while the nerve cord expresses predominantly isoform II. Isoform III is expressed in the brain at a lower level. The level of gene expression also changes considerably during development. Manse-AT immunoreactivity has been detected in the brain, frontal ganglion, subesophageal ganglion, thoracic and abdominal ganglia, and midgut endocrine cells of *M. sexta*, although the pattern of distribution differs between larvae and pharate adults [2]. There is close correlation between Manse-AT-immunoreactive cells and Manse-AT mRNA-containing cells localized by *in situ* hybridization [2]. In the kissing bug *Triatoma infestans* (Hemiptera), an AT-like peptide is suggested to be produced by the malpighian tubules [4].

Tissue Content

AT contents in the brain, corpus cardiacum−corpus allatum (CC−CA) complex, and ventral nerve cord of *M. sexta* adults are 1.2, 0.01, and 1.7 pmol/insect, respectively [5].

Receptors

Structure

The AT receptor (ATR) was first identified in the silkmoth *Bombyx mori* as a G protein-coupled receptor that can react with Manse-AT *in vitro* when expressed in cultured cells. cDNAs for its ortholog have been cloned in some insects, including *M. sexta*, where the predicted receptor protein has 457 aa residues and 7 transmembrane domains [6]. This receptor shows sequence similarity to orexin receptors of vertebrates, although there is no orexin-like peptide found in insects and other invertebrates. The orexin-like receptors

Y. Takei, H. Ando, & K. Tsutsui (Eds): Handbook of Hormones. DOI: http://dx.doi.org/10.1016/B978-0-12-801028-0.00057-X

Table 57.1 Comparison of the Amino Acid Sequences of ATs and AT-like Peptides

Manse—AT (moth)	− − −GFK−NVEMMTARGF−NH$_2$
Locmi−AG−Myo−I (locust)	− − −GFK−NVALSTARGF−NH$_2$
Aedas−ATL (mosquito)	− −APFR−NSEMMTARGF−NH$_2$
Fusfe−FEP4 (shellfish)	− − −GFRMNSSNRVAHGF−NH$_2$
Lymst−MATP (snail)	− − −GFRANSASRVAHGY−NH$_2$
Eisfo−ETP (earthworm)	− − −GFKDGAADRISHGF−NH$_2$
Manse−ATL−I	− −GTFKPNSNILIARGY−NH$_2$
Manse−ATL−II	GTPTFK−SPTVGIARDF−NH$_2$
Manse−ATL−III	− −PWFNPKSKLLVSTRF−NH$_2$

Conserved aa residues are shaded.

(i.e., "ATRs") are also found in the genomes of several insects from different orders, including Diptera, Coleoptera, Hymenoptera, and Lepidoptera [6], and some of them have been characterized for the spatio-temporal pattern of gene expression and ligand specificity. Various tissues express the ATR genes in a stage-specific and a species-specific manner. Expression in the CA is high in *M. sexta* and the mosquito *Aedes aegypti* [7], but low in *B. mori*, in which the CC shows a much higher level of gene expression [8]. The binding affinities of AT and AT-like peptides to ATR have been examined in *M. sexta* by using ATR-expressing cell lines [6]. The affinity of Manse-AT-like peptide-I is much higher than that of AT, suggesting the possibility that the identified receptor is for Manse-AT-like peptide-I rather than for AT.

Signal Transduction Pathway

The action of AT on the CA is suggested to be mediated by Ca^{2+} and phosphoinositide signaling [9].

Biological Functions

Target Cells/Tissues and Functions

Manse-AT stimulates JH biosynthesis by adult female CA from many lepidopteran insects, and larval CA from some lepidopteran species [2]. Manse-AT is also reported to stimulate the CA of non-lepidopteran insects, such as the honey bee and blow fly, although no allatotropic peptide has been isolated from these insects [2]. The expression of ATR in the CA observed in some insects readily explains the action of AT on the CA, but its expression in the CC rather than the CA in *B. mori* does not. In this species, since the ATR-expressing cells in the CC produce short neuropeptide-F (sNPF) and this peptide inhibits JH synthetic activity of the CA *in vitro*, it is proposed that ATR may indirectly mediate AT activity on the CA by suppressing sNPF action on the CA [8]. Besides the action on the CA, Manse-AT accelerates heart rate, stimulates oscillation of the ventral diaphragm, regulates foregut movement,

inhibits ion transport across the midgut epithelium, and controls the release of digestive enzymes in the midgut in lepidopteran insects [2]. Furthermore, Manse-AT is suggested to be involved in the photic entrainment of the circadian clock in the cockroach [2]. Manse-ATL peptides also show some of these biological activities. AT family peptides from hemimetabolous insects such as locust myotropin also show cardiostimulatory and myoregulatory activity [2]. The effects of AT on non-insect invertebrates have been examined using synthetic mosquito AT. This AT showed myoregulatory activity on the turbellarians and hydra [10]. A widespread role of AT in myoregulation and the presence of AT in organisms that do not have JH suggest that the primary role of AT is myoregulation and that the control of JH synthesis is the secondary function.

References

1. Kataoka H, Toschi A, Li JP, Carney RL, et al. Identification of an allatotropin from adult *Manduca sexta*. *Science*. 1989;243:1481−1483.
2. Elekonich MM, Horodyski FM. Insect allatotropins belong to a family of structurally-related myoactive peptides present in several invertebrate phyla. *Peptides*. 2003;24:1623−1632.
3. Horodyski FM, Bhatt SR, Lee K-Y. Alternative splicing of transcripts expressed by the *Manduca sexta* allatotropin (Mas-AT) gene is regulated in a tissue-specific manner. *Peptides*. 2001;22:263−269.
4. Santini MS, Ronderos JR. Allatotropin-like peptide in Malpighian tubules: insect renal tubules as an autonomous endocrine organ. *Gen Comp Endocrinol*. 2009;160:243−249.
5. Veenstra JA, Hagedorn HH. Sensitive enzyme immunoassay for *Manduca* allatotropin and the existence of an allatotropin-immunoreactive peptide in *Periplaneta americana*. *Arch Insect Biochem Physiol*. 1993;23:99-09
6. Horodyski M, Verlinden H, Filkin N, et al. Isolation and functional characterization of an allatotropin receptor from *Manduca sexta*. *Insect Biochem Mol Biol*. 2011;41:804−814.
7. Nouzova M, Brockhoff A, Mayoral JG, et al. Functional characterization of an allatotropin receptor expressed in the corpora allata of mosquitoes. *Peptides*. 2012;34:201−208.
8. Yamanaka N, Yamamot S, Zitnan D, et al. Neuropeptide receptor transcriptome reveals unidentified neuroendocrine pathways. *PLoS One*. 2008;3:e3048.
9. Rachinsky A, Tobe SS. Role of second messengers in the regulation of juvenile hormone production in insects, with particular emphasis on calcium and phosphoinositide signaling. *Arch Insect Biochem Physiol*. 1996;33:259−282.
10. Alzugaray ME, Adami ML, Diambra LA, et al. Allatotropin: an ancestral myotropic neuropeptide involved in feeding. *PLoS One*. 2013;8:e720.

Supplemental Information Available on Companion Website

- Accession numbers of the cDNAs for allatotropins and allatotropin receptors/E-Tables 57.1, 57.2

CHAPTER 58

Allatostatin-A

Akira Mizoguchi

Abbreviation: AST-A
Additional names: A-type allatostatin (A-type AST), cockroach allatostatin, FGLamide

One of three types of allatostatins, AST-A inhibits juvenile hormone biosynthesis by the corpora allata in some groups of insects and exhibits myoinhibitory actions in a wide variety of arthropods.

Discovery

AST-A (AST herein) was first identified in 1989 as four similar neuropeptides that inhibit juvenile hormone (JH) biosynthesis by the corpora allata (CA) of the cockroach *Diploptera punctata*. Subsequently, molecular cloning of the gene for its precursor revealed that the four peptides are derived from a single precursor protein together with additional nine similar peptides. Homologous peptides have been identified in many other insects and in the crustaceans [1−3]. Furthermore, evidence for the existence of AST-like peptides in the nervous system has been reported in other invertebrates, including mollusks, annelida, cnidaria, flatworms, and nematodes [1−3].

Structure

Structural Features

This family of peptides is characterized by the presence of a highly conserved pentapeptide sequence at the C-terminus [1] (Table 58.1).

Primary Structure

Most peptides of this family consist of 6−18 aa residues with a common C-terminal motif, (Y/F)XFG(L/I) amide, which is essential for their biologic activity [1]. So far, hundreds of distinct AST peptides have been identified across the animal kingdom. One estimate suggests that, within Arthropoda, a total of 431 AST sequences have been identified, of which 233 are different sequences and 168 are specific to a species. There is little overlap between hemimetabolous and holometabolous insects, and between insects and crustaceans [4].

Synthesis and Release

Gene, mRNA, and Precursor

There is a single copy of the AST gene per haploid genome [5]. In each species, varying numbers of AST peptides are generated from a single precursor protein by endoproteolytic cleavage at dibasic sites (E-Figure 58.1). The C-terminus of predicted AST peptides in the precursor sequence is all glycine, suggesting that these peptides are amidated in mature forms [1,2]. The number of AST peptides derived from the precursor is 2−14 in insects and up to 35 (or more) in crustaceans [4,5]. Some long AST peptides have a dibasic cleavage site within the sequence, thus potentially generating a shorter peptide [1,4].

Distribution of mRNA and Peptide

In cockroaches, crickets, termites, and locusts, where ASTs inhibit JH biosynthesis, the CA is innervated by AST-immunoreactive neurons, which originate from the pars lateralis in the brain and from the subesophageal ganglion. In contrast, in the insects where ASTs do not inhibit the CA activity, little or no AST immunoreactivity is detected in the CA [2,4]. Detailed observation of AST distribution throughout the body has been reported in the cockroaches, where AST-immunoreactive neurons distribute widely in the central nervous system and stomatogastric nervous system [1,2]. AST-immunoreactive axons from medial neurosecretory cells in the protocerebrum and cells from the tritocerebrum enter the stomatogastric nervous system and innervate the foregut and midgut. The hindgut is innervated by neurons originating from the last abdominal ganglion. Endocrine cells in the midgut, and hemocytes, also produce ASTs [1,2]. In Lepidoptera, too, AST-immunoreactive neurons distribute widely in the central and peripheral nervous systems as well as midgut endocrine cells. In the tobacco hornworm *Manduca sexta*, large AST-immunoreactive neurons are located in the frontal ganglion and terminal abdominal ganglion, innervating the foregut and hindgut, respectively [2]. ASTs are also widespread in the central and stomatogastric nervous system of crustaceans. In the prawn *Macrobrachium rosenbergii*, expression of the AST gene was detected in the central nervous system and in the gut [5]. All these observations demonstrate that ASTs are so-called brain−gut peptides [2].

Tissue Content

The AST content in the brain of the cockroach *Diploptera punctata* is 1−4 pmol. AST concentration in the hemolymph varies within a range of 0−2.5 nM over reproductive cycles [6]. The half life of ASTs in the hemolymph varies largely between molecular species, ranging from 22 minutes for Dippu-AST 7 and 153 minutes for Dippu-AST 5 [7].

Y. Takei, H. Ando, & K. Tsutsui (Eds): Handbook of Hormones. DOI: http://dx.doi.org/10.1016/B978-0-12-801028-0.00058-1

Table 58.1 *D. punctata* AST Peptides*

AST1	LYDFGL–NH$_2$
2	AYSYVSEYKRLPVYNFGL–NH$_2$
3	SKMYGFGL–NH$_2$
4	DGRMYSFGL–NH$_2$
5	DRLYSFGL–NH$_2$
6	ARPYSFGL–NH$_2$
7	APSGAQRLYGFGL–NH$_2$
8	GGSLYSFGL–NH$_2$
9	GDGRLYAFGL–NH$_2$
10	PVNSGRSSRFNFGL–NH$_2$
11	YPQEHRFSFGL–NH$_2$
12	PFNFGL–NH$_2$
13	IPMYDFGI–NH$_2$

*Highly conserved residues are shaded. Underlining indicates a potential cleavage site.

Receptors

Structure

Two G protein-coupled receptors were first identified in the fruit fly *Drosophila melanogaster* as AST receptors by using heterologous functional assay [3,4]. These receptors, DAR-1 and -2, consist of 394 and 357 aa residues, respectively, and are structurally related to the mammalian somatostatin/galanin/opioid receptor family [4]. The sequence identity between DAR-1 and -2 is 47% [5]. Later, homologous receptors were identified and characterized in cockroaches and the silkworm [5]. The cockroach AST receptor (Dippu-AstR) shows 32% and 27% overall identity with the rat galanin receptor type-1 and the rat somatostatin receptor type-2, respectively [8]. The Dippu-AstR gene is highly expressed in the brain and to a lesser extent in the gut, testes, and ovaries [8].

Signal Transduction Pathway

cAMP, Ca^{2+}, and phosphoinositide are suggested to be involved in the signal transduction of AST action on the CA [2]. It is proposed that inhibition of JH biosynthesis by AST occurs in the early step(s) of the JH synthesis pathway [2].

Biological Functions

Target Cells/Tissues and Functions

ASTs inhibit JH biosynthesis in specific groups of insects, including cockroaches, crickets, and termites, at a concentration of 10^{-9}–10^{-8} M. In *D. punctata*, all the members of innate ASTs show CA-inhibiting activity *in vitro*, although EC$_{50}$ values differ considerably between the AST molecular species [9]. In crustaceans, the mandibular organ, which is homologous to the CA of insects, produces methyl farnesoate (MF; the equivalent of JH in insects). It has been demonstrated in the adult crayfish that cockroach ASTs stimulate, but do not inhibit, the production of MF by the mandibular organ [5]. Insect ASTs also exert various actions [1–3], including inhibition of muscle contraction in the foregut, midgut, hindgut, oviduct, and heart, inhibition of vitellogenin release from the fat body (cockroach), stimulation of carbohydrate-metabolizing enzyme activity in the midgut, and promotion of adipokinetic hormone release from the corpora cardiaca, some of which are species-specific. These multiple actions of ASTs often involve a myoinhibitory reaction, suggesting that regulation of JH biosynthesis may be a secondary role for these peptides and their role(s) may be primarily myoregulatory [2,3]. In these actions, AST may act as a neurotransmitter, hormone, or paracrine factor. The presence of AST-immunoreactive interneurons in the central nervous system suggests that ASTs also serve as a neuromodulators [1].

Phenotype in Gene-Modified Animals

Injection of dsDippu-AstR RNA into mated female cockroachs leads to a significant increase in JH release from the CA [7]. Genetically based manipulation of AST-neuron activity inhibits feeding behavior in adult *Drosophila* [10].

References

1. Bendena WG, Donly BC, Tobe SS. Allatostatins: a growing family of neuropeptides with structural and functional diversity. *Ann NY Acad Sci.* 1999;891:311–329.
2. Hoffmann KH, Meyering-Vos M, Lorenz MW. Allatostatins and allatotropins: Is the regulation of corpora allata activity their primary function? *Eur J Endomol.* 1999;96:255–266.
3. Gäde G. Allatoregulatory peptides – molecules with multiple functions. *Invert Rep Dev.* 2002;41:127–135.
4. Weaver RJ, Audsley N. Neuropeptide regulators of juvenile hormone synthesis. *Ann NY Acad Sci.* 2009;1163:316–329.
5. Stay B, Tobe SS. The role of allatostatins in juvenile hormone synthesis in insects and crustaceans. *Annu Rev Entomol.* 2007;52:277–299.
6. Yu CG, Stay B, Joshi S. Allatostatin content of brain, corpora allata and hemolymph at different developmental stages of the cockroach, *Diploptera puctata*: Quantitation by ELISA and bioassay. *J Insect Physiol.* 1993;39:111–122.
7. Garside CS, Hayes TK, Tobe SS. Degradation of Dip-allatostatins by hemolymph from the cockroach, *Diploptera punctata*. *Peptides.* 1997;18:17–25.
8. Lungchukiet P, Donly BC, Zhang J, et al. Molecular cloning and characterization of an allatostatin-like receptor in the cockroach *Diploptera punctata*. *Peptides.* 2008;29:276–285.
9. Tobe SS, Zhang JR, Bowser PRF, et al. Biological activities of the allatostatin family of peptides in the cockroach, *Diploptera punctata*, and potential interactions with receptors. *J Insect Physiol.* 2000;46:231–242.
10. Hergarden AC, Tayler TD, Anderson DJ. Allatostatin-A neurons inhibit feeding behavior in adult *Drosophila*. *Proc Natl Acad Sci USA*. 2012;109:3967–3972.

Supplemental Information Available on Companion Website

- Accession numbers of the cDNAs for allatostatin-A and allatostatin-A receptor/E-Tables 58.1 and 58.2
- Amino acid sequence of *D. punctata* AST precursor/E-Figure 58.1

CHAPTER 59

Adipokinetic Hormone

Shinji Nagata

Abbreviation: AKH
Additional names: red pigment concentrating hormone (RPCH), hypertrehalosemic hormone (HrH)

Adipokinetic hormone (AKH) mobilizes carbohydrate and lipids from storage into the hemolymph for energy demands such as a long flight. Thus, AKH is a functional counterpart of vertebrate glucagon. Red pigment concentrating hormone (RPCH), structurally related to AKH, controls the granule of the crustacean chromatophore, eventually changing the body color.

Discovery

AKH was first discovered as a hormone mobilizing the energy supply from lipid stores for a long flight in the migratory locust *Locusta migratoria* [1]. The peptide was isolated from *L. migratoria* by its biological activity in mobilizing lipid into the hemolymph [2]. By contrast, RPCH, which is a crustacean peptide similar to insect AKHs, was isolated from the northern shrimp *Pandalus borealis* by pigment concentration [3].

Structure

Structural Features

Most AKH/RPCHs are octapeptides with an N-terminal pyroGlu residue, C-terminal amidation, and two conserved aromatic amino acids at the fourth and eighth residues. Sequences of AKH and RPCH are highly conserved over arthropod species (Table 59.1).

Structure and Subtype

AKH has been identified in more than 40 insect species, so far [4]. In most insect species, several AKH subtypes are present. In *L. migratoria*, two structurally related subtypes (AKH-II and -III) have been identified [5]. Some AKH subtypes are exceptionally nonapeptides and decapeptides. The number of subtypes differs according to species. No subtypes of crustacean RPCH are known.

Synthesis and Release

Gene, mRNA, and Precursor

An mRNA encoding AKH/RPCH produces a precursor peptide composed of 60−70 amino acids. The precursor peptide contains a signal sequence, a biologically active peptide, and a predomain.

Distribution of mRNA

AKH mRNA is produced in the corpora cardiaca (CC) associated with the insect brain. RPCH mRNA is produced by the X-organ in the eyestalks of crustaceans.

Regulation of Synthesis and Release

AKH is released from the secretary granule in the CC. AKH secretion is modulated by nutritional status, including the need for extra loading of lipid and carbohydrate for long migratory flights [6]. AKH is released from the CC by a high concentration of potassium *in vitro* [6,7].

Receptors

Structure and Subtype

Receptors for AKH/RPCH are GPCRs that have been characterized in several arthropod species [8]. All annotated AKH/RPCH receptors exhibit 60−90% similarity with one another. Most AKH/RPCH receptors have no subtypes. In the mosquito *Aedes aegypti*, however, two AKHR subtypes are confirmed by alternative splicing, having different C-terminal portions, including phosphorylation sites and a possible internalization domain. Phylogenetic analyses have revealed the presence of some receptor genes related to the AKH/RPCH receptor family. For example, a novel paralog of receptors for AKH and corazonin has been found in several insect species, including *B. mori*, *Anopheles gambiae*, and *Tribolium castaneum*. The undetermined ligand for those paralogous receptors is designated ACP (AKH/corazonin-related peptide). In addition, phylogenetic analyses have shown that vertebrate GnRHRs are also distant ancient receptors. In fact, some reports have illustrated the structural similarity between AKH and GnRH.

Signal Transduction Pathway

It has been reported that cAMP and inositol 1,4,5-triphosphate are the second messengers of the AKH signal in *L. migratoria* and *Pachnoda sinuata*, respectively. Consequent activation of glycogen phosphorylase drives trehalose production after intracellular calcium elevation (Figure 59.1). Hormone-sensitive lipase is also activated to produce fatty acid and diacyl glycerol (DAG) from the stored triacyl glycerol (TAG) in *Manduca sexta*, *P. sinuata*, and *L. migratoria*.

Y. Takei, H. Ando, & K. Tsutsui (Eds): Handbook of Hormones. DOI: http://dx.doi.org/10.1016/B978-0-12-801028-0.00059-3

Table 59.1 Alignment of AKHs and RPCH

Gryllus bimaculatus	AKH	pEVNFSTW—NH$_2$
Drosophila melanogaster	AKH	pELTFSTGW—NH$_2$
Locusta migratoria	AKH	pELNFTPNWGT—NH$_2$
	AKH-II	pELNFSAGW—NH$_2$
	AKH-III	pELNFTPWW—NH$_2$
Pandalus borealis	RPCH	pELTFSTGW—NH$_2$

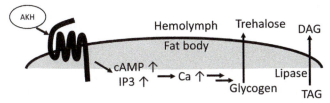

Figure 59.1 Schematic diagram of the AKH signaling pathway.

Biological Functions

Target Cells and Function

AKH stimulates the fat body, where lipid and carbohydrate are stored, eventually resulting in lipid and carbohydrate mobilization into the hemolymph. In *L. migratoria*, abdominal injection of synthetic AKH results in a drastic increase in hemolymph trehalose and lipid in a couple of hours. The increase in hemolymph lipid is accompanied by activation of the lipid transfer system. Therefore, AKH is thought to be a functional counterpart of mammalian glucagon. According to the effects of the increase in hemolymph carbohydrate and lipid, locomotor and flight muscles are concomitantly activated [6]. RPCH, however, influences the concentration of pigment chromatophore, causing a drastic body color changes in the crustacean species.

Phenotype in Gene-Modified Animals

Knock-in flies, in which AKH is ectopically overexpressed in the fat body, show hypertrehalosemia, while AKH-deficient flies suffer from the opposite phenotype in which hemolymph trehalose levels decrease and storage lipid in the fat body accumulates. In addition, hyperlocomotion is observed by AKH overexpressed flies, consistent with the effects of AKH

injection. In *D. melanogaster*, an AKH receptor-deficient strain shows a similar phenotype to AKH-deficient flies [9]. AKH receptor null mutant flies have reinforced starvation resistance, resulting in the activation of hormone-sensitive lipase. In the two-spotted cricket *Gryllus bimaculatus*, AKH receptor knockdown by RNAi increases feeding frequency, as observed in AKH receptor null mutant flies [10]. AKHR knockdown crickets also show reduced locomotor activity.

References

1. Mayer RJ, Candy DJ. Control of haemolymph lipid concentration during locust flight – an adipokinetic hormone from corpora cardiaca. *J Insect Physiol.* 1969;15:611–620.
2. Stone JV, Mordue W, Gatley KE, et al. Structure of locust adipokinetic hormone, a neurohormone that regulates lipid utilization during flight. *Nature.* 1979;263:207–211.
3. Joseffson L. Invertebrate neuropeptide hormones. *Int J Pept Protein Res.* 1983;21:459–470.
4. Gäde G. Peptides of the adipokinetic hormone/red pigment-concentrating hormone family: a new take on biodiversity. *Ann NY Acad Sci.* 2009;1163:125–136.
5. Oudejans R, Kooiman FP, Heerma W, et al. Isolation and structure elucidation of a novel adipokinetic hormone (Lom-AKH-III) from the glandular lobes of the corpus cardiacum of the migratory locust, *Locusta migratoria. Eur J Biochem.* 1991;195:351–359.
6. Goldworthy GJ. The endocrine control of flight metabolism in locusts. *Adv Insect Physiol.* 1983;17:149–204.
7. Gäde G, Auerswald L. Mode of action of neuropeptides from the adipokinetic hormone family. *Gen Comp Endocrinol.* 2003;132:10–20.
8. Grönke S, Muller G, Hirsch J, et al. Dual lipolytic control of body fat storage and mobilization in *Drosophila. PLoS Biol.* 2007;5: e137.
9. Hansen KK, Stafflinger E, Schneider M, et al. Discovery of a novel insect neuropeptide signaling system closely related to the insect adipokinetic hormone and corazonin hormonal systems. *J Biol Chem.* 2010;285:10736–10747.
10. Konuma T, Morooka N, Nagasawa H, et al. Knockdown of the adipokinetic hormone receptor increases feeding frequency in the two-spotted cricket *Gryllus bimaculatus. Endocrinology.* 2012;153:3111–3122.

Supplemental Information Available on Companion Website

- Accession Numbers of AKH and AKH receptor/E-Table 59.1

CHAPTER 60

Neuroparsin

Yoshiaki Tanaka

Abbreviation: NP
Additional names: ovary ecdysteroidogenic hormone (OEH)

Neuroparsins (NPs) and ovary ecdysteroidogenic hormone (OEH) are large peptides that belong to the most cysteine-rich family of neurohormones. NPs are insect gonad-inhibiting hormones in locusts, whereas OEH has been identified as a gonadotropin in the female mosquito.

Discovery

Neuroparsin (NP) was first isolated from the corpora cardiaca (CC) of the migratory locust *Locusta migratoria*, based on its inhibitory action on vitellogenesis [1]. Ovary ecdysteroidogenic hormone (OEH) was initially identified in the yellow fever mosquito *Aedes aegypti* as a neurohormone that stimulates the ovaries to secrete ecdysteroids [2].

Structure

Structural Features

NPs/OEH are large peptides (about 8 kDa in locusts) that belong to the most cysteine-rich family of neurohormones (12−14 cycteines per peptide) [1]. Although the overall sequence similarity among these peptides is variable, the spacing pattern of 11 Cys residues is well conserved [3].

Structure and Subtype

NP family peptides have been identified in a variety of insects and crustaceans, although most of them have not been functionally characterized. The homology of overall sequences is low except for the conserved Cys residues among NP family peptides. Hemimetabolous insects and the mosquito have multiple isoforms of NP, which could be generated due to alternative splicing of the NP gene, presence of multiple NP genes [4,5], or different processing of the precursor peptide [2]. Some insect peptides showing pronounced sequence similarity to the N-terminal domain of vertebrate and mollusk insulin-like growth factor binding proteins (IGFBPs) are proposed to be additional members of the NP family [4]. However, these IGFBP-like peptides are likely different from NP because the spacing patterns of Cys residues are somewhat different (Figure 60.1) and because both the NP and IGFBP-like peptides occur in the red flour beetle *Tribolium castaneum* and the honey bee *Apis mellifera* [5]. Only a IGFBP-like peptide is found in the silkworm *Bombyx mori* (Figure 60.1).

Synthesis and Release

Gene, mRNA, and Precursor

Each NP/OEH gene encodes a transcript of the NP/OEH precursor. Mature peptides are processed by removal of the signal peptide and cleavage at the C-terminus by a dipeptidase. NP family peptides are variable in Diptera. NP family peptides occur in the mosquito and are coded in the genome of the fruit fly *Drosophila ananassae*, but the fruit fly *Drosophila auraria* lacks a long sequence including 4 of the 12 conserved Cys residues (Figure 60.1). Moreover, no NP gene can be detected in the genome of the fruit fly *Drosophila melanogaster* [5]. As *D. auraria* stands phylogenetically between *D. ananassae* and *D. melanogaster*, the NP gene appears to have been lost in the *melanogaster* subgroup of *Drosophila* during evolution.

Distribution of mRNA

NP mRNAs are detected in some neural tissues, including the brain [6], corpora cardiaca, and subesophageal ganglion of insects, and the X-organ, brain, and subesophageal and thoracic ganglia of the spiny lobster *Jasus lalandii* [7]. NP mRNAs are also detected in non-neural tissues, such as the fat body, male accessory glands, leg muscle, flight muscle, reproductive organs, and antennae of insects [8], and in the heart and ovary of *J. lalandii* [7].

Regulation of Synthesis and Release

Regulation of NP gene expression is strongly phase-dependent. An mRNA for the desert locust *Schistocerca gregaria* NP precursor 3 is much more abundantly detected in the fat bodies of gregarious locusts than in those of solitary animals [8]. A blood meal stimulates the release of OEH from the medial neurosecretory cells in the brain of mosquitoes.

Receptors

Unidentified as yet.

Biological Functions

Target Cells/Tissues and Functions

NP inhibits vitellogenesis, and might play a role in the context of phase transition in locusts [1]. These actions appear to be independent of juvenile hormone (JH), as NP has no effect on JH biosynthesis by the corpora allata. NP also shows an

Y. Takei, H. Ando, & K. Tsutsui (Eds): Handbook of Hormones. DOI: http://dx.doi.org/10.1016/B978-0-12-801028-0.00060-X

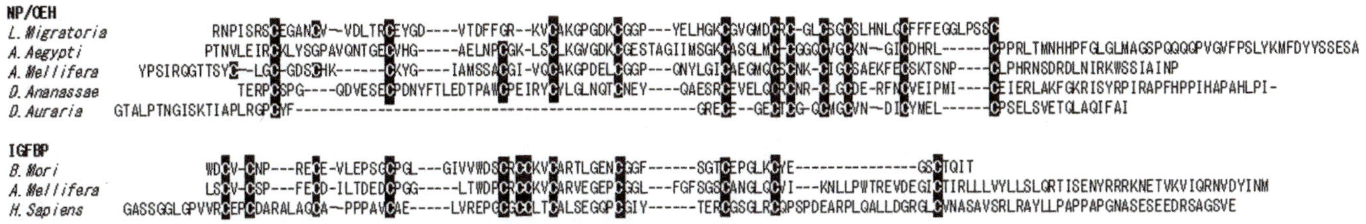

Figure 60.1 Comparison of conserved Cys residues between NP/OEH and IGFBPs. Conserved Cys residues are indicated as black oblongs.

antidiuretic effect [9] and neuritogenic effects in the locusts [10]. OEH stimulates vitellogenesis by virtue of its stimulation of ecdysone synthesis by the ovary in *A. aegypti* [2].

References

1. Girardie J, Girardie A, Huet JC, et al. Amino acid sequence of locust neuroparsins. *FEBS Lett*. 1989;245:4–8.
2. Brown MR, Graf R, Swiderek KM. Identification of a steroidogenic neurohormone in female mosquitoes. *J Biol Chem*. 1998;273:3967–3971.
3. Dhara A, Eum JH, Robertson A, et al. Ovary ecdysteroidogenic hormone functions independently of the insulin receptor in the yellow fever mosquito, *Aedes aegypti*. *Insect Biochem Mol Biol*. 2013;43:1100–1108.
4. Badisco L, Claeys I, Van Loy T, et al. Neuroparsins, a family of conserved arthropod neuropeptides. *Gen Comp Endocrinol*. 2007;153:64–71.
5. Veenstra JA. What the loss of the hormone neuroparsin in the melanogaster subgroup of *Drosophila* can tell us about its function. *Insect Biochem Mol Biol*. 2010;40:354–361.
6. Lagueux M, Kromer E, Girardie J. Cloning of a Locusta cDNA encoding neuroparsin A. *Insect Biochem Mol Biol*. 1992;22: 511–516.
7. Marco HG, Anders L, Gäde G. cDNA cloning and transcript distribution of two novel members of the neuroparsin peptide family in a hemipteran insect (*Nezara viridula*) and a decapod crustacean (*Jasus lalandii*). *Peptides*. 2014;53:97–105.
8. Claeys I, Simonet G, Breugelmans B, et al. Quantitative real-time RT-PCR analysis in desert locusts reveals phase dependent differences in neuroparsin transcript levels. *Insect Mol Biol*. 2005;14:415–422.
9. Fournier B, Herault JP, Proux J. Antidiuretic factor from the nervous corpora cardiaca of the migratory locust: improvement of an existing *in vitro* bioassay. *Gen Comp Endocrinol*. 1987;68:49–56.
10. Vanhems E, Delbos M, Girardie J. Insulin and neuroparsin promote neurite outgrowth in cultured locust CNS. *Eur J Neurosci*. 1990;2:776–782.

Supplemental Information Available on Companion Website

- Primary structures of NP/OEH/IGFBPs /E-Figure 60.1
- Accession numbers of genes and cDNAs for preproNP/OEH/IGFBPs/E-Table 60.1

CHAPTER 61

Ovary Maturating Parsin

Yoshiaki Tanaka

Abbreviation: OMP

Ovary maturating parsin (OMP) was the first gonadotropic factor to be identified in insects. OMP stimulates oocyte growth, probably by inducing ovarian ecdysone production in the locusts. OMPs likely only occur in the order of Orthoptera.

Discovery

Ovary maturating parsin (OMP) was first identified as a gonadotropic factor from the corpora cardiaca of the migratory locust *Locusta migratoria* [1].

Structure

Structural Features

OMPs consist of 48–65 amino acid residues and include a high concentration of alanine, but are devoid of cysteine, methionine, lysine, and threonine.

Structure and Subtype

OMP molecules of the desert locust *Schistocerca gregaria* (Scg OMPs) comprise four isoforms (Table 61.1) [2]. Two long isoforms (OMP1, -2) correspond to *L. migratoria* OMP (Lom OMP), differing from each other only by the presence or absence of a tripeptide insertion (Pro−Ala−Ala, PAA) at position 21. Two short isoforms (OMP3, -4) are deprived of N-terminal 13 residues of the long isoforms. OMPs probably only occur in the order Orthoptera, and more specifically in the family of Acrididae [3].

Gene, mRNA, and Precursor

Two cDNAs coding for highly similar OMP-precursor proteins (OMP-DH-L and OMP-DH-S; Figure 61.1) have been cloned, each of which also encodes the same 46-aa peptide representing a member of the CRF-like diuretic hormone (DH)

peptide family in *S. gregaria* [4]. OMP-DH-L contains OMP1 and OMP3, and OMP-DH-S contains OMP2 and OMP4. Two short forms possibly result from differential processing of precursor proteins.

Distribution of mRNA

OMP-DH precursor transcripts occur throughout the central nervous system (CNS), with the highest level in the brain [4]. *Scg*-OMP-DH transcripts are also detected in the malpighian tubules, midgut, and hindgut. No difference in the transcript level was observed between the two different time points in either sex.

Receptors

Unidentified as yet.

Biological Functions

Target Cells/Tissues and Functions

OMP stimulates oocyte growth [5], probably by inducing ovarian ecdysone production [6].

Phenotype in Gene-Modified Animals

Injection with OMP-DH dsRNA causes a significantly larger oocyte size, higher ecdysteroid titers in the hemolymph and ovary, and a higher food intake in *S. gregaria* [4]. As injection of DH results in the opposite effects, the observed effects by RNAi are probably caused by reducing the DH signaling system, with the effect of reduced OMP signaling possibly being overruled by the reduction of DH signaling.

References

1. Girardie J, Richard O, Huet JC, et al. Physical characterization and sequence identification of the ovary maturating parsin. A new neurohormone purified from the nervous corpora cardiaca of the African locust (*Locusta migratoria migratorioides*). *Eur J Biochem.* 1991;202:1121−1126.

Table 61.1 Comparison of Amino Acid Sequences of OMPs*

		1	10	20	30	40	50	60
L. migratoria	OMP	YYEAPPDGRHLLLQPAPAAPVAPAAPASWPHQQRRQALDEFAAAAAAAADAQF QDEEEDGGRRV
S. gregaria	OMP1	YYEAPPDGQRLLLQAAPAAAPAAPAAASWPHQQRRQAI DEFAAAAAAAADAQY QDEEEDGARRV
S. gregaria	OMP2	YYEAPPDGQRLLLQAAPAAA-----PAAASWPHQQRRQAI DEFAAAAAAAADAQY QDEEEDGARRV
S. gregaria	OMP3	QAAPAAAPAAPAAASWPHQQRRQAI DEFAAAAAAAADAQY QDEEEDGARRV
S. gregaria	OMP4	QAAPAAA-----PAAASWPHQQRRQAI DEFAAAAAAAADAQY QDEEEDGARRV

*Outlining indicates a tripeptide insertion.

Y. Takei, H. Ando, & K. Tsutsui (Eds): Handbook of Hormones. DOI: http://dx.doi.org/10.1016/B978-0-12-801028-0.00061-1

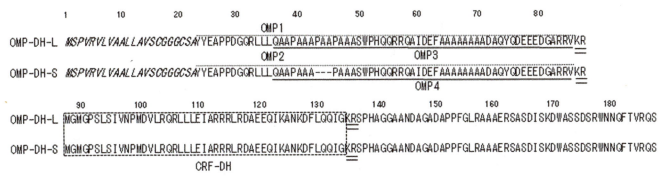

Figure 61.1 **OMP-DH precursors of *S. gregaria.*** CRF-DHs are outlined; signal peptides are indicated in italics. =, cleavage site.

2. Girardie J, Huet JC, Atay-Kadiri Z, et al. Isolation, sequence determination, physical and physiological characterization of the neuroparsins and ovary maturing parsins of *Schistocerca gregaria*. *Insect Biochem Mol Biol*. 1998;28:641–650.
3. Richard O, Tamarelle M, Girardie J, et al. Restricted occurrence of *Locusta migratoria* ovary maturing parsin in the brain–corpora cardiaca complex of various insect species. *Histochemistry*. 1994;102:233–239.
4. Van Wielendaele P, Dillen S, Marchal E, et al. CRF-like diuretic hormone negatively affects both feeding and reproduction in the desert locust, *Schistocerca gregaria*. *PLoS One*. 2012;7:e31425.
5. Girardie J, Richard O, Girardie A. Time-dependent variations in the activity of a novel ovary maturing neurohormone from the nervous corpora cardiaca during oögenesis in the locust, *Locusta migratoria migratorioides*. *J Insect Physiol*. 1992;38:215–221.
6. Girardie J, Girardie A. Lom OMP, a putative ecdysiotropic factor for the ovary in *Locusta migratoria*. *J Insect Physiol*. 1996;42:215–221.

Supplemental Information Available on Companion Website

- Accession numbers of genes and cDNAs for preproOMPs/E-Table 61.1

Regulation of Myo/Cardio-Activities

CHAPTER 62

LF Peptides

Toshio Takahashi

History

Members of the LF peptide group have been isolated from the sea anemone *Anthopleura elegantissima* [1], the freshwater polyp *Hydra magnipapillata* [2], and the ascidian *Ciona intestinalis* [3].

Structure

Structural Features

LF peptides contain Leu—Phe at the C-termini. Head activator, one of the LF peptide family isolated from the sea anemone *A. elegantissima*, is an undecapeptide with pyroglutamate at the N-terminus (Table 62.1) [1]. LF-like peptide isolated from *H. magnipapillata*, Hym-323, is a 16-amino acid peptide with Lys—Phe at the C-terminus (Table 62.1) [2]. Hym-323 shares some homology with the head activator. Recently, mass spectrometry-based peptidomic analysis detected eight LF peptides from *C. intestinalis* [3]. The *Ciona* peptides (Ci-LFs) have a variety of structures from five to nine residues in length (Table 62.1).

Distribution of LF peptides

Head activator itself has been identified in *Hydra*, as well as in rat and human hypothalamus and intestine [4—6]. Hym-323

is a novel, hydra-specific LF-like peptide [2]. Ci-LF peptides are also ascidian-specific peptides [3].

Biological Functions

As head activator increases the rate of head regeneration and bud formation, and increases the number of regenerated tentacles of *Hydra*, the peptide has been labeled as a morphogen. Head activator also functions as a mitogen and an inducer of neuron differentiation. Both *in situ* hybridization analysis and immunohistochemistry showed the presence of Hym-323 in both ectodermal and endodermal epithelial cells throughout the body, with the exception of the basal disk and the head region in *Hydra*, indicating that Hym-323 is an epitheliopeptide [7]. Hym-323 enhances the capacity of *Hydra* for ectopic foot formation from tissues of the body column [7].

In situ hybridization demonstrates that *ci-lf* is localized to various neurons in the brain ganglion, providing evidence that Ci-LF peptides serve as neuropeptides [3]. Functional analysis of these peptides is also currently in progress.

References

1. Schaller HC, Bodenmueller H. Isolation and amino acid sequence of a morphogenetic peptide from hydra. *Proc Natl Acad Sci USA*. 1981;78:7000—7004.
2. Takahashi T, Muneoka Y, Lohmann Y, et al. Systematic isolation of peptide signal molecules regulating development in hydra: LWamide and PW families. *Proc Natl Acad Sci USA*. 1997;94:1241—1246.
3. Kawada T, Ogasawara M, Sekiguchi T, et al. Peptidomic analysis of the central nervous system of the protochordate, *Ciona intestinalis*: homolog and prototypes of vertebrate peptides and novel peptides. *Endocrinology*. 2011;152:2416—2427.
4. Schaller HC. A neurohormone from hydra is also present in the rat brain. *J Neurochem*. 1975;25:187—188.
5. Bodenmueller H, Schaller HC, Darai G. Human hypothalamus and intestine contain a hydra-neuropeptide. *Neurosci Lett*. 1980;16:71—74.
6. Bodenmueller H, Schaller HC. Conserved amino acid sequence of neuropeptide, the head activator, from coelenterates to humans. *Nature*. 1981;293:579—580.
7. Harafuji N, Takahashi T, Hatta M, et al. Enhancement of foot formation in *Hydra* by a novel epitheliopeptide, Hym-323. *Development*. 2001;128:437—446.

Table 62.1 LF Peptides in Hydra, Sea Anemone, and Ascidian

Hydra magnipapillata	
Head activator	pE*PPGGSKVILF
Hym-323	KWVQGKPTGEVKQIKF
Anthopleura elegantissima	
Head activator	pE*PPGGSKVILF
Ciona intestinalis	
Ci-LF-1	FQSLF
Ci-LF-2	YPGFQGLF
Ci-LF-3	HNPHLPDLF
Ci-LF-4	YNSMGLF
Ci-LF-5	SPGMLGLF
Ci-LF-6	SDARLQGLF
Ci-LF-7	YPNFQGLF
Ci-LF-8	GNLFSLF

pE*, pyroglutamic acid.

Y. Takei, H. Ando, & K. Tsutsui (Eds): Handbook of Hormones. DOI: http://dx.doi.org/10.1016/B978-0-12-801028-0.00062-3

Head Activator

Toshio Takahashi

Head activator is a neuropeptide which regulates pattern-forming processes in Hydra magnipapillata.

Discovery

Head activator was first isolated from the sea anemone *Anthopleura elegantissima* [1]. An identical peptide was also purified and used to determine the sequence in *Hydra magnipapillata*, as well as in the hypothalamus and intestine of rats and humans [2–4].

Structure

Structural Features

Head activator is an undecapeptide characterized by the N-terminal pyroglutamate residue and the X–Pro–Pro sequence (Table 62A.1). These characteristics are thought to render the peptide resistant to aminopeptidases.

Synthesis and Release

Distribution of Head Activator

Head activator has been identified in *Hydra*, as well as in the hypothalamus and intestine of rats and humans, where the peptide appears to have a role in neuronal differentiation [2–4]. It has been suggested that head activator acts as a neuropeptide [5], although direct evidence is lacking. Recently, the *Hydra* Genome Project was completed [6]; however, the *Hydra* genome sequence does not include a putative head activator gene, and therefore the nature of this factor is uncertain.

Biological Functions

Pattern Formation

Hydra is one of the simplest multicellular animals, having a simple body plan with a single body axis (Figure 62A.1). Gierer and Meinhardt proposed a reaction–diffusion model that involves two counteracting factors, head activator and head inhibitor [7]. When a certain threshold ratio of head activator to head inhibitor is attained, a head forms. Head activator increases the rate of head regeneration and bud formation, and increases the number of regenerated tentacles. Thus, this peptide was initially labeled a morphogen. Head activator also functions as a mitogen and an inducer of neuronal differentiation [8,9]. It has been reported that the

Table 62A.1 Structure of Head Activators

Anthopleura elegantissima	
Head activator	pE*PPGGSKVILF
Hydra magnipapillata	
Head activator	pE*PPGGSKVILF

pE*, pyroglutamic acid.

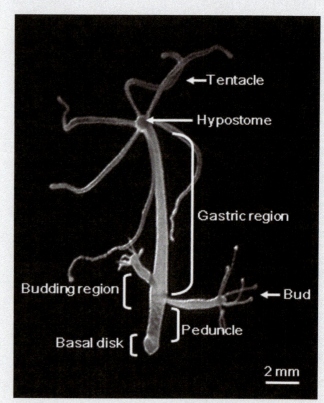

Figure 62A.1 Simple body plan of *H. magnipapillata*.

morphogenetic function of head activator can be explained by its mitogenetic activity [10]. Thus, it seems doubtful that head activator is a morphogen. However, it does have the capacity to activate mitosis and neuronal differentiation in *Hydra* and cultured mammalian cells. The structure of head inhibitor remains to be determined.

Y. Takei, H. Ando, & K. Tsutsui (Eds): Handbook of Hormones. DOI: http://dx.doi.org/10.1016/B978-0-12-801028-0.00215-4

References

1. Schaller HC, Bodenmueller H. Isolation and amino acid sequence of a morphogenetic peptide from hydra. *Proc Natl Acad Sci USA*. 1981;78:7000−7004.
2. Schaller HC. A neurohormone from hydra is also present in the rat brain. *J Neurochem*. 1975;25:187−188.
3. Bodenmueller H, Schaller HC, Darai G. Human hypothalamus and intestine contain a hydra-neuropeptide. *Neurosci Lett*. 1980;16:71−74.
4. Bodenmueller H, Schaller HC. Conserved amino acid sequence of neuropeptide, the head activator, from coelenterates to humans. *Nature*. 1981;293:579−580.
5. Schaller HC, Geirer A. Distribution of the head-activating substance in hydra and its localization in membranous particles in nerve cells. *J Embryol Exp Morph*. 1973;29:39−52.
6. Chapman JA, Kirkness EF, Simakov O, et al. The dynamic genome of *Hydra. Nature*. 2010;464:592−596.
7. Gierer A, Meinhardt H. A theory of biological pattern formation. *Kybernetik*. 1972;12:30−39.
8. Schaller HC. Action of head activator as a growth hormone in *Hydra. Cell Differ*. 1976;5:1−11.
9. Schaller HC. Action of head activator on the determination of interstitial cells in *Hydra. Cell Differ*. 1976;5:13−20.
10. Javois LC, Tombe VK. Head activator dose not qualitatively alter head morphology in regenerates of *Hydra oligactis. Roux's Arch Dev Biol*. 1991;199:402−408.

Hym-323

Toshio Takahashi

Hym-323 is produced in epithelial cells and enhances foot formation by increasing foot activation potential in Hydra.

Discovery

Hym-323 was identified during the course of a systematic screening of peptide signaling molecules in *Hydra magnipapillata* [1].

Structure

Structural Features

Hym-323 is composed of 16 amino acid residues with a free C-terminus (Table 62B.1). Hym-323 shares identical amino acid residues in 5 out of 10 positions in its C-terminal region with head activator (Table 62B.1).

Synthesis and Release

cDNA of the Hym-323-Encoding Gene

The Hym-323 preprohormone contains a single copy of Hym-323 at its C-terminus (Figure 62B.1) [2]. However, the prepro-hormone lacks a signal peptide. Although the mechanism of its secretion is unknown, some proteins without a signal peptide are secreted extracellularly via a chaperone-like protein [3]. Also, a large number of peptides secreted independently of the classical ER–Golgi vesicular pathway have been reported in a large-scale peptidomic analysis of the mouse brain [4]. The Hym-323-encoding sequence is preceded by a Thr residue at the N-terminus (Figure 62B.1). Thr has been shown to be a cleavage site in cnidarians [5]. Thus, the Thr residue could act as a cleavage site leading to the production of mature Hym-323.

Table 62B.1 Structure of Hym-323

Hydra magnipapillata	
Hym-323	KWVQGKPTGEVKQIKF
Head activator	pE*PPGGSKVILF

pE*, pyroglutamic acid.

Distribution of mRNA and Peptide

In situ hybridization analysis and immunohistochemistry with an antibody specific to Hym-323 show the presence of Hym-323 in both ectodermal and endodermal epithelial cells throughout the *Hydra* body, with the exception of the basal disk and head region, indicating that Hym-323 is an epitheliopeptide [2]. During foot regeneration, the peptide is not present during basal disk cell formation [2]. Since epithelial cells continually undergo mitosis, and as the body column can always regenerate a foot, the presence of the Hym-323 peptide in the epithelial cells of the body column will be maintained at an appropriate level.

Biological Functions

Target Cells/Tissues and Functions

The *Hydra* foot, including the peduncle and basal disk, is considered to be composed mainly of terminally differentiated cells. Epithelial stem cells of the ectoderm can undergo differentiation into peduncular and basal disk cells of the foot. Hym-323 treatment enhances the capacity of *Hydra* for ectopic foot formation from tissues of the body column by grafting [2]. For example, epithelial tissue from the body column of a donor hydra that was treated with Hym-323 peptide was grafted onto the body column of a host hydra and ectopic foot formation was assayed (Figure 62B.2). The result indicates that the potential for activation of foot formation is enhanced in the peptide-treated tissue as compared with the untreated control tissue. Hym-323 significantly enhances foot regeneration of epithelial *Hydra* lacking nerve cells, strongly suggesting that Hym-323 acts directly on epithelial cells. Collectively, upon initiation of foot formation, the stored Hym-323 peptide is released from epithelial cells and Hym-323 concentration exceeds the activation threshold, resulting in Hym-323 binding to a receptor on epithelial cells in an autocrine manner. Hym-323 autocrine signaling increases the foot-forming potential of the tissue, eventually enhancing foot formation.

The Target Gene of Hym-323

The target gene of Hym-323 is an astacin matrix metalloprotease termed foot activator-responsive matrix metalloprotease (*Farm1*) [6]. *Farm1* is normally expressed in epithelial cells in the gastric region, and is absent from apical and basal tissues. Treatment of *Hydra* with Hym-323 induces immediate downregulation of this gene. Thus, *Farm1* is a transcriptional target of positional signals that specify foot differentiation and appears to elicit a potent effect on basal patterning processes.

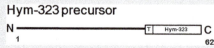

Figure 62B.1 Schematic presentation of the Hym-323 preprohormone.

Y. Takei, H. Ando, & K. Tsutsui (Eds): Handbook of Hormones. DOI: http://dx.doi.org/10.1016/B978-0-12-801028-0.00216-6

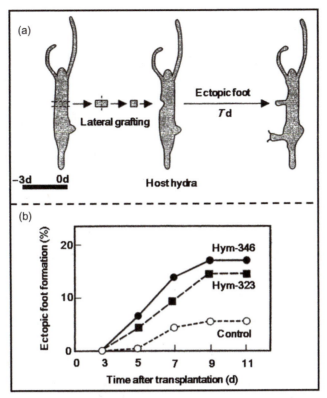

Figure 62B.2 A schematic outline of lateral transplantation. The other peptide, Hym-346 which is a counterpart of pedibin, also enhances foot formation.

References

1. Takahashi T, Muneoka Y, Lohmann Y, et al. Systematic isolation of peptide signal molecules regulating development in hydra: LWamide and PW families. *Proc Natl Acad Sci USA*. 1997;94:1241−1246.
2. Harafuji N, Takahashi T, Hatta M, et al. Enhancement of foot formation in *Hydra* by a novel epitheliopeptide, Hym-323. *Development*. 2001;128:437−446.
3. Suzuki Y, Kobayashi M, Miyashita H, et al. Isolation of a small vasohibin-binding protein (SVBP) and its role in vasohibin secretion. *J Cell Sci*. 2010;123:3094−3101.
4. Fricker LD. Analysis of mouse brain peptides using mass spectrometry-based peptidomics: implications for novel functions ranging from non-classical neuropeptides to microproteins. *Mol Biosyst*. 2010;6:1355−1365.
5. Darmer D, Hauser F, Nothacker H-P, et al. Three different prohormones yield a variety of Hydra-RFamide (Arg−Phe−NH$_2$) neuropeptides in *Hydra magnipapillata*. *Biochem J*. 1998;332:403−412.
6. Kumpfmeuller G, Rybakine V, Takahashi T, et al. Identification of an astacin matrix metalloprotease as target gene for *Hydra* foot activator peptides. *Dev Genes Evol*. 1999;209:601−607.

Supplemental Information Available on Companion Website

- Accession number of the Hym-323 cDNA/E-Table 62B.1
- Primary structure of preproHym-323/E-Figure 62B.1

CHAPTER 63

Invertebrate Kinins

Yoshiaki Tanaka

Abbreviation: K, LK
Additional names: insect kinin, leucokinin, myokinin, drosokinin

Invertebrate kinins are myotropic peptides found in a wide variety of invertebrates, regulating diuresis, feeding, and other processes in insects. The kinin family is characterized by an amidated C-terminal pentapeptide motif (Phe—X—X—Trp—Gly—NH$_2$). Both kinin and kinin receptor are lost in several insect genomes.

Discovery

Invertebrate kinins were first isolated from the Madeira cockroach *Leucophaea maderae* as peptides that induce hindgut contraction [1].

Structure

Structural Features

The kinin family is characterized by an amidated C-terminal pentapeptide motif (Phe—X—X—Trp—Gly—NH$_2$), which is the minimal peptide sequence required for biological activity [2].

Structure and Subtype

Kinins have been identified in insects, Acari, and crustaceans, but appear to be lost in the genomes of some insects, including the red flour beetle *Tribolium castaneum*, the wasp *Nasonia vitripennis*, and many of the ants [3]. The length of mature peptides is variable among species, and the C-terminal glycine is substituted by serine in the kinin of the pond snail *Lymnaea stagnalis* [4] and by alanine in some of the shrimp kinins (Table 63.1).

Synthesis and Release

Gene, mRNA, and Precursor

The preprokinin proteins encoded by kinin mRNAs contain variable numbers of kinin peptides (E-Figure 63.1). The pre-propeptide of the fruit fly *Drosophila melanogaster* contains only 1 kinin, whereas the prepropeptide of the kissing bug *Rhodnius prolixus* contains 12 kinins [5]. Transcripts encoding insect kinin-like peptides have been predicted in other Ecdysozoa species, such as nematodes and a water bear [5].

Distribution of mRNA

Kinin mRNA is expressed in the central nervous system (CNS) of insects. In *D. melanogaster*, kinin mRNA is expressed in the lateral horn neurons and anterior neurons of the brain, subesophageal neurons, and abdominal ganglion neurons [6]. Kinin-containing presynaptic terminals are found close to the kinin receptor-expressing neurons in the brain and ventral ganglia [7]. In *R. prolixus*, kinin mRNA is expressed in the CNS of 4–6 weeks-unfed fifth instars, with prominent expression observed in the posterior-lateral neurosecretory cells of metathoracic ganglia [5].

Receptors

Structure and Subtypes

The kinin receptor is a GPCR, and the first known receptor was identified from *L. stagnalis* by a functional intracellular calcium assay [4]. Kinin receptors have also been identified in many insects, Acari, and crustaceans (E-Figures 63.2 and 63.3), although orthologs of the kinin receptors are not found in *T. castaneum*, *N. vitripennis*, or any of the ants [3].

Distribution of Receptor

Kinin receptors are expressed in the malpighian tubules, CNS, foregut, posterior midgut, hindgut, and ovaries in *D. melanogaster* [7,8].

Signal Transduction Pathway

The kinin receptor behaves as a multiligand receptor and responds functionally to different kinin isoforms in the yellow fever mosquito *Aedes aegypti* [9]. Activation of kinin receptors leads to an increased production of inositol 1,4,5-triphosphate (IP3), causing the release of calcium from intracellular stores through IP3 receptors [10].

Biological Functions

Target Cells/Tissues and Functions

Kinins stimulate hindgut contractions, but their effects on the foregut and oviduct are significantly lower than on the hindgut, and the heart does not respond to any of the kinins in insects [2]. Kinins are also associated with diuretic activity in insects. The kinins of the moth *Heliothis zea* are more potent stimulators of fluid secretion from malpighian tubules than the endogenous diuretic peptides in the tobacco hornworm *Manduca sexta* [2]. Additionally, these peptides may play a role in regulating food intake in *D. melanogaster* [7] and may also serve as neuromodulators or neurotransmitters in the insect CNS [2].

Y. Takei, H. Ando, & K. Tsutsui (Eds): Handbook of Hormones. DOI: http://dx.doi.org/10.1016/B978-0-12-801028-0.00063-5

Table 63.1 Structure of Invertebrate Kinins

Insecta	*L. maderae*	Leucokinin-1	DPAFNSWG—NH₂
	L. migratoria	Locustakinin	AFSSWG—NH₂
	A. aegypti	Aedeskinin-1	NSKYVSKQKFYSWG—NH₂
Crustacea	*P. vannamei*	Pev-kinin1	ASFSPWG—NH₂
		Pev-kinin5	pEAFSPWA—NH₂
		Pev-kinin6	AFSPWA—NH₂
Gastropoda	*L. stagnalis*	Lymnokinin	PSFHSWS—NH₂

Phenotype in Gene-Modified Animals

RNAi knockdown of the kinin receptor reduces fluid excretion *in vivo* by 50% during post-peak diuresis of the mosquito [8]. Mutations in the kinin and kinin receptor genes result in altered feeding phenotypes in *D. melanogaster* adults — i.e., an increase in meal size with a compensatory reduction in meal frequency [7].

References

1. Holman GM, Cook BJ, Nachman RJ. Primary structure and synthesis of a blocked myotropic neuropeptide isolated from the cockroach, *Leucophaea maderae*. *Comp Biochem Physiol C.* 1986;85:219—224.
2. Torfs P, Nieto J, Veelaert D, et al. The kinin peptide family in invertebrates. *Ann NY Acad Sci.* 1999;897:361—373.
3. Veenstra JA, Rodriguez L, Weaver RJ. Allatotropin, leucokinin and AKH in honey bees and other Hymenoptera. *Peptides.* 2012;35:122—130.
4. Cox KJA, Tensen CP, Van der Schors RC, et al. Cloning, characterization, and expression of a G-protein-coupled receptor from *Lymnaea stagnalis* and identification of a leucokinin-like peptide, PSFHSWSamide, as its endogenous ligand. *J Neurosci.* 1997;17:1197—1205.
5. Bhatt G, da Silva R, Nachman R, et al. The molecular characterization of the kinin transcript and the physiological effects of kinins in the blood-gorging insect, *Rhodnius prolixus*. *Peptides.* 2014;53:148—158.
6. de Haro M, Al-Ramahi I, Benito-Sipos J, et al. Detailed analysis of leucokinin-expressing neurons and their candidate functions in the *Drosophila* nervous system. *Cell Tissue Res.* 2010;339:321—336.
7. Al-Anzi B, Armand E, Nagamei P, et al. Mosquito *Aedes aegypti* (L.) leucokinin receptor is critical for *in vivo* fluid excretion post blood feeding. *Curr Biol.* 2010;20:969—978.
8. Kersch CN, Pietrantonio PV. The leucokinin pathway and its neurons regulate meal size in *Drosophila*. *FEBS Lett.* 2011;585:3507—3512.
9. Pietrantonio PV, Jagge C, Taneja-Bageshwar S, et al. The mosquito *Aedes aegypti* (L.) leucokinin receptor is a multiligand receptor for the three *Aedes* kinins. *Insect Mol Biol.* 2005;14:55—67.
10. Radford JC, Davies SA, Dow JA. Systematic G-protein-coupled receptor analysis in *Drosophila melanogaster* identifies a leucokinin receptor with novel roles. *J Biol Chem.* 2002;277:38810—38817.

Supplemental Information Available on Companion Website

- Primary structures of prepro-invertebrate kinins/E-Figure 63.1
- Alignment of kinin receptors/E-Figure 63.2
- Phylogenetic tree of kinin receptors/E-Figure 63.3
- Accession number of genes and cDNAs for prepro-invertebrate kinins and kinin receptors/E-Table 63.1

CHAPTER 64

FMRFamides

Shinji Nagata

Abbreviation: FMRFa
Additional names: FMRFamide-related peptide (FaRP)

FMRFamide (FMRFa) was originally isolated as a cardioacce-leratory peptide from mollusks. To date, a number of biologically active FMRFa and related peptides have been found in various forms.

Discovery

FMRFa was originally discovered as a factor with cardioacce-leratory activities from the extract of mollusk ganglia [1]. Since the discovery of FMRFa, peptides with extended length at the N-terminal portion have been reported. The extended forms and related peptides are together designated FMRFamide-related peptides (FaRPs), consequently generating a large family of FaRPs.

Structure

Structural Features

FaRPs share the C-terminal RF-NH_2. In *D. melanogaster*, FMRFa is a heptapeptide with a C-terminal FMRF−NH_2 sequence. In addition to FMRFa, several extended forms and other related peptides (FaRPs) are produced from a prepropeptide; these have different lengths and do not always have a conserved C-terminal FMRF−NH_2 but similar sequence domain such as FMLF-NH_2 (Table 64.1) [2].

Structure and Subtype

In *D. melanogaster*, a preproFMRFa is composed of 10 copies of FMRFa and FaRPs. These 10 peptides have different aa lengths, ranging from 7 to 10. Three peptides of the 10 have a substitution within the C-terminal FMRF−NH_2 (Table 64.1) [3]. The number of FaRP subtypes in a single species varies according to the species [4]. The structural characteristics of FaRPs sometimes extensively define the broad range of peptides,

Table 64.1 Alignment of *D. melanogaster* FMRFa and FaRPs (Extended Forms of FMRFa) Derived from the Prepropeptide

FMRFa-1	SVQDMPMHF−NH_2
FMRFa-2	DPKQD−FMRF−NH_2
FMRFa-3	TPAED−FMRF−NH_2
FMRFa-4	SDNFMRF−NH_2
FMRFa-5	SPKQDNFMRF−NH_2
FMRFa-6	PDNFMRF−NH_2
FMRFa-7	SAPQD−FVRF−NH_2
FMRFa-8	MDSNFIRF−NH_2

including sulfakinin, myosuppressin, and other similar peptides with C-terminal RF−NH_2. In this chapter, only those peptides derived from FMRFa prepropeptides are described; other related peptides are described in the appropriate chapters (e.g., Chapter 46, Sulfakinin; Chapter 66, Myosuppressin).

Synthesis and Release

Gene, mRNA, and Precursor

An mRNA encoding *D. melanogaster* FMRFa produces a deduced prepropeptide composed of 347 amino acids. The prepropeptide is composed of a signal sequence, and 10 copies of FMRFa and extended FaRPs (because two identical peptides overlap, structurally there are eight peptides).

Distribution of mRNA

FMRFa in *D. melanogaster* and the blowfly is distributed in the six large neurosecretory cells (Tv neurons) in the thoracic ganglia and in the thoracic endocrinal cells [4,5]. While expression sites are widely observed throughout the CNS, FaRPs are observed not only in the endocrinal secretary cells but also in the central interneurons − for example, in the optic lobes in the brain, and the dorsal and caudal protocer-ebrum in the midbrain.

Receptors

Structure and Subtype

In *D. melanogaster*, a receptor for FMRFa (CG2114) has been identified as a GPCR [6]. In the silkworm *Bombyx mori*, the FaRP receptor exhibits high similarity with the *D. melanoga-ster* FMRFa receptor. All subtypes of FaRPs stimulate one receptor in one species.

Signal Transduction Pathway

In *D. melanogaster*, stimulation of FMRFa introduces signals involving intracellular CaMKII activity, leading eventually to an increase in muscle activity [7].

Biological Functions

Target Cells and Function

The predominant effect of FaRPs is the reduction of spontane-ous muscle contractions, such as in the intestinal muscle and the heart rate. Among FaRP subtypes, some FaRPs function in gut contraction and other subtypes do not. Such minor physi-ological discrimination of FaRPs results in the wide range of functions [8]. For example, FaRPs exert acceleratory effects on

Y. Takei, H. Ando, & K. Tsutsui (Eds): Handbook of Hormones. DOI: http://dx.doi.org/10.1016/B978-0-12-801028-0.00064-7

the heart rate, inhibitory effects on gut motility, and the modulation of synaptic activity. Furthermore, the ecdysis sequential behavior is also regulated by FaRPs, and is controlled downstream of the eclosion hormone (EH) and ecdysis-triggering hormone (ETH) [9]. Coincidentally, the fact that Tv neurons in the FaRP expression site are activated by ETH indicates a function of FaRPs during ecdysis.

References

1. Price DA, Greenberg MJ. Purification and characterization of a cardioexcitatory neuropeptide from the central ganglia of a bivalve mollusk. *Science*. 1977;7:261−281.
2. Nässel DR, Winther AM. *Drosophila* neuropeptides in regulation of physiology and behavior. *Prog Neurobiol*. 2010;92:42−104.
3. Santama N, Benjamin RR. Gene expression and function of FMRFamide-related neuropeptides in the snail *Lymnaea*. *Microsc Res Tech*. 2000;49:547−556.
4. Nichol R. Signaling pathways and physiological functions of *Drosophila melanogaster* FMRFamide-related peptides. *Annu Rev Entomol*. 2003;48:485−503.
5. Wgute J, Hurteau T, Punsal P. Neuropeptide-FMRFamide-like immunoreactivity in *Drosophila*: development and distribution. *J Comp Neurol*. 1986;247:430−438.
6. Cazamali G, Grimmelikhuijzen CJ. Molecular cloning and functional expression of the first insect FMRFamide receptor. *Proc Natl Acad Sci USA*. 2002;99:12073−12078.
7. Dunn TW, Mercier AJ. Synaptic modulation by a *Drosophila* neuropeptide is motor neuron-specific and requires CaMKII activity. *Peptides*. 2005;26:269−276.
8. Orchard I, Lange AB, Bendena WG. FMRFamide-related peptides: a multifunctional family of structurally related neuropeptides in insects. *Adv Insect Physiol*. 2001;28:267−329.
9. Kim YJ, Zitnan D, Galizia CG, et al. A command chemical triggers an innate behavior by sequential activation of multiple peptidergic ensembles. *Curr Biol*. 2006;16:1395−1407.

Supplemental Information Available on Companion Website

- Accession numbers of an FMRFa preproprotein and its receptor in *D. melanogaster*/E-Table 64.1

CHAPTER 65

Myoinhibiting Peptide

Yoshiaki Tanaka

Abbreviations: MIP, AST-B, PTSP
Additional names: allatostatin-B, prothoracicostatic peptide, $W(X_6)$Wamide

Myoinbiting peptide is characterized by two tryptophan (W) residues at the C-terminus denoted as the $W(X_6)$Wamide or $W(X_7)$Wamide motif. MIP receptor (MIPR) is a multiligand receptor and can be activated by both MIPs and sex peptide (SP) of the fruit fly Drosophila melanogaster.

Discovery

The myoinhibiting peptide (MIP) was first isolated from extracts of the brain−corpora cardiaca−corpora allata−subesophageal ganglion complexes of the migratory locust *Locusta migratoria*, based on its ability to inhibit the spontaneous contractions of the hindgut and oviduct [1]. This peptide is also described as prothoracicostatic peptide (PTSP) [2] or allatostatin-B (AST-B) [3] based on different physiological effects in specific insects.

Structure

Structural Features

MIPs contain 9−12 amino acids and are characterized by two tryptophan (W) residues at the C-terminus denoted as the $Trp−(X_6)−Trp−NH_2$ ($W(X_6)$Wamide) motif (Table 65.1).

Structural Feature and Subtype

The members of this family of peptides with a $W(X_6)$Wamide motif have since been identified in a number of insect species as well as in crustacean and molluskan species (E-Figure 65.1) [4]. Both MIPs and homologous peptides with a $W(X_7)$ Wamide motif are found in the hemipteran insects and the water flea *Daphnia pulex* (Table 65.1) [4]. No members of this family have been reported in genome or peptidomic searches of Hymenoptera.

Table 65.1 Comparison of MIPs

L. migratoria MIP	AWQDL−NAGW−NH₂
M. sexta MIP-I/*B. mori* PTSP-1	AWQDL−NSAW−NH₂
G. bimaculatus AST-B1	GWQDL−NGGW−NH₂
R. prolixus MIP-7	AWNSL−HGGW−NH₂
R. prolixus MIP-1	AWKDLQSSGW−NH₂

Synthesis and Release

Gene and mRNAs

MIP mRNA encodes multiple copies of MIPs. *MIP* transcripts in the hemipteran insects and *D. pulex* code for both $W(X_6)$ Wamide and $W(X_7)$Wamide peptides [4].

Distribution of mRNA

MIP transcripts are detected in numerous cells throughout the central nervous system (CNS) [5,6]. In the silkworm *Bombyx mori*, prominent signals are observed in a pair of medial neurosecretory cells of the brain, the abdominal ganglia, frontal ganglion, and terminal ganglion [6]. Transcripts are also detected in the midgut endocrine cells in the fruit fly *Drosophila melanogaster* [5], and the peripheral endocrine organ, epiproctodeal gland (EPG), in *B. mori* [6].

Regulation of Synthesis and Release

Expression of *MIP* in the EPG of *B. mori* is regulated by the levels of hemolymph ecdysteroid, and correlates with the expression of its receptor in the PG, which is upregulated by the decline of the ecdysteroid titer in *B. mori* [6].

Receptors

Structure and Subtype

A functional receptor for MIPs was originally identified in *B. mori* using a heterologous expression system [6]. The identified receptor was the GPCR which had been previously reported as the *Bombyx* ortholog of *Drosophila* sex peptide (SP) receptor [7]. This receptor can be activated by both MIPs and SP, but MIPs are more potent agonists for MIPR than SP itself [8]. MIPR is also activated by MIPs with a $W(X_7)$Wamide motif (Table 65.2) [4]. These three peptides are structurally quite different except for the two Trp residues. Thus, it appears that only the two Trp residues are necessary for total receptor activation. Moreover, MIPR allows for some degree of freedom in the spacing of the two Trp residues [8].

Both MIPs and MIPR are found in most invertebrates, including Ecdysozoa and Lophotrochozoa (E-Figures 65.2 and 65.3). By contrast, SP is only detected in the genomes of a limited number of *Drosophila* species. These findings suggest that MIPs are the ancestral ligands for MIPR, and that SP later hijacked MIPR during the evolution of *Drosophila* species [8].

Distribution of mRNA

The receptor is highly expressed in the prothoracic glands (PGs) during each larval molting period and pupal molting

Y. Takei, H. Ando, & K. Tsutsui (Eds): Handbook of Hormones. DOI: http://dx.doi.org/10.1016/B978-0-12-801028-0.00065-9

Table 65.2 MIP Receptor Activation by MIPs and SP [4]

Peptide	Sequence	EC$_{50}$ of *D. melanogaster* MIPR Activation (nM)
R. prolixus MIP-7 (W(X$_6$)Wamide)	AWNSLHGG--W-NH$_2$	0.59
R. prolixus MIP-4 (W(X$_7$)Wamide)	AWSDLQSSG-W-NH$_2$	53
D. melanogaster SP	MKTLALFLVLVCVLGLVQAW-EWPWNRKPTKFPIPSPNPRD-KWCRLNLGPAWGGRC	15

period, but only weakly expressed during each intermolt period, in *B. mori* [6]. *MIPR* (i.e., *SPR*) is expressed in the female's reproductive tract and CNS in adult *D. melanogaster* [7], but MIPs are not required for post-mating behavior at this stage [8].

Signal Transduction Pathway

MIP inhibits ecdysteroidogenesis in the PG by blocking dihydropyridine-sensitive Ca($^{2+}$) channels [9].

Biological Functions

Target Cells/Tissues and Functions

MIPs show diverse effects in insects. They suppress visceral muscle contractions in the tobacco hornworm *Manduca sexta* and *L. migratoria* [1]. These peptides suppress both production of juvenile hormone by the corpora allata, and ecdysteroidogenesis by the ovary in the cricket *Gryllus bimaculatus* [3]. In *B. mori*, MIPs are released from EPG during each larval and pupal ecdysis, and suppress activation of the PG to finely regulate the ecdysteroid titer in hemolymph [2]. MIPs act together with crustacean cardioactive peptide (CCAP) to initiate the ecdysis motor pattern in the CNS of insects [10].

Phenotype in Gene-Modified Animals

Knockdown of the MIP gene in the CNS does not affect the post-mating response in the female adult of *D. melanogaster* [8].

References

1. Schoofs L, Holman GM, Hayes TK, et al. Isolation, identification and synthesis of locustamyoinhibiting peptide (LOM-MIP), a novel biologically active neuropeptide from *Locusta migratoria*. *Regul Pept*. 1991;36:111–119.
2. Hua YJ, Tanaka Y, Nakamura K, et al. Identification of a prothoracicostatic peptide in the larval brain of the silkworm, *Bombyx mori*. *J Biol Chem*. 1999;274:31169–31173.
3. Lorenz MW, Kellner R, Hoffmann KH. A family of neuropeptides that inhibit juvenile hormone biosynthesis in the cricket, *Gryllus bimaculatus*. *J Biol Chem*. 1995;270:21103–21108.
4. Lange AB, Alim U, Vandersmissen HP, et al. The distribution and physiological effects of the myoinhibiting peptides in the kissing bug, *Rhodnius prolixus*. *Front Neurosci*. 2012;6:98.
5. Williamson M, Lenz C, Winther AM. Molecular cloning, genomic organization, and expression of a B-type (cricket-type) allatostatin preprohormone from *Drosophila melanogaster*. *Biochem Biophys Res Commun*. 2001;281:544–550.
6. Yamanaka N, Hua YJ, Roller L, et al. *Bombyx* prothoracicostatic peptides activate the sex peptide receptor to regulate ecdysteroid biosynthesis. *Proc Natl Acad Sci USA*. 2010;107:2060–2065.
7. Yapici N, Kim YJ, Ribeiro C, et al. A receptor that mediates the post-mating switch in *Drosophila* reproductive behaviour. *Nature*. 2008;451:33–37.
8. Kim YJ, Bartalska K, Audsley N, et al. MIPs are ancestral ligands for the sex peptide receptor. *Proc Natl Acad Sci USA*. 2010;107:6520–6525.
9. Dedos SG, Birkenbeil H. Inhibition of cAMP signalling cascade-mediated Ca^{2+} influx by a prothoracicostatic peptide (Mas-MIP I) via dihydropyridine-sensitive Ca^{2+} channels in the prothoracic glands of the silkworm, *Bombyx mori*. *Insect Biochem Mol Biol*. 2003;33:219–228.
10. Kim YJ, Zitnan D, Cho KH, et al. Central peptidergic ensembles associated with organization of an innate behavior. *Proc Natl Acad Sci USA*. 2006;103:14211–14216.

Supplemental Information Available on Companion Website

- Primary structures of prepro-MIPs/E-Figure 65.1
- Alignment of MIPRs/E-Figure 65.2
- Phylogenetic tree of MIPRs/E-Figure 65.3
- Accession number of genes and cDNAs for prepro-MIP s and MIPRs/E-Table 65.1

CHAPTER 66

Myosuppressin

Yoshiaki Tanaka

Abbreviation: MS
Additional names: FLRFamide, F10

Myosuppressin (MS) shows inhibitory activities against various kinds of visceral muscles of arthropods, and is thought to be one of the core sets of neuropeptides found among insect species. MS receptors are not evolutionarily related to the insect FMRFamide receptors.

Discovery

Myosuppressin (MS) was first isolated as a myoinhibiting peptide from the Madeira cockroach *Leucophaea maderae*, based on its ability to reduce the frequency of hindgut contractions [1].

Structure

Structural Features

MSs are decapeptides and have the general conserved sequence X_1−Asp−X_3−X_4−His−X_6−Phe−Leu−Arg−Phe−NH_2 (XDXXHXFLRFamide), where X_1 can be pGlu, Pro, Thr, or Ala; X_3 can be Val or Leu; X_4 can be Asp, Gly, or Val; and X_6 can be Val, Ser, or Ile [2]. MSs appear to be one of the core sets of neuropeptides found in all sequenced insect genomes.

Structure and Subtype

The MS peptides are represented by the consensus structure XDXXHXFLRFamide, but MS in the kissing bug *Rhodnius prolixus* [3] has a unique amino acid sequence with methionine[8] instead of leucine (Table 66.1). Two extended myosuppressins, F24 and F39, have been isolated from the midgut of parasitized tobacco hornworm *Manduca sexta* (Table 66.1) [4]. The sequence of F24 is identical to the C-terminal 24 amino acids of F39. Moreover, the C-terminal 10-aa sequence of F24 and F39 is identical to that of MS, suggesting that they are generated from the same precursor.

Synthesis and Release

Gene, Precursor, and mRNAs

The mRNA encoding MS produces the precursor peptide containing one copy of MS near the C-terminus. Both MS and extended MSs are derived from a single gene, and the extended MSs are generated by atypical cleavage at the N-terminal portion of the precursor peptides in the midgut cells of *M. sexta* under the parasitized condition [4]. In the copepod crustacean *Calanus finmarchicus*, multiple genes coding for one copy of myosuppressin prepropeptide have been identified [5].

Distribution of mRNA

MS mRNA is expressed in the neurosecretory cells (NSCs) of the brain [4,6], frontal ganglion, neurons/NSCs of all ganglia of the ventral nerve cord, corpora cardiaca, link neuron 1 (L1) that are located at connections between the peripheral nerves, and midgut endocrine cells of insects [3,4].

Regulation of Synthesis and Release

In the silkworm *Bombyx mori*, MS is mainly produced in the brain and secreted in the feeding period of the final larval instar [6]. In the nerve cord of *M. sexta*, mRNA levels fluctuate during development and increase during pupal and adult ecdysis [4].

Receptors

Structure and Subtype

Two GPCRs were first identified as MS receptors (MSRs) in *D. melanogaster* [7]. These receptors, DMSR-1 and DMSR-2, are highly homologous (65% overall amino acid residue identity; 71% identity in the transmembrane region). MSRs have been functionally characterized in the mosquito *Anopheles gambiae* and *B. mori* [6]. Moreover, MSR orthologs have been found in many insects, mites, and crustaceans. Although MSs are often grouped together with insect FMRFamides under the name FaRPs (FMRFamide-related peptides), MSRs are not activated by the neuropeptides that resemble MS only in their C-terminal moiety, such as *Anopheles* FMRFamide-3 (Pro−Asp−Arg−Asn−Phe−Leu−Arg−Phe−NH_2) and other FMRFamide peptides [7]. This observation suggests that the N-terminal portion of MS in addition to the FLRFamide motif is important for the activation of MSR, and MSRs are not evolutionarily related to the insect FMRFamide receptors.

Distribution of mRNA

In the adult of *D. melanogaster*, DMSR-1 is highly expressed in the head but virtually absent in the body, whereas DMSR-2 is present in both the head and body [7]. In *B. mori*, *MSR* is expressed in the prothoracic gland in addition to the midgut, hindgut, and malpighian tubule [6].

Signal Transduction Pathway

MS inhibits both the basal and PTTH-stimulated cAMP accumulation in the prothoracic gland of *B. mori* in a dose-dependent manner [6].

Y. Takei, H. Ando, & K. Tsutsui (Eds): Handbook of Hormones. DOI: http://dx.doi.org/10.1016/B978-0-12-801028-0.00066-0

Table 66.1 MS and MS-Like Peptides in the Arthropods

Species	Sequence
D. melanogaster	TDVDHVFLRF–NH$_2$
A. gambiae	TDVDHVFLRF–NH$_2$
T. castaneum	pEDVDHVFLRF–NH$_2$
A. mellifera	pEDVVHFLRF–NH$_2$
B. mori	pEDVVHSFLRF–NH$_2$
M. sexta	pEDVVHSFLRF–NH$_2$
R. prolixus	pEDIDHVFMRF–NH$_2$
S. gregaria	PDVDHVFLRF–NH$_2$
D. pulex	pEDVDHVFLRF–NH$_2$
P. clarkii	pEDLDHVFLRF–NH$_2$
C. finmarchicus	IDSGHIFLRF–NH$_2$
C. finmarchicus	pEPDPDHVFLRF–NH$_2$
C. finmarchicus	DPDHVFLRF–NH$_2$
C. finmarchicus	DPQHLFLRF–NH$_2$
C. finmarchicus	NDPDHVFLRF–NH$_2$
C. finmarchicus	TKDPDHVFLRF–NH$_2$
C. finmarchicus	AKPDHVFLRF–NH$_2$
M. sexta (F39)	YAEAEQVPEYQALVRDYPLLDSGMKRQDVVHSFLRF–NH$_2$
M. sexta (F24)	VRDYPLLDSGMKRQDVVHSFLRF–NH$_2$
M. sexta (F7G)	GNSFLRF–NH$_2$
M. sexta (F7D)	DPSFLRF–NH$_2$
A. gambiae (FMRFamide-3)	PDRNFLRF–NH$_2$
C. finmarchicus (FMRFamide)	NFLRF–NH$_2$

Biological Functions

Target Cells/Tissues and Functions

MS shows inhibitory activities against visceral muscles of various organs, including the gut, heart, oviduct, malpighian tubule, and salivary gland in insects [2]. MS also inhibits feeding activity in the cockroach *Blattella germanica* and in the tobacco cutworm *Spodoptera littoralis* [8]. Moreover, MS inhibits the release of adipokinetic hormones from the corpora cardiaca of the migratory locust *Locusta migratoria* [9], and appears to act as a prothoracicostatic hormone in *B. mori* [6].

Phenotype in Gene-Modified Animals

In the oothecae-carrying females of *B. germanica*, knockdown of MS increases the frequency of heartbeat [10].

References

1. Holman GM, Cook BJ, Nachman RJ. Primary structure and synthesis of a blocked myotropic neuropeptide isolated from the cockroach, *Leucophaea maderae*. *Comp Biochem Physiol C.* 1986;85:219–224.
2. Orchard I, Lange AB, Bendena WG. Molecular characterization of the inhibitory myotropic peptide leucomyosuppressin. *Adv Insect Physiol.* 2001;28:267–329.
3. Lee D, Taufique H, da Silva R, Lange AB. An unusual myosuppressin from the blood-feeding bug *Rhodnius prolixus*. *J Exp Biol.* 2012;215:2088–2095.
4. Lu D, Lee KY, Horodyski FM, et al. Molecular characterization and cell-specific expression of a *Manduca sexta* FLRFamide gene. *J Comp Neurol.* 2002;446:377–396.
5. Christie AE, Roncalli V, Wu LS, et al. Peptidergic signaling in *Calanus finmarchicus* (Crustacea, Copepoda): *in silico* identification of putative peptide hormones and their receptors using a de novo assembled transcriptome. *Gen Comp Endocrinol.* 2013;187:117–135.
6. Yamanaka N, Hua YJ, Mizoguchi A, et al. Identification of a novel prothoracicostatic hormone and its receptor in the silkworm *Bombyx mori*. *J Biol Chem.* 2005;280:14684–14690.
7. Egerod K, Reynisson E, Hauser F, et al. Molecular cloning and functional expression of the first two specific insect myosuppressin receptors. *Proc Natl Acad Sci USA.* 2003;100:9808–9813.
8. Audsley N, Weaver RJ. Neuropeptides associated with the regulation of feeding in insects. *Gen Comp Endocrinol.* 2009;162:93–104.
9. Vullings HG, Ten Voorde SE, Passier PC, et al. A possible role of SchistoFLRFamide in inhibition of adipokinetic hormone release from locust corpora cardiaca. *J Neurocytol.* 1998;27:901–913.
10. Maestro JL, Tobe SS, Belle X. Leucomyosuppressin modulates cardiac rhythm in the cockroach *Blattella germanica*. *J Insect Physiol.* 2011;57:1677–1681.

Supplemental Information Available on Companion Website

- Primary structures of preproMSs/E-Figure 66.1
- Alignment of MSRs/E-Figure 66.2
- Phylogenetic tree of MSRs/E-Figure 66.3
- Accession number of genes and cDNAs for preproMSs and MSRs/E-Table 66.1

CHAPTER 67

Proctolin

Yoshiaki Tanaka

Abbreviation: Proct

Proctolin was the first bioactive peptide to be isolated from insect tissues and structurally characterized. Proctolin shows diverse activities, including a myotropic effect, regulation of JH synthesis, and behavioral effects.

Discovery

Proctolin was first identified in 1975 from the whole body of the American cockroach *Periplaneta americana*, based upon its ability to stimulate hindgut contractions [1].

Structure

Structural Features

Proctolin is almost invariably present in the form of the pentapeptide Arg−Tyr−Leu−Pro−Thr (Figure 67.1).

Structure and Subtype

Two alternative forms, Ala−Tyr−Leu−Pro−Thr and Arg−Tyr−Leu−Met−Thr, along with typical proctolin, have been reported in the Colorado potato beetle *Leptinotarsa decemlineata* [2] and the water flea *Daphnia pulex* [3], respectively. Proctolin appears to be unique to arthropods, and there is no convincing evidence for its presence in any other invertebrate or vertebrate [4].

Synthesis and Release

Gene, mRNA, and Precursor

The gene (CG7105, *Proct*) encoding a proctolin precursor protein was first identified in the fruit fly *Drosophila melanogaster* [5]. The predicted protein encoded by *Proct* has a single copy of the Arg−Tyr−Leu−Pro−Thr sequence that directly follows the predicted signal peptidase cleavage point flanked by a potential peptide cleaving site (Figure 67.1), and similar precursor structures are conserved in other insects and in non-insect arthropods (E-Figure 67.1). The proctolin gene has been found in a variety of arthropods [4]. However, the genes for proctolin and its receptor appear to be absent from the genomes of Hymenoptera, Lepidoptera and the yellow fever mosquito *Aedes aegypti* [4].

Distribution of mRNA

Proctolin mRNA is expressed in the brain and ventral nerve cord of *D. melanogaster* [5], and in the testes of fifth instars and reproductive tissues of adults of the kissing bug *Rhodnius prolixus* [4].

Receptors

Structure and Subtype

Drosophila CG6986 encoding a GPCR has been identified as a proctolin receptor independently by two research groups [6]. An interesting feature of the proctolin receptors of *Drosophila* species is the large extracellular loop between TM4 and TM5, which is very unusual for a GPCR (E-Figure 67.2). However, such extra-large loops are not found in the orthologous receptors of other species.

Distribution of mRNA

Proctolin receptor is only weakly expressed in embryos, larvae, pupae, and the thorax and abdomen of adult flies, but is strongly expressed in the adult heads of *D. melanogaster* [6].

Signal Transduction Pathway

Proctolin can increase intracellular calcium concentrations by modulating voltage-dependent or -independent channels in insect muscles [7], and induces a decrease in intracellular

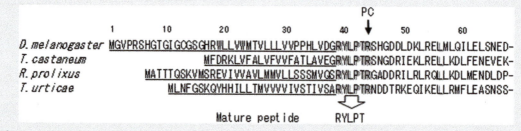

Figure 67.1 The structure of preproproctolin. Signal peptides are underlined. PC, cleavage site by a proprotein convertase.

Y. Takei, H. Ando, & K. Tsutsui (Eds): Handbook of Hormones. DOI: http://dx.doi.org/10.1016/B978-0-12-801028-0.00067-2

cGMP concentration via PKC-dependent inhibition of guanylate cyclase in the skeletal muscle of the marine isopod *Idotea emarginata* [8].

Biological Functions

Target Cells/Tissues and Functions

Proctolin displays potent myotropic activity in visceral, cardiac, and skeletal muscle from a range of insect species [4]. In addition to the well-known role of proctolin as a myotropic agent, proctolin may act as a releasing factor for adipokinetic hormones (AKHs) and juvenile hormone (JH) in the locust [9]. However, it does not regulate JH release in the cockroach. Moreover, proctolin plays a modulatory role in a neuronal circuit controlling sexual behavior in the bow-winged grasshopper *Chorthippus biguttulus* [10].

Phenotype in Gene-Modified Animals

Pupae overexpressing *Proct* displayed a 14% increase in heart rate in *D. melanogaster* [10].

References

1. Starratt AN, Brown BE. Structure of the pentapeptide proctolin, a proposed neurotransmitter in insects. *Life Sci.* 1975;17:1253–1256.
2. Spittaels K, Vankeerberghen A, Torrekens S, et al. Isolation of Ala¹-proctolin, the first natural analogue of proctolin, from the brain of the Colorado potato beetle. *Mol Cell Endocrinol.* 1995;110:119–124.
3. Dircksen H, Neupert S, Predel R, et al. Genomics, transcriptomics, and peptidomics of *Daphnia pulex* neuropeptides and protein hormones. *J Proteome Res.* 2011;10:4478–4504.
4. Orchard I, Lee do H, da Silva R, et al. The proctolin gene and biological effects of proctolin in the blood-feeding bug, *Rhodnius prolixus. Front Endocrinol.* 2011;2:59.
5. Taylor CA, Winther AM, Siviter RJ, et al. Identification of a proctolin preprohormone gene (Proct) of *Drosophila melanogaster*: expression and predicted prohormone processing. *J Neurobiol.* 2004;58:379–391.
6. Isaac RE, Taylor CA, Hamasaka Y, et al. Proctolin in the post-genomic era: new insights and challenges. *Invert Neurosci.* 2004;5:51–64.
7. Wegener C, Nässel DR. Peptide-induced Ca(2+) movements in a tonic insect muscle: effects of proctolin and periviscerokinin-2. *J Neurophysiol.* 2000;84:3056–3066.
8. Philipp B, Rogalla N, Kreissl S. The neuropeptide proctolin potentiates contractions and reduces cGMP concentration via a PKC-dependent pathway. *J Exp Biol.* 2006;209:531–540.
9. Clark L, Zhang JR, Tobe S, et al. Proctolin: a possible releasing factor in the corpus cardiacum/corpus allatum of the locust. *Peptides.* 2006;27:559–566.
10. Vezenkov SR, Danalev DL. From molecule to sexual behavior: the role of the neuropentapeptide proctolin in acoustic communication in the male grasshopper *Chorthippus biguttulus. Eur J Pharmacol.* 2009;619:57–60.

Supplemental Information Available on Companion Website

- Primary structures of preproproctolins/E-Figure 67.1
- Alignment of proctolin receptors/E-Figure 67.2
- Phylogenetic tree of proctolin receptors/E-Figure 67.3
- Accession numbers of genes and cDNAs for preproproctolins and proctolin receptors/E-Table 67.1

CHAPTER 68

Orcokinins

Yoshiaki Tanaka

Abbreviation: OK

Orcokinins were first identified as crustacean myotropic peptides and more recently they have been reported in many insects and other arthropods. Orcokinins play diverse physiological roles in various target tissues.

Discovery

Orcokinin was originally isolated from approximately 1,200 abdominal nerve cords of the crayfish *Orconectes limosus* as a potent hindgut myostimulating factor [1].

Structure

Structural Features

Orcokinins are characterized by a conserved N-terminal motif Asn−Phe−Asp−Glu−Ile−Asp−Arg (NFDEIDR). The C-terminal sequences are divergent among the isoforms.

Primary Structure and Subtype

Orcokinins occur in various decapod crustaceans, insects, and arachnids (Table 68.1). The substitution of amino acid residues within the conserved N-terminal motif occurs in several insects and arthropods [2−5]. C-terminal amidated orcokinins are identified in the malaria mosquito *Anopheles gambiae* and the spider mite *Tetranychus urticae* (Table 68.1).

Synthesis and Release

Gene and mRNA

Orcokinin mRNA encodes one or more isoforms of orcokinin (E-Figure 68.1) [6], and alternative splicing of a single orcokinin gene gives rise to multiple transcripts in several insects (Figure 68.1) [7]. Orcokinin peptides having a conserved N-terminal motif (orcokinin-A) are encoded in the transcript A (OKA), and unique peptides (orcokinin-B) having the consensus sequence X_1−Asp−X−X_1−Gly−Gly−Gly−X_2−Leu−X_3 (X_1 = Ile or Leu; X_2 = Asn or Val; X_3 = Ile, Leu, or Val) are encoded in the transcript-B (OKB, Figure 68.1). However, both orcokinin-A and -B are encoded in the same transcripts (OKA + B) in the water flea *Daphnia pulex*, the cattle tick *Rhipicephalus microplus*, and the silkworm *Bombyx mori* (Figure 68.1). Orcokinin-B is not found in the pea aphid *Acyrthosiphon pisum*, *O. limosus*, American lobster *Homarus americanus*, and the spider mite *T. urticae*, whereas only orcokinin-B is found in the red flour beetle *Tribolium castaneum* (Table 68.1) [7]. Orcokinin-B is so named as it

Table 68.1 Orcokinins in the Arthropods

Species	Orcokinin-A	Orcokinin-B
A. gambiae	NFDEIDRFARFN−NH$_2$	NQRSLDSIGGGNLL
	NFDEIDRFN−NH$_2$	ASLDSIGGGNLL
D. melanogaster	NFDEIDKASASFSILNQLV	GLDSIGGGHLI
B. mori	NFDEIDRSSLNTFV	NLDSLGGGHLI
	NFDEIDRSSMPFPYAI	NLDPLGGGNLV
T. castaneum		SLDRIGGGNLV−NH$_2$
		SLDGIGGGNLV−NH$_2$
A. mellifera	NLDEIDRVGWSGFV	NLDQIGGGNLL
		NLDSIGGGNLV
A. pisum	NFDEIDRSGFDRFV	
	NFDEIDRSAFNSFV	
N. lugens	NMDEMDRSGFSSFI	NLDGIGGGNLL
	NFDELDRAGFDSFI	
	NFDEMDRAGFNSFV	
R. prolixus	FDPLSSAYAAD	EGGTLDSLGGGHLI
	NFDEIDRSGFNSFI	DFLDGLGGGHLL
	NFDEIDRSGFDGFV	EFLDPLGGGHLI
	NFDEIDRVGFGSFI	GIDSIGGGHLL
		GLDSIGGGHLV
T. urticae	NFDEIDRSELSGFA−NH$_2$	
	NFDEIDRSAFVGFN	
	NFDEIDRSGLAGFR−NH$_2$	
	NFDEIDRSSFTGFTG−NH$_2$	
	NFDEIDRTNLNGFR−NH$_2$	
	NYDEIDNSGFV−NH$_2$	
	NFDEIDRSAFSGFT	
R. microplus	NFDEIDRNGFEGFT	SLDKIGGGEYI
	NFDEIDRTGFEGFY	
D. pulex	NLDEIDRSNFGTFA	
	NLDEIDRSDFGRFV	
H. americanus	NFDEIDRSGFGFN	
	NFDEIDRSGFGFH	
	NFDEIDRSGFGFV	
P. clarkii	NFDEIDRSGFGFN	
	NFDEIDRSGFGFV	
	NFDEIDRSGFGFA	
	NFDEIDRSGFGFH	

originated from alternative splicing of the orcokinin gene [7]; however, it is not structurally related to orcokinin-A and its biological functions remain to be analyzed.

Distribution of mRNA

Orcokinin mRNAs are expressed in numerous neurons of the CNS and midgut endocrine cells [6,7], but OKA expression is restricted to the CNS in the kissing bug *Rhodnius prolixus* [7]. In *B. mori*, orcokinin is co-expressed with *Bombyx* FMRFa (BRFa) in one pair of the neurosecretory cells of the prothoracic ganglion, and orcokinin peptides are delivered to the surface of the prothoracic gland (PG) through the neurons projecting to the PGs [6]. In *O. limosus*, orcokinins are co-localized with orcomyotropin in approximately 80−90

Y. Takei, H. Ando, & K. Tsutsui (Eds): Handbook of Hormones. DOI: http://dx.doi.org/10.1016/B978-0-12-801028-0.00068-4

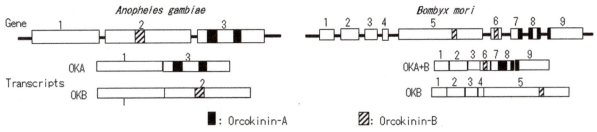

Figure 68.1 Alternative splicing of orcokinin gene in *Anopheles gambiae* and *Bombyx mori.* The numbers indicate the exon number.

neurons of the terminal abdominal ganglion that have been shown to innervate the entire hindgut muscularis via the intestinal nerve [8].

Biological Functions

Target Cells/Tissues and Functions

Orcokinins play diverse physiological roles in various target tissues. In crustacea, orcokinins have an effect on hindgut myoactivity in *O. limosus* [1], and regulate the pyloric rhythm of adult *H. americanus* [9]. In insects, orcokinins regulate the circadian locomotor activity in the Madeira cockroach *Leucophaea maderae* [10] and are involved in the neuronal regulation of ecdysterodogenesis by the prothoracic gland in *B. mori* [7].

References

1. Stangier J, Hilbich C, Burdzik S. Orcokinin: a novel myotropic peptide from the nervous system of the crayfish, *Orconectes limosus*. *Peptides*. 1992;13:859–864.
2. Liu F, Baggerman G, D'Hertog W, et al. *In silico* identification of new secretory peptide genes in *Drosophila melanogaster*. *Mol Cell Proteomics*. 2006;5:510–522.
3. Hummon AB, Richmond TA, Verleyen P, et al. From the genome to the proteome: uncovering peptides in the *Apis* brain. *Science*. 2006;314:647–649.
4. Veenstra JA, Rombauts S, Grbić M. *In silico* cloning of genes encoding neuropeptides, neurohormones and their putative G-protein coupled receptors in a spider mite. *Insect Biochem Mol Biol*. 2012;42:277–295.
5. Tanaka Y, Suetsugu Y, Yamamoto K. Transcriptome analysis of neuropeptides and G-protein coupled receptors (GPCRs) for neuropeptides in the brown planthopper *Nilaparvata lugens*. *Peptides*. 2014;53:125–133.
6. Yamanaka N, Roller L, Žitňan D, et al. *Bombyx* orcokinins are brain-gut peptides involved in the neuronal regulation of ecdysteroidogenesis. *J Comp Neurol*. 2011;519:238–246.
7. Sterkel M, Oliveira PL, Urlaub H. OKB, a novel family of brain-gut neuropeptides from insects. *Insect Biochem Mol Biol*. 2012;42:466–473.
8. Dircksen H, Burdzik S, Sauter A, et al. Two orcokinins and the novel octapeptide orcomyotropin in the hindgut of the crayfish *Orconectes limosus*: identified myostimulatory neuropeptides originating together in neurones of the terminal abdominal ganglion. *J Exp Biol*. 2000;203:2807–2818.
9. Li L, Pulver SR, Kelley WP, et al. Orcokinin peptides in developing and adult crustacean stomatogastric nervous systems and pericardial organs. *J Comp Neurol*. 2002;444:227–244.
10. Wei H, Stengl M. Light affects the branching pattern of peptidergic circadian pacemaker neurons in the brain of the cockroach *Leucophaea maderae*. *J Biol Rhythms*. 2011;26:507–517.

Supplemental Information Available on Companion Website

- Primary structures of prepro-orcokinins/E-Figure 68.1
- Accession numbers of genes and cDNAs for prepro-orcokinins /E-Table 68.1

CHAPTER 69

Crustacean Cardioactive Peptide

Dušan Žitňan and Ivana Daubnerová

Abbreviation: CCAP

The most conserved neuropeptide found in arthropods, CCAP has also been identified in insects and crustaceans. It is expressed in a conserved network of segmental neurons that control ecdysis.

Discovery

Crustacean cardioactive peptide (CCAP) was originally identified from the central nervous system (CNS) of the crab *Carcinus maenas*, based on its heart stimulatory activity [1]. An identical neuropeptide and the gene encoding a single copy of CCAP were later identified or annotated in many other insects and crustaceans [2].

Structure

Structural Features

CCAP is a cyclic amidated peptide composed of 9 amino acids, including two cysteines that form a disulfide bridge. Amidation is crucial for biological activity (Figure 69.1).

Primary Structure

CCAP is the most conserved arthropod neuropeptide. An identical aa sequence, $PFCNAFTGC-NH_2$, has been found in all examined insects and crustaceans.

Properties

Mr 958; soluble in water, physiological saline, 50−70% methanol or ethanol.

Synthesis and Release

Gene, mRNA, and Precursor

In the silkworm *Bombyx mori* the *ccap* gene contains five exons, and mRNA is composed of 651 nt and encodes a simple 126-aa precursor (Figure 69.1). The first 21 aa of the precursor represent a signal peptide, followed by 19 aa and a proteolytic cleavage site KKR, the sequence of neuropeptide CCAP, and another amidation and cleavage site GRKR. No other bioactive peptides seem to be encoded by the gene.

Distribution of Peptide and mRNA

The *ccap* gene is expressed in several interneurons of the brain and in a network of paired neurons in each ventral ganglion. This network is composed of neurosecretory cells (named 27 or L_1) and interneurons (IN-704), and is conserved in all studied insects (network 27/704). Developmental studies in moths showed that in addition to the network 27/704, CCAP expression appears in late larval and pupal stages in lateral motoneurons (MN1), medial neurosecretory cells (M_4), and some other neurons. After adult eclosion, the entire network of 27/704 neurons undergoes programmed cell death [3,4]. In moths, CCAP is also produced by peripheral lateral link neurons in each abdominal segment [4]. This neuropeptide was detected in similar neurons in the CNS of several crustaceans, spiders, and an opilionid.

Hemolymph Concentration

Considerable increases of CCAP levels were detected in the hemolymph during ecdysis of crustaceans [5].

Regulation of Synthesis and Release

Expression of the *ccap* gene is probably constitutive. CCAP is apparently released at ecdysis of insects and crustaceans [2,3,5]. In moths, its release is induced by ecdysis-triggering hormone (ETH) and eclosion hormone (EH) [2,3,6]. In flies, CCAP secretion is probably under the control of ETH, EH, and bursicon [7,8]. In the cockroach *Periplaneta americana*, CCAP expression in the midgut endocrine cells is induced by intake of protein nutrients (starch and casein) [9].

Receptors

Structure and Subtype

The CCAP receptor belongs to a family of GPCRs. The first CCAP receptor (CG6111) was identified in the fruit fly *Drosophila melanogaster* [10]. The mRNA contains 2,986 nt with seven exons (617, 393, 136, 198, 261, 286, 1,095−nt), and the receptor protein is composed of 494 amino acids. Two closely related orthologs of CCAP receptor (A26, A30) have been found in *B. mori*. Moderate expression levels of the A26 gene have been detected in peripheral tissues (fat body, muscles, midgut, malpighian tubules, and epidermis), while A30 gene expression is restricted to the brain and midgut.

Biological Functions

Target Cells/Tissues and Functions

CCAP acts on the heart to elicit increased frequency and amplitude of the heartbeat in crustaceans and insects [1−3]. Its main central role is to induce the specific ecdysis behavior in moths [2,3]. Its function in flies is less clear. CCAP null mutants do not show any obvious defects at ecdysis, and develop normally. However, physiological and genetic

Y. Takei, H. Ando, & K. Tsutsui (Eds): Handbook of Hormones. DOI: http://dx.doi.org/10.1016/B978-0-12-801028-0.00069-6

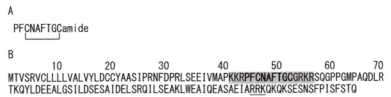

Figure 69.1 CCAP sequence depicting a disulfide bond (A) and CCAP precursor of *Manduca sexta* (B). The CCAP sequence is in bold; cleavage/amidation sequences are shaded.

experiments indicate that CCAP acts together with myoinhibitory peptides and the bursicon subunit (pburs) to control this behavior [7,8]. In feeding cockroaches, CCAP is released from midgut endocrine cells to stimulate α-amylase and protease activity in the intestine [9].

Phenotype in Gene-Modified Animals

In *D. melanogaster*, ablation of CCAP neurons (that also produce myoinhibiting peptides and bursicon) had little effect on larval and adult ecdyses, but caused lethal defects at pupal ecdysis. The majority of pupating flies showed a failure of abdominal inflation. Surviving animals also exhibited abnormal circadian timing of adult eclosion. As mentioned above, however, CCAP null mutants develop normally, indicating that a mixture of different peptides controls the ecdysis behavior [8].

References

1. Stangier J, Hilbich C, Beyruether K, Keller R. Unusual crustacean cardioactive peptide ccap from pericardial organs of the shore crab *Carcinus maenas*. *Proc Natl Acad Sci USA*. 1987;84: 575−579.
2. Žitňan D, Adams ME. Neuroendocrine regulation of ecdysis. In: Gilbert LI, ed. *Insect Endocrinology*. Amsterdam: Elsevier;2012:253−309.
3. Ewer J, Reynolds S. Neuropeptide control of molting in insects. In: Pfaff DW, Arnold AP, Fahrbach SE, et al., eds. *Hormones, Brain, and Behavior*, Vol 3. San Diego: Academic Press;2002:1−92.
4. Davis NT, Homberg U, Dircksen H, et al. Crustacean cardioactive peptide-immunoreactive neurons in the hawkmoth *Manduca sexta* and changes in their immunoreactivity during postembryonic development. *J Comp Neurol*. 1993;338:612−627.
5. Phlippen MK, Webster SG, Chung JS, Dircksen H. Ecdysis of decapod crustaceans is associated with a dramatic release of crustacean cardioactive peptide into the haemolymph. *J Exp Biol*. 2000;203:521−536.
6. Kim Y-J, Žitňan D, Cho K-H, et al. Central peptidergic ensembles associated with organization of an innate behavior. *Proc Natl Acad Sci USA*. 2006;103:14211−14216.
7. Kim Y-J, Žitňan D, Galizia G, et al. A command chemical triggers an innate behavior by sequential activation of multiple peptidergic ensembles. *Curr Biol*. 2006;16:1395−1407.
8. Lahr EC, Dean D, Ewer J. Genetic analysis of ecdysis behavior in *Drosophila* reveals partially overlapping functions of two unrelated neuropeptides. *J Neurosci*. 2012;32:6819−6829.
9. Sakai T, Satake H, Takeda M. Nutrient-induced alpha-amylase and protease activity is regulated by crustacean cardioactive peptide (CCAP) in the cockroach midgut. *Peptides*. 2006;27: 2157−2164.
10. Park Y, Kim YJ, Adams ME. Identification of G protein-coupled receptors for *Drosophila* PRXamide peptides, CCAP, corazonin, and AKH supports a theory of ligand-receptor coevolution. *Proc Natl Acad Sci USA*. 2002;99:11423−11428.

Supplemental Information Available on Companion Website

- Accession numbers of gene and mRNAs for CCAP and its receptor in insects and crustaceans/E-Tables 69.1−69.3

CHAPTER 70

Cardioacceleratory Peptide 2b

Shinji Nagata

Additional names/Abbreviations: CAP2b/CAPA

Cardioacceleratory peptide 2b (CAP2b) was originally identified as a cardioacceleratory peptide in the moth. This peptide is a pleiotropic factor with activities including insect diuresis.

Discovery

Cardioacceleratory peptide 2b (CAP2b) has been identified from a crude extract of the abdominal nerve cord of the tobacco hornworm *Manduca sexta*. Two significant cardioaccelatory fractions from the extract were identified as CAP1 and CAP2 [1]. In the CAP2 fraction, CAP2b was identified as a separate fraction from CAP2a and CAP2c [2,3]. In *Drosophila melanogaster*, a CAP2b candidate peptide has been assigned to the gene "*capability*." Thus, *D. melanogaster* CAP2b was designated CAPA peptide [4].

Structure

Structural Features

Original CAP2b isolated from *M. sexta* as a cardioaccessory peptide is a heptapeptide, pELYAFPRV-NH$_2$. All CAP2b and CAPA peptides are composed of 8–15 amino acids with a C-terminal amidation (Table 70.1). CAP2b/CAPA is widely conserved among arthropods. Although there are several structurally similar peptides, including diapause hormone (−PRLamide) and pyrokinin (−PRLamide), only limited functional relationships between CAP2b/CAPA peptides and those similar peptides have been observed.

Structure and Subtype

In one species, several subtypes of CAP2b/CAPA are present. In *D. melanogaster*, CAPA has two subtypes (CAPA-1 and CAPA-2) produced from one preproprotein (Figure 70.1). In the preproprotein, there is a third CAPA peptide (CAPA-3), which had originally been identified as a different biologically active peptide, pyrokinin. Similarly, in the blood-sucking bug *Rhodnius prolixus* two subtypes are encoded in a precursor (CAPA-α), one of which is a PK gene. Interestingly, *R. prolixus* has another paralogous precursor gene (CAPA-β) that

has two CAPA subtype peptides. As CAPA-α2 is identical to CAPA-β2, this species has three CAPA peptides.

Synthesis and Release

Gene, mRNA, and Precursor

A mRNA encoding a CAP2b/CAPA precursor is composed of 60–70 aa. In *D. melanogaster*, three subtypes, CAPA-1, CAPA-2, and CAPA-3, are produced after cleavage of signal sequences and cleavages at dibasic sites [5].

Distribution of mRNA

CAPA mRNA is produced in the central nervous system. In *D. melanogaster* adults, the CAPA gene is expressed in six neurosecretory cells of the abdominal ganglion and two large nervous cells in the subesophageal ganglion.

Receptors

Structure and Subtype

The first receptor for CAPA (capaR) has been identified as a class of GPCR from *D. melanogaster* [6,7]. capaRs have been identified from other species, including dipteran species such as the mosquito *Aedes aegypti*, and the hymenopteran *R. prolixus*. capaR is specifically expressed in larval and adult malpighian tubules, where it plays a role in insect diuresis.

Signal Transduction Pathway

CAPA-1 elevates intracellular cGMP and Ca^{2+} ions in the renal tubules. In addition, involvement of CAPA in the peripheral calcium channel and V-ATPase (or other ATPase) has been proposed in *D. melanogaster* and *Aedes gambiae* [8]. CAP2b/CAPA stimulation leads to the activation of NOS, eventually producing intracellular cGMP in the tubules [7].

Biological Functions

Target Cells and Function

CAP2b/CAPA targets the malpighian tubules, where the fluid secretion takes place in insect diuresis to maintain water and ion homeostasis. The predominant function is to modulate fluid secretion in the tubules [6–8].

Phenotype in Gene-Modified Animals

Knockdown flies, in which *capaR* is ectopically downregulated in the malpighian tubules, show no response to CAPA-1 in terms of intracellular Ca^{2+} flux. Interestingly, stress resistance, including desiccation, increased in the capaR RNAi

Table 70.1 Alignment of CAP2b and CAPA Peptides

Manduca sexta	CAP2b	pELYAFPRVa
Drosophila melanogaster	CAPA-1	GANMGLYAFPRVa
	CAPA-2	ASGLVAFPRVa
	CAPA-3	TGPSASSGLWFGPRLa
Periplaneta americana	Pyrokinin	SESEVPGMWFGPRLa

Y. Takei, H. Ando, & K. Tsutsui (Eds): Handbook of Hormones. DOI: http://dx.doi.org/10.1016/B978-0-12-801028-0.00070-2

```
 1        10          20          30          40          50          60
MKSMLVHIVL VIFIIAEFST AETDJDKNRR GANMGLYAFP RVGRSDPSLA NSLRDGLEAG
        Signal peptide                     CAPA-1
         70          80          90          100         110         120
VLDGIYGDAS QEDYNEADFQ KKASGLVAFP RVGRGDAELR KWAHLLALQQ VLDKRTGPSA
                                 CAPA-2
         130         140         150
SSGLWFGPRL GKRSVDAKSF ADISKGQKEL N*
      CAPA-3
```

Figure 70.1 Amino acid sequence of _D. melanogaster_ CAPA preproprotein. Mature peptides are boxed in yellow. The signal peptide is underlined.

flies. This suggests that CAPA signaling is strongly related to homeostasis in water balance via control of fluid secretion.

References

1. Tublitz NJ, Vate M, Davies SA, et al. Insect cardioactive peptides in _Manduca sexta_: a comparison of the biochemical and molecular characteristics of cardioactive peptides in larvae and adults. _J Exp Biol_. 1992;165:265–272.
2. Cheung CC, Loi PK, Sylwester AW, et al. Primary structure of a cardioactive neuropeptide from the tobacco hawkmoth, _Manduca sexta_. _FEBS Lett_. 1992;313:165–168.
3. Huesmann GR, Cheung CC, Loi PK, et al. Amino acid sequence of CAP2b, an insect cardioacceleratory peptide from the tobacco hawkmoth _Manduca sexta_. _FEBS Lett_. 1995;371:311–314.
4. Davies SA, Huesmann GR, Maddrell SHP, et al. CAP2b, a cardioacceleratory peptide, is present in _Drosophila_ and stimulates tubule fluid secretion via cGMP. _Am J Physiol_. 1995;269:R1321–R1326.
5. Kean L, Cazenave W, Costes L, et al. Two nitridergic peptides are encoded by the gene capability in _Drosophila melanogaster. Am J Physiol Regul Int Comp Physiol_. 2002;282:R1297–R1307.
6. Davies SA, Cabrero P, Povsic M, et al. Signaling by _Drosophila_ capa neuropeptides. _Gen Comp Endocrinol_. 2013;188:60–66.
7. Predel R, Wegener C. Biology of the CAPA peptides in insects. _Cell Mol Life Sci_. 2006;63:2477–2490.
8. Terhzaz S, Cabrero P, Robben JH, et al. Mechanism and function of _Drosophila_ capa GPCR: a desiccation stress-responsive receptor with functional homology to human neuromedinU receptor. _PLoS One_. 2012;7:e29897.

Supplemental Information Available on Companion Website

- Accession numbers of CAP2b, CAPA, and capaR/E-Table 70.1

CHAPTER 71

Achatina Cardio-Excitatory Peptide-1

Fumihiro Morishita

Abbreviations: ACEP-1

Achatina cardioexitatory peptide-1 (ACEP-1) is an undeca-peptide identified from the African giant snail Achatina fulica. *Structurally related peptides are known in a sea hare,* Aplysia californica, *and a freshwater snail,* Lymnaea stagnalis. *AG protein-coupled receptor has been identified in* Lymnaea.

Discovery

ACEP-1 was originally isolated from the peptidic extract of atria of *Achatina fulica*, through the combination of fraction-ation with reversed-phase or cation-exchange HPLC and bio-assay on the ventricle of the animal [1]. In *Aplysia*, the *L5-67* gene specifically expressed in the left upper quadrant (LUQ) neurons of the abdominal ganglia encoded a precursor pro-tein of an ACEP-1 related peptide [2,3]. The peptide was named as LUQIN [2]. In the freshwater snail *L. stagnalis*, ACEP-1 related peptide, LyCEP, was purified from the ganglia of the animal as the endogenous ligand to an orphan GPCR, GRL106 [4].

Structure

Structural Features

ACEP-1 is an undecapeptide, while all of the other related peptides are decapeptides. All of them share the identical C-terminal WRPQGRF-NH$_2$ structure (Table 71.1).

Synthesis and Release

Gene, mRNA, and Precursor

Precursor proteins of ACEP-1 related peptides so far known consist of 94−124 amino acids [2,4,5]. On every precursor, a single copy of the peptide sequence was located just C-terminal to the signal peptide. A related precursor predicted from EST data of the limpet *Lottia gigantea* was also listed as LgCAP [6].

Distribution of mRNA

In situ hybridization revealed that the LUQIN precursor gene (*L5-67*) is highly expressed in the identifiable neurons L2, L3, L4, and L6 in the abdominal ganglion of *Aplysia* [3]. A similar technique localized the ACEP-1 precursor gene expressing

neurons in the central nervous system of *A. fulica*, although characterization of those neurons is not complete.

Tissue Content

Tissue content of the LUQIN precursor was estimated in the central nervous system of *Aplysia*, based on the immunoreac-tivity to anti-LUQIN precursor antibody as follows (fmol/g tis-sue): abdominal ganglion, $14,371 \pm 95$; pleural ganglia, 350.7 ± 39.0; pedal ganglia, 107.4 ± 10.0; cerebral ganglion, 993.9 ± 89 [3]. Immuno-staining with specific antibody dem-onstrated the distribution of ACEP-1 and LUQIN in the cardio-vascular regions of *Achatina* and *Aplysia*, as well as some neurons in the central nervous system [3,7].

Receptors

Structure and Subtypes

Receptor for LyCEP was cloned as an orphan G protein-coupled receptor, GRL106 [4]. This receptor gene is expressed in some caudo-dorsal cells (CDCs) in the cerebral ganglion of *L. stagnalis*. The amino acid sequence of GRL106 is partly similar to those of neuropeptide Y and tachykinin receptors of *Drosophila*.

Signal Transduction Pathway

LyCEP can elicit Ca^{2+} influx into the *Xenopus* oocyte trans-fected with the LyCEP-receptor (GRL106) gene [4]. However, the endogenous signal transduction pathway in *Lymnaea* is unknown.

Biological Functions

Target Cells/Tissues and Functions

ACEP-1 and LyCEP have cardio-excitatory action on *A. fulica* and *L. stagnalis*, respectively [1,4,6]. A major difference in the action of the two peptides is that ACEP-1 augments the beat-ing amplitude of the ventricle but not the auricle of *Achatina*, while LyCEP augments the beating frequency of the auricle of *Lymnaea*. ACEP-1 increases the phasic contractions of the penis retractor muscle and buccal muscle that was elicited by electrical stimulation of the muscles [1]. It also induces spon-taneous discharge of buccal B4 neurons. In the cerebral gan-glion of *Lymnaea*, the LyCEP receptor gene is expressed in some CDCs that contain the egg-laying hormone (ELH) [4]. LyCEP seems to decrease the secretion of ELH from CDCs

Y. Takei, H. Ando, & K. Tsutsui (Eds): Handbook of Hormones. DOI: http://dx.doi.org/10.1016/B978-0-12-801028-0.00071-4

Table 71.1 Structures of ACEP-1 and its Related Peptides*

Species	Name	Structure	Mr
Achatina fulica	ACEP-1	SGQSWRPQGRF–NH$_2$	1234.34
Lymnaea stagnalis	LyCEP	TPHWRPQGRF–NH$_2$	1281.44
Aplysia californica	LUQIN	APSWRPQGRF–NH$_2$	1201.35
Lottia gigantea	LgCAP	APQWRPQGRF–NH$_2$	1242.41

*ACEP-1 and related peptides found in mollusks are indicated. C-terminal amino acids shared by all of ACEP-related peptides are highlighted.

because CDCs are surrounded by LyCEP-containing nerve-endings, and application of LyCEP to CDCs reduces the electrical discharge of the cells.

References

1. Fujimoto K, Ohta N, Yoshida M, et al. A novel cardio-excitatory peptide isolated from the atria of the African giant snail, *Achatina fulica*. *Biochem Biophys Res Commun*. 1990;167:777–783.
2. Aloyz RS, DesGroseillers L. Processing of the L5-67 precursor peptide and characterization of LUQIN in the LUQ neurons of *Aplysia californica*. *Peptides*. 1995;16:331–338.
3. Giardino ND, Aloyz RS, Zollinger M, et al. L5-67 and LUQ-1 peptide precursors of *Aplysia californica*: distribution and localization of immunoreactivity in the central nervous system and in peripheral tissues. *J Comp Neurol*. 1996;374:230–245.
4. Tensen CP, Cox KJ, Smit AB, et al. The *Lymnaea* cardioexcitatory peptide (LyCEP) receptor: a G-protein-coupled receptor for a novel member of the RFamide neuropeptide family. *J Neurosci*. 1998;18:9812–9821.
5. Satake H, Takuwa K, Minakata H. Characterization of cDNA and expression of mRNA encoding an *Achatina* cardioexcitatory RFamide peptide. *Peptides*. 1999;20:1295–1302.
6. Veenstra JA. Neurohormones and neuropeptides encoded by the genome of *Lottia gigantea*, with reference to other mollusks and insects. *Gen Comp Endocrinol*. 2010;167:86–103.
7. Fujiwara-Sakata M, Kobayashi M. Localization of FMRFamide- and ACEP-1-like immunoreactivities in the nervous system and heart of a pulmonate mollusc, *Achatina fulica*. *Cell Tissue Res*. 1994;278:451–460.

Supplemental Information Available on Companion Website

- Accession numbers of precursor proteins of ACEP-1 and related peptides/E-Table 71.1
- Amino acid sequences of precursor proteins of ACEP-1 and related peptides/E-Figure 71.1
- Structure of precursor protein for the LyCEP receptor (GRL106)/E-Figure 71.2
- Alignment of amino acid sequences of invertebrate GPCR/E-Figure 71.3

CHAPTER 72

Fulicins

Fumihiro Morishita

Additional name: fulicin gene-related peptides (FGRPs), FGRP-9 (fulyal)

Fulicin is a peptapeptide isolated from the circum-esophageal ganglia of the African giant snail Achatina fulica. The structural feature of fulicin is D-configuration of an asparagine (Asn) residue at the second position. On the fulicin precursor, fulicin and nine fulicin gene-related peptides (FGRPs) are encoded. Interestingly, FGRP-9 (fulyal) has D-Ala².Thus, fulicin precursor is the source of two distinct D-amino acids containing neuropeptide.

Discovery

Fulicin is a D-amino acid-containing pentapeptide isolated from the circum-esophageal ganglia of the pulmonate mollusk *Achatina fulica* by the combination of HPLC fractionation and bioassay using isolated penis retractor muscle [1]. When the cDNA encoding fulicin precursor was cloned, it was predicted that the precursor included nine structurally related peptides, the fulicin gene-related peptides (FGRPs) [2]. Structural analysis of FGRPs purified from *Achatina* atria revealed that FGRP-9, but not other FGRPs, has D-alanine in the N-terminal second position, and is named fulyal [3].

Structure

Structural Features

Structural features of fulicin and fulyal are the D-configuration of asparagine for fulicin and that of alanine for fulyal. When fulicin was purified, structural analysis by amino acid sequencing and mass spectrometry suggested its structure as being Phe−Asn−Glu−Phe−Val−NH₂. However, the synthetic peptide did not co-elute with the native peptide from the anion-exchange column, and its bioactivity was much less than that of the native one. By comparing the elution profiles of synthetic peptides with D-amino acid to that of the native one, the D-configuration of the asparagine residue was determined. Another structural feature is that both of the peptides have a C-terminal amide.

Structure and Subtype

On the precursor protein of fulicin, nine structurally related peptides, FGRPs, are also encoded (Figure 72.1a). All of the fulicin and nine FGRPs were isolated from the *Achatina* atria by the HPLC-fractionation and bioassay. Structural analysis of the isolated peptides revealed that FGRP-9, but not other FGRPs, has D-alanine. Thus, FGRP-9 was named fulyal. In addition, all the FGRPs have a C-terminal amide. In *Aplysia californica*, precursor cDNA encoding FGRP has been cloned (Figure 72.1b) [4]. Predicted structures of *Aplysia* FGRPs (AcFGRPs) are listed in Table 72.1.

Biological Functions

Known bioactivities of fulicin on *Achatina* tissues include: (1) potentiation of electrically induced contraction of the penis retractor muscle; (2) attenuation of electrically induced contraction of the radula retractor muscle; (3) enhancement of the beating amplitude of the ventricle; (4) contraction of the oviduct; and (5) initiation of the bursting discharge of B1 and

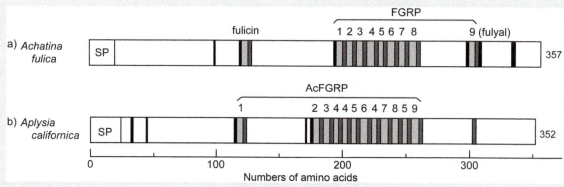

Figure 72.1 Scale models showing the distribution of the predicted peptides on the precursor protein. Localizations of fulicin and FGRPs on the precursors of *A. fulica* (a) and *A. californica* (b) are indicated. Light-gray bars represent peptide coding regions, dark-gray bars represent amidation signals, and black bars represent dibasic or monobasic cleavage sites. The number on the right side of each model indicates the total number of amino acids. SP, signal peptide.

Y. Takei, H. Ando, & K. Tsutsui (Eds): Handbook of Hormones. DOI: http://dx.doi.org/10.1016/B978-0-12-801028-0.00072-6

Table 72.1 Structures of Fulicin and Related Peptides Encoded on the Precursor Proteins of *Achatina* and *Aplysia**

Achatina fulica		Aplysia californica	
Peptide Name	Structure	Peptide Name	Structure
fulicin	FNEFV—NH$_2$		
FGRP-1	GYEFV—NH$_2$	AcFGRP-1	FTEFL—NH$_2$
FGRP-2	SYNFV—NH$_2$	AcFGRP-2	QWEFV—NH$_2$
FGRP-3	TYNFL—NH$_2$	AcFGRP-3	PYDFV—NH$_2$
FGRP-4	SPYYFL—NH$_2$	AcFGRP-4	YDFV—NH$_2$
FGRP-5	YNPL—NH$_2$	AcFGRP-5	YDFL—NH$_2$
FGRP-6	SPYNFI—NH$_2$	AcFGRP-6	NPYEFI—NH$_2$
FGRP-7	SYNFV—NH$_2$	AcFGRP-7	RPYDFL—NH$_2$
FGRP-8	SPYNFV—NH$_2$	AcFGRP-8	SYDFL—NH$_2$
FGRP-9 (fulyal)	YAEFL—NH$_2$	AcFGRP-9	NPYEFV—NH$_2$

*Note that the asparagine of fulicin and alanine of FGRP-1 (fulyal) are D-amino acids.

B4 neurons in the buccal ganglion [1,5]. A physiological test of fulicin analogs on the penis retractor muscle of *Achatina* revealed that D-asparagine[2] is essential for the bioactivity of fulicin, although it could be replaced by D-threonine, D-gluta-mine, or D-glutamic acid without a reduction in bioactivity [6]. Fulyal is myoactive on the penis retractor muscle and oviduct [5]. By contrast, FGRPs 1—8 have little action on these tissues.

Gene, mRNA, and Precursor

The predicted fulicin precursor protein consists of 357 amino acids, including a 20-aa signal peptide, a single copy of fuli-cin, 8 FGRPs that are aligned in tandem on the precursor, and a single copy of fulyal (Figure 72.1a). Therefore, the fulicin precursor gene encodes two structurally related but distinct D-amino acid-containing neuropeptides. cDNA encoding FGRP-like peptide precursor has been cloned in the sea hare

A. californica (Figure 72.1b). However, the presence of D-amino acid in the matured form of FGRP-like peptide has not yet been confirmed.

References

1. Ohta N, Kubota I, Takao T, et al. Fulicin, a novel neuropeptide containing a D-amino acid residue isolated from the ganglia of *Achatina fulica*. *Biochem Biophys Res Commun*. 1991;178:486—493.
2. Yasuda-Kamatani Y, Nakamura M, Minakata H, et al. A novel cDNA sequence encoding the precursor of the D-amino acid-containing neuropeptide fulicin and multiple alpha-amidated neuropeptides from *Achatina fulica*. *J Neurochem*. 1995;64:2248—2255.
3. Yasuda-Kamatani Y, Kobayashi M, Yasuda A, et al. A novel D-amino acid-containing peptide, fulyal, coexists with fulicin gene-related peptides in *Achatina* atria. *Peptides*. 1997;18:347—354.
4. Moroz LL, Edwards JR, Puthanveettil SV, et al. Neuronal transcriptome of *Aplysia*: neuronal compartments and circuitry. *Cell*. 2006;127:1453—1467.
5. Fujisawa Y, Masuda K, Minakata H. Fulicin regulates the female reproductive organs of the snail, *Achatina fulica*. *Peptides*. 2000;21:1203—1208.
6. Fujita K, Minakata H, Nomoto K, et al. Structure-activity relations of fulicin, a peptide containing a D-amino acid residue. *Peptides*. 1995;16:565—568.

Supplemental Information Available on Companion Website

- Accession number of fulicin precursor cDNA and protein/ E-Table 72.1
- Amino-acid sequence of fulicin precursors predicted from the nucleotide sequence of cloned fulicin-precursor cDNA/E-Figure 72.1

CHAPTER 73

Buccalins

Fumihiro Morishita

Buccalin was originally identified in the buccal ganglia of the sea hare Aplysia californica and buccalins are now recognized as a family of peptides consisting of as many as 19 different but structurally related peptides. Buccalins are released, along with ACh, from the buccal B15/B16 neurons innervating the accessory radula closer muscle of the buccal mass, and regulate feeding behavior of Aplysia.

Discovery

Buccalin A was isolated from the cholinergic neurons B15/B16 of the buccal ganglion of the gastropod Mollusca *Aplysia californica* [1]. Subsequently, two different buccalins, named buccalins B and C, were identified from the same neurons [2,3].

Structure

Structural Features

Buccalins include 19 kinds of different peptides, consisting of 10−12 amino acids (Table 73.1). The common features shared by all the buccalins are asparagine at the ninth position from the C-terminus, and the C-terminal amide. In addition, some buccalins share the N-terminal−GMD sequence (buccalins A, I, J, M, O, and Q), while others share the C-terminal −GGL−NH_2 (buccalins A, B, E, J, M) or −PGL−NH_2 (buccalins I, K, L, O, and R) sequence.

Structure and Subtype

In addition to buccalins A−S, a buccalin-related peptide has been isolated from a spindle snail, *Fusinus ferrugineus* [4] (Table 73.1).

Synthesis and Release

Gene, mRNA, and Precursor

The total length of cloned cDNA encoding the buccalin precursor is 2,276 bp [2]. The predicted precursor protein consists of 505 amino acids, with the N-terminal 25 amino acids predicted to be the signal peptide. The total number of encoded buccalin peptides is 29. Buccalin precursors have been predicted in the limpet *Lottia gigantea* and the nematode *Caenorhabditis elegans*, by translating the nucleotide sequences of cloned cDNA. The structure of each peptide is listed in E-Table 73.1, while localizations of buccalin-related peptides on the precursors are indicated schematically in Figure 73.1. cDNA fragments encoding a part of the buccalin precursor are also found in EST data for several mollusks such as the oyster *Crassostrea gigas*, and freshwater snails *Biomphalaria glabrata* and *Lymnaea stagnalis*. Buccalin precursor-like protein was also found in the EST data of a rotifer *Philodina roseola* (E-Table 73.2).

Synthesis and Release

It is generally accepted that, when a particular neuron releases both the low-molecular-weight neurotransmitter and the neuropeptide as the signal mediator, the neurotransmitter is released at a low firing rate while the neuropeptide is released at a higher firing rate. However, this is not the case for B15/B16 neurons in the buccal ganglion of *A. californica* [5,6]. The neurotransmitter of the neurons is acetylcholine (ACh). B15/B16 neurons also synthesize and release buccalin, together with another neuropeptide, small cardioactive peptide b (SCPb). Sensitive radio-immunoassay for the neuropeptides revealed that these neurons co-release buccalin and SCPb, even when the firing rate is low (7.5 Hz). This co-release of neuropeptides depends on Ca^{2+} in the external bathing medium. There is evidence that buccalin reduces SCPb- and ACh-release from B15/B16 neurons, while serotonin reduces buccalin release. It is also recognized that the amount of buccalin release per action potential increases as the frequency of discharge is increased, probably via an accumulation of Ca^{2+} in the nerve ending. These phenomena seem to be characteristic of B15/B16 neurons, because other buccalin-releasing neurons such as B1/B2 have not shown such changes. Since buccalin has been detected in the cerebro-pedal and cerebro-buccal connectives with MALDI-TOF-MS, inter-ganglionic axonal transport of buccalin is suggested [7].

Biological Functions

Target Cells and Function

An important physiological action of buccalin is the regulation of feeding behavior of *Aplysia* [8,9]. B15/16 neurons in the buccal ganglion of *A. californica*, which release buccalin, project onto the accessory radula closer (ARC) muscle in the buccal mass. Released buccalin reduces contraction of the ARC muscle induced by electrical stimulation of B15/B16, not by application of ACh to the muscle. Thus, buccalin may reduce ACh release by acting on the presynaptic element. It is suggested that presynaptic inhibition includes modification of activity of the N-type Ca^{2+} channel in the nerve terminal [10].

Y. Takei, H. Ando, & K. Tsutsui (Eds): Handbook of Hormones. DOI: http://dx.doi.org/10.1016/B978-0-12-801028-0.00073-8

Table 73.1 Structures of Buccalins Found in Gastropod Mollusks*

Species & Peptide Name	Structure
Aplysia californica	
Buccalin A	GMDSLAFSGGL—NH₂
B	GLDRYGFVGGL—NH₂
C	GFDHYGFTGGI—NH₂
D	PNVDPYSYLPSV—NH₂
E	AFDHYGFTGGL—NH₂
F	IDHFGFVGGL—NH₂
G	QIDPLGFSGGI—NH₂
H	YDSFQYSAGL—NH₂
I	GMDSFTFAPGL—NH₂
J	GMDSLAFAGGL—NH₂
K	MDGFAFAPGL—NH₂
L	MDSFAFAPGL—NH₂
M	GMDHFAFTGGL—NH₂
N	GLDAYSFTGAL—NH₂
O	GMDDFAFSPGL—NH₂
P	MDSFMFGSRL—NH₂
Q	GMDRFSFSGHL—NH₂
R	MDQFSFGPGL—NH₂
S	QLDPMLFSGRL—NH₂
Fusinus ferrugineus	
	RMNSMMFGPQL—NH₂

*Buccalins identified in *Aplysia* and *Fusinus* are shown. Amino acids shared by several buccalins are shaded. Note that the precursor protein for buccalin has not been identified in *Fusinus*.

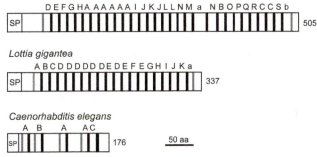

Aplysia californica

Lottia gigantea

Caenorhabditis elegans

Figure 73.1 Scale drawings showing the distribution of buccalin and related peptides in respective precursor proteins. Buccalin and related peptide precursors of *Aplysia californica*, *Lottia gigantea*, and *Caenorhabidits elegans* are indicated. Black columns indicate dibasic cleavage sites with Gly on their N-terminal side (amidation signals), while gray columns indicate dibasic cleavage sites without Gly. Letters on each drawing match to those of the peptide numbering in E-Table 73.1. Numbers on the right side of each drawing represent the total number of amino acids of each precursor. Scale bar: 50 amino acids; SP, signal peptide.

References

1. Cropper EC, Miller MW, Tenenbaum R, et al. Structure and action of buccalin: a modulatory neuropeptide localized to an identified small cardioactive peptide-containing cholinergic motor neuron of *Aplysia californica*. *Proc Natl Acad Sci USA*. 1988;85:6177–6181.
2. Miller MW, Beushausen S, Cropper EC, et al. The buccalin-related neuropeptides: isolation and characterization of an *Aplysia* cDNA clone encoding a family of peptide cotransmitters. *J Neurosci*. 1993;13:3346–3357.
3. Vilim FS, Cropper EC, Rosen SC. Structure, localization, and action of buccalin B: a bioactive peptide from *Aplysia*. *Peptides*. 1994;15:956–969.
4. Kanda T, Takabatake I, Fujisawa Y, et al. Biological activities of ganglion extracts from a prosobranch mollusc, *Fusinus ferrugineus*. *Hiroshima J Med Sci*. 1989;38:109–116.
5. Vilim FS, Cropper EC, Price DA, et al. Release of peptide cotransmitters in *Aplysia*: regulation and functional implications. *J Neurosci*. 1996;16:8105–8114.
6. Vilim FS, Price DA, Lesser W, et al. Costorage and corelease of modulatory peptide cotransmitters with partially antagonistic actions on the accessory radula closer muscle of *Aplysia californica*. *J Neurosci*. 1996;16:8092–8104.
7. Li L, Moroz TP, Garden RW, et al. Mass spectrometric survey of interganglionically transported peptides in *Aplysia*. *Peptides*. 1998;19:1425–1433.
8. Cropper EC, Lloyd PE, Reed W, et al. Multiple neuropeptides in cholinergic motor neurons of *Aplysia*: evidence for modulation intrinsic to the motor circuit. *Proc Natl Acad Sci USA*. 1987;84:3486–3490.
9. Baux G, Fossier P, Trudeau LE, Tauc L. Presynaptic receptors for FMRFamide, histamine and buccalin regulate acetylcholine release at a neuro-neuronal synapse of *Aplysia* by modulating N-type Ca²⁺ channels. *J Physiol Paris*. 1992;86:3–13.
10. Baux G, Fossier P, Trudeau LE, Tauc L. *J Physiol Paris*. 1992;86:3–13.

Supplemental Information Available on Companion Website

- EST clones encoding part of the precursor protein for buccalin-related peptide/E-Table 73.1
- Accession numbers for cDNA of precursors of buccalin and related peptides/E-Table 73.2
- Amino acid sequence of buccalin precursors of several invertebrates, predicted from precursor cDNAs/E-Figure 73.1

CHAPTER 74

Eisenia Inhibitory Pentapeptides

Fumihiro Morishita and Hiroyuki Minakata

Additional name/Abbreviation: EIPP

EIPPs inhibit tension and frequency of the spontaneous contractions of the anterior gut (crop-gizzard) of the oligochaete annelid Eisenia foetida. Pentapeptides having a similar structural motif (a basic amino acid residue at the second position and a C-terminal −FV−NH$_2$ or −FL−NH$_2$ structure) have been found from mollusks and an echiuroid worm.

Discovery

Three EIPPs (EIPP-1, -2, and -3) have been isolated from the gut of the oligochaete annelid *Eisenia foetida*, as inhibitory regulators of the rhythmic spontaneous contractions of the gut [1]. The amino acid sequences of EIPPS are similar to molluskan and echiuroid peptides previously identified [1].

Structure

Structural Features

EIPPs are pentapeptides with a similar sequence motif, namely a basic amino acid (Arg, His, or Lys) at the second position and a −FV−NH$_2$ or −FL−NH$_2$ at the C-terminal region [1].

Structure and Subtype

A pentapeptide (MIP-RP) is found in the anterior retractor muscle of *Mytilus edulis* [2]. Five pentapeptides are present in the central nervous system of *Aplysia kurodai* and *A. californica* [3]. Three peptides are found in the ventral nerve cords of *Urechis unicinctus* (Table 74.1) [4].

Table 74.1 Structures of EIPP-Related Peptides*

Animals		Names	Structures
Annelida	*Eisenia foetida*	EIPP-1	SRLFV−NH$_2$
		EIPP-2	SHLFV−NH$_2$
		EIPP-3	VHLFV−NH$_2$
Molluska	*Mytilus edulis*	MIP-RP	MRYFV−NH$_2$
	Aplysia kurodai	ApPP	PRQFV−NH$_2$
	A. californica	ApPP	PRQFV−NH$_2$
			AREFV−NH$_2$
			VRDFV−NH$_2$
			VREFV−NH$_2$
			IREFV−NH$_2$
Echiura	*Urechis unicinctus*		ARYFL−NH$_2$
			AKYFL−NH$_2$

*EIPP-related peptides found in Annelida, Mollusca, and Echiura are indicated. MIP-RP, *Mytilus* inhibitory peptide related peptide.

Synthesis and Release

Gene, mRNA, and Precursor

A cDNA encoding precursor protein of *Aplysia* pentapeptide (ApPP) has been cloned on a random primed *A. californica* ganglionic cDNA lambda library [3]. The precursor cDNA contains a 2,586-bp open reading frame encoding an 862-aa precursor, containing 33 copies of ApPP, and single copies of AREFV−NH$_2$, VRDFV−NH$_2$, and IREFV−NH$_2$ [3]. mRNAs of EIPPs, *Mytilus* pentapepide, and *Urechis* pentapeptide have not been cloned.

Distribution of mRNA

ApPP precursor mRNA is distributed throughout the central ganglia of *Aplysia*. The relative abundance is abdominal ≫ pedal > pleural > buccal ganglia [3].

Biological Functions

Target Cells and Function

Known biological activities of EIPP and EIPP-related peptides are largely inhibitory. In *E. foetida*, EIPP inhibits spontaneous rhythmic contraction of the esophagus, crop, gizzard, and intestine [1]. In *Aplysia*, ApPP inhibits spontaneous contraction of the gut with EC$_{50}$ around 1.6×10^{-7} M, and electrically induced contraction of the anterior aorta with an EC$_{50}$ around 4×10^{-7} M [4]. It also reduces the frequency of discharge of B31/32 and B4 neurons in the buccal ganglia, which constitute the neural circuit for the feeding central pattern generator.

Localization of ApPP is well characterized in *Aplysia* [4]. Immunostaining and *in situ* hybridization visualized that ApPP-containing neurons formed some clusters in the pedal and abdominal ganglia. In other ganglia, ApPP-containing neurons are relatively few and scattered. In the peripheral nervous system, dense ramifications of ApPP-containing nerve processes were found on the digestive tract and vasculature, including the gill, kidney, and blood vessels. ApPP may act as a modulator within the feeding system as well as in the cardiovascular system of *Aplysia* [4].

References

1. Ukena K, Oumi T, Matsushima O, et al. Inhibitory pantapeptides isolated from the gut of the earthworm, *Eisenia foetida*. *Comp Biochem Physiol*. 1996;114A:245−249.
2. Fujisawa Y, Takahashi T, Ikeda T, et al. Further identification of bioactive peptides in the anterior byssus retractor muscle of *Mytilus*: two contractile and three inhibitory peptides. *Comp Biochem Physiol*. 1993;106C:261−267.

Y. Takei, H. Ando, & K. Tsutsui (Eds): Handbook of Hormones. DOI: http://dx.doi.org/10.1016/B978-0-12-801028-0.00074-X

3. Ikeda T, Kubota I, Miki W, et al. Structures and actions of 20 novel neuropeptides isolated from the ventral nerve cords of an echiuroid worm, *Urechis unicinctus*. In: Yanaihara N, ed. *Peptide Chemistry 1992*. Leiden: ESCOM B.V.; 1993:583–585.

4. Furukawa Y, Nakamaru K, Sasaki K, et al. PRQFVamide, a novel pentapeptide identified from the CNS and gut of *Aplysia*. *J Neurophysiol*. 2003;89:3114–3127.

Supplemental Information Available on Companion Website

- Accession number of cDNA encoding precursor protein of *Aplysia* PRQFVamide/E-Table 74.1

CHAPTER 75

GGNG Peptides

Fumihiro Morishita and Hiroyuki Minakata

Additional names: earthworm excitatory peptide (EEP), leech excitatory peptide (LEP), polychaete excitatory peptide (PEP)

GGNG peptides are myoactive peptides isolated from several species of annelids. Two types of peptides have been isolated: earthworm excitatory peptide (EEP) and leech excitatory peptide (LEP), which are highly homologous with each other but act specifically on each species.

Discovery

GGNG peptides (EEP-1, EEP-2, and EEP-3) have been isolated from the gut tissue or the whole body of the two species of earthworm *Eisenia foetida* and *Pheretima vittata* [1]. The name, GGNG peptides, comes from the fact that EEP-1 to -3 share the C-terminal GGNG sequence. Highly homologous peptides have also been isolated from other animals. These peptides include LEP from the leeches *Whitmania pigra* and *Hirudo nipponia* [2], polychaete excitatory peptide (PEP) from the marine worm *Perinereis vancaurica* [3], and *Thais* excitatory peptide (TEP) from the marine snail *Thais clavigera* [4]. Although the C-terminal structure of those peptides is GGN−NH$_2$, and not GGNG, they still share the GGNG sequence on their precursor protein. These peptides show excitatory effects on spontaneous contraction of the gut.

Structure

Structural Features

GGNG peptides consist of 17 or 18 amino acid residues with a unique C-terminal sequence, −GGNG, and an intramolecular disulfide bridge.

Structure and Subtype

PEP has also been characterized from the whole body of a marine polychaete *Perinereis vancaurica*, using an immunoblot assay [3]. Two GGNG peptides (TEP-1 and TEP-2) were isolated from the marine snail *Thais clavigera*, with a bioassay using an isolated preparation of esophagus [4]. Unlike other GGNG peptides, these molluskan peptides have two Trp residues within the ring structure (Table 75.1). Genome analysis has revealed that GGNG peptides are present in the polychaete worm *Capitella teleta*, and in the mollusk *Lottia gigantea* [5,6]. It has been proposed that crustacean cardioactive

Table 75.1 Structures of GGNG Peptides and CCH*

Animals	Species	Names	Structures	
Annelida	*Eisenia foetida*	EEP-1	APKCSGRWAI	HSCGGGNG
		EEP-2	GKCAGQWAI	HACAGGNG
	Pheretima vittata	EEP-3	RPKCAGRWAI	HSCGGGNG
	Whitmania pigra	LEP	AKCEGEWAI	HSCLGGN−NH$_2$
	Hirudo nipponia	LEP	AKCEGEWAI	HSCLGGN−NH$_2$
	Perinereis vancaurica	PEP	KCTGPWAI	HACGGGN−NH$_2$
Molluska	*Thais clavigera*	TEP-1	KCSGKWAI	HACWGGN−NH$_2$
		TEP-2	KCYGKWAM	HACWGGN−NH$_2$
Arthropoda	*Bombyx mori*	CCH	GCSA−FG−	HSCFGGH−NH$_2$

*Structures of GGNG peptides are indicated with insect CCH. Note that Cys residues on each GGNG peptide are connected by intramolecular disulfide bonds. Amino acids shared by GGNG peptides are highlighted. Two gaps (−) have been placed in the structure of CCH in order to align with other GGNG peptides.

hormone (CCH) peptide of the silk moth *Bombyx mori,* is one of the GGNG peptides of insects [6].

Synthesis and Release

Gene, mRNA, and Precursor

The total length (684 bp) of mRNA encoding EEP-2 produces a precursor protein (227 aa) which is composed of a signal peptide, EEP-2, and a −Lys−Arg sequence [7]. The leech cDNA includes an open reading frame (472 bp) and the translated amino acid sequence contains precursor protein (111 amino acid residues) of LEP with a signal peptide, several proteolytic cleavage sites, and LEP peptides flanked by mono- and dibasic residues. Alignment of amino acid sequences of precursor proteins for GGNG peptides is shown in Figure 75.1.

Distribution of mRNA

Expression of the LEP-precursor gene has been detected in some neurons in the supra- and subesophageal ganglia of *W. pigra* by *in situ* hybridization [8].

Receptors

Structure and Subtype

The structure of the EEP receptor has not yet been clarified, but EEP binds a specific receptor with a K_d of 4.9 ± 1.2 nM and a B_{max} of 15.9 ± 2.0 fmol/mg wet tissue of the crop gizzard of *E. foetida* [9].

Y. Takei, H. Ando, & K. Tsutsui (Eds): Handbook of Hormones. DOI: http://dx.doi.org/10.1016/B978-0-12-801028-0.00075-1

```
LEP     1    M---------  --KLLFLVLC  LSL----LHV  TRGVERRYRL  GSDEGRIERR
EEP-2   1    M-RRP-----  VTNIMLLVIC  YLW----CHI  QSTSGG----  ----------
PEP     1    MDRTG-----  IWYLLLVSIL  YIWTGQKVS-  -----G----  ----------
LgGGN   1    MEVRCTTLFA  VTSLLYLTIS  VSL----VS-  -----G----  ----------
CCH     1    MA--------  --QICLAVSI  AVL----LMM  SQGVSAKRG-  ----------

LEP    36    WRLRSDETVR  GTRAKCEGEW  AIHACLGGNG  KRSQQRRNID  SPT-------
EEP-2  27    ----------  ----KCAGQW  AIHACAGGNG  KRSDIDDLVR  ATNLGTDGEI
PEP    26    ----------  ----KCTGPW  AIHACGGGNG  KRSDV-----  ----------
LgGGN  27    ----------  ----KCSGRW  AIHACFGGNG  KRSD------  ----------
CCH    26    ----------  -----CSAFG  --HSCFGGHG  KRS-------  ----------

LEP    79    ----------  ----------  ----------  ---------E  EKDNLLL---
EEP-2  63    SETGLQEARG  DERGGGGGRE  GDE-------  ----------  ----------
PEP    47    ----------  -EAGGLGDRE  KESRISLQRI  LRGFGEYSNE  DEDSTHFGDP
LgGGN  47    ----------  ----------  ----------  ----------  ----------
CCH    42    ----------  ----------  ----------  ----------  -------GEP

LEP    87    ----------  ----------  -----NKQFR  SVLPTLSARL  P---------
EEP-2  86    ----------  ----------  ----------  ----------  --FRGHREG
PEP    86    ---DLSNYVE  DVLSQQPDLA  YIQKENRKQQ  SLLRKLVAML  K---------
LgGGNG 47    -PSLTDNT--  ----------  -----ENSRQE TLLRQIL--L  PQTYEYHKSN
CCH    45    APMDMANQ--  ----------  ----DMMVRH  QLGQEET--P  PHPGYPHSSY

LEP   103    ----------  ----------  ----------  ----------  -FRVEEKSYY
EEP-2  93    DDLRSELSEG  RQLLEDKQLL  DIVNRG----  ----------  -----HPPRS
PEP   124    ----------  ----------  ----------  ----------  -LRQ------
LgGGN  78    DALLQD----  ----------  DDINTY----  ----------  -----EKSHE
CCH    77    NVLQPG----  ----------  DDIIPIRDGG  VYDHDAAARD  VMKY------

LEP   112    ----------  ----------  ----------  ----------  ----------
EEP-2 124    PTPLESALSR  LRSLLRTTEL  DVDRREEEEK  PEQWQEEELP  RHRSTDRNSR
PEP   127    ---------N  LQNLQN----  ----------  ----------  ----------
LgGGN  95    ESQRMRDLRA  LNILLKTLMM  EQKVRTENSV  MA--------  ----------
CCH   107    ---------K  LRNIFKHW-M  DNYRRSQNT-  ----------  ----------

LEP   112    ----------  ----------  ----------  ----------  ----------
EEP-2 174    YNIKDSLTML  RRILDDIEGR  FDSQPVSNSE  RKTRGYADDV  GNDSYSLQKR
PEP   134    ----------  ----------  ----------  ----------  ----------
LgGGN 127    ----------  ----------  ----------  ----------  ----------
CCH   126    ----------  ----------  ----------  ----------  -NDEYFLETI

LEP   112    ----
EEP-2 224    SADQ
PEP   134    ----
LgGGN 127    ----
CCH   135    ----
```

Figure 75.1 Alignment of GGNG-peptide precursors. Amino acid sequences of precursors of LEP, EEP-2, PEP, LgGGN, and CCH are aligned by Edialign.

Biological Functions

Target Cells and Function

EEP-1, -2, and -3 augment spontaneous repetitive contraction of the gut of *E. foetida* in a concentration dependent fashion within the range of 10^{-10} to 10^{-8} M [1]; LEP induces tonic contraction in the gut of *H. nipponia* [2]. In the leech *W. pigra*, LEP seems to modulate motility of the penial complex [8]. PEP induces repetitive contraction of the esophagus of *P. vancaurica* [3]. In *T. clavigera*, TEP is myoactive on the penial complex and the esophagus [4]. It has been demonstrated that each of the two Trp residues of TEP is essential for bioactivity. This result supports the fact that side chains of Trp protrude to the outside, to the backbone ring structure of the peptide [10].

References

1. Oumi T, Ukena K, Matsushima O, et al. The GGNG peptides: novel myoactive peptides isolated from the gut and the whole body of the earthworms. *Biochem Biophys Res Commun*. 1995;216: 1072–1078.
2. Minakata H, Fujita T, Kawano T, et al. The leech excitatory peptide, a member of the GGNG peptide family: isolation and comparison with the earthworm GGNG peptides. *FEBS Lett*. 1997;410:437–442.
3. Matsushima O, Takahama H, Ono Y, et al. A novel GGNG-related neuropeptide from the polychaete *Perinereis vancaurica*. *Peptides*. 2002;23:1379–1390.
4. Morishita F, Minakata H, Takeshige K, et al. Novel excitatory neuropeptides isolated from a prosobranch gastropod, *Thais clavigera*: The molluscan counterpart of the annelidan GGNG peptides. *Peptides*. 2006;27:483–492.

5. Veenstra JA. Neuropeptide evolution: neurohormones and neuro-peptides predicted from the genomes of *Capitella teleta* and *Helobdella robusta*. *Gen Comp Endocrinol*. 2011;171:160–175.

6. Veenstra JA. Neurohormones and neuropeptides encoded by the genome of *Lottia gigantea*, with reference to other mollusks and insects. *Gen Comp Endocrinol*. 2010;167:86–103.

7. Minakata H, Ikeda T, Nagahama T, et al. Comparison of precursor structures of the GGNG peptides derived from the earthworm *Eisenia foetida* and the leech *Hirudo nipponia*. *Biosci Biotechnol Biochem*. 1999;63:443–445.

8. Nagahama T, Ukena K, Oumi T, et al. Localization of leech excit-atory peptide, a member of the GGNG peptides, in the central nervous system of a leech (*Whitmania pigra*) by immunohis-tochemistry and *in situ* hybridization. *Cell Tissue Res*. 1999;297:155–162.

9. Niida T, Nagahama T, Oumi T, et al. Characterization of binding of the annelidan myoactive peptides, GGNG peptides, to tissues of the earthworm, *Eisenia foetida*. *J Exp Zool*. 1997;279:562–570.

10. Horiuch Y, Nose T, Abe Y, et al. Structural analysis of excitatory neuropeptides TEP-1 and TEP-2 isolated from the prosobranch gastropod *Thais clavigera*. *Peptide Sci*. 2008;2007:273–276.

Supplemental Information Available on Companion Website

- Accession numbers of cDNA for GGNG precursor pro-teins/E-Table 75.1

Regulation of Behaviors

CHAPTER 76

Eclosion Hormone

Dušan Žitňan and Ivana Daubnerová

Abbreviation: EH

Eclosion hormone is a brain neuropeptide that controls the release of ecdysis triggering hormone (ETH) from Inka cells and activates a network of 27/704 neurons in the ventral nerve cord responsible for ecdysis behavior and post-ecdysis processes in moths and probably other insects.

Discovery

Initial experiments with transplanted brains of adult moths indicated the presence of a humoral factor released from a neurohemal organ complex, the corpora cardiaca−corpora allata (CC−CA), during adult eclosion. Further work showed that the CC−CA extracts elicit a premature ecdysis sequence when injected into pharate larvae, pharate pupae, and pharate adults of moths. The aa sequence of EH was identified from the CC−CA or the head extracts of the moths *Bombyx mori* and *Manduca sexta* [1]. Subsequently, cDNAs or genes encoding EH have been cloned or deduced from available databases of flies and other arthropods [2].

Structure

Structural Features

Sequences of EH in arthropods are highly conserved with six cysteine residues (Table 76.1). EH of *M. sexta* is 62-aa peptide which has three disulfide bonds, between Cys^{14}−Cys^{38}, Cys^{18}−Cys^{34}, and Cys^{21}−Cys^{49}.

Primary Structure

Closely related peptides occur in insects, crabs, and tardigrades. The position and number of cysteines are highly conserved [2].

Properties

Mr 6,813; soluble in water, physiological saline, 50−70% methanol or ethanol.

Synthesis and Release

Gene, mRNA, and Precursor

The *eh* gene of *B. mori* contains two exons (101 and 616 bp) and the mRNA (718 nt) encodes an 88-aa prepropeptide composed of a signal sequence (26 aa) immediately followed by the EH sequence (62 aa).

Distribution of mRNA

In moths, expression of EH is restricted to two pairs of ventromedial neurosecretory cells (VM; type V) that project their axons through the ventral nerve cord and terminate in the neurohemal proctodeal nerves on the hindgut surface. These cells develop additional axonal projections into the CC after pupation. In the fruit fly *Drosophila melanogaster* there is only a single pair of EH-producing neurons, while hemimetabolous insects contain two to six pairs of VM cells in the brain [3].

Hemolymph Concentration

Hemolymph concentrations are unknown, but biologically active concentrations are very low (2−10 pM) [3].

Regulation of Synthesis and Release

Expression of the *eh* gene is probably constitutive. EH release is induced by ecdysis-triggering hormone (ETH) action on its receptor (ETHR-A) in the VM cells [4,5]. EH release during pre-ecdysis of moths and flies is indicated by a loss of immunoreactivity in VM axons and the appearance of cGMP in Inka cells and neuronal network 27/704 [3,6].

Receptors

Structure and Subtype

The receptor for EH that belongs to a family of guanylate cyclase receptors has been identified in the fruit fly *Bactrocera dorsalis*. There are two isoforms (BdmGC-1 and BdmGC-1B) sharing the same conserved domains and putative N-glycosylation sites. A high-affinity receptor (BdmGC-1) lacks a 46-aa insertion in the extracellular domain and contains the additional C-terminal tail. Low concentrations of EH elicit massive cGMP increase in HEK cells expressing BdmGC-1, while higher EH levels are required for activation of the B-isoform [7]. BdmGC-1 and its orthologs in *D. melanogaster* and *B. mori* are expressed in Inka cells and another adjacent exocrine cell of the epitracheal gland [3].

Signal Transduction Pathway

EH acts through guanylate cyclase receptor to elicit cGMP elevation and mobilization of Ca^{2+} from intracellular stores in target cells. cGMP activates a protein kinase (PKG) for phosphorylation of downstream substrate protein(s) [7,8].

Y. Takei, H. Ando, & K. Tsutsui (Eds): Handbook of Hormones. DOI: http://dx.doi.org/10.1016/B978-0-12-801028-0.00076-3

Table 76.1 Comparison of EH Sequences in Arthropods*

Bombyx mori	SPAIASSYDAMEICIENCAQCKKMFGPWFEGSLCAESCIKARGKDIPECESFASISPFLNKL
Manduca sexta	NPAIATGYDPMEICIENCAQCKKMLGAWFEGPLCAESCIKFKGKLIPECEDFASIAPFLNKL
Helicoverpa armigera	NPAIATGYDPMEVCIENCAQCKKMLGAWFEGPLCAESCITFKGKLIPECENFASISPFLNKL
Apis mellifera	NAEVRSGYIDDGVCIRNCAQCKKMFGPYFLGQKCADSCFKNKGKLIPDCEDEDSIQPFLQAL
Tribolium castaneum	SSLLVVDANPIGVCIRNCAQCKKMFGPYFEGQLCADACVKFKGKIIPDCEDITSIAPFLNKF
Aedes aegypti	NPQLDILGGYDMLSVCINNCAQCKRMFGEFFEGRLCAEACIQFKGKMVPDCEDINSIAPFLTKLN
Drosophila melanogaster	FDSMGGIDFVQVCLNNCVQCKTMLGDYFQGQTCALSCLKFKGKAIPDCEDIASIAPFLNAE
Crab *Callinectes*	ASAAVAANRKVSICIKNCGQCKKMYTDYFNGGLCGDFCLQTEGRFIPDCNRPDILIPFFLQRLE
Tardigrade *Hypsibius*	PIASKGMVGNISICIANCAQCHDIVGEVFDHRKCSRDCVQRRGTFIPDCTSPVAIREYLNL

*Homologous sequences are shaded. Note that the position of cysteines (bold, shaded) is highly conserved.

Biological Functions

Target Cells/Tissues and Functions

There are central and peripheral roles of EH in moths. Centrally released EH activates a network of neurons (27/704) through elevation of cGMP that are specific for regulation of the ecdysis and post-ecdysis processes. Peripherally released EH elicits cGMP production in endocrine Inka cells and causes massive ETH secretion to activate additional neurons controlling ecdysis [1–3]. EH probably controls tracheal inflation and may act on Verson's glands during ecdysis [9].

Phenotype in Gene-Modified Animals

Ablation of VM neurons producing EH in flies causes defects in tracheal inflation, ecdysis behaviors, and postemergence expansion and hardening of the adult cuticle [10].

References

1. Truman JW. Hormonal control of insect ecdysis: endocrine cascades for coordinating behavior with physiology. *Vitam Horm.* 2005;73:1–30.
2. Žitňan D, Kim YJ, Zitnanova I, et al. Complex steroid–peptide–receptor cascade controls insect ecdysis. *Gen Comp Endocrinol.* 2007;153:88–96.
3. Žitňan D, Adams ME. Neuroendocrine regulation of ecdysis. In: Gilbert LI, ed. *Insect Endocrinology.* San Diego: Academic Press; 2012:253–309.
4. Kim Y-J, Žitňan D, Cho K-H, et al. Central peptidergic ensembles associated with organization of an innate behavior. *Proc Natl Acad Sci USA.* 2006;103:14211–14216.
5. Kim Y-J, Žitňan D, Galizia G, et al. A command chemical triggers an innate behavior by sequential activation of multiple peptidergic ensembles. *Curr Biol.* 2006;16:1395–1407.
6. Gammie SC, Truman JW. Eclosion hormone provides a link between ecdysis-triggering hormone and crustacean cardioactive peptide in the neuroendocrine cascade that controls ecdysis behavior. *J Exp Biol.* 1999;202:343–352.
7. Chang JC, Yang RB, Adams ME, et al. Receptor guanylyl cyclases in Inka cells targeted by eclosion hormone. *Proc Natl Acad Sci USA.* 2009;106:13371–13376.
8. Kingan TG, Cardullo RA, Adams ME. Signal transduction in eclosion hormone-induced secretion of ecdysis-triggering hormone. *J Biol Chem.* 2001;276:25136–25142.
9. Ewer J, Reynolds S. Neuropeptide control of molting in insects. In: Pfaff DW, Arnold AP, Fahrbach SE, et al., eds. *Hormones, and Brain Behavior.* San Diego: Academic Press; 2002:1–92.
10. McNabb SL, Baker JD, Agapite J, et al. Disruption of a behavioral sequence by targeted death of peptidergic neurons in *Drosophila*. *Neuron.* 1997;19:813–823.

Supplemental Information Available on Companion Website

- Accession numbers of genes and mRNAs for EH and EH receptor in insects, a crab, and a fly/E-Tables 76.1, 76.2

CHAPTER 77

Ecdysis Triggering Hormone

Dušan Žitňan and Ivana Daubnerová

Abbreviation: ETH

One or two closely related peptide hormones produced and secreted by endocrine Inka cells of the epitracheal glands in insects, ETHs are major regulators of the ecdysis sequence and associated processes in insects and possibly other arthropods.

Discovery

ETH was originally isolated and identified from the epitracheal gland extracts of the tobacco hornworm *Manduca sexta* [1]. Later a related peptide, pre-ecdysis triggering hormone (PETH), was identified from these glands, and similar peptides (ETH1, ETH2) have been isolated from tracheal extracts of several other insects (E-Table 77.1) [2,3]. Precursors or genes encoding ETH and related peptides have been deduced from available databases of many insects, a crustacean, and a tick [3].

Structure

Structural Features

Linear amidated peptides are composed of 11–34 amino acids. Amidation is crucial for biological activity (Table 77.1).

Primary Structure

ETHs are identified in many insects and deduced from genomes of a crustacean (*Daphnia pulex*) and a tick (*Ixodes scapularis*) (E-Table 77.1). Most ETHs share a conserved C-terminus (KSVPRXamide; X = I, L, V, M), but have a variable N-terminus [3]. Some neuropeptides of the CAPA/pyrokinin/PBAN family share PRLamide or PRVamide C-termini with some ETH sequences, but they are produced by neurons in the ventral nerve cord and their function is completely different.

Properties

PETH, Mr 1,240; ETH, Mr 2,960. Soluble in water, physiological saline, 50–70% methanol or ethanol.

Synthesis and Release

Gene, mRNA, and Precursor

The *M. sexta eth* gene is composed of three exons and two introns. The mRNA codes for a 114-aa prepropeptide, including a 22-aa signal peptide immediately followed by PETH (11 aa), ETH (26 aa), and an unrelated ETH-associated peptide (47 aa). The upstream regulatory region contains direct repeats of the ecdysone receptor response element AGGTCA separated by two base pairs followed by a TATA box transcription initiation site at positions 969–982 bp. In addition, several putative Broad Complex response elements have been observed in this promoter region (E-Figure 77.1) [2].

Distribution of mRNA

The *eth* gene is exclusively expressed in Inka cells attached to the tracheal system. In all hemimetabolous and most holometabolous insects, a large number of small Inka cells are scattered all over the tracheal tubes. Only some beetles and hymenopterans, and all lepidopterans and dipterans contain eight or nine pairs of large Inka cells which are part of the segmental epitracheal glands (E-Figure 77.2) [3].

Tissue Content

One Inka cell of *M. sexta* contains 2 pmol PETH/ETH in pharate fifth instar larvae and 8–10 pmol PETH/ETH in pharate pupae [2,3].

Hemolymph Concentration

Concentration in hemolymph is 20–35 nM PETH/ETH during larval ecdysis of *M. sexta* [2].

Regulation of Synthesis and Release

Expression of the *eth* gene and synthesis of ETH are controlled by increased ecdysteroid levels that act through the ecdysone receptor and the basic-leucine zipper (bZIP) gene *cryptocephal* (*crc*) [2,4]. During increased ecdysteroid levels, Inka cells are not able to secrete ETH. Competence of Inka cells to release ETH is induced by ecdysteroid decline and expression of the nuclear transcription factor βFTZ-F1 [5]. ETH secretion is regulated by subsequent actions of the neuropeptides corazonin and eclosion hormone [6,7].

Receptors

Structure and Subtype

The ETH receptor belongs to a family of G protein-coupled receptors. It was originally identified in the fruit fly *Drosophila melanogaster* (GC5911), and the gene contains five exons (E-Figure 77.3). It has two subtypes produced by alternative splicing (ETHR-A and ETHR-B). Each subtype is differentially expressed in various peptidergic and aminergic neurons of the central nervous system (CNS), frontal ganglion, and corpora allata. Expression of *ETHR* is regulated by increased ecdysteroid levels. A similar receptor gene has been identified and characterized in *M. sexta*, and closely related

Y. Takei, H. Ando, & K. Tsutsui (Eds): Handbook of Hormones. DOI: http://dx.doi.org/10.1016/B978-0-12-801028-0.00077-5

Table 77.1 Amino Acid Sequences of PETH and ETH in
M. sexta*

M. sexta PETH	SFIKPN−NVPRV−NH$_2$
M. sexta ETH	SNEAISPFDQGMMGYVIKTNKNIPRM−NH$_2$

*Identical aa residues are shaded.

orthologs of ETHRs have been found in numerous insects,
the crustacean *D. pulex*, and the tick *I. scapularis* [3,8,9].

Biological Functions

Target Cells/Tissues and Functions

ETH is a crucial peripheral signal for activation of the ecdysis
sequence in insects. Its major targets are specific neurons in
the CNS expressing ETH receptors. Subsequent activation of
individual neuronal networks results in initiation of different
behaviors (pre-ecdysis, ecdysis, post-ecdysis) for shedding
the old cuticle and tracheal system [3,8,9]. Recent data indicate that ETH also stimulates the activity of corpora allata to
produce juvenile hormones during larval ecdyses.

Phenotype in Gene-Modified Animals

Drosophila mutants with deleted *eth* gene fail to inflate the
tracheal system, show severe behavioral defects during the
first larval ecdysis, and die. This indicates that ETH is a necessary endocrine signal for regulation of the ecdysis sequence
in insects [10].

References

1. Žitňan D, Kingan TG, Adams ME. Identification of ecdysis-
   triggering hormone from an epitracheal endocrine system.
   Science. 1996;271:88−91.
2. Žitňan D, Ross LS, Žitňanová I, et al. Steroid induction of a peptide hormone gene leads to orchestration of a defined behavioral
   sequence. *Neuron*. 1999;23:523−535.
3. Žitňan D, Adams ME. Neuroendocrine regulation of ecdysis. In:
   Gilbert LI, ed. *Insect Endocrinology*. San Diego: Academic Press;
   2012:253−309.
4. Gauthier SA, Hewes RA. Transcriptional regulation of neuropeptide and peptide hormone expression by the *Drosophila* dimmed
   and cryptocephal genes. *J Exp Biol*. 2006;209:1803−1815.
5. Cho K-H, Daubnerová I, Park Y, et al. Secretory competence in a
   gateway endocrine cell conferred by the nuclear receptor βFTZ-
   F1 enables stage-specific ecdysone responses throughout development in *Drosophila*. *Dev Biol*. 2014;385:253−262.
6. Kim YJ, Cho KH, Zitnanova I, et al. Corazonin receptor signaling
   in ecdysis initiation. *Proc Natl Acad Sci USA*.
   2004;101:6704−6709.
7. Kingan TG, Cardullo RA, Adams ME. Signal transduction in eclosion hormone-induced secretion of ecdysis-triggering hormone. *J
   Biol Chem*. 2001;276:25136−25142.
8. Kim Y-J, Žitňan D, Cho K-H, et al. Central peptidergic ensembles
   associated with organization of an innate behavior. *Proc Natl
   Acad Sci USA*. 2006;103:14211−14216.
9. Kim Y-J, Žitňan D, Galizia G, et al. A command chemical triggers
   an innate behavior by sequential activation of multiple peptidergic ensembles. *Curr Biol*. 2006;16:1395−1407.
10. Park Y, Filippov V, Gill SS. Deletion of the ecdysis-triggering hor-
    mone gene leads to lethal ecdysis deficiency. *Development*.
    2002;129:493−503.

Supplemental Information Available on Companion Website

- Comparison of ETHs from diverse insects, the water flea
  and the tick/E-Table 77.1
- Organization of *Manduca* and *Drosophila eth* genes/E-
  Figure 77.1
- Variability of Inka cells in different insects and possibly
  homologous cells in the tick/E-Figure 77.2
- Organization of the ETH receptor gene in *Drosophila*/E-
  Figure 77.3
- Accession numbers of mRNAs for ETH and ETH receptor/
  E-Tables 77.2 and 77.3

CHAPTER 78

Sex Peptide

Yusuke Tsukamoto

Abbreviation: SP
Additional names: accessory gland protein 70A, ACP70A, paragonial substance

Sex peptide (SP) is one of the male accessory gland proteins transferred to females in the seminal fluid during mating. SP elicits post-mating responses (PMRs) in mated females and virgin females injected with SP. PMRs include stimulating ovulation, reducing receptivity for courting males, and several behaviors related to reproductively.

Discovery

SP was first discovered as a hormone inducing oviposition in virgin female *Drosophila* species [1]. The peptide has been isolated from the male accessory gland in the fruit fly *Drosophila melanogaster* as a biologically active peptide capable of reducing receptivity and stimulating ovulation in virgin females [2].

Structure

Structural Features

SP contains 36 aa residues and is composed of three functional regions [3]; an N-terminal tryptophan-rich region, a central region containing five hydroxyproline residues, and a C-terminal region enclosed by a disulfide bridge (Figure 78.1). The C-terminal region contains the $W(X)_8W$ domain, which is similar to that in myoinhibiting peptide (MIP, synonymous with allatostatin type B or PTSP).

Primary Structure and Subtype

SP has been identified only in the Drosophilidae. In some Drosophilidae species, a closely related ortholog, ductus ejaculatorius peptide 99B (DUP99B) [4] composed of 31 aa residues, has been identified. The C-terminal regions of SP and DUP99B are similar to each other, while DUP99B does not have the $W(X)_8W$ domain at its C-terminal region (Table 78.1).

Synthesis and Release

Gene, mRNA, and Precursor

An mRNA encoding SP produces a 54- to 60-aa residue precursor peptide.

Distribution of mRNA

SP mRNA is produced in the male accessory gland, the origin of seminal fluid.

Regulation of Synthesis and Release

SP is released into the female seminal vesicle together with seminal fluid during ejaculation. Only SP can induce transient PMRs; long-term responses are required for other seminal fluid proteins. For example, *D. melanogaster* CG17575, which encodes a cysteine-rich secretory protein, is necessary for accumulating lectins (CG1652 and CG1656). These lectins are thought to activate a protease (CG9997). In the presence of all these factors, SP is able to elicit long-term responses in females [5].

Receptors

Structure and Subtype

Although SP has been identified only in the Drosophilidae, receptors that can respond to SP have been identified from several insect species, as a class of GPCR (CG16752 in *D. melanogaster*) [6]. All annotated SP receptors (SPRs) exhibit 60−90% similarity with one another.

SPR is expressed in the female reproductive organs and the central nervous system (CNS) of both males and females through the larval to adult stages. SP and *D. melanogaster* myoinhibiting peptides (MIPs) activate *D. melanogaster* SPR expressed in CHO cells. That MIP can also activate SPR has been observed in several insect species; therefore, MIP is considered to be an ancestral ligand against SPR. (See MIPR, allatostatin receptor, PTSP receptor).

Signal Transduction Pathway

Activated SPR in a set of six to eight sensory neurons of the female reproductive tract induces PMRs. These *spr*-expressing neurons which co-express both *fruitless* (*fru*) and *pickpocket* (*ppk*) are necessary and sufficient for inducing PMRs [7,8]. The *spr* neurons project to the CNS. Furthermore, a sex-determination gene, *doublesex*, is also critical for PMRs involved in SP signaling [9]. Inhibition of protein kinase A (PKA) activity in *fru* and *ppk* neurons, possibly resulting from signals from $G_{\alpha i}$, induces oviposition in females and reduces the receptivity of males [8].

Biological Functions

Endocrinological and Behavioral Functions

Three regions within the SP contribute to separate biological functions. The N-terminal tryptophan-rich region supports sperm-binding and JH synthesis. Owing to sperm binding, SP is proteolytically reinforced in the female uterus, consequently causing long-term responses. The central region,

Y. Takei, H. Ando, & K. Tsutsui (Eds): Handbook of Hormones. DOI: http://dx.doi.org/10.1016/B978-0-12-801028-0.00078-7

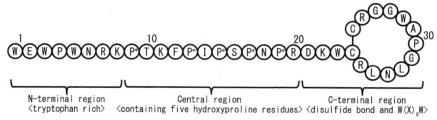

Figure 78.1 **Structure of SP.** P* indicates hydroxyproline residue.

Table 78.1 Alignment of Amino Acid Sequences of SP, DUP99B, and DmMIP1*

SP	WEWPWNRKP*TKFP*IP*SP*NP*RDKWCRLNLGPAWGGRC
DUP99B	pEDRN DTEW IQSQ KDRE KWCRLNLGPYL GGRC
DmMIP1	AWQSLQSS W—NH₂

P* indicates hydroxyproline residue.

which has five hydroxyproline residues, induces the innate immune response. The hydroxyproline motif is considered to be a chemical mimic of sugar components of the bacterial cell wall, and thus induces the innate system via pattern recognition receptors. The C-terminal region, which has a disulfide bond and $W(X)_8W$ motif, is the predominant PMRs inducible region. Eventually, SP contributes to a repertoire of PMRs, stimulating oviposition, reducing receptivity, increasing food intake, switching dietary preference, shortening longevity, altering sleep patterns, and concentrating fecal deposits.

References

1. Chen PS, Bühler R. Paragonial substance (sex peptide) and other free ninhydrin-positive components in male and female adults of *Drosophila melanogaster. J Insect Physiol.* 1970;16:615−627.
2. Chen PS, Stumm-Zollinger E, Aigaki T, et al. A male accessory gland peptide that regulates reproductive behavior of female *D. melanogaster. Cell.* 1988;54:291−298.
3. Moehle K, Freund A, Kubli E, et al. NMR studies of the solution conformation of the sex peptide from *Drosophila melanogaster. FEBS Lett.* 2011;585:1197−1202.
4. Saudan P, Hauck K, Soller M, et al. Ductus ejaculatorius peptide 99B (DUP99B), a novel *Drosophila melanogaster* sex-peptide pheromone. *Eur J Biochem.* 2002;269:989−997.
5. Ravi Ram K, Wolfner MF. A network of interactions among seminal proteins underlies the long-term postmating response in *Drosophila. Proc Natl Acad Sci USA.* 2009;106:15384−15389.
6. Yapici N, Kim YJ, Ribeiro C, et al. A receptor that mediates the post-mating switch in *Drosophila* reproductive behavior. *Nature.* 2008;451:33−37.
7. Häsemeyer M, Yapici N, Heberlein U, et al. Sensory neurons in the *Drosophila* genital tract regulate female reproductive behavior. *Neuron.* 2009;61:511−518.
8. Yang CH, Rumpf S, Xiang Y, et al. Control of the postmating behavioral switch in *Drosophila* females by internal sensory neurons. *Neuron.* 2009;61:519−526.
9. Rezaval C, Pavlou HJ, Dornan AJ, et al. Neural circuitry underlying *Drosophila* female postmating behavioral responses. *Curr Biol.* 2012;22:1155−1165.

Supplemental Information Available on Companion Website

- Accession numbers of SP preproprotein and SP receptor/ E-Table 78.1

CHAPTER 79

APWGamide

Shinji Nagata

Abbreviation: APWGa

APWGamide (APWGa) controls molluskan male reproductive behavior. Male mollusks possess this peptide more in the nerve cells, particularly in the reproductive accessory organ, in comparison with females.

Discovery

APWGa was originally isolated from the CNS of molluskan *Fusinus ferrugineus* by assaying inhibitory effects on the twitch contraction [1].

Structure

Structural Features

APGWa is a tetrapeptide with C-terminal amidation. Similar peptides have been identified from other molluscan species. All related peptides are composed of four amino acids and share a C-terminal PWG−NH$_2$ [2].

Structure and Subtype

In some mollusks, such as *Mytilus edulis*, an N-terminal aa residue of APGWa is substituted (Table 79.1). The squid *Sepia officinalis* has a dipeptide GW−NH$_2$ [3].

Synthesis and Release

Gene, mRNA, and Precursor

An APGWa precursor mRNA encodes a protein of 190−220 amino acids, providing 6−10 copies of APGWa peptides after cleavage at dibasic sites (E-Figure 79.1) [2]. The number of mature bioactive APGWa peptides derived from a prepropotein differs among species [4].

Distribution of mRNA

Immunoreactive cells are observed in the CNS, especially in the brain [5]. In addition, the reproductive accessory organ includes immunoreactive cells. Interestingly, the accessory organ in males includes more APGWa-immunopositive cells than that in females, consistent with the fact that this peptide functions in the male reproductive process in mollusks [6].

Biological Functions

Target Cells and Function

APGWa plays a role in the reproduction system of cephalopods such as the octopus. The oviducts, neurons and fibers are located at the posterior olfactory lobule and subpenduncolate area, innervating reproduction-related neurons that are immunoreactive sites for an antiserum against APGWa. Thus, APGWa plays in a pivotal role in reproductive behavior. As it has also been indicated that APGWa modulates the central pattern generator for the ingesting locomotor muscle, APGWa controls feeding behavior in mollusks.

References

1. Kuroki Y, Kanda T, Kubota I, et al. A molluscan neuropeptide related to the crustacean hormone, RPCH. *Biochem Biophys Res Commun.* 1990;167:273−279.
2. Favrel P, Mathieu M. Molecular cloning of a cDNA encoding the precursor of Ala−Pro−Gly−Trp amide-related neuropeptides from the bivalve mollusc *Mytilus edulis. Neurosci Lett.* 1996;205:210−214.
3. Sirinupong P, Suwanjarat J, van Minnen J. Distribution of APGWamide-immunoreactivity in the brain and reproductive organs of adult pygmy squid, *Idiosepius pygmaeus. Invert Neurosci.* 2011;11:97−102.
4. Di Cristo C, Di Cosmo A. Neuropeptidergic control of *Octopus* oviducal gland. *Peptides.* 2007;28:163−168.
5. Croll RP, Minnern J. Distribution of the peptide Ala−Pro−Gly−Trp−NH$_2$ (APGWamide) in the nervous system and periphery of the snail *Lymnaea stagnalis* as revealed by immunocytochemistry and *in situ* hybridization. *J Comp Neurol.* 1992;324:567−574.
6. Oberdorster E, Romano J, McClellan-Green P. The neuropeptide APGWamide as a penis morphogenic factor (PMF) in gastropod mollusks. *Integr Comp Biol.* 2005;45:28−32.

Supplemental Information Available on Companion Website

- Amino acid sequence of *Aplysia* APGWa preproprotein (AAD00568.1)/E-Figure 79.1
- Accession numbers for cDNA of APGWa precursor/E-Table 79.1

Table 79.1 Alignment of APGWa and its Related Peptides

Fusinus ferrugineus	APGW−NH$_2$
Lymnaea stagnalis	APGW−NH$_2$
Aplysia	APGW−NH$_2$
Mytilus edulis	RPGW−NH$_2$
	KPGW−NH$_2$
	TPGW−NH$_2$
Sepia officinalis	GW−NH$_2$

Y. Takei, H. Ando, & K. Tsutsui (Eds): Handbook of Hormones. DOI: http://dx.doi.org/10.1016/B978-0-12-801028-0.00079-9

CHAPTER 80

SIFamide

Shinji Nagata

Additional name/Abbreviation: SIFa

SIFamide (SIFa) was originally identified as a myoactive factor in the crude extract from various species of flies. SIFa is a possible factor coordinating sexual behavior, from courtship to mating.

Discovery

SIFa was identified as a peptide with a C-terminal LF—NH$_2$ from the crude extract of *Neobellieria bullata* by its myostimulatory activity on *Leucophaea maderae* hindgut contraction [1]. The corrected SIFa sequence (not leucine but isoleucine residue) was reported after purification using 350,000 *N. bullata* adults [2].

Structure

Structural Features

SIFa is composed of 12 amino acids with C-terminal amidation. The C-terminal Y(or F)RKP(or F)PFNGSIF—NH$_2$ sequence of SIFa is highly conserved over arthropods (Table 80.1) [3,4].

Structure and Subtype

Although no subtypes are present, two truncated forms are found in some species, including *Drosophila melanogaster*.

Synthesis and Release

Gene, mRNA, and Precursor

A precursor peptide mRNA encodes approximately 70–90 amino acids, comprising a signal sequence, a mature SIFa, and a prepeptide domain.

Distribution of mRNA

SIFa mRNA is expressed by the two and four cells in the pars intercerebralis of the flies *N. bullata* and *D. melanogaster*, respectively. SIFa is also expressed in the extensive arborizations of the CNS. In the crayfish *Procambarus clarkii*, expression sites are similar to those of insect species. No immunoreactivity was observed in the insect gut, while immunoreactivity was observed in the gut of crustacean species [3,4].

Receptors

Structure and Subtype

The SIFa receptor has been identified by deorphanization of *D. melanogaster* GPCRs [5]. Genomic information indicates that the SIFa receptor is extensively conserved over insect species.

Biological Functions

Target Cells and Function

In *D. melanogaster*, the SIFa receptor is expressed in the brain and CNS in adult flies [5]. SIFa plays a role in the regulation of sex-specific behavior, including courtship behavior.

Phenotype in Gene-Modified Animals

Using targeted RNAi, SIFa-deficient *D. melanogaster* was shown to lack sex-specific sexual behavior, especially in courtship. This indicates that SIFa contributes to sexual behavior [6].

References

1. Fonagy A, Schoofs L, Proost P, et al. Isolation, primary structure and synthesis of neomyosuppressin, a myoinhibiting neuropeptide from the grey flesh fly, *Neobellieria bullata*. *Com Biochem Physiol C*. 1992;102:239–245.
2. Janssen J, Schoofs L, Spittaels K, et al. Isolation of NEB-LFamide, a novel myotropic neuropeptide from the grey fleshfly. *Mol Cell Endocrinol*. 1996;117:157–165.
3. Verleyen P, Huybrechts G, Sas F, et al. Neuropeptidomics of the grey flesh fly, *Neobellieria bullata*. *Biochem Biophys Res Commun*. 2004;316:763–770.
4. Verleyen P, Huybrechts J, Schoofs L. SIFamide illustrates the rapid evolution in Arthropod neuropeptide research. *Gen Comp Endocrinol*. 2009;162:27–35.
5. Jorgensen LM, Hauser F, Cazzamali G, et al. Molecular identification of the first SIFamide receptor. *Biochem Biophys Res Commun*. 2006;340:696–701.
6. Terhzaz S, Rosay P, Goodwin SF, Veenstra JA. The neuropeptide SIFamide modulates sexual behavior in *Drosophila*. *Biochem Biophys Res Commun*. 2007;352:305–310.

Supplemental Information Available on Companion Website

- Accession numbers of SIFa and SIFa receptor/E-Table 80.1

Table 80.1 Alignment of Sequences of SIFa

Species	Sequence
Neobellieria bullata	AYRKPPFNGSIF-NH$_2$
Drosophila melanogaster	AYRKPPFNGSIF-NH$_2$
Tribolium castaneum	TYRKPPFNGSIF-NH$_2$
Bombyx mori	TYRKPPFNGSIF-NH$_2$

Y. Takei, H. Ando, & K. Tsutsui (Eds): Handbook of Hormones. DOI: http://dx.doi.org/10.1016/B978-0-12-801028-0.00080-5

CHAPTER 81

Egg-Laying Hormone

Fumihiro Morishita and Hiroyuki Minakata

Abbreviation: ELH
Additional names: caudo-dorsal cell hormone, CDCH

Egg-laying hormone (ELH) is a peptide hormone which controls a series of complex but stereotyped egg-laying behaviors of the sea slug Aplysia californica. *In the ELH precursor several bioactive peptides are also found, which work together to induce egg laying.*

Discovery

Egg-laying hormone (ELH) was purified from the bag cell, a pair of clusters of neurosecretory cells located on the rostral side of the abdominal ganglion of *Aplysia californica*. The peptide triggers a series of egg-laying behaviors of the animal [1].

Structure

Structural Features

The ELH of *Aplysia californica* consists of 36 amino acids, and the C-terminal Lys is amidated.

Structure and Subtype

Egg-releasing hormone (ERH) and califin A, B, and C are ELH-related peptides isolated from the atrial gland (an exocrine organ that secrets materials into the genital duct) of *A. californica* (Table 81.1). Califin A, B, and C consist of two subunits connected by a disulfide bond. The N-terminal 18 amino acid residues of the longer subunits of califins are identical to those of ELH, and the shorter subunit has similarity to the acidic peptide found in the C-terminal region of the ELH precursor (see E-Figure 81.1 and 81.2). Califins are different in an amino acid residue in the C-terminal region of the longer subunit. ERH consists of 34 aa, and the N-terminal region of ERH has similarity to that of ELH but it does not have a C-terminal amide. In the freshwater snail *Lymnaea stagnalis*, the caudo-dorsal cell hormone (CDCH) that promotes egg-laying was identified from the caudo-dorsal cells [2]. As does ELH, CDCH consists of 36 aa and has a C-terminal amide, and ELH and CDCH share 18 of the same amino acid residues. ELH-related peptides have also been identified in the abalone *Haliotis rubra* and the rhynchobdellid leech *Theromyzon tessulatum*.

Synthesis and Release

Gene, mRNA, and Precursor

The ELH gene consists of three exons and one intron (DDBJ:M57962) [2]. Exon 1 (40 bp) and Exon 2 (149 bp) are connected directly to the genome, while Exons 2 and Exon 3 are separated by 5 kbp of intron (E-Figure 81.1). Either Exon 1 or Exon 2 is connected to Exon 3 before maturation of mRNA. However, the open-reading frame for ELH is derived from Exon 3. On the precursor protein, ELH is located near the C-terminus. Some bioactive peptides that coordinate with ELH are generated from the middle region of the precursor. These peptides are named α-, β-, δ-, and γ-bag cell peptides (BCPs). Localization of bioactive peptides on the CDCH precursor in *L. stagnalis* is quite similar to that of the ELH precursor.

In the atrial gland of *Aplysia*, ELH gene transcripts are derived from a variant Exon 3. The main difference in the variant is the deletion of 240 bp of nucleotides that encode β-, δ-, and γ-BCPs of ELH in the bag cell. Interestingly, the precursor protein derived from the variant Exon 3 includes the atrial gland peptide (AGP) A or B in the middle region, while it includes one of the califins in the C-terminal region (E-Table 81.2). These peptides are also involved in the egg-laying of *Aplysia* (see below).

Distribution of mRNA

The ELH gene expresses in the bag cell, while AGP-A and -B genes express in the atrial gland. The two genes also express in the ganglia, including the cerebral and pedal ganglia; however, the expression level is relatively low.

Regulation of Synthesis and Release

The ELH precursor protein consists of 271 aa, including an N-terminal signal peptide. Prohormone in the trans-Golgi network is cleaved to the N-terminal half that includes BCPs, and the C-terminal half that includes ELH; each processing intermediate is then packaged into the different vesicles [3]. Thus, BCPs and ELH can be processed and released, separately.

The bag cell shows a unique property of discharge when it releases ELH (E-Figure 81.3). When the neural input from the cerebral ganglion has fully excited the bag cell, the cell continues to discharge action potentials even after the cessation of neural input (after discharge). Stimulation of the bag cell activates both the adenylate cyclase/cAMP pathway and the phospholipase (PLC)/diacylglycerol (DAG)/inositol triphosphate (IP3) pathway in the cell. cAMP reduces K^+-conductance of

Y. Takei, H. Ando, & K. Tsutsui (Eds): Handbook of Hormones. DOI: http://dx.doi.org/10.1016/B978-0-12-801028-0.00081-7

Table 81.1 Structures of ELH-Related Peptides*

		Sequence
A. californica	ELH	I SI NQDLK AI T D ML L TEQI R ERQRYLAD LR QRLLEK-NH2
	ERH	I SI VSLFK AI T D ML L TEQIY ANYFSTPR LR FYPI
	Califin-A (L)	I SI NQDLK AI T D ML L TEQIQ ARR RCLDA LR QRLLDL-NH2
	Califin-B (L)	I SI NQDLK AI T D ML L TEQIQ ARQRCLDA LR QRLLDL-NH2
	Califin-C (L)	I SI NQDLK AI T D ML L TEQIQ ARR RCLA A LR QRLLDL-NH2
L. stagnalis	CDCH I	L SI TNDLR AI A D SY L YDQHK LRER QEEN LR RRFLEL-NH2
	CDCH II	SI TNDLR AI A D SY L YDQHK LREQQEEN LR RRFYELSLRP YPDNL
H. rubra	Abalone ELH	L SI TNDLR AI A D SY L YDQHK LRER QEEN LR RRFLRL-NH2
T. tessulatum	LELH	GSGVSNGG TE M I QL S HIRER QRYWAQDN LR RRFLEK-NH2
A. californica	Califin-A (S)	DSDVSLFNGD LLPNGRCS
	acidic peptide	SSGVSLLTSN KDEEQRELLKAISNLLD

*Amino acids identical to ELH are highlighted dark gray, while those identical to CDCH I are highlighted light gray. Amino acids shared by both groups are boxed. Note that cysteine residues in long (L) and short (S) subunits of califins are connected by a disulfide bond.

the cell membrane through the protein kinase (PK)-A dependent phosphorylation of the membrane protein. Such modification augments excitability of the bag cell and maintains discharge of the cell [4]. On the other hand, activated PLC elevates intracellular levels of IP3 and DAG. IP3 induces Ca^{2+}-mobilization from the intracellular Ca^{2+} store, while DAG modifies phosphorylation of the Ca^{2+}-channel by Ca^{2+}/PK-C [5]. These modifications induce a long-lasting elevation of the Ca^{2+} level in the cell, even after the discharge of the cell is terminated. cAMP and IP3 can also augment ELH release in different ways. cAMP stimulates ELH synthesis through acceleration of the processing of the ELH precursor [6], and also stimulates packaging of ELH into secretory granules and axonal transport of the secretory vesicle to the releasing site [7]. IP3 and PK-C, in turn, promote exocytotic release of ELH, as the intracellular Ca^{2+} is elevated [8]. These modifications, triggered by the discharge of the cell, result in prolonged ELH release, even after the discharge is terminated. By contrast, a neuropeptide, FMRFamide [8], and a neurotransmitter, 5-HT, are known to terminate after discharge of the bag cell.

The neural trigger that initiates ELH release from the bag cell is not fully understood. However, some neurons in the cerebral and right pleural ganglia are known to contain ELH and BCPs (E-Figure 81.4), and stimulation of those neurons induces discharge of the bag cell [9]. Thus, it is likely that some sensory input, such as pheromones or chemoreception, activates the ELH/BCP cells in the ganglia, and that then the signal is transmitted to the bag cell to induce ELH release.

Receptors

Structure and Subtype

No information is available regarding the ELH receptor.

Signal Transduction Pathway

The signal transduction pathway mediating the actions of ELH on egg-laying behavior is largely unknown.

Biological Functions

Target Cells and Function

ELH induces egg-laying when it is injected into *Aplysia* and other gastropods [9]. It also induces ovulation from the isolated ovotestis of *Aplysia*. However, coordination of multiple peptides encoded on the ELH precursor is essential for egg-laying behavior [8]. For example, ELH augments the bursting frequency of R15 neurons that regulates the motility of the reproductive tract. α- and β-BCP regulate the neural activities of the left upper quadrant (LUQ) cell in the abdominal ganglion. Moreover, these peptides have a self-excitatory action on the bag cell [10]. This positive cycle may also contribute to the prolonged ELH release. Injection of extract from the atrial gland induces

egg-laying in *Aplysia*, and there is evidence that atrial gland hormones, such as califin and AGP-A/B, mediate the response. For example, AGP-A and -B can trigger the discharge of the bag cell, and califin modulates neural activities of LUQ cells. Besides the regulatory action on egg-laying, ELH may be involved in regulation of the feeding behavior of *Aplysia*. This peptide hormone augments the bursting activity of B16 in the buccal ganglion, which innervates the accessory radula closer muscle.

References

1. Chiu AY, Hunkapiller MW, Heller E, et al. Purification and primary structure of the neuropeptide egg-laying hormone of *Aplysia californica*. *Proc Natl Acad Sci USA*. 1979;76:6656−6660.
2. Mahon AC, Nambu JR, Taussig R, et al. Structure and expression of the egg-laying hormone gene family in *Aplysia*. *J Neurosci*. 1985;5:1872−1880.
3. Sossin WS, Sweet-Cordero A, Scheller RH. Dale's hypothesis revisited: different neuropeptides derived from a common prohormone are targeted to different processes. *Proc Natl Acad Sci USA*. 1990;87:4845−4848.
4. Strong JA, Kaczmarek LK. Multiple components of delayed potassium current in peptidergic neurons of *Aplysia*: modulation by an activator of adenylate cyclase. *J Neurosci*. 1986;6:814−822.
5. Gardam KE, Magoski NS. Regulation of cation channel voltage and Ca^{2+} dependence by multiple modulators. *J Neurophysiol*. 2009;102:259−271.
6. Azhderian EM, Kaczmarek LK. Cyclic AMP regulates processing of neuropeptide precursor in bag cell neurons of *Aplysia*. *J Mol Neurosci*. 1990;2:61−70.
7. Azhderian EM, Hefner D, Lin CH, et al. Cyclic AMP modulates fast axonal transport in *Aplysia* bag cell neurons by increasing the probability of single organelle movement. *Neuron*. 1994;12:1223−1233.
8. Roubos EW, van de Van AM, ter Maat A. Quantitative ultrastructural tannic acid study of the relationship between electrical activity and peptide secretion by the bag cell neurons of *Aplysia californica*. *Neurosci Lett*. 1990;111:1−6.
9. Morishita F, Furukawa Y, Matsushima O, Minakata H. Regulatory actions of neuropeptides and peptide hormones on the reproduction of molluscs. *Can J Zool*. 2010;88:825−845.
10. Bernheim SM, Mayeri E. Complex behavior induced by egg-laying hormone in *Aplysia*. *J Comp Physiol A*. 1995;176:131−136.

Supplemental Information Available on Companion Website

- List of accession numbers for precursors of ELH and related peptides/E-Table 81.1
- List of structures of ELH and related peptides/E-Table 81.2
- Gene structure of *Aplysia* ELH genome/E-Figure 81.1
- Structure of precursor proteins for ELH, atrial gland peptides, ELH related peptides, and CDCH/E-Figure 81.2
- Signal transduction cascade that induces ELH-release to the bag cell/E-Figure 81.3
- Regulation of egg-laying by the ELH/BCP system/E-Figure 81.4

SUBSECTION II.2.4

Other Hormones and Neuropeptides

CHAPTER 82

Growth Blocking Peptide

Yoichi Hayakawa

Abbreviation: GBP

Additional names: juvenile hormone esterase activity repressive factor, ENF peptide, paralytic peptide (PP), plasmatocyte spreading peptide (PSP), cardioactive peptide, insect cytokine

Insect cytokine GBP was first isolated from Lepidoptera Mythimna (Pseudaletia) separata. *It exerts multifunctional effects, such as larval growth regulation, paralysis induction, plasmatocyte spreading induction, cell proliferation, and cardioacceleration.*

Discovery

Presence of GBP in the hemolymph of *M. separata* larvae parasitized by parasitoid wasp *Cotesia kariyai* was reported in 1990, and it was isolated and sequenced in 1991 [1].

Structure

Structural Features

M. separata proGBP consists of 152 aa residues with bioactive mature GBP at the C-terminus. *M. separata* GBP consists of 23 aa residues with an intramolecular ring structure of 13 aa residues flanked by two Cys residues and with N-terminal and C-terminal extensions [2]. NMR analysis showed that GBP consists of disordered N- and C-termini and a well-defined core stabilized by a disulfide bridge between two Cys residues, hydrophobic interactions, and a short β-hairpin (Figure 82.1) [3].

Primary Structure

Sequence identity is low in the pro-segment (proGBP). The mature GBP sequence is relatively conserved in the same order of insects, but is variable across different orders.

Conserved Motif of Orthologs and Homologs

Until *Drosophila* GBP was identified in 2012 [4], GBP and its orthologs with a conserved primary structure had been found only in Lepidoptera. Although they were referred to as the "ENF peptide family" based on the conserved N-terminal amino acid sequence, a similar family peptide without the ENF N-terminus has been recently identified even in Lepidoptera [2]. Furthermore, GBP functional homologs and their homologous genes have been isolated or identified in insect species spanning five orders — Lepidoptera, Diptera, Coleoptera, Hymenoptera, and Hemiptera — but they do not conserve a significant homology. However, it was found that most of these GBP homologs contain the conserved motif C

$(X_2)G(X_{4,6})G(X_{1,2})C(K/R)$, which shares a certain similarity with the motif in the mammalian EGF family (Table 82.1) [4].

Properties

M. separata GBP: Mr 2,494, pI 8.14. Freely soluble in water and ethanol.

Synthesis and Release

Gene and mRNA

The fat body expresses 5′-terminal truncated forms of the GBP transcript while the brain produces the 5′-terminal elongated form in *M. separata* larvae. Brain GBP gene consists of two exons separated by a single intron which contains a GATA motif (TGATAA) and two cognates (KGATAW) that have all been reported to be important for expression in the insect fat body [5].

Distribution of mRNA

GBP mRNA is detected most abundantly in the fat body, and is also found in the nervous tissues, brain, and ventral ganglia [2]. In the brain, GBP mRNA is expressed mainly in the glial cell layer [6]. In the fat body, GBP mRNA is expressed in most cells [2].

Plasma Concentration

GBP concentrations in the plasma vary between several nM and several 10s of nM, depending on the developmental stages and physiological conditions; a temporal increase is observed after parasitization by a parasitoid wasp and during the time of every molt [2].

Regulation of Synthesis and Release

ProGBP synthesized in the fat body is released into the hemocoel, in which its concentrations are normally kept much higher than those of active GBP. ProGBP is proteolyzed by the processing enzyme (a serine protease) after activation from the inactive proenzyme by various stressors, such as parasitization, wounding, and environmental changes [2]. ProGBP processing enzyme has not been purified yet. The exposure of hemocytes to external pathogens, such as bacteria, enhances the release of proGBP from hemocytes [7].

Receptors

Structure

Neither a GBP receptor nor other orthologous (homologous) peptide receptors have been isolated. Competitive binding assays using ^{125}I-GBP demonstrated the specific binding to cultured cells such as human keratinocytes ($K_d = 0.12$ nM) and insect SF9 cells ($K_d = 0.25$ nM) [8].

Y. Takei, H. Ando, & K. Tsutsui (Eds): Handbook of Hormones. DOI: http://dx.doi.org/10.1016/B978-0-12-801028-0.00082-9

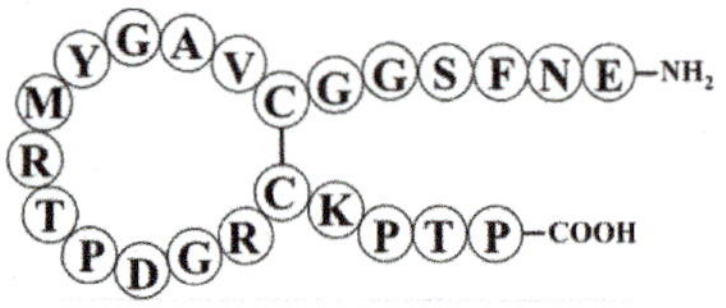

Figure 82.1 Structure of *Mythimna (Pseudaletia) separata* GBP.

Table 82.1 Comparison of GBP Homologous Peptide Sequences in Representative Species of Five Insect Orders*

Mythimna separata GBP	-----EN--FSGGCVAGYMRTPDGRCKPTF--------
Mamestra brassicae GBP	-----EN-FAGGCLTGFMRTPDGRCKPTF--------
Spodoptera litura GBP	-----EN-FAAGCATGYQRTADGRCKPTF--------
Bombyx mori PPI	-----EN-FVGGCATGFKRTADGRCKPTF--------
Bombyx mori PP II	-----DN-FKGECATGFKRTADGRCRPIF--------
Manduca sexta PSP	-----EN-FAGGCATGFLRTADGRCKPTF--------
Zophobas atratus GBP	-----RIIS-AGSNCPAGQRADSAGNCREEW--------
LOC661094	-----RVIA-VGANCPPGFRGDGKGNCREEY--------
(*T. castaneum*)	
CG15917	-----ILLE-TTQKCKPGFELFG-KRCRKPA--------
(*D. melanogaster*)	
GG20666 (*D. erecta*)	-----ILLE-TTRKCKPGFELFG-KRCRKPA--------
GE11651 (*D. yakuba*)	-----ILLE-TTQKCKPGYQLFG-KRCRKPA--------
AAEL008441-PA	-----RQIFV-
(*A. aegypti*)	APVVCPSGQKPDHRGRCRPVWSM--------
LOC100649983	-----GARFRCPTGQQRDHLGKCRDVFVVPLKNDV
(*B. terrestris*)	
SINV_01554 (*S. invicta*)	-----N-AKLGCPSGEKLDPRGMCRKVL--------
RPRC004573-RA	-----RAID-AKELCESGFQKDPSGKCRGVFGL--------
(*R. prolixus*)	
AGAP008923-PB	----ALFD-APIVCPDGTVLDHKGVCRRPMGR--------
(*A. gambiae*)	
XP_001842206	--VFGLFD-APYVCPPGQSADLKGKCKERF--------
(*C. quinquefasciatus*)	
LOC100575717 (*A. pisum*)	-----KP-DGGQCPRGYSITSNGNCKPSFTG--------
Lucilia cuprina GBP	-----TILS-APSNCQ-------QTDFKGRCL--------

*GBP orthologs and homologs have been identified in Lepidoptera, Diptera, Coleoptera, Hymenoptera, and Hemiptera.

Signal Transduction Pathway

GBP exerts innate cellular and humoral immune responses. GBP signaling in mediating cellular immune responses requires the adaptor protein P77 with a molecular mass of 77 kDa containing SH2/SH3 domain binding motifs and an immunoreceptor tyrosine-based activation motif (ITAM)-like domain in the cytoplasmic region of the C-terminus [7]. Soon after binding of GBP to GBP receptor, the tyrosine residues of P77 are phosphorylated, which induces tyrosine phosphorylation of the integrin β-chain and subsequent activation of cellular responses of plasmatocytes, plausibly via a drastic change in organization of the actin cytoskeleton. GBP signaling in mediating humoral responses, such as antimicrobial peptide (AMP) expression, passes through the Imd/JNK pathway: recruitment of the adaptor protein Imd to the GBP receptor activated by GBP initiates intracellular signaling to enhance AMP expression via the JNK pathway [9]. It has been also reported that GBP induces Ca^{2+} mobilization via classical activation of phospholipase C in *Drosophila* S3 cells, which induces cell spreading (an initiating event in cellular innate immunity) via activation of Pvr/ERK signaling [10].

Biological Functions

Since the first report on *M. separata* GBP in 1990 [1], various orthologs have been identified as peptides with different functions, as described above in relation to Lepidoptera. GBP and other orthologs have been recognized as multifunctional cytokines since it was demonstrated in the year 2000 that both GBP and plasmatocyte spreading peptide (PSP) induce larval growth retardation, plasmatocyte spreading, and larval

paralysis [6]. GBP secreted from the subesophageal body plays an essential role in formation of the procephalic domain during early embryogenesis [2]. Furthermore, GBP enhances *dopa decarboxylase* (*DDC*) expression in integuments, nervous tissues, and hemocytes of *M. separata* larvae, which elevates dopamine concentrations in the hemolymph and nervous tissues. The GBP—dopamine system has been proposed to regulate induction of insect diapause as well as larval growth [2].

Phenotype in Gene-Modified Animals

GBP RNAi knockdown made *Drosophila* larvae more sensitive to Gram-negative bacteria [9]. On the other hand, overexpressing the GBP gene enhanced expression levels of *AMP* such as *Metchnikowin* and *Diptericin* in *Drosophila* larvae. Plasmatocytes prepared from P77 RNAi larvae of *M. separata* were not activated by GBP. Furthermore, GBP overexpression and knockdown during embryogenesis produced severe abnormal phenotypes of the head structure in *M. separata*.

References

1. Hayakawa Y. Growth-blocking peptide: an insect biogenic peptide that prevents the onset of metamorphosis. *J Insect Physiol.* 1995;41:1—6.
2. Hayakawa Y. Insect cytokine growth-blocking peptide (GBP) regulates insect development. *Appl Entomol Zool.* 2006;41:545—554.
3. Aizawa T, Fujitani N, Hayakawa Y, et al. Solution structure of an insect growth factor, growth-blocking peptide. *J Biol Chem.* 1999;274:1887—1890.
4. Matsumoto H, Tsuzuki S, Date-Ito A, et al. Characteristics common to a cytokine family spanning five orders of insects. *Insect Biochem Mol Biochem.* 2012;42:446—454.
5. Hayakawa Y, Noguchi H. Growth-blocking peptide expressed in the insect nervous system: cloning and functional characterization. *Eur J Biochem.* 1998;253:810—816.
6. Hayakawa Y, Ohnishi A, Mizoguchi A, et al. Distribution of growth-blocking peptide in the insect central nervous tissue. *Tissue Cell Res.* 2000;300:459—464.
7. Oda Y, Matsumoto H, Kurakake M, et al. Adaptor protein is essential for insect cytokine signaling in hemocytes. *Proc Natl Acad Sci USA.* 2010;107:15862—15867.
8. Ohnishi A, Oda Y, Hayakawa Y. Characterization of receptors of insect cytokine, growth-blocking peptide, in human keratinocyte and insect Sf9 Cells. *J Biol Chem.* 2001;276:37974—37979.
9. Tsuzuki S, Ochiai M, Matsumoto H, et al. *Drosophila* growth-blocking peptide-like factor mediates acute immune reactions during infectious and non-infectious stress. *Sci Rep.* 2012;2:210.
10. Tsuzuki S, Matsumoto H, Furihata S, et al. Switching between humoral and cellular immune responses in *Drosophila* is guided by the cytokine GBP. *Nat Commun.* 2014;5:4628.

Supplemental Information Available on Companion Website

- Three-dimensional structure of *Mythimna (Pseudaletia) separata* GBP/E-Figure 82.1
- Sequence alignment of ENF peptides found in Lepidoptera/E-Figure 82.2
- Precursor and mature GBP homolog sequences of various insects/E-Figure 82.3
- Phylogenetic analysis of the GBP precursor amino acid sequences/E-Figure 82.4
- Primary structure of GBP signaling adaptor protein P77 of *Mythimna (Pseudaletia) separata*/E-Figure 82.5
- Primary structure of GBP signaling adaptor protein P77 of various insects/E-Figure 82.6
- Accession numbers of peptides, cDNAs, and genes for GBP, GBP ortholog, GBP homolog, and GBP receptor adaptor protein/E-Table 82.1

CHAPTER 83

Yamamarin

Koichi Suzuki

Additional names: diapause-maintaining peptide, growth-suppressing pentapeptide

Yamamarin provides a clue to the solution of the consequence of the metabolic and cell proliferation arrests throughout diapause.

Discovery

Yamamarin was first proposed as a repressive factor (RF) which is secreted from the mesothorax throughout diapause in pharate first-instar larvae of the wild silkmoth *Antheraea yamamai* [1]. An RF candidate was isolated in the form of a novel pentapeptide which may function to maintain diapause; however, the pattern of its production, its source, and its exact role in diapausing insects remain to be determined [2]. This peptide is called yamamarin after its source, *A. yamamai* Figure 83.1.

Structure

Structural Features

Yamamarin consists of 5 aa residues with C-terminal amidation and was determined from a fraction of 1,500 diapausing individuals (about 7.5 g), using the purification method. When the molecular mass and N-terminal amino acid sequence of this fraction were determined by MALDI-TOF MS and automated Edman degradation, another yamamarin form in the silkmoth was confirmed, with the following amino acid sequence: Asp−Ile−Leu−Arg−Gly−COOH [3]. The two forms of yamamarin are present in diapausing pharate first-instar larvae, but only the C-terminal amidated form is biologically active [2,3]. The chemical structure of the C-terminal part (−Arg−Gly−NH_2) of yamamarin is also essential to its activity [4]. Yamamarin cDNA has been registered with GenBank Accession No. P84863.

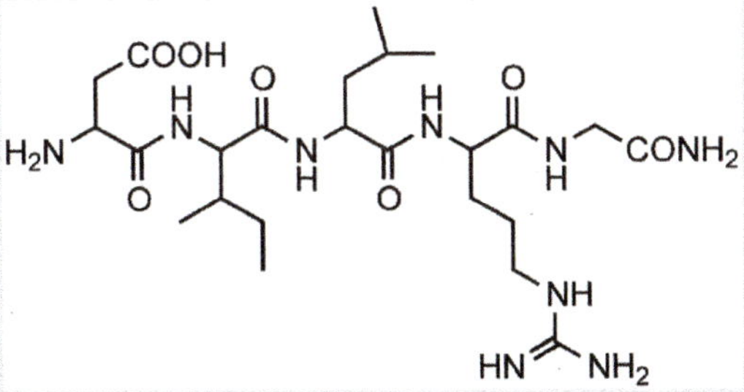

Figure 83.1 Structure of yamamarin.

Synthetic Yamamarin Products

The structures of yamamarin and its acetyl-C (C2), octanoyl-C (C8), and palmitoyl-C (C16) conjugates are synthesized. The C2 and C8 peptides show virtually no activity, but the C16 peptide is over twenty times more effective than yamamarin [3] (Figure 83.2).

Biological Functions

Although yamamarin is isolated as a developmental inhibitor from the diapausing pharate-instar larvae of the silkmoth, the compounds show attractive potential for the development of biopharmaceuticals and pest management. Synthetic yamamarin exhibits suppressive activity against rat hepatoma cells, and reversible cell-cycle arrest rather than apoptotic cell death [2]. C16-yamamarin also causes reversible growth arrest in rat hepatoma cells and *Drosophila* Schneider S2 cells [3,5]. In addition, this peptide compound suppresses murine leukemic cell lines expressing the human gene *Bcr/ABl*, and a farnesoyl peptide induces embryonic diapause in *Bombyx mori* [5]. With respect to the organelles and molecular mechanisms in states similar to diapause, C16-yamamarin inhibits mitochondrial respiration and expresses the remarkable upregulation and downregulation genes of some functional categories [5]. Moreover, yamamarin has revealed several other biological activities: a strong cardio-inhibitory effect, in cardiotropic testing on a semi-isolated heart of *Tenebrio molitor* [6]; *in vitro* growth inhibition of the plant pathogens, *Phoma narcissi* and *Botrytis* tulipae [7]; *in vitro* antiviral activity against human herpes virus type 1 in Vero cells [8]; and an *in vivo* antagonistic effect against insect peptide (poneratoxin)-induced analgesia [9]. It is unknown whether yamamarin is a novel insect hormone, and the gene and mRNA encoding the peptide and receptors are yet to be isolated. However, yamamarin may open up new vistas in the development of pharamaceutical agents (Figure 83.2).

Y. Takei, H. Ando, & K. Tsutsui (Eds): Handbook of Hormones. DOI: http://dx.doi.org/10.1016/B978-0-12-801028-0.00083-0

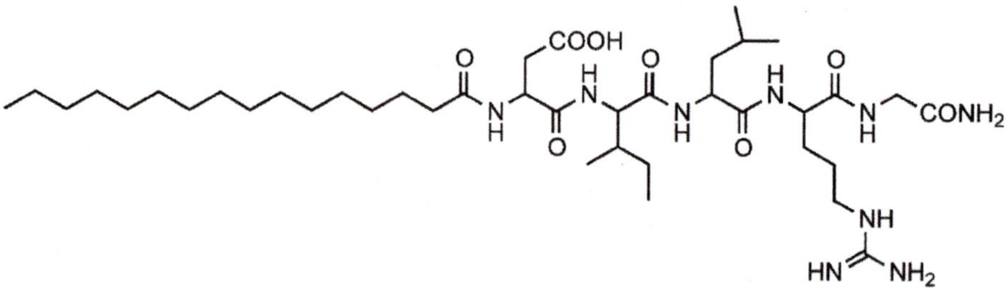

Figure 83.2 Structure of C16-yamamarin.

References

1. Suzuki K, Minagawa T, Kumagai T, et al. Control mechanism of diapause of the pharate first-instar larvae of the silkworm *Antheraea yamamai*. *J Insect Physiol*. 1990;36:855–860.
2. Yang P, Abe S, Zhao Y, et al. Growth suppression of rat hepatoma cells by a pentapeptide from *Antheraea yamamai*. *J Insect Biotech Sericol*. 2004;73:7–13.
3. Yang P, Abe S, Sato Y, et al. A palmitoyl conjugate of an insect pentapeptide causes growth arrest in mammalian cells and mimics the action of diapause hormone. *J Insect Biotech Sericol*. 2007;76:63–69.
4. Kamiya M, Oyauchi K, Sato Y, et al. Structure–activity relationship of a novel pentapeptide with cancer cell growth-inhibitory activity. *J Pept Sci*. 2010;16:242–248.
5. Sato Y, Yang P, An Y, et al. A palmitoyl conjugate of insect pentapeptide Yamamarin arrests cell proliferation and respiration. *Peptides*. 2010;31:827–833.
6. Szymanowska-Dziubasik K, Marciniak P, Rosiński G, et al. Synthesis, cardiositmulatory, and cardioinhibitory effects of selected insect peptides on *Tenebrio molitor*. *J Pept Sci*. 2008;14:708–713.
7. Kuczer M, Dzuibasik K, Łuczak M, et al. The search for new biological activities for selected insect peptides. *Pestycydy*. 2008;1–2:5–11.
8. Kuczer M, Dziubasik K, Midak-Siewirska A, et al. Studies of insect peptides alloferon, *Any*-GS and their analogues. Synthesis and anti-herpes activity. *J Pept Sci*. 2010;16:186–189.
9. Rykaczewska-Czerwińska M, Oleś P, Konopińska D, et al. Pentapeptide Any-GS blocks antinociceptive effect of poneratoxin in rats. *J Agric Sci Tech A*. 2012;2:702–708.

CHAPTER 84

CCHamide

Shinji Nagata

Additional name/Abbreviation: CCHa

CCHamide (CCHa) was found to be a peptide having weak activity on parturition in the tsetse fly Glossina morsitans. *Recent studies suggest a contribution of CCHa to nutritional conditioning and olfactory sense modification.*

Discovery

CCHa is an *in silico* peptide with homology to CCM, having weak activity on parturition in the tsetse fly *Glossina morsitans* [1]. Subsequent transcriptomic analyses in the silkworm *Bombyx mori* confirmed the presence of this CCM ortholog [2], and the peptide was designated CCHa after two intramolecular cysteine residues and two histidine residues.

Structure

Structural Features

CCHa is composed of 13 amino acids forming a cyclic peptide via an S—S bridge of two intramolecular cysteine residues. Additionally, the C-terminal amidated histidine residue is conserved (Figure 84.1).

Structure and Subtype

All transcriptomic and genomic DNA researches in arthropods have revealed the presence of CCHa. In all arthropods, CCHa has two subtypes, CCHa-1 and CCHa-2. The names of the subtypes (CCHa-1 or CCHa-2) are related to their similarity to aa sequences of *Drosophila melanogaster* CCHa-1 and CCHa-2: SCLEYGHSCWGAH—NH$_2$ and GCQAYGHVCYGGH—NH$_2$, respectively [3]. Exceptionally, some crustacean species have only one CCHa.

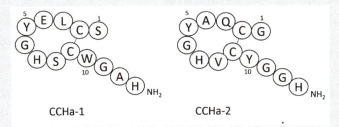

Figure 84.1 Schematic diagram of *D. melanogaster* CCHa-1 and CCHa-2.

Synthesis and Release

Gene, mRNA, and Precursor

There are two CCHamide peptides (CCHa-1 and CCHa-2). CCH-1 is encoded by a different cDNA from that encoding CCH-2.

Distribution of mRNA

Expression of CCHa mRNAs is mainly observed in the brain, CNS, and gut [3]. Interestingly, CCHa-1 and CCHa-2 are expressed in different locations in the gut. In *D. melanogaster*, immunostaining indicates that CCHa-1 is somewhat highly expressed near the anterior and posterior borders of the larval midgut, while CCHa-1 and CCHa-2 are expressed in similar intestinal endocrine cells.

Receptors

Structure and Subtype

The receptors for CCHa-1 and CCHa-2 have been identified in *D. melanogaster* in the course of de-orphanization of *D. melanogaster* GPCRs [3]. There are two subtypes of receptors with high sequence homology. Interestingly, these two homologous GPCRs are activated only by CCHa-1 or CCHa-2 independently, indicating that CCHa-1 and CCHa-2 have individual sensitive receptors [3].

Biological Functions

Target Cells and Function

As CCHa receptors express in the gut and nerve cells, CCHa is thought to contribute to feeding behavior and nutritional sensing [4]. The expression levels of CCHa-1 receptor and CCHa-2 receptor are differently modulated, like the expression levels of their ligands differing according to growth and location. *D. melanogaster* CCHa-2 is not highly expressed compared with CCHa-1 expression levels in the brain, and CCHa-2R shows a very low expression level compared to that of CCHa-1R in the gut. As there is intestinal expression of this peptide, several researches have demonstrated that CCHa plays a role in appetite and food consumption [5].

Phenotype in Gene-Modified Animals

In *D. melanogaster*, CCHa-1R deficient flies have impaired olfactory responses under starvation conditions, indicating that CCHa-1 governs starvation-derived olfactory modification. This phenotype in CCHa-1R deficient flies is demonstrated to be involved with the short neuropeptide F (sNPF) receptor [6].

Y. Takei, H. Ando, & K. Tsutsui (Eds): Handbook of Hormones. DOI: http://dx.doi.org/10.1016/B978-0-12-801028-0.00084-2

References

1. Zdarek J, Nachman RJ, Denlinger DL. Parturition hormone in the tsetse *Glossina morsitans*: activity in reproductive tissues from other species and response of tsetse to identified neuropeptides and other neuroactive compounds. *J Insect Physiol.* 2000;46:213−219.
2. Roller L, Yamanaka N, Watanabe K, et al. The unique evolution of neuropeptide genes in the silkworm *Bombyx mori. Insect Biochem Mol Biol.* 2008;38:1147−1157.
3. Hansen KK, Hauser F, Williumson M, et al. The *Drosophila* genes CG14593 and CG30106 code for G-protein-coupled receptors specifically activated by the neuropeptides CCHamide-1 and CCHamide-2. *Biochem Biophys Res Commun.* 2011;404:184−189.
4. Li S, Torre-Muruzabal T, Sogaard KC, et al. Expression patterns of the *Drosophila* neuropeptide CCHamide-2 and its receptor may suggest hormonal signaling from the gut to the brain. *PLoS One.* 2013;8:e76131.
5. Ida T, Takahashi T, Tominaga H, et al. Isolation of the bioactive peptides CCHamide-1 and CCHamide-2 from *Drosophila* and their putative role in appetite regulation as ligands for G protein-coupled receptors. *Front Endocrinol.* 2012;3:e177.
6. Farhan A, Gulati J, Grobe-Wilde E, et al. The CCHamide 1 receptor modulates sensory perception and olfactory behavior in starved *Drosophila. Sci Rep.* 2013;3:e2765.

Supplemental Information Available on Companion Website

- Accession numbers of CCHa and CCHa receptor/E-Table 84.1

CHAPTER 85

Corazonin

Dušan Žitňan and Ivana Daubnerová

Abbreviation: CRZ

A highly conserved neuropeptide produced by the central nervous system in insects, crustaceans, and ticks, CRZ has been implicated in controlling different species- and stage-specific functions.

Discovery

Corazonin was isolated as a cardioactive peptide from the corpora cardiaca (CC) of the cockroach *Periplaneta americana*, and subsequently identified in many arthropods. This neuropeptide controls cuticle melanization in locusts and crayfish, initiation of the ecdysis sequence in moths, and modulation of stress responses and male reproduction in flies [1–5].

Structure

Structural Features

Corazonin is a very conserved N-terminally blocked undecapeptide amidated at the C-terminus. The aa sequence is conserved in all examined arthropods (insects, crustaceans, and ticks) (Table 85.1). Corazonin is distantly related to vertebrate gonadotropin-releasing hormone (GnRH) and GnRH-like peptide from mollusks [1].

Primary Structure

The sequence pETFQYSRGWTN–NH_2 is very conserved in most arthropods examined. Only locusts and a honey bee contain conserved substitutions of 1–2 aa [2].

Properties

Mr 1,350–1,370; soluble in water, physiological saline, 50–70% methanol or ethanol.

Synthesis and Release

Gene, mRNA, and Precursor

The *crz* gene of the silkworm *Bombyx mori* contains three exons (141, 217, and 487 bp). The mRNA encodes 101-aa precursor protein composed of a signal peptide, corazonin, and a longer unrelated corazonin precursor-related peptide (CPRP) (Figure 85.1). The corazonin mRNA in the fruit fly *Drosophila melanogaster* encodes a 154-aa precursor protein composed of a signal peptide, corazonin, and CPRP. Organization of the precursor in other insects is also very similar. The CPRP sequence is very variable even in related species, and its function (if any) is unknown [2]. Corazonin precursors are similar to the preprohormones for adipokinetic hormone (AKH), each containing a short amidated peptide (AKH) and a longer adipokinetic preprohormone-related peptide (APRP) [2].

Distribution of mRNA

The *crz* gene is expressed in the brain lateral neurosecretory cells and segmental paired neurons in the abdominal ganglia in different insects [6]. It is also present in the central nervous system, pericardial organs and eyestalks of crustaceans, and in the synganglion of ticks [7].

Hemolymph Concentration

Concentrations of 30–80 pM have been detected 20–30 minutes prior to the onset of pre-ecdysis in pharate fifth-instar larvae of the tobacco hornworm *Manduca sexta* [3].

Regulation of Synthesis and Release

Corazonin expression appears to be constitutive. In locusts, external environmental conditions such as high population density and low temperature seem to elicit corazonin secretion from the CC [2,4]. In *D. melanogaster*, corazonin is centrally released from the abdominal interneurons during mating [5].

Receptors

Structure and Subtype

The first corazonin receptor (CRZR) was identified as GPCR in *D. melanogaster* (CG10698). The transcript of 3,230 nt contains six exons (355, 946, 184, 206, 152, 1,387 nt). The open reading frame is composed of 1,740 nt that encode a 579-aa receptor protein [8]. Using *in vitro* expression systems or bioinformatics, orthologous receptors have been identified in representatives of moths (A21), mosquitoes, honey bee, and other insects. Corazonin and its receptor are apparently absent in beetles and the aphid *Acyrthosiphon pisum* [2]. CRZR cDNA in *M. sexta* encodes a protein of 436 amino acids with 7 putative transmembrane domains [3]. It is strongly expressed in Inka cells of *M. sexta* and *B. mori* [3,9]. CRZR is also expressed in a small group of male-specific neurons in the posterior abdominal ganglion of *D. melanogaster* [5]. CRZR shows high sequence homology to the GnRH receptor [1,2].

Y. Takei, H. Ando, & K. Tsutsui (Eds): Handbook of Hormones. DOI: http://dx.doi.org/10.1016/B978-0-12-801028-0.00085-4

Table 85.1 Comparison of Corazonin, Adipokinetic Hormone (AKH), and Gonadotropin-Releasing Hormone (GnRH)

Most arthropod crz	pE—TFQYSRGWTN—NH₂
Locust crz	pE—TFQYSHGWTN—NH₂
Honeybee crz	pE—TFTYSHGWTN—NH₂
B. mori AKH1	pELTFT—SS—WG—NH₂
B. mori AKH2	pELTFT——PGWGQ—NH₂
B. mori AKH3	pEITF——SRDWSG—NH₂
Aplysia (mollusk)	
GnRH-like peptide	pENYHF—SNGWYAG—NH₂
Homo sapiens GnRH	pE——HW—SYGLRP—NH₂

```
          10        20        30        40        50        60
Gam MATNITMFLIVITLTSVAAQTFQYSRGWTNGKRDGHKTEDIRDLTNNLERILSPCQMNKL
Drm MLRLLLLPLFLFTLSMCMGQTFQYSRGWTNGKRSFNAASPL--LANGHLHRASELGLTDL
          70        80        90        100       110
Gam KYVLEGKPLNERLLGPCDTSKTRSTTNPSDTNTSAVKTPCSTHFNKHCYSFSY
Drm -YDLQDWSSDRRLER
```

Figure 85.1 Sequences of corazonon protein precursors of the greater wax moth *Galleria mellonella* and *D. melanogaster*. Identical aa residues are shaded and corazonin sequences underlined.

Biological Functions

Target Cells/Tissues and Function

Corazonin strimulates heartbeat in *P. americana in vitro*, but is inactive *in vivo* and in other insects [1]. Corazonin controls eye pigment migration, melanization of the cuticle, and a shift to the gregarious phase in locusts [2,4]. In the crustacean *Procambarus clarkii* it regulates chromatophore migration in the integument [2]. In moths (*M. sexta* and *B. mori*) it acts on its receptor in Inka cells to elicit the secretion of pre-ecdysis triggering hormone/ecdysis triggering hormone (PETH/ETH) and initiate the ecdysis sequence [3]. In *B. mori* it also suppresses the spinning behavior and production of silk [1,2]. In *D. melanogaster* it mediates metabolic, osmotic, and oxidative stress accompanying locomotor activity [10]. In male flies, corazonin from abdominal interneurons acts on its receptor in a small cluster of posterior serotoninergic neurons to control activity of the accessory glands and sperm ejaculation during mating [5].

Phenotype in Gene-Modified Animals

Ablation of corazonin neurons extends the lifespan in *D. melanogaster* exposed to starvation, osmotic, and oxidative stress. It also increases locomotion and dopamine levels in male flies [10]. Silencing of corazonin or CRZR neurons in male flies elicits infertility and blocks sperm and seminal fluid ejaculation. Conversely, activation of these neurons causes premature ejaculation [5].

References

1. Veenstra JA. Does corazonin signal nutritional stress in insects? *Insect Biochem Mol Biol*. 2009;39:755–762.
2. Boerjan B, Verleyen P, Huybrechts J. In search for a common denominator for the diverse functions of arthropod corazonin: a role in the physiology of stress? *Gen Comp Endocrinol*. 2010;166:222–233.
3. Kim YJ, Cho KH, Žitňanová I. Corazonin receptor signaling in ecdysis initiation. *Proc Natl Acad Sci USA*. 2004;101:6704–6709.
4. Tanaka S. Corazonin and locus phase polyphenism. *Appl Entomol Zool*. 2006;41:179–193.
5. Tayler TD, Pacheco DA, Hergarden AC. A neuropeptide circuit that coordinates sperm transfer and copulation duration in *Drosophila. Proc Natl Acad Sci USA*. 2012;109:20697–20702.
6. Roller L, Tanaka Y, Tanaka S. Corazonin and corazonin-like substances in the central nervous system of the Pterygote and Apterygote insects. *Cell Tissue Res*. 2003;312:393–406.
7. Šimo L, Slovák M, Park Y, Žitňan D. Identification of a complex peptidergic neuroendocrine network in the hard tick, *Rhipicephalus appendiculatus. Cell Tissue Res*. 2009;335:639–655.
8. Park Y, Kim YJ, Adams ME. Identification of G protein-coupled receptors for *Drosophila* PRXamide peptides, CCAP, corazonin, and AKH supports a theory of ligand–receptor coevolution. *Proc Natl Acad Sci USA*. 2002;99:11423–11428.
9. Žitňan D, Adams ME. Neuroendocrine regulation of ecdysis. In: Gilbert LI, ed. *Insect Endocrinology*. London: Academic Press; 2012:253–309.
10. Zhao Y, Bretz CA, Hawksworth SA, et al. Corazonin neurons function in sexually dimorphic circuitry that shape behavioral responses to stress in *Drosophila. PLoS One*. 2010;5:e9141.

Supplemental Information Available on Companion Website

- Accession numbers of mRNAs for corazonin and its receptor in insects/E-Tables 85.1, 85.2

CHAPTER 86

Trypsin-Modulating Oostatic Factor

Yoshiaki Tanaka

Abbreviation: TMOF

Trypsin-modulating oostatic factor (TMOF) inhibits trypsin biosynthesis in the midgut and indirectly inhibits oocyte maturation. TMOFs are isolated from the gray flesh fly and the mosquito, but these two TMOFs show no significant sequence similarity.

Discovery

The factor inhibiting oocyte maturation was first isolated from the ovaries of the mosquito *Aedes aegypti* [1]. Because this peptide also inhibits proteolytic enzyme biosynthesis in the midgut, it was named trypsin-modulating oostatic factor (Aea-TMOF). TMOF has also been isolated from the ovaries of the gray flesh fly *Neobellieria bullata* (Neb-TMOF) [2].

Structure

Structural Features

Aea-TMOF consists of 10 aa residues and contains as many as 7 proline residues, with 6 of them located in a row at the C-terminus of the peptide chain (Table 86.1) [1]. Neb-TMOF is a hexapeptide, and has no structural similarity with Aea-TMOF (Table 86.1) [2]. However, computer modeling demonstrates that, in both molecules, an aromatic amino acid (Tyr^1 for Aea-TMOF and His^6 for Neb-TMOF) sticks out of the molecular axis [2].

Structure and Subtype

No homology has been found with any insect neuropeptides identified so far. However, Aea-TMOF resembles the primary structure of oligoproline-rich regions within a bacterial surface protein [3].

Synthesis and Release

Distribution of Peptide

In adult females, the ovary appears to be the only site of synthesis of Neb-TMOF and of its precursor [4]. Aea-TMOF is produced by the epithelium ovary cells and is transported through hemolymph to the endothelial alimentary tract cells of the midgut [5].

Table 86.1 Structure of TMOFs

A. aegypti	YDPAPPPPP
N. bullata	NPTNLH

Receptors

No receptors have yet been identified.

Biological Functions

Target Cells/Tissues and Functions

Both TMOFs inhibit *de novo* biosynthesis of trypsin in the midgut cells [2,5]. This results in a lack of free amino acids for vitellogenin production in the fat body and, consequently, a termination of oocyte growth [6]. Neb-TMOF is also a very potent inhibitor of ecdysone biosynthesis by the ring gland [7]. Neb-TMOF exerts a strong gonado-inhibiting effect on ovary development and oocyte maturation in the meal worm *Tenebrio molitor* [8].

References

1. Borovsky D, Carlson DA, Griffin PR, et al. Mosquito oostatic factor: a novel decapeptide modulating trypsin-like enzyme biosynthesis in the midgut. *FASEB J.* 1990;4:3015−3020.
2. Bylemans D, Borovsky D, Hunt DF, et al. Sequencing and characterization of trypsin modulating oostatic factor (TMOF) from the ovaries of the grey fleshfly, *Neobellieria (Sarcophaga) bullata.* *Regul Pept.* 1994;50:61−72.
3. Southwick FS, Purich DL. Inhibition of *Listeria* locomotion by mosquito oostatic factor, a natural oligoproline peptide uncoupler of profilin action. *Infect Immun.* 1995;63:182−190.
4. Bylemans D, Verhaert P, Janssen I, et al. Immunolocalization of the oostatic and prothoracicostatic peptide, Neb-TMOF, in adults of the fleshfly, *Neobellieria bullata.* *Gen Comp Endocrinol.* 1996;103:273−280.
5. Borovsky D, Mahmood F. Feeding the mosquito *Aedes aegypti* with TMOF and its analogs; effect on trypsin biosynthesis and egg development. *Regul Pept.* 1995;57:273−281.
6. Zhu W, Vandingenen A, Huybrechts R, et al. *In vitro* degradation of the Neb-Trypsin modulating oostatic factor (Neb-TMOF) in gut luminal content and hemolymph of the grey fleshfly, *Neobellieria bullata.* *Insect Biochem Mol Biol.* 2001;31:87−95.
7. Hua YJ, Bylemans D, De Loof A, et al. Inhibition of ecdysone biosynthesis in flies by a hexapeptide isolated from vitellogenic ovaries. *Mol Cell Endocrinol.* 1994;104:R1−R4.
8. Kuczer M, Wasielewski O, Skonieczna M. Insect oostatic and gonadotropic peptides: synthesis and new biological activities in *Tenebrio molitor* and *Zophobas atratus.* *Pestycydy/Pesticides.* 2004;3−4:25−31.

Supplemental Information Available on Companion Website

- Accession number of TMOF/E-Table 86.1

Y. Takei, H. Ando, & K. Tsutsui (Eds): Handbook of Hormones. DOI: http://dx.doi.org/10.1016/B978-0-12-801028-0.00086-6

CHAPTER 87

Colloostatin

Yoshiaki Tanaka

Additional name: Neb-colloostatin

Colloostatin is an oostatic factor. The name is derived from sequence similarities with parts of several known collagens and its oostatic activity.

Discovery

Colloostatin was discovered as a novel factor with oostatic activity during the purification of trypsin modulating oostatic factor (TMOF) from the ovaries of the gray flesh fly *Neobellieria bullata* [1].

Structure

Structural Features

Colloostatin consists of 19 amino acids (Ser−Ile−Val−Pro−Leu−Gly−Leu−Pro−Val−Pro−Ile−Gly−Pro−Ile−Val−Val−Gly−Pro−Arg−OH), and exhibits structural similarity to known vertebrate and invertebrate collagens [1]. Like collagen, the peptide chain of colloostatin contains characteristic repeats of tripeptide amino acid sequences (Gly−X−X or Val−X−X; Table 87.1).

Structure and Subtype

No colloostatin-like peptide has been identified in other insects. However, the largest structural analogy is observed between colloostatin and the 468−480 sequence of preprocollagen α1 (IV) present in *D. melanogaster* (Table 87.1) [2].

Synthesis and Release

Distribution of Peptide

This peptide is synthesized in the ovaries of the adult *N. bullata* [1].

Table 87.1 Comparison of Colloostatin and Preprocollagen

N. bullata	Colloostatin	SIVPLGLPVPIGPIVVGPR
D. melanogaster	Preprocollagen α 1 (IV)	−SIGPIGHPGPPGPEGQKG−
H. sapiens	Preprocollagen α 1 (VIII)	−−IGPMGIPGPQGGPPGPHG−

Receptors

No receptor has yet been identified.

Biological Functions

Target Cells/Tissues and Functions

Colloostatin inhibits yolk uptake by previtellogenic oocytes as do *Aea*-TMOF and *Neb*-TMOF but, unlike TMOFs, it inhibits neither trypsin biosynthesis in the gut nor ecdysone biosynthesis by larval ring glands [1]. Although no colloostatin-like peptide has been identified from the mealworm *Tenebrio molitor*, colloostatin interferes with vitellogenin production by the fat body as well as with vitellogenin uptake by oocytes through modification of the patency of ovaries [3]. Moreover, injection of colloostatin at physiological concentrations results in significant hemocytotoxicity with a marked increase in apoptotic activity in *T. molitor* hemocytes [4].

References

1. Bylemans D, Proost P, Samijn B, et al. Neb-colloostatin, the second folliculostatin of the grey fleshfly, *Neobellieria bullata*. *Eur J Biochem*. 1995;228:45−49.
2. Kuczer M, Rosiński G, Konopińska D. Insect gonadotropic peptide hormones: some recent developments. *J Pept Sci*. 2007;13:16−26.
3. Wasielewski O, Rosinski G. Gonadoinhibitory effects of Neb-colloostatin and Neb-TMOF on ovarian development in the mealworm, *Tenebrio molitor* L. *Arch Insect Biochem Physiol*. 2007;64:131−141.
4. Czarniewska E, Mrówczyńska L, Kuczer M, et al. The pro-apoptotic action of the peptide hormone Neb-colloostatin on insect haemocytes. *J Exp Biol*. 2012;215:4308−4313.

Supplemental Information Available on Companion Website

- Accession number of colloostatin/E-Table 87.1

Y. Takei, H. Ando, & K. Tsutsui (Eds): Handbook of Hormones. DOI: http://dx.doi.org/10.1016/B978-0-12-801028-0.00087-8

CHAPTER 88

Pigment Dispersing Hormone

Masatoshi Iga

Abbreviation: PDH
Additional names: pigment dispersing factor (PDF), light-adapting hormone (LAH), distal retinal pigment hormone (DRPH).

Pigment dispersing hormone (PDH) regulates the color change of shielding pigments in the compound eye and the epithelial chromatophoral pigment in Crustacea. PDH is also an important factor for regulating circadian rhythms.

Discovery

The first PDH was isolated from the eyestalks of the shrimp *Pandalus borealis* as a light-adapting hormone, and its amino acid sequence was determined in 1976 [1]. The insect ortholog of PDH, pigment dispersing factor (PDF), was first identified from the lubber grasshopper *Romalea microptera* in 1987 [2].

Structure

Structural Features

PDH/PDF is a peptide hormone widely conserved in crustaceans, insects, and nematodes (Table 88.1, E-Table 88.1), while no homologs are present in vertebrates [3]. The identified and deduced crustacean PDHs and insect PDFs are highly homologous octadecapeptides with C-terminal amidation. The crustacean PDHs were classified into α-PDFs and β-PDHs, and the most significant difference was observed at the third and fourth amino acid residues: α-PDH (glycine and methionine) and β-PDH (glutamic acid and leucine).

Primary Structure

β-PDH has been identified in many crustaceans to date, whereas α-PDH has only been identified in *Pandalus* and *Macrobrachium*. Only *P. jordani* has both α- and β-PDH, and several species have multiple copies of β-PDH. Insect PDFs show high similarity to crustacean β-PDH. The silkworm *Bombyx mori* is the only exception, in that the third amino acid residue is not glutamic acid but aspartic acid. PDF-like peptides are present in nematodes, but they have different characters compared to crustacean PDHs and insect PDFs [3]. Nematode PDF-like peptides consist of 20−22 aa residues with low similarity to PDH/PDF.

Synthesis and Release

Gene, mRNA, and Precursor

The mature PDH/PDF peptide is generated from a precursor peptide consisting of 43−182 amino acids. The precursor contains a signal peptide, a PDH/PDF-associated peptide, and a PDH/PDF peptide, except for the cricket *Gryllus bimaculatus*. The mature peptide is cleaved at a mono and/or dibasic cleavage site.

Distribution of mRNA

Immunological studies indicate that PDH is predominantly produced in several cells of the eyestalks and median protocerebrum in crustaceans [3]. In insects, PDF is predominantly expressed in the central nervous system, especially in the brain and abdominal ganglia [4]. However, the presence of abdominal PDF neurons depends on the species. In nematodes, PDF-like peptides are predominantly produced in the nervous system, lectal cells, and posterior arcade cells [5].

Regulation of Synthesis and Release

PDH is released from the sinus glands in crustaceans. Since PDH secretion is stimulated by photoreception, regulation of PDH synthesis and/or release is possibly related to the visual system. According to reports of immunological studies, PDF is released from several neurons in the brain and abdominal ganglia in insects. Three transcription factors (VRILLE, PAR-domain protein 1, and KAYAK-α) have been reported as the transcriptional regulators of PDF in *Drosophila melanogaster* [3,6]. These factors are important components for regulating the circadian rhythm. Nematode PDF-like peptides are mainly expressed in neurons, whereas it is unclear whether the neurons secrete the peptides or not. Involvement of ATF-2 and CES-2 (VRILLE and PAR-domain protein 1 homologs) is reported in the transcriptional regulation of PDF-like peptides in *Caenorhabditis elegans* [5].

Receptors

Structure and Subtype

The receptor for PDH/PDF (PDHR/PDFR) has been identified as a class B (secretin type) GPCR from several species (E-Table 88.2). To date, 2 PDHRs and 11 PDFRs (including PDFR-like receptors) have been identified/annotated with 36−98% similarity. Most species have a single PDH/PDF receptor, whereas the existence of splice variants has been reported in *D. melanogaster* (three variants) and *C. elegans* (three variants). *D. melanogaster* PDFR exhibits weak affinity to diuretic hormone 31 (DH31), whereas *B. mori* PDFR, neuropeptide GPCR-B2 (BNGR-B2), exhibits no response to DH31 [7,8].

Y. Takei, H. Ando, & K. Tsutsui (Eds): Handbook of Hormones. DOI: http://dx.doi.org/10.1016/B978-0-12-801028-0.00088-X

Table 88.1 Amino Acid Sequence Alignment of PDHs and PDFs

α-PDH	*Pandalus borealis*	NSGMINSIL	GIPRVMTEA–NH₂
	Pandalus jordani-II	NSGMINSIL	GIPRVMTEA–NH₂
	Pandalus jordani-III	NSGMINSIL	GIPKVMADA–NH₂
	Macrobrachium rosenbergii	NSGMINSIL	GIPKVMAEA–NH₂
β-PDH	*Uca pugilator*	NSELINSIL	GLPKVMNDA–NH₂
	Penaeus aztecus	NSELINSLL	GIPKVMNDA–NH₂
	Pandalus jordani-I	NSELINSLL	GIPKVMNDA–NH₂
	Daphnia pulex	NSELINSLL	GLPRFMKVV–NH₂
PDF	*Romalea microptera*	NSEIINSLL	GLPKLLNDA–NH₂
	Drosophila melanogaster	NSELINSLL	SLPKNMNDA–NH₂
	Bombyx mori	NADLINSLL	ALPKDMNDA–NH₂
PDF-like	*Caenorhabditis elegans*-1a	SNAELINGLI	GMDLGKLSAV–NH₂
peptide	*Caenorhabditis elegans*-1b	SNAELINGLL	SMNLNKLSGA–NH₂
	Caenorhabditis elegans-2	NNAEVVNHIL	KNFGALDRLG DV–NH₂

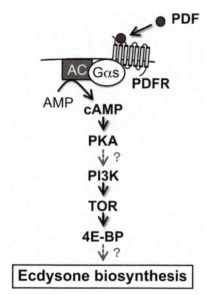

Figure 88.1 PDF signaling model in ecdysone synthesis.

PDF → PDFR

AC / Gαs

AMP → cAMP

PKA

PI3K ?

TOR

4E-BP ?

Ecdysone biosynthesis

Signal Transduction Pathway

PDF-induced intracellular cAMP and Ca^{2+} increase has been reported in insects and nematodes. The relationship between cAMP and Ca^{2+} is not clear; one or both of them might act as the second messenger(s). In contrast, downstream signaling of the second messenger is not fully understood. Recently, the PDF signaling pathway in ecdysone (insect molting hormone) biosynthesis was partially revealed in *B. mori*, and PKA, PI3K, target of rapamycin (TOR), and eukaryotic translation initiation factor 4E (eIF4E)-binding protein (4E-BP) were involved in the downstream signaling of cAMP (Figure 88.1) [8].

Biological Functions

Target Cells and Function

PDH regulates dispersion of the retinal screening pigment and integumental chromatophores, and the phototransduction process and circadian entrainment in crustaceans (Figure 88.2). Insect PDFs can induce pigment dispersion of crustacean chromatophores, whereas insects have no integumental chromatophores. Thus, other functions of PDF in insects were long expected. Currently, a wide variety of PDF-induced physiological phenomena have been reported (Figure 88.2). Regulation of circadian rhythms, locomotor

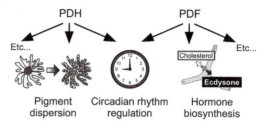

Figure 88.2 PDH/PDF function overview.

activity, courtship behavior, and male sex pheromone production has been reported in *D. melanogaster*. Most of these functions were observed in adults. Conversely, PDF function in the larvae was recently reported in *B. mori*. The PDF affects the prothoracic glands and the ecdysteroidogenic organ in insects, and induces ecdysone biosynthesis [8]. In nematodes, PDF-like peptides affect locomotor behavior [5,9]. In addition to neurohormonal functions, PDHs/PDFs possibly act as neuromodulators and/or neurotransmitters in the nervous system.

Phenotype in Gene-Modified Animals

PDF gene: In *D. melanogaster*, *pdf*-null mutants show abnormal circadian locomotor rhythms and geotaxis [7]. Furthermore, *pdf*-null mutants show low production of male sex pheromones in the oenocytes, which affects mating behavior [10]. In the nematode *C. elegans*, *pdf1*-null mutants and the *pdf2*-overexpressing transgenic line show locomotor defects [9].

PDFR gene: *pdfr*-null mutants show similar phenotypes to those observed in PDF-deficient mutants.

References

1. Fernlund P. Structure of a light-adapting hormone from the shrimp, *Pandalus borealis*. *Biochim Biophys Acta*. 1976;439:17–25.
2. Rao KR, Mohrherr CJ, Riehm JP, et al. Primary structure of an analog of crustacean pigment-dispersing hormone from the lubber grasshopper *Romalea microptera*. *J Biol Chem*. 1987;262:2672–2675.
3. Meelkop E, Temmerman L, Schoofs L, et al. Signalling through pigment dispersing hormone-like peptides in invertebrates. *Prog Neurobiol*. 2011;93:125–147.
4. Helfrich-Förster C. Development of pigment-dispersing hormone-immunoreactive neurons in the nervous system of *Drosophila melanogaster*. *J Comp Neurol*. 1997;380:335–354.
5. Janssen T, Husson SJ, Meelkop E, et al. Discovery and characterization of a conserved pigment dispersing factor-like neuropeptide pathway in *Caenorhabditis elegans*. *J Neurochem*. 2009;111:228–241.
6. Ling J, Dubruille R, Emery P. KAYAK-α modulates circadian transcriptional feedback loops in *Drosophila* pacemaker neurons. *J Neurosci*. 2012;32:16959–16970.
7. Mertens I, Vandingenen A, Johnson EC, et al. PDF receptor signaling in *Drosophila* contributes to both circadian and geotactic behaviors. *Neuron*. 2005;48:213–219.
8. Iga M, Nakaoka Y, Suzuki Y, et al. Pigment dispersing factor regulates ecdysone biosynthesis via *Bombyx* neuropeptide G protein coupled receptor-B2 in the prothoracic glands of *Bombyx mori*. *PLoS One*. 2014;9:e103239.
9. Janssen T, Husson SJ, Lindemans M, et al. Functional characterization of three G protein-coupled receptors for pigment dispersing factors in *Caenorhabditis elegans*. *J Biol Chem*. 2008;283:15241–15249.
10. Krupp JJ, Billeter JC, Wong A, et al. Pigment-dispersing factor modulates pheromone production in clock cells that influence mating in *Drosophila*. *Neuron*. 2013;79:54–68.

Supplemental Information Available on Companion Website

- Accession numbers of PDH and PDF/E-Table 88.1
- Accession numbers of PDHR and PDFR/E-Table 88.2

CHAPTER 89

GLWamide

Toshio Takahashi

Additional names: metamorphosin A (MMA)

The GLWamide family of peptides comprises neuropeptides important for regulating the developmental and physiological processes in cnidarians.

Discovery

The GLWamide family peptides were first isolated as a neuropeptide, termed metamorphosin A (MMA), from the anthozoan *Anthopleura elegantissima* [1]. Subsequently, seven GLWamide family members were independently isolated from *Hydra magnipapillata* [2,3]. A cDNA from the colonial hydroid *Hydractinia echinata* has two GLWamide family peptides [4]. Other GLWamide family peptides are predicted in two other anthozoans: *Actinia equina*, with eight GLWamides, included in MMA; and *Anemonia sulcata*, with five GLWamides, included in MMA [4].

Structure

Structural Features

GLWamides have characteristic structural features in their N- and C-terminal regions. For example, most of the peptides share a GLW−NH$_2$ motif at their C-termini (Table 89.1). In *Hydra*, seven of the GLWamide peptides were found to have a Pro residue at the second position (X−Pro), or at the second and third positions (X−Pro−Pro), of their N-terminal regions (Table 89.1). MMA has a pyroGlu residue at the N-terminus (Table 89.1). Both of these N-terminal structures confer resistance to aminopeptidase digestion [5]. Two further peptides, possibly encoded in the preprohormones of *Actinia* and *Anemonia*, are likely to be processed into GMW−NH$_2$ (Ae-MWamide) and GIW−NH$_2$ (As-IWamide) at their C-termini (Table 89.1). These two may not belong to the GLWamide family, as it has been reported that substitution of the Leu in GLWamide with Met or Ile results in complete abolition of contractile activity in the retractor muscle of *Anthopleura fuscoviridis* [6].

Possible Distribution of the GLWamide Family

Using an antibody against GLW−NH$_2$, GLWamide-like immunoreactivities have been detected in a variety of animals (Figure 89.1). In all cases, neurons were observed to be immunoreactive. While this does not necessarily mean that GLWamide-like peptides are present in these species, it does suggest that, like the RFamide family, GLWamide-like peptides are widely distributed throughout the animal kingdom. In mammals, GLWamide-like immunoreactivity has been observed in the cell bodies of neurons and thin varicose fibers in some regions of the rat brain, suggesting that the rat nervous system contains as yet unidentified GLWamide-like peptides [7].

Biological Functions

Induction of Metamorphosis in Planula Larvae

MMA can induce the metamorphosis of planula larvae of *H. echinata* [1]. All of the GLWamides from *Hydra* also have the ability to induce metamorphosis of *Hydractinia* planula larvae into polyps [2]. Moreover, the shortest possible predicted peptide PPGLW−NH$_2$, as well as (A)KPPGLW−NH$_2$, which are predicted from the cDNA structure of the precursor protein of *Hydractinia*, induce metamorphosis in a dose-dependent

Table 89.1 The GLWamide Family Peptides in Cnidaria

Anthopleura elegantissima	
MMA	pE*QPGLW-NH$_2$
Hydra magnipapillata	
Hym-53	NPYPGLW-NH$_2$
Hym-54	GPMTGLW-NH$_2$
Hym-248	EPLPIGLW-NH$_2$
Hym-249	KPIPGLW-NH$_2$
Hym-331	GPPPGLW-NH$_2$
Hym-338	GPP^{hp}**GLW-NH$_2$
Hym-370	KPNAYKGKLPIGLW-NH$_2$
Hydractinia echinata	
He-LWamide I	pERPPGLW-NH$_2$
He-LWamide I	KPPGLW-NH$_2$
Actinia equina	
Ae-LWamide I	pEQHGLW-NH$_2$
Ae-LWamide II	pENPGLW-NH$_2$
Ae-LWamide III	pEPGLW-NH$_2$
Ae-LWamide IV	PEKAGLW-NH$_2$
Ae-LWamide V	pEIGLW-NH$_2$
Ae-LWamide VI	RSRIGLW-NH$_2$
Ae-MWamide	pEDLDIGMW-NH$_2$
MMA	pEQPGLW-NH$_2$
Anemonia sulcata	
As-LWamide I	pEQAGLW-NH$_2$
As-LWamide II	pEHPGLW-NH$_2$
As-IWamide	pEERIGIW-NH$_2$
Ae-LWamide II	pENPGLW-NH$_2$
MMA	pEQPGLW-NH$_2$

pE*; pyroglutamic acid.
^{hp}P**; hydroxyproline.

Y. Takei, H. Ando, & K. Tsutsui (Eds): Handbook of Hormones. DOI: http://dx.doi.org/10.1016/B978-0-12-801028-0.00089-1

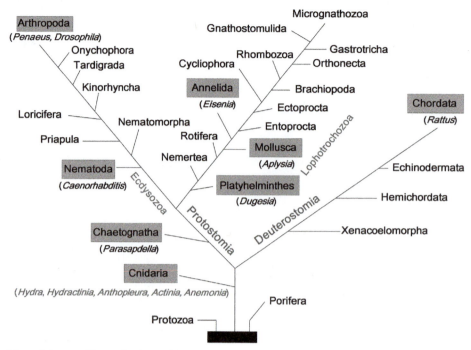

Figure 89.1 **Phyla with anti-GLWamide antibody-positive neurons.** Immunoreactive phyla are shaded gray.

manner [4]. Furthermore, a GLWamide family member in *Hydra*, Hym-248, is capable of inducing metamorphosis of acroporid larvae into polyps at high rates (approximately 100%), with *Acropora* planulae responding to the peptide in a dose-dependent manner [8].

Muscle Contraction

In adult *Hydra*, GLWamides induce detachment of the bud from a parental polyp [2]. In this process, a circular muscle, or sphincter muscle, is formed in the basal disk late in bud development. Contraction of the sphincter muscle results in constriction of the basal disk of the bud, thereby pinching off the bud from the parent. Hym-248 induces not only bud detachment but also elongation of the body column in epithelial hydra, which consists mainly of epithelial cells and lacks multipotent interstitial stem cells and their derivatives, including neurons [3]. In *Hydra*, the muscle processes of ecto-dermal and endodermal epithelial cells run in perpendicular directions; ectoderm processes run along the body axis and endoderm processes circumferentially. Hym-248 causes contraction of endodermal muscles. All GLWamides also induce contraction of the retractor muscle in *A. fuscoviridis* [3].

Planula Migration

Planula larvae have the ability to migrate toward light. This pattern of periodic migration is regulated by KPPGLW—NH$_2$ [9]. The peptide stimulates migration primarily by prolonging the active periods. Since sensory neurons containing GLWamides are present in planula larvae, planula migration and metamorphosis are regulated by the release of endoge-nous neuropeptides in response to environmental cues.

Oocyte Maturation and Spawning

Oocyte maturation and subsequent spawning in hydrozoan jellyfish are generally triggered by the light–dark cycle. The peptide, Hym-53, is effective in inducing oocyte maturation and spawning in the hydrozoan jellyfish *Cytaeis uchidae* [10]. This indicates that a neuropeptide signal transduction path-way is involved in mediating this process in jellyfish.

References

1. Leitz T, Morand K, Mann M. Metamorphosin A: a novel peptide controlling development of the lower metazoan *Hydractinia echinata* (Coelenterata, Hydrozoa). *Dev Biol.* 1994;163:440−446.
2. Takahashi T, Muneoka Y, Lohmann Y, et al. Systematic isolation of peptide signal molecules regulating development in hydra: LWamide and PW families. *Proc Natl Acad Sci USA.* 1997;94: 1241−1246.
3. Takahashi T, Kobayakawa Y, Muneoka Y, et al. Identification of a new member of the GLWamide peptide family: physiological activity and cellular localization in cnidarian polyps. *Comp Biochem Physiol Part B.* 2003;135:309−324.
4. Gajewski M, Schmutzler C, Plickert G. Structure of neuropeptide precursors in cnidaria. *Ann NY Acad Sci.* 1998;839:311−315.
5. Carstensen K, Rinehart KL, McFarlane ID, et al. Isolation of Leu−Pro−Pro−Gly−Pro−Leu−Pro−Arg−Pro−NH$_2$ (Antho-RPamide), an N-terminally protected, biologically active neuro-peptide from sea anemones. *Peptides.* 1992;13:851−857.
6. Takahashi T, Ohtani M, Muneoka Y, et al. Structure−activity rela-tion of LWamide peptides synthesized with a multipeptide synthe-sizer. In: Kitada C, ed. *Peptide Chemistry.* 1997. Osaka: Protein Research Foundation; 1996:193−196.
7. Hamaguchi-Hanada K, Fujisawa Y, Koizumi O, et al. Immunohistochemical evidence for the existence of novel mam-malian neuropeptides related to the Hydra GLW-amide neuro-peptide family. *Cell Tissue Res.* 2009;337:15−25.
8. Iwao K, Fujisawa T, Hatta M. A cnidarian neuropeptide of the GLWamide family induces metamorphosis of reef-building corals in the genus *Acropora. Coral Reefs.* 2002;21:127−129.
9. Katsukura Y, Ando H, David CN, et al. Control of planula migration by LWamide and RFamide neuropeptides in *Hydractinia echinata. J Exp Biol.* 2004;207:1803−1810.
10. Takeda N, Nakajima Y, Koizumi O, et al. Neuropeptides trigger oocyte maturation and subsequent spawning in the hydrozoan jellyfish *Cytaeis uchidae. Mol Reprod Dev.* 2013;80:223−232.

Supplemental Information Available on Companion Website

- Accession numbers of cDNA for GLWamides/E-Table 89.1

CHAPTER 90

Hym-176

Toshio Takahashi

A novel neuropeptide, Hym-176, induces contraction of the ectodermal muscle in Hydra.

Discovery

Hym-176, a novel myoactive neuropeptide, was isolated from *Hydra magnipapillata* during the course of the Hydra Peptide Project [1].

Structure

Structural Features

Hym-176 peptide is a decapeptide characterized by an N-terminal X—Pro sequence and amidation at the C-terminus (Table 90.1). The gene encoding Hym-176 contains another peptide, Hym-357 [2]. Hym-357 shares identical amino acid residues in 5 out of 11 positions with Hym-176 (Table 90.1). Hym-357 was also identified in the Hydra Peptide Project.

Synthesis and Release

cDNA of the Hym-176 Encoding Gene

A full-length cDNA encoding the preprohormone of Hym-176 has been cloned [2]. The preprohormone, which contains a typical signal sequence, contains one copy of the Hym-176 precursor peptide and one copy of the Hym-357 precursor peptide. The Hym-176 precursor sequence is flanked at the N-terminus by Tyr. N-terminal processing likely takes place at the C-terminal side of Tyr [3]. At the C-terminus, there is a possible amidation signal, Gly—Arg, where the basic residue Arg likely serves as a substrate for processing enzymes and Gly serves as an amide donor for amidation enzymes [3].

Distribution of mRNA and Peptide

Immunohistochemistry and *in situ* hybridization analysis revealed that *Hym-176* mRNA and the peptide are expressed in neurons (Figure 90.1), hence Hym-176 can be categorized as a neuropeptide [2]. The Hym-176-encoding gene is expressed intensely in peduncle neurons, and less intensely in neurons scattered throughout the gastric region [2].

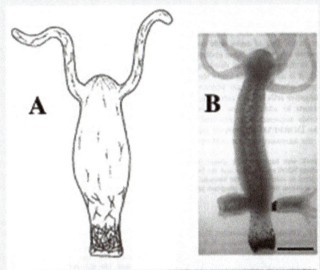
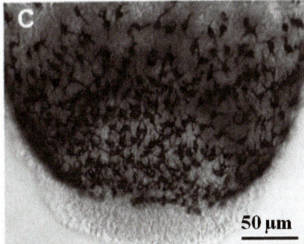

Figure 90.1 Distribution and localization of Hym-176 and its transcripts in a *Hydra* body.

Table 90.1 Structures of Hym-176 and Hym-357

Hym-176 Gene Family	
Hym-176	APFIFPGPKV-NH₂
Hym-357	KPAFLFKGYKP-NH₂

Y. Takei, H. Ando, & K. Tsutsui (Eds): Handbook of Hormones. DOI: http://dx.doi.org/10.1016/B978-0-12-801028-0.00090-8

Biological Functions

Target Cells/Tissues and Functions

Hym-176 specifically and reversibly induces contraction of the ectodermal muscle of the body column, particularly in the peduncle region of epithelial *Hydra*, which lacks all cells of intestinal cell lineage [1]. The expression pattern of *Hym-176* mRNA correlates with the myoactive function of the peptide, suggesting localized function by a localized peptide. Hym-357 strongly induces both tentacle and body contraction in normal *Hydra*, but has no effect on epithelial *Hydra* [4]. This suggests that the peptide acts indirectly on epithelial muscle cells via nerve cells, which in turn release neurotransmitters to directly induce muscle contraction. The receptors for Hym-176 and Hym-357 have not yet been identified.

References

1. Yum S, Takahashi T, Koizumi O, et al. A novel neuropeptide, Hym-176, induces contraction of the ectodermal muscle in *Hydra*. *Biochem Biophys Res Commun*. 1998;248:584−590.

2. Yum S, Takahashi T, Hatta M, et al. The structure and expression of a preprohormone of a neuropeptide, Hym-176, in *Hydra magnipapillata*. *FEBS Lett*. 1998;439:31−34.

3. Grimmelikhuijzen CJP, Leviev I, Carstensen K. Peptide in the nervous systems of cnidarians: structure, function, and biosynthesis. *Int Rev Cytol*. 1996;167:37−89.

4. Fujisawa T. *Hydra* peptide project 1993−2007. *Dev Growth Differ*. 2008;50(Suppl. 1):257−268.

Supplemental Information Available on Companion Website

- Accession number of Hym-176 cDNA/E-Table 90.1

CHAPTER 91

Hym-301

Toshio Takahashi

Additional names: epitheliopeptide

Hym-301 is produced by epithelial cells in Hydra and is therefore also termed epitheliopeptide.

Discovery

Hym-301 is a peptide that was discovered as part of a project aimed at isolating novel peptides from *Hydra magnipapillata* (the *Hydra* Peptide Project) [1].

Structure

Structural Features

Hym-301 is composed of 14 amino acid residues with one intra-molecular disulfide bond, N-terminal X–Pro–Pro, and C-terminal amidation (Figure 91.1) [1]. The N-terminal and C-terminal structures confer resistance to aminopeptidase digestion [2].

Synthesis and Release

cDNA of the Hym-301-Encoding Gene

The Hym-301 precursor protein is a typical preprohormone as described for neuropeptide precursor proteins (Figure 91.2). The coding region contains a signal sequence in the N-terminal region and a single copy of the unprocessed Hym-301 near the C-terminus [3]. A typical dibasic amino acid (Lys–Lys) processing site occurs at both ends of the peptide [3]. It has been shown that the presence of the amino acid Thr is often important for precursor protein processing in cnidarians [4]; thus it is plausible that the removal of Glu and Thr at the N-terminal of the peptide would occur after cleavage of the Lys–Lys site. At the C-terminus of the unprocessed Hym-301, there is a typical sequence for amidation (Gly–Lys–Lys) [3]. Using *H. magnipapillata* ESTs and its genome as a resource, three additional novel genes related to the Hym-301-encoding gene have been identified [5]. Hym-301 homologous genes have not yet been identified in other animals, indicating that this novel gene family is likely to have evolved as taxon-specific.

Distribution of mRNA and Peptide

In situ hybridization analysis has revealed the presence of the Hym-301-encoding gene in epithelial cells, especially in *Hydra* ectodermal epithelial cells, indicating that Hym-301 is an epitheliopeptide [3]. Further examination shows that the ectoderm expressing the Hym-301- encoding gene is localized within the tentacle zone. *Hym-301* is also expressed at the apical end of the body column. *H. magnipapillata* undergoes *de novo* head formation during bud formation and head regeneration. The expression of *Hym-301* during both budding and regeneration indicates that *Hym-301* is expressed relatively early during head formation [3]. Expression of the three novel genes related to the Hym-301- encoding gene is variable. Two genes are expressed in the head, whereas the other is expressed along the body column [5]. It is generally accepted that neuropeptides are localized at the granular core of dense-cored vesicles. A fusion protein of the Hym-301 peptide precursor and green fluorescent protein (GFP) is localized in secretory vesicle-like structures beneath the cell membrane [5]. Furthermore, an immunogold precipitation method (Figure 91.3) was used to obtain direct evidence of Hym-301 intracellular localization; the peptide is stored in vesicles located adjacent to the external surface of ectodermal epithelial cells [6].

Biological Functions

Target Cells/Tissues and Functions

Hym-301 plays a role in head formation, namely in determining the number of tentacles formed [3]. Analysis of the role of Hym-301 in head formation was performed by assaying the effect of the peptide on tentacle formation during head regeneration and by assaying the effect of knockdown of *Hym-301* gene expression by RNAi [3]. Treatment with the peptide results in an increase in the number of tentacles formed. The introduction of *Hym-301* dsRNA into developing buds, in order to block the activity of endogenous *Hym-301*, transiently reduced the number of tentacles formed and concomitantly decreased the levels of *Hym-301* RNA in the developing buds. Khalturin and co-workers [5] produced transgenic polyps that overexpress *Hym-301* in all ectodermal epithelial cells. They revealed that Hym-301 overexpression affects both the speed of tentacle regeneration and the pattern in which tentacles arise [5]. Addition of the peptide significantly increases the number of tentacles formed in the epithelial *Hydra* compared to the control *Hydra*. Epithelial *Hydra* is a preparation of normal *Hydra* devoid of cells of the interstitial cell lineage. Thus, Hym-301 acts directly on the epithelial cells. However, the cellular receptor for Hym-301 remains to be identified.

Tissue Manipulation

Expression of *Hym-301* is restricted to the head region. Thus, to examine whether the expression of *Hym-301* is under the control of a positional gradient, *Hydra* was treated

Y. Takei, H. Ando, & K. Tsutsui (Eds): Handbook of Hormones. DOI: http://dx.doi.org/10.1016/B978-0-12-801028-0.00091-X

Hydra magnipapillata

Hym-301   KPPRRCYLNGYCSP-NH$_2$

Figure 91.1 Hym-301 peptide in *Hydra magnipapillata*.

Hym-301 precursor

N▬■▬▬▬▬▬KK│ET│Hym-301│GKK▬C
  1   18   96

Figure 91.2 Schematic representation of the Hym-301 preprohormone.

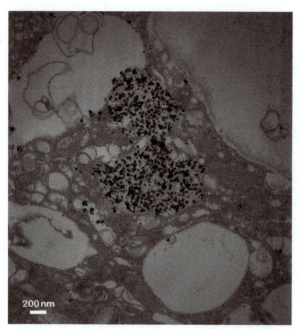

200 nm

Figure 91.3 Immunoelectron microscopy using an anti-Hym-301 antibody.

with 2-mM LiCl. At this concentration, LiCl is known to distalize *Hydra* tissues—i.e., the head region spreads widely, and ectopic head and tentacle formation is induced in the body column [7]. Following 24-hour LiCl treatment, *Hym-301* gene expression spreads throughout the body toward the foot region [8]. Thus, *Hym-301* is under the control of positional signals. Hym-301 may be involved in the Wnt pathway in *Hydra*. Supporting evidence for this possibility emerges from the observation that LiCl treatment inhibits GSK-3β, an important component of the Wnt pathway, resulting in the spread of *Hym-301* gene expression toward the body column.

References

1. Takahashi T, Muneoka Y, Lohmann Y, et al. Systematic isolation of peptide signal molecules regulating development in hydra: LWamide and PW families. *Proc Natl Acad Sci USA*. 1997;94: 1241–1246.
2. Carstensen K, Rinehart KL, McFarlane ID, et al. Isolation of Leu–Pro–Pro–Gly–Pro–Leu–Pro–Arg–Pro–NH$_2$ (Antho-RPamide), an N-terminally protected, biologically active neuropeptide from sea anemones. *Peptides*. 1992;13:851–857.
3. Takahashi T, Hatta M, Yum S, et al. Hym-301, a novel peptide, regulates the number of tentacles formed in *Hydra*. *Development*. 2005;132:2225–2234.
4. Darmer D, Hausen F, Nothacker H-P, et al. Three different prohormones yield a variety of Hydra-RFamide (Arg–Phe–NH$_2$) neuropeptides in *Hydra magnipapillata*. *Biochem J*. 1998;332: 403–412.
5. Khalturin K, Anton-Erxleben F, Sassmann S, et al. A novel gene family controls species-specific morphological traits in *Hydra*. *PLoS Biol*. 2008;6:2436–2449.
6. Takaku Y, Shimizu H, Takahashi T, et al. Subcellular localization of the epitheliopeptide, Hym-301, in *Hydra*. *Cell Tissue Res*. 2013; 351:419–424.
7. Hassel M, Bieller A. Stepwise transfer from high to low lithium concentrations increases the head-forming potential in *Hydra vulgaris* and possibly activates the PI cycle. *Dev Biol*. 1996;177: 439–448.
8. Takahashi T, Fujisawa T. Important roles for epithelial cell peptides in hydra development. *BioEssays*. 2009;31:610–619.

Supplemental Information Available on Companion Website

- Accession number of Hym-301 cDNA/E-Table 91.1
- Primary structure of preproHym-301/E-Figure 91.1

CHAPTER 92

Leech Osmoregulatory Factor

Fumihiro Morishita and Hiroyuki Minakata

Abbreviation: LORF

LORF is an osmoregulatory peptide isolated from the central nervous systems of the leeches Erpobdella octoculata *and* Theromyzon tessulatum. *LORF is also present in the rat brain, coupling to NO release in leech, rat, and human tissues. It is likely that LORF is generated from hemerythrin through cleavage by aspartyl protease.*

Discovery

LORF has been isolated from the central nervous systems of the leeches *Erpobdella octoculata* and *Theromyzon tessulatum*, with both ELISA and dot-blotting assay for oxytocin [1,2]. An identical peptide has been also isolated from rat brain [3].

Structure

Structural Features

Although LORF is immunoreactive against oxytocin antiserum, its structure is different from that of oxytocin [1]. LORF is an octapeptide (Ile−Pro−Glu−Pro−Tyr−Val−Trp−Asp−OH) identified from two annelids, *Erpobdella octoculata* and *Theromyzon tessulatum*. The same peptide was also found in a mammal, *Rattus norvegicus*. This is an interesting fact,

considering the evolutionary distance between annelid and mammal.

Structure and Subtype

In *T. tessulatum*, LORF with a C-terminal amide (Ile−Pro−Glu−Pro−Tyr−Val−Trp−Asp−NH$_2$) has also been identified [2].

Synthesis and Release

Gene, mRNA, and Precursor

The precursor protein mRNA of LORF has yet to be clarified, but the peptide sequence is identical to the N-terminal part of the respiratory pigments myohemerythrin (14 kDa) of the sipunculid *Themiste zostericola* and hemerythrin of *Hirudo medicinalis*, while it is highly homologous to those of hemerythrin and a yolk protein, ovohemerythrin (14 kDa) of the leech *T. tessulatum* (Table 92.1) [1]. If degradation of those proteins liberates LORF, processing of the LORF precursor is different from that of other neuropeptides, because LORF sequences are not flanked by the dibasic or monobasic cleavage sites. It is likely that an aspartyl protease may play an important role in the synthesis of LORF [4].

Table 92.1 Structures of Hemerythrin and Related Proteins of Sipuncula and Annelida*

```
                      10         20         30         40         50
Myohemerythrin  -GWEIPEPYVWDESFRVFYEQLDEEHKKIFKGIFDCIR---DNSAPNLAT
Hemerythrin-1   MGFEIPEPYVWDESFKVFYENLDEEHKGLFQAIFNLSKSPADNGA--LKH
Hemerythrin-2   MVFEIPEPYQWDETFEVFYEKLDEEHKGLFKGIKDLSDSPACSET--LEK
Ovohemerythrin  --YDIPEPFRWDESFKVFYE

                      60         70         80         90        100
Myohemerythrin  LVKVTTNHFTHEEAMMDAAKYSEVVPHKKMHKDFLEKIGGLSAPVDAKNV
Hemerythrin-1   LSKVIDEHFSHEEDMMKKASYSDFDNHKKAHVDFQASLGGLSSPVANDKV
Hemerythrin-2   LVKLIEDHFTDEEEMMKSKSYEDLDSHKKIHSDFVETLKGVKAPVSEENI

                     110        120
Myohemerythrin  DYCKEWNVNHIKGTDFKYKGKL
Hemerythrin-1   HWAKDWLVNHIKGTDFLYKGKL
Hemerythrin-2   KMAKEWLVNHIKGTDFKYKGKL
```

*Amino acid sequence of myohemerythrin of a sipunculid, *Themiste zostericola*, was aligned with hemerythrin-related peptides of Annelida, namely two hemerythrins of *Hirudo medicinalis* and *Theromyzon tessulatum* and ovohemerythrin of *T. tessulatum*. Note that only the N-terminal partial sequence of ovohemerythrin is currently known.

Y. Takei, H. Ando, & K. Tsutsui (Eds): Handbook of Hormones. DOI: http://dx.doi.org/10.1016/B978-0-12-801028-0.00092-1

Distribution of mRNA

Localization of LORF in some neurons on the segmental ganglia of *E. octoculata and T. tessulatum* has been demonstrated by immunostaining [1]. LORF is packaged in the electron-dense secretory granules (~ 100 nm in diameter) in the cell body of neurons in the subesophageal ganglion [2].

Biological Functions

Target Cells and Functions

Injection of LORF or LORF$-$NH$_2$ into the leech provokes an increase in body mass, reflecting an uptake of water; thus, the two peptides exert an antidiuretic effect [2]. Measurement of transepitherial conductance using leech skin suggests that inhibition of Na$^+$ conductance is involved in the antidiuretic action of LORF [1].

It was reported that application of LORF at 10^{-5} M or 10^{-6} M to the human saphenous vein augmented nitric oxide (NO) release from the tissue, which was inhibited by a NO synthase inhibitor, N-nitro-L-arginine [3].

References

1. Salzet M, Bulet P, Weber W-M, et al. Structural characterization of a novel neuropeptide from the central nervous system of the leech *Erpobdella octoculata*. The leech osmoregulator factor. *J Biol Chem*. 1996;271:7237−7243.
2. Salzet M, Vandenbulcke F, Verger-Bocquet M. Structural characterization of osmoregulator peptides from the brain of the leech *Theromyzon tessulatum*: IPEPYVWD and IPEPYVWD-amide. *Mol Brain Res*. 1996;43:301−310.
3. Salzet M, Salzet B, Sautiere P, et al. Isolation and characterization of a leech neuropeptide in rat brains: coupling to nitric oxide release in leech, rat and human tissues. *Mol Brain Res*. 1998;55:173−179.
4. Salzet M. Annelids endocrinology. *Can J Zool*. 2001;79:175−191.

Supplemental Information Available on Companion Website

Accession numbers for hemerythrin of leech and sipunculid/E-Table 92.1

PART III

Lipophilic Hormones in Vertebrates

CHAPTER 93

Thyroid Hormones

Kiyoshi Yamauchi

History

In 1895 Adolf Magnus-Levy found that the feeding of thyroid gland extract to patients increased their metabolic rate [1]. The active compound was isolated from the thyroid glands by Edward Calvin Kendall in 1914 and was named thyroxine (T_4) [2]. T_4 was synthesized in 1927. Later, 3,3′,5-triiodothyronine (T_3), which is physiologically more active than T_4, was isolated from plasma. In 1966, T_3 was found to stimulate DNA-dependent RNA polymerase activity. Nuclear thyroid hormone receptor (TR) cDNAs were cloned in 1986. Thyronamines, decarboxylated and deiodinated metabolites of thyroid hormones (THs), were shown to be potential ligands for a trace amine-associated receptor 1 (TAAR1), a new class of G protein-coupled receptors (GPCRs), in 2004 [3]. A plasma membrane receptor for THs was identified to be on integrin $\alpha_v\beta_3$ in 2005 [4].

Structure

Structural Features

THs are iodothyronines that are derivatives of amino acid tyrosine. T_4 is a major TH that has four iodine atoms while T_3 is less abundant and has three iodine atoms (Figure 93.1, *left*). Other iodothyronines are also present in human plasma: 3,3′,5′-triiodothyronine (rT_3), diiodothyronines (T_2), and monoiodothyronines (T_1). 3,3′,5,5′-Tetraiodothyroacetic acid (Tetrac) and 3,3′,5-triiodothyroacetic acid (Triac) are acetic acid analogs of THs (Figure 93.1, *middle*). 3,5-Diiodo- and 3-monoiodo-thyronamines (T_2AM and T_1AM) and thyronamine (T_0AM) are decarboxylated analogs of iodothyronines (Figure 93.1, *right*).

Iodotyrosines and Iodothyronines in Organisms

THs have the same structure in all vertebrates. However, a variety of iodotyrosines or iodothyronines are present in algae and animals: iodotyrosines in algae, sponges, corals, starfish, mollusks, annelids, crustaceans, and insects; and T_3 and T_4 in echinoderms, mollusks, urochordates, and cephalochordates. In amphioxus, Triac is a ligand for TR [5]. Reference ranges of serum THs and thyrotropin (TSH) in humans are shown in E-Table 93.1.

Receptors

Structure and Subtypes

In jawed vertebrates, there are two subtypes of thyroid hormone receptors (TRα and TRβ) [6] (E-Figures 93.1 and 93.2), which belong to the nuclear receptor superfamily and are named NR1A1 and NR1A2 according to nuclear hormone receptor nomenclature. Like other members of this superfamily, TRs have an A/B domain for ligand-independent transactivation (AF1), a C domain for DNA-binding domain (DBD) that contains ~70 amino acids forming two zinc fingers, a D domain serving as a hinge region, and an E domain for ligand-binding, dimerization, and ligand-dependent transactivation (AF2) (Figure 93.2). TRs have approximately one order of magnitude higher affinity for T_3 ($K_d = 2 \times 10^{-10}$ M) than for T_4. Plasma membrane receptors are involved in non-genomic actions of iodothyronines and iodothyronamines. The extracellular domain of integrin $\alpha_v\beta_3$ acts as a membrane receptor for THs [4], which is the initiation site for T_4-induced activation of intracellular signaling cascades. The receptor for T_1AM, TAAR1 [3], has a structural similarity to other members of TAARs for tyramine, tryptamine, β-phenylethylamine, and octopamine [7].

Signal Transduction Pathway

The major signaling pathway of T_3 is mediated by nuclear TRs, which are ligand-dependent transcription factors. TRs bind to TH-response elements (TREs) in the target genes as a homodimer or heterodimer with retinoid X receptors (RXRs). T_3 directly upregulates gene expression. Unliganded TRs act as repressors by associating with a corepressor complex, whereas liganded TRs act as activators by recruiting a coactivator complex [8]. Unlike TH effects via nuclear TR activation, where its effects manifest over several hours to days, biological effects of iodothyronines or iodothyronamines via membrane receptors are rapidly exerted within seconds or minutes. The TH membrane receptor on the integrin $\alpha_v\beta_3$ activates MAPK (ERK1/2) [3], through which nuclear factors such as TRβ1, estrogen receptor α (ERα), STAT1α, STAT3, and p53 are phosphorylated [9]. The iodothyronamine membrane receptor stimulates cAMP accumulation [3]. Signaling via nuclear TRs is conserved in all vertebrates, although there are only a few studies on signaling via membrane receptors in vertebrates other than mammals.

Biological Functions

THs exert biological effects in most tissues. In general, THs control metabolism (basic metabolic rate, lipid and carbohydrate metabolism, body heat production, and oxygen consumption), growth in children and young animals, and development (fetal and neonatal brain, and animal metamorphosis). THs affect the cardiovascular, nervous, and reproductive systems (E-Table 93.2).

Y. Takei, H. Ando, & K. Tsutsui (Eds): Handbook of Hormones. DOI: http://dx.doi.org/10.1016/B978-0-12-801028-0.00093-3

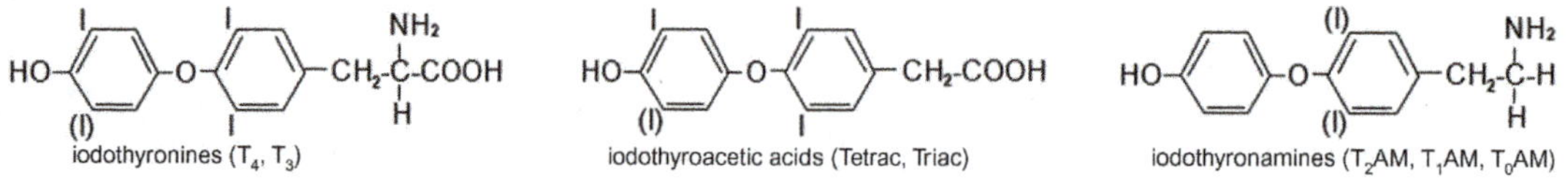

Figure 93.1 Structural formulae of iodothyronines, iodothyroacetic acids and iodothyronamines.

N-[A/B | C | D/E]-C

Figure 93.2 Domain structure of nuclear TRs. In an isoform of human TRα (TRα1), A/B, C, and D/E domains consist of 1–51, 52–119, and 120–410 amino acid residues, respectively.

Target Cells/Tissues and Functions

1. Genomic actions: Direct TH response genes have TREs in regulatory regions, and some of these encode factors involved in the transcription or signaling pathway (S14 in lipogenic tissues of rats, growth hormone in rat pituitary, TSHα and β subunits in the pituitary of mammals) or other important proteins in specific tissues (myelin basic protein in developing brain of rats). These proteins indirectly regulate TH-response genes.

2. Non-genomic actions: Examples of TH actions via membrane receptor on integrin $\alpha_v\beta_3$ are activation of Na^+ channel bursting in rabbit ventriculocytes, inward Na^+ current in cat atrial myocytes and neonatal rat cardiac myocytes, and Na^+/H^+ antiporter in rat skeletal myoblasts [9].

References

1. Magnus-Levy A. Uber den respiratorischen Gaswechsel unter dem Einfluss der Thyroidea sowie unter verschiedenen physiologischen Zustanden. *Berl Klin Wochenschr.* 1895;32:650–652.
2. Kendall EC. The isolation in crystalline form of the compound containing iodin, which occurs in the thyroid: its chemical nature and physiologic activity. *J Am Med Assoc.* 1915;64:2042–2043.
3. Scanlan TS, Suchland KL, Hart ME, et al. 3-Iodothyronamine is an endogenous and rapid-acting derivative of thyroid hormone. *Nat Med.* 2004;10:638–642.
4. Bergh JJ, Lin HY, Lansing L, et al. Integrin $\alpha_v\beta_3$ contains cell surface receptor site for thyroid hormone that is linked to activation of mitogen-activated protein kinase and induction of angiogenesis. *Endocrinology.* 2005;146:2864–2871.
5. Paris M, Escriva H, Schubert M, et al. Amphioxus postembryonic development reveals the homology of chordate metamorphosis. *Curr Biol.* 2008;18:825–830.
6. Wu W, Niles EG, LoVerde PT. Thyroid hormone receptor orthologues from invertebrate species with emphasis on *Schistosoma mansoni. BMC Evol Biol.* 2007;7:150.
7. Borowsky B, Adham N, Jones KA. Trace amines: identification of a family of mammalian G protein-coupled receptors. *Proc Natl Acad Sci USA.* 2001;98:8966–8971.
8. Kato S, Yokoyama A, Fujiki R. Nuclear receptor coregulators merge transcriptional coregulation with epigenetic regulation. *Trends Biochem Sci.* 2011;36:272–281.
9. Davis PJ, Davis FB, Cody V. Membrane receptors mediating thyroid hormone action. *Trends Endocrinol Metab.* 2005;16:429–435.
10. Agency for Toxic Substances and Disease Registry (ATSDR). Toxicological profile for perchlorates. Atlanta, GA, USA; US Department of Health and Human Services, Public Health Service, 2008.
11. Murk AJ, Rijntjes E, Blaauboer BJ, et al. Mechanism-based testing strategy using *in vitro* approaches for identification of thyroid hormone disrupting chemicals. *Toxicol In Vitro.* 2013;27:1320–1346.

Supplemental Information Available on Companion Website

- Alignment of amino acid sequences of DNA-binding and ligand-binding domains of TRs/E-Figure 93.1
- Molecular phylogenetic tree of TRs/E-Figure 93.2
- Reference ranges for serum THs and TSH in humans/E-Table 93.1
- Physiological and pathological processes related to the thyroid hormone system/E-Table 93.2

3,3′,5-Triiodothyronine

Kiyoshi Yamauchi

Abbreviation: T_3
Additional names: liothyronine, liothyronin, tresitope
IUPAC Name: (2S)-2-amino-3-[4-(4-hydroxy-3-iodo-phen-oxy)-3,5-diiodo-phenyl]propanoic acid
CAS No. 6893-02-3
T_3 is an active form of thyroid hormone (TH), which plays an important role in body control, including growth and development, metabolism, body temperature, and heart rate.

Discovery

Jack Gross and Rosalind Pitt-Rivers identified T_3 in human plasma as a more active component than thyroxine (T_4), in 1952 [1]. Rat liver deiodinase 1 was described as the first deiodinase enzyme for THs in 1976 [2].

Structure

Structural Features

T_3 is a derivative of amino acid tyrosine. T_3 has one iodine atom in the phenolic ring and two in the tyrosyl ring.

Properties

Molecular formula, $C_{15}H_{12}I_3NO_4$; Mr 650.97; physical state: solid, odorless, and tasteless; water solubility: 3.96 mg/l at 37°C.

Synthesis and Release

Synthesis

Eighty percent of circulating T_3 is generated from T_4 by deiodinases 1 and 2 (DIO1 and DIO2) in peripheral tissues (E-Table 93A.1). These enzymes are selenocysteine-dependent membrane proteins. However, T_4 is also converted to various metabolites in addition to T_3 by specifc enzymes [3] as shown in Figure 93A.1. Daily production rates of T_3 are 48 nmol/day/70 kg, respectively [4].

Gene and mRNA

DIO1 and *DIO2* are located on 1p33−p32 and 14q24.2−24.3, respectively.

Receptors

Structure and Subtype

Nuclear TH receptors (TRs) function as T_3-dependent transcription factors [5]. TRs are encoded by two genes: *THRA* for TRα and *THRB* for TRβ, which are located on 17q11.2 and 3p24.2, respectively, in humans. By alternative splicing and usage of internal ATG as a translational initiation site, these genes produce variable isoforms in rats and mice (E-Figure 93A.1), five of which are T_3-binding isoforms: α1, p43, β1, β2, and β3 (Figure 93A.2) [6]. The isoform p43 acts as a TR in the mitochondrial matrix. The others lack TH binding activity due to the replacement of the critical TH binding region in the C-terminal region of TRα1, and act as TR antagonists. These isoforms are differentially expressed, showing isoform-specific physiological roles with some level of redundancy. Variable actions of THs are generated by the diversity of TH-response elements (TREs) (Figure 93A.3) and heterodimeric partners of TRs as well as different isoforms of TRs. The core TREs sequence is the hexanucleotide half-site (A/G)GGT(C/A/G)A. TRE half-sites are present in pairs: direct repeat, inverted repeat, and everted repeat. Retinoid X receptor (RXR), vitamin D3 receptor (VDR), and peroxisome proliferator-activated receptor (PPAR) are known as heterodimer partners. In mitochondria, the truncated form of TRα1, p43, binds to one of the highly related TREs in the D-loop region.

Signal Transduction Pathways

After T_3 binding to TRs on TREs in TH-induced target genes, major conformational changes in the helix 12 in the ligand-binding domain occurs. This induces a coregulator switch from a corepressor complex to a coactivator complex. Major corepressors are nuclear receptor corepressor (N-CoR) and silencing mediator of retinoic acid and TR (SMRT), which can associate with histone deacetylase 3 (HDAC3) to form a complex resulting in a compact state of chromatin via HDAC activity. Coactivators, including steroid receptor coactivators (SRC/p160) family and TR-associated proteins (TRAP), have histone acetyltransferase (HAT) activity or can recruit HATs to create a relaxed state of chromatin. This facilitates the recruitment of general transcription factors and RNA polymerase II. Phosphorylation of TRβ1 at serine 142 by MAPK (ERK1/2) also accelerates the coregulator switch.

Agonists

T_3 and Triac are more potent agonists than T_4. A derivative of Triac, GC-1, is a synthetic TRβ-selective agonist [7]. Tetrabromobisphenol A has weak agonist activity.

Y. Takei, H. Ando, & K. Tsutsui (Eds): Handbook of Hormones. DOI: http://dx.doi.org/10.1016/B978-0-12-801028-0.00217-8

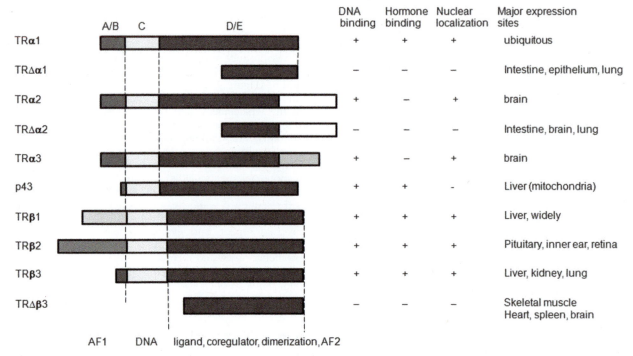

Figure 93A.1 T_3 synthesis and various metabolic pathways of T_4.

	DNA binding	Hormone binding	Nuclear localization	Major expression sites
TRα1	+	+	+	ubiquitous
TRΔα1	−	−	−	Intestine, epithelium, lung
TRα2	+	−	+	brain
TRΔα2	−	−	−	Intestine, brain, lung
TRα3	+	−	+	brain
p43	+	+	−	Liver (mitochondria)
TRβ1	+	+	+	Liver, widely
TRβ2	+	+	+	Pituitary, inner ear, retina
TRβ3	+	+	+	Liver, kidney, lung
TRΔβ3	−	−	−	Skeletal muscle Heart, spleen, brain

Figure 93A.2 Mammalian TR subtypes, and their domain structure and expression sites.

	Example
Direct repeat (DR+4): AGGTCAn₄AGGTCA	rat malic enzyme promoter
Inverted repeat (Pal): AGGTCATGACCT	rat GH promoter
Everted repeat: TGACCTn₆AGGTCA	chicken lysozyme promoter

Figure 93A.3 Three types of TREs.

Antagonists

A derivative of Triac, NH-3, is a synthetic antagonist [7]. Tetrabromobisphenol A shows weak antagonist activity.

Biological Functions

Target Cells/Tissues and Functions

THs control normal growth and development in bone and central nervous system, and regulate lipids in adipose tissue. THs also increase the metabolic rate in most metabolically active tissues in mammals by increasing absorption of carbohydrates from intestine, protein breakdown in muscle, O_2 dissociation from hemoglobin, and O_2 consumption [4]. These TH effects can be divided into two mechanistically different actions: genomic actions mediated by nuclear TRs in the nucleus and nongenomic actions that occur at the plasma membrane, in cytoplasm, and at subcellular organelles. T_3-regulated hepatic and HepG2 genes, which are determined by microarray analyses, are shown in E-Table 93A.2 and E-Figure 93A.2.

Phenotype in Gene-Modified Animals

Mice lacking TRα1 have abnormal heart function and decreased body temperature. Mice lacking both TRα1 and TRα2 have impaired postnatal development and decreased postnatal survival, whereas mice lacking the TRβ gene have mild dysfunction of the pituitary–thyroid axis, and a deficit in auditory function and eye development. Knock-in mice harboring a C-terminal frameshifted TRβ have dysfunction of the pituitary-thyroid axis and the nervous system, abnormal regulation of cholesterol, neurological growth retardation, hearing loss, and thyrotoxic skeletal phenotype [5,6]. In an amphibian, transgenic *Xenopus laevis* overexpressing dominant negative TRα prevents coactivator recruitment. This affects proliferation of the jaw and brain, resorption of the gills and tail, and remodeling of the intestinal tract [8].

Pathophysiological Implications

Clinical Implications

A close relationship is found between more than 347 family members with TH resistance (Refetoff syndrome) and 124 different mutations in the TRβ gene. Symptoms of this disorder are increased TH with non-suppressible thyrotropin, goiter, short stature, decreased weight, tachycardia, hearing loss, attention deficit hyperactivity disorder, and dyslexia, but most patients are heterozygous and are euthyroid. Mutant proteins act as a dominant negative form of TRβ. In contrast, there have been no reports on TRα mutation in humans [5].

Use for Diagnosis and Treatment

The free or total T_4 and/or T_3 levels are routinely measured in the diagnosis of thyroid diseases, including hyperthyroidism, thyroiditis, and hypothyroidism. Patients with TH resistance are usually identified with elevated levels of free or total T_4 and/or T_3 in association with normal or slightly elevated thyrotropin [5]. The most common cause is TRβ gene mutations. However, mutations in monocarboxylate transporter 8 responsible for T_3 uptake and in selenocysteine insertion sequence binding protein 2 that is required for a translational step of selenium-containing deiodinase transcripts have also been associated with this condition. T_3 mimetic eprotirome (KB2115) and sobetirome (GC-1) effectively decrease plasma low-density lipoprotein cholesterol and stimulate bile acid synthesis in humans.

References

1. Gross J, Pitt-Rivers R. The identification of 3:5:3′-L-triiodothyronine in human plasma. *Lancet*. 1952;1:439—441.
2. Visser TJ, Does-Tobé I, Docter R, et al. Subcellular localization of a rat liver enzyme converting thyroxine into tri-iodothyronine and possible involvement of essential thiol groups. *Biochem J*. 1976;157:479—482.
3. Wu SY, Green WL, Huang WS, et al. Alternate pathways of thyroid hormone metabolism. *Thyroid*. 2005;15:943—958.
4. Braverman LE, Utiger RD, (eds). *Werner and Ingbar's The Thyroid: A fundamental and clinical text*. 7th ed. Philadelphia: Lippincott-Raven; 1996.
5. Cheng SY, Leonard JL, Davis PJ. Molecular aspects of thyroid hormone actions. *Endocr Rev*. 2010;31:139—170.
6. Flamant F, Samarut J. Thyroid hormone receptors: lessons from knockout and knock-in mutant mice. *Trends Endocrinol Metab*. 2003;14:85—90.
7. Webb P, Nguyen NH, Chiellini G, et al. Design of thyroid hormone receptor antagonists from first principles. *J Steroid Biochem Mol Biol*. 2002;83:59—73.
8. Furlow JD, Neff ES. A developmental switch induced by thyroid hormone: *Xenopus laevis* metamorphosis. *Trends Endocrinol Metab*. 2006;17:38—45.
9. Huang Y-H, Tsai M-M, Lin K-H, et al. Thyroid hormone dependent regulation of target genes and their physiological significance. *Chan Gung Med J*. 2008;31:325—333.
10. Bianco AC, Kim BW. Deiodinases: implications of the local control of thyroid hormone action. *J Clin Invest*. 2006;116:2571—2579.
11. Feng X, Jiang Y, Meltzer P, et al. Thyroid hormone regulation of hepatic genes in vivo detected by complementary DNA microarray. *Mol Endocrinol*. 2000;14:947—955.

Supplemental Information Available on Companion Website

- TRα and TRβ gene structures in the mouse genome and TR isoforms/E-Figure 93A.1
- Pie chart diagram of expression profiling data of 149 upregulated genes from microarray analysis in HepG2-TRα cells treated with 100 nM T3/E-Figure 93A.2
- Human iodothyronine deiodinases and their properties/E-Table 93A.1
- List of hepatic genes regulated by T3 determined by microarray analyses/E-Table 93A.2

Thyroxine

Kiyoshi Yamauchi

Abbreviation: T_4

Additional names: 3,3′,5,5′-tetraiodothyronine, levothyroxine, synthroid, thyroxin, Thyrax, Thyratabs, Thyreoideum, Thyroxinal, thyroxine iodine

IUPAC Name: *(2S)-2-amino-3-[4-(4-hydroxy-3,5-diiodophenoxy)-3,5-diiodophenyl]propanoic acid*

CAS No. 51-48-9

T_4 is the quantitatively major thyroid hormone (TH), derived from the thyroid gland, but less potent than 3, 3′, 5-triiodothyronine (T_3) in many genomic actions of THs. Although T_4 has been thought to be a prohormone of the active form T_3, it turns out that T_4 functions as a hormone in some non-genomic actions.

Discovery

Edward Calvin Kendall first isolated T_4 in crystalline form in 1914 [1]. Charles Robert Harington determined its chemical formula in 1926 [2], and synthesized it in 1927.

Structure

Structural Features

T_4 has two iodine atoms in the phenolic (outer) ring and two in the tyrosyl (inner) ring.

Properties

Molecular formula, $C_{15}H_{11}I_4NO_4$; Mr 776.87; physical state, white needle-like crystals, odorless, and tasteless; water solubility, 0.105 mg/l at 25°C.

Synthesis and Release

Synthesis

T_4 and a small proportion of T_3 are synthesized by iodination of specific sites on tyrosine residues, subsequent oxidative coupling between the two iodotyrosines on 660 kDa thyroglobulin (TG) molecules at the luminal surface of the thyroid follicle cells, and then processing in lysosome to be released into the bloodstream (Figure 93B.1). Daily production rates of T_4 are 130 nmol/day/70 kg, respectively [3].

Gene and mRNA

The human *TG* gene, located on 8q24, consists of 48 exons and has binding sites for TTF-1 (NKX2A/TITF1), TTF-2 (FOXE2/TITF2), and PAX8 in the promoter region. *TG* mRNA has 8,450 bp and encodes 2,768 residues. The expression site is restricted to the thyroid follicular cells. *TG* transcription is controlled by thyrotropin [3].

Tissue and Plasma Concentrations

Concentrations of T_4 and T_3 in human serum are 86 ng/ml and 1.35 ng/ml, respectively [3]. Plasma T_3 concentrations in other species are comparable to that in humans; however, plasma T_4 concentrations in humans and *Petromyzon marinus* (sea lamprey) are one order of magnitude higher than those in other species (Table 93B.1) [3–6].

Receptors

Structure and Subtype

T_4 can act via nuclear TH receptors (TRs), but its potency is one order of magnitude lower than that of T_3. Therefore, T_3 is the natural ligand of nuclear TRs. Plasma membrane TR for THs was found on integrin $\alpha_v\beta_3$, a heterodimeric transmembrane glycoprotein, belonging to the integrin family. In general, integrins transduce signals between the cell interior and the extracellular matrix (EMC). Integrin $\alpha_v\beta_3$ acts as a receptor for cytotactin, fibronectin, laminin, matrix metalloproteinase-2, osteopontin, osteomodulin, prothrombin, thrombospondin, vitronectin, and von Willebrand factor. An arginine-glycine-aspartic acid (RGD) peptide recognition site on the β_3 chain is essential for the binding of EMC proteins that have an RGD peptide, and also is involved in T_4 binding (Figure 93B.2) [7]. The TH binding domain comprises two binding sites: one is for T_3 and the other is for T_3 and T_4.

Signal Transduction Pathways

In plasma membrane TR on integrin $\alpha_v\beta_3$, T_4 administration induces mitogen-activated protein kinase (MAPK; ERK1/2) activation (Figure 93B.2) and MAPK-dependent nuclear accumulation of phosphorylated ERK1/2 in CV-1 cells that lack nuclear TRs [7]. MAPK-dependent phosphorylation of the TRβ1 DNA-binding domain (DBD) causes shedding of corepressor proteins and recruits coactivator proteins in CV-1 cells transfected with wild-type nuclear TRβ1, altering transcriptional activity of nuclear TR [8]. The involvement of protein kinase C α (PKCα), phospholipase C (PLC), intracellular [Ca^{2+}], AMP-activated protein kinase (AMPK) and Akt/protein kinase B, and phosphoinositide 3-kinase (PI3-K) mediated signaling, are reported as other signaling pathways via plasma membrane TR on the integrin $\alpha_v\beta_3$, nuclear TR in cytoplasm [9], or unidentified initiation sites.

Y. Takei, H. Ando, & K. Tsutsui (Eds): Handbook of Hormones. DOI: http://dx.doi.org/10.1016/B978-0-12-801028-0.00218-X

$H_2O_2 + I^-$

| Iodination by thyroperoxidase |

in thyroid follicle

Tyrosine residues in thyroglobulin molecule

| Coupling & hydrolysis |

in lysosome

L-Thyroxine (T_4)

3,3',5-triiodothyronine (T_3)

Release into bloodstream

Figure 93B.1 Synthetic pathway of T_4 in the thyroid gland.

Table 93B.1 Plasma Thyroid Hormone Levels in Various Vertebrate Species

Class/order	Species	Stages	T_3 (ng/ml)	T_4 (ng/ml)
Agnatha/Cyclostoma	*Petromyzon marinus* [4]	mature female	0.3–1.2	54–108
Osteichthyes/Teleostei	*Oreochromis mossambicus* [5]	during oogenesis	2.7–4.8	3.4–6.5
Amphibia/Anura	*Rana catesbeiana* [6]	metamorphic stages	0.75	5
Mammalia/Primates	*Homo sapiens* [3]	adults	1.35	86

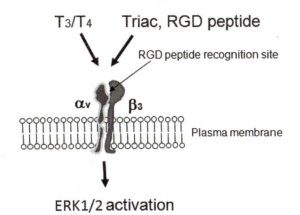

Figure 93B.2 Plasma membrane TR on integrin $\alpha_v\beta_3$.

Agonists

For purified integrin $\alpha_v\beta_3$, T_4 is a more potent agonist than T_3 [7]. EC_{50} for T_4 is 371 pM [7]. GC-1 also acts as an agonist.

Antagonists

For plasma membrane TR on the integrin $\alpha_v\beta_3$, Tetrac, Triac, SB-273005, and RGD-containing peptide are antagonists [9].

Biological Functions

Target Cells/Tissues and Functions

T_4 actions via plasma membrane TR on the integrin $\alpha_v\beta_3$ are detected in mouse glioma (cell proliferation by MAPK) and chicken embryonic chorioallantoic membrane (angiogenesis by MAPK and bFGF2). T_4 also acts on the inward flux of Na^+ in rat neonatal myocardiocytes and activates cell excitability, on actin polymerization via extranuclear TRΔα1 in rat cultured astrocytes and alters cell attachment, and on membrane Ca^{2+}-ATPase activity in human and rat erythrocytes [9].

Phenotype in Gene-Modified Animals

Deiodinase 3 (Dio3) is upregulated by TH (see E-Table 93A.1 in Chapter 93A, 3,3',5-Triiodothyronine). Transgenic *Xenopus laevis* tadpoles overexpressing a GFP-Dio3 fusion protein show resistance to exogenous TH and retardation in their development. Gill and tail resorption is also delayed (Figure 93B.3) [10].

Pathophysiological Implications

Clinical Implications

Plasma membrane TR on the integrin $\alpha_v\beta_3$ stimulates cancer cell proliferation and angiogenesis. The inhibitory effect of

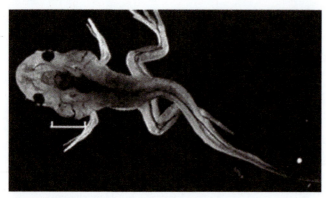

Figure 93B.3 Transgenic tadpole overexpressing a GFP-Dio3 fusion protein. *Cited from [10], with permission.*

nanoparticular Tetrac on angiogenesis in tumors opens routes toward targeted therapies. In a rat model system of experimental myocardial infarction, TH administration increases abundance of new blood vessels [9]. 3,5,-Diiodothyropropionic acid is a potent TR ligand that does not require cellular uptake by the plasma membrane T_3 transporter, monocarboxylate transporter 8. Therefore, 3,5,-Diiodothyropropionic acid might be an effective alternative to TH treatment for those patients with X-linked Allan-Herndon-Dudley syndrome (due to mutations in the *SLC16A2* gene encoding monocarboxylate transporter 8).

Use for Diagnosis and Treatment

The free or total T_4 and/or T_3 levels are routinely measured in the diagnosis of thyroid diseases, including hyperthyroidism, thyroiditis, and hypothyroidism. Anti-thyroid medicine such as carbimazole (used in the UK), methimazole (used in the US), propylthiouracil, and radioactive iodine are used in treatments for hyperthyroidism. Patients with hypothyroidism are treated with a synthetic long-acting form of T_4, known as levothyroxine. Triac and D-T_4 have been used as cholesterol- and lipid-lowering drugs.

References

1. Kendall EC. The isolation in crystalline form of the compound containing iodin, which occurs in the thyroid: its chemical nature and physiologic activity. *J Am Med Assoc.* 1915;64:2042–2043.
2. Harington CR. Chemistry of thyroxine: isolation of thyroxine from the thyroid gland. *Biochem J.* 1926;20:293–299.
3. Braverman LE, Utiger RD (eds). *Werner and Ingbar's The Thyroid: A fundamental and clinical text.* 7th ed. Philadelphia: Lippincott-Raven; 1996.
4. Norris DO, Carr JA. *Vertebrate Endocrinology.* 5th ed. Amsterdam: Academic Press; 2013.
5. Weber GM, Okimoto DK, Richman III NH, et al. Patterns of thyroxine and triiodothyronine in serum and follicle-bound oocytes of the tilapia, *Oreochromis mossambicus*, during oogenesis. *Gen Comp Endocrinol.* 1992;85:392–404.
6. White BA, Nicoll CS. Hormonal control of amphibian metamorphosis. In: Gilbert LI, Frieden E, eds. *Metamorphosis: A Problem in Developmental Biology.* 2nd ed. NY: Plenum Press; 1981.
7. Bergh JJ, Lin HY, Lansing L, et al. Integrin alphaVbeta3 contains a cell surface receptor site for thyroid hormone that is linked to activation of mitogen-activated protein kinase and induction of angiogenesis. *Endocrinology.* 2005;146:2864–2871.
8. Lin HY, Zhang S, West BL, et al. Identification of the putative MAP kinase docking site in the thyroid hormone receptor-beta1 DNA-binding domain: functional consequences of mutations at the docking site. *Biochemistry.* 2003;42:7571–7579.
9. Cheng SY, Leonard JL, Davis PJ. Molecular aspects of thyroid hormone actions. *Endocr Rev.* 2010;31:139–170.
10. Huang H, Marsh-Armstrong N, Brown DD. Metamorphosis is inhibited in transgenic *Xenopus laevis* tadpoles that overexpress type III deiodinase. *Proc Natl Acad Sci USA.* 1999;96:962–967.
11. Davis PJ, Davis FB, Cody V. Membrane receptors mediating thyroid hormone action. *Trends Endocrinol Metab.* 2005;16:429–435.

Supplemental Information Available on Companion Website

- Non-genomic actions of thyroid hormones/E-Table 93B.1

Thyronamines

Kiyoshi Yamauchi

Abbreviation: TAM, T_xAM,

Additional names: (1) thyronamine (T_0AM), IUPAC name, 4-[4-(2-aminoethyl)phenoxy]phenol; CAS No. 500-78-7; (2) 3-monoiodothyronamine (3-T_1AM)

IUPAC name: 2-[4-(4-hydroxyphenoxy)-3-iodophenyl]ethylazanium for 3-T_1AM

CAS No. 712349-95-6

TAMs are decarboxylated and deiodinated metabolites of iodothyronines (THs), a type of trace amines. T_0AM and 3-T_1AM are detected in various tissues and plasma, and have hypothermic and metabolic suppressive effects which may serve to fine-tune or antagonize TH effects.

Discovery

Although there has been some literature regarding the synthesis and biological effects of TAMs since the 1930s, their physiological roles are still not clear. In 2004 [1], TAMs were rediscovered as potential ligands for a new class of G protein-coupled receptor (GPCR) superfamily, which is now referred to as trace amine-associated receptor (TAAR).

Structure

Structural Features

There are nine TAMs depending on the number and the position of iodine atoms [2]. TAMs are designated T_xAMs, with x indicating the number of iodine atoms per molecule (Figure 93C.1). The 3D structure of T_3AM resembles that of T_3, with a twist-skewed diphenyl ether portion.

Properties

For T_0AM, the molecular formula is $C_{14}H_{15}NO_2$; Mr 229.27. For T_1AM, the molecular formula is $C_{14}H_{15}INO_2$; Mr 356.18.

Synthesis and Release

Biosynthesis and Metabolism

TAMs appear to share many of the metabolic reactions of thyronines (see Figure 93A.1 in Subchapter 93A, 3,3′,5-Triiodothyronine). From the structural resemblance, precursors of TAMs are thought to be iodothyronines including THs. There is currently no firm evidence to show what the decarboxylase for iodothyronines is. At least human aromatic L-amino acid decarboxylase fails to catalyze TH decarboxylation. TAMs are demonstrated to be substrates of all mouse deiodinases 1 to 3 (Dio1, Dio2, and Dio3 in Figure 93C.2) [4]. T_0AM, 3-T_1AM, and T_3AM are preferred substrates of human

liver sulfotransferases (SULT1A3, SULT1A1, and SULT1E1), and 3-T_1AM is also oxidatively deaminated by amine oxidases such as monoamine oxidase and semicarbazide-sensitive amine oxidase to generate thyroacetic acids.

Tissue Distribution

Only T_0AM and 3-T_1AM are detected in blood, heart, liver, adipose tissue, thyroid, and brain of adult male C57BL/6 mice, and in brain of Long-Evans rats and guinea pigs [1].

Tissue Content

For T_1AM, Long-Evans rat brain, sub-pmol/g [1]; Wistar rat (female) heart, 68 pmol/g with variation from 1 to 120 pmol/g [2]; human thyroid, skeletal muscle, adipose tissue, and prostate, ~ 60 fmol/ml [2].

Plasma Concentration

For T_1AM (fmol/ml), Djungarian hamster 6 [4]; human 60 [2].

Receptors

Possible Candidates

A receptor for TAMs has not been unequivocally identified, although several models have been proposed. A candidate receptor is trace amine-associated receptor 1 (TAAR1, E-Figure 93C.1), based on the fact that TAMs stimulated the intracellular accumulation of cAMP in HEK-293 cells transfected with rat TAAR1, with EC_{50} of 14 nM [1]. Ligands for TAARs include β-phenylethylamine, tyramine, octopamine, and tryptamine,

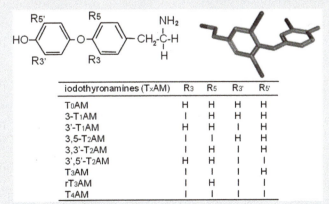

iodothyronamines (T_xAM)	R3	R5	R3′	R5′
T_0AM	H	H	H	H
3-T_1AM	I	H	H	H
3′-T_1AM	H	H	I	H
3,5-T_2AM	I	I	H	H
3,3′-T_2AM	I	H	I	H
3′,5′-T_2AM	H	H	I	I
T_3AM	I	I	I	H
rT_3AM	I	H	I	I
T_4AM	I	I	I	I

Figure 93C.1 Structure and nomenclature of TAMs. R, variable residue; H, hydrogen; I, iodine. 3D structures are obtained from PubChem Compound, CID: 165262 [3].

Y. Takei, H. Ando, & K. Tsutsui (Eds): Handbook of Hormones. DOI: http://dx.doi.org/10.1016/B978-0-12-801028-0.00219-1

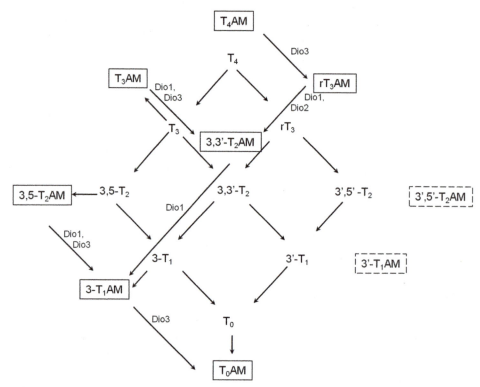

Figure 93C.2 Possible pathways for TAM biosynthesis in mice. T_x and T_xAM, iodothyronines and iodothyronamines, respectively, with x indicating the number of iodine atoms per molecule; Dio1 to Dio3, deiodinases 1 to 3. Dashed lines, T_xAMs that are excluded as precursors of biosynthesis of $3\text{-}T_1AM$ and T_0AM. *Modified from Piehl et al. [2], with permission.*

besides TAMs (E-Figure 93C.2). Another candidate receptor is α_{2A} adrenergic receptor (ARA2A), which shows a high affinity for $3\text{-}T_1AM$ *in vitro* in COS7 cells transfected with human or mouse *ARA2A/Ara2a* gene. The receptor antagonist yohimbine inhibits the hyperglycemic effect of $3\text{-}T_1AM$ [6]. Recent studies propose that TAAR1 and ARA2A might play a concerted role in TAM actions. On the insulin secretion of pancreatic β cells, $3\text{-}T_1AM$ stimulates it via TAAR1 but inhibits it via ARA2A.

Signal Transduction Pathway

Although $3\text{-}T_1AM$ might signal via tyrosine phosphorylation and dephosphorylation, Ca^{2+} and protein kinase C (PKC) might also be involved in TAM signalling [7].

Biological Functions

Target Cells/Tissues and Functions

Although physiological roles of TAMs are not clearly addressed, TAM administration shows some biological effects in mice, rats, or Djungarian hamsters: metabolic depression, hypothermia (~10°C drop), negative chronotropy, negative inotropy, hyperglycemia, reduction of the respiratory quotient (RQ, from 0.90 to 0.7 in hamsters), ketonuria, and reduction of fat mass (Table 93C.1). Among them, negative chronotropy and negative inotropy are direct effects of TAMs [8]. T_1AM is more potent than T_0AM. T_0AM is a negative inotropic agent but not a negative chronotropic agent [7]. Intracerebroventricular infusion indicated TAMs acts more profoundly in the central nervous system on hepatic glucose output and metabolism [8]. In heterogeneous expression systems and primary rat brain synaptosomes, $3\text{-}T_1AM$ acts as an inhibitor of the dopamine transporter, vesicular monoamine transporter 2, and norepinephrine transporter [9].

Table 93C.1   Effects of TAMs in Mammals

Species	Tissues	Functions	Ref.
mice		hypothermia due to a decrease in metabolic rate	[1]
	heart	negative chronotropy, negative inotropy	[1]
mice, hamsters		decreases in metabolic rate, RQ, and fat mass; hyperthermia;	[5]
		ketonuria, changes from carbohydrate to lipid utilization	[5]
rat	pituitary-thyroid	decreases in TSH, T_3, and T_4 levels, and metabolic rate; hyperglycemia,	[8]
	pancreatic islets	hyperglucagonemia, and hypoinsulinemia	[8]

Phenotype of Gene-Modified Animals

In *Ara2a* knockout mice, $3\text{-}T_1AM$ fails to induce hyperglycemia.

Pathophysiological Implications

Clinical Implications

The induction of hypothermia by $3\text{-}T_1AM$ may have potential neuroprotective benefit in the case of ischemic injury such as stroke [10]. Pretreatment of 3-T1AM is associated with reduced infarct size on cardiac ischemic injury [11].

Treatment

$3\text{-}T_1AM$ pretreatment in adult mice subjected to experimentally induced stroke, suggests that $3\text{-}T_1AM$ can be used as an antecedent treatment to induce neuroprotection and cardioprotection against subsequent ischemia [10,11].

References

1. Scanlan TS, Suchland KL, Hart ME, et al. 3-Iodothyronamine is an endogenous and rapid-acting derivative of thyroid hormone. *Nat Med.* 2004;10:638−642.
2. Piehl S, Hoefig CS, Scanlan TS, et al. Thyronamines — past, present, and future. *Endocr Rev.* 2011;32:64−80.
3. National Center for Biotechnology Information. PubChem Compound Database; CID = 165262. <http://pubchem.ncbi.nlm.nih.gov/summary/summary.cgi?cid = 165262>; Accessed 30.03.15.
4. Piehl S, Heberer T, Balizs G, et al. Thyronamines are isozyme-specific substrates of deiodinases. *Endocrinology.* 2008;149:3037−3045.
5. Braulke LJ, Klingenspor M, DeBarber A, et al. 3-Iodothyronamine: a novel hormone controlling the balance between glucose and lipid utilisation. *J Comp Physiol B.* 2008;178:167−177.
6. Regard JB, Kataoka H, Cano DA, et al. Probing cell type-specific functions of Gi in vivo identifies GPCR regulators of insulin secretion. *J Clin Invest.* 2007;117:4034−4043.
7. Chiellini G, Frascarelli S, Ghelardoni S, et al. Cardiac effects of 3-iodothyronamine: a new aminergic system modulating cardiac function. *FASEB J.* 2007;21:1597−1608.
8. Klieverik LP, Foppen E, Ackermans MT, et al. Central effects of thyronamines on glucose metabolism in rats. *J Endocrinol.* 2009;201:377−386.
9. Snead AN, Santos MS, Seal RP, et al. Thyronamines inhibit plasma membrane and vesicular monoamine transport. *ACS Chem Biol.* 2007;2:390−398.
10. Doyle KP, Suchland KL, Ciesielski TM, et al. Novel thyroxine derivatives, thyronamine and 3-iodothyronamine, induce transient hypothermia and marked neuroprotection against stroke injury. *Stroke.* 2007;38:2569−2576.
11. Zucchi R, Ghelardoni S, Chiellini G. Cardiac effects of thyronamines. *Heart Fail Rev.* 2010;15:171−176.
12. Borowsky B, Adham N, Jones KA, et al. Trace amines: Identification of a family of mammalian G protein-coupled receptors. *Proc Natl Acad Sci USA.* 2001;98:8966−8971.

Supplemental Information Available on Companion Website

- A phylogenetic tree for trace amine-associated receptors/E-Figure 93C.1
- Structures of trace amines and classical biogenic amines/E-Figure 93C.2

CHAPTER 94

Gonadal Steroids

Yukiko Ogino, Tomomi Sato, and Taisen Iguchi

History

Vertebrate steroid hormones are gonadal steroids such as progestins, estrogens, and androgens mainly produced in the ovary and testis, and corticoids such as glucocorticoids and mineralocorticoids produced in the adrenal cortex. Chemical structures of steroid hormones were identified from the 1930s to the 1950s after crystallization. Buteandt and co-workers identified the chemical structure of androsterone in 1931, and later the same research group identified the chemical structures of estrone and progesterone. Kendall and colleagues identified the chemical structure of corticosterone in 1937.

Functions

Naturally produced gonadal steroids are estrogens, progestogens, and androgens in vertebrates. Estrogens and progestins are mainly produced in the ovary and androgens are produced in the testis. Estrogens stimulate proliferation of the uterine endometrium, induce LH surge for ovulation as a positive feedback, and inhibit pituitary hormone release via negative feedback. Estrogens are important for female sex differentiation in reptiles, amphibians, and fish species. Progestins function in the maintenance of pregnancy. Androgens are important for differentiation and development of genital organs, spermatogenesis, and development of secondary sex characteristics.

Synthetic Mechanisms

All vertebrate steroid hormones, corticosteroids (21 carbons or C21), progestogens (C21), androgens (C19), and estrogens (C18) are synthesized from cholesterol (C27) by steroidogenic enzymes located in the endoplasmic reticulum and mitochondria. Cholesterol is converted to pregnenolone by cytochrome CYP11A1 in mitochondria, and then converted to progesterone by 3β-hydroxysteroid dehydrogenase (3β-HSD) or metabolized to 17α-hydroxypregnenolone, then to dehydroepiandrosterone (DHEA), which is converted to androst-4-en-3,17-dione. Progesterone is converted to 17α-hydroxyprogesterone, then to androst-4-en-3,17-dione. Androst-4-en-3,17-dione is converted to testosterone (T), which is converted to 5α-dihydrotestosterone (DHT) by 5α-reductase. Androstenedione converted from DHEA, is converted to T by 17β-HSD or to estrone (E1) by aromatase (p450aro). Both T and E1 are converted to estradiol-17β (E2) by P450aro and 17β-HSD, respectively. E2 is metabolized to estriol (E3). Steroidogenic acute regulating protein (StAR) has been found to facilitate the transfer of cholesterol from the outer mitochondrial membrane to the inner membrane where P450scc (side-chain cleaving enzyme) is located (Figure 94.1).

Quantification Method

Antibodies against estrogens, progestins, and androgens are obtainable from reagent companies in general. RIA, a radioimmunoassay with specific antibody, is used routinely to measure many steroids in nanogram and even picogram quantities. HPLC (high-performance liquid chromatography) or TLC (thin-layer chromatography) is used for separation and determination. Currently, LC/MS/MS analysis is also applied for determination of endogenous steroids in biomatrices.

Receptors

Estrogens, androgens, and progestogens bind to nuclear receptors: the estrogen receptor (ER), androgen receptor (AR), and progesterone receptor (PR), respectively. Human ER [1,2], human AR [3], and chicken PR [4] were cloned in 1986, 1988, and 1986, respectively. The ancestral steroid receptor is believed to have been similar to the estrogen-related receptor (ERR) in the lamprey. ER evolved from the ERR. Following the emergence of the ER gene, a corticoid receptor and a receptor for 3-ketogonadal steroids (androgens and progestins) were generated. Following the duplication of the corticoid receptor, the glucocorticoid receptor and mineralocorticoid receptor genes were produced, whereas the 3-ketogonadal steroid receptor was duplicated into PR and AR genes [5]. The ER gene duplicated and diverged into two distinct forms, ERα and ERβ, in mammals, avians, reptiles, amphibians, and ancient fish species. In teleost fish species, three ER subtypes, ERα, ERβ1, and ERβ2, are present. In AR, two AR subtypes, ARα and ARβ, are present in the teleost [6]. The ER, AR, and PR share structural similarities, containing three functional domains (N-terminal A/B transactivation domain; DNA-binding domain; ligand-binding domain). Membrane ER and PR have been reported.

Phenotype of Receptor Gene Disrupted Animals

ERα knockout female mice, generated in 1993, are infertile due to the inability to maintain efficient levels of gonadotropins and lack of sensitivity to E2 treatment [7]. ERβ knockout female mice, generated in 2003, exhibit reduced ovulation rates [8]; however, ERβ knockout male mice are normal. After superovulation, the ovary of ERα knockout and ERβ knockout mice exhibited a reduced number of corpora lutea and a number of unruptured antral follicles. ERα and ERβ double

Y. Takei, H. Ando, & K. Tsutsui (Eds): Handbook of Hormones. DOI: http://dx.doi.org/10.1016/B978-0-12-801028-0.00094-5

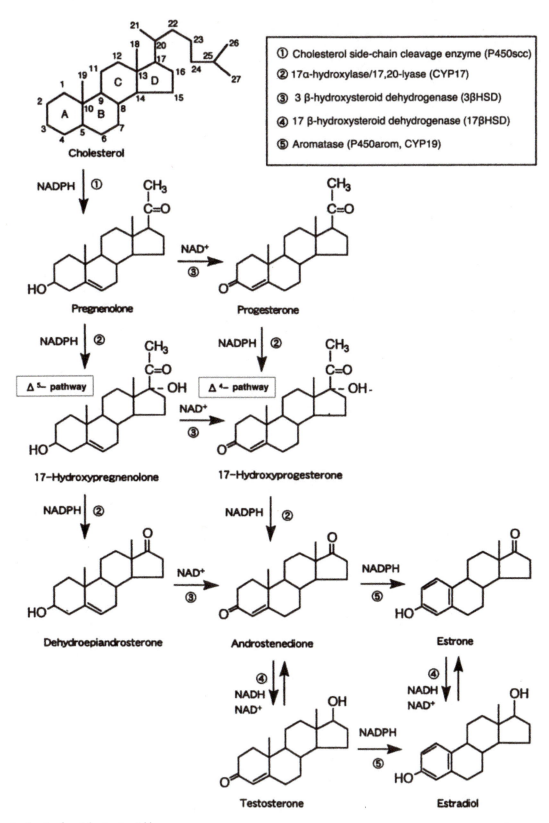

① Cholesterol side-chain cleavage enzyme (P450scc)
② 17α-hydroxylase/17,20-lyase (CYP17)
③ 3 β-hydroxysteroid dehydrogenase (3βHSD)
④ 17 β-hydroxysteroid dehydrogenase (17βHSD)
⑤ Aromatase (P450arom, CYP19)

Figure 94.1 Synthesis of vertebrate steroid hormones.

knockout female mice generated in 1999 are infertile due to a lack of ovulation [9]. AR knockout mice were generated in 2002 [10]. Thereafter, cell specific AR knockout mice models were created. The ubiquitous AR knockout male mice have female-like external genitalia, small undescended testes, a blind ending vagina, and absent accessory sex organs; and germ cell development was severely disrupted, which was similar to human complete androgen insensitivity syndrome

(CAIS). The female AR knockout mice had considerable reproductive defects, with decreased fertility, defective follicular development, reduced ovulation, and premature ovarian failure. PR knockout female mice generated in 1995 are infertile due to a failure of ovulation [10].

References

1. Green GL, Gilna P, Waterfield M, et al. Sequence and expression of human estrogen receptor complementary DNA. *Science.* 1986;231:1150−1154.
2. Green S, Walter P, Kumar V, et al. Human oestrogen receptor cDNA: sequence, expression and homology to v-erb-A. *Nature.* 1986;320:134−139.
3. Ham J, Thomson A, Needham M, et al. Characterization of response elements for androgens, glucocorticoids and progestins in mouse mammary tumour virus. *Nucleic Acids Res.* 1988;16:5263−5276.
4. Conneely OM, Sullivan WP, Toft DO, et al. Molecular cloning of the chicken progesterone receptor. *Science.* 1986;233:767−770.
5. Thornton JW. Evolution of vertebrate steroid receptors from an ancestral estrogen receptor by ligand exploitation and serial genome expansions. *Proc Natl Acad Sci USA.* 2001;98:5671−5676.
6. Ogino Y, Katoh H, Kuraku S, et al. Evolutionary history and functional characterization of androgen receptor genes in jawed vertebrates. *Endocrinology.* 2009;150:5415−5427.
7. Lubahn DB, Moyer JS, Golding TS, et al. Alteration of reproductive function but not prenatal sexual development after insertional disruption of the mouse estrogen receptor gene. *Proc Natl Acad Sci USA.* 1993;90:11162−11166.
8. Dupont S, Krust A, Gransmuller A, et al. Effect of single and compound knockouts of estrogen receptors alpha (ERalpha) and beta (ERbeta) on mouse reproductive phenotypes. *Development.* 2000;127:4277−4291.
9. Couse JF, Hewitt SC, Bunch DO, et al. Postnatal sex reversal of the ovaries in mice lacking estrogen receptors alpha and beta. *Science.* 1999;286:2328−2331.
10. De Gendt K, Verhoeven G. Tissue- and cell-specific functions of the androgen receptor revealed through conditional knockout models in mice. *Mol Cell Endocrinol.* 2012;352:13−25.

Progesterone

Tomomi Sato, Shinichi Miyagawa, and Taisen Iguchi

Abbreviation: P4
Additional names: pregn-4-ene-3,20-dione, 4-pregnene-3,20-dione, Δ^4-pregnene-3,20-dione, luteohormone

Progesterone (P4) is a steroid hormone and an intermediate of sex steroids produced by all steroidogenic tissues. P4 maintains pregnancy and the secretory condition of the uterine endometrium during the luteal phase, and inhibits the release of gonadotropins.

Discovery

In 1934, progesterone was isolated from the corpus luteum of a pregnant sow's ovaries.

Structure

Structural Features

P4 is a C21 steroid hormone, with a double bond in the A ring and a ketone on carbon 3 (Figure 94A.1).

Properties

Molecular formula: $C_{21}H_{30}O_2$, Mr 314.46. Two crystalline forms of equal physiological activity exist. Soluble in alcohol, acetone, dioxane, and concentrated H_2SO_4. Insoluble in water.

Synthesis and Release

Gene, mRNA, and Precursor

Progesterone is converted from pregnenolone by Δ^5,3β-hydroxysteroid dehydrogenase (3β-HSD) enzyme in the granulosa cells and luteal cells of the ovary, Leydig cells of the testis, zona fasciculata and zona glomerulosa of the adrenal gland, and the placenta (E-Figure 94A.1). The synthesis of P4 varies in different mammalian species. Chemical names and properties of progestogens are found in E-Table 94A.1.

Tissue Content

Plasma Concentration (ng/ml (nmol/l))

Human (male) mature 0−0.48 (0−1.5), (female) mature follicular phase 0−5.3 (0−17), luteal phase 6−21 (19−67), early pregnancy 40 (127), before parturition 80−250 (254−795).

Regulation of Synthesis and Release

The synthesis and release of progesterone in the theca and luteal cells of the ovary are regulated by LH.

Receptors

Structure and Subtype

The progesterone receptor (PR) is a member of the nuclear receptor superfamily of transcription factors (E-Figure 94A.2). Human PR has two natural isoforms, PRA and PRB, which are transcribed from the same gene located on chromosome 11q22−q23. The PRA consists of 769 amino acids (aa) residues and the molecular weight is 94,000, while PRB consists of 933 aa. residues and the molecular weight is 114,000 (E-Figure 94A.3). The cDNA of human PR was identified in 1987 [1], and mouse PR was subsequently isolated in 1991 [2]. PRs are organized into six regions of varying homology denoted A−F. The N-terminal A/B domain is important for stimulating transcription from certain progesterone-responsive promoters without ligand. It contains a highly variable transcriptional activation function-1 (TAF-1) domain situated near the N terminus. The C domain (the DNA-binding domain, DBD), composed of two zinc fingers, binds to progesterone response elements (PREs) in the target DNA [3]. The D domain is a hinge region containing the nuclear localization signal. The E domain, containing the ligand-binding domain (LBD) and a highly conserved TAF-2 domain, transactivates the gene expression in response to the ligand [3]. Membrane-associated, progesterone-specific receptors are also isolated and these receptors can mediate non-genomic responses to progesterone [4].

Signal Transduction Pathway

The PR is located in either the cytoplasm or the nucleus, bound by chaperone proteins (hsp90, hsp70, and hsp59) when P4 is absent. Once P4 has bound to PR, the PR homodimerizes and binds to specific DNA sequences (PREs) within the promotor region of target genes to initiate transcription. PR can also be activated, travel to the nucleus, and activate the target genes in the absence of ligand [5].

Agonist

19-Nortestosterone derivatives (norethisterone levonorgestrel, desogestrel, gestodene) combined with estrogen are used as oral contraceptives for preventing ovulation (E-Figure 94A.4).

Antagonist

Mifepristone (RU486) binds to PR and antagonizes progesterone to inhibit myometrial contractility. It provides a means of nonsurgical abortion in the early stages of a pregnancy (E-Figure 94A.4).

Y. Takei, H. Ando, & K. Tsutsui (Eds): Handbook of Hormones. DOI: http://dx.doi.org/10.1016/B978-0-12-801028-0.00220-8

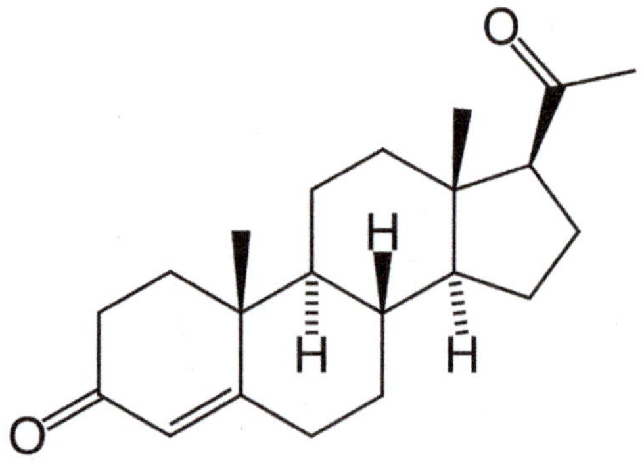

Figure 94A.1 Structure of P4.

Table 94A.1 Functions of Progesterones

Species	Function	Ref.
Mouse	Inhibition of epithelial cell proliferation in the uterus before treatment with estradiol	[6,7]
Mouse	Sensitization of the stromal cells to respond to estradiol with increased mitosis	[8]
Mouse	Induction of mitosis in the epithelium and stroma of the uterus	[8]
Chicken	Synthesis of a specific protein, avidin, in the minced oviduct from estrogen-treated chickens	[9]
Xenopus	Induction of maturation in oocytes via a membrane-bound receptor coupled to heterotrimeric G-proteins	[10]

Biological Functions

Target Cells/Tissues and Functions

PR is expressed specifically in granulosa cells of preovulatory follicles of the ovary and is required specifically for LH-dependent follicular rupture leading to ovulation (Table 94A.1) [6–10]. In the luteal phase, progesterone from the luteal cells prevents ovulation and the overproliferation of the endometrial tissue. PR is expressed in the epithelial, stromal, and myometrial cells of the uterus, controlled by estrogen. In the mammary gland, PR is found in the epithelial and stromal cells. PR is co-localized with ERα in the hypothalamus, medial preoptic area, and arcuate nucleus. Although only estradiol can induce the LH surge, estradiol induces the expression of PRA in the hypothalamus, which can activate kisspeptin neurons in the anteroventral periventricular nucleus [11].

Phenotype of Gene-Modified Animals

PR knockout female mice are infertile due to a failure of ovulation despite exposure to superovulatory levels of gonadotropins [12]. PRA knockout female mice exhibit severely reduced ovulation rates and implantation defects, while PRB is required for normal proliferative responses of the mammary gland to progesterone [13,14]. In the uterus of PR knockout mice, hyperplasia of the epithelium due to the inhibitory action of estrogen is observed, whereas the stromal cells are hypocellular by the absence of a stromal proliferative signal after estrogen and progesterone treatment. PR knockout male mice exhibit lower FSH levels, higher inhibin levels, larger testes, and a significantly increased sperm count in the cauda epididymis [15].

Pathophysiological Implications

Clinical Implications

Progesterone is an intermediate in steroidogenesis, and converting enzyme (3β-HSD) deficiency is both rare and fatal.

Use for Diagnosis and Treatment

Progesterone has been approved for the treatment of irregular and anovulatory menstrual cycles. Progesterone combined with estrogen is used for hormonal contraception and the prevention of endometrial hyperplasia in postmenopausal hormonal replacement therapy. It is also used without estrogen as a progestin-only contraceptive agent and in alternative treatment for early stages of endometrial cancer (at high dose) in premenopausal women.

References

1. Misrahi M, Atger M, d'Auriol L, et al. Complete amino acid sequence of the human progesterone receptor deduced from cloned cDNA. *Biochem Biophys Res Commun*. 1987;143:740–748.
2. Schott DR, Shyamala G, Schneider W, et al. Molecular cloning, sequence analyses, and expression of complementary DNA encoding murine progesterone receptor. *Biochemistry*. 1991;30:7014–7020.
3. Tsai M-J, O'Malley BW. Molecular mechanisms of action of steroid/thyroid receptor superfamily members. *Annu Rev Biochem*. 1994;63:451–486.
4. Bramley T. Non-genomic progesterone receptors in the mammalian ovary: some unresolved issues. *Reproduction*. 2003;125:3–15.
5. Li X, O'Malley BW. Unfolding the action of progesterone receptors. *J Biol Chem*. 2003;278:39261–39264.
6. Martin L, Das RM, Finn CA, et al. The inhibition by progesterone of uterine epithelial proliferation in the mouse. *J Endocrinol*. 1973;57:549–554.
7. Das RM, Martin L. Progesterone inhibition of uterine epithelial proliferation. *J Endocrinol*. 1973;59:205–206.
8. Martin L, Finn CA. Duration of progesterone treatment required for a stromal response to oestradiol-17β in the uterus of the mouse. *J Endocrinol*. 1969;44:279–280.
9. O'Malley BW, McGuire WL. Studies on the mechanism of action of progesterone in regulation of the synthesis of specific protein. *J Clin Invest*. 1968;47:654–664.
10. Masui Y. Relative roles of the pituitary, follicle cells, and progesterone in the induction of oocyte maturation in *Rana pipiens*. *J Exp Zool*. 1967;166:365–375.
11. Sinchak K, Wagner EJ. Estradiol signaling in the regulation of reproduction and energy balance. *Front Neuroendocrinol*. 2012;33:342–363.
12. Lydon JP, DeMayo FJ, Funk CR, et al. Mice lacking progesterone receptor exhibit pleiotropic reproductive abnormalities. *Genes Dev*. 1995;9:2266–2278.
13. Mulac-Jericevic B, Mullinax RA, DeMayo FJ, et al. Subgroup of reproductive functions of progesterone mediated by progesterone receptor-B isoform. *Science*. 2000;289:1751–1754.
14. Ismail PM, Amato P, Soyal SM, et al. Progesterone involvement in breast development and tumorigenesis—as revealed by progesterone receptor "knockout" and "knockin" mouse models. *Steroids*. 2003;68:779–787.
15. Lue Y, Wang C, Lydon JP, et al. Functional role of progestin and the progesterone receptor in the suppression of spermatogenesis in rodents. *Andrology*. 2013;1:308–317.

Supplemental Information Available on Companion Website

- Synthesis of steroid hormones from cholesterol/E-Figure 94A.1
- The structure of the steroid receptor family/E-Figure 94A.2
- The structure of the human progesterone receptor/E-Figure 94A.3
- Agonists and antagonists of progesterone/E-Figure 94A.4
- Chemical names and characters of progestogens/E-Table 94A.1

17,20β-Dihydroxy-4-pregnen-3-one

Yukiko Ogino, Shinichi Miyagawa, and Taisen Iguchi

Additional names: 17α,20β-Dihydroxy-4-pregnen-3-one, 17α,20β-dihydroxyprogesterone (17α,20β-DP), 17α,20β-DHP (DHP), 4-pregnen-17α,20β-diol-3-one, maturation inducing hormone (MIH), maturation inducing steroid (MIS)

17,20β-Dihydroxy-4-pregnen-3-one (17α,20β-DP) was identified as a maturation-inducing hormone (MIH) in several teleost fish species [1,2]. Among various C21 steroids, 17α,20β-DP is the most effective steroid in the induction of germinal vesicle breakdown (GVBD) during oocyte maturation in many fish species [2].

Discovery

In the 1980s, 17α,20β-DP was identified as an MIH in several teleost fish species [1].

Structure OF 17,20 β-Dihydroxy-4-Pregnen-3-One

Structural Features

17,20β-Dihydroxy-4-pregnen-3-one is a C21 steroid secreted by the ovarian follicular cells. See Figure 94B.1.

Properties

Molecular formula: $C_{21}H_{32}O_3$, Mr 332. Melting point: 206−209°C; soluble in alcohol, acetone and other organic solvents. Appearance: white crystalline powder.

Synthesis and Release

The hydroxylase activity of CYP17 (P450c17) and 20β-hydroxysteroid dehydrogenase (20β-HSD) is required for the production of 17α,20β-DP [3].

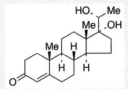

Figure 94B.1 Structure of 17,20β-dihydroxy-4-pregnen-3-one.

Plasma Concentration (ng/ml)

Amago salmon (vitellogenic females) ≤ 2; amago salmon (mature and ovulated females) 50−70; amago salmon (mature males during the period of active spermiation) ≥ 8 [4,5].

Regulation of Synthesis and Release

Pituitary gonadotropin stimulates thecal cells to produce 17α-hydroxyprogesterone, which is converted within the follicular cells to 17α,20β-DP [3].

Receptors

Structure and Subtype

Specific binding of 17α,20β-DP to plasma membrane of the rainbow trout oocyte was demonstrated in 1993 [6]. In several teleost fish species, G protein-coupled membrane-bound MIH receptors have been shown to mediate non-genomic actions of 17α,20β-DP [1]. Nuclear 17α,20β-DP receptor (DPR) was identified from the testis in Japanese eel, which showed 17α,20β-DP-dependent activation of transcription [7].

Biological Functions

Target Cells/Tissues and Functions

17α,20β-DP is produced in the ovarian follicle under the influence of pituitary gonadotropin. 17α,20β-DP induces oocyte maturation by stimulating the *de novo* synthesis of cyclin B, a regulatory subunit of maturation-promoting factor (MPF) [8]. The testicular production of 17α,20β-DP is responsible for the acquisition of sperm motility in salmonid fish [9]. In goldfish, 17,20β-DP released into the water during ovulation stimulates, in the male, gonadotropin release and milt production prior to spawning [10].

Pathophysiological Implications

Clinical Implications

None.

Use for Diagnosis and Treatment

None.

Y. Takei, H. Ando, & K. Tsutsui (Eds): Handbook of Hormones. DOI: http://dx.doi.org/10.1016/B978-0-12-801028-0.00221-X

References

1. Nagahama Y. 17α,20β-Dihydroxy-4-pregnen-3-one, a maturation-inducing hormone in fish oocytes: mechanisms of synthesis and action. *Steroids.* 1997;62:190−196.
2. Miwa T, Yoshizaki G, Naka H, et al. Ovarian steroid synthesis during oocyte maturation and ovulation in Japanese catfish (*Silurus asotus*). *Aquaculture.* 2001;198:179−191.
3. Nagahama Y, Yamashita M. Regulation of oocyte maturation in fish. *Devel Growth Differ.* 2008;50:S195−S219.
4. Young G, Crim LW, Kagawa H, et al. Plasma 17α,20β-dihydroxy-4-pregnen-3-one levels during sexual maturation of amago salmon (*Oncorhynchus rhodurus*): correlation with plasma gonadotropin and *in vitro* production by ovarian follicles. *Gen Comp Endocrinol.* 1983;51:96−105.
5. Ueda H, Young G, Crim LW, et al. 17α,20β-Dihydroxy-4-pregnen-3-one: plasma levels during sexual maturation and in vitro production by the testes of amago salmon (*Oncorhynchus rhodurus*) and rainbow trout (*Salmo gairdneri*). *Gen Comp Endocrinol.* 1983;51:106−112.
6. Yoshikuni M, Shibata N, Nagahama Y. Specific binding of [³H] 17α, 20β-dihydroxy-4-pregnen-3-one to oocyte cortices of rainbow trout (*Oncorhynchus mykiss*). *Fish Physiol Biochem.* 1993;11:15−24.
7. Todo T, Ikeuchi T, Kobayashi T, et al. Characterization of a testicular 17α, 20β-dihydroxy-4-pregnen-3-one (a spermiation-inducing steroid in fish) receptor from a teleost, Japanese eel (*Anguilla japonica*). *FEBS Lett.* 2000;465:12−17.
8. Katsu Y, Yamashita M, Nagahama Y. Translational regulation of cyclin B mRNA by 17α,20β-dihydroxy-4-pregnen-3-one (maturation-inducing hormone) during oocyte maturation in a teleost fish, the goldfish (*Carassius auratus*). *Mol Cell Endocrinol.* 1999;158:79−85.
9. Miura T, Yamauchi K, Takahashi H, et al. The role of hormones in the acquisition of sperm motility in salmonid fish. *J Exp Zool.* 1992;261:359−363.
10. Stacey NE, Sorensen PW, Van der Kraak GJ, et al. Direct evidence that 17α,20β-dihydroxy-4-pregnen-3-one functions as a goldfish primer pheromone: preovulatory release is closely associated with male endocrine responses. *Gen Comp Endocrinol.* 1989;75:62−70.

Supplemental Information Available on Companion Website

- 17,20β-dihydroxy-4-pregnen-3-one synthetic pathway/E-Figure 94B.1

17,20β,21-Trihydroxy-4-pregnen-3-one

Yukiko Ogino, Shinichi Miyagawa, and Taisen Iguchi

Additional names: 20β-S,17,20β,21-P,20β-dihydro-11-deoxy-cortisol, maturation inducing steroid (MIS)

17,20β,21-Trihydroxy-4-pregnen-3-one (20β-S) is a major maturation-inducing steroid (MIS) produced in the ovaries of teleost fish including the Atlantic croaker, spotted seatrout, and bambooleaf wrasse. It stimulates germinal vesicle break-down (GVBD) in oocytes during the final maturation processes [1−4].

Discovery

In the 1980s, 20β-S was identified as a major MIS or maturation-inducing hormone (MIH) produced by the ovary of the Atlantic croaker [1].

Structure

Structural Features

The C21 steroid is secreted by ovarian follicular cells under-going final oocyte maturation. (See Figure 94C.1.)

Properties

Molecular formula: $C_{21}H_{32}O_4$, Mr 348. Melting point: 196−198°C, Soluble in alcohol, acetone, and other organic solvents. Appearance: white crystalline powder.

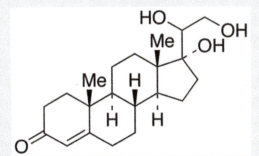

Figure 94C.1   Structure of 17,20β,21-trihydroxy-4-pregnen-3-one.

Synthesis and Release

Following the production of 11-deoxycortisol from 17-hydroxyprogesterone by cytochrome P450c21 (steroid 21-hydroxylase), 20β-hydroxysteroid dehydrogenase (20β-HSD) converts 11-deoxycortisol to 20β-S.

Plasma Concentration

Serum 20β-S level in bambooleaf wrasse (mature female), 735 ± 240 pg/ml [3]. Changes in serum levels of 20β-S during ovarian development are documented in the bambooleaf wrasse [3].

Regulation of Synthesis and Release

Gonadotropin stimulates 20β-S production in Atlantic croaker ovarian follicles undergoing final maturation *in vitro* [5].

Receptors

Structure and Subtype

The maturation-inducing action of 20β-S is through the binding to membrane receptors in the oocyte plasma membrane [6]. A progestin membrane receptor characterized in spotted sea trout (*Cynoscion nebulosus*) ovaries mediates the induction of oocyte meiotic maturation by 20β-S [7]. 20β-S also induces oocyte hydration and ovulation through nuclear progestin receptors [8].

Biological Functions

Target Cells/Tissues and Functions

20β-S is produced by the ovarian follicular cells, under the regulation of pituitary gonadotropin. 20β-S is a highly effec-tive steroid at inducing germinal vesicle breakdown (GVBD). 20β-S-induced GVBD is dependent on activation of the phos-phatidylinositol 3-kinase/Akt in the Atlantic croaker [9]. 20β-S also stimulates sperm hypermotility in the flounder [10].

Pathophysiological Implications

Clinical Implications

None.

Use for Diagnosis and Treatment

None.

Y. Takei, H. Ando, & K. Tsutsui (Eds): Handbook of Hormones. DOI: http://dx.doi.org/10.1016/B978-0-12-801028-0.00222-1

References

1. Trant JM, Thomas P, Shackleton CH. Identification of 17α,20β, 21-trihydroxy-4-pregnen-3-one as the major ovarian steroid produced by the teleost *Micropogonias undulatus* during final oocyte maturation. *Steroids*. 1986;47:89—99.
2. Thomas P, Trant JM. Evidence that 17α,20β,21-trihydroxy-4-pregnen-3-one is a maturation-inducing steroid in spotted seatrout. *Fish Physiol Biochem*. 1989;7:185—191.
3. Matsuyama M, Ohta K, Morita S, et al. Circulating levels and *in vitro* production of two maturation-inducing hormones in teleost: 17α,20β-dihydroxy-4-pregnen-3-one and 17α,20β,21-trihydroxy-4-pregnen-3-one, in a daily spawning wrasse, *Pseudolabrus japonicus*. *Fish Physiol Biochem*. 1998;19:1—11.
4. Ohta K, Yamaguchi S, Yamaguchi A, et al. Biosynthesis of steroids in ovarian follicles of red seabream, *Pagrus major* (Sparidae, Teleostei) during final oocyte maturation and the relative effectiveness of steroid metabolites for germinal vesicle breakdown *in vitro*. *Comp Biochem Physiol B Biochem Mol Biol*. 2002;133:45—54.
5. Patino R, Thomas P. Gonadotropin stimulates 17α,20β,21-trihydroxy-4-pregnen-3-one production from endogenous substrates in Atlantic croaker ovarian follicles undergoing final maturation *in vitro*. *Gen Comp Endocrinol*. 1990;78:474—478.
6. Thomas P, Zhu Y, Pace M. Progestin membrane receptors involved in the meiotic maturation of teleost oocytes: a review with some new findings. *Steroids*. 2002;67:511—517.
7. Zhu Y, Rice CD, Pang Y, et al. Cloning, expression, and characterization of a membrane progestin receptor and evidence it is an intermediary in meiotic maturation of fish oocytes. *Proc Natl Acad Sci USA*. 2003;100:2231—2236.
8. Pinter J, Thomas P. Characterization of a progestogen receptor in the ovary of the spotted seatrout, *Cynoscion nebulosus*. *Biol Reprod*. 1995;52:667—675.
9. Pace MC, Thomas P. Steroid-induced oocyte maturation in Atlantic croaker (*Micropogonias undulatus*) is dependent on activation of the phosphatidylinositol 3-kinase/Akt signal transduction pathway. *Biol Reprod*. 2005;73:988—996.
10. Tubbs C, Tan W, Shi B, et al. Identification of 17,20β,21-trihydroxy-4-pregnen-3-one (20β-S) receptor binding and membrane progestin receptor α on southern flounder sperm (*Paralichthys lethostigma*) and their likely role in 20β-S stimulation of sperm hypermotility. *Gen Comp Endocrinol*. 2011;170:629—639.

Supplemental Information Available on Companion Website

- Progestin synthesis in teleost ovary during the oocyte maturation/E-Figure 94C.1

Dehydroepiandrosterone

Yukiko Ogino, Shinichi Miyagawa, and Taisen Iguchi

Abbreviation: DHEA

Additional names: androstenolone, dehydroisoandrosterone, diandrone, psicosterone, trans-dehydroandrosterone, prasterone, diandron

Dehydroepiandrosterone (DHEA) and its sulfate ester, DHEAS, are the most abundant steroids circulating in human blood. They are mainly produced in the adrenals, and are converted into potent androgens and/or estrogens in peripheral tissues. DHEA is also produced in the brain, acting as a neurosteroid.

Discovery

DHEA was first discovered in human urine in 1931 and was isolated from human blood in 1954 [1].

Structure

Structural Features

DHEA is a major C19 steroid produced mainly by the adrenal cortex (Figure 94D.1).

Properties

Molecular formula: $C_{19}H_{28}O_2$, Mr 288. Melting point: 140−141°C.

Synthesis and Release

The adrenal cortex is the main site of production of DHEA, which is converted to active sex hormones in peripheral target tissues. DHEA is also produced in small quantities in the testis, ovary, and brain. Following the conversion of cholesterol to pregnenolone by CYP11A (P450scc, cholesterol side-chain cleavage), pregnenolone is converted to DHEA by CYP17 (P450c17, 17α-hydroxylase, 17,20-lyase). Most DHEA is sulfated (DHEA-sulfate, DHEAS) before secretion. DHEA can be converted to androstenedione by 3β-hydroxysteroid dehydrogenase (3β-HSD), and subsequently to active androgens and estrogens.

Plasma Concentration

In adult humans, plasma concentrations of DHEA and DHEAS are in the range of 12 and 2,000 nM, respectively [2]. DHEA-S levels are high in newborn babies, and then quickly drop. They rise again during puberty [3]. Circulating serum levels of DHEA and DHEAS decline progressively and markedly with aging [4].

Regulation of Synthesis and Release

DHEA is produced in the zona reticulata and zona fasciculata of the adrenal cortex under the stimulation of adrenocorticotropic hormone (ACTH) secreted by the pituitary gland. DHEA is implicated in the stress response [5].

Biological Functions

Target Cells/Tissues and Functions

DHEA and DHEAS are transformed into androgens and/or estrogens in peripheral target tissues such as gonads, brain, bone, breast, skin, lymph nodes, and adipose tissue. Transformation of DHEA and DHEAS depends on the level of expression of the various steroidogenic and metabolizing enzymes in each of these peripheral tissues.

Pathophysiological Implications

Clinical Implications

High levels of DHEA and DHEAS in children can cause early puberty (adrenarche) [3]. The phenotypic result of adrenarche is pubarche or the development of axillary and pubic hair that occurs at approximately 8 years of age [6]. The production of DHEA and DHEAS during fetal development provides a substrate for the massive rise in estrogen biosynthesis during gestation and may play a role in neural development [7].

Use for Diagnosis and Treatment

As compared to DHEA, serum DHEAS concentrations do not have a circadian rhythm because the plasma half-life of DHEAS is much longer. Serum DHEAS concentrations are clinically useful in the differential diagnosis of some adrenal disorders, including adrenal adenoma, adrenal carcinoma, Cushing's syndrome, and congenital adrenal hyperplasia.

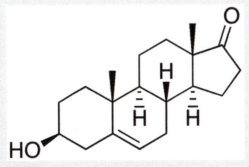

Figure 94D.1   Structure of dehydroepiandrosterone.

Y. Takei, H. Ando, & K. Tsutsui (Eds): Handbook of Hormones. DOI: http://dx.doi.org/10.1016/B978-0-12-801028-0.00223-3

DHEA has been known as an inhibitor of carcinogenesis in various organs such as the mammary gland and prostate [8,9]. DHEAS treatment is used as an anti-aging and anti-obesity hormone therapy [10]. Indeed, DHEA has many effects, including vasodilatory, anti-aging, anti-inflammatory, anti-atherosclerotic, and antidepressant effects. However, the clinical benefits and safety of DHEA are not well established.

References

1. Migeon CJ, Plager JE. Identification and isolation of dehydro-isoandrosterone from peripheral human plasma. *J. Biol Chem.* 1954;209:767−772.
2. Lavallee B, Provost PR, Roy R, et al. Dehydroepiandrosterone-fatty acid esters in human plasma: formation, transport and delivery to steroid target tissues. *J Endocrinol.* 1996;150(Suppl):S119−S124.
3. Havelock JC, Auchus RJ, Rainey WE. The rise in adrenal androgen biosynthesis: adrenarche. *Semin Reprod Med.* 2004;22:337−347.
4. Orentreich N, Brind JL, Rizer RL, et al. Age changes and sex differences in serum dehydroepiandrosterone sulfate concentrations throughout adulthood. *J Clin Endocrinol Metab.* 1984;59:551−555.
5. Zinder O, Dar DE. Neuroactive steroids: their mechanism of action and their function in the stress response. *Acta Physiol Scand.* 1999;167:181−188.
6. Auchus RJ, Rainey WE. Adrenarche − physiology, biochemistry and human disease. *Clin Endocrinol.* 2004;60:288−296.
7. Compagnone NA, Mellon SH. Dehydroepiandrosterone: a potential signalling molecule for neocortical organization during development. *Proc Natl Acad Sci USA.* 1998;95:4678−4683.
8. Shilkaitis A, Green A, Punj V, et al. Dehydroepiandrosterone inhibits the progression phase of mammary carcinogenesis by inducing cellular senescence via a p16-dependent but p53-independent mechanism. *Breast Cancer Res.* 2005;7:R1132−R1140.
9. Rao KV, Johnson WD, Bosland MC, et al. Chemoprevention of rat prostate carcinogenesis by early and delayed administration of dehydroepiandrosterone. *Cancer Res.* 1999;59:3084−3089.
10. Gomez-Santos C, Larque E, Granero E, et al. Dehydroepiandrosterone-sulphate replacement improves the human plasma fatty acid profile in plasma of obese women. *Steroids.* 2011;76:1425−1432.

Testosterone/ Dihydrotestosterone

Yukiko Ogino, Shinichi Miyagawa, and Taisen Iguchi

Abbreviation: T/DHT

Additional names: 17β-Hydroxy-4-androsten-3-one/5α-dihydrotestosterone (5α-DHT), androstanolone, atanolone, 5α-androstan-17β-ol-3-one, 17β-hydroxy-5α-androstan-3-one, andractim, anaboleen, anabolex, androlone, neodrol, proteina, protona

Androgens play important roles in male sexual differentiation, development, reproductive development, and function. Testosterone (T) is a predominant androgen mainly secreted by Leydig cells of the testis in vertebrates. It can be converted into its active metabolites, 5α-dihydrotestosterone (DHT) by 5α-reductase and/or 17β-estradiol by P450arom, in peripheral tissues. The production of testosterone is stimulated by luteinizing hormone (LH) from the pituitary gland. In most vertebrates, T and DHT are required for masculinization.

Discovery

The biological effects of the testis and T have been known for many years from testicular transplantation experiments. John Hunter transplanted testes into capons in 1786. Berthold's classic study on domesticated roosters in 1849 demonstrated that testicular secretions are necessary for the normal expression of aggressive behavior. Following his observations, in 1889, Brown-Séquard, aged 72, reported dramatic rejuvenating effects after self-administering testicular extracts of dogs and guinea pigs. In 1935, Ernest Laqueur isolated the molecule from bull testes, naming the compound T, and then androgen therapy was started.

Structure of Testosterone

Structural Features

A 19-carbon steroid, secreted by the testis. Aromatization of the A ring converts it into estradiol-17β. Reduction of the δ4 double bond can convert it into DHT (Figure 94E.1).

Properties

Molecular formula: $C_{19}H_{28}O_2$, Mr 288. Melting point: 155°C. Appearance: white crystalline powder, or colorless or yellowish-white crystals.

Structure of DHT

Structural Features

A biologically active metabolite of T, DHT is generated by the 5α-reduction of T. The structural difference between DHT and T is that T has a 4,5 double-bond on the A ring (Figure 94E.1).

Properties

Molecular formula: $C_{19}H_{30}O_2$, Mr 290. Melting point: 181°C. Appearance: white crystalline powder.

Synthesis and Release

Pregnenolone can be converted to T by two alternative routes, referred to as the Δ4- and Δ5-pathways, based on whether the steroid intermediates are 3-keto, Δ4 steroids (Δ4) or 3 hydroxy, Δ5 steroids (Δ5). In mammals, T is secreted primarily in testicular Leydig cells and ovarian follicle theca cells, although small amounts are also secreted by the adrenal gland. Ovarian T is required for ovarian estrogen biosynthesis. The synthesis of T from cholesterol involves four enzymes, cytochrome P450 side-chain cleavage enzyme (P450scc), cytochrome P450 17α-hydroxylase/17,20-lyase (CYP17), 3β-hydroxysteroid dehydrogenase (3β-HSD), and 17β-hydroxysteroid dehydrogenase (17β-HSD). Most circulating DHT is derived from the peripheral conversion of T by 5α-reductase type 1 and type 2. The synthesis of DHT also occurs from 5α-androstanedione, instead of via T [1]. DHT has much greater affinity to the androgen receptor (AR) than does T.

Plasma Concentration (ng/ml)

Testosterone

Human (male) mature 3–10 (10–35 nmol/l), human (female) mature < 1 (<3.5 nmol/l), human (before puberty) 0.05–0.2 ng/ml (0.17–0.69 nmol/l) [2], human (testicular venous blood) 700–800 (2,430–2,778 nmol/l) [3]. Chum salmon (male) mature 25–40 (87–139 nmol/l), chum salmon (female) mature 5–20 (17–69 nmol/l) [4].

DHT

Human (male) mature 0.3–0.9 (1.0–3.1 nmol/l), human (female) mature 0.02–0.25 (0.07–0.86 nmol/l) [2].

Y. Takei, H. Ando, & K. Tsutsui (Eds): Handbook of Hormones. DOI: http://dx.doi.org/10.1016/B978-0-12-801028-0.00224-5

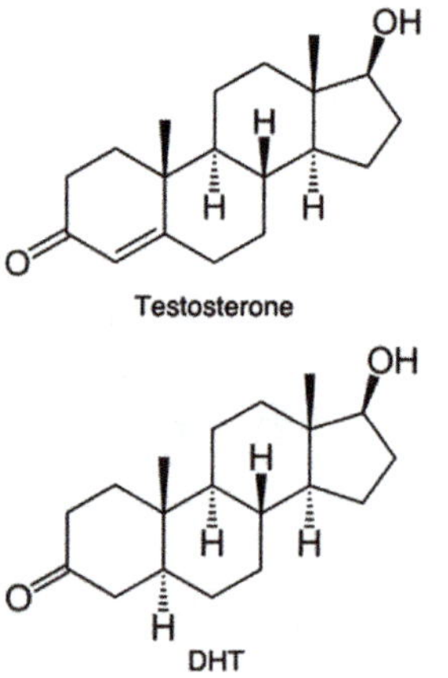

Figure 94E.1 Structure of T and DHT.

Regulation of Synthesis and Release

In the testes, LH binds to LH receptors on Leydig cells, stimulating synthesis and secretion of T. In the ovary, LH stimulates production of T in theca cells, which is converted into estrogen by adjacent granulosa cells. The pituitary secretion of LH is controlled by negative feedback from T.

Receptors

Structure and Subtype

Androgen receptor (AR, alternative name NR3C4) belongs to the nuclear receptor (NR) superfamily. The human AR gene is located in the X chromosome at Xq11−12 and is encoded in eight exons. Amino-terminal domain (NTD) is encoded by the first exon, the DNA binding domain (DBD) containing two zinc fingers by Exons 2 and 3, the hinge region by part of Exon 4, and the ligand binding domain (LBD) by Exons 4−8. Two main transactivation function (AF) domains, AF1 in the NTD and AF2 in the LBD, have been identified. Two pentapeptide regions of the AR NTD, ^{23}FQNLF^{27} (FXXLF) and ^{433}WHTLF^{437} (WXXLF), mediate binding of the N terminus to the C-terminal region of AR. AR is generally known to be expressed in the cytoplasm and translocated into the nucleus upon ligand stimulation. The liganded-AR subsequently transactivates the target genes by binding to a specific DNA sequence, the androgen-response element (ARE) [5].

Agonists

R1881, also known as methyltrienolone, mibolerone and 17α-methyltestosterone, are known as synthetic agonists of the AR.

Antagonists

Cyproterone acetate, hydroxyflutamide, vinclozolin, *p,p'*-DDE, linuron, megestrol acetate, bicalutamide, and nilutamide are known as anti-androgenic chemicals. A large number of chemicals that bind to AR are summarized by Fang et al. [6].

5α-Reductase Inhibitor

Finasteride.

Biological Functions

Target Cells/Tissues and Functions

T regulates the differentiation of the Wolffian ducts into the epididymis, vas deferens, and seminal vesicle. T's actions in many tissues are mediated through its active metabolites, estradiol-17β and DHT. For instance, the sexual differentiation of the brain and trabecular bone resorption requires its aromatization to estradiol. DHT is needed for prostate development and growth, masculinization of external genitalia, and male patterns of facial and body hair growth. Male rodents are androgenized by a perinatal T surge and females are feminized by the lack of perinatal T. This perinatal androgen surge is critical for organizing male and female sexual dimorphisms in rodents. T and DHT can both exert anabolic effects on the muscle.

Phenotype of Gene-Modified Animals

The first animal model of androgen insensitivity, an X-linked gene for Tfm (testicular feminized male) in the mouse, was described in 1970. A transgenic mouse model with a ubiquitous knockout of AR was first generated in 2002 [7]. Thereafter, cell specific AR knockout mouse models were created [8]. The ubiquitous AR knockout male mice have female-like external genitalia, small undescended testes, a blind ending vagina, and absent accessory sex organs, and germ cell development is severely disrupted, which is similar to human complete androgen insensitivity syndrome (CAIS) [8]. The female AR knockout mice had considerable reproductive defects, with decreased fertility, defective follicular development, reduced ovulation, and premature ovarian failure [9].

Pathophysiological Implications

Clinical Implications

Diseases such as androgen insensitivity syndrome (AIS), prostate cancer, Kennedy's disease, and infertility can be caused by mutations in the AR. Mutations of the *AR* gene in patients with AIS and/or prostate cancer are listed in the *AR* mutation database [10]. Mutations that completely eliminate the function of the AR cause CAIS. Some patients with CAIS present with male hermaphroditism characterized by female external genitalia, a blind vaginal pouch, and well-developed breasts. Partial AIS patients with genetic changes that reduce the AR's activity may have a male phenotype and milder abnormalities such as hypospadias, gynecomastia, and infertility. Alterations in the *AR* gene are also associated with an increased risk of androgenetic alopecia. An abnormal length of the polyglutamide tract (CAG and GCG repeats) in exon 1 of the *AR* has been associated with spinal and bulbar muscular atrophy (Kennedy's disease). Women with polycystic ovary syndrome (PCOS), in which ovaries produce excessive amounts of androgens, experience irregular menstrual cycles and decreased fertility.

Use for Diagnosis and Treatment

AR ligands are widely utilized in a variety of clinical applications (i.e., agonists are used for hypogonadism, while antagonists, such as bicalutamide and hydroxyflutamide, are employed for prostate cancer therapy).

References

1. Sharifi N. The 5α-androstanedione pathway to dihydrotestosterone in castration-resistant prostate cancer. *J Investig Med.* 2012;60:504−507.
2. O'Malley BW, Strott CA, O'Malley BW, Strott CA. Steroid hormones: metabolism and mechanism of action. In: Yen SSC, Jaffe RB, Barbieri RL, eds. *Reproductive Endocrinology.* 4th edition. Philadelphia: W.B. Saunders Company; 1994:110−133.
3. Maddocks S, Hargreave TB, Reddie K, et al. Intratesticular hormone levels and the route of secretion of hormones from the testis of the rat, guinea pig, monkey and human. *Int J Androl.* 1993;16:272−278.

4. Asahina K, Kobayashi T, Soeda H. Changes in plasma electrolyte and hormone concentrations during homing migration of chum salmon *Oncorhynchus keta* with special reference to the development of nuptial color. *Nippon Suisan Gakkaishi.* 1991;57:599−605.

5. Ham J, Thomson A, Needham M, et al. Characterization of response elements for androgens, glucocorticoids and progestins in mouse mammary tumour virus. *Nucleic Acids Res.* 1988;16:5263−5276.

6. Fang H, Tong W, Branham WS, et al. Study of 202 natural, synthetic, and environmental chemicals for binding to the androgen receptor. *Chem Res Toxicol.* 2003;16:1338−1358.

7. Yeh S, Tsai MY, Xu Q, et al. Generation and characterization of androgen receptor knockout (ARKO) mice: an *in vivo* model for the study of androgen functions in selective tissues. *Proc Natl Acad Sci USA.* 2002;99:13498−13503.

8. De Gendt K, Verhoeven G. Tissue- and cell-specific functions of the androgen receptor revealed through conditional knockout models in mice. *Mol Cell Endocrinol.* 2012;352:13−25.

9. Sen A, Prizant H, Light A, et al. Androgens regulate ovarian follicular development by increasing follicle stimulating hormone receptor and microRNA-125b expression. *Proc Natl Acad Sci USA.* 2014;111:3008−3013.

10. Gottlieb B, Beitel LK, Nadarajah A, et al. The androgen receptor gene mutations database: 2012 update. *Hum Mutat.* 2012;33:887−894.

Supplemental Information Available on Companion Website

- Gene and primary structure of human AR/E-Figure 94E.1
- Steroid hormone synthetic pathways/E-Figure 94E.2
- DHT synthetic pathways/E-Figure 94E.3
- Chemical structure of AR agonists and antagonists/E-Figure 94E.4
- Accession numbers of AR cDNAs/E-Table 94E.1

11-Ketotestosterone

Yukiko Ogino, Shinichi Miyagawa, and Taisen Iguchi

Additional names: ketotestosterone, 11-oxotestosterone, 17β-hydroxy-4-androstene-3,11-dione
Abbreviation: 11-KT, 11KT

Androgens play important roles in male sexual differentiation, development, and reproductive development and function. 11-Ketotestosterone (11-KT) is the primary and most potent androgen in teleost fish. 11-KT stimulates the development of secondary sexual characteristics and expression of male reproductive behavior in various teleost fishes.

Discovery

11-KT was first identified in the plasma of the salmon *Oncorhynchus nerka* [1].

Structure of 11-KT

Structural Features

11-KT is an oxidized form of testosterone that contains a keto group at position-11 (Figure 94F.1).

Properties

Molecular formula: $C_{19}H_{26}O_3$, Mr: 302 g/mol. Melting point: 181−184°C. Appearance: white crystalline powder.

Synthesis and Release

In some teleosts, it was reported that 11-KT is synthesized in the gonads [2,3]. 11-KT is synthesized from testosterone by the action of two enzymes, 11β-hydroxylase (P45011β) and 11β-hydroxysteroid dehydrogenase (11β-HSD).

Plasma Concentration (ng/ml)

Coho salmon (male) spawning season, 80−150 ng/ml (65−97 nmol/l), coho salmon (female) spawning season 4−12 ng/ml (13−40 nmol/l) [4].

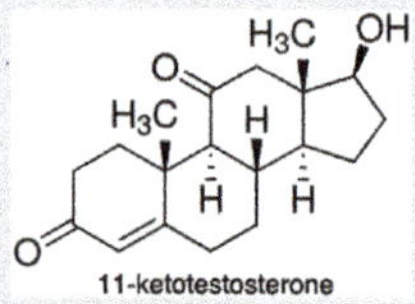

Figure 94F.1 Structure of 11-KT.

Regulation of Synthesis and Release

Gonadotropins stimulate the production of 11-KT [5].

Receptors

Structure and Subtype

Androgen receptor (AR, alternative name NR3C4) belongs to the nuclear receptor (NR) superfamily. In teleost fish, two distinct paralogous copies of ARs, ARα and ARβ, have been identified from several species [6,7]. ARβ is generally known to be expressed in the cytoplasm and translocated into the nucleus upon ligand stimulation, while ARα is known to be translocated into the nucleus without ligand stimulation [7]. The liganded-AR subsequently transactivates the target genes by binding to a specific DNA sequence, the androgen-response element (ARE) [7].

Agonist

R1881, also known as methyltrienolone, mibolerone, and 17α-methyltestosterone, is known as a synthetic agonist of the AR.

Antagonist

Cyproterone acetate, hydroxyflutamide, vinclozolin, p,p'-DDE, linuron, megestrol acetate, bicalutamide (Casodex), and nilutamide are known as anti-androgenic chemicals. A large number of chemicals that bind to AR are summarized by Fang et al. [8].

Biological Functions

Target Cells/Tissues and Functions

11-KT has important roles in spermatogenesis, male secondary sexual characteristics, behavior, and nuptial coloration [9,10].

References

1. Idler DR, Schmidt PJ, Ronald AP. Isolation and identification of 11-ketotestosterone in salmon plasma. *Can J Biochem Physiol*. 1960;38:1053−1058.
2. Idler DR, Macnab HC. The biosynthesis of 11-ketotestosterone and 11-β-hydroxytestosterone by Atlantic salmon tissues *in vitro*. *Can J Biochem*. 1967;45:581−589.
3. Alam MA, Bhandari RK, Kobayashi Y, et al. Changes in androgen-producing cell size and circulating 11-ketotestosterone level during female-male sex change of honeycomb grouper *Epinephelus merra*. *Mol Reprod Dev*. 2006;73:206−214.
4. Fitzpatrick MS, Van der Kraak G, Schreck CB. Profiles of plasma sex steroids and gonadotropin in coho salmon, *Oncorhynchus kisutch*, during final maturation. *Gen Comp Endocrinol*. 1986;62:437−451.

Y. Takei, H. Ando, & K. Tsutsui (Eds): Handbook of Hormones. DOI: http://dx.doi.org/10.1016/B978-0-12-801028-0.00225-7

5. Planas JV, Swanson P. Maturation-associated changes in the response of the salmon testis to the steroidogenic actions of gonadotropins (GTH I and GTH II) *in vitro. Biol Reprod.* 1995;52:697−704.

6. Douard V, Brunet F, Boussau B, et al. The fate of the duplicated androgen receptor in fishes: a late neofunctionalization event? *BMC Evol Biol.* 2008;8:336.

7. Ogino Y, Katoh H, Kuraku S, et al. Evolutionary history and functional characterization of androgen receptor genes in jawed vertebrates. *Endocrinology.* 2009;150:5415−5427.

8. Fang H, Tong W, Branham WS, et al. Study of 202 natural, synthetic, and environmental chemicals for binding to the androgen receptor. *Chem Res Toxicol.* 2003;16:1338−1358.

9. Miura T, Yamauchi K, Takahashi H, et al. Hormonal induction of all stages of spermatogenesis *in vitro* in the male Japanese eel (*Anguilla japonica*). *Proc Natl Acad Sci USA.* 1991;88:5774−5778.

10. Borg B. Androgens in teleost fishes. *Comp Biochem Physiol.* 1994;109C:219−245.

Supplemental Information Available on Companion Website

- 11-KT synthetic pathway/E-Figure 94F.1

Estradiol-17β

Tomomi Sato, Shinichi Miyagawa, and Taisen Iguchi

Abbreviation: E2
Additional names: 1,3,5(10)-estratriene-3,17β-diol, 17β-estradiol

Estradiol-17β (E2) is a steroid hormone mainly produced in the ovary. It is also produced in the testis but in low concentrations. E2 stimulates proliferation of the uterine endometrium, induces the LH surge for ovulation as a positive feedback, and inhibits pituitary hormone release via negative feedback.

Discovery

In 1936, E2 was isolated from sow ovaries [1].

Structure

Structural Features

A C18 steroid hormones, with a phenolic A ring and a hydroxyl group on carbon 17 in the β-conformation (Figure 94G.1).

Properties

Molecular formula: $C_{18}H_{24}O_2$, Mr 272.38. Soluble in alcohol, acetone, dioxane, and other organic solvents. Almost insoluble in water.

Synthesis and Release

Gene, mRNA, and Precursor

Estradiol-17β is converted from testosterone by aromatase (P450arom encoded by the CYP19 gene) or from estrone by 17β-hydroxysteroid dehydrogenase (17β-HSD) in the granulosa cells and luteal cells of the ovary, Leydig cells of the testis, adrenal gland, placenta (human), brain (mouse), and adipose tissue (E-Figures 94G.1 and 94G.2) [2]. Two oxygen molecules are utilized for oxidation of the C19 methyl group and the third oxidative reaction is considered to be a peroxidative attack on the C19 methyl group combined with 1β-hydrogen elimination, yielding formic acid and the phenolic A ring. Chemical names and characters of estrogens are found in E-Table 94G.1.

Plasma Concentration

See Table 94G.1 [3–6] for plasma concentrations.

Regulation of Synthesis and Release

The synthesis and release of E2 in the granulosa cells of the ovary are considered a "two cell model"; androstenedione is synthesized in the theca interna by LH stimulation and aromatized in the granulosa cells under the influence of FSH. Similarly, testosterone produced in the thecal cell is converted into E2 in the granulosa cells in teleosts [7]. In the chicken ovary, the granulosa cells produce progesterone and the thecal cells convert them into testosterone and E2 [8]. Equilin and equilenin in pregnant mare urine are ring B unsaturated estrogens in the placenta, which are not synthesized from cholesterol via standard steroidogenesis (E-Figure 94G.3) [9].

Receptors

Structure and Subtype

Estrogen receptor (ER) has two distinct forms, ERα and ERβ in mammals, avians, reptilians, amphibians, and ancient fish species, and has three ER subtypes, ERα, ERβ1, and ERβ2, in teleosts (E-Figure 94G.4). The cDNA of human ERα was identified in 1986, and then rat ERα was isolated in 1987 [10]. In 1996, another type of ER was identified, called ERβ [11]. ERs are members of the nuclear receptor superfamily of transcription factors. The ERα gene is located on chromosome 6q25.1, the protein consists of 595 aa residues, and its molecular weight is 66,000. The ERβ gene is located in 14q23.2, the protein consists of 530 aa, and its molecular weight is 60,000~64,000. The structure of ERα and ERβ is categorized into six regions of varying homology denoted A–F (E-Figure 94G.5). The N-terminal A/B domain is important for stimulating transcription from certain estrogen-responsive promoters without ligand (ligand-independent activation function, AF-1). The C domain (the DNA-binding domain, DBD), composed of two zinc fingers, binds to estrogen response elements (ERE) in the target DNA. The consensus ERE is GGTCAnnnTGACC (where n is any nucleotide). The D domain is a hinge region containing the nuclear localization signal. The E domain, containing the ligand-binding domain (LBD) and the ligand-dependent transactivation function (AF-2), transactivates the gene expression in response to the ligand. The specific function of the C terminus in the F domain has not yet been fully clarified. The central DNA-binding regions in ERα and ERβ are similar structures (98% identity), whereas their activation domains are largely different (less than 15% homology in their N termini). Splice variants have been found both in ERα and ERβ; however, all the variants have not been identified at the protein level. In addition, the G protein-coupled ER (GPER-1, formerly GPR30), a 7-transmembrane spanning G protein-coupled receptor, can mediate rapid cellular effects by E2 via activation of ERK1/2 and adenylyl cyclase [12].

Y. Takei, H. Ando, & K. Tsutsui (Eds): Handbook of Hormones. DOI: http://dx.doi.org/10.1016/B978-0-12-801028-0.00226-9

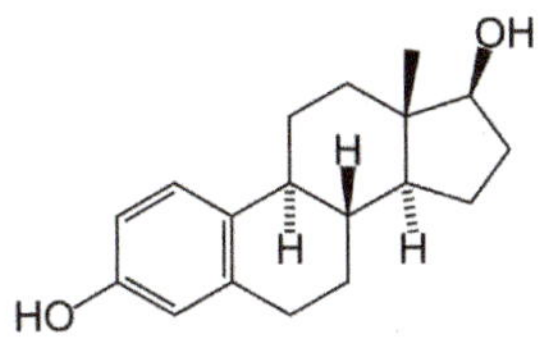

Figure 94G.1 Structure of E2.

Table 94G.1 Plasma Concentrations of E2 in Mammals and Non-Mammals

Species	Sex and Reproductive State	Concentrations (pg/ml (pmol/l))	Ref.
Human	(male) mature	10−57 (37−210)	[3]
	(female) mature follicular phase	10−98 (37−360)	
	(female) intermediate phase	170−770 (625−2830)	
	(female) luteal phase	190−340 (699−1250)	
	(female) postmenopause	10−38 (37−140)	
Salmon	(male) mature	250−350 (919−1287)	[4]
	(female)	14−16 ng/ml (51−59 nmol/l)	
Goldfish	(female) spawning season	2−6 ng/ml (7−22 nmol/l)	[5]
Red sea bream	(female) spawning season	200−1200 (735−4412)	[6]

Signal Transduction Pathway

ERs are located in either the cytoplasm or nucleus bound by chaperone proteins when E2 is absent. Once E2 has bound to ERs, the ER homodimerizes and binds to specific DNA sequences (EREs) within the promotor region of target genes to initiate transcription. ERα also forms a heterodimer with ERβ. The ligand-ERs complex recruits transcriptional comodulators including members of the steroid receptor complex (SRC)/p160 family and the thyroid receptor-associated protein (TRAP220) complex.

Agonists

Diethylstilbestrol, hexestrol, mestranol, 17α-ethinylestradiol (EE2). EE2 combined with progestins is used as oral contraceptives for prevention of ovulation (E-Figure 94G.6).

Antagonists

Tamoxifen and raloxifene act as antagonists or agonists depends on the tissues and organs, called selective ER modulators (SERMs). Tamoxifen and raloxifene compete with estrogen to bind to ER and inhibit the proliferation of uterine and mammary epithelial cells and uterine cancer cells. Therefore, they are used as hormonal therapy for patients who exhibit ER positive breast cancer. The chance of getting endometrial cancer is also increased by tamoxifen treatment (E-Figure 94G.6).

Biological Functions

Target Cells/Tissues and Functions

In avians, reptiles, amphibians, and fish species, E2 stimulates synthesis of vitellogenin, the lipoprotein yolk precursor, in the liver. In the late stage of yolk protein accumulation, activity of aromatase and synthesis and release of E2 are reduced. E2 stimulates choriogenin synthesis, the precursor of egg envelope protein, in the liver of fish species including medaka fish [13]. Ovalbumin and conalbumin, major types of albumin found in egg white, are synthesized under the stimulation of E2 and progesterone in the oviduct of chicken. In mammals, ERα is expressed in liver, uterus, mammary gland, pituitary, hypothalamus, ovary, cervix, and vagina, whereas

Table 94G.2 Functions of Estrogens

Species	Function	Ref.
Mammals	Stimulation of epithelial cell differentiation and proliferation in the oviduct, uterus, and vagina	
Mammals	Development of mammary glands during the last third of the pregnancy	
Mammals	Induction of negative feedback effects on GnRH neurons via Kiss1 neurons in the arcuate nucleus and positive feedback effects in the preoptic periventricular nucleus	[14−16]
Mammals	Regulation of bone homeostasis and bone turnover, and maintenance of bone mass	[17]
Xenopus	Stimulation of the vitellogenin gene transcription and vitellogenesis in hepatic cells	[18]
All vertebrates	Posttranslational glycosylation and phosphorylation of vitellogenins in the endoplasmic reticulum and Golgi complex, and secretion to the circulating plasma. Cleavage into lipovitellins (Lv), phosvitin (Pv), and beta component (β′)	[19]
Japanese eels	Stimulation of spermatogonial stem cell renewal via eel spermatogenesis related substance 34 (eSRS34)	[20,21]
Lizards	Induction of spermatogonial proliferation through ERKs phosphorylation	[22]

ERβ is found in the ovary, lung and prostate. E2 is synthesized in the granulosa cell of the growing follicle in the follicular phase. It peaks on about day 14 of the menstrual cycle and stimulates accelerated pulsatile GnRH release leading to the LH surge and ovulation. E2 stimulates differentiation and proliferation of the endometrium in the proliferative phase, and maintains the proliferated uterine endometrium and increases hyperemia during the secretory phase (Table 94G.2) [14−22].

Phenotype of Gene-Modified Animals

ERα knockout female mice are infertile due to inefficient levels of gonadotropins and lack of sensitivity to E2 treatment. ERβ knockout female mice exhibit reduced ovulation rates; however, ERβ knockout male mice are normal and fertile. After superovulation, the ovaries of ERα and ERβ knockout mice exhibit a reduced number of corpora lutea and a number of unruptured antral follicles. ERα and ERβ double knockout female mice are infertile due to a lack of ovulation and exhibit sex-reversed somatic cells similar to what is found in aromatase knockout mice. Mammary glands of ERα knockout and ERαβ knockout female mice exhibit no pubertal development [8]. ERα knockout and aromatase knockout male mice are infertile because of the failure of concentrating epididymal sperm through regulation of fluid absorption in the epididymis and maintaining normal adult spermatogenesis. GPER-1 knockout female mice exhibit hyperglycemia, impaired glucose tolerance, and altered bone growth [23].

Pathophysiological Implications
Clinical Implications

A null mutation in ERα causes estrogen unresponsiveness. Mutations in the aromatase gene produce severe estrogen deficiency in women and severe osteopenia in men.

Use for Diagnosis and Treatment

Mainly for postmenopausal HRT (hormone replacement therapy), estrogenic derivatives are available as drugs administered by oral, transdermal, and injection routes. Also, in adolescent patients and those of reproductive age, such drugs are used in conditions and diseases involving low and no endogenous estrogens.

References

1. MacCorquodale DW, Thayer SA, Doisy EA. The isolation of the principal estrogenic substance of liquor folliculi. *J Biol Chem.* 1936;115:435–448.
2. Payne AH, Hales DB. Overview of steroidogenic enzymes in the pathway from cholesterol to active steroid hormones. *Endocr Rev.* 2004;25:947–970.
3. Desjardins C. Endocrine regulation of reproductive development and function in the male. *J Animal Sci.* 1978;47(Suppl 2):56–79.
4. Maddocks S, Hargreave TB, Reddie K, et al. Intratesticular hormone levels and the route of secretion of hormones from the testis of the rat, guinea pig, monkey and human. *Int J Androl.* 1993;16:227–228.
5. Asahina K, Kobashi T, Soeda H. Changes in plasma electrolyte and hormone concentrations during homing migration of chum salmon *Oncorhynchus keta* with special reference to the development of nuptial color. *Nippon Suisan Gakkaishi.* 1991;57:599–605.
6. Matsuyama M, Adachi S, Nagahama Y, et al. Diurnal rhythm of oocyte development and plasma steroid hormone levels in the female red sea bream, *Pagrus major*, during the spawning season. *Aquaculture.* 1988;73:359–372.
7. Nagahama Y. Gonadotropin action on gametogenesis and steroidogenesis in teleost gonads. *Zool Sci.* 1987;4:209–222.
8. Huang ES, Kao KJ, Nalbandov AV. Synthesis of sex steroids by cellular components of chicken follicles. *Biol Reprod.* 1979; 20:454–461.
9. Bhavnani BR. Oestrogen biosynthesis in the pregnant mare. *J Endocrinol.* 1981;89(Suppl):19–32.
10. Koike S, Sakai M, Muramatsu M. Molecular cloning and characterization of rat estrogen receptor cDNA. *Nucleic Acids Res.* 1987;15:2499–2513.
11. Kuiper GG, Enmark E, Pelto-Huikko M, et al. Cloning of a novel receptor expressed in rat prostate and ovary. *Proc Natl Acad Sci USA.* 1996;93:5925–5930.
12. Couse JF, Korach KS. Estrogen receptor null mice: what have we learned and where will they lead us? *Endocr Rev.* 1999;20:358–417.
13. Murata K, Sugiyama H, Yasumasu S, et al. Cloning of cDNA and estrogen-induced hepatic gene expression for choriogenin H, a precursor protein of the fish egg envelope (chorion). *Proc Natl Acad Sci USA.* 1997;94:2050–2055.
14. Dungan Lemko HM, Elias CF. Kiss of the mutant mouse: how genetically altered mice advanced our understanding of kisspeptin's role in reproductive physiology. *Endocrinology.* 2012; 153:5119–5129.
15. George JT, Seminara SB. Kisspeptin and the hypothalamic control of reproduction: lessons from the human. *Endocrinology.* 2012;153:5130–5136.
16. Pinilla L, Aguilar E, Dieguez C, et al. Kisspeptins and reproduction: physiological roles and regulatory mechanisms. *Physiol Rev.* 2012;92:1235–1316.
17. Nelson ER, Wardell SE, McDonnell DP. The molecular mechanisms underlying the pharmacological actions of estrogens, SERMs and oxysterols: implications for the treatment and prevention of osteoporosis. *Bone.* 2013;53:42–50.
18. Skipper JK, Hamilton TH. Regulation by estrogen of the vitellogenin gene. *Proc Natl Acad Sci USA.* 1977;74:2384–2388.
19. Finn RN. Vertebrate yolk complexes and the functional implications of phosvitins and other subdomains in vitellogenins. *Biol Reprod.* 2007;76:926–935.
20. Miura T, Miura C, Ohta T, et al. Estradiol-17beta stimulates the renewal of spermatogonial stem cells in males. *Biochem Biophys Res Commun.* 1999;264:230–234.
21. Miura T, Ohta T, Miura CI, et al. Complementary deoxyribonucleic acid cloning of spermatogonial stem cell renewal factor. *Endocrinology.* 2003;144:5504–5510.
22. Chieffi P, Colucci D'Amato L, Guarino F, et al. 17beta-Estradiol induces spermatogonial proliferation through mitogen-activated protein kinase (extracellular signal-regulated kinase 1/2) activity in the lizard (*Podarcis s. sicula*). *Mol Reprod Dev.* 2002; 61:218–225.
23. Mårtensson UE, Salehi SA, Windahl S, et al. Deletion of the G protein-coupled receptor 30 impairs glucose tolerance, reduces bone growth, increases blood pressure, and eliminates estradiol-stimulated insulin release in female mice. *Endocrinology.* 2009;150:687–698.

Supplemental Information Available on Companion Website

- Synthesis of steroid hormones from cholesterol/E-Figure 94G.1
- Synthesis of estrogens and androgens from DHEA/E-Figure 94G.2
- Equine estrogens/E-Figure 94G.3
- Human estrogen receptors/E-Figure 94G.4
- The structure of the steroid receptor family/E-Figure 94G.5
- Agonists and antagonists of estrogen/E-Figure 94G.6
- Chemical names and characters of estrogens/E-Table 94G.1

Estrone

Tomomi Sato, Shinichi Miyagawa, and Taisen Iguchi

Abbreviation: E1
Additional names/Abbreviation: 3-hydroxy-1,3,5(10)-estra-triene-17-one, 1,3,5-estratriene-3-ol-17-one

Estrone is a steroid hormone produced in the ovary, testis, and placenta, and has weak estrogenic activity.

Discovery

In 1930, estrone was crystallized from pregnant-human urine [1].

Structure

Structural Features

Estrone is a C18 steroid hormone. It is a metabolite of estradiol-17β, and has a phenolic A ring and a ketone on carbon 17 (Figure 94H.1).

Properties

Molecular formula, $C_{18}H_{22}O_2$; Mr 270.37; soluble in alcohol, acetone, dioxane, and other organic solvents. Almost insoluble in water.

Synthesis and Release

Gene, mRNA, and Precursor

Converted from androstenedione by aromatase (P450arom encoded by the CYP19 gene) or from estradiol-17β by 17β-hydroxysteroid dehydrogenase (17β-HSD) in granulosa cells and luteal cells of the ovary, Leydig cells of the testis, and placenta (mouse) (Figure 94H.2).

Tissue Content

Plasma Concentration (pg/ml (pmol/l))

Human (male) mature 10−68 (37−250), (female) mature follicular phase 30−108 (110−400), luteal phase 84−178 (310−660), postmenopause 6−62 (22−230).

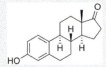

Figure 94H.1 Estrone structure.

Regulation of Synthesis and Release

Several isozymes of 17β-HSDs have been identified, and 17β-HSD type 2 is associated with the final step of biosynthesis of estrone [2]. The synthesis of estrone in rodent placenta is catalyzed by 17β-HSD type 2, thereby inactivating estradiol-17β to estrone [3]. 17β-HSD type 2 is found in the breast, uterus, prostate, liver, and kidney in humans [4]. 17β-HSD types 1 and 7 catalyze the conversion of estrone to estradiol-17β in granulosa cells and corpora lutea of ovary in mice [5], respectively. The human placenta expresses 17β-HSD type 1 enzyme [6]. Estrone can be converted to estriol via 16-hydroxyestrone [7]. In postmenopausal women, estrone is mainly synthesized in adipose tissue from adrenal dehydro-epiandrosterone [8].

Receptors

Structure and Subtype

Estrone can bind to both types of estrogen receptors (ERs), ERα and ERβ [9]. Estrone shows 1.7 times and 2.7 times lower affinity to ERα and ERβ compared to estradiol-17β, respectively.

Signal Transduction Pathway

See Subchapter 94G, Estradiol-17β.

Agonist

Estrogenic activity of estrone is 3.5% that of estradiol-17β in the reporter gene assay using the medaka estrogen receptor α [10].

Antagonist

See Subchapter 94G, Estradiol-17β.

Biological Functions

Target Cells/Tissues and Functions

See Subchapter 94G, Estradiol-17β.

Pathophysiological Implications

Clinical Implications

See Subchapter 94G, Estradiol-17β.

Y. Takei, H. Ando, & K. Tsutsui (Eds): Handbook of Hormones. DOI: http://dx.doi.org/10.1016/B978-0-12-801028-0.00227-0

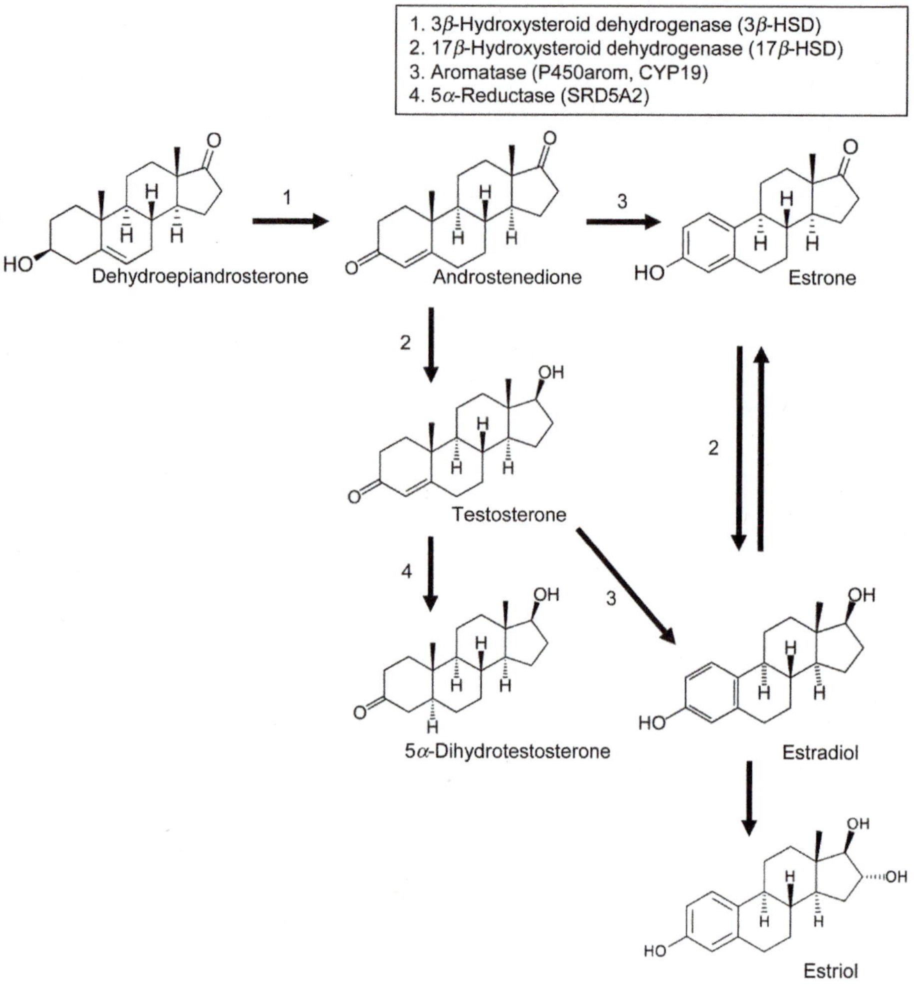

Figure 94H.2 Synthesis of estrogens and androgens from dehydroepiandrosterone.

Use for Diagnosis and Treatment

As an alternative choice for estrogenic agents, conjugated estrone derivatives from equine origin are available as an oral clinical drug.

References

1. Doisy EA, Thayer SA, Levin L, et al. A new tri-atomic alcohol from the urine of pregnant women. *Proc Soc Exp Biol Med (Maywood)*. 1930;28:88—89.
2. Wu L, Einstein M, Geissler WM, et al. Expression cloning and characterization of human 17 beta-hydroxysteroid dehydrogenase type 2, a microsomal enzyme possessing 20 alpha-hydroxysteroid dehydrogenase activity. *J Biol Chem*. 1993;268:12964—12969.
3. Mustonen M, Poutanen M, Chotteau-Lelievre A, et al. Ontogeny of 17beta-hydroxysteroid dehydrogenase type 2 mRNA expression in the developing mouse placenta and fetus. *Mol Cell Endocrinol*. 1997;134:33—40.
4. Vihko P, Isomaa V, Ghosh D. Structure and function of 17beta-hydroxysteroid dehydrogenase type 1 and type 2. *Mol Cell Endocrinol*. 2001;171:71—76.
5. Nokelainen P, Peltoketo H, Vihko R, et al. Expression cloning of a novel estrogenic mouse 17 beta-hydroxysteroid dehydrogenase/17-ketosteroid reductase (m17HSD7), previously described as a prolactin receptor-associated protein (PRAP) in rat. *Mol Endocrinol*. 1998;12:1048—1059.
6. Fournet-Dulguerov N, MacLusky NJ, Leranth CZ, et al. Immunohistochemical localization of aromatase cytochrome P-450 and estradiol dehydrogenase in the syncytiotrophoblast of the human placenta. *J Clin Endocrinol Metab*. 1987;65:757—764.
7. Martucci CP, Fishman J. P450 enzymes of estrogen metabolism. *Pharmacol Ther*. 1993;57:237—257.
8. Geisler J, Lønning PE. Aromatase inhibition: translation into a successful therapeutic approach. *Clin Cancer Res*. 2005;11:2809—2821 (Review).
9. Kuiper GGJM, Carlsson B, Grandien K, et al. Comparison of the ligand binding specificity and transcript tissue distribution of estrogen receptors alpha and beta. *Endocrinology*. 1997;138:863—870.
10. Lange A, Katsu Y, Miyagawa S, et al. Comparative responsiveness to natural and synthetic estrogens of fish species commonly used in the laboratory and field monitoring. *Aquat Toxicol*. 2012;109:250—258.

CHAPTER 95

Corticosteroids

Yoshinao Katsu and Taisen Iguchi

History

Corticosteroids, glucocorticoids, and mineralocorticoids are a class of chemicals that includes steroid hormones naturally produced in the adrenal cortex of vertebrates. The adrenal cortex of adult mammals may be subdivided by means of histological criteria into three well-defined regions: zona glomerulosa, zona fasciculata, and zona reticularis. These regions are arranged as concentric shells surrounding the adrenal medulla. The cells of the outermost region of the adrenal cortex, the zona glomerulosa, are smaller and more rounded and contain less lipid than those of the more central zona fasciculata. The zona fasciculata is responsible for synthesis of glucocorticoids, and the zona glomerulosa is responsible for synthesis of mineralocorticoids [1,2]. Edward Calvin Kendall, Tadeus Reichstein, and Philip Showalter Hench were awarded the Nobel Prize in 1950 for their work on steroids of the adrenal cortex.

Structure and Biosynthesis

Structural Features and Biosynthetic Pathways

All vertebrate steroid hormones including corticosteroids are synthesized from cholesterol by steroidogenic enzymes located in the endoplasmic reticulum and mitochondria (Figure 95.1 and E-Figure 95.1) [2]. The steroidogenic enzymes, 3β-hydroxysteroid dehydrogenase (3β-HSD) and five different cytochrome P450s, CYPs (CYP11A1, CYP17, CYP21, CYP11B1, and CYP11B2) are necessary for corticosteroid production from cholesterol. CYP11A1 is a side chain cleavage enzyme, CYP17 is a steroid 17α-hydroxylase and 17,20-lyase, CYP21 is a steroid 21-hydroxylase, CYP11B1 is a steroid 11β-hydroxylase, and CYP11B2 is an aldosterone synthase. Cholesterol is converted to pregnenolone by cytochrome CYP11A1 in mitochondria, and then converted to deoxycortisol, deoxycorticosterone, and dehydroepiandrosterone or androstenedione by enzymes in the endoplasmic reticulum. Deoxycortisol and deoxycorticosterone move to mitochondria and are converted to cortisol and corticosterone by CYP11B. Further, corticosterone is converted to aldosterone by CYP11B1 or CYP11B2 localized in the mitochondria. Corticosteroid production in the adrenal cortex is regulated by the pituitary hormone adrenocorticotropin (ACTH), which is released from the pituitary gland, and ACTH production is regulated by hypothalamic corticotropin-releasing hormone (CRH). Aldosterone secretion is promoted by angiotensin II and potassium ions with ACTH in the blood [2].

Receptors

The glucocorticoids and mineralocorticoids bind to nuclear receptors belonging to subfamily 3C: the glucocorticoid receptor (GR) and mineralocorticoid receptor (MR), respectively. GR was cloned in 1985 and MR shortly thereafter [3]. Individually and in combination, these two corticosteroid receptors play pivotal roles in some of the most fundamental aspects of physiology such as the stress response, metabolism, immune function, electrolyte homeostasis, growth, and development. Multiple signaling pathways have been established in these receptors, and several common mechanisms have been revealed [4]. The GR and MR share structural similarities, containing three functional domains [A/B N-terminal transactivation domain; DNA-binding domain (DBD); ligand-binding domain (LBD)]. Compared with the human GR, the sequence identity of the A/B domain of the human MR is 38%, the DBD is 94%, and the LBD is 57%, respectively [5]. The crystal structures for the LBD of the GR and MR have been made available through the work of several groups [5,6] (E-Figures 95.1–95.6, E-Tables 95.1 and 95.2). The main signaling pathway is via direct DNA binding and transcriptional regulation of responsive genes (E-Figure 95.2). Reflecting the high degree of homology within their DBD, GR and MR bind to common nuclear hormone response elements (HREs), with a consensus 15-nucleotide sequence of AGAACAnnnTGTTCT as a homodimer [3,7].

Biological Functions

Target Cells/Tissues and Functions

Corticosteroids are implicated in stress response, carbohydrate metabolism, protein catabolism, retention of sodium in the kidney, and regulation of inflammation. Corticosteroids also are involved in bone development, blood electrolyte levels, and behavior [6].

Phenotype of Gene-Modified Animals

Homozygous MR-deficient mice have a normal prenatal development, and during the first week of life, these animals develop symptoms of pseudohypoaldosteronism, lose weight, and eventually die at around day 10 due to kidney failure. The other, GR null, mice die within hours after birth because of respiratory failure. They have atelectatic lungs, impaired liver function, impaired hypothalamus-pituitary-adrenal (HPA) axis, increased plasma levels of ACTH and corticosterone, and enlarged adrenal glands that produce no adrenaline [3,8].

Y. Takei, H. Ando, & K. Tsutsui (Eds): Handbook of Hormones. DOI: http://dx.doi.org/10.1016/B978-0-12-801028-0.00095-7

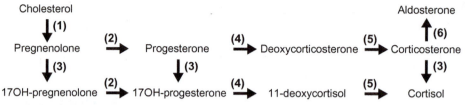

Figure 95.1 Synthesis of corticosteroids from cholesterol. The steps (1–5) in the biosynthetic pathway are catalyzed by: CYP11A1 (1), 3β-HSD (2), CYP17 (3), CYP21(4), and CYP11B1 (5). Step (6) is catalyzed by CYP11B1 or CYP11B2.

References

1. Mesiano S, Jaffe RB. Developmental and functional biology of the primate fetal adrenal cortex. *Endocrine Rev.* 1997;18:378–403.
2. Miller WL. Molecular biology of steroid hormone synthesis. *Endocrine Rev.* 1988;9:295–318.
3. Mangelsdorf DJ, Thummel C, Beato M, et al. The nuclear receptor superfamily: the second decade. *Cell.* 1995;83:835–839.
4. Bledsoe RK, Montana VG, Stanley TB, et al. Crystal structure of the glucocorticoid receptor ligand binding domain reveals a novel mode of receptor dimerization and coactivator recognition. *Cell.* 2002;110:93–105.
5. Huyet J, Pinon GM, Fay MR, et al. Structural determinants of ligand binding to the mineralocorticoid receptor. *Mol Cell Endocrinol.* 2012;350:187–195.
6. Lu NZ, Wardell SE, Burnstein KL, et al. International union of pharmacology. LXV. The pharmacology and classification of the nuclear receptor superfamily: glucocorticoid, mineralocorticoid, progesterone, and androgen receptors. *Pharmacol Rev.* 2006;58:782–797.
7. Ratman D, Vanden Berghe W, Dejager L, et al. How glucocorticoid receptors modulate the activity of other transcription factors: a scope beyond tethering. *Mol Cell Endocrinol.* 2013;380:41–54.
8. Berger S, Bleich M, Schmid W, et al. Mineralocorticoid receptor knockout mice: pathophysiology of Na$^+$ metabolism. *Proc Natl Acad Sci USA.* 1998;95:9424–9429.

Supplemental Information Available on Companion Website

- Synthesis pathway for corticosteroids/E-Figure 95.1
- Receptor of glucocorticoids/E-Figure 95.2
- Gene, mRNA, structures of GR /E-Figure 95.3
- Gene, mRNA, structures of MR/E-Figure 95.4
- Primary structure of GR of various animals/E-Figure 95.5
- Primary structure of MR of various animals/E-Figure 95.6
- Accession numbers of gene and cDNA for vertebrate GR and MR/E-Table 95.1 and 95.2

Corticosterone

Yoshinao Katsu and Taisen Iguchi

Abbreviation: B or CORT
Additional names: 11β,21-dihydroxy-4-pregnene-3,20-dione, 11β,21-dihydroxyprogesterone, Kendall's compound B, Reichstein's substance H, 4-pregnene-11β,21-diol-3,20-dione

Corticosterone is one of the glucocorticoids isolated from the adrenal gland; however, the glycogen retention and the anti-inflammatory action are weaker than those of cortisol in humans. It is synthesized in the zona fasciculata of the adrenal cortex in many animals, and converted into aldosterone via 18-hydroxycorticosterone in the zona glomerulosa. Rats, mice, birds, reptiles, and amphibians have no 17α-hydroxylase activity in the adrenal cortex, and corticosterone (B) is the only glucocorticoid. In humans, B is important as an intermediate product of the reaction to produce aldosterone from pregnenolone.

Discovery

In 1937, B was isolated and identified from adrenal gland extracts. Further, B has been crystallized and biological activity has been also examined using adrenalectomized dogs (Figure 95A.1) [1].

Structure

Structural Feature

Pregnenolone can be modified to deoxycorticosterone by two enzyme activities caused by 3β-hydroxysteroid dehydrogenase and 21α-hydroxylase at the smooth endoplasmic reticulum. B is converted from deoxycorticosterone by 11β-hydroxylase at the mitochiondria in the zona fasciculata of the adrenal gland. Thus, B is important mainly as an intermediate in the steroidogenic pathway from pregnenolone to aldosterone in humans.

Properties

Mr 346.46. Melting point: 179°C. Water solubility: 0.199 mg/ml.

Synthesis and Release

Gene, mRNA, and Precursor

The human CYP11B1 gene, located on chromosome 8 (8q21), consists of nine exons. Human CYP11B1 mRNA has 3,534 bp that encode a protein of 503 aa residues (E-Figures 95A.1 and 95A.2, and E-Table 95A.1).

Blood Concentration

Serum concentrations in humans are 0.38–8.42 ng/ml for males and 0.21–8.48 ng/ml for females [2]. In amphibians, reptiles, and birds, the major corticosteroids synthesized by adrenocortical tissue are aldosterone and B as stress responses (Table 95A.1) [2].

Regulation of Synthesis and Release

Rats, mice, birds, reptiles, and amphibians have no 17α-hydroxylase activity that catalyzes the conversion of B to cortisol in the adrenal cortex. B is the only glucocorticoid in these animals. Further, several studies show that stress increases the concentration of B in amphibians, reptiles, birds, and mammals [3].

Biological Functions

Physiological Actions

Several studies indicate that endogenous B is a causal factor to behavioral response to stress [4]. In amphibians, stressors cause elevation in plasma B levels. Administration of either aldosterone or B induces hyperglycemia, but it is likely that only B is a physiological glucocorticoid. Seasonal rhythms of B levels that show highest levels during breeding have been reported in plasma samples from several amphibian species. Although high levels of B can inhibit reproduction, B serves as a competitor to androgens. In reptiles, the seasonal pattern of B secretion generally shows a peak during the breeding season, although some species show no breeding secretion increase and a few exhibit a secretion decrease with breeding [4]. Cortisol and B are glucocorticoids, but their physiological functions are different. Both steroids play a basic role in the functional zonation of the adrenal gland. However, these steroids have different effects on adrenal steroidogenesis; cortisol inhibits 18-hydroxydeoxycorticosterone and aldosterone production in tissue culture of adrenals, whereas B inhibits cortisol production and stimulates the increase of androgen production. Further, B plays different roles in the growth of undifferentiated and differentiated fetal rat adrenocortical cells; a low concentration supports the proliferation of undifferentiated zona glomerulosa cells, and a high concentration inhibits the proliferation of differentiated zona fasciculata cells. Son et al. (2014) reported that B directly induces the transcription of the gonadotropin-inhibitory hormone (GnIH) gene, which inhibits gonadotropin release from the anterior pituitary, by recruitment of glucocorticoid receptor (GR) to its promoter [5].

Y. Takei, H. Ando, & K. Tsutsui (Eds): Handbook of Hormones. DOI: http://dx.doi.org/10.1016/B978-0-12-801028-0.00228-2

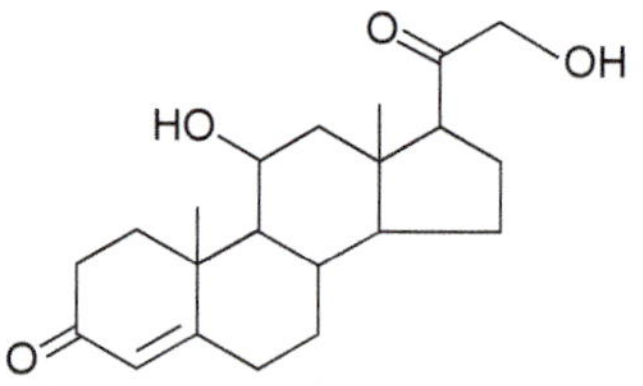

Figure 95A.1 Structure of corticosterone.

Table 95A.1 Basal Plasma Corticosterone and Stress in Selected Amphibians, Reptiles, and Birds

Species	Basal Levels (ng/ml)	Stressed Levels (ng/ml)
Amphibian		
American bullfrog	2.0−16.0	180.0
red-legged salamander	12.5	22.8
Reptiles		
green sea turtle	0.3	7.0−15.0
Australian crocodile	4.0	~40.0
Birds		
starling	~10.0	~40.0
white crowned sparrow	~15.0	~27.0

Based on Norris and Carr (2013) [2].

Mechanisms of Actions

During periods of stress, elevated B concentrations inhibit reproductive behaviors in amphibians, and the B response is rapid. Genomic actions of B via nuclear GR require a period of hours to induce changes in transcriptional regulation; however, B administration inhibits reproductive behaviors including clasping within a few minutes, suggesting that the rapid effect of B involves cell surface receptors [6]. In general, B acts via the GR. For example, the effects of B on blood pressure regulation within the hindbrain are mediated in part by the GR, but are also likely to involve MR-mediated effects. Relative glucocorticoid and mineralocorticoid activities of B, cortisol, and aldosterone are shown in Table 95A.2.

Pharmacological Implications

Clinical Implications

The concentration of B increases by 17α-hydroxylase deficiency and aldosterone synthase deficiency among congenital adrenal cortex hyperplasias [7]. Patients with these diseases show a decrease of blood cortisol and an increase of adrenocorticotropic hormone (ACTH), and as a result, blood corticosterone shows high concentrations. Symptoms of 17α-hydroxylase deficiency often come out after puberty with high blood pressure and hypogonadism. When such symptoms appear, 17α-hydroxylase deficiency is suspected and B levels are measured [7]. B decreases in 3β-hydroxysteroid dehydrogenase deficiency, and 11β-hydroxylase deficiency [8]. As production of B as well as cortisol are ACTH dependent, high concentration levels lead to Cushing's disease and ectopic ACTH-producing tumors, while low concentration levels lead to pan-hypopituitarism and isolated ACTH

Table 95A.2 Relative Activities of Corticosteroids

Steroid	MR[a]	GR[a]	Na⁺[b]	GLY[c]	AI[d]
Corticosterone	0.2	2.0	0.004	0.36	0.3
Cortisol	0.1	1.0	0.001	1.0	1.0
Aldosterone	1.0	1.0	1.0	0.15	−

[a]: Based on receptor competition binding assay
[b]: Based on change in urinary Na^+/K^+ with bioassay
[c]: Based on glycogen deposition in the adrenalectomized rat liver
[d]: based on anti-inflammatory activity
Based on Norris and Carr (2013) [2].

deficiency. Therefore, the clinical significance is low. Even a primary adrenal insufficiency such as Addison's disease shows low B concentration, similarly to cortisol, due to increasing ACTH. Further, B deficiency induces the decrease of RNA levels and enzyme activities of P450scc and 17β-HSD, and it causes the impairment of testosterone production in Leydig cells [9].

Use for Diagnosis and Treatment

As stated above, corticosterone is helpful for diagnosis of certain diseases. However, unlike cortisol, it is not used for clinical treatment.

References

1. Mason HL, Hoehn WM, Mckenzie BF, et al. Chemical studies of the suprarenal cortex: III. The structures of compounds A, B, and H. *J Biol Chem.* 1937;120:719−741.
2. Norris DO, Carr JA. *Vertebrate Endocrinology.* 5th edn. New York: Academic Press; 2013.
3. Cockrem JF. Individual variation in glucocorticoid stress responses in animals. *Gen Comp Endocrinol.* 2013;181:45−58.
4. Lutterschmidt DI, Mason RT. Endocrine mechanisms mediating temperature-induced reproductive behavior in red-sided garter snakes (*Thamnophis sirtalis parietalis*). *J Exp Biol.* 2009;212:3108−3118.
5. Son YL, Ubuka T, Narihiro M, et al. Molecular basis for the activation of gonadotropin-inhibitory hormone gene transcription by corticosterone. *Endocrinology.* 2014;155:1817−1826.
6. Orchinik M, Moore FL, Rose JD. Mechanistic and functional studies of rapid corticosteroids actions. *Ann N Y Acad Sci.* 1994;746:101−112.
7. Yanase T, Simpson ER, Waterman MR. 17alpha-hydroxylase/17,20-lyase deficiency: from clinical investigation to molecular definition. *Endocr Rev.* 1991;12:91−108.
8. Simard J, Ricketts M, Gingras S, et al. Molecular biology of the 3beta-hydroxysteroid dehydrogenase/delta5-delta4 isomerase gene family. *Endocr Rev.* 2005;26:525−582.
9. Parthasarathy C, Balasubramanian K. Effects of corticosterone deficiency and its replacement on Leydig cell steroidogenesis. *J Cell Biochem.* 2008;104:1671−1683.

Supplemental Information Available on Companion Website

- Gene, mRNA, and precursor structure of human CYP11B1/E-Figure 95A.1
- Primary structure of CYP11B1 of various animals/E-Figure 95A.2
- Accession numbers of gene and cDNA for CYP11B1/E-Table 95A.1

18-Hydroxycorticosterone

Yoshinao Katsu and Taisen Iguchi

Abbreviation: 18-OHB
Additional names: 11β,18,21-trihydroxypregn-4-ene-3,20-dione

A derivative of corticosterone, 18-hydroxycorticosterone has a weak electrolyte action. It serves as an intermediate in the synthesis of aldosterone by aldosterone synthase in the zona glomerulosa.

Discovery

In 1960, 18-hydroxycorticosterone (18-OHB) was isolated from the adrenal tissue of the bullfrog, and the corresponding lactone from pig adrenals (Figure 95B.1) [1].

Structure

Structural Features

Corticosterone is converted to aldosterone, and it is a major biosynthetic route to mineralocorticoid. The terminal steps in the biosynthesis of aldosterone from corticosterone in the human adrenal cortex are mediated by the enzymes CYP11B1 (11β-hydroxylase) and CYP11B2 (aldosterone synthase), respectively. CYP11B1 catalyzes 11β-hydroxylation of 11-deoxycorticosterone to corticosterone in both the zona glomerulosa and zona fasciculata/reticularis. CYP11B2 is expressed exclusively in the zona glomerulosa, where it catalyzes the 11β-hydroxylation of deoxycorticosterone to corticosterone, and further, 18-hydroxylase (CYP11B2) acts on corticosterone to form 18-OHB. 18-OHB secretion moves closely and parallel to that of aldosterone in bodies of healthy and diseased subjects. 18-Oxidation activity by 18-hydroxydehydrogenase (CYP11B2) catalyzes the conversion of 18-OHB to its metabolite aldosterone (Figure 95B.2) [2].

Properties

Mr 362.46. Water solubility: 0.15 g/l.

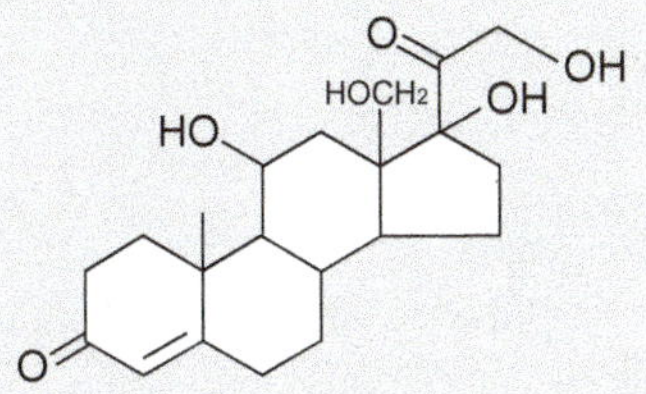

Figure 95B.1 Structure of 18-OHB.

Synthesis and Release

Gene, mRNA, and Precursor

The human CYP11B2 gene, located on chromosome 8 (8q21–q22), consists of nine exons (E-Figure 95B.1). Human CYP11B2 mRNA has 2,935 bp that encode a protein of 503 aa residues (E-Figure 95B.2 and E-Table 95B.1).

Tissue and Plasma Concentrations

Recumbent position: Adult male: 10.3 ± 4.2 ng/100 ml; adult female: 12.4 ± 4.5 ng/100 ml [4]. There is no difference between the sexes or the periods of the menstrual cycle. Further, adrenocorticotropic hormone (ACTH) stimulation increases the plasma concentration of 18-OHB, while dexamethasone markedly decreases the levels of this steroid [3].

Regulation of Synthesis and Release

Serum concentration of 18-OHB increases with increased aldosterone synthesis due to sodium depletion and angiotensin II infusion. ACTH also has a powerful influence over 18-OHB secretion. ACTH is able to stimulate 18-OHB secretion, although it may not be necessary for maintaining its normal basal level. It is unknown whether 18-OHB is the main aldosterone precursor *in vivo*, since it is reported that the conversion of 18-OHB to aldosterone is approximately one-twentieth that of corticosterone to aldosterone [4].

Biological Functions

Physiological Actions

Very little is known about the physiological function of 18-OHB in vertebrates. 18-OHB is an intermediate precursor in the aldosterone biosynthetic pathway and has low affinity for the mineralocorticoid receptor (MR). 18-OHB binds poorly (0.2% compared to aldosterone) to renal MRs, and lacks the salt-retaining action in humans. Using adrenalectomized rats, a physiological dose of 18-OHB has been found to have positive effects on the brush border Na^+/K^+ exchanger, as with corticosterone [5]. High concentration of 18-OHB has an ability to stimulate electrogenic Na^+ absorption, although its calculated K_m value is outside the physiological range for a corticosteroid [6].

Pathophysiological Implications

Clinical Implications

In patients with Cushing's syndrome, the plasma level of 18-OHB is high in cases of adrenocortical hyperplasia and

Y. Takei, H. Ando, & K. Tsutsui (Eds): Handbook of Hormones. DOI: http://dx.doi.org/10.1016/B978-0-12-801028-0.00229-4

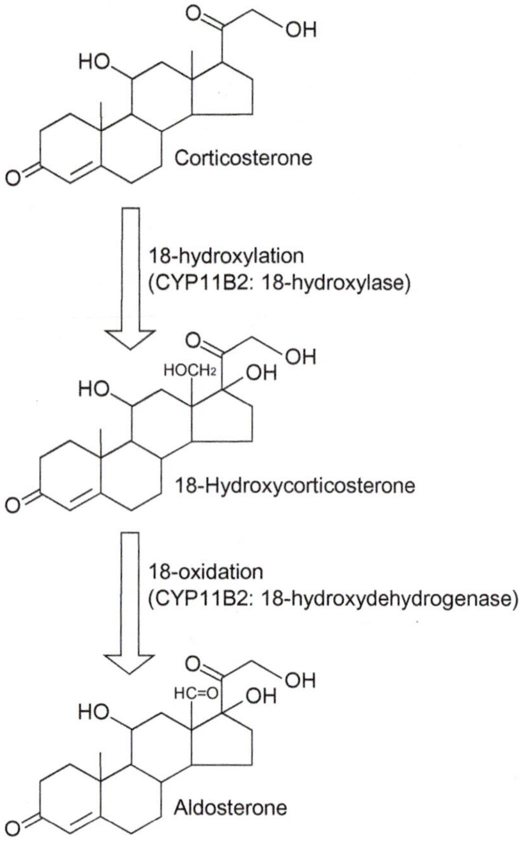

Figure 95B.2 Synthesis of 18-OHB.

adrenocortical adenoma, and also increases in primary aldosteronism, idiopathic hyperaldosteronism, and congenital 17α-hydroxylase deficiency. On the other hand, 18-OHB levels decrease in Addison's disease and the salt-losing type of congenital 21α-hydroxylase deficiency [7]. Congenital hypoaldosteronism is subdivided into two types according to the levels of aldosterone and its precursors: one type shows elevated serum levels of corticosterone and low levels of 18-OHB and aldosterone, while the other shows high levels of 18-OHB. The former type is generated by loss of both 18-hydroxylation and 19-oxidation enzyme activities, and the latter is associated with deletion of the ability to convert 18-OHB to aldosterone. Further, Witzgall et al. (1982) reported that the increase of 18-OHB was significantly greater with the administration of furosemide in patients with low-renin essential hypertension than in those with normal renin essential hypertension, suggesting that this steroid seems to be a good marker for this phenomenon [8].

Use for Diagnosis and Treatment

None.

References

1. Ulick S, Nicolis GL, Vetter KK. Relationship of 18-hydroxycorticosterone to aldosterone. In: Baulieu EE, Robel P, eds. *Aldosterone A Symposium.*. Oxford: Blackwell; 1964:3−17.
2. White PC. Aldosterone synthase deficiency and related disorders. *Mol Cell Endocrinol.* 2004;217:81−87.
3. Ojima M, Kambegawa A. *In vitro* and *in vivo* studies on 18-hydroxy-11-deoxycorticosterone and 18-hydroxycorticosterone in normal subjects and in those with various adrenocortical disorders. *Folia Endocrinol Jpn.* 1979;55:994−1006.
4. Ulick S. Correction of the nomenclature and mechanism of the aldosterone biosynthetic defects. *J Clin Endocrinol Metab.* 1996;81:1299−1300.
5. Igarreta P, Lantos CP, Calvo JC, et al. Dose-dependence of the effects of corticosterone and the non-glucocorticoid, 18-hydroxycorticosterone, on the brush-border Na + /H + exchanger. *Endocr Res.* 1998;24:601−605.
6. Grotjohann I, Schulzke JD, Fromm M. Electrogenic Na$^+$ transport in rat late distal colon by natural and synthetic glucocorticosteroids. *Am J Physiol.* 1999;276(2 Pt 1):G491−G498.
7. De Groot L.J., ed. Endotext: The free complete source for clinical endocrinology. MDText.com. 2015. Available from: <www.endotext.org>
8. Witzgall H, Thayil G, Weber PC. Rapid increase of mineralocorticoids after furosemide in low-renin essential hypertension: evidence for 18-hydroxycorticosterone to be a better marker than aldosterone. *Klin Wochenschr.* 1982;60:847−852.

Supplemental Information Available on Companion Website

- Gene, mRNA, and precursor structure of human CYP11B2/E-Figure 95B.1
- Primary structure of CYP11B2 of various animals/E-Figure 95B.2
- Accession numbers of gene and cDNA for CYP11B2/E-Table 95B.1

1α-Hydroxycorticosterone

Yoshinao Katsu and Taisen Iguchi

Abbreviation: 1α-OHB
Additional names: 4-pregnen-1α,11β,21-triol-3,20-dione; 1α,11β,21-trihydroxypregnen-4-ene-3,20-dione

1α-Hydroxycorticosterone (1α-OHB) is a physiologically important corticosteroid synthesized only in elasmobranchs. It is synthesized in the interrenal gland in copious amounts relative to other endogenous steroids. 1α-Hydroxylation is necessary for 1α-OHB synthesis and its reaction is detected in steroidogenic tissues of elasmobranchs, but the underlying gene has not been identified.

Discovery

In 1966, a novel corticosteroid, 1α,11β,21-trihydroxypregnen-4-ene-3,20-dione was isolated from the plasma of two species of ray, and this steroid was later confirmed to be 1α-OHB (Figure 95C.1) [1]. Further, 1α-OHB was found to be the dominant corticosteroid produced by the interrenal tissue in elasmobranchs [2]. In 1972, 1α-OHB was synthesized biochemically and a radioimmunoassay (RIA) was developed for the measurement of the steroid from significantly smaller volumes of plasma.

Structure

Structural Feature

Normal biosynthetic pathway of adrenocortical steroids is used for 1α-OHB production until corticosterone synthesis (Figure 95C.2). In the elasmobranchs, corticosterone is further hydroxylated to form the bioactive 1α-OHB. Synthesized deoxycorticosterone is translocated to the mitochondria where the enzyme CYP11B is responsible for the α-hydroxylation of deoxycorticosterone to corticosterone. The final step in the synthesis of 1α-OHB is the hydroxylation of C1 in corticosterone [3]. Corresponding genes for hydroxylation at position 1 for vitamin D3 are characterized in humans [4].

Properties

Mr 362.46.

Synthesis and Release

Gene, mRNA, and Precursor

A functional equivalent gene for the hydroxylation of corticosterone has not been identified in elasmobranchs.

Regulation of Synthesis and Release

Many genes encoding steroidogenic enzymes have been identified in elasmobranchs; however, there are a few key enzymes unidentified which have to be examined including 1α-hydroxylase, and CYP11B, which acts on the process of the conversion of deoxycorticosterone to corticosterone [2]. In mammals, the major mineralocorticoid is aldosterone. However, fish do not synthesize and release aldosterone. Corticosteroids may play dual roles regulating both mineralocorticoid and glucocorticoid responses in fish. The glucocorticoid response in fish is triggered by release of corticotrophin-releasing hormone (CRH) in the hypothalamus. In fact, injection of CRH induced an increase in the circulating levels of 1α-OHB in the dogfish; further, adenocorticotropic hormone (ACTH) has been shown to stimulate 1α-OHB production both *in vitro* and *in vivo* [5].

Tissue and Plasma Concentration

The presence of 1α-OHB as a major plasma corticosteroid has been confirmed in the ray, skate, and dogfish, and *in vitro* incubation has demonstrated conversion of corticosterone to 1α-OHB in interrenal tissues in the ray. Plasma levels of 1α-OHB have been reported in several elasmobranchs (Table 95C.1) [2,6,7]. For quantitation of 1α-OHB, HPLC fractionation is recommended, because it is difficult to distinguish 1α-OHB from corticosterone accurately with the ELISA method.

Biological Functions

Physiological Actions

1α-OHB has a mineralocorticoid-like role in the regulation of salt and osmolyte balance. Administration of 1α-OHB to rats produced significant sodium retention, and it has been shown to stimulate sodium transport across the isolated toad bladder. The increase in plasma sodium is coupled with a significant increase in plasma 1α-OHB concentration, suggesting that 1α-OHB possesses a mineralocorticoid role in renal function. A mineralocorticoid function of 1α-OHB in elasmobranchs has been demonstrated in key tissues involved in osmoregulation such as the gill, kidney, and rectal gland of the skate [9]. The evidence for 1α-OHB's glucocorticoid-like effects is speculative. ACTH clearly influences corticosteroidogenesis in elasmobranchs, implying that 1α-OHB is the principal glucocorticoid in elasmobranchs. There is no evidence of the direct action of 1α-OHB on glucose metabolism in elasmobranchs. Salinity transfer studies showed an increase in circulating levels of 1α-OHB, but it is clear there is a need for a definitive

Y. Takei, H. Ando, & K. Tsutsui (Eds): Handbook of Hormones. DOI: http://dx.doi.org/10.1016/B978-0-12-801028-0.00230-0

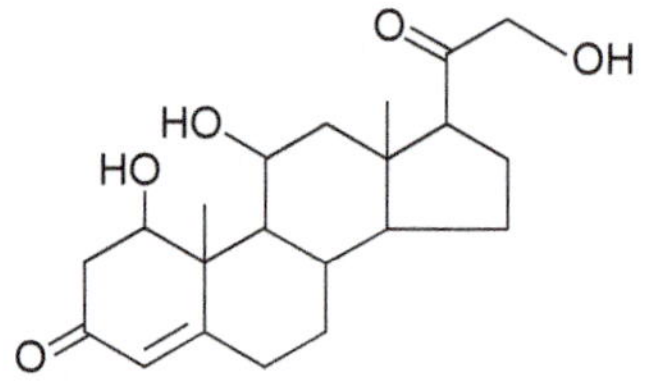

Figure 95C.1 Structure of 1α-OHB.

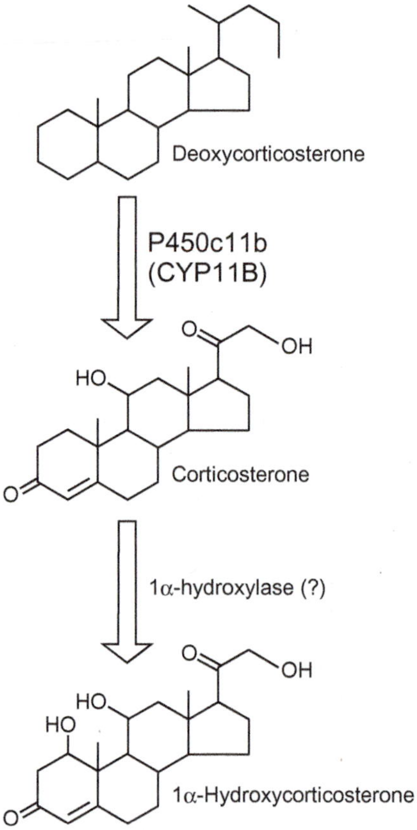

Figure 95C.2 Synthesis of 1α-OHB.

measurement of 1α-OHB in the plasma of elasmobranchs following a stressful event.

Mechanisms of Action

The mineralocorticoid receptor (MR) and glucocorticoid receptor (GR) have been isolated from elasmobranchs (E-Figures 95C.1 and 95C.2, and E-Table 95C.1); however, the molecular basis for 1α-OHB action in elasmobranchs, particularly its potential interactions with MR and GR, remains unclear. Carroll et al. (2008) showed that using an *in vitro* reporter gene assay, MR serves as a high-sensitivity and GR as a low-sensitivity receptor (Table 95C.2) [8,10]. MR is a high-sensitivity receptor that responds to low nanomolar concentrations of 1α-OHB including other corticosteroids, while two to four orders of higher magnitude concentrations are needed to achieve half-maximal activation of GR. Endogenous concentrations of 1α-OHB in elasmobranchs are expected to activate the MR [11].

Table 95C.1 Plasma 1α-OHB Levels Reported for Elasmobranchs

Species	1α-OHB (ng/ml)
Raja clavata	<0.8−139.0
Raja radiata	2−43
Raja ocellata	5−20
Scyliorhinus caninus	1−7.8

References: Anderson WG (2012) [2], Pankhurst NW (2011) [6].

Table 95C.2 Ligand Sensitivities of Skate MR and GR

	EC$_{50}$ (nM)	
Ligand	MR	GR
Corticosterone	0.09	58.4
Deoxycorticosterone	0.03	305.57
1α-Hydroxycorticosterone	3.84	947.35
Cortisol	1.04	138.92
Aldosterone	0.07	11.0

Reference: Carroll SM et al. (2008) [10].

Pathophysiological Implications

None.

References

1. Idler DR, Truscott B. 1α-hydroxycorticosterone from cartilaginous fish: a new adrenal steroid in blood. *J Fish Res Bd Canada*. 1966;23:615−619.
2. Anderson WG. The endocrinology of 1α-hydroxycorticosterone in elasmobranch fish: a review. *Comp Biochem Physiol, Part A*. 2012;162:73−80.
3. Wishart DS, Jewison T, Guo AC, et al. HMDB 3.6 − The Human Metabolome Database. *Nucleic Acids Res*. 2015;. Available from: <http://www.hmdb.ca/>
4. Bury NR, Sturm A. Evolution of the corticosteroid receptor signaling pathway in fish. *Gen Comp Endocrinol*. 2007;153:47−56.
5. Nunez S, Trant JM. Regulation of interrenal gland steroidogenesis in the Atlantic stingray (*Dasyatis sabina*). *J Exp Zool*. 1999;284:517−525.
6. Pankhurst NW. The endocrinology of stress in fish: an environmental perspective. *Gen Comp Endocrinol*. 2011;170:265−275.
7. Anderson WG. The endocrinology of 1α-hydroxycorticosterone in elasmobranch fish: a review. *Comp Biochem Physiol*. 2012;162:73−80.
8. Carroll ST, Bridgham JT, Thornton JW. Evolution of hormone signaling in elasmobranchs by exploitation of promiscuous receptors. *Mol Biol Evol*. 2008;25:2643−2652.
9. Idler DR, Kane KM. Cytosol receptor glycoprotein for 1 alpha-hydroxycorticosterone in tissues of an elasmobranch fish (*Raja ocellata*). *Gen Comp Endocrinol*. 1980;42:259−266.
10. Carroll SM, Bridgham JT, Thornton JW. Evolution of hormone signaling in elasmobranchs by exploitation of promiscuous receptors. *Mol Biol Evol*. 2008;25:2643−2652.
11. Norris DO, Carr JA. *Vertebrate Endocrinology*. 5th ed. New York: Academic Press; 2013.

Supplemental Information Available on Companion Website

- Primary structure of Chondrichthyes glucocorticoid receptor/E-Figure 95C.1
- Primary structure of little skate GR and MR/E-Figure 95C.2
- Accession numbers of gene and cDNA for Chondrichthyes GR and MR/E-Table 95C.1

Cortisol

Yoshinao Katsu and Taisen Iguchi

Abbreviation: F
Additional names: hydrocortisone; 4-pregnene-11β,17α,21-triol-3,20-dione; 11β,17α,21-trihydroxypregn-4-ene-3,20-dione; Kendall's compound F; Reichstein's substance M

One of the adrenal steroid hormones, cortisol is the mammalian principal glucocorticoid hormone, being essential for life. It influences protein and carbohydrate metabolism under certain conditions, and its major physiological role appears to be influencing peripheral utilization of glucose and maintaining blood pressure, immune function, and the body's anti-inflammatory processes.

Discovery

During the 1930s, Edward Calvin Kendall and colleagues isolated six different compounds from the adrenal gland, naming them compounds A through F (F is cortisol), and analyzed their structures (Figure 95D.1) [1].

Structure

Structural Features

Cortisol (F) is synthesized from pregnenolone derived from cholesterol by removal of the side chain at C20 catalyzed by cytochrome P450scc (CYP11A1) at the mitochondria [2]. Newly synthesized pregnenolone moves into the endoplasmic reticulum, where it is converted to 17-hydroxypregnenolone by CYP17 or to progesterone by 3β-hydroxysteroid dehydrogenase (3β-HSD). 17-Hydroxyprogesterone is synthesized from 17-hydroxypregnenolone by 3β-HSD, and progesterone by CYP17. 17-Hydroxyprogesterone is converted to 11-deoxycortisol by 21α-hydroxylase activity of CYP21, and then 11-deoxycortisol moves to the mitochondria from the endoplasmic reticulum. For the final step of F biosynthesis, 11-deoxycortisol undergoes 11β-hydroxylation that is catalyzed by CYP11B1 or CYP11B2 [3].

Property

Mr 362.46. Water solubility: 0.32 mg/ml. DMSO solubility: 100 mM. Ethanol solubility: 25 mM.

Synthesis and Release

Gene, mRNA, and Precursor

The human CYP17A gene, located on chromosome 10 (10q24.3), consists of eight exons. Human CYP17A mRNA has 1,869 bp that encodes a protein of 508 aa residues (E-Figures 95D.1 and 95D.2, E-Table 95D.1).

Tissue and Plasma Concentration

F levels are regulated by adrenocorticotropic hormone (ACTH), which responds to corticotropin-releasing hormone (CRH). CRH is released in a cyclic fashion by the hypothalamus, resulting in diurnal peaks (6−8 a.m.; 7−25 μg/dl) and nadirs (11 p.m.; 2−14 μg/dl) in the plasma cortisol level [4]. The majority of F circulates and binds to cortisol-binding globulin (CBG) and albumin. Normally, <5% of circulating cortisol is free (unbound), while the other 95% of F is bound to either CBG (80%) or albumin (15%). The unbound F is the physiologically active form [5].

Regulation of Synthesis and Release

F is produced in the animal body in the zona fasciculata of the adrenal gland [5]. The major synthetic pathways for F were described previously. The release of F is controlled by the hypothalamus-pituitary-adrenal (HPA) axis [6,7]. Synthesis and secretion of F are regulated by the pituitary hormone, ACTH, which is regulated by the hypothalamic hormone, CRH. ACTH stimulates the concentration of cAMP, and increases the concentration of cholesterol in the inner mitochondrial membrane via regulation of steroidogenic acute regulatory protein (StAR), which is related to mitochondrial cholesterol import. The cholesterol is converted to pregnenolone, catalyzed by P450scc (CYP11A1), and then the reaction of steroid synthesis advances.

Biological Functions

Physiological Actions

F has widespread actions: it reduces inflammation, contributes to the maintenance of constant blood pressure, suppresses the immune system, helps the body to manage stress, and increases blood sugar through gluconeogenesis. The relationship between F and the immune system is complex but necessary for the development and maintenance of normal human immunity. In addition, F is critical for the functions of the central nervous system, and digestive, hematopoietic, renal, and reproductive systems [5].

Mechanisms of Action

F binds to an intracellular receptor, called the glucocorticoid receptor (GR) within target cells. GR, being expressed in almost all tissues, is a transcriptional regulatory factor consisting of three different domains: an N-terminal domain containing transactivation functions, a DNA-binding domain, and a

Y. Takei, H. Ando, & K. Tsutsui (Eds): Handbook of Hormones. DOI: http://dx.doi.org/10.1016/B978-0-12-801028-0.00231-2

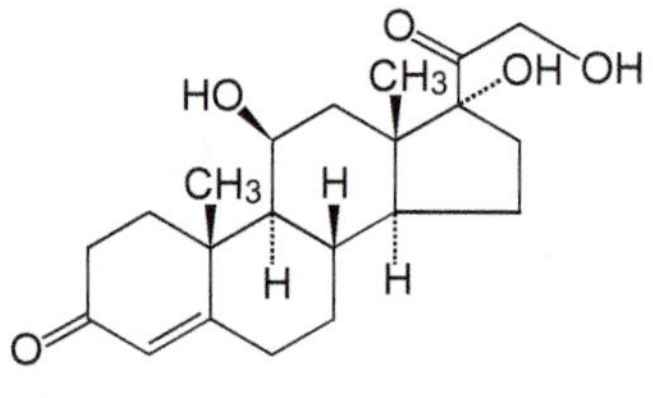

Figure 95D.1 Structure of F.

ligand-recognition domain. Cortisol passes through the plasma membrane and binds to GR with high affinity within the cytoplasm. After dissociation from the interacted chaperone protein(s), GR-F complex tranaslocates into the nucleus, and then it binds to a specific DNA motif (hormone-response element, HRE) as a homodimer. The GR-F complex interacted with specific DNA induces or inhibits gene transcription [6]. While the genomic action of F is becoming better understood, little is known of the rapid non-genomic action of F. However, F may have an effect on prolactin release as a non-genomic function in fish cell culture systems [7].

Pathophysiological Implications

Clinical Implications

Increase in F production (glucocorticoid hypersecretion or hyperadrenocorticism) is associated with Cushing's syndrome, showing excessive levels of F in the blood. Cushing's syndrome results from chronic exposure to excessive circulating levels of F. Normally, F secretion shows a circadian rhythm; however, the loss of the circadian rhythm, together with loss of the negative feedback mechanism of the HPA axis, results in chronic exposure to excessive circulating F levels that gives rise to the clinical state of endogenous Cushing's syndrome [8]. Decreased F production is associated with adrenal insufficiency and ACTH deficiency (congenital adrenal hyperplasia, CAH). In adrenal insufficiency, the F level is insufficient in the blood. Addison's disease is characterized by an absence of F resulting in hypersecretion of ACTH. A patient with Addison's disease is usually hypoglycemic due to the lack of F and reduced capacity for gluconeogenesis. CAH is a genetic defect in gene(s) for steroidogenic enzymes involved in corticosteroid synthesis; it causes reduced glucocorticoid synthesis and hence excessive ACTH secretion due to the absence of feedback. The most common form of CAH is caused by the absence of the P450c21 (21-hydroxylase activity), which blocks synthesis of corticosterone, F, and aldosterone. A newborn female with CAH may be diagnosed as of ambiguous sex or as a male depending on the extent of masculinization of the external genitalia, due to the adrenal androgen production caused by the elevated levels of ACTH [9].

Use for Diagnosis and Treatment

The concentrations of serum cortisol and urinary free cortisol are routinely measured in the diagnosis of many diseases, which have been mentioned above. As well as hydrocortisone, the synthetic glucocorticoids such as prednisolone and dexamethasone are commonly used to treat a variety of patients with adrenal insufficiency and inflammatory, allergic, and immunological disorders.

References

1. Mason HL, Hoehn WM, Kendall EC. Chemical studies of the suprarenal cortex. IV. Structures of compounds C, D, E, F, and G. *J Biol Chem*. 1938;124:459–474.
2. Mesiano S, Jaffe RB. Developmental and functional biology of the primate fetal adrenal cortex. *Endocr Rev*. 1997;18:378–403.
3. Miller WL, Auchus RJ. The molecular biology, biochemistry, and physiology of human steroidogenesis and its disorders. *Endocr Rev*. 2011;32:81–151.
4. Buckley TM, Schatzberg AF. On the interactions of the hypothalamic-pituitary-adrenal (HPA) axis and sleep: normal HPA axis activity and circadian rhythm, exemplary sleep disorders. *J Clin Endocrinol Metab*. 2005;90:3106–3114.
5. Lu NZ, Wardell SE, Burnstein KL, et al. International Union of Pharmacology. LXV. The pharmacology and classification of the nuclear receptor superfamily: glucocorticoid, mineralocorticoid, progesterone, and androgen receptors. *Pharmacol Rev*. 2006;58:782–797.
6. Stahn C, Löwenberg M, Hommes DW, et al. Molecular mechanisms of glucocorticoid action and selective glucocorticoid receptor agonists. *Mol Cell Endocrinol*. 2007;275:71–78.
7. Borski RJ, Hyde GN, Fruchtman S. Signal transduction mechanisms mediating rapid, nongenomic effects of cortisol on prolactin release. *Steroids*. 2002;67:539–548.
8. Lacroix A, Baldacchino V, Bourdeau I, et al. Cushing's syndrome variants secondary to aberrant hormone receptors. *Trends Endocrinol Metab*. 2004;15:375–382.
9. Erichsen MM, Husebye ES, Michelsen TM, et al. Sexuality and fertility in women with Addison's disease. *J Clin Endocrinol Metab*. 2010;95:4354–4360.

Supplemental Information Available on Companion Website

- Gene, mRNA, and precursor structure of human CYP17/E-Figure 95D.1
- Primary structure of CYP17 of various animals/E-Figure 95D.2
- Accession numbers of gene and cDNA for CYP17/E-Table 95D.1

Aldosterone

Yoshinao Katsu and Taisen Iguchi

Abbreviation: A or ALDO
Additional names: 11β,21-dihydroxypregn-4-ene-3,18,20-trione; 18-aldocorticosterone; 11β,21-dihydroxy-3,20-dioxo-4-pregnen-18-al; Reichstein X

Aldosterone is a steroid hormone produced by the adrenal cortex in the adrenal gland to regulate sodium and potassium balance in the blood. It specifically regulates electrolyte and water balance by increasing the renal retention of sodium and the excretion of potassium. It is synthesized from cholesterol by aldosterone (ALDO) synthase in the zona glomerulosa of the adrenal gland. Aldosterone is the principal mammalian mineralocorticoid.

Discovery

Simpson and Tait (1953) developed a bioassay with high sensitivity for mineralocorticoid activity, and crystallized ALDO (electrocortin) from beef adrenal glands.

Structure

Structural Features

By 1954 the structure of ALDO had been reported (Figure 95E.1) [1]. ALDO is produced from 18-hydroxycorticosterone by ALDO synthase (CYP11B) activity. The production of ALDO is regulated at the two critical enzyme steps: (1) in its biosynthetic pathway (the conversion of cholesterol to pregnenolone by cholesterol side chain cleavage enzyme) and (2) the conversion of corticosterone to ALDO by ALDO synthase. Steroidogenic acute regulatory protein (StAR) controls the transport of cholesterol to the inner mitochondrial membrane where P450scc (CYP11A1) is located. CYP11A1 converts cholesterol to pregnenolone. Pregnenolone can be modified to deoxycorticosterone by two enzyme activities − 3β-hydroxysteroid dehydrogenase (3β-HSD) and 21-hydroxylase at the smooth endoplasmic reticulum. Deoxycorticosterone is converted to corticosterone by 11β-hydroxylase at the mitochondria in the zona fasciculata of the adrenal gland. In the zona glomerulosa of the adrenal gland, corticosterone is modified to 18-hydroxycorticosterone and then to ALDO. This synthetic cascade is regulated by ALDO synthase, CYP11B2 [2]. This ALDO synthase is absent in other sections of the adrenal gland, and the teleost fishes probably lack ALDO synthase CYP11B2, suggesting that the dominant corticosteroid is cortisol [3].

Properties

Mr 360.44. Melting point: 166.5°C. Water solubility: 0.0512 mg/ml. Experimental solubility: soluble in ethanol, chloroform, glycols, vegetable oils, acetone, ether, and methanol. Insoluble in water.

Synthesis and Release

Gene, mRNA, and Precursor

The human CYP11B2 gene, located on chromosome 8 (8q21−q22), consists of nine exons. Human CYP11B2 mRNA has 2,935 bp that encode a protein of 503 aa residues (E-Figures 95E.1 and 95E.2, and E-Table 95E.1).

Tissue and Plasma Concentration

Reference values are 50−800 pg/ml for children (standing), 30−350 pg/ml for children (lying down), 70−300 pg/ml for adults (standing), and 30−160 pg/ml for adults (lying down) [4].

Regulation of Synthesis and Release

A variety of factors modify ALDO secretion, such as the levels of pituitary adrenocorticotropic hormone (ACTH) and kidney angiotensin II, and the plasma concentration of potassium ions. ACTH acts as an enhancer of StAR synthesis via the activation of cAMP-dependent protein kinase A. Angiotensin II, which is converted from angiotensin I by angiotensin-converting enzyme, is involved in the regulation of ALDO [5]. Angiotensin II stimulates synthesis and release of ALDO from zona glomerulosa (renin−angiotensin system). High levels of serum potassium directly stimulate aldosterone secretion from the zona glomerulosa. An increase in dietary potassium induces an increase in ALDO secretion whereas potassium depletion causes a decrease. Some animal studies show the physiological function for potassium is the regulation of ALDO secretion [5]. Thus, alterations in potassium balance as well as acute increments in serum potassium can stimulate the production of ALDO [6]. On the other hand, atrial natriuretic peptide (ANP) is an inhibitory factor of ALDO secretion. ANP secretion is increased in response to sodium and/or water loading, and it in turn inhibits ALDO secretion [7]. In human, the dosage of ANP induces the reduction of plasma renin and ALDO concentrations and inhibits angiotensin II-induced ALDO secretion.

Y. Takei, H. Ando, & K. Tsutsui (Eds): Handbook of Hormones. DOI: http://dx.doi.org/10.1016/B978-0-12-801028-0.00232-4

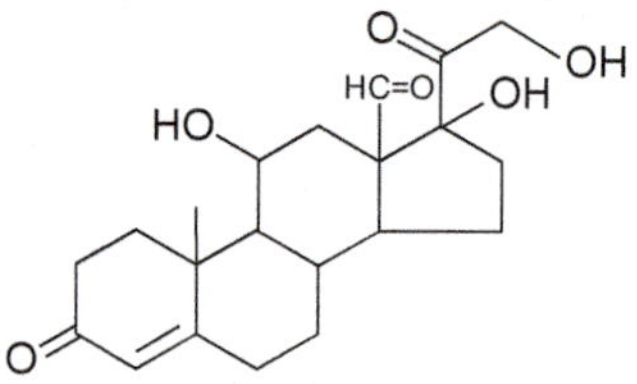

Figure 95E.1   Structure of ALDO.

Biological Functions

Physiological Actions

ALDO achieves its physiological effects by controlling the transcription of specific genes attached to the mineralocorticoid receptor (MR) in the target cell. ALDO has the most powerful mineralocorticoid activity, corresponding to controlling sodium homeostasis. ALDO regulates the transcription of the epithelial sodium channel and Na^+/K^+-ATPase subunit genes. Consequently, ALDO promotes the reabsorption of sodium from renal tubules. Thus, ALDO is a critical regulator of serum sodium and electrolytes [6].

Mechanisms of Actions

ALDO achieves the typical steroidal pattern of action on target cells after binding to the cytoplasmic receptor, MR. This mechanism is called "genomic action" of ALDO. The ALDO-MR complex binds to a specific hormone response element located at the gene promoter region, and leads to its gene specific transcription. Human MR consists of 984 amino acids, and has 57% amino acid identity with the human glucocorticoid receptor (GR) in the ligand-binding domain, and 94% in the DNA-binding domain. The non-genomic pathway as well as the genomic pathway also affects the ALDO mediated biological function at the cellular level via the non-classical receptor, MR. This non-genomic pathway, less well characterized to date, involves numerous signaling pathways in kidney, colon, vascular wall, and heart [8].

Pathophysiological Implications

Clinical Implications

Hyperaldosteronism is characterized by low blood potassium and high blood sodium. Elevated plasma sodium levels bring about water retention and may produce hypertension. Primary hyperaldosteronism is induced by the overproduction of ALDO by the adrenal gland; secondary hyperaldosteronism is due to overactivity of the renin—angiotensin system. Treatment with ALDO inhibitors can reduce these symptoms until the source of the excessive ALDO is removed. Hypoaldosteronism induces a reduction of potassium excretion. Further water retention is impaired, and sodium is lost in the urine.

Use for Diagnosis and Treatment

The concentrations of serum aldosterone are routinely measured in the diagnosis of many diseases associated with hyperaldosteronism and hypoaldosteronism. Aldosterone itself is not used for clinical treatment. Fludrocortisone, a synthetic mineralocorticoid, is administered to some patients in relation to mineralocorticoid deficiency.

References

1. Simpson SA, Tait JF, Wettstein A, et al. Aldosteron. Isolierung und Eigenschaften. Über Bestandteile der Nebennierenrinde und verwandte Stoffe. 91. Mitteilung. *Helv Chim Acta.* 1954;37:1163—1200.
2. Lisurek M, Bernhardt R. Modulation of aldosterone and cortisol synthesis on the molecular level. *Mol Cell Endocrinol.* 2004;215:149—159.
3. Bury NR, Sturm A. Evolution of the corticosteroid receptor signaling pathway in fish. *Gen Comp Endocrinol.* 2007;153:47—56.
4. Fischbach FT, Dunning III MB, eds. *Manual of Laboratory and Diagnostic Tests.* 8th ed. Philadelphia: Lippincott Williams and Wilkins; 2009.
5. Stocco DM, Clark BJ. The role of the steroidogenic acute regulatory protein in steroidogenesis. *Steroids.* 1997;62:29—36.
6. Amasheh S, Epple HJ, Mankertz J, et al. Differential regulation of EnaC by aldosterone in rat early and late distal colon. *Ann NY Acad Sci.* 2000;915:92—94.
7. Spät A, Hunyady L. Control of aldosterone secretion: a model for convergence in cellular signaling pathways. *Physiol Rev.* 2004; 84:489—539.
8. Funder JW. The nongenomic actions of aldosterone. *Endocrine Rev.* 2005;26:313—321.

Supplemental Information Available on Companion Website

- Gene, mRNA, and precursor structure of human CYP11B2/E-Figure 95E.1
- Primary structure of CYP11B2 of various animals/E-Figure 95E.2
- Accession numbers of gene and cDNA for CYP11B2/E-Table 95E.1

CHAPTER 96

Neurosteroids

Kazuyoshi Tsutsui and Shogo Haraguchi

History

The central and peripheral nervous systems have the capacity to synthesize steroids from cholesterol, the so-called neurosteroids [1–5]. *De novo* formation of neurosteroids in the brain was originally demonstrated in mammals, and subsequently in birds, amphibians, and fish [1–5]. Thus, *de novo* synthesis of neurosteroids from cholesterol in the brain appears to be conserved across vertebrate species.

Structure and Biosynthesis

Structural Features and Biosynthetic Pathways

The brain possesses several kinds of steroidogenic enzymes, such as cytochrome P450 side-chain cleavage enzyme (P450scc; *Cyp11a*), hydroxysteroid sulfotransferase (HST; *Sult2*), cytochrome P450 7α-hydroxylase (cytochrome P450$_{7\alpha}$; *Cyp7b*), 3α-hydroxysteroid dehydrogenase/Δ^5-Δ^4-isomerase (3α-HSD; *Hsd3a*) and 3β-hydroxysteroid dehydrogenase/Δ^5-Δ^4-isomerase (3β-HSD; *Hsd3b*), cytochrome P450 21-hydroxylase (P450$_{c21}$; *Cyp21*), cytochrome P450 11β-hydroxylase (P450$_{c11}$; *Cyp11b*), 5α-reductase (*Srd5a*) or 5β-reductase (*Srd5b*; only birds), cytochrome P450 17α-hydroxylase/c17,20-lyase (P450$_{17\alpha,lyase}$; *Cyp17*), 17β-hydroxysteroid dehydrogenase (17β-HSD; *Hsd17b*), and cytochrome P450 aromatase (P450arom; *Cyp19*), and produces pregnenolone, pregnenolone sulfate, 7α-hydroxypregnenolone, dehydroepiandrosterone, progesterone, corticosterone, allopregnanolone or epipregnanolone (only birds), androstenedione, testosterone, and estradiol-17β from cholesterol (Figure 96.1) [1–6]. Similar biosynthetic pathways of neurosteroids are also evident in the pineal gland [7–9].

Molecular Evolution of Family Members

Each neurosteroid has a common structural feature in vertebrates. However, only birds possess epipregnanolone instead of allopregnanolone in the brain [5]. In contrast to the structural identities of neurosteroids, the amino acid sequences of steroidogenic enzymes show diversity in vertebrates.

Neurosteroidogenic Cells

It was first accepted that glial cells, such as oligodendrocytes and astrocytes, are the site for neurosteroid formation in the brains of vertebrates [1]. Subsequently, it was demonstrated that neurons located in the brain also produce neurosteroids in vertebrates [2,4–6]. The Purkinje cell, a cerebellar neuron, expresses steroidogenic enzymes and actively produces neurosteroids *de novo* in rats and other vertebrates [2,4]. In the rat hippocampus, pyramidal neurons in the CA1–CA3 regions as well as granule cells in the dentate gyrus express steroidogenic enzymes [2,4]. Steroidogenic enzymes are also expressed in neurons in the retinal ganglion, sensory neurons in the dorsal root ganglia, and motor neurons in the spinal cord of rat [2,4]. Recently, the pineal gland has been shown to be a major neurosteroidogenic site that actively produces neurosteroids *de novo* in birds [7–9].

Regulation of Biosynthesis

Recent studies have shown that the production of neurosteroids is finely regulated by neurotransmitters, such as glutamate, γ-aminobutyric acid (GABA) and endozepines [6], neuropeptides, such as vasotocin, mesotocin, neuropeptide Y (NPY) and gonadotropin-inhibitory hormone (GnIH) [6,10], melatonin, a pineal hormone [9], prolactin, an adenohypophysial hormone [9], and glucocorticoid, an adrenal steroid hormone [9].

Receptors

Structure and Subtypes

Neurosteroids exhibit a large array of biological activities in the brain either through a conventional genomic action or through interaction with membrane receptors. In particular, neurosteroids act as allosteric modulators of the GABA$_A$ receptor (Figure 96.2), N-methyl-D-aspartic acid (NMDA) receptor (Figure 96.2), kainate receptor, α-amino-3-hydroxy-5-methyl-4-isoxazolepropionate acid (AMPA) receptor, σ receptor, glycine receptor, serotonin receptor, nicotinic receptor, and muscarinic receptor [6]. Neurosteroids may directly activate the G protein-coupled membrane receptor or indirectly modulate the binding of neuropeptides to their receptor [6].

Biological Functions

Target Cells/Tissues and Functions

Neurosteroids are implicated in differentiation, proliferation, activity, and survival of nerve cells. Neurosteroids are also involved in the control of behavioral, neuroendocrine, and metabolic processes, such as regulation of food intake, locomotor activity, sexual activity, aggressiveness, cognition, arousal, stress, depression, anxiety, sleep, and blood pressure [6,9].

Y. Takei, H. Ando, & K. Tsutsui (Eds): Handbook of Hormones. DOI: http://dx.doi.org/10.1016/B978-0-12-801028-0.00096-9

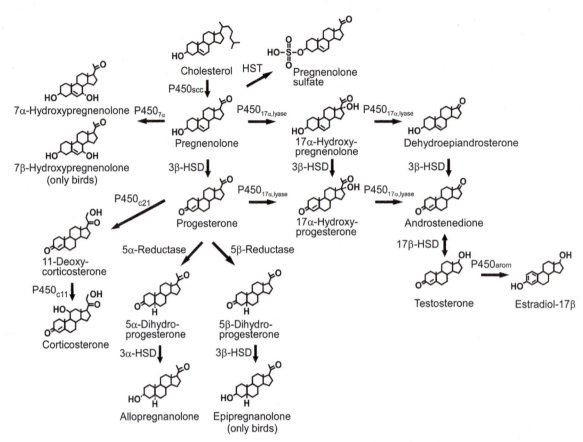

Figure 96.1   Structural features of neurosteroids and neurosteroidogenic pathways in the brain of vertebrates.

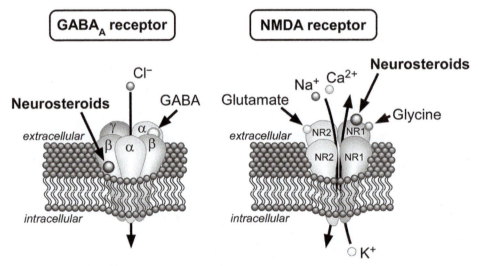

Figure 96.2   Neurosteroid actions mediated by GABA$_A$ receptor and NMDA receptor.

References

1. Baulieu EE. Neurosteroids: of the nervous system, by the nervous system, for the nervous system. *Recent Prog Horm Res.* 1997;52:1–32.
2. Tsutsui K, Ukena K, Usui M, et al. Novel brain function: biosynthesis and actions of neurosteroids in neurons. *Neurosci Res.* 2000;36:261–273.
3. Mellon SH, Vaudry H. Biosynthesis of neurosteroids and regulation of their synthesis. *Int Rev Neurobiol.* 2001;46:33–78.
4. Tsutsui K, Mellon SH. Neurosteroids in the brain neuron: biosynthesis, action and medicinal impact on neurodegenerative disease. *Central Nerv Syst Agents Med Chem.* 2006;6:73–82.
5. Tsutsui K, Matsunaga M, Miyabara H, et al. Neurosteroid biosynthesis in the quail brain. *J Exp Zool.* 2006;305A:733–742.
6. Do Rego JL, Seong JY, Burel D, et al. Neurosteroid biosynthesis: enzymatic pathways and neuroendocrine regulation by neurotransmitters and neuropeptides. *Front Neuroendocrinol.* 2009;30:259–301.

7. Hatori M, Hirota T, Iitsuka M, et al. Light-dependent and circadian clock-regulated activation of sterol regulatory element-binding protein, X-box-binding protein 1, and heat shock factor pathways. *Proc Natl Acad Sci USA*. 2011;108:4864—4869.

8. Haraguchi S, Hara S, Ubuka T, et al. Possible role of pineal allopregnanolone in Purkinje cell survival. *Proc Natl Acad Sci USA*. 2012;109:21110—21115.

9. Tsutsui K, Haraguchi S, Fukada Y, et al. Brain and pineal 7α-hydroxypregnenolone stimulating locomotor activity: identification, mode of action and regulation of biosynthesis. *Front Neuroendocrinol*. 2013;34:179—189.

10. Ubuka T, Haraguchi S, Tobari Y, et al. Hypothalamic inhibition of socio-sexual behaviour by increasing neuroestrogen synthesis. *Nat Commun*. 2014;5:3061.

Supplemental Information Available on Companion Website

- Primary structure of P450scc of various animals/E-Figure 96.1
- Primary structure of 3β-HSD of various animals/E-Figure 96.2
- Primary structure of $P450_{c21}$ of various animals/E-Figure 96.3
- Primary structure of $P450_{17\alpha,lyase}$ of various animals/E-Figure 96.4
- Primary structure of P450arom of various animals/E-Figure 96.5
- Primary structure of $GABA_A$ receptor subunits of various animals/E-Figures 96.6—96.8
- Primary structure of NMDA receptor subunits of various animals/E-Figures 96.9 and 96.10
- Accession numbers of gene and cDNA for neurosteroidogenic enzymes/E-Tables 96.1—96.5
- Accession numbers of gene and cDNA for $GABA_A$ receptor and NMDA receptor/E-Tables 96.6—96.10

Pregnenolone Sulfate

Kazuyoshi Tsutsui and Shogo Haraguchi

Additional names/abbreviation: 5-pregnen-3β-ol-20-one sulfate, pregnenolone monosulfate.

Pregnenolone sulfate is a neuroactive metabolite of pregnenolone. It has significant regulatory effects on the release of many important neurotransmitters, such as glutamate, GABA, acetylcholine, norepinephrine, and dopamine.

Discovery

The presence of high amounts of pregnenolone sulfate in the rat brain was reported and it was found that the concentration of this sulfated steroid was not affected by adrenalectomy and castration [1].

Structure

Structural Features

The synthesis of pregnenolone sulfate is regulated by a specific cytosolic enzyme hydroxysteroid sulfotransferase (HST) (Figure 96A.1), which catalyzes the transfer of a sulfate moiety from 3′-phosphoadenosine 5′-phosphosulfate on the 3-hydroxyl group of non-conjugated steroids [2]. HST was cloned and characterized [3]. The presence of the mRNA encoding the HST variant SULT2A1 in the brain was shown [3]. The presence of the corresponding enzyme in the brain was confirmed using an antibody against rat recombinant SULT2A1 [3].

Properties

Mr 396.54.

Synthesis and Release

Gene, mRNA, and Precursor

Human HST gene (*SULT2A1*), located on chromosome 19 (19q13.3), consists of six exons. Human HST mRNA has 1,987 bp that encode a protein of 285 aa residues.

Distribution of mRNA

In vertebrates, HST mRNA is highly expressed in the brain. In amphibians, immunohistochemical studies using antibodies against HST have shown the presence of the enzyme in two populations of frog hypothalamic neurons: in the anterior preoptic area and the dorsal part of the magnocellular preoptic nucleus [2]. A dense network of varicose HST-fibers is located in the telencephalon at the border between the medial septum and the nucleus accumbens, a formation that is homologous to the septum of mammals [2]. HST-immunoreactive processes originating from diencephalic cell bodies also project towards the mesencephalon [2].

Tissue and Plasma Concentrations

The pregnenolone sulfate concentration in the anterior and posterior brain of rat is approximately 15.8 and 5.7 ng/g tissue respectively [1]. That in the brain of quail is approximately 1.8 ng/g tissue [4]. The pregnenolone sulfate concentration in the male *Xenopus* brain (0.162 nmol/g tissue) is significantly higher than that in the testis (0.078 nmol/g tissue) and plasma (0.035 pmol/ml) [5]. Also in the female *Xenopus*, the concentration of pregnenolone sulfate in the brain (0.087 nmol/g tissue) is significantly higher than that in the ovary (0.021 nmol/g tissue) and plasma (0.014 pmol/ml) [5].

Regulation of Synthesis and Release

Neuropeptide Y inhibits the biosynthesis of pregnenolone sulfate in the frog hypothalamus through activation of Y_1 receptors [6]. Gonadotropin-releasing hormone (GnRH) stimulates the biosynthesis of pregnenolone sulfate in the frog hypothalamus [7].

Receptors

Structure and Subtype

It has been proposed that the receptor mechanism of the effect of pregnenolone sulfate is through activation of the σ-1 receptor. The evidence supporting this statement is that the σ-1 receptor antagonists, haloperidol and BD-1063, can block the pregnenolone sulfate-induced increase in miniature excitatory postsynaptic current (mEPSC) frequency and the σ-1 receptor agonist pentazocine can mimic the effect of pregnenolone sulfate [8]. However, a lack of involvement of the σ-1 receptor in the mechanism of pregnenolone sulfate-induced increase in mEPSC frequency was also observed [8]. In addition, the roles of N-methyl-D-aspartic acid (NMDA) receptors in the effect of pregnenolone sulfate in cultured hippocampus neurons and neonatal hippocampus slices are completely different: the NMDA receptor is not required for the presynaptic actions of pregnenolone sulfate in cultured hippocampal neurons, but it is required in neonatal slices [8]. Pregnenolone sulfate has inhibitory actions on the GABA$_A$ receptor [9]. Pregnenolone sulfate acts as a noncompetitive antagonist of the GABA$_A$ receptor by interacting with a site that is distinct from the interacting site of neurosteroids, such as allopregnanolone [9]. Pregnenolone sulfate's negative modulatory action on the GABA$_A$ receptor occurs through a

Y. Takei, H. Ando, & K. Tsutsui (Eds): Handbook of Hormones. DOI: http://dx.doi.org/10.1016/B978-0-12-801028-0.00233-6

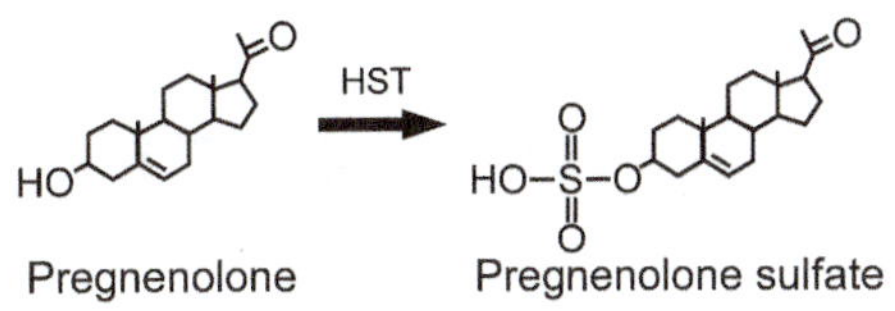

Figure 96A.1 Pregnenolone sulfate synthetic pathway.

reduction in channel opening frequency, although the precise mechanism is not understood well [9].

Signal Transduction Pathway

The downstream mechanism of the activation of the σ-1 receptor by pregnenolone sulfate is thought to involve $G_{i/o}$ proteins and intracellular Ca^{2+} because the $G_{i/o}$ protein inhibitor pertussis toxin and the Ca^{2+} chelator BAPTA can block the effect of pregnenolone sulfate [8].

Antagonist

The σ-1 receptor antagonists haloperidol and BD-1063 block the effect of pregnenolone sulfate [8].

Biological Functions

Target Cells/Tissues and Functions

Pregnenolone sulfate has multiple important effects on brain functions, such as cognitive enhancing, promnesic, antistress, and antidepressant effects [8]. Pregnenolone sulfate also has significant regulatory effects on the release of many important neurotransmitters, such as glutamate, GABA, acetylcholine, norepinephrine, and dopamine [8].

Pathophysiological Implications

Clinical Implications

Central acetylcholine neurotransmission is known to be involved in the regulation of cognitive processes [10] and is affected in normal aging and severely altered in human neurodegenerative pathologies like Alzheimer's disease [10]. The level of acetylcholine cortical release induced by infusion of pregnenolone sulfate was correlated to the spatial memory performance of animals [10]. Infusion of pregnenolone sulfate

stimulated the release of acetylcholine, suggesting that this effect could partly explain the memory enhancing properties of this neurosteroid [10].

References

1. Corpéchot C, Synguelakis M, Talha S, et al. Pregnenolone and its sulfate ester in the rat brain. *Brain Res*. 1983;270:119—125.
2. Do Rego JL, Seong JY, Burel D, et al. Neurosteroid biosynthesis: enzymatic pathways and neuroendocrine regulation by neurotransmitters and neuropeptides. *Front Neuroendocrinol*. 2009;30:259—301.
3. Shimada M, Yoshinari K, Tanabe E, et al. Identification of ST2A1 as a rat brain neurosteroid sulfotransferase mRNA. *Brain Res*. 2001;920:222—225.
4. Tsutsui K, Yamazaki T. Avian neurosteroids. II. Localization of a cytochrome P450scc-like substance in the quail brain. *Brain Res*. 1995;678:1—9.
5. Takase M, Ukena K, Yamazaki T, et al. Pregnenolone, pregnenolone sulfate, and cytochrome P450 side-chain cleavage enzyme in the amphibian brain and their seasonal changes. *Endocrinology*. 1999;140:1936—1944.
6. Beaujean D, Do Rego JL, Galas L, et al. Neuropeptide Y inhibits the biosynthesis of sulfated neurosteroids in the hypothalamus through activation of Y1 receptors. *Endocrinology*. 2002;143:1950—1963.
7. Burel D, Li JH, Do Rego JL, et al. Gonadotropin-releasing hormone stimulates the biosynthesis of pregnenolone sulfate and dehydroepiandrosterone sulfate in the hypothalamus. *Endocrinology*. 2013;154:2114—2128.
8. Zheng P. Neuroactive steroid regulation of neurotransmitter release in the CNS: action, mechanism and possible significance. *Prog Neurobiol*. 2009;89:134—152.
9. Reddy DS. Neurosteroids: endogenous role in the human brain and therapeutic potentials. *Prog Brain Res*. 2010;186:113—137.
10. Mayo W, George O, Darbra S, et al. Individual differences in cognitive aging: implication of pregnenolone sulfate. *Prog Neurobiol*. 2003;71:43—48.

Supplemental Information Available on Companion Website

- Gene, mRNA and precursor structures of human HST/E-Figure 96A.1
- Primary structure of HST of various animals/E-Figure 96A.2
- Accession numbers of gene and cDNA for HST/E-Table 96A.1

7α-Hydroxypregnenolone

Kazuyoshi Tsutsui and Shogo Haraguchi

Additional names/abbreviation: 5-pregnen-3β,7α-diol-20-one, 7α-OH PREG

7α-Hydroxypregnenolone is a neuroactive metabolite of pregnenolone. It acts as an important neuromodulator to increase locomotor activity.

Discovery

7α-Hydroxylation of pregnenolone was found in the rat brain [1]. Subsequently, 7α-hydroxypregnenolone formation was demonstrated in the brain of various vertebrates [2–6] and in the pineal gland of birds [6–8].

Structure

Structural Features

In the brain, 7α-hydroxypregnenolone is synthesized from pregnenolone by the action of cytochrome P450 7α-hydroxylase ($P450_{7\alpha}$) in vertebrates (Figure 96B.1) [2–5]. $P450_{7\alpha}$ was cloned and its physiological role was characterized [2,4,5]. Only birds produce both 7α-hydroxypregnenolone and its stereoisomer 7β-hydroxypregnenolone in the brain (Figure 96B.1) [4,6].

Properties

Mr of 7α-hydroxypregnenolone is 332.48 and its melting point is 197°C.

Synthesis and Release

Gene, mRNA, and Precursor

The human $P450_{7\alpha}$ gene (*CYP7B1*), located on chromosome 8 (8q21.3), consists of six exons. Human $P450_{7\alpha}$ mRNA has 2,395 bp that encode a protein of 506 aa residues.

Distribution of mRNA

$P450_{7\alpha}$ mRNA is expressed in the brain [2,4–6] and the pineal gland [6–8]. In the quail brain, $P450_{7\alpha}$ mRNA is detected in the nucleus preopticus medialis (POM), nucleus paraventricularis magnocellularis (PVN), nucleus ventromedialis hypothalami (VMN), nucleus dorsolateralis anterior thalami (DLA), and nucleus lateralis anterior thalami (LA) by *in situ* hybridization [4]. In the newt brain, $P450_{7\alpha}$-like immunoreactivity is detected in the lateral pallium, dorsal pallium, anterior preoptic area (POA), magnocellular preoptic nucleus (Mg), tegmental area (TA), and central gray [5].

Tissue and Plasma Concentrations

In quail, the concentration of 7α-hydroxypregnenolone in the male diencephalon changed during a 24-h period of observation, with a maximal level at 11:00 a.m. (approximately 1.8 ng/g tissue) when locomotor activity of males was high [4,6]. In contrast to males, 7α-hydroxypregnenolone concentration is constantly low (approximately 0.1 ng/g tissue) in females during the same observeation period [4,6]. In newt, 7α-hydroxypregnenolone concentration in the brain is approximately 0.6 μg/g tissue in the breeding period and 0.1 μg/g tissue in the nonbreeding period [3,5].

Regulation of Synthesis and Release

Melatonin derived from the pineal gland and eyes regulates 7α-hydroxypregnenolone synthesis in the quail brain, thus inducing diurnal locomotor changes [4,6]. The combination of pinealectomy (Px) and orbital enucleation (Ex) increases the production and concentration of 7α-hydroxypregnenolone in quail [4,6]. Melatonin administration to Px/Ex quail decreases the production and concentration of 7α-hydroxypregnenolone in the brain [4,6]. The inhibitory effect of melatonin on 7α-hydroxypregnenolone synthesis in the quail brain is abrogated by luzindole, a melatonin receptor antagonist [4,6]. Prolactin regulates 7α-hydroxypregnenolone synthesis in the newt brain, and this may be related to seasonal locomotor changes [5,6]. It is well known that locomotor activity of male newts increases during the breeding period [3,5,6]. Prolactin acts on $P450_{7\alpha}$ neurons to stimulate 7α-hydroxypregnenolone synthesis, causing seasonal locomotor changes of newt [5,6]. 7α-Hydroxypregnenolone synthesis is also regulated by stress via the action of glucocorticoids in the brain of newts [6,9]. A 30-min restraint stress increases 7α-hydroxypregnenolone synthesis in the brain tissue concomitant with the increase in plasma corticosterone concentrations in newts [6,9]. Decreasing plasma corticosterone concentrations by hypophysectomy decreases 7α-hydroxypregnenolone synthesis in the diencephalon including the dorsomedial hypothalamus (DMH), whereas administration of corticosterone to newts increases 7α-hydroxypregnenolone synthesis [6,9]. Acute stress increases the synthesis of 7α-hydroxypregnenolone via corticosterone action in the DMH, and 7α-hydroxypregnenolone activates serotonergic neurons in the DMH that may coordinate behavioral responses to stress in newts [6,9].

Y. Takei, H. Ando, & K. Tsutsui (Eds): Handbook of Hormones. DOI: http://dx.doi.org/10.1016/B978-0-12-801028-0.00234-8

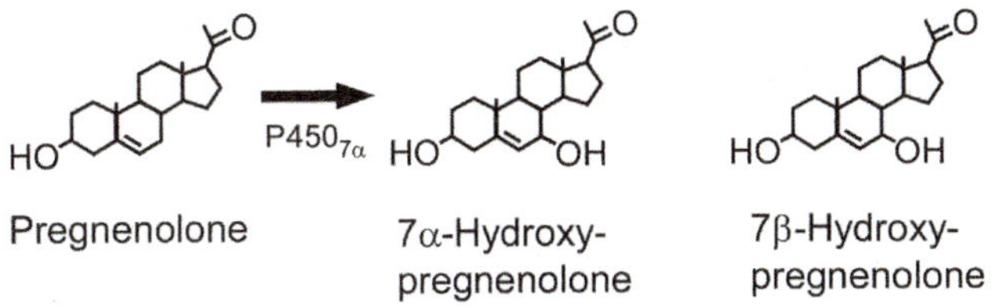

Figure 96B.1 7α-Hydroxypregnenolone synthetic pathway.

Receptors

Structure and Subtype

The receptor for 7α-hydroxypregnenolone is not known. 7α-Hydroxypregnenolone acutely increases locomotor activity in newts [3,6] and quail [4,6,7], strongly suggesting that this neurosteroid acts through a non-genomic mechanism rather than a genomic mechanism via a membrane type receptor.

Biological Functions

Target Cells/Tissues and Functions

Dopamine neurons are located in the medial brain region, specifically in the TA and posterior tuberal nucleus, and they project to the rostral brain region including the striatum [3,6]. 7α-Hydroxypregnenolone increases the concentration of dopamine in the telencephalic region including the striatum [3,6], which is known to be involved in the regulation of locomotor behavior in vertebrates. Thus, there is potential for direct regulation of dopamine release by the action of 7α-hydroxypregnenolone synthesized in the TA. Acute stress increases the synthesis of 7α-hydroxypregnenolone through corticosterone action in newts [6,9]. 7α-Hydroxypregnenolone increases serotonin concentrations in the diencephalon including the DMH under stress [6,9]. The pineal gland, an endocrine organ located close to the brain, is an important site of production of neurosteroids *de novo* from cholesterol in birds [6–8]. 7α-Hydroxypregnenolone is a major pineal neurosteroid that stimulates locomotor activity in juvenile chickens, connecting light-induced gene expression with locomotion [6,7].

Pathophysiological Implications

Clinical Implications

7α-Hydroxypregnenolone administration to memory-impaired aged rats for 11 days enhanced spatial memory retention in the eight-arm radial-arm version of the water maze [10].

References

1. Akwa Y, Morfin RF, Robel P, et al. Neurosteroid metabolism. 7α-Hydroxylation of dehydroepiandrosterone and pregnenolone by rat brain microsomes. *Biochem J.* 1992;288:959–964.
2. Stapleton G, Steel M, Richardson M, et al. A novel cytochrome P450 expressed primarily in brain. *J Biol Chem.* 1995;270:29739–29745.
3. Matsunaga M, Ukena K, Baulieu EE, et al. 7α-Hydroxypregnenolone acts as a neuronal activator to stimulate locomotor activity of breeding newts by means of the dopaminergic system. *Proc Natl Acad Sci USA.* 2004;101:17282–17287.
4. Tsutsui K, Inoue K, Miyabara H, et al. 7α-Hydroxypregnenolone mediates melatonin action underlying diurnal locomotor rhythms. *J Neurosci.* 2008;28:2158–2167.
5. Haraguchi S, Koyama T, Hasunuma I, et al. Prolactin increases the synthesis of 7α-hydroxypregnenolone, a key factor for induction of locomotor activity, in breeding male newts. *Endocrinology.* 2010;151:2211–2222.
6. Tsutsui K, Haraguchi S, Fukada Y, et al. Brain and pineal 7α-hydroxypregnenolone stimulating locomotor activity: identification, mode of action and regulation of biosynthesis. *Front Neuroendocrinol.* 2013;34:179–189.
7. Hatori M, Hirota T, Iitsuka M, et al. Light-dependent and circadian clock-regulated activation of sterol regulatory element-binding protein, X-box-binding protein 1, and heat shock factor pathways. *Proc Natl Acad Sci USA.* 2011;108:4864–4869.
8. Haraguchi S, Hara S, Ubuka T, et al. Possible role of pineal allopregnanolone in Purkinje cell survival. *Proc Natl Acad Sci USA.* 2012;109:21110–21115.
9. Haraguchi S, Koyama T, Hasunuma I, et al. Acute stress increases the synthesis of 7α-hydroxypregnenolone, a new key neurosteroid stimulating locomotor activity, through corticosterone action in newts. *Endocrinology.* 2012;153:794–805.
10. Yau JL, Noble J, Graham M, et al. Central administration of a cytochrome P450-7B product 7α-hydroxypregnenolone improves spatial memory retention in cognitively impaired aged rats. *J Neurosci.* 2006;26:11034–11040.

Supplemental Information Available on Companion Website

- Gene, mRNA and precursor structures of human P450$_{7\alpha}$//E-Figure 96B.1
- Primary structure of P450$_{7\alpha}$ of various animals/E-Figure 96B.2
- Accession numbers of gene and cDNA for P450$_{7\alpha}$/ E-Table 96B.1

Allopregnanolone

Kazuyoshi Tsutsui and Shogo Haraguchi

Additional names/abbreviation: 3α-hydroxy-5α-pregnan-20-one; 3α,5α-tetrahydroprogesterone; THP; ALLO

Allopregnanolone is a neuroactive metabolite of progesterone. It is a potent positive allosteric modulator of the action of GABA on the GABA$_A$ receptor.

Discovery

Presence of a progestational compound in the rabbit adrenal gland was reported in 1933, and it was isolated in 1938 from ox adrenal gland [1]. Allopregnanolone formation was demonstrated in the brain of various vertebrates [2–5] and the pineal gland of birds [6].

Structure

Structural Features

In the brain, allopregnanolone is synthesized from progesterone by the sequential actions of two enzymes [2–5], 5α-reductase type I (5α-RI), which transforms progesterone into 5α-dihydroprogesterone, and 3α-hydroxysteroid dehydrogenase (3α-HSD), which converts 5α-dihydroprogesterone into allopregnanolone (Figure 96C.1). Both enzymes have been cloned and characterized.

Properties

Mr of allopregnanolone is 318.49 and its melting point is 168°C.

Synthesis and Release

Gene, mRNA, and Precursor

Human 5α-RI gene (*SRD5A1*), located on chromosome 5 (5p15), consists of five exons. Human 5α-RI mRNA has 2,285 bp that encode a protein of 259 aa residues. Human 3α-HSD gene (*AKR1C4*), located on chromosome 10 (10p15.1), consists of nine exons. Human 3α-HSD mRNA has 1,192 bp that encode a protein of 323 aa residues.

Distribution of mRNA

5α-RI mRNA is highly expressed in the brain, pineal gland, prostate, and testis [3,6]. In the human brain, 5α-RI mRNA is detected in the temporal cortex, subcortical white matter, and hippocampal tissue obtained from patients with chronic temporal lobe epilepsy, as well as in the cerebellum, hypothalamus, and pons from post-mortem

brains [3,5]. In the human brain, 3α-HSD mRNA is widely expressed. In particular, 3α-HSD mRNA is found in the frontotemporal lobes, subcortical white matter, putamen, cerebellum, medulla, and spinal cord [3,5].

Tissue and Plasma Concentrations

Allopregnanolone concentration in rat hippocampus or diencephalon is approximately 40 or 10 ng/g tissue [7]. Plasma concentration of allopregnanolone is approximately 3 ng/ml [7].

Regulation of Synthesis and Release

Allopregnanolone synthesis is stimulated by swim stress [7]. In swim stress models, increased allopregnanolone levels are associated with decreased dopamine and norepinephrine levels in the prefrontal cortex, suggesting that allopregnanolone could influence the mesolimbocortical dopamine pathway [7]. *In vivo* treatment of rats with the dopamine D4 antagonist clozapine induces a rapid increase in the concentration of allopregnanolone in the cerebral cortex and striatum [8].

Receptors

Structure and Subtype

Allopregnanolone rapidly alters neuronal excitability through direct interaction with the GABA$_A$ receptor (Figure 96C.2), which is the main receptor for the inhibitory neurotransmitter GABA [3,5,6]. The GABA$_A$ receptor is an ionotropic receptor and a ligand-gated ion channel (Figure 96C.2). Allopregnanolone is a potent positive allosteric modulator of the action of GABA on the GABA$_A$ receptor (Figure 96C.2).

Signal Transduction Pathway

Activation of the GABA$_A$ receptor leads to an influx of chloride ions and to a hyperpolarization of the membrane that dampens the excitability (Figure 96C.2).

Agonist

Epipregnanolone, an isoform of allopregnanolone, acts as a partial agonist acting on the allopregnanolone modulatory site of the GABA$_A$ receptor in avian brain.

Antagonists

The GABA$_A$ receptor antagonists bicuculline and SR95531 reduce the effect of allopregnanolone.

Y. Takei, H. Ando, & K. Tsutsui (Eds): Handbook of Hormones. DOI: http://dx.doi.org/10.1016/B978-0-12-801028-0.00235-X

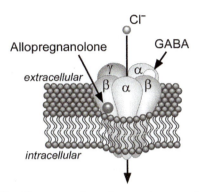

Figure 96C.1 Allopregnanolone synthetic pathway.

Progesterone — 5α-RI → 5α-Dihydro-progesterone — 3α-HSD → Allopregnanolone

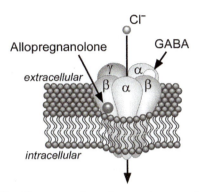

Figure 96C.2 Allopregnanolone is a potent positive allosteric GABA_A receptor modulator.

Biological Functions

Target Cells/Tissues and Functions

Allopregnanolone mediates its effects through modulation of the GABA_A receptor [3,5,6]. Allopregnanolone is reported to modulate the GABA-ergic function by increasing the GABA_A receptor opening frequency and duration at concentrations in the nanomolar range. A balance between unbinding, desensitization, and reopening of the desensitized GABA_A receptor underlies delay of inhibitory postsynaptic currents. Allopregnanolone slows the rate of recovery of the GABA_A receptor from desensitization and possibly increases the rate of entry into fast desensitized states. Allopregnanolone exerts neurogenetic, neuroprotective, antidepressant, and anxiolytic effects. Reduced levels of allopregnanolone are found to be associated with major depression, anxiety disorders, premenstrual dysphoric disorder, and Alzheimer's disease. Allopregnanolone is produced in much greater quantities in the pineal gland compared with the brain, and pineal allopregnanolone acts on Purkinje cells to prevent apoptosis in juvenile quail [6].

Phenotype in Gene-Modified Animals

Niemann-Pick type C (NP-C) disease is a fetal, autosomal recessive, childhood neurodegenerative disease characterized by accumulation of cholesterol and other lipids in the viscera and central nervous system and patterned Purkinje cell death in the cerebellum [9]. The activities of 5α-RI and 3α-HSD and allopregnanolone synthesis were substantially decreased in all principal brain regions in the postnatal NP-C mouse [9]. Neonatal administration of allopregnanolone delays the onset of neurological symptoms, and increases Purkinje and granule cell survival [9].

Pathophysiological Implications

Clinical Implications

Decreased production of allopregnanolone leads to NP-C; thus allopregnanolone treatment may be useful in ameliorating progression of the disease [9]. In patients with Alzheimer's disease, the level of allopregnanolone in the temporal cortex was significantly lower than that in controls, in contrast to pregnenolone and dehydroepiandrosterone, the concentrations of which were increased [10]. This may be explained by altered regulation of the neurosteroid biosynthetic pathway, which blocks allopregnanolone formation.

References

1. Beall D, Reichstein T. Isolation of progesterone and allopregnanolone from the adrenal. *Nature.* 1938;142:479.
2. Agís-Balboa RC, Pinna G, Zhubi A, et al. Characterization of brain neurons that express enzymes mediating neurosteroid biosynthesis. *Proc Natl Acad Sci USA.* 2006;103:14602–14607.
3. Tsutsui K, Haraguchi S. Biosynthesis and biological action of pineal allopregnanolone. *Front Cell Neurosci.* 2014;8:118.
4. Tsutsui K. Neurosteroids in the Purkinje cell: biosynthesis, mode of action and functional significance. *Mol Neurobiol.* 2008;37:116–125.
5. Do Rego JL, Seong JY, Burel D, et al. Neurosteroid biosynthesis: enzymatic pathways and neuroendocrine regulation by neurotransmitters and neuropeptides. *Front Neuroendocrinol.* 2009;30:259–301.
6. Haraguchi S, Hara S, Ubuka T, et al. Possible role of pineal allopregnanolone in Purkinje cell survival. *Proc Natl Acad Sci USA.* 2012;109:21110–21115.
7. Frye CA, Paris JJ, Walf AA, et al. Effects and mechanisms of 3α,5α,-THP on emotion, motivation, and reward functions involving pregnane Xenobiotic receptor. *Front Neurosci.* 2011;5:136.
8. Barbaccia ML, Affricano D, Purdy RH, et al. Clozapine, but not haloperidol, increases brain concentrations of neuroactive steroids in the rat. *Neuropsychopharmacology.* 2001;25:489–497.
9. Griffin LD, Gong W, Verot L, et al. Niemann-Pick type C disease involves disrupted neurosteroidogenesis and responds to allopregnanolone. *Nat Med.* 2004;10:704–711.
10. Naylor JC, Kilts JD, Hulette CM, et al. The function of very long chain polyunsaturated fatty acids in the pineal gland. *Biochim Biophys Acta.* 2010;1801:95–99.

Supplemental Information Available on Companion Website

- Gene, mRNA and precursor structures of human 5α-RI and 3α-HSD/E-Figures 96C.1 and 96C.2
- Primary structure of 5α-RI and 3α-HSD of various animals/E-Figures 96C.3 and 96C.4
- Accession numbers of gene and cDNA for 5α-RI and 3α-HSD/E-Tables 96C.1 and 96C.2

CHAPTER 97

Vitamin D Derivatives

Yoshihiko Ohyama and Toshimasa Shinki

History

Vitamin D was discovered in cod liver oil as the anti-rickets factor in the early 20th century and several vitamin D compounds (D_1: mixture; D_2 to D_7: different in the side chains) were isolated. Their structures were determined in the 1930s. Due to their biological activities, at present vitamin D refers to vitamin D_2 and vitamin D_3. Vitamin D_3 is the form of vitamin D that is synthesized by vertebrates, whereas vitamin D_2 is the naturally occurring form of vitamin D in plants.

Although vitamin D is classified as a vitamin, it is in fact a prohormone. In the late 1960s to early 1970s, several metabolites of vitamin D_3 were discovered [1,2]. Finally, it was shown that vitamin D_3 synthesized in the body is converted to its active form (1,25-dihydroxyvitamin D_3) by cytochrome P450 enzymes (CYPs) (vitamin D_3 25-hydroxylase and 25-hydroxyvitamin D_3 1α-hydroxylase) (Figure 97.1) and reveals its biological activities via a dedicated receptor [1,2].

Structure

Structural Features

Vitamin D has a secosteroid structure in which a bond (C9–C10) in ring B of the steroid structure is broken. Vitamin D_3 and vitamin D_2 are produced by the photochemical reaction of 7-dehydrocholestrol and ergosterol with ultraviolet light B (naturally with sunlight), and subsequent heat isomerization, respectively. These two chemical reactions (not enzymatic reactions) are essential for vitamin D synthesis. In human, these reactions of 7-dehydrocholestrol occur in the skin. The structural difference between vitamin D_3 and vitamin D_2 is in their side chains (Figure 97.2).

Evolutional Aspects of Vitamin D

Vitamin D is regularly found in phytoplankton and zooplankton. Therefore, vitamin D already existed for millions of years as inactive products before the emergence of vertebrates.

Receptors

Vitamin D receptor (VDR) is expressed in the kidney, intestine, and bone in avian as well as mammalian species. In the other organs (parathyroid gland, skin, pancreas, placenta, pituitary, ovary, testis, mammary gland, and heart), VDRs are also found at relatively low expression levels. VDR was purified in the early 1980s and first cloned from chickens in 1987. VDR is a single gene product, but chicken has alternative start sites, producing two isoforms.

VDR has two functional domains (DNA-binding domain and ligand-binding domain) separated by a hinge region (Figure 97.3). The DNA binding domain consists of two sets of a zinc-coordinated finger structure. The ligand-binding domain obtained by trypsin-digestion from VDR can bind 1,25-dihydroxyvitamin D_3, indicating that the domain can independently achieve its function similarly to other nuclear receptors. The overall sequence similarity of rat, mouse, and avian receptors to that of human VDR is 79, 86, and 66%, respectively [3]. Similarity of the DNA binding domains shows a level above 95% [3].

VDR is a member (NR1I1) of the nuclear receptor superfamily and acts as a transcription factor [4,5]. VDR specifically binds to 1,25-dihydroxyvitamin D and then forms a heterodimer with the retinoid-X receptor. The complex binds the vitamin D responsive element (VDRE) in target gene promoters and modulates gene expression. The VDRE in osteocalcin, osteopontin, 24-hydroxylase genes has been well characterized and the consensus sequence consists of two imperfect repeats of AGGTCA separated by three nonspecified nucleotides [6].

Biological Functions

Target Cells/Tissues and Functions

One of the most important roles of vitamin D is to maintain calcium balance by enhancing calcium absorption in the intestine, calcium mobilization in the bone, and calcium reabsorption in the kidney. Vitamin D deficiencies cause rickets in which the skeleton is poorly mineralized and deformed. Vitamin D is also known to be involved in cell proliferation and differentiation.

1,25-Dihydroxyvitamin D is the sole vitamin D that can activate VDR. Vitamin D-dependent rickets type I is an autosomal recessive disorder that does not produce 1,25-dihydroxyvitamin D_3 because of mutations in the 25-hydroxyvitamin D_3 1α-hydroxylase gene. Vitamin D-dependent rickets type II is caused by mutations in the VDR gene that alter the function of the receptor, leading to complete or partial resistance to 1,25-dihydroxyvitamin D_3.

The vitamin D endocrine system is characterized by VDR, specific vitamin D metabolizing CYPs, and plasma transport proteins. Vitamin D has clear effects on calcium and bone homeostasis of mammals and birds. Vitamin D deficiency in amphibians, reptiles, birds, and mammals engenders rickets [7]. Many fishes (excepting zebrafish) show no biological effect of vitamin D on calcium homeostasis. Although VDR is present in early fish such as lampreys, the vitamin D endocrine system is unlikely to regulate calcium homeostasis [7,8]. Therefore, it seems that the vitamin D endocrine system

Y. Takei, H. Ando, & K. Tsutsui (Eds): Handbook of Hormones. DOI: http://dx.doi.org/10.1016/B978-0-12-801028-0.00097-0

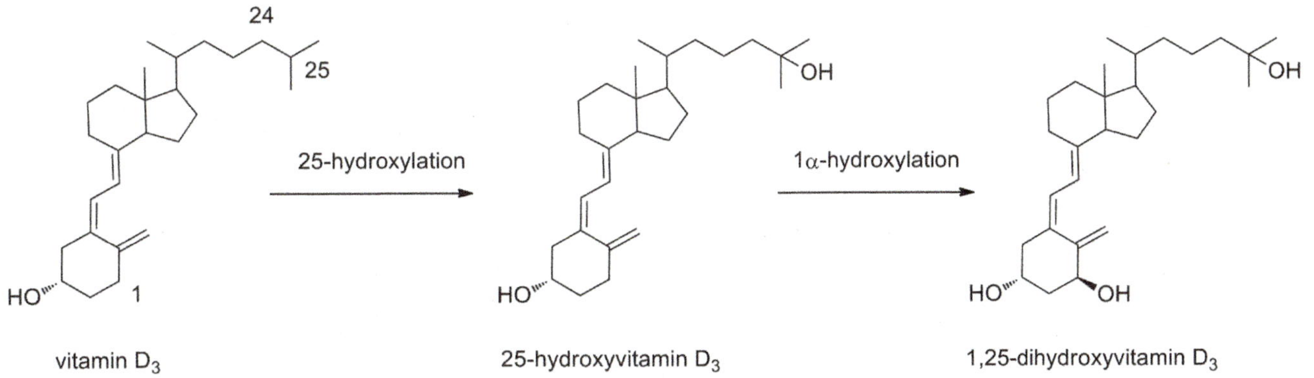

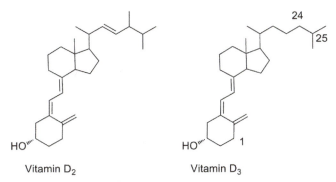

Figure 97.1 Activation of the vitamin D molecule.

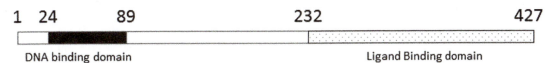

Figure 97.2 Nutritional forms of vitamin D.

Figure 97.3 Functional domain of human VDR.

started during the evolution of fish first for catabolic degradation and later for the calcium conserving mechanism.

References

1. DeLuca HF. Overview of general physiologic features and functions of vitamin D. *Am J Clin Nutr*. 2004;80:1689S–1696SS.
2. DeLuca HF. Historical perspective. In: Feldman D, Pike JW, Glorieux FH, eds. *Vitamin D,* 2nd ed. San Diego: Elsevier, Academic Press; 2005:3–12, Chapter 1.
3. Pike JE, Shevde NK. The vitamin D receptor. In: Feldman D, Pike JW, Glorieux FH, eds. *Vitamin D,* 2nd ed. San Diego: Elsevier, Academic Press; 2005:167–191, Chapter 11.
4. Laudet V, Auwerx J, Gustafsson J-A, et al. A unified nomenclature system for the nuclear receptor superfamily. *Cell.* 1999;97:161–163.
5. Germain P, Staels B, Dacquet C. Overview of nomenclature of nuclear receptors. *Pharmacological Reviews*. 2006;58:685–704.
6. Umesono K, Murakami KK, Thompson CC, et al. Direct repeats as selective response elements for the thyroid hormone, retinoic acid, and vitamin D_3 receptors. *Cell.* 1991;65:1255–1266.
7. Bouillon R, Suda T. Vitamin D: calcium and bone homeostasis during evolution. *Bonekey Rep.* 2014;3:article number 480.
8. Lock EJ, Ørnsrud R, Aksnes L, et al. The vitamin D receptor and its ligand 1α,25-dihydroxyvitamin D_3 in Atlantic salmon (*Salmo salar*). *J Endocrinol*. 2007;193:495–471.

Supplemental Information Available on Companion Website

- Phylogenetic tree of the human VDR and related nuclear receptors /E-Figure 97.1
- Phylogenetic tree of the vertebrate VDRs/E-Figure 97.2
- Properties of VDRs/ E-Table 97.1

Calcitriol

Yoshihiko Ohyama and Toshimasa Shinki

Abbreviation: none
Additional names: 1α,25-dihydroxycholecalciferol; 1α,25-dihydroxyvitamin D_3; 1α,25$(OH)_2D_3$

Calcitriol is a calciotropic hormone secreted by renal proximal tubular cells. It stimulates intestinal calcium absorption and bone mineralization. The drug form is used for treatment of osteoporosis and renal failure.

Discovery

In 1971, the purification and identification of 1α,25-dihydroxyvitamin D_3 were reported for vitamin D-deficient intestine from chicks given radiolabeled vitamin D_3 [1]. The configuration of the hydroxyl group at position 1 was deduced by co-chromatography with chemically synthesized 1α- and 1β-dihydroxyvitamin D_3 compounds.

Structure

Structural Features

Calcitriol is a secosteroid hormone essential for bone and mineral homeostasis.

Properties

Molecular formula: $C_{27}H_{44}O_3$; Mr 416.6. White crystalline powder, MP 111−115°C. UV max (abs. ethanol): 264 nm (ε 19,000). $[\alpha]_D^{25}$ +48 (methanol). Slightly soluble in methanol, ethanol, ethylacetate, and THF. Air- and light-sensitive.

Synthesis and Release

Metabolism

Vitamin D is a secosteroid whose actions are dependent on specific metabolic steps catalyzed by cytochrome P450 enzymes (CYPs). The first of these steps occurs in the liver and involves the enzyme vitamin D_3 25-hydroxylase (CYP2R1), which catalyzes synthesis of 25-hydroxyvitamin D_3 (25$(OH)D_3$), the main circulating form of vitamin D. The 25$(OH)D_3$ molecule then acts as the substrate for 25$(OH)D_3$ 1α-hydroxylase (CYP27B1), which catalyzes synthesis of 1α,25-dihydroxyvitamin D_3, an active form of vitamin D_3 [2]. 25$(OH)D_3$ and 1α,25-dihydroxyvitamin D_3 receive hydroxylation by CYP24 at the position of C-24 as degradation steps (Figure 97A.1).

Gene and mRNA

The human 25-hydroxyvitamin D_3 1α-hydroxylase gene, CYP27B1, location chromosome 12q14.1, consists of nine exons [3]. Human 25-hydroxyvitamin D_3 1α-hydroxylase mRNA has 2,503 bp that encode a protein of 508 aa residues. The protein has a predicted topology that is similar to mitochondrial P450, with an N-terminal mitochondrial signal sequence and conserved ferrodoxin- and heme-binding sites. The rat CYP27B1 gene is mapped to chromosome 7. Rat CYP27B1 mRNA has 2,426 bp that encode a protein of 501 aa residues [4]. The mouse CYP27B1 gene is mapped to chromosome 10. Mouse CYP27B1 mRNA has 2,440 bp that encode a protein of 507 aa residues.

Regulation

Calcitriol is produced in the proximal tubular cells in the kidney. The synthesis of 1α,25$(OH)_2D_3$ by renal 1α-hydroxylase (CYP27B1) is tightly regulated by the levels of plasma 1α,25$(OH)_2D_3$, calcium and parathyroid hormone (PTH). The renal CYP27B1 mRNA expression is strongly upregulated by PTH [4]. The promoter has three potential cAMP-responsive element sites, two AP-1 sites, and two vitamin D-response element (VDRE) sites [5]. These results indicate that the mechanism to induce CYP27B1 gene transcription might occur through activation of the adenylate cyclase system.

Plasma Concentration

1α,25-Dihydroxyvitamin D_3 concentration (pg/ml): rat (female) mature 30; rat pregnant 90; human (adult) 40; human pregnant 80.

Receptors

Structure and Subtype

The initial discovery of the vitamin D receptor (VDR) was accomplished through biochemical examination of chick intestinal mucosa [6]. Calcitriol binds VDR with high affinity ($K_d = 10^{-10}$ M). VDR exhibits species-specific molecular weights that range between 48,000 and 60,000 and binds both calcitriol and DNA. The human VDR gene, located on chromosome 12 (12q13.11), consists of 14 exons [7]. The main VDR transcript is 4.8 kbp and encodes the 427 aa residues. VDR, like all steroid hormone receptors, is a zinc-finger transcription factor; the zinc atoms are attached to four cysteines. VDR functions as a heterodimer with the retinoid X receptor (RXR) [2].

Y. Takei, H. Ando, & K. Tsutsui (Eds): Handbook of Hormones. DOI: http://dx.doi.org/10.1016/B978-0-12-801028-0.00236-1

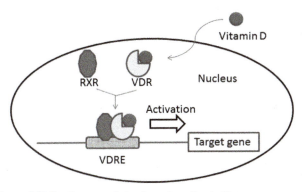

25-hydroxyvitamin D_3 → (25-hydroxylation, CYP2R1) → 25-hydroxyvitamin D_3 → (1α-hydroxylation, CYP27B1) → 1,25-dihydroxyvitamin D_3

(24-hydroxylation, CYP24) → 24,25-dihydroxyvitamin D_3

(24-hydroxylation, CYP24) → 1,24,25-trihydroxyvitamin D_3

Figure 97A.1 Synthesis and degradation of $1,25(OH)D_3$.

Figure 97A.2 Gene regulation by active vitamin D.

Signal Transduction Pathway

The biological effects of calcitriol are mediated by VDR, a ligand-dependent transcription factor that belongs to the superfamily of nuclear hormone receptors. In various vertebrate species, calcitriol mediates its biological effects by binding to VDR, which is principally located in the nuclei of target cells. VDR acts as a ligand-dependent transcription factor that binds to vitamin D response element(s) (VDRE) as a VDR:RXR heterodimer (Figure 97A.2) [2]. The binding of the calcitriol-VDR complex to VDRE allows VDR to act as a transcription factor that modulates the gene expression of calbindin, osteocalcin, and PTH.

Biological Function

Target Cells/Tissues and Functions

VDR is present in cells of the intestinal epithelium, renal tubules, parathyroid gland cells, epidermis, mammary epithelium, pancreatic cells, pituitary gland, bone cells, spleen cells, lymph node, and germ tissues. VDR is a member of the nuclear receptor superfamily and plays a central role in the biological functions of vitamin D. VDR regulates the expression of numerous genes involved in calcium and phosphorus homeostasis, cellular proliferation and differentiation, and immune response, largely in a ligand-dependent manner.

Phenotype in Gene-Modified Animals

CYP27B1 knockout mice exhibit an identical phenotype to the human disease hereditary vitamin D-dependent rickets type I (VDDR-I) [8]. VDR knockout mice exhibit an identical phenotype to the human disease hereditary vitamin D-dependent rickets type II (VDDR-II) [9].

Pathophysiological Implications

Clinical Implications

VDDR-I is an autosomal recessive disorder. Patients with VDDR-I have inactivating mutations in the CYP27B1 gene, which encodes the enzyme $25(OH)D_3$ 1α-hydroxylase. VDDR-I is characterized by early onset of skeletal disease and severe hypocalcemia. Patients exhibit muscle weakness and rickets. VDDR-II is a rare autosomal recessive disorder. The disorder is characterized by end organ hyporesponsiveness to vitamin D. VDDR-II is characterized by the development of hypocalcemia in infancy, accompanied by rickets, osteomalacia, and secondary hyperparathyroidism. Alopecia is observed in some kindred with VDDR-II.

Use for Treatment

Calcitriol is used in the treatment of patients with osteoporosis and renal osteodystrophy. Maxacalcitol ointment is used to treat plaque psoriasis. Maxacalcitol injection is used for the treatment of secondary hyperparathyroidism. Calcipotriol ointment is used to treat plaque psoriasis. Eldecalcitol is used in the treatment of patients with osteoporosis.

References

1. Holick MF, Schnoes HK, DeLuca HF, et al. Isolation and identification of 1,25-dihydroxycholecalciferol. A metabolite of vitamin D active in intestine. *Biochemistry*. 1971;10:2799−2804.
2. DeLuca HF. History of the discovery of vitamin D and its active metabolites. *Bonekey Rep*. 2014;3:479.
3. Monkawa T, Yoshida T, Wakino S, et al. Molecular cloning of cDNA and genomic DNA for human 25-hydroxyvitamin D_3 1 α-hydroxylase. *Biochem Biophys Res Commun*. 1997;239:527−533.
4. Shinki T, Shimada H, Wakino S, et al. Cloning and expression of rat 25-hydroxyvitamin D_3-1α-hydroxylase cDNA. *Proc Natl Acad Sci USA*. 1997;94:12920−12925.
5. Brenza HL, Kimmel-Jehan C, Jehan F, et al. Parathyroid hormone activation of the 25-hydroxyvitamin D_3-1α-hydroxylase gene promoter. *Proc Natl Acad Sci USA*. 1998;95:1387−1391.
6. Haussler MR, Haussler CA, Bartik L, et al. Vitamin D receptor: molecular signaling and actions of nutritional ligands in disease prevention. *Nutr Rev*. 2008;66:S98−S112.
7. Crofts LA, Hancock MS, Morrison NA, et al. Multiple promoters direct the tissue-specific expression of novel N-terminal variant human vitamin D receptor gene transcripts. *Proc Natl Acad Sci USA*. 1998;95:10529−10534.
8. Panda DK, Miao D, Tremblay ML, et al. Targeted ablation of the 25-hydroxyvitamin D 1α-hydroxylase enzyme: evidence for skeletal, reproductive, and immune dysfunction. *Proc Natl Acad Sci USA*. 2001;98:7498−7503.
9. Li YC, Pirro AE, Amling M, et al. Targeted ablation of the vitamin D receptor: an animal model of vitamin D-dependent rickets type II with alopecia. *Proc Natl Acad Sci USA*. 1997;94:9831−9835.

Supplemental Information Available on Companion Website

- Primary structure of VDR of various animals/E-Figure 97A.1
- Primary structure of CYP27B1 of various animals/E-Figure 97A.2
- Primary structure of CYP24 of various animals/E-Figure 97A.3
- Vitamin D analogs/E-Figure 97A.4
- Distribution of VDR/E-Table 97A.1
- Vitamin D response elements/E-Table 97A.2
- Accession numbers of gene and cDNA for VDR/E-Table 97A.3
- Accession numbers of gene and cDNA for CYP27B1/E-Table 97A.4
- Accession numbers of gene and cDNA for CYP24/E-Table 97A.5

Cholecalciferol

Yoshihiko Ohyama and Toshimasa Shinki

Abbreviation: none
Additional names: vitamin D_3, $(3\beta,5Z,7E)$-9,10-secocholesta-5,7,10(19)-trien-3-ol

Vitamin D was discovered as an anti-rachitic factor and is the precursor of calcitriol involved in calcium and bone homeostasis.

Discovery

Presence of an anti-rachitic factor was reported in the 1910s. The chemical structure of vitamin D_3 was determined in 1936.

Structure

Structural Features

Vitamin D_3 has the seco-steroid structure, with a broken C9—10 bond of the B ring in 7-dehydrocholesterol. Vitamin D_3 is found in vertebrates as well as in plankton.

Properties

Molecular formula: $C_{27}H_{44}O$, Mr 384.6. White crystal (fine needles); MP 84—88°C. UV λmax 265 nm. Insoluble in water. Soluble in organic solvents: ether, alcohol, etc. Oxidized by moist air. Heat and light labile.

Synthesis and Release

Metabolism

Vitamin D_3 is only produced via the photochemical reaction of 7-dehydrocholestrol with ultraviolet light B (naturally with sunlight) and subsequent thermal isomerization (Figure 97B.1).

These two chemical reactions (not enzymatic reactions) are essential for vitamin D synthesis. In humans, these reactions occur in the skin. Vitamin D_3 is converted to calcidiol, 25-hydroxyvitamin D_3 ($25(OH)D_3$) in the liver. $25(OH)D_3$ is the main circulating form of vitamin D [1,2].

The 25-hydroxylation reaction is catalyzed by several cytochrome P450 enzymes (CYPs). The most important CYP in human and mouse is CYP2R1, because a genetic defect of CYP2R1 shows abnormality in calcium metabolism [3]. However, CYP27A1 (mammals), CYP3A4 (human), CYP2D25 (pig), CYP2C11 (male rat), and CYP2J3 (female rat) are also reported to catalyze the 25-hydroxylation reaction [4,5]. The existence of multiple enzymes provides possibilities of redundancy and compensation for synthesis of $25(OH)D_3$. 25 $(OH)_2D_3$ circulates in the blood, forming a complex with a specific vitamin D binding protein, DBP (mammals and birds). $25(OH)D_3$ is finally converted to the biologically active form of vitamin D_3, calcitriol, by CYP27B1 (see Subchapter 97A, Calcitriol).

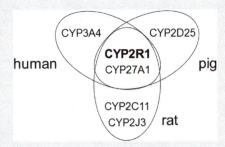

Figure 97B.2 CYPs catalyzing vitamin D_3 25-hydroxylation.

Figure 97B.1 Synthesis and activation of vitamin D_3.

7-dehydrocholesterol   previtamin D_3 vitamin D_3 25-hydroxyvitamin D_3

Y. Takei, H. Ando, & K. Tsutsui (Eds): Handbook of Hormones. DOI: http://dx.doi.org/10.1016/B978-0-12-801028-0.00237-3

Gene and mRNA

The human CYP2R1 gene is located on chromosome 11 (11p15) and consists of five exons. The mouse CYP2R1 gene is located on chromosome 7 and consists of five exons. Mouse CYP2R1 mRNAs are about 1.6 and 1.1 kbp. The protein of human and mouse contains 501 aa residues. CYP2R1 is highly conserved across vertebrate species from pufferfish to human [6]. The human CYP27A1 gene, located on chromosome 2 (2q35), consists of nine exons. Human CYP27A1 mRNA is approximately 1.8 to 2.2 kbp and encodes 531 aa residues (a 33-amino acid mitochondrial signal sequence followed by a mature protein of 498 aa residues). The mouse CYP27A1 gene locates on chromosome 1 and consists of nine exons. Mouse CYP27A1 mRNA encodes a protein of 533 aa residues (with signal sequence). The CYP27A1 gene is conserved in chimpanzee, rhesus monkey, dog, cow, mouse, rat, chicken, frog, zebrafish, fruit fly, and mosquito.

Distribution of mRNA

CYP2R1 mRNA is highly expressed in the liver and testis, and lower expression is detected in numerous other tissues. CYP27A1 mRNA is found in the liver, kidney, intestine, ovary, lung, and skin.

Regulation

The concentration of vitamin D depends on synthesis in the skin and dietary intake. The levels of $25(OH)D_3$, the main circulation form of vitamin D_3, seem to reflect intake of vitamin D_3, thus serving as an indicator of vitamin D status. There is no clear data showing whether 25-hydroxylase activates are coordinated physiologically. The following 1α-hydroxylation reaction is controlled strictly by physiological factors.

Plasma Concentration

Vitamin D_3, 1–3 ng/ml (human); $25(OH)D_3$, 20–60 ng/ml (human). $1,25(OH)_2D_3$, 20–60 pg/ml (human).

Receptors

Structure and Subtype

The vitamin D receptor (VDR) is a member of the nuclear receptor superfamily and designated as NR1I1 (nuclear receptor subfamily 1, group I, member 1). VDR binds to calcitriol with high affinity ($K_d = 10^{-10}$ M) but not cholecalciferol itself. The human VDR gene, located on chromosome 12 (12q13.11), consists of 14 exons (eight coding exons 2–9 and a variable number of untranslated 5′ exons 1A–1F). The major VDR transcript is 4.8 kb and encodes 427 aa residues. VDR, like all steroid hormone receptors, is a zinc-finger transcription factor; four cysteines each tetrahedrally coordinate two zinc ions.

Signal Transduction Pathway

Cholecalciferol does not control gene expression via VDR. After two sequential hydroxylation reactions of vitamin D_3, the product of $1,25(OH)_2D_3$ acquires affinity to VDR. $1,25(OH)_2D_3$-binding VDR forms a heterodimer with the retinoid X receptor (RXR), and then binds to vitamin D response element(s) [VDRE(s): a direct repeat of the AGGTCA motif separated by three nucleotides] that are identified in promoters of many target genes, such as calbindin, osteocalcin, and CYP24 (see Subchapter 97A, Calcitriol).

Biological Function

Target Cells/Tissues and Functions

One of the most important roles of vitamin D is to maintain calcium concentration in blood by promoting calcium absorption in the intestine and maintaining calcium and phosphate levels for bone formation. Vitamin D deficiency causes rickets in amphibians, reptiles, birds, and mammals [7]. Vitamin D_2 that is synthesized from ergosterol shows similar activity to vitamin D_3 in human and rat. Vitamin D_2 exhibits about 1–2% biopotency of vitamin D_3 in chicken. This difference is caused by the function of vitamin D binding protein in blood between human and chicken.

Phenotype in Gene-Modified Animals

CYP2R1 knockout mice had greater than 50% reduction in serum 25-hydroxyvitamin D_3 indicating that CYP2R1 is a major contributor to 25-hydroxylation of vitamin D *in vivo*. However, the calcitriol level in the serum remained unchanged. CYP27A1 is known to work not only as vitamin D 25-hydroxylase but also as sterol 27-hydroxylase, since it hydroxylates cholesterol and bile acid intermediates at C-27. CYP27A1 knockout mice exhibited normal serum $25(OH)D_3$ level despite significant reduction in bile acid production, suggesting a minor contribution of CYP27A1 in $25(OH)D_3$ synthesis.

Pathophysiological Implications

Clinical Implications

A Nigerian patient with a homozygous mutation (Leu 99 Pro) in the CYP2R1 gene had abnormally low plasma levels of 25-hydroxyvitamin D_3 and classic symptoms of vitamin D deficiency, including skeletal abnormalities, hypocalcemia, and hypophosphatemia [8]. Later, a handful of patients from Nigeria and Saudi Arabia had $25(OH)D_3$ deficiency, showing relevance of a mutation in CYP2R1 gene. Mutations in the CYP27A1 gene cause cerebrotendinous xanthomatosis (CTX), a rare inborn disorder of bile acid and cholesterol metabolism. In most cases, CTX patients do not exhibit abnormalities in vitamin D status.

Use for Diagnosis and Treatment

The measurement of serum $25(OH)D_3$ is clinically important in diagnosis of some diseases related to vitamin D status, because $25(OH)D_3$ is the main circulating form of vitamin D. It is reported that vitamin D can decrease the risk of many chronic disorders, including cancers, autoimmune diseases, infectious diseases, and cardiovascular diseases. However, vitamin D and its metabolites are presently used for the treatment of plaque psoriasis and some illnesses relating to calcium and phosphorus homeostasis, such as vitamin D deficiency, osteoporosis, and secondary hyperparathyroidism. Calcidiol, which does not require hepatic 25-hydroxylation, is available for patients with liver disease. Calcitriol, which does not require renal 1αhydroxylation, is useful for patients with chronic renal failure or vitamin D-dependent rickets type 1.

References

1. DeLuca HF. Overview of general physiologic features and functions of vitamin D. *Am J Clin Nutr*. 2004;80:1689S–1696S.
2. Horst RL, Reinhardt TA, Reddy GS. Vitamin D metabolism. In: Feldman D, Pike JW, Glorieux FH, eds. *Vitamin D*. 2nd ed. San Diego: Elsevier, Academic Press; 2005:15–36, Chapter 2.

3. Zhu JG, Ochalek JT, Kaufmann M, et al. CYP2R1 is a major, but not exclusive, contributor to 25-hydroxyvitamin D production in vivo. *Proc Natl Acad Sci USA*. 2013;110:15650−15655.
4. Ohyama Y, Yamasaki T. Eight cytochrome P450s catalyze vitamin D metabolism. *Front Biosci*. 2004;9:3007−3018.
5. Zhu J, DeLuca HF. Vitamin D 25-hydroxylase − Four decades of searching, are we there yet? *Arch Biochem Biophys*. 2012;523:30−36.
6. Nelson DR. Comparison of P450s from human and fugu: 420 million years of vertebrate P450 evolution. *Arch Biochem Biophys*. 2003;409:18−24.
7. Bouillon R, Suda T. Vitamin D: calcium and bone homeostasis during evolution. *Bonekey Rep*. 2014;3:480.
8. Cheng JB, Levine MA, Bell NH, et al. Genetic evidence that the human CYP2R1 enzyme is a key vitamin D 25-hydroxylase. *Proc Natl Acad Sci USA*. 2004;101:7711−7715.

Supplemental Information Available on Companion Website

- Primary structure of CYP2R1 of various animals/E-Figure 97B.1
- Primary structure of CYP27A1 of various animals/E-Figure 97B.2
- Accession numbers of gene and cDNA for CYP2R1/E-Table 97B.1
- Accession numbers of gene and cDNA for CYP27A1/E-Table 97B.2
- Accession numbers of gene and 25-hydroxylase of various animals/E-Table 97B.3

PART IV

Lipophilic Hormones in Invertebrates

CHAPTER 98

Ecdysteroids

Yoshiaki Nakagawa and Haruyuki Sonobe

History

The generic term "ecdysteroid" was originally proposed by Goodwin and colleagues in 1978. Ecdysteroids are involved in the control of many physiological events in arthropods, such as molting, metamorphosis and reproduction. The first ecdysteroid to be isolated was ecdysone, from silkworm pupae in 1954, and to date about 500 ecdysteroids have been isolated from animal and plant sources (http://ecdybase.org/). Among them, 20-hydroxyecdysone (20E), the principal molting hormone in arthropods, has also been identified in non-arthropod invertebrates such as nematodes, mollusks, and annelids. Zooecdysteroids and phytoecdysteroids refer to the ecdysteroids that have been isolated only from animals or plants, respectively, because numerous compounds are common to both groups. The action of ecdysteroids on gene expression was originally modeled by Ashburner in 1972.

Structure

Features of the steroid skeleton are *cis* A/B ring fusion (5β-H), which is different from the plant steroid hormone (brassinosteroids), a 7-ene-6-one chromophore, and *trans* C/D ring fusion (14α-OH or -H), as shown in Figure 98.1.

Molecular Evolution of Family Members

Although A/B ring fusions are different between ecdysteroids and brassinosteroids, the hybrid compounds with the side chain of ponasterone A and steroid moiety of brassinosteroids exhibited ecdysone-like activity [1]. For this hybrid compound, the 22*R*-isomer had 100-fold greater binding affinity to the receptor than the corresponding 22*S*-isomer. However, interestingly for a compound without a 20-OH group, it is essential to have an *S*-configuration at the 22-position [2]. The binding affinity was enhanced 250-fold by conversion of the A/B ring fusion from *trans* to *cis* [3]. This structure—activity relationship information was used for *in silico* screening of ecdyone agonists [4].

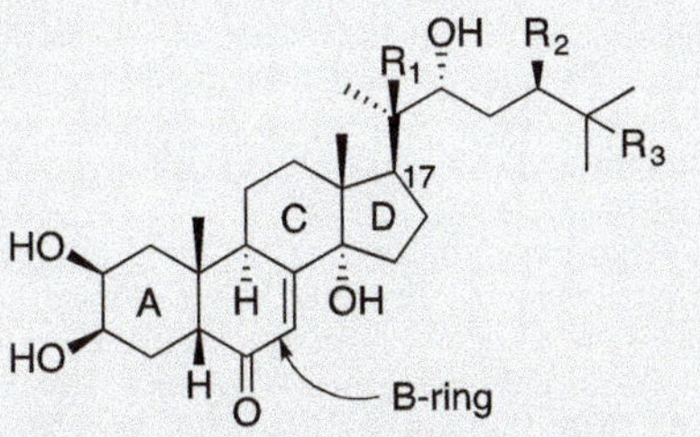

Receptors

Structure and Subtypes

Ecdysteroids bind to their intracellular receptors, ecdysone receptors (EcRs), that migrate to the nucleus and transactivate the genes with the collaboration of ultraspiracle (USP) [5]. USP is the homolog of retinoid X receptor (RXR). EcR and USP (RXR) are categorized in the nuclear receptor superfamily, and their basic structure includes A/B (transactivation), C (DNA binding), D (hindge), E (ligand binding), and F domains. The F domains of most EcRs except for that of Diptera are very short. DNA binding domains are highly homologous among insect species, but ligand-binding domains are varied among insect orders. Three-dimensional structures of the ligand binding domains of some EcRs have been solved by X-ray crystal structure analysis. Recently, in insects and crustaceans, it has been suggested that ecdysteroids may interact with transmembrane receptors, G protein-coupled receptors (GPCRs), and take part in the activation of various second messenger pathways. Recently, an insect GPCR interacting with ecdysteroids (*Drosophila* GPCR, CG18314, or DmDopEcR) has been identified [6].

Biological Functions

Target Cells/Tissues and Functions

In insects, ecdysteroids are synthesized mainly by the prothoracic gland (PG) and ovary—egg system. Ecdysteroid synthesis in the PG is stimulated by prothoracicotropic hormone, which is produced in brain neurosecretory cells. Almost the entire insect is the target of PG-derived ecdysteroids, which are involved in, for example, control of choriogenesis, stimulation of the growth and development of imaginal discs, initiation of the breakdown of larval structures during metamorphosis, and elicitation of the deposition of cuticle by the epidermis [7]. On the other hand, ecdysteroids in the ovary—egg system participate in embryonic development at an early embryonic stage when the PGs have not yet differentiated [8]. In the crustacean, ecdysteroids are synthesized by the Y-organ. Ecdysteroid synthesis in the Y-organ is negatively controlled by molt-inhibiting hormone, which is produced in the X-organ. Ecdysteroids in crustaceans are involved in the molt cycle,

Ecdysone (R$_1$=R$_2$=H, R$_3$=OH)
20-Hydroxyecdysone (R$_1$=OH, R$_2$=H, R$_3$=OH)
Ponasterone A (R$_1$=OH, R$_2$=H, R$_3$=H)
Makisterone A (R$_1$=OH, R$_2$=CH$_3$, R$_3$=OH)

Figure 98.1 Chemical structures of ecdysteroids.

Y. Takei, H. Ando, & K. Tsutsui (Eds): Handbook of Hormones. DOI: http://dx.doi.org/10.1016/B978-0-12-801028-0.00098-2

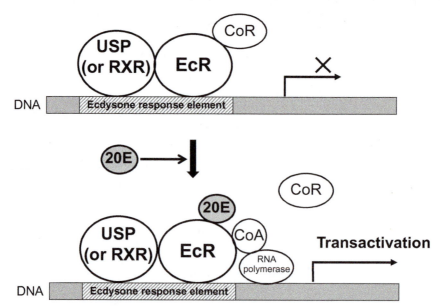

Figure 98.2 Orchestration of hormone and receptors and action on DNA.

including, for example, proliferation of epidermal cells, secretion of new cuticle, withdrawal and storage of calcium salts from the cuticle, construction of new exoskelton, and growth of a regenerating limb bud [9].

Agonists/Antagonists

In 1988, diacylhydrazine (DAH)-type compound (RH-5849) was first discovered as a non-steroidal ecdysone agonist. By the modification of benzene rings, tebufenozide (RH-5992) was launched as an insecticide for agriculture. Most such compounds are specifically toxic against Lepidoptera but weak or inactive against other insect orders. Currently, five DAHs (tebufenozide, methoxyfenozide, chromafenozide, halofenozide, and fufenozide) are on the market. Other compounds with different chemical structures are also reported to be ecdysonergic [10]. Three-dimensional structures of the ligand binding domains of EcRs with non-steroidal compounds are solved by X-ray crystal analysis.

References

1. Watanabe B, Nakagawa Y, Ogura T, et al. Stereoselective synthesis of (22R)- and (22S)-castasterone/ponasterone A hybrid compounds and evaluation of their molting hormone activity. *Steroids.* 2004;69:483−493.
2. Voigt B, Whiting P, Dinan L. The ecdysteroid agonist/antagonist and brassinosteroid-like activities of synthetic brassinosteroid/ecdysteroid hybrid molecules. *Cell Mol Life Sci.* 2001;58:1133−1140.
3. Arai H, Watanabe B, Nakagawa Y, et al. Synthesis of ponasterone A derivatives with various steroid skeleton moieties and evaluation of their binding to the ecdysone receptor of Kc cells. *Steroids.* 2008;73:1452−1464.
4. Harada T, Nakagawa Y, Ogura T, et al. Virtual screening for ligands of the insect molting hormone receptor. *J Chem Inf Model.* 2011;51:296−305.
5. Hill RJ, Graham LD, Turner KA, et al. Structure and function of ecdysone receptors − interactions with ecdysteroids and synthetic agonists. *Adv Insect Physiol.* 2012;43:299−351.
6. Evans PD, Bayliss A, Reale V. GPCR-mediated rapid, nongenomic actions of steroids: comparisons between DmDopEcR and GPER1 (GPR30). *Gen Com Endocrinol.* 2014;195:157−163.

7. Gilbert LI, Rybczynski R, Tobe SS. Endocrine cascade in insect metamorphosis. In: Gilbert LI, Tata JR, Atkinson BG, eds. *Metamorphosis.* New York, NY: Academic Press; 1996:59−107.
8. Sonobe H, Ito Y. Phosphoconjugation and dephosphorylation reactions of steroid hormone in insects. *Mol Cell Endocrinol.* 2009;307:25−35.
9. Hopkins PM. Crustacean ecdysteroids and their receptors. In: Smagghe G, ed. *Ecdysone:* Structures and Functions. New York, NY: Springer; 2009:73−97.
10. Dinan L, Nakagawa Y, Hormann RE. Structure-activity relationships of ecdysteroids and non-steroidal ecdysone agonists. *Adv Insect Physiol.* 2012;43:251−298.
11. Minakuchi C, Nakagawa Y, Kamimura M, et al. Binding affinity of non-steroidal ecdysone agonists against the ecdysone receptor complex (EcR/USP) determines the strength of their molting hormonal activity. *Eur J Biochem.* 2003;270:4095−4104.
12. Swevers L, Cherbas L, Cherbas P, et al. *Bombyx* EcR (BmEcR) and *Bombyx* USP (BmCF1) combine to form a functional ecdysone receptor. *Insect Biochem Mol Biol.* 1996;26:217−221.
13. Yao T-P, Forman BM, Jiang Z, et al. Functional ecdysone receptor is the product of *EcR* and *ultraspiracle* genes. *Nature.* 1993;366:476−479.
14. Rauch P, Grebe M, Elke C, et al. Ecdysteroid receptor and ultraspiracle from *Chironomus tentans* (Insecta) are phosphoproteins and are regulated differently by molting hormone. *Insect Biochem Mol Biol.* 1998;28:265−275.
15. Ogura T, Minakuchi C, Nakagawa Y, et al. Molecular cloning, expression analysis and functional confirmation of ecdysone receptor and ultraspiracle from the Colorado potato beetle *Leptinotarsa decemlineata.* *FEBS J.* 2005;272:4114−4128.
16. Nakagawa Y, Sakai A, Magata F, et al. Molecular cloning of the ecdysone receptor and the retinoid X receptor from the scorpion *Liocheles australasiae.* *FEBS J.* 2007;274:6191−6203.
17. Minakuchi C, Nakagawa Y, Miyagawa H. Validity analysis of a receptor binding assay for ecdysone agonists using cultured intact insect cells. *J Pestic Sci.* 2003;28:55−57.
18. Minakuchi C, Nakagawa Y, Kamimura M, et al. Measurement of receptor-binding activity of non-steroidal ecdysone agonists using *in vitro* expressed receptor proteins (EcR/USP complex) of *Chilo suppressalis* and *Drosophila melanogaster.* ACS Symposium Series 892. In: Clark J, Ohkawa H, eds. *New Discoveries in Agrochemicals.* Hawaii: American Chemical Society; 2005:191−200.

19. Minakuchi C, Ogura T, Miyagawa H, et al. Effects of the structures of ecdysone receptor (EcR) and ultraspiracle (USP) on the ligand-binding activity of the EcR/USP heterodimer. *J Pestic Sci.* 2007;32:379–384.

20. Nakagawa Y, Henrich V. Arthropod nuclear receptors and their role in molting. *FEBS J.* 2009;276:6128–6157.

21. Nakagawa Y, Minakuchi C, Ueno T. Inhibition of [^{3}H]ponasterone A binding by ecdysone agonists in the intact Sf-9 cell line. *Steroids.* 2000;65:537–542.

22. Nakagawa Y, Minakuchi C, Takahashi K, et al. Inhibition of [^{3}H] ponasterone A binding by ecdysone agonists in the intact Kc cell line. *Insect Biochem Mol Biol.* 2002;32:175–180.

23. Ogura T, Nakagawa Y, Minakuchi C, et al. QSAR for binding affinity of substituted dibenzoylhydrazines to intact Sf-9 cells. *J Pestic Sci.* 2005;30:1–6.

Supplemental Information Available on Companion Website

- Alignment of primary sequences of representative insect EcR-As/E-Figure 98.1
- Alignment of the primary sequences of ligand binding domains for various EcRs/E-Figure 98.2
- Alignment of the primary sequences of representative USPs/E-Figure 98.3
- Phylogenetic trees constructed from primary sequences of EcR/E-Figure 98.4
- Phylogenetic trees constructed from primary sequences of USP (RXR)/E-Figure 98.5
- Structures of representative diacylhydrazines/E-Figure 98.6
- Structures of representative non-steroidal ecdysone agonists other than diacylhydrazines/E-Figure 98.7
- Dissociation constants of ponasterone A for the *in vitro* translated EcR/USP (RXR)/E-Table 98.1
- Molting hormonal activity in terms of IC$_{50}$ (μM) for ecdysteroids and non-steroidal ecdysone-like compounds/E-Table 98.2

20-Hydroxyecdysone

Yoshiaki Nakagawa and Haruyuki Sonobe

Abbreviation: 20E
Additional names: β-ecdysone, crustecdysone, ecdysterone, isoinokosterone, polypodine A, molting hormone/20E

20-Hydroxyecdysone (20E) is an important steroid hormone forming a central dogma of insect and crustacean endocrinology. 20E exerts biological effects on target tissues through a nuclear receptor (EcR/USP complex) or seven-transmembrane-spanning G protein-coupled receptor (GPCR).

Discovery

20-Hydroxyecdysone (20E) was isolated and identified from the seawater crayfish *Jasus lalandii* and the silkworm *Bombyx mori* in 1966. 20E has also been identified in the Japanese scorpion *Liocheles australasiae* [1]. It has been established that 20E is the principal molting hormone of all arthropods.

Structure

Structural Features

Since 20E is biosynthesized from cholesterol, the common name based on cholesterol is often used in the naming of 20E − i.e., 2β,3β,14α,20R,22R,25-hexa-hydroxy-5β-cholest-7-en-6-one. The A/B ring fusion is *cis*, and the C/D ring fusion is *trans* (Figure 98A.1). Total synthesis of 20E was performed by Kametani and co-workers [2].

Properties

Molecular formula, $C_{27}H_{44}O_7$; Mr, 480.63. UV absorption maximum at 243 nm: $\varepsilon = 10,300$. Melting point, 241−242.5°C (http://ecdybase/), $[\alpha]_D^{20} + 61.8°C$ (MeOH). Soluble in water and ethanol; stable at 5°C, but unstable in alkaline solution.

Synthesis and Release

Ecdysone is synthesized in insect prothoracic glands and crustacean Y-organs, secreted to hemolymph, and oxidized to 20E in peripheral tissues such as the fat body. Ecdysone is synthesized from cholesterol (C27) and other plant steroids (C28) such as stigmasterol, β-sitosterol, and campesterol. The first step is the conversion of cholesterol to 7-dehydrocholesterol (7dC), which is mediated by 7,8-dehydrogenase encoded by *neverland*. Conversion of 7dC to the Δ^4-diketol constitutes the so-called Black Box, because no intermediates have been identified. However, it is reported that the Halloween genes *spook* (CYP307A1), *spookier* (CYP307A2), and *spookiest*

(CYP307B1) are involved in the Black Box reaction(s) in the fruit fly *Drosophila melanogaster*. Furthermore, *non-molting glossy* in *B. mori* or its ortholog *shroud* in *D. melanogaster*, which encodes a short-chain dehydrogenase/reductase, take part in Black Box reactions [3,4]. However, the exact functions of these enzymes are still not clear. Subsequently, 5β-reduction and 3β-reduction steps convert the Δ^4-diketol to the 5β-ketodiol. The products of Halloween genes *phantom* (CYP306A1), *disembodied* (CYP302A1), *shadow* (CYP315A1), and *shade* (CYP314A1) sequentially convert 5β-ketodiol to 2,22-dideoxyecdysone, 2-deoxyecdysone, ecdysone, and 20E (Figure 98A.2) [3]. The main pathway for 20E biosynthesis in crustaceans is considered to be similar to that in insects.

Regulation of Synthesis and Release

In insects, prothoracicotropic hormone (PTTH) stimulates the prothoracic glands, leading to the production of ecdysone [5]. In the crustacean, molt inhibiting hormone (MIH) negatively regulates the production of ecdysone by the Y-organ [6]. The molecular mechanisms by which PTTH and MIH regulate 20E synthesis still remain to be solved, because Black Box reactions are thought to contain the rate-limiting step in 20E biosynthesis [3,4].

Receptors

Structure and Subtype

20E binds to ecdysone receptors (EcRs) and transactivates target genes in collaboration with ultraspiracle (USP), which is a homolog of the retinoic acid X receptor (RXR). Genes encoding EcR and USP (RXR) have been cloned from various molting animals, and the primary sequences of EcR and USP (RXR) deduced [7]. EcR and USP (RXR) belong to the nuclear receptor superfamily, and are constructed from five domains (A/B, C, D, E, F). 20E binds to the E domain (ligand binding domain; LBD) of EcR. Although C (DNA binding) domains are highly conserved among insects and other arthropods, E domains vary somewhat from order to order. A phylogenetic tree has been constructed from the primary sequences of EcR or EcR-LBD. The primary structures of USP (RXR) are also used to construct phylogenetic trees to categorize insect orders and other animals.

Molecular Mechanism of Receptor

The EcR/USP heterodimer is able to bind to the ecdysone response element (EcRE) of target genes, but the genes are not activated in the absence of 20E [8]. During the intermolt

Y. Takei, H. Ando, & K. Tsutsui (Eds): Handbook of Hormones. DOI: http://dx.doi.org/10.1016/B978-0-12-801028-0.00238-5

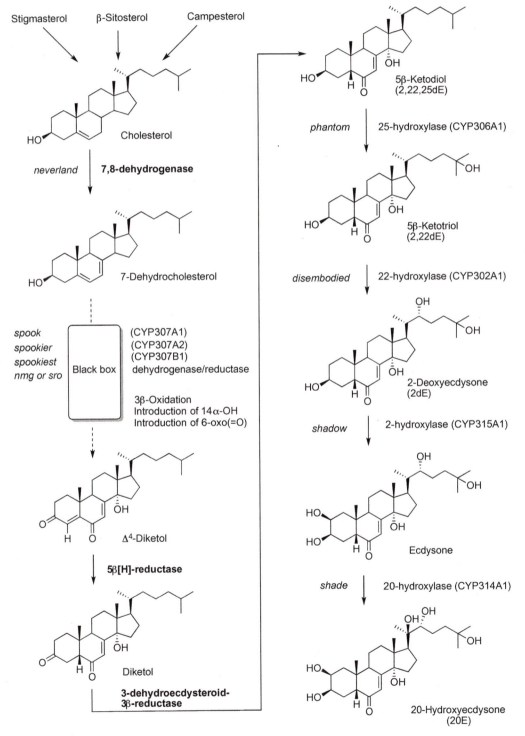

Figure 98A.1 Chemical structure of 20E.

period, co-repressor (CoR) binds to EcR to prevent gene activation. When 20E or its agonists bind to EcR, CoR is released from EcR and co-activator (CoA) is recruited; then RNA polymerase is set to synthesize RNA. These processes are known as constituting the genomic function of 20E. In addition to the genomic action, it is becoming clear that 20E brings about changes in second messenger levels, e.g., intracellular calcium and cyclic AMP, and the activation of MAP kinase, via specific trans-membrane receptor GPCRs [9,10]. These mechanisms are designated as non-genomic actions of 20E. A

Figure 98A.2 Biosynthesis of 20E.

Drosophila GPCR (DmDopEcR), which can be activated by both ecdysteroids and dopamine, has been isolated (CG18314) [10].

Biological Functions

Target Cells and Functions

It takes a long time (several hours to several days) to see the genomic effects of 20E, such as morphological changes, metabolic changes, and secretory activity of salivary glands. However, non-genomic actions of 20E are very rapid, taking place without any changes in gene expression – for example, a Na^+/K^+ pump in the hypodermis of the crayfish *Orconectes limosus*; puff formation in the midge *Chironomus thummi* by changes in the Na^+/K^+ ratio; induction of programmed cell death in the anterior silk gland of *B. mori*; and metamorphosis of the insect nervous system, including neuronal maturation, remodeling, and cell death. [9,10].

Phenotype in Gene-Modified Animals

The strength of the affinity of ligand molecules with EcRs governs the morphological effects. Mutation of specific amino acid residues in the E domain reduces ligand binding affinity. Halloween gene mutants show some morphological abnormalities at the embryonic stage owing to the 20E deficiency. The *non-molting glossy* mutant of *B. mori* shows low ecdysteroid levels and developmentally arrests at the first or second larval instar.

Agonists/Antagonists

See Chapter 98, Ecdysteroids.

Ecdysteroid Inactivation and Storage

In insects, major reactions contributing to the inactivation of ecdysteroids (ecdysone and 20-hydroxyecdysone) are (1) epimerization, (2) ecdysonic acid formation, (3) phosphorylation, (4) fatty acylation, and (5) glucosylation, summarized below.

1. Reactions involved in the epimerization of the C-3 hydroxy group of ecdysteroids are catalyzed by ecdysone oxidase and NAD(P)H-dependent 3-dehydroecdysteroid (3DE) 3α-reductase. The intermediate of these reactions, 3DE, is reduced back to ecdysone by an NAD(P)H-dependent 3DE 3β-reductase (E-Figure 98A.1). The reduction of 3DE to ecdysone catalyzed by 3DE 3β-reductase is suggested to be an important regulatory step in 20E production. The molecular cloning and characterization of the cDNAs encoding the three enzymes mentioned above have been reported [11–13].
2. Ecdysonic acid (26-oic acid) is formed via the following inactivation pathway: ecdysteroid → 26-hydroxyecdysteroid → ecdysteroid 26-aldehyde → ecdysteroid 26-oic acid. Properties of enzymes that catalyze these three reactions remain to be elucidated [14].
3. Ecdysteroids are synthesized in ovaries and eggs as well as prothoracic glands. The bulk of ecdysteroids synthesized in the ovaries exist as physiologically inactive phosphate conjugates. Among them, 2- and 3-phosphate esters are inactive products, whereas 22-phosphate esters are a storage form of ecdysteroids before the prothoracic glands exert their influence on the embryos. The biochemical characteristics and physiological significance of two enzymes, ecdysteroid 22-kinase (EcKinase) and ecdysteroid-phosphate phosphatase (EPPase) (E-Figure 98A.2), in the ovary–egg system have been reported [15–17].
4. Apolar conjugates (linoleate, palmitate, oleate esters) of ecdysteroids have been detected in various species of ticks, spiders, scorpions, myriapods, and crustaceans, in addition to insects. It has been demonstrated that an acyl-CoA: ecdysone acyltransferase takes part in the formation of ecdysteroid 22-fatty acyl ester [18]. However, the property of this enzyme is not clear.
5. There is little information regarding the isolation and identification of ecdysteroid glucoside conjugate in insects except for the demonstration of 26-hydroxyecdysone 22-glucoside in late embryos of the tobacco hornworm, *Manduca sexta*. However, when insects were infected with the nuclear polyhedrosis virus, glucosylated ecdysteroids such as ecdysone 22-glucoside and 20-hydroxyecdysone 22-glucoside were produced. It has been reported that the virus has a gene, *egt*, that encodes ecdysteroid UDP-glucosyl transferase (EGT). EGT decreases 20-hydroxyecdysone titer in the infected larvae by catalyzing the glucosylation of ecdysteroids, and leading to interference in the molting of host larvae [19].

References

1. Miyashita M, Matsushita K, Nakamura S, et al. LC/MS/MS identification of 20-hydroxyecdysone in a scorpion (*Liocheles australasiae*) and its binding affinity to in vitro-translated molting hormone receptors. *Insect Biochem Mol Biol*. 2011; 41:932–937.
2. Kametani T, Tsubuki M, Nemoto H. Total synthesis of steroid hormones. Efficient stereocontrolled synthesis of 17-methoxy-6-oxo-D-homo-18-nor-5β-androsta-2,13,15,17-tetraene. *J Org Chem*. 1980;45:4391–4398.
3. Gilbert LI, Rewitz KF. The function and evolution of the Halloween genes: The pathway to the arthropod molting hormone. In: Smagghe G, ed. *Ecdysone: Structures and Functions*. New York, NY: Springer; 2009:231–269.
4. Niwa R, Namiki T, Ito K, et al. *Non-molting glossy/shroud* encodes a short-chain dehydrogenase/reductase that functions in the "Black Box" of the ecdysteroid biosynthesis pathway. *Development*. 2010;137:1991–1999.
5. Lin TJ, Gu SH. *In vitro* and *in vivo* stimulation of extracellular signal-regulated kinase (ERK) by the prothoracicotropic hormone in prothoracic gland cells and its developmental regulation in the silkworm, *Bombyx mori*. *J Insect Physiol*. 2007;53:622–631.
6. Nakatsuji T, Sonobe H, Watson D. Molt-inhibiting hormone-mediated regulation of ecdysteroid synthesis in Y-organs of the crayfish (*Procambarus clarkii*): Involvement of cyclic GMP and cyclic nucleotide phosphodiesterase. *Mol Cell Endocrinol*. 2006;253:76–82.
7. Nakagawa Y, Henrich V. Arthropod nuclear receptors and their role in molting. *FEBS J*. 2009;276:6128–6157.
8. Yao T-P, Forman BM, Jiang Z, et al. Functional ecdysone receptor is the product of EcR and Ultraspiracle genes. *Nature*. 1993; 366:476–479.
9. Iga M, Sakurai S. Genomic and nongenomic actions of 20-hydroxyecdysone in programmed cell death. In: Smagghe G, ed. *Ecdysone: Structures and Functions*. New York, NY: Springer; 2009:411–423.
10. Evans PD, Bayliss A, Reale V. GPCR-mediated rapid, non-genomic actions of steroids: comparisons between DmDopEcR and GPER1 (GPR30). *Gen Com Endocrinol*. 2014;195:157–163.
11. Chen J, Turner PH, Rees HH. Molecular cloning and characterization of hemolymph 3-dehydroecdysone 3β-reductase from the cotton leafworm, *Spodoptera littoralis*. *J Biol Chem*. 1999; 274:10551–10556.
12. Takeuchi H, Chen J, O'Reilly DR, et al. Regulation of ecdysteroid signaling: molecular cloning, characterization and expression of 3-dehydroecdysone 3α-reductase, a novel eukaryotic member of the short-chain dehydrogenases/reductases superfamily from the cotton leafworm, *Spodoptera littoralis*. *Biochem J*. 2000; 349:239–245.
13. Takeuchi H, Rigden DJ, Ebrahimi B, et al. Regulation of ecdysteroid signalling during *Drosophila* development: identification,

characterization and modelling of ecdysone oxidase, an enzyme involved in control of ligand concentration. *Biochem J.* 2005; 389:637−645.

14. Lafont R, Dauphin-Villemant C, Warren JT, et al. Ecdysteroid chemistry and biochemistry. In: Gilbert LI, Iatrou K, Gill SS, eds. *Comprehensive Molecular Insect Science*. Amsterdam: Elsevier BV; 2005:125−195.

15. Yamada R, Sonobe H. Purification, kinetic characterization, and molecular cloning of a novel enzyme ecdysteroid-phosphate phosphatase. *J Biol Chem.* 2003;278:26365−26373.

16. Sonobe H, Ohira T, Ieki K, Ito Y, et al. Purification, kinetic characterization, and molecular cloning of a novel enzyme, ecdysteroid 22-kinase. *J Biol Chem.* 2006;281:29513−29524.

17. Sonobe H, Ito Y. Phosphoconjugation and dephosphorylation reactions of steroid hormone in insects. *Mol Cell Endocrinol.* 2009;307:25−35.

18. Slinger AJ, Isaac RE. Acyl-CoA:ecdysone acyltransferase activity from the ovary of *P. Americana*. *Insect Biochem.* 1988;18:779−784.

19. O'Reilly DR. Baculovirus encoded ecdysteroid UDP-glucosyl-transferases. *Insect Biochem Mol Biol.* 1995;25:541−550.

Supplemental Information Available on Companion Website

- Ecdysteroid inactivation and storage/E-Figures 98A.1, 98A.2
- Accession numbers for nuclear receptors and enzymes involved in 20E biosynthesis and metabolism/E-Tables 98A.1, 98A.2

CHAPTER 99

Juvenile Hormone

Tetsuro Shinoda

Abbreviation: JH

JH represents acyclic sesquiterpenoid hormones produced by the corpora allata in insects. JH regulates various aspects of insect physiology, including metamorphosis, reproduction, diapause, and polyphenisms. JH analogs are used as insect growth regulators.

Discovery

In the 1930s, Vincent Wigglesworth discovered that the corpus allatum (CA) produces juvenile hormone (JH), which prevents metamorphosis in the blood-sucking bug *Rhodnius prolixus*. The chemical structure of JH was first determined (for JH I) by Röller and his colleagues in 1967 [1].

Structure

Structural Features

JHs are derivatives of methyl esters of farnesoic acid (FA) or its ethyl branched homologs (Figure 99.1). All JHs except methyl farnesoate (MF) have an epoxide group distal to the ester group. JH III, which is found in many insect species, is the simplest form. JH 0, JH I, JH II, and iso-JH 0 are found only in Lepidoptera, and have one to three ethyl branches. JH III bisepoxide (JHB$_3$) of higher Diptera and JH III skipped bisepoxide (JHSB$_3$) of Hemiptera have a second epoxide group [2].

Properties

JHs are lipophilic and highly soluble in ethanol and acetone. JH I can be dissolved in aqueous solution up to a concentration of $2.5-3.0 \times 10^{-5}$ M, and be kept at up to 10^{-3} M in solution with bovine serum albumin. JH is strongly absorbed by plastic materials [3].

Synthesis, Release, and Degradation

Biosynthetic Pathway

JH is synthesized sequentially through the common mevalonate pathway and JH-specific pathway [4]. In the mevalonate pathway, acetyl-CoA is converted to farnesyl pyrophosphate (FPP). In the JH-specific pathway, FPP is converted to FA by phosphatase and by dehydrogenase/oxydase. Finally, FA is catalyzed to JH III by epoxidase (CYP15) and JH acid methyltransferase (JHAMT). The order of the last two steps differs between species. Lepidoptera can utilize propionyl-CoA as well as acetyl-CoA as starting materials, thereby producing ethyl-branched JHs through the same pathway.

Regulation of Synthesis and Release

JH synthesis in the CA is under complex control by various neuroendocrine and neuronal factors. Allatostatins (AST-A, -B, and -C) and allatotropin (AT) represent neuropeptides with inhibitory and stimulatory roles in JH synthesis, respectively. Other neuropeptides such as sNPF and ETH, neurotransmitters such as dopamine and glutamate, and ecdysteroid and nutrition also influence JH synthesis [5]. The relevance of these factors to JH synthesis varies with developmental stages and insect species.

Catabolism and Transport

JH is catabolized mainly by two enzymes, JH esterase (JHE), present in hemolymph, and JH epoxide hydrolase (JHEH), present in tissues. The relative importance of these enzymes in JH degradation varies with species, tissues, and developmental stages. JH is delivered from the CA to target tissues by hemolymph proteins that bind tightly to JH and protect it from degradation by non-specific enzymes. Hexamerins and lipophorins represent high molecular weight JH transporters. Low molecular weight hemolymph JH binding proteins (hJHBPs) of about 30 kDa are present in Lepidoptera and Diptera [2].

Receptors

Structure

Both a membrane-bound receptor(s) and a nuclear receptor(s) are involved in JH actions, but to date only one of the latter has been molecularly identified. The identified JH receptor is a member of the basic helix−loop−helix (bHLH)/Per−Arnt−Sim (PAS) family proteins, encoded in *methoprene tolerant* (*Met*) and its paralog *Germ-cell expressed* (*gce*) in the fruit fly *Drosophila melanogaster*. The silkworm *Bombyx mori* also has two *Met* genes, but the red flower beetle *Tribolium castaneum* and the mosquito *Aedes aegypti* have only one. JH III binds Met protein of *D. melanogaster* and *T. castaneum* with high affinity ($K_d = 5.3$ and 2.9 nM, respectively). Computational modeling has shown the presence of a hydrophobic-ligand binding pocket in the PAS-B domain that is critical for JH binding [6].

Signal Transduction Pathway

In the presence of JH, Met forms a functional complex with another bHLH-PAS protein, steroid receptor co-activator (SRC). The Met/SRC/JH complex interacts with JH response elements (JHREs) in the promoter of JH inducible genes such as *early trypsin* (*ET*) and *Krüppel homolog 1* (*Kr-h1*) and activates their transcription [6].

Y. Takei, H. Ando, & K. Tsutsui (Eds): Handbook of Hormones. DOI: http://dx.doi.org/10.1016/B978-0-12-801028-0.00099-4

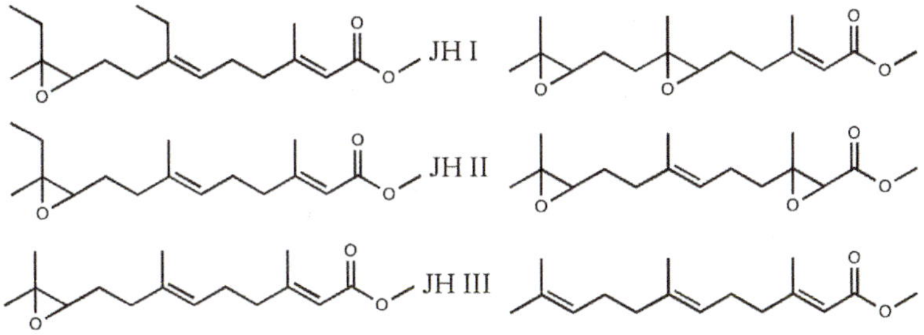

Figure 99.1 Structures of major insect juvenile hormones.

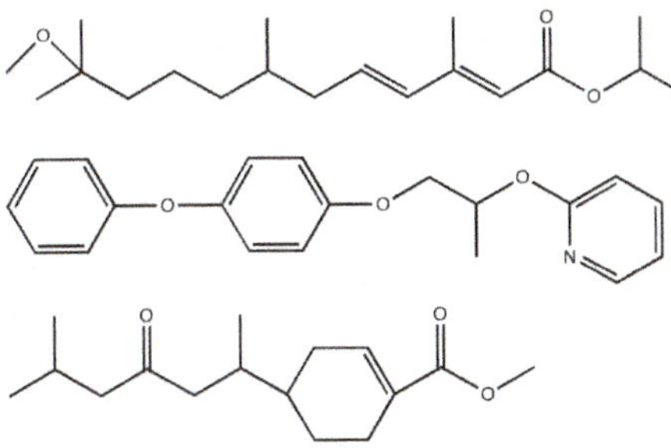

Figure 99.2 Structure of synthetic and natural JH agonists.

Agonists

Numerous JH agonists have been synthesized, and some of them are practically applied for control of insect pests. Methoprene and pyriproxyfen represent agonists with JH-like and non-JH-like structure, respectively (Figure 99.2). Juvabione is a natural JH agonist isolated from the balsam fir [7].

Antagonists

Several anti-JH compounds causing precocious metamorphosis have been reported [4]; however, a *bona fide* JH antagonist that targets the JH receptor is as yet unknown.

Biological Functions

Target Cells/Tissues and Functions

JH is highly pleiotropic and critical for the regulation of development, reproduction, polyphenisms, diapause, and behavior in insects. JH modifies the action of ecdysteroid during molting, and inhibits metamorphosis in both hemimetabolous and holometabolous insects. JH affects virtually all tissues during metamorphosis, among which the epidermis and imaginal discs are the most conspicuous targets [8]. JH promotes reproductive maturation in many insects, and induces synthesis of vitellogenin in the fat body and its uptake into the ovaries in female adults [9]. Adult diapause or reproductive diapause is caused by the lack of JH production in the CA, while larval diapause is caused by elevated levels of JH. Caste polyphenisms observed in social insects such as bees, ants, and termites are regulated by the levels of JH during immature stages. JH is an important determinant of phase polyphenisms in non-social insects, such as solitary or migratory phases in locusts and winged or apterous forms of aphids [10]. In the context of the abovementioned physiological events, JH has profound effects on the nervous system, and changes various insect behaviors.

Phenotype in Gene-Modified Animals

The single mutant of JH receptor *Met* or *gce* is fully viable, but the double mutant dies during the larval—pupal transition in *D. melanogaster*. *Met* RNAi in early stage larvae causes precocious metamorphosis in both holometabolous and hemimetabolous insects. RNAi of the key JH biosynthesis enzyme, *JHAMT*, causes precocious metamorphosis in *T. castaneum*, while ubiquitous expression of *JHAMT* results in pharate adult lethality in *D. melanogaster*. Overexpression of JHE from the embryonic stage results in precocious pupation after the third stadium in *B. mori* [6].

References

1. Gilbert LI, Granger NA, Roe RM. The juvenile hormones: historical facts and speculations on future research directions. *Insect Biochem Mol Biol*. 2000;30:617—644.
2. Goodman W.G., Cusson M. The juvenile hormones. In: *Insect Endocrinology*. Academic Press, New York, NY. 2012: 310—365.
3. Giese C, Spindler KD, Emmerich H. The solubility of insect juvenile hormone in aqueous solutions and its adsorption by glassware and plastics. *Z. Naturforsch*. 1977;32c:158—160.
4. Cusson M, Sen SE, Shinoda T. Juvenile hormone biosynthetic enzymes as targets for insecticide discovery. In: Ishaaya I, Palli SR, Horowitz AR, eds. *Advanced Technologies for Managing Insect Pests*. New York, NY: Springer; 2013:31—55.
5. Hiruma K, Kaneko Y. Hormonal regulation of insect metamorphosis with special reference to juvenile hormone biosynthesis. *Curr Top Dev Biol*. 2013;103:73—100.
6. Jindra M, Palli SR, Riddiford LM. The juvenile hormone signaling pathway in insect development. *Annu Rev Entomol*. 2013;58:181—204.
7. Sláma K. Insect hormones: more than 50 years after the discovery of insect juvenile hormone analogues (JHA, juvenoids). *Terr Arthropod Rev*. 2013;6:257—333.
8. Riddiford LM. Cellular and molecular actions of juvenile hormone I. General considerations and premetamorphic actions. *Adv. Insect Physiol*. 1994;24:213—274.
9. Wyatt GR, Davey KG. Cellular and molecular actions of juvenile hormone. II. Roles of juvenile hormone in adult insects. *Adv. Insect Physiol*. 1996;26:1—155.
10. Hartfelder K. Insect juvenile hormone: from "status quo" to high society. *Braz J Med Biol Res*. 2000;33:157—177.

Supplemental Information Available on Companion Website

- List of accession numbers of cDNA for insect *Met*/E-Table 99.1
- List of accession numbers of cDNA for insect *SRC*/E-Table 99.2
- Comparison of amino acid sequences of insect JH receptor, *methoprene tolerant (Met)*/E-Figure 99.1
- Comparison of amino acid sequences insect *SRC*/E-Figure 99.2
- Molecular phylogenetic tree of insect and crustacean *Met*/E-Figure 99.3
- Molecular phylogenetic tree of insect and crustacean *SRC*/E-Figure 99.4

CHAPTER 100

Methyl Farnesoate

Tetsuro Shinoda

Abbreviation: MF
Additional names: (E, E)-methyl farnesoate, farnesoic acid methyl ester

MF is a crustacean hormone produced by the mandibular organs. It has multiple functions in crustaceans, including the regulation of larval development, molting, reproduction, metamorphosis, polyphenisms, and sex determination. MF has also been found in several insect species and a species of mite.

Discovery

In 1987, Laufer and his colleagues first identified methyl farnesoate (MF) from the hemolymph of the spider crab *Libinia emarginata* as a juvenile hormone (JH)-like factor produced by the mandibular organs (MOs). Since then, MF has been found in more than 35 crustacean species [1]. MF is also found in several insect species, and a species of mite [2,3].

Structure

Structural Features

MF is a methyl ester of farnesoic acid (FA) or an unepoxidized form of JH III (Figure 100.1).

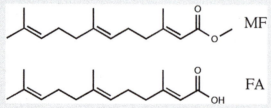

Figure 100.1 Chemical structures of methyl farnesoate and farnesoic acid.

Properties

MF is lipophilic, and highly soluble in hexane and ethanol.

Synthesis and Release

Biosynthetic Pathway

MF is synthesized by and secreted from the MOs of crustaceans. Like insect JH, MF is synthesized from acetyl-CoA through the mevalonate pathway and MF-specific pathway [4]. The *farnesoic acid O-methyltransferase (FAMeT)* gene has been considered as encoding the enzyme that catalyzes FA to MF in crustaceans [5]. In contrast, in insects MF is synthesized from FA by an enzyme encoded in *JHAMT*, which belongs to a gene family distinct from *FAMeT* [6]. Recently, *JHAMT* homologs were also found in the waterflea *Daphnia pulex* [7] and a species of mite [3]. Therefore, the relevance of *FAMeT* and *JHAMT* in MF synthesis in crustaceans and other arthropods needs to be reassessed.

Regulation of Synthesis and Release

MF biosynthesis by crustacean MOs is negatively regulated by neuropeptides, i.e., mandibular organ inhibitory hormones (MOIHs), which are produced by the X-organ/sinus gland complex in the eyestalk. MOIHs have been isolated from *L. emarginata* and the edible crab *Cancer pagurus* and identified to be members of the crustacean hyperglycemic hormone (CHH) family of neuropeptides. MOIH inhibits FAMeT or JHAMT activity at the final step of MF biosynthesis in the MOs. MF is transported to target tissues by MF binding proteins in the hemolymph, and is metabolized by a specific esterase present in the hepatopancreas and gonads [4].

Receptors

Structure

Similar to insect JH receptor, a complex of methoprene tolerant (Met) and steroid receptor co-activator (SRC) functions as an MF receptor in crustaceans. The Met and SRC of *D. pulex* and *Daphnia magna* form a heterodimer in the presence of juvenoids, including MF [8]. The effect of MF on the dimerization is about 10-fold higher than that of JH III. An amino acid substitution in the PAS-B domain of *Daphnia* Met in part accounts for the difference in sensitivity between MF and JH III [8]. Ultraspiracle (USP), an ortholog of vertebrate RXR, binds to MF with high affinity (K_d in the nanomolar range) and is proposed as working as an MF receptor in the fruit fly *Drosophila melanogaster* [2].

Signal Transduction Pathway

MF is thought to induce the expression of hemoglobin genes and sex-determining genes through MF receptors in daphnids [9]. However, the MF signal transduction pathway downstream of the MF receptor is as yet unknown in any crustacean species.

Agonists

JH III, methoprene, pyriproxyfen, and fenoxycarb work as MF agonists in the induction of male offspring in *Daphnia* [8].

Antagonists

Specific antagonists for the MF receptor have not yet been identified.

Y. Takei, H. Ando, & K. Tsutsui (Eds): Handbook of Hormones. DOI: http://dx.doi.org/10.1016/B978-0-12-801028-0.00100-8

Biological Functions

Target Cells/Tissues and Functions

MF has multiple physiological functions, including the regulation of metabolism, molting, reproduction, behavior, and morphogenesis in crustaceans [4]. It stimulates general protein synthesis and ecdysteroid secretion by the Y-organ, and influences the molting cycle [10]. MF plays a key role in the regulation of crustacean reproduction; it enhances egg production in females, and affects reproductive morphology and behavior in males [1]. Expression of the vitellogenin (Vg) gene is induced by MF in the hepatopancreas, although FA is more potent than MF for this activity [1]. MF has either a stimulatory or an inhibitory role in metamorphosis, depending on the species and developmental stages [10]. MF functions as a sex determinant and induces the production of male offspring in daphnids and other cladoceran species [9], and is also involved in predator-induced polyphenisms in *Daphnia* [7].

Phenotype in Gene-Modified Animals

RNAi mediated knockdown of the MF receptor component, *Met* or *SRC*, causes embryonic death in *D. magna* [8].

References

1. Nagaraju GP. Reproductive regulators in decapod crustaceans: an overview. *J Exp Biol.* 2011;214:3–16.
2. Jones D, Jones G, Teal PE. Sesquiterpene action, and morphogenetic signaling through the ortholog of retinoid X receptor, in higher Diptera. *Gen Comp Endocrinol.* 2013;194:326–335.
3. Grbic M, Van Leeuwen T, Clark RM, et al. The genome of *Tetranychus urticae* reveals herbivorous pest adaptations. *Nature.* 2011;479:487–492.
4. Nagaraju GPC. Is methyl farnesoate a crustacean hormone? *Aquaculture.* 2007;272:39–54.
5. Hui JH, Hayward A, Bendena WG, et al. Evolution and functional divergence of enzymes involved in sesquiterpenoid hormone biosynthesis in crustaceans and insects. *Peptides.* 2010;31:451–455.
6. Shinoda T, Itoyama K. Juvenile hormone acid methyltransferase: a key regulatory enzyme for insect metamorphosis. *Proc Natl Acad Sci USA.* 2003;100:11986–11991.
7. Miyakawa H, Imai M, Sugimoto N, et al. Gene up-regulation in response to predator kairomones in the water flea, *Daphnia pulex. BMC Dev Biol.* 2010;10:45.
8. Miyakawa H, Toyota K, Hirakawa I, et al. A mutation in the receptor Methoprene-tolerant alters juvenile hormone response in insects and crustaceans. *Nat Commun.* 2013;4:1856.
9. Eads BD, Andrews J, Colbourne JK. Ecological genomics in *Daphnia*: stress responses and environmental sex determination. *Heredity (Edinb).* 2008;100:184–190.
10. Homola E, Chang ES. Methyl farnesoate: Crustacean juvenile hormone in search of functions. *Comp Biochem & Physiol B: Comp Biochem.* 1997;117:347–356.

Supplemental Information Available on Companion Website

- List of accession numbers of cDNA for crustacean *Met*/ E-Table 100.1
- List of accession numbers of cDNA for crustacean *SRC*/ E-Table 100.2
- Comparison of amino acid sequences of crustacean MF receptor, *methoprene tolerant* (*Met*)/ E-Figure 100.1
- Comparison of amino acid sequences of crustacean *SRC*/ E-Figure 100.2
- Molecular phylogenetic tree of insect and crustacean *Met*/ E-Figure 100.3
- Molecular phylogenetic tree of insect and crustacean *SRC*/ E-Figure 100.4

PART V

Endocrine Disrupting Chemicals

CHAPTER 101

Endocrine Disruptors

Shinichi Miyagawa, Tomomi Sato, and Taisen Iguchi

Abbreviation: EDs
Additional names: endocrine disrupting chemicals (EDCs), environmental hormones

History

In 1991 Theo Colborn organized a meeting at Wingspread, Wisconsin, USA, concerning environmental chemicals, and its outcome was published in 1992 [1]. Based on the outcome, Colborn and colleagues wrote a book, *Our Stolen Future* [2], and discussed concerns regarding the health effects of environmental chemicals with hormone-like activities (endocrine disruptors, endocrine disrupting chemicals (EDCs)) on wildlife and humans. This book stimulated international organizations such as WHO (the World Health Organization) and OECD (the Organization for Economic Cooperation and Development), and the EU countries, the USA, and Japan to start their own research programs to tackle endocrine disruptor issues.

The National Research Council, USA, published *Hormonally Active Agents in the Environment* in 1999 [3]. In 1992, WHO and IPCS (International Programme on Chemical Safety) provided definitions relating to an EDC, such as "An endocrine disruptor is an exogenous substance or mixture that alters function(s) of the endocrine system and consequently causes adverse health effects in an intact organism, or its progeny, or (sub)populations," and "A potential endocrine disruptor is an exogenous substance or mixture that possesses properties that might be expected to lead to endocrine disruption in an intact organism, or its progeny, or (sub)populations." Based on these definitions, WHO/IPCS evaluated potential correlations of environmental chemicals and adverse effects on wildlife and humans, using published papers [4]. In 2013, WHO and UNEP (the United Nations Environment Programme) published a follow-up of EDCs [5]. In 2010, the EU also published a report on endocrine disruptor issues [6]. OECD has established screening and testing methods for chemicals having hormonal activities using four fish species (fathead minnow, zebrafish, medaka, stickleback), an amphibian (*Xenopus laevis*), an avian (Japanese quail), mammals (rats), and invertebrates (copepods, *Daphnia magna*, chironomids, and snails). *In vitro* screening methods have also been prepared for detecting chemicals having (anti) estrogenic, (anti)androgenic, or (anti)thyroid hormone activities, and *in silico* and docking models for estrogenic and androgenic chemicals.

Historical studies for estrogens and estrogenic chemicals were summarized by Hertz in 1985 [7,8]. Cyclical growth of uterus and vaginal cornification were found in the guinea pig [9] and rat [10], and then estrone was identified in ovaries as a stimulant of uterine cell proliferation and vaginal cornification by Allen and Doisy in 1923 [11]. Estrogens were found to be able to induce neoplastic growth in numerous species of animals, including the squirrel monkey, and in multiple tissue sites [12]. Estrogenic substances had been found in the diet and in the environment as early as the 1920s, when Zondek demonstrated biologically active estrogens in the flowers of the willow tree. By 1946, frequent abortions and male infertility in sheep exposed to a variant form of clover had been recorded in New Zealand, and the active compound in the clover was found to be genistein [13]. Various phytoestrogens were summarized by Biggers in 1959 [14]. Infertility of mouse colonies caused by contamination with stilbestrol was also found in the 1950s. Stilbestrol contamination in the vitamin capsule caused nodular breast enlargement in a 4-year-old girl and her 6-year-old brother [15]. In NIH, methoxychlor used to combat ectoparasite infestation was discovered to have potent estrogenic activity [16] and induced spontaneous uterotrophic effects in untreated weanling rats [17]. In 1928 Zondek [18] found very high estrogenic activity in the urine of pregnant mares, and later equilenin, equilin, etc., were isolated from horse urine. In 1938, Dodds *et al.* [19] synthesized diethylstilbestrol (DES) as a potent estrogen, which has been used for prevention of abortion [8,20]. Herbst *et al.* [21] found that DES exposure *in utero* induced vaginal adenocarcinoma in women. In 1962, Takasugi *et al.* [22] demonstrated that the estrogen exposure in perinatal female mice induced persistent vaginal epithelial proliferation and vaginal cancer [23].

Biological Functions

Estrogenic effects of industrial chemicals such as bisphenol-A, nonylphenol, and octylphenol have been identified using cell proliferation of estrogen-responsive human breast cancer cells in *in vitro* and *in vivo* induction of vitellogenin (egg yolk precursor protein) in male fish [2]. o,p'-DDT was known to have estrogenic activity in the 1950s [7]. 17α-Ethinylestradiol, synthesized for contraceptive pills and equilins used for hormone replacement therapy, have been found in river water in various countries, including the UK [5,6]. These chemicals may be affecting reproduction and sex differentiation of fish species [24,25].

Recently, anti-androgenic activity in many pesticides has been reported using the reporter gene assay system [6]. Anti-androgenic activities were found in effluents of waste water treatment works in the UK [26]. Chemicals having anti-thyroid hormone activity have also been reported, using amphibian metamorphosis assay [5,6].

Y. Takei, H. Ando, & K. Tsutsui (Eds): Handbook of Hormones. DOI: http://dx.doi.org/10.1016/B978-0-12-801028-0.00101-X

Receptors and Structures

Most chemicals that have estrogenic activity have been identified using estrogen-receptor based transactivation assays or estrogen receptor binding assays, using estrogen receptors from various animal species. Many chemicals with estrogenic activity show structural similarity in their phenol ring. For screening of anti-androgenic chemicals and (anti)thyroid hormone activities, androgen-receptor based transactivation assays and thyroid hormone receptor based transactivation assays are also used. No chemical with both androgenic activity and thyroid hormone activity has been found to date, except as a pharmaceutical.

Representative chemicals showing estrogenic activity (nonylphenol, octylphenol, bisphenol-A, *o,p'*-DDT, 17α-ethinylestradiol, equilin, equilenin), anti-estrogenic activity (TCDD, PCBs), anti-androgenic activity (DDE, vinclozolin, polybrominated diphenyl ether), and anti-thyroid hormone activity (tetrabromobisphenol-A, perchlorate) are explained in detail in the following subchapters.

References

1. Colborn T, Clement C, eds. *Chemically-Induced Alterations in Sexual and Functional Development: The Wildlife/Human Connection.* Princeton, NJ: Princeton Sci Pub Co.; 1992, 403 pp.
2. Colborn T, Dumanoski D, Myers JP. *Our Stolen Future.* New York, NY: Dutton; 1996, 306 pp.
3. National Research Council. *Hormonally Active Agents in the Environment.* Washington, DC: National Academy Press; 1999, 430 pp.
4. Damstra T, Barlow S, Bergman Å, Kavlock R, van der Kraak G, eds. *Global Assessment of the State-of-the-Science of Endocrine Disruptors.* Geneva: WHO/IPCS/EDC/02.2; 2002, 180 pp.
5. Bergman Å, Heindel JJ, Jobling S, Kidd K, Zoeller RT, eds. *State of the Science of Endocrine Disrupting Chemicals – 2012.* Geneva: WHO/UNEP; 2013, 260 pp.
6. Kortenkamp A, Martin O, Faust M et al. *State of the Art Assessment of Endocrine Disrupters.* Final Report, Project Contract Number 070307/2009/550687/SER/D3, 135 pp.
7. McLachlan JA, ed. *Estrogens in the Environment. II: Influence on Development.* New York, NY: Elsevier; 1985, 435 pp.
8. Hertz R. The estrogen problem-retrospect and prospect. In: McLachlan JA, ed. *Estrogens in the Environment II – Influences on Development.* New York, NY: Elsevier; 1985:1–11.
9. Stockard CR, Papanicolaou GN. A rhythmical "heat period" in the guinea-pig. *Science.* 1917;46:42–44.
10. Long JA, Evans HM. The oestrous cycle in the rat and its associated phenomena. *Mem Univ Calif.* 1922;6:1–148.
11. Allen E, Doisy EA. An ovarian hormone; preliminary report of its localization, extraction and partial purification and action in test animals. *J Am Med Assoc.* 1923;81:819.
12. Herts R. A review of the evidence linking estrogens and cancer in animals. *Pediatrics.* 1978;62:1138–1142.
13. Bennetts HW. Metaplasia in the sex organs of castrated male sheep maintained on early subterranean clover pastures. *Aust Vet J.* 1946;22:70.
14. Biggers JD. Plant phenols possessing oestrogenic activity. In: Fairbairn JW, ed. *The Pharmacology of Plant Phenolics.* New York, NY: Academic Press; 1959:51–69.
15. Herts R. Accidental ingestion of estrogens by children. *Pediatrics.* 1958;21:203–206.
16. Tullner WW. Uterotrophic action of the insecticide methoxychlor. *Science.* 1961;133:647.
17. Evans JM, Young JP, Hertz R, et al. The ability of the human liver to reduce the biologic activity of injected estrogens. *J Clin Endocrinol Metab.* 1952;12:495–501.
18. Zondek B. Darstellung des weiblichen Sexualhormones aus dem Harn, insbesondere dem Harn von Schwangeren. *Klinische Wochenschrift.* 1928;7:485–486.
19. Dodds EC, Goldberg L, Lawson W, et al. Estrogenic activity of certain synthetic compounds. *Nature.* 1938;141:247–248.
20. Smith OW. Diethylstilbestrol in the prevention and treatment of complications of pregnancy. *Am J Obstet Gynecol.* 1948;56:821.
21. Herbst AL, Ulfelder H, Poskanzer DC. Adenocarcinoma of the vagina. Association of maternal stilbestrol therapy with tumor appearance in young women. *N Engl J Med.* 1971;284:878–881.
22. Takasugi N, Bern HA, DeOme KB. Persistent vaginal cornification in mice. *Science.* 1963;138:438–439.
23. Herbst AL, Bern HA, eds. *Developmental Effects of Diethylstilbestrol (DES) in Pregnancy.* Stratton, New York, NY: Thieme; 1981, 203 pp.
24. Lange A, Paull GC, Hamilton PB, et al. Implications of persistent exposure to treated wastewater effluent for breeding in wild roach (*Rutilus rutilus*) populations. *Environ Sci Technol.* 2011;12:1673–1679.
25. Tyler CR, Filby AL, Bickley LK, et al. Environmental health impacts of equine estrogens derived from hormone replacement therapy. *Environ Sci Technol.* 2009;43:3897–3904.
26. Hill EM, Evans KL, Horwood J, et al. Profiles and some initial identifications of (anti)androgenic compounds in fish exposed to wastewater treatment works effluents. *Environ Sci Technol.* 2010;44:1137–1143.

Nonylphenol

Shinichi Miyagawa, Tomomi Sato, and Taisen Iguchi

Abbreviation: NP
Additional names: *p*-nonylphenol, 4-nonylphenol

Nonylphenol (NP) is a subset of the alkylphenols. NP is a precursor of nonylphenol ethoxylates, which are widely used as industrial surfactants and in a variety of pesticides and consumer products. NP can bind to estrogen receptors and elicit estrogenic action in vivo *and* in vitro.

Discovery

NP was first synthesized in the 1940s.

Structure

Structural Features

NP features both polar and hydrophobic subunits, the phenol and the hydrocarbon tail. Technically, nonylphenol contains approximately 20 isomers and predominantly *para*-substituted nonylphenol (Figure 101A.1).

Properties

A pale yellow viscous liquid, molecular formula $C_{15}H_{24}O$, Mr 220.34, density 0.95 g/ml (20°C), melting point −8°C. Poorly soluble in water but soluble in alcohol.

Production

NP is produced by acid-catalyzed alkylation of phenol with a mixture of nonenes.

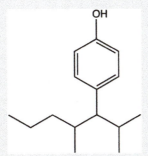

Figure 101A.1   Structure of NP (branched).

Application

NP is subjected to ethoxylation to give nonylphenol ethoxylates, which are widely used as industrial surfactants and as lubricant additives, polymer stabilizers, antioxidants, and agricultural chemicals.

Biological and Pathophysiological Implications

Background

NP has estrogenic activity by binding to estrogen receptors (ERs) in both *in vitro* and *in vivo* assay systems. In 1991, Soto *et al.* reported that polystyrene tubes used in routine laboratory procedures released a substance with estrogenic properties and identified the substance as a nonylphenol. They showed that NP induced cell proliferation and progesterone receptor gene expression in MCF-7 cells, and activated cell proliferation in ovariectomized rat endometrium [1].

Effects on Mammals

NP affected both male (decreased epididymal sperm density and testicular spermatid head counts) and female (increased estrus cycle length and decreased ovarian weights) rats in multigeneration tests [2].

Environment

In the environment, NP originates from degradation of nonylphenol ethoxylates, and the principal source is treated wastewater effluent [3,4]. Sumpter and Jobling suggested that NP contained in effluent from sewage treatment works is associated with vitellogenesis in male rainbow trout [5]. NP has greater binding affinity for fish ER than for human ER, and induces the formation of testis—ova (oocytes in the testis) of Japanese medaka at environmentally relevant concentrations [6]. *In vitro* reporter gene assay revealed that relative potency of NP to estradiol-17β for medaka estrogen receptor α is 0.5% [7]. In *in vivo* experiments, nonylphenol was found to induce vitellogenins (female egg yolk proteins) in the male liver at 20.3 μg/l in rainbow trout, and at 11.6 μg/l in medaka [8,9].

Safety Standards and Regulatory Compliance

NP and nonylphenol ethoxylates have been restricted in the European Union, as a hazard to humans and aquatic animals

Y. Takei, H. Ando, & K. Tsutsui (Eds): Handbook of Hormones. DOI: http://dx.doi.org/10.1016/B978-0-12-801028-0.00239-7

[10]. Nonylphenol ethoxylates are being replaced by other surfactants, mainly alcohol ethoxylates, in most European, Canadian, and Japanese markets.

References

1. Soto AM, Justicia H, Wray JW, et al. *p*-Nonyl-phenol: an estrogenic xenobiotic released from "modified" polystyrene. *Environ Health Perspect*. 1991;92:167–173.
2. National Institute of Environmental Health Services. Reproductive toxicity of nonylphenol (CA 84852-15-3) administered by Gavage to Sprague–Dawley rats. Study RACB94021. NTIS P97-210900. National Technical Information Services, Springfield, VA, USA; 1998.
3. Langford KH, Lester JN. Fate and behaviour of endocrine disrupters in wastewater treatment processes. In: Brikett JW, Lester JN, eds. *Endocrine Disrupters in Wastewater and Sludge Treatment Processes*. Boca Raton, FL, USA: CRC Press Inc; 2002.
4. Giger W, Brunner PH, Schaffner C. 4-Nonylphenol in sewage sludge: accumulation of toxic metabolites from nonionic surfactants. *Science*. 1984;225:623–625.
5. Sumpter JP, Jobling S. Vitellogenesis as a biomarker for estrogenic contamination of the aquatic environment. *Environ Health Perspect*. 1995;103(Suppl 7):173–178.
6. Japan Environment Agency. The Report on the Estrogenic Action of 4-Nonylphenol on the Fishes. Tokyo, Japan; 2001.
7. Miyagawa S, Anke L, Hirakawa I, et al. Differing species responsiveness of estrogenic contaminants in fish is conferred by the ligand binding domain of the estrogen receptor. *Environ Sci Toxicol*. 2014;48:5254–5263.
8. Jobling S, Sheahan D, Osborne JA, et al. Inhibition of testicular growth in rainbow trout (*Oncorhynchus mykiss*) exposed to estrogenic alkylphenolic chemicals. *Environ Toxicol Chem*. 1996;15:194–202.
9. Seki M, Yokota H, Maeda M, et al. Effects of 4-nonylphenol and 4-*tert*-octylphenol on sex differentiation and vitellogenin induction in medaka (*Oryzias latipes*). *Environ Toxicol Chem*. 2003;22:1507–1516.
10. Soares A, Guieysse B, Jefferson B, et al. Nonylphenol in the environment: a critical review on occurrence, fate, toxicity and treatment in wastewaters. *Environ Int*. 2008;34:1033–1049.

Octylphenol

Shinichi Miyagawa, Tomomi Sato, and Taisen Iguchi

Abbreviation: OP
Additional names: 4-(1,1,3,3-tetramethylbutyl)phenol, 4-*tert*-octylphenol

Octylphenol (OP) is used as an intermediate in the production of phenolic resins and in the manufacture of octylphenol ethoxylates. OP is an endocrine disruptor because of its estrogenic activity and adverse effect in fish species.

Structure

Structural Features

The octyl group is located at the 2-, 3-, or 4-position of the benzene ring. 4-*tert*-octylphenol is the most commercially important isomer (Figure 101B.1).

Properties

Molecular formula $C_{14}H_{22}O$, Mr 206.32, density 0.961 g/ml (25°C), melting point 79−82°C. Poorly soluble in water.

Production

OP is produced by the reaction of phenol and *tert*-octene (di-isobutene) in the presence of an ion-exchange resin or boron trifluoride complex in a batch reactor, or a fixed bed ion-exchange resin in a continuous process.

Application

OP is used as an intermediate in the production of phenolic resins and in the manufacture of octylphenol ethoxylates. Phenolic resins are used in rubber processing, to make tyres, and in inks.

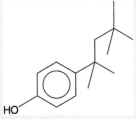

Figure 101B.1   Structure of 4-*tert*-octylphenol.

Biological and Pathophysiological Implications

Background

OP is mainly released during manufacturing, and during the use and disposal of products containing it. Environmental OP is also derived from the breakdown of octylphenol ethoxylates in the environment. 4-*tert*-Octylphenol is a high production volume chemical, and is the most likely immediate replacement for nonylphenol. However, there is the possibility of adverse environmental effects for several endpoints in fish studies, with risks similar to those predicted for nonylphenol [2]. This is to be expected, in view of the similarities in the environmental fate, behavior, and toxicity of the two substances [1].

Estrogenic Activities

Alkylphenols act as weak estrogens. OP stimulates cell proliferation in MCF-7 cells [2]. Estrogenic activity of 4-*tert*-octylphenol in mammals was also evident in rodent bioassays such as the uterotrophic assay [3]. Neonatal exposure of 4-*tert*-octylphenol in rats led to reduced numbers of corpora lutea with higher numbers of preantral and atretic follicles, and persistent estrus [4].

Effects on Fish

4-*tert*-Octylphenol also influences the endocrine system in fish species, with effects including vitellogenin induction, presence of testis−ova, increased testicular degeneration, interstitial (Leydig) cell hyperplasia/hypertrophy and decrease in secondary sex characteristics in males, and a female-biased phenotypic sex ratio during the sexual development of medaka (*Oryzias latipes*) [5,6]. *In vitro* reporter gene assay revealed that the potency of 4-*tert*-octylphenol relative to estradiol-17β for medaka estrogen receptor α is 0.5% [7]. OP is capable of induction of vitellogenesis (female proteins) in male medaka at 11.4 ng/ml [6].

References

1. Chemicals and Nanotechnologies Division, Defra. 4-tert-Octylphenol Risk Reduction Strategy and Analysis of Advantages and Drawbacks. London UK; 2006.
2. Andersen HR, Andersson AM, Arnold SF, et al. Comparison of short-term estrogenicity tests for identification of hormone-disrupting chemicals. *Environ Health Perspect*. 1999;107(Suppl 1):89−108.

Y. Takei, H. Ando, & K. Tsutsui (Eds): Handbook of Hormones. DOI: http://dx.doi.org/10.1016/B978-0-12-801028-0.00240-3

3. Laws SC, Carey SA, Ferrell JM, et al. Estrogenic activity of octylphenol, nonylphenol, bisphenol A and methoxychlor in rats. *Toxicol Sci.* 2000;54:154−167.

4. Willoughby KN, Sarkar AJ, Boyadjieva NI, et al. Neonatally administered *tert*-octylphenol affects onset of puberty and reproductive development in female rats. *Endocrine.* 2005;26:161−168.

5. European Chemicals Agency. Substance name: 4-(1,1,3,3-tetramethylbutyl)phenol, 4-*tert*-octylphenol. 2011. ECHA, Helsinki, Finland. <http://echa.europa.eu/documents/10162/f88ff7a0-59f1-44c1-ac22-295c4d14dbf3>.

6. Seki M, Yokota H, Maeda M, et al. Effects of 4-nonylphenol and 4-*tert*-octylphenol on sex differentiation and vitellogenin induction in medaka (*Oryzias latipes*). *Environ Toxicol Chem.* 2003;22:1507−1516.

7. Miyagawa S, Anke L, Hirakawa I, et al. Differing species responsiveness of estrogenic contaminants in fish is conferred by the ligand binding domain of the estrogen receptor. *Environ Sci Toxicol.* 2014;48:5254−5263.

Bisphenol A

Shinichi Miyagawa, Tomomi Sato, and Taisen Iguchi

Abbreviation: BPA
Additional names: 2,2-bis(4-hydroxyphenyl)propane4,4-(1-methylethlidene)bisphenol

Bisphenol A (BPA) is a chemical produced in large quantities for use primarily in the production of polycarbonate plastics and epoxy resins. It is a representative environmental estrogen and also affects a wide array of biological processes, including the metabolic, thyroid hormone, and androgen systems.

Discovery

BPA was first synthesized in the 1890s. In the early 1930s, Dodds and Lawson discovered that BPA was estrogenic, but less effective than estradiol-17β (E$_2$) [1]. Since Dodds discovered diethylstilbestrol as an effective synthetic estrogen, BPA has never been used as a drug.

Structure

Structural Features

BPA is a carbon-based synthetic compound containing two 4-hydroxyphenyl rings (Figure 101C.1).

Properties

A white solid, molecular formula $C_{15}H_{16}O_2$, Mr 228.29, density 1.20 g/ml (25°C), melting point 158−159°C. Poorly soluble in water.

Production

BPA is synthesized by the condensation of acetone with two equivalents of phenol.

Applications

BPA is used to make certain polycarbonate plastic and epoxy resins. Polycarbonate plastic is utilized in reusable food and drink containers, including baby milk and water bottles, and in tableware and water pipes. The inner walls of cans and the lids of glass jars and bottles for foods and beverages are lined with epoxy resins as a protective coating. BPA is also found in some polyvinyl chloride plastics and paper products.

Biological and Pathophysiological Implications

Background and Estrogenic Effects

In mammalian uterotrophic assay, BPA was found to be about 10,000-fold less potent than E$_2$ [2]. In *in vitro* reporter gene assay, the relative potency of BPA to E$_2$ for medaka estrogen receptor α is 0.08% [3]. Thus, BPA is a weak estrogen exhibiting a relatively short half-life, but can build up to concentrations of concern by continuous exposure to it. Developmental animals (i.e., fetuses and young animals) may be more susceptible to BPA (and other endocrine disruptors) exposure than adults.

Low-Dose Effects

Some effects of BPA are observed at extremely low concentrations, and higher concentrations often do not result in the same effects. This dose−response relationship is often referred to as an inverted U response. In 1997, adverse effects of low-dose exposure of laboratory animals to BPA were first shown as an increase in size of the fetal mouse prostate [4]. Some examples of low-dose effects of BPA in rodents include increased postnatal growth, early onset of sexual maturation in females, decrease in daily sperm production and fertility in males, stimulation of mammary gland development in female offspring, altered immune function, decrease in antioxidant enzymes, changes in the brain (including an increase in progesterone receptor mRNA levels), and behavioral effects (including hyperactivity, increase in aggressiveness, and decreased maternal behavior) [5].

Obesity

BPA may increase the risk for obesity, acting via fat cells and brain to regulate adipose tissue deposition and food intake in rodents [3]. BPA also affects β cells in the pancreas, and induces increase of insulin resistance and glucose intolerance [6].

Other Hormonal Effects

BPA acts as an androgen receptor antagonist and can block testosterone synthesis [5]. It also affects thyroid hormone systems [5]. In addition, BPA binds to orphan receptor estrogen-related receptor γ with relatively high affinity [7]. BPA disrupts microtubule functions, resulting in a high degree of aneuploidy [8].

Y. Takei, H. Ando, & K. Tsutsui (Eds): Handbook of Hormones. DOI: http://dx.doi.org/10.1016/B978-0-12-801028-0.00241-5

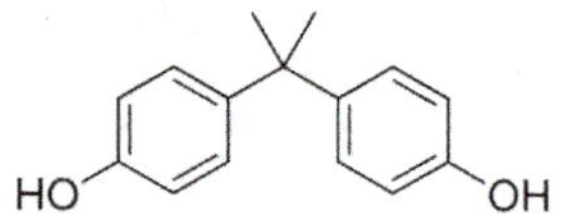

Figure 101C.1 Structure of BPA.

Safety Standards and Regulatory Compliance

The European Union and Canada have banned the use of BPA in baby bottles. In 2010, Environment Canada declared BPA to be a toxic substance. In 2015, the French government banned the use of BPA in materials which directly affect food.

References

1. Dodds EC, Lawson W. Synthetic strogenic agents without the phenanthrene nucleus. *Nature*. 1936;137:996.
2. Milligan SR, Balasubramanian AV, Kalita JC. Relative potency of xenobiotic estrogens in an acute *in vivo* mammalian assay. *Environ Health Perspect*. 1998;106:23−26.
3. Miyagawa S, Lange A, Hirakawa I, et al. Differing species responsiveness of estrogenic contaminants in fish is conferred by the ligand binding domain of the estrogen receptor. *Environ Sci Toxicol*. 2014;48:5254−5263.
4. vom Saal FS, Timms BG, Montano MM, et al. Prostate enlargement in mice due to fetal exposure to low doses of estradiol or diethylstilbestrol and opposite effects at high doses. *Proc Natl Acad Sci USA*. 1997;94:2056−2061.
5. vom Saal FS, Hughes C. An extensive new literature concerning low-dose effects of bisphenol A shows the need for a new risk assessment. *Environ Health Perspect*. 2005;113:926−933.
6. Alonso-Magdalena P, Vieira E, Soriano S, et al. Bisphenol A exposure during pregnancy disrupts glucose homeostasis in mothers and adult male offspring. *Environ Health Perspect*. 2010;118:1243−1250.
7. Matsushima A, Kakuta Y, Teramoto T, et al. Structural evidence for endocrine disruptor bisphenol A binding to human nuclear receptor ERR gamma. *J Biochem*. 2007;142:517−524.
8. Hunt PA, Koehler KE, Susiarjo M, et al. Bisphenol a exposure causes meiotic aneuploidy in the female mouse. *Curr Biol*. 2003;13:546−553.

Dichlorodiphenyltrichloroethane

Shinichi Miyagawa, Tomomi Sato, and Taisen Iguchi

Abbreviation: DDT

Dichlorodiphenyltrichloroethane (DDT) is a persistent organic pollutant with high bioaccumulation properties. In addition to toxic effects, DDT and its derivatives have estrogenic and/ or anti-androgenic properties.

Discovery

DDT was first synthesized in 1873, and its insecticidal action was discovered by Müller in 1939. Müller was awarded the Nobel Prize in Physiology or Medicine in 1948 for his discovery of the high efficiency of DDT as a contact poison against several arthropods.

Structure

Structural Features

DDT is an organochloride (Figure 101D.1).

Properties

A white, crystalline powder, molecular formula $C_{14}H_9Cl_5$, Mr 354.49. Poorly soluble in water.

The term DDT/DDTs is often used to refer to a family of isomers (*p,p'*-DDT and *o,p'*-DDT) and their breakdown products (*p,p'*-dichlorodiphenyldichloroethylene (DDE), *o,p'*-DDE, *p,p'*-dichlorodiphenyldichloroethane (DDD), and *o,p'*-DDD).

Production

DDT is produced by the reaction of chloral with chlorobenzene in the presence of sulfuric acid as a catalyst.

Applications

DDT was used to control insects during World War II, and then as an agricultural insecticide. Almost all uses of DDT were banned in most developed countries in the 1970s–1980s. In some countries, DDT was applied to the inside walls of homes to kill or repel mosquitoes. Technical grade DDT (generally used as an insecticide) contains 65–80% (active ingredient) of *p,p'*-DDT, 15–21% of *o,p'*-DDT, and up to 4% of *p,p'*-DDD.

Biological and Pathophysiological Implications

Health and Environmental Hazards

In 1962, the book *Silent Spring*, by Rachel Carson, warned that DDT and other pesticides were a threat to wildlife. DDT and its derivatives cannot be broken down and eliminated by organisms and they are very persistent in both the environment and organisms. Bioaccumulation of DDT and DDE was especially pronounced in some predator birds, causing eggshell thinning and resulting in severe population declines [1]. In addition, DDT and DDE possibly cause lower testosterone and demasculinization in polar bears and alligators, intersex in fish and amphibians, adrenal hyperplasia in seals, and suppression of thyroid hormone function in marine mammals, birds, and amphibians [2].

Persistent Organic Pollutants

DDT is one of the POPs (persistent organic pollutants). Once released into the environment, they (1) remain intact for exceptionally long periods of time; (2) become widely distributed throughout the environment as a result of natural processes involving soil, water and, most notably, air; (3) accumulate in the fatty tissue of living organisms, including humans, and are found at higher concentrations at higher levels in the food chain; (4) are toxic to both humans and wildlife. Some POPs, including DDT and its derivatives, are also considered to be endocrine disruptors, which, by altering the hormonal system, can damage the reproductive systems of exposed individuals as well as their offspring.

Hormone Activities

Regarding steroid hormone disruption effects, *p,p'*-DDT, DDT's main component, has little or no androgenic or estrogenic activity. *o,p'*-DDT has estrogenic activity, and the relative potency of *o,p'*-DDT with respect to estradiol-17β for medaka estrogen receptor α is 0.16% [3]. The DDT metabolite, DDE, can act as an anti-androgen but not as an estrogen [4–7]. Of the DDT metabolites, *p,p'*-DDE is the most potent anti-androgen.

Carcinogenicity

The carcinogenicity of DDT and its metabolites is evident in rodents. DDT was found to increase incidences of liver and lung tumors, and lymphomas, in mice [8].

Y. Takei, H. Ando, & K. Tsutsui (Eds): Handbook of Hormones. DOI: http://dx.doi.org/10.1016/B978-0-12-801028-0.00242-7

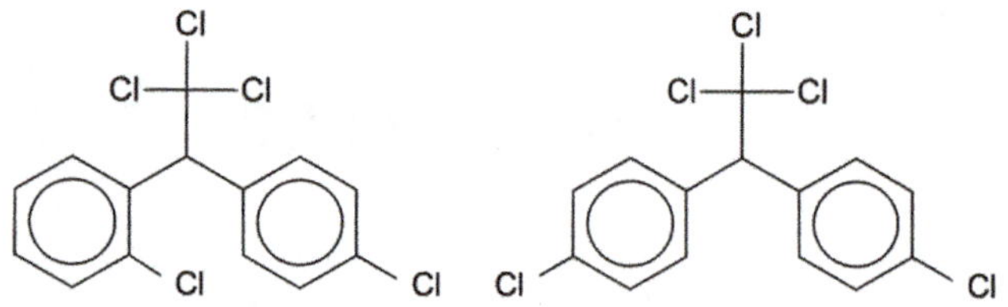

Figure 101D.1   Structures of *o,p'*-DDT (left) and *p,p'*-DDT (right).

Safety Standards and Regulatory Compliance

DDT was included in the Stockholm Convention on Persistent Organic Pollutants, which banned its production and use worldwide. The Stockholm Convention has, however, given an exemption from the ban for the production and public health use of DDT for indoor applications for vector control use, pending acceptable alternatives.

References

1. Vos JG, Dybing E, Greim HA, et al. Health effects of endocrine-disrupting chemicals on wildlife, with special reference to the European situation. *Crit Rev Toxicol.* 2000;30:71−133.
2. World Health Organization & United Nations Environment Programme. State of the Science of Endocrine Disrupting Chemicals-2012. In: Bergman ÅK, Heindel JJ, Jobling S, Kidd KA, Zoeller RT, eds. *Inter-Organization Programme for the Sound Management of Chemicals.* Geneva, Switzerland: WHO Press; 2012.
3. Miyagawa S, Lange I, Hirakawa I, et al. Differing species responsiveness of estrogenic contaminants in fish is conferred by the ligand binding domain of the estrogen receptor. *Environ Sci Technol.* 2014;48:5254−5263.
4. Bitman J, Cecil HC, Harris SJ, et al. Estrogenic activity of *o,p'*-DDT in the mammalian uterus and avian oviduct. *Science.* 1968;162:371−372.
5. Davis DL, Bradlow HL, Wolff M, et al. Medical hypothesis: xenoestrogens as preventable causes of breast cancer. *Environ Health Perspect.* 1993;101:372−377.
6. Kelce WR, Stone CR, Laws SC, et al. Persistent DDT metabolite *p,p'*-DDE is a potent androgen receptor antagonist. *Nature.* 1995;375:581−585.
7. Xu LC, Sun H, Chen JF, et al. Androgen receptor activities of *p,p'*-DDE, fenvalerate and phoxim detected by androgen receptor reporter gene assay. *Toxicol Lett.* 2006;160:151−157.
8. Turusov V, Rakitsky V, Tomatis L. Dichlorodiphenyltrichloroethane (DDT): ubiquity, persistence, and risks. *Environ Health Perspect.* 2002;110:125−128.

Supplemental Information Available on Companion Website

- Listed substances in the Stockholm Convention on Persistent Organic Pollutants/E-Table 101D.1

17α-Ethinylestradiol

Shinichi Miyagawa, Tomomi Sato, and Taisen Iguchi

Abbreviation: EE$_2$
Additional name: 19-nor-17α-pregna-1,3,5(10)-trien-20-yne-3,17-diol

17α-Ethinylestradiol (EE$_2$), commonly used as pharmaceutical, is one of the most potent estrogenic compounds identified in the aquatic environment. EE$_2$ has the potential to disrupt reproductive processes in fish at relatively low (ng/l) concentrations in such an environment.

Discovery

EE$_2$ was first synthesized in 1938, by Inhoffen and Hohlweg [1].

Structure

Structural Features

EE$_2$ is an alkylated estradiol with a 17α-ethinyl substitution (Figure 101E.1).

Properties

A white solid, molecular formula $C_{20}H_{24}O_2$, Mr 296.40.

EE$_2$ is hormonally effective by activating the estrogen receptor. EE$_2$ is more resistant to degradation compared with estradiol-17β, which is readily inactivated by the liver.

Production

EE$_2$ is prepared from estrone by ethinylation with ethyne, sodium, and sodium amide.

Application

EE$_2$ is a synthetic estrogen and effective in activating estrogen receptors. EE$_2$ is used in many formulations of combined oral contraceptive pills. It is generally prescribed with progestogen.

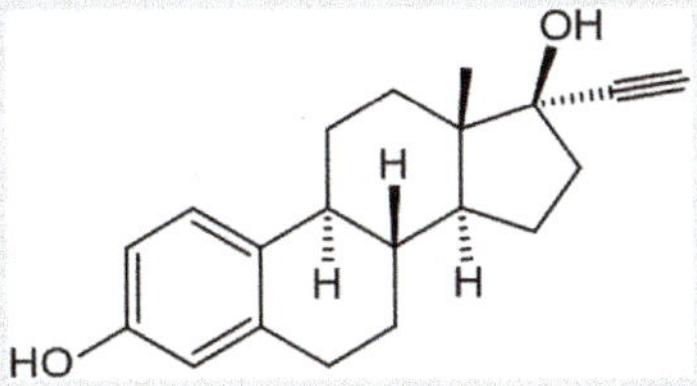

Biological and Pathophysiological Implications

Effects on Aquatic Animals

EE$_2$ is released into the environment from the urine and feces of people who take it as a medication. EE$_2$ is implicated among the major and potent estrogenic contaminants in effluents with high persistence and a tendency to bioconcentrate in organisms [2]. Several fish species have been exposed to environmental concentrations of EE$_2$ (up to 10 ng/l) to ascertain any effect on reproduction [3]. Long-term exposure to environmentally relevant concentrations of EE$_2$ can affect fish reproductive physiology by disturbing development and reproduction, hence possibly decreasing fertility in wild fish species, and risking a long-term reduction in population size [4]. EE$_2$ has been demonstrated to induce testis—ova and sex reversal at environmentally relevant concentrations in several fishes (for example, medaka, *Oryzias latipes*, at 20 ng/l) and amphibians (tropical clawed frog, *Silurana tropicalis*, at 30 ng/l) [5,6].

References

1. Inhoffen HH, Hohlweg W. Neue per os wirksame weibliche Keimdruesenhormonderivate: 17-Aethinyl-oestradiol und Pregnen-in-on-3-ol-17. *Naturwissenschaften*. 1938;26:96.
2. Larsson DGJ, Adolfsson-Erici M, Parkkonen J, et al. Ethinylestradiol — an undesired fish contraceptive? *Aquat Toxicol*. 1999;45:91—97.
3. Soffker M, Tyler CR. Endocrine disrupting chemicals and sexual behaviors in fish — a critical review on effects and possible consequences. *Crit Rev Toxicol*. 2012;42:653—668.
4. Kidd KA, Blanchfield PJ, Mills KH, et al. Collapse of a fish population after exposure to a synthetic estrogen. *Proc Natl Acad Sci USA*. 2007;104:8897—8901.
5. Hirakawa I, Miyagawa S, Mitsui N, et al. Developmental disorders and altered gene expression in the tropical clawed frog (*Silurana tropicalis*) exposed to 17α-ethinylestradiol. *J Appl Toxicol*. 2013;33:1001—1010.
6. Hirakawa I, Miyagawa S, Katsu Y, et al. Gene expression profiles in the testis associated with testis-ova in adult Japanese medaka (*Oryzias latipes*) exposed to 17α-ethinylestradiol. *Chemosphere*. 2012;87:668—674.

Figure 101E.1 Structure of EE$_2$.

Y. Takei, H. Ando, & K. Tsutsui (Eds): Handbook of Hormones. DOI: http://dx.doi.org/10.1016/B978-0-12-801028-0.00243-9

Equilin

Shinichi Miyagawa, Tomomi Sato, and Taisen Iguchi

Additional names: 1,3,5(10),7-estratetraen-3-ol-17-one; 3-hydroxy-1,3,5(10),7-estratetraen-17-one; 7-dehydroestrone

Equilin is an equine estrogen, and is one of the components of commonly used pharmaceuticals for hormone replacement therapy in the USA.

Discovery

Equilin was originally prepared from the urine of pregnant mares by Girard *et al.* [1].

Structure

Structural Features

Equilin, an estrogenic steroid produced by horses, has a total of four double bonds in the A- and B-rings (Figure 101F.1).

Properties

A white solid, molecular formula $C_{18}H_{20}O_2$, Mr 268.35. Melting point 238−240°C.

Production

Equilin is derived from the urine of horses.

Biological and Pathophysiological Implications

Application

Equilin is one of the major components of estrogens for hormone replacement therapy, along with estrone and equilenin (another equilin estrogen; 517-09-9). These estrogens are administered as salts of their sulfate esters. There is an increased risk of endometrial cancer in women, as well as strokes, heart attacks, blood clots, and breast cancer. Equilin (and equilenin)

is not normally present in women, and is the only human estrogen replacement drug that is derived from animals.

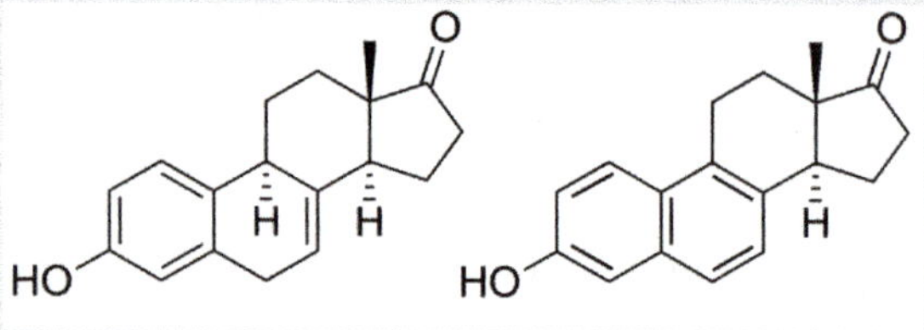

Environmental Estrogen

It is known that effluent from sewage treatment works contains contaminants that can be estrogenic to aquatic animals. Estrogens used in hormone replacement therapy and/or oral contraceptives are implicated as some of the causative compounds. Equilin and its metabolite 4-dihydroequilin were detected at nanogram per liter (pg/ml) concentrations in the influent and effluent [2]. Equilin has been found to be concentrated in the bile of fishes, and to increase the blood levels of the egg yolk protein vitellogenin [2].

Estrogenic Activity

Relative potency with regard to estradiol-17β in the T47D-KBluc assay is 50% [3]. *In vitro* reporter gene assay revealed that relative potency of equilin with regard to estradiol-17β for several fish estrogen receptors α is in the range of 3−5% [4].

Other Effects

It has been reported that equine estrogen metabolites react readily with DNA *in vitro*, resulting in the formation of a number of DNA adducts [5,6].

References

1. Girard A, Sandulesco G, Friedenson A. Sur une nouvelle hormone sexuelle cristallisee retiree de l'urine de juments. *Compt Rend Acad Sci.* 1932;194:909.
2. Tyler CR, Filby AL, Bickley LK, et al. Environmental health impacts of equine estrogens derived from hormone replacement therapy. *Environ Sci Technol.* 2009;43:3897−3904.
3. Bermudez DS, Gray LE, Wilson VS. Modelling defined mixtures of environmental oestrogens found in domestic animal and sewage treatment effluents using an *in vitro* oestrogen-mediated transcriptional activation assay (T47D-KBluc). *Int J Androl.* 2012;35:397−406.
4. Miyagawa S, Lange I, Hirakawa I, et al. Differing species responsiveness of estrogenic contaminants in fish is conferred by the ligand binding domain of the estrogen receptor. *Environ Sci Technol.* 2014;48:5254−5263.
5. Bolton JL, Pisha E, Zhang F, et al. Role of quinoids in estrogen carcinogenesis. *Chem Res Toxicol.* 1998;11:1113−1127.
6. Shen L, Qiu S, Chen Y, et al. Alkylation of 2′-deoxynucleosides and DNA by the Premarin metabolite 4-hydroxyequilenin semiquinone radical. *Chem Res Toxicol.* 1998;11:94−101.

Figure 101F.1   Structure of equilin (left) and equilenin (right).

Y. Takei, H. Ando, & K. Tsutsui (Eds): Handbook of Hormones. DOI: http://dx.doi.org/10.1016/B978-0-12-801028-0.00244-0

2,3,7,8-Tetrachlorodibenzo-*p*-dioxin/Polychlorinated Biphenyls

Yukiko Ogino, Shinichi Miyagawa, and Taisen Iguchi

Abbreviations: TCDD, 2,3,7,8-TCDD, dioxin/PCBs

2,3,7,8-Tetrachlorodibenzo-p-dioxin (TCDD) and polychlorinated biphenyls (PCBs) have a range of toxic effects on reproduction. TCDDs are formed as a side product in organic synthesis and burning. PCBs are man-made chlorinated hydrocarbons. These chemicals are still found in the environment, and mainly accumulate in the fatty tissue of animals.

Discovery

The toxic effects of TCDDs became well known from the contamination caused by Agent Orange, a herbicide used in the Malayan Emergency and the Vietnam War. The toxicity associated with PCBs was recognized due to industrial incidents. Numerous papers on the toxicity of PCBs were published before 1940.

Structure of TCDD

Structural Features

"Dioxins" are used to specify polychlorinated dibenzo-*p*-dioxins (PCDDs), dibenzo-furans (F), and dioxin-like polychlorinated biphenyls (PCBs). TCDD is a polychlorinated dibenzo-*p*-dioxin that is considered the most toxic of the congeners. The dibenzo-*p*-dioxin contains two benzene rings joined by two oxygen bridges (Figure 101G.1). The name dioxin formally refers to the central dioxygenated ring. In TCDD, chlorine atoms are attached to two benzene rings at positions 2,3,7, and 8.

Properties

A solid chemical compound with no color or odor, molecular formula $C_{12}H_4Cl_4O_2$, Mr 321.97, melting point 305°C. Poorly soluble in water: 0.2 ng/ml at 25°C.

Production

TCDD is not intentionally produced commercially except as a pure chemical for scientific research. TCDD is formed by the condensation of two molecules of sodium 2,4,5-trichlorophenate at high temperatures. TCDDs may be formed during the chlorine bleaching process used in the manufacture of pulp and paper mills. They are also produced as side products of burning.

Structure of PCBs

Structural Features

PCBs are aromatic, synthetic chemicals. They consist of the biphenyl structure with two linked benzene rings in which some or all of the hydrogen atoms have been substituted by chlorine atoms. PCBs are a family of 209 congeners of structurally similar organic chemicals. Co-planar PCBs, which have a fairly rigid structure with the two phenyl rings in the same plane, are considered to be most toxic. Three coplanar PCBs, namely 3,3′,4,4′-tetrachlorobiphenyl, 3,3′,4,4′,5-pentachlorobiphenyl, and 3,3′,4,4′,5,5′-hexachlorobiphenyl, exhibit dioxin-like effects. Due to their similar toxicological properties, they are referred to as dioxin-like PCBs.

Properties

Liquid at room temperature, molecular formula $C_{12}H_{10-n}Cl_n$ (where *n* ranges from 1 to 10). High flash point of 170–380°C, density 1.182–1.566 kg/l. Low water solubility, readily soluble in organic solvents.

Production

PCBs are formed by electrophilic chlorination of biphenyl with chlorine gas. PCBs were commercially produced worldwide on a large scale between the 1930s and 1980s.

Biological and Pathophysiological Implications

Applications

The only present use for TCDD is in scientific research. Formerly, PCBs were widely used as electric fluids in transformers and capacitors, as pesticide extenders, adhesives, dedusting agents, cutting oils, flame retardants, heat transfer fluids, hydraulic lubricants, sealants, and paints, and in carbonless copy paper, due to their non-flammability, chemical stability, high boiling point, and electrical insulating properties.

Receptors

The aryl hydrocarbon receptor (AhR) plays a role in the toxicity of TCDD and its congeners, such as PCBs. [1] AhR repressor (AhRR) inhibits AhR function by competing with AhR for dimerizing with AhR nuclear translocator (Arnt) and binding to the xenobiotic responsive element (XRE) sequence [1].

Y. Takei, H. Ando, & K. Tsutsui (Eds): Handbook of Hormones. DOI: http://dx.doi.org/10.1016/B978-0-12-801028-0.00245-2

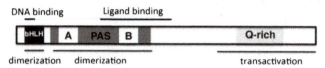

Figure 101G.1 Structure of 2,3,7,8-TCDD and PCBs. Polychlorinated biphenyls (PCBs) comprise a class of 209 individual compounds. The possible positions of chlorine atoms on the benzene rings are shown by numbers assigned to the carbon atoms.

Figure 101G.2 The domain structure of AhR.

Structure and Subtype

AhR, AhRR, and Arnt belong to a superfamily of the basic helix−loop−helix (bHLH)− Per−AhR/Arnt−Sim homology sequence (PAS) transcription factor superfamily (Figure 101G.2) [1]. Recently, the crystal structure of the PAS-A domain of AhR was reported [2].

Signal Transduction Pathway

AhR is activated by ligand binding. The liganded AhR is translocated from the cytoplasm to the nucleus, and then the complex switches its partner protein from Hsp90 to Arnt. AhR/Arnt heterodimer binds to XRE/DRE (dioxin response element) sequences in the promoter and enhances transcription of the target genes [3].

Health and Environmental Hazards

TCDD and PCBs have been demonstrated to alter endocrine, immune, and nervous system functions, and to have adverse effects on the reproduction and development of animals, including humans. The most characteristic symptoms of severe acute TCDD toxicity are chloracne and porphyria [4]. Acute exposure to TCDD may also cause transient hepatotoxicity, peripheral and central neurotoxicity, vomiting, and diarrhea. Because of the long-term persistence of TCDD in the human body, chronic effects appear as atherosclerosis, hypertension, diabetes, tumor promotion, and signs of neural system damage, including neuropsychological impairment [5]. PCBs have properties similar to those of TCDD if they can exist in a planar configuration (dioxin-like PCBs). After maternal exposure to PCBs, decreased embryonic growth, delayed implantation, and increased abortion rates have been observed [6]. AhR knockout mice are resistant to TCDD-induced teratogenesis such as cleft palate and hydronephrosis, and thymic atrophy [7].

Hormonal Effects

The biological activities of TCDD and PCBs have been reported to include both estrogenic and anti-estrogenic effects, and therefore pose a risk to the perinatal development of the female reproductive tract [8,9]. A relationship between these toxicities and the developing reproductive system of male offspring has been suggested.

Species Differences

Marked interspecies variability exists in the acute toxicity of TCDD, with the guinea pig having an oral LD_{50} dose (0.6 μg/kg body weight) about 10,000-fold greater than that of the hamster (5 mg/kg body weight) [10], and even among rat strains there may be a 1,000-fold difference.

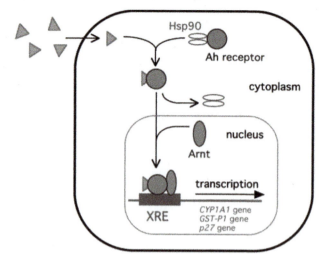

Figure 101G.3 Activation of the AhR pathway.

Safety Standards and Regulatory Compliance

The production of PCBs was banned towards the end of the 1970s in most countries. The US Environmental Protection Agency prohibited PCB production in 1977. PCB production, use, and importation were banned in Japan in 1972.

References

1. Mimura J, Ema M, Sogawa K, et al. Identification of a novel mechanism of regulation of Ah (dioxin) receptor function. *Genes Dev.* 1999;13:20−25.
2. Wu D, Potluri N, Kim Y, Rastinejad F. Structure and dimerization properties of the aryl hydrocarbon receptor PAS-A domain. *Mol Cell Biol.* 2013;33:4346−4356.
3. Fujisawa-Sehara A, Sogawa K, Yamane M, Fujii-Kuriyama Y. Characterization of xenobiotic responsive elements upstream from the drug-metabolizing cytochrome P-450c gene: a similarity to glucocorticoid regulatory elements. *Nucleic Acids Res.* 1987;15:4179−4191.
4. Geusau A, Abraham K, Geissler K, et al. Severe 2,3,7,8-tetrachlorodibenzo-*p*-dioxin (TCDD) intoxication: clinical and laboratory effects. *Environ Health Perspect.* 2001;109:865−869.
5. Pelclova D, Urban P, Preiss J, et al. Adverse health effects in humans exposed to 2,3,7,8-tetrachlorodibenzo-*p*-dioxin (TCDD). *Rev Environ Health.* 2006;21:119−138.
6. Kholkute SD, Rodriguez J, Dukelow WR. Reproductive toxicity of Aroclor-1254: effects on oocyte, spermatozoa, *in vitro* fertilization, and embryo development in the mouse. *Reprod Toxicol.* 1994;8:487−493.
7. Safe S, Lee SO, Jin UH. Role of the aryl hydrocarbon receptor in carcinogenesis and potential as a drug target. *Toxicol Sci.* 2013;135:1−16.
8. Buchanan DL, Sato T, Peterson RE, Cooke PS. Antiestrogenic effects of 2,3,7,8-tetrachlorodibenzo-*p*-dioxin in mouse uterus: critical role of the aryl hydrocarbon receptor in stromal tissue. *Toxicol Sci.* 2000;57:302−311.
9. Ma R, Sassoon DA. PCBs exert an estrogenic effect through repression of the Wnt7a signaling pathway in the female reproductive tract. *Environ Health Perspect.* 2006;114:898−904.
10. Kociba RJ, Schwetz BA. Toxicity of 2,3,7,8-tetrachlorodibenzo-*p*-dioxin (TCDD). *Drug Metab Rev.* 1982;13:387−406.

Supplemental Information Available on Companion Website

• Accession numbers of AhR cDNAs/E-Table 101G.1

1,1-Dichloro-2,2-bis (*p*-chlorophenyl)ethylene

Tomomi Sato, Shinichi Miyagawa, and Taisen Iguchi

Abbreviation: DDE
Additional names: Dichlorodiphenyldichloroethane; 1,1-dichloro-2,2-bis(*p*-chlorophenyl)ethylene; 1,1'-(2,2-dichloro-ethylidene)bis(4-chlorobenzene); 4,4'-DDE; *p,p'*-DDE

DDE is a metabolite of 1,1,1-trichloro-2,2-bis(p-chlorophenyl) ethane (DDT), and an anti-androgen.

Discovery

DDE is one of the metabolites of DDT, which was first synthesized in 1874, and was used as an insecticide from 1939 to 1972 in the USA.

Structure

Structural Features

DDE is a chlorinated pesticide, and a breakdown product of DDT by the loss of hydrogen chloride (dehydrohalogenation). DDE in the atmosphere may be broken down by reactions caused by the sun; however, DDE in the soil lasts for a very long time.

Properties

Molecular formula $C_{14}H_8Cl_4$, Mr 318.03. Soluble in lipids and other organic solvents, also soluble in in water at 0.12 µg/ml at 25°C [1].

Synthesis and Release

Tissue and Plasma Concentrations

Tissue: Tissue contents in milk fat and adipose tissue in humans are shown in Table 101H.1 [2–6].
  Plasma: Human, 6.93–7.29 (ng/ml) [7], 2.3–16.9 mg/ml [8].

Receptors

Structure and Subtype

DDE can inhibit androgen binding to the androgen receptor (AR), but has no affinity for the estrogen receptor (ER). IC_{50} values of DDE for AR binding, ER binding, and 5α-reductase enzymatic activity are 5 µM, >1,000 µM, and >1,000 µM, respectively [9].

Target Cells/Tissues and Functions

DDE inhibits androgen-induced gene expression *in vitro* [9]. DDE treatment reduces androgen-dependent organ weights in adult rats and the anogenital distance of male pups at birth by *in utero* effects (Table 101H.2) [9–13]. Prenatal DDE treatment results in histological and ultrastructural changes in fetal testis and adrenal glands of rats [14]. High concentrations of DDE inhibit 5α-reductase activity in prostate homogenates [15].

Table 101H.1 Tissue Content of DDE in Humans

Tissues	Content (mg/kg)	Reference
Milk fat	129	[2]
	0.34	[3]
	0.594	[4]
	5.302	[4]
Adipose (abdominal)	4.36	[5]
	970	[6]

Table 101H.2 Effects of DDE Observed in the Reproductive System of Animals [10]

Species (sex)	Observation	Reference(s)
Rat (male offspring, adult)	Nipple retention	[9,11,12]
	Hypospadias	[11]
	Reduced accessory sex organ weights	[9,11]
Rat (male offspring)	Reduced anogenital distance	[9,12]
Rat (male pubertal)	Delayed preputial separation	[9]
Alligators (juvenile male)	Abnormally small penis, poorly organized testis, decreased plasma testosterone levels	[13]
Alligators (juvenile female)	Increased plasma estradiol levels, abnormal ovarian morphology with large number of polyovular follicles and polynuclear oocytes	[13]

References

1. Toxicological profile for DDT, DDE, and DDD. US Department of Health and Human Services, Public Health Service, Agency for Toxic Substances and Disease Registry September 2002.
2. Norén K, Meironyté D. Certain organochlorine and organobromine contaminants in Swedish human milk in perspective of past 20–30 years. *Chemosphere*. 2000;40:1111–1123.
3. Dewailly E, Ayotte P, Laliberté C, et al. Polychlorinated biphenyl (PCB) and dichlorodiphenyl dichloroethylene (DDE) concentrations in the breast milk of women in Quebec. *Am J Public Health*. 1996;86:1241–1246.

Y. Takei, H. Ando, & K. Tsutsui (Eds): Handbook of Hormones. DOI: http://dx.doi.org/10.1016/B978-0-12-801028-0.00246-4

4. Torres-Arreola L, López-Carrillo L, Torres-Sánchez L, et al. Levels of dichloro-diphenyl-trichloroethane (DDT) metabolites in maternal milk and their determinant factors. *Arch Environ Health.* 1999;54:124–129.

5. Waliszewski SM, Aguirre AA, Infanzon RM, et al. Organochlorine pesticide levels in maternal adipose tissue, maternal blood serum, umbilical blood serum, and milk from inhabitants of Veracruz, Mexico. *Arch Environ Contam Toxicol.* 2001;40:432–438.

6. Archibeque-Engle SL, Tessari JD, Winn DT, et al. Comparison of organochlorine pesticide and polychlorinated biphenyl residues in human breast adipose tissue and serum. *J Toxicol Environ Health.* 1997;52:285–293.

7. Gammon MD, Wolff MS, Neugut AI, et al. Temporal variation in chlorinated hydrocarbons in healthy women. *Cancer Epidemiol Biomarkers Prev.* 1997;6:327–332.

8. Schmid K, Lederer P, Goen T, et al. Internal exposure to hazardous substances of persons from various continents: investigations on exposure to different organochlorine compounds. *Int Arch Occup Environ Health.* 1997;69:399–406.

9. Kelce WR, Stone CR, Laws SC, et al. Persistent DDT metabolite *p,p′*-DDE is a potent androgen receptor antagonist. *Nature.* 1995;375:581–585.

10. World Health Organization and United Nations Environment Programme. *Endocrine Disrupters and Child Health.* Geneva, Switzerland: WHO/UNEP; 2012.

11. Wolf Jr C, Lambright C, Mann P, et al. Administration of potentially antiandrogenic pesticides (procymidone, linuron, iprodione, chlozolinate, *p,p′*-DDE, and ketoconazole) and toxic substances (dibutyl- and diethylhexyl phthalate, PCB 169, and ethane dimethane sulphonate) during sexual differentiation produces diverse profiles of reproductive malformations in the male rat. *Toxicol Ind Health.* 1999;15:94–118.

12. You L, Casanova M, Archibeque-Engle S, et al. Impaired male sexual development in perinatal Sprague-Dawley and Long-Evans hooded rats exposed *in utero* and lactationally to *p,p′*-DDE. *Toxicol Sci.* 1998;45:162–173.

13. Guillette Jr LJ, Gross TS, Masson GR, et al. Developmental abnormalities of the gonad and abnormal sex hormone concentrations in juvenile alligators from contaminated and control lakes in Florida. *Environ Health Perspect.* 1994;102:680–688.

14. Adamsson A, Salonen V, Paranko J, et al. Effects of maternal exposure to di-isononylphthalate (DINP) and 1,1-dichloro-2,2-bis (*p*-chlorophenyl)ethylene (*p,p′*-DDE) on steroidogenesis in the fetal rat testis and adrenal gland. *Reprod Tox.* 2009;28:66–74.

15. Lo S, King I, Alléra A, Klingmüller D. Effects of various pesticides on human 5α-reductase activity in prostate and LNCaP cells. *Toxicol In Vitro.* 2007;21:502–508.

Vinclozolin

Tomomi Sato, Shinichi Miyagawa, and Taisen Iguchi

Additional names: 3-(3,5-dichlorophenyl)-5-ethenyl-5-methyl-2,4-oxazolidinedione, 3-(3,5-dichlorophenyl)-5-methyl-5-vinyloxazolidine-2,4-dione, Ronilan®, Curalan®, Vorlan®

Vinclozolin is a dicarboximide fungicide, and anti-androgen.

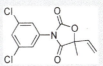

Figure 101I.1   Structure of vinclozolin.

Discovery

Vinclozolin was registered in the USA in 1984.

Structure

Structural Features

Vinclozolin is a dicarboximide fungicide that is degraded to two hydrolytic products: 2-(3,5-dichloropheniyl)-carbamoyloxy-2-methyl-3-butenoic acid (M1) and 3′,5′-dichloro-2-hydroxy-2-methylbut-3-enanilide (M2) (Figure 101I.1) [1].

Properties

Molecular forumula $C_{12}H_9Cl_2NO_3$, Mr 286.11. Soluble in acetone, benzene, chloroform, and ethyl acetate.

Receptors

Structure and Subtype

Vinclozolin and its metabolites (M1 and M2) can inhibit androgen binding to the androgen receptor (AR) ($K_i > 700\,\mu M$ [vinclozolin], $K_i = 92\,\mu M$ [M1] and $9.7\,\mu M$ [M2]) but have no affinity for the estrogen receptor [1].

Biological Functions

Target Cells/Tissues and Functions

Vinclozolin inhibits androgen-induced gene expression *in vivo* (Table 101I.1) [1–9]. Vinclozolin treatment reduces androgen-dependent organ weights in adult rats, and anogenital distance of male pups at birth, by *in utero* exposure [6]. Embryonic exposure to vinclozolin also modifies the developing testis and prostate transcriptome including DNA methyltransferases [10]. The metabolites of vinclozolin (M1 and M2) block AR binding to the androgen response element (ARE) and inhibit dihydrotestosterone (DHT)-induced transcriptional activation [11]. Conversely, 10 μM M2 promotes AR binding to ARE, and acts as an agonist in the absence of DHT [11].

Table 101I.1   Effects of vinclozolin Observed in the Reproductive System of Animals [2]

Species (sex)	Observation	References
Rat (male)	Hypospadias with cleft phallus Reduced anogenital distance Decreased testis weight Cryptorchidism Increased number of apoptotic germ cells in testis, reduced elongated spermatid content per testis Nipple retention Glandular atrophy and chronic inflammation of prostate Epididymal granulomas Agenesis of prostate, decreased fertility, decreased serum testosterone levels Low ejaculated sperm count Reduction of erections during the *ex copula* penile reflex test, increase in seminal emissions during the *ex copula* penile reflex tests	[3–9]
Rat (male), mouse (male)	Reduced accessory sex organ weights Decreased sperm number and daily sperm production Abnormal morphology of seminiferous tubules	[4–9]
Rat (male adult)	Reduced secretion and chronic inflammation of seminal vesicles, chronic inflammation of epididymis, spermatogenic granuloma	[7]
Mouse (male)	Increased sperm head abnormalities	[5]

References

1. Kelce WR, Monosson E, Gamcsik MP, et al. Environmental hormone disruptors: evidence that vinclozolin developmental toxicity is mediated by antiandrogenic metabolites. *Toxicol Appl Pharmacol*. 1994;126:276–285.
2. World Health Organization and United Nations Environment Programme. *Endocrine Disrupters and Child Health*. Geneva, Switzerland: WHO/UNEP; 2012.
3. Colbert NK, Pelletier NC, Cote JM, et al. Perinatal exposure to low levels of the environmental antiandrogen vinclozolin alters sex-differentiated social play and sexual behaviors in the rat. *Environ Health Perspect*. 2005;113:700–707.

Y. Takei, H. Ando, & K. Tsutsui (Eds): Handbook of Hormones. DOI: http://dx.doi.org/10.1016/B978-0-12-801028-0.00247-6

4. Cowin PA, Gold E, Aleksova J, et al. Vinclozolin exposure in utero induces postpubertal prostatitis and reduces sperm production via a reversible hormone-regulated mechanism. *Endocrinology.* 2010;151:783–792.

5. Elzeinova F, Novakova V, Buckiova D, et al. Effect of low dose of vinclozolin on reproductive tract development and sperm parameters in CD1 outbred mice. *Reprod Toxicol.* 2008;26:231–238.

6. Gray Jr LE, Ostby J, Kelce WR. Developmental effects of an environmental antiandrogen: the fungicide vinclozolin alters sex differentiation of the male rat. *Toxicol App Pharmacol.* 1994;129:46–52.

7. Hellwig J, van Ravenzwaay B, Mayer M, Gembardt C. Pre- and postnatal oral toxicity of vinclozolin in Wistar and Long-Evans rats. *Reg Toxicol Pharmcol.* 2000;32:42–50.

8. Ostby J, Monosson E, Kelce WR, et al. Environmental antiandrogens: low doses of the fungicide vinclozolin alter sexual differentiation of the male rat. *Toxicol Ind Health.* 1999;15:48–64.

9. Wolf Jr C, Lambright C, Mann P, et al. Administration of potentially antiandrogenic pesticides (procymidone, linuron, iprodione, chlozolinate, *p,p*′-DDE, and ketoconazole) and toxic substances (dibutyl- and diethylhexyl phthalate, PCB 169, and ethane dimethane sulphonate) during sexual differentiation produces diverse profiles of reproductive malformations in the male rat. *Toxicol Ind Health.* 1999;15:94–118.

10. Anway MD, Rekow SS, Skinner MK. Transgenerational epigenetic programming of the embryonic testis transcriptome. *Genomics.* 2008;91:30–40.

11. Wong C, Kelce WR, Sar M, et al. Androgen receptor antagonist versus agonist activities of the fungicide vinclozolin relative to hydroxyflutamide. *J Biol Chem.* 1995;270:19998–20003.

Polybrominated Diphenyl Ether

Tomomi Sato, Shinichi Miyagawa, and Taisen Iguchi

Abbreviation: PBDE

PBDEs are organobromine compounds used as flame retardants, and show anti-androgenic and anti-thyroid activities.

Discovery

The production of PBDEs and their commercial use started in the 1970s.

Structure

Structural Features

Organobromine compounds consist of 209 possible congeners (PBDE = $C_{12}H_{(10-x)}Br_xO$). Three commercial PBDE products (i.e., penta-, octa-, and decabromodiphenyl ethers) are available; there are 10 homologous groups of PBDE congeners, and each group contains one or more isomers based on the number of bromine substituents.

Properties

Molecular formula $C_{12}H_{(10-x)}Br_xO$. Soluble in acetone and benzene.

Synthesis and Release

Tissue and Plasma Concentrations

Tissue: Human, adipose tissue, 1.36–459 (ng/g lipid weight); breast milk, 0.18–18.4.

Plasma: BDE-47, human blood 0.63–95 (ng/g lipid weight).

Receptors

Structure and Subtype

PBDE congeners DE-71, BDE-47, BDE-99, and BDE-100 can inhibit androgen binding to the cytosolic androgen receptor (AR) in the rat ventral prostate [1].

Biological Functions

Target Cells/Tissues and Functions

Reduced serum tyroxine levels and thyroid follicular cell hyperplasia were observed in rats and mice orally exposed to lower brominated PBDE mixtures [2–4]. PBDE congeners BDE-19 and -10 inhibit AR binding [1]. BDE-99 treatment during pregnancy reduced estradiol-17β and testosterone levels

Table 101J.1 Effects of PBDE Observed in the Reproductive System of Animals [7]

Species (sex)	Observation	References
Rat (male and female)	Delayed onset of puberty Reduced seminal vesicle and ventral prostate weights Delay in the age of vaginal opening	[8]
Rat (male)	Reduced sperm counts	[9]
Mouse (male)	Affected anogenital distance and testicular histopathology; sperm-head abnormalities	[10]
Rat (female)	Ultrastructural alterations in ovaries Altered folliculogenesis in ovaries	[11,12]

in male offspring in rats [5], and affected progesterone and estrogen receptor (ER)-α and -β levels in the uterus of female offspring [6]. Table 101J.1 summarizes the observed effects of endocrine disrupters on the reproductive system of rats and mice [7–12]. PBDEs and their metabolites are agonistic to ERα and ERβ, and/or antagonistic to ERα, ERβ, AR, glucocorticoid receptor and thyroid hormone receptor (TR)α1 and TRβ1 in Chinese hamster ovary cells [13].

References

1. Stoker TE, Cooper RL, Lambright CS, et al. *In vivo* and *in vitro* anti-androgenic effects of DE-71, a commercial polybrominated diphenyl ether (PBDE) mixture. *Toxicol Appl Pharmacol.* 2005;207:78–88.
2. Zhou T, Ross DG, DeVito MJ, Crofton KM. Effects of short-term *in vivo* exposure to polybrominated diphenyl ethers on thyroid hormones and hepatic enzyme activities in weanling rats. *Toxicol Sci.* 2001;61:76–82.
3. Zhou T, Taylor MM, DeVito MJ, Crofton KM. Developmental exposure to brominated diphenyl ethers results in thyroid hormone disruption. *Toxicol Sci.* 2002;66:105–116.
4. Darnerud PO. Brominated flame retardants as possible endocrine disrupters. *Int J Androl.* 2008;31:152–160.
5. Lilienthal H, Hack A, Roth-Härer A, et al. Effects of developmental exposure to 2,2,4,4,5-pentabromodiphenyl ether (PBDE-99) on sex steroids, sexual development, and sexually dimorphic behavior in rats. *Environ Health Perspect.* 2006;114:194–201.
6. Ceccatelli R, Faass O, Schlumpf M, et al. Gene expression and estrogen sensitivity in rat uterus after developmental exposure to the polybrominated diphenylether PBDE 99 and PCB. *Toxicology.* 2006;220:104–116.
7. US Department of Health and Human Services. *Public Health Service Agency for Toxic Substances and Disease Registry. Toxicological profile for DDT, DDE, and DDD.* September. Atlanta, GA: ATSDR; 2002.
8. Stoker TE, Laws SC, Crofton KM, et al. Assessment of DE-71, a commercial polybrominated diphenyl ether (PBDE) mixture, in the EDSP male and female pubertal protocols. *Toxicol Sci.* 2004;78:144–155.

Y. Takei, H. Ando, & K. Tsutsui (Eds): Handbook of Hormones. DOI: http://dx.doi.org/10.1016/B978-0-12-801028-0.00248-8

9. Kuriyama SN, Talsness CE, Grote K, Chahoud I. Developmental exposure to low dose PBDE 99: effects on male fertility and neurobehavior in rat offspring. *Environ Health Perspect*. 2005;113: 149–154.

10. Tseng LH, Hsu PC, Lee CW, et al. Developmental exposure to decabrominated diphenyl ether (BDE-209): effects on sperm oxidative stress and chromatin DNA damage in mouse offspring. *Environ Toxicol*. 2013;28:380–389.

11. Talsness CE, Shakibaei M, Kuriyama SN, et al. Ultrastructural changes observed in rat ovaries following in utero and lactational exposure to low doses of a polybrominated flame retardant. *Toxicol Lett*. 2005;157:189–202.

12. Talsness CE, Kuriyama SN, Sterner-Kock A, et al. In utero and lactational exposures to low doses of polybrominated diphenyl ether-47 alter the reproductive system and thyroid gland of female rat offspring. *Environ Health Perspect*. 2008;116:308–314.

13. Kojima H, Takeuchi S, Uramaru N, et al. Nuclear hormone receptor activity of polybrominated diphenyl ethers and their hydroxylated and methoxylated metabolites in transactivation assays using Chinese hamster ovary cells. *Environ Health Perspect*. 2009;117:1210–1218.

CHAPTER 102

Anti-Thyroid Hormone Active Chemicals

Kiyoshi Yamauchi

History

Thyroid-related disorders and diseases in humans and wildlife have gradually risen throughout the world with increasing exposure to man-made chemicals over the past several decades. Despite the fact that production of persistent organic pollutants (POPs) such as polychlorinated biphenyls (PCBs) was banned in the late 1970s, POPs are still found in the environment and the tissues of humans and animals. Although there is a large number of chemicals suspected to be potential thyroid disrupting chemicals (TDCs) (E-Table 102.1) [1,2], it is unclear whether all of these chemicals alter functions of the thyroid system. Observations of experimental animals indicate that some chemicals affect the vertebrate thyroid system at a specific developmental window.

Hypothalamic–Pituitary–Thyroid Axis

The thyroid system in vertebrates consists of the HPT axis (Figure 102.1) [3], and is remarkably evolutionarily conserved among vertebrates. The HPT axis begins to mature during the second half of gestation in humans, or during the second postnatal week in rodents [4]. Maternal thyroid hormones (TH) influence fetal development before the establishment of the HPT axis. Thus, the sensitive window for TDCs is the fetal and neonatal periods. THs negatively control thyrotropin-releasing hormone secretion from the hypothalamus and thyrotropin secretion from the pituitary. Thyroxine (T_4), secreted from the thyroid gland, is peripherally converted to 3,3′,5-triiodothyronine (T_3). Chemical-induced thyroid disruption is defined as a change in hormone production, transport, function, or metabolism resulting in impaired homeostasis, by targeting a site or sites of the HPT axis, including peripheral tissues (sites 1–11 in Figure 102.1) [5]. However, once the HPT axis is established at a developmental stage, the HPT axis is likely to acquire an ability to compensate the chemical-induced changes in TH levels by negative feedback loops.

Structure

Structures of some chemicals that are suspected to be TDCs are shown in Figure 102.2. Their structures resemble those of THs.

Biological Functions and Pathophysiological Implications

Human Studies

There has been a trend towards increasing incidence of thyroid diseases and disorders (goiter, congenital and adult hypothyroidism, autoimmune thyroiditis, hyperthyroidism, and thyroid cancer) and neurodevelopmental problems (attention deficit hyperactivity disorder [ADHD], autism, and learning disability) with increasing exposure to pollutants over the past several decades. Exposure to TDCs is one suspicious factor [5]. Children whose mothers ate the fish most contaminated with PCBs showed delayed neuromuscular and neurological development [5]. Although PCBs are epidemiologically linked to thyroid diseases and disorders, there is no convincing evidence that PCBs interfere with thyroid function in humans [5].

Experimental Animal Studies

In rodents with experimentally induced hypothyroidism at specific developmental stages, several disorders in nervous system development are found at later stages: cochlear development (permanent hearing loss in adults), structural abnormalities in the brain (myelination), reduction in cell density of the corpus callosum, incorrect positioning of neurons, and a decrease in the numbers of myelin-forming oligodendrocytes. PCB and polybrominated diphenyl ether (PBDE) exposure causes a reduction of total and/or free serum T_4 in a congener-dependent manner. Some congeners are TH antagonists while others are TH agonists (E-Figure 102.1) [6]. Hydroxylated PCBs, metabolites of PCBs, are also potent as TDCs, some of which have affinity for TRs. Amphibian metamorphosis is a TH-dependent developmental process. Using this system, PCBs, PBDEs, organochlorine pesticides, perchlorate, bisphenol A (BPA), tetrabromobisphenol A (TBBPA), and nonylphenol have been shown to have thyroid disrupting activity.

Wildlife Studies

Positive relationships have been identified between body burden of POPs and thyroid-related diseases and disorders (enlarged thyroid glands, interfollicular fibrosis, other thyroid abnormalities, TH transport, and metabolism) in marine animals (seals in the Wadden Sea; sea lions in Año Nuevo

Y. Takei, H. Ando, & K. Tsutsui (Eds): Handbook of Hormones. DOI: http://dx.doi.org/10.1016/B978-0-12-801028-0.00102-1

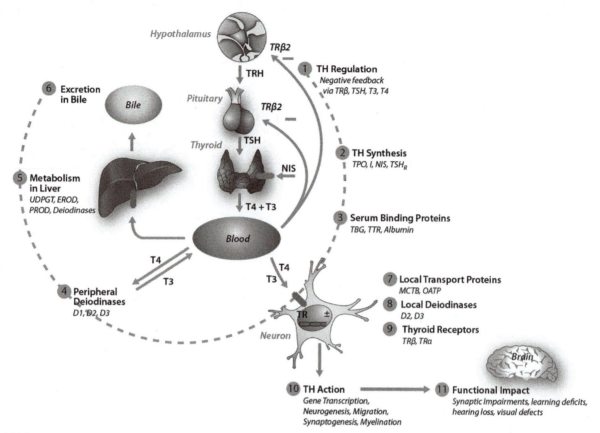

Figure 102.1 **Hypothalamic—pituitary—thyroid (HPT) axis and possible action sites of thyroid disrupting chemicals.** *Reproduced from Gilbert et al., 2012 [3], with permission.*

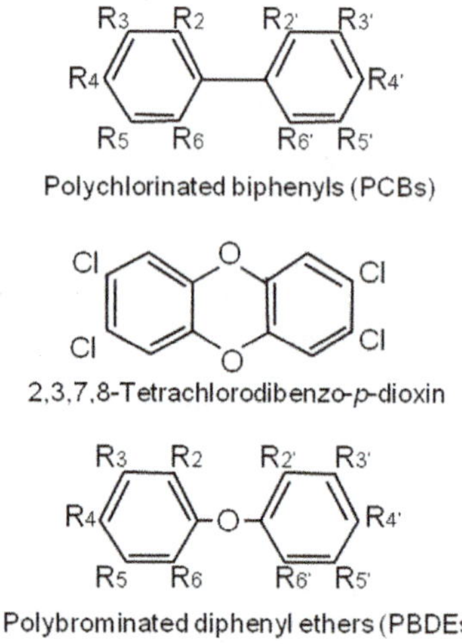

Polychlorinated biphenyls (PCBs)

2,3,7,8-Tetrachlorodibenzo-*p*-dioxin

Polybrominated diphenyl ethers (PBDEs)

Figure 102.2 Structural formulae of suspicious thyroid disrupting chemicals. R = Cl or H for PCBs; R = Br or H for PDDEs.

Island, CA; beluga whales in the St Lawrence estuary; polar bears in Spitzbergen; waterbirds in the Great Lakes, Barents Sea, and Tokyo Bay; salmonid fish in the Great Lakes during the 1970s—1980s; and mummichogs in New Jersey and San Francisco Bay) [5]. Chemical metabolism and subsequent exposure differs among vertebrates, resulting in species-specific effects of TDCs.

References

1. Boas M, Feldt-Rasmussen U, Skakkebaek NE, et al. Environmental chemicals and thyroid function. *Eur J Endocrinol.* 2006;154:599—611.
2. Woodruff TJ. Bridging epidemiology and model organisms to increase understanding of endocrine disrupting chemicals and human health effects. *J Steroid Biochem Mol Biol.* 2011;127:108—117.
3. Gilbert ME, Rovet J, Chen Z, et al. Developmental thyroid hormone disruption: prevalence, environmental contaminants and neurodevelopmental consequences. *Neurotoxicology.* 2012;33:842—852.
4. Howdeshell KL. A model of the development of the brain as a construct of the thyroid system. *Environ Health Perspect.* 2002;110:337—348.
5. Bergman Å, Heindel JJ, Jobling S, et al., eds. *State of the Science of Endocrine Disrupting Chemicals — 2012.* Geneva, Switzerland: UNEP and WHO Press; 2013.
6. Diamanti-Kandarakis E, Bourguignon JP, Giudice LC, et al. Endocrine-disrupting chemicals: an Endocrine Society scientific statement. *Endocr Rev.* 2009;30:293—342.
7. Ulbrich B, Stahlmann R. Developmental toxicity of polychlorinated biphenyls (PCBs): a systematic review of experimental data. *Arch Toxicol.* 2004;78:252—268.

Supplemental Information Available on Companion Website

- Target sites and effects of thyroid disrupting chemicals or suspected thyroid disrupting chemicals/E-Table 102.1 [1,2]
- Various effects of different PCB congeners on behavioral parameters, brain biochemistry or electrophysiology/E-Figure 102.1 [7]

Tetrabromobisphenol A

Kiyoshi Yamauchi

Abbreviation: TBBPA
Additional names: 2,2′,6,6′-tetrabromo-4,4′-isopropylidene-diphenol, 2,2′,6,6′-tetrabromobisphenol A, 2,2-bis(3,5-dibromo-4-hydroxyphenyl)propane, 4,4′-isopropylidenebis(2,6-dibromophenol), Bromdian, Tetrabromodian
IUPAC name: 2,6-dibromo-4-[2-(3,5-dibromo-4-hydroxyphenyl)propan-2-yl]phenol, CAS No. 79-94-7

TBBPA, a brominated derivative of bisphenol A, is the highest-selling brominated flame retardant. From its structural resemblance to thyroid hormone (TH), TBBPA interferes with TH binding to the plasma protein transthyretin (TTR) and with TH action through TH receptors (TRs).

Discovery

Brominated flame retardants have been routinely used for several decades to prevent the ignition of combustible organic materials. Meerts and colleagues [1] first reported that TBBPA affects TH binding to human transthyretin *in vitro* in 2000.

Structure

Structural Features

TBBPA has four bromine atoms, at the 2, 2′, 6, and 6′ positions of bisphenol A. The structural formula and 3D structure of TBBPA are shown in Figure 102A.1.

Properties

A crystalline or powdered white solid, molecular formula $C_{15}H_{12}Br_4O_2$, Mr 543.9. Water solubility 0.063 mg/l at 21°C [2]; $LogK_{ow}$ 5.9 [2]; pK_a 7.5 and 8.5.

Usage

TBBPA is a flame retardant bound to epoxy and polycarbonate resins in printed circuit boards and casings used in

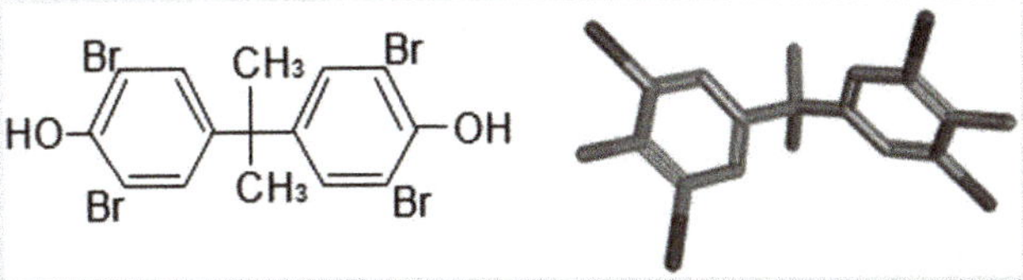

Figure 102A.1 Structural formula and 3D-structure of TBBPA.
3D-structures are obtained from PubChem Compound (http://pubchem.ncbi.nlm.nih.gov, CID: 6618).

personal computers, printers, fax machines, and copiers; and an additive flame retardant of acrylonitrile-butadiene-styrene resin and polystyrene that contains dimethyl TBBPA in automotive parts, pipes and fittings, domestic and office appliances, packaging, electrical and electronic equipment enclosures, furniture, and construction materials [3].

Annual Consumption

Consumption in Asia was 89,400 tonnes in 2001, which was 56% of the total market demand for that year [4]. TBBPA is mainly manufactured in Israel, the USA, Jordan, Japan, and China.

Synthesis and Release

Origins of Environmental and Biota TBBPA

The origins of environmental and biota TBBPA are anthropogenic, primarily from waste streams or effluents of manufacturing plants using TBBPA. TBBPA is also detected in sewage sludge and landfill leachate [5]. Levels of TBBPA in humans are shown in Table 102A.1 [6–15].

Bioaccumulation

Bioaccumulation is low, with peak concentrations of C^{14}-TBBPA within the first hour in all tissues (fat > liver, sciatic nerve, muscle, and adrenals) in the rat. Approximately 95% of the dose is excreted in the feces as a parent compound in the first 72 hours [4]. *Xenopus* metabolizes >94% of C^{14}-TBBPA within 8 hours, with sulfated conjugates as major metabolites. However, perturbation of TH signaling by TBBPA is likely due to the rapid direct action of the parent compound.

Excretion (Half-Life)

Half-lives are <1 day in fish, and <5 days in oysters [4].

Biological and Pathophysiological Implications

Action Mechanisms

TBBPA targets transthyretin and TR *in vitro*. The detailed thyroid disruption effects of TBBPA are summarized in E-Table 102A.1. TBBPA also has adverse effects on reproductive and immune functions, although there is still controversy regarding these effects. The European Union (EU) has concluded that TBBPA shows no risk to human health or the environment. The possible main action mechanisms of TBBPA in the thyroid system are as follows.

Y. Takei, H. Ando, & K. Tsutsui (Eds): Handbook of Hormones. DOI: http://dx.doi.org/10.1016/B978-0-12-801028-0.00249-X

Table 102A.1 TBBPA Levels in Human Serum and Milk

Country	Sample	Level (mg/kg fat)	Reference
1. Serum			
Sweden	Computer technicians	<0.5−1.8	[6]
	Electrics dismantlers	1.1−3.8	[7]
	Computer technicians	<1.0−3.4	[8]
Norway	Electrics dismantlers	0.64−1.8	[9]
Belgium		nd−0.186	[10]
Canada	Northern Quebec	<0.01−0.48	[11]
Japan	Adults (men and women, 37−49 years)	nd∼3.7	[12]
2. Milk			
Germany		0.29−0.94	[13]
France		0.06−37.3	[14]
China	Values in 75% of samples	<1	[15]

1. *Binding to serum TH binding protein transthyretin.* In *in vitro* competitive binding assays, TBBPA binding to transthyretin is more potent than TH binding [1]. It is theorized that TBBPA may decrease serum TH concentrations by displacing TH from transthyretin, leading to increased clearance. However, the consequences of TBBPA binding to transthyretin *in vivo* are not completely clear [16]. As transthyretin may mediate the maternal to fetal transport of T_4 through the placenta, TBBPA bound to transthyretin may be transported to the fetal compartment and affect fetal brain development [17]. Plasma T_4 levels are increased in European flounder (*Platichthys flesus*) exposed to TBBPA (0.1−0.8 µM for 105 days).

2. *Binding to TRs and TR activation.* In vitro competitive binding to TRs shows that TBBPA is 10^4 times less potent than T_3. TBBPA exerts anti-TH effects at $10^{-6}−10^{-5}$ M in a TH-responsive reporter assay using the CHO-K1 cell line [18]. TBBPA (10 and 100 nM) increases gelatinase B mRNA levels in the tail and TRα mRNA levels in the brain of the Pacific tree frog. TBBPA (∼10 µM) affects regulation of transcription by both apo- and holo-TRα1 *in vitro*. TBBPA (10−50 µM) shows TH-like activity on stimulation of GH3 cell growth, which is inhibited by the anti-estrogen ICI 182780 [3].

3. *Transcriptome and proteome analysis.* In the liver of zebrafish, TBBPA (0.75 and 1.5 µM for 14 days) interferes with the thyroid system and vitamin A homeostasis, induces oxidative and general stress responses, and alters cellular metabolism [19].

In Vitro *and* In Vivo *Assays for Detection of Thyroid System Disruption Activity of TBBPA*

In Vitro *Cell-Based Assays*

In vitro assays include the competitive TTR binding assay (∼10 times more potent than TH); the competitive TR binding assay (effective at submicromolar concentrations); the TH-response reporter gene assay in *Xenopus*, Chinese hamster, rat, and yeast cells (weak agonist and antagonist activities at 0.1−1.0 µM); a T-screen (rat GH3 cell growth assay, effective at 10 µM); a yeast two-hybrid assay (effective at $10^{-6}−10^{-5}$ M); and a cultured frog tadpole tail fin biopsy assay (C-fin assay; no effects at 0.01−1.0 µM) [20].

In Vivo *or* Ex Vivo *Assays*

In vivo or *ex vivo* assays include amphibian *Xenopus laevis* metamorphosis assays (inhibition of larval development at 920 nM for 21 days; weak agonist and antagonist activities on TH-responsive gene activation at 920 nM for 1−3 days).

References

1. Meerts IA, van Zanden JJ, Luijks EA, et al. Potent competitive interactions of some brominated flame retardants and related compounds with human transthyretin in vitro. *Toxicol Sci.* 2000;56:95−104.
2. EU (European Union), *Risk Assessment Report: Tetrabromobisphenol-A (TBBP-A) CAS No: 79-94-7. EINECS: 201-236-9, Part II Human Health* 2006; EU Joint Research Center, EUR 22161 EN, 4th Priority List, vol. 63.
3. Talsness CE, Andrade AJ, Kuriyama SN, et al. Components of plastic: experimental studies in animals and relevance for human health. *Phil Trans R Soc B.* 2009;364:2079−2096.
4. Birnbaum LS, Staskal DF. Brominated flame retardants: cause for concern? *Environ Health Perspect.* 2004;112:9−17.
5. Environment Canada. *2013 Screening Assessment Report, Phenol,4,4′-(1-methylethylidene)bis[2,6-dibromo-.* Environment Canada, Health Canada.
6. Hagmar L, Jakobsson K, Thuresso K, et al. Computer technicians are occupationally exposed to polybrominated diphenyl ethers and tetrabromobisphenol A. *Organohalogen Comp.* 2000;47:202−205.
7. Hagmar L, Sjödin A, Höglund P, et al. Biological half-lives of polybrominated diphenyl ethers and tetrabromobisphenol A in exposed workers. *Organohalogen Comp.* 2000;47:198−201.
8. Jakobsson K, Thuresson K, Rylander L, et al. Exposure to polybrominated diphenyl ethers and tetrabromobisphenol A among computer technicians. *Chemosphere.* 2002;46:709−716.
9. Thomsen C, Leknes H, Lundanes E, et al. A new method for determination of halogenated flame retardants in human milk using solid-phase extraction. *J Anal Toxicol.* 2002;26:129−137.
10. Kiciński M, Viaene MK, Den Hond E, et al. Neurobehavioral function and low-level exposure to brominated flame retardants in adolescents: a cross-sectional study. *Environ Health.* 2012;11:86.
11. Dallaire R, Ayotte P, Pereg D, et al. Determinants of plasma concentrations of perfluorooctanesulfonate and brominated organic compounds in Nunavik Inuit adults (Canada). *Environ Sci Technol.* 2009;43:5130−5136.
12. Nagayama J, Tsuji H, Takasuga T. Comparison between brominated flame retardants and dioxins or organochlorine compounds in blood levels of Japanese adults. *Organohalogen Comp.* 2000;48:27−30.
13. Kemmlein S. Polybromierte Flammschutzmittel: Entwicklung eines Analyseverfahrens und Untersuchung und Bewertung der Belastungssituation ausgewählter Umweltkompartimente. Fachbereich 06. *Vorgelegt von Diplom-Chemikerin Berlin.* 2000.
14. Cariou R, Antignac JP, Zalko D, et al. Exposure assessment of French women and their newborns to tetrabromobisphenol-A: occurrence measurements in maternal adipose tissue, serum, breast milk and cord serum. *Chemosphere.* 2008;73:1036−1041.
15. Shi ZX, Wu YN, Li JG, et al. Dietary exposure assessment of Chinese adults and nursing infants to tetrabromobisphenol-A and hexabromocyclododecanes: occurrence measurements in foods and human milk. 2009b. *Environ Sci Technol.* 2009;43:4314−4319.
16. Diamanti-Kandarakis E, Bourguignon JP, Giudice LC, et al. Endocrine-disrupting chemicals: an Endocrine Society scientific statement. *Endocr Rev.* 2009;30:293−342.

17. Boas M, Feldt-Rasmussen U, Skakkebaek NE, et al. Environmental chemicals and thyroid function. *Eur J Endocrinol.* 2006;154:599—611.

18. Kitamura S, Kato T, Iida M, et al. Anti-thyroid hormonal activity of tetrabromobisphenol A, a flame retardant, and related compounds: affinity to the mammalian thyroid hormone receptor, and effect on tadpole metamorphosis. *Life Sci.* 2005;76:1589—1601.

19. De Wit M, Keil D, Remmerie N, et al. Molecular targets of TBBPA in zebrafish analysed through integration of genomic and proteomic approaches. *Chemosphere.* 2008;74:96—105.

20. OECD, *In vitro & ex vivo assays for identification of modulators of thyroid hormone signaling.* Draft, 2013 <www.oecd.org/env/ehs/testing/Thyroid_scoping_Part1_WNT_comm%20_2_201213-clean.pdf>.

21. van der Ven LT, Kuil T, Verhoef A, et al. Endocrine effects of tetrabromobisphenol-A (TBBPA) in Wistar rats as tested in a one-generation study and a subacute toxicity study. *Toxicology.* 2008;245:76—89.

22. Choi JS, Lee YJ, Kim TH, et al. Molecular mechanism of tetrabromobisphenol A (TBBPA)-induced target organ toxicity in Sprague-Dawley male rats. *Toxicol Res.* 2011;27:61—70.

23. Jagnytsch O, Opitz R, Lutz I, et al. Effects of tetrabromobisphenol A on larval development and thyroid hormone-regulated biomarkers of the amphibian *Xenopus laevis. Environ Res.* 2006;101:340—348.

24. Veldhoen N, Boggs A, Walzak K, et al. Exposure to tetrabromobisphenol-A alters TH associated gene expression and tadpole metamorphosis in the Pacific tree frog *Pseudacris regilla. Aquat Toxicol.* 2006;78:292—302.

25. Chan WK, Chan KM. Disruption of the hypothalamic-pituitary-thyroid axis in zebrafish embryo-larvae following waterborne exposure to BDE-47, TBBPA and BPA. *Aquat Toxicol.* 2012;108:106—111.

26. Kuiper RV, van den Brandhof EJ, Leonards PEG, et al. Toxicity of tetrabromobisphenol A (TBBPA) in zebrafish (*Danio rerio*) in a partial life-cycle test. *Arch Toxicol.* 2007;81:1—9.

27. Hamers T, Kamstra JH, Sonneveld E, et al. *In vitro* profiling of the enodocrine-disrupting potency of brominated flame retardants. *Toxicol Sci.* 2006;92:157—173.

28. Kudo Y, Yamauchi K, Fukazawa H, et al. In vitro and in vivo analysis of the thyroid system-disrupting activities of brominated phenolic and phenol compounds in *Xenopus laevis. Toxicol Sci.* 2006;92:87—95.

29. Kitamura S, Jinno N, Ohta S, et al. Thyroid hormonal activity of the flame retardants tetrabromobisphenol A and tetrachlorobisphenol A. *Biochem Biophys Res Commun.* 2002;293:554—559.

30. Ghisari M, Bonefeld-Jorgensen EC. Impact of environmental chemicals on the thyroid hormone function in pituitary rat GH3 cells. *Mol Cell Endocrinol.* 2005;244:31—41.

31. Jugan ML, Lévy-Bimbot M, Pomérance M, et al. A new bioluminescent cellular assay to measure the transcriptional effects of chemicals that modulate the alpha-1 thyroid hormone receptor. *Toxicol in Vitro.* 2007;21:1197—1205.

32. Sun H, Shen O, Wang X, et al. Anti-thyroid hormone activity of bisphenol A, tetrabromobisphenol A and tetrachlorobisphenol A in an improved reporter gene assay. *Toxicol in Vitro.* 2009;23:950—954.

33. Freitas J, Cano P, Craig-Veit C, et al. Detection of thyroid hormone receptor disruptors by a novel stable in vitro reporter gene assay. *Toxicol in Vitro.* 2011;25:257—266.

34. Hinther A, Domanski D, Vawda S, et al. C-fin: a cultured frog tadpole tail fin biopsy approach for detection of thyroid hormone-disrupting chemicals. *Environ Toxicol Chem.* 2010;29:380—388.

35. Terasaki M, Kosaka K, Kunikane S, et al. Assessment of thyroid hormone activity of halogenated bisphenol A using a yeast two-hybrid assay. *Chemosphere.* 2011;84:1527—1530.

Supplemental Information Available on Companion Website

- Thyroid system disruption effects of TBBPA/E-Table 102A.1 [1,18,21—35]

Perchlorate

Kiyoshi Yamauchi

Additional names: ClO_4^-, CAS No. 14797-73-0

Perchlorate interferes with iodide uptake to the thyroid follicular cells, and acts as a goitrogen.

Discovery

Perchlorate has been known to inhibit the transport of iodide into the thyroid gland, and has been used clinically for the treatment of thyrotoxicosis since the 1950s. Industrially, perchlorate has been produced in large amounts for rocket fuels and military applications [1]. In the 1990s it was discovered that perchlorate remains unreacted in the environment, including soil, plants, and animals located near many military installations and rocket manufacturing facilities, at potentially thyroid-disrupting levels [2].

Structure

Properties

Colorless, odorless, molecular formula ClO_4^-, Mr 99.451. Stable at room or normal temperature. Perchlorates are soluble in water (249 g/l for ammonium perchlorate), and do not volatilize from soil or water surfaces.

Usage

Perchlorate is used as an oxidant in solid rocket propellants, ordnance, fireworks, and airbag deployment systems [1]. Perchlorates are also detected as impurities in consumer products such as bleach [2].

Annual Consumption

Annual consumption of perchlorate is several million kilograms/year.

Synthesis and Release

Origins of Environmental Perchlorates

Environmental perchlorates originate from natural and anthropogenic sources. Five perchlorates are manufactured in large amounts, as magnesium, potassium, ammonium, sodium, and lithium salts.

Environmental Levels

The maximum perchlorate concentration in some wells in several facilities with perchlorate release in the USA was ≥ 1 mg/l [2], and in drinking water wells it was $4-260\,\mu g/l$ in

California in 1997; in various water sources in the USA, the concentration was $0.06-6.8\,\mu g/l$ in 2005.

Exposure Routes

Exposure to perchlorates is via food (especially leafy green vegetables and milk) or drinking water, but not through the skin.

Biodegradation

Biodegradation occurs via a wide variety of microorganisms under anaerobic conditions, and by plants.

Reference Dose

A concentration of $0.7\,\mu g/kg$ per day in drinking water is the no-observed-effect level (NOEL) proposed by the National Academy of Science, USA [3]. This is equivalent to 24.5 ppb in drinking water (drinking water equivalent level, DWEL).

Biological and Pathophysiological Implications

Targets of Perchlorates

Perchlorates affect iodide uptake into the thyroid follicular cells by interfering with the function of the sodium/iodide symporter (NIS) and the chloride/iodide transporter pendrin (PDS) (Figure 102B.1 [4]).

NIS (or Solute Carrier Family 5 Member 5 [SLC5A5])

NIS transports two sodium cations (Na^+) for each iodide anion (I^-) into the cell (E-Figures 102B.1, 102B.2). NIS-mediated uptake of iodide into the follicular cells of the thyroid gland is the first step in the synthesis of thyroid hormones (THs). The ability of perchlorate to inhibit thyroidal iodide uptake is assumed to induce hypothyroidism. The anion selectivity of NIS is $TcO_4^- \geq ClO_4^- > ReO_4^- > SCN^- > BF_4^- > I^- > NO_3^- > Br^- > Cl^-$ [5]. All these chemicals act as competitive inhibitors of NIS. The relative potencies of ClO_4^-, SCN^-, and NO_3^- are 1, 1/20 to 1/10, and 1/550 to 1/240, respectively [6], on the molar concentration basis. The inhibitory effect of a mixture of NIS inhibitors on I^- uptake is expressed as a perchlorate equivalent concentration (PEC) [6].

Pendrin

Perchlorates also interact with pendrin (PDS), a member of the solute carrier (SLC) family, another iodide transporter, which is a sodium-independent chloride/iodide transporter

Y. Takei, H. Ando, & K. Tsutsui (Eds): Handbook of Hormones. DOI: http://dx.doi.org/10.1016/B978-0-12-801028-0.00250-6

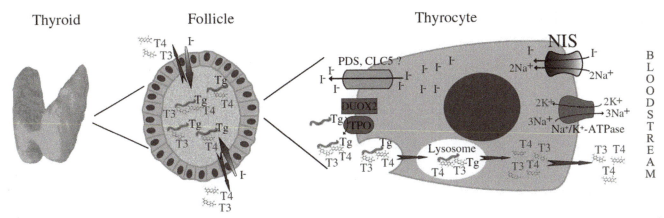

Figure 102B.1 Sodium/iodide symporter (NIS) and TH synthesis in the thyroid gland. NIS actively transports iodide into thyrocytes. The iodide ions cross the cells and are organified (covalently linked) inside the thyroid follicles by thyroid peroxidase (TPO) into thyroglobulin (Tg) tyrosine residues. These iodotyrosines are then coupled to form thyroid hormones. After endocytosis, the iodinated Tg is proteolyzed, and the THs are released into the bloodstream. *Figure reproduced from Darrouzet et al., 2014 [4], with permission.*

encoded by the *SLC26A4* gene. Pendrin is located on the apical membrane of the thyroid follicular cells [1].

Health Effects on the Thyroid System

Human Studies

It remains controversial whether environmentally occurring levels of perchlorates have any effects on human health (E-Table 102B.1) [7]. The effects of perchlorates may depend on gender, the length and developmental stages of exposure, and how much iodine is consumed [1,2].

Experimental Studies

When pregnant rats were treated with low amounts of perchlorate (≥0.009 mg/kg per day), it altered the thyroid gland (hypertrophy and hyperplasia) and serum THs and thyrotropin (TSH) levels in newborn animals.

Wildlife Studies

Few studies of the bioaccumulation in fish and aquatic organisms or food chain bioaccumulation have been carried out. Since I$^-$ levels in fish plasma are much higher than normal human levels, fish may not suffer from I$^-$ deficiency [8].

In Vitro *and* In Vivo *Assays for Detection of Perchlorate or Other Chemicals that Target NIS*

In Vitro *Cell-Based Assays* [8,9]

Assays include: the Na$^+$/I$^-$ uptake assay using radioactive iodide in transiently transfected rat FRTL-5 cells; the Na$^+$/I$^-$ uptake inhibitory assay (transient transfection of human embryonic kidney 293-derived cells; radioactive/colorimetric readouts); and the non-radioactive Na$^+$/I$^-$ uptake assay based on the Sandell-Kolthoff reaction using FRTL-5 cells.

In Vivo *or* Ex Vivo *Assays* [8]

Assays include the modified perchlorate discharge test using thyroid scintigraphy; and amphibian metamorphic climax and prometamorphic assays (developmental stage, morphological changes, TH levels, thyroid histology), although these are not NIS inhibitor-specific.

Clinical Implications

There are some mutations in the NIS gene (*SLC5A5*) that cause hypothyroidism and thyroid dyshormonogenesis. Anti-NIS antibodies have been found in thyroid autoimmune

diseases. Defect of the pendrin gene (*PDS, SLC26A4*) results in Pendred syndrome associated with profound sensorineural hearing loss and a thyroid condition called goiter.

Use for Diagnosis and Treatment

Therapeutic applications include dosage of more than 400 mg/day for hyperthyroidism and for amiodarone-induced thyrotoxicosis, while diagnostic applications include impairments in the synthesis of THs in the thyroid [2]. Side effects include skin rashes, nausea, and vomiting [2,5]. This treatment started in the USA, followed by the UK. In Germany, perchlorate is used in conjunction with thionamides for prophylaxis against iodine excess derived from coronary angiography [5].

References

1. Bergman À, Heindel JJ, Jobling S, et al. State of the Science of Endocrine Disrupting Chemicals-2012. Geneva, Switzerland: UNEP and WHO Press; 2013.
2. ATSDR (Agency for Toxic Substances and Disease Registry). *Toxicological Profile for Perchlorates.* Atlanta, GA, USA: ATSDR; 2008.
3. NRC (National Research Council). *Health Implications of Perchlorate Ingestion.* Washington DC: National Academic Press; 2005.
4. Darrouzet E, Lindenthal S, Marcellin D, et al. The sodium/iodide symporter: state of the art of its molecular characterization. *Biochim Biophys Acta.* 2014;1838:244−253.
5. Wolff J. Perchlorate and the thyroid gland. *Pharmacol Rev.* 1998;50:89−105.
6. De Groef B, Decallonne BR, Van der Geyten S, et al. Perchlorate versus other environmental sodium/iodide symporter inhibitors: potential thyroid-related health effects. *Eur J Endocrinol.* 2006;155:17−25.
7. Hartoft-Nielsen ML, Boas M, Bliddal S, et al. Do thyroid disrupting chemicals influence foetal development during pregnancy? *J Thyroid Res.* 2011;2011:342189.
8. EPA. *Draft detailed review paper on thyroid hormone disruption assays.* 2005; Contract No. 68-W-01-023 US EPA Washington, DC.
9. OECD. *In Vitro & Ex Vivo Assays for Identification of Modulators of Thyroid Hormone Signaling.* OECD; 2013.

Supplemental Information Available on Companion Website

- Structure of NIS/E-Figure 102B.1
- NIS family/E-Figure 102B.2
- Effects of perchlorate on the human thyroid system/E-Table 102B.1

PART VI

Gasotransmitters

CHAPTER 103

Gasotransmitter Family

John Alexander Donald

History

Nitric oxide (NO), carbon monoxide (CO), and hydrogen sulfide (H_2S) are often grouped together as a family of signaling molecules called gasotransmitters. The term "gasotransmitter" is now considered somewhat misleading, as the molecules are gases at Standard Temperature and Pressure but act as solutes in the aqueous environment of the intracellular and extracellular fluids [1]. However, the term is routinely used to classify NO, CO, and H_2S as a group of regulatory molecules, and it will be used here and in the subsequent subchapters for each specific gas for simplicity. NO, CO, and H_2S are all naturally occurring atmospheric gases that were present in the prebiotic world, and there is evidence that these gases were involved in the origin of life when atmospheric O_2 was low. In recent times, these gases were thought of as byproducts of industrialization that were considered pollutants prior to their biological role being discovered. They are all toxic to humans at high concentrations, but it is at low concentrations that they function as important signaling molecules in biological systems. The general concept of gases as biological signaling molecules began when it was finally shown in 1987 that NO was the labile factor released from the vascular endothelium of mammals to mediate acetylcholine-induced vasodilation, thus showing for the first time that a gas was an endogenously produced signaling molecule in biological systems [2]. This discovery led to Robert Furchgott, Louis Ignarro, and Ferid Murad receiving the 1998 Nobel Prize in Physiology or Medicine. In the mid-1990s, CO was discovered to be a putative signaling molecule involved in neurotransmission, and then it was proposed at the turn of the century that H_2S was also an endogenously produced gasotransmitter and a regulator of neuronal function [1]. Research into the biological role of gasotransmitters has grown rapidly, and it is now recognized that they are not classical signaling molecules but are regulators of cellular function through complex chemical interactions with each other (including metabolites) and target proteins (Figure 103.1) [1,3]. Of the gasotransmitters, NO can now be considered an endocrine molecule in the form of circulating nitrite (NO_2^-), which can be reduced to NO in cells and blood by nitrite reductases [4]. An endocrine role for H_2S is unlikely, based on the measurements of sulfide in the blood, and CO will circulate tightly bound to hemoglobin [5]. Each of the gasotransmitters is now being developed as a therapeutic through the use of engineered molecules that can be introduced into the body and release the gas in a controlled manner.

Structure

Structural Features

As the name implies, the gasotransmitters are low molecular weight molecules (NO, 30 Da; CO, 28 Da; H_2S, 34 Da), which are hydrophilic and lipophilic. NO, CO, and H_2S are not stored, and NO and H_2S in particular have a short half-life. Each gasotransmitter has a dependence on O_2 as it is required for NO and CO synthesis and H_2S metabolism, respectively. NO, CO, and H_2S have different chemical properties. NO is a radical with an unpaired electron, and CO is biologically inactive. H_2S in solution is a weak acid that partially dissociates to form an equilibrium between H_2S and HS^-. H^+ and HS^- may further dissociate to $2H$ and S^{2-}, but this is not favored in biological fluids. Both NO and H_2S can participate in redox reactions, and can react with each other and their respective metabolites. The three gases can form coordination chemistries with heme, with NO and CO generally interacting with iron (II) and H_2S with iron (III), respectively [1].

Receptors

Structure and Signal Transduction Pathways

NO, CO, and H_2S can interact with a large number of target proteins through coordination chemistries with prosthetic metal groups or protein modifications such as *S*-nitrosylation and *S*-sulfhydration (Figure 103.1) [3]. Soluble guanylyl cyclase (sGC) is a heme protein receptor that is involved in the classical signaling pathway for NO and CO, as both ligands bind to sGC to generate the second messenger cGMP; H_2S does not bind to sGC [6]. The ability of CO to bind to sGC is dependent on NO concentration, as NO binds to sGC with much greater affinity than CO. This is a good example of the interdependence of the gasotransmitters on their functionality. H_2S in involved in cGMP signaling by inhibition of cGMP phosphodiesterases that augments intracellular cGMP concentration [7].

Biological Functions

Target Cells/Tissues and Functions

Gasotransmitters can affect the function of all cells that express target proteins, such as those summarized in Figure 103.1. Thus, they are involved in the control of a vast array of physiological functions, including regulation of the

Y. Takei, H. Ando, & K. Tsutsui (Eds): Handbook of Hormones. DOI: http://dx.doi.org/10.1016/B978-0-12-801028-0.00103-3

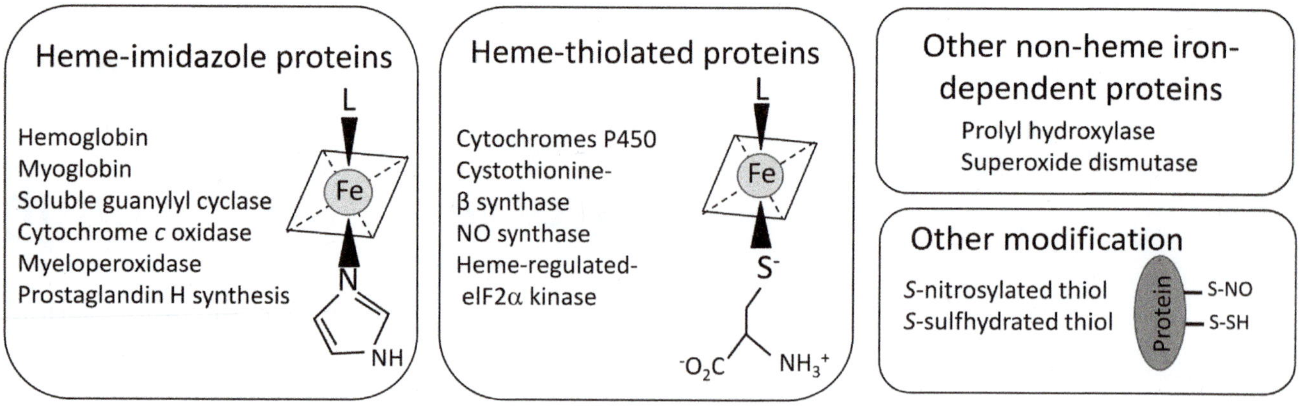

Heme-imidazole proteins

Hemoglobin
Myoglobin
Soluble guanylyl cyclase
Cytochrome *c* oxidase
Myeloperoxidase
Prostaglandin H synthesis

Heme-thiolated proteins

Cytochromes P450
Cystothionine-
β synthase
NO synthase
Heme-regulated-
eIF2α kinase

Other non-heme iron-dependent proteins

Prolyl hydroxylase
Superoxide dismutase

Other modification

S-nitrosylated thiol
S-sulfhydrated thiol

— S-NO
— S-SH

Figure 103.1 Summary of gas targets classified by the nature of protein binding. NO, CO, and H$_2$S can interact with many proteins that regulate cellular function. The nature of the protein binding can be broadly divided into two groups: (1) coordination chemistry of the gas with prosthetic metal complexes, and (2) binding to a specific region of the protein function such as the cysteine thiol group. *Figure modified from Kajimura et al., 2012 [3].*

cardiovascular, nervous, gastrointestinal, excretory, and immune systems [6,8,9]. Gasotransmitters are involved in regulating many cellular functions, including cytoprotection, apoptosis, proliferation, inflammation, and gene transcription [3,6,8,9]. In addition, they regulate fundamental intracellular functions such as cellular respiration and ATP synthesis, for which NO and CO are inhibitory [3,6,10]. H$_2$S is bifunctional, as it increases ATP production at low concentrations but decreases it at high concentrations [8,10]. Both H$_2$S and CO are strongly implicated in O$_2$ sensing, as their respective intracellular concentrations are dependent on the O$_2$ concentration.

References

1. Feelisch M, Olson K. Embracing sulfide and CO to understand nitric oxide biology. *Nitric Oxide.* 2013;35:2−4.
2. Palmer RM, Ferrige AG, Moncada S. Nitric oxide release accounts for the biological activity of endothelium-derived relaxing factor. *Nature.* 1987;327:524−526.
3. Kajimura M, Nakanishi T, Takenouchi T. Mechanisms of nitrite bioactivation. *Respir Physiol Neurobiol.* 2012;184:139−148.
4. Kim-Shapiro DB, Gladwin MT. Is hydrogen sulfide a circulating "gasotransmitter" in vertebrate blood? *Nitric Oxide.* 2014;38:58−68.
5. Olson KR. Gas biology: tiny molecules controlling metabolic systems. *Biochim Biophs Acta.* 2009;1787:856−863.
6. Gao Y. The multiple actions of NO. *Pflugers Arch Eur J Physiol.* 2010;459:829−839.
7. Bucci M, Papapetropoulos A, Vellecco V, et al. cGMP-dependent protein kinase contributes to hydrogen sulfide-stimulated vasorelaxation. *PLoS One.* 2012;7:e53319.
8. Wang R. Physiological implications of hydrogen sulphide: a whiff exploration that blossomed. *Physiol Rev.* 2012;92:791−896.
9. Kolluru GK, Shen X, Bir SC, et al. Heme oxygenase-1/carbon monoxide: from basic science to therapeutic applications. *Nitric Oxide.* 2013;35:5−20.
10. Ryter SW, Alam J, Choi AM. Hydrogen sulfide chemical biology: Pathophysiological roles and detection. *Physiol Rev.* 2006;86:583−650.

Nitric Oxide

John Alexander Donald

Abbreviation: NO
Additional names: nitrogen monoxide, nitrogen (II) oxide

Nitric oxide was the first gasotransmitter discovered as an important signaling molecule in the cardiovascular, nervous, and immune systems

Discovery

Like the other gasotransmitters, nitric oxide (NO) is naturally produced in the atmosphere, and is a by-product of human activity that is produced during combustion of substances in automobile engines and power plants. About 40 years ago it was discovered that the NO donors, nitroprusside and nitroglycerin, stimulated the production of the second messenger, guanosine 3':5'-cyclic monophosphate (cGMP) via a soluble guanylyl cyclase (sGC) receptor. Later, Furchgott and Zawadzki (1980) demonstrated that relaxation of the rabbit aorta by the neurotransmitter acetylcholine (ACh) required the presence of an intact endothelium, and that ACh stimulated the release of a labile substance called endothelium-derived relaxing factor (EDRF) that appeared to signal via cGMP [1]. In 1987, Palmer and colleagues then demonstrated that the EDRF was pharmacologically identical to NO, which initiated a new and vast research area on the biological role of NO. NO is now known to be a critical regulator of many biological function in animals [2]. In the 1990s it was discovered that NO is generated by enzymes called nitric oxide synthases (NOS), of which there are three isoforms in mammals. It is now known that NO can be generated by reduction of nitrite (NO_2^-) by NO_2^- reductases, and that this source of NO is very important during hypoxia.

Structure

Primary Structure, Structural Features, and Properties

NO is a 30-Da free radical that is a colorless gas with a melting point of 163.6°C. NO has one unpaired electron and is paramagnetic. It can react with transition metals to form metal nitrosyls, which is an important aspect of its signaling properties. NO has a short half-life and is rapidly metabolized. NO has poor solubility in water, but as a gas it is lipophilic and freely crosses the lipid bilayer of the plasma membrane, which is important for paracrine regulation of physiological processes.

Synthesis, Release, and Metabolism

NO can be generated enzymatically by NOS or by chemical reduction of NO_2^- [3]. In tetrapods, there are three NOS isoforms called NOS1 (neuronal or nNOS), NOS2 (inducible or iNOS), and NOS3 (endothelial or eNOS) [4]. NOS1 and NOS3 are also known as constitutive NOS enzymes. In the presence of L-arginine and O_2, all NOS enzymes catalyze the reaction that generates NO and L-citrulline (Figure 103A.1) [3]; accordingly, NO generated by NOS is O_2 dependent [4]. NO can also be generated by the reduction of NO_2^- that is facilitated by NO_2^- reductases such as heme globins (e.g., hemoglobin and myoglobin), molybdenum-containing enzymes (e.g., xanthine oxidase), NOS, and various components of the mitochondrial electron transport chain [5]. NO_2^- reduction to NO is important in hypoxia when NOS generated NO is limited by low O_2. Thus, the role of NO_2^- in NO homeostasis during hypoxia is important. The major pathway for NO metabolism is via oxidation to NO_2^- and nitrate (NO_3^-), which may proceed via autooxidation or be catalyzed by a number of factors. NO_2^- can then be recycled back to NO by NO_2^- reduction (Figure 103A.1). NO also reacts with superoxide (O_2^-) to form the radical oxygen species peroxynitrite ($ONOO^-$) [6].

Gene, mRNA, and Protein

NO is not a gene product. In tetrapods, there are three NOS genes encoding NOS1, NOS2, and NOS3, respectively [4]. Most fishes do not possess the *nos3* gene, and this is reflected physiologically as there is an absence of endothelial NO signaling in blood vessels [3]. Amphibians possess a *nos3* gene but also do not have endothelial NO signaling [3]. NOS genes are broadly expressed in many tissues. The *nos3* gene is expressed in the endothelium of amniotic vertebrates, while the *nos1* gene is widely expressed in the nervous system of all vertebrates where NO is an important neurotransmitter [3]. The expression of the *nos2* gene is associated with elements of the immune sytem of vertebrates. NOS proteins share a similar molecular structure, with an N-terminal oxygenase domain and a C-terminal reductase domain [4]. The NOS1 protein has a distinctive N-terminal PDZ domain that is not present in NOS2 and NOS3. Mammalian NOS proteins have the following molecular weights: NOS1, 155 kDa; NOS2, 130 kDa; and NOS3, 140 kDa [4]. There are many reductase genes that encode proteins involved in NO_2^- reduction.

Tissue and Plasma Concentrations

NO has a very short half-life and NO_2^- concentrations are an indirect indicator of NO production, which can be measured

Y. Takei, H. Ando, & K. Tsutsui (Eds): Handbook of Hormones. DOI: http://dx.doi.org/10.1016/B978-0-12-801028-0.00251-8

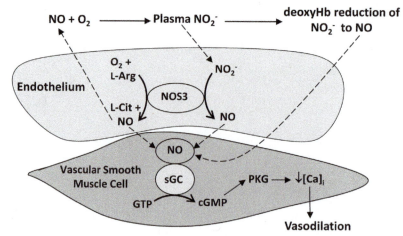

Figure 103A.1 Schematic diagram showing NO signaling and recycling in blood vessels of mammals. NO is generated by NOS3 and diffuses into neighboring vascular smooth muscle cells, where it binds to sGC and activates the cGMP signaling pathway. This leads to a reduction in intracellular Ca^{+2} and vasodilation. In addition, NO reacts with O_2 to generate NO_2^- that can then be reduced back to NO by deoxy Hb in the blood. In this way, NO_2^- provides a circulating reservoir of NO. NO_2^- can also be reduced to NO by NOS3 in endothelial cells and endogenous NO_2^- reductase in vascular smooth muscle cells. *Modified from Olson et al., 2012 [3].*

to determine NO production. It is now considered that NO_2^- is a stable endocrine reservoir of NO that can circulate in the plasma at nanomolar concentrations compared with micromolar concentrations in tissues, and be reduced to NO by NO_2^- reductases [5]. The majority of NO_2^- comes from the oxidation of NO generated by NOS, with a smaller amount (30%) obtained from the diet or reduction of NO_3^- [5].

Receptors

Structure and Signal Transduction Pathways

NO signaling can occur by binding to sGC that generates the second messenger cGMP, which activates protein kinase G to affect downstream processes such as vasodilation (Figure 103A.1) [7,8]. NO can also interact with components of the mitochondrial electron transport chain to regulate cellular respiration and reduce damage by reactive oxygen species. The regulation of gene expression and protein function can be altered by posttranslation modification of proteins by nitrosylation, through the addition of a nitrosyl ion (NO^-) to a metal to form iron-nitrosyl (FeNO) compounds or to a thiol to form S-nitroso (SNO) compounds [7,8].

Agonists

NO can be provided directly as NO gas, but this can be problematic due to its short half-life. As a replacement, compounds that can release NO, called NO donors, are used in experimental and therapeutic settings. There are various classes of NO donors; some release NO directly, such as sodium nitroprusside, and others require metabolism, such as organic nitrates and nitrite esters [9]. Reduction of NO_2^- is also a mechanism for providing a source of NO.

Antagonists

The primary antagonists of receptor-mediated NO signaling are componds that block sGC, such as ODQ (1H-[1,2,4]oxadiazolo[4,3,–a]quinoxalin-1-one). In addition, synthesis of NO by NOS can be antagonized with compounds such as nitroarginine (L-NNA) that are generic inhibitors of NOS. NOS

inhibition is routinely used to demonstrate NO signaling in tissues. There is also a broad range of compounds that can provide selective inhibition of the different NOS isoforms and thus NO production; there is a greater range of inhibitors for NOS1 and NOS2, respectively, than for NOS3.

Biological Functions

Target Cells/Tissues and Functions

NO mediates its effects as a paracrine regulator and a neurotransmitter. NO can affect all cells that express sGC receptors or proteins that can interact with NO. It can regulate a vast array of physiological functions, including muscle contractility, metabolism, platelet aggregation, neuronal behavior, and immune responses. NO signaling via cGMP is involved in regulation of the cardiovascular and gastrointestinal systems. NO relaxes smooth muscle primarily via cGMP and PKG by decreasing Ca^{2+} flux into the cells or by modulation of ion channels. In the nervous system, nerves releasing NO are referred to as nitrergic nerves. NO is a neurotransmitter in both the central and peripheral nervous systems, and provides nitrergic nerve regulation of many effectors. In the autonomic nervous system, NO is a co-transmitter with ACh in many parasympathetic postganglionic nitrergic nerves projecting to craniosacral targets such as the cerebral vasculature and the genitalia. NO can regulate metabolism by stimulating glucose uptake, glycolysis, and AMP-activated protein kinase activity, and can inhibit mitochondrial respiration by binding to and inhibiting the activity of cytochrome c oxidase. NO can directly affect the behavior of many proteins (e.g., ion channels) and enzymes by direct nitrosylation [7,8].

Phenotype in Gene-Modified Animals

NOS3 knockout mice have elevated blood pressure that is about 20 mmHg higher than in wild-type mice, indicating a tonic role for endothelial NO in blood pressure control [10]. Furthermore, the aorta from NOS3 knockout mice does not relax in response to ACh, demonstrating that NOS3 is essential for endothelium-dependent NO vasodilation [10].

Pathophysiological Implications

Clinical Implications and Use for Diagnosis and Treatment

Dysfunction of NO production (NOS) or signaling is implicated in many diseases of most organ systems in the body [6]. In addition, the role of NO metabolites such as NO_2^- and peroxynitrite in disease states is becoming increasingly recognized [6]. Classical NO therapy involves the use of NO donors in the treatment of cardiovascular disease, particularly the symptoms of ischemic heart disease and pulmonary hypertension [9]. In addition, phosphodiesterase inhibitors such as sildenafil are used to prolong cGMP-mediated vasodilation in the penis for the treatment of erectile dysfunction, as the penile vasculature is richly supplied with nitrergic nerves. Exhaled NO is used as a marker of respiratory diseases such as asthma and chronic obstructive pulmonary disease, and there is ongoing research on the therapeutic use of inhaled NO in the treatment of acute respiratory distress syndrome. Currently, new classes of non-steroidal anti-inflammatory drugs that release NO (called NO-NSAIDs) are being developed to provide a more effective means of providing analgesia or anti-inflammatory treatment. The addition of NO to the NSAID compound provides cytoprotection in the gastrointestinal system by reducing mucosal damage and bleeding [9].

References

1. Furchgott RF, Zawadzki JV. The obligatory role of endothelial-cells in the relaxation of arterial smooth-muscle by acetylcholine. *Nature.* 1980;288:373–376.
2. Palmer RM, Ferrige AG, Moncada S. Nitric oxide release accounts for the biological activity of endothelium-derived relaxing factor. *Nature.* 1987;327:524–526.
3. Olson KR, Donald JA, Dombkowski RA, et al. Evolutionary and comparative aspects of nitric oxide, carbon monoxide and hydrogen sulfide. *Respir Physiol Neurobiol.* 2012;184:117–129.
4. Daff S. NO synthase: structures and mechanisms. *Nitric Oxide.* 2010;23:1–11.
5. Shiva S. Nitrite: a physiological store of nitric oxide and modulator of mitochondrial function. *Redox Biol.* 2013;1:40–44.
6. Pacher P, Beckman JS, Liaudet L. Nitric oxide and peroxynitrite in health and disease. *Physiol Rev.* 2007;87:315–424.
7. Gao Y. The multiple actions of NO. *Pflugers Arch Eur J Physiol.* 2010;459:829–839.
8. Martínez-Ruiz A, Cadenas S, Lamas S. Nitric oxide signaling: classical, less classical, and nonclassical mechanisms. *Free Radical Biol Med.* 2011;51:17–29.
9. Ignarro LJ, Napoli C, Loscalzo J. Nitric oxide donors and cardiovascular agents modulating the bioactivity of nitric oxide – An overview. *Circ Res.* 2002;90:21–28.
10. Huang PL. Mouse models of nitric oxide synthase deficiency. *J Am Soc Nephrol.* 2000;11:S120–S123.
11. Hall AV, Antoniou H, Wang Y, et al. Structural organisation of the human neuronal nitric oxide synthase gene (NOS1). *J Biol Chem.* 1994;269:33082–33090.
12. Chartrain NA, Geller DA, Koty PP, et al. Molecular cloning, structure, and chromosomal localisation of the human inducible nitric oxide synthase gene. *J Biol Chem.* 1994;269:6765–6772.
13. Nadaud S, Bonnardeaux A, Lathrop M, et al. Gene structure, polymorphism and mapping of the endothelial nitric oxide synthase gene. *Biochem Biophys Res Comm.* 1994;198:1027–1033.
14. Donald JA, Forgan LG, Cameron MS. The evolution of nitric oxide signalling in vertebrate blood vessels. *J Comp Physiol.* 2015;185:153–171.
15. Alderton WK, Cooper CE, Knowles G. Nitric oxide synthases: structure, function and inhibition. *Biochem J.* 2001;357:593–615.

Supplemental Information Available on Companion Website

- Gene structure of human NOS1, NOS2, and NOS3/E-Figure 103A.1
- Phylogenetic tree of vertebrate nitric oxide synthase proteins/E-Figure 103A.2
- Accession numbers of full-length vertebrate NOS proteins/E-Table 103A.1
- Protein domain analysis of human NOS proteins/E-Figure 103A.3

Carbon Monoxide

John Alexander Donald

Abbreviation: CO
Additional names: carbonous oxide, carbon (II) oxide

Carbon monoxide is the second gasotransmitter that was found to have a physiological role in neurotransmission, cardiovascular regulation, and oxygen sensing

Discovery

Carbon monoxide (CO) is a naturally occurring atmospheric gas and a chemical by-product of industrialization. It is toxic to humans at high concentrations, as it binds hemoglobin 200–300 times more strongly than does O_2 to form carboxyhemoglobin (CoHb), thus reducing O_2 storage and transport in the blood to a concentration where death may result from asphyxiation. It can also have toxic effects in cells due to interaction with many intracellular protein targets. In a biological context, it was discovered in the 1960s that CO can be endogenously produced in the body by heme oxygenase (HO) metabolism of heme, thus initiating the idea that CO at non-toxic concentrations is a regulator of biological function. Since the 1990s CO has been recognized as a physiologically important signaling molecule, and is now considered as one of the three gasotransmitters with NO and H_2S. CO can signal in cells to regulate many physiological functions, particularly in the nervous and cardiovascular systems [1–3].

Structure

Primary Structure, Structural Features, and Properties

CO is a stable oxide of carbon that is produced when there is partial oxidation of carbon-containing compounds. It is a chemically stable 28-Da molecule with carbon and oxygen atoms connected by a triple bond. It is a colorless and odorless gas at temperatures above $-190°C$, and chemical reduction requires temperatures above $100°C$. CO is poorly soluble in water, and does not react with water without considerable energy input. As a gas, CO is lipophilic and freely crosses the lipid bilayer of the plasma membrane, which is important for physiological regulation. CO does not contain free electrons and is a relatively stable neutral molecule; therefore, it is not as reactive as the other gasotransmitters [1,2].

Synthesis, Release, and Metabolism

Most CO production is generated by the catabolism of heme by HOs, which are common to all multicellular and unicellular organisms (Figure 103B.1) [3].

Heme is an important cofactor in many proteins, such as hemoglobin and myoglobin, that bind oxygen. Accordingly, intracellular heme levels are controlled such that CO production is rate-limited by the availability of heme, and therefore indirectly by heme biosynthesis. There are three isoforms of HO: inducible HO-1 and two constitutive forms, namely HO-2 and HO-3. Heme is metabolized by HO in a tightly regulated manner to produce CO, iron, and biliverdin, the latter being rapidly converted to bilirubin (Figure 103B.1). In addition to HO metabolism of heme, minor amounts of CO can be produced via lipid peroxidation, oxidation of phenols, and cytochrome b5 reduction. The endogenous production rate of CO has been measured at $16.4\ \mu mol/h$. Endogenously generated CO can circulate in the blood bound to hemoglobin, or be sequestered in cells bound to heme-containing proteins. The vast majority of endogenously produced CO is exhaled via the lungs. Oxidation of CO has not been shown in vertebrates, but it does occur reversibly in many microbes via CO dehydrogenases [1–6].

Gene, mRNA, and Protein

This section will focus on HO, as its regulation is critical in CO production. The human HO-1 gene consists of five exons and four introns, and the promotor sequence contains consensus binding sites for stress-responsive transcription factors, such as nuclear factor-κB, activator protein-1 (AP-1), AP-2, and Sp1 c-myc/max, among others. There is evidence that transcriptional regulation of HO-1 varies between different animal models. The HO-1 gene is most abundantly expressed in the spleen for red blood cell degradation, with lower levels occurring in other tissues. Generally, HO-1 expression can be regulated by many endogenous and exogenous factors that are associated with cellular stress. The HO-2 gene has a structure similar to that of HO-1 and is constitutively expressed in the testis at high levels, followed by the liver, nervous system, endothelium, and smooth muscle. The HO-3 protein is believed to result from differential splicing of HO-2. The HO-1 and HO-2 proteins have a molecular mass of 32 kD and 36 kD, respectively, and have a homology of approximately 40%. The HO-2 protein contains an additional N-terminal segment that is not found in HO-1, and both HO-1 and HO-2

Y. Takei, H. Ando, & K. Tsutsui (Eds): Handbook of Hormones. DOI: http://dx.doi.org/10.1016/B978-0-12-801028-0.00252-X

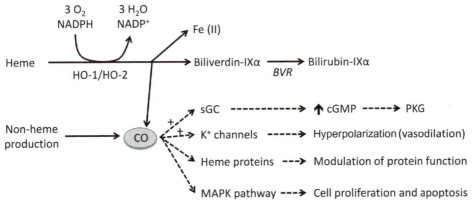

Figure 103B.1 **The production of CO and examples of some cellular mechanisms.** Heme is catabolized by inducible HO-1 or constitutive HO-2 into iron (II), biliverdin-IX, and CO in an oxygen- and NADPH-dependent manner. Biliverdin-IXα is then converted to bilirubin-IXα by biliverdin reductase (BVR). There is also non-heme CO production. CO can elicit physiological regulation by different pathways, including soluble guanylyl cyclase (sGC) and cGMP signaling, direct activation of K^+ channels leading to membrane hyperpolarization, binding to heme-containing proteins, and modulation of the MAPK signaling pathway associated with control of cell proliferation and apoptosis. *Figure modified from Olson* et al., 2012 [3].

share a heme-regulatory domain that is essential for heme metabolism. The HO proteins are membrane-bound, with both HO-1 and HO-2 anchored to the endoplasmic reticulum [1–9].

Receptors

Structure and Signal Transduction Pathways

There is no specific receptor for CO, as it can interact with a vast array of heme-containing proteins (Figure 103B.1). The binding of CO to the heme in proteins causes a conformational change that regulates their activity. Some key heme-containing proteins include transcription factors that regulate a number of functions, such as cellular proliferation and circadian rhythm, cystathione β-synthase that is involved in H_2S synthesis, potassium channels involved in the regulation of smooth muscle function, nitric oxide synthase, and soluble guanylyl cyclase (sGC). sGC is a heme-containing heterodimeric receptor protein that catalyzes the conversion of GTP to the important second messenger cGMP; sGC is also the receptor for NO. CO can bind to the distal side of the heme iron of sGC to increase its catalytic activity, albeit to a much lower degree than NO. Therefore, the interaction of CO with sGC is indirectly regulated by NO availability, which is an important consideration in assessing the function of CO [1–9].

Agonists

In experimental and clinical scenarios, CO gas and CO-releasing molecules (called CORMs) are used to provide CO in body fluids and tissues in an amount that is below toxic concentrations [9]. CORMs are transition metal carbonyls that release CO, and are the predominant means to provide CO for physiological experimentation [1,2,9].

Antagonists

As there is no specific receptor for CO, antagonism of CO signaling can be achieved indirectly as follows. Synthesis of CO can be blocked using HO inhibitors such as metalloporphyrins that include chromium mesoporphyrin, manganese protoporphyrin, manganese mesoporphyrin, zinc protoporphyrin, and tin mesoporphyrin. The potency of metalloporpyrins in inhibiting HO is dependent on the species of metal cation linked to the porphyrin ring. CO signaling via sGC can be

inhibited using specific sGC inhibitors such as ODQ (1H-[1,2,4]oxadiazolo[4,3–a]quinoxalin-1-one) [1–9].

Biological Functions

Target Cells/Tissues and Functions

CO will target all cells, as intracellular heme-containing proteins are ubiquitous in cells. Therefore, a vast array of biological processes could be regulated by CO via multiple pathways and cellular targets such as the cGMP and MAPK signaling pathways, direct regulation of potassium channels, and modulation of proteins involved in inflammation, mitochondrial function, and cytoprotection (Figure 103B.1). Accordingly, CO is involved in many physiological functions, including regulation of the cardiovascular and nervous systems [1–9].

Phenotype in Gene-Modified Animals

HO-knockout mice are unlikely to survive, but if they do they exhibit a complex phenotype generally associated with oxidative stress, and have enlarged organs and a shortened lifespan [1,5,7]. The cellular effects of HO deficiency can be ameliorated with applied CO via CO donors [1,5,7]. HO-1 deficiency in humans is linked to growth retardation and blood disorders such as anemia and abnormal coagulation, and results in early death [7].

Pathophysiological Implications

Clinical Implications and Use for Diagnosis and Treatment

The most reliable indicator of CO exposure is the measurement of COHb in the blood, as CO binds Hb with high affinity. The analysis of COHb is used in both clinical and forensic settings. As the lung is the sole site of CO elimination, the measurement of CO in the exhaled breath can predict endogenous CO production and indicate dysfunction of the HO/CO system in disease states. For example, patients with diabetes or asthma have elevated levels of CO in the breath. The therapeutic use of CO via CO gas or CORMs is receiving considerable attention for the treatment of cardiovascular disease (e.g., pulmonary hypertension) and inflammation, and in the area of organ transplantation [1,3,6–9].

References

1. Wu L, Wang R. Carbon monoxide: Endogenous production, physiological functions, and pharmacological applications. *Pharmacol Rev.* 2005;57:585–630.
2. Heinemann SH, Toshinori H, Westerhausen M, et al. Carbon monoxide – physiology, detection and controlled release. *Chem Comm.* 2014;50:3644–3660.
3. Olson KR, Donald JA, Dombkowski RA, et al. Evolutionary and comparative aspects of nitric oxide, carbon monoxide and hydrogen sulfide. *Respir Physiol Neurobiol.* 2012;184:117–129.
4. Ryter SW, Alam J, Choi AM. Heme oxygenase-1/carbon monoxide: from basic science to therapeutic applications. *Physiol Rev.* 2006;86:583–650.
5. Dennery PA. Signaling function of heme oxygenase proteins. *Antioxid Redox Signal.* 2014;20:1743–1753.
6. Kajimura M, Nakanishi T, Takenouchi T, et al. Gas biology: tiny molecules controlling metabolic systems. *Respir Physiol Neurobiol.* 2012;184:139–148.
7. Wegiel B, Nemeth Z, Correa-Costa M, et al. Heme oxygenase-1: a metabolic Nike. *Antioxid Redox Signal.* 2014;20:1709–1722.
8. Rochette L, Cottin Y, Zeller M, et al. Carbon monoxide: mechanisms of action and potential clinical implications. *Pharmacol Therap.* 2013;137:133–152.
9. Motterlini R, Otterbein LE. The therapeutic potential of carbon monoxide. *Nat Rev Drug Discov.* 2010;9:728–743.
10. Shibahara S, Sato M, Muller RM, et al. Structural organisation of the human heme oxygenase gene and the function of its promoter. *Eur J Biochem.* 1989;179:557–563.

Supplemental Information Available on Companion Website

- Gene structure of human heme oxygenases/ E-Figure 103B.1
- Phylogenetic tree of vertebrate heme oxygenase proteins/ E-Figure 103B.2
- Accession numbers of full-length vertebrate heme oxygenase proteins/ E-Table 103B.1

Hydrogen Sulfide

Leonard G. Forgan and John Alexander Donald

Abbreviation: H_2S
Additional names: sulfane (IUPAC), hydrogen sulfide, hydrosulfuric acid, dihydrogen monosulfide, dihydrogen sulfide, sulfurated hydrogen, sulfur hydride

Hydrogen sulfide is important in the regulation of many cellular processes, and is strongly implicated in oxygen sensing.

Discovery

Hydrogen sulfide (H_2S) is a naturally occurring compound that is replete throughout many physical and biological systems. It is acutely toxic at high concentrations, the toxicity attributed to inhibition of mitochondrial cytochrome c oxidase (CCO), carbonic anhydrase (CA), monoamine oxidase, Na^+/K^+-ATPase and cholinesterases [1]. It was not until 1996 that the physiological signaling properties of H_2S were recognized. Originally identified as a neuromodulator [2], a plethora of studies have since been published demonstrating the involvement of H_2S in the cardiovascular system, and nearly all other organ systems examined to date [3].

Structure

Primary Structure, Structural Features, and Properties

H_2S is a 34.08-Da molecule that has a molecular charge of zero and tetrahedral electron pair geometry with two pairs of bonding electrons and two lone pairs. The molecule is slightly polar and a strong reductant. H_2S is both hydrophilic and lipophilic. In solution, H_2S is a weak acid [$H_2S \leftrightarrow H^+ + HS^-$, pK_a 6.76, 37°C] [4]. In intracellular fluid, nearly equal amounts of H_2S and HS^- are present, but in the extracellular fluid there is approximately a 1:4 ratio of H_2S to HS^- at 37°C and pH 7.4 [4].

Synthesis, Release, and Metabolism

H_2S synthesis occurs via enzyme-catalyzed reactions, some of which involve multiple steps. These processes have been comprehensively described [3,5,6]. Current evidence suggests that most H_2S is synthesized from L-cysteine by cystathionine β-synthase (CBS), cystathionine γ-lyase (CSE), or cysteine aminotransferase (CAT) and 3-mercaptopyruvate sulfurtransferase (3-MST), which act in tandem (Figure 103C.1A). However, H_2S is also produced by cysteine lyase (CLY) (Figure 103C.1B), and non-enzymatically in the thiosulfate cycle (Figure 103C.1A) and via thiols (Figure 103C.1C).

Carbonyl sulfide has been proposed as a substrate for H_2S synthesis, with CA as the catalyst (Figure 103C.1D).

Catabolism of H_2S occurs in mitochondria. Membrane-bound sulfide:quinone oxidoreductase (SQOR) simultaneously reduces elemental sulfur and cysteine disulfide to form persulfide groups (H_2S_2) at one of the SQOR cysteines. Sulfur dioxygenase oxidizes a persulfide to sulfite (H_2SO_4), with O_2 and H_2O consumed in the process. Most thiosulfate ($S_2O_3^{2-}$) is further metabolized to sulfate by thiosulfate-dithiol sulfurtransferase and sulfite oxidase. Electrons from H_2S flow into the electron transport chain (ETC) via the quinone pool to O_2 at CCO.

Gene, mRNA, and Protein

Human CBS is encoded by the *cbs* gene found on chromosome 21 (21q22.3), producing a 1,656-bp mRNA encoding a 551-aa protein (61 kDa). Human CSE is encoded by the *cth* gene found on chromosome 1 (1p31.1), producing a 1218-bp mRNA and 405-aa protein (45 kDa). Human 3-MST is encoded by the *mpst* gene found on chromosome 22 (22q13.1), producing a 894-bp mRNA and a 297-aa protein (33 kDa). Comparative gene and protein data from representative vertebrate animals are shown in E-Table 103C.1. 3D protein folding, putative conserved domains, and posttranslational modifications predictions for the three human biosynthetic enzymes are shown in E-Figure 103C.1 and E-Table 103C.2.

Distribution of Protein and mRNA

CBS is a major source of H_2S in the brain and nervous system, where mRNA and protein have been localized to neurons and astrocytes [4]. CSE mRNA is also expressed in the brain, although not abundantly [4]. CSE is a major source of H_2S in the vasculature, but recent evidence suggests that the CBS protein is also expressed in vascular smooth muscle (VSM) and endothelial cells [3,4]. Interestingly, both CBS and CSE proteins have been identified in the serum proteome [3]. 3-MST has been found in the vascular endothelium and in VSM cells, with CAT only in VSM [3]. 3-MST mRNA has been localized in the brain of *cbs* knockout mice. Many other sites of H_2S production have been identified [4].

Tissue and Plasma Concentrations

Numerous studies report H_2S involvement in a multitude of physiological processes at concentrations as high as 100 mM. Similarly, H_2S tissue and plasma concentrations of between 10 and 300 μM have been reported [4]. However, more accurate measurements in recent years using H_2S sensors have demonstrated that: (1) sulfide exists at extremely low concentrations under normoxia (<1 μM), (2) production is

Y. Takei, H. Ando, & K. Tsutsui (Eds): Handbook of Hormones. DOI: http://dx.doi.org/10.1016/B978-0-12-801028-0.00253-1

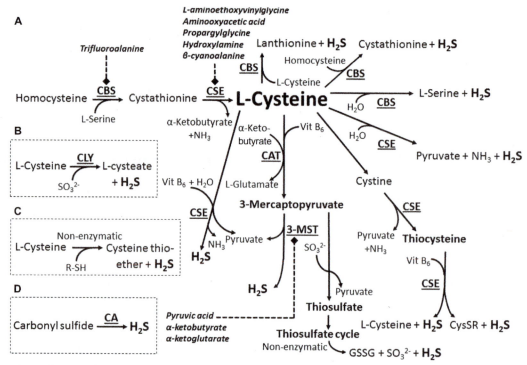

Figure 103C.1 **Biosynthesis and inhibition of H₂S production.** *Figure redrawn from [3,4,6].*

inversely related to O_2 tension, and (3) mitochondria rapidly catabolize H_2S [3]. Furthermore, the presence of an oxidant (e.g., $Fe^{3+}Hb$ in RBCs) rapidly oxidizes H_2S [3]. All these data suggest that H_2S is unlikely to act as an endocrine molecule and instead favor an autocrine and/or paracrine mechanism of signal transduction [6].

Receptors

Since the recognition of H_2S as a signaling molecule, a seemingly endless number of intracellular targets have been reported, although no specific receptor has been identified [1,3,4].

Signal Transduction Pathway

The list of signal transduction pathways affected by H_2S is vast. Present evidence suggests that H_2S participates in redox chemistry, the regulation of transcription factors, S-sulfydration of proteins, and mitochondrial function [3,4]. In the cardiovascular system, specific activation of K_{ATP} channels has been reported in VSM cells in addition to inhibition of cGMP phosphodiesterase that enhances vasodilation [4]. K_{Ca}, Cl^-, and Ca^{2+} channels have also been implicated as cellular targets of H_2S [4]. Vascular endothelial cells generate H_2S through CSE via stimulation of muscarinic cholinergic receptors in the vascular endothelium, which increase intracellular Ca^{2+} and activate Ca^{2+}-dependent calmodulin [4]. It is a concentration-dependent stimulus for and inhibitor of ATP production by acting as an electron donor and inorganic source of energy, and through the inhibition of CCO in mitochondria, respectively, which potentially affects every cell type [7].

Agonists

Commonly used H_2S donors include the inorganic salts sodium sulfide (Na_2S) and sodium hydrosulfide (NaHS) that cause a transient spike in H_2S concentration when dissolved. Slow-release donors have been synthesized, including

numerous water-soluble molecules such as GYY4137, and anethole trithione hydroxide (ADT-OH) complexes [3,4]. Naturally occurring compounds are useful H_2S donors, including diallyl disulfide and diallyl trisulfide from garlic, and sulforaphane from broccoli [3,4]. Applied cysteine analogs thereof such as S-propyl cysteine, and other sulfur-donating molecules like 3-mercaptopyruvate, have a stimulatory effect on endogenous H_2S production [3].

Antagonists

The primary inhibitors available for H_2S research are for either CSE or CBS. D,L-propargylglycine (IC_{50} 40 μM), β-cyanoalanine (IC_{50} 14 μM), and L-aminoethoxyvinylglycine (IC_{50} 1.0 μM) are specific inhibitors of CSE [8]. Aminooxyacetic acid has an 8.5-fold greater selectivity for CSE over CBS (IC_{50} 1 vs. 8.5) [8]. Hydroxylamine is a general inhibitor of pyridoxyl 5′-phosphate-dependent enzymes and has a 57-fold greater selectivity for CSE over CBS (IC_{50} 4.8 vs. 278) [8]. Trifluoroalanine has a four-fold greater IC_{50} for CBS than for CSE (IC_{50} 66 vs. 289) [8]. 3-MST is inhibited by α-keto acids: pyruvic acid IC_{50} 13 mM), α-ketobutyrate (IC_{50} 14 mM), α-ketoglutarate (IC_{50} 10 mM) [9].

Biological Functions

Target Cells/Tissues and Functions

A year after recognizing its effects in the nervous system (1997), Kimura and colleagues identified the vasorelaxant properties of H_2S [2]. In the vasculature, H_2S acts on both endothelial cells and VSM to induce dilation, making H_2S another endothelium-derived relaxing factor and possibly an endothelium-derived hyperpolarizing factor [4]. Both negative chronotropic and inotropic effects have been reported in the heart [4]. In a number of tissues, H_2S is cytoprotective by mediating oxidative damage through interactions with the mitochondrial ETC, among other mechanisms including S-sulfhydration [2,4]. H_2S is also important in respiratory, renal, hepatic, endocrine, gastrointestinal, and reproductive tissues [4]. It is proposed that H$_2$S is an "oxygen sensor" in key tissues in vertebrates, including the vasculature, carotid bodies, and fish gill chemoreceptors [6].

Phenotype in Gene-Modified Animals

Homozygous knockout of *cbs* is fatal in mice (4-week lifespan). In heterozygous knockout mice endogenous H$_2$S production is still significant, but these animals have hyperhomocysteinemia [4]. Knockout of *cth* eliminates most H$_2$S production in the cardiovascular system, resulting in hypertension and severe disruption of endothelium-dependent relaxation of small resistance arteries [4]. Effects on other organ systems in these mice are relatively poorly characterized.

Pathophysiological Implications

Clinical Implications

Despite the myriad of physiological effects reportedly caused by H$_2$S, relatively few clinical conditions are directly attributable to H$_2$S metabolism [3]. Nonetheless, perturbations in endogenous production, or experimental manipulations, have implicated H$_2$S in angiogenesis, apoptosis, asthma, other respiratory diseases, atherosclerosis, cancer, wound healing, diabetes, erectile dysfunction, hypertension, neurodegenerative diseases, and oxidative stress associated with ischemia–reperfusion and injury in myocardial and liver tissues [4].

Use for Diagnosis and Treatment

As the fundamentals of H$_2$S biology are still being elucidated, the therapeutic potential of H$_2$S is yet to be realized. Nonetheless, numerous H$_2$S donating compounds (e.g., ADT-OH) are already available or under development, many as chimeras with existing drugs such as NSAIDs (e.g., S-diclofenac, S-aspirin, and ATB-343, an indomethacin analog) [4,10].

References

1. Liu YH, Lu M, Hu LF, et al. Hydrogen sulfide in the mammalian cardiovascular system. *Antioxid Redox Signal*. 2012;17:141–185.
2. Kimura H. Hydrogen sulfide: its production, release and functions. *Amino Acids*. 2011;41:113–121.
3. Olson KR. The therapeutic potential of hydrogen sulfide: separating hype from hope. *Am J Physiol*. 2011;301:R297–R312.
4. Wang R. Physiological implications of hydrogen sulfide: a whiff exploration that blossomed. *Physiol Rev*. 2012;92:791–896.
5. Li Q, Lancaster JR. Chemical foundations of hydrogen sulfide biology. *Nitric Oxide*. 2013;35:21–34.
6. Olson KR, Donald JA, Dombkowski RA, et al. Evolutionary and comparative aspects of nitric oxide, carbon monoxide and hydrogen sulfide. *Respir Physiol Neurobiol*. 2012;184:117–129.
7. Módis K, Coletta C, Erdélyi K, et al. Intramitochondrial hydrogen sulfide production by 3-mercaptopyruvate sulfurtransferase maintains mitochondrial electron flow and supports cellular bioenergetics. *FASEB J*. 2012;27:601–611.
8. Asimakopoulou A, Panopoulos P, Chasapis CT, et al. Selectivity of commonly used pharmacological inhibitors for cystathionine beta synthase (CBS) and cystathionine gamma lyase (CSE). *Br J Pharmacol*. 2013;169:922–932.
9. Porter DW, Baskin SI. The effect of three α-keto acids on 3-mercaptopyruvate sulfurtransferase activity. *J Biochem Toxicol*. 1996;11:45–50.
10. Kolluru GK, Shen X, Bir SC, et al. Hydrogen sulfide chemical biology: pathophysiological roles and detection. *Nitric Oxide*. 2013;35:5–20.

Supplemental Information Available on Companion Website

- 3D structures of human H$_2$S biosynthetic enzymes showing putative domains/E-Figure 103C.1
- Gene and protein data from representative vertebrate animals/E-Table 103C.1
- Post-translational modification predictions of human H$_2$S biosynthetic enzymes/E-Table 103C.2

Note: Page numbers followed by "*f*" and "*t*" refer to figures and tables, respectively.

Index

Index

Index

Index

Index